Log on.

Tune in.

Succeed.

KT-547-365

Your steps to success.

STEP 1: Register

All you need to get started is a valid email address and the access code below. To register, simply:

1. Go to www.geneticsplace.com
2. Click the appropriate book cover.
   Cover must match the textbook edition being used for your class.
3. Click **"Register"** under **"First-Time User?"**
4. Leave **"No, I Am a New User"** selected.
5. Using a coin, scratch off the silver coating below to reveal your access code.
   Do not use a knife or other sharp object, which can damage the code.
6. Enter your access code in lowercase or uppercase, without the dashes.
7. Follow the on-screen instructions to complete registration.
   During registration, you will establish a personal login name and password to use for logging into the website. You will also be sent a registration confirmation email that contains your login name and password.

Your Access Code is:

Note: If there is no silver foil covering the access code, it may already have been redeemed, and therefore may no longer be valid. In that case, you can purchase access online using a major credit card. To do so, go to www.geneticsplace.com, click the cover of your textbook, click **"Buy Now"**, and follow the on-screen instructions.

STEP 2: Log in

1. Go to www.geneticsplace.com and click the appropriate book cover.
2. Under **"Established User?"** enter the login name and password that you created during registration. *If unsure of this information, refer to your registration confirmation email.*
3. Click **"Log In"**.

STEP 3: (Optional) Join a class

Instructors have the option of creating an online class for you to use with this website. If your instructor decides to do this, you'll need to complete the following steps using the Class ID your instructor provides you. By "joining a class," you enable your instructor to view the scored results of your work on the website in his or her online gradebook.

To join a class:

1. Log into the website. For instructions, see "STEP 2: Log in."
2. Click **"Join a Class"** near the top left.
3. Enter your instructor's **"Class ID"** and then click **"Next"**.
4. At the Confirm Class page you will see your instructor's name and class information. If this information is correct, click Next.
5. Click **"Enter Class Now"** from the Class Confirmation page.
• To confirm your enrollment in the class, check for your instructor and class name at the top right of the page. You will be sent a class enrollment confirmation email.
• As you complete activities on the website from now through the class end date, your results will post to your instructor's gradebook, in addition to appearing in your personal view of the Results Reporter.

To log into the class later, follow the instructions under "STEP 2: Log in."

Got technical questions?

Visit http://247.aw.com. Email technical support is available 24/7.

SITE REQUIREMENTS

For the latest updates on Site Requirements, go to www.geneticsplace.com, choose your text cover, and click Site Reqs.

WINDOWS
366 MHz CPU
Windows 98, 2000, XP
Browser: Internet Explorer 5.0, 5.5 or 6.0; Netscape 6.2.3 or 7.0

MACINTOSH
OS 9.2.2, 10.2.4, 10.3.2
Browsers: Internet Explorer 5.1 and 5.2; Netscape 6.2.3; Safari 1.2*

* The gradebook and class manager features of this website have been partially tested on Mac OS X 10.3.2 with Safari 1.2. A future update will fully support this configuration.

Register and log in

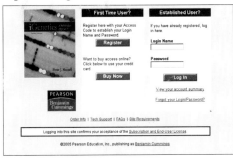

Join a class

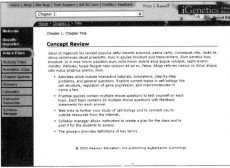

Important: Please read the Subscription and End-User License agreement, accessible from the book website's login page, before using the Genetics Place website and CD-ROM. By using the website or CD-ROM, you indicate that you have read, understood, and accepted the terms of this agreement.

iGenetics *A Molecular Approach*

Peter J. Russell

Reed College

PEARSON

Benjamin
Cummings

San Francisco Boston New York
Capetown Hong Kong London Madrid Mexico City
Montreal Munich Paris Singapore Sydney Tokyo Toronto

Sponsoring Editors: Jim Smith and Susan Winslow
Project Editor: Susan Minarcin
Publishing Assistant: Cinnamon Hearst
Instructional Media Designer: Margy Kuntz, Science Technologies
Developmental Editor: Jean Sims Fernango
Managing Editor: Erin Gregg
Designer, Cover: Gary Hespenheide
Illustrations and Photographs: Laura Murray Productions, Electronic
  Publishing Services Inc., NYC; Steve McEntee
Project Management and Composition: Progressive Information
  Technologies
Senior Production Supervisor: Corinne Benson
Senior Marketing Manager: Scott Dustan

About the Cover:
The front cover shows a color enhanced DNA Fingerprint. The image was photographed by S. Miller, Cumstom Medical Stock Photo.

ISBN 0-321-31207-4

If you purchased this book within the United States or Canada you should be aware that it has been wrong-
fully imported without the approval of the Publisher or the Author.

www.aw-bc.com

www.geneticsplace.com   1 2 3 4 5 6 7 8 9 10—QWD—08 07 06 05

Brief Contents

Detailed Contents *v*

Preface *xiii*

chapter 1 **Genetics: An Introduction** *1*

chapter 2 **DNA: The Genetic Material** *13*

chapter 3 **DNA Replication** *43*

chapter 4 **Gene Function** *67*

chapter 5 **Gene Expression: Transcription** *87*

chapter 6 **Gene Expression: Translation** *111*

chapter 7 **DNA Mutation, DNA Repair, and Transposable Elements** *133*

chapter 8 **Recombinant DNA Technology** *175*

chapter 9 **Applications of Recombinant DNA Technology** *211*

chapter 10 **Genomics** *243*

chapter 11 **Mendelian Genetics** *271*

chapter 12 **Chromosomal Basis of Inheritance** *299*

chapter 13 **Extensions of Mendelian Genetic Principles** *339*

chapter 14 **Quantitative Genetics** *373*

chapter 15 **Gene Mapping in Eukaryotes** *405*

chapter 16 **Advanced Gene Mapping in Eukaryotes** *429*

chapter 17 **Variations in Chromosome Structure and Number** *453*

chapter 18 **Genetics of Bacteria and Bacteriophages** *481*

chapter 19 **Regulation of Gene Expression in Bacteria and Bacteriophages** *515*

chapter 20 **Regulation of Gene Expression in Eukaryotes** *543*

chapter 21 **Genetic Analysis of Development** *573*

chapter 22 **Genetics of Cancer** *605*

chapter 23 **Non-Mendelian Inheritance** *631*

chapter 24 **Population Genetics** *657*

chapter 25 **Molecular Evolution** *705*

Glossary *729*

Suggested Readings *749*

Solutions to Selected Questions and Problems *763*

Credits *811*

Index *815*

+

C1 ENFE

<R>

Detailed Contents

Preface *xiii*

CHAPTER *1*

Genetics: An Introduction *1*
Classical and Modern Genetics *2*
Basic Concepts of Genetics *2*
DNA, Genes, and Chromosomes *2*
Transmission of Genetic Information *3*
Expression of Genetic Information *4*
Sources of Genetic Variation *5*
Geneticists and Genetic Research *5*
The Subdisciplines of Genetics *6*
Basic and Applied Research *6*
Genetic Databases and Maps *7*
Organisms for Genetics Research *9*
Summary *12*

CHAPTER *2*

DNA: The Genetic Material *13*
The Search for the Genetic Material *14*
Griffith's Transformation Experiment *15*
Avery's Transformation Experiments *16*
The Hershey–Chase Bacteriophage
  Experiments *16*
The Discovery of RNA as Viral Genetic
  Material *19*
The Composition and Structure of DNA and RNA *20*
The DNA Double Helix *22*
Different DNA Structures *24*
DNA in the Cell *26*
RNA Structure *26*
The Organization of DNA in Chromosomes *27*
Viral Chromosomes *27*
Prokaryotic Chromosomes *28*
Eukaryotic Chromosomes *29*
Unique-Sequence and Repetitive-Sequence DNA *36*
Summary *37*

Analytical Approaches to Solving Genetics
  Problems *38*
Questions and Problems *39*

CHAPTER *3*

DNA Replication *43*
Semiconservative DNA Replication *44*
The Meselson-Stahl Experiment *45*
Semiconservative DNA Replication
  in Eukaryotes *47*
**DNA Polymerases, the DNA Replicating
  Enzymes** *47*
DNA Polymerase I *47*
Roles of DNA Polymerases *48*
Molecular Model of DNA Replication *50*
Initiation of Replication *51*
Semidiscontinuous DNA Replication *52*
Replication of Circular DNA and the Supercoiling
  Problem *54*
Rolling Circle Replication *54*
DNA Replication in Eukaryotes *57*
Replicons *57*
Initiation of Replication *58*
Eukaryotic Replication Enzymes *58*
Replicating the Ends of Chromosomes *59*
Assembling Newly Replicated DNA
  into Nucleosomes *60*
Summary *62*
Analytical Approaches to Solving Genetics Problems *62*
Questions and Problems *63*

CHAPTER *4*

Gene Function *67*
Gene Control of Enzyme Structure *68*
Garrod's Hypothesis of Inborn Errors
  of Metabolism *68*
The One-Gene–One-Enzyme Hypothesis *69*

Genetically Based Enzyme Deficiencies in Humans *73*
Phenylketonuria *73*
Albinism *75*
Lesch–Nyhan Syndrome *75*
Tay–Sachs Disease *75*
Gene Control of Protein Structure *76*
Sickle-Cell Anemia *77*
Other Hemoglobin Mutants *78*
Cystic Fibrosis *79*
Genetic Counseling *79*
Carrier Detection *80*
Fetal Analysis *80*
Summary *81*
Analytical Approaches to Solving Genetics Problems *82*
Questions and Problems *83*

CHAPTER 5

Gene Expression: Transcription *87*
Gene Expression—The Central Dogma: An Overview *88*
The Transcription Process *88*
RNA Synthesis *88*
Initiation of Transcription at Promoters *89*
Elongation and Termination of an RNA Chain *92*
Transcription in Eukaryotes *93*
Eukaryotic RNA Polymerases *93*
Transcription of Protein-Coding Genes by RNA Polymerase II *93*
Eukaryotic mRNAs *95*
Transcription of Other Genes *102*
Summary *106*
Analytical Approaches to Solving Genetics Problems *108*
Questions and Problems *108*

CHAPTER 6

Gene Expression: Translation *111*
Proteins *112*
Chemical Structure of Proteins *112*
Molecular Structure of Proteins *112*
The Nature of the Genetic Code *115*
The Genetic Code Is a Triplet Code *115*
Deciphering the Genetic Code *117*
Characteristics of the Genetic Code *118*
Translation: The Process of Protein Synthesis *119*
The mRNA Codon Recognizes the tRNA Anticodon *119*

Adding an Amino Acid to tRNA *119*
Initiation of Translation *120*
Elongation of the Polypeptide Chain *123*
Termination of Translation `126*
Protein Sorting in the Cell *127*
Summary *128*
Analytical Approaches to Solving Genetics Problems *128*
Questions and Problems *129*

CHAPTER 7

DNA Mutation, DNA Repair, and Transposable Elements *133*
DNA Mutation *134*
Adaptation versus Mutation *134*
Mutations Defined *136*
Spontaneous and Induced Mutations *140*
Detecting Mutations *148*
Repair of DNA Damage *149*
Direct Reversal of DNA Damage *150*
Base Excision Repair *150*
Repair Involving Excision of Nucleotides *150*
Human Genetic Diseases Resulting from DNA Replication and Repair Mutations *153*
Transposable Elements *153*
General Features of Transposable Elements *153*
Transposable Elements in Prokaryotes *155*
Transposable Elements in Eukaryotes *157*
Summary *165*
Analytical Approaches to Solving Genetics Problems *166*
Questions and Problems *168*

CHAPTER 8

Recombinant DNA Technology *175*
DNA Cloning *176*
Restriction Enzymes *176*
Cloning Vectors and DNA Cloning *179*
Recombinant DNA Libraries *183*
Genomic Libraries *184*
Chromosome Libraries *185*
cDNA Libraries *185*
Finding a Specific Clone in a DNA Library *187*
Screening a cDNA Library *187*
Screening a Genomic Library *189*
Identifying Genes in Libraries by Complementation of Mutations *190*
Using Heterologous Probes to Identify Specific DNA Sequences in Libraries *190*
Using Oligonucleotide Probes to Identify Genes or cDNAs in Libraries *190*

Molecular Analysis of Cloned DNA *192*
Restriction Mapping *192*
Southern Blot Analysis of Sequences
  in the Genome *195*
Northern Blot Analysis of RNA *197*

DNA Sequencing *197*

The Polymerase Chain Reaction (PCR) *200*
PCR Steps *200*
Advantages and Limitations of PCR *202*
Applications of PCR *202*
RT-PCR and mRNA Quantification *203*

Summary *203*

Analytical Approaches to Solving Genetics
  Problems *204*

Questions and Problems *204*

CHAPTER *9*

Applications of Recombinant DNA Technology *211*

Site-Specific Mutagenesis of DNA *212*

Analysis of DNA Polymorphisms in Genomes *213*
DNA Polymorphism Defined *213*
Classes of DNA Polymorphisms *214*

**DNA Molecular Testing for Human Genetic Disease
  Mutations** *218*
Concept of DNA Molecular Testing *218*
Purposes of Human Genetic Testing *219*
Examples of DNA Molecular Testing *219*
Availability of DNA Molecular Testing *222*

Isolation of Human Genes *222*
Cloning the Cystic Fibrosis Gene *222*

DNA Typing *224*
DNA Typing in a Paternity Case *224*
Crime Scene Investigation: DNA Forensics *225*
Other Applications of DNA Typing *227*

Analysis of Expression of Individual Genes *228*
Regulation of Transcription: Glucose Repression
  of the Yeast *GAL1* Gene *228*
Alternative Pre-mRNA Splicing: A Role in Sexual
  Behavior in *Drosophila* *228*

Analysis of Protein–Protein Interactions *229*

Gene Therapy *231*

Biotechnology: Commercial Products *232*

Genetic Engineering of Plants *233*
Transformation of Plant Cells *233*
Applications to Plant Genetic Engineering *234*

Summary *236*

Analytical Approaches to Solving Genetics
  Problems *236*

Questions and Problems *238*

CHAPTER *10*

Genomics *243*

Structural Genomics *245*
Sequencing Genomes *245*
Whole-Genome Shotgun Sequencing
  of Genomes *249*
Selected Examples of Genomes Sequenced *251*
Bacterial Genomes *251*
Eukaryotic Genomes *252*
Insights from Genome Analysis: Genome Sizes
  and Gene Densities *255*

Functional Genomics *257*
Identifying Genes in DNA Sequences *258*
Sequence Similarity Searches to Assign
  Gene Function *258*
Assigning Gene Function Experimentally *259*
Describing Patterns of Gene Expression *260*

Comparative Genomics *264*

Ethics and the Human Genome Project *264*

Summary *265*

Analytical Approaches to Solving Genetics Problems *266*

Questions and Problems *266*

CHAPTER *11*

Mendelian Genetics *271*

Genotype and Phenotype *272*

Mendel's Experimental Design *272*

**Monohybrid Crosses and Mendel's Principle
  of Segregation** *275*
The Principle of Segregation *278*
Representing Crosses with a Branch Diagram *278*
Confirming the Principle of Segregation: The Use
  of Testcrosses *279*
The Wrinkled-Pea Phenotype *280*

**Dihybrid Crosses and Mendel's Principle
  of Independent Assortment** *282*
The Principle of Independent Assortment *282*
Branch Diagram of Dihybrid Crosses *284*
Trihybrid Crosses *285*

The "Rediscovery" of Mendel's Principles *286*

**Statistical Analysis of Genetic Data:
  The Chi-Square Test** *286*

Mendelian Genetics in Humans *288*
Pedigree Analysis *288*
Examples of Human Genetic Traits *289*

Summary *291*

Analytical Approaches to Solving Genetics Problems *291*

Questions and Problems *293*

CHAPTER 12

Chromosomal Basis of Inheritance 299

Chromosomes and Cellular Reproduction 300
Eukaryotic Chromosomes 300
Mitosis 302
Meiosis 306

Chromosome Theory of Inheritance 312
Sex Chromosomes 312
Sex Linkage 314
Nondisjunction of X Chromosomes 317

Sex Chromosomes and Sex Determination 320
Genotypic Sex Determination 320
Genic Sex Determination 326

Analysis of Sex-Linked Traits in Humans 326
X-Linked Recessive Inheritance 327
X-Linked Dominant Inheritance 327
Y-Linked Inheritance 327

Summary 329

Analytical Approaches to Solving Genetics Problems 330

Questions and Problems 331

CHAPTER 13

Extensions of Mendelian Genetic Principles 339

Determining the Number of Genes for Mutations with the Same Phenotype 340

Multiple Alleles 341
ABO Blood Groups 341
Drosophila Eye Color 344
Relating Multiple Alleles to Molecular Genetics 344

Modifications of Dominance Relationships 346
Incomplete Dominance 346
Codominance 346
Molecular Explanations of Incomplete Dominance and Codominance 347

Gene Interactions and Modified Mendelian Ratios 348
Gene Interactions That Produce New Phenotypes 348
Epistasis 349

Essential Genes and Lethal Alleles 356

Gene Expression and the Environment 357
Penetrance and Expressivity 358
Effects of the Environment 359
Nature versus Nurture 362

Summary 363

Analytical Approaches to Solving Genetics Problems 364

Questions and Problems 366

CHAPTER 14

Quantitative Genetics 373

The Nature of Continuous Traits 374
Questions Studied in Quantitative Genetics 375

The Inheritance of Continuous Traits 375
Polygene Hypothesis for Quantitative Inheritance 375
Polygene Hypothesis for Wheat Kernel Color 375

Statistical Tools 377
Samples and Populations 377
Distributions 377
The Mean 378
The Variance and the Standard Deviation 378
Correlation 380
Regression 381
Analysis of Variance 382

Quantitative Genetic Analysis 383
Inheritance of Ear Length in Corn 383

Heritability 385
Components of the Phenotypic Variance 385
Broad-Sense and Narrow-Sense Heritability 387
Understanding Heritability 388
How Heritability Is Calculated 389

Response to Selection 390
Estimating the Response to Selection 390
Genetic Correlations 392

Quantitative Trait Loci 394

Summary 397

Analytical Approaches to Solving Genetics Problems 398

Questions and Problems 400

CHAPTER 15

Gene Mapping in Eukaryotes 405

Early Studies of Genetic Linkage: Morgan's Experiments with *Drosophila* 406

Gene Recombination and the Role of Chromosomal Exchange 408
Corn Experiments 408
Drosophila Experiments 408

Constructing Genetic Maps 412
Detecting Linkage through Testcrosses 412
Gene Mapping with Two-Point Testcrosses 414
Generating a Genetic Map 414
Gene Mapping with Three-Point Testcrosses 415
Calculating Accurate Map Distances 420

Summary 421

Analytical Approaches to Solving Genetics Problems 421

Questions and Problems 423

CHAPTER *16*

Advanced Gene Mapping in Eukaryotes *429*

Tetrad Analysis in Certain Haploid Eukaryotes *430*

Using Random-Spore Analysis to Map Genes in Haploid Eukaryotes *431*

Calculating Gene–Centromere Distance in Organisms by Using Ordered Tetrads *431*

Using Tetrad Analysis to Map Two Linked Genes *433*

Mitotic Recombination *436*

Discovery of Mitotic Recombination *436*

Mitotic Recombination in the Fungus *Apergillus nidulans* *438*

Retinoblastoma, a Human Tumor That Can Be Caused by Mitotic Recombination *442*

Mapping Human Genes *443*

Mapping Human Genes by Recombination Analysis *443*

lod Score Method for Analyzing Linkage of Human Genes *444*

High-Density Genetic Maps of the Human Genome *444*

Physical Mapping of Human Genes *445*

Summary *446*

Analytical Approaches to Solving Genetics Problems *446*

Questions and Problems *448*

CHAPTER *17*

Variations in Chromosome Structure and Number *453*

Types of Chromosomal Mutations *454*

Variations in Chromosome Structure *454*

Deletion *455*

Duplication *457*

Inversion *459*

Translocation *460*

Chromosomal Mutations and Human Tumors *462*

Position Effect *465*

Fragile Sites and Fragile X Syndrome *465*

Variations in Chromosome Number *466*

Changes in One or a Few Chromosomes *467*

Changes in Complete Sets of Chromosomes *472*

Summary *474*

Analytical Approaches to Solving Genetics Problems *474*

Questions and Problems *475*

CHAPTER *18*

Genetics of Bacteria and Bacteriophages *481*

Genetic Analysis of Bacteria *482*

Gene Mapping in Bacteria by Conjugation *484*

Discovery of Conjugation in *E. coli* *484*

The Sex Factor *F* *484*

High-Frequency Recombination Strains of *E. coli* *485*

F′ Factors *487*

Using Conjugation to Map Bacterial Genes *487*

Circularity of the *E. coli* Map *489*

Genetic Mapping in Bacteria by Transformation *490*

Genetic Mapping in Bacteria by Transduction *492*

Bacteriophages *492*

Transduction Mapping of Bacterial Chromosomes *493*

Mapping Bacteriophage Genes *498*

Fine-Structure Analysis of a Bacteriophage Gene *499*

Recombination Analysis of *rII* Mutants *500*

Deletion Mapping *501*

Defining Genes by Complementation (*Cis-Trans*) Tests *503*

Summary *506*

Analytical Approaches to Solving Genetics Problems *506*

Questions and Problems *508*

CHAPTER *19*

Regulation of Gene Expression in Bacteria and Bacteriophages *515*

The *lac* Operon of *E. coli* *516*

Lactose as a Carbon Source for *E. coli* *516*

Experimental Evidence for the Regulation of *lac* Genes *517*

Jacob and Monod's Operon Model for the Regulation of *lac* Genes *520*

Positive Control of the *lac* Operon *525*

Molecular Details of *lac* Operon Regulation *526*

The *trp* Operon of *E. coli* *528*

Gene Organization of the Tryptophan Biosynthesis Genes *529*

Regulation of the *trp* Operon *529*

Regulation of Gene Expression in Phage Lambda *533*

Early Transcription Events *533*

The Lysogenic Pathway *534*

The Lytic Pathway *534*

Summary *536*

Analytical Approaches to Solving Genetics Problems *537*

Questions and Problems *538*

CHAPTER 20

Regulation of Gene Expression in Eukaryotes *543*

Operons in Eukaryotes *544*

Levels of Control of Gene Expression in Eukaryotes *546*

Control of Transcription Initiation *546*
Chromatin Remodeling *547*
Activation of Transcription by Activators and Coactivators *549*
Blocking Transcription with Repressors *550*
Combinatorial Gene Regulation *550*
Case Study: Regulation of Galactose Utilization in Yeast *551*
Regulation of Gene Expression by Steroid Hormones *553*

Gene Silencing and Genomic Imprinting *557*

Posttranscriptional Control *561*
RNA Processing Control *561*
mRNA Transport Control *563*
mRNA Translation Control *563*
mRNA Degradation Control *564*
Protein Degradation Control *564*

RNA Interference: A Mechanism for Silencing Gene Expression *565*

Summary *567*

Analytical Approaches to Solving Genetics Problems *568*

Questions and Problems *568*

CHAPTER 21

Genetic Analysis of Development *573*

Basic Events of Development *574*

Model Organisms for the Genetic Analysis of Development *575*

Development Results from Differential Gene Expression *577*
Constancy of DNA in the Genome During Development *577*
Examples of Differential Gene Activity During Development *579*
Exception to the Constancy of Genomic DNA During Development: DNA Loss in Antibody-Producing Cells *581*

Case Study: Sex Determination and Dosage Compensation in Mammals and *Drosophila* *585*
Sex Determination in Mammals *585*
Dosage Compensation Mechanism for X-Linked Genes in Mammals *586*
Sex Determination in *Drosophila* *586*
Dosage Compensation in *Drosophila* *590*

Case Study: Genetic Regulation of the Development of the *Drosophila* Body Plan *591*
Drosophila Developmental Stages *591*
Embryonic Development *591*
Microarray Analysis of *Drosophila* Development *598*

Summary *600*

Analytical Approaches to Solving Genetics Problems *601*

Questions and Problems *601*

CHAPTER 22

Genetics of Cancer *605*

Relationship of the Cell Cycle to Cancer *607*
Molecular Control of the Cell Cycle *607*
Regulation of Cell Division in Normal Cells *607*

Cancers Are Genetic Diseases *609*

Genes and Cancer *610*
Oncogenes *610*
Tumor Suppressor Genes *618*
Mutator Genes *624*

Telomere Shortening, Telomerase, and Human Cancer *624*

The Multistep Nature of Cancer *625*

Chemicals and Radiation as Carcinogens *626*
Chemical Carcinogens *626*
Radiation *626*

Summary *627*

Analytical Approaches to Solving Genetics Problems *627*

Questions and Problems *628*

CHAPTER 23

Non-Mendelian Inheritance *631*

Origin of Mitochondria and Chloroplasts *632*

Organization of Extranuclear Genomes *632*
Mitochondrial Genome *632*
Chloroplast Genome *636*

Rules of Non-Mendelian Inheritance *639*

Examples of Non-Mendelian Inheritance *639*
Shoot Variegation in the Four O'Clock *639*
The [poky] Mutant of *Neurospora* *641*
Yeast *petite* Mutants *643*
Non-Mendelian Inheritance in *Chlamydomonas* *645*
Human Genetic Diseases and Mitochondrial DNA Defects *647*
Cytoplasmic Male Sterility and Hybrid Seed Production *647*
Exceptions to Maternal Inheritance *648*
Infectious Heredity: Killer Yeast *649*

Contrasts to Non-Mendelian Inheritance *650*

Maternal Effect *650*

Summary *651*

Analytical Approaches to Solving Genetics Problems *652*

Questions and Problems *653*

CHAPTER *24*

Population Genetics *657*

Genetic Structure of Populations *659*

Genotype Frequencies *659*
Allele Frequencies *660*

The Hardy-Weinberg Law *663*

Assumptions of the Hardy-Weinberg Law *664*
Predictions of the Hardy-Weinberg Law *664*
Derivation of the Hardy-Weinberg Law *665*
Extensions of the Hardy-Weinberg Law
   to Loci with More Than Two Alleles *667*
Extensions of the Hardy-Weinberg Law to
   Sex-Linked Alleles *667*
Testing for Hardy-Weinberg Proportions *667*
Using the Hardy-Weinberg Law to Estimate
   Allelic Frequencies *668*

Genetic Variation in Space and Time *669*

Genetic Variation in Natural Populations *670*

Measuring Genetic Variation at the Protein Level *671*
Measuring Genetic Variation at the DNA Level *673*

**Forces That Change Gene Frequencies in
   Populations** *676*

Mutation *677*
Random Genetic Drift *679*
Migration *685*
Hardy-Weinberg and Natural Selection *687*
Balance Between Mutation and Selection *694*
Assortative Mating *695*
Inbreeding *695*

**Summary of the Effects of Evolutionary Forces
   on the Genetic Structure of a Population** *696*

Changes in Allelic Frequency Within
   a Population *696*
Genetic Divergence Among Populations *696*
Increases and Decreases in Genetic Variation
   Within Populations *697*

The Role of Genetics in Conservation Biology *697*

Speciation *697*

Barriers to Gene Flow *698*
Genetic Basis for Speciation *698*

Summary *699*

Analytical Approaches to Solving Genetics
   Problems *699*

Questions and Problems *700*

CHAPTER *25*

Molecular Evolution *705*

Patterns and Modes of Substitutions *706*

Nucleotide Substitutions in DNA Sequences *706*
Rates of Nucleotide Substitutions *708*
Variation in Evolutionary Rates Between
   Genes *711*
Rates of Evolution in Mitochondrial DNA *713*
Molecular Clocks *714*

Molecular Phylogeny *715*

Phylogenetic Trees *716*
Reconstruction Methods *718*
Phylogenetic Trees on a Grand Scale *721*

Acquisition and Origins of New Functions *723*

Multigene Families *723*
Gene Duplication and Gene Conversion *724*
Arabidopsis Genome Results *724*

Summary *725*

Analytical Approaches to Solving Genetics
   Problems *725*

Questions and Problems *726*

Glossary *729*

Suggested Readings *749*

Solutions to Selected Questions and Problems *763*

Credits *811*

Index *815*

Preface

An Approach to Teaching Genetics

The structure of DNA was first described in 1953, and since that time genetics has become one of the most exciting and ground-breaking sciences. Our understanding of gene structure and function has progressed rapidly since molecular techniques were developed to clone or amplify genes, and rapid methods for sequencing DNA became available. In recent years, the sequencing of the genomes of a number of organisms has changed the scope of experiments performed by geneticists. For example, we can study a genome's worth of genes now in one experiment, allowing us to obtain a deeper understanding of gene expression.

I have taught genetics for over 30 years, while maintaining a research program involving undergraduates. Students learn genetics best if they are given a balanced approach that integrates their understanding of the abstract nature of genes (from the transmission genetics part) with the molecular nature of genes (from the molecular genetics part). My goal, in this edition, has been to provide students with a clear and logical presentation of the material, in combination with an experimental theme throughout, which makes clear to students how we know what we know. It is my hope that you will find my approach as helpful to you in successfully teaching this course, as have so many colleagues who have used past editions.

The general features of *iGenetics: A Molecular Approach 2e* are as follows:

Modern Coverage. The field of genetics has grown rapidly in recent years. In creating this text I have worked with experts in the field to ensure that we present these exciting developments with the highest degree of accuracy. The book covers all major areas of genetics, balancing classical and molecular aspects to give students an integrated view of genetic principles. The classical genetics material tends to be abstract and more intuitive, while the molecular genetics is more factual and conceptual. Teaching genetics, therefore, requires teaching these two styles, as well as conveying the necessary information, and the modern coverage reflects that. The molecular material, which is

the material that changes most rapidly in genetics, is current and presented at a suitable level for students.

Experimental Approach. Research is the foundation of our present knowledge of genetics. Through the presentation of research experiments throughout, students learn about the formulation and study of scientific questions in a way that will be of value in their study of genetics and, more generally, in all areas of science. The amount of information that students must learn is constantly growing, making it critical that students don't simply memorize facts but rather learn how to learn. In my classroom and in this text I emphasize basic principles, but I place them in the meaningful context of classic and modern experiments. Thus, in observing the process of science, students learn for themselves the type of rigorous thinking that leads to the formulation of hypotheses and experimental questions and, thence, to the generation of new knowledge.

Classic Principles. Our present understanding of genes is built on a foundation of classic experiments, a number of them leading to Nobel Prizes. These classic experiments are presented so that students can appreciate how ideas about genetic processes have developed to our present-day understanding. These experiments include:

- Griffith's Transformation Experiment
- Avery and His Colleague's Transformation Experiment
- Hershey and Chase's Bacteriophage Experiment
- Meselson and Stahl's DNA Replication Experiment
- Beadle and Tatum's One-Gene–One-Enzyme Hypothesis Experiments
- Mendel's Experiments on Gene Segregation
- Thomas Hunt Morgan's Experiments on Gene Linkage
- Seymour Benzer's Experiments on the Fine Structure of the Gene
- Jacob and Monod's Experiments on the *lac* Operon

Human Applications. The impact of modern genetics on our daily lives cannot be understated. Gene therapy, gene mapping, genetic disorders, genetic screening, genetic engineering, and the human genome: these topics directly impact human lives. By illustrating important concepts with numerous examples of applications from human genetics, students are attracted by a natural curiosity to learn about themselves and our species. For instance, there are discussions about specific genetic diseases (in Chapter 4 on Gene Function, for example), about DNA analysis approaches used to detect human gene mutations and in forensics (in Chapter 9), and about the Human Genome Project (Chapter 10). Human genes mentioned in the text are keyed to the OMIM (Online Mendelian Inheritance in Man) online database of human genes and genetic disorders at http://www3.ncbi.nlm.nih.gov/Omim/, where the most up-to-date information is available about the genes.

Using Media to Teach Genetics. Media for this textbook include interactive activities to allow students to self-assess their understanding of key chapter concepts, and animations to provide a dynamic representation of processes that are difficult to visualize from a static figure. I was involved in the development of most of these pieces, ensuring that their look and quality match that of the textbook.

- Twenty-six interactive activities called *iActivities*—one per chapter—have been designed to promote interactive problem solving. Found on the *iGenetics* CD-ROM, these activities are based on case studies presented at the beginnings of the chapters. An example from Chapter 18 is the analysis of DNA microarray results for a fictional patient with breast cancer to determine gene expression differences and then determine which drugs would be useful for treating her cancer. I have worked closely with the development teams for most *iActivities* to help ensure accuracy and quality. A brief description of the *iActivity* appears in the text at the beginning of each chapter, and then, at the point in the chapter at which it is appropriate to use the media there is a message urging students to try it.
- Fifty-nine narrated animations on the *iGenetics* CD-ROM help students visualize challenging concepts or complex processes, such as meiosis, gene mapping, DNA replication, translation, restriction mapping, and DNA molecular testing for human genetic disease mutations. As with the *iActivities*, I have worked closely with the development teams for most of the animations: outlining topics, editing the storyboards, helping describe the steps for the artists, and working closely with the animators until the animations were complete. We have made a special effort to base the animations on the text figures so that students do not have to think about the processes in a different graphic format. These animations are of very high quality, showing a level of detail not typical of animations that are supplements to texts. A media flag with the title of the animation appears next to the discussion of that topic in the chapter.

Accuracy. An intense developmental effort, along with numerous third party reviews of both text and media, ensure the highest degree of accuracy. In addition, we established a **Problem Review Board,** whose members provided us with comments and suggestions to ensure that the content, accuracy, and appropriate level of our problems has been achieved.

Organization

This text focuses on a molecular first presentation of materials. After the introductory chapter, a core set of nine chapters covers the molecular details of gene structure and function, and the cloning and manipulation of DNA, before the Mendelian genetics, gene segregation, and gene mapping principles are developed. However, the chapters can readily be used in any sequence to fit the needs of individual instructors.

Changes from iGenetics 1e
- All molecular material in the book was updated where necessary.
- One chapter (Chapter 7) brings together a discussion of DNA mutation, DNA repair, and transposable elements in the context of changes affecting gene function
- Mitosis and meiosis are now covered in Chapter 12 on the chromosomal basis of inheritance.
- A section on using the complementation test to determine the number of genes for mutations with the same phenotype is added to the chapter on extensions of Mendelian genetics principles (Chapter 13).
- Quantitative Genetics (Chapter 14) is now presented immediately after the core chapters on gene segregation principles.
- The chapter on Gene Mapping in Eukaryotes has been divided into two chapters, one on general principles of gene mapping (Chapter 15: Gene Mapping in Eukaryotes), and one on more specialized approaches to gene mapping, such as tetrad analysis, mitotic recombination, and mapping human genes (Chapter 16: Advanced Gene Mapping in Eukaryotes).
- The Recombinant DNA Technology chapter (Chapter 8) was revised to streamline the discussion of cloning vectors, and to amplify the presentation of PCR (including describing real time reverse transcriptase-PCR).
- The chapter on Applications of Recombinant DNA Technology (Chapter 9) was reconceived and many parts were rewritten. The chapter is now organized with a flow from DNA to RNA to protein analysis. The chapter starts with a description of site-specific mutagenesis (previously in the mutation chapter). The largely rewritten DNA polymorphism section follows. Polymorphisms are introduced in the context of genome projects, and foreshadows the involvement

of polymorphisms in DNA testing and forensics that is covered later in the chapter. The DNA forensics material is expanded compared with 1e, and now includes case studies.

- **The chapter on Genomics (Chapter 10)** has a streamlined presentation of structural genomics, which includes a discussion of the analysis of complete genomes. Material was added about the mouse and rat genomes. A new section on insights from genome analysis discusses what we are learning about genome organization and gene densities from genome projects. Coverage of comparative genomics remains in **Chapter 25, Molecular Evolution.**

- The chapter in the previous edition on **Regulation of Gene Expression in Eukaryotes** was divided into two chapters for this edition: **Regulations of Gene Expression in Eukaryotes (Chapter 20)**, and **Genetic Analysis of Development (Chapter 21)**. The Regulation chapter was reorganized and rewritten. It now includes, for instance, a discussion of chromatin remodeling, and a presentation of RNA interference in a new section on Gene Silencing and Genome Imprinting. The Development chapter is newly constructed from some material in the previous Regulation chapter and new material.

- The chapter on **Genetics of Cancer (Chapter 22)** now has the material on cyclins and cyclin-dependent kinases in the section on the Relationship of the Cell Cycle to Cancer, making a more logical connection between molecular events controlling the cell cycle and the consequences of mutations affecting key genes for the cell cycle. The coverage of signal transduction has been expanded in the section on the Regulation of Cell Division in Normal Cells. Presentation of the functions of the retinoblastoma and p53 tumor-suppressor proteins has been updated and refined.

- **Coverage of cytoplasmic male sterility was added to the chapter on Non-Mendelian Inheritance (Chapter 23).**

Coverage

The four major areas of genetics—transmission genetics, molecular genetics, population genetics, and quantitative genetics—are covered in 25 chapters.

Chapter 1 is an introductory chapter designed to summarize the main branches of genetics, explain the basic concepts of genetics (the molecular nature of genes, the transmission of genetic information, the expression of genes, and the sources of genetic variation), describe what geneticists do and what their areas of research encompass, and introduce genetic databases and maps.

Chapters 2 through 7 are core chapters covering genes and their functions. In **Chapter 2,** we cover the structure of DNA, and the details of DNA structure and organization

in prokaryotic and eukaryotic chromosomes. We cover DNA replication in prokaryotes and eukaryotes and recombination between DNA molecules in **Chapter 3.** In **Chapter 4,** we examine some aspects of gene function, such as the genetic control of the structure and function of proteins and enzymes and the role of genes in directing and controlling biochemical pathways. A number of examples of human genetic deseases that result from enzyme deficiencies are described to reinforce the concepts, The discussion of gene function in **Chapter 4** enables students to understand the important concept that genes specify proteins and enzymes, setting them up for the next two chapters, in which gene expression is discussed. In **Chapter 5,** we discuss transcription, and in **Chapter 6,** we describe the structure of proteins, the evidence for the nature of the genetic code, and the process of translation in both prokaryotes and eukaryotes. Then, the ways in which genetic material can change or be changed are presented in **Chapter 7.** Topics include the processes of gene mutation, some of the mechanisms that repair damage to DNA, some of the procedures used to screen for particular types of mutants, and the structures and movements of transposable gentic elements in prokaryotes and eukaryotes.

Gene manipulation and genome analysis is described in the following three chapters. In **Chapter 8,** we discuss recombinant DNA technology and other molecular techniques that are essential tools of most areas of modern genetics. There are descriptions of the use of recombinant DNA technology to clone and characterize genes and to manipulate DNA. Then, in **Chapter 9** we discuss the applications of recombinant DNA technology in analyzing DNA, RNA and protein, including the types of DNA polymorphisms in genomes, the diagnosis of human diseases, the isolation of human genes, forensics (DNA typing), gene therapy, the development of commercial products, and the genetic engineering of plants. In **Chapter 10** we discuss genomics, the analysis of genomes, focusing on the Human Genome Project for mapping and sequencing the complete genomes of humans and other selected organisms. The chapter outlines the methods for sequencing complete genomes, discusses the features of a number of genomes that have been sequenced, and goes into the types of research scientists are engaging in to detail the global expression of genes in cells at the RNA and protein levels.

Chapters 11 through 18 are core chapters covering the principles of gene segregation analysis. Chapters 11 and 12 present the basic principles of genetics in relation to Mendel's laws. **Chapter 11** is focused on Mendel's contributions to our understanding of the principles of heredity, and **Chapter 12** covers mitosis and meiosis in the context of animal and plant life cycles, the experimental evidence for the relationship between genes and chromosomes, and methods of sex determination. Mendelian genetics in humans is introduced in Chapter 11 with a focus on pedigree analysis and autosomal traits. The topic is continued in Chapter 12 with respect to sex-linked genes. The exceptions to and extensions of Mendelian analysis (such as the existence of multiple alleles, the modification of dominance relationships, gene interactions and modified Mendelian ratios, essential genes and lethal alleles, and the relationship between genotype and phenotype) are described in **Chapter 13**. In **Chapter 14**, "Quantitative Genetics," we consider the heredity of traits in groups of individuals that are determined by many genes simultaneously. In this chapter we also discuss heritability; the relative extent to which a characteristic is determined by genes or by the environment. Discussions of the application of molecular tools to this area of genetics is also included. In Chapters 15 and 16, gene mapping in eukaryotes is presented. In **Chapter 15**, we describe how the order of and distance between the genes on eukaryotic chromosomes are determined in genetic experiments designed to quantify the crossovers that occur during meiosis. **Chapter 16** focusses on the more specialized analysis of genes by tetrad analysis, primarily in fungal systems, the phenomenon of recombination in mitosis, and gene mapping in humans. Chromosomal mutations—changes in normal chromosome structure or chromosome number—are discussed in **Chapter 17**. Chromosomal mutations in eukaryotes and human disease syndromes that result from chromosomal mutations, including triplet repeat mutations, are emphasized. In **Chapter 18**, we discuss the ways of mapping genes in bacteria and in bacteriophages, which take advantage of the processes of conjugation, transformation, and transduction. Fine structure analysis of bacteriophage genes concludes this chapter.

Gene regulation is covered in the following two chapters. **Chapter 19** focuses on the regulation of gene expression in prokaryotes. In this chapter, we discuss the operon as a unit of gene regulation, the current molecular details in the regulation of gene expression in bacterial operons, and regulation of genes in bacteriophages. **Chapter 20** focuses on the regulation of gene expression in eukaryotes, explaining how eukaryotic gene expression is regulated, stressing molecular changes that accompany gene regulation and short-term gene regulation in simple and complex eukaryotes.

Chapter 21 discusses genetic analysis of development. The chapter describes basic events in development, and the evidence that development results from differential gene expression, before illustrating gene regulation principles at work in case studies of well-characterized developmental processes, namely sex determination and dosage compensation, and the development of the *Drosophila* body plan. Next, in **Chapter 22** we discuss the relationship of the cell cycle to cancer and the various types of genes that, when mutated, play a role in the development of cancer.

In **Chapter 23**, we address the organization and genetics of extranuclear genomes of mitochondria and chloroplasts. We cover the current molecular information about the organization of genes within the extranuclear genomes and the classic genetic experiments that are used to study non-Mendelian inheritance.

In **Chapter 24**, we present the basic principles in population genetics, extending our studies of heredity from the individual organism to a population of organisms. This chapter includes an integrated discussion of the developing area of conservation genetics. **Chapter 25** discusses evolution at the molecular level of DNA and protein sequences. The study of molecular evolution uses the theoretical foundation of population genetics to address two essentially different sets of questions: how DNA and protein molecules evolve and how genes and organisms are evolutionarily related.

Pedagogical Features

Because the field of genetics is complex, making the study of it potentially difficult, we have incorporated a number of special pedagogical features to assist students and to enhance their understanding and appreciation of genetic principles:

- Each chapter opens with an outline of its contents and a section called "**Principal Points**." Principal Points are short summaries that alert students to the key concepts they will encounter in the material to come.
- Throughout each chapter, strategically placed "**Keynote**" summaries emphasize important ideas and critical points that allow students to check their progress.
- Important **terms and concepts**—highlighted in bold—are defined where they are introduced in the text. For easy reference, they are also compiled in a glossary at the back of the book.
- Each chapter closes with a "**Summary**," further reinforcing the major points that have been discussed.
- With the exception of the introductory Chapter 1, all chapters conclude with a section titled "**Analytical Approaches to Solving Genetics Problems.**" Genetics principles have always been best taught with a problem-solving approach. However, beginning students often do not acquire the necessary experience with basic concepts that would enable them to attack assigned problems methodically. In the Analytical Approaches sections, typical genetic problems are talked through in step-by-step detail to help students understand how to tackle a genetics problem by applying fundamental principles.

- The problem sets that close the chapters include approximately **600 questions and problems** designed to give students further practice in solving genetics problems. The problems for each chapter represent a range of topics and difficulty. The answers to questions marked by an asterisk (*) can be found at the back of the book, and answers to all questions are available in a separate supplement, *Study Guide and Solutions Manual*.
- Some chapters include **boxes covering special topics** related to chapter coverage. Some of these boxed topics are *Equilibrium Density Gradient Centrifugation* (Chapter 3), *Labeling DNA* (Chapter 8), *Elementary Principles of Probability* (Chapter 11); *Genetic Terminology* (Chapter 11), and *Hardy, Weinberg, and the History of Their Contribution to Population Genetics* (Chapter 24).
- Suggested readings and selected websites for the material in each chapter are listed at the back of the book.
- Special care has been taken to provide an extensive, accurate, and well cross-referenced index.

Supplements

For Students

Study Guide and Solutions Manual for *iGenetics: A Molecular Approach 2e* (0-8053-4694-5)
Prepared by Bruce Chase of the University of Nebraska, the *Study Guide and Solutions Manual* has detailed solutions for all the problems in the text and contains the following features for each chapter: chapter outline of text material, key terms, suggestions for analytical approaches, problem-solving strategies; and 1,000 additional questions for practice and review. It also includes a review of important terms and concepts; a "Thinking Analytically" section, which provides guidance and tips on solving problems and avoiding common pitfalls; additional questions for practice and review; and questions that relate to chapter-specific animations and *iActivities*.

The Genetics Place (www.geneticsplace.com)
This online learning environment offers interactive learning activities, practice quizzes, links to related websites, a syllabus manager, and a glossary. A free one-year subscription to the Genetics Place is available and included with each copy of the text. This online learning environment offers 25 activities, each with follow-up questions, 59 animations, and practice quizzes that can be graded and the results entered directly into the instructor's grade book. A **syllabus manager** and a **grade book function** are provided as instructor resources.

For Instructors

Instructor's Guide to Text and Media for *iGenetics: A Molecular Approach 2e* (0-8053-4689-9)
Written by Rebecca Ferrell of the Metropolitan State College of Denver, this guide presents sample lecture outlines, teaching tips for the text, and media tips for using and assigning the media component in class.

Instructor's Resource CD ROM for *iGenetics: A Molecular Approach 2e* (0-8053-4690-2)
This cross-platform CD-ROM features all illustrations, photos, and tables from the text, with each available in high-resolution JPEG and PowerPoint® formats.

Transparency Acetates for *iGenetics: A Molecular Approach 2e* (0-8053-4691-0)
These high-quality acetates include 175 static figures, plus 25 selected figures that are layered for step-by-step lecture presentation.

Printed Test Bank (0-8053-4693-7)
Prepared by James Costa at Western Carolina University and Katherine Matthews at Western Carolina University, the test bank includes more than 1,200 multiple-choice, matching, true/false, and short answer questions. The entire set of questions is also available on a cross-platform CD-ROM that allows you to generate tests with a user-friendly interface on which you can edit, sort, and even add your own questions.

Acknowledgments

Publishing a textbook and all its supplements is a team effort. I have been very fortunate to have some very talented individuals working with me on this project. I thank Jean Sims Fornango for her work as developmental editor. For their help in honing the text, I thank all of the reviewers and class testers involved in this edition; their names are listed on the following pages.

I would like to thank the following individuals for their talents and efforts in crafting some of the chapters in the text: Dr. Kevin Livingstone (Trinity University) for his revision of Chapter 14, "Quantitative Genetics"; Dr. Kent Holsinger (University of Connecticut) for his revision of Chapter 24, "Population Genetics"; and Dr. Dan E. Krane (Wright State University) for revising Chapter 25, "Molecular Evolution."

I would also like to thank Bruce Chase (University of Nebraska, Omaha) for his work on the end-of-chapter questions, including his contribution of many new problem sets, and for his excellent work on putting together the *Study Guide and Solutions Manual*. I thank Barbara

Littlewood for generating an excellent, comprehensive index for this book. I also thank Laura Murray, Art Editor, for her attention to quality in the art program, and Ruth Steyn for editing the Glossary.

I want to acknowledge a number of talented individuals who worked with me to develop the material found on the *iGenetics: A Molecular Approach 2e* CD-ROM: Margy Kuntz, who did an excellent job researching this subject matter and then authoring most of the highly creative and rich *iActivities,* all of which are designed to enhance critical thinking in genetics; Dr. Todd Kelson (Ricks College; animation storyboards); Dr. Hai Kinal (Springfield College, animation storyboards); Dr. Robert Rothman (Rochester Institute of Technology; animation storyboards); Steve McEntee (*iActivity* art development, art style for the animations and text art); Kristin Mount (animations); Richard Sheppard (animations); Eric Stickney (animations); and James Costa (Western Carolina University; CD-ROM quiz questions). In addition, I thank Dr. James Caras, Principal, Jon Harmon, Content Developer, and the rest of the Science Technologies staff for developing and producing additional *iActivities* and *Animations* for the CD-ROM.

I am grateful to the literary executor of the late Sir Ronald A. Fisher, F.R.S.; to Dr. Frank Yates, F.R.S.; and to Longman Group Ltd. London, for permission to reprint Table IV from their book *Statistical Table for Biological, Agricultural and Medical Research* (Sixth Edition, 1974).

I would like to thank Corinne Benson, Senior Production Supervisor at Benjamin Cummings, as well as Crystal Clifton and the staff at Progressive Publishing Alternatives for their handling of the production phase of the book.

Finally, I wish to thank the editorial and marketing staff at Benjamin Cummings who helped to make *iGenetics: A Molecular Approach 2e,* a reality. In particular, I thank Jim Smith, Acquisitions Editor; Frank Ruggirello, Vice President and Editorial Director; Kay Ueno, Director of Development; and Scott Dustan, Senior Marketing Manager. I am especially grateful to Susan Minarcin, Project Editor, for her excellent management of all aspects of the production of the book; her tireless efforts have made this textbook and its supplements of the highest quality.

Peter J. Russell

Reviewers

George Bajszar
University of Colorado, Colorado Springs

Ruth Ballard
California State University, Sacramento

Anna Berkovitz
Purdue University

Joanne Brock
Kennesaw State University

Paul J. Bottino
University of Maryland

Patrick Calie
Eastern Kentucky University

Clarissa Cheney
Pomona College

Richard Cheney
Christopher Newport University

Claire Chronmiller
University of Virginia

James Costa
Western Carolina University

Frank Doe
University of Dallas

David Durcia
University of Oklahoma

Larry Eckroat
Pennsylvania State University at Erie

Bert Ely
University of South Carolina

Russ Feirer
St. Norbert College

Wayne Forrester
Indiana University

Elaine Freund
Pomona College

David Fromson
California State University, Fullerton

Gail Gasparich
Towson State University

Peter Gegenheimer
University of Kansas

Vaughn Gehle
Southwest Minnesota State

Richard C. Gethmann
University of Maryland, Baltimore County

Elliot Goldstein
Arizona State University

Pamela Gregory
Jacksonville State University

Pamela Hanratty
Indiana University

Ernie Hannig
University of Texas, Dallas

David Haymer
University of Hawaii

Mary Healy
Springfield College

Robert Hinrichsen
Indiana University

Margaret Hollingsworth
State University of New York, Buffalo

Lynne Hunter
University of Pittsburgh

Gregg Jongeward
University of the Pacific

Todd Kelson
Ricks College

Alexander Lai
Oklahoma State University

Sandy Latourelle
Plattsburg State University

Michael Lentz
University of North Florida

Hai Kanal
Springfield College

Larry Kline
State University of New York, Brockport

Alan Leonard
Florida Institute of Technology

Mark J. M. Magbanua
University of California at Davis

Karen Malatesta
Princeton University

Russell Malmburg
University of Georgia, Athens

Patrick H. Masson
University of Wisconsin, Madison

Steven McCommas
Southern Illinois University

David McCullough
Wartburg College

Denis McGuire
St. Cloud State University

Kim McKim
Rutgers University

John Merruam
University of California, Los Angeles

Stan Metzenberg
University of California, Northridge

Dwight Moore
Emporia State University

Roderick Morgan
Grand Valley State University

Muriel Nesbit
University of California, San Diego

Brent Nelson
Auburn University

James M. Pipas
University of Pittsburgh

Diane Robbins
University of Michigan Medical School

Harry Roy
Rensselaer Polytechnic Institute

Thomas Rudge
Ohio State University

Rey Antonio L. Sia
State University of New York

Randy Small
University of Tennessee, Knoxville

William Steinhart
Bowdoin College

Millard Sussman
University of Wisconsin, Madison

Sara Tolsma
Northwestern University

Melina Wales
Texas A&M University

Robert West
University of Colorado

Matthew White
Ohio University

Media Reviewers

Mary D. Healey
Springfield College

Sidney R. Kushner
University of Georgia

Gayle LoPiccolo
Montgomery College

Maria Orive
University of Kansas

Kajan Ratnakumar
Desplan Laboratory, New York University

Class Testers

Richard Cheney
Christopher Newport University

Carol Chihara
University of San Francisco

Candi Coffin
Briar Cliff College

Alix Dardin
The Citadel

Betsy Dyer
Wheaton College

Larry Eckroat
Pennsylvania State University, Erie

Victor Fet
Marshall University

Stephanie Fore
Truman State University

Jenna Hellack
University of Central Okalahoma

Debora Hetinger
Texas Lutheran University

Sarwar Jahangir
Wabash College

J. Michael Jones
Culver-Stockton College

Gregory Jongeward
University of the Pacific

David Kass
Eastern Michigan University

Todd Kleson
Ricks College

Hai Kanai
Springfield College

Keith Kline
Minnesota State University, Mankato

Larry Kline
State University of New York, Brockport

Sandy Latourelle
Plattsburg State University

Carl Lucian
Indiana University of Pennsylvania

Steve McCommas
Illinois State University

Virginia McDonough
Hope College

Phillip Meneely
Haverford College

Grant Mitmann
Montana Tech

Michelle Morek
Brescia University

Peggy Redshaw
Austin College

Lisa Sardinia
Pacific University

Todd Stanislav
Xavier University

Kathleen Truman
Franklin and Marshall College

Lisa Urry
Mills College

Carol Weaver
Union University

David Weber
Illinois State University

Problem Review Board

Colleen M. Belk
University of Minnesota, Duluth

Nancy Elwess
Plattsburg State University

James M. Ford
Stanford University School of Medicine

Richard C. Gethmann
University of Maryland, Baltimore County

Gail Gasparich
Towson University

Mary D. Healey
Springfield College

Lynne Hunter
University of Pennsylvania

Sidney R. Kushner
University of Georgia

Sandy Latourelle
Plattsburg State University

1

Genetics:
An Introduction

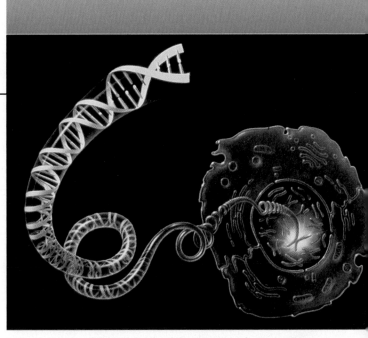

Stylized diagram of the relationship between DNA, chromosomes, and the cell.

iActivity

i DO YOU HAVE YOUR MOTHER'S EYES AND MUSICAL ability? Your father's nose and hand clasping pattern? How can you determine which of these traits you have inherited from your parents? Or what traits you might pass on to your children? The answers lie in the tiny segments of DNA known as genes and the way in which those genes are transmitted from parent to child. In this chapter, you will have a chance to review the basic concepts of genetics and to learn more about what geneticists do and how they work. Then, in the iActivity, you will discover some of the different methods geneticists might use to determine whether a specific trait is inherited.

PRINCIPAL POINTS

- Genetics often is divided into four major subdisciplines: transmission genetics, which deals with the transmission of genes from generation to generation; molecular genetics, which deals with the structure and function of genes at the molecular level; population genetics, which deals with heredity in groups of individuals for traits that are determined by one or a few genes; and quantitative genetics, which deals with heredity of traits in groups of individuals wherein the traits are determined by many genes simultaneously. Depending whether the goal is to obtain a fundamental understanding of genetic phenomena or exploit discoveries, genetic research is considered to be basic or applied, respectively.

- Eukaryotes are organisms in which the genetic material is located in a membrane-bound nucleus within the cells. The genetic material is distributed among several linear chromosomes. Prokaryotes, by contrast, lack a membrane-bound nucleus.

Welcome to the study of **genetics,** the science of heredity. Genetics is concerned primarily with understanding biological properties that are transmitted from parent to offspring. The subject matter of genetics includes heredity, the molecular nature of the genetic material, the ways in which genes (which determine the characteristics of organisms) control life functions, and the distribution and behavior of genes in populations.

Genetics is central to biology because gene activity underlies all life processes, from cell structure and function to reproduction. Learning what genes are, how genes are transmitted from generation to generation, how genes are expressed, and how gene expression is regulated is the focus of this book. Genetics is expanding so rapidly that it is not possible to describe everything we know about it between these covers. The important principles and concepts are presented carefully and thoroughly; readers who want to go further may consult the list of suggested readings at the end of the text or search for research papers via the Internet—for example, by searching the PubMed database supported by the National Library of Medicine, National Institutes of Health, at http://www.ncbi.nlm.nih.gov/PubMed/.

It is assumed that you understand the general features of genetics. This chapter presents a brief introduction to genetics to provide you with a framework for your subsequent studies of genes.

Classical and Modern Genetics

Humans recognized long ago that offspring tend to resemble their parents. Humans have also performed breeding experiments with animals and plants for centuries: classic genetic engineering, if you will. However, the principles of heredity were not understood until the mid-nineteenth century when Gregor Mendel analyzed quantitatively the results of crossing pea plants that varied in easily observable characteristics. The importance of his findings was not recognized in Mendel's lifetime, but when the principles of heredity were rediscovered at the turn of the twentieth century, it was realized that Mendel's work constituted the foundation of modern genetics.

Since the turn of the twentieth century, genetics has been an increasingly powerful tool for studying biological processes. An important approach used by many geneticists is to isolate mutants affecting a particular biological process and then, by comparing the mutants with normal strains, to obtain an understanding of the process. Such research has gone in many directions, such as analyzing heredity in populations, analyzing evolutionary processes, identifying the genes that control the steps in a process, mapping the genes involved, determining the products of the genes, and analyzing the molecular features of the genes, including the regulation of the genes' expression.

Research in genetics was revolutionized in 1972, when Paul Berg constructed the first recombinant DNA molecule in vitro and, in 1973, when Herbert Boyer and Stanley Cohen cloned a recombinant DNA molecule for the first time. Kary Mullis's development in 1986 of the polymerase chain reaction (PCR) to amplify specific segments of DNA spawned a second revolution. Recombinant DNA technology, PCR, and other molecular technologies are leading to an ever-increasing number of exciting discoveries that are furthering our knowledge of basic biological functions and will lead to improvements in the quality of human life. Already, the complete genomic DNA sequences have been determined for a number of organisms, including humans, thus heralding the exciting field of *genomics*. As scientists analyze the data, we can expect major contributions to our biological knowledge. For instance, we will know about every gene in the human genome: where they are in the genome, their sequences, and their regulatory sequences. Such knowledge undoubtedly will lead to a better understanding of human genetic diseases and contribute to their cures. We can imagine that the science-fiction scenario of each of us carrying our DNA genome sequence on a chip may become reality. However, knowledge about our genomes will raise social and ethical concerns that must be resolved carefully. For example, if a person has a genome sequence that includes a gene with the potential to cause a life-shortening disease, to what extent should that information be private? These questions notwithstanding, this is a very exciting time to be a student of genetics.

Basic Concepts of Genetics

In this section, we summarize some of the important concepts and processes of genetics so that they are clear to you as we develop the field of genetics in more detail.

DNA, Genes, and Chromosomes

The genetic material of all living organisms, both eukaryotes and prokaryotes, is **DNA (deoxyribonucleic acid).** (See Chapter 2.) DNA is also the genetic material of many viruses that infect prokaryotes and eukaryotes. A number of other viruses have **RNA (ribonucleic acid)** as genetic material. The full DNA sequence of an organism or the full DNA or RNA sequence of a virus is called its **genome.**

DNA is made up of two strands (also called chains); each strand consists of building blocks called **nucleotides,** each of which consists of the sugar deoxyribose, a phosphate group, and a **base.** The arrangement of the nucleotides in the chains forms a double helix (Figure 1.1). There are four bases in DNA: adenine (A), guanine (G), cytosine (C), and thymine (T). In RNA, uracil (U) occurs

Figure 1.1

DNA. (a) Three-dimensional molecular model of DNA.
(b) Stylized diagram of the DNA double helix.

a) Molecular model **b) Stylized diagram**

Axis of helix

O

P

H

Base pairs
(C and N)

|← 1 nm →|

A = T
T = A

C ≡ G
G ≡ C

T = A

0.34 nm

A = T

G ≡ C

C ≡ G

T = A

A = T

≡ G

G ≡ C

G ≡ C

A = T

3.4 nm

C

Base pairs

Backbones

Base pairs

Backbones

in place of thymine. The sequence of these bases within a strand determines the genetic information stored in that strand. **Genes,** which Mendel called factors, are specific sequences of nucleotides. These sequences describe the traits that are passed on from parent to offspring.

In the cell, the genetic material is organized into structures called **chromosomes** (See Chapter 2.) *Chromosome* means "colored body" and is so named because these threadlike structures are visible under the light microscope only after they are stained with dyes. Many, but not all, prokaryotes have a single, usually circular chromosome. In eukaryotes, the nucleus contains a number of linear chromosomes, each consisting of a single DNA molecule complexed with histone proteins. Different organisms have different numbers of chromosomes.

Outside the nucleus, eukaryotes have DNA in the mitochondria—(in animals and plants)—and in chloroplasts—(in plants). (See Chapter 23.) This extranuclear DNA differs in structure between species, but in many cases it is circular.

Transmission of Genetic Information

Genes determine such things as what we look like, how we grow, and what genetic diseases we may have. But how are

such traits passed on from parents to offspring? As mentioned earlier, Gregor Mendel was the first to determine the laws of heredity. (See Chapter 11.) Mendel performed a series of careful breeding experiments with the garden pea. In brief, he picked strains of peas that differed in particular characteristics (also called *traits*). For example, the pea seeds were either smooth or wrinkled, and the flowers were either purple or white (Figure 1.2). Then he made genetic crosses, counted the number of times the traits appeared in the progeny, and interpreted the results. (This basic experimental design is still used in gene transmission studies today.) From this kind of data, Mendel was able to conclude that inherited characteristics are determined by factors, or genes, and that each organism contains two copies of each gene: one inherited from its mother and one from its father. Mendel also hypothesized that alternative versions of genes, what we now call **alleles,** account for variations in inherited characters. For example, the gene for pea seed color exists in two versions: one for yellow (allele *Y*) and the other for green (allele *y*).

An organism having a pair of identical alleles for a trait is said to be **homozygous** for the trait (e.g., *YY* or *yy* for the seed color trait), whereas an organism having two different alleles for a gene is said to be **heterozygous** (e.g., *Yy* for the seed color trait). The complete genetic makeup of an organism is its **genotype.** All the observable properties an organism has are its **phenotype.** The genotype alone is not responsible for the phenotype; rather, the genotype interacts with the environment—the external environment and the internal environment—to produce the phenotype. Hence, two individuals with identical genotypes (identical twins, for example) are not necessarily identical in phenotype.

Mendel considered the factors that controlled the phenotypes he studied in abstract terms. He understood that individuals contain two (paired) factors and that each is contributed by one parent. Further, he correctly deduced that the factors segregated randomly into the gametes (Mendel's first law, the principle of

Figure 1.2

Example of easily distinguishable alternative traits: purple-flowered (left) vs. white-flowered (right) pea plants.

segregation) and that, during gamete formation, the two factors controlling one trait assorted independently from the two factors controlling another trait (Mendel's second law, the principle of independent assortment.) (See Chapter 11.) Recall the thinking behind the principle of segregation, as we look briefly at the following cross of two homozygotes:

P (parental) generation	YY	×	yy
Parental phenotype	yellow seeds		green seeds
Haploid gametes	Y		y

Fusion of the gametes produces the following:

F₁ generation	Yy
F₁ phenotype	Yellow seeds because Y allele is **dominant** to y allele; can also say that y is **recessive** to Y

When F₁ plants are crossed:

F₁ × F₁   Yy × Yy

F₁ gametes   Y   y   Y   y

The F₂ generation is produced by random fusion of the gametes as follows:

F₂ generation

	Y	y
Y	YY	Yy
y	Yy	yy

The result is a genotypic ratio of 1 YY : 2 Yy : 1 yy; because Y is dominant to y, the phenotypic ratio is 3 yellow : 1 green. We will learn more about Mendel's experiments and the principles of heredity he deduced in Chapter 11.

It was almost two decades after Mendel's death (in 1884) that the material basis of gene segregation from generation to generation was shown. In 1902, Walter

Sutton and Theodor Boveri proposed the chromosome theory of heredity, which posited that genes are on chromosomes and that the segregation patterns of genes can be explained entirely by the segregation of chromosomes from generation to generation. (See Chapter 12.) This is generally accepted today, with gene segregation paralleling the segregation of chromosomes in meiosis.

Expression of Genetic Information

Genes control all aspects of the life of an organism, encoding the products that are responsible for development, reproduction and so forth. The process by which a gene produces its product and the product carries out its function is known as **gene expression.** But what are the products? And how are these products related to specific traits?

In 1941, George Beadle and Edward Tatum, working with mutations of the fungus *Neurospora crassa* ("new-ROSS-pore-a crass-a") (orange bread mold) that imposed nutritional requirements on the organism, showed that there was a firm relationship between genes and enzymes (a special subset of proteins; see Chapter 4). That is, cells function as a result of many, interacting biochemical pathways, each step of which is catalyzed by an enzyme (or more than one enzyme). The scientists' studies of mutants elegantly showed that each step in the biochemical pathways they studied was under genetic control. Because the steps are catalyzed by enzymes, Beadle and Tatum proposed the **one gene–one enzyme hypothesis.** This hypothesis has been modified to the *one gene–one polypeptide hypothesis,* because not all proteins are enzymes and not all proteins consist of only one polypeptide.

Beadle and Tatum showed that genes provide the instructions for making specific proteins. This process consists of two main steps: transcription (see Chapter 5) and translation (see Chapter 6). In **transcription,** the DNA separates locally into single strands, and an enzyme—**RNA polymerase**—makes an RNA copy of one of the strands of the DNA molecule (Figure 1.3). In prokaryotes, three major classes of RNA are made: **messenger RNA (mRNA), transfer RNA (tRNA),** and **ribosomal RNA (rRNA).** Eukaryotes encode these three

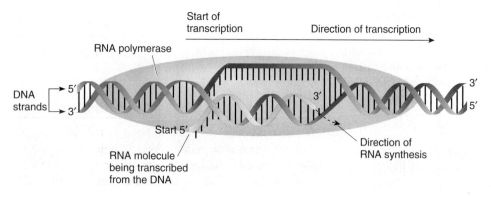

Start of transcription

Direction of transcription

RNA polymerase

DNA strands — 5′ 3′

3′

3′

5′

Start 5′

RNA molecule being transcribed from the DNA

Direction of RNA synthesis

Figure 1.3

Transcription. The DNA separates locally into single strands, and RNA polymerase makes an RNA copy of one of the DNA strands.

classes of RNA and **small nuclear RNA (snRNA).** The RNA molecules are essential for cell function in both prokaryotes and eukaryotes. The mRNAs specify the amino acid sequences of proteins, which are important structural and functional components of cells.

The process by which the base sequence information in mRNA is converted into an amino acid sequence in proteins is called **translation.** Translation occurs on **ribosomes** (Figure 1.4)—large complexes of rRNA molecules and proteins. The base sequence information that specifies the amino acid sequence of a protein is called the **genetic code.** Each amino acid is specified by a three-nucleotide sequence of the mRNA; this sequence is called a **codon.** Other DNA sequences specify where the RNA copy is to stop.

At any one time, only some of the genes in a particular genome are active and, in complex multicellular organisms, only a specific set of genes are active in each tissue and organ. How is all this accomplished? We do not have anywhere near a complete understanding yet, but at the general level, this is the result of a finely tuned array of gene regulation signals determining which genes are active and which are inactive.

Our comprehension of the regulation of gene expression began in bacteria, when, in 1961, François Jacob and Jacques Monod proposed an **operon** model to explain the coordinated regulation of expression of genes that encode enzymes needed to metabolize lactose (a sugar; see Chapter 19). In their model, a genetic switch is involved. When the switch is set one way, transcription of the genes encoding the enzymes is blocked, and when the switch is set in the other way, transcription of the genes can take place. This model has been shown to be a general description of many gene systems in bacteria and their viruses, as well as of some genes in the nematode worm, *Caenorhabditis elegans* ("see-no-rab-DYE-tiss elly-gans") (*C. elegans*).

Genetic switches also regulate gene expression in eukaryotes, but those switches mostly are different from, and typically more complex than, those identified in bacteria. (See Chapter 19.) Moreover, many mechanisms are eukaryote specific. Unquestionably, although much has been learned about gene regulation in eukaryotes, much remains to be learned.

Sources of Genetic Variation

Genetics has shown us that many of the differences between organisms are the result of differences in the genes they carry. These differences have resulted from the evolutionary process of **mutation** (a change in the genetic material; see Chapter 7), **recombination** (the exchange of genetic material between chromosomes; see Chapters 15 and 16), and **selection** (the favoring of particular combinations of genes in a given environment; see Chapter 24).

A mutation is any heritable alteration in the genetic material. Mutations can occur spontaneously or be induced experimentally by the use of mutagenic agents such as radiation or chemicals. As we will learn in Chapter 7, living cells have a variety of systems that repair damage to the genetic material, which is what mutagens do. Thus, mutations are changes in the genetic material that become permanently incorporated into the genetic code because they remain uncorrected by the repair systems.

Recombination, the exchange of genetic material between chromosomes, is brought about by enzymes that cut and rejoin DNA molecules. In eukaryotes, recombination is a common event in meiosis when the physical exchange of homologous chromosomes occurs by crossing-over. (See Chapter 12.) Uncommonly, crossing-over also may occur in mitosis. In prokaryotes, which do not undergo meiosis or mitosis, recombination still occurs by crossing-over whenever two DNA molecules with identical or nearly identical sequences become aligned.

Another source of genetic variation is selection. Selection was discovered by Charles Darwin in the mid-nineteenth century. The main consequence of selection is a change in the frequencies of genes affecting the trait or traits on which selection acts. As a result of selection, different genotypes contribute alleles to the next generation in a proportion that relates to the selective advantage they have, rather than in proportion to their number in the population.

These three mechanisms of mutation, recombination, and selection individually and collectively produce new genetic variations that are essential for evolution.

Figure 1.4

A ribosome, the organelle on which translation of mRNA (protein synthesis) takes place. Two views of three-dimensional models of the *E. coli* ribosome. The large subunit is red, and the small subunit is yellow.

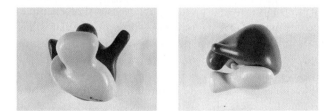

Geneticists and Genetic Research

The material presented in this book is the result of an incredible amount of research done by geneticists working in many areas of biology. Geneticists use the methods of science in their studies. As researchers, geneticists typically use the **hypothetico-deductive method of investigation.** This consists of making *observations*, forming *hypotheses* to explain the observations, making experimental *predictions* based on the hypotheses, and finally *testing* the predictions. The last step provides new observations, producing a cycle that leads to a refinement of the hypotheses and perhaps, eventually, to the establishment of a theory that attempts to explain a set of observations.

As in all other areas of scientific research, the exact path a research project will follow cannot be predicted precisely. In part, the unpredictability of research makes it exciting and motivates the scientists engaged in it. The discoveries that have revolutionized genetics were not planned; they developed out of research in which basic genetic principles were being examined. Barbara McClintock's work on the inheritance of patches of color on corn kernels is an excellent example. (See Chapter 7.) After accumulating a large amount of data from genetic crosses, she hypothesized that the appearance of colored patches was the result of the movement (transposition) of a DNA segment from one place to another in the genome. Only many years later were these DNA segments—called *transposons* or *transposable elements*—isolated and characterized in detail. (A more complete discussion of this discovery and of Barbara McClintock's life is presented in Chapter 7.) We know now that transposons are ubiquitous, playing a role not only in the evolution of species, but also in some human diseases.

The Subdisciplines of Genetics

Geneticists often divide genetics into four major subdisciplines. **Transmission genetics** (sometimes called classical genetics) is the subdiscipline dealing with how genes and genetic traits are transmitted from generation to generation and how genes recombine. Analyzing the pattern of trait transmission in a human pedigree or in crosses of experimental organisms is an example of a transmission genetics study. **Molecular genetics** is the subdiscipline dealing with the molecular structure and function of genes. Analyzing the molecular events involved in gene expression is an example of a molecular genetics study. **Population genetics** is the subdiscipline that studies heredity in groups of individuals for traits that are determined by one or only a few genes. Analyzing the frequency of a disease-causing gene in the human population is an example of a population genetics study. **Quantitative genetics** also considers the heredity of traits in groups of individuals, but the traits of concern are determined by many genes simultaneously. Analyzing the fruit weight and crop yield in agricultural plants are examples of quantitative genetics studies. Although these subdisciplines help us think about genes from different perspectives, there are no sharp boundaries between them. Increasingly, for example, population and quantitative geneticists analyze

molecular data to determine gene frequencies in large groups. Historically, transmission genetics developed first, followed by population genetics and quantitative genetics, and then molecular genetics.

Through the products they encode, genes influence all aspects of an organism's life. Understanding transmission genetics, population genetics, and quantitative genetics will help you understand such things as population biology, ecology, evolution, and animal behavior. Similarly, understanding molecular genetics is useful when you study such topics as neurobiology, cell biology, developmental biology, animal physiology, plant physiology, immunology, and, of course, the structure and function of genomes.

Basic and Applied Research

Genetics research, and research in general, may be either basic or applied. In **basic research,** experiments are done to gain an understanding of fundamental phenomena, whether or not the knowledge gained leads to any immediate applications. In **applied research,** experiments are done with an eye toward overcoming specific problems in society or exploiting discoveries. Basic research was responsible for most of the facts we discuss in this book. For example, we know how the expression of many prokaryotic and eukaryotic genes is regulated as a result of basic research on model organisms such as the bacterium *Escherichia coli* (*E. coli*), the yeast *Saccharomyces cerevisiae* ("sack-a-row-MY-seas serry-VEE-see-eye"), and the fruit fly *Drosophila melanogaster* ("dra-SOFF-ee-la muh-LANO-gas-ter"). The knowledge obtained from basic research is used largely to fuel more basic research.

Applied research is done with different goals in mind. In agriculture, applied genetics has contributed significantly to improvements in animals bred for food (such as reducing the amount of fat in beef and pork) and in crop plants (such as increasing the amount of protein in soybeans). A number of diseases are caused by genetic defects, and great strides are being made in understanding the molecular bases of some of those diseases. For example, drawing on knowledge gained from basic research, applied genetic research involves developing rapid diagnostic tests for genetic diseases and producing new pharmaceuticals for treating diseases.

There is no sharp dividing line between basic and applied research. Indeed, in both areas, researchers use similar techniques and depend on the accumulated body of information when building hypotheses. For example, **recombinant DNA technology**—procedures that allow molecular biologists to splice a DNA fragment from one organism into DNA from another organism and to clone (make many identical copies of) the new recombinant DNA molecule—has had a profound effect on both basic and applied research. (See Chapters 8, 9, and 10.) Many biotechnology companies owe their existence to recombinant DNA technology as they seek to clone and manipulate genes in developing their products. In the area of plant breeding, recombinant DNA technology has made it easier to introduce traits such as disease resistance from noncultivated species into cultivated species. Such crop improvement traditionally was achieved by using conventional breeding experiments. In animal breeding, recombinant DNA technology is being used in the beef, dairy, and poultry industries, for example, to increase the amount of lean meat, the amount of milk, and the number of eggs. In medicine, the results are equally impressive. Recombinant DNA technology is being used to produce many antibiotics, hormones, and other medically important agents such as clotting factor and human insulin (marketed under the name Humulin; Figure 1.5) and to diagnose and treat a number of human genetic diseases. In forensics, *DNA typing* (also called *DNA fingerprinting* or *DNA profiling*) is being used in paternity cases, criminal cases, and anthropological studies. In short, the science of genetics is currently in an exciting and dramatic growth phase, and there is still much to discover.

>
> Go to the iActivity *A Question of Taste* on your CD-ROM, and investigate different ways to determine whether the ability to taste a chemical called phenylthiocarbamide (PTC) is inherited.

KEYNOTE

> Genetics can be divided into four major subdisciplines: transmission genetics, molecular genetics, population genetics, and quantitative genetics. Depending on whether the goal is to obtain a fundamental understanding of genetic phenomena or to exploit discoveries, genetic research is considered to be basic or applied, respectively.

Figure 1.5

Example of a product developed as a result of recombinant DNA technology. Humulin, human insulin for insulin-dependent diabetics.

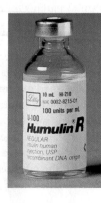

Genetic Databases and Maps

In this section, we talk about two important resources for genetic research: genetic databases and genetic maps. Genetic databases have become much more sophisticated and expansive as computer analysis tools have been developed and Internet access to databases has become routine. Constructing genetic maps has been part of genetic analysis for about 100 years.

Genetic Databases. The amount of information about genetics has increased dramatically. No longer can we learn everything about genetics by going to a college or university library; the computer now plays a major role. For example, a useful way to look for genetic information through the Internet is by entering key words into search engines such as Google (http://www.google.com). Typically, a vast number of hits are listed, some useful and some not. Therefore, it is difficult to summarize all genetic databases that are useful in this section. You must search for yourself and be critical about what you find. However, we can consider a set of important and extremely useful genetic databases at the National Center for Biotechnology Information (NCBI) website (http://www.ncbi.nlm.nih.gov). NCBI was created in 1988 as a national resource for molecular biology information. Its role is to "create public databases, conduct research in computational biology, develop software tools for analyzing genome data, and disseminate biomedical information—all for the better understanding of molecular processes affecting human health and disease."

Some of the search tools available at the NCBI site are as follows:

- PubMed is used to access literature citations and abstracts and provides links to sites with electronic versions of research journal articles, which sometimes can be viewed or you must pay a one-time fee or obtain a free subscription. You search PubMed by entering terms, author names or journal titles. It is highly recommended that you use PubMed to find research articles on genetic topics that interest you.

- OMIM (Online Mendelian Inheritance in Man) is a database of human genes and genetic disorders authored and edited by Dr. Victor A. McKusick and his colleagues. You search OMIM by entering terms in a text box search window; the result is a list of linked pages, each with specific OMIM entry number. The pages have detailed information about the gene or genetic disorder specified in the original search, including genetic, biochemical, and molecular data, along with an up-to-date list of references. We refer to OMIM entries throughout the book each time we discuss a human gene or genetic disease and give the OMIM entry number.

- BLAST is a tool used to compare a nucleotide sequence or protein sequence with all sequences in the database to find possible matches. This is useful, for example, if you have sequenced a new gene and want to find out whether anything similar has been sequenced previously. Moreover, genes with related functions may be listed in the databases, allowing you to focus your research on the function of the gene you are studying.

- GenBank is the National Institutes of Health (NIH) genetic-sequence database. This database is an annotated collection of all publicly available DNA sequences (more than 39 billion bases as of April 2004). You search GenBank by entering terms in the search window. For example, if you are interested in the human disease cystic fibrosis, enter the term "cystic fibrosis" into the search window, and you will find all sequences that have been entered into GenBank that include those two words in the annotations.

- Entrez is a system for searching several linked databases. The particular database is chosen from a pull-down menu. The databases include PubMed; Nucleotide, for the GenBank DNA and RNA sequences database; Protein, for protein amino acid sequences; Structure, for three-dimensional macromolecular structures; Genome, for complete genome assemblies; OMIM, the Online Mendelian Inheritance in Man human gene database; and PopSet, population study datasets. The database can be selected from the hot links, or a pull-down menu choice on the main Entrez page will guide your search terms appropriately. For example, if you are interested in nucleotide sequences related to the human disease cystic fibrosis, you would select "Nucleotide" in the pull-down menu and enter "cystic fibrosis" in the search window. A list of relevant sequence entries will be returned.

A powerful feature of the NCBI databases is that they are linked, enabling the user to move smoothly between them and hence integrate the knowledge obtained in each of them. For example, a literature citation found in PubMed will have links to sequences in nucleotide and protein databases.

Genetic Maps.   Since 1902, much effort has been made to construct **genetic maps** (Figure 1.6) for the commonly used experimental organisms in genetics. Genetic maps are like road maps that show the relative locations of towns along a road; that is, they show the arrangements of genes along the chromosomes and the genetic distances between the genes. The position of a gene on the map (and of the chromosome to which the map relates) is called a **locus** or **gene locus**. The genetic distances between genes on the same chromosome are calculated from the results of genetic crosses by counting the frequency of recombination—that is, the percentage of the time among the progeny that the arrangement of

Figure 1.6

Example of a genetic map, illustrating some of the genes on chromosome 2 of the fruit fly, *Drosophila melanogaster*. The numerical values represent the positions of the genes from the chromosome end (top) measured in map units. (From *Principles of Genetics* by Robert Tamarin. Copyright © 1996. Reproduced with permission of The McGraw-Hill Companies.)

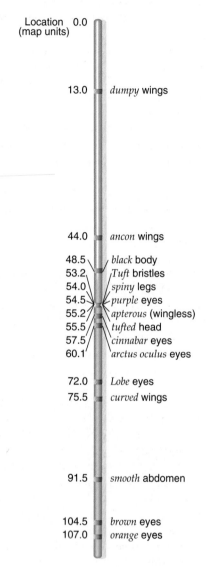

alleles in the two original parents switches (i.e., recombines; see Chapter 15). The unit of genetic distance is the **map unit.**

The goal of constructing genetic maps has been to obtain an understanding of the organization of genes along the chromosomes (e.g., to inform us whether genes with related functions are on the same chromosome and, if they are, whether they are close to each other). Genetic maps have also proved very useful in efforts to clone and sequence particular genes of interest and, more recently, as part of genome projects, the efforts to obtain the complete sequences of genomes.

Organisms for Genetics Research

The principles of heredity were first established in the nineteenth century by Gregor Mendel's experiments with the garden pea. Since Mendel's time, many organisms have been used in genetic experiments. Among the qualities that make an organism a particularly good model for genetic experimentation are the following:

- The genetic history of the organism involved is well known.
- The organism has a short life cycle so that a large number of generations occur within a short time. In this way, researchers can obtain data readily over many generations. Fruit flies, for example, produce offspring in 10 to 14 days.
- A mating produces a large number of offspring.
- The organism should be easy to handle. For example, hundreds of fruit flies can be kept easily in small bottles.
- Most importantly, genetic variation must exist between the individuals in the population so that the inheritance of traits can be studied. The more marked the variations, the easier the genetic analysis.

Both eukaryotes and prokaryotes are used in genetics research. **Eukaryotes** (meaning "true nucleus") are organisms with cells within which the genetic material (DNA) is located in the **nucleus** (a discrete structure bounded by a nuclear envelope). Eukaryotes can be unicellular or multicellular. In genetics today, a great deal of research is done with six eukaryotes (Figure 1.7a–f): *Saccharomyces cerevisiae* (budding yeast), *Drosophila melanogaster* (fruit fly), *Caenorhabditis elegans* (a nematode worm), *Arabidopsis thaliana* ("a-rab-ee-DOP-sis thal-ee-AH-na," a small weed of the mustard family), *Mus musculus* ("muss MUSS-cue-lus," a mouse), and *Homo sapiens* ("homo SAY-pee-ens," human). Humans are included not because they meet the criteria for an organism well suited for genetic experimentation, but because ultimately we want to understand as much as we can about human genes and their expression. With this understanding we will be able to combat genetic diseases and gain fundamental knowledge about our species' development and evolution.

Over the years, research with the following seven eukaryotes has also contributed significantly to our understanding of genetics (Figure 1.7g–m): *Neurospora crassa* (orange bread mold), *Tetrahymena* ("tetra-hi-me-na," a protozoan), *Paramecium* ("para-ME-see-um," a protozoan), *Chlamydomonas reinhardtii* ("clammy-da-MOAN-as rhine-HEART-ee-eye," a green alga), *Pisum sativum* ("PEA-zum sa-TIE-vum," garden pea), *Zea mays* (corn), and *Gallus* (chicken). Of these, *Tetrahymena, Paramecium, Chlamydomonas,* and *Saccharomyces* are unicellular organisms, and the rest are multicellular.

You probably learned about many features of eukaryotic cells in your introductory biology course. Figure 1.8 shows a generalized higher plant cell and a generalized animal cell. Surrounding the cytoplasm of both plant cells and animal cells is the *plasma membrane*. Plant cells, but not animal cells, have a rigid cell wall outside the plasma membrane. The nucleus of eukaryotic cells contains DNA complexed with proteins and organized into a number of linear structures called chromosomes. The nucleus is separated from the rest of the cell—the cytoplasm and associated organelles—by the double membrane of the nuclear envelope. The membrane is selectively permeable and has pores about 20 to 80 (nm = nanometer = 10^{-9} meter) in diameter that allow certain materials to move between the nucleus and the cytoplasm. For example, ribosomes, which function in the cytoplasm to synthesize proteins, are assembled within the nucleus in the **nucleolus** and reach the cytoplasm through the pores.

The cytoplasm of eukaryotic cells contains many different materials and organelles. Of special interest to geneticists are the *centrioles,* the *endoplasmic reticulum* (ER), *ribosomes, mitochondria,* and *chloroplasts.* Centrioles (also called basal bodies) are found in the cytoplasm of nearly all animal cells (see Figure 1.8b), but not in most plant cells. In animal cells, a pair of centrioles is located at the center of the centrosome, a region of undifferentiated cytoplasm which organizes the spindle fibers that are involved in chromosome segregation in mitosis and meiosis (discussed in Chapter 12.)

The ER is a double-membrane system, continuous with the nuclear envelope, that runs throughout the cell. Rough ER has ribosomes attached to it, giving it a rough appearance, whereas smooth ER does not. Ribosomes bound to rough ER synthesize proteins to be secreted by the cell or to be localized in the cell membrane or particular organelles within the cell. The synthesis of proteins other than those distributed via the ER is performed by ribosomes that are free in the cytoplasm.

Mitochondria (singular: mitochondrion; see Figure 1.8) are large organelles surrounded by a double membrane—the inner membrane is highly convoluted. Mitochondria play a crucial role in processing energy for the cell. They also contain DNA that encodes some of the proteins that function in the mitochondrion and some components of the mitochondrial protein synthesis machinery.

Many plant cells contain chloroplasts, large, triple-membraned, chlorophyll-containing organelles involved in photosynthesis. (See Figure 1.8a.) Chloroplasts also contain DNA that encodes some of the proteins that function in the chloroplast and some components of the chloroplast protein synthesis machinery.

In contrast to eukaryotes, **prokaryotes** (meaning "prenuclear") do not have a nuclear envelope surrounding their DNA (Figure 1.9); this is the major distinguishing feature of prokaryotes. Included in the prokaryotes are all the bacteria, which are spherical, rod-shaped, or spiral-shaped organisms; most are single-celled, although a few are multicellular and filamentous. The shape of a bacterium is maintained by a rigid cell wall located outside the cell membrane. Prokaryotes are divided into two evolutionarily distinct

Figure 1.7

Eukaryotic organisms that have contributed significantly to our knowledge of genetics. (a) *Saccharomyces cerevisiae* (a budding yeast). (b) *Drosophila melanogaster* (fruit fly). (c) *Caenorhabditis elegans* (a nematode). (d) *Arabidopsis thaliana* (Thale cress, a member of the mustard family). (e) *Mus musculus* (a mouse). (f) *Homo sapiens* (human). (g) *Neurospora crassa* (orange bread mold). (h) *Tetrahymena* (a protozoan). (i) *Paramecium* (a protozoan). (j) *Chlamydomonas reinhardtii* (a green alga). (k) *Pisum sativum* (a garden pea). (l) *Zea mays* (corn). (m) *Gallus* (chicken).

a)

b)

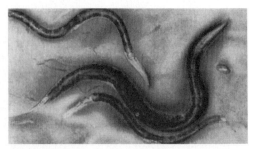

c)

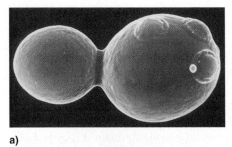

d)

e)

f)

g)

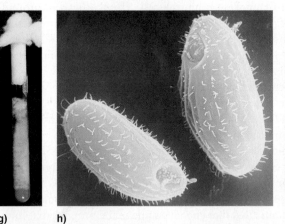

h)

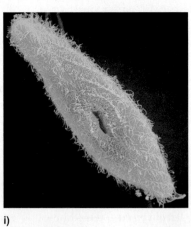

i)

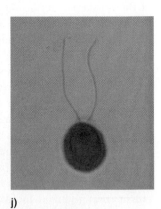

j)

k)

l)

m)

Figure 1.8

Eukaryotic cells. Cutaway diagrams of (**a**) a generalized higher plant cell and (**b**) a generalized animal cell, showing the main organizational features and the principal organelles in each.

a) Plant cell

b) Animal cell

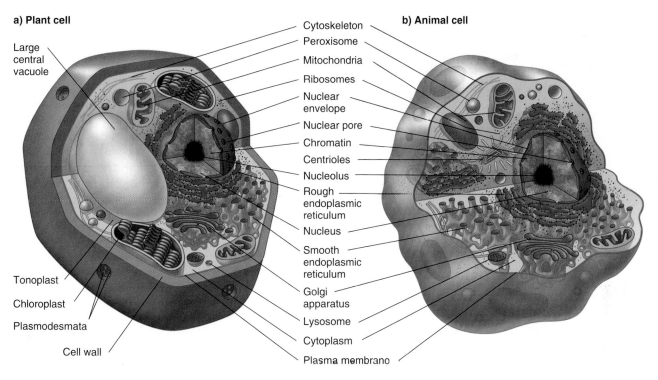

Large central vacuole

Cytoskeleton
Peroxisome
Mitochondria
Ribosomes
Nuclear envelope
Nuclear pore
Chromatin
Centrioles
Nucleolus
Rough endoplasmic reticulum
Nucleus
Smooth endoplasmic reticulum
Golgi apparatus
Lysosome
Cytoplasm
Plasma membrane

Tonoplast

Chloroplast

Plasmodesmata

Cell wall

Figure 1.9

Cutaway diagram of a generalized prokaryotic cell.

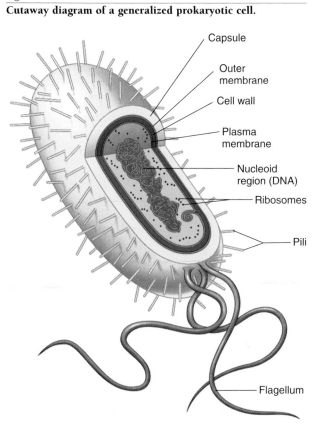

Capsule

Outer membrane

Cell wall

Plasma membrane

Nucleoid region (DNA)

Ribosomes

Pili

Flagellum

groups: the Bacteria and the Archaea. The Bacteria are the common varieties found in living organisms (naturally or by infection), in soil, and in water. Archaea are the prokaryotes found often in much more inhospitable conditions, such as hot springs, salt marshes, methane-rich marshes, or the ocean depths, where bacteria do not thrive. Archaea are also found under typical conditions, such as water and soil. Bacteria generally vary in size from about 100 nm in diameter to 10 μm in diameter and 60 μm in length. The largest species, the spherical *Thiomargarita namibiensis*, can reach ¾ mm in diameter at which point it is visible to the naked eye (about the size of a *Drosophila* eye).

In most cases, the prokaryotes studied in genetics are members of the Bacteria group. The most intensely studied is *E. coli* (Figure 1.10), a rod-shaped bacterium common in human intestines. Studies of *E. coli* have significantly advanced our understanding of the regulation of gene expression and the development of molecular biology. *E. coli* is also used extensively in recombinant DNA experiments.

K E Y N O T E

Eukaryotes are organisms that have cells in which the genetic material is located in a membrane-bound nucleus. The genetic material is distributed among several linear chromosomes. Prokaryotes, by contrast, lack a membrane-bound nucleus.

Figure 1.10

Colorized scanning electron micrograph of *Escherichia coli*, a rod-shaped eubacterium common in human intestines.

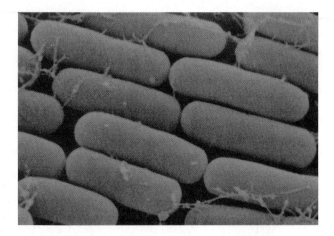

Summary

This chapter introduced the subject of genetics briefly. Genetics often is divided into four main branches: transmission genetics, molecular genetics, population genetics, and quantitative genetics. Geneticists investigate all aspects of genes, and, depending on whether the goal is to obtain fundamental knowledge or to exploit discoveries, genetic research is considered to be basic or applied, respectively.

This chapter also introduced some of the organisms that have been fruitful subjects of genetic studies. As the simplest cellular organisms, prokaryotes lack a membrane-bound nucleus and have a single chromosome located in a nucleoid region. Eukaryotic organisms may be single-celled or multicellular and have several chromosomes located within a membrane-bound nucleus.

2

DNA: The Genetic Material

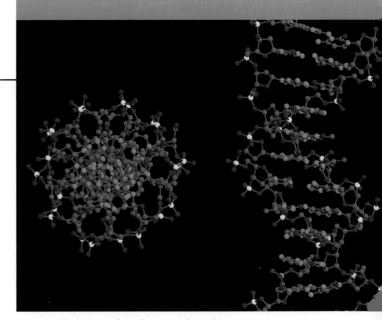

DNA double helix. Left, end view; right, side view.

PRINCIPAL POINTS

- Organisms contain genetic material that governs an individual's characteristics and that is transferred from parent to progeny.

- Deoxyribonucleic acid (DNA) is the genetic material of all living organisms and some viruses. Ribonucleic acid (RNA) is the genetic material only of certain viruses.

- DNA and RNA are macromolecules composed of smaller building blocks called nucleotides. Each nucleotide consists of a five-carbon sugar (deoxyribose in DNA, ribose in RNA) to which are attached a nitrogenous base and a phosphate group. In DNA, the four possible bases are adenine, guanine, cytosine, and thymine; in RNA, the four possible bases are adenine, guanine, cytosine, and uracil.

- According to Watson and Crick's model, the DNA molecule consists of two polynucleotide (polymers of nucleotides) chains joined by hydrogen bonds between pairs of bases (adenine [A] and thymine [T]; and guanine [G] and cytosine [C]) in a double helix.

- The genetic material of viruses may be linear or circular double-stranded DNA, single-stranded DNA, double-

stranded RNA, or single-stranded RNA, depending on the virus. The genomes of some viruses are organized into a single chromosome, whereas others have a segmented genome.

- The genetic material of prokaryotes is double-stranded DNA localized into one or a few chromosomes.

- A bacterial chromosome is compacted by supercoiling of the DNA helix to produce looped domains.

- The complete set of metaphase chromosomes in a eukaryotic cell is called its karyotype.

- The nuclear chromosomes of eukaryotes are complexes of DNA and histone and nonhistone chromosomal proteins. Each chromatid of a chromosome consists of one linear, unbroken, double-stranded DNA molecule running throughout its length; the DNA is variously coiled and folded. The histones are constant from cell to cell within an organism, whereas the nonhistones vary significantly between cell types.

- The large amount of DNA present in the eukaryotic chromosome is compacted by its association with histones in nucleosomes and by higher levels of folding of the nucleosomes into chromatin fibers. Highly condensed chromosomes consist of a large number of looped domains of 30-nm chromatin fibers spirally attached to a protein scaffold. The more condensed a region of a chromosome is, the less likely it is that the genes in that region will be active.

- The centromere region of each eukaryotic chromosome is responsible for the accurate segregation of the replicated chromosome to the daughter cells during mitosis and meiosis. The DNA sequences of centromeres vary a little within an organism and extensively between organisms.

- The ends of eukaryotic chromosomes—the telomeres—often are associated with the nuclear envelope. Telomeres consist of simple, short, tandemly repeated sequences that are species specific.

- Prokaryotic genomes consist mostly of unique DNA sequences. They have only a few repeated sequences and genes. Eukaryotes have both unique and repetitive sequences in the genome. The spectrum of complexity of repetitive DNA sequences among eukaryotes is extensive.

i IMAGINE THAT YOU ARE HANDED A SEALED BLACK box and are told that it contains the secret of life. Determining the chemical composition, molecular structure, and function of the thing inside the box will allow you to save lives, feed the hungry, solve crimes, and even create new life-forms. What's inside the box? What tools and techniques could you use to find out?

In this chapter, you will discover how scientists identified the contents of this "black box" and, in doing so, unraveled the "secret of life." Later in the chapter, you can apply what you've learned by trying the iActivity, in which you use many of the same tools and techniques to determine the genetic nature of a virus that is ravaging rice plants in Asia.

Simple observation shows that a lot of variation exists between individuals of a given species. For example, there are numerous breeds of dogs, including Bernese mountain dogs, dalmatians, pointers, dachshunds, Pomeranians, and so on. Even though all dogs belong to the same species, *Canis familiaris*, each breed has characteristic sizes, shapes, colors, and behaviors. Similarly, individual humans vary in eye color, height, skin color, and hair color, even though all humans belong to the species *Homo sapiens*. The differences between individuals within and among species are mainly the result of differences in the DNA sequences that constitute their genes. The genetic information coded in DNA is responsible for determining the structure, function, and development of the cell and the organism.

In the next several chapters, we explore the molecular structure and function of genetic material—both **deoxyribonucleic acid (DNA)** and **ribonucleic acid (RNA)**—and examine the molecular mechanisms by which genetic information is transmitted from generation to generation. You will see exactly what the genetic message is, and you will learn how DNA stores and expresses that message. We begin by recounting how scientists discovered the nature and structure of the genetic material. These discoveries led to an explosion of knowledge about the molecular aspects of biology.

The Search for the Genetic Material

Long before DNA and RNA were known to carry genetic information, scientists realized that living organisms contain some substance—a genetic material—that is responsible for the characteristics that are passed on from parent to child. Geneticists knew that the material responsible for hereditary information must have three principal characteristics:

1. It must contain, in a stable form, *the information* about an organism's cell structure, function, development, and reproduction.

2. It must *replicate accurately*, so that progeny cells have the same genetic information as the parental cell.

3. It must be capable of *change*. Without change, organisms would be incapable of variation and adaptation, and evolution could not occur.

The Swiss biochemist Friedrich Miescher is credited with the discovery, in 1869, of nucleic acid. He isolated a substance from cells of pus in used bandages. At first he believed the substance to be protein, but chemical tests indicated that it contained carbon, hydrogen, oxygen, nitrogen, and phosphorus, the last of which was not known to be a component of proteins. Miescher then searched for the same substance in other sources. He found it in the nucleus of all the samples he studied and, therefore, he called the substance *nuclein*.

In the early 1900s, experiments showed that **chromosomes**—the threadlike structures found in nuclei—are carriers of hereditary information. Chemical analysis over the next 40 years revealed that chromosomes are composed of protein and **nucleic acids,** a class of compounds that includes DNA and RNA. At first, many scientists believed that the genetic material must be protein. They reasoned that proteins have a great capacity for storing information because they are composed of 20 different amino acids. By contrast, DNA, with its four nucleotides, was thought to be too simple a molecule to account for the variation found in living organisms. However, beginning in the late 1920s, a series of experiments led to the definitive identification of DNA as the genetic material.

Figure 2.1

Electron micrograph of the bacterium *Streptococcus pneumoniae*.

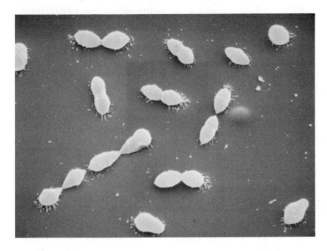

Griffith's Transformation Experiment

In 1928, Frederick Griffith, a British medical officer, was working with *Streptococcus pneumoniae* (also called pneumococcus), a bacterium that causes pneumonia (Figure 2.1). Griffith used two strains of the bacterium:

the *S* strain, which produces smooth, shiny colonies and is highly infectious (virulent); and the *R* strain, which produces rough colonies and is harmless (avirulent). Although this distinction was not known at the time, each cell of the *S* strain is surrounded by a polysaccharide (complex sugar) coat, or capsule, that results in the strain's infectious properties and is responsible for its smooth, shiny appearance. The *R* strain, which is a *mutation* of the *S* strain, lacks the polysaccharide coat.

Several variations of the *S* strain have differences in the chemical composition of the polysaccharide coat. Griffith worked with two varieties, known as the *IIS* and *IIIS* strains. Occasionally, *S*-type cells can mutate, or change, into *R*-type cells, and vice versa. The mutations are type specific; that is, if a *IIS* cell mutates into an *R* cell, then that *R* cell can mutate back only into a *IIS* cell, not a *IIIS* cell.

Griffith injected mice with different strains of the bacterium and observed whether the mice remained healthy or died (Figure 2.2). When mice were injected with *IIR* bacteria (*R* bacteria derived by mutation from *IIS* bacteria), the *IIR* bacteria did not affect the mice. When mice were injected with living *IIIS* bacteria, the mice died, and living *IIIS* bacteria could be isolated from their blood. However, if the *IIIS* bacteria were killed by heat

Figure 2.2

Griffith's transformation experiment. Mice injected with type *IIIS* pneumococcus died, whereas mice injected with either type *IIR* or heat-killed type *IIIS* bacteria survived. When injected with a mixture of living type *IIR* and heat-killed type *IIIS* bacteria, however, the mice died.

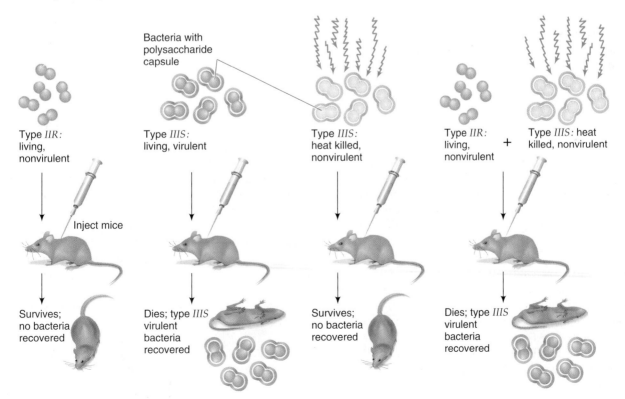

before injection, the mice survived. These experiments showed that the bacteria had to be alive and had to have the polysaccharide coat to be infectious and kill the mice.

In his key experiment, Griffith injected mice with a mixture of living *IIR* bacteria and heat-killed *IIIS* bacteria. Surprisingly, the mice died, and living *S* bacteria were present in the blood. These bacteria were all of type *IIIS* and therefore could not have arisen by mutation of the *R* bacteria, because mutation would have produced *IIS* colonies. Griffith concluded that some *IIR* bacteria had somehow been transformed into smooth, infectious *IIIS* cells by interaction with the dead *IIIS* cells. Griffith believed that the unknown agent responsible for the change in genetic material was a protein. He called the agent the **transforming principle.** (See Chapter 18 for a discussion of bacterial transformation.)

Avery's Transformation Experiments

In the 1930s and 1940s, American biologist Oswald T. Avery, along with his colleagues Colin M. MacLeod and Maclyn McCarty, followed up on Griffith's experiments.

animation

a DNA as Genetic Material: The Avery Experiment

The scientists tried to identify the transforming principle by studying the transformation of *R*-type bacteria to *S*-type bacteria in the test tube.

They lysed (broke open) *IIIS* cells with a detergent and used a centrifuge to separate the cellular components—the cell extract—from the cellular debris. They incubated the extract with a culture of living *IIR* bacteria and then plated cells on a medium in a Petri dish. Colonies of *IIIS* bacteria

appeared on the plate, showing that the extract contained the transforming principle. Avery and his colleagues knew that one of the macromolecular components in the extract—polysaccharides, proteins, RNA, or DNA—must be the transforming principle. To determine which, they carried out a series of enzyme treatments to degrade components one at a time, testing after each step to see if transformation still occurred. After degrading polysaccharides and then proteins, transformation of *IIR* to *IIIS* still occurred, indicating that the transforming principle was either RNA or DNA. Avery and his colleagues then used specific **nucleases**—enzymes that degrade nucleic acids—to determine whether DNA or RNA was the transforming principle (Figure 2.3). When they treated the nucleic acids with **ribonuclease (RNase),** which degrades RNA, but not DNA, the transforming activity was still present. However, when they used **deoxyribonuclease (DNase),** which degrades only DNA, no transformation followed. These results strongly suggested that DNA was the genetic material. Although Avery's work was important, it was criticized at the time because the nucleic acids isolated from the bacteria were contaminated by proteins and a number of scientists at the time believed strongly that protein was the genetic material.

The Hershey–Chase Bacteriophage Experiments

In 1953, Alfred D. Hershey and Martha Chase published a paper that provided more evidence that DNA was the genetic material. They were studying a bacteriophage called

animation

a DNA as Genetic Material: The Hershey–Chase Experiment

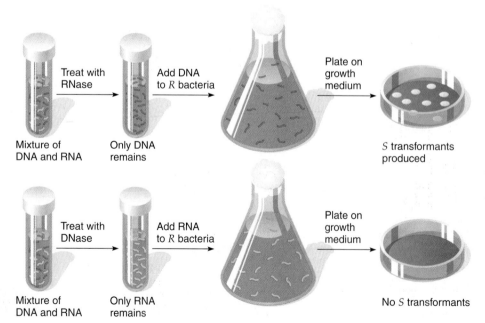

Treat with RNase → Add DNA to *R* bacteria → Plate on growth medium → *S* transformants produced

Mixture of DNA and RNA → Only DNA remains

Treat with DNase → Add RNA to *R* bacteria → Plate on growth medium → No *S* transformants

Mixture of DNA and RNA → Only RNA remains

Figure 2.3

Experiment which showed that DNA, not RNA, was the transforming principle. When the nucleic acid mixture of DNA and RNA was treated with ribonuclease (RNase), *S* transformants still resulted. However, when the DNA and RNA mixture was treated with deoxyribonuclease (DNase), no *S* transformants resulted. (*Note:* *R* colonies resulting from untransformed cells were present on both plates; they have been omitted from the drawings.)

Figure 2.4

Electron micrograph and diagram of bacteriophage T2 (1 nm = 10⁻⁹m).

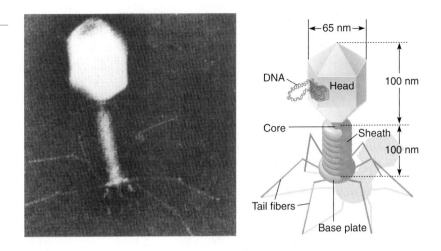

T2 (Figure 2.4). **Bacteriophages** (also called **phages**) are viruses that attack bacteria. Like all viruses, the T2 phage cannot reproduce on its own. Instead, it replicates by invading a living cell and using the cell's molecular machinery to make more viruses. The host cell then ruptures, releasing 100–200 progeny phages. This process is known as the **lytic cycle** (Figure 2.5). Recall that T2 replicates by invading an *E. coli* cell and using its molecular machinery to make progeny phages. The host cell then lyses (breaks open), releasing 100–200 progeny phages. The suspension of released progeny phages is called a **phage lysate.**

Hershey and Chase knew that T2 consisted of only DNA and protein and believed that the DNA was the genetic material. To prove this, they grew cells of *E. coli* in media containing either a radioactive isotope of phosphorus (^{32}P) or a radioactive isotope of sulfur (^{35}S) (Figure 2.6a). They used these isotopes because DNA

Figure 2.5

Lytic life cycle of a virulent phage, such as T2.

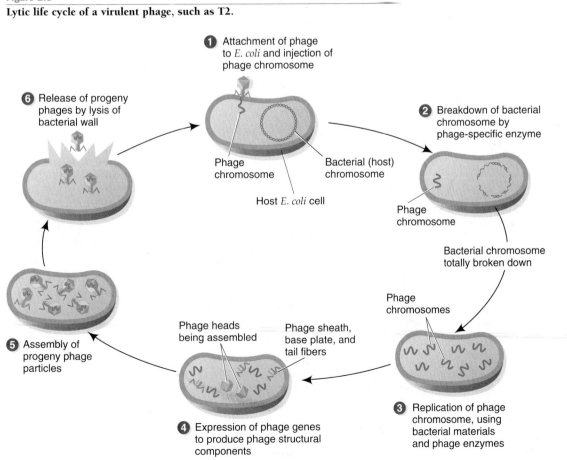

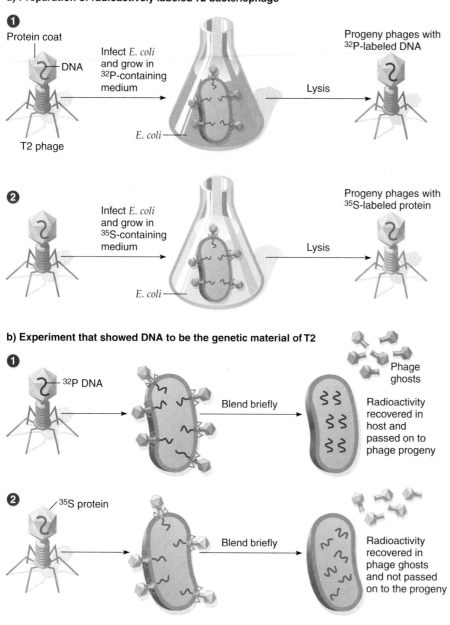

a) Preparation of radioactively labeled T2 bacteriophage

1

Protein coat

DNA

T2 phage

Infect *E. coli* and grow in ^{32}P-containing medium

E. coli

Lysis

Progeny phages with ^{32}P-labeled DNA

2

Infect *E. coli* and grow in ^{35}S-containing medium

E. coli

Lysis

Progeny phages with ^{35}S-labeled protein

b) Experiment that showed DNA to be the genetic material of T2

1

^{32}P DNA

Blend briefly

Phage ghosts

Radioactivity recovered in host and passed on to phage progeny

2

^{35}S protein

Blend briefly

Radioactivity recovered in phage ghosts and not passed on to the progeny

Figure 2.6

Hershey–Chase experiment.
(a) The production of T2 phages with (1) ^{32}P-labeled DNA or (2) ^{35}S-labeled protein. **(b)** The experimental evidence showing that DNA is the genetic material in T2: (1) The ^{32}P is found within the bacteria and appears in progeny phages, whereas (2) the ^{35}S is not found within the bacteria and is released with the phage ghosts.

contains phosphorus, but no sulfur, and protein contains sulfur, but no phosphorus. They infected the bacteria with T2 and collected the progeny phages. At this point, Hershey and Chase had two batches of T2; one had the proteins radioactively labeled with ^{35}S, and the other had the DNA labeled with ^{32}P.

They then infected *E. coli* with the two types of radioactively labeled T2 (Figure 2.6b). When the infecting phage was ^{32}P labeled, most of the radioactivity was found within the bacteria soon after infection. Very little was found in protein parts of the phage (the *phage ghosts*) released from the cell surface after the cells were agitated in a kitchen blender. After lysis, some of the ^{32}P was

found in the progeny phages. In contrast, after *E. coli* were infected with ^{35}S-labeled T2, almost none of the radioactivity appeared within the cell or in the progeny phage particles, while most of the radioactivity was in the phage ghosts.

Because genes serve as the blueprint for making the progeny virus particles, it was also presumed that the blueprint must get into the bacterial cell for new phage particles to be built. Therefore, reasoned Hershey and Chase, because *it was DNA and not protein that entered the cell*, as evidenced by the presence of ^{32}P and the absence of ^{35}S, *DNA must be the material responsible for the function and reproduction of phage T2.* The protein, they hypothesized, provided a structural framework that contains both the DNA and the specialized structures required to inject the DNA into the bacterial cell.

Alfred Hershey shared the 1969 Nobel Prize in Physiology or Medicine for his "discoveries concerning the genetic structure of viruses."

The Discovery of RNA as Viral Genetic Material

Most of the organisms and viruses discussed in this book (such as a human, *Drosophila*, yeast, *E. coli*, and phage T2) have DNA as their genetic material. However, some bacterial viruses (for example, Qβ), some animal viruses (for instance, poliovirus), and some plant viruses (such as, tobacco mosaic virus) have RNA as their genetic material. No known prokaryotic or eukaryotic organism has RNA as its genetic material. The classic experiment described next showed that RNA is the genetic material

of the *tobacco mosaic virus* (TMV), a virus that causes lesions on the leaves of the tobacco plant.

Like phage T2, TMV contains two chemical components: RNA and protein, each in a spiral (helical) configuration (Figure 2.7). The protein surrounds the RNA core, protects it from attack by nucleases, and, along with the RNA core, functions in the infection of the plant cells. Many varieties of TMV are known, differing in the plants they infect and in the extent to which they affect those plants.

In 1956, A. Gierer and G. Schramm showed that when tobacco plants were inoculated with the purified RNA of TMV (that is, RNA without the protein coat), they developed typical virus-induced lesions. No lesions were produced when the RNA had been degraded by treatment with ribonuclease and then injected into the plant. These results indicated strongly that RNA was the genetic material of TMV.

Figure 2.7

Demonstration that RNA is the genetic material in tobacco mosaic virus (TMV).
A TMV particle consists of a helical RNA core surrounded by a helical arrangement of protein subunits. Hybrid particles were made from the protein subunits of one TMV strain and the RNA of a different TMV strain. Tobacco leaves were infected with the reconstituted hybrid viruses, and the progeny viruses isolated from the resulting leaf lesions were analyzed. The progeny viruses always had protein subunits specified by the RNA component; that is, the character of the protein coat cannot be transmitted from the hybrid particles to their progeny.

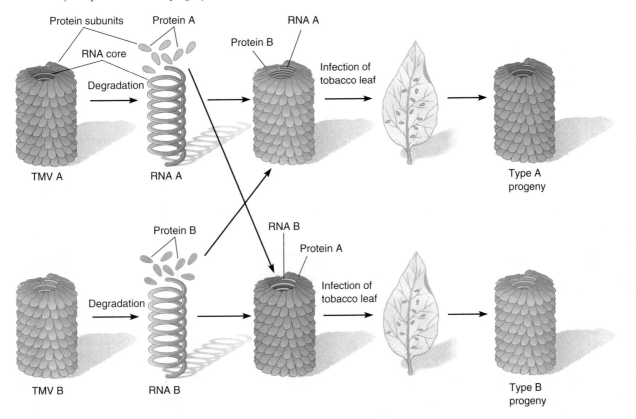

In 1957, Heinz Fraenkel-Conrat and B. Singer confirmed Gierer and Schramm's conclusions. They isolated the RNA and protein components of two distinct TMV strains and reconstituted the RNA of one type with the protein of the other type and vice versa. (See Figure 2.7.) They then infected the tobacco leaves with the two hybrid viruses. The progeny viruses isolated from the resulting lesions were of the type specified by the RNA, not by the protein. These results conclusively showed that RNA is the genetic material of TMV.

KEYNOTE

> A series of experiments proved that the genetic material consists of one of two types of nucleic acids: DNA or RNA. Of the two, DNA is the genetic material of all living organisms and of some viruses, and RNA is the genetic material of the remaining viruses.

The Composition and Structure of DNA and RNA

What is the molecular structure of DNA? DNA and RNA are *polymers*—large molecules that consist of many similar smaller molecules, called *monomers*, linked together. The monomers that make up DNA and RNA are called **nucleotides.** Each nucleotide consists of three distinct parts: a **pentose** (five-carbon) **sugar,** a **nitrogenous** (nitrogen-containing) **base,** and a **phosphate group.**

Figure 2.8

Structures of deoxyribose and ribose, the pentose sugars of DNA and RNA, respectively. The difference between the two sugars is highlighted.

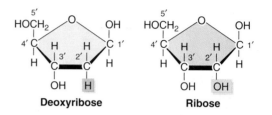

The pentose sugar in DNA is **deoxyribose,** and the sugar in RNA is **ribose** (Figure 2.8). The two sugars differ by the chemical groups attached to the 2′ carbon: a hydrogen atom (H) in deoxyribose and a hydroxyl group (OH) in ribose. (The carbon atoms in the pentose sugar are numbered 1′ to 5′ to distinguish them from the numbered carbon and nitrogen atoms in the rings of the bases.)

There are two classes of nitrogenous bases: the **purines,** which are nine-membered, double-ringed structures, and the **pyrimidines,** which are six-membered, single-ringed structures. There are two purines—**adenine (A)** and **guanine (G)**—and three different pyrimidines—**thymine (T), cytosine (C),** and **uracil (U).** The chemical structures of the five bases are shown in Figure 2.9. (The carbons and nitrogens of the purine rings are numbered 1 to 9, and those of the pyrimidines are numbered 1 to 6.) Both DNA and RNA contain adenine, guanine, and cytosine; however, thymine is found only in DNA, and uracil is found only in RNA.

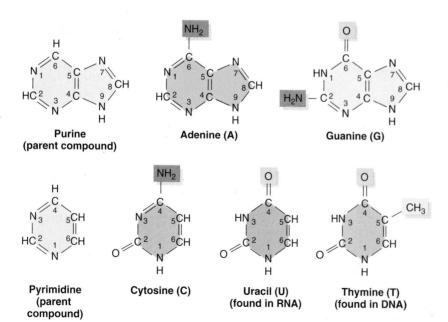

Figure 2.9

Structures of the nitrogenous bases in DNA and RNA. The parent compounds are purine (top left) and pyrimidine (bottom left). Differences between the bases are highlighted.

In DNA and RNA, bases are always attached to the 1′ carbon of the pentose sugar by a covalent bond. The purine bases are bonded at the 9 nitrogen, and the pyrimidines bond at the 1 nitrogen. The combination of a sugar and a base is called a **nucleoside.** Addition of a phosphate group to a nucleoside yields a **nucleoside phosphate,** also called a nucleotide. The phosphate group (PO_4^{2-}) is attached to the 5′ carbon of the sugar in both DNA and RNA. Examples of a DNA nucleotide (a **deoxyribonucleotide**) and an RNA nucleotide (a **ribonucleotide**) are shown in Figure 2.10a. A complete list of the names of the bases, nucleosides, and nucleotides is in Table 2.1.

To form **polynucleotides** of either DNA or RNA, nucleotides are linked together by a covalent bond between the phosphate group of one nucleotide and the 3′ carbon of the sugar of another nucleotide. These 5′-3′ phosphate linkages are called **phosphodiester bonds.** A short polynucleotide chain is diagrammed in Figure 2.10b. The phosphodiester bonds are relatively strong, so the repeated sugar–phosphate–sugar–phosphate backbone of DNA and RNA is a stable structure.

To understand how a polynucleotide chain is synthesized (discussed in Chapter 3), you must be aware of one more feature: The two ends of the chain are not the same; that is, the chain has a 5′ carbon (with a phosphate group on it) at one end and a 3′ carbon (with a hydroxyl group on it) at the other end, as shown in Figure 2.10b. This asymmetry is called the *polarity* of the chain.

Figure 2.10

Chemical structures of DNA and RNA. (a) Basic structures of DNA and RNA nucleosides (sugar plus base) and nucleotides (sugar, plus base, plus phosphate group), the fundamental building blocks of DNA and RNA molecules. Here, the phosphate groups are orange, the sugars are red, and the bases are brown. (b) A segment of a polynucleotide chain, in this case a single strand of DNA. The deoxyribose sugars are linked by phosphodiester bonds (shaded) between the 3′ carbon of one sugar and the 5′ carbon of the next sugar.

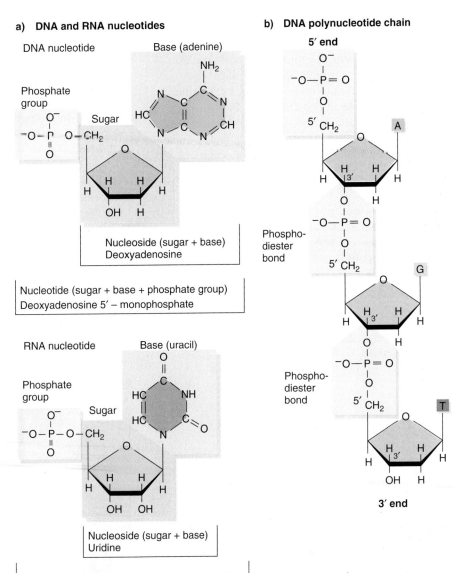

Table 2.1   **Names of the Base, Nucleoside, and Nucleotide Components Found in DNA and RNA**

		Base: Purines (Pu)		Base: Pyrimidines (Py)		
		Adenine (A)	Guanine (G)	Cytosine (C)	Thymine (T) (deoxyribose only)	Uracil (U) (ribose only)
DNA	Nucleoside: deoxyribose + base	Deoxyadenosine (dA)	Deoxyguanosine (dG)	Deoxycytidine (dC)	Thymidine (dT)	
	Nucleotide: deoxyribose + base + phosphate group	Deoxyadenylic acid or deoxyadenosine monophosphate (dAMP)	Deoxyguanylic acid or deoxyguanosine monophosphate (dGMP)	Deoxycytidylic acid or deoxycytidine monophosphate (dCMP)	Thymidylic acid or thymidine monophosphate (TMP)	
RNA	Nucleoside: ribose + base	Adenosine (A)	Guanosine (G)	Cytidine (C)		Uridine (U)
	Nucleotide: ribose + base + phosphate group	Adenylic acid or adenosine monophosphate (AMP)	Guanylic acid or guanosine monophosphate (GMP)	Cytidylic acid or cytidine monophosphate (CMP)		Uridylic acid or uridine monophosphate (UMP)

K E Y N O T E

> DNA and RNA occur in nature as macromolecules composed of smaller building blocks called nucleotides. Each nucleotide consists of a 5-carbon sugar (deoxyribose in DNA, ribose in RNA) to which is attached a phosphate group and one of four nitrogenous bases: adenine, guanine, cytosine, and thymine (in DNA) or adenine, guanine, cytosine, and uracil (in RNA).

The DNA Double Helix

In 1953, James D. Watson and Francis H. C. Crick (Figure 2.11) published a very important paper in which they proposed a model for the physical and chemical structure of the DNA molecule. The model they devised, which fit all the known data on the composition of the DNA molecule, is the now-famous double helix model for DNA. Unquestionably, the determination of the structure of DNA was a momentous occasion in biology, leading directly to the transformation in our understanding of all aspects of the life sciences.

At the time of Watson and Crick's discovery, DNA was known to be composed of nucleotides. However, it was not known how the nucleotides formed the structure of DNA. Watson and Crick thought that understanding the structure of DNA would help determine how DNA acts as the genetic basis for living organisms. The data they used to help generate their model came primarily from two sources: base composition studies

Figure 2.11

James Watson (left) and Francis Crick (right) in 1993 at a 40th-anniversary celebration of their discovery of the structure of DNA and in 1953 with the model of DNA structure.

conducted by Erwin Chargaff and X-ray diffraction studies conducted by Rosalind Franklin and Maurice H. F. Wilkins.

Base Composition Studies. By chemical treatment, Erwin Chargaff had hydrolyzed the DNA of a number of organisms and had quantified the purines and pyrimidines released. His studies showed that, in all double-stranded DNAs, 50 percent of the bases were purines and 50 percent were pyrimidines. More important, the amount of adenine (A) was equal to that of thymine (T), and the amount of guanine (G) was equal to that of cytosine (C). These equivalencies have become known as Chargaff's rules. In comparisons of double-stranded DNAs from different organisms, the A/T ratio is 1 and the G/C ratio is 1, but the (A + T)/(G + C) ratio (typically denoted %GC) varies. Because the amount of purines equals the amount of pyrimidines, the (A + G)/(C + T) ratio is 1. (See Table 2.2.)

X-Ray Diffraction Studies. Rosalind Franklin, working with Maurice H. F. Wilkins (Figure 2.12a), studied isolated fibers of DNA by X-ray diffraction, a technique in which a beam of parallel X rays is aimed at molecules. The beam is diffracted (broken up) by the atoms in a pattern that is characteristic of the atomic weight and the spatial arrangement of the molecules. The diffracted X rays are recorded on a photographic plate (Figure 2.12b). Analyzing the photographs, Franklin obtained information about the molecule's atomic structure. In particular, she concluded that DNA is a helical structure with two distinctive regularities of 0.34 nm and 3.4 nm along the axis of the molecule (1 nanometer [nm] = 10^{-9} meter = 10 angstrom units [Å]; 1 Å = 10^{-10} meter).

Watson and Crick's Model. Watson and Crick used Franklin's data to build three-dimensional models of the structure of DNA. Figure 2.13a shows a three-dimensional model of the DNA molecule, and Figure 2.13b is a diagram of the same molecule, showing the arrangement of the sugar–phosphate backbone and base pairs in a stylized way. Figure 2.13c shows the chemical structure of double-stranded DNA.

Watson and Crick's double-helical model of DNA has the following main features:

1. The DNA molecule consists of two polynucleotide chains wound around each other in a right-handed double helix; that is, viewed on end (from either end), the two strands wind around each other in a clockwise (right-handed) fashion.

2. The two chains are **antiparallel** (show *opposite polarity*); that is, the two strands are oriented in opposite directions, with one strand oriented in the 5′-to-3′ way and the other strand oriented 3′ to 5′. More simply if the 5′ end is the "head" of the chain and the 3′ end is the "tail," *antiparallel* means that the head of one chain is against the tail of the other chain and vice versa.

3. The sugar-phosphate backbones are on the outsides of the double helix, with the bases oriented toward the central axis. (See Figure 2.13.) The bases of both chains are flat structures oriented perpendicularly to the long axis of the DNA; that is, the bases are stacked like pennies on top of one another, following the twist of the helix.

4. The bases in each of the two polynucleotide chains are bonded together by hydrogen bonds, which are relatively weak chemical bonds. The specific pairings observed are A bonded with T (two hydrogen bonds, Figure 2.14a) and G bonded with C (three hydrogen bonds, Figure 2.14b). The hydrogen bonds make it relatively easy to separate the two strands of the DNA—for example, by heating. The A-T and G-C base pairs are the only ones that can fit the physical dimensions of the helical model, and their arrangement is in accord with Chargaff's rules. The specific A-T and G-C pairs are called **complementary base pairs,** so the nucleotide sequence in one strand dictates the nucleotide sequence of the other. For instance, if one chain has the sequence 5′-TATTCCGA-3′, then the opposite, antiparallel chain must bear the sequence 3′-ATAAGGCT-5′.

5. The base pairs are 0.34 nm apart in the DNA helix. A complete (360°) turn of the helix takes 3.4 nm; therefore, there are 10 base pairs per turn. The external diameter of the helix is 2 nm.

Table 2.2	Base Compositions of DNAs from Various Organisms						
	Percentage of Base in DNA				Ratios		
DNA origin	A	T	G	C	A/T	G/C	(A + T)/(G + C)
Human (sperm)	31.0	31.5	19.1	18.4	0.98	1.03	1.67
Corn (*Zea mays*)	25.6	25.3	24.5	24.6	1.01	1.00	1.04
Drosophila	27.3	27.6	22.5	22.5	0.99	1.00	1.22
Euglena nucleus	22.6	24.4	27.7	25.8	0.93	1.07	0.88
Escherichia coli	26.1	23.9	24.9	25.1	1.09	0.99	1.00

Figure 2.12

X-ray diffraction analysis of DNA. (a) Rosalind Franklin and Maurice H. F. Wilkins (photographed in 1962, the year he received the Nobel Prize shared with Watson and Crick). (b) The X-ray diffraction pattern of DNA that Watson and Crick used in developing their double-helix model. The dark areas that form an X shape in the center of the photograph indicate the helical nature of DNA. The dark crescents at the top and bottom of the photograph indicate the 0.34-nm distance between the base pairs.

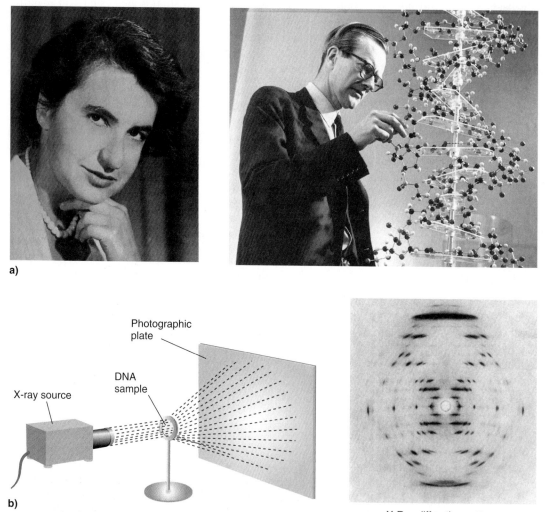

a)

b)

X-Ray diffraction pattern

6. Because of the way the bases bond with each other, the two sugar–phosphate backbones of the double helix are not equally spaced from one another along the helical axis. This unequal spacing results in grooves of unequal size between the backbones; one groove is called the *major* (wider) *groove*, the other the *minor* (narrower) *groove*. (See Figure 2.13a.) The edges of the base pairs are exposed in the grooves, and both grooves are large enough to allow protein molecules to make contact with the bases.

For their "discoveries concerning the molecular structure of nucleic acids and its significance for information transfer in living material," the 1962 Nobel Prize in Physiology or Medicine was awarded to Francis Crick, James Watson, and Maurice Wilkins. What was Rosalind Franklin's contribution to the discovery? This has been the subject of some debate, and it is likely that we will never know whether she would have shared the prize. She died in 1962, and Nobel Prizes are never awarded posthumously.

Different DNA Structures

In recent years, methods have been developed to synthesize short DNA molecules (called **oligomers;** *oligo* means "few") of defined sequences. The pure DNA oligomers

Figure 2.13

Molecular structure of DNA. (a) Three-dimensional molecular model of DNA as prepared by Watson and Crick. (b) Stylized representation of the DNA double helix. (c) Chemical structure of a segment of double-stranded DNA.

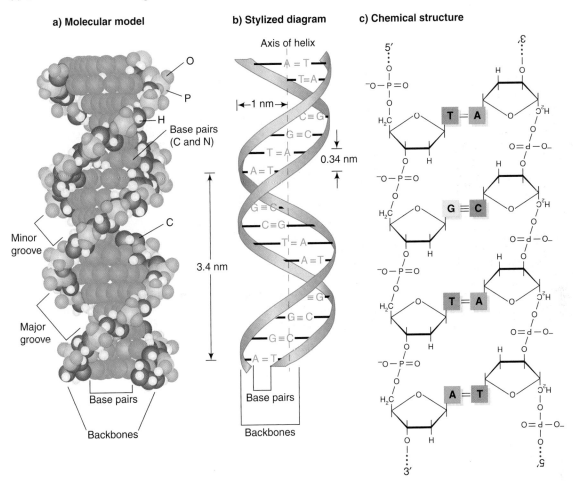

can be crystallized and analyzed by X-ray diffraction. The various approaches have revealed that DNA can exist in several different forms—most notably, the A-, B-, and Z-DNA forms (Figure 2.15). Some of the key structural features of each DNA type are listed in Table 2.3.

A-DNA and B-DNA. A-DNA and B-DNA are right-handed double helices with 10.9 and 10.0 base pairs, respectively, per 360° turn of the helix. The bases are inclined away from the perpendicular to the helix axis (which would be 0°) by 13° in A-DNA and 2° in B-DNA.

B-DNA is the form of DNA seen under conditions of high humidity. Watson and Crick's double-helical model was based on B-DNA analyzed in hydrated fibers outside of the cell. A-DNA is seen only in conditions of low humidity. The A-DNA double helix is short and wide with a narrow, very deep major groove and a wide, shallow minor groove. (Think of these descriptions in terms of canyons: Narrow and wide describe the distance from rim to rim, and shallow and deep describe the distance from the rim down to the bottom of the canyon.) The B-DNA double helix is thinner and longer for the same number of base pairs than A-DNA, with a wide major groove and a narrow minor groove; both grooves are of similar depths.

Z-DNA. Z-DNA is a left-handed helix with a zigzag sugar–phosphate backbone. (The latter property gave this form of DNA its "Z" designation.) Z-DNA has 12.0 base pairs per complete helical turn. The bases are inclined away from the perpendicular to the helix axis by 8.8 degrees. The Z-DNA helix is thin and elongated with a deep, minor groove. The major groove is very near the surface of the helix, so it is not distinct.

Figure 2.14

Structures of the complementary base pairs found in DNA.
In both cases, a purine pairs with a pyrimidine: (a) The
adenine–thymine bases, which pair through two hydrogen bonds.
(b) The guanine–cytosine bases, which pair through three hydrogen bonds.

You must now determine the molecular composition and structure of a virus infecting the rice
crops of Asia. Go to the iActivity *Cracking a
Viral Code* on your CD-ROM.

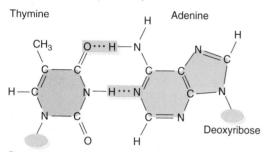

**a) Adenine–thymine base
    (Double hydrogen bond)**

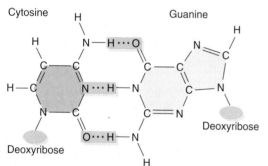

**b) Guanine–cytosine base
    (Triple hydrogen bond)**

DNA in the Cell

In solution, as in the cell, DNA is closest in form to B-DNA,
although it is a little more twisted, with about 10.4 base
pairs per complete helical turn. A-DNA is found only when
the DNA is dehydrated, so it is unlikely that any lengthy
sections of A-DNA exist within cells. Whether Z-DNA
exists in living cells has long been a topic of debate among
scientists. In those organisms where there is some evidence
for Z-DNA, its physiological role is still unknown.

RNA Structure

As has already been mentioned, RNA has a structure similar to that of DNA. To reiterate, the RNA molecule is
a polymer of RNA nucleotides (ribonucleotides) in which
the sugar is ribose. Three of the four bases in RNA are the
same as those in DNA: A, G, and C. The distinctive base
in RNA is uracil, and that in DNA is thymine.

In the cell, the various functional forms of RNA—
messenger RNA (mRNA), transfer RNA (tRNA), ribosomal RNA (rRNA), and small nuclear RNA (snRNA)—are
single-stranded molecules. However, you should not
think of these molecules as stiff, linear rods. Rather, wherever bases can pair together, they will do so. This means
that a single-stranded RNA molecule will fold up on itself

Figure 2.15

Space-filling models of different forms of DNA. (a) A-DNA. (b) B-DNA. (c) Z-DNA.

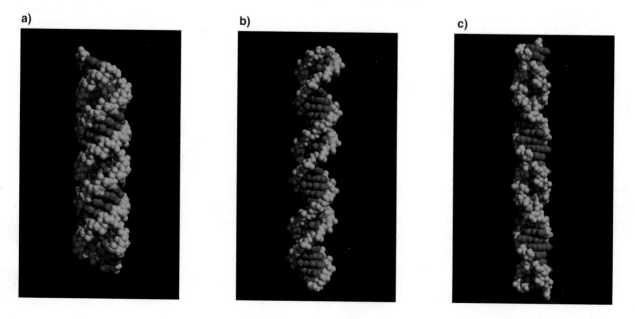

Table 2.3	Properties of A-DNA, B-DNA, and Z-DNA		
Property	**A-DNA**	**B-DNA**	**Z-DNA**
Helix direction	Right handed	Right handed	Left handed
Base pairs per helix turn	10.9	10.0	12.0
Overall morphology	Short and wide	Longer and thinner	Elongated and thin
Major groove	Extremely narrow and very deep	Wide and of intermediate depth	Flattened out on helix surface
Minor groove	Very wide and shallow	Narrow and of intermediate depth	Extremely narrow and very deep
Helix axis location	Major groove	Through base pairs	Minor groove
Helix diameter	2.2 nm	2.0 nm	1.8 nm

to produce regions of double-stranded RNA separated by short segments of unpaired RNA. This configuration is called the secondary structure of the molecule.

A number of viruses have RNA as their genome. (See next section.) In some cases, the RNA genome is single stranded; in others, it is double-stranded. Double-stranded RNA has a structure similar to that of double-stranded DNA, with antiparallel strands, the sugar–phosphate backbones on the outside of the helical molecule, and complementary base pairs formed by hydrogen bonding in the middle of the helix.

The Organization of DNA in Chromosomes

The DNA in a cell is organized into physical structures called chromosomes. The full DNA sequence of an organism's haploid set of chromosomes is its **genome.** In prokaryotes, the genome is usually, but not always, a single circular chromosome. In eukaryotes, the genome is typically distributed among a number of chromosomes in the cell nucleus. (Eukaryotes also contain a mitochondrial genome and, in plants, a chloroplast genome, as discussed in Chapter 23.) To understand the process by which the information within a gene is accessed (Chapter 5), it is important to understand how DNA is organized in chromosomes. In the sections that follow, we discuss the organization of DNA molecules in chromosomes of viruses, prokaryotes, and eukaryotes.

Viral Chromosomes

Viruses are fragments of nucleic acid surrounded by proteins. Viruses infect all types of living organisms. Depending on the virus, the genetic material may be double-stranded DNA, single-stranded DNA, double-stranded RNA, or single-stranded RNA, and it may be circular or linear. The genomes of some viruses are organized into a single chromosome, whereas other viruses have a segmented genome: The genome is distributed

among a number of DNA molecules. In this section, we look at the chromosome structure of three representative bacteriophages of genetic significance: T-even phages, ΦX174 ("fi-X-one-seventy-four"), and lambda (λ).

T2, T4, and T6—the T-even bacteriophages—have genomes consisting of double-stranded DNA. The T4 genome is 168,900 bp (168 kb, where 1 kb = 1 kilobase = 1,000 base pairs) in size. These virulent phages have similar structures (see Figure 2.4) and contain a single, linear DNA chromosome surrounded by a protein coat. The chromosome is packaged within the head of the virus.

The virulent phage ΦX174 is a DNA phage that infects *E. coli.* The ΦX174 phage is smaller than the T-even phages and has a single chromosome that contains 5,386 nucleotides. Unlike the T-even phages, the ΦX174 phage does not have a contractile sheath, a baseplate, or tail fibers. Instead, the phage is an icosahedron consisting of protein subunits surrounding the genetic material (Figure 2.16). In 1959, Robert Sinsheimer found that the DNA of ΦX174 has a ratio of bases of 25A:33T:24G:18C. This ratio does not match the A = T and G = C expected of double-stranded DNA, and the reason is that the chromosome is single stranded rather than double stranded. In addition, the

Figure 2.16

Electron micrograph of bacteriophage ΦX174.

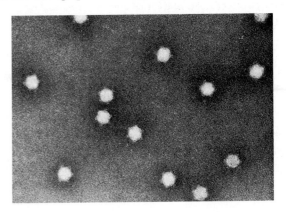

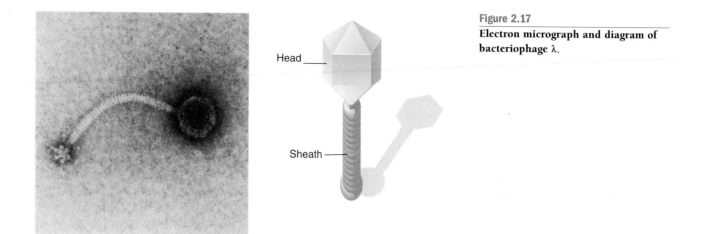

Figure 2.17

Electron micrograph and diagram of bacteriophage λ.

Head

Sheath

chromosome is resistant to digestion by an exonuclease—an enzyme that removes nucleotides from the ends of linear molecules. The reason here is that the chromosome is circular rather than linear.

The λ (lambda) bacteriophages are similar in structure to the T-even phages. (See Figure 2.17.) Unlike the T-even phages, however, the λ phage chromosome structure changes between linear and circular forms. Within the phage particle, the λ chromosome is linear, double-stranded DNA, and the two ends of the DNA molecule have 12-nucleotide, single-stranded segments that are complementary (Figure 2.18). When a λ phage infects a cell, the linear chromosome converts to a circular chromosome. The complementary ends of the molecule (called *sticky ends*) pair together to produce the circular form. (See Figure 2.18.) When the phage reproduces, the chromosome is converted back to the linear form. (A more detailed discussion of this process appears in Chapter 3.)

Prokaryotic Chromosomes

Most prokaryotes contain a single, double-stranded, circular DNA chromosome. The remaining prokaryotes

have genomes that consist of one or more chromosomes that may be circular or linear. In the latter cases, there is typically a main chromosome and one or more smaller chromosomes. When a minor chromosome is dispensable to the life of the cell, it is more correctly called a plasmid. For example, among the bacteria, *Borrelia burgdorferi,* the causative agent of Lyme disease, has a 0.91-Mb (1 Mb = 1 megabase = 1 million base pairs) linear chromosome and at least 17 small linear and circular chromosomes with a combined size of 0.53 Mb. *Agrobacterium tumefaciens,* the causative agent of crown gall disease in some plants, has a 3.0-Mb circular chromosome and a 2.1-Mb linear chromosome. Among the archaea, chromosome organization also varies, although no linear chromosomes have yet been found. For example, *Methanococcus jannaschii* has 1.66-Mb, 58-kb, and 16-kb circular chromosomes, and *Archaeoglobus fulgidus* has a single 2.2-Mb circular chromosome.

In bacteria and archaea, the chromosome is arranged in a dense clump in a region of the cell known as the **nucleoid.** Unlike eukaryotic nuclei, bacterial nuclei have no membrane between the nucleoid region and the rest of the cell.

Figure 2.18

λ chromosome structure varies at stages of lytic infection of E. coli. Parts of the λ chromosome, showing the nucleotide sequence of the two single-stranded, complementary ("sticky") ends and the chromosome circularizing after infection by pairing of the ends, with the single-stranded gaps filled in to produce a covalently closed circle.

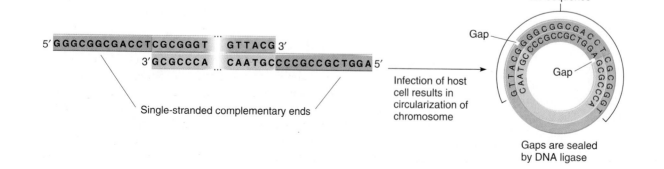

5′ GGGCGGCGACCTCGCGGGT ... GTTACG 3′
3′ GCGCCCA ... CAATGCCCCGCCGCTGGA 5′

Single-stranded complementary ends

Infection of host cell results in circularization of chromosome

cos sequence

Gap

Gap

Gaps are sealed by DNA ligase

Figure 2.19

Chromosome released from a lysed *E. coli* cell.

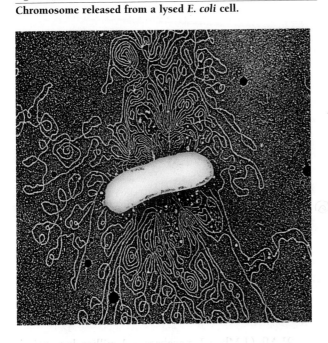

If an *E. coli* cell is lysed (broken open) gently, its DNA is released in a highly folded state (Figure 2.19). The double-stranded DNA is present as a single, 4.6-Mb circular chromosome, approximately 1,100 µm long. The length of DNA in the *E. coli* chromosome is approximately 1,000 times the length of the *E. coli* cell. The DNA fits into the nucleoid region of the cell in part because it is **supercoiled**; that is, the double helix has been twisted in space about its own axis.

animation

DNA Super-coiling

To understand supercoiling, consider a 208-base-pair linear piece of cellular DNA (Figure 2.20a). Such a molecule has two free ends, and because it has 10.4 base pairs per helical turn, there are 20 helical turns in this molecule. If we simply join the two ends, we have produced a circular DNA molecule that is *relaxed* (Figure 2.20b). Alternatively, if we first untwist one end of the linear DNA molecule by two turns (Figure 2.20c) and then join the two ends, the circular DNA molecule produced will have 18 helical turns and a small unwound region (Figure 2.20d). Such a structure is not energetically favored and will switch to a structure with 20 helical turns and two superhelical turns—a supercoiled form of DNA (Figure 2.20e). The very act of supercoiling produces tension in the DNA molecule. Therefore, if a break—called a nick—is introduced into the sugar–phosphate backbone of a supercoiled circular DNA molecule, the molecule spontaneously untwists and produces a relaxed DNA circle. Although not intuitively obvious, supercoiling can also occur in a linear DNA molecule. That is, if we twist a length of rope on one end without holding the other end, the rope just spins in the air and remains linear (relaxed). However, with a large, linear DNA

molecule, supercoiling occurs in localized regions and the ends behave as if they are fixed.

Figure 2.21 pictures supercoiled circular DNA to show how much more compact a supercoiled molecule is compared with a nonsupercoiled (relaxed) molecule. There are two types of supercoiling: *negative supercoiling* and *positive supercoiling*. To visualize supercoiling, think of the DNA double helix as a spiral staircase that turns in a clockwise direction. If you untwist the spiral staircase by one complete turn, *you have the same number of stairs to climb, but you have one less 360° turn to make*; this is a negative supercoil. If, instead, you twist the spiral staircase by one more complete turn, *you have the same number of stairs to climb, but now there is one more 360° turn to make*; this is a positive supercoil. Either type of supercoiling causes the DNA to become more compact. The amount and type of DNA supercoiling is controlled by **topoisomerases**—enzymes that are found in all organisms.

Chromosomes also become compacted because the DNA is organized into **looped domains.** In *E. coli*, each domain consists of a loop of about 40 kb of negatively supercoiled DNA, so there are approximately 100 domains for the entire genome. The ends of each domain are held—presumably by proteins—such that the supercoiled DNA state in one domain is not affected by events that influence the supercoiling of DNA in the other domains (Figure 2.22). The compaction achieved by organizing into looped domains is about tenfold.

KEYNOTE

Viral genomes may be double-stranded or single-stranded DNA, double-stranded or single-stranded RNA, and circular or linear. The genomes of some viruses are organized into a single chromosome, whereas other viruses have a segmented genome. The genetic material of bacteria and archaea is double-stranded DNA localized into one or a few chromosomes. The *E. coli* chromosome is circular and is organized into about 100 independent looped domains of supercoiled DNA.

Eukaryotic Chromosomes

A fundamental difference between prokaryotes and eukaryotes is that most prokaryotes have a single type of chromosome (sometimes present in more than one copy in the cell), whereas eukaryotes have several chromosomes. More specifically, each eukaryotic species has a characteristic number of chromosomes. There are 46 human chromosomes, a number stemming from the fact that humans are diploid (2N) organisms, possessing one haploid (N) set of chromosomes (23 chromosomes) from the egg and another haploid set from the sperm.

The total amount of DNA in the haploid genome of a species is known as the species' **C value.** Table 2.4 lists

Figure 2.20

Illustration of DNA supercoiling. (a) Linear DNA with 20 helical turns (208 base pairs). (b) Relaxed circular DNA produced by joining the two ends of the linear molecule of (a). (c) The linear DNA molecule of (a) unwound from one end by two helical turns. (d) A possible circular DNA molecule produced by joining the two ends of the linear molecule of (c). The circular molecule has 18 helical turns and a short unwound region. (e) The more energetically favored form of (d), a supercoiled DNA with 20 helical turns and two superhelical turns.

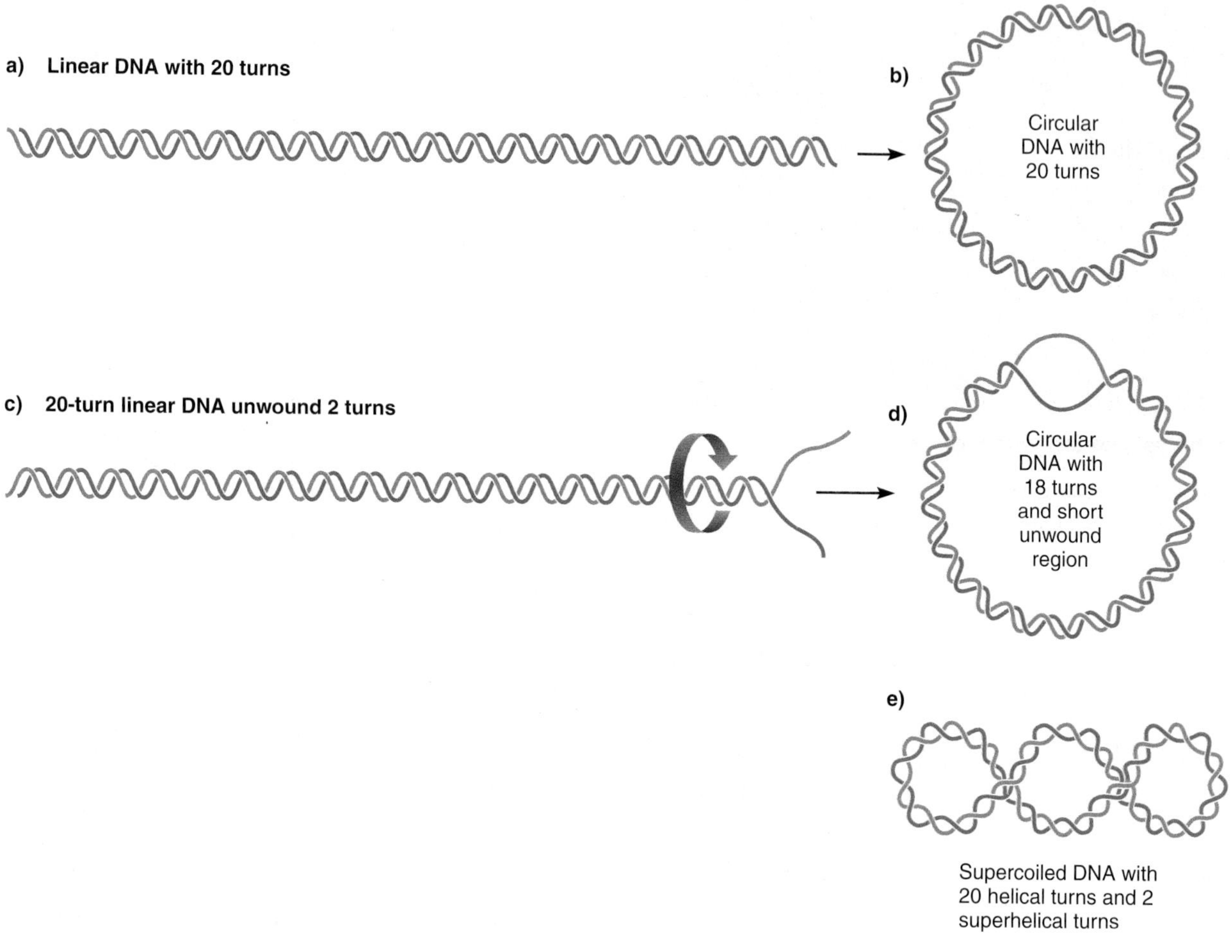

a) **Linear DNA with 20 turns**

b) Circular DNA with 20 turns

c) **20-turn linear DNA unwound 2 turns**

d) Circular DNA with 18 turns and short unwound region

e) Supercoiled DNA with 20 helical turns and 2 superhelical turns

the C values for some selected species. C-value data show that the amount of DNA found among organisms varies widely, and there may or may not be significant variation in the amount between related organisms. For example, mammals, birds, and reptiles show little variation, both across each other and among species within each class, whereas amphibians, insects, and plants vary over a wide range, often tenfold or more. There is also no direct relationship between the C value and the structural or organizational complexity of the organism, a situation called the *C value paradox*. At least one reason for this absence of a direct link is variation in the amount of repetitive sequence DNA in the genome (pp. 36–37).

As we will learn in Chapter 12 (see pp. 302–304 and Figure 12.4), eukaryotic cells reproduce in a cell cycle consisting of four phases: G_1, S, G_2, and M. During G_1 phase, each chromosome is a single structure. During S phase, the chromosomes duplicate to produce two sister chromatids joined by the duplicated, but not-yet-separated, centromeres. This state remains during G_2, and then, during M phase (mitosis), the centromeres separate and the sister chromatids become known as daughter chromosomes. We must keep this cycle clear when we think about chromosomes. Each eukaryotic chromosome in G_1 consists of one linear, double-stranded DNA molecule running throughout its length and complexed with about twice as much protein by weight as DNA. The duplicated chromosomes at other stages with sister chromatids have one linear, double-stranded DNA molecule running the length of each chromatid. To complete our discussion, we must learn another term: *chromatin*. **Chromatin** is the stainable material in a cell nucleus: DNA and proteins. More broadly,

Figure 2.21

Electron micrographs of a circular DNA molecule, showing relaxed and supercoiled states. (a) Relaxed (nonsupercoiled) DNA. (b) Supercoiled DNA. Both molecules are shown at the same magnification.

a)

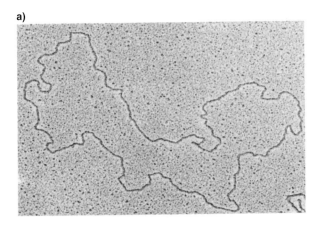

b)

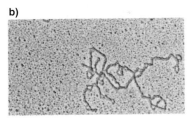

the term is commonly used in descriptions of chromosome structure and function. The fundamental structure of chromatin is essentially identical in all eukaryotes.

Chromatin Structure. Two major types of proteins are associated with DNA in chromatin: **histones** and **nonhistones.**

Figure 2.22

Model for the structure of a bacterial chromosome. The chromosome is organized into looped domains, the bases of which are anchored in an unknown way.

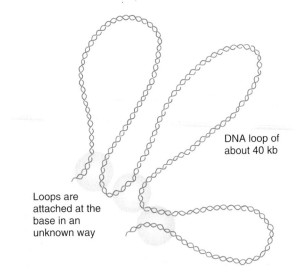

DNA loop of about 40 kb

Loops are attached at the base in an unknown way

Table 2.4	Haploid DNA Content, or C Value, of Selected Species
Species	**C Value (bp)**
Viruses and Phages	
λ (bacteriophage)	48,502[a]
T4 (bacteriophage)	168,900
Feline leukemia virus (cat virus)	8,448[a]
Simian virus 40 (SV40)	5,243[a]
Human immunodeficiency virus-1 (HIV-1, causative agent of AIDS)	9,750
Measles virus (human virus)	15,894[a]
Bacteria	
Bacillus subtilis	4,214,814[a]
Borrelia burgdorferi (Lyme disease spirochete)	910,724[a]
Escherichia coli	4,639,221[a]
Heliobacter pylori (bacterium that causes stomach ulcers)	1,667,867[a]
Neisseria meningitis	2,272,351[a]
Archaea	
Methanococcus jannaschii	1,664,970[a]
Eukarya	
Saccharomyces cerevisiae (budding yeast; Brewer's yeast)	13,105,020[a]
Schizosaccharomyces pombe (fission yeast)	14,000,000
Lilium formosanum (lily)	36,000,000,000
Zea mays (maize, corn)	5,000,000,000
Amoeba proteus (amoeba)	290,000,000,000
Drosophila melanogaster (fruit fly)	180,000,000
Caenorhabditis elegans (nematode)	97,000,000
Danio rerio (zebrafish)	1,900,000,000
Xenopus laevis (African clawed frog)	3,100,000,000
Mus musculus (mouse)	3,454,200,000
Rattus rattus (rat)	3,093,900,000
Canis familiaris (dog)	3,355,500,000
Equus caballus (horse)	3,311,000,000
Homo sapiens (humans)	3,400,000,000

[a] These C values derive from the complete genome sequence, all others are estimates based on other measurements.

Both types of proteins play an important role in determining the physical structure of the chromosome.

The histones are the most abundant proteins in chromatin. They are small basic proteins—they contain large amounts of arginine and lysine—with a net positive charge that facilitates their binding to the negatively charged DNA. Five main types of histones are associated with eukaryotic DNA: H1, H2A, H2B, H3, and H4. Weight for weight, there is an equal amount of histone and DNA in chromatin.

The amounts and proportions of histones relative to DNA are constant from cell to cell in all eukaryotic organisms. The amino-acid sequences of H2A, H2B, H3, and H4 histones are highly conserved (very similar), evolutionarily speaking, even between distantly related species. Evolutionary conservation of these sequences is a strong indicator that histones perform the same basic role in organizing the DNA in the chromosomes of all eukaryotes.

The histones play a crucial role in chromatin packing as well, as we will see. A human cell, for example, has more than 700 times as much DNA as does *E. coli*. Without the compacting of the 6×10^9 base pairs of DNA in the diploid genome, the DNA of the chromosomes of a single human cell would be more than 2 meters long (about 6.5 feet) if the molecules were placed end to end. Several levels of packing enable chromosomes that would be several millimeters or even centimeters long to fit into a nucleus that is a few micrometers in diameter.

Nonhistones are all the proteins associated with DNA, apart from the histones. Nonhistones are far less abundant than histones. Examples of nonhistones are DNA binding proteins that play a role in the processes of DNA replication, DNA repair, transcription (including gene regulation), and recombination. Many nonhistones are acidic proteins—proteins with a net negative charge—and are likely to bind to positively charged histones in the chromatin. Each eukaryotic cell has many different nonhistones in the nucleus, but, in contrast to the histones, the nonhistone proteins differ markedly in number and type from cell type to cell type within an organism, at different times in the same cell type, and from organism to organism.

With the electron microscope, we can see different chromatin structures, especially at different stages of the cell cycle. The least compact form seen is the **10-nm chromatin fiber,** which has a characteristic "beads-on-a-string" morphology; the beads have a diameter of about 10 nm (Figure 2.23). The beads are **nucleosomes,** which are the basic structural units of eukaryotic chromatin. A nucleosome is about 11 nm in diameter and consists of a core of eight histone proteins—two each of H2A, H2B, H3, and H4 (Figure 2.24a)—around which a 147-bp segment of DNA is

Figure 2.23

Electron micrograph of unraveled chromatin, showing the nucleosomes in a "beads-on-a-string" morphology.

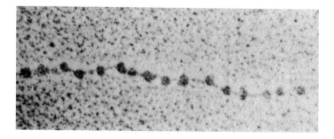

Figure 2.24

Basic eukaryotic chromosome structure. (a) Histone core for the nucleosome. (b) Diagram of nucleosomes in "beads-on-a-string" chromatin. A nucleosome is 11 nm × 5.7 nm. (c) Chromosome condensation brought about by the binding of histone H1.

a) Histone core

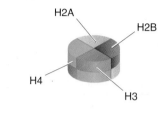

b) Basic nucleosome structure in "beads-on-a-string" chromatin

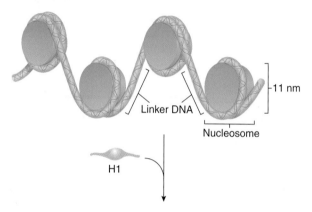

c) Chromatin condensation by H1 binding

wound about 1.65 times (Figure 2.24b). This configuration serves to compact the DNA by a factor of about six.

Individual nucleosomes are connected by strands of *linker DNA*. (See Figures 2.23 and 2.24b.) The amount of linker DNA varies within and among organisms. Human linker DNA, for example, has a length of 38–53 bp, giving a total amount of 185–200 bp of DNA per nucleosome.

The next level of chromatin condensation is brought about by histone H1. A single molecule of H1 binds both to the linker DNA at one end of the nucleosome and to the middle of the DNA segment wrapped around core histones. The binding of H1 causes the nucleosomal DNA to assume a more regular appearance (Figure 2.24c) and

Figure 2.25

The 30-nm chromatin fiber. (a) Electron micrograph. (b) Solenoid model for the packaging of nucleosomes into the 30-nm chromatin fiber. (H1 is not shown.)

a)

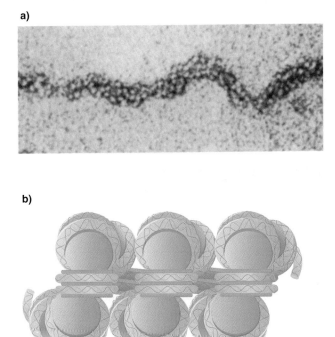

b)

Figure 2.26

Electron micrograph of a metaphase chromosome depleted of histones. The chromosome maintains its general shape by a nonhistone protein scaffold from which loops of DNA protrude.

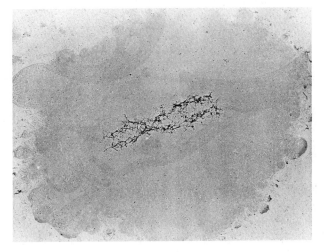

the nucleosomes themselves to compact into a structure about 30 nm in diameter called the **30-nm chromatin fiber** (Figure 2.25a). One possible model for the 30-nm fiber—the *solenoid model*—has the nucleosomes spiraling helically, with about six nucleosomes per complete turn (Figure 2.25b).

Chromatin packing beyond the 30-nm chromatin filaments is less well understood. Current models derive from 1970s-vintage electron micrographs of metaphase chromosomes depleted of histones (Figure 2.26). The photos show 30–90-kb loops of DNA attached to a

proteinaceous "scaffold" with the characteristic X shape of the paired sister chromatids. If the histones are not removed, loops of chromatin—looped domains—are seen, consisting of 180–300 nucleosomes organized into 30-nm fibers. An average human chromosome has approximately 2,000 looped domains.

Current information indicates that each loop is held together at its base by nonhistone proteins that are part of the *chromosome scaffold*, but the details of this structure are not known (Figure 2.27a). Particular stretches of DNA called *scaffold-associated regions*, or *SARs*, bind to the nonhistone proteins to determine the loops. It is simplest to think of these loops as being arranged in a spiral fashion around the central chromosome scaffold (Figure 2.27b). In cross section, the loops would be seen to radiate out from the center like the petals of a flower. With about 15 loops per turn, we can account for the 700-nm diameter of the cylindrical chromatid arms of a

Figure 2.27

Looped domains in metaphase chromosomes. (a) Fiber loops 30 nm attached at scaffold-associated regions to the chromosome scaffold by nonhistone proteins. (b) Schematic of a section of the metaphase chromosome. Shown is the spiraling of looped domains.

a)   **b)**

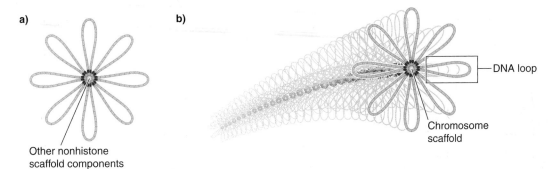

Other nonhistone
scaffold components

DNA loop

Chromosome
scaffold

metaphase chromosome. Overall, this packing produces a chromosome that is about 10,000 times shorter, and about 400 times thicker, than naked DNA.

KEYNOTE

> The nuclear chromosomes of eukaryotes are complexes of DNA, histone proteins, and nonhistone chromosomal proteins. Each chromosome consists of one linear, unbroken, double-stranded DNA molecule running throughout the length of the chromosome. There are five main types of histones (H1, H2A, H2B, H3, and H4), which are constant from cell to cell within an organism. Nonhistones, of which there are a large number, vary significantly between cell types, both within and among organisms as well as with time in the same cell type. The large amount of DNA present in the eukaryotic chromosome is compacted by its association with histones in nucleosomes and by higher levels of folding of the nucleosomes into chromatin fibers. Each chromosome contains a large number of looped domains of 30-nm chromatin fibers attached to a protein scaffold. The functional state of the chromosome is related to the extent of coiling: The more condensed a part of a chromosome is, the less likely it is that the genes in that region will be active.

Euchromatin and Heterochromatin.

The degree of DNA packing changes throughout the cell cycle. The most dispersed is when the chromosomes are about to duplicate (beginning of S phase of the cell cycle), and the most highly condensed is within mitosis and meiosis.

Historically, two forms of chromatin have been defined, each on the basis of chromosome-staining properties. **Euchromatin** is the chromosomes or regions of chromosomes that show the normal cycle of chromosome condensation and decondensation in the cell cycle. Most of the genome of an active cell is euchromatic. Visually, euchromatin undergoes a change in intensity of staining ranging from the darkest in the middle of mitosis (metaphase stage) to the lightest in the S phase. Typically, (1) euchromatin is actively transcribed, meaning that the genes within it can be expressed, and (2) euchromatic regions are devoid of repetitive sequences.

Heterochromatin, by contrast, is the chromosomes or chromosomal regions that usually remain condensed—more darkly staining than euchromatin—throughout the cell cycle, even in interphase. Heterochromatic DNA often replicates later than the rest of the DNA in the S phase. Genes within heterochromatic DNA are usually transcriptionally inactive. There are two types of heterochromatin. **Constitutive heterochromatin** is present in all cells at identical positions on both homologous chromosomes of a pair. This form of heterochromatin consists mostly of repetitive DNA and is exemplified by the centromere regions.

Facultative heterochromatin, by contrast, varies in state in different cell types, at different developmental stages, or, sometimes, from one homologous chromosome to another. This form of heterochromatin represents condensed, and therefore inactivated, segments of euchromatin. A well-known example of facultative heterochromatin is the *Barr body,* an inactivated X chromosome in somatic cells of XX mammalian females. (See Chapter 12, p. 322.)

KEYNOTE

> DNA in the eukaryotic chromatin is compacted by its association with histones in nucleosomes and by higher levels of folding of the nucleosomes into chromatin fibers. Each chromosome contains a large number of looped domains of 30-nm chromatin fibers attached to a protein scaffold. The functional state of the chromosome is related to the extent of coiling: The more condensed a part of a chromosome is, the less likely it is that the genes in that region will be active.

Centromeric and Telomeric DNA.

All eukaryotic chromosomes have two areas of special function: the centromere and the telomere. We will learn in Chapter 12 that the behavior of chromosomes in mitosis and meiosis depends on the *kinetochores* that form on the centromeres. The telomeres also have a special role, in this case in the replication of the DNA in the chromosome. The DNA sequences that make up these regions are responsible for their function.

Centromeres are the DNA sequences found near the point of attachment of mitotic or meiotic spindle fibers. The centromere region of each chromosome is responsible for accurately segregating the replicated chromosomes to the daughter cells during mitosis and meiosis. The error frequency for this process is low, but significant, and varies from organism to organism. In yeast, for example, segregation errors occur at a frequency of 1 in 10^5 or less.

The DNA sequences of centromeres have been analyzed extensively in the yeast *Saccharomyces cerevisiae.* These sequences are called **CEN sequences,** after the *cen*tromere. Although each yeast centromere has the same function, the *CEN* regions, though similar, are not identical to one another in nucleotide sequence and organization. The common core centromere region in each yeast chromosome consists of 112 to 120 base pairs that can be grouped into three sequence domains (centromere DNA elements, or CDEs; Figure 2.28). CDEII, a 78- to 86-bp region, more than 90 percent of which is composed of A-T base pairs, is the largest domain. To one side is CDEI, which has an 8-bp sequence [R T C A C R T G, where R is a purine (i.e., either A or G)], and to the other side is CDEIII, a 26-bp sequence domain that is also AT rich. Centromere sequences have been determined for a number of other organisms as

Figure 2.28

Consensus sequence for centromeres of the yeast *Saccharomyces cerevisiae*. R = a purine. Base pairs that appear in 15 to 16 of the 16 centromeres are highly conserved and are indicated by capital letters. Base pairs found in 10 to 13 of the 16 centromeres are conserved and are indicated by lowercase letters. Nonconserved positions are indicated by dashes.

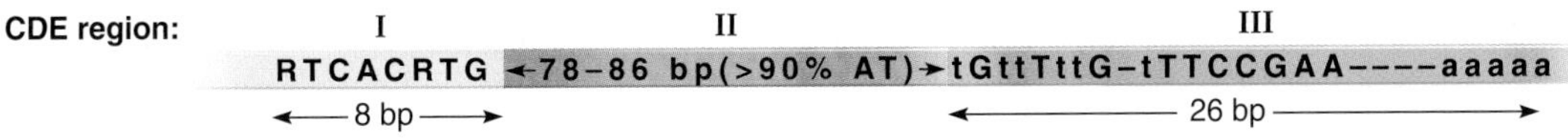

```
CDE region:        I              II                        III
             RTCACRTG ◄-78-86 bp(>90% AT)►tGttTttG-tTTCCGAA----aaaaa
             ◄───8 bp───►                 ◄────── 26 bp ───────►
```

well and are different both from those of yeast and from each other. The centromeres of the fission yeast *Schizosaccharomyces pombe*, for example, are 40–80 kb long, with complex arrangements of several repeated sequences. Human centromeres are even longer, ranging from 240 kb to several million base pairs. Thus, although centromeres carry out the same function in all eukaryotes, there is no common sequence that is responsible for that function.

In yeast, mutational studies have led to the identification of various proteins that interact with the centromere and with the spindle microtubule in the kinetochore structure. A much-simplified model of the yeast kinetochore structure based on the mutational data is shown in Figure 2.29. In this model, Cbf1 (centromere binding factor 1) binds to CDEI, and a complex of proteins making up Cbf3 binds to CDEIII. The longer CDEII sequence may be wrapped once around a histone octamer. Perhaps through a linker protein, Cbf3 binds to the end of a microtubule; other proteins are bound around the end of the microtubule.

A telomere is required for replication and stability of the chromosome. Telomeres are characteristically heterochromatic. In most organisms that have been examined, the telomeres are positioned just inside the nuclear envelope and often are found associated with each other as well as with the nuclear envelope.

All telomeres in a given species share a common sequence. Most telomeric sequences may be divided into two types:

1. **Simple telomeric sequences** are at the extreme ends of the chromosomal DNA molecules. Simple telomeric sequences are the essential functional components of telomeric regions, in that they are sufficient to supply a chromosomal end with stability. These sequences are species specific and consist of a series of simple DNA sequences repeated one after the other (called *tandemly repeated DNA sequences*). In the ciliate *Tetrahymena*, for example, reading the sequence toward the end of one DNA strand, we see that the repeated sequence is 5'-TTGGGG-3' (Figure 2.30a), and in humans the repeated sequence is 5'-TTAGGG-3'. Note that the DNA is not double stranded all the way out to the end. In a new model, the telomere DNA loops back on itself, forming a *t-loop* (Figure 2.30b). The single-stranded end invades the double-stranded telomeric

sequences, causing a *displacement loop*, or *D-loop*, to form. In the next chapter, you will see the role these telomere sequences have in the replication of eukaryotic chromosomes.

2. **Telomere-associated sequences** are regions internal to the simple telomeric sequences. These sequences often contain repeated, but still complex, DNA sequences extending many thousands of base pairs in from the chromosome end. The significance of such sequences is not yet known.

Whereas the telomeres of most eukaryotes contain short, simple, repeated sequences, the telomeres of *Drosophila* are quite different. *Drosophila* telomeres consist of *transposons*—DNA sequences that can move to other locations in the genome.

Figure 2.29

Hypothetical model of the kinetochore of yeast, showing the relationship of the spindle fiber microtubule to the centromere DNA elements. (Adapted with permission from Pluta et al., *Science* 270 (1995):1591–94. Copyright © 1995 American Association for the Advancement of Science.)

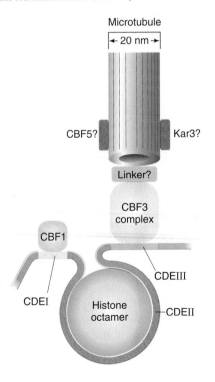

Figure 2.30

Telomeres. (a) Simple telomeric sequences at the ends of *Tetrahymena* chromosomes. (b) Model of telomere structure in which the telomere DNA loops back to form a t-loop. The single-stranded end invades the double-stranded telomeric sequences to produce a displacement loop (D-loop).

a) *Tetrahymena* **simple telomeric sequences**

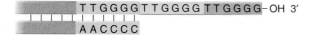

b) t-loop model for telomeres

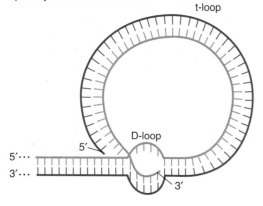

Unique-Sequence and Repetitive-Sequence DNA

Now that you know about the basic structure of DNA and its organization in chromosomes, we can discuss the distribution of certain sequences in the genomes of prokaryotes and eukaryotes. From molecular analyses, geneticists have found that some sequences are present only once in the genome, whereas other sequences are repeated. For convenience, these sequences are grouped into three categories: **unique-sequence DNA** (present in one to a few copies in the genome), **moderately repetitive DNA** (present in a few to about 10^5 copies in the genome), and **highly repetitive DNA** (present in about 10^5 to 10^7 copies in the genome). In prokaryotes, with the exception of the ribosomal RNA genes, transfer RNA genes, and a few other sequences, all of the genome is present as unique-sequence DNA. Eukaryotic genomes, by contrast, consist of both unique-sequence and repetitive-sequence DNA, with the latter typically being quite complex in number of types, number of copies, and distribution. To date, we have sketchy information about the distribution of the various classes of sequences in the genome. However, as the complete DNA sequences of more and more eukaryotic genomes are determined, we will develop a precise understanding of the molecular organization of genomes.

Unique-Sequence DNA. Unique sequences, sometimes called single-copy sequences, are defined as sequences that are present as single copies in the genome. (Thus, there are two copies per diploid cell.) Actually, in current usage, the term usually applies to sequences that have one to a few copies per genome. Most of the genes we know about—those which code for proteins in the cell—are in the unique-sequence class of DNA. But not all unique-sequence material contains protein-coding sequences. In humans, unique sequences are estimated to make up roughly 55–60 percent of the genome.

Repetitive-Sequence DNA. Both moderately repetitive and highly repetitive DNA sequences are sequences that appear many times within a genome. These sequences can be arranged within the genome in one of two ways: distributed at irregular intervals (known as **dispersed repeated DNA** or **interspersed repeated DNA**) or clustered together so that the sequence repeats many times in a row (known as **tandemly repeated DNA**).

The distribution of dispersed repeated sequences has been studied in a variety of organisms. The picture that has emerged is one of families of repeated sequences interspersed through the genome with unique-sequence DNA. Each family consists of a set of related sequences characteristic of the family. Often, small numbers of families have very high copy numbers and make up most of the dispersed repeated sequences in the genome. Two general interspersion patterns are encountered. In one, the families have sequences about 1,000–7,000 bp or more long; these sequences are called **long interspersed elements (LINEs).** In the other, the families have sequences 100–400 bp long; these sequences are called **short interspersed elements (SINEs).** All eukaryotic organisms have LINEs and SINEs, although the relative proportions vary widely. *Drosophila* and birds, for example, have mostly LINEs, whereas humans and frogs have mostly SINEs. LINEs and SINEs represent a significant proportion of all the moderately repetitive DNA in the genome.

Mammalian diploid genomes have about 500,000 copies of the LINE-1 (L1) family of repeated sequences, representing about 15% of the genome. Other LINE families may be present also, but they are much less abundant than LINE-1. Full-length LINE-1 family members are 6–7 kb long, although most are truncated elements of about 1–2 kb. The full-length LINE-1 elements are transposons, meaning that they encode the enzymes for movement of the elements in the genome.

The SINEs pattern of repeated sequences is found in a diverse array of eukaryotic species, including mammals, amphibians, and sea urchins. Each species with SINEs has its own characteristic array of SINE families. A well-studied SINE family is the *Alu* family of certain primates. This family is named for the cleavage site for the restriction enzyme

*Alu*I ("Al-you-one"), typically found in the repeated sequence. In humans, the *Alu* family is the most abundant SINE family in the genome, consisting of 200- to 300-bp sequences repeated as many as a million times and making up about 9 percent of the total haploid DNA. One *Alu* repeat is located every 5,000 bp in the genome, on average. The SINEs are also transposons, but they do not encode the enzymes they need for movement. They can move, however, if those enzymes are supplied by an active LINE transposon.

Unlike dispersed-sequence DNA, tandemly repeated DNA is arranged one after the other in the genome in a head-to-tail organization. Tandemly repeated DNA is quite common in eukaryotic genomes, in some cases in short sequences 1–10 bp long and in other cases associated with genes and in much longer sequences. The simple telomeric sequences shown in Figure 2.30a exemplify the former type of tandemly repeated DNA. Examples of the latter are the genes for ribosomal RNA (rRNA; see Chapter 5) and transfer RNA (tRNA; see Chapter 5), which are tandemly repeated in one or more clusters in most eukaryotes. However, the greatest amount of tandemly repeated DNA is associated with centromeres and telomeres. At each centromere, hundreds to thousands of copies of simple, short tandemly repeated sequences (highly repetitive sequences) are found. Indeed, a significant proportion of the eukaryotic genome may consist of the highly repeated sequences found at centromeres: 8 percent in the mouse, about 50 percent in the kangaroo rat, and about 5–10 percent in humans. (See Chapter 18 for a more detailed discussion of nongenic tandemly repeated DNA, as well as a description of what we have learned from genome sequencing about the organization of genes and repeated sequences in the genomes of human and other organisms of genetic importance.)

KEYNOTE

> Prokaryotic genomes consist mostly of unique-sequence DNA, with only a few sequences and genes repeated. Eukaryotes have both unique and repetitive sequences in the genome, with an extensive, complex spectrum of the repetitive sequences among species. Some of the repetitive sequences are genes, but most are not.

Summary

In this chapter, you have learned that DNA or RNA is the genetic material. All prokaryotic and eukaryotic organisms and most viruses have DNA as their genetic material, whereas some viruses have RNA as their genetic material. The form of the genetic material varies among organisms and viruses. In living organisms the DNA is always double stranded, whereas in viruses the genetic material may be double- or single-stranded DNA or RNA, depending on the virus.

Chemical analysis has revealed that DNA and RNA are macromolecules composed of building blocks called nucleotides. Each nucleotide consists of a five-carbon sugar (deoxyribose in DNA, ribose in RNA) to which is attached one of four nitrogenous bases and a phosphate group. In DNA, the four nitrogenous bases are adenine, guanine, cytosine, and thymine; in RNA, the four bases are adenine, guanine, cytosine, and uracil. From chemical and physical analysis, it was determined that the DNA molecule consists of two polynucleotide chains joined by hydrogen bonds between pairs of bases (A and T; and G and C) in a double helix. The diameter of the helix is 2 nm, and there are 10 base pairs in each complete turn of the helix (3.4 nm).

A number of different types of double-helical DNA have been identified by X-ray diffraction analysis. The three major types are the right-handed A- and B-DNAs and the left-handed Z-DNA. The common form found in cells is B-DNA (the form analyzed by Watson and Crick). A-DNA probably does not exist in cells. Z-DNA may exist in cells, but its function is unknown.

Examples of chromosome organization in bacteriophages were discussed, including the linear double-stranded DNA chromosomes in T-even phages; circular, single-stranded DNA in ΦX174; and the changing structure of the λ double-stranded DNA chromosome. The chromosomes of prokaryotic organisms consist of circular, double-stranded DNA molecules that are complexed with a number of proteins. The chromosome is condensed within the cell by supercoiling of the DNA helix and the formation of looped domains of supercoiled DNA.

A distinguishing feature of eukaryotic chromosomes is that DNA is distributed among a number of them. The DNA in the genome specifies an organism's structure and function. The amount of DNA in a genome (made up of a prokaryotic chromosome or of the haploid set of chromosomes in eukaryotes) is called the C value. There is no direct relationship between the C value and the structural or organizational complexity of an organism.

As regards the organization of DNA in eukaryotic chromosomes, the chromosomes of eukaryotes are complexes of DNA with histone and nonhistone chromosomal proteins. Each unreplicated chromosome consists of one linear, double-stranded DNA molecule that runs through its length. As a class, the histones are constant from cell to cell within an organism and have been evolutionarily conserved. Nonhistone chromosomal proteins, by contrast, vary significantly among cell types and among organisms. DNA in chromatin is highly condensed by its association with histones to form nucleosomes and by the several higher levels of folding of nucleosomes into chromatin fibers. Like prokaryotic chromosomes, eukaryotic chromosomes are organized into a large number of looped domains. The loops are attached to a protein scaffold. The most highly condensed structure is seen in metaphase chromosomes, the least condensed in interphase chromosomes. The factors controlling the transitions between different levels of chromosome folding are not well understood.

Two specialized eukaryotic chromosome structures are centromeres and telomeres. Significant progress has been made in defining the sequences of centromeres and telomeres; for example, the DNA sequences of centromeres are complex and are species specific, and the DNA sequences of telomeres are simple, short, tandemly repeated sequences that are also species specific.

Molecular analysis has provided information about the distribution of DNA sequences in genomes. Such analysis has revealed that prokaryotic and viral genomes consist mostly of unique-sequence DNA, with only a few repeated sequences. In contrast, the genomes of eukaryotes contain both unique-sequence DNA and repeated-sequence DNA, and there is a wide spectrum of complexity of the repeated DNA sequences. Many genes are found in unique-sequence DNA, but not all unique-sequence DNA contains genes.

Repeated DNA sequences may be arranged tandemly or interspersed with unique-sequence DNA in the eukaryotic genome. Although some tandemly repeated DNA is composed of gene families, the greatest amounts are associated, not with genes, but with centromeres and telomeres. At each centromere, for example, hundreds to thousands of copies of simple, short, tandemly repeated sequences may be found. Dispersed repeated sequences include both gene sequences and nongene sequences. Dispersed repeated gene sequences typically are present in low numbers (fewer than 50), whereas dispersed nongene sequences may be present in hundreds of thousands of copies.

Eukaryotic species have characteristic families of repeated sequences interspersed with unique-sequence DNA throughout the genome. The two general interspersion families are LINEs (long interspersed elements) and SINEs (short interspersed repeats). LINEs and SINEs are found in all eukaryotes, although the relative proportions vary widely. Both of these families of repeated elements are transposons, meaning that they can move in the genome. Only full-length LINEs encode enzymes for the movement, and those enzymes can move truncated LINEs or SINEs, neither of which encode their own enzymes for movement.

Analytical Approaches to Solving Genetics Problems

The most practical way to reinforce genetics principles is to solve genetics problems. In this and all subsequent chapters, we discuss how to approach genetics problems by presenting examples of such problems and discussing their answers. The problems use familiar and unfamiliar examples and pose questions designed to get you to think analytically.

Q2.1 The linear chromosome of phage T2 is 52 μm long. The chromosome consists of double-stranded DNA, with 0.34 nm between each base pair. How many base pairs does a chromosome of T2 contain?

A2.1 This question involves the careful conversion of different units of measurement. The first step is to put the lengths in the same units: 52 μm is 52 millionths of a meter, or $52,000 \times 10^9$ m, or 52,000 nm. One base occupies 0.34 nm in the double helix, so the number of base pairs in the chromosome of T2 is 52,000 divided by 0.34, or 152,941 base pairs.

The human genome contains 3×10^9 bp of DNA, for a total length of about 1 meter, distributed among 23 chromosomes. The average length of the double helix in a human chromosome is 3.8 cm, which is 3.8 hundredths of a meter, or 38 million nm—much longer than the T2 chromosome! There are more than 111.7 million base pairs in the average human chromosome.

Q2.2 The following table lists the relative percentages of bases of nucleic acids isolated from different species:

Species	Adenine	Guanine	Thymine	Cytosine	Uracil
(i)	21	29	21	29	0
(ii)	29	21	29	21	0
(iii)	21	21	29	29	0
(iv)	21	29	0	29	21
(v)	21	29	0	21	29

For each species, what type of nucleic acid is involved? Is it double or single stranded? Explain your answer.

A2.2 This question focuses on the base-pairing rules and the difference between DNA and RNA. In analyzing the data, we should determine first whether the nucleic acid is RNA or DNA and then whether it is double or single stranded. If the nucleic acid has thymine, it is DNA; if it has uracil, it is RNA. Thus, species (i), (ii), and (iii) must have DNA as their genetic material, and species (iv) and (v) must have RNA as their genetic material.

Next, we must analyze the data for strandedness. Double-stranded DNA must have equal percentages of A and T and of G and C. Similarly, double-stranded RNA must have equal percentages of A and U and of G and C. Therefore, species (i) and (ii) have double-stranded DNA, whereas species (iii) must have single-stranded DNA, because the base-pairing rules are violated, with A = G and T = C, but A ≠ T and G ≠ C. As for the RNA-containing species, (iv) contains double-stranded RNA, because A = U and G = C, and (v) must contain single-stranded RNA.

Q2.3 Here are four characteristics of one 5'-3' strand of a particular long, double-stranded DNA molecule:

i. Thirty-five percent of the adenine-containing nucleotides (As) have guanine-containing nucleotides (Gs) on their 3′ sides.

ii. Thirty percent of the As have Ts as their 3′ neighbors.

iii. Twenty-five percent of the As have Cs as their 3′ neighbors.

iv. Ten percent of the As have As as their 3′ neighbors.

Use the preceding information to answer the following questions as completely as possible, explaining your reasoning in each case:

a. In the complementary DNA strand, what will be the frequencies of the various bases on the 3′ side of A?

b. In the complementary strand, what will be the frequencies of the various bases on the 3′ side of T?

c. In the complementary strand, what will be the frequency of each kind of base on the 5′ side of T?

d. Why is the percentage of A not equal to the percentage of T (and the percentage of C not equal to the percentage of G) among the 3′ neighbors of A in the 5′-to-3′ DNA strand described?

A2.3

a. This question cannot be answered without more information. Although we know that the As neighbored by Ts in the original strand will correspond to As neighbored by Ts in the complementary strand, there will be additional As in the complementary strand about whose neighbors we know nothing.

b. This question cannot be answered. All the As in the original strand correspond to Ts in the complementary strand, but we know only about the 5′ neighbors of these Ts, not the 3′ neighbors.

c. On the original strand, 35 percent were 5′-AG-3′, so on the complementary strand, 35 percent of the sequences will be 3′-TC-5′. Thus, 35 percent of the bases on the 5′ side of T will be C. Similarly, on the original strand, 30 percent were 5′-AT-3′, 25 percent were 5′-AC-3′, and 10 percent were 5′-AA-3′, meaning that, on the complementary strand, 30 percent of the sequences were 3′-TA-5′, 25 percent were 3′-TG-5′, and 10 percent were 3′-TT-5′. So 30 percent of the bases on the 5′ side of T will be A, 25 percent will be G, and 10 percent will be T.

d. The A = T and G = C rule applies only when one is considering both strands of a double-stranded DNA. Here, we are considering only the original single strand of DNA.

Q2.4 When double-stranded DNA is heated to 100°C, the two strands separate because the hydrogen bonds between the strands break. Depending on the conditions, when the solution is cooled, the two strands can find each other and re-form the double helix, a process

called renaturation or reannealing. Consider the DNA double helix:

```
┬┬┬┬┬┬┬┬┬┬┬┬┬┬
G C G C G C G C G C G C G C
C G C G C G C G C G C G C G
┴┴┴┴┴┴┴┴┴┴┴┴┴┴
```

If this DNA is heated to 100°C and then cooled, what might be the structure of the single strands if the two strands never find one another?

A2.4 This question serves two purposes. First, it reinforces certain information about double-stranded DNA, and, second, it poses a problem that can be solved by simple logic.

We can analyze the base sequences themselves to see whether there is anything special about them and avoid an answer of "Nothing significant happens." The DNA is a 14-base-pair segment of alternating G-C and C-G base pairs. By examining just one of the strands, we can see that there is an axis of symmetry at the midpoint such that it is possible for the single strand to form a double-stranded DNA molecule by intrastrand (within-strand) base pairing. The result is a double-stranded hairpin structure, as shown in the following diagram (from the top strand; the other strand will also form a hairpin structure):

Questions and Problems

2.1 Griffith's experiment injecting a mixture of dead and live bacteria into mice demonstrated that (choose the correct answer):

a. DNA is double stranded.

b. mRNA of eukaryotes differs from mRNA of prokaryotes.

c. a factor was capable of transforming one bacterial cell type to another.

d. bacteria can recover from heat treatment if live helper cells are present.

*2.2 In the 1920s, while working with *Streptococcus pneumoniae* (the agent that causes pneumonia), Griffith injected mice with different types of bacteria. For each of the following bacteria types injected, indicate whether the mice lived or died:

a. type *IIR*

b. type *IIIS*

c. heat-killed *IIIS*

d. type *IIR* + heat-killed *IIIS*

2.3 Several years after Griffith described the transforming principle, Avery, MacLeod, and McCarty investigated the same phenomenon.

a. List the steps they used to show that DNA from dead *S. pneumoniae* cells was responsible for the change from a nonvirulent to a virulent state.

b. Did their work confirm or disconfirm Griffith's work, and how?

c. What was the role of the enzymes used in their experiments?

*2.4 Hershey and Chase showed that when phages were labeled with ^{32}P and ^{35}S, the ^{35}S remained outside the cell and could be removed without affecting the course of infection, whereas the ^{32}P entered the cell and could be recovered in progeny phages. What distribution of isotopes would you expect to see if parental phages were labeled with isotopes of

a. C?

b. N?

c. H?

Explain your answer.

2.5 What is the evidence that the genetic material of tobacco mosaic virus (TMV) is RNA?

2.6 The X-ray diffraction data obtained by Rosalind Franklin suggested (choose the correct answer)

a. DNA is a helix with a pattern that repeats every 3.4 nm

b. purines are hydrogen bonded to pyrimidines.

c. DNA is a left-handed helix.

d. DNA is organized into nucleosomes.

2.7 What evidence do we have that, in the helical form of the DNA molecule, the base pairs are composed of one purine and one pyrimidine?

2.8 What exactly is a deoxyribonucleotide made up of, and how many different deoxyribonucleotides are there in DNA? Describe the structure of DNA and describe the bonding mechanism of the molecule (i.e., the kind of bonds on the sides of the "ladder" and the kind of bonds hold the two complementary strands together). Base pairing in DNA consists of purine–pyrimidine pairs, so why can't A-C and G-T pairs form?

*2.9 What is the base sequence of the DNA strand that would be complementary to the following single-stranded DNA molecules?

a. 5'-AGTTACCTGATCGTA-3'

b. 5'-TTCTCAAGAATTCCA-3'

*2.10 Describe the bonding properties of G-C and T-A. Which would be the hardest to break? Why?

2.11 The double-helix model of DNA, as suggested by Watson and Crick, was based on DNA data gathered by other researchers. The facts fell into the following two general categories:

a. chemical composition

b. physical structure

Give two examples of each.

*2.12 For double-stranded DNA, which of the following base ratios always equals 1?

a. $(A + T)/(G + C)$

b. $(A + G)/(C + T)$

c. C/G

d. $(G + T)/(A + C)$

e. A/G

2.13 If the ratio of $(A + T)$ to $(G + C)$ in a particular DNA is 1.0. Does this ratio indicate that the DNA is probably composed of two complementary strands of DNA or a single strand of DNA, or is more information necessary?

2.14 The percentage of cytosine in a double-stranded DNA is 17. What is the percentage of adenine in that DNA?

*2.15 A double-stranded DNA polynucleotide contains 80 thymidylic acid and 110 deoxyguanylic acid residues. What is the total nucleotide number in this DNA fragment?

*2.16 Analysis of DNA from a bacterial virus indicates that it contains 33 percent A, 26 percent T, 18 percent G, and 23 percent C. Interpret these data.

*2.17 The following are melting temperatures for different double-stranded DNA molecules:

a. 73°C   **d.** 78°C

b. 69°C   **e.** 82°C

c. 84°C

Arrange these molecules from lower to higher content of G-C pairs.

2.18 What is a DNA oligomer?

2.19 The genetic material of bacteriophage ΦX174 is single-stranded DNA. What base equalities or inequalities might we expect for single-stranded DNA?

2.20 Through X-ray diffraction analysis of crystallized DNA oligomers, different forms of DNA have been identified. These forms include A-DNA, B-DNA, and Z-DNA, and each has unique molecular attributes.

a. Which form is the most common form in living cells?

b. Z-DNA has an unusual conformation resulting in more base pairs per helical turn than B-DNA has. What is the conformation? Does this molecule have any function in living cells?

c. Which of the given forms is never found in living cells?

2.21 If a virus particle contains 200,000 bp of double-stranded DNA, how many complete 360° turns occur in its genome?

*2.22 A double-stranded DNA molecule is 100,000 bp (100 kb) long.
a. How many nucleotides does it contain?
b. How many complete turns are there in the molecule?
c. How long is the DNA molecule?

2.23 Different organisms have vastly different amounts of genetic material. *E. coli* has about 4.6 million bp of DNA in one circular chromosome, the haploid budding yeast (*S. cerevisiae*) has 12,067,280 bp of DNA in 16 chromosomes, and the gametes of humans have about 2.75 billion bp of DNA in 23 chromosomes.
a. For each of these organism's cells, if all of the DNA were B-DNA, what would be the average length of a chromosome in the cell?
b. On average, how many complete turns would be in each chromosome?
c. Would your answers to (a) and (b) be significantly different if the DNA were composed of, say, 20 percent Z-DNA and 80 percent B-DNA?
d. What implications do your answers to these questions have for the packaging of DNA in cells?

*2.24 If nucleotides were arranged at random in a piece of single-stranded RNA 10^6 nucleotides long, and if the base composition of this RNA was 20 percent A, 25 percent C, 25 percent U, and 30 percent G, how many times would you expect the specific sequence 5'-GUUA-3' to occur?

*2.25 Two double-stranded DNA molecules from a population of T2 phages were denatured to single strands by heat treatment. The result was the following four single-stranded DNAs:

1 TAGCTCC →   3 GCTCCTA →
    and
2 ← ATCGAGG   4 ← CGAGGAT

These separated strands were then allowed to renature. Diagram the structures of the renatured molecules most likely to appear when (a) strand 2 renatures with strand 3 and (b) strand 3 renatures with strand 4. Label the strands and indicate sequences and polarity.

2.26 Define topoisomerases, and list the functions of these enzymes.

2.27 What is the relationship between cellular DNA content and the structural or organizational complexity of the organism?

2.28 In a particular eukaryotic chromosome (choose the best answer),
a. heterochromatin and euchromatin are regions where genes make functional gene products (that is, where genes are active).
b. heterochromatin is active, but euchromatin is inactive.
c. heterochromatin is inactive, but euchromatin is active.
d. both heterochromatin and euchromatin are inactive.

*2.29 Compare and contrast eukaryotic chromosomes and bacterial chromosomes with respect to the following features:
a. centromeres   f. nonhistone protein scaffolds
b. hexose sugars   g. DNA
c. amino acids   h. nucleosomes
d. supercoiling   i. circular chromosome
e. telomeres   j. looping

2.30 Discuss the structure of nucleosomes, their hierarchical packaging, and the composition of a core particle of a nucleosome. Also, discuss the functions of nucleosomes.

*2.31 Set up the following "rope trick": Start with a belt (representing a DNA molecule; imagine the phosphodiester backbones lying along the top and bottom edges of the belt) and a soda can. Holding the belt buckle at the bottom of the can, wrap the belt flat against the side of the can, counterclockwise three times around the can. Now remove the "core" soda can, and, holding the ends of the belt, pull the ends of the belt taut. After some reflection, answer the following questions:
a. Did you make a left- or a right-handed helix?
b. How many helical turns were present in the coiled belt before it was pulled taut?
c. How many helical turns were present in the coiled belt after it was pulled taut?
d. Why does the belt appear more twisted when pulled taut?
e. About what percentage of the length of the belt was decreased by this packaging?
f. Is the DNA of a linear chromosome that is coiled around histones supercoiled?
g. Why are topoisomerases necessary to package linear chromosomes?

*2.32 What are the main molecular features of yeast centromeres?

2.33 Telomeres are unique repeated sequences. Where on the DNA strand are they found? Do they serve a function?

*2.34 Would you expect to find most protein-coding genes in unique-sequence DNA, in moderately repetitive DNA, or in highly repetitive DNA?

2.35 Both histone and nonhistone proteins are essential for DNA packaging in eukaryotic cells. However, these classes of proteins are fundamentally dissimilar in a number of ways. Describe how they differ in terms of

a. their protein characteristics
b. their presence and abundance in cells
c. their interactions with DNA
d. their role in DNA packaging

*2.36 Rearrangements at the end of 16p (the short arm of chromosome 16) underlie a variety of common human genetic disorders, including β-thalassemia (a defect in hemoglobin metabolism caused by mutations in the β-globin gene), mental retardation, and the adult form of polycystic kidney disease. The availability of approximately 285-kilobase (kb) pairs of DNA sequence at the end of human chromosome 16p has allowed very detailed analysis of the structure of this chromosome region. The first functional gene lies about 44 kb from the region of simple telomeric sequences and about 8 kb from the telomere-associated sequences. Analysis of sequences proximal (nearer the centromere) to the first gene reveals a sinusoidal variation in GC content, with GC-rich regions associated with gene-rich areas and AT-rich regions associated with *Alu*-dense areas. The β-globin gene lies about 130 kb from the telomere-associated sequences.

a. Discuss these findings in light of the current view of telomere structure and function as presented in the text.

b. What new information have the preceding data revealed about the distribution of SINEs in the terminus of 16p? (SINEs and LINEs are, respectively, short and long interspersed nuclear elements.)

3

DNA Replication

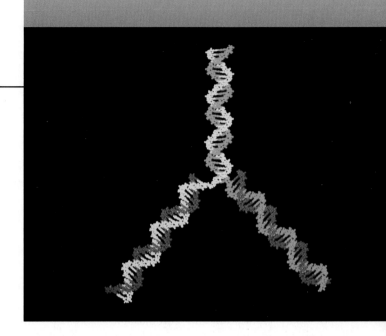

PRINCIPAL POINTS

- DNA replication in prokaryotes and eukaryotes occurs by a semiconservative mechanism in which the two strands of a DNA double helix are separated and a new complementary strand of DNA is synthesized on each of the two parental template strands. This mechanism ensures that genetic information will be copied faithfully at each cell division.

- The enzymes called DNA polymerases catalyze the synthesis of DNA. Using deoxyribonucleoside 5′-triphosphate (dNTP) precursors, all DNA polymerases make new strands in the 5′-to-3′ direction.

- DNA polymerases cannot initiate the synthesis of a new DNA strand. All new DNA strands need a short primer of RNA, the synthesis of which is catalyzed by the enzyme DNA primase.

- DNA replication in *E. coli* requires two DNA polymerases and several other enzymes and proteins. In both prokaryotes and eukaryotes, the synthesis of DNA is

continuous on one template strand and discontinuous on the other template strand, a process called semidiscontinuous replication.

- In eukaryotes, DNA replication occurs in the S phase of the cell cycle and is biochemically and molecularly similar to replication in prokaryotes. To enable long chromosomes to replicate efficiently, DNA replication is initiated at many sites (origins) along the chromosomes and proceeds bidirectionally.

- Special enzymes—telomerases—replicate the ends of chromosomes in eukaryotes. A telomerase is a complex of proteins and RNA. The RNA acts as a template for the synthesis of the complementary telomere repeat of the chromosome. In mammals, telomerase activity is limited to immortal cells (such as tumor cells). The absence of telomerase activity in nontumor cells results in a progressive shortening of chromosome ends as the cell divides, thereby limiting the number of cell divisions before the cell dies.

ⓘ A BASIC PROPERTY OF GENETIC MATERIAL IS ITS ability to replicate in a precise way so that the genetic information encoded in the nucleotides can be transmitted from each cell to all of its progeny. James Watson and Francis Crick recognized that the complementary relationship between DNA strands probably would be the basis for DNA replication. However, even after scientists confirmed this fact five years after Watson and Crick developed their model, many questions about the mechanics of DNA replication remained. In this chapter, you will learn about the steps and enzymes involved in the replication of prokaryotic and eukaryotic DNA molecules. Then, in the iActivity, you will have a chance to investigate the specifics of DNA replication in *E. coli.*

Your goal in the chapter is to learn about the mechanisms of DNA replication and chromosome duplication in prokaryotes and eukaryotes and about some of the enzymes and other proteins needed for replication. Some of these enzymes are also involved in the repair of damage to DNA, a topic we discuss in Chapter 7.

Semiconservative DNA Replication

When Watson and Crick proposed their double-helix model for DNA in 1953, they realized that DNA replication would be straightforward if their model was correct. That is, if the DNA molecule were unwound and the two strands separated, each strand would be a template for the synthesis of a new, complementary strand of DNA that would remain bound to the parental strand. This model for DNA replication is known as the **semiconservative model,** because each progeny molecule retains one of the parental strands (Figure 3.1a).

At the time, two other models for DNA replication were proposed: the **conservative model** (Figure 3.1b) and the **dispersive model** (Figure 3.1c). In the conservative model, the two parental strands of DNA remain together or pair again after replication and, as a whole, serve as a template for the synthesis of new progeny DNA double helices. Thus, one of the two progeny DNA molecules actually is the parental double-stranded DNA molecule, and the other consists of new material. In the dispersive model, the parental double helix is cleaved into double-stranded DNA segments that act as templates for the synthesis of new double-stranded DNA segments. Somehow, the segments reassemble into complete DNA double helices, with parental and progeny DNA segments interspersed. Thus, although the two progeny DNAs are identical with respect to their base-pair sequence, double-stranded parental DNA has become dispersed throughout both progeny molecules. It is hard to imagine how the DNA sequences of chromosomes could be kept the same with such a

Figure 3.1

Three models for DNA replication. (a) Semiconservative model (the correct model). (b) Conservative model. (c) Dispersive model. The parental strands are shown in taupe, and the newly synthesized strands are shown in red.

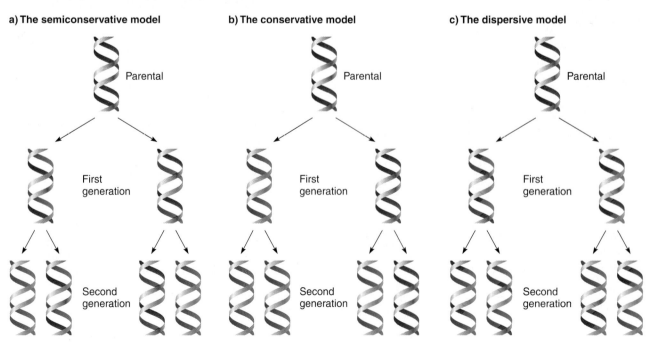

a) **The semiconservative model**

Parental

First generation

Second generation

b) **The conservative model**

Parental

First generation

Second generation

c) **The dispersive model**

Parental

First generation

Second generation

mechanism. We include this model only for historical completeness.

The Meselson–Stahl Experiment

In 1958, Matthew Meselson and Frank Stahl obtained experimental evidence that the semiconservative replication model was correct. Meselson and Stahl grew

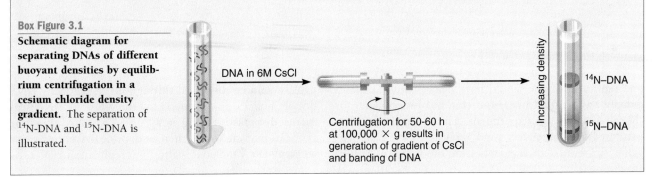

animation

The Meselson–Stahl Experiment

E. coli in a medium in which the only nitrogen source was $^{15}NH_4Cl$ (ammonium chloride) (Figure 3.2). In this compound, the normal isotope of nitrogen, ^{14}N, is replaced with ^{15}N, the heavy isotope. (*Note:* Density is weight divided by volume, so ^{15}N, with one extra neutron in its nucleus, is $^1/_{14}$ denser than ^{14}N.) As a result, all the bacteria's nitrogen-containing compounds, including its DNA, contained ^{15}N instead of ^{14}N. ^{15}N DNA can be separated from ^{14}N DNA by using equilibrium density gradient centrifugation (described in Box 3.1). Briefly, in this technique, through high-speed centrifugation, a solution of cesium chloride (CsCl) forms a density gradient, with the lightest material at the top of the tube and the densest material at the bottom. If DNA is present in the solution, it forms a band at a position where its buoyant density is the same as that of the surrounding cesium chloride.

Next, the ^{15}N-labeled bacteria were transferred to a medium containing nitrogen in the normal ^{14}N form, and the bacteria were allowed to reproduce for several gener-

ations. During this time, samples of *E. coli* were taken and the DNA was extracted and analyzed in CsCl density gradients. (See Figure 3.2.) After one replication cycle (one generation) in the ^{14}N medium, all of the DNA had a density that was exactly intermediate between that of ^{15}N DNA and that of ^{14}N DNA. After two replication cycles, half the DNA was of the intermediate density and half was of the density of DNA containing entirely ^{14}N. These observations, presented in Figure 3.2, and those obtained from subsequent replication cycles were exactly what the semiconservative model predicted.

If the conservative model for DNA replication were correct, after one replication cycle there would have been two bands of DNA: one in the heavy-density position of the gradient containing parental DNA molecules with both strands labeled with ^{15}N, the other in the light-density position containing progeny DNA molecules with both strands labeled with ^{14}N. (See Figure 3.1b.) The heavy parental DNA band would have been seen at each subsequent replication cycle, in the amount found at the start of the experiment. All new DNA molecules would than have had both strands labeled with ^{14}N. Therefore, the amount of DNA in the light-density position would have increased with each replication cycle. In the conservative model of DNA replication, then, the most significant prediction was that *at no time would any DNA of intermediate density be found*. The fact that intermediate-density DNA *was* found ruled out the conservative model.

If the dispersive model for DNA replication were correct, then all DNA present in the ^{14}N medium after

---

Box 3.1 | Equilibrium Density Gradient Centrifugation

In equilibrium density gradient centrifugation, a concentrated solution of cesium chloride (CsCl) is centrifuged at high speed to produce a linear concentration gradient of the CsCl. The actual densities of CsCl at the extremes of the gradient are related to the CsCl concentration that is centrifuged.

For example, to examine DNA of density 1.70 g/cm³ (a typical density for DNA), one makes a gradient which spans that density—for example, from 1.60 to 1.80 g/cm³. If DNA

is mixed with the CsCl and the mixture is centrifuged, the DNA comes to equilibrium at the point in the gradient where its buoyant density equals the density of the surrounding CsCl. (See the accompanying figure.) The DNA is said to have *banded* in the gradient. If DNAs that have different densities are present, as is the case with ^{15}N DNA and ^{14}N DNA, they band (come to equilibrium) in different positions. The DNA is detected in the gradient by its ultraviolet absorption.

Box Figure 3.1

Schematic diagram for separating DNAs of different buoyant densities by equilibrium centrifugation in a cesium chloride density gradient. The separation of ^{14}N-DNA and ^{15}N-DNA is illustrated.

DNA in 6M CsCl

Centrifugation for 50-60 h at 100,000 × g results in generation of gradient of CsCl and banding of DNA

Increasing density

^{14}N–DNA

^{15}N–DNA

Figure 3.2

The Meselson–Stahl experiment. The demonstration of semiconservative replication in *E. coli*. Cells were grown in a ^{15}N-containing medium for several replication cycles and then were transferred to a ^{14}N-containing medium. At various times over several replication cycles, samples were taken; the DNA was extracted and analyzed by CsCl equilibrium density gradient centrifugation. Shown in the figure are a schematic interpretation of the DNA composition after various replication cycles, photographs of the DNA bands, and densitometric scans of the bands.

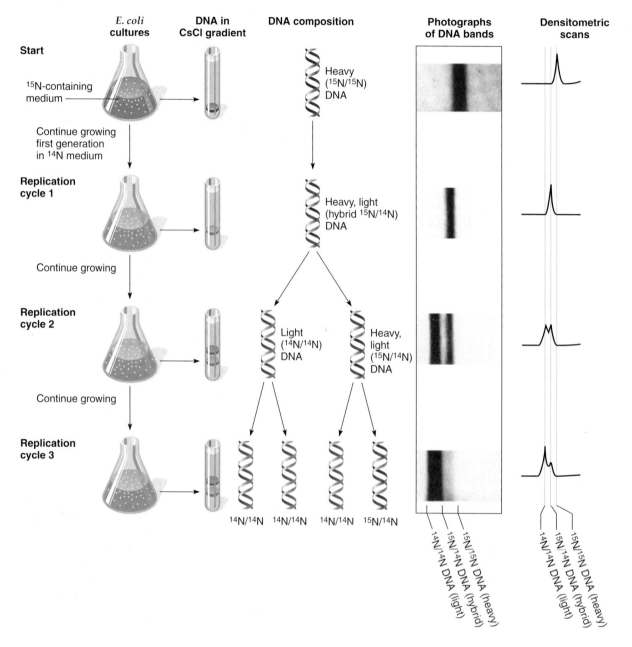

one replication cycle would have been of intermediate density (see Figure 3.1c), and this was indeed seen in the Meselson–Stahl experiment. The dispersive model predicted that, after a second replication cycle in the same medium, DNA segments from the first replication cycle would be dispersed throughout the progeny DNA double helices produced. Thus, the ^{15}N–^{15}N DNA segments dispersed among new ^{14}N–^{14}N DNA after one replication cycle would then be distributed among twice as many DNA molecules after two replication cycles. As a result, the DNA molecules would be found in one band located halfway between the intermediate-density and light-density band positions. With subsequent replication cycles, there would continue to be one band, and it would become lighter in density with each replication cycle. The results of the Meselson–Stahl experiment did not bear out this prediction, so the dispersive model was ruled out.

Semiconservative DNA Replication in Eukaryotes

DNA replication in eukaryotes also occurs by a semiconservative mechanism. One way in which the process can be visualized is by staining replicating chromosomes in cells growing in tissue culture with 5-bromodeoxyuridine (BUdR). BUdR is a *base analog,* with a structure similar to that of the normal base thymine in DNA. During replication, wherever a T is called for in the new strand, BUdR can be incorporated instead. Counterintuitively, mitotic chromosomes stained with a fluorescent dye and Giemsa stain become *lighter* as the amount of BUdR incorporated *increases.*

To examine DNA replication, Chinese hamster ovary (CHO) cells in culture were treated with BUdR. After two cycles of replication, mitotic chromosomes were stained, with the result shown in Figure 3.3. Sister chromatids are visible, one darkly stained and one lightly stained. The dark–light staining pattern brings to mind the costumes of harlequins, so the chromosomes are called *harlequin chromosomes.* The interpretation of the pattern is as follows: The darker staining chromatids have less BUdR in their DNA than the lighter staining chromatids. That is, if semiconservative replication occurs, then, after one replication cycle, two progeny DNAs will be produced, each with one T strand and one BUdR strand. Following that, a second replication cycle in the presence of BUdR will produce one sister chromatid consisting of DNA with one T-containing strand and one BUdR-containing strand (darkly staining) and another sister chromatid consisting of DNA with two BUdR-containing strands (lightly staining). This is the staining pattern that was observed, showing that semiconservative DNA replication also is the correct model in eukaryotes.

KEYNOTE

DNA replication in *E. coli* and other prokaryotes and in eukaryotes occurs by a semiconservative mechanism in which the strands of a DNA double helix separate and a new complementary strand of DNA is synthesized on each of the two parental template strands. Semiconservative replication results in two double-stranded DNA molecules, each of which has one strand from the parent molecule and one newly synthesized strand.

Figure 3.3

Visualization of semiconservative DNA replication in eukaryotes. Shown are harlequin chromosomes in Chinese hamster ovary cells that have been allowed to go through two rounds of DNA replication in the presence of the base analog 5-bromodeoxyuridine, followed by staining. Arrows indicate the sites where crossing-over has occurred.

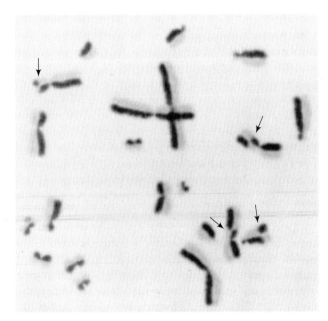

DNA Polymerases, the DNA Replicating Enzymes

In 1955, Arthur Kornberg and his colleagues were the first to identify the enzymes necessary for DNA replication. Their work focused on bacteria, because it was assumed that bacterial replication machinery would be less complex than that of eukaryotes. Kornberg shared the 1959 Nobel Prize in Physiology or Medicine for his "discovery of the mechanisms in the biological synthesis of deoxyribonucleic acid."

DNA Polymerase I

Kornberg's approach was to identify all the ingredients needed to synthesize *E. coli* DNA in vitro. The first successful DNA synthesis was accomplished in a reaction mixture containing DNA fragments—a mixture of four deoxyribonucleoside 5′-triphosphate precursors (dATP, dGTP, dTTP, and dCTP, collectively abbreviated dNTP, for *deoxyribonucleoside triphosphate*)—and an *E. coli* lysate (cells of the bacteria, broken open to release their contents). To measure the minute quantities of DNA expected to be synthesized in the reaction, Kornberg used radioactively labeled dNTPs.

Kornberg analyzed the lysate and isolated an enzyme that was capable of DNA synthesis. This enzyme was originally called the *Kornberg enzyme,* but is now called **DNA polymerase I** (or **DNA Pol I;** by definition, enzymes that catalyze DNA synthesis are called **DNA polymerases**).

With DNA Pol I isolated, more detailed information could be obtained about DNA synthesis in vitro. Researchers found that four components were needed. If any one of the following four components was omitted, DNA synthesis would not occur:

1. All four dNTPs. (If any one dNTP is missing, no synthesis occurs.) These molecules are the precursors for the nucleotide (phosphate-sugar-base) building blocks of DNA described in Chapter 10 (p. 258).
2. A fragment of DNA to act as a template.
3. DNA Pol I.
4. Magnesium ions (Mg^{2+}), needed for optimal DNA polymerase activity.

Subsequent experiments showed that the fragments of DNA acted as a template for the synthesis of the new DNA; that is, the new DNA made in vitro was a faithful base-pair–for–base-pair copy of the original DNA.

Roles of DNA Polymerases

All DNA polymerases from prokaryotes and eukaryotes catalyze the polymerization of nucleotide precursors (dNTPs) into a DNA chain (Figure 3.4a). The same reaction in shorthand notation is shown in Figure 3.4b. The reaction has three main features:

animation

a **DNA Biosynthesis: How a New DNA Strand Is Made**

1. At the growing end of the DNA chain, DNA polymerase catalyzes the formation of a phosphodiester bond between the 3'-OH group of the deoxyribose on the last nucleotide and the 5'-phosphate of the dNTP precursor. The energy for the formation of the phosphodiester bond comes from the release of two of three phosphates from the dNTP. The important concept here is that *the lengthening DNA chain acts as a primer in the reaction*—a preexisting polynucleotide chain to which a new nucleotide can be added at the free 3'-OH.
2. At each step in lengthening the new DNA chain, DNA polymerase finds the correct precursor dNTP that can form a complementary base pair with the

nucleotide on the template strand of DNA. Nucleotides are added rapidly—for example, 850 per second in *E. coli* and 60–90 per second in human tissue culture cells. The process does not occur with 100 percent accuracy, but the error frequency is very low.

3. The direction of synthesis of the new DNA chain is only from 5' to 3', because of the properties of DNA polymerase.

One of the best understood systems of DNA replication is that of *E. coli.* For several years after the discovery of DNA polymerase I, scientists believed that that enzyme was the only DNA replication enzyme in *E. coli.* However, genetic studies disproved that hypothesis. One way to study the action of an enzyme in vivo is to induce a mutation in the gene that codes for the enzyme. In this way, the phenotypic consequences of the mutation can be compared with the wild-type phenotype. The first DNA Pol I mutant, *polA1,* was isolated in 1969 by Peter DeLucia and John Cairns. This mutant shows less than 1 percent of normal polymerizing activity and near-normal 5'→3' exonuclease activity. DNA polymerase was expected to be essential to cell function, so a mutation in the gene for that enzyme was expected to be lethal or at least crippling. Unexpectedly, however, *E. coli* cells carrying the *polA1* mutation grew and divided normally. Nonetheless, *polA1* mutants have a high mutation rate when they are exposed to ultraviolet (UV) light and chemical mutagens, a property which was interpreted to mean that DNA polymerase I has an important function in repairing damaged (chemically changed) DNA.

To study the consequences of mutations in genes coding for essential proteins and enzymes, geneticists find it easiest to work with **temperature-sensitive mutants**—mutants that function normally until the temperature is raised past some threshold level, at which time some temperature-sensitive defect is manifested. At *E. coli*'s normal growth temperature of 37°C, temperature-sensitive *polAex1* mutant strains produce DNA Pol I with normal activity. In vitro at 42°C, however, the temperature-sensitive DNA Pol I has near-normal polymerizing activity, but is defective in 5'→3' exonuclease activity (the progressive removal of nucleotides from a free 5' end toward the 3' end). A 42°C temperature-sensitive *polAex1* mutants die (are lethal) showing that 5'→3' exonuclease activity of DNA Pol I is essential to DNA replication. Taken together, the results of studies of the *polA1* and *polAex1* DNA Pol I mutants indicated that there must be other DNA-polymerizing enzymes in the cell. With improvements in preparing cell extracts and in enzyme assay techniques, two new *E. coli* DNA polymerases were identified. Martin Gefter, Rolf Knippers, and C. C. Richardson, all working independently, discovered DNA Pol II in 1970, and Tom Kornberg and

Figure 3.4

DNA chain elongation catalyzed by DNA polymerase. (a) Mechanism at molecular level. (b) The same mechanism, using a shorthand method to represent DNA.

a) Mechanism of DNA elongation

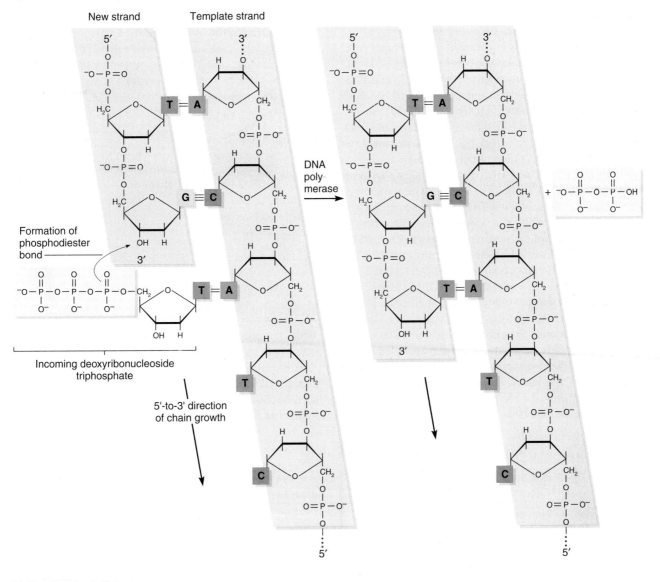

b) Shorthand notation

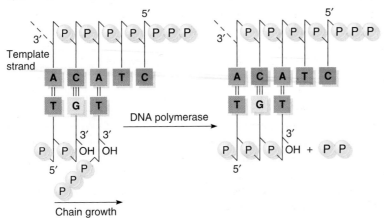

Gefter, working together, discovered DNA Pol III in 1971. Since that time, two other *E. coli* DNA polymerases—DNA Pol IV and DNA Pol V—have been discovered. DNA Pol II (*polB* gene), DNA Pol IV (*dinB* gene), and DNA Pol V (*umuDC* gene) are polymerases that function in DNA repair, while DNA Pol I and DNA Pol III are the polymerases necessary for replication.

DNA polymerase I is encoded by the *polA* gene and consists of one polypeptide. The core DNA polymerase III contains the catalytic functions of the enzyme and consists of three polypeptides: α (alpha, encoded by the *dnaE* gene), ε (epsilon, encoded by the *dnaQ* gene), and θ (theta, encoded by the *holE* gene). The complete DNA Pol III enzyme, called the DNA Pol III holoenzyme, contains an additional six different polypeptides.

Both DNA Pol I and DNA Pol III replicate DNA in the 5′→3′ direction in the cell. Both enzymes also have 3′→5′ exonuclease activity: They can remove nucleotides from the 3′ end of a DNA chain. This enzyme activity is part of an error correction mechanism. If an incorrect base is inserted by DNA polymerase (an event that occurs at a frequency of about 10^{-6} for both DNA polymerase I and DNA polymerase III), in many cases the error is recognized immediately. Then, the 3′→5′ exonuclease activity excises the erroneous nucleotide from the new strand. This process resembles that of the backspace or delete key on a computer keyboard. After excision, the DNA polymerase resumes motion in the forward direction and inserts the correct character. Thus, in DNA replication, 3′→5′ exonuclease activity is a **proofreading** mechanism that helps keep the frequency

of errors very low. With proofreading, the frequency of replication errors by DNA polymerase I or III is reduced to less than 10^{-9}. The 3′→5′ exonuclease activity of DNA Pol I is a domain of the single polypeptide enzyme, while, for DNA Pol III, it is the function of the ε subunit, stimulated by the θ subunit.

Uniquely of the two, DNA Pol I has 3′→5′ exonuclease activity and can remove either DNA or RNA nucleotides from the 5′ end of a nucleic acid strand. This activity is important in DNA replication and is examined later in the chapter.

KEYNOTE

> The enzymes that catalyze the synthesis of DNA are called DNA polymerases. All known DNA polymerases synthesize DNA in the 5′→3′ direction. Polymerases may also have other activities, such as proofreading, or removing nucleotides from a strand in the 5′→3′ direction.

Molecular Model of DNA Replication

Table 3.1 presents the functions of some of the *E. coli* DNA replication genes and key DNA sequences involved in replication. A number of the genes were identified by mutational analysis. In this section, we discuss a molecular model of DNA replication involving these genes and sequences.

Table 3.1 Functions of Some of the Genes and DNA Sequences Involved in DNA Replication in *E. coli*

Gene Product or Function	Gene
DNA polymerase I	*polA*
DNA polymerase III	*dnaE, dnaQ, dnaX, dnaN, dnaD holA→E*
Initiator protein; binds to *oriC*	*dnaA*
IHF protein (DNA binding protein); binds to *oriC*	*himA*
FIS protein (DNA binding protein); binds to *oriC*	*fis*
Helicase and activator of primase	*dnaB*
Complexes with *dnaB* protein and delivers it to DNA	*dnaC*
Primase; makes RNA primer for extension by DNA polymerase III	*dnaG*
Single-stranded binding (SSB) proteins; bind to unwound single-stranded arms of replication forks	*ssb*
DNA ligase; seals single-stranded gaps	*lig*
Gyrase (type II topoisomerase); replication swivel to avoid tangling of DNA as replication fork advances	*gyrA, gyrB*
Origin of chromosomal replication	*oriC*
Terminus of chromosomal replication	*ter*
TBP (*ter* binding protein); stalls replication forks	*tus*

Initiation of Replication

The initiation of replication is directed by a DNA sequence called the **replicator.** The replicator usually includes the **origin of replication,** the specific region where the DNA double helix denatures into single strands and within which replication commences. The locally denatured segment of DNA is called a **replication bubble.** The segments of untwisted single strands on which the new strands are made (in accordance with complementary base-pairing rules) are called the **template strands.**

When DNA untwists to expose the two single-stranded template strands for DNA replication, a Y-shaped structure called a **replication fork** is formed. A replication fork moves in the direction of untwisting of the DNA. When DNA untwists in the middle of a DNA molecule, as in a circular chromosome, there are two replication forks—like two Ys joined together at their tops. In many (but not all) cases, each replication fork is active, so DNA replication proceeds *bidirectionally.*

An outline of the initiation of replication in *E. coli* is shown in Figure 3.5. The *E. coli* replicator is *oriC,* which spans 245 bp and contains a cluster of three copies of a 13-bp AT-rich sequence and four copies of a 9-bp sequence. An AT-rich region is relatively easy to denature to single strands and is characteristic of replicators in all organisms. For the initiation of replication, an **initiator protein** or proteins bind to the replicator and stimulate the local denaturing at the AT-rich region. The *E. coli* initiator protein is DnaA (*dnaA* gene), which binds to the 9-bp regions in multiple copies, leading to the denaturing of the region with the 13-bp sequences. **DNA helicases** (the *dnaB* gene) are recruited and loaded onto the DNA. The helicases begin untwisting the DNA in both directions from the origin of replication. The energy for the untwisting comes from the hydrolysis of ATP—a reaction that causes a change in the shape of the helicase, enabling the enzyme to move along a single strand of DNA. By repeated ATP hydrolysis, the helicase can move along the single strand and untwist any double-stranded DNA it encounters.

Next, each DNA helicase recruits the enzyme **DNA primase** (a product of the *dnaG* gene), forming a complex called the **primosome.** DNA primase is important in DNA replication because no known DNA polymerases can initiate the synthesis of a DNA strand; they can only add nucleotides to a preexisting strand. The DNA primase on each template strand is activated by its associated DNA helicase and synthesizes a short **RNA primer** (about 5–10 nucleotides) to which new nucleotides can be added by DNA polymerase. The RNA primer is removed later and replaced with DNA, we will return to discuss this event further. At this point, the bidirectional replication of DNA has just begun.

Figure 3.5

Initiation of replication in *E. coli*. The DnaA initiator protein binds to *oriC* (the replicator) and stimulates denaturation of the DNA. DNA helicases are recruited and begin to untwist the DNA to form two head-to-head replication forks.

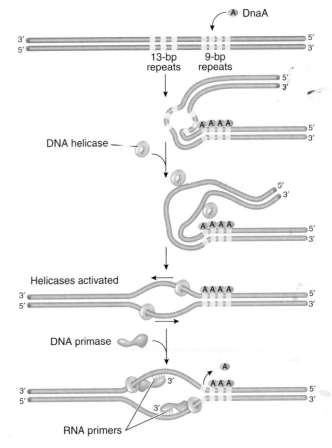

We must be clear about the difference between a *template* and a *primer* with respect to DNA replication. A template strand is the one on which the new strand is synthesized according to complementary base-pairing rules. A primer is a short segment of nucleotides bound to the template strand. The primer acts as a substrate for DNA polymerase, which extends the primer as a new DNA strand, the sequence of which is complementary to the template strand.

KEYNOTE

The initiation of DNA synthesis first involves the denaturation of double-stranded DNA at an origin of replication, catalyzed by DNA helicase. Next, DNA primase binds to the helicase and the denatured DNA and synthesizes a short RNA primer. The RNA primer is extended by DNA polymerase as new DNA is made. Later, the RNA primer is removed.

Semidiscontinuous DNA Replication

The foregoing discussion of the initiation of replication considered the production of two replication forks when DNA denatures at an origin. The replication events are identical with each replication fork, so we will focus on just one fork in the discussion that follows.

animation

a Molecular Model of DNA Replication

As we just discussed (see Figure 3.5), helicase untwists the DNA to produce single-stranded template strands. **Single-strand DNA-binding (SSB) proteins** bind to the single-stranded DNA, stabilizing it (Figure 3.6) and preventing it from reannealing. In *E. coli*, the SSB protein (*ssb* gene) is a tetramer of four identical subunits that binds to a 32-nucleotide segment of DNA. More than 200 of the proteins bind to each replication fork.

The RNA primer made by DNA primase in Figure 3.5 is at the 5′ end of the new DNA strand being synthesized on the template strand at the bottom of Figure 3.6. The DNA primase at the fork synthesizes another RNA primer, this one on the top template DNA strand (Figure 3.6a). These RNA primers are lengthened by DNA polymerase III, which synthesizes new DNA complementary to the template strands while simultaneously displacing the bound SSB proteins (Figure 3.6a). Recall that DNA polymerases can make DNA only in the 5′→3′ direction; however, the two DNA strands are of opposite polarity. To maintain the 5′→3′ polarity of DNA synthesis on each template, and to maintain one overall direction of replication fork movement, DNA is made in opposite directions on the two template strands. (See Figure 3.6a.) The new strand being made in the 5′→3′ direction (in the *same* direction as the movement of the replication fork) is called the **leading strand,** and the new strand being made in the direction *opposite* that of the movement of the replication fork is called the **lagging strand.** The leading strand needs a single RNA primer for its synthesis, whereas the lagging strand needs a series of primers, as we will see.

As the replication fork moves, helicase untwists more DNA (Figure 3.6b). DNA gyrase (a form of topoisomerase) relaxes the tension produced in the DNA ahead of the replication fork. This tension could be considerable, because the replication fork rotates at about 3,000 rpm. On the leading-strand template (the bottom strand in Figure 3.6), the leading strand is synthesized continuously toward the replication fork. Because DNA synthesis can proceed only in the 5′→3′ direction, however, lagging-strand synthesis has gone as far as it can. For DNA replication to continue on the lagging-strand template (the top strand in Figure 3.6), a new initiation of DNA synthesis must occur. An RNA primer is made by the DNA primase at the replication fork. (See Figure 3.6b.) DNA polymerase III then adds DNA to the RNA primer to make another DNA fragment. Since the leading strand is

being made continuously, whereas the lagging strand can be made only in pieces, or discontinuously, DNA replication as a whole occurs in a **semidiscontinuous** manner.

The fragments of lagging strand made in the process just described are called **Okazaki fragments,** after their discoverers, Reiji and Tuneko Okazaki and colleagues. Experimentally, the Okazakis added a radioactive DNA precursor (^{3}H-thymidine) to cultures of *E. coli* for 0.5 percent of a generation time. Next, they added a large amount of nonradioactive thymidine to prevent the incorporation of any more of the radioactive precursor into the DNA. At various times, they extracted the DNA and determined the size of the newly labeled molecules. At intervals soon after the labeling period, most of the radioactive ^{3}H-thymidine was present in low-molecular-weight DNA about 100 to 1,000 nucleotides long. As time increased, a greater and greater proportion of the labeled molecules was found in DNA of a high molecular weight. These results indicated that DNA replication normally involves the synthesis of short DNA segments—the now-named Okazaki fragments—that are subsequently linked together.

In Figure 3.6c, the process repeats itself: Helicase untwists the DNA, DNA is continuously synthesized on the leading-strand template, and DNA is discontinuously synthesized on the lagging-strand template, with new RNA primers being synthesized every 1,000–2,000 nucleotides along the template. Eventually, the unconnected Okazaki fragments on the lagging-strand template are joined into a continuous DNA strand. Joining them requires the activities of DNA polymerase I and **DNA ligase.** Consider two adjacent Okazaki fragments. The 3′ end of the newer fragment is adjacent, but not joined, to the primer at the 5′ end of the previously made fragment. DNA polymerase III leaves the DNA, and DNA polymerase I continues the 5′→3′ synthesis of the fragment, simultaneously removing the primer section of the older fragment by its 5′→3′ exonuclease activity (Figure 3.6d). When DNA polymerase I has replaced all the RNA primer nucleotides with DNA nucleotides, a single-stranded nick is left between the two fragments. The fragments are joined by DNA ligase to produce a longer DNA strand (Figure 3.6e). The catalytic reaction of DNA ligase is diagrammed in Figure 3.7. The entire sequence of events is repeated until all the DNA is replicated.

In sum, DNA replication in *E. coli* is a complicated process, and what we have shown is a simplified version of it. In fact, the key replication proteins are closely associated to form a **replisome.** Figure 3.8 shows the lagging-strand DNA, folded so that its DNA polymerase III is complexed with the DNA polymerase III on the leading strand. These are two copies of the core enzyme described earlier (see p. 50), held together by the six other polypeptides to form the DNA Pol III holoenzyme. Only

Figure 3.6

Model for the events occurring around a single replication fork of the E. coli chromosome. (a) Initiation. (b) Further untwisting and elongation of the new DNA strands. (c) Further untwisting and continued DNA synthesis. (d) Removal of the primer by DNA polymerase I. (e) Joining of adjacent DNA fragments by the action of DNA ligase. Green = RNA; red = new DNA.

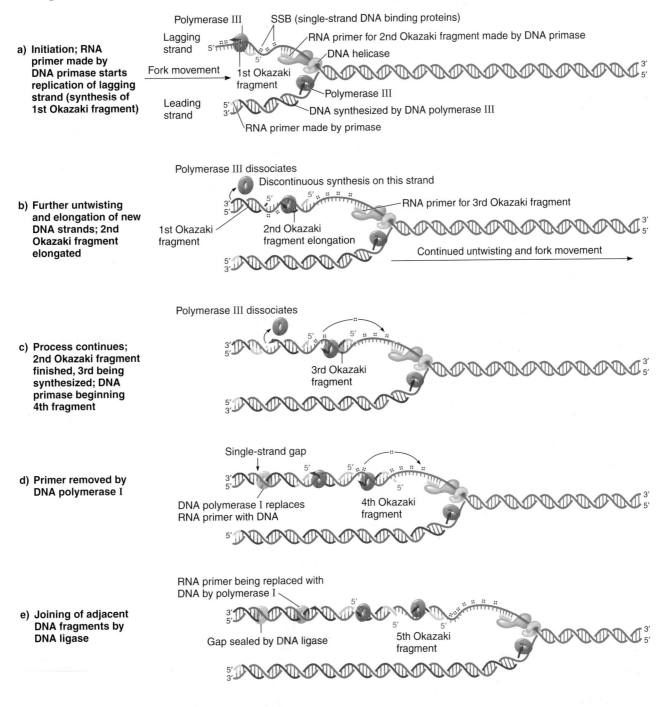

the core enzymes are shown in the figure, for simplicity. The folding of the lagging-strand template brings the 3′ end of each completed Okazaki fragment near the site where the next Okazaki fragment will start. The primase stays near the replication fork, synthesizing new RNA primers intermittently. Similarly, because the lagging-strand polymerase is complexed with the other replication proteins at the fork, that polymerase can be reused continually at the same replication fork, synthesizing a string of Okazaki fragments as it moves with the rest of the

Figure 3.7

Action of DNA ligase in sealing the nick between adjacent DNA fragments (e.g., Okazaki fragments) to form a longer, covalently continuous chain. The DNA ligase catalyzes the formation of a phosphodiester bond between the 3′-OH and the 5′-phosphate groups on either side of a nick, sealing the nick.

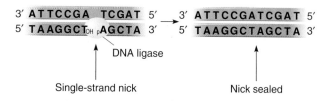

replication machine. That is, the complex of replication proteins that forms at the replication fork moves mostly as a unit along the DNA and enables new DNA to be synthesized efficiently on both the leading-strand and lagging-strand templates. Finally, with **bidirectional replication,** we can visualize two replisomes, each the mirror image of the other, on replication forks that are moving in opposite directions (Figure 3.9).

iActivity

Identify some of the specific elements and processes needed for DNA replication in the iActivity *Unraveling DNA Replication* on your CD-ROM.

Replication of Circular DNA and the Supercoiling Problem

In *E. coli*, the parental DNA strands remain in a circular form throughout the replication cycle. This is true of many, but not all, circular DNA molecules. During replication, these circular DNA molecules exhibit a theta-like (θ) shape, because of a replicating bubble's initiation at the replication origin. (See Figure 3.9.)

As the two DNA strands in a circular chromosome untwist during replication, positive supercoils form elsewhere in the molecule. For the replication fork to move, then, the chromosome ahead of the fork must rotate. Given a rate of movement of the replication fork of 500 nucleotides per second, at 10 base pairs per turn, the helix ahead of the fork must rotate at 50 revolutions per second, or 3,000 rpm.

The supercoiling problem is solved by the action of topoisomerases (see Chapter 2, pp. 29–30)—enzymes that introduce negative supercoils into DNA or that convert negatively supercoiled DNA into relaxed DNA. Topoisomerases play an important role in replication by preventing excessively supercoiled DNA from forming and thereby allowing both parental strands to remain intact during the replication cycle as the replication fork migrates. That is, the unreplicated part of the theta structure ahead of the replication fork repeatedly has negative supercoils introduced into it by the action of DNA gyrase (a type II topoisomerase), relieving the positive supercoiling that occurs as DNA is untwisted during replication.

Rolling Circle Replication

A rolling circle model of replication (Figure 3.10) applies to the replication of several viral DNAs, such as bacteriophages ΦX174 (see Figure 2.16) and λ (see Figure 2.17). The starting point is a circular, double-stranded DNA molecule. In the case of ΦX174, this is produced when a complementary strand is synthesized with the use of the circular, single-stranded genomic DNA as a template. In the case of λ, the circular molecule is produced when the short, complementary, single-stranded ends of the linear double-stranded DNA genome (see Figure 3.11, p. 56) pair together after infecting the bacterial cell.

The first step in rolling circle replication is the generation of a specific cut (nick) in one of the two strands at the origin of replication. The 5′ end of the cut strand is then displaced from the circular molecule, creating a replication fork and leaving a single-stranded

Figure 3.8

Model for the replisome, the complex of key replication proteins, with the DNA at the replication fork. The DNA polymerase III on the lagging-strand template (top of figure) is just finishing the synthesis of an Okazaki fragment.

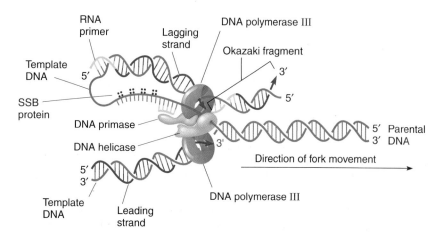

Figure 3.9

Bidirectional replication of circular DNA molecules.

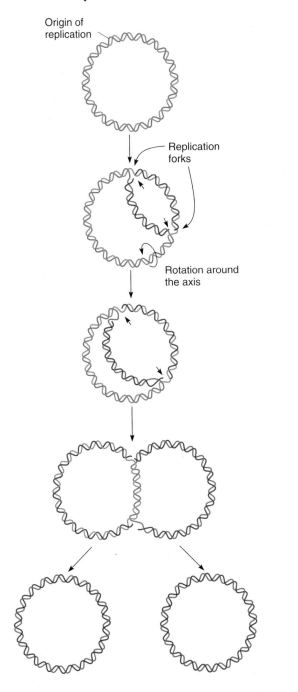

Figure 3.10

The replication process of double-stranded circular DNA molecules through the rolling circle mechanism. The active force that unwinds the 5′ tail is the movement of the replisome propelled by its helicase components.

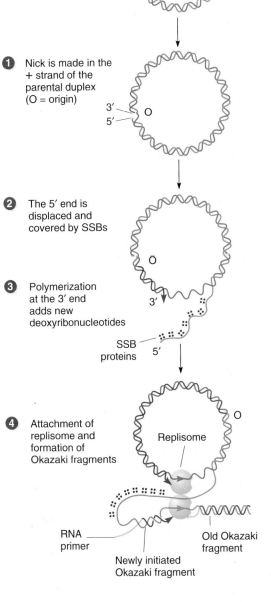

1 Nick is made in the + strand of the parental duplex (O = origin)

2 The 5′ end is displaced and covered by SSBs

3 Polymerization at the 3′ end adds new deoxyribonucleotides

SSB proteins

4 Attachment of replisome and formation of Okazaki fragments

Replisome

RNA primer

Newly initiated Okazaki fragment

Old Okazaki fragment

stretch of DNA that serves as a template for the addition of deoxyribonucleotides to the free 3′ end by DNA polymerase III, using the intact circular DNA as a template. This new DNA synthesis occurs continuously as the 5′ cut end is displaced from the circular molecule; the intact circular DNA acts as the leading-strand template.

The 5′ end of the cut DNA strand is rolled out as a free "tongue" of increasing length as replication proceeds. This single-stranded DNA tongue becomes covered with SSB proteins. New DNA is synthesized on the displaced DNA in

the 5′→3′ direction, meaning from the circle out toward the end of the displaced DNA. With further displacement, new DNA is synthesized again, beginning at the circle and moving outward along the displaced DNA strand. Thus, synthesis on this strand is discontinuous; that is, the displaced strand is the lagging strand template. (See

Figure 3.6.) As the single-stranded DNA tongue rolls out, DNA synthesis continues on the circular DNA template.

Since the parental DNA circle can continue to roll, it is possible to generate a linear double-stranded DNA molecule that is longer than the circumference of the circle. For example, in the later stages of phage λ DNA replication, linear tongues that are many times the circumference of the original circle are produced by rolling circle replication. These molecules are cut into individual linear λ chromosomes, and those unit-length molecules are then packaged into phage heads.

Let us consider the rolling circle mechanism of DNA replication in the context of the phage λ life cycle.

Recall that phage λ is a temperate phage, meaning that when it infects *E. coli,* it has a choice of entering the *lytic pathway,* which results in cell lysis, or following the *lysogenic pathway* to a quiescent state that does not result in phage reproduction. (The life cycle of phage λ is described in Chapter 18, pp. 492–497 and is diagrammed in Figure 18.12, p. 493.) Regardless of which pathway λ follows when it infects a cell, the first step after infection is the conversion of the linear molecule into a circular molecule by pairing of the complementary "sticky" ends and then sealing of the single-stranded nicks by DNA ligase (Figure 3.11a). The paired ends are called the *cos* sequence. In the lysogenic cycle,

Figure 3.11

λ chromosome structure varies at stages of lytic infection of *E. coli*. (a) Parts of the λ chromosome, showing the nucleotide sequence of the two single-stranded, complementary ("sticky") ends and the chromosome circularizing after infection by pairing of the ends, with the single-stranded nicks filled in to produce a covalently closed circle. (b) Generation of the "sticky" ends of the λ DNA during the lytic cycle. During replication of the λ chromosome, a giant concatameric DNA molecule is produced; it contains tandem repeats of the λ genome. The diagram shows the joining of two adjacent λ chromosomes and the extent of the *cos* sequence. The *cos* sequence is recognized by the *ter* gene product, an endonuclease that makes two cuts at the sites shown by the arrows. These cuts produce a complete λ chromosome from the concatamer.

a) **Linear λ chromosome (~48,000 base pairs) forms circular λ chromosome**

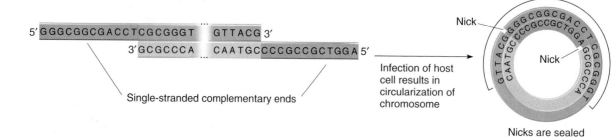

b) **Production of progeny, linear λ chromosomes from concatamers (multiple copies linked end to end at complementary ends)**

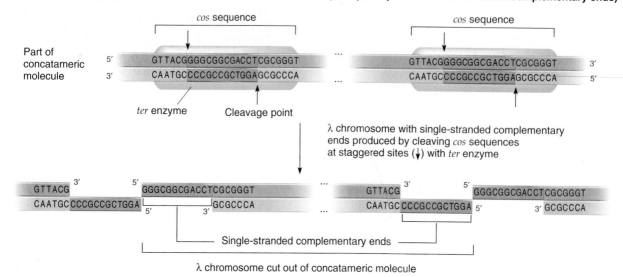

the circular DNA finds a particular site in the *E. coli* chromosome and is integrated into the main chromosome by a crossing-over event.

In the lytic cycle, the rolling circle replication mechanism produces a very long molecule with head-to-tail copies of the λ chromosome. A DNA molecule like this, made up of repeated monomers, is called a *concatamer*. From this concatameric molecule, unit-length progeny phage λ chromosomes are generated as follows: The phage λ chromosome has a gene called *ter* (for terminus-generating activity, Figure 3.11b), the product of which is a DNA endonuclease (an enzyme that digests a nucleic acid chain by cutting somewhere along its length rather than at the termini). The endonuclease recognizes the *cos* sequence. Once *ter* is aligned on the DNA at the *cos* site, the endonuclease makes a staggered cut such that linear λ chromosomes with the correct complementary ("sticky"), 12-base-long, single-stranded ends are produced. The chromosomes are then packaged in the assembled phage heads, and progeny λ phages are assembled and released from the cell when it lyses.

KEYNOTE

> During DNA replication, new DNA is made in the 5′→3′ direction, so chain growth is continuous on one strand and discontinuous (i.e., in segments that are later joined) on the other strand. This semidiscontinuous model is applicable to many other prokaryotic replication systems, each of which differs in the number and properties of the enzymes and proteins needed.

DNA Replication in Eukaryotes

The biochemistry and molecular biology of DNA replication are similar in prokaryotes and eukaryotes. However, an added complication in eukaryotes is that DNA is distributed among many chromosomes rather than just one. In this section, we summarize some of the important aspects of DNA replication in eukaryotes.

Replicons

Each eukaryotic chromosome consists of one linear DNA double helix. For example, there are about 3 billion base pairs of DNA in the haploid human genome (24 chromosomes), and the average chromosome is roughly 10^8 base pairs long. Replication fork movement is much slower in eukaryotes than in *E. coli*, so, if there were only one origin of replication per chromosome, replicating each chromosome would take many days.

Actual measurements show that the chromosomes in eukaryotes replicate much faster than would be the case with only one origin of replication per chromosome. The diploid set of chromosomes in *Drosophila* embryos, for example, replicates in 3 minutes. This is six times faster than the replication of the *E. coli* chromosome, even though there is about 100 times more DNA in *Drosophila* than in *E. coli*.

Eukaryotic chromosomes duplicate rapidly because DNA replication initiates at many origins of replication throughout the genome. At each origin of replication, the DNA denatures (as in *E. coli*), and the replication proceeds bidirectionally. Eventually, each replication fork runs into an adjacent replication fork, initiated at an adjacent origin of replication. In eukaryotes, the stretch of DNA from the origin of replication to the two termini of replication (where adjacent replication forks fuse) on each side of the origin is called a **replicon** or **replication unit** (Figure 3.12). In general, the replicon size is much smaller, and the rate of fork movement much slower, in eukaryotic organisms than in bacteria. For example, the *E. coli* genome consists of one replicon, of size 4.6 Mb (million base pairs, the entire genome size), with a replication fork movement rate of 2.2 Mb per hour. By contrast, eukaryotic replicons are relatively small (an average of 30 kb in adult frogs and 160 kb in yeast), with slower rates of replication fork movement (18 kb per hour in *Drosophila* and ~4 kb per hour in the mustard *Crepis*).

Figure 3.12

Replicating DNA of *Drosophila melanogaster*. (a) Electron micrograph showing replication units (replicons). (b) An interpretation of the electron micrograph shown in (a).

a)

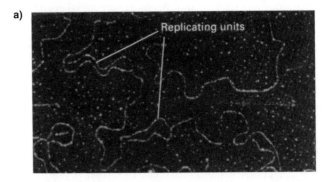

b)

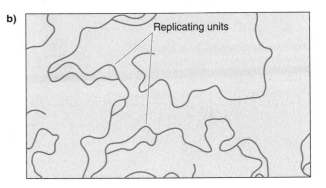

DNA replication does not occur simultaneously in all the replicons in an organism's genome. Instead, there is a cell-specific timing of initiation of replication at the various origins. Figure 3.13 shows one segment of one chromosome in which three replicons always begin replicating at distinct times. When the replication forks fuse at the margins of adjacent replicons, the chromosome has replicated into two sister chromatids. In general, replication of a segment of chromosomal DNA occurs after the synchronous activation of a cluster of origins.

Initiation of Replication

Replicators are less well defined in eukaryotes than in prokaryotes. In the yeast *Saccharomyces cerevisiae*, specific chromosomal sequences have been identified that, when they are included as part of an extrachromosomal, circular DNA molecule, confer upon that molecule the ability to replicate autonomously within the yeast cell. These approximately 100-bp sequences are yeast replicators and are called **autonomously replicating sequences (ARSs)**. Three sequence elements are typically found at yeast replicators in the order A, B1, and B2. Replicators of more complex, multicellular organisms are poorly understood.

The initiator protein in eukaryotes is the multisubunit **origin recognition complex (ORC)**. At yeast replicators, the ORC binds to A and B1, and recruits other replication proteins, among which is the protein needed for DNA unwinding in B2. The origin of replication is between B1 and B2.

DNA replication takes place in a specific stage of the cell cycle. The cell cycle consists of four stages (see Figure 12.4, p. 302): G_1 during which the cell prepares for DNA replication, S during which DNA replication occurs, G_2 during which the cell prepares for cell division, and M, the division of the cell by mitosis. A key issue involves the control of replication initiation at the replicators. No origin of replication must be used more than once in the cell cycle. This is accomplished by a fairly complicated series of events that we will only outline. In essence, the initiation of replication involves two temporally separate steps. The first step is *replicator selection*, in which particular proteins assemble on each replicator to form *prereplicative complexes (pre-RC)*. This selection step occurs in the G_1 phase of the cell cycle and begins with recognition of the replicator by the ORC. Once bound, ORC recruits the other proteins. In contrast to the situation in bacteria, the binding of the initiator to the replicator does not lead immediately to unwinding of the DNA. Rather, the pre-RCs are activated to initiate replication at the origins only after passage of the cell from G_1 to the S phase. The limiting of replication initiation to the S phase is controlled by *cyclin-dependent kinases (Cdks)*, key enzymes that carefully regulate the progression of a cell through the cell cycle. Cdks are needed to activate pre-RCs to initiate replication, but Cdk activity inhibits the formation of new pre-RCs. No active Cdks are present in G_1, allowing the pre-RCs to form. Active Cdks are present in the rest of the cell cycle, so when the cell enters S, the pre-RCs are activated and replication starts. Once a replicator has "fired," a new pre-RC cannot form on it until the next G_1, when active Cdks are again absent.

Eukaryotic Replication Enzymes

As we saw earlier, many of the enzymes and proteins involved in prokaryotic DNA replication have been identified. Less is known about the enzymes and proteins involved in eukaryotic DNA replication. However, it is clear that the steps described for DNA synthesis in prokaryotes also occur in DNA synthesis in eukaryotes—namely, denaturation of the DNA double helix and the semiconservative, semidiscontinuous replication of DNA.

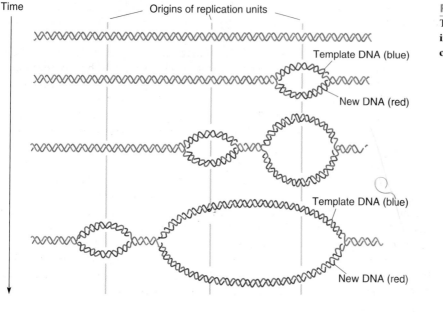

Time

Origins of replication units

Template DNA (blue)

New DNA (red)

Template DNA (blue)

New DNA (red)

Figure 3.13

Temporal ordering of DNA replication initiation events in replication units of eukaryotic chromosomes.

Eukaryotic cells have 15 or more DNA polymerases. Replication of the nuclear DNA requires 3 of these: Pol α (alpha)/primase, Pol δ (delta), and Pol ε (epsilon). Pol α (alpha)/primase initiates new strands in replication by primase making about 10 nucleotides of an RNA primer, which is extended by about 30 nucleotides of DNA by Pol α. The RNA/DNA primers are extended by Pol δ and Pol ε. One of these enzymes synthesizes the leading-strand DNA, and the other synthesizes the lagging-strand DNA, but which synthesizes which is not clear.

Other eukaryotic DNA polymerases replicate mitochondrial or chloroplast DNA or are involved in specific DNA repair processes.

Replicating the Ends of Chromosomes

Because DNA polymerases can synthesize new DNA only by extending a primer, there are special problems in replicating the ends—the telomeres—of eukaryotic chromosomes (Figure 3.14). A parental chromosome (Figure 3.14a) is replicated, resulting in two new DNA molecules, each of which has an RNA primer at the 5′ end of the newly synthesized strand in the telomere region (Figure 3.14b). The RNA primers are removed, leaving a single-stranded stretch of DNA—a gap—at the 5′ end of the new strand. The gap cannot be filled in by DNA polymerase, because that enzyme cannot initiate new DNA synthesis. If nothing were done about these gaps, the chromosomes would get shorter and shorter with each replication cycle.

There is a special mechanism, however, for replicating the ends of chromosomes. Most eukaryotic chromosomes have tandemly repeated, species-specific, simple sequences at their telomeres. (See Chapter 2, pp. 35–36.) Elizabeth Blackburn and Carol W. Greider have shown that an enzyme called **telomerase** maintains chromosome lengths by adding telomere repeats to the chromosome ends. This mechanism does not involve the regular replication machinery.

Figure 3.15 is a simplified diagram of the mechanism that has been deduced for the protozoan *Tetrahymena*. The repeated sequence in *Tetrahymena* is 5′-TTGGGG-3′, reading toward the end of the DNA on the top strand in the figure. Telomerase acts at the stage shown in Figure 3.14c—that is, where a chromosome end has been produced with a gap at the 5′ end of the new DNA (Figure 3.15a). Telomerase is an enzyme made up of both protein and RNA. The RNA component includes a base sequence that is complementary to the telomere repeat unit of the organism in which it is found. Therefore, the telomerase binds specifically to the overhanging telomere repeat at the end of the chromosome (Figure 3.15b). Next, the telomerase catalyzes the synthesis of three nucleotides of new DNA—TTG—using the telomerase RNA as a template (Figure 3.15c). The telomerase then slides toward the end of the chromosome, so that its AAC at the 3′ end of the RNA template now pairs with the newly synthesized TTG on the DNA

Figure 3.14

The problem of replicating completely a linear chromosome in eukaryotes. (a) Schematic diagram of a parent double-stranded DNA molecule representing the full length of a chromosome. (b) After semiconservative replication, new DNA segments hydrogen bonded to the template strands have RNA primers at their 5′ ends. (c) The RNA primers are removed, DNA polymerase fills the resulting gaps, and DNA ligase joins the adjacent fragments. However, at the two telomeres, there are still gaps at the 5′ ends of the new DNA. The gaps result from RNA primer removal, because no new DNA synthesis could fill them in.

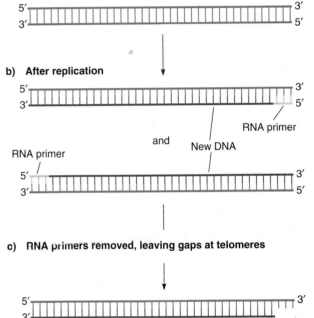

(Figure 3.15d). Telomerase then makes the rest of the TTGGG telomere repeat (Figure 3.15e). The process recurs, to add more telomere repeats. In this way, the chromosome is lengthened by the addition of a number of telomere repeats. Then, by primer synthesis and DNA synthesis catalyzed by DNA polymerase in the conventional way, the former gap is filled in, and the new chromosomal DNA is lengthened (Figure 3.15f). After removal of the RNA primer, a new 5′ gap is left (Figure 3.15g), but any net shortening of the chromosome has been averted.

Introducing into cells mutant telomerase RNA genes with certain of their template bases changed showed that telomerase RNA is used as a template to synthesize new chromosomal telomere repeats. The new repeats made by the mutated telomerase had sequences that were complementary to the altered RNA, rather than the normal

Figure 3.15

Synthesis of telomeric DNA by telomerase. The example is of *Tetrahymena* telomeres. The process is described in the text. (**a**) The starting point is the chromosome end with 5′ gap left after primer removal. (**b**) Binding of telomerase to the overhanging telomere repeat at the end of the chromosome. (**c**) Synthesis of three-nucleotide DNA segment at chromosome end, using the RNA template of telomerase. (**d**) The telomerase moves so that the RNA template can bind to the newly synthesized TTG in a different way. (**e**) Telomerase catalyzes the synthesis of a new telomere repeat, using the RNA template. The process recurs, to add more telomere repeats. (**f**) After telomerase has left, new DNA is made on the template, starting with an RNA primer. (**g**) After the primer is removed, the result is a longer chromosome than at the start, with a new 5′ gap.

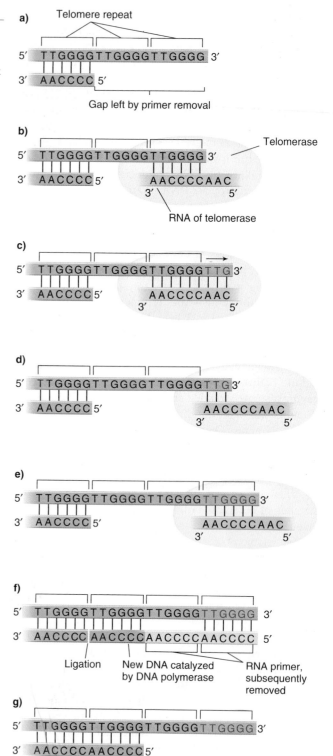

sequence. The synthesis of DNA from an RNA template is called *reverse transcription,* so telomerase is an example of a **reverse transcriptase enzyme.** (The telomerase reverse transcriptase is abbreviated TERT.)

Telomere length, while not identical from chromosome end to chromosome end, is nonetheless regulated to an average length for the organism and cell type. In wild-type yeast, for example, the simple telomere sequences (TG$_{1-3}$, a repeating sequence of one T followed by one to three Gs) occupy an average of about 300 bp. Mutants that affect telomere length have been identified. For example, if the *TLC1* gene (which encodes the telomerase RNA) is deleted or the *EST1* (ever shorter telomeres) gene is mutated, telomeres shorten continuously until the cells die. This phenotype provides evidence that telomerase activity is necessary for long-term cell viability. The product of the *EST1* gene, the protein Est1p, is either a component of the telomere RNA–protein complex or a separate factor that is essential to telomerase function. Mutations of the *TEL1* and *TEL2* genes cause cells to maintain their telomeres at a new, shorter-than-wild-type length, making it clear that telomere length is regulated genetically.

Current evidence suggests many levels of regulation of telomere activity and telomere length. For example, attention is being given to the observation that telomerase activity in mammals is limited to immortal cells (such as tumor cells). The absence of telomerase activity in other cells results in progressive shortening of chromosome ends during successive divisions, because of the failure to replicate those ends, and also results in a limited number of cell divisions before the cell dies.

KEYNOTE

Special enzymes—telomerases—are used to replicate the ends of chromosomes in eukaryotes. A telomerase is a complex of proteins and RNA. The RNA acts as a template for synthesizing the complementary telomere repeat of the chromosome, so telomerase is a type of reverse transcriptase enzyme.

Assembling Newly Replicated DNA into Nucleosomes

Eukaryotic DNA is complexed with histones in nucleosomes, which are the basic units of chromosomes. (See Chapter 2, p. 32.) Recall that there are eight histones in the nucleosome, two each of H2A, H2B, H3, and H4.

Therefore, when the DNA is replicated, the histone complement must be doubled so that all nucleosomes are duplicated. Doubling involves two processes: the synthesis of new histone proteins and the assembly of new nucleosomes.

Most histone synthesis is coordinated with DNA replication. The transcription of the genes for the five histones is initiated near the end of the G_1 phase, just before S. Translation of the histone mRNAs occurs throughout S, producing the histones to be assembled into nucleosomes as the chromosomes are duplicated.

Electron microscopy studies have shown that newly replicated DNA is assembled into nucleosomes almost immediately. Nonetheless, for replication to proceed, nucleosomes must disassemble during the short time when a replication fork passes. New nucleosomes are assembled as follows (Figure 3.16): Each parental histone core of a nucleosome separates into an H3–H4 tetramer (two copies each of H3 and H4) and two copies of an H2A–H2B dimer. The H3–H4 tetramer is transferred directly to one of the two replicated DNA double helices past the fork, whereupon it begins

Figure 3.16

Assembly of new nucleosomes at a replication fork. New nucleosomes are assembled first with the use of either a parental or a new H3–H4 tetramer and then by completing the structure with a pair of H2A–H2B dimers.

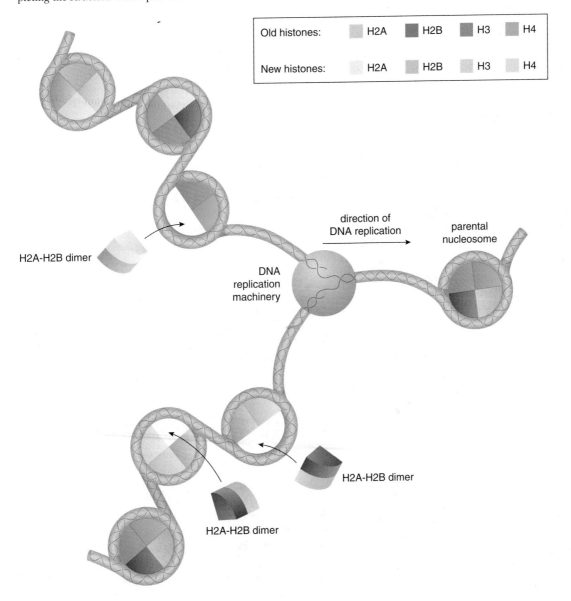

Old histones: H2A H2B H3 H4

New histones: H2A H2B H3 H4

direction of DNA replication

parental nucleosome

H2A-H2B dimer

DNA replication machinery

H2A-H2B dimer

H2A-H2B dimer

nucleosome assembly. The H2A–H2B dimers are released, joining the pool of newly synthesized H2A–H2B's that have assembled. A pool of new H3–H4 tetramers is also present, and one of these tetramers initiates nucleosome assembly on the other DNA double helix past the fork. The rest of the new nucleosomes is assembled from H2A–H2B dimers, which may be parental or new. Thus, a new nucleosome will have either a parental or new H3–H4 tetramer and a pair of H2A–H2B dimers that may be parental–parental, parental–new, or new–new. The assembly process is not spontaneous in the cell, however, requiring proteins known as *histone chaperones* to direct it. Self-assembly of nucleosomes has been shown in vitro, but the conditions are nonphysiological.

Summary

In this chapter, we discussed DNA replication. Many aspects of DNA replication are similar in prokaryotes and eukaryotes: Both types of organism employ a semiconservative and a semidiscontinuous mechanism, synthesis of new DNA in the $5' \rightarrow 3'$ direction, and the use of RNA primers to initiate DNA chains. The enzymes that catalyze DNA synthesis are the DNA polymerases. In *E. coli*, there are three DNA polymerases, two of which are known to be involved in DNA replication along with several other enzymes and proteins. The DNA polymerases have $3' \rightarrow 5'$ exonuclease activity, which permits proofreading to take place during DNA synthesis if an incorrect nucleotide is inserted opposite the template strand. In eukaryotes, two DNA polymerases are involved in nuclear DNA replication. Neither has associated proofreading activity; that function presumably is the property of a separate protein.

In prokaryotes, DNA replication begins at specific chromosomal sites. Such sites are known also for yeast. A prokaryotic chromosome has one initiation site for DNA replication, whereas a eukaryotic chromosome has many initiation sites dividing the chromosome into replication units, or replicons. It is not clear whether the initiation sites that are used are the same sequences from cell generation to cell generation. The existence of replicons means that DNA replication of the entire set of chromosomes in a eukaryotic organism can proceed quickly, in some cases faster than with the single *E. coli* chromosome, despite the presence of orders of magnitude more DNA.

Since eukaryotic chromosomes are linear, there is a special problem of maintaining the lengths of chromosomes, because the removal of RNA primers results in a shorter new DNA strand. This problem is overcome by special enzymes called telomerases that maintain the length of chromosomes. Telomerases are a combination of proteins and RNA. The RNA component acts as a template to guide the synthesis of new telomere repeat units at the chromosome ends.

DNA replication in eukaryotes occurs in the S phase of the cell division cycle. Eukaryotic chromosomes are complexes of DNA with histones and nonhistone chromosome proteins, so not only must DNA be replicated, but the chromosome structure must be duplicated. In particular, the nucleosome organization of chromosomes must be duplicated as the replication forks migrate. Nucleosomes are disassembled to allow the replication fork to pass, and then new nucleosomes are assembled soon after a replication fork passes. Nucleosome assembly is an orderly process directed with the aid of histone chaperones.

Analytical Approaches to Solving Genetics Problems

Q3.1

a. Meselson and Stahl used ^{15}N-labeled DNA to prove that DNA replicates semiconservatively. The method of analysis was cesium chloride equilibrium density gradient centrifugation, in which bacterial DNA labeled in both strands with ^{15}N (the heavy isotope of nitrogen) bands to a different position in the gradient than DNA labeled in both strands with ^{14}N (the normal isotope of nitrogen). Starting with a mixture of ^{15}N-containing and ^{14}N-containing DNAs, then, two bands result after CsCl density gradient centrifugation.

When double-stranded DNA is heated to 100°C, the two strands separate because the hydrogen bonds between the strands break—a process called denaturation. When the solution is cooled slowly, any two complementary single strands will find each other and re-form the double helix—a process called renaturation or reannealing. If the mixture of ^{15}N-containing and ^{14}N-containing DNAs is first heated to 100°C and then cooled slowly before centrifuging, the result is different. In this case, two bands are seen in exactly the same positions as before, and a new third band is seen at a position halfway between the other two. From its position relative to the other two bands, the new band is interpreted to be intermediate in density between the other two bands. Explain the existence of the three bands in the gradient.

b. DNA from *E. coli* containing ^{15}N in both strands is mixed with DNA from another bacterial species, *Bacillus subtilis*, containing ^{14}N in both strands. Two bands are seen after CsCl density gradient centrifugation. If the two DNAs are mixed, heated to 100°C, slowly cooled, and then centrifuged, two bands again result. The bands are in the same positions as in the unheated DNA experiment. Explain these results.

A3.1

a. When DNA is heated to 100°C, it is denatured to single strands. If denatured DNA is allowed to cool slowly, complementary strands renature to produce double-stranded DNA again. Thus, when mixed, denatured ^{15}N–^{15}N DNA and ^{14}N–^{14}N DNA from the same species is cooled slowly, the single strands pair randomly during renaturation so that ^{15}N–^{15}N, ^{14}N–^{14}N, and ^{15}N–^{14}N double-stranded DNA are produced. The latter type of DNA has a density intermediate between those of the two other types, accounting for the third band. Theoretically, if all DNA strands pair randomly, there should be a 1:2:1 distribution of ^{15}N–^{15}N, ^{15}N–^{14}N, and ^{14}N–^{14}N DNAs, and this ratio should be reflected in the relative intensities of the bands.

b. DNAs from different bacterial species have different sequences. In other words, DNA from one species typically is not complementary to DNA from another species. Therefore, only two bands are seen because only the two *E. coli* DNA strands can renature to form ^{15}N–^{15}N DNA and only the two *B. subtilis* DNA strands can renature to form ^{14}N–^{14}N DNA. No ^{15}N–^{14}N hybrid DNA can form, so in this case there is no third band of intermediate density.

Q3.2 What would be the effect on chromosome replication in *E. coli* strains carrying deletions of the following genes?

a. *dnaE*   **d.** *lig*
b. *polA*   **e.** *ssb*
c. *dnaG*   **f.** *oriC*

A3.2 When genes are deleted, the function encoded by those genes is lost. All the genes listed in the question are involved in DNA replication in *E. coli*, and their functions are briefly described in Table 3.1 and discussed further in the text.

a. *dnaE* encodes a subunit of DNA polymerase III, the principal DNA polymerase in *E. coli* that is responsible for elongating DNA chains. A deletion of the *dnaE* gene undoubtedly would lead to a nonfunctional DNA polymerase III. In the absence of DNA polymerase III activity, DNA strands could not be synthesized from RNA primers; therefore, new DNA strands could not be synthesized, and there would be no chromosome replication.

b. *polA* encodes DNA polymerase I, which is used in DNA synthesis to extend DNA chains made by DNA polymerase III while simultaneously excising the RNA primer by 5′-to-3′ exonuclease activity. As discussed in the text, in mutant strains lacking the originally studied DNA polymerase—DNA polymerase I—chromosome replication still occurred. Thus, chromosomes would replicate normally in an *E. coli* strain carrying a deletion of *polA*.

c. *dnaG* encodes DNA primase, the enzyme that synthesizes the RNA primer on the DNA template. Without the synthesis of the short RNA primer, DNA polymerase III cannot initiate DNA synthesis, so chromosome replication will not take place.

d. *lig* encodes DNA ligase, the enzyme that catalyzes the ligation of Okazaki fragments. In a strain carrying a deletion of *lig*, DNA would be synthesized, but stable progeny chromosomes would not result, because the Okazaki fragments could not be ligated together, so the lagging strand synthesized discontinuously on the lagging-strand template would be in fragments.

e. *ssb* encodes the single-strand binding proteins that bind to and stabilize the single-stranded DNA regions produced as the DNA is unwound at the replication fork. In the absence of single-strand binding proteins, DNA replication would be impeded or absent, because the replication bubble could not be kept open.

f. *oriC* is the origin-of-replication region in *E. coli*,—that is, the location at which chromosome replication initiates. Without the origin, the initiator protein cannot bind and no replication bubble can form, so chromosome replication cannot take place.

Questions and Problems

3.1 Describe the Meselson–Stahl experiment, and explain how it showed that DNA replication is semiconservative.

*3.2 In the Meselson–Stahl experiment, ^{15}N-labeled cells were shifted to a ^{14}N medium at what we can designate as generation 0.

a. For the semiconservative model of replication, what proportion of ^{15}N–^{15}N, ^{15}N–^{14}N, and ^{14}N–^{14}N would you expect to find after one, two, three, four, six, and eight replication cycles?

b. Answer (a) in terms of the conservative model of DNA replication.

3.3 A spaceship lands on Earth and with it a sample of extraterrestrial bacteria. You are assigned the task of determining the mechanism of DNA replication in this organism.

You grow the bacteria in an unlabeled medium for several generations and then grow it in the presence of ^{15}N for exactly one generation. You extract the DNA and subject it to CsCl centrifugation. The banding pattern you find is as follows:

Control   Experimental
                    sample
^{15}N–^{15}N ^{14}N–^{14}N

It appears to you that this is evidence that DNA replicates in the semiconservative manner, but you are wrong. Why? What other experiment could you perform (using the same sample and technique of CsCl centrifugation) that would further distinguish between semiconservative and dispersive modes of replication?

*3.4 The elegant Meselson–Stahl experiment was among the first experiments to contribute to what is now a highly detailed understanding of DNA replication. Consider this experiment again in light of current molecular models by answering the following questions:

a. Does the fact that DNA replication is semiconservative mean that it must be semidiscontinuous?
b. Does the fact that DNA replication is semidiscontinuous ensure that it is also semiconservative?
c. Do any properties of known DNA polymerases ensure that DNA is synthesized semiconservatively?

*3.5 List the components necessary to make DNA in vitro, using the enzyme system isolated by Kornberg.

*3.6 How do we know that the Kornberg enzyme is not the main enzyme involved in DNA synthesis for chromosome duplication in the growth of *E. coli*?

3.7 Kornberg isolated DNA polymerase I from *E. coli*. DNA polymerase I has an essential function in DNA replication. What are the functions of the enzyme in DNA replication?

3.8 Suppose you have a DNA molecule with the base sequence TATCA, going from the 5′ to the 3′ end of one of the polynucleotide chains. The building blocks of the DNA are drawn as in the following figure:

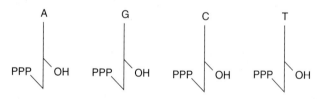

Use this shorthand system to diagram the completed double-stranded DNA molecule, as proposed by Watson and Crick.

*3.9 Base analogs are compounds that resemble the natural bases found in DNA and RNA, but are not normally found in those macromolecules. Base analogs can replace their normal counterparts in DNA during in vitro DNA synthesis. Researchers studied four base analogs for their effects on in vitro DNA synthesis using *E. coli* DNA polymerase. The results were as follows, with the amounts of DNA synthesized expressed as percentages of the DNA synthesized from normal bases only:

Analog	Normal Bases Substituted by the Analog			
	A	**T**	**C**	**G**
A	0	0	0	25
B	0	54	0	0
C	0	0	100	0
D	0	97	0	0

Which bases are analogs of adenine? of thymine? of cytosine? of guanine?

3.10 Concerning DNA replication:
a. Describe (draw) models of continuous, semidiscontinuous, and discontinuous DNA replication.
b. What was the contribution of Reiji and Tuneko Okazaki and colleagues with regard to these replication models?

3.11 The following events, steps, or reactions occur during *E. coli* DNA replication. For each entry in column A, select its match(es) from column B. Each entry in A may have more than one match, and each entry in B can be used more than once.

Column A	Column B
_____ a. Unwinds the double helix	A. Polymerase I
	B. Polymerase III
_____ b. Prevents reassociation of complementary bases	C. Helicase
	D. Primase
_____ c. Is an RNA polymerase	E. Ligase
_____ d. Is a DNA polymerase	F. SSB protein
_____ e. Is the "repair" enzyme	G. Gyrase
_____ f. Is the major elongation enzyme	H. None of these
_____ g. Is a 5′→3′ polymerase	
_____ h. Is a 3′→5′ polymerase	
_____ i. Has 5′→3′ exonuclease function	
_____ j. Has 3′→5′ exonuclease function	
_____ k. Bonds free 3′-OH end of a polynucleotide to a free 5′-monophosphate end of polynucleotide	
_____ l. Bonds 3′-OH end of a polynucleotide to a free 5′ nucleotide triphosphate	
_____ m. Separates daughter molecules and causes supercoiling	

3.12 How long would it take *E. coli* to replicate its entire genome $(4.2 \times 10^6 \text{ bp})$, assuming a replication rate of 1,000 nucleotides per second at each fork with no pauses?

*3.13 A diploid organism has 4.5×10^8 bp in its DNA. The DNA is replicated in 3 minutes. Assuming that all replication forks move at a rate of 10^4 bp per minute, how many replicons (replication units) are present in the organism's genome?

*3.14 Describe the molecular action of the enzyme DNA ligase. What properties would you expect an *E. coli* cell to have if it had a temperature-sensitive mutation in the gene for DNA ligase?

*3.15 Chromosome replication in *E. coli* commences from a constant point, called the origin of replication. It is known that DNA replication is bidirectional. Devise a biochemical experiment to prove that the *E. coli* chromosome replicates bidirectionally. (*Hint*: Assume that the amount of gene product is directly proportional to the number of genes.)

3.16 Reiji Okazaki concluded that both DNA strands could not replicate continuously. What evidence led him to this conclusion?

*3.17 A space probe returns from Jupiter and brings with it a new microorganism for study. It has double-stranded DNA as its genetic material. However, studies of replication of the alien DNA reveal that, although the process is semiconservative, DNA synthesis is continuous on both the leading-strand and the lagging-strand templates. What conclusions can you draw from this result?

3.18 Phage such as λ and T4 are packaged from concatamers.
a. What are concatamers, and what type of DNA replication is responsible for producing concatamers?
b. In what ways does this type of DNA replication differ from that used by *E. coli*?

*3.19 Although λ is replicated into a concatamer, linear unit-length molecules are packaged into phage heads.
a. What enzymatic activity is required to produce linear unit-length molecules, how does it produce molecules that contain a single complete λ genome, and what gene encodes the enzyme involved?
b. What types of ends are produced when this enzyme acts on DNA, and how are these ends important in the λ life cycle?

*3.20 M13 is an *E. coli* bacteriophage whose capsid holds a closed circular DNA molecule with 2,221 T, 1,296 C, 1,315 G, and 1,575 A nucleotides. M13 lacks a gene for DNA polymerase and so must use bacterial DNA polym-

erases for replication. Unlike λ or T4, this phage does not form concatamers during replication and packaging.
a. Suppose the M13 chromosome were replicated in a manner similar to the way the *E. coli* chromosome is replicated, using semidiscontinuous replication from a double-stranded circular DNA template. How would the semidiscontinuous DNA replication mechanism discussed in the text need to be modified?
b. Suppose the M13 chromosome were replicated in a manner similar to the way the λ chromosome is replicated, using rolling circle replication. How would the rolling circle replication mechanism discussed in the text need to be modified?

*3.21 Compare and contrast eukaryotic and prokaryotic DNA polymerases.

3.22 What mechanism do eukaryotic cells employ to keep their chromosomes from replicating more than once per cell cycle?

*3.23 Autoradiography is a technique that allows radioactive areas of chromosomes to be observed under the microscope. The slide is covered with a photographic emulsion, which is exposed by radioactive decay. In regions of exposure, the emulsion forms silver grains on being developed. The tiny silver grains can be seen on top of the (much larger) chromosomes. Devise a method to find out which regions in the human karyotype replicate during the last 30 minutes of the S phase. (Assume a cell cycle in which the cell spends 10 hours in G_1, 9 hours in S, 4 hours in G_2, and 1 hour in M.)

3.24 In typical human fibroblasts in culture, the G_1 period of the cell cycle lasts about 10 hours, S lasts about 9 hours, G_2 takes 4 hours, and M takes 1 hour. Suppose you added radioactive (^{3}H) thymidine to the medium, left it there for 5 minutes, and then washed it out and replaced it with an ordinary medium.
a. What percentage of cells would you expect to become labeled by incorporating the ^{3}H-thymidine into their DNA?
b. How long would you have to wait after removing the ^{3}H medium before you would see labeled metaphase chromosomes?
c. Would one or both chromatids be labeled?
d. How long would you have to wait if you wanted to see metaphase chromosomes containing ^{3}H in the regions of the chromosomes that replicated at the beginning of the S period?

3.25 Suppose you performed the experiment in Question 3.24, but left the radioactive medium on the cells for 16 hours instead of 5 minutes. How would your answers change?

*3.26 In Figure 3.3, semiconservative DNA replication is visualized in eukaryotic cells with the harlequin chromosome-staining technique.

a. Explain what the harlequin chromosome-staining technique is and how it provides evidence for semiconservative DNA replication in eukaryotes.

b. Propose a hypothesis to explain why, in Figure 3.3, some chromatids appear to contain segments of both DNA containing T and DNA containing BUdR, while others appear to consist entirely of DNA with T or DNA with BUdR.

3.27 When the eukaryotic chromosome duplicates, the nucleosome structures must duplicate. Discuss how the synthesis of histones is related to the cell cycle, and discuss new nucleosomes are assembled at replication forks.

*3.28 A mutant *Tetrahymena* has an altered repeated sequence in its telomeric DNA. What change in the telomerase enzyme would produce this phenotype?

3.29 What is the evidence that telomere length is regulated in cells, and what are the consequences of the misregulation of telomere length?

4

Gene Function

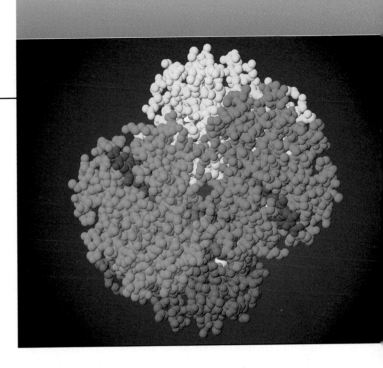

PRINCIPAL POINTS

- There is a specific relationship between genes and enzymes, initially embodied in the one-gene–one-enzyme hypothesis stating that each gene controls the synthesis or activity of a single enzyme. Since some enzymes consist of more than one polypeptide, and genes code for individual polypeptide chains, a more modern name for this hypothesis is "one gene–one polypeptide."

- Many human genetic diseases are caused by deficiencies in enzyme activities. Although some of these diseases are inherited as dominant traits, most are inherited as recessive traits.

- From the study of alterations in proteins other than enzymes, convincing evidence was obtained that genes control the structures of all proteins.

- Genetic counseling consists of an analysis of the risk that prospective parents may produce a child with a genetic defect, together with a presentation to appropri-

ate family members of the available options for avoiding or minimizing those risks. Early detection of a genetic disease is done by carrier detection and fetal analysis.

i WITHIN THE FIRST FEW MINUTES OF LIFE, MOST newborns in the United States are subjected to a battery of tests: Reflexes are tested, respiration and skin color assessed, and blood samples collected and rushed to a lab. Assays of the blood samples help health practitioners determine whether the child has a debilitating or even lethal genetic disease. What are genetic diseases? What is the relationship between genes, enzymes, and genetic disease? How can understanding gene function help prevent or minimize the risk of such diseases?

What do bread mold and certain human genetic disorders have in common? In the iActivity for this chapter, you will use Beadle and Tatum's experimental procedure to learn the answer to that question.

iActivity

In this chapter, we examine gene function. We present some of the classic evidence that genes code for enzymes and for other proteins. In particular, we examine the involvement of certain sets of genes in directing and controlling a particular biochemical pathway: the series of enzyme-catalyzed steps needed to break down or synthesize a particular chemical compound. Instead of thinking about the gene in isolation, we will see that a gene often must work in cooperation with other genes for cells to function properly. The experiments discussed represent the beginnings of molecular genetics, historically speaking, in that their goal was to understand better a gene at the molecular level. In later chapters, we develop our modern understanding of gene structure and function further.

Gene Control of Enzyme Structure

Garrod's Hypothesis of Inborn Errors of Metabolism

In 1902, Archibald Garrod, an English physician, studied *alkaptonuria* (Online Mendelian Inheritance in Man [OMIM], http://www3.ncbi.nlm.nih.gov/Omim/, entry 203500), a human disease characterized by urine that turns black upon exposure to the air and by a tendency to develop arthritis later in life. Because of the urine phenotype, the disease is easily detected soon after birth.

Garrod and geneticist William Bateson concluded that alkaptonuria is a genetically controlled trait because several members of the same families often had alkaptonuria and the disease was much more common among children of marriages involving first cousins than among children of marriages between unrelated partners. This finding was significant because first cousins have many alleles in common, so the chances are greater for recessive alleles to be homozygous in children of first-cousin marriages.

Garrod found that people with alkaptonuria excrete homogentisic acid (HA) in their urine, whereas people without the disease do not; it is the HA in urine that turns it black in air. This result indicated to Garrod that normal people can metabolize HA, but that people with alkaptonuria cannot. In Garrod's terms, the disease is an example of an **inborn error of metabolism;** that is, alkaptonuria is a genetic disease caused by the absence of a particular enzyme necessary for HA metabolism. Figure 4.1 shows part of the phenylalanine–tyrosine metabolic pathway, including the HA-to-maleylacetoacetic acid step, which is not carried out in people with alkaptonuria. The mutation responsible for alkaptonuria is recessive, so only people homozygous for the mutant gene express the defect. Later analysis has pinpointed the location of this gene on chromosome 3. Garrod's work provided the first evidence of a specific relationship between genes and enzymes.

Garrod also studied three other human genetic diseases that affected biochemical processes, and in each case he was able to conclude correctly that a step in a metabolic pathway could not be carried out. An important aspect of Garrod's analysis of these human diseases was his understanding that the position of a block in a metabolic pathway can be determined by the accumulation of the chemical compound (HA in the case of alkaptonuria) that precedes the blocked step. However, the significance of Garrod's work was not appreciated by his contemporaries.

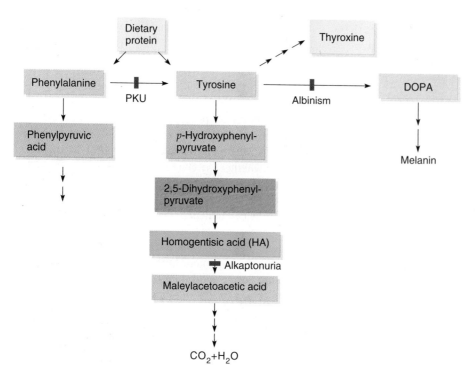

Figure 4.1

Phenylalanine-tyrosine metabolic pathways. People with alkaptonuria cannot metabolize homogentisic acid (HA) to maleylacetoacetic acid, and HA accumulates. People with PKU cannot metabolize phenylalanine to tyrosine, and phenylpyruvic acid accumulates. People with albinism cannot synthesize much melanin from tyrosine.

The One-Gene–One-Enzyme Hypothesis

In 1942, George Beadle and Edward Tatum heralded the beginnings of biochemical genetics, a branch of genetics that combines genetics and biochemistry to explain the nature

 of metabolic pathways. Results of their studies involving the haploid fungus *Neurospora crassa* (orange bread mold) showed a direct relationship between genes and enzymes and led to the *one-gene–one-enzyme hypothesis*, a landmark in the history of genetics. Beadle and Tatum shared one-half of the 1958 Nobel Prize in Physiology or Medicine for their "discovery that genes act by regulating definite chemical events."

ⓐ The One-
Gene–One-
Enzyme
Hypothesis

Isolation of Nutritional Mutants of *Neurospora*.

To understand Beadle and Tatum's experiment, we must understand the life cycle of *Neurospora crassa*, the orange bread mold (Figure 4.2). *Neurospora crassa* is a mycelial-form fungus, meaning that it spreads over its growth medium in a web-

like pattern. The mycelium produces asexual spores called conidia; their orange color gives the fungus its common name. *Neurospora* has important properties that make it useful for genetic and biochemical studies including the fact that it is a haploid organism, so the effects of mutations may be seen directly, and that it has a short life cycle, facilitating study of the segregation of genetic defects.

Neurospora can be propagated vegetatively (asexually) by inoculating pieces of the mycelial growth or the asexual spores (conidia) on a suitable growth medium to give rise to a new mycelium. *Neurospora crassa* can also reproduce by sexual means. There are two mating types, called *A* and *a*. The two mating types look identical and can be distinguished only by the fact that strains of the *A* mating type do not mate with other *A* strains, and *a* strains do not mate with other *a* strains. The sexual cycle is initiated by mixing *A* and *a* mating type strains on nitrogen-limiting medium. Under these conditions, cells of the two mating types fuse, followed by fusion of two haploid nuclei to produce a transient *A/a*

Figure 4.2

Life cycle of the haploid, mycelial-form fungus *Neurospora crassa*. (Parts not to scale.)

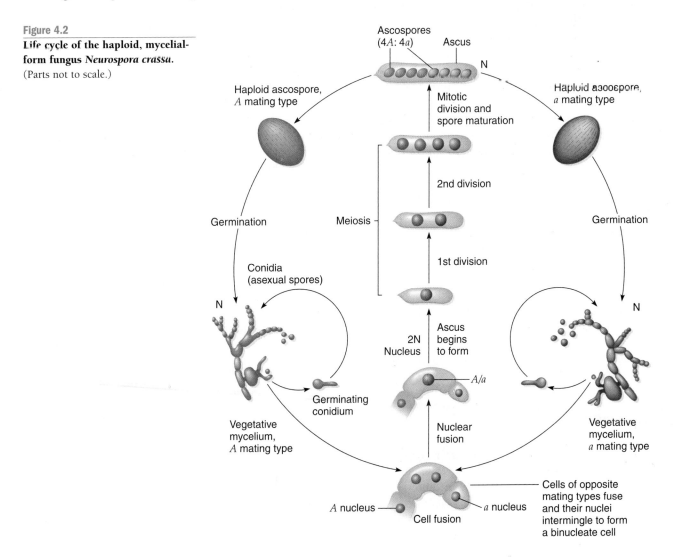

diploid nucleus, which is the only diploid stage of the life cycle. The diploid nucleus immediately undergoes meiosis and produces four haploid nuclei (two *A* and two *a*) within an elongating sac called an *ascus*. A subsequent mitotic division results in a linear arrangement of eight haploid nuclei around which spore walls form to produce eight sexual ascospores (four *A* and four *a*). Each ascus, then, contains all the produces of the initial, single meiosis. When the ascus is ripe, the ascospores (sexual spores) are shot out of the ascus and out of the fruiting body that encloses the ascus to be dispersed by wind currents. Germination of an ascospore begins the formation of a new haploid mycelium.

Important for Beadle and Tatum's experiments is the fact that *Neurospora* has simple growth requirements. Wild-type (prototrophic) *Neurospora* can grow on a simple minimal medium containing only inorganic salts (including a source of nitrogen), an organic carbon source (such as glucose or sucrose), and the vitamin biotin. Beadle and Tatum reasoned that *Neurospora* synthesized the other materials it needed for growth (e.g., amino acids, nucleotides, vitamins, nucleic acids, proteins) from the chemicals present in the minimal medium. They also realized that it should be possible to isolate **nutritional mutants** (also called **auxotrophic mutants** or **auxotrophs**) of *Neurospora* that required nutritional supplements to grow. (A strain that can grow on the minimal medium is called a **prototrophic strain** or a **prototroph.**) Such auxotrophs could be isolated because they would not grow on the minimal medium.

Figure 4.3 shows how Beadle and Tatum isolated and characterized auxotrophic mutants. They treated conidia with X rays to induce genetic mutants and then crossed them with a wild-type strain (a **prototrophic** strain that can grow on minimal medium) of the opposite mating type. By crossing the mutagenized spores with the wild type, they ensured that any auxotrophic mutant they isolated had segregated in a cross and therefore had a genetic basis, rather than a nongenetic reason, for requiring the nutrient.

One progeny spore per ascus from the crosses was allowed to germinate in a nutritionally complete medium that contained all necessary amino acids, purines, pyrimidines, and vitamins, in addition to the sucrose, salts, and biotin found in minimal medium. In complete medium, any strain that could not make one or more necessary compounds from the basic ingredients in minimal medium could still grow by using the compounds supplied in the growth medium. Each culture grown on the complete medium was then tested for growth on minimal medium. The strains that did not grow were the auxotrophs. These mutants, in turn, were tested individually for their ability to grow on minimal medium plus amino acids and on minimal medium plus vitamins. Theoretically, an amino acid auxotroph—a mutant strain that has lost the ability to synthesize a particular amino acid—

would grow on minimal medium plus amino acids, but not on minimal medium plus vitamins or on minimal medium alone. Similarly, vitamin auxotrophs would grow only on minimal medium plus vitamins.

Beadle and Tatum next did a second round of screening to determine which specific substance each auxotrophic strain needed for growth. Suppose an amino acid auxotroph was identified. To determine which of the 20 amino acids was required, the strain was inoculated into 20 tubes, each containing minimal medium plus a different one of the 20 amino acids. In the example shown in Figure 4.3, a tryptophan auxotroph was identified because it grew only in the tube containing minimal medium plus tryptophan. Crosses between each mutant and wild type segregated in the 1:1 ratio of wild-type : tryptophan-requiring phenotypes expected for a single gene defect.

Genetic Dissection of a Biochemical Pathway.   Once Beadle and Tatum had isolated and identified auxotrophic mutants, they investigated the biochemical pathways affected by the mutations. They assumed that *Neurospora* cells, like all other cells, function through the interaction of the products of a very large number of genes. Furthermore, they reasoned that wild-type *Neurospora* converted the constituents of minimal medium into amino acids and other required compounds by a series of reactions that were organized into pathways. In this way, the synthesis of cellular components occurred through a series of small steps, each catalyzed by an enzyme. As an example of the analytical approach Beadle and Tatum used that led to an understanding of the relationship between genes and enzymes, let us consider the genetic dissection of the pathway for the biosynthesis of the amino acid methionine in *Neurospora crassa*.

Starting with a set of methionine auxotrophs— mutants that require the addition of methionine to minimal medium to grow—genetic analysis (complementation tests; see Chapter 13, p. 340) identifies four genes: *met-2*[+], *met-3*[+], *met-5*[+], and *met-8*[+]. A mutation in any one of them gives rise to auxotrophy for methionine. Next, the growth pattern of the four mutant strains on media supplemented with chemicals thought to be intermediates involved in the methionine biosynthetic pathway— O-acetyl homoserine, cystathionine, and homocysteine— is determined, with the results shown in Table 4.1. By definition, all four mutant strains can grow on methionine and none can grow on unsupplemented minimal medium.

The sequence of steps in a pathway can be deduced from the pattern of growth supplementation. The principles are as follows: The farther along in a pathway a mutant strain is blocked, the fewer intermediate compounds permit the strain to grow. If a mutant strain is blocked at early steps, a larger number of intermediates enable the strain to grow. Thus, in these analyses, not only is the pathway deduced, but the steps controlled by

Figure 4.3

Method devised by Beadle and Tatum to isolate auxotrophic mutations in *Neurospora*. Here, the mutant strain isolated is a tryptophan auxotroph.

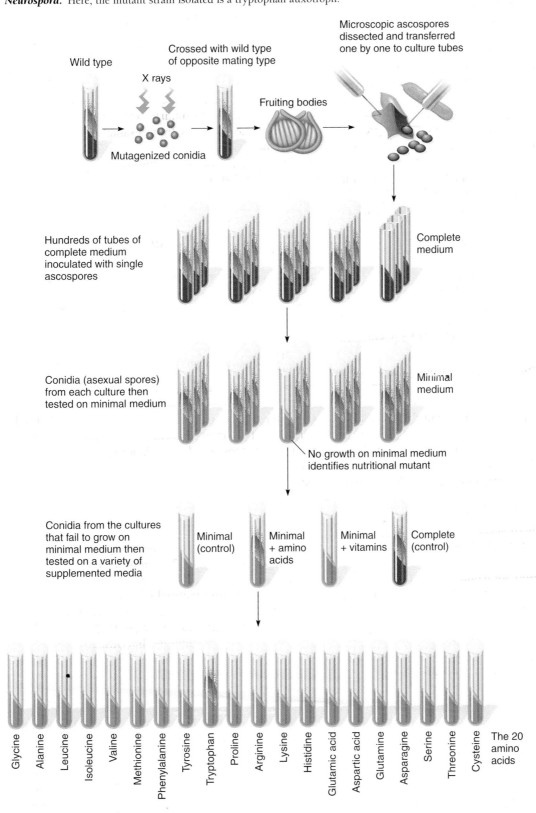

Table 4.1	**Growth Responses of Methionine Auxotrophs**				
	Growth Response on Minimal Medium and . . .				
Mutant Strains	**Nothing**	**O-Acetyl Homoserine**	**Cystathionine**	**Homocysteine**	**Methionine**
Wild type	+	+	+	+	+
met-5	–	+	+	+	+
met-3	–	–	+	+	+
met-2	–	–	–	+	+
met-8	–	–	–	–	+

each gene are determined. In addition, a genetic block in a pathway may lead to an accumulation of the intermediate compound used in the step that is blocked.

From Table 4.1, the *met-8* mutant strain grows when supplemented with methionine, but not when supplemented with any of the intermediates. This means that the *met-8* gene must control the last step in the pathway, which leads to the formation of methionine. The *met-2* mutant strain grows on media supplemented with methionine or homocysteine, so homocysteine must be immediately before methionine in the pathway, and the *met-2* gene must control the synthesis of homocysteine from another chemical. The *met-3* mutant strain grows on media supplemented with methionine, homocysteine, or cystathionine, so cystathionine must precede homocysteine in the pathway, and the *met-3* gene must control the synthesis of cystathionine from another compound. The *met-5* strain grows on media supplemented with either methionine, homocysteine, cystathionine, or *O*-acetyl homoserine, so *O*-acetyl homoserine must precede cystathionine in the pathway, and the *met-5* gene must control the synthesis of *O*-acetyl homoserine from another compound. The methionine biosynthesis pathway involved here (which is part of a larger pathway) is shown in Figure 4.4. Gene *met-5*[+] encodes the enzyme for converting homoserine to

O-acetyl homoserine, so mutants for this gene can grow on a minimal medium plus either *O*-acetyl homoserine, cystathionine, homocysteine, or methionine. Gene *met-3*[+] codes for the enzyme that converts *O*-acetyl homoserine to cystathionine, so a *met-3* mutant strain can grow on a minimal medium plus either cystathionine, homocysteine, or methionine, and so on.

From the results of experiments of this kind, Beadle and Tatum proposed that a specific gene encodes each enzyme. This hypothetical relationship between an organism's genes and the enzymes that catalyze the steps in a biochemical pathway is called the **one-gene–one-enzyme hypothesis.** Gene mutations that result in the loss of enzyme activity lead to the accumulation of precursors in the pathway (and to possible side reactions) and to the absence of the end product of the pathway. With the approach described, then, a biochemical pathway can be dissected genetically; that is, we can determine the sequence of steps in the pathway and relate each step to a specific gene or genes.

However, more than one gene may control each step in a pathway. An enzyme[1] may have two or more different

---

[1]We will see later in the book that some enzymes are RNA molecules, not proteins.

Figure 4.4

Methionine biosynthetic pathway showing four genes in *Neurospora crassa* that code for the enzymes that catalyze each reaction. (The *met-5* and *met-2* genes are on the same chromosome; *met-3* and *met-8* are on two other chromosomes.)

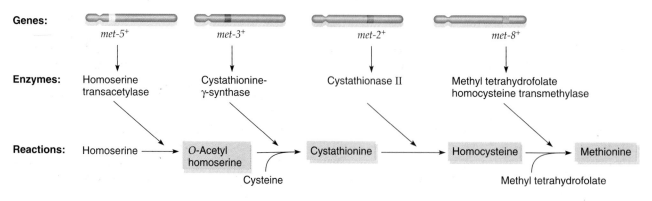

Figure 5.1

Transcription process. The DNA double helix is denatured by RNA polymerase in prokaryotes and by other proteins in eukaryotes. RNA polymerase then catalyzes the synthesis of a single-stranded RNA chain, beginning at the "start of transcription" point. The RNA chain is made in the 5′-to-3′ direction, with only one strand of the DNA used as a template to determine the base sequence.

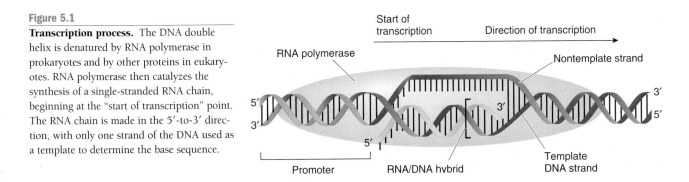

is done by other proteins that bind to the DNA at the start point for transcription.

In transcription, RNA is synthesized in the 5′-to-3′ direction. The 3′-to-5′ DNA strand that is read to make the RNA strand is called the **template strand.** The 5′-to-3′ DNA strand complementary to the template strand and that has the same polarity as the resulting RNA strand is called the *nontemplate strand.*

animation

a **RNA Biosynthesis**

The RNA precursors for transcription are the ribonucleoside triphosphates ATP, GTP, CTP, and UTP, collectively called NTPs (*nucleoside triphosphates*). RNA synthesis occurs by polymerization reactions that are similar to those involved in DNA synthesis (Figure 5.2; DNA polymerization is shown in Figure 3.4). RNA polymerase selects the next nucleotide to be added to the chain by its ability to pair with the exposed base on the DNA template strand. Unlike DNA polymerases, RNA polymerases can initiate new polynucleotide chains. (No primer is needed.)

Recall that RNA chains contain nucleotides with the base uracil instead of thymine and that uracil pairs with adenine. Therefore, where an A nucleotide occurs on the DNA template chain, a U nucleotide is placed in the RNA chain instead of a T. For example, if the template DNA strand reads

3′-ATACTGGAC-5′,

then the RNA chain will be synthesized in the 5′-to-3′ direction and will have the sequence

5′-UAUGACCUG-3′.

KEYNOTE

> Transcription, the process of transcribing the genetic information in DNA into RNA base sequences, exhibits basic similarities in prokaryotes and eukaryotes. The DNA unwinds in a short region next to a gene, and an RNA polymerase catalyzes the synthesis of an RNA molecule in the 5′-to-3′ direction along the 3′-to-5′ template strand of the DNA. Only one strand of the double-stranded DNA is transcribed into an RNA molecule.

Initiation of Transcription at Promoters

In both prokaryotes and eukaryotes, the process of transcription occurs in three steps: initiation, elongation, and termination. Particular mechanisms are used to signal initiation and termination of the RNA chain. In this section, we discuss the initiation of transcription in prokaryotes, focusing on *E. coli*. A prokaryotic gene may be divided into three sequences with respect to its transcription (Figure 5.3):

1. A sequence, called the **promoter,** upstream of the start of the RNA coding sequence. The RNA polymerase interacts with the promoter to begin transcription. The promoter ensures that the initiation of every RNA occurs at the same site.

2. The RNA-coding sequence itself—that is, the DNA sequence transcribed by RNA polymerase into the RNA transcript.

3. A **terminator,** specifying where transcription will stop.

From comparisons of sequences upstream of coding sequences and from studies of the effects of specific base-pair changes at every position upstream of transcription initiation sites, two DNA sequences in most promoters of *E. coli* genes have been shown to be critical for specifying the initiation of transcription. These sequences generally are found at −35 and −10; that is, they are centered at 35 and 10 base pairs upstream from +1 the base pair at which transcription starts. The **consensus sequence** (the sequence found most frequently at each position) for the −35 region (the −35 box) is 5′-TTGACA-3′. The consensus sequence for the −10 region (the **−10 box,** formerly called the **Pribnow box,** after David Pribnow, the researcher who first discovered it) is 5′-TATAAT-3′.

For transcription to begin, a form of RNA polymerase called the *holoenzyme* (or *complete enzyme*) binds to the promoter. This holoenzyme consists of the **core enzyme** form of RNA polymerase (which has four polypeptides—two α, one β, and one β′, bound to another polypeptide called the *sigma factor* (σ)). The sigma factor is essential for recognizing the −35 and −10 regions of the promoter.

The RNA polymerase holoenzyme binds to the promoter in two steps (Figure 5.4). First, it binds loosely to the −35 box (Figure 5.4a). For this step, the promoter is

Figure 5.2

Chemical reaction involved in the RNA polymerase–catalyzed synthesis of RNA on a DNA template strand.

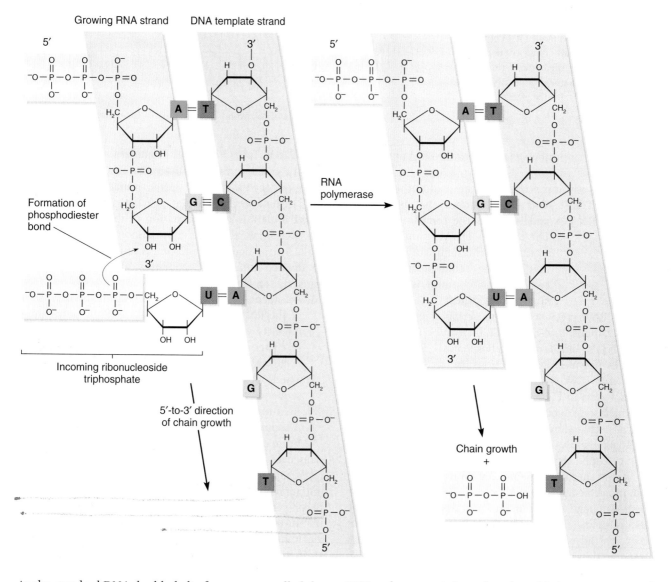

in the standard DNA double-helix form, a state called the *closed promoter complex.* Then the holoenzyme binds more tightly to the −10 box, accompanied by untwisting of the DNA (Figure 5.4b). The untwisted form of the promoter is called the *open promoter complex.* Once the

RNA polymerase is bound at the −10 box, it is oriented properly to begin transcription at the correct nucleotide.

Promoters differ in their actual sequence, so the binding efficiency of RNA polymerase varies. As a result, the rate at which transcription is initiated varies

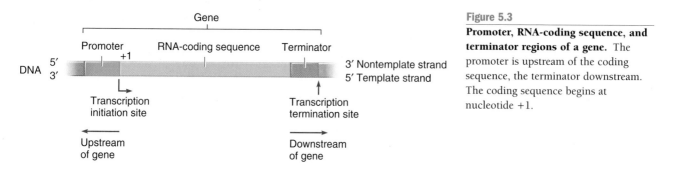

Figure 5.3

Promoter, RNA-coding sequence, and terminator regions of a gene. The promoter is upstream of the coding sequence, the terminator downstream. The coding sequence begins at nucleotide +1.

Figure 5.4

Action of *E. coli* RNA polymerase in the initiation and elongation stages of transcription. **(a)** In initiation, the RNA polymerase holoenzyme first binds loosely to the promoter at the −35 region. **(b)** As initiation continues, RNA polymerase binds more tightly to the promoter at the −10 region, accompanied by a local untwisting of about 17 bp around that region. At this point, the RNA polymerase is correctly oriented to begin transcription at +1. **(c)** After eight to nine nucleotides have been polymerized, the sigma factor dissociates from the core enzyme. **(d)** As the RNA polymerase elongates the new RNA chain, the enzyme untwists the DNA ahead of it; as the double helix re-forms behind the enzyme, the RNA is displaced away from the DNA.

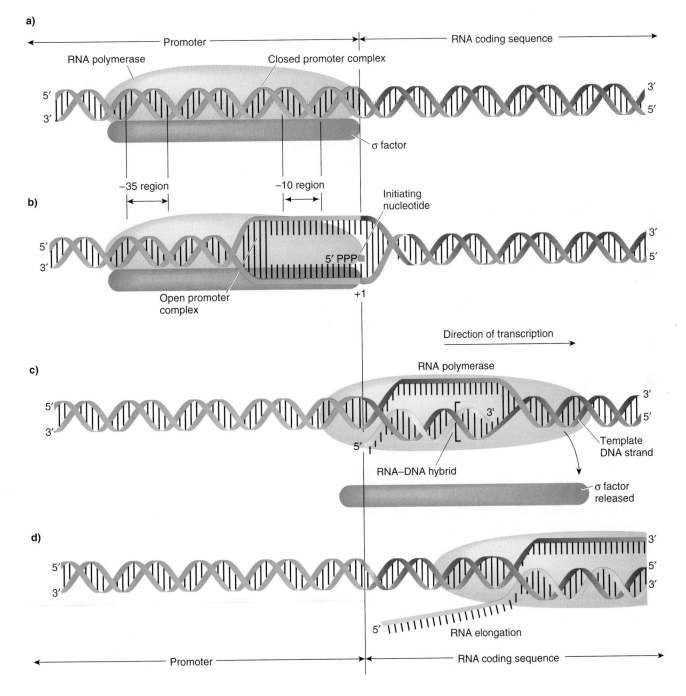

from gene to gene, which explains in part why different genes have different levels of expression. In other words, the relative strengths of promoters relate directly to how similar they are to the consensus sequence. For example, a −10 region sequence of 5′-GATACT-3′ has a lower rate of transcription initiation than 5′-ATAAT-3′, because the ability of the sigma factor component of the RNA polymerase holoenzyme to recognize and bind to the former sequence is lower than it is to the latter.

Several different sigma factors in *E. coli* play important roles in regulating gene expression. Each type of sigma factor binds to the core RNA polymerase and permits the holoenzyme to recognize different promoters. Most promoters have the recognition sequences we have just discussed, and these sequences are recognized by a sigma factor with a molecular weight of 70,000 Da, called σ^{70}. Under conditions of high heat (heat shock) and other forms of stress, a different sigma factor, called σ^{32} (molecular weight 32,000 Da), increases in amount, thereby directing some RNA polymerase molecules to bind to the promoters of genes that encode proteins required to cope with the stress. Such promoters have recognition sequences specific to the σ^{32} factor: CCCCC at −39 and TATAAATA at −15. Under conditions of limiting nitrogen, a third sigma factor, σ^{54} (molecular weight 54,000 Da), is produced. This sigma factor recognizes promoters with the consensus sequences GTGGC at −26 and TTGCA at −14. A fourth sigma factor, σ^{23} (molecular weight 23,000 Da), is made when cells are infected with phage T4. This sigma factor recognizes promoters with the consensus sequence TATAATA at −15. Other sigma factors control the expression of yet other types of genes under various conditions. Multiple sigma factors are also found in bacterial species different from *E. coli*.

Elongation and Termination of an RNA Chain

RNA synthesis takes place in a region of DNA that has separated to form a transcription bubble. Once about 10 RNA nucleotides have been linked together, the sigma factor dissociates from the RNA polymerase core enzyme (Figure 5.4c) and is used again in other transcription initiation reactions. The core enzyme then completes the transcription of the gene.

As the core RNA polymerase moves along, it untwists the DNA double helix ahead of itself and reanneals the DNA behind (Figure 5.4d). Within the untwisted region, RNA nucleotides are base paired to the DNA in a temporary RNA–DNA hybrid. Transcription proceeds at a rate averaging 30 to 50 nucleotides per second. Recent evidence has shown that RNA polymerase has two proofreading activities. One of these is similar to the proofreading by DNA polymerase in which the incorrectly inserted nucleotide is removed by the enzyme's reversing its synthesis reaction, backing up one step, and then replacing the incorrect nucleotide with the correct one in a forward step. The other proofreading activity involves the enzyme moving back one or more nucleotides and then cleaving the RNA at that position before resuming synthesis in the forward direction.

The termination of prokaryotic gene transcription is signaled by *terminator sequences*. One important protein involved in the termination of transcription of some *E. coli* genes is Rho (ρ) The terminators of such genes are called *Rho-dependent terminators* (also, type II terminators). At many other terminators, the core RNA polymerase itself carries out the termination events. Those terminators are called *Rho-independent terminators* (also, type I terminators).

Rho-independent terminators consist of an inverted repeat sequence that is about 16 to 20 base pairs upstream of the transcription termination point, followed by a string of about 4 to 8 A-T base pairs. The RNA polymerase transcribes the terminator sequence, which, because of the inverted repeat arrangement, folds into a hairpin loop structure and is followed by the string of U's transcribed from the string of A-T base pairs (Figure 5.5). In other words, the terminator sequence must be transcribed into the RNA in order to function. (Accordingly, the terminator is actually part of the initial RNA-coding sequence of the gene.) The hairpin structure is thought to cause termination by affecting RNA polymerase's ability to continue elongation.

Rho-dependent terminators lack the A-T string found in Rho-independent terminators, and many cannot form hairpin structures. The model for this type of termination is as follows: Rho has RNA-binding and ATPase domains. Rho binds to the terminator sequence on the RNA as it exits the RNA polymerase. Once bound, Rho's ATPase activity hydrolyzes ATP, and the energy from this reaction separates the RNA from the RNA polymerase and from the DNA template.

Only one type of RNA polymerase is found in prokaryotes, so all classes of genes—protein-coding genes, tRNA genes, and rRNA genes—are transcribed by it.

KEYNOTE

In *E. coli*, the initiation and termination of transcription are signaled by specific sequences that flank the RNA-coding sequence of the gene. The promoter is recognized by the sigma factor component of the RNA polymerase–sigma factor complex. Two types of termination sequences are found, and a particular gene has one or the other. One type of terminator is recognized by the RNA polymerase alone, and the other type is recognized by the enzyme in association with the *rho* factor.

Figure 5.5

Sequence of a Rho-independent terminator and structure of the terminated RNA. The mutations in the stem (yellow section) partially or completely prevent termination.

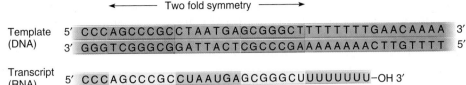

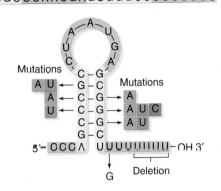

Transcript folded to form termination hairpin

Transcription in Eukaryotes

Transcription is more complicated in eukaryotes than in prokaryotes, because eukaryotes possess three different classes of RNA polymerases and because of the way in which transcripts are processed to their functional forms.

Eukaryotic RNA Polymerases

In eukaryotes, three different RNA polymerases transcribe the genes for the four types of RNAs. **RNA polymerase I,** located exclusively in the nucleolus, catalyzes the synthesis of three of the RNAs found in ribosomes: the 28S, 18S, and 5.8S rRNA molecules. (The S values derive from the rate at which the rRNA molecules sediment during centrifugation and give a very rough indication of molecular sizes.) **RNA polymerase II,** found only in the nucleoplasm of the nucleus, synthesizes messenger RNAs (mRNAs) and some small nuclear RNAs (snRNAs), some of which are involved in RNA processing events. **RNA polymerase III,** found only in the nucleoplasm, synthesizes (1) the transfer RNAs (tRNAs), which bring amino acids to the ribosome, (2) 5S rRNA, a small rRNA molecule found in each ribosome, and (3) the snRNAs not made by RNA polymerase II.

Compared with what we know about the structure and function of the RNA polymerases in *E. coli* and a number of other prokaryotes, less is known about the structure and function of eukaryotic RNA polymerases. What we do know is that all eukaryotic RNA polymerases consist of multiple subunits; thus, several genes encode these enzymes. For example, yeast RNA polymerase II consists of 12 subunits and has a U-shaped structure (Figure 5.6). Interestingly, five of these subunits are also part of yeast RNA polymerase III. Other eukaryotic RNA polymerase II's appear to have similar structures.

KEYNOTE

In *E. coli,* a single RNA polymerase synthesizes mRNA, tRNA, and rRNA. Eukaryotes have three distinct nuclear RNA polymerases, each of which transcribes different gene types: RNA polymerase I transcribes the genes for the 28S, 18S, and 5.8S ribosomal RNAs; RNA polymerase II transcribes mRNA genes and some snRNA genes; and RNA polymerase III transcribes genes for the 5S rRNAs, the tRNAs, and the remaining snRNAs.

Transcription of Protein-Coding Genes by RNA Polymerase II

In this section, we discuss the sequences and molecular events involved in transcribing a protein-coding gene. In eukaryotes, RNA polymerase II transcribes such

Figure 5.6

Three-dimensional structure of RNA polymerase II from yeast.

genes. The product of transcription is a **precursor-mRNA (pre-mRNA)** molecule—a transcript that must be modified, processed, or both to produce the mature, functional mRNA molecule.

Promoters and Enhancers. Promoters of protein-coding genes are analyzed in two principal ways. One way is to examine the effect of mutations that delete or alter base pairs upstream from the starting point of transcription and to see whether those mutants affect transcription. Mutations that significantly affect transcription define important promoter elements. The second way is to compare the DNA sequences upstream of a number of protein-coding genes to see whether there are any regions with similar sequences. The results of these experiments show that the promoters of protein-coding genes encompass about 200 base pairs upstream of the transcription initiation site and contain various sequence elements. We can divide the promoter into two regions: (1) the core promoter; and (2) promoter proximal elements.

The **core promoter** is the set of cis-acting sequence elements needed for the transcription machinery to start RNA synthesis at the correct site. These elements are typically within no more than 50 base pairs upstream of that site. The best-characterized core promoter elements are (1) a short sequence element called *Inr* (initiator), which spans the transcription initiation start site (defined as +1), and (2) the TATA **box**, or TATA **element** (also called the **Goldberg-Hogness box,** after its discoverers), located at about position −30. The TATA box has the seven-nucleotide consensus sequence TATAAAA. The Inr and TATA elements specify where the transcription machinery assembles and determine where transcription will begin. However, in the absence of other elements, transcription will occur only at a low level.

Promoter proximal elements are further upstream from the TATA box, in the area from 50 to 200 nucleotides from the start site of transcription. Examples of these elements are the CAAT (*"cat"*) *box,* named for its consensus sequence and centered at about −75, and the GC **box,** with consensus sequence GGGCGG, centered at about −90. Both the CAAT box and the GC box work in either orientation (meaning with the sequence element oriented either towards or away from the direction of transcription). Mutations in either of these elements (or other proximal promoter elements not mentioned) markedly decrease transcription initiation from the promoter, indicating that they play a role in determining the efficiency of the promoter.

Promoters contain various combinations of core promoter elements and promoter proximal elements that together determine promoter function. The promoter proximal elements are important in determining how and when a gene is expressed. Key to this regulation are tran-scription regulatory proteins called **activators,** which determine the efficiency of transcription initiation. For example, genes that are expressed in all cell types for basic cellular functions—"housekeeping genes"—have promoter proximal elements that are recognized by activators ubiquitous to all cell types. Examples of housekeeping genes are the actin gene and the gene for the enzyme glucose 6-phosphate dehydrogenase. By contrast, genes that are expressed only in particular cell types or at particular times have promoter proximal elements recognized by activators found only in those cell types or at those particular times.

Other sequences—**enhancers**—are required for the maximal transcription of a gene. Enhancers are another type of cis-acting element. By definition, enhancers function either upstream or downstream from the transcription initiation site, although, commonly, they are upstream of the gene they control, sometimes thousands of base pairs away. In other words, enhancers modulate transcription from a distance. Enhancers contain a variety of short sequence elements, some of which are the same as those found in the promoter. Activators also bind to these elements and with other protein complexes. The DNA containing the enhancer is brought close to the promoter DNA to which the transcription machinery is bound, stimulating transcription to the maximal level for the particular gene. (We will discuss activators, promoters, and enhancers and how eukaryotic protein-coding genes are regulated in more detail in Chapter 20.)

Transcription Initiation. Accurate initiation of transcription of a protein-coding gene involves the assembly of RNA polymerase II and a number of other proteins called **general transcription factors** (GTFs) on the core promoter. All three eukaryotic RNA polymerases require GTFs in order to initiate transcription. The GTFs are numbered for the RNA polymerase with which they work and are lettered to reflect their order of discovery. For example, TFIID is the fourth basal transcription factor (D) discovered that works with RNA polymerase II.

For protein-coding genes, the GTFs and RNA polymerase II bind to promoter elements in a particular order in vitro to produce the *complete transcription initiation complex,* also called the *preinitiation complex* (PIC) because it is ready to begin transcription (Figure 5.7). As has been mentioned, the binding of activators to promoter proximal elements and to enhancer elements determines the overall efficiency of transcription initiation at the promoter.

While the in vitro experiments indicate a sequential order of loading of GTFs and RNA polymerase II onto the promoter, the situation is less clear in vivo. Some data indicate that the initiation complex comes to the promoter in a single complex. That issue aside, transcription

Figure 5.7

Assembly of the transcription initiation machinery. First, TFIID binds to the TATA box to form the *initial committed complex*. The multisubunit TFIID has one subunit called the TATA-binding protein (TBP), which recognizes the TATA box sequence, and a number of other proteins called TBP-associated factors (TAFs). In vitro, the TFIID–TATA box complex acts as a binding site for the sequential addition of other transcription factors. Initially, TFIIA and then TFIIB bind, followed by RNA polymerase II and TFIIF, to produce the *minimal transcription initiation complex*. (RNA polymerase II, like all eukaryotic RNA polymerases, cannot directly recognize and bind to promoter elements.) Next, TFIIE and TFIIH bind to produce the *complete transcription initiation complex*, also called the *preinitiation complex* (PIC). TFIIH's helicase-like activity now unwinds the promoter DNA, and transcription is ready to begin.

Assembly of preinitiation complex

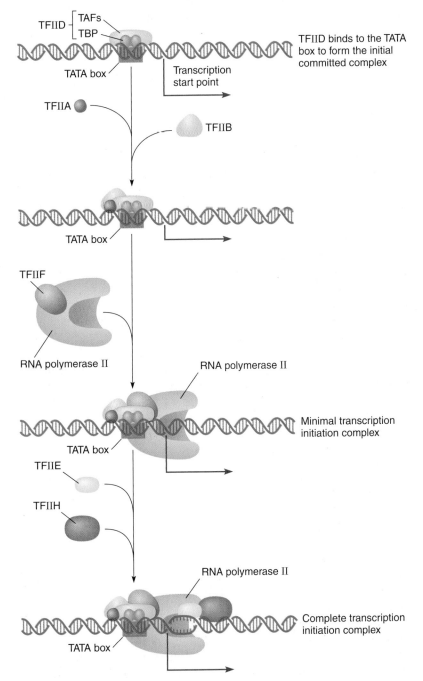

TFIID binds to the TATA box to form the initial committed complex

initiation in vivo is clearly more complicated simply because of the nucleosome organization of chromosomes. (This complication is addressed in Chapter 20.)

iActivity

Investigate how mutations at different regions in the β-globin gene affect mRNA transcription and the production of β-globin in the iActivity, *Investigating Transcription in Beta-Thalassemia Patients,* on your CD-ROM.

Eukaryotic mRNAs

Figure 5.8 shows the general structure of the mature, biologically active mRNA as it exists in both prokaryotic and eukaryotic cells. The mRNA molecule has three main parts. At the 5′ end is a **leader sequence**, or **5′ untranslated region (5′ UTR)**, which varies in length between mRNAs of different genes. Following

animation

a **mRNA Production in Eukaryotes**

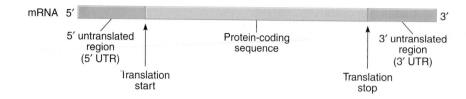

mRNA 5′ ___ 3′

5′ untranslated region (5′ UTR)

Translation start

Protein-coding sequence

Translation stop

3′ untranslated region (3′ UTR)

Figure 5.8

General structure of mRNA found in both prokaryotic and eukaryotic cells.

the 5′ leader sequence is the **coding sequence** of the mRNA—the sequence that specifies the amino acid sequence of a protein during translation. Following the amino acid–coding sequence is a **trailer sequence**, or **3′ untranslated region (3′ UTR)**, which may contain information that signals the stability of the particular mRNA. (See Chapter 20.)

mRNA production is different in prokaryotes and eukaryotes. In prokaryotes (Figure 5.9a), the RNA transcript functions directly as the mRNA molecule; that is, the base pairs of a prokaryotic gene are colinear with the bases of the translated mRNA. In addition, because prokaryotes lack a nucleus, an mRNA begins to be translated on ribosomes before it has been transcribed completely; this process is called *coupled transcription and translation*. In eukaryotes (Figure 5.9b), the RNA transcript (the pre-mRNA) must be modified in the nucleus by a series of events known as *RNA processing* to produce the mature mRNA. Also, the mRNA must migrate from the nucleus to the cytoplasm (where the ribosomes are located) before it can be translated. Thus, a eukaryotic mRNA is always transcribed completely and processed before it is translated.

Figure 5.9

Processes that synthesize functional mRNA in prokaryotes and eukaryotes. (a) In prokaryotes, the mRNA synthesized by RNA polymerase does not have to be processed before it can be translated by ribosomes. Also, because there is no nuclear membrane, mRNA translation can begin while transcription continues, resulting in a coupling of transcription and translation. (b) In eukaryotes, the primary RNA transcript is a precursor-mRNA (pre-mRNA) molecule, which is processed in the nucleus by the addition of a 5′ cap and a 3′ poly(A) tail and the removal of introns. Only when that mRNA is transported to the cytoplasm can translation occur.

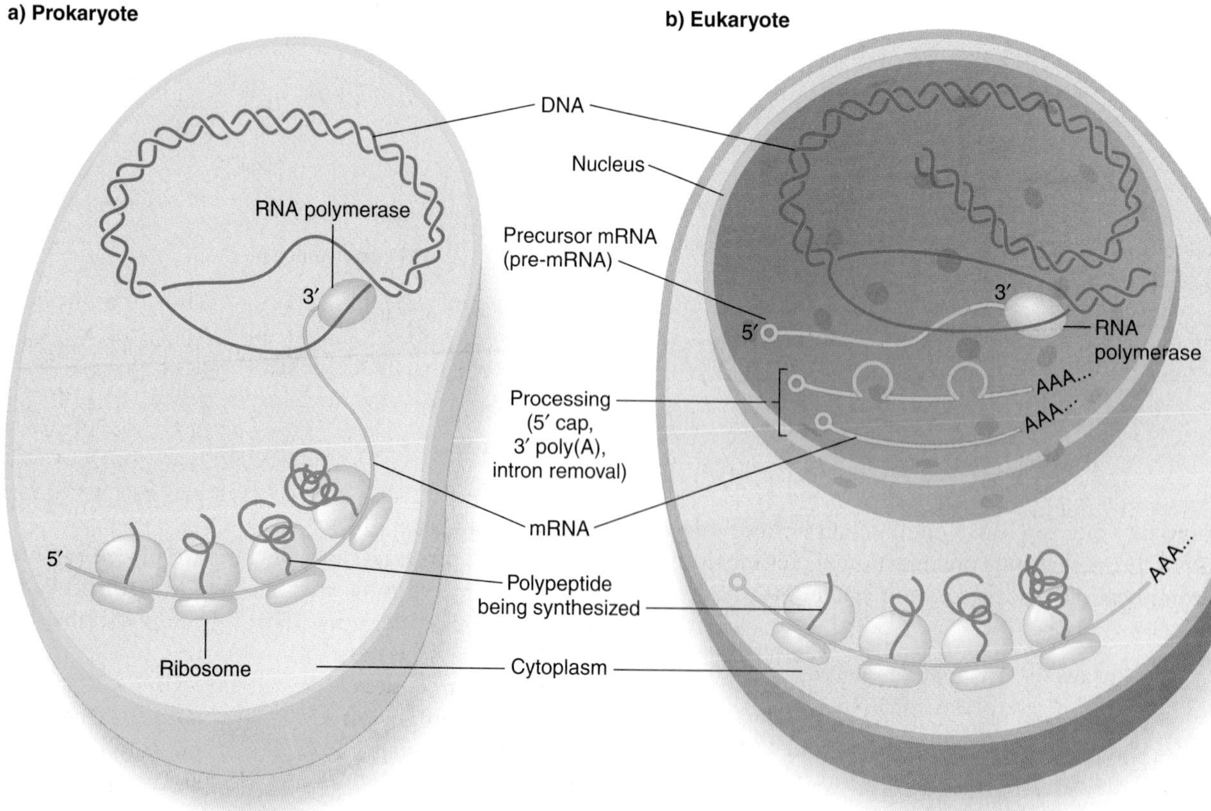

a) Prokaryote

b) Eukaryote

DNA

Nucleus

Precursor mRNA (pre-mRNA)

RNA polymerase

3′

5′

RNA polymerase

3′

5′

AAA...

AAA...

Processing (5′ cap, 3′ poly(A), intron removal)

mRNA

5′

Polypeptide being synthesized

AAA...

Ribosome

Cytoplasm

Another fundamental difference between prokaryotic and eukaryotic mRNAs is that prokaryotic mRNAs often are *polycistronic*, meaning that they contain the amino acid–coding information from more than one gene, whereas eukaryotic mRNAs are always *monocistronic*, meaning that they contain the amino acid–coding information from just one gene.

Production of Mature mRNA in Eukaryotes. Unlike prokaryotic mRNAs, eukaryotic mRNAs are modified at both the 5′ and 3′ ends. In addition, a very exciting discovery in the history of molecular genetics took place in 1977 when Richard Roberts, Philip Sharp, and Susan Berget showed that the genes of certain animal viruses contain internal sequences that are not expressed in the amino acid sequences of the proteins they encode. Subsequently, the same phenomenon was seen in eukaryotes. In fact, in eukaryotes in general, most protein-coding genes have non-amino-acid–coding sequences called **introns** between the other sequences that are present in mRNA: the **exons**. The term *intron* is derived from *intervening sequence*—a sequence that is not translated into an amino acid sequence—and the term *exon* is derived from *expressed sequence*. Exons include the 5′ and 3′ UTRs, as well as the amino acid-coding portions. In the processing of pre-mRNA to the mature mRNA molecule, the introns are removed. The 1993 Nobel Prize in Physiology or Medicine was awarded to Roberts and Sharp for their independent discoveries of split genes.

5′ and 3′ Modifications. Once RNA polymerase II has made about 20 to 30 nucleotides of pre-mRNA, a *capping enzyme* adds a guanine nucleotide—most commonly, 7-methyl guanosine (m^7G)—to the 5′ end by an unusual 5′-to-5′ linkage, as opposed to the usual 5′-to-3′ linkage (Figure 5.10). The process is called **5′ capping.** The sugars of the next two nucleotides are also modified by methylation. The 5′ cap remains throughout processing and is present in the mature mRNA, protecting it against degradation by exonucleases because of the unusual 5′–5′ linkage. The 5′ cap is also important for the binding of the ribosome as an initial step of translation.

Most eukaryotic pre-mRNAs become modified at their 3′ ends by the addition of a sequence of about 50 to 250 adenine nucleotides called a **poly(A) tail.** There is no DNA template for the poly(A) tail, which remains while pre-mRNA is processed to mature mRNA. The poly(A) tail is required for efficient export of the mRNA from the nucleus to the cytoplasm. Once in the cytoplasm, the poly(A) tail protects the 3′ end of the mRNA by buffering coding sequences against early degradation by exonucleases. The poly(A) tail also plays an important role in processes that regulate the stability of mRNA.

Addition of the poly(A) tail marks the 3′ end of the mRNA. That is, there typically is no termination

Figure 5.10

Cap structure at the 5′ end of a eukaryotic mRNA. The cap results from the addition of a guanine nucleotide and two methyl groups.

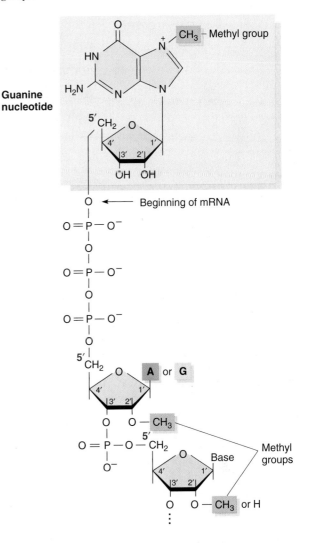

sequence in the DNA to signal the end of transcription of a protein-coding gene. Instead, mRNA transcription continues, in some cases for hundreds or thousands of nucleotides, past a site in the RNA called the **poly(A) site,** which is about 10 to 30 nucleotides downstream of the poly(A) consensus sequence AAUAAA. For poly(A) addition to the RNA, a number of proteins, including CPSF (*cleavage and polyadenylation specificity factor*) protein, CstF (*cleavage stimulation factor*) protein, and two *cleavage factor* proteins (CFI and CFII), bind to and cleave the RNA (Figure 5.11). Then, the enzyme **poly(A) polymerase** (PAP) uses ATP as a substrate and catalyzes the addition of A nucleotides to the 3′ end of the RNA to produce the poly(A) tail. During this process, PAP is bound to CPSF. As the poly(A) tail is synthesized, molecules of poly(A) binding protein II (PABII) bind to it.

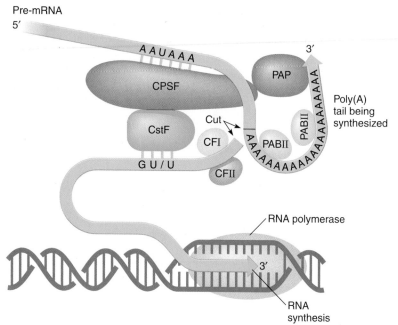

Pre-mRNA
5′

Figure 5.11

Schematic diagram of the 3′ end formation of mRNA and the addition of the poly(A) tail to that end in mammals. In eukaryotes, the formation of the 3′ end of an mRNA is produced by cleavage of the lengthening RNA chain. CPSF binds to the AAUAAA signal, and CstF binds to a GU-rich or U-rich sequence (GU/U) downstream of the poly(A) site. CPSF and CstF also bind to each other, producing a loop in the RNA. CFI and CFII bind to the RNA and cleave it. Poly(A) polymerase then adds the poly(A) tail to which poly(A) binding proteins attach.

Introns. Pre-mRNAs often contain a number of introns. To generate a mature mRNA that can be translated into a complete polypeptide, introns must be excised from each pre-mRNA. The mature mRNA, then, contains, in a contiguous form, the exons that, in the gene, were separated by introns.

At the time introns were discovered, it was known that the nucleus contains a large population of RNA molecules of various sizes, known as **heterogeneous nuclear RNAs (hnRNAs).** It was correctly assumed that hnRNAs include pre-mRNA molecules. In 1978, Philip Leder's group was studying the β-globin gene in cultured mouse cells. This gene encodes the 146-amino-acid β-globin polypeptide that is part of a hemoglobin protein molecule. The researchers isolated a 1.5-kb RNA molecule nuclear hnRNA that was the β-globin pre-mRNA. Like the 0.7-kb mature mRNA, the pre-mRNA has a 5′ cap and a 3′ poly(A) tail. Leder's group demonstrated that the 1.5-kb pre-mRNA is colinear with the gene that encoded it, whereas the 0.7-kb β-globin mRNA is not. The scientists interpreted their results to mean that the β-globin gene has an intron of about 800 nucleotide pairs. Transcription of the gene results in a 1.5-kb pre-mRNA containing both exon and intron sequences. This RNA is found only in the nucleus. The intron sequence is excised by processing events, and the flanking exon sequences are spliced together to produce a mature mRNA. (Subsequent research showed that the β-globin gene contains two introns; the second, smaller intron was not detected in the early research.)

At the time of this discovery, scientists believed that the gene sequence was completely colinear with the amino acid sequence of the encoded protein. Thus, finding that genes could be "in pieces" was most surprising and was one of those highly significant discoveries that changed our thinking about genes. In the years since the discovery of introns, we have learned that most eukaryotic protein-coding genes contain introns. Interestingly, some bacteriophage genes also contain introns. For example, there is a single intron in the bacteriophage T4 thymidylate synthetase gene.

KEYNOTE

The transcripts of protein-coding genes are messenger RNAs or their precursors. These molecules are linear and vary widely in length with the size of the polypeptides they specify and whether they contain introns. Prokaryotic mRNAs are not modified once they are transcribed, whereas most eukaryotic mRNAs are modified by the addition of a cap at the 5′ end and a poly(A) tail at the 3′ end. Many eukaryotic pre-mRNAs contain non–amino acid–coding sequences called introns, which must be excised from the mRNA transcript to make a mature, functional mRNA molecule. The amino acid–coding segments separated by introns are called exons.

Processing of Pre-mRNA to Mature mRNA. Messenger RNA production from genes with introns involves transcription of the gene by RNA polymerase II, addition of the 5′-cap and poly(A) tail to produce the pre-mRNA molecule, and processing of the pre-mRNA in the nucleus to remove the introns and splice the exons together to produce the mature mRNA (Figure 5.12).

animation

a RNA Splicing

Figure 5.12

General sequence of steps in the formation of eukaryotic mRNA. Not all steps are necessary for all mRNAs.

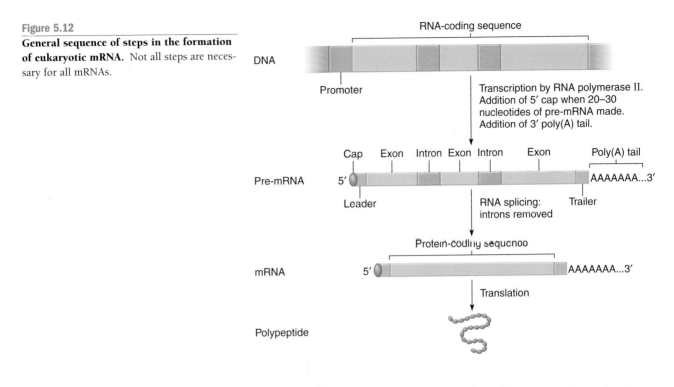

Introns are identifiable because they typically begin with 5′-GU and end with AG-3′, although more than just those nucleotides is needed to specify a junction between an intron and an exon. Introns in pre-mRNAs are removed and exons joined in the nucleus by **mRNA splicing.** The splicing events occur in complexes called **spliceosomes,** which consist of the pre-mRNA bound to **small nuclear ribonucleoprotein particles** (**snRNPs;** also called *snurps*). snRNPs are snRNAs associated with proteins. There are five principal snRNAs involved (named U1, U2, U4, U5, and U6), and each is associated with a number of proteins to form the snRNPs. U4 and U6 snRNAs are found within the same snRNP (U4/U6 snRNP), and the others are found within their own special snRNPs. Each snRNP type is abundant in the nucleus, with at least 10^5 copies per cell.

Figure 5.13 shows a simplified model of splicing for two exons separated by an intron. The steps are as follows:

1. U1 snRNP binds to the 5′ splice junction of the intron. This binding is primarily the result of base pairing of U1 snRNA in the snRNP to the 5′ splice junction.
2. U2 snRNP binds to a sequence called the **branch-point sequence,** which is located upstream of the 3′ splice junction.
3. A U4/U6 snRNP and a U5 snRNP interact, and the combination binds to the U1 and U2 snRNPs, causing the intron to loop, thereby bringing its two ends closer together.
4. U4 snRNP dissociates, resulting in the formation of the active spliceosome.

5. The snRNPs in the spliceosome cleave the intron from exon 1 at the 5′ splice junction, and the now-free 5′ end of the intron is bonded to an A nucleotide in the branch-point sequence. Because of its resemblance to the rope cowboys use, the looped back structure is called an *RNA lariat structure.* The branch point in the RNA that produces the lariat structure involves an unusual 2′–5′ phosphodiester bond formed between the 2′ OH of the adenine nucleotide in the branch-point sequence and the 5′ phosphate of the guanine nucleotide at the end of the intron. The A itself remains in normal 3′–5′ linkage with its adjacent nucleotides of the intron.
6. Next, the intron is excised (still in lariat shape) by cleavage at the 3′ splice junction, and exons 1 and 2 are ligated together. The snRNPs are released at this time. The process is repeated for each intron.

K E Y N O T E

Introns are removed from pre-mRNAs in a series of well-defined steps. Intron removal begins with the cleavage of the pre-mRNA at the 5′ splice junction. The free 5′ end of the intron loops back and bonds to a site upstream of the 3′ splice junction. Cleavage at that junction releases the intron, which is shaped like a lariat. Once the intron is excised, the exons that flanked it are spliced together. The removal of introns from eukaryotic pre-mRNA occurs in the nucleus in complexes called spliceosomes, which consist of several snRNPs bound specifically to each intron.

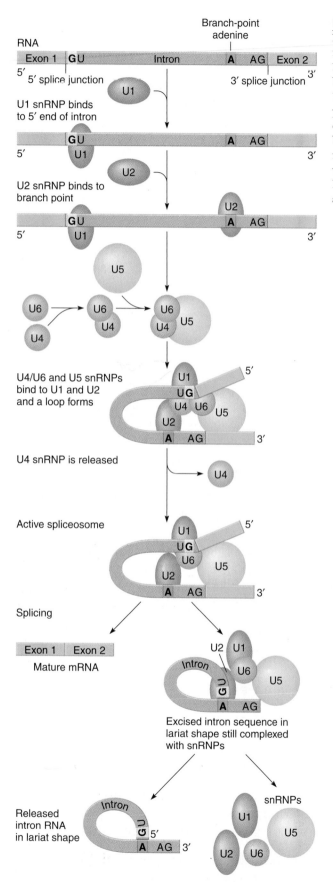

Figure 5.13

Model for intron removal by the spliceosome. At the 5′ end of an intron is the sequence GU, and at the 3′ end is the sequence AC. Near the 3′ end of the intron is an A nucleotide located within the branch-point sequence, which, in mammals, is YNCURAY, where Y = pyrimidine, N = any base, R = purine, and A = adenine, and, in yeast, is UACUAAC. (The italic A in each sequence is where the 5′ end of the intron bonds). With the aid of snRNPs, intron removal begins with a cleavage at the first exon–intron junction. The G at the released 5′ of the intron folds back and forms an unusual 2′–5′ bond with the A of the branch-point sequence. This reaction produces a lariat-shaped intermediate. Cleavage at the 3′ intron–exon junction and ligation of the two exons completes the removal of the intron.

Self-Splicing Introns. In some species of the protozoan *Tetrahymena*, the genes for the 28S ribosomal RNA found in the large ribosomal subunit (see p. 102) are interrupted by a 413-bp intron. The excision of this intron—now called a *group I intron*—was shown to occur by a *protein-independent reaction* in which the RNA intron folds into a secondary structure that promotes its own excision. The process, called **self-splicing,** was unexpectedly discovered in 1982 by Tom Cech and his research group. In 1989, Cech shared the Nobel Prize in Chemistry for his discovery. The self-splicing of the *Tetrahymena* pre-rRNA intron was the first example of what is now called *group I intron self-splicing.* Group I introns are rare. Other self-splicing group I introns have been found in nuclear rRNA genes, in some mitochondrial protein-coding genes, and in some mRNA and tRNA genes in bacteriophages.

Figure 5.14 diagrams the self-splicing reaction for the group I intron in *Tetrahymena* pre-rRNA. The steps are as follows:

1. The pre-rRNA is cleaved at the 5′ splice junction as guanosine is added to the 5′ end of the intron.
2. The intron is cleaved at the 3′ splice junction.
3. The two exons are spliced.
4. The excised intron circularizes to produce a lariat molecule, which is cleaved to produce a circular RNA and a short linear piece of RNA.

The self-splicing activity of the intron RNA sequence cannot be considered an enzyme activity. That is, although the RNA carries out the reaction, it is not regenerated in its original form at the end of the reaction, as is the case with protein enzymes. Modified forms of the *Tetrahymena* intron RNA and of other self-cleaving RNAs that function catalytically have been produced in the lab. These **RNA enzymes** are called **ribozymes;** they can be used experimentally to cleave RNA molecules at specific sequences.

The discovery that RNA can act like a protein was an important landmark in biology and has revolutionized theories about the origin of life. Previous theories proposed that proteins were required for replication of the first nucleic acid molecules. The new theories propose that the first nucleic acid was self-replicating through ribozyme-like activity carried out in the molecule.

Figure 5.14

Self-splicing reaction for the group I intron in *Tetrahymena* pre-rRNA. Cleavage at the 5′ splice junction takes place first, accompanied by the addition of a G nucleotide to the 5′ end of the intron. Cleavage at the 3′ splice junction releases the intron; the two exons are spliced. The intron now circularizes by the 3′ nucleotide bonding to an internal nucleotide, producing a lariat-shaped molecule. Cleavage of the lariat molecule produces a circular RNA molecule and a linear molecule.

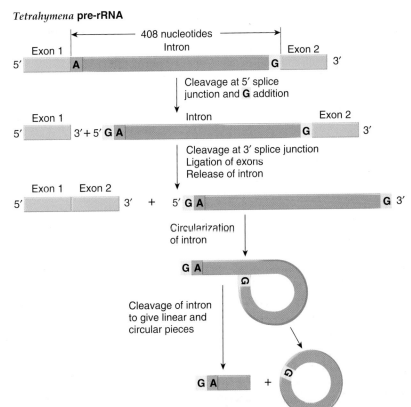

Tetrahymena pre-rRNA
408 nucleotides
Exon 1
Intron
Exon 2
5′
A
G
3′
Cleavage at 5′ splice junction and G addition
Exon 1
Intron
Exon 2
5′
3′+5′ G A
G
3′
Cleavage at 3′ splice junction
Ligation of exons
Release of intron
Exon 1
Exon 2
5′
3′
+
5′ G A
G 3′
Circularization of intron
G A
G
Cleavage of intron to give linear and circular pieces
G A
+
G

KEYNOTE

In some precursor rRNAs, there are introns whose RNA sequences fold into a secondary structure that excises itself in a process called self-splicing. The self-splicing reaction does not involve any proteins.

RNA Editing. **RNA editing** involves the posttranscriptional insertion or deletion of nucleotides or the conversion of one base to another. As a result, the functional RNA molecule has a base sequence that does not match the DNA coding sequence.

RNA editing was discovered in the mid-1980s in some mitochondrial mRNAs of trypanosomes, the protozoa that cause sleeping sickness. For example, the sequences of the *COIII* gene for subunit III of cytochrome oxidase and its mRNA transcripts for the protozoans *Trypanosome brucei* (*Tb*), *Crithridia fasiculata* (*Cf*), and *Leishmania tarentolae* (*Lt*) are shown in Figure 5.15. Although the mRNA sequences are highly similar among the three organisms, only the *Cf* and *Lt* gene sequences are colinear with the mRNAs. Strikingly, the *Tb* gene has a sequence that cannot produce the mRNA it apparently encodes. The differences between the two are U nucleotides in the mRNA that are

Figure 5.15

Comparison of the DNA sequences of the cytochrome oxidase subunit III gene (*COIII*) in the protozoans *Trypanosome brucei (Tb)*, *Crithridia fasiculata (Cf)*, and *Leishmania tarentolae (Lt)*, aligned with the conserved mRNA for *Tb*. The lower-case *u*'s are the U nucleotides added to the transcript by RNA editing. The template T's in *Tb* DNA that are not in the RNA transcript are yellow.

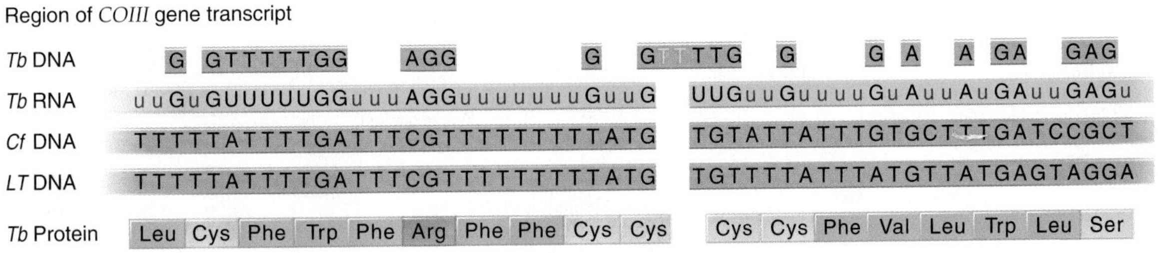

Region of *COIII* gene transcript
Tb DNA G GTTTTTGG AGG G GTTTTG G G A A GA GAG
Tb RNA uuGuGUUUUUGGuuuAGGuuuuuuuGuuG UUGuuGuuuuGuAuuAuGAuuGAGu
Cf DNA TTTTTATTTTGATTTCGTTTTTTTTTATG TGTATTATTTGTGCTTTGATCCGCT
LT DNA TTTTTATTTTGATTTCGTTTTTTTTTATG TGTTTTATTTATGTTATGAGTAGGA
Tb Protein Leu Cys Phe Trp Phe Arg Phe Phe Cys Cys Cys Cys Phe Val Leu Trp Leu Ser

not encoded in the DNA and T nucleotides in the DNA that are not found in the transcript. Once it is made, the transcript of the *Tb COIII* gene is edited to add U nucleotides in the appropriate places and remove the U nucleotides encoded by the T nucleotides in the DNA. As the figure shows, there are extensive insertions of U nucleotides and deletions of several templated T nucleotides. The magnitude of the changes is even more apparent when the whole sequence is examined: More than 50 percent of the mature mRNA consists of posttranscriptionally added U nucleotides. This RNA editing must be accurate in order to reconstitute the appropriate sequence for translation into the correct protein. A special RNA molecule, called a *guide RNA* (gRNA), is involved in the process. The gRNA pairs with the mRNA transcript and is thought to be responsible for cleaving the transcript, templating the missing U nucleotides, and ligating the transcript back together again.

RNA editing is not confined to trypanosomes. In the slime mold *Physarum polycephalum*, single C nucleotides are added posttranscriptionally at many positions of several mitochondrial mRNA transcripts. In higher plants, the sequences of many mitochondrial and chloroplast mRNAs are edited by C-to-U changes. C-to-U editing is also involved in producing an AUG initiation codon from an ACG codon in some chloroplast mRNAs in a number of higher plants. In mammals, C-to-U editing occurs in the nuclear gene-encoded mRNA for apolipoprotein B and results in tissue-specific generation of a stop codon. Also in mammals, A-to-G editing has been shown to occur in the glutamate receptor mRNA, and pyrimidine editing occurs in a number of tRNAs.

Transcription of Other Genes

In this section, we discuss the transcription of non–protein-coding genes.

Ribosomal RNA and Ribosomes. Protein synthesis takes place on ribosomes, of which each cell contains thousands. Ribosomes bind to mRNA and facilitate the binding of the tRNA to the mRNA so that a polypeptide chain can be synthesized.

Figure 5.16

Model of the complete (70S) ribosome of *E. coli*. The small (30S) ribosomal subunit is yellow, and the large (50S) ribosomal subunit is red.

Ribosome Structure. In both prokaryotes and eukaryotes, ribosomes consist of two unequally sized subunits—the large and small ribosomal subunits—each of which consists of a complex between RNA molecules and proteins. Each subunit contains one or more rRNA molecules and a large number of **ribosomal proteins.**

The *E. coli* ribosome has a size of 70S (Figure 5.16), with distinctly shaped subunits of 50S (large subunit) and 30S (small subunit). The 50S subunit has 34 different proteins, a 23S rRNA (2,904 nucleotides), and a 5S rRNA (120 nucleotides). The 30S ribosomal subunit has 20 different proteins and a 16S rRNA (1,542 nucleotides). Transcription of the ribosomal protein genes occurs by the mechanism we have already discussed with regard to protein-coding genes of bacteria.

Eukaryotic ribosomes are larger and more complex than their prokaryotic counterparts, and they vary in size and composition among eukaryotic organisms. Mammalian ribosomes, for example, have a size of 80S and consist of a large 60S subunit and a small 40S subunit (Figure 5.17). The 40S subunit contains 18S rRNA and about 35 ribosomal proteins, and the 60S subunit contains 28S, 5.8S, and 5S rRNAs and about 50 ribosomal proteins. Transcription of the genes that code for

Figure 5.17

Composition of whole ribosomes and of ribosomal subunits in mammalian cells.

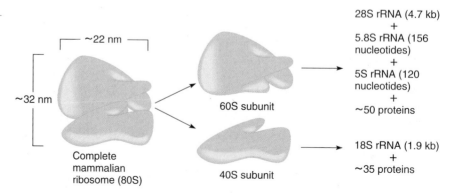

~22 nm

~32 nm

Complete mammalian ribosome (80S)

60S subunit

40S subunit

28S rRNA (4.7 kb)
+
5.8S rRNA (156 nucleotides)
+
5S rRNA (120 nucleotides)
+
~50 proteins

18S rRNA (1.9 kb)
+
~35 proteins

Figure 5.18

rRNA genes and rRNA production in *E. coli*. (a) General organization of an *E. coli* rRNA transcription unit (an *rrn* region). (b) The scheme for the synthesis and processing of a precursor rRNA (p30S) to the mature 16S, 23S, and 5S rRNAs of *E. coli*. (The tRNAs have been omitted from this depiction of the processing scheme.)

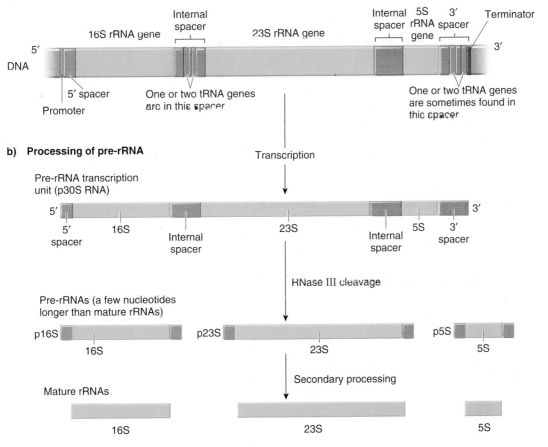

a) *E. coli* rRNA transcription unit

b) Processing of pre-rRNA

ribosomal proteins occurs by the same mechanism as for protein-coding genes of eukaryotes.

Transcription of rRNA Genes. In prokaryotes and eukaryotes, the regions of DNA that contain the genes for rRNA are called **ribosomal DNA (rDNA)** or **rRNA transcription units**. Figure 5.18a shows an *E. coli* rRNA transcription unit, called an *rrn* region. Seven such regions are scattered in the *E. coli* chromosome. Each *rrn* contains one copy each of the 16S, 23S, and 5S rRNA coding sequences, arranged in the order 16S–23S–5S. There are one or two tRNA genes in the spacer region between the 16S and 23S rRNA sequences and another one or two tRNA genes in the 3′ spacer region between the end of the 5S rRNA sequence and the 3′ end of the transcription unit.

The rRNA transcription units are transcribed by RNA polymerase to produce a single **precursor rRNA (pre-rRNA)** molecule—the 30S pre-rRNA (p30S)—which contains a 5′ spacer, the 16S, 23S, and 5S rRNA sequences

(each separated by spacer sequences), and a 3′ spacer (Figure 5.18b). RNase III cleaves p30S to produce the p16S, p23S, and p5S precursors. Other processing enzymes release the tRNAs from the spacers. In normal cells, RNase III cleaves the transcript of the rRNA transcription unit to produce the three rRNA precursors while transcription is still occurring. As a result, the p30S is not seen in normal cells. However, in mutants that have temperature-sensitive RNase III activity, the p30S molecule accumulates at the high temperature. Through studies of these mutants, the pathway of p30S synthesis and processing shown in Figure 5.18b was discovered. The transcript-processing events take place within a complex formed between the rRNA transcript (as it is being transcribed) and ribosomal proteins. In this way, the functional ribosomal subunits are assembled.

Most eukaryotes have many copies of the genes for each of the four rRNA species 18S, 5.8S, 28S, and 5S. The genes for 18S, 5.8S, and 28S rRNAs are found adjacent to

a) rDNA repeat unit

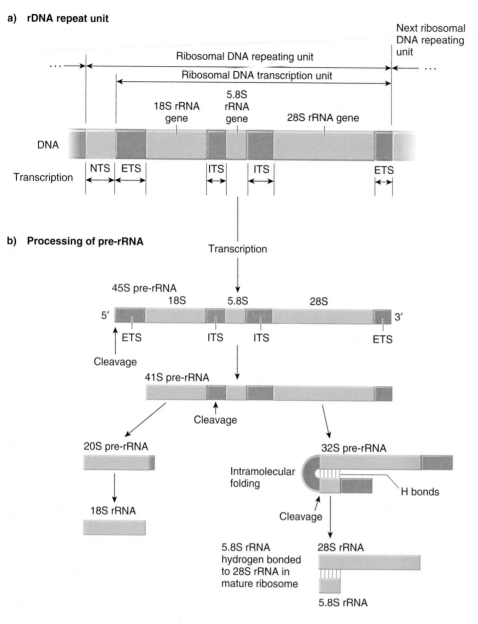

b) Processing of pre-rRNA

Figure 5.19

rRNA genes and rRNA production in eukaryotes. (a) Generalized diagram of a eukaryotic ribosomal DNA repeat unit. The coding sequences for 18S, 5.8S, and 28S rRNAs are indicated in green. NTS = nontranscribed spacer, ETS = external transcribed spacer, ITS = internal transcribed spacer. (b) The transcription and processing of 45S pre-rRNA to produce the mature 18S, 5.8S, and 28S rRNAs.

one another in the order 18S–5.8S–28S, with each set of three genes typically repeated 100 to 1,000 times (depending on the organism), to form tandem arrays called **rDNA repeat units** (Figure 5.19a). Human cells, for example, have 1,250 gene sets. The rDNA repeat units are organized into one or more clusters in the genome, and, as a result of active transcription of 18S, 5.8S, and 28S rRNAs, a nucleolus forms around each cluster. Typically, the multiple nucleoli so formed fuse to form one nucleolus.

Each eukaryotic rDNA repeat unit is transcribed by RNA polymerase I to produce a pre-rRNA molecule (45S pre-rRNA in mammalian cells) (Figure 5.19b). The pre-rRNA contains the 18S, 5.8S, and 28S rRNA sequences, as well as **spacer sequences**—sequences between and flanking the three rRNA sequences. The external transcribed spacers (ETSs) are located upstream of the 18S sequence

and downstream of the 28S sequence. The internal transcribed spacers (ITSs) are located on each side of the 5.8S sequence—that is, between the 18S sequence and the 5.8S sequence and between the 5.8S sequence and the 28S rRNA sequence. Between each rDNA repeat unit is the **nontranscribed spacer (NTS) sequence,** which, as the name indicates, is not transcribed. (See Figure 5.19a.) The promoter is located in the NTS.

To produce the 18S, 5.8S, and 28S rRNAs, the transcribed pre-rRNA is cleaved at specific sites to remove ITS and ETS sequences. For example, Figure 5.19b shows how pre-rRNA is processed in mammalian cells. First, the 5′ ETS sequence is removed, producing a precursor molecule containing all three rRNA sequences. The next cleavage produces the 20S precursor to 18S rRNA, and the 32S precursor is processed to 28S and

5.8S rRNAs. Removal of the ITS sequence generates the 18S rRNA from the 20S precursor.

All the pre-rRNA-processing events take place in complexes formed between the pre-rRNA, 5S rRNA, and ribosomal proteins. The 5S rRNA is produced by transcription of the 5S rRNA genes that are located elsewhere in the genome by RNA polymerase III (described shortly), and the ribosomal proteins are produced by transcription of the ribosomal protein genes by RNA polymerase II and the subsequent translation of the mRNAs in the cytoplasm. As pre-rRNA processing proceeds, the complexes undergo changes in shape, resulting in the formation of the 60S and 40S ribosomal subunits, which are then transported to the cytoplasm, where they function in protein synthesis.

It is important to understand the distinction between an intron and a spacer. The removal of a spacer releases the flanking RNAs, and they remain separate. Intron removal, by contrast, results in the splicing together of the RNA sequences that flanked the intron.

KEYNOTE

Ribosomes consist of two unequally sized subunits, each containing one or more ribosomal RNA molecules and ribosomal proteins. The three prokaryotic rRNAs and three of the four eukaryotic rRNAs are encoded in rRNA transcription units. The fourth eukaryotic rRNA is encoded by separate genes. The transcription of rRNA transcription units by RNA polymerase produces pre-rRNA molecules that are processed to mature rRNAs by the removal of spacer sequences. The processing events occur in complexes of the pre-rRNAs with ribosomal proteins and other proteins and are part of the formation of the mature ribosomal subunits.

Transcription of Genes by RNA Polymerase III. RNA polymerase III transcribes eukaryotic 5S rRNA genes, tRNA genes, and some snRNA genes. 5S rRNA is a 120-nucleotide RNA found in the large subunit of the eukaryotic ribosome. tRNAs are 75- to 90-nucleotide RNAs that function in both prokaryotes and eukaryotes to bring amino acids to the ribosome–mRNA complex, where they are polymerized into protein chains in the translation (protein synthesis) process.

Prokaryotic 5S rRNA genes are part of the rDNA, whereas eukaryotic 5S rRNA genes are found in multiple copies in the genome, usually separate from the rDNA repeat units. Prokaryotic tRNA genes are found in one or at most a few copies in the genome, whereas eukaryotic tRNA genes are repeated many times in the genome. In the South African clawed toad *Xenopus laevis,* for example, there are about 200 copies of each tRNA gene. Each type of tRNA molecule has a different nucleotide sequence, although all tRNAs have the sequence CCA at their 3′ ends.

This sequence is added to the tRNA posttranscriptionally. The differences in nucleotide sequences explain the ability of a particular tRNA molecule to bind a specific amino acid. All tRNA molecules are also extensively modified chemically by enzyme reactions after transcription.

The nucleotide sequences of all tRNAs can be arranged into what is called a cloverleaf (Figure 5.20a). The cloverleaf results from complementary base pairing between different sections of the molecule, producing four base-paired "stems" separated by four loops: I, II, III, and IV. Loop II contains the three-nucleotide **anticodon** sequence, which pairs with a three-nucleotide **codon** sequence in mRNA by complementary base pairing during translation. This codon–anticodon pairing is crucial to the addition of the amino acid specified by the mRNA to the growing polypeptide chain. Figures 5.20b and c show the tertiary structure of phenylalanine tRNA from yeast. All other tRNAs that have been examined exhibit similar upside-down L-shaped structures in which the 3′ end of the tRNA—the end to which the amino acid attaches—is at the end of the L that is opposite from the anticodon loop.

How does RNA polymerase III transcribe the genes under its control? The 5S rRNA genes were the first eukaryotic genes for which the promoter structure and transcription factor requirements were determined. On the basis of the knowledge of upstream promoters for bacterial genes, the expectation was that the 5S rRNA genes would also have an upstream promoter. However, it was discovered that their promoter is *within the transcribed region.* This internal promoter is called the **internal control region (ICR).** The tRNA genes (tDNA), which encode the 75- to 90-nucleotide tRNAs, also have internal promoters, whereas snRNA genes transcribed by RNA polymerase III typically have promoters upstream of the genes.

The internal promoter functions as follows: Transcription factors bind to sequence elements of the ICR, facilitating binding of RNA polymerase III. The enzyme is positioned on the gene so that it initiates transcription at the beginning of the gene, which is upstream of where it is bound.

Transcription of a 5S rRNA gene directly produces the mature 5S rRNA; no sequences need be removed. Transcription of tRNA genes produces **precursor tRNA (pre-tRNA),** which has extra sequences at each end that are subsequently removed to produce the mature tRNA.

Some tRNA genes in certain eukaryotic organisms contain introns. For example, about 10 percent of the 400 tRNA genes in yeast have introns. Depending on the tRNA, the intron is 14 to 60 bp long and is almost always located between the first and second nucleotides 3′ to the anticodon. Removal of the introns is different from pre-mRNA splicing. A specific endonuclease cleaves the pre-tRNA at each end of the intron, and the tRNA pieces are then spliced together by an enzyme called **RNA ligase.**

Figure 5.20

Transfer RNA. (a) Cloverleaf structure of yeast alanine tRNA. Py = pyrimidine. Modified bases: I = inosine, T = ribothymidine, ψ = pseudouridine, D = dihydrouridine, GMe = methylguanosine, GMe₂ = dimethylguanosine, IMe = methylinosine. (b) Schematic of the three-dimensional structure of yeast phenylalanine tRNA, as determined by X-ray diffraction of tRNA crystals. Note the characteristic L-shaped structure. (c) Photograph of a space-filling molecular model of yeast phenylalanine tRNA. The CCA end of the molecule to which the amino acid attaches is at the upper right, and the anticodon loop is at the bottom.

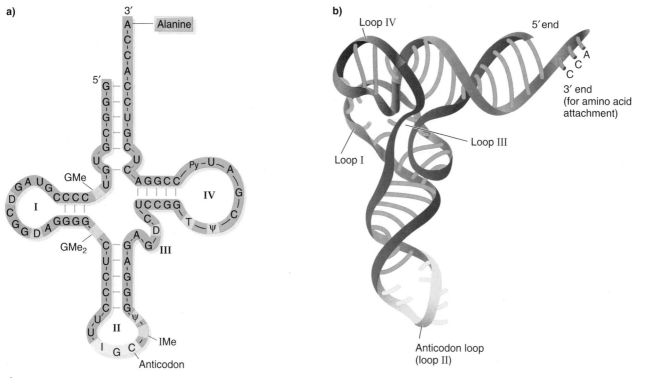

a)

b)

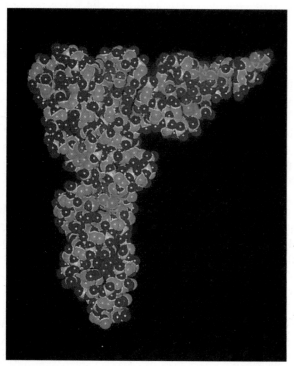

c)

KEYNOTE

RNA polymerase III transcribes 5S rRNA genes (5S rDNA), tRNA genes (tDNA), and some snRNA genes. For 5S rDNA and tDNA, the promoter for RNA polymerase III is located within the gene itself. The internal promoter is called the internal control region. Transcription factors bind to sequence elements in the internal control region and enable RNA polymerase III to bind. Transcription then commences upstream of the promoter, at the beginning of the gene.

Summary

Transcription

When a gene is expressed, the DNA base-pair sequence is transcribed into the base sequence of an RNA molecule. Transcription of four classes of genes produces messenger RNA (mRNA), transfer RNA (tRNA), ribosomal RNA (rRNA), and small nuclear RNA (snRNA). snRNA is found only in eukaryotes, and the other three classes are found in both prokaryotes and eukaryotes. Only mRNA is translated to produce a protein molecule.

When a gene is transcribed by RNA polymerase, just one of the two DNA strands is copied. The direction of RNA synthesis is 5′ to 3′. In addition to possessing coding sequences, genes contain other sequences that are important in regulating transcription, including promoter sequences and terminator sequences. Promoter sequences specify where transcription of the gene is to begin, and terminator sequences specify where transcription is to stop.

Only one type of RNA polymerase is found in bacteria, so all classes of RNA are synthesized by the same enzyme. Consequently, the promoters for all three classes of genes are similar. The promoter is recognized by a complex between the RNA polymerase core enzyme and a protein factor called sigma. Once transcription is initiated correctly, the sigma factor dissociates from the enzyme and is reused in other transcription initiation events. Termination is signaled by one of two possible termination sequences.

In eukaryotes, three different RNA polymerases are located in the nucleus. RNA polymerase I, in the nucleolus, transcribes the 18S, 5.8S, and 28S rRNA sequences. These rRNAs are part of ribosomes. RNA polymerase II, in the nucleoplasm, transcribes protein-coding genes (into mRNA precursors) and some snRNA genes. RNA polymerase III, in the nucleoplasm, transcribes tRNA genes, 5S rRNA genes, and other snRNA genes. The promoters for the three types of RNA polymerase differ, but in each case none of the three eukaryotic RNA polymerases binds directly to the promoter. Instead, the promoter sequences for the genes they transcribe are first recognized by transcription factors, which then bind to the DNA and facilitate binding of the polymerase to the transcription factor–DNA complex so that transcription will be initiated correctly. For 5S rRNA genes and tRNA genes transcribed by RNA polymerase III, the promoter is located within the gene itself.

Although transcription is similar in prokaryotes and eukaryotes, the molecular components of the process are different. Even within eukaryotes, three different promoters have evolved along with three distinct RNA polymerases. Much remains to be learned about the associated transcription factors and regulatory proteins before we have a complete understanding of transcription and its regulation.

RNA Molecules and RNA Processing

We discussed the structure, synthesis, and function of mRNA, tRNA, and rRNA. Each mRNA encodes the amino acid sequence of a polypeptide chain. The nucleotide sequence of the mRNA is translated into the amino acid sequence of the polypeptide chain. Ribosomes consist of two unequal-sized subunits, each of which contains both rRNA and protein molecules. The amino acids that are assembled into proteins are brought to the ribosome bound to tRNA molecules.

mRNAs have three main parts: a 5′ untranslated region (UTR), the amino acid–coding sequence, and the 3′ untranslated region. In prokaryotes, the gene transcript functions directly as the mRNA molecule; in eukaryotes, the RNA transcript must be modified in the nucleus to produce mature mRNA. Modifications include the addition of a 5′ cap and a 3′ poly(A) tail and the removal of any introns. Intron removal is known as RNA splicing and involves specific interactions with snRNPs in structures called spliceosomes. Only when all processing events have been completed is the mRNA functional; at that point, it can be translated in the cytoplasm once it leaves the nucleus.

In the pre-rRNA of *Tetrahymena*, the 28S rRNA sequence is interrupted by an intron that is removed during processing of the pre-rRNA in the nucleolus. The intron is excised by a protein-independent reaction in which the RNA sequence of the intron folds into a secondary structure that promotes its own excision. The process is called self-splicing.

In some organisms, RNA editing inserts or deletes nucleotides or converts one base to another in an RNA posttranscriptionally. As a result, the functional RNA molecule has a base sequence that does not match the DNA coding sequence. Many RNAs that are edited are encoded by the mitochondrial and chloroplast genomes.

Ribosomal RNAs are important structural components of ribosomes. The smaller subunit contains one rRNA molecule: 16S in prokaryotes and 18S in eukaryotes. The prokaryotic larger subunit contains 23S and 5S rRNAs, and the eukaryotic larger subunit contains 28S, 5.8S, and 5S rRNAs.

In prokaryotes, the genes for the 16S, 23S, and 5S rRNAs are transcribed into a single pre-rRNA molecule. Some tRNA genes are found in the spacer regions of each transcription unit. The pre-rRNA molecules are processed to remove the noncoding sequences found at the ends of the molecules and also to remove spacers between the rRNA sequences. At the same time, the tRNAs are released. All the processing events occur while the pre-rRNA is associating with the ribosomal proteins, so that when processing is complete, the functional 50S and 30S subunits have been assembled.

In eukaryotes, the 18S, 5.8S, and 28S rRNA genes constitute a transcription unit, and many such transcription units are organized into a tandem array. The 5S rRNA genes are also present in many copies, but are usually located elsewhere in the genome. The 18S, 5.8S, and 28S sequences are transcribed into pre-rRNA molecules that contain, in addition to the rRNA sequences, spacer sequences at the ends and between the rRNA sequences.

The transcription of tandem arrays of these rRNA genes results in the formation of a nucleolus around each cluster. The 60S and 40S ribosomal subunits are assembled in the nucleolus. That is, the spacers are removed by specific processing events that take place while the rRNA sequences are associated with ribosomal proteins and with

5S rRNA, which is transcribed elsewhere in the nucleus. The completed ribosomal subunits exit the nucleus and participate in protein synthesis in the cytoplasm.

All tRNAs bring amino acids to the ribosomes. Thus, all tRNAs are similar in length (75 to 90 nucleotides) and in secondary and tertiary structure. Introns are present in some tRNA genes of yeast. Removal of these introns involves a mechanism different from the other RNA splicing mechanisms described.

Analytical Approaches to Solving Genetics Problems

Q5.1 If two RNA molecules have complementary base sequences, they can hybridize to form a double-stranded helical structure, just as DNA can. Imagine that, in a particular region of the genome of a certain bacterium, one DNA strand is transcribed to give rise to the mRNA for protein A and the other DNA strand is transcribed to give rise to the mRNA for protein B.
a. Would there be any problem in expressing these genes?
b. What would you see in protein B if a mutation occurred that affected the structure of protein A?

A5.1
a. mRNA A and mRNA B would have complementary sequences, so they might hybridize with each other and not be available for translation.
b. Every mutation in gene A would also be a mutation in gene B, so protein B might also be abnormal.

Q5.2 Compare the following two events in terms of their potential consequences: In event 1, an incorrect nucleotide is inserted into the new DNA strand during replication and is not corrected by the proofreading or repair systems before the next replication. In event 2, an incorrect nucleotide is inserted into an mRNA during transcription.

A5.2 Assuming that it occurred within a gene, event 1 would result in a mutation. The mistake would be inherited by future generations and would affect the structure of all mRNA molecules transcribed from the region; therefore, all molecules of the corresponding protein could be affected.

Event 2 would result in a single aberrant mRNA that could then produce a few aberrant protein molecules. Additional normal protein molecules would exist, because other, normal mRNAs would have been transcribed. The abnormal mRNA would soon be degraded. The mRNA mistake would not be hereditary.

Q5.3 You are given three different RNA samples. Sample I has a homogeneous molecular weight, sample II is produced by the processing of a larger precursor RNA, and sample III has an additional sequence added onto the orig-

inal transcript. For each sample, state whether the RNA could be rRNA, mRNA, or tRNA. If it is not one of those, state what it might be. Note that, for each sample, more than one RNA could apply. Give reasons for your choices.

A5.3 In sample I, rRNA and tRNA species have homogeneous molecular weights, because they carry out specific functions within the cell for which their length and three-dimensional configuration are important. Messenger RNA and heterogeneous nuclear RNA are heterogeneous in length. In sample II, rRNA, tRNA, and some mRNAs are produced by the processing of a larger precursor RNA molecule. In sample III, both tRNA and eukaryotic mRNA may have sequences added after they are transcribed. For tRNA, this sequence is the CCA at the 3′ end; for mRNA, the sequence is the poly(A) tail at the 3′ end.

Questions and Problems

*5.1 Compare DNA and RNA with regard to their structure, function, location, and activity. Also, how do these molecules differ with regard to the polymerases used to synthesize them?

5.2 All base pairs in the genome are replicated during the DNA synthesis phase of the cell cycle, but only some of the base pairs are transcribed into RNA. How is it determined which base pairs of the genome are transcribed into RNA?

*5.3 Discuss the similarities and differences between the E. coli RNA polymerase and eukaryotic RNA polymerases.

5.4 What are the most significant differences between the organization and expression of prokaryotic genes and eukaryotic genes?

5.5 Discuss the molecular events involved in the termination of RNA transcription in prokaryotes. In what ways is this process fundamentally different in eukaryotes?

5.6 More than 100 promoters in prokaryotes have been sequenced. One element of these promoters is sometimes called the Pribnow box, named after the investigator who compared several E. coli and phage promoters and discovered a region they held in common. Discuss the nature of this sequence. (Where is it located and why is it important?) Another consensus sequence appears a short distance from the Pribnow box. Diagram the positions of the two prokaryotic promoter elements relative to the start of transcription for a typical E. coli promoter.

*5.7 An E. coli transcript with the first two nucleotides 5′-AG-3′ is initiated from the following segment of double-stranded DNA:

```
5'  TAGTGTATTGACATGATAGAAGCACTCTTACTATAATCTCAATAGCTACG  3'
3'  ATCACATAACTGTACTATCTTCGTGAGAATGATATTAGAGTTATCGATGC  5'
```

a. Where is the transcription start site?

b. What are the approximate locations of the regions that bind the RNA polymerase homoenzyme?

c. Does transcription elongation proceed towards the right or left?

d. Which DNA strand is the template strand?

e. Which DNA strand is the RNA-coding strand?

5.8 Figure 5.A shows the sequences, given $5' \rightarrow 3'$, that lie upstream a subset of *E. coli* genes transcribed by RNA polymerase and σ^{70}. Carefully examine the sequences in the −10 and −35 regions, and then answer the following questions:

a. The −10 and −35 regions have the consensus sequences 5'-TATAAT-3' and 5'-TTGACA-3', respectively. How many of the genes that are listed have sequences that perfectly match the −10 consensus? How many have perfect matches to the −35 consensus?

b. On the basis of your examination of these sequences, what does the term "consensus sequence" mean?

c. What is the function of these consensus sequences in transcription initiation?

d. More generally, what might you infer about a DNA sequence if it is part of a consensus sequence?

e. None of these promoters have perfect consensus sequences, but some have better matches than others. Speculate what consequence this might have on the efficiency of transcription initiation?

*5.9 The single RNA polymerase of *E. coli* transcribes all of its genes, even though these genes do not all have identical promoters.

a. What different types of promoters are found in the genes of *E. coli*?

b. How is the single RNA polymerase of *E. coli* able to initiate transcription even though it uses different types of promoters?

c. Why might it be to *E. coli*'s advantage to have genes with different types of promoters?

*5.10 Three different RNA polymerases are found in all eukaryotic cells, and each is responsible for synthesizing a different class of RNA molecules. How do the characteristics of the eukaryotic RNA polymerases differ in terms of their cellular location and products?

5.11 *E. coli* RNA polymerase can transcribe all the genes of *E. coli*, but the three eukaryotic RNA polymerases transcribe only specific, nonoverlapping subsets of eukaryotic genes. What mechanisms are used to restrict the transcription of each of the three eukaryotic polymerases to a particular subset of eukaryotic genes?

5.12 Figure 5.3 shows the structure of a prokaryotic gene, including its promoter, RNA-coding sequence, and terminator region. Modify the figure to show the general structures of eukaryotic genes transcribed by

a. RNA polymerase II

b. RNA polymerase III

5.13 A piece of mouse DNA was sequenced as follows (a space is inserted after every 10th base for ease in counting; (...) means a lot of unspecified bases):

```
AGAGGGCGGT CCGTATCGGC CAATCTGCTC ACAGGGCGGA
TTCACACGTT GTTATATAAA TGACTGGGCG TACCCCAGGG
TTCGAGTATT CTATCGTATG GTGCACCTGA CT(...)
GCTCACAAGT ACCACTAAGC(...)
```

What can you see in this sequence to indicate that it might be all or part of a transcription unit?

5.14 How do the structures of mRNA, rRNA, and tRNA differ? Hypothesize a reason for the difference.

Figure 5.A

Gene	−35 Region	−10 Region	Initiation Region
lac	ACCCAGGCTTTACACTTTATGGCTTCCGGCTCGTATGTTGTGTGGAATTGTGAGCGG		
lacI	CCATCGAATGGCGCAAAACCTTTCGCGGTATGGCATGATAGCGCCCGGAAGAGAGTC		
galP2	ATTTATTCCATGTCACACTTTTCGCATCTTTGTTATGCTATGGTTATTTCATACCAT		
araB,A	GGATCCTACCTGACGCTTTTTATCGCAACTCTCTACTGTTTCTCCATACCCGTTTTT		
araC	GCCGTGATTATAGACACTTTTGTTACGCGTTTTTGTCATGGCTTTGGTCCCGCTTTG		
trp	AAATGAGCTGTTGACAATTAATCATCGAACTAGTTAACTAGTACGCAAGTTCACGTA		
bioA	TTCCAAAACGTGTTTTTTGTTGTTAATTCGGTGTAGACTTGTAAACCTAAATCTTTT		
bioB	CATAATCGACTTGTAAACCAAATTGAAAAGATTTAGGTTTACAAGTCTACACCGAAT		
tRNA^{Tyr}	CAACGTAACACTTTACAGCGGCGCGTCATTTGATATGATGCGCCCCGCTTCCCGATA		
rrnD1	CAAAAAAATACTTGTGCAAAAAATTGGGATCCCTATAATGCGCCTCCGTTGAGACGA		
rrnE1	CAATTTTTCTATTGCGGCCTGCGGAGAACTCCCTATAATGCGCCTCCATCGACACGG		
RRNa2	AAAATAAATGCTTGACTCTGTAGCGGGAAGGCGTATTATGCACACCCCGCGCCGCTG		

5.15 Many eukaryotic mRNAs, but not prokaryotic mRNAs, contain introns. Describe how these sequences are removed during the production of mature mRNA.

5.16 How is the mechanism of group I intron removal different from the mechanism used to remove the introns in most eukaryotic mRNAs? Speculate as to why these different mechanisms for intron removal might have evolved and how each might be advantageous to a eukaryotic cell.

5.17 Distinguish between the following terms: *leader sequence, trailer sequence, coding sequence, intron, spacer sequence, nontranscribed spacer sequence, external transcribed spacer sequence,* and *internal transcribed sequence.* Give examples of actual molecules in your answer.

*5.18 Discuss the posttranscriptional modifications and processing events that take place on the primary transcripts of eukaryotic rRNA and protein-coding genes.

5.19 Small RNA molecules such as snRNAs and gRNAs play essential roles in eukaryotic transcript processing.
a. Where are these molecules found in the cell, and what roles do they have in transcript processing?
b. How is the abundance of snRNAs related to their role in transcript processing?

5.20 Describe the organization of the ribosomal DNA repeating unit of a higher eukaryotic cell.

*5.21 Which of the mutations that follow would be likely to be recessive lethal mutations (i.e., mutations causing lethality when they are the only alleles present in a homozygous individual) in humans? Explain your reasoning.
a. deletion of the U1 genes
b. a single base-substitution mutation in the U1 gene that prevented U1 snRNP from binding to the 5'-GU-3' sequence found at the 5' splice junctions of introns
c. deletion within intron 2 of β-globin
d. deletion of four bases at the end of intron 2 and three bases at the beginning of exon 3 in β-globin

*5.22 The following figure shows the transcribed region of a typical eukaryotic protein-coding gene:

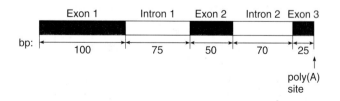

What is the size (in bases) of the fully processed, mature mRNA? Assume a poly(A) tail of 200 As in your calculations.

*5.23 Most human obesity does not follow Mendelian inheritance patterns, because body fat content is determined by a number of interacting genes and environmental variables. Insights into how specific genes function to regulate body fat content have come from studies of mutant, obese mice. In one mutant strain, *tubby (tub)*, obesity is inherited as a recessive trait. A comparison of the DNA sequence of the *tub*+ and *tub* alleles has revealed a single base-pair change: Within the transcribed region, a 5' G-C base pair has been mutated to a T-A base pair. The mutation causes an alteration of the initial 5' base of the first intron. Therefore, in the homozygous *tub/tub* mutant, a longer transcript is found. Propose a molecularly based explanation for how a single base change causes a nonfunctional gene product to be produced, why a longer transcript is found in *tub/tub* mutants, and why the tub mutant is recessive.

5.24 Which of the deletions that follow could occur in a single mutational event in a human? Explain.
a. deletion of 10 copies of the 5S ribosomal RNA genes only
b. deletion of 10 copies of the 18S rRNA genes only
c. simultaneous deletion of 10 copies of the 18S, 5.8S, and 28S rRNA genes only
d. simultaneous deletion of 10 copies each of the 18S, 5.8S, 28S, and 5S rRNA genes

5.25 During DNA replication in a mammalian cell, a mistake occurs: Ten wrong nucleotides are inserted into a 28S rRNA gene. The mistake is not corrected. What will probably be its effect on the cell?

*5.26 Choose the correct answers, noting that each blank may have more than one correct answer and that each answer (1 through 4) could be used more than once.

Answers:
1. eukaryotic mRNAs
2. prokaryotic mRNAs
3. transfer RNAs
4. ribosomal RNAs

_____ **a.** have a cloverleaf structure
_____ **b.** are synthesized by RNA polymerases
_____ **c.** display one anticodon each
_____ **d.** are the template of genetic information during protein synthesis
_____ **e.** contain exons and introns
_____ **f.** are of four types in eukaryotes and only three types in *E. coli*
_____ **g.** are capped on their 5' end and polyadenylated on their 3' end

6

Gene Expression: Translation

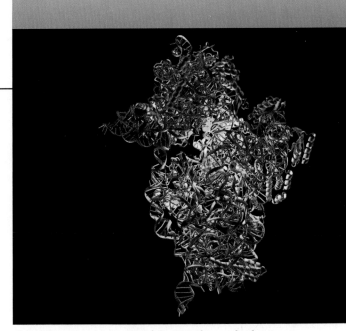

Three-dimensional structure of the 30S ribosomal subunit.

PRINCIPAL POINTS

- A protein consists of one or more subunits called polypeptides, which are composed of smaller building blocks called amino acids. The amino acids are linked together in the polypeptide by peptide bonds.

- The amino-acid sequence of a protein (its primary structure) determines its secondary, tertiary, and quaternary structures and hence its functional state.

- The genetic code is a triplet code in which each three-nucleotide codon in an mRNA specifies one amino acid. Some amino acids are represented by more than one codon. The code is almost universal, and it is read without gaps in successive, nonoverlapping codons.

- The mRNA is translated into a polypeptide chain on ribosomes. Amino acids for polypeptide synthesis come to the ribosome on tRNA molecules. The correct amino-acid sequence is achieved by specific binding of each amino acid to its specific tRNA and by specific binding between the codon of the mRNA and the complementary anticodon of the tRNA.

- In prokaryotes and eukaryotes, AUG (methionine) is the initiator codon for the start of translation. Elongation of

the protein chain involves peptide bond formation between the amino acid on the tRNA in the A site of the ribosome and the growing polypeptide on the tRNA in the adjacent P site. Once the peptide bond has formed, the ribosome translocates one codon along the mRNA in preparation for the next tRNA. The incoming tRNA with its amino acid binds to the next codon occupying the A site.

- Translation continues until a chain-terminating codon (UAG, UAA, or UGA) is reached in the mRNA. These codons are read by release factor proteins, and then the polypeptide is released from the ribosome and the other components of the protein synthesis machinery dissociate.

- In eukaryotes, proteins are found free in the cytoplasm and in various cell compartments, such as the nucleus, mitochondria, chloroplasts, and secretory vesicles. Mechanisms exist that sort proteins to their appropriate cell compartments. For example, proteins that are to be secreted have N-terminal signal sequences that facilitate their entry into the endoplasmic reticulum for later sorting in the Golgi apparatus and beyond.

iActivity

 CHANGING A SINGLE LETTER IN A WORD CAN COMpletely change the meaning of the word. This, in turn, can change the meaning of the sentence containing that word. In living organisms, a sequence of three nucleotide "letters" produces an amino acid "word." The amino acids are strung together to form polypeptide "sentences." In this chapter, you will study the process by which nucleotide "letters" are translated into polypeptide "sentences."

One of the most important applications of human genome research is the use of sequence information to track down the causes of genetic diseases. In the iActivity for this chapter, you will investigate part of the gene responsible for cystic fibrosis, the most common fatal genetic disease in the United States, and try to identify possible causes of the disease.

The information identifying the proteins found in a cell is encoded in the structural genes of the cell's genome. A protein-coding gene is expressed by transcription of the gene to produce an mRNA (discussed in Chapter 5), followed by translation of the mRNA: the conversion of the mRNA base-sequence information into the amino-acid sequence of a polypeptide. The nucleotide information that specifies the amino-acid sequence of a polypeptide is called the genetic code. In this chapter, your goal is to learn how the nucleotide sequence of mRNA is translated into the amino-acid sequence of a polypeptide.

Proteins

Chemical Structure of Proteins

A **protein** is a high-molecular-weight, nitrogen-containing organic compound of complex shape and composition. Each cell type has a characteristic set of proteins that give it its functional properties. A protein consists of one or more macromolecular subunits called **polypeptides,** which are composed of smaller building blocks: the **amino acids.** The sequence of amino acids gives the polypeptide its three-dimensional shape and its properties in the cell.

With the exception of proline, the amino acids have a common structure, shown in Figure 6.1. The structure consists of a central carbon atom (α-carbon), to which is bonded an amino group (NH_2), a carboxyl group (COOH), and a hydrogen atom. Figure 6.1 shows an amino acid at the pH commonly found within cells. At that pH, the NH_2 and COOH groups of free amino acids are in a charged state, $-NH_3^+$ and $-COO^-$, respectively. Also bound to the α-carbon is the R group, which varies from one amino acid to another and gives each amino acid its distinctive properties. Because different polypeptides have different sequences and proportions of amino acids, the arrangement and nature of the R groups give a polypeptide its structural and functional properties.

Figure 6.1

General structural formula for an amino acid.

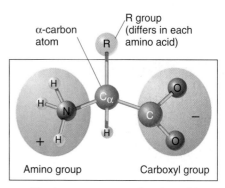

Structures common to all amino acids

Twenty amino acids are used to make proteins in living cells; their names, three-letter and one-letter abbreviations, and chemical structures are shown in Figure 6.2. The 20 amino acids are divided into subgroups on the basis of whether the R group is acidic, basic, neutral and polar, or neutral and nonpolar.

Amino acids of a polypeptide are joined by a **peptide bond**—a covalent bond formed between the carboxyl group of one amino acid and the amino group of an adjacent amino acid (Figure 6.3). A polypeptide, then, is an unbranched molecule that consists of a sequence of many amino acids (usually 100 or more) joined by peptide bonds. Every polypeptide has a free amino group at one end (called the N terminus, or the N-terminal end) and a free carboxyl group at the other end (called the C terminus, or the C-terminal end). The N-terminal end is defined as the beginning of a polypeptide chain.

Molecular Structure of Proteins

Proteins can have four levels of structural organization (Figure 6.4).

1. The *primary structure* of a polypeptide chain is the amino-acid sequence (Figure 6.4a). The amino acid sequence is directly determined by the base-pair sequence of the gene that encodes it.

2. The *secondary structure* of a protein is the regular folding and twisting of a single polypeptide chain into a variety of shapes. A polypeptide's secondary structure is the result of weak bonds, such as electrostatic or hydrogen bonds, between NH and CO groups of amino acids that are near each other on the chain. One type of secondary structure found in regions of many polypeptides is the α-*helix* (Figure 6.4b), a structure discovered by Linus Pauling and Robert Corey in 1951. Note the hydrogen bonding between the NH group of one amino acid (i.e., an NH group that is part of a peptide bond) and the CO group (also part of a peptide bond) of an

Figure 6.2

Structures of the 20 naturally occurring amino acids, organized according to chemical type. Below each amino acid name are its three-letter and one-letter abbreviations.

Acidic

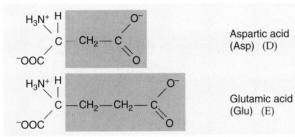

Aspartic acid
(Asp) (D)

Glutamic acid
(Glu) (E)

Basic

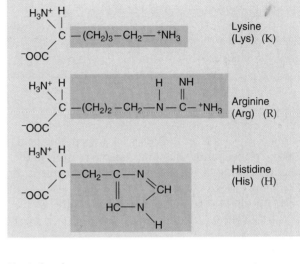

Lysine
(Lys) (K)

Arginine
(Arg) (R)

Histidine
(His) (H)

Neutral, nonpolar

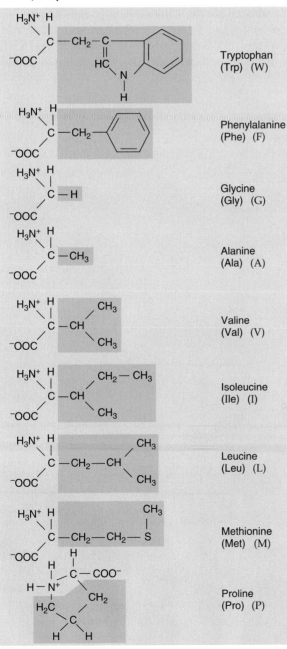

Tryptophan
(Trp) (W)

Phenylalanine
(Phe) (F)

Glycine
(Gly) (G)

Alanine
(Ala) (A)

Valine
(Val) (V)

Isoleucine
(Ile) (I)

Leucine
(Leu) (L)

Methionine
(Met) (M)

Proline
(Pro) (P)

Neutral, polar

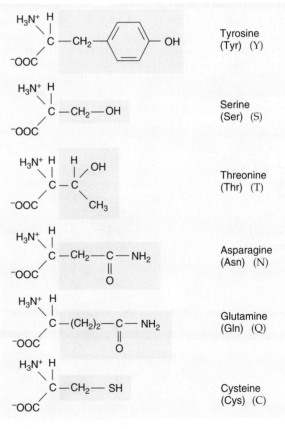

Tyrosine
(Tyr) (Y)

Serine
(Ser) (S)

Threonine
(Thr) (T)

Asparagine
(Asn) (N)

Glutamine
(Gln) (Q)

Cysteine
(Cys) (C)

Figure 6.3

Peptide bond formation.

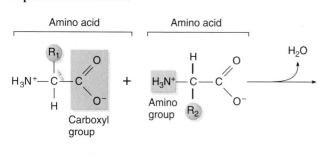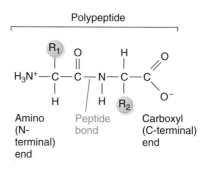

amino acid that is four amino acids away in the chain. The repeated formation of this bonding results in the helical coiling of the chain. The α-helix content of proteins varies.

Another type of secondary structure is the β-pleated sheet (not illustrated). The β-pleated sheet involves a polypeptide chain or chains folded in a zigzag way, with parallel regions or chains linked by hydrogen bonds. Many proteins contain a mixture of α-helical and β-pleated sheet regions.

3. A protein's *tertiary structure* (Figure 6.4c) is the three-dimensional structure of a single polypeptide

Figure 6.4

Four levels of protein structure. (a) Primary, the sequence of amino acids in a polypeptide chain. (b) Secondary, the folding and twisting of a single polypeptide chain into a variety of shapes. Shown is one type of secondary structure: the α-helix. Both structures are stabilized by hydrogen bonds. (c) Tertiary, the specific three-dimensional folding of the polypeptide chain. Shown here is the β polypeptide chain of hemoglobin, a heme-containing polypeptide that carries oxygen in the blood. (d) Quaternary, the aggregate of polypeptide chains. Shown here is hemoglobin; it consists of two α chains, two β chains, and four heme groups.

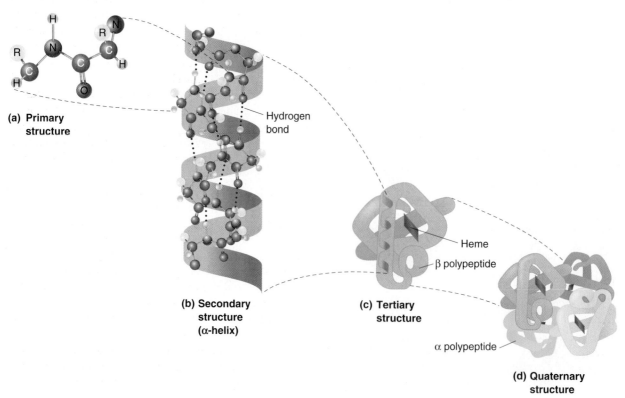

chain. The three-dimensional shape of a polypeptide often is called its *conformation*. Tertiary folding is a direct property of the amino-acid sequence of the chain and thus is related to the distribution of the R groups along the chain. Figure 6.4c shows the tertiary structure of the β polypeptide of hemoglobin. (The 1962 Nobel Prize in Chemistry was awarded to Max Perutz and Sir John Kendrew for their studies of the structures of globular proteins, and the 1972 Nobel Prize in Chemistry was awarded to Christian Anfinsen for his work on ribonuclease, especially concerning the connection between the amino-acid sequence and the biologically active conformation.)

4. The *quaternary structure* is the complex of polypeptide chains in a multisubunit protein, so quaternary structure is found only in proteins having more than one polypeptide chain (Figure 6.4d). Shown in Figure 6.4d is the quaternary structure of a multimeric (many-subunit) protein—the oxygen-carrying protein hemoglobin, which consists of four polypeptide chains (two 141-amino-acid α polypeptides and two 146-amino-acid β polypeptides), each of which is associated with a heme group that is involved in the binding of oxygen. In the quaternary structure of hemoglobin, each α chain is in contact with each β chain, but there is little interaction between the two α chains or between the two β chains.

For many years, it was thought that the amino-acid sequence alone was sufficient to specify how a protein folds into its functional state. We now know that, for many proteins, folding depends on one or more of a family of proteins called *chaperones* (also called *molecular chaperones*). Chaperones act analogously to enzymes, in that they interact with the proteins they help fold, but do not become part of the functional protein produced. A detailed discussion of chaperones is beyond the scope of this book.

KEYNOTE

A protein consists of one or more molecular subunits called polypeptides, which are themselves composed of smaller building blocks, the amino acids, linked together by peptide bonds to form long chains. The primary amino-acid sequence of a protein determines its secondary, tertiary, and quaternary structure and hence its functional state.

The Nature of the Genetic Code

How do nucleotides in the mRNA molecule specify the amino-acid sequence in proteins? With four different nucleotides (A, C, G, U), a three-letter code generates 64 possible codons. If it were a one-letter code, only four

amino acids could be encoded. If it were a two-letter code, then only 16 (4 × 4) amino acids could be encoded. A three-letter code, however, generates 64 (4 × 4 × 4) possible codes, more than enough to code for the 20 amino acids found in living cells. Since there are only 20 different amino acids, the assumption of a three-letter code suggests that some amino acids may be specified by more than one codon, which is in fact the case.

The Genetic Code Is a Triplet Code

The evidence that the **genetic code** is a triplet code—that is, that a set of three nucleotides (a **codon**) in mRNA code form one amino acid in a polypeptide chain—came from genetic experiments done by Francis Crick, Leslie Barnett, Sidney Brenner, and R. Watts-Tobin in the early 1960s. The experiments used bacteriophage T4, which Benzer also had used in his work on the fine structure of the gene. (See Chapter 18, pp. 500–501.) Recall that T4 is a virulent phage, meaning that when it infects *E. coli*, it undergoes the lytic cycle, producing 100 to 200 progeny phages that are released from the cell when the cell lyses. Some mutants of T4 are known to affect the lytic cycle. As discussed in Chapter 18, p. 500, *rII* mutants produce clear plaques on the strain *E. coli B*, whereas the wild type *r*+ strain produces turbid plaques. Furthermore, in contrast to the *r*+ strain, *rII* mutants are unable to reproduce in strain *E. coli K12(λ)*.

Crick and his colleagues began with an *rII* mutant strain that had been produced by treating the *r*+ strain with a *mutagen*—a chemical that causes mutations (discussed in more detail in Chapter 7). They used the mutagen proflavin, which induces mutations by causing the addition or deletion of a base pair in the DNA. When such mutations occur in the amino-acid-coding part of a gene, the mutations are **frameshift mutations.** (See Chapter 7, p. 138.) That is, a series of three-nucleotide "words" is read by the translation machinery to assemble the correct polypeptide chain. If a single base pair is deleted or added in this region, the words after the deletion or addition are now different—they are in another frame—and a different set of amino acids will be specified.

Crick and his colleagues reasoned that if the *rII* mutant phenotypes resulted from either an addition or a deletion, treatment of the *rII* mutant with proflavin could reverse the mutation to the wild-type—*r*+—state. The process of changing a mutant back to the wild-type state is called **reversion,** and the wild type produced in this way is called a *revertant*. So if the original mutation was an addition, it could be corrected by a deletion. The researchers isolated a number of *r*+ revertant strains by plating a population of *rII* phages that had been treated with proflavin onto a lawn of *E. coli K12(λ)*, in which only *r*+ phages can grow. This approach made it very easy to isolate the low number of *r*+ revertants produced by the proflavin treatment.

Some of the revertants resulted from an exact correction of the original mutation; that is, an addition corrected the deletion, or a deletion corrected the addition. A second type of revertant was much more useful for determining the nature of the genetic code. This type of revertant resulted from a second mutation within the *rII* gene very close to, but distinct from, the original mutation site. For example, if the first mutation was a deletion of a single base pair, the reversion of that mutation would involve an addition of a base pair nearby. Figure 6.5a shows a hypothetical segment of DNA. For the purposes of discussion, we have assumed that the code is a triplet code. Thus, the mRNA transcript of the DNA would be read ACG ACG ACG, etc., giving a polypeptide with a string of identical amino acids—threonine—each specified by ACG. This is our starting **reading frame**—that is, the codons (words) that are read sequentially to specify the amino acids. If proflavin treatment causes a deletion of the second AT base pair, the mRNA will now read ACG CGA CGA CGA, etc., giving a polypeptide starting with

the amino acid specified by ACG (threonine), followed by a string of amino acids that are specified by the repeating CGA (arginine) (Figure 6.5b). This mutation is a frameshift mutation because the codons after the deletion are changed. That is, after the ACG, the reading frame of the message is now a string of CGA codons. In that repeated CGA codon sequence, the repeated ACG sequence is still present, with the A as the last letter of the CGA codon and the CG as the first two letters of the CGA. This deletion mutation can revert by adding a base pair nearby. For example, the insertion of a GC base pair after the GC in the third triplet results in an mRNA that is read as ACG CGA CGG ACG ACG, and so on (Figure 6.5c). This gives a polypeptide consisting mostly of the amino acid specified by ACG (threonine), but with two wrong amino acids: those specified by CGA and CGG (both arginine). Thus, the second mutation has restored the reading frame, and a nearly wild-type polypeptide is produced. As long as the incorrect amino acids in the short segment between the mutations do not significantly affect the function of the polypeptide, the double mutant will have a normal or near-normal phenotype.

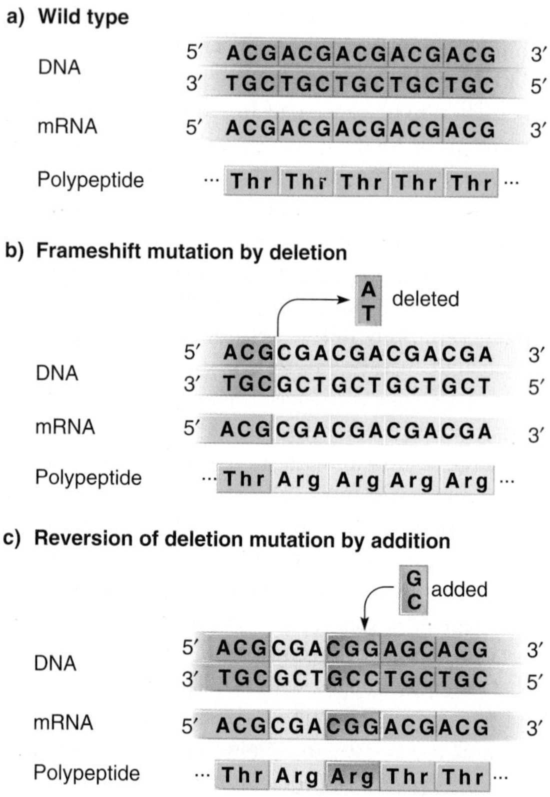

In short, an addition mutation can be reverted by a nearby deletion mutation, and a deletion mutation can be reverted by a nearby addition mutation. We symbolize addition mutations as + mutations and deletion mutations as − mutations. The next step Crick and his colleagues took was to combine genetically distinct *rII* mutations of the same type (either all + or all − mutations)[1] in various numbers to see whether any combinations reverted the *rII* phenotypes. Figure 6.6 is a hypothetical presentation of the type of results they obtained, showing the effects of the mutations just on the mRNA. The figure shows a 30-nucleotide segment of mRNA that codes for 10 different amino acids in the polypeptide. If we add three base pairs at nearby locations in the DNA coding for this mRNA segment, the result will be a 33-nucleotide segment that codes for 11 amino acids, one more than the original. However, the amino acids between the first and third insertions are not the same as the wild-type mRNA. In essence, the reading frame is correct before the first insertion and again after the third insertion. The incorrect amino acids between those points may result in a not-quite wild-type phenotype for the revertant.

Crick and his colleagues found that the combination of three nearby + mutations or three nearby − mutations gave r^+ revertants. No multiple combinations worked, except multiples of three. Therefore, they concluded that the genetic code is a triplet code.

Figure 6.5

Reversion of a deletion frameshift mutation by a nearby addition mutation. (a) Hypothetical segment of normal DNA, mRNA transcript, and polypeptide in the wild type. (b) Effect of a deletion mutation on the amino-acid sequence of a polypeptide. The reading frame is disrupted. (c) Reversion of the deletion mutation by an addition mutation. The reading frame is restored, leaving a short segment of incorrect amino acids.

a) Wild type

DNA
5′ ACGACGACGACGACG 3′
3′ TGCTGCTGCTGCTGC 5′

mRNA
5′ ACGACGACGACGACG 3′

Polypeptide
··· Thr Thr Thr Thr Thr ···

b) Frameshift mutation by deletion

A / T deleted

DNA
5′ ACGCGACGACGACGA 3′
3′ TGCGCTGCTGCTGCT 5′

mRNA
5′ ACGCGACGACGACGA 3′

Polypeptide
··· Thr Arg Arg Arg Arg ···

c) Reversion of deletion mutation by addition

G / C added

DNA
5′ ACGCGACGGAGCACG 3′
3′ TGCGCTGCCTGCTGC 5′

mRNA
5′ ACGCGACGGACGACG 3′

Polypeptide
··· Thr Arg Arg Thr Thr ···

[1]Crick and his colleagues did not know whether an *rII* mutant resulted from a + or a − mutation. But they did know which of their single-mutant *rII* strains were of one sign and which were of the other sign. That is, all mutants of one sign (e.g., +) could be reverted by nearby mutants of the other sign (i.e., −) and vice versa.

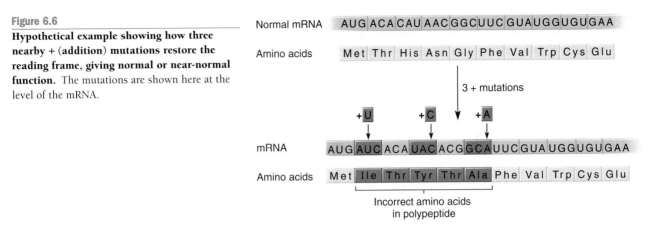

Figure 6.6

Hypothetical example showing how three nearby + (addition) mutations restore the reading frame, giving normal or near-normal function. The mutations are shown here at the level of the mRNA.

Deciphering the Genetic Code

The exact relationship of the 64 codons to the 20 amino acids was determined by experiments done mostly in the laboratories of Marshall Nirenberg and Ghobind Khorana, who shared the 1968 Nobel Prize in Physiology or Medicine with Robert Holley. Essential to these experiments was the use of *cell-free, protein-synthesizing systems* with components isolated and purified from *E. coli*. These systems contain ribosomes, tRNAs with amino acids attached, and all the necessary protein factors for polypeptide synthesis. Radioactively labeled amino acids were used to measure the incorporation of amino acids into new proteins.

In one approach to establishing which codons specify which amino acids, synthetic mRNAs containing one, two, or three different types of bases were made and added to the cell-free protein-synthesizing systems. The polypeptides produced in these systems were then analyzed. When the synthetic mRNA contained only one type of base, the results were unambiguous. Synthetic poly(U) mRNA, for example, directed the synthesis of a polypeptide consisting of a chain of phenylalanines. Since the genetic code is a triplet code, this result indicated that UUU is a codon for phenylalanine. Similarly, a synthetic poly(A) mRNA directed the synthesis of a lysine chain, and poly(C) directed the synthesis of a proline chain, indicating that AAA is a codon for lysine and CCC is a codon for proline. The results from poly(G) were inconclusive because the poly(G) folds up upon itself, so it cannot be translated in vitro.

Synthetic mRNAs made by the random incorporation of two different bases (called *random copolymers*) were also analyzed. When mixed copolymers are made, the bases are incorporated into the synthetic molecule in a random way. Thus, poly(AC) molecules can contain the eight different codons CCC, CCA, CAC, ACC, CAA, ACA, AAC, and AAA. In the cell-free protein-synthesizing system, poly(AC) synthetic mRNAs resulted in the incorporation of asparagine, glutamine, histidine, and threonine into polypeptides, in addition to the lysine expected from

AAA codons and the proline expected from CCC codons. The proportions of asparagine, glutamine, histidine, and threonine incorporated into the polypeptides that were produced depended on the A:C ratio used to make the mRNA and were used to deduce information about the codons that specify the amino acids. For example, because an AC random copolymer containing much more A than C resulted in the incorporation of many more asparagines than histidines, researchers concluded that asparagine is coded by two A's and one C and histidine by two C's and one A. With experiments of this kind, the base composition (but *not* the base sequence) of the codons for a number of amino acids was determined.

Another experimental approach also used copolymers, but these copolymers were synthesized, so that they had a known sequence, not a random one. For example, a repeating copolymer of U and C gives a synthetic mRNA of UCUCUCUCUCUC, etc. When this copolymer was tested in a cell-free protein-synthesizing system, the resulting polypeptide had a repeating amino acid pattern of leucine–serine–leucine–serine. Therefore, UCU and CUC specify leucine and serine, although which coded for which cannot be determined from the result.

Yet another approach used a *ribosome-binding assay,* developed in 1964 by Nirenberg and Philip Leder. This assay depends on the fact that, in the absence of protein synthesis, specific tRNA molecules bind to ribosome-mRNA complexes. For example, when a synthetic mRNA codon, UUU, is mixed with ribosomes, it forms a UUU–ribosome complex, and only a phenylalanine tRNA (the tRNA with an AAA anticodon that brings phenylalanine to an mRNA) binds to the UUU codon. This codon-binding property made it possible to determine the specific relationships between many codons and the amino acids for which they code. Note that in this particular approach, *the specific nucleotide sequence of the codon is determined.* Using the ribosome-binding assay, Nirenberg and Leder resolved many ambiguities that had arisen from other approaches. For example, UCU was found to be a codon for serine, and CUC was found to be a codon for leucine. All in all, about 50 codons were identified with this approach.

In sum, no single approach produced an unambiguous set of codon assignments, but information obtained through all of the approaches enabled 61 codons to be assigned to amino acids; the other 3 codons do not specify amino acids (Figure 6.7). Each codon is written as it appears in mRNA and reads in a 5'-to-3' direction.

Characteristics of the Genetic Code

The characteristics of the genetic code are as follows:

1. *The code is a triplet code.* Each mRNA codon that specifies an amino acid in a polypeptide chain consists of three nucleotides.
2. *The code is comma free; that is, it is continuous.* The mRNA is read continuously, three nucleotides at a time, without skipping any nucleotides of the message.
3. *The code is nonoverlapping.* The mRNA is read in successive groups of three nucleotides.
4. *The code is almost universal.* Almost all organisms share the same genetic language. Therefore, we can

isolate an mRNA from one organism, translate it by using the machinery from another organism, and produce the protein as if it had been translated in the original organism. The code is not completely universal, however. For example, the mitochondria of some organisms, such as mammals, have minor changes in the code, as does the nuclear genome of the protozoan *Tetrahymena*.

5. *The code is degenerate.* With two exceptions (only AUG codes for methionine and only UGG codes for tryptophan), more than one codon occurs for each amino acid. This multiple coding is called the **degeneracy** of the code. There are particular patterns in this degeneracy. (See Figure 6.8). When the first two nucleotides in a codon are identical and the third letter is U or C, the codon always codes for the same amino acid. For example, UUU and UUC specify phenylalanine, and CAU and CAC specify histidine. Also, when the first two nucleotides in a codon are identical and the third letter is A or G, the same amino acid often is specified. For example, UUA and UUG specify leucine, and AAA and AAG specify lysine. In a few cases, when the first two nucleotides in a codon are identical and the base in the third position is U, C, A, or G, the same amino acid often is specified. For example, CUU, CUC, CUA, and CUG all code for leucine.

6. *The code has start and stop signals.* Specific start and stop signals for protein synthesis are contained in the code. In both eukaryotes and prokaryotes, AUG (which codes for methionine) is almost always the start codon for protein synthesis.

Only 61 of the 64 codons specify amino acids; these codons are called *sense codons.* (See Figure 6.7.) The other three codons—UAG (amber), UAA (ochre), and UGA (opal)—do not specify an amino acid, and no tRNAs in normal cells carry the appropriate

Figure 6.7

The genetic code. Of the 64 codons, 61 specify one of the 20 amino acids. The other 3 codons are chain-terminating codons and do not specify any amino acid. AUG, one of the 61 codons that specify an amino acid, is used in the initiation of protein synthesis.

Second letter

	U	C	A	G	
U	UUU Phe (F) / UUC / UUA Leu / UUG (L)	UCU / UCC Ser (S) / UCA / UCG	UAU Tyr (Y) / UAC / UAA Stop / UAG Stop	UGU Cys (C) / UGC / UGA Stop / UGG Trp (W)	U C A G
C	CUU / CUC Leu (L) / CUA / CUG	CCU / CCC Pro (P) / CCA / CCG	CAU His (H) / CAC / CAA Gln (Q) / CAG	CGU / CGC Arg (R) / CGA / CGG	U C A G
A	AUU / AUC Ile (I) / AUA / AUG Met (M)	ACU / ACC Thr (T) / ACA / ACG	AAU Asn (N) / AAC / AAA Lys (K) / AAG	AGU Ser (S) / AGC / AGA Arg (R) / AGG	U C A G
G	GUU / GUC Val (V) / GUA / GUG	GCU / GCC Ala (A) / GCA / GCG	GAU Asp (D) / GAC / GAA Glu (E) / GAG	GGU / GGC Gly (G) / GGA / GGG	U C A G

First letter (left), Third letter (right)

■ = Chain termination codon (stop)
▨ = Initiation codon

Figure 6.8

Example of base-pairing wobble. Two different leucine codons (CUC, CUU) can be read by the same leucine tRNA molecule, contrary to regular base-pairing rules.

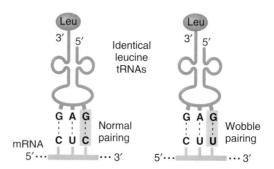

anticodons. (The three-nucleotide anticodon pairs with the codon in the mRNA by complementary base pairing during translation.) These three codons are the **stop codons**, also called **nonsense codons** or **chain-terminating codons**. They are used singly or in tandem groups (UAG UAA, for example) to specify the end of translation of a polypeptide chain. Thus, when we read a particular mRNA sequence, we look for a stop codon located at a multiple of three nucleotides—*in the same reading frame*—from the AUG start codon to determine where the amino-acid-coding sequence for the polypeptide ends. This is called an **open reading frame**.

7. *Wobble occurs in the anticodon.* Since 61 sense codons specify amino acids in mRNA, a total of 61 tRNA molecules could have the appropriate anticodons. According to the **wobble hypothesis** proposed by Francis Crick, the complete set of 61 sense codons can be read by fewer than 61 distinct tRNAs, because of pairing properties of the bases in the anticodon (Table 6.1). Specifically, the base at the 5′ end of the anticodon complementary to the base at the 3′ end of the codon, or the third letter, is not as constrained three dimensionally as the other two bases. This feature allows for less exact base pairing, so that the base at the 5′ end of the anticodon can pair with more than one type of base at the 3′ end of the codon; in other words, it can wobble. As the table shows, no single tRNA molecule can recognize four different codons. But if the tRNA molecule contains the modified purine inosine at the 5′ end of the anticodon, then that tRNA can recognize three different codons. Figure 6.8 gives an example of how a single leucine tRNA can read two different leucine codons by base-pairing wobble.

i

Learn how to use sequencing information to track down part of the gene responsible for cystic fibrosis in the iActivity *Determining Causes of Cystic Fibrosis* on your CD-ROM.

Table 6.1	Wobble in the Genetic Code	
Nucleotide at 5′ End of Anticodon		**Nucleotide at 3′ End of Codon**
G	can pair with	U or C
C	can pair with	G
A	can pair with	U
U	can pair with	A or G
I (inosine)	can pair with	A, U, or C

KEYNOTE

The genetic code is a triplet code in which each codon (a set of three contiguous bases) in an mRNA specifies one amino acid. The code is degenerate: Some amino acids are specified by more than one codon. The genetic code is nonoverlapping and almost universal. Specific codons are used to signify the start and end of protein synthesis.

Translation: The Process of Protein Synthesis

Protein synthesis takes place on ribosomes, where the genetic message encoded in mRNA is translated. The mRNA is translated in the 5′-to-3′ direction, and the polypeptide is made in the N-terminal-to-C-terminal direction. Amino acids are brought to the ribosome bound to tRNA molecules. The correct amino-acid sequence is achieved as a result of (1) the binding of each amino acid to its own specific tRNA and (2) the binding between the codon of the mRNA and the complementary anticodon in the tRNA.

The mRNA Codon Recognizes the tRNA Anticodon

That the mRNA codon recognizes the tRNA anticodon, and not the amino acid carried by the tRNA, was proved by G. von Ehrenstein, B. Weisblum, and S. Benzer. These researchers attached cysteine in vitro to tRNA.Cys. (This terminology indicates the amino acid specified by the anticodon of the tRNA—in this case, cysteine); then they chemically converted the attached cysteine to alanine. The resulting Ala-tRNA.Cys was used in the in vitro synthesis of hemoglobin. In vivo, the α and β chains of hemoglobin each contain one cysteine. When the hemoglobin made in vitro was examined, however, the amino acid alanine was found in both chains at the positions normally occupied by cysteine. This result could only mean that the Ala-tRNA.Cys had read the cysteine codon and had inserted the amino acid it carried—in this case, alanine. Therefore, the researchers concluded that *the specificity of codon recognition lies in the tRNA molecule, not in the amino acid it carries.*

Adding an Amino Acid to tRNA

The correct amino acid is attached to the tRNA by a type of enzyme called an **aminoacyl-tRNA synthetase**. The process is called aminoacylation, or **charging**, and produces an **aminoacyl-tRNA** (or **charged tRNA**). Aminoacylation uses energy from ATP hydrolysis. Since there are 20 different amino acids, there are 20 different aminoacyl-tRNA synthetases. Each aminoacyl-tRNA synthetase

Figure 6.9

Charging of a tRNA molecule by aminoacyl-tRNA synthetase to produce an aminoacyl-tRNA (charged tRNA).

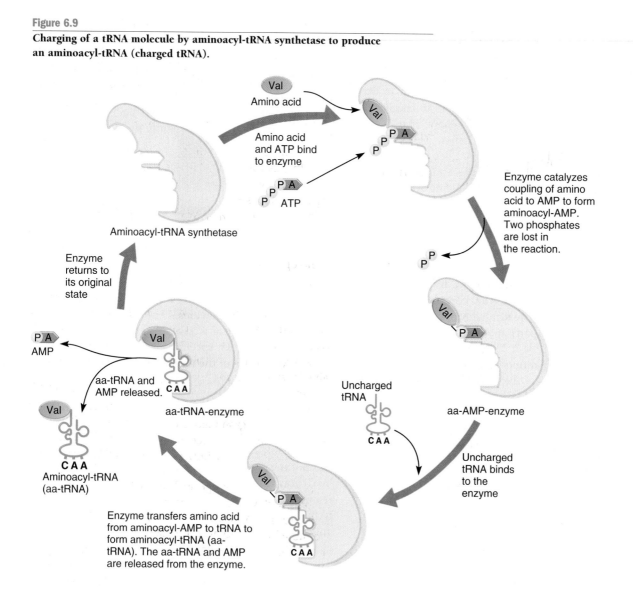

recognizes particular structural features of the tRNA or tRNAs it charges.

Figure 6.9 shows the charging of a tRNA molecule to produce valine-tRNA (Val-tRNA). First, the amino acid and ATP bind to the specific aminoacyl-tRNA synthetase enzyme. The enzyme then catalyzes a reaction in which the ATP loses two phosphates and is coupled to the amino acid as AMP to form aminoacyl-AMP. Next, the tRNA molecule binds to the enzyme, which transfers the amino acid from the aminoacyl-AMP to the tRNA and then displaces the AMP. The aminoacyl-tRNA molecule produced is then released from the enzyme. Chemically, the amino acid attaches at the 3′ end of the tRNA by a covalent linkage between the carboxyl group of the amino acid and the 3′-OH or 2′-OH group of the ribose of the adenine nucleotide found at the end of every tRNA (Figure 6.10).

Initiation of Translation

The three basic stages of protein synthesis—*initiation, elongation,* and *termination*—are similar in prokaryotes and eukaryotes. In the sections that follow, we discuss each of these stages in turn, concentrating on the processes in *E. coli*. In the discussions, we note where significant differences in translation occur between prokaryotes and eukaryotes.

animation

a Initiation of Translation

The initiation of translation is shown in Figure 6.11. Initiation involves an mRNA molecule, a ribosome, a specific initiator tRNA, three different protein **initiation factors,** GTP (guanosine triphosphate), and magnesium ions. In prokaryotes, the first step in translation is the binding of the 30S small ribosomal subunit to the region of the mRNA with the AUG initiation codon. The ribosomal

Figure 6.10

Attachment of an amino acid to a tRNA molecule. In an aminoacyl-tRNA molecule (charged tRNA), the carboxyl group of the amino acid is attached to the 3′-OH or 2′-OH group of the 3′ terminal adenine nucleotide of the tRNA.

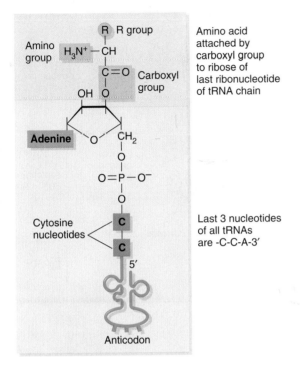

subunit comes to the mRNA bound to three initiation factors (IF1, IF2, and IF3) and to a molecule of GTP and magnesium ions.

The AUG initiation codon is not sufficient to tell the ribosomal subunit where to bind to the mRNA; a sequence upstream (to the 5′ side in the leader of the mRNA) of the AUG called the **ribosome-binding site (RBS)** is also needed. Evidence from the work of John Shine and Lynn Dalgarno indicates that the purine-rich RBS sequence (AGGAG or some similar sequence) and sometimes other nucleotides in this region are complementary to a pyrimidine-rich region (which always contains the sequence UCCUCC) at the 3′ end of 16S rRNA (Figure 6.12). As a result of this work, the mRNA RBS region is also known as the **Shine–Dalgarno sequence.**

Most of the RBSs are 8 to 12 nucleotides upstream from the initiation codon. The model is that the formation of complementary base pairs between the mRNA and 16S rRNA allows the ribosome to locate the true sequence in the mRNA for the initiation of protein synthesis. There is genetic evidence that the model is correct. If the Shine–Dalgarno sequence is mutated so that its possible pairing with the 16S rRNA sequence is significantly diminished or prevented, the particular mRNA involved cannot be translated. Likewise, if the rRNA sequence complementary to the Shine–Dalgarno

sequence is mutated, mRNA translation does not occur. Since it can be argued that the loss of translatability as a result of mutations in one or the other RNA partner could be caused by effects unrelated to the loss of pairing of the two RNA segments, a more elegant experiment was done. That is, mutations were made in the Shine–Dalgarno sequence to abolish pairing with the wild-type rRNA sequence, and compensating mutations were made in the rRNA sequence so that the two mutated sequences could pair. In this case, mRNA translation occurred normally, indicating the importance of the pairing of the two RNA segments. (This type of experiment, in which compensating mutations are made in two sequences that are hypothesized to interact, has been used in a number of other systems to explore the roles of specific interactions in biological functions.)

The next step in the initiation of translation is the binding of the initiator tRNA to the AUG start codon to which the 30S subunit is bound. In both prokaryotes and eukaryotes, the AUG initiator codon specifies methionine. As a result, newly made proteins in both types of organisms begin with methionine. In many cases, the methionine is removed later.

In prokaryotes, the initiator methionine is a modified form of methionine, called **formylmethionine** (fMet), in which a formyl group has been added to the methionine's amino group. The fMet is brought to the ribosome attached to a special initiator tRNA, tRNA.fMet, which has the anticodon 5′-CAU-3′ to bind to the AUG start codon. This tRNA is special because it is involved specifically with the initiation process of protein synthesis. The aminoacylation of tRNA.fMet occurs as follows: First, methionyl-tRNA synthetase catalyzes the addition of methionine to the tRNA. Then the enzyme *transformylase* adds the formyl group to the methionine. The resulting molecule is designated fMet-tRNA.fMet. (This nomenclature indicates that the tRNA is specific for the attachment of fMet and that fMet is attached to it.)

When an AUG codon in an mRNA molecule is encountered at a position other than the start of the amino acid-coding sequence, a different tRNA, called tRNA.Met, is used to insert methionine at that point in the polypeptide chain. This tRNA is charged by the same aminoacyl-tRNA synthetase as is tRNA.fMet to produce Met-tRNA.Met. However, tRNA.Met and tRNA.fMet molecules are coded for by different genes and have different sequences. We will see later in the chapter how the two tRNAs are used differently.

When the fMet-tRNA binds to the start codon of the 30S–mRNA complex, IF3 is released. At this point, the *30S initiation complex*, consisting of the mRNA, 30S subunit, fMet-tRNA, IF1, and IF2, has been formed. Next, the 50S ribosomal subunit binds, leading to GTP hydrolysis and the release of IF1 and IF2. The final complex is called the *70S initiation complex*. The

Figure 6.11

Initiation of protein synthesis in prokaryotes. A 30S ribosomal subunit, complexed with initiation factors and GTP, binds to mRNA and fMet-tRNA to form a 30S initiation complex. Next, the 50S ribosomal subunit binds, forming a 70S initiation complex. During this event, the initiation factors are released and GTP is hydrolyzed.

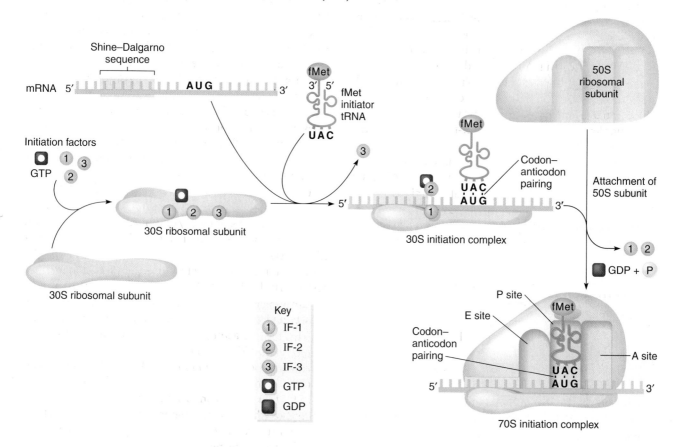

70S ribosome has three binding sites for aminoacyl-tRNA: the exit (E), peptidyl (P), and aminoacyl (A) sites. The fMet-tRNA is bound to the mRNA in the P site, and the E and A sites are vacant. (See Figure 6.11.)

The initiation of translation is similar in eukaryotes. The main differences are that (1) the initiator methionine is unmodified, although a special initiator tRNA brings it to the ribosome, and (2) Shine–Dalgarno sequences are

not found in eukaryotic mRNAs. Instead, the eukaryotic ribosome uses another way to find the AUG initiation codon. First, a *eukaryotic initiator factor eIF-4F*—a multimer of several proteins, including eIF-4E, the *cap-binding protein* (CBP)—binds to the cap at the 5′ end of the mRNA. (See Chapter 5.) Then, a complex of the 40S ribosomal subunit with the initiator Met-tRNA, several eIF proteins, and GTP binds, together with other eIFs,

a) **Sequence at 3′ end of 16S rRNA**

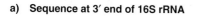

3′   **AUUCCUCCAUAG**   5′

b) **Example of mRNA leader and 16S rRNA pairing**

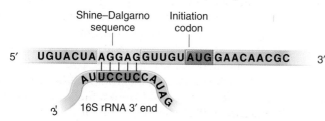

Figure 6.12

Sequences involved in the binding of ribosomes to the mRNA in the initiation of protein synthesis in prokaryotes. (a) Nucleotide sequence at the 3′ end of *E. coli* 16S rRNA. (b) Example of how the 3′ end of 16S rRNA can base pair with the nucleotide sequence 5′ upstream from the AUG initiation codon.

and moves along the mRNA, scanning for the initiator AUG codon. The AUG codon is embedded in a short sequence—called the Kozak sequence, after Marilyn Kozak—which indicates that it is the initiator codon. This process is called the *scanning model* for initiation. The AUG codon is almost always the first AUG codon from the 5′ end of the mRNA, but, to be an initiator codon, it must be in an appropriate sequence context. Once the 40S subunit finds this AUG, it binds to it, and then the 60S ribosomal subunit binds, displacing the eIFs (with the exception of eIF-4F, which is needed for the subsequent initiation of translation), producing the 80S initiation complex with the initiator Met-tRNA bound to the mRNA in the P site of the ribosome.

The poly(A) tail of the eukaryotic mRNA also plays a role in translation. Poly(A) binding protein (PABP; see Figure 5.10) bound to the poly(A) tail also binds to eIF-4G, one of the proteins of eIF-4F at the cap, thereby looping the 3′ end of the mRNA close to the 5′ end. In this way, the poly(A) tail stimulates the initiation of translation.

animation

a Elongation of the Polypeptide Chain

Elongation of the Polypeptide Chain

Figure 6.13 depicts the elongation events—the addition of amino acids to the growing polypeptide chain one by one—as they take place in prokaryotes. This phase has three steps:

1. Aminoacyl-tRNA (charged tRNA) binds to the ribosome.
2. A peptide bond forms.
3. The ribosome moves (translocates) along the mRNA one codon at a time.

Binding of Aminoacyl-tRNA. At the start of elongation, the anticodon of fMet-tRNA is hydrogen bonded to the AUG initiation codon in the P (peptidyl) site of the ribosome (Figure 6.13, part 1). The next codon in the mRNA is in the A (aminoacyl) site. In Figure 6.13, this codon (UCC) specifies serine (Ser).

Next, the appropriate aminoacyl-tRNA (here, Ser-tRNA.Ser) binds to the codon in the A site (Figure 6.13, part 2). This aminoacyl-tRNA is brought to the ribosome bound to the protein elongation factor EF-Tu and a molecule of GTP. When the aminoacyl-tRNA binds to the codon in the A site, GTP hydrolysis releases EF-Tu-GDP. As shown in Figure 6.13, part 2, EF-Tu is recycled. First, a second elongation factor, EF-Ts, binds to EF-Tu and displaces the GDP. Next, GTP binds to the EF-Tu-EF-Ts complex to produce an EF-Tu-GTP complex simultaneously with the release of EF-Ts. An aminoacyl-tRNA binds to the EF-Tu-GTP, and that complex can bind to the A site in the ribosome when the appropriate codon is exposed.

Peptide Bond Formation. The ribosome maintains the two aminoacyl-tRNAs in the P and A sites in the correct positions, so that a peptide bond can form between the two amino acids (Figure 6.13, part 3). Two steps are involved in the formation of this peptide bond (Figure 6.14). First, the bond between the amino acid and the tRNA in the P site is cleaved. In this case, the breakage is between the fMet and its tRNA. Second, the peptide bond is formed between the now-freed fMet and the Ser attached to the tRNA in the A site in a reaction catalyzed by **peptidyl transferase.** For many years, this enzyme activity was thought to be a result of the interaction of a few ribosomal proteins of the 50S ribosomal subunit. However, in 1992, Harry Noller and his colleagues found that when most of the proteins of the 50S ribosomal subunit were removed, leaving only the ribosomal RNA, peptidyl transferase activity could still be measured. In addition, this activity was inhibited by the antibiotics chloramphenicol and carbomycin, both of which are known specifically to inhibit peptidyl transferase activity. Furthermore, when the rRNA was treated with ribonuclease T1, which degrades RNA, but not protein, the peptidyl transferase activity was lost. These results suggested that the 23S rRNA molecule of the large ribosomal subunit is intimately involved with the peptidyl transferase activity and may in fact be that enzyme. In this case, the rRNA would be acting as a ribozyme (catalytic RNA; see Chapter 5, p. 100). Recently, the atomic structure of a bacterial large ribosomal subunit at 2.4 Å resolution has been reported. From the structure, it has been deduced that the peptidyl transferase consists entirely of RNA. Thus, in a reversal of what was once thought, the ribosomal proteins are the structural units that help to organize the rRNA into key ribozyme elements.

Once the peptide bond has formed (see Figure 6.13, part 3), a tRNA without an attached amino acid (an uncharged tRNA) is left in the P site. The tRNA in the A site, now called peptidyl-tRNA, has the first two amino acids of the polypeptide chain attached to it—in this case, fMet-Ser.

Translocation. In the last step in the elongation cycle, **translocation** (Figure 6.13, part 4), the ribosome moves one codon along the mRNA toward the 3′ end. In prokaryotes, translocation requires the activity of another protein elongation factor, EF-G. An EF-G-GTP complex binds to the ribosome, GTP is hydrolyzed, and translocation of the ribosome occurs along with displacement of the uncharged tRNA away from the P site. It is possible that GTP hydrolysis changes the structure of EF-G, which facilitates the translocation event. Translocation is similar in eukaryotes; the elongation factor in this case is eEF-2, which functions similarly to the way EF-G does.

A 50S subunit site called E (for *exit*) is involved in the release of the uncharged tRNA from the *E. coli* ribosome. The uncharged tRNA moves from the P site and then binds

Figure 6.13

Elongation stage of translation in prokaryotes.

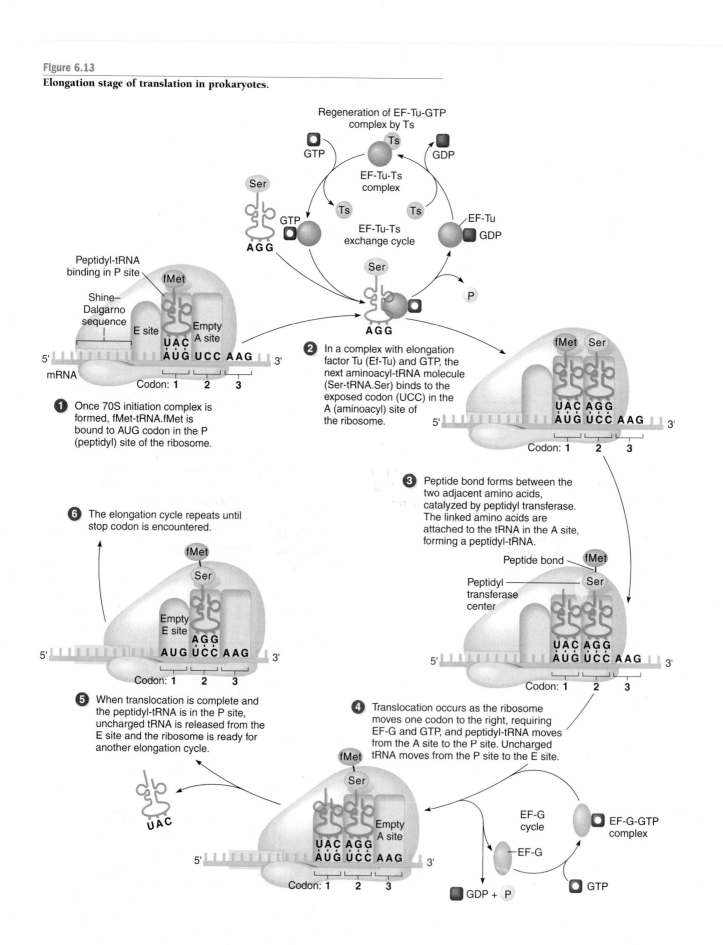

Figure 6.14

The formation of a peptide bond between the first two amino acids (fMet and Ser) of a polypeptide chain is catalyzed on the ribosome by peptidyl transferase.
(a) Adjacent aminoacyl-tRNAs bound to the mRNA at the ribosome. (b) After peptide bond formation, an uncharged tRNA is in the P site and a dipeptidyl-tRNA is in the A site.

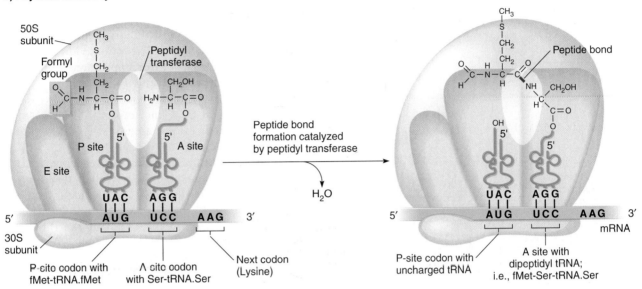

to the E site, effectively blocking the next aminoacyl-tRNA from binding to the A site until translocation is complete and the peptidyl-tRNA is bound correctly in the P site. After translocation, the EF-G is released in a reaction requiring GTP hydrolysis; EF-G is then reused, as shown in Figure 6.13, part 4. During the translocation step, the peptidyl-tRNA remains attached to its codon on the mRNA, and because the ribosome has moved, the peptidyl-tRNA is now located in the P site (hence the name *peptidyl site*). The exact mechanism for the physical translocation of the ribosome is not known.

After translocation is completed, the A site is vacant. An aminoacyl-tRNA with the correct anticodon binds to the newly exposed codon in the A site, reiterating the process already described. The whole process is repeated until translation terminates at a stop codon (Figure 6.13, part 5).

In eukaryotes, the elongation and translocation steps are similar to those in prokaryotes, although there are differences in the number and properties of elongation factors and in the exact sequences of events.

In both prokaryotes and eukaryotes, once the ribosome moves away from the initiation site on the mRNA, another initiation event occurs, and the process repeats until, typically, several ribosomes are translating each mRNA simultaneously. The complex between an mRNA molecule and all the ribosomes that are translating it simultaneously is called a **polyribosome**, or **polysome** (Figure 6.15). Each ribosome in a polysome translates the entire mRNA and produces a single, complete polypeptide. Having several ribosomes translating an mRNA simultaneously enables a large amount of protein to be produced quickly and efficiently.

Figure 6.15

Diagram of a polysome—a number of ribosomes, each translating the same mRNA sequentially.

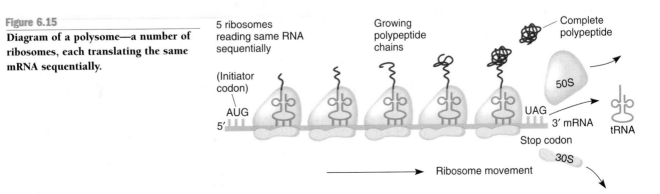

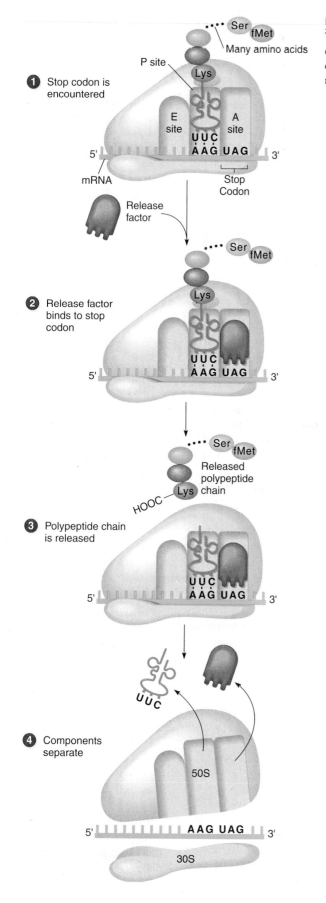

1. Stop codon is encountered

   P site

   Ser ••• fMet
   Many amino acids
   Lys
   E site   A site
   UUC
   AAG UAG
   5′ ———————— 3′
   mRNA
   Stop Codon

   Release factor

2. Release factor binds to stop codon

   Ser ••• fMet
   Lys
   UUC
   AAG UAG
   5′ ———————— 3′

3. Polypeptide chain is released

   Ser ••• fMet
   Released polypeptide chain
   HOOC Lys
   UUC
   AAG UAG
   5′ ———————— 3′

4. Components separate

   UUC
   50S
   AAG UAG
   5′ ———————— 3′
   30S

Figure 6.16

Termination of translation. The ribosome recognizes a chain termination codon (UAG) with the aid of release factors. A release factor reads the stop codon, initiating a series of specific termination events leading to the release of the completed polypeptide.

Termination of Translation

Elongation continues until the polypeptide coded for in the mRNA is completed. The termination of translation is signaled by one of three stop codons (UAG, UAA, and UGA), which are the same in prokaryotes and eukaryotes (Figure 6.16, part 1). The stop codons do not code for any amino acid, so no tRNAs in the cell have anticodons for them. The ribosome recognizes a stop codon with the help of proteins called **termination factors,** or **release factors (RF),** which read the codons (Figure 6.16, part 2) and then initiate a series of specific termination events. *E. coli* has three RFs (RF1, RF2, and RF3), and each is a single polypeptide. Factor RF1 recognizes UAA and UAG, and RF2 recognizes UAA and UGA. Factor RF3, which does not recognize any of the stop codons, stimulates the termination events. In eukaryotes, a single release factor—eukaryotic release factor 1 (eRF1)—recognizes all three stop codons, and eRF3 stimulates the termination events.

The specific termination events triggered by the release factors are (1) release of the polypeptide from the tRNA in the P site of the ribosome in a reaction catalyzed by peptidyl transferase (Figure 6.16, part 3); (2) release of the tRNA from the ribosome (Figure 6.16, part 4); and then (3) dissociation of the two ribosomal subunits and the RF from the mRNA (Figure 6.16, part 4). In both prokaryotes and eukaryotes, the initiating amino acids—fMet and Met, respectively—usually are cleaved from the completed polypeptide.

animation
a **Translation Termination**

KEYNOTE

Translation is a complicated process requiring many protein factors and energy. The AUG (methionine) initiator codon signals the start of translation in prokaryotes and eukaryotes. Elongation proceeds when a peptide bond forms between the amino acid attached to the tRNA in the A site of the ribosome and the growing polypeptide attached to the tRNA in the P site. Translocation occurs when the now-uncharged tRNA in the P site is released from the ribosome and the ribosome moves one codon down the mRNA. Termination occurs as a result of the interaction of a release factor with a chain-terminating codon.

Protein Sorting in the Cell

In prokaryotes and eukaryotes, some proteins may be secreted, and in eukaryotes, some other proteins must be placed in different cell compartments, such as the nucleus, a mitochondrion, a chloroplast, and a lysosome. The sorting of proteins to their appropriate compartments is under genetic control, in that specific "signal" or "leader" sequences on the proteins direct them to the correct organelles. Similarly, in prokaryotes, certain proteins become localized in the membrane and others are secreted.

Let us briefly describe how proteins are secreted from a eukaryotic cell. Such proteins are passaged through the endoplasmic reticulum (ER) and Golgi apparatus. In 1975, Günther Blobel, B. Dobberstein, and colleagues found that secreted proteins and other proteins sorted by the Golgi initially contain extra amino acids at the amino terminal end. Blobel's work led to the **signal hypothesis,** which states that proteins sorted by the Golgi bind to the ER by a hydrophobic amino terminal extension (the **signal sequence**) to the membrane that is subsequently removed and degraded (Figure 6.17). Blobel won the Nobel Prize in Physiology or Medicine in 1999 for this work.

The signal sequence of a protein destined for the ER consists of about 15-to-30 N-terminal amino acids. When the signal sequence is translated and exposed on the ribosome surface, a cytoplasmic **signal recognition particle** (SRP, an RNA-protein complex) binds to the sequence and blocks further translation of the mRNA until the growing polypeptide– SRP–ribosome–mRNA complex reaches and binds to the ER. (See Figure 6.17.) The SRP binds to a **docking protein** in the ER membrane, causing the firm binding of the ribosome to the ER, release of the SRP, and the resumption of translation. The growing polypeptide extends through the ER membrane into the cisternal space of the ER.

Once the signal sequence is fully into the cisternal space of the ER, it is removed from the polypeptide by the action of an enzyme called **signal peptidase.** When the complete polypeptide is entirely within the ER cisternal space, it is typically modified by the addition of specific carbohydrate groups to produce *glycoproteins*, which are then transferred in vesicles to the Golgi apparatus, where most of the sorting occurs. Proteins destined to be secreted, for example, are packaged into secretory storage vesicles, which migrate to the cell surface, where they fuse with the plasma membrane and release their packaged proteins to the outside of the cell.

K E Y N O T E

> Eukaryotic proteins that enter the endoplasmic reticulum have signal sequences at their N-terminal ends which target them to that organelle. The signal sequence first binds to a signal recognition particle, arresting translation. The complex then binds to a docking protein in the outer ER membrane, translation resumes, and the polypeptide is translocated into the cisternal space of the ER. Once in the ER, the signal sequence is removed by signal peptidase. The proteins are then sorted to their final destinations by the Golgi complex.

Figure 6.17

Model for the translocation of proteins into the endoplasmic reticulum in eukaryotes.

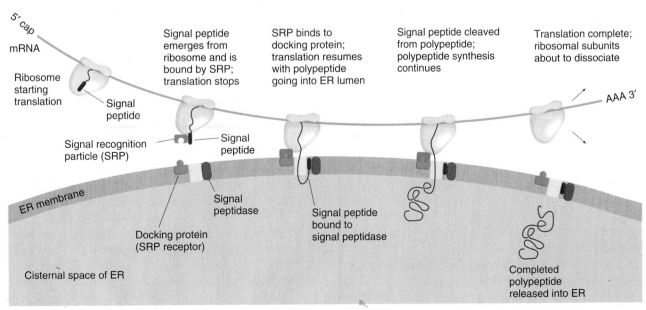

Summary

In this chapter, we discussed features of the genetic code and the translation of an mRNA to produce a protein. We learned that a protein consists of one or more subunits called polypeptides, which are themselves composed of smaller building blocks called amino acids, which in turn are linked together in the polypeptide by peptide bonds. The amino-acid sequence of a protein determines its secondary, tertiary, and quaternary structures and hence its functional state.

RNA consists of 4 different subunits (nucleotides), and protein consists of 20 different subunits (amino acids). The genetic code is a triplet code (meaning that a sequence of three nucleotides specifies one amino acid), is continuous, is nonoverlapping, is almost universal, and is degenerate. Of the 64 codons, 61 specify amino acids, and the other 3 are stop codons. One codon, AUG (for methionine), is almost always used to specify the first amino acid in a protein chain.

Protein synthesis occurs on ribosomes, where the genetic message encoded in mRNA is translated into a sequence of amino acids, brought to the ribosome on charged tRNA molecules. Each tRNA has an anticodon that binds specifically to a codon in the mRNA. As a result, the correct amino-acid sequence is achieved by the specific binding of each amino acid to its own specific tRNA and the specific binding between the codon of the mRNA and the complementary anticodon in the tRNA.

In both prokaryotes and eukaryotes, an AUG codon is the initiator codon for the start of protein synthesis. In prokaryotes, the initiation of protein synthesis requires a sequence upstream of the AUG codon, to which the small ribosomal subunit binds. This sequence is the Shine–Dalgarno sequence, which binds specifically to the 3′ end of the 16S rRNA of the small ribosomal subunit, thereby associating the small subunit with the mRNA. No functionally equivalent sequence occurs in eukaryotic mRNAs; instead, the ribosomes load onto the mRNA at its 5′ end and scan toward the 3′ end, initiating translation at the first AUG codon.

In both prokaryotes and eukaryotes, the initiation of protein synthesis requires protein factors called initiation factors. Bound to the ribosome–mRNA complex during the initiation phase, initiation factors dissociate once the polypeptide chain has been started. During the elongation phase, the polypeptide chain is lengthened one amino acid at a time. Elongation occurs simultaneously with the movement of the ribosome toward the 3′ end of the mRNA one codon at a time. Protein factors called elongation factors play important catalytic roles in this part of translation.

The signal for polypeptide chain growth to stop is the presence of a stop codon (UAG, UAA, or UGA) in the mRNA. No naturally occurring tRNA has an anticodon that can read a stop codon. Instead, specific protein factors called release factors read the stop codon and initiate the events characteristic of the termination of protein synthesis, namely, the release of the completed polypeptide from the ribosome, the release of the tRNA from the ribosome, and the dissociation of the two ribosomal subunits from the mRNA.

In eukaryotes, certain proteins are secreted from the cell, and others become localized in different cell compartments. Specific sequences in the proteins sort them to their appropriate compartments. Proteins destined for the ER, for example, have a signal sequence at their amino ends. A special process recognizes the signal sequence, blocks translation until the growing polypeptide–ribosome–mRNA complex binds to the ER, and then inserts the polypeptide into the cisternal space of the ER, where the signal sequence is removed. Once inside the ER, the protein is modified and then transported to the Golgi apparatus, where it undergoes further modifications before being sorted to its final location.

In sum, protein synthesis is a complex process involving interaction among three major classes of RNA (mRNA, tRNA, and rRNA), and a large number of accessory protein factors that act catalytically in the process. By repeated translation of an mRNA molecule by a string of ribosomes (producing a polysome), a large number of identical protein molecules can be produced. Thus, from a single gene, large quantities of a protein can be produced by two steps: the production of multiple mRNAs from the gene and the production of many protein molecules by repeated translation of each mRNA.

Analytical Approaches to Solving Genetics Problems

Q6.1

a. How many of the 64 codons can be made from the three nucleotides A, U, and G?

b. How many of the 64 codons can be made from the four nucleotides A, U, G, and C with one or more Cs in each codon?

A6.1

a. This question involves probability. There are four bases, so the probability of a cytosine at the first position in a codon is $1/4$. Conversely, the probability of a base other than cytosine in the first position is $(1 - 1/4) = 3/4$. These same probabilities apply to the other two positions in the codon. Therefore, the probability of a codon without a cytosine is $(3/4)^3 = 27/64$.

b. This question involves the relative frequency of codons that have one or more cytosines. We have already calculated the probability of a codon not having a cytosine, so all the remaining codons have one or more cytosines. The answer to this question, therefore, is $(1 - {}^{27}/_{64}) = {}^{37}/_{64}$.

Q6.2 Random copolymers were used in some of the experiments directed toward deciphering the genetic code. For each of the following ribonucleotide mixtures, give the expected codons and their frequencies, and give the expected proportions of the amino acids that would be found in a polypeptide directed by the copolymer in a cell-free protein-synthesizing system:
a. 2 U : 1 C
b. 1 U : 1 C : 2 G

A6.2

a. The probability of a U at any position in a codon is $^2/_3$, and the probability of a C at any position in a codon is $^1/_3$. Thus, the codons, their relative frequencies, and the amino acids for which they code are as follows:

$UUU = (2/3)(2/3)(2/3) = 8/27 = 0.296 = 29.6\%$ Phe
$UUC = (2/3)(2/3)(1/3) = 4/27 = 0.148 = 14.8\%$ Phe
$UCC = (2/3)(1/3)(1/3) = 2/27 = 0.0741 = 7.41\%$ Ser
$UCU = (2/3)(1/3)(2/3) = 4/27 = 0.148 = 14.8\%$ Ser
$CUU = (1/3)(2/3)(2/3) = 4/27 = 0.148 = 14.8\%$ Leu
$CUC = (1/3)(2/3)(1/3) = 2/27 = 0.0741 = 7.41\%$ Leu
$CCU = (1/3)(1/3)(2/3) = 2/27 = 0.0741 = 7.41\%$ Pro
$CCC = (1/3)(1/3)(1/3) = 1/27 = 0.037 = 3.7\%$ Pro

In sum, we have 44.4% Phe, 22.21% Ser, 22.21% Leu, and 11.11% Pro. (The total does not quite add up to 100%, because of rounding.)

b. The probability of a U at any position in a codon is $^1/_4$, the probability of a C at any position in a codon is $^1/_4$, and the probability of a G at any position in a codon is $^1/_2$. Thus, the codons, their relative frequencies, and the amino acids for which they code are as follows:

$UUU = (1/4)(1/4)(1/4) = 1/64 = 1.56\%$ Phe
$UUC = (1/4)(1/4)(1/4) = 1/64 = 1.56\%$ Phe
$UCU = (1/4)(1/4)(1/4) = 1/64 = 1.56\%$ Ser
$UCC = (1/4)(1/4)(1/4) = 1/64 = 1.56\%$ Ser
$CUU = (1/4)(1/4)(1/4) = 1/64 = 1.56\%$ Leu
$CUC = (1/4)(1/4)(1/4) = 1/64 = 1.56\%$ Leu
$CCU = (1/4)(1/4)(1/4) = 1/64 = 1.56\%$ Pro
$CCC = (1/4)(1/4)(1/4) = 1/64 = 1.56\%$ Pro
$UUG = (1/4)(1/4)(1/2) = 2/64 = 3.13\%$ Leu
$UGU = (1/4)(1/2)(1/4) = 2/64 = 3.13\%$ Cys
$UGG = (1/4)(1/2)(1/2) = 4/64 = 6.25\%$ Trp
$GUU = (1/2)(1/4)(1/4) = 2/64 = 3.13\%$ Val
$GUG = (1/2)(1/4)(1/2) = 4/64 = 6.25\%$ Val
$GGU = (1/2)(1/2)(1/4) = 4/64 = 6.25\%$ Gly
$GGG = (1/2)(1/2)(1/2) = 8/64 = 12.5\%$ Gly
$CCG = (1/4)(1/4)(1/2) = 2/64 = 3.13\%$ Pro
$CGC = (1/4)(1/2)(1/4) = 2/64 = 3.13\%$ Arg
$CGG = (1/4)(1/2)(1/2) = 4/64 = 6.25\%$ Arg
$GCC = (1/2)(1/4)(1/4) = 2/64 = 3.13\%$ Ala
$GCG = (1/2)(1/4)(1/2) = 4/64 = 6.25\%$ Ala
$GGC = (1/2)(1/2)(1/4) = 4/64 = 6.25\%$ Gly
$UCG = (1/4)(1/4)(1/2) = 2/64 = 3.13\%$ Ser
$UGC = (1/4)(1/2)(1/4) = 2/64 = 3.13\%$ Cys
$CUG = (1/4)(1/4)(1/2) = 2/64 = 3.13\%$ Leu
$CGU = (1/4)(1/2)(1/4) = 2/64 = 3.13\%$ Arg
$GUC = (1/2)(1/4)(1/4) = 2/64 = 3.13\%$ Val
$GCU = (1/2)(1/4)(1/4) = 2/64 = 3.13\%$ Ala

In sum, 3.12% Phe, 6.25% Ser, 9.38% Leu, 6.25% Pro, 6.26% Cys, 6.25% Trp, 12.51% Val, 25% Gly, 12.51% Arg, 12.51% Ala.

Questions and Problems

*6.1 The form of genetic information used directly in protein synthesis is (choose the correct answer):
a. DNA
b. mRNA
c. rRNA
d. tRNA

6.2 Most genes encode proteins. What exactly is a protein, structurally speaking? List some of the functions of proteins.

*6.3 In each of the following cases stating how a certain protein is treated, indicate what level(s) of protein structure would change as the result of the treatment:
a. Hemoglobin is stored in a hot incubator at 80°C.
b. Egg white (albumin) is boiled.
c. RNase (a single polypeptide enzyme) is heated to 100°C.
d. Meat in your stomach is digested (gastric juices contain proteolytic enzymes).
e. In the β polypeptide chain of hemoglobin, the amino acid valine replaces glutamic acid at the number-six position.

*6.4 Bovine spongiform encephalopathy (BSE) ("mad cow") disease and the human version, Creutzfeldt–Jakob disease (CJD), are characterized by the deposition of amyloid—insoluble, nonfunctional protein deposits—in the brain. In these diseases, amyloid deposits contain an abnormally folded version of the prion protein. Whereas the normal prion protein has lots of alpha-helical regions and is soluble, the abnormally folded version has α-helical regions converted into β-pleated sheets and is insoluble. Curiously, small amounts of the abnormally folded version can trigger the conversion of an α-helix to a β-pleated sheet in the normal protein, making the abnormally folded version infectious.

a. Some cases of CJD may have arisen from ingesting beef having tiny amounts of the abnormally folded protein. What would you expect to find if you examined the primary structure of the prion protein in the affected tissues? What levels of protein structural organization are affected in this form of prion disease?

b. Answer the questions posed in part (a) for cases of CJD in which susceptibility to CJD is inherited due to a rare mutation in the gene for the prion protein.

6.5 The structure and function of the rRNA and protein components of ribosomes have been investigated by separating those components from intact ribosomes and then using reconstitution experiments to determine which of the components are required for specific ribosomal activities.

a. Contrast the components of prokaryotic ribosomes with those of eukaryotic ribosomes.

b. What is the function of ribosomes, what steps are used by ribosomes to carry out that function, and which components of ribosomes are active in each step?

*6.6 What is the evidence that the rRNA component of the ribosome serves more than a structural role?

6.7 The term *genetic code* refers to the set of three-base code words (codons) in mRNA that stand for the 20 amino acids in proteins. What are the characteristics of the code?

6.8 If codons were four bases long, how many codons would exist in a genetic code?

*6.9 Base-pairing wobble occurs in the interaction between the anticodon of the tRNAs and the codons. On a theoretical level, determine the minimum number of tRNAs needed to read the 61 sense codons.

*6.10 Antibiotics have been highly useful in elucidating the steps of protein synthesis. If you have an artificial messenger RNA with the sequence AUGUUUUUUUUUUUUU ..., it will produce the following polypeptide in a cell-free protein-synthesizing system: fMet-Phe-Phe-Phe.... Suppose that, in your search for new antibiotics, you find one called putyermycin, which blocks protein synthesis. When you try it with your artificial mRNA in a cell-free system, the product is fMet-Phe. What step in protein synthesis does putyermycin affect? Why?

6.11 Describe the reactions involved in the aminoacylation (charging) of a tRNA molecule.

6.12 If the initiating codon of an mRNA were altered by mutation, what might be the effect on the transcript?

6.13 What differences are found in the initiation of protein synthesis between prokaryotes and eukaryotes? What differences are found in the termination of protein synthesis between prokaryotes and eukaryotes?

*6.14 In Chapter 5, we saw that eukaryotic mRNAs are posttranscriptionally modified at their 5′ and 3′ ends. What role does each of these modifications play in translation?

6.15 Translation usually initiates at an AUG codon near the 5′ end of an mRNA, but mRNAs often have multiple AUG triplets near their 5′ ends. How is the initiation AUG codon correctly identified in prokaryotes? How is it correctly identified in eukaryotes?

6.16 Energy is required during multiple steps in translation. What are the sources of this energy, which steps require energy for their completion, and what is energy used for during these steps?

*6.17 Random copolymers were used in some of the experiments that revealed the characteristics of the genetic code. For each of the following ribonucleotide mixtures, give the expected codons and their frequencies, and give the expected proportions of the amino acids that would be found in a polypeptide directed by the copolymer in a cell-free protein-synthesizing system:

a. 4 A : 6 C
b. 4 G : 1 C
c. 1 A : 3 U : 1 C
d. 1 A : 1 U : 1 G : 1 C

*6.18 What would the minimum word (codon) size need to be if, instead of four, the number of different bases in mRNA were

a. two
b. three
c. five

6.19 Suppose that, at stage A in the evolution of the genetic code, only the first two nucleotides in the coding triplets led to unique differences and that any nucleotide could occupy the third position. Then, suppose there was a stage B in which differences in meaning arose, depending on whether a purine (A or G) or pyrimidine (C or U) was present at the third position. Without reference to the number of amino acids or the multiplicity of tRNA molecules, how many triplets of different meaning can be constructed out of the code at stage A? at stage B?

*6.20 A gene encodes a polypeptide 30 amino acids long containing an alternating sequence of phenylalanine and tyrosine. What are the sequences of nucleotides corresponding to this sequence in each of the following:

a. the DNA strand that is read to produce the mRNA, assuming that Phe = UUU and Tyr = UAU in mRNA

b. the DNA strand that is not read

c. tRNAs

6.21 A segment of a polypeptide chain is Arg-Gly-Ser-Phe-Val-Asp-Arg. It is encoded by the following segment of DNA:

```
—G G C T A G C T G C T T C C T T G G G G A—
 | | | | | | | | | | | | | | | | | | | | |
—C C G A T C G A C G A A G G A A C C C C T—
```

Which strand is the template strand? Label each strand with its correct polarity (5′ and 3′).

*6.22 Two populations of RNAs are made by the random combination of nucleotides. In population A the RNAs contain only A and G nucleotides (3 A : 1 G), whereas in population B the RNAs contain only A and U nucleotides (3 A : 1 U). In what ways other than amino-acid content will the proteins produced by translating the population A RNAs differ from those produced by translating the population B RNAs?

6.23 In *E. coli*, a particular tRNA normally has the anticodon 5′-GGG-3′, but because of a mutation in the tRNA gene, the tRNA has the anticodon 5′-GGA-3′.
a. What codon would the normal tRNA recognize?
b. What codon would the mutant tRNA recognize?

*6.24 A protein found in *E. coli* normally has the N-terminal amino-acid sequence Met-Val-Ser-Ser-Pro-Met-Gly-Ala-Ala-Met-Ser.... A mutation alters the anticodon of a tRNA from 5′-GAU-3′ to 5′-CAU-3′. What would be the N-terminal amino-acid sequence of this protein in the mutant cell? Explain your reasoning.

6.25 The gene encoding an *E. coli* tRNA containing the anticodon 5′-GUA-3′ mutates so that the anticodon is now 5′-UUA-3′. What will be the effect of this mutation? Explain your reasoning.

*6.26 The following diagram shows the normal sequence of the coding region of an mRNA, along with six mutant versions of the same mRNA:

```
Normal     AUGUUCUCUAAUUAC(...)AUGGGGUGGGUGUAG
Mutant a   AUGUUCUCUAAUUAG(...)AUGGGGUGGGUGUAG
Mutant b   AGGUUCUCUAAUUAC(...)AUGGGGUGGGUGUAG
Mutant c   AUGUUCUCGAAUUAC(...)AUGGGGUGGGUGUAG
Mutant d   AUGUUCUCUAAAUAC(...)AUGGGGUGGGUGUAG
Mutant e   AUGUUCUCUAAUUUC(...)AUGGGGUGGGUGUAG
Mutant f   AUGUUCUCUAAUUAC(...)AUGGGGUGGGUGUGG
```

Indicate what protein would be formed in each case, where (. . .) denotes a multiple of three unspecified bases.

6.27 The following diagram shows the normal sequence of a particular protein, along with several mutant versions of it:

```
Normal     Met-Gly-Glu-Thr-Lys-Val-Val-...-Pro
Mutant 1   Met-Gly
Mutant 2   Met-Gly-Glu-Asp
Mutant 3   Met-Gly-Arg-Leu-Lys
Mutant 4   Met-Arg-Glu-Thr-Lys-Val-Val-...-Pro
```

For each mutant, explain what mutation occurred in the coding sequence of the gene, where (. . .) denotes a multiple of three unspecified bases.

6.28 The N-terminus of a protein has the sequence Met-His-Arg-Arg-Lys-Val-His-Gly-Gly. A molecular biologist wants to synthesize a DNA chain that can encode this portion of the protein. How many DNA sequences can encode this polypeptide?

6.29 In the recessive condition in humans known as sickle-cell anemia, the β-globin polypeptide of hemoglobin is found to be abnormal. The only difference between it and the normal β-globin is that the sixth amino acid from the N-terminal end is valine, whereas the normal β-globin has glutamic acid at this position. Explain how this amino acid substitution occurred in terms of differences in the DNA and the mRNA.

*6.30 Cystic fibrosis is an autosomal recessive disease in which the cystic fibrosis transmembrane conductance regulator (CFTR) protein is abnormal. The transcribed portion of the cystic fibrosis gene spans about 250,000 base pairs of DNA. The CFTR protein, with 1,480 amino acids, is translated from an mRNA of about 6,500 bases. The most common mutation in this gene results in a protein that is missing a phenylalanine at position 508 (ΔF508).
a. Why is the RNA coding sequence of this gene so much larger than the mRNA from which the CFTR protein is translated?
b. About what percentage of the mRNA together makes up 5′ untranslated leader, and 3′ untranslated trailer, sequences?
c. At the DNA level, what alteration would you expect to find in the ΔF508 mutation?
d. What consequences might you expect if the DNA alteration you describe in (c) occurred at random in the protein-coding region of the cystic fibrosis gene?

*6.31 Antibiotics have been useful in determining whether cellular events depend on transcription or translation. For example, actinomycin D is used to block transcription, and cycloheximide (in eukaryotes) is used to block translation. In some cases, though, surprising results are obtained after antibiotics are administered. Adding actinomycin D, for example, may result in an increase, not a decrease, in the activity of a particular enzyme. Discuss how this result might come about.

7

DNA Mutation, DNA Repair, and Transposable Elements

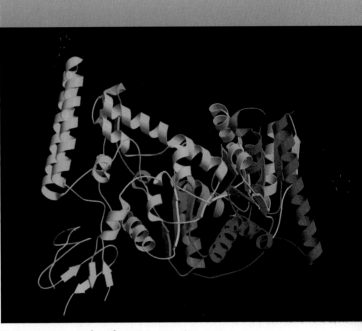

UvrB protein; a nucleotide excision repair enzyme.

PRINCIPAL POINTS

- Changes in heritable traits result from random mutation.

- Mutation is the process by which the sequence of base pairs in a DNA molecule is altered. The alteration can be as simple as a single base-pair substitution, insertion, or deletion or as complex as rearrangement, duplication, or deletion of whole sections of a chromosome. Mutations may occur spontaneously, including by the effects of natural radiation or errors in replication, or they may be induced experimentally by the application of mutagens.

- Mutations at the level of the chromosome are called chromosomal mutations. (See Chapter 17.) Mutations in the sequences of genes and in other DNA sequences at the level of the base pair are called point mutations.

- The consequences to an organism of a mutation in a gene depend on a number of factors, especially the extent to which the amino-acid coding information for a protein is changed.

- By studying mutants that have defects in certain cellular processes, geneticists have made great progress in understanding how those processes take place. A number of screening procedures have been developed to help find mutants of interest after mutagenizing cells or organisms.

- The effects of a gene mutation can be reversed either by reversion of the base-pair sequence to its original state or by a mutation at a site distinct from that of the original mutation. The latter is called a suppressor mutation.

- High-energy radiation may cause damage to genetic material by producing chemicals that interact with DNA or by causing unusual bonds between DNA bases. Mutations result if the genetic damage is not repaired. Ionizing radiation may also break chromosomes.

- Gene mutations may be caused by exposure to a variety of chemicals called chemical mutagens, a number of which exist in the environment and can cause genetic diseases in humans and other organisms.

- In prokaryotes and eukaryotes, a number of enzymes repair different kinds of DNA damage. Not all DNA damage is repaired; therefore, mutations do appear, but at low frequencies. At high dosages of mutagens, repair systems cannot correct all of the damage, and cancer (in the case of eukaryotic, multicellular organisms) or cell death can result.

- Transposable elements are DNA segments that can insert themselves at one or more sites in a genome. Transposable elements in a cell usually are detected by the changes they bring about in the expression and activities of the genes at or near the chromosomal sites into which they integrate.

- In bacteria, two important types of transposable elements are insertion sequence (IS) elements and transposons (Tn), each of which has repeated sequences at its ends and encode proteins, such as transposases, that are responsible for its transposition. Transposons also carry genes that encode other functions, such as drug resistance.

- Many transposable elements in eukaryotes resemble bacterial transposons in both general structure and transposition properties. Transposons may transpose either while leaving a copy behind in the original site or by excision from the chromosome. Transposons integrate at a target site by a precise mechanism, so that the integrated elements are flanked at the insertion site by a short duplication of target-site DNA. Some transposons are autonomous elements that can direct their own transposition, and some are nonautonomous elements that can transpose only when activated by an autonomous element in the same genome. Although most transposons move by means of a DNA-to-DNA mechanism, some eukaryotic transposons move via an RNA intermediate (using a transposon-encoded reverse transcriptase). Such transposons resemble retroviruses in genome organization and other properties.

i A MUTATION IN A GENE CAN LEAD TO A CHANGE IN A phenotype. What types of mutations can occur in our DNA? And what effect do DNA mutations have on our health? In one iActivity for this chapter you are investigating the possible health hazards associated with contaminated ground water. In a second iActivity we will look at another way that DNA can change. In the 1940's Barbara McClintock found that "jumping genes", or transposable elements, can create gene mutations, affect gene expression, and produce various types of chromosome mutations. In this iActivity, you will have the opportunity to explore further how a transposable element in *E. coli* moves from one location to another.

DNA can be changed in a number of ways—through spontaneous changes, errors in the replication process, or the action of particular chemicals or radiation, for example. We considered **chromosomal mutations**—changes involving whole chromosomes or sections of them—in Chapter 17. Another broad type of change in the genetic material is the **point mutation,** a change of one or a few base pairs. Unless it occurs within a gene or in the sequences regulating the gene, a point mutation will not change the phenotype of the organism. Thus, the point mutations that have been of particular interest to geneticists are **gene mutations**—those which affect the function of genes. A gene mutation can alter the phenotype by changing the function of a protein, as illustrated in Figure 7.1. In this chapter, you will learn about some of the mechanisms that cause point mutations, some of the repair systems that can fix genetic damage, and some of the methods used to detect genetic mutants. As you learn about the specifics of point mutations, you should be aware that mutations are a major source of genetic variation in a species and therefore are important elements of the evolutionary process.

Genetic change also can occur when certain genetic elements in the chromosomes of prokaryotes and eukaryotes move from one location to another in the genome. These mobile genetic elements are known as **transposable elements,** because the term reflects the **transposition** (change in position) events associated with them. The discovery of transposable elements was a great surprise that altered our classic picture of genes and genomes and brought to our attention a new phenomenon to consider in developing theories about the evolution of genomes. In this chapter, we learn about the nature of transposable elements and about how they move.

DNA Mutation

Adaptation versus Mutation

Although we currently know that the variation in heritable traits result from mutations, this was not always the case. In the early part of the twentieth century, there were two opposing schools of thought. Some geneticists believed that variation among organisms resulted from

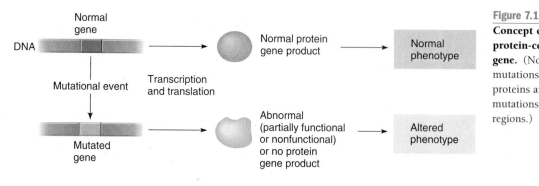

Figure 7.1

Concept of a mutation in the protein-coding region of a gene. (Note that not all mutations lead to altered proteins and that not all mutations are in protein-coding regions.)

random mutations which sometimes happened to be adaptive. Others believed that variations resulted from *adaptation*; that is, they that held the environment induced an adaptive inheritable change. The adaptation theory was based on Lamarckism, which is the doctrine of the inheritance of acquired characteristics. Some observations with bacteria fueled the debate. Wild-type *E. coli*, for example, is sensitive to the virulent bacteriophage T1. If a culture of wild-type *E. coli* started from a single cell is plated in the presence of an excess of phage T1, most of the bacteria are killed. However, a very few survive and produce colonies, because they are resistant to infection by T1. The resistance trait is heritable and results from a change in the bacterial cell surface that prevents the phages from binding to it. Supporters of the adaptation theory argued that the resistance trait arose as a result of the presence of the T1 phage in the environment. In the opposite camp, supporters of the mutation theory argued that mutations occur randomly, so that, at any time in a large enough population of cells, some cells have undergone a mutation that makes them resistant to T1 (in the example at hand), even though they have never been exposed to the bacteriophage. Thus, when T1 is

subsequently added to the culture, the T1-resistant bacteria are selected for.

The acquisition of resistance to T1 was used by Salvador Luria and Max Delbrück in 1943 to demonstrate that the mutation mechanism was correct and the adaptation mechanism was incorrect. The test they designed is known as the *fluctuation test* and is based on the idea that the different mechanisms would give rise to observably different proportions of resistant bacteria within a sample of cultures.

For example, consider a dividing population of wild-type *E. coli* that started with a single cell (Figure 7.2). Assume that phage T1 is added at generation 4, when there are 16 cells. (This number is for illustration; in the actual experiment, the number of cells was much higher.) If the adaptation theory is correct, a certain proportion of the generation-4 cells will be induced *at that time* to become resistant to T1 (Figure 7.2a). Most importantly, *that proportion will be the same for all identical cultures, because adaptation would not commence until T1 was added.* However, if the mutation theory is correct, then the number of generation-4 cells that are resistant to T1 depends on when in the culturing process the random mutational event

Figure 7.2

Representation of a dividing population of T1 phage-sensitive wild-type *E. coli*. At generation 4, T1 phage is added. (a) If the adaptation theory is correct, cells mutate only when T1 phage is added, so the proportions of resistant cells in duplicate cultures are the same. (b) If the mutation theory is correct, cells mutate independently of when T1 phage is added, so the proportions of resistant cells in duplicate cultures are different. *Left:* If one cell mutates to become resistant to T1 phage infection at generation 3, then 2 of the 16 cells at generation 4 are resistant to T1. *Right:* If one cell mutates to become resistant to T1 phage infection at generation 1, then 8 of the 16 cells at generation 4 are resistant to T1.

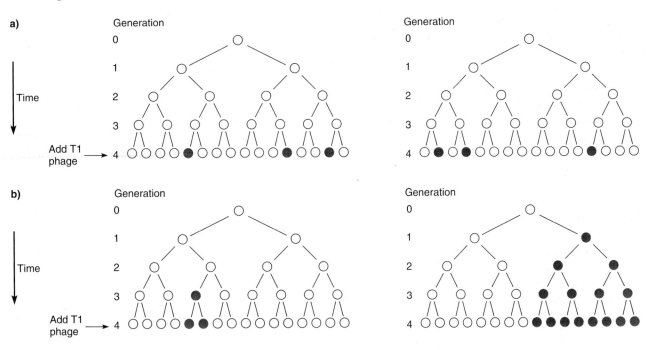

occurred that confers resistance to T1. If the mutational event occurs in generation 3 in our example, then 2 of the 16 cells in generation 4 will be T1 resistant (Figure 7.2b). However, if the mutational event occurs instead at generation 1, then 8 of the 16 generation-4 cells will be T1 resistant (Figure 7.2b). The key point is that if the mutation theory is correct, there should be a *fluctuation in the number of T1-resistant cells in generation 4 because the mutation to T1 resistance occurred randomly in the population and did not require the presence of T1.*

Luria and Delbrück observed a large range in the number of resistant colonies among identical cultures. This high degree of fluctuation in the number of resistant bacteria was taken as proof that resistance resulted from random mutation rather than adaptation.

KEYNOTE

Heritable adaptive traits result from random mutation, rather than by adaptation as a result of induction by environmental influences.

Mutations Defined

Mutation is the process by which the sequence of base pairs in a DNA molecule is altered. A mutation, then, is a change in either a DNA base pair or a chromosome.

A cell with a mutation is a mutant cell. If a mutation happens to occur in a somatic cell (in multicellular organisms), the mutant characteristic affects only the individual in which the mutation occurs and is not passed on to the succeeding generation. This type of mutation is called a **somatic mutation.** In contrast, mutations in the germ line of sexually reproducing organisms may be transmitted by the gametes to the next generation, producing an individual with the mutation in both its somatic and its germ-line cells. Such mutations are called **germ-line mutations.**

Two different terms are used to give a quantitative measure of the occurrence of mutations. The **mutation rate** is the probability of a particular kind of mutation as a function of time, such as the number of mutations per nucleotide pair per generation or the number per gene per generation. The **mutation frequency** is the number of occurrences of a particular kind of mutation, expressed as the proportion of cells or individuals in a population, such as the number of mutations per 100,000 organisms or the number per 1 million gametes.

Types of Point Mutations. Point mutations—those mutations which change only one or a few base pairs—can be divided into two general categories: base-pair substitutions and base-pair insertions or deletions. A **base-pair substitution mutation** involves a change in the DNA such that one base pair is replaced by another. There are two general types of base-pair substitution mutations. A **transition mutation**

(Figure 7.3a) is a mutation from one purine–pyrimidine base pair to the other purine–pyrimidine base pair. The four types of transition mutations are AT to GC, GC to AT, TA to CG, and CG to TA. A **transversion mutation** (Figure 7.3b) is a mutation from a purine–pyrimidine base pair to a pyrimidine– purine base pair. The eight types of transversion mutations are AT to TA, TA to AT, GC to CG, CG to GC, AT to CG, CG to AT, GC to TA, and CG to AT.

Base-pair substitutions in protein-coding genes can also be defined according to their effects on amino-acid sequences in proteins. Depending on how a base-pair substitution is translated via the genetic code, the mutations can result in no change to the protein, an insignificant change, or a noticeable change. (You can explore the effects of these mutations further with the Creating DNA Mutations learning activity.)

A **missense mutation** (Figure 7.3c) is a gene mutation in which a base-pair change in the DNA causes a change in an mRNA codon so that a different amino acid is inserted into the polypeptide. A phenotypic change may or may not result, depending on the amino-acid change involved. In Figure 7.3c, an AT-to-GC transition mutation changes the DNA from $\frac{5'\text{-AAA-}3'}{3'\text{-TTT-}5'}$ to $\frac{5'\text{-GAA-}3'}{3'\text{-CTT-}5'}$ altering the mRNA codon from 5′-AAA-3′ (lysine) to 5′-GAA-3′ (glutamic acid).

A **nonsense mutation** (Figure 7.3d) is a gene mutation in which a base-pair change in the DNA changes an mRNA codon for an amino acid to a stop (nonsense) codon (UAG, UAA, or UGA). For example, in Figure 7.3d, an AT-to-TA transversion mutation changes the DNA from $\frac{5'\text{-AAA-}3'}{3'\text{-TTT-}5'}$ to $\frac{5'\text{-TAA-}3'}{3'\text{-ATT-}5'}$ and this in turn changes the mRNA codon from 5′-AAA-3′ (lysine) to 5′-UAA-3′, which is a nonsense codon. A nonsense mutation causes premature chain termination, so instead of complete polypeptides, shorter-than-normal polypeptide fragments (often nonfunctional) are released from the ribosomes (Figure 7.4).

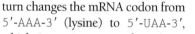

animation

ⓐ Nonsense Mutations and Nonsense Suppressor Mutations

A **neutral mutation** (Figure 7.3e) is a base-pair change in a gene that changes a codon in the mRNA such that the resulting amino-acid substitution produces no detectable change in the *function* of the protein translated from that message. A neutral mutation is a subset of missense mutations in which the new codon codes for a different amino acid that is chemically equivalent to the original and therefore does not affect the protein's function. Consequently, the phenotype does not change. In Figure 7.3e, an AT-to-GC transition mutation changes the codon from 5′-AAA-3′ to 5′-AGA-3′, which substitutes the basic amino acid arginine for the basic amino acid lysine. Because arginine

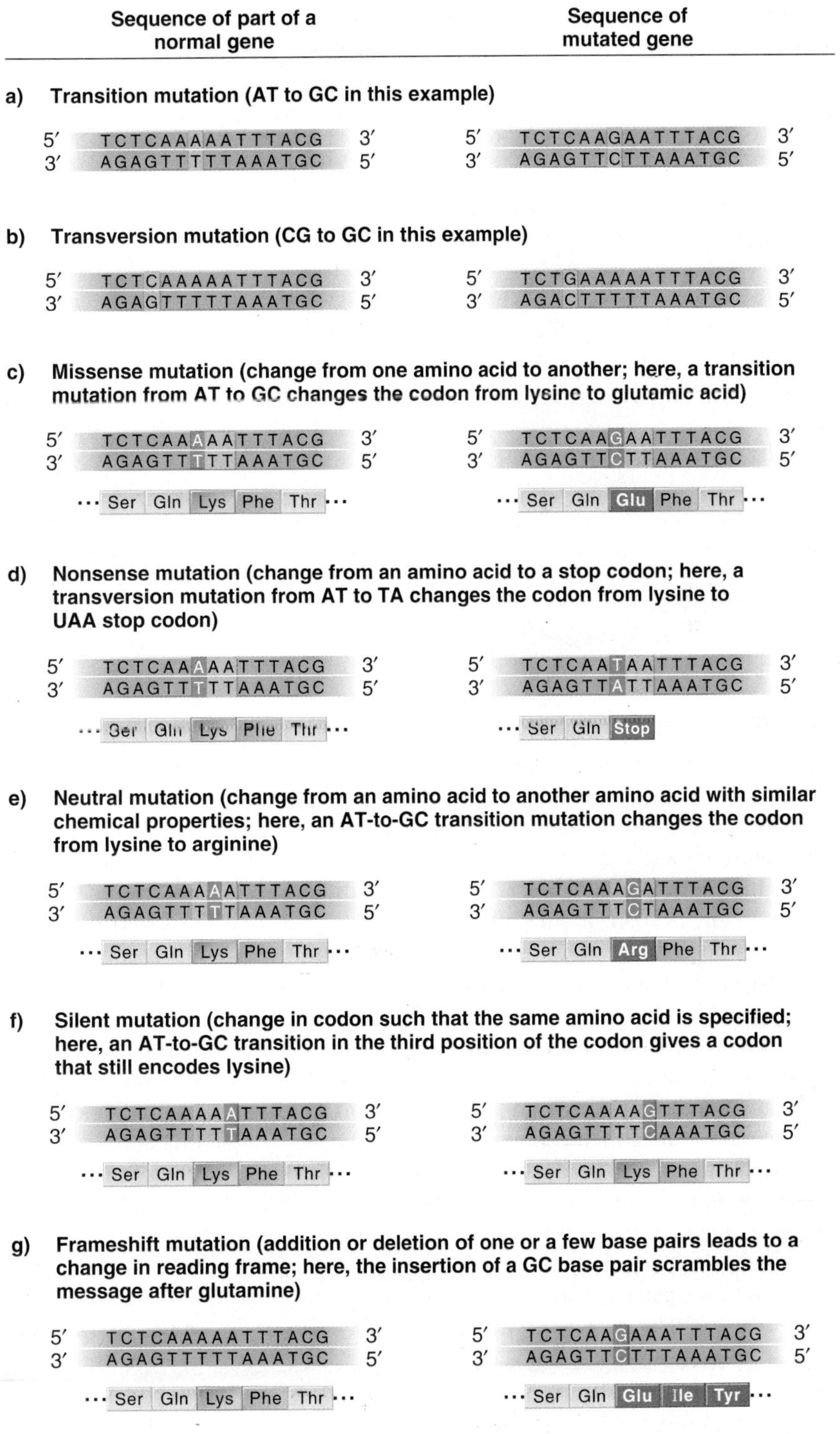

Figure 7.3

Types of base-pair substitution mutations. Transcription of the segment shown produces an mRNA with the sequence 5′...UCUCAAAAAAUUUACG...3′, which encodes ...-Ser-Gln-Lys-Phe-Thr-...

Sequence of part of a normal gene	Sequence of mutated gene

a) Transition mutation (AT to GC in this example)

```
5′  TCTCAAAAATTTACG  3′        5′  TCTCAAGAATTTACG  3′
3′  AGAGTTTTTAAATGC  5′        3′  AGAGTTCTTAAATGC  5′
```

b) Transversion mutation (CG to GC in this example)

```
5′  TCTCAAAAATTTACG  3′        5′  TCTGAAAAATTTACG  3′
3′  AGAGTTTTTAAATGC  5′        3′  AGACTTTTTAAATGC  5′
```

c) Missense mutation (change from one amino acid to another; here, a transition mutation from AT to GC changes the codon from lysine to glutamic acid)

```
5′  TCTCAAAAATTTACG  3′        5′  TCTCAAGAATTTACG  3′
3′  AGAGTTTTTAAATGC  5′        3′  AGAGTTCTTAAATGC  5′

 ··· Ser Gln Lys Phe Thr ···      ··· Ser Gln Glu Phe Thr ···
```

d) Nonsense mutation (change from an amino acid to a stop codon; here, a transversion mutation from AT to TA changes the codon from lysine to UAA stop codon)

```
5′  TCTCAAAAATTTACG  3′        5′  TCTCAATAATTTACG  3′
3′  AGAGTTTTTAAATGC  5′        3′  AGAGTTATTAAATGC  5′

 ··· Ser Gln Lys Phe Thr ···      ··· Ser Gln Stop
```

e) Neutral mutation (change from an amino acid to another amino acid with similar chemical properties; here, an AT-to-GC transition mutation changes the codon from lysine to arginine)

```
5′  TCTCAAAAATTTACG  3′        5′  TCTCAAAGATTTACG  3′
3′  AGAGTTTTTAAATGC  5′        3′  AGAGTTTCTAAATGC  5′

 ··· Ser Gln Lys Phe Thr ···      ··· Ser Gln Arg Phe Thr ···
```

f) Silent mutation (change in codon such that the same amino acid is specified; here, an AT-to-GC transition in the third position of the codon gives a codon that still encodes lysine)

```
5′  TCTCAAAAATTTACG  3′        5′  TCTCAAAAGTTTACG  3′
3′  AGAGTTTTTAAATGC  5′        3′  AGAGTTTTCAAATGC  5′

 ··· Ser Gln Lys Phe Thr ···      ··· Ser Gln Lys Phe Thr ···
```

g) Frameshift mutation (addition or deletion of one or a few base pairs leads to a change in reading frame; here, the insertion of a GC base pair scrambles the message after glutamine)

```
5′  TCTCAAAAATTTACG  3′        5′  TCTCAAGAAATTTACG  3′
3′  AGAGTTTTTAAATGC  5′        3′  AGAGTTCTTTAAATGC  5′

 ··· Ser Gln Lys Phe Thr ···      ··· Ser Gln Glu Ile Tyr ···
```

and lysine have similar properties, the protein's function may not be altered significantly.

A **silent mutation** (Figure 7.3f) is also a subset of missense mutations that occurs when a base-pair change in a gene alters a codon in the mRNA such that the *same* amino acid is inserted in the protein. In this case, the protein obviously has a wild-type function. For example, in Figure 7.3f, a silent mutation results from an AT-to-GC

transition mutation that changes the codon from 5′-AAA-3′ to 5′-AAG-3′, both of which specify lysine.

If one or more base pairs are added to or deleted from a protein-coding gene, the reading frame of an mRNA can change downstream of the mutation. An addition or deletion of one base pair, for example, shifts the mRNA's downstream reading frame by one base, so that incorrect amino acids are added to the polypeptide chain

Figure 7.4

A nonsense mutation and its effect on translation.

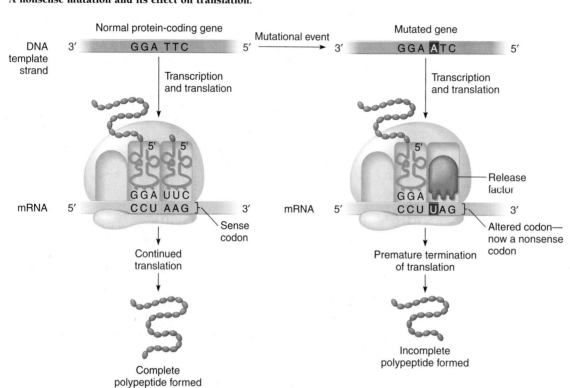

after the mutation site. This type of mutation, called a **frameshift mutation** (Figure 7.3g), usually results in a nonfunctional protein. Frameshift mutations may generate new stop codons, resulting in a shortened protein; they may result in a read-through of the normal stop codon, resulting in longer-than-normal proteins; or they may result in a complete alteration of the amino-acid sequence of a protein. In Figure 7.3g, an insertion of a GC base pair scrambles the message after the codon specifying glutamine. Since each codon consists of three bases, a frameshift mutation is produced by the insertion or deletion of any number of base pairs in the DNA that is not divisible by three. Frameshift mutations were instrumental in scientists' determining that the genetic code is a triplet code. (See Chapter 6, pp. 115–116.)

KEYNOTE

> Mutation is the process by which the sequence of base pairs in a DNA molecule is altered. Mutations that affect a single base pair of DNA are called base-pair substitution mutations or point mutations. Mutations in the sequences of genes are called gene mutations.

Reverse Mutations and Suppressor Mutations. There are two classes of point mutations in terms of their effects on the phenotype: (1) **Forward mutations** change

the genotype from wild type to mutant; and (2) **reverse mutations** (or **reversions** or **back mutations**) change the genotype from mutant to wild type or to partially wild type. Reversion of a nonsense mutation, for instance, occurs when a base-pair change results in a change of the mRNA nonsense codon to a codon for an amino acid. If this reversion is back to the wild-type amino acid, the mutation is a **true reversion.** If the reversion is to some other amino acid, the mutation is a **partial reversion,** and complete or partial function may be restored, depending on the change. Reversion of missense mutations occurs in the same way.

The effects of a mutation may be diminished or abolished by a **suppressor mutation**—a mutation at a different site from that of the original mutation. A suppressor mutation masks or compensates for the effects of the initial mutation, but it does not reverse the original mutation.

Suppressor mutations may occur within the same gene in which the original mutations occurred, but at a different site (called **intragenic** [*intra* = within] **suppressors**), or they may occur in a different gene (called **intergenic** [*inter* = between] **suppressors**). Both intragenic and intergenic suppressors operate to allow the production of functional or partially functional copies of the protein that was initially rendered inactive by the original mutation. However, the mechanisms of the two suppressors are completely different.

Intragenic suppressors act by altering a different nucleotide in the same codon in which the original mutation occurred or by altering a nucleotide in a different codon. An example of the latter is the suppression of a base-pair addition frameshift mutation by a nearby base-pair deletion. (See Chapter 6, Figure 6.5)

Intergenic suppression occurs as a result of a second mutation in another gene. Genes that cause the suppression of mutations in other genes are called **suppressor genes.** Many suppressor genes work by changing the way the mRNA is read. Each suppressor gene can suppress the effects of only one type of nonsense, missense, or frameshift mutation; therefore, suppressor genes can suppress just a small proportion of the point mutations that theoretically can occur within a gene. By contrast, a given suppressor gene suppresses all mutations to which it is specific—whatever gene the mutation is in—if the resulting amino-acid substitution is compatible with the functionality of the affected protein.

Suppressor genes often encode tRNAs. In the case of nonsense suppressors, particular tRNA genes can mutate so that (in contrast to what occurs with wild-type tRNAs) their anticodons recognize a chain-terminating codon and put an amino acid into the chain. Thus, instead of polypeptide chain synthesis being stopped prematurely as a result of a nonsense mutation, the altered (suppressor) tRNA inserts an amino acid at that position, and full or partial function of the polypeptide is restored. This suppression process is not very efficient, but efficient enough for sufficient functional polypeptides to be produced to reverse the phenotype.

There are three classes of nonsense suppressors, one for each of the stop codons UAG, UAA, and UGA. For example, if a gene for a tyrosine tRNA (which has the anticodon 3'-AUG-5') is mutated so that the tRNA has the anticodon 3'-AUC-5', the mutated suppressor tRNA (which still carries tyrosine) reads the nonsense codon 5'-UAG-3'. So, instead of chain termination occurring, tyrosine is inserted at that point in the polypeptide. The process is shown in Figure 7.5.

But we now have a dilemma: If this tRNA.Tyr gene has mutated so that the tRNA's anticodon can read a nonsense codon, it can no longer read the original codon which specifies the amino acid it carries. The solution comes from the discovery that nonsense suppressor tRNA genes typically are produced by mutations of tRNA genes that are redundant in the genome. In other words, two or more different genes code for the *same* species of tRNA.Tyr. If there is a mutation in one of these redundant genes (usually one that codes for minor amounts of

Figure 7.5

Mechanism of action of an intergenic nonsense-suppressor mutation that results from the mutation of a tRNA gene. In this example, a tRNA.Tyr gene has mutated so that the tRNA's anticodon is changed from 3'-AUG-5' to 3'-AUC-5', which can read a UAG nonsense codon, inserting tyrosine in the polypeptide chain at that codon.

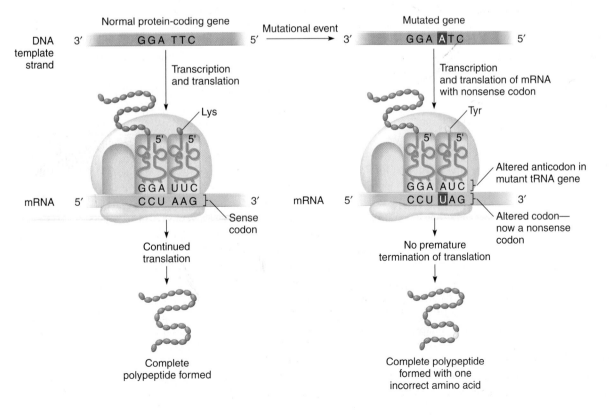

tRNA), so that the tRNA reads, say, UAG, then the other gene(s) specifying the same tRNA produce(s) a tRNA molecule that reads the normal Tyr codon.

Another dilemma is that the mutated tRNA.Tyr gene will also read normal stop codons, leading to longer-than-normal polypeptides. The inefficiency of nonsense suppression by the aforementioned suppressor tRNAs means that most polypeptides are of the correct length, and rarely, if at all, do the small numbers of longer-than-normal polypeptides have phenotypic consequences.

KEYNOTE

> Reverse mutations are mutations that cause the genotype to change from mutant to wild type. A suppressor mutation is a mutation at a second site that completely or partially restores a function which was lost because of a primary mutation. Intragenic suppressors are suppressor mutations that occur within the same gene as that in which the original mutation occurred, but at a different site. Intergenic suppressors are suppressor mutations that occur in a suppressor gene—a gene different from that with the original mutation.

Spontaneous and Induced Mutations

Mutagenesis, the creation of mutations, can occur spontaneously or can be induced. **Spontaneous mutations** are naturally occurring mutations. **Induced mutations** occur when an organism is exposed either deliberately or accidentally to a physical or chemical agent, known as a **mutagen,** that interacts with DNA to cause a mutation. Induced mutations typically occur at a much higher frequency than spontaneous mutations.

Spontaneous Mutations. All types of point mutations can occur spontaneously. Spontaneous mutations can occur during DNA replication, as well as during the G_1 and G_2 phases of the cell cycle. Spontaneous mutations can also result from the movement of transposable genetic elements, a topic we discuss later in the chapter.

In humans, the spontaneous mutation rate for individual genes varies between 10^{-4} and 4×10^{-6} per gene per generation. For eukaryotes in general, the spontaneous mutation rate is 10^{-4} to 10^{-6} per gene per generation, for bacteria and phages, 10^{-5} to 10^{-7} per gene per generation. (The spontaneous mutation frequencies at specific loci for various organisms are presented in Table 24.6.) Furthermore, note that *the rates and frequencies just mentioned represent the mutations that become fixed in DNA.* Most spontaneous errors are corrected by cellular repair systems, which we discuss later in this chapter; only some remain uncorrected as permanent changes.

DNA Replication Errors. *Base-pair substitution mutations*—point mutations involving a change from one base pair to

another—can occur if mismatched base pairs form during DNA replication. Chemically, each base can exist in alternative states, called **tautomers.** When a base changes state, it is said to have undergone a *tautomeric shift.* In DNA, the *keto* form of each base is usually found and is responsible for the normal Watson–Crick base pairing of T with A and C with G (Figure 7.6a). However, non-Watson–Crick base pairing can result if a base is in a rare tautomeric state, called the *enol* form. Figure 7.6b and Figure 7.6c respectively show mismatched base pairs that can occur if purines are in their rare tautomeric states or if pyrimidines are in their rare tautomeric states.

Figure 7.7 illustrates how a mutation can be produced as a result of a mismatch caused by a base shifting to a rare tautomeric. In this example, the rare form of T forms a mismatched base pair with G in the template strand of the DNA. If this mismatch escapes detection by the replication proofreading machinery and by other repair systems, a GC-to-AT transition mutation will result in the subsequent replication cycle.

Small additions and deletions also can occur spontaneously during replication (Figure 7.8). The mechanism for introducing such errors involves the displacement—looping out—of bases from either the template or the growing DNA strand, generally in regions where a run of the same base is present. If DNA loops out from the template strand, DNA polymerase skips the looped-out base or bases, and a deletion mutation results; if DNA polymerase synthesizes an untemplated base or bases, the new DNA loops out from the template, and an addition mutation results. If the additions or deletions occur in the coding region of a structural gene, the errors generate frameshift mutations. (An exception is that three additions or three deletions occurring close together restore the reading frame.)

Spontaneous Chemical Changes. Depurination and deamination of particular bases are two of the most common chemical events that produce spontaneous mutations. These events create lesions, which are damaged sites in the DNA. In **depurination,** a purine, either adenine or guanine, is removed from the DNA when the bond breaks between the base and the deoxyribose sugar, resulting in an *apurinic site.* A mammalian cell typically loses thousands of purines in an average cell generation period. If such lesions are not repaired, there is no base to specify a complementary base during DNA replication, and the DNA polymerase may stall or dissociate from the DNA.

Deamination is the removal of an amino group from a base. For example, the deamination of cytosine produces uracil (Figure 7.9a), which is not a normal base in DNA, although it is a normal base in RNA. A repair system described later in the chapter replaces most of the uracils in DNA, thereby minimizing the mutational consequences of cytosine deamination. However, if the uracil is not replaced, an adenine will be incorporated into the

Figure 7.6

Normal Watson–Crick and non–Watson–Crick base pairing in DNA.
(a) Normal Watson–Crick base pairing between the normal forms of T and A and of C and G. (b) Non–Watson–Crick base pairing between normal forms of pyrimidines and rare tautomeric forms of purines. (c) Non–Watson–Crick base pairing between rare tautomeric forms of pyrimidines and normal forms of purines.

a) Normal Watson-Crick base pairing

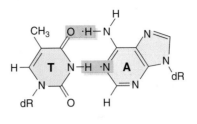
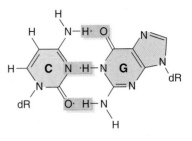

b) Non-Watson-Crick base pairing between normal pyrimidines and rare purines

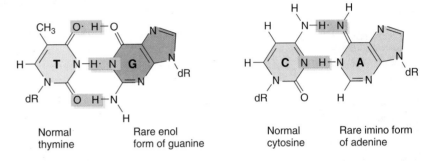

Normal thymine   Rare enol form of guanine   Normal cytosine   Rare imino form of adenine

c) Non-Watson-Crick base pairing between rare pyrimidines and normal purines

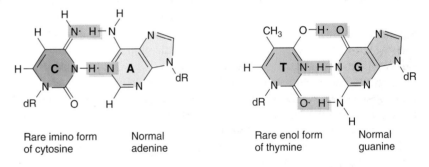

Rare imino form of cytosine   Normal adenine   Rare enol form of thymine   Normal guanine

new DNA strand opposite it during replication. Ultimately, this results in the conversion of a CG base pair to a TA base pair—that is, a transition mutation.

DNA of both prokaryotes and eukaryotes contains small amounts of the modified base 5-methylcytosine (5^mC) (Figure 7.9b) in place of the normal base cytosine. Deamination of 5^mC produces thymine (Figure 7.9b), a normal nucleotide in DNA, so there are no repair mechanisms that can detect and correct such a mutation. As a consequence, deamination of 5^mC results in CG-to-TA transitions. Because significant proportions of other kinds of mutations are corrected by repair mechanisms, but 5^mC deamination mutations are not, locations of 5^mC in the genome often appear as *mutational hot spots*—that is, nucleotides where a higher-than-average frequency of mutation occurs.

Induced Mutations. Mutations can be induced by exposing organisms to physical mutagens, such as radiation, or chemical mutagens. Deliberately induced mutations have played, and continue to play, an important role in the study of mutations. Since the rate of spontaneous muta-

tion is so low, geneticists use mutagens to increase the frequency of mutation so that a significant number of organisms have mutations in the gene being studied.

Radiation. Both X rays and ultraviolet light (UV) induce mutations. Hermann Joseph Müller received the 1946 Nobel Prize in Physiology or Medicine for "the discovery of the production of mutations by means of X-ray irradiation." A form of ionizing radiation, X rays penetrate tissue and collide with molecules, knocking electrons out of orbits, thereby creating ions. The ions can break covalent bonds, including those in the sugar–phosphate backbone of DNA. In fact, ionizing radiation is the leading cause of gross chromosomal mutations in humans.

High dosages of ionizing radiation kill cells—hence their use in treating some forms of cancer. At certain low levels of ionizing radiation, point mutations are commonly produced; at these levels, there is a linear relationship between the rate of point mutations and the radiation dosage. Importantly, for many organisms, including humans, the effects of ionizing radiation doses are cumulative. That is, if a particular dose of

Figure 7.7
Production of a mutation as a result of a mismatch caused by wobble base pairing. The details are explained in the text.

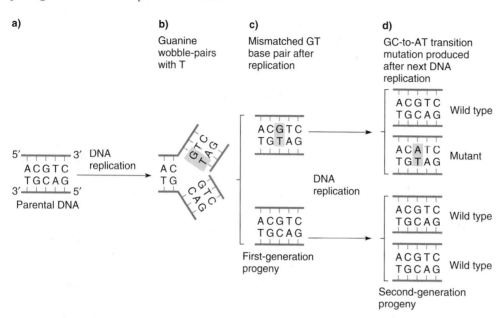

radiation results in a certain number of point mutations, the same number of point mutations will be induced whether the radiation dose is received over a short or over a long period of time. Therefore, people need to be concerned about their exposure to ionizing radiation such as X rays.

UV rays have insufficient energy to induce ionizations in the cell. Instead, UV light causes mutations because the purine and pyrimidine bases in DNA absorb light strongly in the ultraviolet range (254 to 260 nm). At this wavelength, UV light induces point mutations primarily by causing photochemical (light-induced chemical) changes in the DNA. One of the effects of UV radiation on DNA is the formation of abnormal chemical bonds between adjacent pyrimidine molecules in the same strand, or between pyrimidines on the opposite strands, of the double helix. This bonding is induced mostly between adjacent thymines, forming what

Figure 7.8
Spontaneous generation of addition and deletion mutants by DNA looping-out errors during replication.

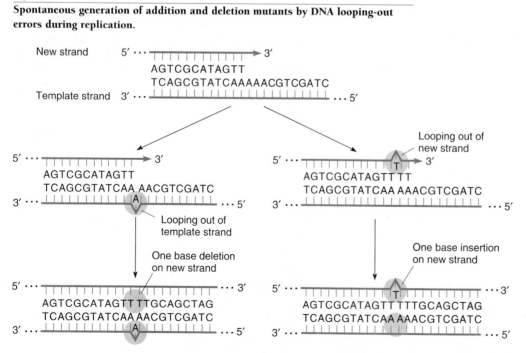

Figure 7.9

(a) Deamination of cytosine to uracil. (b) Deamination of 5-methylcytosine (5ᵐC) to thymine.

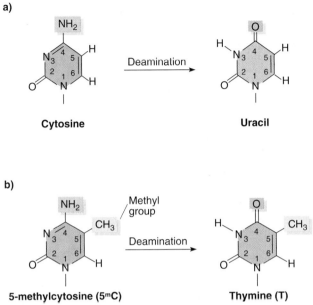

a)

Cytosine → Deamination → Uracil

b)

5-methylcytosine (5ᵐC) → Deamination → Thymine (T)

are called **thymine dimers** (Figure 7.10), usually designated T̂T. (ĈC, ĈT, and T̂C pyrimidine dimers are also produced by UV radiation, but in much lesser amounts.) This unusual pairing produces a bulge in the DNA strand and disrupts the normal pairing of T's with corresponding A's on the opposite strand. Since replication cannot proceed past the lesion, if enough pyrimidine dimers remain unrepaired in the cell, it may die.

KEYNOTE

Radiation may cause genetic damage by producing chemicals that affect the DNA (as in the case of X rays) or by causing the formation of unusual bonds between DNA bases, such as thymine dimers (as in the case of ultraviolet light). If radiation-induced genetic damage is not repaired, mutations or cell death may result. Radiation may also break chromosomes.

Chemical Mutagens. Chemical mutagens include both naturally occurring chemicals and synthetic substances. These mutagens can be grouped into different classes on the basis of their mechanism of action. Here we discuss base analogs, base-modifying agents, and intercalating agents and explain how they induce mutations. Mutations induced by base analogs and intercalating agents depend on replication, whereas base-modifying agents can induce mutations at any state of the cell cycle.

Base analogs are bases that are similar to the bases normally found in DNA. Like normal bases, base analogs exist in normal and rare tautomeric states. In each of the two states, the base analog pairs with a different normal base in DNA. Because base analogs are so similar to the normal nitrogen bases, they may be incorporated into DNA in place of the normal bases.

A common base analog mutagen is 5-bromouracil (5BU), which has a bromine residue instead of the methyl group of thymine. In its normal state, 5BU resembles thymine and pairs only with adenine in DNA (Figure 7.11a). In its rare state, it pairs only with guanine (Figure 7.12b). 5BU induces mutations by switching between its two chemical states once the base analog has been incorporated into the DNA (Figure 7.12c).

animation
Mutagenic Effects of 5BU

If 5BU is incorporated in its more common normal state, it pairs with adenine. If it changes into its rare state during replication, it pairs with guanine instead. In the next round of replication, the 5BU-G base pair is resolved into a C-G base pair instead of the T-A base pair. By this process, a transition mutation is produced, from TA to CG. 5BU can also induce a mutation from CG to TA if it is first incorporated into DNA in its rare state and then switches to the normal state during replication. (See Figure 7.12c.) Thus, 5BU-induced mutations can be reverted by a second treatment of 5BU.

Not all base analogs are mutagens. For example, AZT (azidothymidine), one of the approved drugs given to patients with AIDS, is an analog of thymidine, but it is not a mutagen, because it does not cause base-pair changes.

Figure 7.10

Production of thymine dimers by ultraviolet light irradiation. The two components of the dimer are covalently linked in such a way that the DNA double helix is distorted at that position.

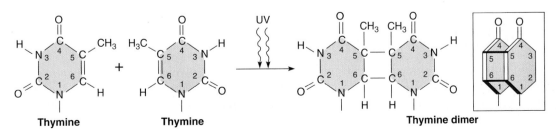

Thymine + Thymine → (UV) → Thymine dimer

Figure 7.11

Mutagenic effects of the base analog 5-bromouracil (5BU). (a) In its normal state, 5BU pairs with adenine. (b) In its rare state, 5BU (indicated by white letters on magenta) pairs with guanine. (c) The two possible mutation mechanisms. 5BU induces transition mutations when it incorporates into DNA in one state and then shifts to its alternate state during the next round of DNA replication.

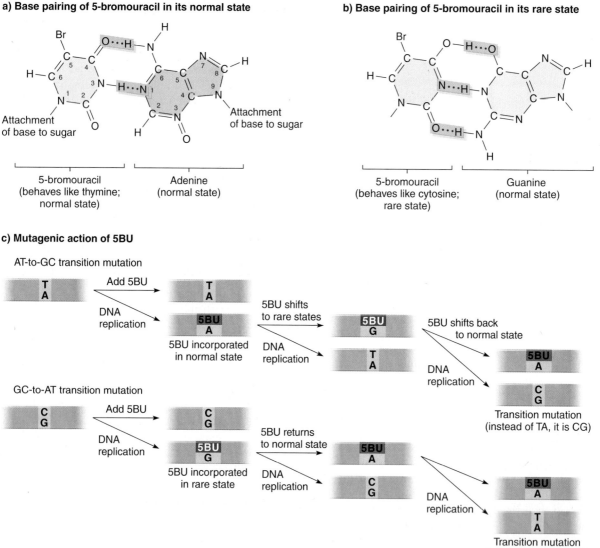

a) Base pairing of 5-bromouracil in its normal state

5-bromouracil
(behaves like thymine;
normal state)

Adenine
(normal state)

b) Base pairing of 5-bromouracil in its rare state

5-bromouracil
(behaves like cytosine;
rare state)

Guanine
(normal state)

c) Mutagenic action of 5BU

AT-to-GC transition mutation

GC-to-AT transition mutation

Base-modifying agents are chemicals that act as mutagens by modifying the chemical structure and properties of bases. Figure 7.12 shows the action of three types of mutagens that work in this way: a deaminating agent, a hydroxylating agent, and an alkylating agent.

Nitrous acid, HNO_2 (Figure 7.12a), is a deaminating agent that removes amino groups ($-NH_2$) from the bases guanine, cytosine, and adenine. Treatment of guanine with nitrous acid produces xanthine, but because this purine base has the same pairing properties as guanine, no mutation results (Figure 7.12a, part 1). Treatment of cytosine with nitrous acid produces uracil (Figure 7.12a, part 2), which pairs with adenine to produce a CG-to-TA transition mutation during replication. Likewise, nitrous acid modifies adenine to produce hypoxanthine, a base that pairs with cytosine rather than thymine, which results in an AT-to-GC transition mutation (Figure 7.12a, part 3). A nitrous acid–induced mutation can be reverted by a second treatment with nitrous acid.

Hydroxylamine (NH_2OH) is a hydroxylating mutagen that reacts specifically with cytosine, modifying it by adding a hydroxyl group (OH) so that it can pair solely with adenine instead of with guanine (Figure 7.12b). Mutations induced by hydroxylamine can only be CG-to-TA transitions, so hydroxylamine-induced mutations cannot be reversed by a second treatment with this chemical.

Figure 7.12

Action of three base-modifying agents. (**a**) Nitrous acid (HNO_2) modifies (**1**) guanine, (**2**) cytosine, and (**3**) adenine. The cytosine and adenine modifications result in mutations, but the guanine modification does not. (**b**) Hydroxylamine (NH_2OH) reacts only with cytosine. (**c**) Methylmethane sulfonate (MMS), an alkylating agent, alkylates guanine. (*Note*: dR = deoxyribose.)

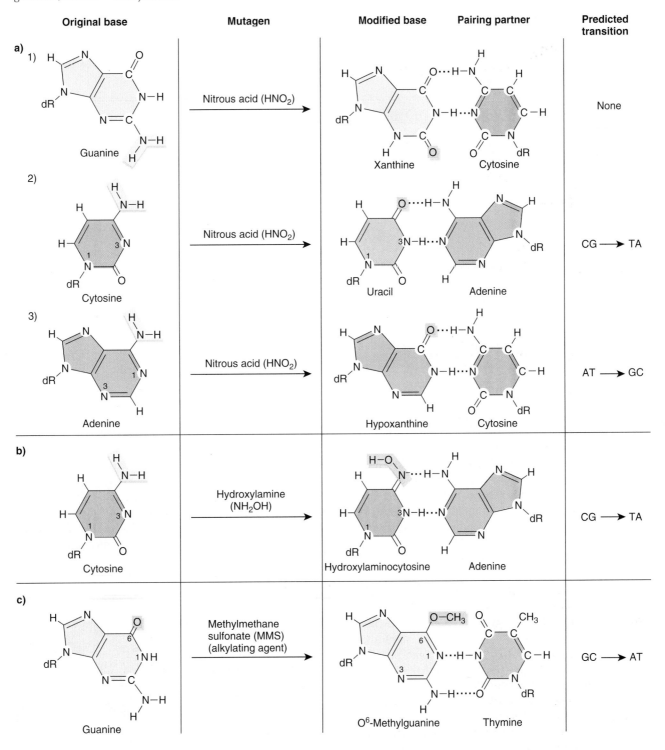

| Original base | Mutagen | Modified base | Pairing partner | Predicted transition |

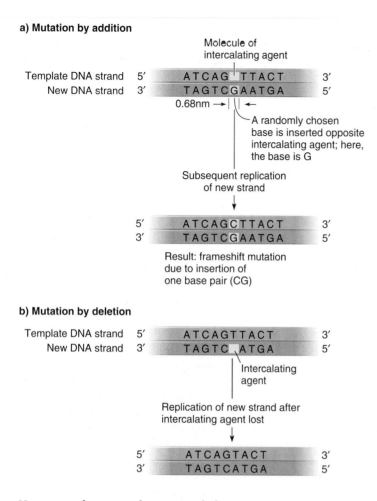

a) Mutation by addition

Molecule of
intercalating agent

Template DNA strand 5′ A T C A G   T T A C T 3′
New DNA strand 3′ T A G T C G A A T G A 5′

0.68nm →

A randomly chosen
base is inserted opposite
intercalating agent; here,
the base is G

Subsequent replication
of new strand

5′ A T C A G C T T A C T 3′
3′ T A G T C G A A T G A 5′

Result: frameshift mutation
due to insertion of
one base pair (CG)

b) Mutation by deletion

Template DNA strand 5′ A T C A G T T A C T 3′
New DNA strand 3′ T A G T C   A T G A 5′

Intercalating
agent

Replication of new strand after
intercalating agent lost

5′ A T C A G T A C T 3′
3′ T A G T C A T G A 5′

Figure 7.13

Intercalating mutations. (a) Frameshift mutation by addition, when agent inserts itself into template strand. (b) Frameshift mutation by deletion, when agent inserts itself into newly synthesizing strand.

However, they can be reverted by treatment with other mutagens (such as 5BU and nitrous acid) that cause TA-to-CG transition mutations.

Methylmethane sulfonate (MMS) is one of a diverse group of alkylating agents that introduce alkyl groups (e.g., $-CH_3$, $-CH_2CH_3$) onto the bases at a number of locations (Figure 7.12c). Most mutations caused by alkylating agents result from the addition of an alkyl group to the 6-oxygen of guanine to produce O^6-alkylguanine. For example, after treatment with MMS, some guanines are methylated to produce O^6-methylguanine. The methylated guanine pairs with thymine rather than cytosine, giving GC-to-AT transitions (Figure 7.12c).

Intercalating agents—such as proflavin, acridine, and ethidium bromide (commonly used to stain DNA in gel electrophoresis experiments)—insert (*intercalate*) themselves between adjacent bases in one or both strands of the DNA double helix, causing the helix to relax (Figure 7.13). If the intercalating agent inserts itself between adjacent base pairs of the DNA strand that is the template for new DNA synthesis (Figure 7.13a), an extra base (chosen at random; G in the figure) must be inserted into the new DNA strand opposite the intercalating agent. After one more round of replication, during which the intercalating agent is lost, the overall result is a base-pair addition

mutation. (CG is added in Figure 7.13a.) If the intercalating agent inserts itself into the new DNA strand in place of a base (Figure 7.13b), then when that DNA double helix replicates after the intercalating agent is lost, the result is a base-pair deletion mutation. (TA is lost in Figure 7.13b.)

If a base-pair addition or base-pair deletion point mutation occurs in a protein-coding gene, the result is a frameshift mutation. Since intercalating agents can cause either additions or deletions, frameshift mutations induced by intercalating agents can be reverted by a second treatment with those same agents.

K E Y N O T E

Mutations can be produced by exposure to chemical mutagens. If the genetic damage caused by the mutagen is not repaired, mutations result. Chemical mutagens act in a variety of ways, such as by substituting for normal bases during DNA replication, modifying the bases chemically, and intercalating themselves between adjacent bases during replication.

Site-Specific In Vitro Mutagenesis of DNA. Spontaneous and induced mutations occur not only in specific genes, but are scattered randomly throughout the genome. However, most geneticists want to study the effects of mutations in particular genes. With recombinant DNA technology, we can clone genes and produce large amounts of DNA for analysis and manipulation. This means that it is now possible to mutate a gene at specific positions in the test tube by **site-specific mutagenesis** and then introduce the mutated gene back into the cell and investigate the phenotypic changes produced by the mutation in vivo. Such techniques enable geneticists to study, for example, genes with unknown function and specific sequences involved in regulating a gene's expression. (We will discuss examples of making site-specific mutants later in the book, after we have discussed recombinant DNA techniques.)

Chemical Mutagens in the Environment. Every day, we are exposed to a wide variety of chemicals in our environment, such as drugs, cosmetics, food additives, pesticides, and industrial compounds. Many of these chemicals can have mutagenic effects, including genetic diseases and cancer. Some banned chemical warfare agents (e.g., mustard gas) also are mutagens—hence the concern about knowing the location of such compounds and the safety concerns regarding their long-term storage.

The Ames Test: A Screen for Potential Carcinogens. As we will learn in Chapter 22, a number of chemicals induce mutations that result in tumorous or cancerous growth.

animation

a Ames Test Protocol

These chemical agents are a subclass of mutagens called chemical **carcinogens.** The mutations typically are base-pair substitutions that produce missense or nonsense

mutations or base-pair additions or deletions that produce frameshift mutations. Directly testing chemicals for their ability to cause tumors in animals is time consuming and expensive. However, the fact that most chemical carcinogens are mutagens led Bruce Ames to develop a simple, inexpensive, indirect assay for mutagens. The **Ames test** assays the ability of chemicals to revert mutant strains of the bacterium *Salmonella typhimurium* to the wild type.

The design of the Ames test is as follows (Figure 7.14): Approximately 10^8 cells of tester bacteria that are auxotrophic for histidine (*his* mutants) are spread with or without a mixture of rat, mouse, or hamster liver enzymes on a culture plate lacking histidine. (Recall that an auxotrophic mutant cannot make a particular molecule that is essential for growth, and it requires a particular nutrient in order to grow. Histidine (*his*) auxotrophs, then, require the presence of histidine in the growth medium; normal (*his*⁺) individuals do not.) An array of tester bacterial strains is available that allows the detection of base-pair substitution mutations and frameshift mutations in the test. The liver enzymes, called the S9 extract, are used because many chemicals are not mutagenic themselves, but are converted to mutagens (and carcinogens) in the liver and other tissues by enzymatic detoxification pathways. A filter disk impregnated with the test chemical is then placed on the plate, which is incubated overnight and scored for colony formation. Control plates are also set up that lack the chemical being tested.

After the incubation period, the control plates will have a few colonies resulting from spontaneous reversion of the *his* strain. A similar result is seen with chemicals

Figure 7.14

The Ames test for assaying the potential mutagenicity of chemicals.

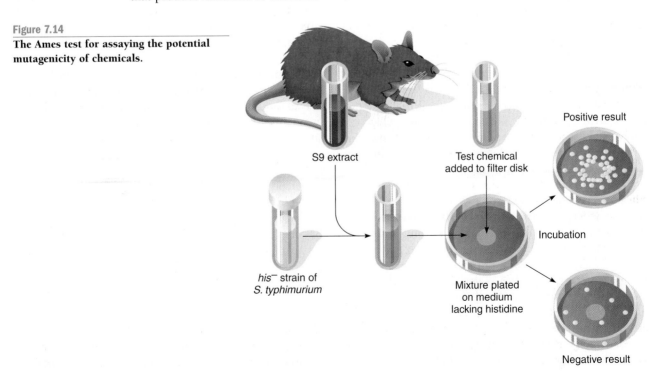

S9 extract

Test chemical added to filter disk

Positive result

his⁻ strain of *S. typhimurium*

Mixture plated on medium lacking histidine

Incubation

Negative result

that are not mutagenic in the Ames test (Figure 7.14). A positive result in the Ames test is observed as an increased number of revertants near the test chemical disk (Figure 7.14).

The Ames test is so straightforward that it is used routinely in many laboratories around the world. To date, the test has identified a large number of mutagens, including many industrial and agricultural chemicals. In general, the Ames test is an excellent indicator of whether a chemical is a carcinogen, but some carcinogenic chemicals assay negative in the test. For example, Ziram, which is used as an agricultural fungicide, gives a positive Ames test for both base substitution and frameshift reversion when S9 extract is present, but a negative test when S9 extract is absent. Thus, this chemical presumably is turned into a mutagen by metabolic enzymes. In contrast, nitrobenzene is negative in the Ames test with or without the S9 extract. Most nitrobenzene is used to manufacture aniline, which is used in the manufacture of polyurethane. Styrene, used in producing polystyrene polymers and resins, similarly tests negative with or without the S9 extract, yet animal tests indicate that it is a carcinogen. Because of results like this, the Ames test is not the sole test relied upon to make the determination as to whether a compound is mutagenic.

Finally, the Ames test can be quantified by using different amounts of chemicals to produce a dose–response curve. With this approach, the relative mutagenicity of different chemicals can be compared.

iActivity

Now it is your turn to investigate the health problems plaguing the inhabitants of Russellville. Conduct your own Ames Test in the iActivity *A Toxic Town* on your CD-ROM.

Detecting Mutations

Geneticists have made great progress over the years in understanding how normal processes take place by studying mutants that have defects in those processes. Researchers have used mutagens to induce mutations at a greater rate than the rate at which spontaneous mutations occur. However, mutagens change base pairs at random, without regard to the positions of the base pairs in the genetic material. Once mutations have been induced, then, they must be detected if they are to be studied. Mutations of haploid organisms are readily detectable because there is only one copy of the genome. In a diploid experimental organism such as *Drosophila*, dominant mutations are also readily detectable, and sex-linked recessive mutations can be detected because they are expressed in half of the sons of a mutated, heterozygous female. However, autosomal recessive mutations can be detected only if the mutation is homozygous.

The detection of mutations in humans is much more difficult than in *Drosophila*, because geneticists cannot make controlled crosses. Dominant mutations can be readily detected, of course, but other types of mutations may be revealed only by pedigree analysis or by direct biochemical or molecular probing.

Fortunately, for some organisms of genetic interest—particularly microorganisms—selection and screening procedures historically have helped geneticists isolate mutants of interest from a heterogeneous mixture in a mutagenized population. Brief descriptions of some of these procedures follow.

Visible Mutations. **Visible mutations** affect the morphology or physical appearance of an organism. Examples of visible mutations are eye-color and wing-shape mutants of *Drosophila*, coat-color mutants of animals (such as albino organisms), colony-size mutants of yeast, and plaque morphology mutants of bacteriophages. Since visible mutations, by definition, are readily apparent, screening is done by inspection.

Nutritional Mutations. An **auxotrophic (nutritional) mutation** affects an organism's ability to make a particular molecule essential for growth. Auxotrophic mutations are most readily detected in microorganisms such as *E. coli* and yeast that grow on simple and defined growth media from which they synthesize the molecules essential to their growth. A number of selection and screening procedures are available to isolate auxotrophic mutants.

One simple procedure called **replica plating** can be used to screen for auxotrophic mutants of any microorganism that grows in discrete colonies on a solid medium (Figure 7.15). In replica plating, samples from a culture of a mutagenized or an unmutagenized colony-forming organism or cell type are plated onto a medium containing the nutrients appropriate for the mutants desired. For example, to isolate arginine auxotrophs, we would plate the culture on a master plate of minimal medium plus arginine. (See Figure 7.15.) On this medium, wild-type and arginine auxotrophs grow, but no other auxotrophs grow. After incubation, colonies are seen on the plate. The pattern of the colonies is transferred onto sterile velveteen cloth, and replicas of the colony pattern on the cloth are then made by gently pressing new plates onto the velveteen. If the new plate contains minimal medium, the wild-type colonies can grow, but the arginine auxotrophs cannot. By comparing the patterns on the original minimal medium plus arginine master plate with those on the minimal medium replica plate, researchers can readily identify the potential arginine auxotrophs. They can then be picked from the original master plate and cultured for further study.

Conditional Mutations. The products of many genes—DNA polymerases and RNA polymerases, for example—are

Figure 7.15

Replica-plating technique to screen for mutant strains of a colony-forming microorganism.

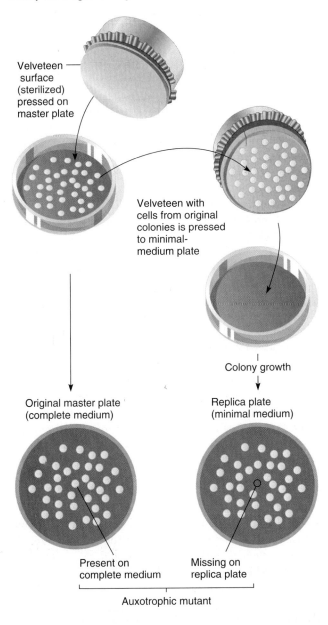

Velveteen surface (sterilized) pressed on master plate

Velveteen with cells from original colonies is pressed to minimal-medium plate

Colony growth

Original master plate (complete medium)

Replica plate (minimal medium)

Present on complete medium

Missing on replica plate

Auxotrophic mutant

important for the growth and division of cells, and most mutations in these genes are lethal. The structure and function of such genes can be studied by inducing **conditional mutations,** which reduce the activity of gene products only under certain conditions. A common type of conditional mutation is a temperature-sensitive mutation. In yeast, for instance, many **temperature-sensitive mutants** that grow normally at 23°C but grow very slowly or not at all at 36°C can be isolated. Heat sensitivity typically results from a missense mutation causing a change in the amino-acid sequence of a protein so that, at the higher temperature, the protein assumes a shape that is nonfunctional.

Essentially the same procedures are used to screen for heat-sensitive mutations of microorganisms as for auxotrophic mutations. For example, replica plating can be used to screen for temperature-sensitive mutants when the replica plate is incubated at a higher temperature than the master plate. That is, such mutants grow on the master plate, but not on the replica plate.

Resistance Mutations. In microorganisms such as *E. coli* and yeast, mutations can be induced for resistance to particular viruses, chemicals, or drugs. For example, in *E. coli*, mutants resistant to phage T1 have been induced (recall the discussion at the beginning of this chapter), and there are mutants that are resistant to antibiotics such as streptomycin. In yeast, for example, some mutants are resistant to antibiotics such as nystatin.

Selecting resistance mutants is straightforward. To isolate azide-resistant mutants of *E. coli*, for example, mutagenized cells are plated on a medium containing azide, and the colonies that grow are resistant to azide. Similarly, antibiotic-resistant *E. coli* mutants can be selected by plating on antibiotic-containing medium.

K E Y N O T E

A number of screening procedures have been developed to isolate mutants of interest from a heterogeneous mixture of cells in a mutagenized population of cells.

Repair of DNA Damage

Spontaneous and induced mutations are damages to the DNA. Especially with high doses of mutagens, the mutational damage can be considerable. What we see as mutations are DNA alterations that are not corrected by various DNA damage repair systems. The simple equation is "mutations = DNA damage − DNA repair." Both prokaryotic and eukaryotic cells have a number of enzyme-based repair systems that deal with damage to DNA. If the repair systems cannot correct all the lesions, the result is a mutant cell (or organism) or, if too many mutations remain, death of the cell (or organism). Life as we know it results from a delicate balance between accuracy in the transmission of DNA to daughter cells and progeny organisms and the occasional mutation that affords raw material for evolution. DNA repair systems provide that delicate balance.

We can group repair systems into different categories on the basis of the way they operate. Some systems correct damaged areas by reversing the damage. This type of repair is called *direct correction* or *direct reversal*. Other systems excise the damaged areas and then repair the gap by new DNA synthesis. Selected repair systems are described in this section.

Direct Reversal of DNA Damage

Mismatch Repair by DNA Polymerase Proofreading. The frequency of base-pair substitution mutations in bacterial genes ranges from 10^{-7} to 10^{-11} errors per generation. However, DNA polymerase makes errors in inserting nucleotides while it is synthesizing the new DNA strand, perhaps at a frequency of 10^{-5} errors per generation. Most of the difference between the two figures given is accounted for by 3′-to-5′ exonuclease proofreading activity of the DNA polymerase in both prokaryotes and eukaryotes. (See Chapter 3, p. 50.) That is, when an incorrect nucleotide is inserted, the mismatched base pair often is detected by the polymerase. The DNA synthesis stalls and cannot proceed until the wrong nucleotide is removed and the correct one is put in its place. Once this occurs, the polymerase moves forward again, resuming its DNA synthesis activity.

The importance of the 3′-to-5′ exonuclease activity of DNA polymerase for maintaining a low mutation rate is shown nicely by the existence of *mutator* mutations in *E. coli*. Strains carrying mutator mutations exhibit a much higher than normal mutation frequency for all genes. These mutations have been shown to affect proteins whose normal functions are required for accurate DNA replication. For example, the *mutD* mutator gene of *E. coli* encodes the ε (epsilon) subunit of DNA polymerase III, the primary replication enzyme of *E. coli*. The *mutD* mutants are defective in 3′-to-5′ proofreading activity, so that many incorrectly inserted nucleotides are left unrepaired.

Repair of UV-Induced Pyrimidine Dimers. Through **photoreactivation,** or **light repair,** UV light-induced thymine (or other pyrimidine) dimers (see Figure 7.10, p. 143) can be reverted directly to the original form by exposure to near-UV light in the wavelength range from 320 to 370 nm. Photoreactivation occurs when an enzyme called *photolyase* (encoded by the *phr* gene) is activated by a photon of light and splits the dimers apart. Strains with mutations in the *phr* gene are defective in light repair. Photolyase has been found in prokaryotes and in simple eukaryotes, but not in humans.

Repair of Alkylation Damage. Alkylating agents transfer alkyl groups (usually methyl or ethyl groups) onto the bases at various locations, such as the oxygen of carbon-6 in guanine. The mutagen MMS methylates guanine, for example. (See Figure 7.12c, p. 145.) In *E. coli*, this alkylation damage can be repaired by an enzyme called O^6-*methylguanine methyltransferase*, encoded by the *ada* gene. The enzyme recognizes the O^6-methylguanine in the DNA and removes the methyl group, thereby changing the base back to its original form. A similar specific system exists to repair alkylated thymine. Mutations of the genes encoding these repair enzymes result in a much higher rate of spontaneous mutations, as would be expected.

Base Excision Repair

Damaged bases are most commonly repaired by removing the base or the nucleotide involved and then inserting the correct base or nucleotide. In **base excision repair,** a repair glycosylase enzyme recognizes the damaged base and removes it from the DNA by cleaving the bond between the base and the deoxyribose sugar. Other enzymes then cleave the sugar–phosphate backbone before and after the now baseless sugar, releasing the sugar and leaving a gap in the DNA chain. The gap is filled with the correct nucleotide by a repair DNA polymerase and DNA ligase, with the opposite DNA strand used as the template.

Repair Involving Excision of Nucleotides

Excision Repair. In 1964, two groups of scientists—R. P. Boyce and P. Howard-Flanders, and R. Setlow and W. Carrier—isolated some UV-sensitive mutants of *E. coli* that, after UV irradiation, showed a higher-than-normal rate of induced mutation in the dark. These mutants were called *uvrA* mutants. (*uvr* means "UV repair.") The *uvrA* mutants can repair thymine dimers only with the input of light; that is, they have a normal photoreactivation repair system. However, the wild-type organism, genotypically *uvrA*⁺, can repair thymine dimers in the dark. Because the normal photoreactive repair system cannot operate in the dark, the investigators hypothesized that there must be another light-independent repair system. They called this system the **dark repair, excision repair system,** now typically referred to as the **nucleotide excision repair (NER)** system.

The NER system in *E. coli* corrects not only thymine dimers, but also other serious damage-induced distortions of the DNA helix. The system works as diagrammed in Figure 7.16 and involves four proteins—UvrA, UvrB, UvrC, and UvrD—that are encoded by the genes *uvrA, uvrB, uvrC,* and *uvrD.* A complex of two UvrA proteins and one UvrB protein slides along the DNA. When the complex recognizes a pyrimidine dimer or another serious distortion in the DNA, the UvrA subunits dissociate and a UvrC protein binds to the UvrB protein at the lesion. The resulting UvrBC protein bound to the lesion makes one cut about four nucleotides to the 3′ side in the damaged DNA strand (done by UvrB) and about seven nucleotides to the 5′ side of the lesion (done by UvrC). UvrB is then released and UvrD binds to the 5′ cut. UvrD is a helicase and unwinds the region between the cuts, releasing the short single-stranded segment. DNA polymerase I fills in

Figure 7.16

Nucleotide excision repair (NER) of pyrimidine dimer and other damage-induced distortions of DNA.

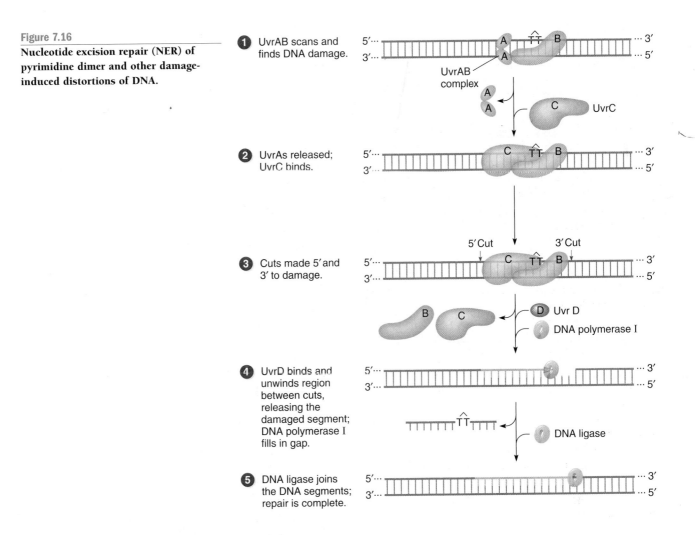

① UvrAB scans and finds DNA damage.

② UvrAs released; UvrC binds.

③ Cuts made 5′ and 3′ to damage.

④ UvrD binds and unwinds region between cuts, releasing the damaged segment; DNA polymerase I fills in gap.

⑤ DNA ligase joins the DNA segments; repair is complete.

the gap in the 5′-to-3′ direction, and DNA ligase seals the final gap. Excision repair systems have been found in most organisms that have been studied. In yeast and mammalian systems, about 12 genes encode proteins involved in nucleotide excision repair.

Repair by Methyl-Directed Mismatch Repair. Despite proofreading by DNA polymerase, a number of mismatched base pairs remain uncorrected after replication has been completed. In the next round of replication, these errors will become fixed as mutations if they are not repaired.

Many mismatched base pairs left after DNA replication can be corrected by another system of repair called **methyl-directed mismatch repair.** This system recognizes mismatched base pairs, excises the incorrect bases, and then carries out repair synthesis. In *E. coli,* the products of three genes—*mutS, mutL,* and *mutH*—are involved in the initial stages of mismatch repair (Figure 7.17). First, the *mutS*-encoded protein, MutS, binds to the mismatch. Then the repair system determines which base is the correct one (the base on the parental DNA strand) and which is the erroneous one (the base on the new

DNA strand). In *E. coli,* the two strands are distinguished by methylation of the A nucleotide in the sequence GATC. This sequence has an axis of symmetry; that is, the same sequence is present 5′-to-3′ on both DNA strands, to give 5′-GATC-3′. Both A nucleotides in the sequence usually 3′-CTAG-5′. are methylated. However, after replication, the parental DNA strand has a methylated A in the GATC sequence, whereas the A in the GATC of the *newly replicated DNA strand* is not methylated until a short time after its synthesis. Therefore, the MutS protein bound to the mismatch forms a complex with the *mutL-* and *mutH-* encoded proteins, MutL and MutH, to bring the unmethylated GATC sequence close to the mismatch. The MutH protein then nicks the unmethylated DNA strand at the GATC site, the mismatch is removed by an exonuclease, and the gap is repaired by DNA polymerase III and ligase.

Mismatch repair also takes place in eukaryotes. However, it is unclear how the new DNA strand is distinguished from the parental DNA strand. In humans, four genes, respectively named *hMSH2, hMLH1, hPMS1,* and *hPMS2,* have been identified; *hMSH2* is homologous to

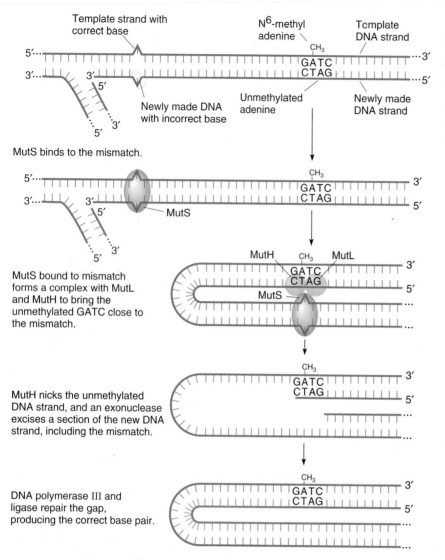

Template strand with correct base

N⁶-methyl adenine

Template DNA strand

Newly made DNA with incorrect base

Unmethylated adenine

Newly made DNA strand

MutS binds to the mismatch.

MutS

MutS bound to mismatch forms a complex with MutL and MutH to bring the unmethylated GATC close to the mismatch.

MutH   MutL

MutS

MutH nicks the unmethylated DNA strand, and an exonuclease excises a section of the new DNA strand, including the mismatch.

DNA polymerase III and ligase repair the gap, producing the correct base pair.

Figure 7.17

Mechanism of mismatch repair. The mismatch correction enzyme recognizes which strand the base mismatch is on by reading the methylation state of a nearby GATC sequence. If the sequence is unmethylated, a segment of that DNA strand containing the mismatch is excised and new DNA is inserted.

E. coli mutS, and the other three genes have homologies to *E. coli mutL.* The genes are known as *mutator genes,* because loss of function of such a gene results in an increased accumulation of mutations in the genome. Mutations in any one of the four human mismatch repair genes confer a phenotype of hereditary predisposition to a form of colon cancer called hereditary nonpolyposis colon cancer (HNPCC: OMIM 120435). The role of mutator genes in cancer is described in Chapter 22, p. 624.

Translesion DNA Synthesis and the SOS Response. As we have learned, some DNA damage involves lesions that block the replication machinery from proceeding past those points. Unrepaired, DNA damage of this kind can be lethal. Fortunately, there is a last-resort process called **translesion DNA synthesis** that allows replication to continue past the lesions. The process involves a special class of DNA polymerases that are synthesized only in response to DNA damage. In *E. coli,* such DNA damage activates a complex system called the *SOS response.* (The name "SOS" comes from the fact that the system is induced as a last-resort, emergency response to mutational damage.) The SOS response allows the cell to survive otherwise lethal events, although often at the expense of generating new mutations.

In *E. coli,* two genes are key to controlling the SOS system: *lexA* and *recA.* Cells with mutant *recA* and *lexA* genes have their SOS response permanently turned on. The SOS response works as follows: In the uninduced state, when there is no DNA damage, the *lexA*-encoded protein, LexA, represses the transcription of about 17 genes whose protein products are involved in repairing and dealing with various kinds of DNA damage. When there has been sufficient damage to DNA, the *recA*-encoded protein, RecA, is activated. Activated RecA stimulates the LexA protein to cleave itself, which in turn relieves the repression of the DNA repair genes. As a result, the DNA repair genes are transcribed, and DNA repair proceeds. After the DNA damage is dealt with, RecA is inactivated, and newly synthesized LexA protein again represses the DNA repair genes.

Among the gene products made during the SOS response is the DNA polymerase for translesion DNA synthesis. This polymerase continues replication over and past the lesion, although it does so by incorporating one or more nucleotides that are not specified by the template strand into the new DNA across from the lesion. These nucleotides may not match the wild-type template sequence; therefore, the SOS response itself is a mutagenic system, because mutations will be introduced into the DNA as a result of its activation. However, such mutations are less harmful than the potentially lethal alternative caused by incompletely replicated DNA.

KEYNOTE

Mutations constitute damage to the DNA. Both prokaryotes and eukaryotes have a number of repair systems that deal with different kinds of DNA damage. All the systems use enzymes to make the correction. Without such repair systems, lesions would accumulate and be lethal to the cell or organism. Not all lesions are repaired, and mutations do appear, but at low frequencies. At high doses of mutagens, repair systems are unable to correct all of the damage, and cell death may result.

Human Genetic Diseases Resulting from DNA Replication and Repair Mutations

We have already discussed a large number of human genetic diseases that result from gene mutations. In particular, Chapter 4 examined genetically based enzyme deficiencies and protein alterations in humans. Genes associated with human traits can be found by searching the OMIM Web site (http://www3.ncbi.nlm.nih.gov/OMIM/).

Some human genetic diseases are attributed to defects in DNA replication or repair; examples are listed in Table 7.1. For instance, *xeroderma pigmentosum* (XP) (OMIM 278700; Figure 7.18) is caused by homozygosity for a recessive mutation in a repair gene. Individuals with this lethal affliction are photosensitive, and portions of their skin that have been exposed to light show intense pigmentation, freckling, and warty growths that can become malignant. Those afflicted are deficient in excision repair of damage caused by ultraviolet light, X rays, gamma radiation, or chemical treatment. Thus, individuals with xeroderma pigmentosum are unable to repair radiation damage to DNA and often die as a result of malignancies that arise from the damage.

Transposable Elements

In this section, we learn about the nature of transposable elements and about the genetic changes they cause.

Figure 7.18

An individual with xeroderma pigmentosum.

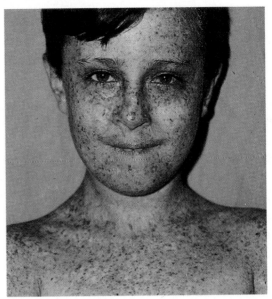

General Features of Transposable Elements

Transposable elements are normal, ubiquitous components of the genomes of prokaryotes and eukaryotes. Transposable elements fall into two general classes with respect to how they move. One class encodes proteins that move the DNA element directly to a new position or replicate the DNA to produce a new element that integrates elsewhere in the genome. Transposable elements of this class are found in both prokaryotes and eukaryotes. Members of the other class are related to retroviruses (see Chapter 22), in that they encode a reverse transcriptase for making DNA copies of their RNA transcripts, which subsequently integrate at new sites in the genome. Transposable elements of this class are found only in eukaryotes.

In prokaryotes, transposable elements can move to new positions on the same chromosome (because there is only one chromosome) or onto plasmids or phage chromosomes; in eukaryotes, transposable elements may move to new positions within the same chromosome or to a different chromosome. In both prokaryotes and eukaryotes, transposable elements insert into new chromosome locations with which they have no sequence homology; therefore, transposition is a process different from homologous recombination and is called *nonhomologous recombination*. Transposable elements are important because of the genetic changes they cause, and that is a reason they have been studied intensively. For example, they can produce mutations by inserting into genes, they can increase or decrease gene expression by inserting into gene regulatory sequences (such as by disrupting promoter function or stimulating a gene's expression through the activity of promoters on the element), and they can produce various

Table 7.1 Some Examples of Naturally Occurring Human Cell Mutants That Are Defective in DNA Replication or Repair

Disease and Mode of Inheritance	Symptoms	Functions Affected	Chromosome Location[a] and OMIM number
Xeroderma pigmentosum (XP) —autosomal recessive	Sensitivity to sunlight, with skin freckling and cancerous growths on skin; lethal at early age as a result of the malignancies	Repair of DNA damaged by UV irradiation or chemicals	9q34.1 —278700
Ataxia-telangiectasia (AT) —autosomal recessive	Muscle coordination defect; propensity for respiratory infection; progressive spinal muscular atrophy in significant proportion of patients in second or third decade of life; marked hypersensitivity to ionizing radiation; cancer prone; high frequency of chromosome breaks leading to translocations and inversions	Repair replication of DNA	11q22.3 —208900
Fanconi anemia (FA) —autosomal recessive	Aplastic anemia;[b] pigmentary changes in skin; malformations of heart, kidney, and limbs; leukemia is a fatal complication; genital abnormalities common in males; spontaneous chromosome breakage	Repair replication of DNA, UV-induced pyrimidine dimers, and chemical adducts not excised from DNA; a repair exonuclease, DNA ligase, and transport of DNA repair enzymes have been hypothesized to be defective in patients with FA	16q24.3 —227650
Bloom syndrome (BS) —autosomal recessive	Pre- and postnatal growth deficiency; sun-sensitive skin disorder; predisposition to malignancies; chromosome instability; diabetes mellitus often develops in second or third decade of life	Elongation of DNA chains intermediate in replication; candidate gene is homologous to *E. coli* helicase Q	15q26.1 —210900
Cockayne syndrome (CS) —autosomal recessive	Dwarfism; precociously senile appearance; optic atrophy; deafness; sensitivity to sunlight; mental retardation; disproportionately long limbs; knee contractures produce bowlegged appearance; early death	Precise molecular defect is unknown, but may involve transcription-coupled repair	5 —216400
Hereditary nonpolyposis colon cancer (HNPCC) —autosomal dominant	Inherited predisposition to non-polyp-forming colorectal cancer	Defect in mismatch repair develops when the remaining wild-type allele of the inherited mutant allele becomes mutated; homozygosity for mutations in any one of four genes (*hMSH2*, *hMLH1*, *hPMS1*, and *hPMS2*, known as mutator genes) has been shown to give rise to HNPCC	2p22-p21 —114500

[a]If multiple complementation groups exist, the location of the most common defect is given.
[b]Individuals with aplastic anemia make no or very few red blood cells.

kinds of chromosomal mutations because of the mechanics of transposition. In fact, transposable elements have made important contributions to the evolution of the genomes of both prokaryotes and eukaryotes through the chromosome rearrangements they have caused.

The frequency of transposition varies with the particular element, but typically is very low. Were it to be high, the genetic changes the transpositions would cause would likely kill the cell.

Transposable Elements in Prokaryotes

Two examples of transposable elements in prokaryotes are insertion sequence (IS) elements and transposons (Tn).

Insertion Sequences.   An **insertion sequence (IS), or IS element,** is the simplest transposable element found in prokaryotes. An IS element contains only genes required to mobilize the element and insert it into a chromosome at a new location. IS elements are normal constituents of bacterial chromosomes and plasmids.

IS elements were first identified in *E. coli* as a result of their effects on the expression of three genes that control the metabolism of the sugar galactose. Some muta-

animation
Insertion
Sequences in
Prokaryotes

tions affecting the expression of these genes did not have properties typical of point mutations or deletions, but rather had an insertion of an approximately 800-bp DNA segment into a gene. This particular DNA segment is now called *insertion sequence 1, or IS1* (Figure 7.19), and the insertion of IS1 into the genome is an example of a *transposition event.*

E. coli contains a number of IS elements (e.g., IS1, IS2, and IS10R), each present in up to 30 copies per genome and each with a characteristic length and unique nucleotide sequence. IS1 (see Figure 7.19), for instance, is 768 bp long and is present in 4 to 19 copies on the *E. coli* chromosome. Among prokaryotes as a whole, the IS elements range in size from 768 bp to more than 5,000 bp and are found in most cells.

Figure 7.19

The insertion sequence (IS) transposable element IS1. The 768-bp IS element has inverted repeat (IR) sequences at the ends. Shown below the element are the sequences for the 23-bp terminal inverted repeats (IR).

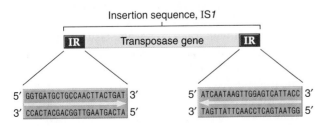

Insertion sequence, IS1

All IS elements end with perfect or nearly perfect terminal inverted repeats (IR's) of 9 to 41 bp. This means that essentially the same sequence is found at each end of an IS, but in opposite orientations. The inverted repeats of IS1, for example, consist of 23 bp with not quite identical sequences. (See Figure 7.19.)

When IS elements integrate at random points along the chromosome, they often cause mutations by disrupting either the coding sequence of a gene or a gene's regulatory region. Promoters within the IS elements themselves may also have effects by altering the expression of nearby genes. In addition, the presence of an IS element in the chromosome can cause mutations such as deletions and inversions in the adjacent DNA. Finally, deletion and insertion events can also occur as a result of crossing-over between duplicated IS elements in the genome.

The transposition of an IS element requires an enzyme encoded by the IS element called **transposase.** The IR sequences are essential to the transposition process, because the transposase recognizes that those sequences initiate transposition. The frequency of transposition is characteristic of each IS element and ranges from 10^5 to 10^7 per generation.

Figure 7.20 shows how an IS element inserts into a new location in a chromosome. Insertion takes place at a *target site* with which the element has no sequence homology. First, a staggered cut is made in the target site and the IS element is then inserted, becoming joined to the single-stranded ends. DNA polymerase and DNA ligase fill in the gaps, producing an integrated IS element with two direct repeats of the target-site sequence flanking the IS element. In this case, "direct" means that the two sequences are repeated in the same orientation. (See Figure 7.20.) The direct repeats are called *target-site duplications,* and they vary with the IS element, but tend to be small (4 to 13 bp).

Transposons.   Like an IS element, a **transposon (Tn)** contains genes for the insertion of the DNA segment into the chromosome and mobilization of the element to other locations on the chromosome. A transposon is more complex than an IS element, however, in that it contains additional genes.

There are two types of prokaryotic transposons: composite transposons and noncomposite transposons (Figure 7.21). *Composite transposons* (Figure 7.21a), exemplified by Tn10, are complex transposons with a central region containing genes (for example, genes that confer resistance to antibiotics), flanked on both sides by IS elements (also called *IS modules*). Composite transposons may be thousands of base pairs long. The IS elements are both of the same type and are called IS*L* (for "left") and IS*R* (for "right"). Depending on the transposon, IS*L* and IS*R* may be in the same or inverted orientation relative to each other. Because the IS's themselves have terminal inverted

Figure 7.20

Process of integration of an IS element into chromosomal DNA. As a result of the integration event, the target site becomes duplicated, producing direct target repeats. Thus, the integrated IS element is characterized by its inverted repeat (IR) sequences, flanked by direct target-site duplications. Integration involves making staggered cuts in the host target site. After insertion of the IS, the gaps that result are filled in with DNA polymerase and DNA ligase. (*Note:* The base sequences given for the IR are for illustration only and are neither the actual sequences found nor their actual length.)

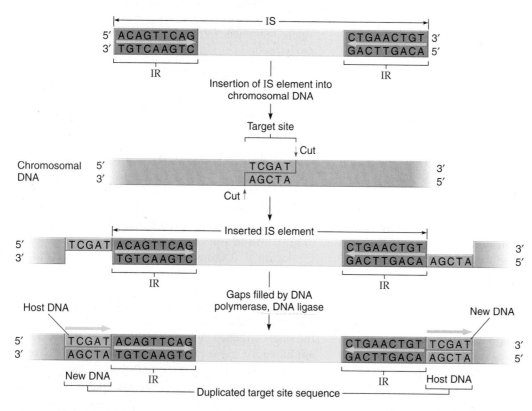

repeats, the composite transposons also have terminal inverted repeats.

Transposition of composite transposons occurs because of the function of the IS elements they contain. One or both IS elements supply the transposase, which recognizes that the inverted repeats of the IS elements at the two ends of the transposon initiate transposition (as with the transposition of IS elements). Transposition of Tn*10* is rare, occurring once in 10^7 cell generations. Like IS elements, composite transposons produce target site duplications after transposition, 9-bp long in the case of Tn*10*.

Like composite transposons, *noncomposite transposons* (Figure 7.21b), exemplified by Tn*3*, contain genes such as those which confer drug resistance, but they do not terminate with IS elements. However, at their ends they do have repeated sequences that are required for transposition. Enzymes for transposition are encoded by genes in the central region of noncomposite transposons. Transposase catalyzes the insertion of a transposon into new sites, and resolvase is an enzyme involved in the particular recombinational events associated with transposition. Again like composite transposons, noncomposite

transposons cause target-site duplications when they move. For example, Tn*3* produces a 5-bp target-site duplication when it inserts into the genome.

A number of detailed models describe the transposition of transposons. Figure 7.22 shows a *cointegration* model involving the transposition of a transposon from one genome to another (for example, from a plasmid to a bacterial chromosome or vice versa). Similar events can occur between two locations on the same chromosome. First, the donor DNA containing the transposable element fuses with the recipient DNA. Because of the way this occurs, the transposable element becomes duplicated, with one copy located at each junction between donor and recipient DNA. The fused product is called a *cointegrate*. Next, recombination between the duplicated transposable elements resolves the cointegrate into two products, each with one copy of the element. Since the transposable element becomes duplicated, the process is called *replicative transposition*. Tn*3* and related noncomposite transposons move by replicative transposition.

A second type of transposition mechanism involves the movement of a transposable element from one location

than the spontaneous reversion frequency for a regular point mutation; therefore, the allele produced by an autonomous element is called a *mutable allele.*

By contrast, mutant alleles resulting from the insertion of a nonautonomous element in a gene are *stable*, because the element is unable to transpose out of the locus by itself. However, if the autonomous element of its family is also either already present in, or introduced into, the genome, the autonomous element can provide the enzymes needed for transposition, and the nonautonomous element can now transpose around the genome.

McClintock's Study of Transposons in Plants. A number of different genes must function together for the synthesis of red anthocyanin pigment, which gives the corn kernel a purple color. Mutation of any one of these genes causes a kernel to be unpigmented. McClintock studied kernels that, rather than being either of a solid color or colorless, had spots of purple pigment on an otherwise white kernel (Figure 7.23). She knew that the phenotype was the result of an unstable mutation. From her careful genetic and cyto-logical studies, she concluded that the spotted phenotype was not the result of any conventional kind of mutation (such as a point mutation), but rather the result of a controlling element, which we now know is a transposon.

The explanation for the spotted kernels is as follows: If the corn plant carries a wild-type *C* gene, the kernel is purple; *c* (colorless) mutations block the production of the purple pigment, so the kernel is colorless. During ker-nel development, revertants of the mutation occur, leading to a spot of purple pigment. The earlier in development the reversion occurs, the larger is the purple spot. McClin-tock determined that the original *c* (colorless) mutation resulted from a "mobile controlling element" (in modern terms, a transposon), called *Ds* for "dissociation," being inserted into the *C* gene (Figure 7.24a and b). We now know this insertion to take place by the transposition of a

nonautonomous transposon. Another mobile controlling element, called *Ac* for "activator" (which we now know is an autonomous element), is required for transposition of *Ds* into the gene. *Ac* can also result in *Ds* transposing (excising perfectly) out of the *c* gene, giving a wild-type revertant with a purple spot (Figure 7.24c). The *Ac-Ds* system is discussed in the next section.

The remarkable fact of McClintock's conclusion was that, at the time, there was no precedent for the existence of transposable genetic elements; indeed, the genome was thought to be static with regard to gene locations. Only much more recently have transposable genetic elements been widely identified and studied, and only in 1983 was direct evidence obtained for the movable genetic ele-ments proposed by McClintock.

The Ac-Ds Transposable Elements in Corn. The *Ac-Ds* family of controlling elements has been studied in detail. The autonomous *Ac* element is 4,563 bp long, with short termi-nal inverted repeats, and a single gene encoding the trans-posase. Upon insertion into the genome, an 8-bp direct duplication of the target site is generated. *Ds* elements are heterogeneous in length and sequence, but all have the same terminal IR's as *Ac* elements, since most have been generated from *Ac* by the deletion of segments or by more complex sequence rearrangements. As a result of either of those processes, no transposase can be produced—hence the trans-position-defective phenotype of the various *Ds* elements.

Transposition of the *Ac* element occurs only during chromosome replication and is by the cut-and-paste (conservative) transposition mechanism (Figure 7.25). Consider a chromosome with one copy of *Ac* at a site called the *donor site*. When the chromosome region con-taining *Ac* replicates, two copies of *Ac* result, one on each progeny chromatid. There are two possible results of *Ac* transposition, depending on whether it occurs to a repli-cated or an unreplicated chromosome site.

Figure 7.23

Corn kernels, some of which show spots of pigment produced by cells in which a transposable genetic element had transposed out of a pigment-producing gene, thereby allowing the gene's function to be restored. The cells in the white areas of the kernel lack pigment because a pigment-producing gene continues to be inactivated by the presence of a transposable element within that gene.

Box 7.1 Barbara McClintock (1902–1992)

Barbara McClintock's remarkable life spanned the history of genetics in the twentieth century. She was born in Hartford, Connecticut, to Sara Handy McClintock, an accomplished pianist, as well as a poet and painter, and Thomas Henry McClintock, a physician. Both parents were quite unconventional in their attitudes toward rearing children: They were interested in what their children would and could be rather than what they should be.

During her high school years, Barbara discovered science, and she loved to learn and figure things out. After high school, Barbara attended Cornell University, where she flourished both socially and intellectually. She enjoyed her social life, but her comfort with solitude and the tremendous joy she experienced in knowing, learning, and understanding things were to be the defining themes of her life. The decisions she made during her university years were consistent with her adamant individuality and self-containment. In Barbara's junior year, after a particularly exciting undergraduate course in genetics, her professor invited her to take a graduate course in genetics. After that, she was treated much like a graduate student. By the time she had finished her undergraduate course work, there was no question in her mind: She had to continue her studies of genetics.

At Cornell, genetics was taught in the plant-breeding department, which at the time did not take female graduate students. To circumvent this obstacle, McClintock registered in the botany department with a major in cytology and a minor in genetics and zoology. She began to work as a paid assistant to Lowell Randolph, a cytologist. McClintock and Randolph did not get along well and soon dissolved their working relationship, but as McClintock's colleague and lifelong friend Marcus Rhoades later wrote, "Their brief association was momentous because it led to the birth of maize cytogenetics." McClintock discovered that metaphase or late-prophase chromosomes in the first microspore mitosis were far better for cytological discrimination than were root-tip chromosomes. In a few weeks, she prepared detailed drawings of the maize chromosomes, which she published in *Science*.

This was McClintock's first major contribution to maize genetics, and it laid the groundwork for a veritable explosion of discoveries that connected the behavior of chromosomes to the genetic properties of an organism, defining the new field of cytogenetics. McClintock was awarded a Ph.D. in 1927 and appointed an instructor at Cornell, where she continued to work with maize. The Cornell maize genetics group was small and included Professor R. A. Emerson, the founder of maize genetics, McClintock, George Beadle, C. R. Burnham, Marcus Rhoades, and Lowell Randolph, together with a few graduate students. By all accounts, McClintock was the intellectual driving force of this talented group.

In 1929, a new graduate student, Harriet Creighton, joined the group and was guided by McClintock. Their work showed, for the first time, that genetic recombination is a reflection of the physical exchange of chromosome segments. (See Chapter 15, p. 409.) A paper on their work, published in 1931,

Barbara McClintock in 1947.

was perhaps McClintock's first seminal contribution to the science of genetics.

Although McClintock's fame was growing, she had no permanent position. Cornell had no female professors in fields other than home economics, so her prospects were dismal. She had already attained international recognition, but as a woman, she had little hope of securing a permanent academic position at a major research university. R. A. Emerson obtained a grant from the Rockefeller Foundation to support her work for two years, allowing her to continue to work independently. McClintock was discouraged and resentful of the disparity between her prospects and those of her male counterparts. Her extraordinary talents and accomplishments were widely appreciated, but she was also seen as difficult by many of her colleagues, in large part because of her quick mind and intolerance of second-rate work and thinking.

In 1936, Lewis Stadler convinced the University of Missouri to offer McClintock an assistant professorship. She accepted the position and began to follow the behavior of maize chromosomes that had been broken by X irradiation. However, soon after her arrival at Missouri, she understood that hers was a special appointment. She found herself excluded from regular academic activities, including faculty meetings. In 1941, she took a leave of absence from Missouri and departed with no intention of returning. She wrote to her friend Marcus Rhoades, who was planning to go to Cold Spring Harbor, New York, for the summer to grow his corn. An invitation for McClintock was arranged through Milislav Demerec (a member, and later the director, of the genetics department at the Carnegie Institution of Washington, then the dominant research laboratory at Cold Spring Harbor), who offered her a year's research appointment. Though hesitant to commit herself, McClintock accepted. When Demerec later offered her an appointment as a permanent member of the

Box 7.1   Barbara McClintock (cont.)

research staff, McClintock accepted, still unsure whether she would stay. Her dislike of making commitments was a given; she insisted that she would never have become a scientist in today's world of grants, because she could not have committed herself to a written research plan. It was the unexpected that fascinated her, and she was always ready to pursue an observation that didn't fit. Nevertheless, McClintock did stay at Carnegie until 1967.

At Carnegie, McClintock continued her studies on the behavior of broken chromosomes. She was elected to the National Academy of Sciences in 1944 and to the presidency of the Genetics Society of America in 1945. In those same two years, McClintock reported observing "an interesting type of chromosomal behavior" involving the repeated loss of one of the broken chromosomes from cells during development. What struck her as odd was that, in this particular stock, it was always chromosome 9 that broke, and it always broke at the same place. McClintock called the unstable chromosome site Dissociation (*Ds*), because "the most readily recognizable consequence of its actions is this dissociation." She quickly established that the *Ds* locus would "undergo dissociation mutations only when a particular dominant factor is present." She named the factor Activator (*Ac*), because it activated chromosome breakage at *Ds*. She also reached the extraordinary conclusion that *Ac* not only was required for *Ds*-mediated chromosome breakage, but also could destabilize previously stable mutations. But more than that, and unprecedentedly, the chromosome-breaking *Ds* locus could "change its position in the chromosome," a phenomenon she called *transposition*. Moreover, she had evidence that the *Ac* locus was required for the transposition of *Ds* and that, like the *Ds* locus, the *Ac* locus was mobile.

Within several years, McClintock had established beyond a doubt that both the *Ac* and *Ds* loci were capable not only of changing their positions on the genetic map, but also of inserting into loci to cause unstable mutations. She presented a paper on her work at the Cold Spring Harbor Symposium of 1951. The reaction to her presentation ranged from perplexed to hostile. Later she published several papers in refereed journals, but from the paucity of requests, for reprints, she inferred an equally cool reaction on the part of the larger biological community to the astonishing news that genes could move.

McClintock's work had taken her far outside the scientific mainstream, and in a profound sense she had lost her ability to communicate with her colleagues. By her own admission, McClintock had neither a gift for written exposition nor a talent for explaining complex phenomena in simple terms. But more important factors underlay her isolation: The very notion that genes can move contradicted the assumption of the regular relationships between genes that serves as a foundation for the construction of linkage maps and the physical mapping of genes onto chromosomes. The concept that genetic elements can move would undoubtedly have met with resistance regardless of its author and presentation.

McClintock was deeply frustrated by her failure to communicate, but her fascination with the unfolding story of transposition was sufficient to keep her working at the highest level of physical and mental intensity she could sustain. By the time of her formal retirement, she had accumulated a rich store of knowledge about the genetic behavior of two markedly different transposable-element families—and beginning about the time her active fieldwork ended, transposable genetic elements began to surface in one experimental organism after another.

These later discoveries came in an altogether different age. In the two decades between McClintock's original genetic discovery of transposition and its rediscovery, genetics had undergone as profound a change as the cytogenetic revolution that had occurred in the second and third decades of the century. The genetic material had been identified as DNA, the manner in which information is encoded in the genes had been deciphered, and methods had been devised to isolate and study individual genes. Genes were no longer abstract entities known only by the consequences of their alteration or loss; they were real bits of nucleic acids that could be isolated, visualized, subtly altered, and reintroduced into living organisms.

By the time the maize transposable elements were cloned and their molecular analysis initiated, the importance of McClintock's discovery of transposition was widely recognized, and her public recognition was growing. For example, she received the National Medal of Science in 1970, she was named Prize Fellow Laureate of the MacArthur Foundation and received the Lasker Basic Medical Research Award in 1981, and in 1982 she shared the Horwitz Prize. Finally, in 1983, 35 years after the publication of the first evidence for transposition, McClintock was awarded the Nobel Prize in Physiology or Medicine.

McClintock was sure she would die at 90, and a few months after her ninetieth birthday she was gone, drifting away from life gently, like a leaf from an autumn tree. What Barbara McClintock was and what she left behind are eloquently expressed in a few short lines written many years earlier by her friend and champion, Marcus Rhoades, whose death preceded hers by a few months:

One of the remarkable things about Barbara McClintock's surpassingly beautiful investigations is that they came solely from her own labors. Without technical help of any kind she has by virtue of her boundless energy, her complete devotion to science, her originality and ingenuity, and her quick and high intelligence made a series of significant discoveries unparalleled in the history of cytogenetics. A skilled experimentalist, a master at interpreting cytological detail, a brilliant theoretician, she has had an illuminating and pervasive role in the development of cytology and genetics.

Adapted by permission of Nina Fedoroff and by courtesy of the National Academy of Sciences, Washington, DC.

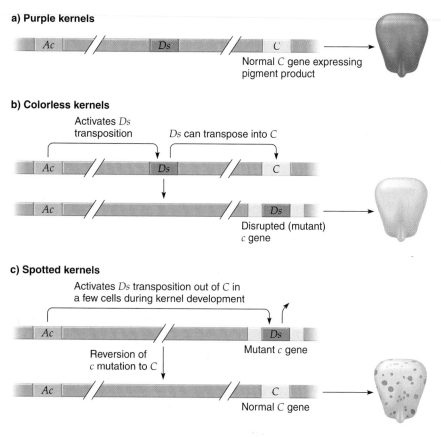

a) Purple kernels

Normal *C* gene expressing
pigment product

b) Colorless kernels

Activates *Ds*
transposition

Ds can transpose into *C*

Disrupted (mutant)
c gene

c) Spotted kernels

Activates *Ds* transposition out of *C* in
a few cells during kernel development

Mutant *c* gene

Reversion of
c mutation to *C*

Normal *C* gene

Figure 7.24

Kernel color and transposon effects in corn. (a) Purple kernels result from the active *C* gene. (b) Colorless kernels can result when the *Ac* transposon activates *Ds* transposition and *Ds* inserts into *C*, producing a mutation. (c) Spotted kernels result by reversion of the *c* mutation during kernel development when *Ac* activates *Ds* transposition out of the *C* gene.

First, let us consider transposition to a replicated chromosome site (Figure 7.25a). If one of the two *Ac* elements transposes to such a site, an empty donor site is left on one chromatid, and an *Ac* element remains in the homologous donor site on the other chromatid. The transposing *Ac* element inserts into a new, already replicated recipient site, which is often on the same chromosome. In Figure 7.25a, the site is shown on the same chromatid as the parental *Ac* element. Thus, in the case of transposition to an already replicated site, there is no net increase in the number of *Ac* elements.

Figure 7.25b shows the transposition of one *Ac* element to an unreplicated chromosome site. As in the first case, one of the two *Ac* elements transposes, leaving an empty donor site on one chromatid and an *Ac* element in the homologous donor site on the other chromatid. But now the transposing element inserts into a nearby recipient site that has yet to be replicated. When that region of the chromosome replicates, the result will be a copy of the transposed *Ac* element on both chromatids, in addition to the one original copy of the *Ac* element at the donor site on one chromatid. Thus, in the case of transposition to an unreplicated recipient site, there is a net increase in the number of *Ac* elements.

The transposition of most *Ds* elements occurs in the same way as *Ac* transposition, using transposase supplied by an *Ac* element in the genome.

KEYNOTE

The mechanism of transposition of plant transposons is similar to that of bacterial IS elements or transposons. Transposons integrate at a target site by a precise mechanism, so that the integrated elements are flanked at the insertion site by a short duplication of target-site DNA of a characteristic length. Many plant transposons occur in families, the autonomous elements of which are able to direct their own transposition and the nonautonomous elements of which are able to transpose only when activated by an autonomous element in the same genome. Most nonautonomous elements are derived from autonomous elements by internal deletions or complex sequence rearrangements.

***Ty* Elements in Yeast.** A *Ty* element is about 5.9 kb long and includes two directly repeated terminal sequences called *long terminal repeats* (LTR) or deltas (δ) (Figure 7.26). Each delta contains a promoter and sequences recognized by transposing enzymes. The *Ty* elements encode a single, 5,700-nucleotide mRNA that begins at the promoter in the delta at the 5′ end of the element. (See Figure 7.26.) The mRNA transcript contains two open reading frames (ORFs), designated *TyA* and *TyB*, that encode two different proteins required for transposition. The number of copies of this element varies

Figure 7.25

The *Ac* transposition mechanism. (a) Transposition to an already replicated recipient site results in no net increase in the number of *Ac* elements in the genome. (b) Transposition to an unreplicated recipient site results in a net increase in the number of *Ac* elements when the region of the chromosome containing the transposed element is replicated.

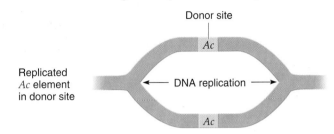

a) Transposition to an already replicated recipient site

b) Transposition to an unreplicated recipient site

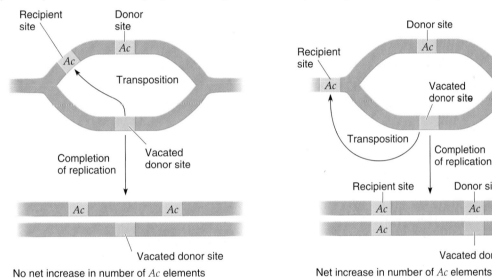

No net increase in number of *Ac* elements

Net increase in number of *Ac* elements

between strains, although, on average, a strain contains about 35.

The organization of yeast *Ty* elements is similar to that of **retroviruses**—single-stranded RNA viruses that replicate via double-stranded DNA intermediates. That is, when a retrovirus infects a cell, its RNA genome is copied by viral reverse transcriptase, producing a double-stranded DNA. The DNA integrates into the host's chromosome, where it can be transcribed to produce progeny RNA viral genomes and mRNAs for viral proteins. HIV, the virus responsible for AIDS in humans, is a retrovirus. As a result of the similarity to retroviruses, *Ty* elements were hypothesized to transpose, not by a DNA-to-DNA mechanism, but by making an RNA copy of the integrated DNA sequence and then creating a new *Ty* element by reverse transcription. The new element would then integrate at a new chromosome location.

Evidence substantiating the hypothesis was obtained through experiments in which DNA manipulation techniques modified *Ty* elements so that they had special features which enabled their transposition to be monitored

easily. One compelling piece of evidence came from experiments in which an intron was placed into the *Ty* element (there are no introns in normal *Ty* elements) and the element was monitored from its initial placement through the transposition event. At the new location, the *Ty* element no longer had the intron sequence, a fact which had to mean that transposition occurred via an RNA intermediate. The

Figure 7.26

The *Ty* transposable element of yeast.

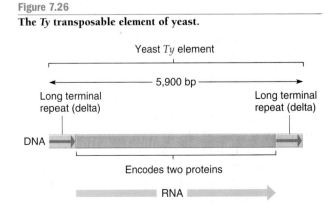

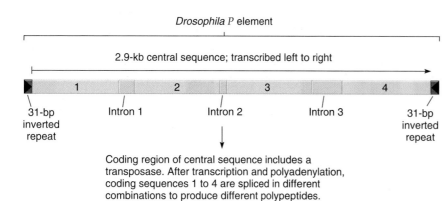

Figure 7.27

Structure of the autonomous *P* transposable element found in *Drosophila melanogaster*.

intron had been removed by the usual splicing processes that excise introns from pre-mRNAs.

Subsequently, it was shown that *Ty* elements encode a reverse transcriptase. Moreover, *Ty* viruslike particles containing *Ty* RNA and reverse transcriptase activity have been identified in yeast cells. Because of their similarity to retroviruses in this regard, *Ty* elements are called **retrotransposons** and the transposition process is called **retrotransposition.**

***Drosophila* Transposons.** A number of classes of transposons have been identified in *Drosophila*. In this organism, it is estimated that about 15 percent of the genome is mobile—a remarkable percentage.

The *P element* is an example of a family of transposons in *Drosophila*. *P* elements vary in length from 500 to 2,900 bp, and each has terminal inverted repeats. The shorter *P* elements are nonautonomous elements, while the longest *P* elements are autonomous elements that encode a transposase needed for transposition of all the *P* elements (Figure 7.27). Insertion of a *P* element into a new site results in a direct repeat of the target site.

P elements are important vectors for transferring genes into the germ line of *Drosophila* embryos, allowing genetic manipulation of the organism. Figure 7.28 illustrates an experiment by G. M. Rubin and A. G. Spradling in which the wild-type *rosy*⁺ (*ry*⁺) gene was introduced into a strain homozygous for a mutant *rosy* allele. The wild-type *rosy* gene was introduced into the middle of a *P* element by recombinant DNA techniques and cloned in a plasmid vector. (See Chapter 8, pp. 181–182.) The plasmids were then microinjected into *rosy* embryos in the regions that would become the germ-line cells. *P* element-encoded transposase then catalyzed the movement of the *P* element, along with the wild-type *rosy* gene it contained, to the *Drosophila* genome in some of the germ-line cells. When the flies that developed from these embryos produced gametes, they contained the wild-type *rosy* gene, so descendants of those flies had normal eye color. In principle, any gene can be transferred into the fly's genome in this way.

Figure 7.28

Illustration of the use of *P* elements to introduce genes into the *Drosophila* genome.

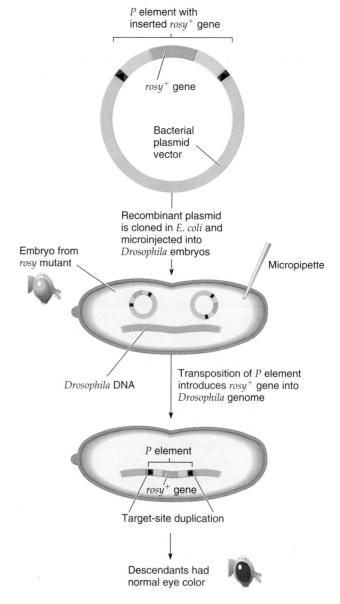

KEYNOTE

> Transposable genetic elements in eukaryotes typically are transposons. Transposons can transpose to new sites while leaving a copy behind in the original site, or they can excise themselves from the chromosome. When the excision is imperfect, deletions can occur, and by various recombination events, other chromosomal rearrangements, such as inversions and duplications, can occur. Whereas most transposons move by using a DNA-to-DNA mechanism, some eukaryotic transposons, such as yeast *Ty* elements, transpose via an RNA intermediate (using a transposon-encoded reverse transcriptase) and so resemble retroviruses.

Human Retrotransposons. In Chapter 2, we discussed the different repetitive classes of DNA sequences found in the genome. Of relevance here are the LINEs (long interspersed sequences) and SINEs (short interspersed sequences) found in the moderately repetitive class of sequences. **LINEs** are repeated sequences more than 5,000 bp long, interspersed among unique-sequence DNA up to approximately 35,000 bp long. **SINEs** are 100- to 400-bp repeated sequences interspersed between unique-sequence DNA 1,000 to 2,000 bp long. Both LINEs and SINEs occur in DNA families whose members are related by sequence.

LINEs and SINEs are retrotransposons. Full-length LINEs are autonomous elements that encode the enzymes for their own retrotransposition and for that of LINEs with internal deletions—nonautonomous derivatives. Those enzymes are also required for the transposition of SINEs, which are nonautonomous elements.

About 20 percent of the human genome consists of LINEs, with one-quarter of them being L1, the best-studied LINE. The maximum length of L1 elements is 6,500 bp, although only about 3,500 of them in the genome are of that full length, the rest having internal deletions of various length (much as corn *Ds* elements have). The full-length L1 elements contain a large open reading frame that is homologous to known reverse transcriptases. When the yeast *Ty* element reverse transcriptase gene was replaced with the putative reverse transcriptase gene from L1, the *Ty* element was able to transpose. Point mutations introduced into the sequence abolished the enzyme activity, indicating that the L1 sequence can indeed make a functional reverse transcriptase. Thus, like corn *Ac* elements, full-length L1 elements (and full-length LINEs of other families) are autonomous elements. L1 and other LINEs do not have LTRs, so they are not closely related to the retrotransposons we have already discussed. Therefore, while transposition is via an RNA intermediate, the mechanism is different. Interestingly, in 1991, two unrelated cases of hemophilia (OMIM 306700) in children were shown to result from insertions of an L1 element into the factor VIII gene, the product of which is required for normal blood clotting. Molecular analysis showed that the insertion was not present in either set of parents, leading to the conclusion that the L1 element had newly transposed. More generally, these results show that L1 elements in humans can transpose and that they can cause disease by inserting into genes, a process called *insertional mutagenesis*.

SINEs are also retrotransposons, but none of them encodes enzymes needed for transposition. Hence, they are nonautonomous elements, and they depend upon the enzymes encoded by LINEs for their transposition. In humans, a very abundant SINE family is the *Alu family*. The repeated sequence in this family is about 300 bp long and is repeated 300,000 to 500,000 times in the genome, amounting to up to 3 percent of the total genomic DNA. The name for the family comes from the fact that the sequence contains a restriction site for the enzyme *AluI* ("Al-you-one").

Evidence that Alu sequences can transpose has come from the study of a young male patient with neurofibromatosis (OMIM 162200), a genetic disease caused by an autosomal dominant mutation. Individuals with neurofibromatosis develop tumorlike growths (neurofibromas) over the body. (See Chapter 13, p. 359.) DNA analysis showed that an Alu sequence was present in one of the introns of the neurofibromatosis gene of the patient. RNA transcripts from this gene are longer than those from normal individuals. The presence of the Alu sequence in the intron disrupts the processing of the transcript, causing one exon to be lost completely from the mature mRNA. As a result, the protein encoded is 800 amino acids shorter than normal and is nonfunctional. Neither parent of the patient has neurofibromatosis, and neither has an Alu sequence in the neurofibromatosis gene. Individual members of the Alu family are not identical in sequence, having diverged over evolutionary time. This divergence made it possible to track down the same Alu sequence in the patient's parents. The analysis showed that an Alu sequence probably inserted into the neurofibromatosis gene by retrotransposition in the germ line of the father.

Summary

In this chapter, we saw that genetic damage can occur to DNA spontaneously through replication errors or through exposure to radiation or chemical mutagens. If the genetic damage is not repaired, mutations result, and if there has been too much damage, cell death may result. We also saw that the movement of transposable elements can result in changes in gene expression (if the element transposes into a gene or into a promoter of a gene), or in chromosomal mutations.

DNA Mutation

Mutations occur spontaneously at low rates. Mutation rates can be increased through the use of mutagens such as radiation and certain chemicals. Chemical mutagens work in a number of different ways, such as by acting as base analogs, by modifying bases, or by intercalating into the DNA. Intercalating results in frameshift mutations, and the other processes result in base-pair substitution mutations. The Ames test can indicate whether chemicals (such as environmental or commercial chemicals) have the potential to cause mutations in humans. A large number of potential human carcinogens have been found in this way.

Repair of DNA Damage

Cells of all organisms possess a number of repair mechanisms that correct most damage to DNA. DNA damage that is not repaired is called a mutation. The collective array of repair enzymes reduces mutation rates for spontaneous errors by several orders of magnitude. However, such repair mechanisms cannot cope with the extensive amount of DNA damage that arises from high dosages of chemical mutagens or UV irradiation, whether from the environment or applied experimentally, so, typically, many mutations result.

We also considered some examples of how a mutagenized population of cells can be screened for particular mutants of interest.

Transposable Elements

Bacteria and eukaryotic cells contain a variety of transposable elements that have the property of moving from one site to another in the genome. We discussed two important types of transposable elements in bacteria: insertion sequence (IS) elements and transposons (Tn's). The simplest type of transposable element is an IS element, which typically consists of terminal inverted repeat sequences flanking a coding region, the products of which provide transposition activity. Tn elements are more complex in that they contain other genes. There are two types of prokaryotic transposons. Composite transposons consist of a central region flanked on both sides by IS elements. The central region contains genes such as those which confer drug resistance. The IS elements contain the genes encoding the proteins required for transposition. Noncomposite transposons consist of a central region containing

genes (again, such as those which confer drug resistance), but they do not end with IS elements. Instead, short repeated sequences that are required for transposition are found at their ends. In these transposons, the transposition functions are encoded by genes in the central region.

Transposable elements in eukaryotes resemble bacterial transposons in general structure and transposition properties. Corn transposons often occur as families, with each family containing an autonomous element (an element capable of transposing by itself) and one or more nonautonomous elements (elements that can transpose only if the autonomous element of the family is also present in the genome). Interestingly, and perhaps uniquely, the timing and frequency of transposition of corn transposons are developmentally regulated. Some eukaryotic transposons, such as yeast *Ty* elements, and human LINE and SINE family members transpose via an RNA intermediate rather than by a DNA-to-DNA mechanism. These types of transposons resemble retroviruses in genome organization and other properties and are therefore called retrotransposons.

The presence of bacterial transposons (IS and Tn elements) and eukaryotic transposons in a cell usually is detected by the changes they bring about in the expression and activities of the genes at or near the chromosomal sites into which they integrate. Gene expression can be increased or decreased if the element inserts into a promoter or some other regulatory sequence, mutant alleles of a gene can be produced if an element inserts within the coding sequence of the gene, and various chromosome rearrangements or chromosome breakage events can occur as a result of the mechanics of transposition. In addition, gene transcription can be turned on next to a transposon because of the action of promoters in the transposons themselves. Often, mutant alleles produced by the insertion of a transposable element are unstable because they revert when the transposable element undergoes a new transposition.

Analytical Approaches to Solving Genetics Problems

Q7.1 Five strains of *E. coli* containing base-substitution mutations that affect the tryptophan synthetase A polypeptide have been isolated. Figure 7.A shows the

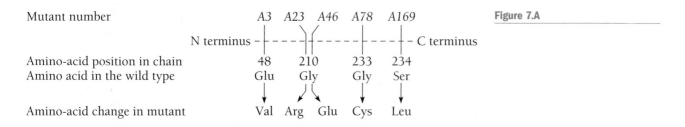

Figure 7.A

changes produced in the protein itself in the indicated mutant strains. In addition, *A23* can be further mutated to insert Ile, Thr, Ser, or the wild-type Gly into position 210.

In the following questions, assume that only a single base change can occur at each step:

a. Using the genetic code (see Figure 6.7, p. 118), explain how the two mutations *A23* and *A46* can result in two different amino acids being inserted at position 210. Give the nucleotide sequence of the wild-type gene at that position and of the two mutants.

b. Can mutants *A23* and *A46* recombine? Why or why not? If recombination can occur, what would be the result?

c. From what you can infer of the nucleotide sequence in the wild-type gene, indicate, for the codons specifying amino acids 48, 210, 233, and 234, whether a nonsense mutant could be generated by a single nucleotide substitution in the gene.

A7.1

a. There are no simple ways to answer questions like this one. The best approach is to scrutinize the genetic-code dictionary and use a pencil and paper to try to define the codon changes that are compatible with all the data. The number of amino-acid changes in position 210 of the polypeptide is helpful in this case. The wild-type amino acid is Gly, and the codons for Gly are GGU, GGC, GGA, and GGG. The *A23* mutant has Arg at position 210, and the arginine codons are AGA, AGG, GGU, GGC, GGA and GGG. Any Arg codon could be generated by a single base change. We have to look at the amino acids at 210 generated by further mutations of *A23*. In the case of Ile, the codons are AUU, AUC, and AUA. The *only* way to get from Gly to Arg in one base change and then to Ile in a subsequent single base change is GGA (Gly) → AGA (Arg) → AUA (Ile). Is this change compatible with the other mutational changes from *A23*? There are four possible Thr codons—ACU, ACC, ACA, and ACG—so a mutation from AGA (Arg) to ACA (Thr) would fit. There are six possible Ser codons—UCU, UCC, UCA, UCG, AGU, and AGC—so a mutation from AGA to either AGU or AGC would fit.

As regards the *A46* mutant, the possible codons for Glu are GGA and GAG. Given that the wild-type codon is GGA (Glu), the only possible single base change that gives Glu is if the Glu codon in the mutant is GAA. So the answer to the question is that the wild-type sequence at position 210 is GGA, the sequence in the *A23* mutant is AGA, and the sequence in the *A46* mutant is GAA. In other words, the *A23*

and *A46* mutations are in different bases of the codon.

b. The answer to this question follows from the answer deduced in part a. Mutants *A23* and *A46* can recombine because the mutations in the two mutant strains are in different base pairs. The results of a single recombination event (at the DNA level) between the first and second base of the codon in AGA × GAA are a wild-type GGA codon (Gly) and a double mutant AAA codon (Lys). Recombination can also occur between the second and third bases of the codon, but the products are AGA and GAA—that is, identical to the parents.

c. Amino acid 48 had a Glu-to-Val change. This change must have involved GAA to GUA or GAG to GUG. In either case, the Glu codon can mutate with a single base-pair change to a nonsense codon, UAA or UAG, respectively.

Amino acid 210 in the wild type has a GGA codon, as we have already discussed. This gene could mutate to the UGA nonsense codon with a single base-pair change.

Amino acid 233 had a Gly-to-Cys change. This change must have involved either GGU to UGU or GGC to UGC. In either case, the Gly codon cannot mutate to a nonsense codon with one base-pair change.

Amino acid 234 had a Ser-to-Leu change. This change was either UCA to UUA or UCG to UUG. If the Ser codon was UCA, it could be changed to AGA in one step, but if the Ser codon was UCG, it cannot change to a nonsense codon in one step.

Q7.2 The chemically induced mutations *a*, *b*, and *c* show specific reversion patterns when subjected to treatment by the following mutagens: 2-aminopurine (AP), 5-bromouracil (BU), proflavin (pro), and hydroxylamine (HA). AP is a base-analog mutagen that induces mainly AT-to-GC changes and can cause GC-to-AT changes also. BU is a base-analog mutagen that induces mainly GC-to-AT changes and can cause AT-to-GC changes also. PRO is an intercalating agent that can cause a single base-pair addition or deletion with no specificity. HA is a base-modifying agent that modifies cytosine, causing one-way transitions from GC to AT. The reversion patterns are shown in the following table:

	Mutagens Tested in Reversion Studies			
Mutation	**AP**	**BU**	**PRO**	**HA**
a	−	−	+	−
b	+	+	+	+
c	+	+	+	−

(Note: "+" indicates that many reversions to wild type were found; "−" indicates that no reversions or very few reversions to wild type were found.)

For each original mutation (a^+ to a, b^+ to b, etc.), indicate the probable base-pair change (AT to GC, deletion of GC, etc.) and the mutagen that was probably used to induce the original change.

A7.2 This question tests your knowledge of the base-pair changes that can be induced by the various mutagens used.

Mutagen AP induces mainly AT-to-GC changes and can cause GC-to-AT changes also. Thus, AP-induced mutations can be reverted by AP.

Base-analog mutagen BU induces mainly GC-to-AT changes and can cause AT-to-GC changes, so BU-induced mutations can be reverted by BU.

Proflavin causes single base-pair deletions or additions, so proflavin-induced changes can be reverted by a second treatment with proflavin.

Mutagen HA causes one-way transitions from GC to AT, so HA-induced mutations cannot be reverted by HA.

With these mutagen specificities in mind, we can answer the questions about each mutation in turn.

Mutation a^+ to a: The a mutation was reverted only by proflavin, indicating that it was a deletion or an addition (a frameshift mutation). Therefore, the original mutation was induced by an intercalating agent such as proflavin, because that is the only class of mutagen that can cause an addition or a deletion.

Mutation b^+ to b: The b mutation was reverted by AP, BU, or HA. A key here is that HA causes only GC-to-AT changes. Therefore, b must be GC, and the original b^+ must have been AT. Thus, the mutational change of b^+ to b must have been caused by treatment with AP or BU, because these are the only two mutagens in the list that can induce that change.

Mutation c^+ to c: The c mutation was reverted only by AP and BU. Since it could not be reverted by HA, c must be AT and c^+ must be GC. The mutational change from c^+ to c therefore involved a GC-to-AT transition and could have resulted from treatment with AP, BU, or HA.

Q7.3 Imagine that you are a corn geneticist and are interested in a gene you call *zma*, which is involved in the formation of the tiny hairlike structures on the upper surfaces of leaves. You have a cDNA clone of this gene. In a particular strain of corn that contains many copies of *Ac* and *Ds*, but no other transposable elements, you observe a mutation of the *zma* gene. You want to figure out whether this mutation involves the insertion of a transposable element into the *zma* gene. How would you proceed? Suggest at least two approaches, and say how your expectations for an inserted transposable element would differ from your expectations for an ordinary gene mutation.

A7.3 One approach would be to make a detailed examination of leaf surfaces in mutant plants. Since there are many copies of *Ac* in the strain, if a transposable element has inserted into *zma*, it should be able to leave again, so that the mutation of *zma* would be unstable. The leaf surfaces should then show a patchy distribution of regions with and regions without the hairlike structures. A simple point mutation would be expected to be more stable.

A second approach would be to digest the DNA from mutant plants and the DNA from normal plants with a particular restriction endonuclease, run the digested DNA on a gel, prepare a Southern blot, and probe the blot using the cDNA. If a transposable element has inserted into the *zma* gene in the mutant plants, then the probe should bind to fragments of different molecular weight in mutant, compared with normal, DNA. This would not be the case if a simple point mutation had occurred.

Questions and Problems

*7.1 Mutations are (choose the correct answer)
a. caused by genetic recombination
b. heritable changes in genetic information
c. caused by faulty transcription of the genetic code
d. usually, but not always, beneficial to the development of the individuals in which they occur

7.2 Answer true or false: Mutations occur more frequently if there is a need for them.

*7.3 Which of the following is *not* a class of mutation?
a. frameshift
b. missense
c. transition
d. transversion
e. none of the above (meaning that all are classes of mutation)

*7.4 Ultraviolet light usually causes mutations by a mechanism involving (choose the correct answer)
a. one-strand breakage in DNA
b. light-induced change of thymine to alkylated guanine
c. induction of thymine dimers and their persistence or imperfect repair
d. inversion of DNA segments
e. deletion of DNA segments
f. all of the above

7.5 The amino-acid sequence shown in the following table was obtained from the central region of a particular polypeptide chain in the wild-type and several mutant bacterial strains:

	Codon								
	1	2	3	4	5	6	7	8	9
a. Wild type:	... Phe	Leu	Pro	Thr	Val	Thr	Thr	Arg	Trp
b. Mutant 1:	... Phe	Leu	His	His	Gly	Asp	Asp	Thr	Val
c. Mutant 2:	... Phe	Leu	Pro	Thr	Met	Thr	Thr	Arg	Trp
d. Mutant 3:	... Phe	Leu	Pro	Thr	Val	Thr	Thr	Arg	
e. Mutant 4:	... Phe	Pro	Pro	Arg					
f. Mutant 5:	... Phe	Leu	Pro	Ser	Val	Thr	Thr	Arg	Trp

For each mutant, say what change has occurred at the DNA level, whether the change is a base-pair substitution mutation (transversion or transition, missense or nonsense) or a frameshift mutation, and in which codon the mutation occurred. (Refer to the codon dictionary in Figure 6.7, p. 118.)

*7.6 In mutant strain X of *E. coli*, a leucine tRNA that recognizes the codon 5′-CUG-3′ in normal cells has been altered so that it now recognizes the codon 5′-GUG-3′. A missense mutation that affects amino acid 10 of a particular protein is suppressed in mutant X cells.

a. What are the anticodons of the two Leu tRNAs, and what mutational event has occurred in mutant X cells?

b. What amino acid would normally be present at position 10 of the protein (without the missense mutation)?

c. What amino acid is put in at position 10 if the missense mutation is not suppressed (i.e., in normal cells)?

d. What amino acid is inserted at position 10 if the missense mutation is suppressed (i.e., in mutant X cells)?

7.7 A researcher using a model eukaryotic experimental system has identified a temperature-sensitive mutation, *rpIIA^ts*, in a gene that encodes a protein subunit of RNA polymerase II. This mutation is a missense mutation. Mutants have a recessive lethal phenotype at the higher, restrictive temperature, but grow at the lower, permissive (normal) temperature. To identify genes whose products interact with the subunit of RNA polymerase II, the researcher designs a screen to isolate mutations that will act as dominant suppressors of the temperature-sensitive recessive lethal mutation.

a. Explain how a new mutation in an interacting protein could suppress the lethality of the temperature-sensitive original mutation.

b. In addition to mutations in interacting proteins, what other type of suppressor mutations might be found?

c. Outline how the researcher might select for the new suppressor mutations.

d. Do you expect the frequency of suppressor mutations to be similar to, much greater than, or much less than the frequency of new mutations at a typical eukaryotic gene?

e. How might this approach be used generally to identify genes whose products interact to control transcription?

*7.8 The mutant *lacZ-1* was induced by treating *E. coli* cells with acridine, whereas *lacZ-2* was induced with 5BU. What kinds of mutants are these likely to be? Explain. How could you confirm your predictions by studying the structure of the β-galactosidase in these cells?

*7.9

a. The sequence of nucleotides in an mRNA is

5′-AUGACCCAUUGGUCUCGUUAG-3′

Assuming that ribosomes could translate this mRNA, how many amino acids long would you expect the resulting polypeptide chain to be?

b. Hydroxylamine is a mutagen that results in the replacement of an AT base pair for a GC base pair in the DNA; that is, it induces a transition mutation. When hydroxylamine was applied to the organism that made the mRNA molecule shown in part (a), a strain was isolated in which a mutation occurred at the 11th position of the DNA that coded for the mRNA. How many amino acids long would you expect the polypeptide made by this mutant to be? Why?

7.10 In a series of 94,075 babies born in a particular hospital in Copenhagen, 10 were achondroplastic dwarfs (an autosomal dominant condition). Two of these 10 had an achondroplastic parent. The other 8 achondroplastic babies each had two normal parents. What is the apparent mutation rate at the achondroplasia locus?

*7.11 Three of the codons in the genetic code are chain-terminating codons for which no naturally occurring tRNAs exist. Just like any other codons in the DNA, though, these codons can change as a result of base-pair changes in the DNA. Confining yourself to single base-pair changes at a time, determine which amino acids could be inserted into a polypeptide by mutation of these chain-terminating codons: (a) UAG, (b) UAA, and (c) UGA. (The genetic code is listed in Figure 6.7, p. 118.)

7.12 The amino-acid substitutions in the following figure occur in the α and β chains of human hemoglobin.

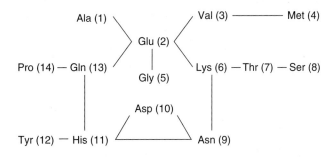

Those amino acids connected by lines are related by single-nucleotide changes. Propose the most likely codon or codons for each of the numbered amino acids. (Refer to the genetic code in Figure 6.7, p. 118.)

*7.13 Charles Yanofsky studied the tryptophan synthetase of *E. coli* in an attempt to identify the base sequence specifying this protein. The wild type gave a protein with a glycine in position 38. Yanofsky isolated two *trp* mutants: *A23* and *A46*. Mutant *A23* had Arg instead of Gly at position 38, and mutant *A46* had Glu at position 38. Mutant *A23* was plated on minimal medium, and four spontaneous revertants to prototrophy were obtained. The tryptophan synthetase from each of the four revertants was isolated, and the amino acids at position 38 were identified. Revertant 1 had Ile, revertant 2 had Thr, revertant 3 had Ser, and revertant 4 had Gly. In a similar fashion, three revertants from *A46* were recovered, and the tryptophan synthetase from each was isolated and studied. At position 38, revertant 1 had Gly, revertant 2 had Ala, and revertant 3 had Val. A summary of these data is given in the following figure:

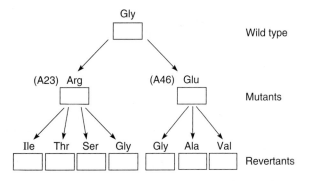

Using the genetic code in Figure 6.7, p. 118, deduce the codons for the wild type, for the mutants *A23* and *A46*, and for the revertants, and place each designation in the space provided in the figure.

7.14 Consider an enzyme chewase from a theoretical microorganism. In the wild-type cell, chewase has the following sequence of amino acids at positions 39 to 47 (reading from the amino end) in the polypeptide chain:

-Met-Phe-Ala-Asn-His-Lys-Ser-Val-Gly-
   39   40   41   42   43   44   45   46   47

A mutant organism that lacks chewase activity was obtained. The mutant was induced by a mutagen known to cause single base-pair insertions or deletions. Instead of making the complete chewase chain, the mutant made a short polypeptide chain only 45 amino acids long. The first 38 amino acids were in the same sequence as the first 38 of the normal chewase, but the last seven amino acids were as follows:

-Met-Leu-Leu-Thr-Ile-Arg-Val
   39   40   41   42   43   44   45

A partial revertant of the mutant was induced by treating it with the same mutagen. The revertant that made a partly active chewase has the following sequence of amino acids at positions 39 to 47 in its amino acid chain:

-Met-Leu-Leu-Thr-Ile-Arg-Gly-Val-Gly-
   39   40   41   42   43   44   45   46   47

Using the genetic code given in Figure 14.7, p. 382, deduce the nucleotide sequences for the mRNA molecules that specify this region of the protein in each of the three strains.

7.15 DNA polymerases from different organisms differ in the fidelity of their nucleotide insertion; however, even the best DNA polymerases make mistakes, usually mismatches. If such mismatches are not corrected, they can become fixed as mutations after the next round of replication.
   a. How does DNA polymerase attempt to correct mismatches during DNA replication?
   b. What mechanism is used to repair such mismatches if they escape detection by DNA polymerase?
   c. How is the mismatched base in the newly synthesized strand distinguished from the correct base in the template strand?

7.16 Two mechanisms in *E. coli* were described for the repair of thymine dimer formation after exposure to ultraviolet light: photoreactivation and excision (dark) repair. Compare these mechanisms, indicating how each achieves repair.

7.17 DNA damage by mutagens has serious consequences for DNA replication. Without specific base pairing, the replication enzymes cannot specify a complementary strand, and gaps are left after the passing of a replication fork.

a. What response has *E. coli* developed to large amounts of DNA damage by mutagens? How is this response coordinately controlled?

b. Why is the response itself a mutagenic system?

c. What effects would loss-of-function mutations in *recA* or *lexA* have on *E. coli*'s response?

*7.18 After a culture of *E. coli* cells was treated with the chemical 5-bromouracil, it was noted that the frequency of mutants was much higher than normal. Mutant colonies were then isolated, grown, and treated with nitrous acid; some of the mutant strains reverted to wild type.

a. In terms of the Watson–Crick model, diagram a series of steps by which 5BU may have produced the mutants.

b. Assuming that the revertants were not caused by suppressor mutations, indicate the steps by which nitrous acid may have produced the back mutations.

7.19 A single, hypothetical strand of DNA is composed of the following base sequence, where A indicates adenine, T indicates thymine, G indicates guanine, C denotes cytosine, U denotes uracil, BU is 5-bromouracil, 2AP is 2-amino-purine, BU-enol is a tautomer of 5BU, 2AP-imino is a rare tautomer of 2AP, HX is hypoxanthine, X is xanthine, and 5′ and 3′ are the numbers of the free, OH-containing carbons on the deoxyribose part of the terminal nucleotides:

    5′-T-HX-U-A-G-BU-enol-2AP-C-BU-X-2AP-imino-3′

a. Opposite the bases of the hypothetical strand, and using the shorthand of the base sequence, indicate the sequence of bases on a complementary strand of DNA.

b. Indicate the direction of replication of the new strand by drawing an arrow next to the new strand of DNA from part (a).

c. When postmeiotic germ cells of a higher organism are exposed to a chemical mutagen before fertilization, the resulting offspring expressing an induced mutation are almost always mosaics for wild-type and mutant tissue. Give at least one reason that these mosaics, and not so-called complete or whole-body mutants, are found in the progeny of treated individuals.

The following information applies to Problems 7.20 through 7.24: A solution of single-stranded DNA is used as the template in a series of reaction mixtures and has the base sequence

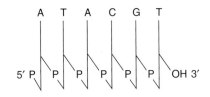

where A = adenine, G = guanine, C = cytosine, T = thymine, H = hypoxanthine, and HNO_2 = nitrous acid. Use the shorthand system shown in the sequence, and draw the products expected from the reaction mixtures. Assume that a primer is available in each case.

7.20 The DNA template + DNA polymerase + dATP + dGTP + dCTP + dTTP + Mg^{2+}.

*7.21 The DNA template + DNA polymerase + dATP + dGMP + dCTP + dTTP + Mg^{2+}.

7.22 The DNA template + DNA polymerase + dATP + dHTP + dGMP + dTTP + Mg^{2+}.

*7.23 The DNA template is pretreated with HNO_2 + DNA polymerase + dATP + dGTP + dCTP + dTTP + Mg^{2+}.

7.24 The DNA template + DNA polymerase + dATP + dGMP + dHTP + dCTP + dTTP + Mg^{2+}.

7.25 A strong experimental approach to determining the mode of action of mutagens is to examine the revertibility of the products of one mutagen by other mutagens. The following table presents data on the revertibility of *rII* mutations in phage T2 by various mutagens (" + " indicates majority of mutants reverted, "−" indicates almost no reversion; BU = 5-bromouracil, AP = 2-aminopurine, NA = nitrous acid, and HA = hydroxylamine):

Mutation Induced by	Proportion of Mutations Reverted by				Base-pair Substitution Inferred
	BU	AP	NA	HA	
BU	+			−	
AP		−	+		
NA	+	+		+	
HA			+	−	GC → AT

Fill in the empty spaces.

7.26

a. Nitrous acid deaminates adenine to form hypoxanthine, which forms two hydrogen bonds with cytosine during DNA replication. After a wild-type strain of bacteria is treated with nitrous acid, a mutant is recovered that is caused by an amino-acid substitution in a protein: wild-type methionine (Met) has been replaced with valine (Val) in the mutant. What is the simplest explanation for this observation?

b. Hydroxylamine adds a hydroxyl (OH) group to cytosine, causing it to pair with adenine. Could mutant organisms like those in part (a) be back mutated (returned to normal) using hydroxylamine? Explain.

*7.27 A wild-type strain of bacteria produces a protein with the amino acid proline (Pro) at one site. Treatment of the strain with nitrous acid, which deaminates C to make it U, produces two different mutants. One mutant has a substitution of serine (Ser), the other has a substitution of leucine (Leu), at the site.

Treatment of the two mutants with nitrous acid now produces new mutant strains, each with phenylalanine (Phe) at the site. Treatment of these new Phe-carrying mutants with nitrous acid then produces no change. The results are summarized in the following figure:

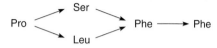

Using the appropriate codons, show how it is possible for nitrous acid to produce these changes and why further treatment has no influence. (Assume that only single-nucleotide changes occur at each step.)

*7.28 Three *ara* mutants of *E. coli* were induced by mutagen X. The ability of other mutagens to cause the reverse change (*ara* to *ara*⁺) was tested, with the results shown in Figure 7.B.

Assume that all *ara*⁺ cells are true revertants. What base changes were probably involved in forming the three original mutations? What kinds of mutations are caused by mutagen X?

7.29 As genes have been cloned for a number of human diseases caused by defects in DNA repair and replication, striking evolutionary parallels have been found between human and bacterial DNA repair systems. Discuss the features of DNA repair systems that appear to be shared in these two types of organism.

7.30 Distinguish between prokaryotic insertion elements and transposons. How do composite transposons differ from noncomposite transposons?

7.31 What properties do bacterial and eukaryotic transposable elements have in common?

7.32 An IS element became inserted into the *lacZ* gene of *E. coli*. Later, a small deletion occurred that removed 40 base pairs near the left border of the IS element. The deletion removed 10 *lacZ* base pairs, including the left copy of the

target site, and the 30 leftmost base pairs of the IS element. What will be the consequence of this deletion?

7.33 Although the detailed mechanisms by which transposable elements transpose differ widely, some features underlying transposition are shared. Examine the shared and different features by answering the following questions:
a. Use an example to illustrate different transposition mechanisms that require
 i. DNA replication of the element
 ii. no DNA replication of the element
 iii. an RNA intermediate
b. What evidence is there that the inverted or direct terminal repeat sequences found in transposable elements are essential for transposition?
c. Do all transposable elements generate a target-site duplication after insertion?

7.34 In addition to single gene mutations caused by the insertion of transposable elements, the frequency of chromosomal aberrations such as deletions or inversions can be increased when transposable elements are present. How?

7.35 A geneticist was studying glucose metabolism in yeast and deduced both the normal structure of the enzyme glucose-6-phosphatase (G6Pase) and the DNA sequence of its coding region. She was using a wild-type strain called A to study another enzyme for many generations when she noticed that a morphologically peculiar mutant had arisen from one of the strain A cultures. She grew the mutant up into a large stock and found that the defect in this mutant involved a markedly reduced G6Pase activity. She isolated the G6Pase protein from these mutant cells and found that it was present in normal amounts, but had an abnormal structure. The N-terminal 70 percent of the protein was normal. The C-terminal 30 percent was present, but altered in sequence by a frameshift reflecting the insertion of 1 base pair, and the N-terminal 70 percent and the C-terminal 30 percent were separated by 111 new amino acids unrelated to normal G6Pase. These amino acids represented predominantly the AT-rich codons (Phe, Leu, Asn, Lys, Ile, and Tyr). There were also two extra amino acids added at the C-terminal end. Explain these results.

*7.36 Consider two theoretical yeast transposons, A and B. Each contains an intron, and each transposes to a new

Figure 7.B

	Frequency of *ara*⁺ Cells among Total Cells after Treatment				
	Mutagen				
Mutant	**None**	**BU**	**AP**	**HA**	**Frameshift**
ara-1	1.5×10^{-8}	5×10^{-5}	1.3×10^{-4}	1.3×10^{-8}	1.6×10^{-8}
ara-2	2×10^{-7}	2×10^{-4}	6×10^{-5}	3×10^{-5}	1.6×10^{-7}
ara-3	6×10^{-7}	10^{-5}	9×10^{-6}	5×10^{-6}	6.5×10^{-7}

location in the yeast genome. Suppose you then examine the transposons for the presence of the intron. In the new locations, you find that A has no intron, but B does. From these facts, what can you conclude about the mechanisms of transposon movement for A and B?

7.37 After the discovery that *P* elements could be used to develop transformation vectors in *Drosophila melanogaster*, attempts were made to use them for the development of germ-line transformation in several different insect species.

Charalambos Savakis and his colleagues successfully used a different transposable element found in *Drosophila*—the *Minos* element—to develop germ-line transformation in that organism and in the medfly, *Ceratitus capitata*, a major agricultural pest present in Mediterranean climates.

a. What is the value of developing a transformation vector for an insect pest?

b. What basic information about the *Minos* element would need to be gathered before it could be used for germ-line transformation?

8

Recombinant DNA Technology

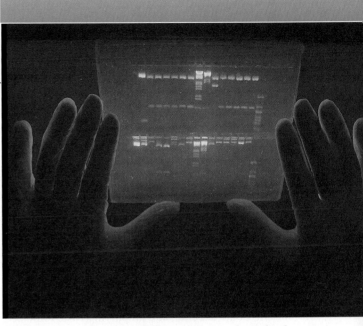

DNA fragments separated by gel electrophoresis and visualized under UV light.

PRINCIPAL POINTS

- Cloning in the molecular biology sense (as opposed to cloning whole organisms) is the making of many copies of a segment of DNA, such as a gene. Cloning makes it possible to generate large amounts of pure DNA for manipulation. DNA is cloned by inserting a DNA fragment into a cloning vector (a DNA molecule capable of replication in a host organism) to make a recombinant DNA molecule and then introducing that molecule into a host cell in which it will replicate. Restriction enzymes and DNA ligase are essential to cloning. Restriction enzymes recognize specific nucleotide pair sequences in DNA (restriction sites) and cleave at a specific point within or close to the sequence. DNA ligase joins cleaved DNA fragments.

- Different kinds of cloning vectors have been developed; plasmids are the most common ones in use. Cloning vectors typically replicate within one or more host organisms, have restriction sites into which foreign DNA can be inserted, and have one or more selectable markers to use in detecting cells that contain the vectors. Shuttle vectors also have these properties, but can replicate in more than one type of host. Yeast artificial chromosomes (YACs) and

bacterial artificial chromosomes (BACs) enable DNA fragments several hundred kilobase pairs long to be cloned in yeast and *E. coli*, respectively.

- Genomic libraries are collections of clones that contain at least one copy of every DNA sequence in an organism's genome. Genomic libraries are useful for isolating specific genes and studying the organization of the genome.

- Individual chromosomes can be isolated and chromosome-specific libraries made from them. If a gene has been localized to a specific chromosome by genetic means, a chromosome-specific library makes it easier to isolate a clone of the gene.

- DNA copies, called complementary DNAs (cDNAs), can be made from mRNA molecules isolated from eukaryotic cells. The cDNAs can be cloned.

- Specific sequences in genomic libraries and cDNA libraries can be identified with a number of approaches, including the use of specific DNA or cDNA probes, specific antibodies, and complementation of mutations.

- Genes and cloned DNA sequences can be analyzed to determine the arrangement and specific locations of restriction sites, a process called restriction mapping. Using recombinant DNA procedures, scientists can analyze gene transcripts to determine tissue specificity and the level of gene expression.

- Methods have been developed for determining the sequence of a cloned piece of DNA. Having the sequence of a gene helps researchers understand the function of the gene. Having the sequence of a genome helps researchers understand the organization of the genome and helps them search for all the genes it contains.

- The polymerase chain reaction (PCR) is a procedure for amplifying a specific segment of DNA. PCR is a highly sensitive procedure: Even DNA from a single cell can be amplified. There are many applications for PCR, including generating specific DNA segments for cloning or sequencing, and amplifying DNA to detect specific genetic defects.

(i) GENETIC ENGINEERING IS ONE OF THE MOST IMPORTANT and controversial aspects of genetic research today. Knowing how to modify and manipulate genes within or between species allows scientists to clone whole organisms such as sheep, create new medicines, produce disease-resistant plants, diagnose human diseases, and even track criminals. What is genetic engineering? What is a clone? What is recombinant DNA? How are clones and recombinant DNA produced?

In this chapter, you will learn the answers to these and other questions. Then you can apply what you've learned by trying the iActivity, in which you use recombinant DNA techniques to create a genetically modified brewing yeast for beer.

The field of molecular genetics changed radically in the 1970s when procedures were developed that enabled researchers to construct recombinant DNA molecules and to clone (make many copies of) those molecules. **Cloning** generates large amounts of pure DNA, such as genes, which can then be manipulated in various ways, including mapping, sequencing, mutating, and transforming cells. Using recombinant DNA technology to manipulate genes for genetic analysis or for developing products or other applications is called **genetic engineering.** For his fundamental studies of the biochemistry of nucleic acids, with particular regard to recombinant DNA, Paul Berg shared the 1980 Nobel Prize in Chemistry. Your goal in this chapter is to learn about DNA cloning and some common molecular procedures by which DNA is analyzed and manipulated.

DNA Cloning

In brief, DNA is cloned molecularly by following these steps:

1. Isolate DNA from an organism.
2. Cut the DNA into pieces with a restriction enzyme, and insert (ligate) each piece individually into a **cloning vector** cut with the same restriction enzyme, thereby making **recombinant DNA** molecules. A cloning vector is an artificially constructed DNA molecule that is capable of replication in a host organism, such as a bacterium.
3. Introduce (transform) the recombinant DNA molecules into a host such as *E. coli*. Replication of the recombinant DNA molecule (**molecular cloning**) occurs in the host cell, producing many identical copies called *clones*. As the host organism reproduces, the recombinant DNA molecules are passed on to all the progeny, giving rise to a population of cells carrying the cloned sequences.

There are many reasons for cloning DNA. For example, suppose we want to study the gene for a particular human protein to determine its DNA sequence and see how its expression is regulated. Each human cell contains only two copies of that gene, making it almost impossible to isolate enough copies of the gene for analysis. By contrast, an essentially unlimited number of copies of the gene can be produced by cloning. Similarly, if we can clone a gene selectively, we can design experiments to manipulate the gene (for example, to change its DNA sequence) and study its function or to synthesize large amounts of the gene's products.

In this section, we describe how DNA can be cloned.

Restriction Enzymes

A **restriction enzyme** (or **restriction endonuclease**) recognizes a specific base-pair sequence in DNA called a *restriction site* and cleaves the DNA (hydrolyzes the phosphodiester backbones) within or near that sequence. All restriction enzymes cut DNA between the 3′ carbon and the phosphate moiety of the phosphodiester bond, so fragments produced by restriction enzyme digestion have 5′ phosphates and 3′ hydroxyls. Restriction enzymes are used both to produce a pool of DNA fragments to be cloned and to analyze the positions of restriction sites in a piece of cloned DNA or in a segment of DNA in the genome. (See pp. 192–195.)

General Properties of Restriction Enzymes. Most restriction enzymes are found naturally in bacteria, although one has been found in the green alga *Chlorella*. In bacteria, restriction enzymes protect the host organism against viruses by cutting up—restricting—invading viral DNA. The bacterium modifies its own restriction sites (by methylation) so that its own DNA is protected from the restriction enzyme(s) it makes. For the discovery of restriction enzymes and their application to problems of molecular genetics, Werner Arber, Daniel Nathans, and Hamilton O. Smith received the 1978 Nobel Prize in Physiology or Medicine.

More than 400 different restriction enzymes have been isolated. They are named for the organisms from which they are isolated. Conventionally, a three-letter system is used. Commonly the first letter is that of the genus, and the second and third letters are from the species name. The letters are italicized or underlined, followed by roman numerals. Letters sometimes are added to signify a particular bacterial strain from which the enzymes were obtained. For example, *Eco*RI is from *E. coli* strain RY13, and *Hind*III is from *Haemophilus influenzae* strain Rd. The names are pronounced in ways that follow no set pattern. For example, *Bam*HI is "bam-H-one," *Bgl*II is "bagel-two," *Eco*RI is "echo-R-one" or "eeko-R-one," *Hind*III is "hin-D-three," *Hha*I is "ha-ha-one," and *Hpa*II is "hepa-two."

Many restriction sites have an axis of symmetry through their midpoint. Figure 8.1 shows this symmetry for the *Eco*RI restriction site: The base sequence from 5′ to 3′ on one DNA strand is the same as the base sequence from 5′ to 3′ on the complementary DNA strand. Thus, the sequences are said to have *twofold rotational symmetry*. A number of restriction sites are shown in Table 8.1. The most commonly used restriction enzymes recognize four base pairs (for example, *Hha*I) or six nucleotide pairs (for example, *Bam*HI or *Eco*RI). Some enzymes recognize eight-nucleotide pair sequences (for instance, *Not*I ["not-one"]). Other classes of enzymes do not fit our model, in that the restriction site is not symmetrical about the center. *Hin*fI ("hin-f-one"), for example, recognizes a five-nucleotide pair sequence in which there is symmetry in the two base pairs on either side of the central base pair, but the central base pair is obviously asymmetrical within the sequence. *Bst*XI ("b-s-t-x-one") is representative of a number of restriction enzymes with a nonspecific spacer region between symmetrical sequences. (See Table 8.1.)

Frequency of Occurrence of Restriction Sites in DNA. Since each restriction enzyme cuts DNA at an enzyme-specific sequence, the number of cuts the enzyme makes in a particular DNA molecule depends on the number of times that particular restriction site occurs. When we cut copies of the same genome with a particular restriction enzyme, the DNA is cleaved at the enzyme's specific restriction sites, which are distributed throughout the genome. Although doing this produces millions of fragments of different sizes, all of the identical chromosomal DNAs in the multiple genome copies will be cut at identical recognition sequences.

On the basis of probability principles, it can be shown that the frequency of a short nucleotide pair sequence in the genome will theoretically be greater than the frequency of a long nucleotide pair sequence, so an enzyme that recognizes a four-nucleotide pair sequence will cut a DNA molecule more frequently than one that recognizes a six-nucleotide pair sequence, and the latter enzyme will cut more frequently than one that recognizes an eight-nucleotide pair sequence.

Now, consider DNA with 50 percent GC and a random distribution of nucleotide pairs. For that DNA, there is an equal chance of finding one of the four possible nucleotide pairs $\frac{G}{C}$, $\frac{C}{G}$, $\frac{A}{T}$ and $\frac{T}{A}$ at any one position. The restriction enzyme *Hpa*II recognizes the sequence 5′-GGCC-3′. The probability of this sequence occurring 3′-CCGG-5′ in DNA is computed as follows:

1st nucleotide pair: $\frac{G}{C}$, probability = ¼

2nd nucleotide pair: $\frac{G}{C}$, probability = ¼

3rd nucleotide pair: $\frac{C}{G}$, probability = ¼

4th nucleotide pair: $\frac{C}{G}$, probability = ¼

The probability of finding any one of the nucleotide pairs is independent of the probability of finding one of the other nucleotide pairs. Therefore, the probability of finding the *Hpa*II restriction site in DNA with a random distribution of nucleotide pairs is ¼ × ¼ × ¼ × ¼ = ¹⁄₂₅₆. In short, the recognition sequence for *Hpa*II occurs, on average, once every 256 base pairs in such a piece of DNA.

Figure 8.1

Restriction site in DNA, showing symmetry of the sequence around the center point. The sequence is a palindrome, reading the same from left to right (5′-to-3′) on the top strand (GAATTC, here) as it does from right to left (5′-to-3′) on the bottom strand. Shown is the restriction site for *Eco*RI.

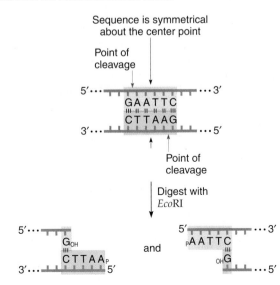

Table 8.1 **Characteristics of Some Restriction Enzymes**

	Enzyme Name	Pronunciation	Organism in Which Enzyme Is Found	Recognition Sequence and Position of Cut[a]
Enzymes with 6-bp Recognition Sequences	BamHI	"bam-H-one"	Bacillus amyloliquefaciens H	5'-G↓GATCC-3' 3'-CCTAG↑G-5'
	BglII	"bagel-two"	Bacillus globigi	A↓GATCT TCTAG↑A
	EcoRI	"echo-R-one"	E. coli RY13	G↓AATTC CTTAA↑G
	HaeII	"hay-two"	Haemophilus aegypticus	RGCGC↓Y Y↑CGCGR
	HindIII	"hin-D-three"	Haemophilus influenzae R_d	A↓AGCTT TTCGA↑A
	PstI	"P-S-T-one"	Providencia stuartii	CTGCA↓G G↑ACGTC
	SalI	"sal-one"	Streptomyces albus	G↓TCGAC CAGCT↑G
	SmaI	"sma-one"	Serratia marcescens	CCC↓GGG GGG↑CCC
Enzymes with 4-bp Recognition Sequences	HaeIII	"hay-three"	Haemophilus aesypticus	GG↓CC CC↑GG
	HhaI	"ha-ha-one"	Haemophilus haemolyticus	GCG↓C C↑GCG
	HpaII	"hepa-two"	Haemophilus parainfluenzae	C↓CGC GGC↑C
	Sau3A	"sow-three-A"	Staphylococcus aureus 3A	↓GATC CTAG↑
Enzyme with 8-bp Recognition Sequence	NotI	"not-one"	Nocardia otitidis-caviarum	GC↓GGCCGC CGCCGG↑CG
Enzyme with Recognition Sequence That is Not Symmetrical	BstXI	"b-s-t-x-one"	Bacillus stearothermophilus	CCANNNNN↓NTGG GGTN↑NNNNNACC

[a]In this column the two strands of DNA are shown with the sites of cleavage indicated by arrows. Since there is an axis of twofold rotational symmetry in each recognition sequence, the DNA molecules resulting from the cleavage are symmetrical. Key: R = purine; Y = pyrimidine; N = any base.

In general, the probability of occurrence of a restriction site in randomly distributed base pairs with 50 percent GC content is given by the formula $(1/4)^n$, where n is the number of nucleotide pairs in the recognition sequence. The values produced are given in Table 8.2. In practice, however, genomes usually do not have exactly 50 percent GC, nor are the base pairs randomly distributed, so a range of sizes of fragments results when genomic DNA is cut with a restriction enzyme. For example, the sequence 5'-CG-3' is very rare in mammalian DNA, so HpaII restriction sites are uncommon and DNA fragments generated by cleavage with HpaII are large.

Restriction Sites and DNA Cloning. One major class of restriction enzymes recognizes a specific DNA sequence and then cuts within that sequence. Restriction enzymes in

Table 8.2	Occurrence of Restriction Sites for Restriction Enzymes in DNA with Randomly Distributed Nucleotide Pairs	
Nucleotide Pairs in Restriction Site	**Probability of Occurrence**	
4	$(1/4)^4 = 1$ in 256 bp	
5	$(1/4)^5 = 1$ in 1,024 bp	
6	$(1/4)^6 = 1$ in 4,096 bp	
8	$(1/4)^8 = 1$ in 65,476 bp	
n	$(1/4)^n$	

this class cut DNA in different general ways. As Table 8.1 indicates, some enzymes, such as *Sma*I ("sma-one"), cut both strands of DNA between the same two base pairs, to produce DNA fragments with blunt ends (Figure 8.2a). Other enzymes, such as *Bam*HI, make staggered cuts in the symmetrical nucleotide pair sequence, generating DNA fragments with **sticky ends,** either 5 overhanging ends, as in the case of cleavage with *Bam*HI (Figure 8.2b) or *Eco*RI, or 3′ overhanging ends, as in the case of cleavage with *Pst*I ("P-S-T-one"; Figure 8.2c). Another class of restriction enzymes has members that recognize a specific DNA sequence and then cut the two strands of DNA outside of that sequence.

Figure 8.2

Examples of how restriction enzymes cleave DNA. (a) *Sma*I results in blunt ends. (b) *Bam*HI results in 5′ overhanging ("sticky") ends. (c) *Pst*I results in 3′ overhanging ("sticky") ends.

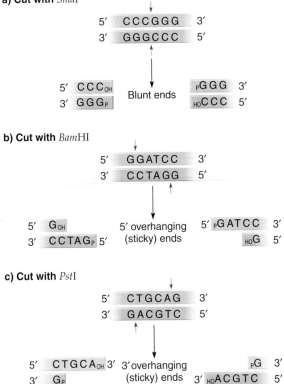

Restriction enzymes that produce sticky ends are of particular value in cloning DNA because every DNA fragment generated by cutting a piece of DNA with the same restriction enzyme has the same base sequence at the two staggered ends. That is, if the ends of two pieces of DNA produced by the action of the same restriction enzyme (such as *Eco*RI)—for example, a cloning vector and a chromosomal DNA fragment—come together in solution, base pairing occurs, and the two single-stranded DNA ends are said to *anneal* (Figure 8.3). The two DNAs can then be covalently linked (ligated) using DNA ligase, to produce a whole DNA molecule with reconstituted restriction sites. (Recall from our discussion of DNA replication that DNA ligase seals nicks in a DNA strand by forming a phosphodiester bond when the two nucleotides have a free 5′ phosphate and a free 3′ hydroxyl group, respectively. (See Figure 3.7 and pp. 52–54.) Even DNA fragments with blunt ends can be ligated together by DNA ligase at high concentrations of the enzyme. The ligation of two DNA fragments is the principle behind the formation of recombinant DNA molecules.

KEYNOTE

Genes are cloned by inserting DNA from an organism into a cloning vector to make a recombinant DNA molecule and then introducing that molecule into a host cell in which it will replicate. Essential to cloning are restriction enzymes, which recognize specific nucleotide pair sequences in DNA (restriction sites) and cleave at a specific point within the sequence.

Cloning Vectors and DNA Cloning

Several types of vectors have been used to clone DNA. Among these vectors are plasmids, bacteriophages (e.g., λ and certain single-stranded DNA species), cosmids (vectors with features of both plasmid and bacteriophage vectors), and artificial chromosomes. The vector types differ in the molecular properties they have and in the maximum size of DNA that can be cloned into each. Each type of vector has been specially constructed in the laboratory. We will focus on plasmid and artificial chromosome vectors in this section.

Plasmid Cloning Vectors. Bacterial **plasmids** are extrachromosomal elements that replicate autonomously within cells. (See Chapter 18.) Their DNA is circular and double stranded and carries sequences (origins, or *ori*) required for plasmid replication and for the plasmid's other functions. Plasmid cloning vectors are derivatives of these natural plas-

animation

a DNA Cloning in a Plasmid Vector

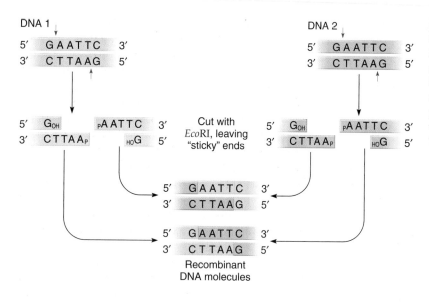

DNA 1

5′ G A A T T C 3′
3′ C T T A A G 5′

5′ G$_{OH}$ 3′ $_P$A A T T C 3′
3′ C T T A A$_P$ 5′ $_{HO}$G 5′

Cut with
*Eco*RI, leaving
"sticky" ends

DNA 2

5′ G A A T T C 3′
3′ C T T A A G 5′

5′ G$_{OH}$ 3′ $_P$A A T T C 3′
3′ C T T A A$_P$ 5′ $_{HO}$G 5′

5′ G A A T T C 3′
3′ C T T A A G 5′

5′ G A A T T C 3′
3′ C T T A A G 5′

Recombinant
DNA molecules

Figure 8.3

Cleavage of DNA by the restriction enzyme *Eco*RI. *Eco*RI makes staggered, symmetrical cuts in DNA, leaving "sticky" ends. A DNA fragment with a sticky end produced by *Eco*RI digestion can bind by complementary base pairing (anneal) to any other DNA fragment with a sticky end produced by *Eco*RI cleavage. The gaps can then be sealed by DNA ligase.

mids and are "engineered" to have features useful for cloning DNA. We focus here on features of *E. coli* plasmid cloning vectors.

An *E. coli* plasmid cloning vector must have three features:

1. An *ori* (origin of DNA replication) sequence, needed for the plasmid to replicate in *E. coli*.

2. A dominant *selectable marker*, so that *E. coli* cells with the plasmid can be distinguished easily from cells that lack the plasmid. Usually, the selectable marker is a gene for resistance to an antibiotic, such as the *amp*^{R} gene for ampicillin resistance or the *tet*^{R} gene for tetracycline resistance. When plasmids carrying antibiotic-resistance genes are added to a population of plasmid-free, and therefore antibiotic-sensitive, *E. coli*, the cells that take up the plasmid can be selected for by culturing the cells on a solid medium containing the appropriate antibiotic; only bacteria with the plasmid will grow on the medium.

3. *One or more unique restriction enzyme cleavage sites*—sites present just once in the vector—for the insertion of the DNA fragments to be cloned. Cloning most commonly involves cutting the plasmid at one of the unique sites with the appropriate restriction enzyme and inserting into that site a piece of DNA that has been cut with the same enzyme.

As an example, Figure 8.4 diagrams the plasmid cloning vector pUC19 ("puck-19"). This 2,686-bp vector has the following features that make it useful for cloning DNA in *E. coli*:

1. It has a high copy number, approaching 100 copies per cell, so many copies of a cloned piece of DNA can be generated readily.

2. It has the *amp*^{R} selectable marker.

Figure 8.4

The plasmid cloning vector pUC19. This plasmid has an origin of replication (*ori*), an *amp*^{R} selectable marker, and a polylinker located within part of the β-galactosidase gene *lacZ*$^+$.

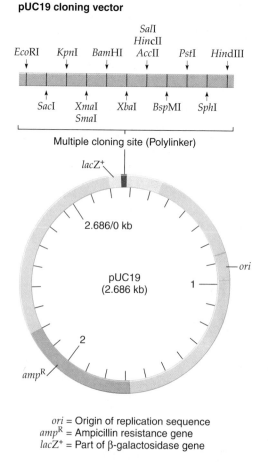

pUC19 cloning vector

*Sal*I
*Hinc*II
*Eco*RI *Kpn*I *Bam*HI *Acc*II *Pst*I *Hind*III

*Sac*I *Xma*I *Xba*I *Bsp*MI *Sph*I
*Sma*I

Multiple cloning site (Polylinker)

lacZ$^+$

2.686/0 kb

pUC19
(2.686 kb)

ori

1

2

amp^R

ori = Origin of replication sequence
amp^R = Ampicillin resistance gene
lacZ$^+$ = Part of β-galactosidase gene

3. It has a number of unique restriction sites clustered in one region, called a **multiple cloning site** or **polylinker.**

4. The multiple cloning site is inserted into part of the *E. coli* β-galactosidase (*lacZ*⁺) gene. (See Figure 8.4.) This part of the *lacZ* gene is from the 5′ end and encodes about two dozen of the N-terminal amino acids of the enzyme called the *alpha fragment*. Then, pUC19, as well as plasmids similarly constructed with such a *lacZ* gene fragment, is usually introduced into an *E. coli* strain that has a *lacZ* gene lacking a short segment from its 5′ end. Expression of the bacterium's gene produces a truncated β-galactosidase—the *omega fragment*—that is inactive. However, if both the alpha fragment and the omega fragment are expressed simultaneously—as is the case when pUC19 is in the cell—functional β-galactosidase is produced by the association of the two fragments. This process is called *alpha complementation*. By contrast, when a piece of DNA is cloned into the polylinker, the *lacZ* fragment on the plasmid is disrupted and no functional alpha fragment can be produced; hence, no alpha complementation can occur in the *E. coli* cell. Therefore, the presence or absence of β-galactosidase indicates whether the plasmid introduced into *E. coli* is the pUC19 vector (enzyme present) or pUC19 with an inserted DNA fragment (enzyme absent). The chemical X-gal—a substrate for β-galactosidase—is included in the medium on which the cells are plated as an indicator for β-galactosidase activity in cells of a colony. If

functional enzyme is present (vector alone), the colony turns blue, whereas if nonfunctional β-galactosidase is made (vector with inserted DNA), the colony is white. This protocol is called *blue–white colony screening*.

Figure 8.5 illustrates how a piece of DNA can be inserted into a plasmid cloning vector such as pUC19. In the first step, pUC19 is cut with a restriction enzyme that has a site in the polylinker. Next, the piece of DNA to be cloned is generated by cutting high-molecular-weight DNA with the same restriction enzyme. Since restriction sites are nonrandomly arranged in DNA, fragments of various sizes are produced. The DNA fragments are mixed with the cut vector in the presence of DNA ligase; in some cases, a fragment becomes inserted between the two cut ends of the plasmid, and DNA ligase joins the two molecules covalently. The resulting recombinant DNA plasmid is introduced into an *E. coli* host by transformation, which may be performed in either of two ways: by incubating the recombinant DNA plasmids with *E. coli* cells treated chemically to take up DNA or by *electroporation*, a method in which an electric shock is delivered to the cells, causing temporary disruptions of the cell membrane to let the DNA enter. Transformed cells are plated onto ampicillin-containing medium. The resulting ampicillin-resistant colonies contain cells transformed by plasmids, and blue–white colony screening distinguishes the recombinant DNA clones from plasmids that were recircularized without an inserted DNA fragment.

Recircularization of a restriction enzyme-digested vector is common because it is a reaction involving only

Figure 8.5

Insertion of a piece of DNA into the plasmid cloning vector pUC19 to produce a recombinant DNA molecule. The vector pUC19 contains several unique restriction enzyme sites localized in a polylinker that are convenient for constructing recombinant DNA molecules. The insertion of a DNA fragment into the polylinker disrupts part of the β-galactosidase (*lacZ*⁺) gene, leading to nonfunctional β-galactosidase in *E. coli*. The blue–white color selection test described in the text can be used to select for vectors with or without inserts.

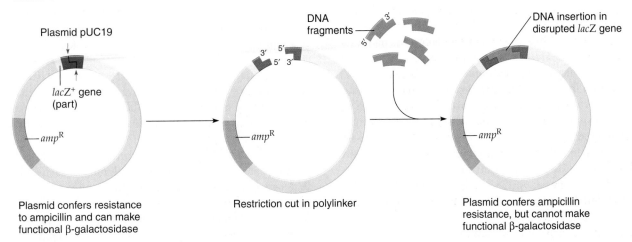

Plasmid pUC19

lacZ⁺ gene (part)

*amp*ᴿ

Plasmid confers resistance to ampicillin and can make functional β-galactosidase

DNA fragments

3′ 5′
5′ 3′

*amp*ᴿ

Restriction cut in polylinker

DNA insertion in disrupted *lacZ* gene

*amp*ᴿ

Plasmid confers ampicillin resistance, but cannot make functional β-galactosidase one DNA molecule. Vector recircularization can be minimized by treating the digested plasmid with alkaline phosphatase to remove the 5′ phosphates, leaving a 5′-OH group at the end of the DNA. Such treated molecules are not substrates for DNA ligase, so they will not recircularize. If this procedure is used before the ligation reaction, the proportion of blue colonies among transformants is reduced drastically, making identification of the desired clones more efficient. Alkaline phosphatase treatment is done routinely for ligations for which blue–white colony screening is not available.

Cloning experiments do not always involve cutting the vector and the DNA to be cloned with a single restriction enzyme. Many times, the experimental design involves using two restriction enzymes. In this case, recircularization of the vector cannot occur because the two ends of the DNA cannot be joined by DNA ligase.

Another difference between cloning strategies that use one restriction enzyme and those which use two concerns the orientation of the insert in the cloning vector. For a one-enzyme strategy, there are two possible orientations for the insert, because the two ends of the insert are the same. For a two-enzyme strategy, there is only one possible orientation of the insert in the clone, because the two ends of the insert are different.

Similar plasmid cloning vectors with variants on the features of pUC19 exist. For example, there are plasmid vectors with different arrays of unique restriction sites in the polylinker and with phage promoters flanking the polylinker–*lacZ* region. The phage promoters, such as those for T7, T3, or SP6 DNA-dependent RNA polymerases, are used to make RNA copies of the cloned DNA sequences in vitro in the presence of the appropriate polymerase. If the RNA is labeled, either radioactively (for example, by the incorporation of ^{32}P-NTPs in the reaction mixture) or nonradioactively, it can be used as a probe—called a *riboprobe*—in further analytical techniques.

Plasmid cloning vectors have been developed for a large variety of prokaryotic and eukaryotic organisms. Their general features are as we have discussed, although in some cases the sequences that are needed for replication in the organism of interest are not known, so the plasmids cannot replicate. Instead, either they integrate into the genome, or the genes they contain are expressed transiently until the plasmid is degraded by cellular enzymes.

DNA fragments of 5 to 10 kb are efficiently cloned in *E. coli* plasmid cloning vectors. Plasmids carrying larger DNA fragments often are unstable and tend to lose most of the inserted DNA. The cloning capacity of plasmid cloning vectors for other organisms varies.

Plasmid Shuttle Vectors. The cloning vectors we have discussed thus far are used mostly to clone DNA within *E. coli* cells. We have also mentioned vectors for intro-

ducing recombinant DNA molecules into other organisms. There are vectors that can be used to transform mammalian cells in culture, as well as vectors to transform other animal cells, plant cells, and yeast cells. Often, these are **shuttle vectors**—that is, cloning vectors that can be introduced into two or more different host organisms. For example, some shuttle vectors can be transformed into, and replicate in, *E. coli* (selected for by using antibiotic resistance) and can also be transformed into yeast (selected for by a nutritional marker, such as the *URA3* gene conferring uracil-independent growth on a *ura3* mutant yeast cell). Different types of yeast-*E. coli* shuttle vectors have been developed, some of which replicate to a high copy number in the nucleus, some of which replicate freely as single copies in the nucleus, and some of which integrate into a yeast nuclear chromosome, replicating when that chromosome replicates.

Expression Vectors. An **expression vector** is a cloning vector containing the regulatory sequences necessary to allow the transcription and translation of a cloned gene or genes. Expression vectors are used to produce the protein encoded by a cloned gene in the transformed host. For example, the biotechnology industry produces pharmaceutically active proteins with the use of expression vectors and the appropriate host. Expression vectors are essentially derivatives of the plasmid cloning vectors used in the host. The modifications include the addition of a promoter specific to the host to allow transcription of the cloned gene and, if appropriate for the host, a transcription termination signal. The cloned gene may also have to be modified if it is to be expressed across the prokaryotic–eukaryotic boundary. That is, different mechanisms are used in prokaryotes and eukaryotes for the translation machinery to identify the start codon. (See Chapter 6, pp. 120–121 and 122–123.) For example, to express a eukaryotic gene in *E. coli* requires the addition of a Shine–Dalgarno sequence at a position upstream of the start codon. (See Chapter 6, p. 121.)

Artificial Chromosomes. Artificial chromosomes are cloning vectors that can accommodate very large pieces of DNA, producing recombinant DNA molecules resembling small chromosomes. We consider two examples here.

Yeast artificial chromosomes. **Yeast artificial chromosomes** (YACs, or "yaks") are cloning vectors that enable artificial chromosomes to be made and cloned in yeast cells. A YAC (shown in its linear form) has the following features (Figure 8.6):

1. A yeast telomere (*TEL*) at each end. (Recall that all eukaryotic chromosomes have a telomere at each end.)
2. A yeast centromere sequence (*CEN*) allowing regulated segregation during mitosis.
3. A selectable marker on each arm for detecting the YAC in yeast (for example, *TRP1* and *URA3* for tryptophan

Figure 8.6

Example of a yeast artificial chromosome (YAC) cloning vector. A YAC vector contains a yeast telomere (*TEL*) at each end, a yeast centromere sequence (*CEN*), a yeast selectable marker for each arm (here, *TRP1* and *URA3*), a sequence that allows autonomous replication in yeast (*ARS*), and restriction sites for cloning.

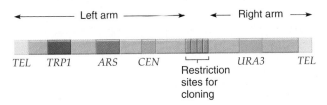

coli. BACs are vectors containing the origin of replication of a natural plasmid called the *F* factor (see Chapter 18, pp. 485), a multiple cloning site, a selectable marker, and, often, some other features. Although YACs can accommodate larger DNA inserts, BACs have the advantage that they can be manipulated like regular bacterial plasmids. Once transformed into *E. coli*, the *F* factor origin of replication keeps the copy number of the plasmid at one per cell. Unlike YACs, they do not undergo rearrangements in the host; therefore, they have essentially superseded YACs in physical mapping studies of genomes.

Because of their ability to accommodate large DNA inserts, BACs form the basis of vectors used to study gene regulation in vertebrates such as mice and zebra fish. That is, the promoter and regulatory sequences of many vertebrate genes are known to often span a large section of DNA. Therefore, a gene and a large segment of DNA upstream of the gene can be cloned in a BAC and the clone transformed into the organism. The hope is that the clone has all the sequences present for normal regulation of the gene, making the study of that regulation feasible.

and uracil independence in *trp1* and *ura3* mutant strains, respectively)

4. An origin-of-replication sequence—*ARS* (autonomously replicating sequence)—that allows the vector to replicate in a yeast cell.

5. Restriction sites unique to the YAC that can be used for inserting foreign DNA.

YAC vectors can accommodate DNA fragments that are several hundred kilobase pairs long—much longer than the fragments that can be cloned in the plasmid vectors we have discussed. Therefore, YAC vectors have been used to clone very large DNA fragments (between 0.2 and 2.0 Mb {Mb = megabase = 1,000,000 bp})—for example, in creating physical maps of large genomes such as the human genome. However, a disadvantage of these very large YAC-based clones is that they frequently undergo rearrangements in the host, making the assembly of a genome sequence very difficult if not impossible.

YAC vectors are propagated in *E. coli* as circular plasmids; in this form, the two telomeres are situated end to end. For cloning experiments, a circular YAC is cut with one restriction enzyme that cuts in the multiple cloning site and another restriction enzyme that cuts between the two *TEL*s. In this way, the left and right arms are produced. High-molecular-weight DNA is ligated to the two arms, and the recombinant molecules are transformed into yeast. By selecting for both *TRP1* and *URA3*, it can be ensured that the transformants have both the left and right arms.

i Better beer through science? Go to the iActivity *Building a Better Beer* on your CD-ROM, and discover how genetically modified yeasts can improve your brew.

Bacterial artificial chromosomes. Bacterial artificial chromosomes (BACs, or "backs") are useful for cloning large DNA fragments up to about 300 kb in *E.*

KEYNOTE

Many different kinds of vectors have been developed to construct and clone recombinant DNA molecules. Most vectors replicate within their host organism. Those which do not replicate extrachromosomally integrate into the genome and are replicated when the genome replicates. Cloning vectors also have unique restriction sites for inserting foreign DNA fragments, as well as having one or more dominant selectable markers. The choice of the vector to use depends on the experimental goal.

Recombinant DNA Libraries

Often, researchers want to study a particular gene or DNA fragment. When genomic DNA is isolated from an organism and cut with a restriction enzyme, and the population of DNA fragments is cloned in a vector, we have a **genomic library**—a collection of clones containing at least one copy of every DNA sequence in the genome.

Genomic libraries have been made for many organisms, including humans (see the discussion of the Human Genome Project in Chapter 16) and many viruses. Related to genomic libraries are **chromosome libraries**, which are collections of clones of fragments of individual chromosomes, and **complementary DNA (cDNA) libraries**, which are collections of clones of DNA copies of mRNAs isolated from cells. The sections that follow describe these different types of libraries.

Genomic Libraries

A genomic library is a collection of clones that theoretically contains at least one copy of every DNA sequence in the genome. (The word "theoretically" is used because, practically speaking, not all of the sequences in the genome can be cloned.) The genomic library can be used to isolate and study a particular clone, such as that for a gene of interest. We will see how this can be done later; in this section, we focus on the construction of genomic libraries of eukaryotic DNA.

Genomic libraries are made by using the basic cloning procedures already described. A restriction enzyme is used to cut up the genomic DNA, and a vector is chosen so that the entire genome is represented in a manageable number of clones. There are three general ways to produce genomic libraries:

1. Digest genomic DNA completely with a restriction enzyme, and clone the resulting DNA fragments in a cloning vector. A drawback of this technique is that if the specific gene the researcher wants to study contains one or more restriction sites for the enzyme, the gene will be split into two or more fragments when the DNA is digested by the restriction enzyme. As a result, the gene would then be cloned in two or more pieces. Another drawback is that the average size of the fragment produced by the digestion of eukaryotic DNA with restriction enzymes is small (about 4 kb for restriction enzymes that have six-base-pair recognition sequences; see Table 8.2). Not only are many genes (especially those in mammals) larger than 4 kb, but also, an entire genomic library would have to contain a very large number of recombinant DNA molecules, and screening for a specific gene would be laborious.

2. The problems of genes split into multiple fragments and the large number of recombinant DNA molecules can be minimized by cloning longer DNA fragments in an appropriate vector. Longer DNA fragments can be generated by mechanically shearing high-molecular-weight (usually 100- to 150-kb) DNA. For example, the DNA can be passed through a syringe needle to produce a population of overlapping DNA fragments of large size. However, because the ends of the resulting fragments will not have been generated by cutting with restriction enzymes, additional enzymatic manipulations will be necessary to add appropriate ends to the molecules for insertion into vector cloning sites.

3. Another approach to producing DNA fragments of appropriate size for constructing a genomic library is to perform a partial digestion of the DNA with restriction enzymes that recognize frequently occurring six- or four-base-pair recognition sequences (Figure 8.7a). *Partial digestion* means that only a portion of the available restriction sites is actually cut with the enzyme. This is achieved by limiting the amount of

Figure 8.7

Use of partial digestion with a restriction enzyme to produce DNA fragments of appropriate size for constructing a genomic library.

a) Partial digestion of DNA by a restriction enzyme (for example *Sau*3A) generates a series of overlapping fragments, each with identical 5′ GATC sticky ends

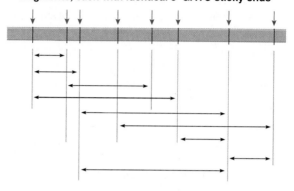

b) Resulting fragments may be inserted into *Bam*HI site of plasmid cloning vector

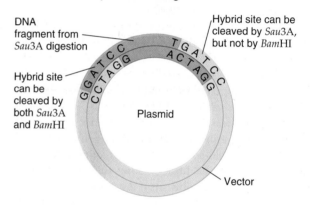

DNA fragment from *Sau*3A digestion

Hybrid site can be cleaved by both *Sau*3A and *Bam*HI

Hybrid site can be cleaved by *Sau*3A, but not by *Bam*HI

Plasmid

Vector

the enzyme used or the time of incubation with the DNA. The ideal result of partial digestion is a population of overlapping fragments representing the entire genome. Fragments of the desired size for cloning are then collected—for example, with the use of agarose gel electrophoresis. (See pp. 192–194.) Those fragments can be cloned directly because their ends were produced by restriction enzyme digestion. For example, if the DNA is digested with the enzyme *Sau*3A, which has the recognition sequence

5′-GATC-3′
3′-CTAG-5′,

the ends are complementary to the ends produced by the digestion of a cloning vector with *Bam*HI, which has the recognition sequence

5′-GGATCC-3′
3′-CCTAGG-5′ (Figure 8.7b).

That is, in

$$\overset{\downarrow}{\text{5′-GATC-3′}}$$
$$\underset{\uparrow}{\text{3′-CTAG-5′}}$$

*Sau*3A cuts to the left of the upper G and to the right of the lower G, giving a 5′ overhang with the sequence 5′-GATC...3′ as follows:

```
5′-                and     5′ GATC-3′
3′-CTAG 5′                         -5′
```

Similarly, in the sequence

```
            ↓
   5′-GGATCC-3′
   3′-CCTAGG-5′
            ↑
```

*Bam*HI cuts between the two G nucleotides, also giving a 5′ overhang with the sequence 5′-GATC...3′ as follows:

```
5′-G               and     5′ GATCC-3′
3′-CCTAG 5′                       G-5′
```

The *Sau*3A and *Bam*HI "sticky" ends can pair to produce a hybrid recognition site.[1]

The recombinant DNA molecules produced by ligating the *Sau*3A-cut fragments and the *Bam*HI-cut vectors together are then introduced into *E. coli*, where the molecules are cloned.

The aim of the methods just described is to produce a library of recombinant molecules that is as complete as possible. However, not all sequences of the eukaryotic genome are equally represented in such a library. For example, if the restriction sites in a particular region are very far apart or extremely close together, the chance of obtaining a fragment of a size that can be cloned is small. In addition, some regions of eukaryotic chromosomes may contain sequences that affect the ability of vectors containing them to replicate in *E. coli*; these sequences would then be lost from the library.

How many clones are needed for the library to contain all sequences in the genome? The number of clones needed to include all sequences in the genome depends on the size of the genome being cloned and the average size of the DNA fragments inserted into the vector. The probability of having at least one copy of any DNA sequence in the genomic library can be calculated from the formula

$$N = \frac{\ln(1 - P)}{\ln(1 - f)}$$

where N is the necessary number of recombinant DNA molecules, P is the probability desired, f is the fractional proportion of the genome in a single recombinant DNA molecule (that is, f is the average size, in kilobase pairs, of the fragments used to make the library, divided by the size of the genome, in kilobase pairs), and ln is the natural logarithm. For example, for a 99 percent chance that a particular yeast DNA fragment is represented in a genomic library of 15-kb fragments, where the yeast genome size is about 12,000 kb, 3,682 recombinant DNA molecules would be needed. For the approximately 3,000,000-kb human genome, more than 920,000 clones would be needed—hence the use of YAC or BAC vectors for making libraries of large genomes. Whatever the genome or vector, to have confidence that all genomic sequences are represented, one must make a library with several times more than the calculated minimum number of clones.

KEYNOTE

A genomic library is a collection of clones that contains at least one copy of every DNA sequence in an organism's genome. Like libraries with books, genomic libraries are great resources of information; in this case, the information is about the genome.

Chromosome Libraries

It is very time consuming to screen for a sequence of interest in a genomic library made from an organism with a large genome. One approach to reducing the screening time is to make libraries of the individual chromosomes in the genome. In humans, this gives 24 different libraries, 1 each for the 22 autosomes, the X, and the Y. Then, if a gene has been localized to a chromosome by genetic means, researchers can focus their attention on the library of that chromosome when they search for its DNA sequence.

Individual chromosomes can be separated if their morphologies and sizes are distinct enough, as is the case for human chromosomes. In one procedure, *flow cytometry*, chromosomes from cells in mitosis are stained with a fluorescent dye and passed through a laser beam connected to a light detector. This system sorts the chromosomes on the basis of differences in dye binding and the resulting light scattering. Once the chromosomes have been sorted and collected from a number of cells, a library of each chromosome type can be made in the manner just described.

cDNA Libraries

DNA copies, called **complementary DNA (cDNA)**, can be made from all mRNA molecules present in a population of eukaryotic cells at a particular time or from a particular tissue. These cDNA molecules can then be cloned to produce a cDNA library. Since a cDNA library reflects the gene activity of the cell type at the time the mRNAs

---

[1] Since the hybrid site contains a 5′-GATC-3′ sequence, it can be cleaved by *Sau*3A. However, whether it can be cleaved by *Bam*HI depends on the base pair "inside" the cloned *Sau*3A-digested fragment. If it is a CG nucleotide pair, then the hybrid site is

```
5′-GGATCC-3′
3′-CCTAGG-5′
```

namely, the recognition site for *Bam*HI. This is the case with the left-hand hybrid site in Figure 16.7b. If any other nucleotide pair is next along the *Sau*3A fragment, the hybrid site is not a *Bam*HI cleavage site (for example, the right-hand hybrid site in the figure).

are isolated, cDNA libraries are useful, for example, for comparing gene activities in different cell types of the same organism or of the same cell type at different times, as in cell differentiation during development.

The clones in the cDNA library represent the mature mRNAs found in the cell. In eukaryotes, mature mRNAs are processed molecules, so the sequences obtained are not equivalent to genomic clones. In particular, intron sequences are present in genomic clones, but not in cDNA clones; hence, cDNA clones are typically smaller than the equivalent gene clone. For any mRNA, cDNA clones can be useful for subsequently isolating the gene that codes for that mRNA. The gene clone can provide more information than can the cDNA clone—for example, on the presence and arrangement of introns and on the regulatory sequences that control expression of the gene.

cDNA libraries are readily made from eukaryotic mRNAs, because, uniquely among RNAs, eukaryotic mRNAs contain a poly(A) tail. (See Chapter 5, p. 97.) These poly(A)+ mRNAs can be purified from a mixture of cellular RNAs by passing the RNA molecules over a column to which short chains of deoxythymidylic acid, called *oligo(dT) chains*, have been attached. As the RNA molecules pass through the column, the poly(A) tails on the mRNA molecules base pair to the oligo(dT) chains. As a result, the mRNAs are captured on the column while the other RNAs pass through. The captured mRNAs are then released and collected—for example, by decreasing the ionic strength of the buffer passing through the column so that the hydrogen bonds are disrupted. This method results in significant enrichment of poly(A)+ mRNAs in the mixed RNA population, to about 50 percent versus approximately 3 percent in the cell.

Figure 8.8 shows how a cDNA molecule can be made from the mRNA molecules. Key to this synthesis is the presence of the 3′ poly(A) tails on the mRNAs. After

Figure 8.8

The synthesis of double-stranded complementary DNA (cDNA) from a polyadeny-lated mRNA, using reverse transcriptase, RNase H, DNA polymerase I, and DNA ligase.

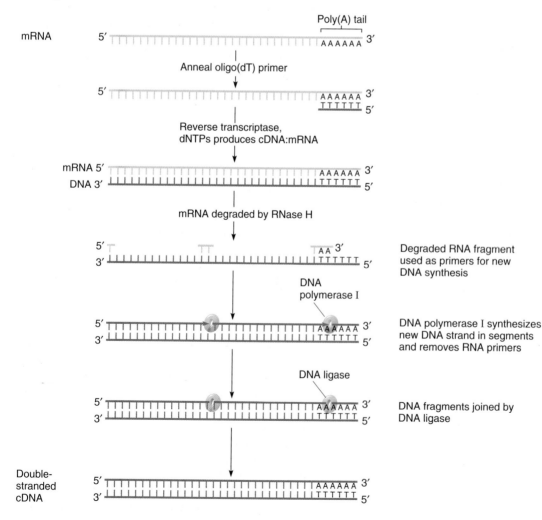

the mRNA has been isolated, the first step in cDNA synthesis is annealing a short oligo(dT) primer to the poly(A) tail. The primer is extended by **reverse transcriptase** (RNA-dependent DNA polymerase) to make a DNA copy of the mRNA strand. The result is a DNA–mRNA double-stranded molecule. Next, RNase H ("R-N-aze H," a type of ribonuclease), DNA polymerase I, and DNA ligase are used to synthesize the second DNA strand. RNase H partially degrades the RNA strand in the hybrid DNA–mRNA; DNA polymerase I makes new DNA fragments, using the partially degraded RNA fragments on the single-stranded DNA as primers; and, finally, DNA ligase ligates the new DNA fragments together to make a complete chain. The result is a double-stranded cDNA molecule that is a faithful DNA copy of the starting mRNA.

How do we clone cDNA molecules? Figure 8.9 illustrates the cloning of cDNA by using a **restriction site linker,** or **linker,** which is a short, double-stranded piece of DNA (oligodeoxyribonucleotide) about 8 to 12 nucleotide pairs long that includes a restriction site—in this case, the site for *Bam*HI. Both the cDNA molecules and the linkers have blunt ends, and they can be ligated

together at high concentrations of T4 DNA ligase. Sticky ends are produced in the cDNA molecule by cleaving the cDNA (with linkers now at each end) with *Bam*HI. The resulting DNA is inserted into a cloning vector that has also been cleaved with *Bam*HI, and the recombinant DNA molecule produced is transformed into an *E. coli* host cell for cloning.

A problem with using linkers for cloning cDNAs is that there may be a restriction site within the cDNA for the enzyme used to cleave the linkers, which would mean that the cDNA would also be cut when the linkers are cut, resulting in cloning the cDNA in pieces. To get around this potential problem, an *adapter* can be added to the cDNA that already has a sticky end on it suitable for cloning, so that the cDNA is never digested with a restriction enzyme. For example, if we anneal together 5'-GATCCAGAC-3' with 5'-GTCTG-3' to form the adapter

5'-GATCCAGAC-3'
GTCTG-5'

and ligate it to a cDNA, the blunt end of the adapter will covalently attach to the blunt end of the cDNA, leaving the 5' overhang GATC at each end. The overhang will base pair with a vector digested with *Bam*HI (see Figure 8.9), and the cDNA will be cloned in one piece.

cDNA molecules can also be cloned by blunt-end cloning. That is, the cDNA molecules have blunt ends, so they can be inserted into a vector that has been cut with a restriction enzyme, such as *Sma*I, (see Table 8.1) that generates blunt ends.

Figure 8.9

The cloning of cDNA, using *Bam*HI linkers.

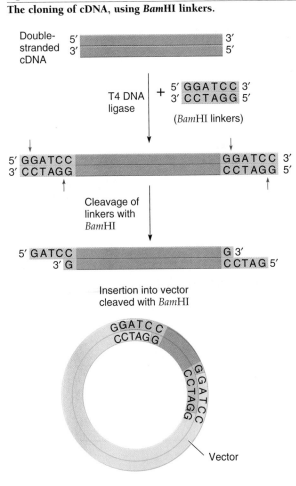

KEYNOTE

> DNA copies, called complementary DNA or cDNA, can be made of the population of mRNAs purified from a cell. First, a primer and the enzyme reverse transcriptase are used to make a single-stranded DNA copy of the mRNA; then, RNase H, DNA polymeralse I, and DNA ligase are used to make a double-stranded DNA copy called cDNA that can be spliced into cloning vectors and cloned.

Finding a Specific Clone in a DNA Library

Unlike libraries of books, clone libraries have no catalog, so they must be searched through—screened—to find a clone of interest. Fortunately, a number of screening procedures have been developed, and we discuss some of them in this section.

Screening a cDNA Library

We can screen a cDNA library in a number of ways to identify a cDNA clone we are interested in studying, including

Figure 8.10

Screening for specific cDNA plasmids in a cDNA library by using an antibody probe.

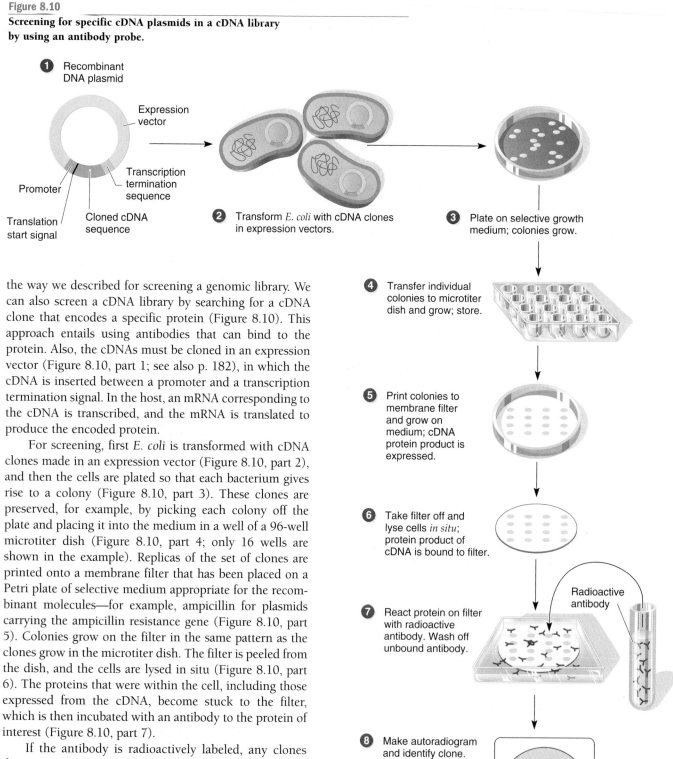

① Recombinant DNA plasmid

Expression vector

Promoter

Translation start signal

Cloned cDNA sequence

Transcription termination sequence

② Transform *E. coli* with cDNA clones in expression vectors.

③ Plate on selective growth medium; colonies grow.

④ Transfer individual colonies to microtiter dish and grow; store.

⑤ Print colonies to membrane filter and grow on medium; cDNA protein product is expressed.

⑥ Take filter off and lyse cells *in situ*; protein product of cDNA is bound to filter.

⑦ React protein on filter with radioactive antibody. Wash off unbound antibody.

Radioactive antibody

⑧ Make autoradiogram and identify clone.

the way we described for screening a genomic library. We can also screen a cDNA library by searching for a cDNA clone that encodes a specific protein (Figure 8.10). This approach entails using antibodies that can bind to the protein. Also, the cDNAs must be cloned in an expression vector (Figure 8.10, part 1; see also p. 182), in which the cDNA is inserted between a promoter and a transcription termination signal. In the host, an mRNA corresponding to the cDNA is transcribed, and the mRNA is translated to produce the encoded protein.

For screening, first *E. coli* is transformed with cDNA clones made in an expression vector (Figure 8.10, part 2), and then the cells are plated so that each bacterium gives rise to a colony (Figure 8.10, part 3). These clones are preserved, for example, by picking each colony off the plate and placing it into the medium in a well of a 96-well microtiter dish (Figure 8.10, part 4; only 16 wells are shown in the example). Replicas of the set of clones are printed onto a membrane filter that has been placed on a Petri plate of selective medium appropriate for the recombinant molecules—for example, ampicillin for plasmids carrying the ampicillin resistance gene (Figure 8.10, part 5). Colonies grow on the filter in the same pattern as the clones grow in the microtiter dish. The filter is peeled from the dish, and the cells are lysed in situ (Figure 8.10, part 6). The proteins that were within the cell, including those expressed from the cDNA, become stuck to the filter, which is then incubated with an antibody to the protein of interest (Figure 8.10, part 7).

If the antibody is radioactively labeled, any clones that expressed the protein of interest can be identified by placing the dried filter against X-ray film, leaving it in the dark for a period (from 1 hour to overnight) to produce an *autoradiogram* (Figure 8.10, part 8). The process is called *autoradiography*. When the film is developed, dark spots are seen wherever the radioactive probe is bound to

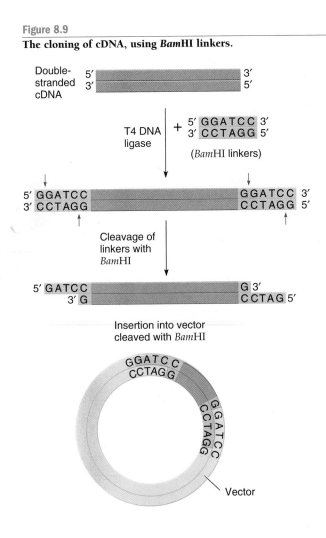

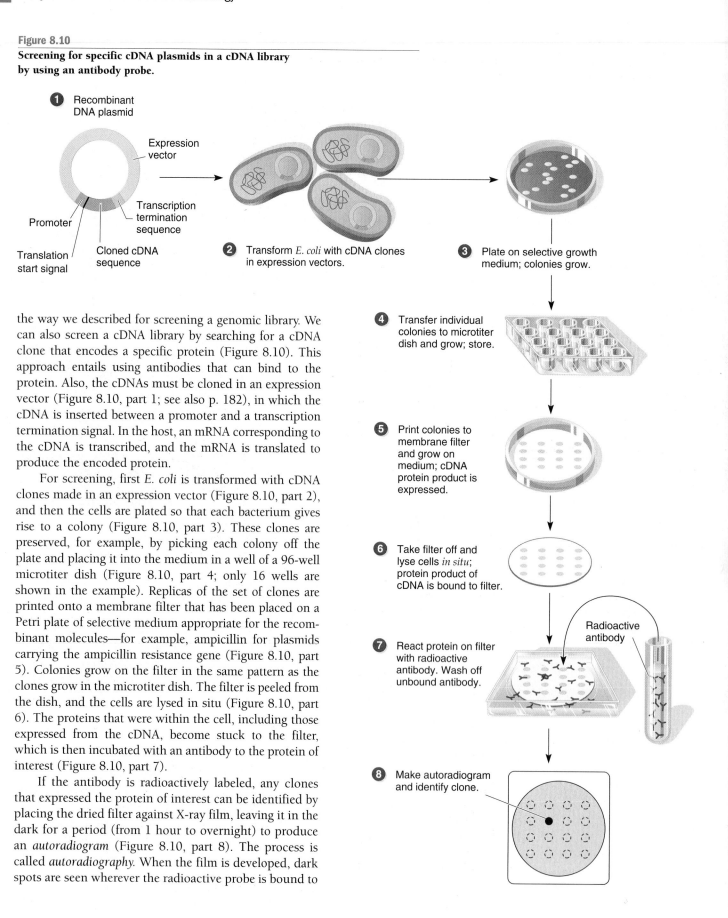

Figure 8.11

Using DNA probes to screen plasmid genomic libraries for specific DNA sequences.

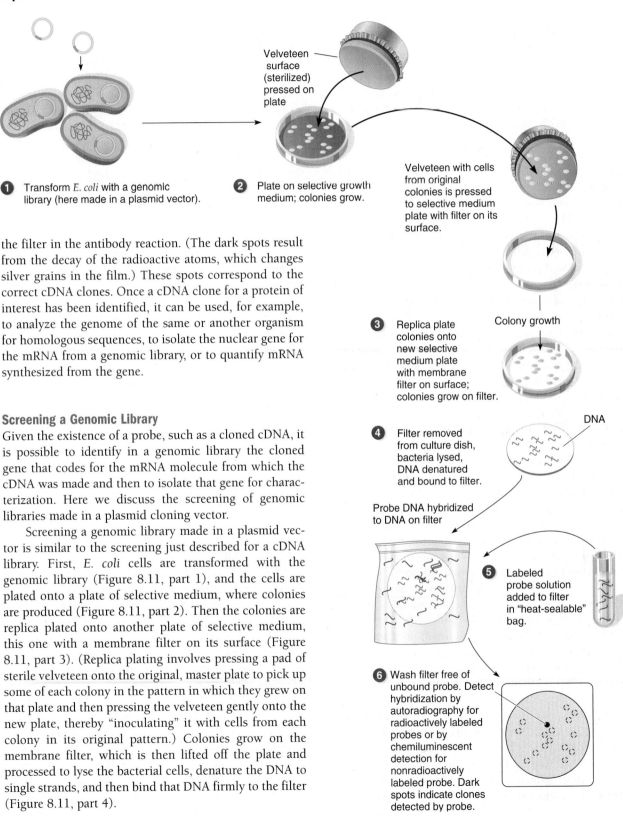

① Transform *E. coli* with a genomic library (here made in a plasmid vector).

② Plate on selective growth medium; colonies grow.

Velveteen surface (sterilized) pressed on plate

Velveteen with cells from original colonies is pressed to selective medium plate with filter on its surface.

③ Replica plate colonies onto new selective medium plate with membrane filter on surface; colonies grow on filter.

Colony growth

④ Filter removed from culture dish, bacteria lysed, DNA denatured and bound to filter.

DNA

Probe DNA hybridized to DNA on filter

⑤ Labeled probe solution added to filter in "heat-sealable" bag.

⑥ Wash filter free of unbound probe. Detect hybridization by autoradiography for radioactively labeled probes or by chemiluminescent detection for nonradioactively labeled probe. Dark spots indicate clones detected by probe.

the filter in the antibody reaction. (The dark spots result from the decay of the radioactive atoms, which changes silver grains in the film.) These spots correspond to the correct cDNA clones. Once a cDNA clone for a protein of interest has been identified, it can be used, for example, to analyze the genome of the same or another organism for homologous sequences, to isolate the nuclear gene for the mRNA from a genomic library, or to quantify mRNA synthesized from the gene.

Screening a Genomic Library

Given the existence of a probe, such as a cloned cDNA, it is possible to identify in a genomic library the cloned gene that codes for the mRNA molecule from which the cDNA was made and then to isolate that gene for characterization. Here we discuss the screening of genomic libraries made in a plasmid cloning vector.

Screening a genomic library made in a plasmid vector is similar to the screening just described for a cDNA library. First, *E. coli* cells are transformed with the genomic library (Figure 8.11, part 1), and the cells are plated onto a plate of selective medium, where colonies are produced (Figure 8.11, part 2). Then the colonies are replica plated onto another plate of selective medium, this one with a membrane filter on its surface (Figure 8.11, part 3). (Replica plating involves pressing a pad of sterile velveteen onto the original, master plate to pick up some of each colony in the pattern in which they grew on that plate and then pressing the velveteen gently onto the new plate, thereby "inoculating" it with cells from each colony in its original pattern.) Colonies grow on the membrane filter, which is then lifted off the plate and processed to lyse the bacterial cells, denature the DNA to single strands, and then bind that DNA firmly to the filter (Figure 8.11, part 4).

Next, the filter is placed in a heat-sealable plastic bag and incubated with the cDNA probe (Figure 8.11, part 5), which has been labeled radioactively or nonradioactively (Box 8.1). To prepare the labeled DNA for use as a probe, the DNA is denatured by boiling and then quickly cooled on ice to produce single-stranded DNA molecules. These labeled molecules are added to the membrane filters to which the denatured (single-stranded) DNA from each colony has been bound. The labeled molecules diffuse over the filter, and, with time, they will find the filter-bound DNA with which they can pair by complementary base pairing. By this hydrogen bonding, DNA–DNA hybrids form between the probe and the colony DNA. If the cDNA probe is derived from the mRNA for β-globin, for example, that probe will hybridize with the filter-bound DNA that encodes the β-globin mRNA (that is, the genomic β-globin gene). After the hybridization step, the filters are washed to remove unbound probe and are subjected to the appropriate detection procedure, depending upon whether the probe was radioactive or nonradioactive: autoradiography for a radioactive probe, chemiluminescent or colorimetric detection for a nonradioactive probe (Figure 8.11, part 6). From the positions of the spots on the film or filter, the locations of the bacterial colony or colonies on the original plate can be determined and the clones of interest isolated for further characterization.

Identifying Genes in Libraries by Complementation of Mutations

For organisms in which genetic systems of analysis have been well developed and for which there are well-defined mutations, it is possible to clone genes by complementation of those mutations. In brief, this approach relies on the expression of the wild-type gene introduced into the cell by transformation overcoming the defect of a mutant form of the gene in the genome. (Complementation is discussed in more detail in Chapter 13, pp. 340–341.) This can be done with the yeast Saccharomyces cerevisiae, for example, which is easy to manipulate genetically and for which efficient integrative and replicative transformation systems using E. coli–yeast shuttle vectors are available.

To clone a yeast gene by complementation, first a genomic library is made of DNA fragments from the wild-type yeast strain in a yeast–E. coli shuttle vector. The library is used to transform a host yeast strain carrying two mutations: one to allow transformants to be selected (ura3, for example) and another in the gene for which the wild-type gene clone is sought. Consider the cloning of the ARG1 gene, the wild-type gene for an enzyme needed for the biosynthesis of arginine (Figure 8.12), by the complementation of an arg1 mutation. A yeast

strain carrying the arg1 mutation has an inactive enzyme for arginine biosynthesis and therefore needs arginine to grow. A genomic library is made by using DNA from a wild-type (ARG1) yeast strain (Figure 8.12, parts 1 and 2). When a population of ura3 arg1 yeast cells is transformed with the genomic library prepared in the shuttle vector (Figure 8.12, part 3), some cells receive plasmids containing the normal (ARG1) gene for the arginine biosynthesis enzyme. The plasmid's ARG1 gene is expressed, enabling the cell to grow on minimal medium—that is, in the absence of arginine—despite the presence of a defective arg1 gene in the cell's genome (Figure 8.12, part 4). The ARG1 gene is said to overcome the functional defect of the arg1 mutation by complementation of that mutation (Figure 8.12, parts 5 and 6). The plasmid is then isolated from the cells, and the cloned gene is characterized.

Using Heterologous Probes to Identify Specific DNA Sequences in Libraries

cDNA probes can be used to identify and isolate specific genes, and a very large number of genes have been cloned from both prokaryotes and eukaryotes with those probes. It is also possible to identify specific genes in a genomic library by using clones of equivalent genes from other organisms as probes. For example, a mouse probe could be used to probe a human genomic library. Such probes generally are called heterologous probes, and their effectiveness depends on a good degree of homology between the probes and the genes. For that reason, the greatest success with this approach has come with highly conserved genes or with probes from a species closely related to the organism from which a particular gene is to be isolated.

Using Oligonucleotide Probes to Identify Genes or cDNAs in Libraries

A number of genes have been isolated from libraries with the use of synthetically made oligonucleotide probes. In this method, at least some of the amino-acid sequence must be known for the protein encoded by the gene. Also, ideally, the amino-acid substitutions associated with specific mutations have been identified. In that case, it may be possible that a consensus sequence (the most common nucleotide at each position) can be determined from previously cloned versions of the gene that are available in GenBank (http://www.ncbi.nlm.nih.gov), a computer database into which sequences are deposited and made available to researchers worldwide. Then, because the genetic code is universal, oligonucleotides about 20 nucleotides long can be designed that, if translated, would give the known amino-acid sequence. Because of the degeneracy of the genetic code—up to six different

Box 8.1   Labeling DNA

DNA can be labeled either radioactively or nonradioactively. Typically, it has been more common to label DNA radioactively, but with increasing regulations pertaining to the disposal of radioactive material and the health risks of exposure to radioactive compounds, great strides have been made in developing nonradioactive DNA labeling methods which produce probes that are as sensitive as radioactive probes in seeking out the target DNA. Thus, it is now possible to detect as little as 0.1 picogram (0.1×10^{-12} g) of DNA with either radioactive or nonradioactive probes. We now discuss briefly some methods for preparing radioactively labeled and nonradioactively labeled DNA probes.

Radioactive Labeling of DNA

A DNA probe can be labeled radioactively by the *random-primer method* (Box Figure 8.1). In this approach, the DNA is denatured to single strands by boiling and quick cooling on ice. DNA primers six nucleotides long (hexanucleotides), synthetically made by the random incorporation of nucleotides, are annealed to the DNA. The *hexanucleotide random primers* pair with complementary sequences in the DNA, and such pairing occurs at many locations because all possible hexanucleotide sequences are present. The primers are elongated by the Klenow fragment of DNA polymerase I, which uses radioactively labeled precursors (dNTPs). (The Klenow fragment, named for the person who discovered it, lacks 5′-to-3 exonuclease activity, which would otherwise remove the short primers, but still has the 3′-to-5′ proofreading activity.) Typically, the label is ^{32}P, located in the phosphate group that is attached to the 5′ carbon of the deoxyribose sugar. This phosphate group is called the α-*phosphate*, because it is the first in the chain of three; the α-phosphate is used in forming the phosphodiester bonds of the sugar–phosphate backbone.

   After the radioactive DNA probe is applied in an experiment, detection depends on the properties of the radioactive isotope. For example, if a ^{32}P-labeled probe has hybridized with a target DNA sequence on a membrane filter, the filter is placed against a piece of X-ray film and the sandwich is placed in the dark. Every location on the filter where there is ^{32}P (a spot, band, etc.) is detected as a black region on the X-ray film after it is developed. This process is called *autoradiography*, and the resulting picture of radioactive signals is called an *autoradiogram*.

Nonradioactive Labeling of DNA

Random-primer labeling also can be used to prepare nonradioactively labeled DNA probes. The difference from preparing radioactively labeled DNA is that a special DNA precursor molecule, rather than a ^{32}P-labeled precursor, is used. For example, in one of many labeling systems, digoxigenin-dUTP (DIG-dUTP) is added to the dATP, dCTP, dGTP, and dTTP precursor mixture. Digoxigenin is a steroid, and it is linked to dUTP (deoxyuridine 5′-triphosphate). During DNA synthesis, DIG-dUTP can be incorporated opposite to A nucleotides on the template DNA strand.

The nonradioactively labeled DNA can be used in experiments in the same way as is radioactively labeled DNA. Detection is different, however. Once the DIG-dUTP-labeled probe has bound to target DNA on a filter, for example, an anti-DIG-AP conjugate is added. The anti-DIG part of the conjugate is an antibody that reacts specifically with DIG, and the AP part of the conjugate is the enzyme alkaline phosphatase. Wherever the DIG-labeled DNA is hybridized to target DNA on the filter, the anti-DIG-AP conjugate binds to form a DNA-DIG–anti-DIG-AP complex. The location of the probe-target hybrid is then visualized by substrates that react with the alkaline phosphatase. To achieve sensitivity matching that of radioactively labeled probes, a chemiluminescent substrate is used. Such a substrate produces light in a reaction catalyzed by alkaline phosphatase, and detection involves exposing X-ray film much like making an autoradiogram. If great sensitivity is *not* necessary, colorimetric substrates for the enzyme are used. In this case, spots or bands develop directly on the filter as purple or blue regions as the enzyme reaction proceeds.

Box Figure 8.1

Random primer method of radioactively labeling DNA.

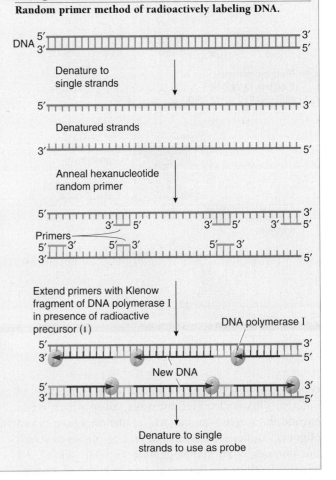

Figure 8.12

Example of cloning a gene by complementation of mutations: cloning of the yeast *ARG1* gene.

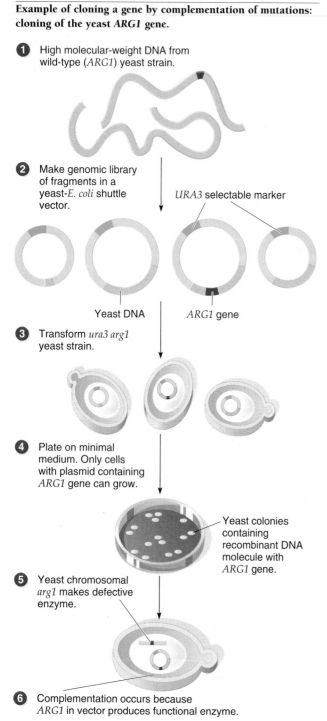

1 High molecular-weight DNA from wild-type (*ARG1*) yeast strain.

2 Make genomic library of fragments in a yeast-*E. coli* shuttle vector.

URA3 selectable marker

Yeast DNA   *ARG1* gene

3 Transform *ura3 arg1* yeast strain.

4 Plate on minimal medium. Only cells with plasmid containing *ARG1* gene can grow.

Yeast colonies containing recombinant DNA molecule with *ARG1* gene.

5 Yeast chromosomal *arg1* makes defective enzyme.

6 Complementation occurs because *ARG1* in vector produces functional enzyme.

codons can specify a given amino acid—a number of different oligonucleotides are made, all of which could encode the targeted amino-acid sequence. These mixed oligonucleotides are labeled and used as probes to search the libraries, with the hope that at least one of the oligonucleotides will detect the gene or cDNA of interest. If the probe, known as a *guessmer*, is labeled radioactively,

detection is by autoradiography, whereas if the probe is labeled nonradioactively, detection is by colorimetry or chemiluminescence. (See Box 8.1.) Though not successful all of the time, oligonucleotide-based library screening has been extremely fruitful and has allowed many genes to be cloned for which previous genetic information was lacking.

KEYNOTE

Specific sequences in cDNA libraries and genomic libraries can be identified via a number of approaches, including the use of specific antibodies, cDNA probes, complementation of mutations, heterologous probes, and oligonucleotide probes.

Molecular Analysis of Cloned DNA

Cloned DNA sequences are resources for experiments designed to answer many kinds of biological questions. Three examples of the use of the sequences are given in this section: restriction mapping, Southern blotting, and northern blotting.

Restriction Mapping

Because cloned DNA sequences represent a homogeneous population of DNA molecules, restriction enzymes cleave cloned DNA into a relatively small number of discretely sized fragments that can be visualized by means of agarose gel electrophoresis and ethidium bromide staining. With the gels that are used, the number and positions of restriction sites for each enzyme can be mapped without the need for hybridization with a labeled probe and subsequent detection. The process is called **restriction mapping,** and the resulting map, analogous to a linkage map with its arrangement of genes is called a *restriction map*. Restriction maps have a number of uses: guiding a researcher in making clones of subsections of the gene or the cDNA, confirming that a particular cloning experiment produced the correct recombinant DNA molecule, and comparing a cDNA with its gene are three examples. Restriction maps were also constructed in early stages of genome sequencing projects as part of efforts to produce comprehensive physical maps of chromosomes.

Let us assume that we have cloned a 5.0-kb region of DNA and want to construct a restriction map of it (Figure 8.13, part 1). One sample of the DNA is digested with *Eco*RI, a second sample is digested with *Bam*HI, and a third sample is digested with both *Eco*RI and *Bam*HI. The DNA restriction fragments of each reaction are

animation

a   Restriction Mapping

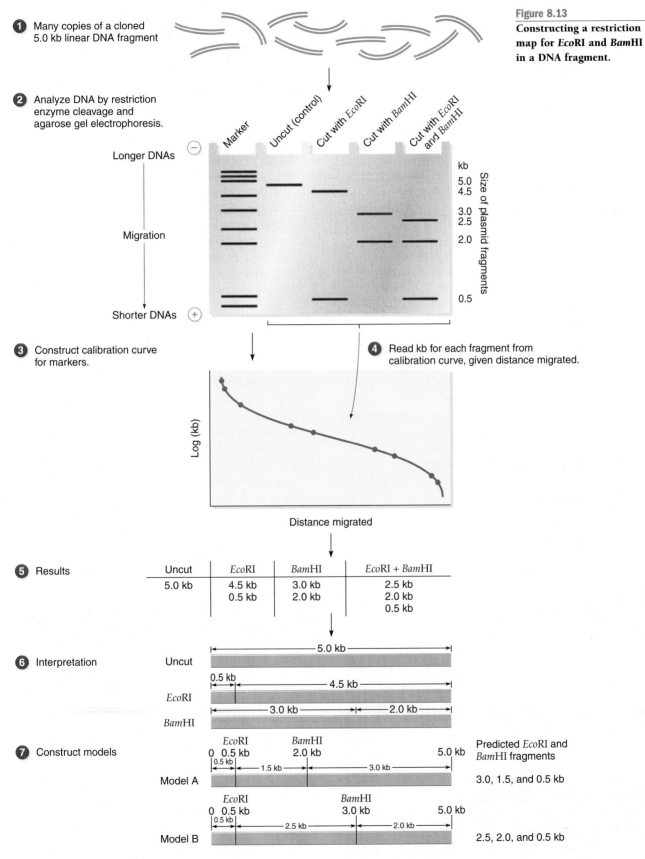

Figure 8.13

Constructing a restriction map for *Eco*RI and *Bam*HI in a DNA fragment.

1 Many copies of a cloned 5.0 kb linear DNA fragment

2 Analyze DNA by restriction enzyme cleavage and agarose gel electrophoresis.

Longer DNAs (−)

Migration

Shorter DNAs (+)

Marker | Uncut (control) | Cut with *Eco*RI | Cut with *Bam*HI | Cut with *Eco*RI and *Bam*HI

kb
5.0
4.5
3.0
2.5
2.0

0.5

Size of plasmid fragments

3 Construct calibration curve for markers.

4 Read kb for each fragment from calibration curve, given distance migrated.

Log (kb)

Distance migrated

5 Results

Uncut	*Eco*RI	*Bam*HI	*Eco*RI + *Bam*HI
5.0 kb	4.5 kb	3.0 kb	2.5 kb
	0.5 kb	2.0 kb	2.0 kb
			0.5 kb

6 Interpretation

Uncut — 5.0 kb

*Eco*RI — 0.5 kb | 4.5 kb

*Bam*HI — 3.0 kb | 2.0 kb

7 Construct models

*Eco*RI 0 0.5 kb | *Bam*HI 2.0 kb | 5.0 kb

Model A — 0.5 kb | 1.5 kb | 3.0 kb

Predicted *Eco*RI and *Bam*HI fragments

3.0, 1.5, and 0.5 kb

*Eco*RI 0 0.5 kb | *Bam*HI 3.0 kb | 5.0 kb

Model B — 0.5 kb | 2.5 kb | 2.0 kb

2.5, 2.0, and 0.5 kb

8 Conclusion

*Eco*RI and *Bam*HI data indicate that model B is correct.

separated according to their molecular sizes by agarose gel electrophoresis; controls are a sample of the same DNA uncut with any enzyme and DNA fragments of known size—DNA fragment size markers, or simply size markers—so that the sizes of the unknown DNA fragments can be computed (Figure 8.13, part 2). A photograph of an agarose gel electrophoresis apparatus is shown in Figure 8.14. The gel is a rectangular, horizontal slab of agarose (a firm, gelatinous material) that has a matrix of pores through which DNA passes in an electric field. Each gel is made by boiling a buffered agarose solution, pouring it into a mold, and allowing it to cool in the mold. A toothed comb is used to form discrete wells in the gels so that different samples can be analyzed simultaneously.

DNA is negatively charged because of its phosphates, so DNA migrates toward the positive pole in an electric field. Migration is in a straight line from the well (called a lane). Because small DNA fragments can move more easily through the pores in the gel, small DNA fragments move through the gel more rapidly than large fragments and, therefore, migrate further in a set time.

After electrophoresis, the DNA is stained with ethidium bromide. The DNA complexed with ethidium bromide fluoresces under ultraviolet light. The opening photograph of the chapter shows a gel with fluorescing DNA bands. The gel is photographed and, from the photograph, the distance each DNA band migrated from the well can be measured. The molecular size of each DNA band in the size markers is known, so a calibration curve can be drawn of DNA size (in log kb) and

Figure 8.14

An agarose gel electrophoresis apparatus (right) and power supply (left).

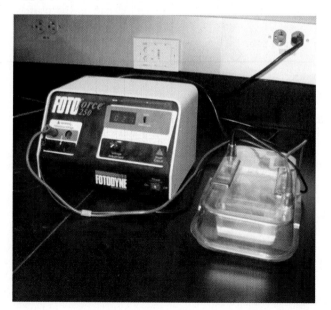

migration distance (in mm) (Figure 8.13, part 3). The migration distances for the DNA bands from the uncut and cut DNA are then used together with the calibration curve to determine the molecular sizes of the DNA fragments in the bands (Figure 8.13, part 4). For our example, the results are shown in Figure 8.13, part 5.

The results are analyzed as follows (Figure 8.13, parts 6–8):

1. When the 5.0-kb DNA is cut with *Eco*RI, 4.5-kb and 0.5-kb DNA fragments are obtained, indicating that there is one restriction site for *Eco*RI in the DNA located 0.5 kb from one end of the molecule (Figure 8.13, part 6).

2. When the same DNA is cut with *Bam*HI, 3.0-kb and 2.0-kb DNA fragments are produced. Logic similar to that used in item 1 then shows that there is one restriction site for *Bam*HI located 2.0 kb from one end of the molecule (Figure 8.13, part 6).

3. At this point, we know there is one restriction site for each enzyme, but we do not know the relationship between the two. However, we can make two models. (See Figure 8.13, part 7.) In model A, the *Eco*RI site is 0.5 kb from one end, and the *Bam*HI site is 2.0 kb from that same end. In model B, the *Eco*RI site is 0.5 kb from one end, and the *Bam*HI site is 3.0 kb from that end (that is, 2.0 kb from the other end). Model A predicts that cutting with *Eco*RI and *Bam*HI will produce three fragments of 0.5, 1.5, and 3.0 kb (going from left to right along the DNA), and model B predicts that cutting with both enzymes will produce three fragments of 0.5, 2.5, and 2.0 kb. The actual data show three fragments with sizes 2.5, 2.0, and 0.5 kb, validating model B (Figure 8.13, part 8).

In real situations, restriction mapping often involves data that are much more complicated (for example, involving more restriction enzymes and a number of sites for each enzyme). Analysis can be done with an entire plasmid (which is circular) or with the cloned sequence or part of it cut out of the plasmid and purified by *preparative agarose gel electrophoresis*. In this technique, large quantities of DNA are cut with appropriate restriction enzymes, and the fragments are separated by agarose gel electrophoresis. After staining with ethidium bromide, the bands can be visualized under ultraviolet light. Then the bands can be physically cut out of the gel and the DNA extracted for analysis.

A common application of restriction mapping is confirming that a particular clone which has been constructed is correct. Suppose we want to clone a 2.0-kb *Eco*RI-*Eco*RI DNA fragment (a piece of DNA with an *Eco*RI restriction site at each end) into the *Eco*RI site of the vector pUC19 (see Figure 8.4) and that we have made a restriction map of the fragment. The fragment

Figure 8.15

Example of restriction mapping to confirm that a plasmid has been constructed correctly. The *Eco*RI–*Eco*RI fragment can insert into the pUC19 vector in two alternative orientations. By cutting with *Aat*II, the site for which is located asymmetrically in the cloned fragment, the orientation of the clones can be distinguished by the restriction fragment sizes.

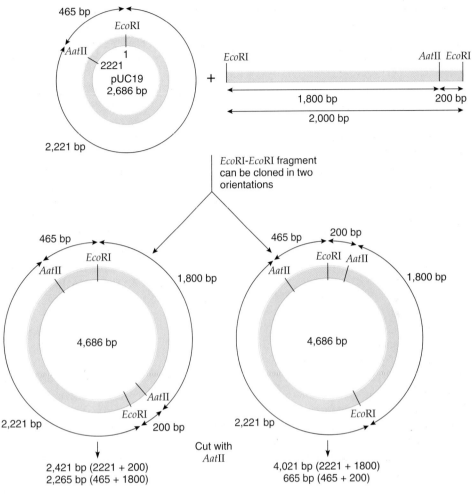

can be inserted into the vector in two possible ways (Figure 8.15). Perhaps we need the fragment to be in only one of the two orientations for our studies. How can we distinguish the two types of clones? The answer is that we scrutinize the restriction maps of the fragment and the vector to find a restriction enzyme or enzymes that will give different sizes of DNA fragments when the clones are digested. Usually, we look for an enzyme that cuts the DNA at one or more sites in the vector and at an asymmetric position in the cloned fragment. In our example, there is an *Aat*II ("a-a-t-two") restriction site in the vector 465 bp away from the *Eco*RI cloning site, and there is an *Aat*II site 200 bp from one end of the cloned fragment. (See Figure 8.15.) Digestion with *Aat*II will give two fragments, the sizes of which will be different, depending on the orientation of the insert. Specifically, the fragments will be either 4,021 bp and 665 bp, or 2,421 bp and 2,265 bp. These clearly distinguishable alternatives then make it easy to screen the clones for the desired one.

Southern Blot Analysis of Sequences in the Genome

As part of the analysis of genes, it can be helpful to determine the arrangement and specific locations of restriction sites in the genome. This information is useful, for example, for comparing homologous genes in different species, analyzing intron organization, planning experiments to clone parts of a gene (such as its promoter or controlling sequences) into a vector, or screening individuals for restriction-site differences associated with disease genes (see the next chapter). The arrangement of restriction sites in a genome can be analyzed directly with a gene probe, with a cDNA probe, or by using as a probe the same gene cloned from a closely related organism. The process of analysis is as follows:

1. Samples of genomic DNA are cut with different restriction enzymes (Figure 8.16, parts 1 and 2), each of which produces DNA fragments of different lengths, depending on the locations of the restriction sites.

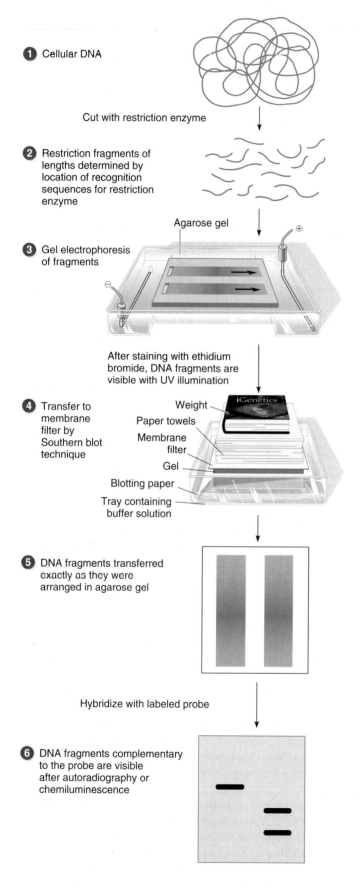

1. Cellular DNA

   Cut with restriction enzyme

2. Restriction fragments of lengths determined by location of recognition sequences for restriction enzyme

3. Gel electrophoresis of fragments

   Agarose gel

   After staining with ethidium bromide, DNA fragments are visible with UV illumination

4. Transfer to membrane filter by Southern blot technique

   Weight
   Paper towels
   Membrane filter
   Gel
   Blotting paper
   Tray containing buffer solution

5. DNA fragments transferred exactly as they were arranged in agarose gel

   Hybridize with labeled probe

6. DNA fragments complementary to the probe are visible after autoradiography or chemiluminescence

Figure 8.16

Southern blot procedure for analyzing cellular DNA for the presence of sequences complementary to a labeled probe, such as a cDNA molecule made from an isolated mRNA molecule. The hybrids, shown as three bands in this theoretical example, are visualized by autoradiography or chemiluminescence.

2. The DNA restriction fragments are separated by size through agarose gel electrophoresis (Figure 8.16, part 3). After electrophoresis, the DNA is stained with ethidium bromide so that it can be seen under ultraviolet light. When genomic DNA is digested with a restriction enzyme, the result is a continuous smear of fluorescence down most of the length of the gel lane, because the enzyme produces many fragments ranging in size from large to small.

3. The DNA fragments are transferred to a membrane filter (Figure 8.16, part 4). In brief, the gel is soaked in an alkaline solution to denature the double-stranded DNA into single strands. The gel is neutralized and placed on a piece of blotting paper on a glass plate. The ends of the paper are in a container of buffer and act as wicks. A piece of membrane filter is laid down so that it covers the gel. Sheets of blotting paper (or paper towels) and a weight are stacked on top of the filter. The blotting paper wicks up the buffer solution in the bottom tray so that the solution passes through the gel and the membrane filter and, finally, into the stack of blotting paper. During this process, the DNA fragments are picked up by the buffer and transferred from the gel to the membrane filter, to which they bind because of the filter's chemical properties. The fragments on the filter are arranged in exactly the same way as they were in the gel (Figure 8.16, part 5).

4. A labeled probe and buffer is added to the membrane filter; it hybridizes to any complementary DNA fragment(s) (Figure 8.16, part 6). Detection of the probe is carried out in a way that is appropriate for whether the probe is radioactive or nonradioactive, to determine the positions of the hybrids (Figure 8.16, part 6). If DNA size markers are separated in a different lane in the agarose gel electrophoresis process, the sizes of the genomic restriction fragments that hybridized with the probe can be calculated. From the fragment sizes obtained, a restriction map can be generated to show the relative positions of the restriction sites. Suppose, for example, that using only *Bam*HI produces a DNA fragment of 3 kb that hybridizes with the labeled probe. If a combination of *Bam*HI and *Pst*I is then used and produces two DNA fragments, one of 1 kb and the other of 2 kb, we would deduce that the 3-kb *Bam*HI fragment contains a *Pst*I restriction site 1 kb from one end and

2 kb from the other end. Further analysis with other enzymes, individually and combined, enables the researcher to construct a map of all the enzyme sites relative to all other sites.

The whole process of separating DNA fragments by agarose gel electrophoresis, transferring (blotting) the fragments onto a filter, and hybridizing with a labelled complementary probe is called **Southern blot analysis** or **Southern blotting** (named after its inventor, Edward Southern). Applications of Southern blot analysis will be described in the next chapter.

Northern Blot Analysis of RNA

A technique that is similar to Southern blot analysis— called **northern blot analysis** or **northern blotting**—is for the study of RNA rather than DNA. (The name does not denote a geographical locale, but indicates that the technique is related to Southern blot analysis.) In northern blot analysis, RNA extracted from cells or a tissue is separated by size with gel electrophoresis, and the RNA molecules are transferred and bound to a filter in a procedure which is essentially identical to that used in Southern blot analysis. After hybridization with a labeled probe and use of the appropriate detection system, bands show the locations of RNA fragments that were complementary to the probe. Given appropriate RNA size markers, the sizes of the RNA fragments identified with the probe can be determined.

Northern blot analysis is useful for revealing the size or sizes of the mRNA encoded by a gene. In some cases, a number of different mRNA species encoded by the same gene have been identified in this way, suggesting that different promoter sites or different terminator sites are used or that alternative mRNA processing can occur. Northern blot analysis can also be used to investigate whether an mRNA is present in a cell type or tissue and how much of it is present. This type of experiment is useful for determining levels of gene activity—for instance, during development, in different cell types of an organism, or in cells before and after they are subjected to various physiological stimuli.

KEYNOTE

Cloned genes and other DNA sequences often are analyzed to determine the arrangement and specific locations of restriction sites. The analytical process involves cleavage of the DNA with restriction enzymes, followed by separation of the resulting DNA fragments by agarose gel electrophoresis. The sizes of the DNA fragments are calculated, enabling restriction maps to be constructed. The many DNA fragments produced by cleaving genomic DNA show a wide range of sizes,

resulting in a continuous smear of DNA fragments in the gel. In this case, specific gene fragments can be visualized only by transferring them to a membrane filter by Southern blotting, hybridizing a specific labeled probe with the DNA fragments, and detecting the hybrids. A similar procedure—northern blotting—is used to analyze the sizes and quantities of RNAs isolated from a cell.

DNA Sequencing

DNA fragments can be analyzed to determine the nucleotide sequence of the DNA and to determine the distribution and location of restriction sites. The latter is the most detailed information one can obtain about a DNA fragment. The information is useful, for example, in computer database analyses for identifying gene sequences and regulatory sequences within the fragment and for comparing the sequences of homologous genes from different organisms. Furthermore, the DNA sequence of a protein-coding gene can be translated by computer to provide information about the properties of the protein for which it codes. Such information can be helpful to an investigator who wants to isolate and study an unknown protein product of a gene for which a clone is available. Walter Gilbert and Frederick Sanger shared one-half of the 1980 Nobel Prize in Chemistry for their "contributions concerning the determination of base sequences in nucleic acids."

Dideoxy Sequencing. The most commonly used method of DNA sequencing, called **dideoxy sequencing** (developed by Fred Sanger in the 1970s), involves the extension of a short primer by DNA polymerase. Both linear DNA and circular DNA can be sequenced with the dideoxy DNA sequencing method. Linear DNA fragments can be generated, for example, by cutting plasmid DNA with one or more restriction enzymes.

animation

a Dideoxy DNA Sequencing

In dideoxy DNA sequencing, the DNA is first denatured to single strands by heat treatment. Next, for each pair of DNA strands produced, a short oligonucleotide primer is annealed to one of the strands (Figure 8.17). The oligonucleotide is designed so that its 3′ end is next to the DNA sequence of interest. The oligonucleotide acts as a *primer* for DNA synthesis, and the 5′-to-3′ orientation chosen ensures that the DNA made is a complementary copy of the DNA sequence of interest. (See Figure 8.17.) Consider, for example, DNA fragments cloned into the plasmid cloning vector pUC19. (See Figure 8.4.) With a pair of oligonucleotide primers complementary to the DNA flanking the multiple cloning site, any DNA insert

Figure 8.17

Dideoxy DNA sequencing of a theoretical DNA fragment.
(The "number of nucleotides" in the gel analysis refers to nucleotides added to the primer during new DNA synthesis.)

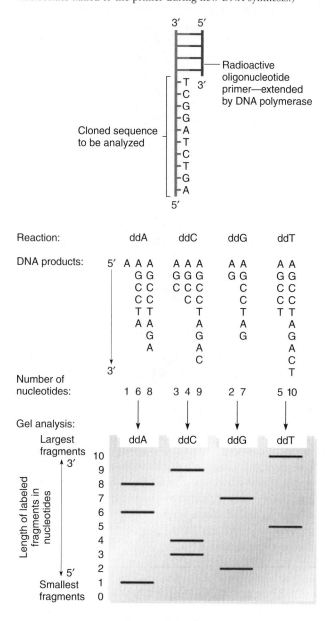

Sequence deduced from banding pattern of autoradiogram made from gel:

5′-A-G-C-C-T-A-G-A-C-T 3′

For each sequencing experiment, four separate reactions are set up. Each reaction contains the single-stranded DNA to be sequenced, the primer annealed to that DNA, DNA polymerase, the four normal deoxynucleotide precursors (dATP, dTTP, dCTP, and dGTP), and a small amount of a modified nucleotide precursor called a **dideoxynucleotide** (ddNTP; Figure 8.18). A dideoxynucleotide differs from a normal deoxynucleotide in that it has a 3′-H rather than a 3-OH on the deoxyribose sugar. Either the primer or one or more of the normal precursors are labeled so that newly synthesized DNA can be detected easily; in Figure 8.17, the primer is labeled.

The four reactions differ in which dideoxynucleotide is present—that is, whether it has A, T, G, or C as the base. Generally, the dideoxy precursor is present in about one one-hundredth the amount of the normal precursor, so that some DNA synthesis occurs in the dideoxy sequencing reactions. When the primer is extended, DNA polymerase occasionally inserts a dideoxynucleotide instead of the normal deoxynucleotide. Once that happens, no further DNA synthesis can occur, because the absence of a 3′-OH prevents the formation of a phosphodiester bond with an incoming DNA precursor.

For example, if an A is specified by the DNA template strand, a dideoxy A nucleotide (ddA), rather than the normal A nucleotide, could be incorporated into the reaction mixture, and elongation of the chain would stop. In a large population of molecules in the same DNA synthesis reaction, new DNA chains stop at all possible positions where the nucleotide is required, because of the incorporation of the dideoxynucleotide. In the ddA reaction, the many different-sized chains produced all end with ddA; in the ddG reaction, all chains end with ddG; and so on (see Figure

Figure 8.18

A dideoxynucleotide (ddNTP) DNA precursor.

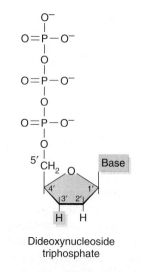

Dideoxynucleoside triphosphate

(Normal DNA precursor has OH at 3′ position)

can be sequenced from each end. In fact, most plasmid cloning vectors have the same sequences flanking their multiple cloning sites, so that *universal sequencing primers* can be used to sequence any cloned insert in these vectors. Any other DNA fragment can be sequenced if some sequence information is available from which a primer can be made.

8.17). In short, each DNA chain that is synthesized starts from the same point and ends at the base determined by the dideoxy nucleotide incorporated.

The DNA chains in each reaction mixture are separated by polyacrylamide gel electrophoresis, and the locations of the DNA bands are revealed by autoradiography (in the case of radioactive DNA sequencing experiments). The DNA sequence of the newly synthesized strand is determined from the autoradiogram by reading up the *sequencing ladder* from the bottom to the top to give the sequence in 5′-to-3′ orientation. In the example, the band that moved the farthest ended with ddA, the band that moved the second farthest ended with ddG, and so on. The complete sequence determined is 5′-AGCCTAGACT-3′; this sequence is *complementary* to that on the template strand. (See Figure 8.17.) An example of a dideoxy sequencing gel result is shown in Figure 8.19. With such a gel, the DNA sequence is recorded manually by the researcher.

KEYNOTE

Methods have been developed for determining the sequence of a cloned piece of DNA. A commonly used method, the dideoxy procedure, uses enzymatic synthesis of a new DNA chain on a cloned template DNA strand. With this procedure, the synthesis of new strands is stopped by the incorporation of a dideoxy analog of the normal deoxyribonucleotide. When four different dideoxy analogs are used, the new strands stop at all possible nucleotide positions, thereby allowing the complete DNA sequence to be determined.

Automated procedures are now routinely used for DNA sequencing. The procedures involve only one reaction containing the four dideoxyribonucleotides, each labeled with a different fluorescent dye. The DNA fragments synthesized are separated by electrophoresis in a single gel lane, which is scanned by a laser device that excites the fluorescent labels and determines which one is present at each position. The output is a series of colored peaks corresponding to each nucleotide position in the sequence (Figure 8.20) and is converted automatically to a sequence of bases by computer. Automated procedures are of great utility to research teams in determining the complete sequences of various genomes.

Analysis of DNA Sequences. Sequences determined by any sequencing method are entered into computer databases. Computer programs are used to analyze DNA sequences for restriction-site locations, for example, and to compare a variety of sequences, homologous regions, transcription regulatory sequences, and so on. Programs can search DNA sequences for possible protein-coding regions by

Figure 8.19

Autoradiogram of a dideoxy sequencing gel. The letters over the lanes (A, C, G, and T) correspond to the particular dideoxy nucleotide used in the sequencing reaction analyzed in a given lane.

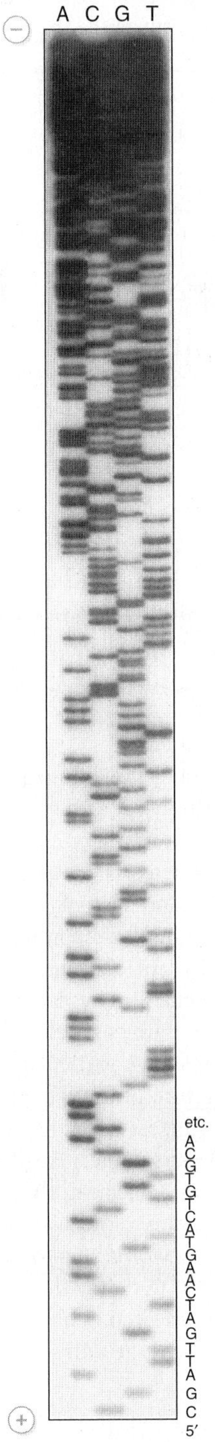

looking for an initiator codon in frame with a stop codon. This region is called an **open reading frame,** or **ORF.** (Note that the discovery of a possible ORF does not mean that that DNA sequence encodes a protein in the cell: Many experiments would have to be done to see whether

Figure 8.20
Results of automated DNA sequence analysis using fluorescent dyes. The procedure is described in the text. The automated sequencer generates the curves shown in
the figure from the fluorescing bands on the gel. The colors are generated by the
machine and indicate the four bases: A is green, G is black, C is blue, and T is red.
Where bands cannot be distinguished clearly, an N is listed.

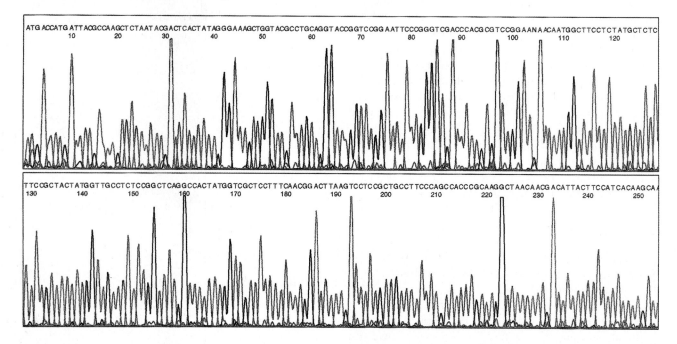

that is the case. Nor does the failure to detect an ORF necessarily mean that the DNA sequence has no function; it
may represent a regulatory region.) Other programs can be
used to translate a cloned DNA sequence into a theoretical
amino-acid sequence and to make predictions about the
structure and function of the protein. This is possible
because the sequences of all sequenced proteins are in
computer databases, enabling researchers to make detailed
and rapid comparisons by computer. For example, many
DNA-binding proteins have particular secondary structure
motifs, and the detection of the sequence for such a motif
in a gene sequence suggests that the gene encodes a DNA
binding protein. Because many regulatory proteins are
DNA-binding proteins, detecting the sequence for a certain secondary structure motif would be of significant
interest. Such a finding would then suggest the direction of
future experimentation to define the actual function of the
gene product. Increasingly, much of this computer analysis of sequences can be done with the use of the Internet.

The Polymerase Chain Reaction (PCR)

Generating large numbers of identical copies of DNA by
the construction and cloning of recombinant DNA
molecules was made possible in the 1970s. Recombinant
DNA techniques revolutionized molecular genetics by

making it possible to analyze genes and their functions in
new ways. However, clMoning DNA is time consuming.
In the mid-1980s, the **polymerase chain reaction (PCR)**
was developed, and this has resulted in a new revolution
in gene analysis. In a process called *amplification*, PCR
produces an extremely large number of copies of a specific DNA sequence from a DNA mixture without having
to clone the sequence in a host organism. The amplified
PCR products are called *amplimers*. PCR has become one
of the most important tools in modern molecular biology.
Kary Mullis, who developed the technique, shared the
Nobel Prize in Chemistry in 1993. (The other recipient,
M. Smith, received the prize for other work.)

PCR Steps

PCR begins with the double-stranded DNA containing
the sequence to be amplified and a pair of oligonucleotide primers which flank that DNA (Figure 8.21). The
primers usually are 20 or more
nucleotides long and are made synthetically, so a limitation of PCR is
that information must be available
about the sequence of interest. In
brief, PCR is done as follows:

animation

a The Polymerase
Chain Reaction
(PCR)

1. Denature the double-stranded DNA to single strands
by heating at 94–95°C (Figure 8.21, part 1).

Figure 8.21

The polymerase chain reaction (PCR) for selective amplification of DNA sequences.

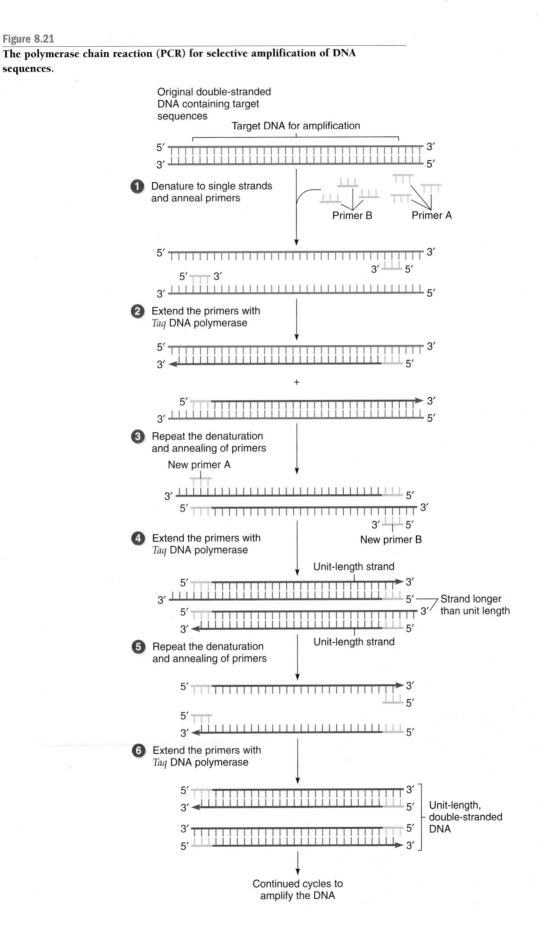

Original double-stranded DNA containing target sequences

Target DNA for amplification

1 Denature to single strands and anneal primers

Primer B   Primer A

2 Extend the primers with *Taq* DNA polymerase

+

3 Repeat the denaturation and annealing of primers

New primer A

4 Extend the primers with *Taq* DNA polymerase

Unit-length strand

Strand longer than unit length

New primer B

Unit-length strand

5 Repeat the denaturation and annealing of primers

6 Extend the primers with *Taq* DNA polymerase

Unit-length, double-stranded DNA

Continued cycles to amplify the DNA

2. Cool the solution, and anneal the primers (A and B in the figure) at 37–65°C, depending on how well the base sequences of the primers complement the base sequence of the DNA. The two primers are designed so that they anneal to the opposite strands of the template DNA flanking the sequence to be amplified. As a result, the 3′ ends of the primers face each other.

3. Extend the primers with DNA polymerase at 70–75°C (Figure 8.21, part 2). For this, a special heat-resistant DNA polymerase, such as *Taq* ("tack") *DNA polymerase*, is used. (This particular enzyme is the DNA polymerase of a thermophilic bacterium, *Thermus aquaticus*.)

4. Repeat the heating cycle to denature the DNA to single strands, and cool the solution to anneal the primers again (Figure 8.21, part 3). (Further amplification of the original strands is omitted in the remainder of the figure.)

5. Repeat the extension of the primer with *Taq* DNA polymerase (Figure 8.21, part 4). In each of the two double-stranded molecules produced in the figure, one strand is of unit length; that is, it is the length of DNA between the 5′ end of primer A and the 5′ end of primer B—the length of the target DNA. The other strand in both molecules is longer than unit length.

6. Repeat the denaturation of DNA and the annealing of new primers (Figure 8.21, part 5). (For simplification, the further amplification of those strands which are longer than unit length is omitted in the rest of the figure.)

7. Repeat the extension of the primer with *Taq* DNA polymerase (Figure 8.21, part 6). This produces unit-length, double-stranded DNA. Note that it took three cycles to produce the two molecules of unit-length DNA. Repeated denaturation, annealing, and extension cycles result in a geometric increase in the amount of unit-length DNA.

With PCR, the amount of new DNA generated increases geometrically. Starting with 1 molecule of DNA, 1 cycle of PCR produces 2 molecules, 2 cycles produce 4 molecules, and 3 cycles produce 8 molecules, 2 of which are the target DNA. A further 10 cycles produce 1,024 copies (2^{10}) of the target DNA, and in 20 cycles there will be 1,048,576 copies (2^{20}) of the target DNA. The procedure is rapid, each cycle taking only a few minutes in a *thermal cycler*, a machine that automatically cycles the reaction through programmed temperature changes.

Advantages and Limitations of PCR

PCR is a powerful technique for amplifying segments of DNA. Such amplification is similar to cloning DNA using vectors. However, PCR is a much more sensitive and quicker technique than cloning. Specifically, PCR can produce millions of copies of a DNA segment, starting with just one DNA molecule, in only a few hours. By contrast, cloning requires a significant amount of starting DNA for restriction digestion, and then at least a week is needed to go through all the cloning steps. There are two major limitations of PCR, however. First, PCR requires the use of specific primers, so sequence information on the DNA to be amplified must be available in order for primers to be designed. Second, the length of DNA that can be amplified by PCR is limited by the enzyme and surrounding conditions to about 20 kb. A further issue with PCR is that *Taq* polymerase has no proofreading activity; accordingly, base-pair mismatches that occur during replication go uncorrected in this in vitro procedure. Since PCR involves a geometric increase in the number of DNA molecules, the lack of error correction is more or less serious depending on when in the amplification process an error is introduced. That is, on the one hand, if an error is introduced in the first round of PCR, then all derivative DNA molecules will have the error. On the other hand, an increasingly lower fraction of the DNA molecules will have the error the later in the rounds of PCR it is introduced. Some alternative enzymes are available for PCR that do have proofreading activity, and these enzymes significantly decrease the error frequency. One such enzyme is Vent polymerase, which was originally extracted from a bacterium growing around high-temperature deep-sea oceanic vents. Finally, the excellent sensitivity of PCR is a liability in some applications. Because PCR can produce many copies from a single DNA molecule, great care has to be taken that the right DNA molecules are amplified. In forensic applications, for example, it is crucial that DNA used for evidence have no chance of being contaminated by DNA from the investigators or researchers.

Applications of PCR

PCR has many applications, including amplifying DNA for cloning or subcloning (moving part of a cloned sequence to a new vector), amplifying DNA from genomic DNA preparations for sequencing without cloning, mapping DNA segments, the diagnosis of disease, the sex determination of embryos, forensics, and studies of molecular evolution. In diagnosing disease, for example, PCR can be used to detect bacterial pathogens or viral pathogens such as HIV (human immunodeficiency virus, the causative agent of AIDS) and hepatitis B virus. PCR can also be used in diagnosing genetic diseases, a topic discussed in Chapter 9.

PCR is useful for subcloning a segment of cloned DNA. This application follows from the discussion of cloning a yeast gene by complementation earlier in the chapter. (See pp. 190–192 and Figure 8.12.) The concept presented was that a yeast genomic library can be used to identify a particular wild-type gene by complementation of a mutation. Experimentally, a clone in the library is identified because it confers a wild-type phenotype on the mutant cell it transforms. In the specific example discussed, the wild-type *ARG1* yeast gene was identified. Let us now refine the analysis. The plasmid clone that complements the *arg1* mutant must contain the *ARG1* gene. The plasmid is extracted from the yeast, and the sequence of the cloned fragment is determined. If there is only one gene in the fragment, then, of course, it must be the *ARG1* gene. However, if there is more than one gene, further steps are needed to identify the *ARG1* gene. Since we have determined the sequence of the cloned fragment, we can design PCR primers and amplify each gene individually. The amplified genes can then be cloned separately into a vector, just as in the case of the construction of the genomic library in Figure 8.12. Now each cloned gene can be tested separately for its ability to complement the *arg1* mutant gene, and in this way, the *ARG1* gene is found.

In forensics, PCR can be used, for example, to amplify trace amounts of DNA in samples such as hair, blood, or semen collected from a crime scene. The amplified DNA can be analyzed and compared with DNA from a victim and a suspect, and the results can be used to implicate or exonerate suspects in the crime. This analysis, called *DNA typing, DNA fingerprinting,* or *DNA profiling,* is discussed in more detail in Chapter 9 (pp. 224–227). Another interesting application of PCR is amplifying ancient DNA for analysis, using samples of tissues preserved hundreds or thousands of years ago, such as 440,000-year-old mammoths. PCR makes it possible to amplify selected DNA sequences and then analyze the sequences of those DNA molecules for comparison with contemporary DNA samples. These analyses enable us to make evolutionary comparisons between ancient forebears and present-day descendants.

RT-PCR and mRNA Quantification

Like regular PCR, **RT-PCR—reverse transcriptase PCR**—is a highly sensitive technique for detecting and quantifying RNA. There are two steps to the technique. In the first, cDNA is synthesized from RNA using a primer (oligo(dT), for example, for mRNA) and the enzyme reverse transcriptase (RT). (See Figure 8.8.) In the second step, the specific cDNA made is amplified by PCR (Figure 8.21).

RT-PCR is used either for testing for the presence of a particular RNA or for quantifying the amount of an RNA. For example, some viruses have RNA genomes, and theoretically, RT-PCR could be used to detect whether an individual has been infected by the virus. Such a test has been developed for HIV, measles, and mumps virus.

The use of RT-PCR to quantify mRNA is as follows: With conventional RT-PCR, the amplified DNA produced after a set number of PCR cycles is analyzed by agarose gel electrophoresis. The DNA is visualized by staining with ethidium bromide and UV illumination. The intensity of the fluorescence is a measure of DNA concentration, and this is related to the starting amount of mRNA in the sample tested. If mRNA levels are being measured in different cell types or under different conditions, a control is run of an mRNA species that is known not to change. However, a limitation of this method is that ethidium bromide is not a very sensitive stain for DNA, so the results are only semi-quantitative. More accurate for mRNA quantification is northern blot analysis, which involves no amplification, but does require many manipulations and particular care in the hybridization step.

A relatively new version of RT-PCR—*real-time RT-PCR*—allows for much more accurate quantification of mRNA levels. In this technique, the reverse transcriptase step is carried out as usual to produce the first-strand cDNA copy of the mRNA, but the DNA amplification step is then done in the presence of SYBR green, a highly sensitive fluorescent dye that stains double-stranded DNA. As the DNA is amplified by PCR, the staining is measured in real time (hence the name of the technique) by means of a special thermal cycler that uses laser detection of the fluorescence. The rate of production of stained amplified DNA compared with controls of known amounts of starting mRNA allows the amount of mRNA in the experimental sample to be quantified. RT-PCR is becoming extensively used to quantify mRNA levels for many genes in a wide range of cells and tissues in numerous organisms.

KEYNOTE

The polymerase chain reaction (PCR) uses specific oligonucleotide primers to amplify a particular segment of DNA many thousandfold in an automated procedure. PCR has many applications in research and in the commercial arena, including generating specific DNA segments for cloning or sequencing, amplifying DNA to detect specific genetic defects, and amplifying DNA for fingerprinting in crime scene investigations.

Summary

In this chapter, we have discussed some of the procedures involved in recombinant DNA technology and DNA manipulation. Collectively, these procedures are also called genetic engineering. We have seen how it is possible to cut DNA at specific sites by using restriction enzymes, how DNA can be cloned into specially constructed vectors, and how the cloned DNA can be analyzed in various ways.

Through the construction of genomic libraries and cDNA libraries and the application of screening procedures to those libraries, a large number of genes from a wide variety of organisms have been cloned and identified.

Restriction mapping analysis has provided detailed molecular maps of genes and chromosomes analogous to the genetic maps constructed on the basis of recombination analysis. DNA sequencing methods have given us an enormous amount of information about the DNA organization of genes, both coding sequences and regulatory sequences. The amount of DNA sequence information is growing rapidly, and computer databases of such sequences are available for researchers to analyze. For example, when a new gene is sequenced, we can determine whether it has any sequences in common with genes already in the database.

In the mid-1980s, a new technique called polymerase chain reaction (PCR) was developed. Given some sequence information about a DNA fragment, synthetic oligonucleotide primers can be made and used to amplify the DNA fragment from the genome in a repeated cycle of DNA denaturation (separation of strands), annealing of the primers, and extension of the primers with a special DNA polymerase. Numerous applications have been found for PCR, including cloning rare pieces of DNA, preparing DNA for sequencing without cloning, and diagnosing genetic diseases. A derivative of PCR, RT-PCR, makes it possible to make DNA copies of an mRNA rapidly. RT-PCR is used, for example, to generate DNA from mRNA for cloning and to quantify mRNA.

Analytical Approaches to Solving Genetics Problems

Q8.1 A piece of DNA 900 bp long is cloned and then cut out of the vector for analysis. Digestion of this linear piece of DNA with three different restriction enzymes singly and in all possible pairs of enzymes gave the following restriction fragment size data:

Enzymes	Restriction Fragment Sizes
*Eco*RI	200 bp, 700 bp
*Hind*III	300 bp, 600 bp
*Bam*HI	50 bp, 350 bp, 500 bp
*Eco*RI + *Hind*III	100 bp, 200 bp, 600 bp
*Eco*RI + *Bam*HI	50 bp, 150 bp, 200 bp, 500 bp
*Hind*III + *Bam*HI	50 bp, 100 bp, 250 bp, 500 bp

Construct a restriction map from these data.

A8.1 The approach to this kind of problem is to consider a pair of enzymes and to analyze the data from the single and double digestions. First, consider the *Eco*RI

and *Hind*III data. Cutting with *Eco*RI produces two fragments, one of 200 bp and the other of 700 bp, and cutting with *Hind*III also produces two fragments, one of 300 bp and the other of 600 bp. Thus, we know that both restriction sites are asymmetrically located along the linear DNA fragment, with the *Eco*RI site 200 bp from an end and the *Hind*III site 300 bp from an end. When we consider the *Eco*RI + *Hind*III data, we can determine the positions of these two restriction sites relative to one another. For example, on the one hand, if the *Eco*RI site is 200 bp from the end of the fragment, and the *Hind*III site is 300 bp from that same end, then we would predict that cutting with both enzymes would produce three fragments, of sizes 200 bp (from one end to the *Eco*RI site), 100 bp (from the *Eco*RI site to the *Hind*III site), and 600 bp (from the *Hind*III site to the other end). On the other hand, if the *Eco*RI site is 200 bp from one end of the fragment and the *Hind*III site is 300 bp from the other end, then cutting with both enzymes would produce three fragments, of sizes 200 bp (from one end to the *Eco*RI site), 400 bp (from the *Eco*RI site to the *Hind*III site), and 300 bp (from the *Hind*III site to the other end). The actual data support the first model.

Now we randomly pick another pair of enzymes: *Hind*III and *Bam*HI. (We could have picked *Eco*RI and *Bam*HI.) Cutting with *Hind*III produces fragments of 300 bp and 600 bp, as we have seen, and cutting with *Bam*HI produces three fragments, of sizes 50 bp, 350 bp, and 500 bp, indicating that there are two *Bam*HI sites in the DNA fragment. Again, the double-digestion products are useful in locating the sites. Double digestion with *Hind*III and *Bam*HI produces four fragments, of 50 bp, 100 bp, 250 bp, and 500 bp. The simplest interpretation of the data is that the 300-bp *Hind*III fragment is cut into the 50-bp and 250-bp fragments by *Bam*HI and that the 600-bp *Hind*III fragment is cut into the 100-bp and 500-bp fragments by *Bam*HI. Thus, the restriction map shown in the following figure can be drawn:

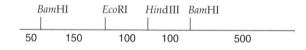

The *Bam*HI + *Eco*RI data are compatible with this model.

Questions and Problems

8.1 The ability to clone and manipulate DNA fragments provides a set of tools for molecular biologists to investigate the structure and function of our genes and their protein products. What are the basic elements of research in recombinant DNA technology?

*8.2 The ability of complementary nucleotides to base pair using hydrogen bonding, and the ability to selectively disrupt or retain accurate base pairing by treatment with chemicals (e.g., alkaline conditions) and/or heat is critical to many methods used to produce and analyze recombinant DNA. Give three examples of methods that rely on complementary base pairing, and explain what role complementary base pairing plays in each of these methods.

8.3 Restriction endonucleases are naturally found in bacteria. What purposes do they serve?

*8.4 A new restriction endonuclease is isolated from a bacterium. This enzyme cuts DNA into fragments that average 4,096 base pairs long. Like many other known restriction enzymes, the new one recognizes a sequence in DNA that has twofold rotational symmetry. From the information given, how many base pairs of DNA constitute the recognition sequence for the new enzyme?

8.5 An endonuclease called *Avr*II ("a-v-r-two") cut DNA whenever it finds the sequence 5'-CCTAGG-3' / 3'-GGATCC-5'.

a. About how many cuts would *Avr*II make in the human genome, which contains about 3×10^9 base pairs of DNA, and in which 40 percent of the base pairs are G-C?
b. On average, how far apart (in terms of base pairs) will two *Avr*II sites be in the human genome?
c. In the cellular slime mold *Dictyostelium discoidium*, about 80 percent of the base pairs in regions between genes are A-T. On average, how far apart (in terms of base pairs) will two *Avr*II sites be in these regions?

8.6 About 40 percent of the base pairs in human DNA are G-C. On average, how far apart (in terms of base pairs) will the following sequences be?

a. two *Bam*HI sites   c. two *Not*I sites
b. two *Eco*RI sites   d. two *Hae*III sites

*8.7 *E. coli*, like all bacterial cells, has its own restriction endonucleases that could interfere with the propagation of foreign DNA in plasmid vectors. For example, wild-type *E. coli* has a gene, *hsdR*, that encodes a restriction endonuclease that cleaves DNA that is not methylated at certain A residues. Why is it important to inactivate this enzyme by mutating the *hsdR* gene in strains of *E. coli* that will be used to propagate plasmids containing recombinant DNA?

8.8 There are many varieties of cloning vectors that are used to propagate cloned DNA. One type of cloning vector used in *E. coli* is a plasmid vector. What features does a plasmid vector have that makes it useful for constructing and cloning recombinant DNA molecules?

8.9 *E. coli* is a commonly used host for propagating DNA sequences cloned into plasmid vectors. Wild-type *E. coli* turns out to be an unsuitable host, however: Not only are the plasmid vectors "engineered," but so is the host bacterium. For example, nearly all strains of *E. coli* used for propagating recombinant DNA molecules carry mutations in the *recA* gene. The wild-type *recA* gene encodes a protein that is central to DNA recombination and DNA repair. Mutations in *recA* eliminate general recombination in *E. coli*, and render *E. coli* sensitive to UV light. How might a *recA* mutation make an *E. coli* cell a better host for propagating a plasmid carrying recombinant DNA? [Hint: What type of events involving recombinant plasmids and the *E. coli* chromosome will *recA* mutations prevent?] What additional advantage might there be to using *recA* mutants, considering that some of the *E. coli* cells harboring a recombinant plasmid could be accidentally released into the environment?

8.10 Much effort has been spent on developing cloning vectors that replicate in organisms other than *E. coli*.

a. Describe several different reasons one might want to clone DNA in an organism other than *E. coli*.
b. What is a shuttle vector, and why is it used?
c. Describe the salient features of a vector that could be used for cloning DNA in yeast.

8.11 What is a cDNA library, and from what cellular material is it derived? How is a cDNA library used in cloning particular genes?

*8.12 Suppose you have cloned a eukaryotic cDNA and want to express the protein it encodes in *E. coli*. What type of vector would you use, and what features must this vector have? How would this vector need to be modified to express the protein in a mammalian tissue culture cell?

*8.13 Suppose you wanted to produce human insulin (a peptide hormone) by cloning. Assume that you could do this by inserting the human insulin gene into a bacterial host where, given the appropriate conditions, the human gene would be transcribed and then translated into human insulin. Which would be better to use as your source of the gene: human genomic insulin DNA or a cDNA copy of this gene? Explain your choice.

*8.14 You have inserted human insulin cDNA in the cloning vector pUC19 and transformed the clone into *E. coli*, but insulin was not expressed. Propose several hypotheses to explain why not.

8.15 Three students are working as a team to construct a plasmid library from *Neurospora* genomic DNA. They want the library to have, on average, about 4-kb inserts. Each student proposes a different strategy for constructing the library, as follows:

  Mike: Cleave the DNA with a restriction enzyme that recognizes a 6-bp site, which appears about once every 4096 bp on average, and leaves sticky, overhanging ends. Ligate this DNA into the plasmid vector cut with the same enzyme, and transform the ligation products into bacterial cells.

  Marisol: Partially digest the DNA with a restriction enzyme that cuts DNA very frequently, say once every 256 bp, and which also leaves sticky overhanging ends. Select DNA that is about 4 kb in size (e.g., purify fragments this size after the products of the digest are resolved by gel electrophoresis) and then ligate this DNA into a plasmid vector cleaved with a restriction enzyme that leaves the same sticky overhangs, and transform the ligation products into bacterial cells.

  Hesham: Irradiate the DNA with ionizing radiation, which will cause double-stranded breaks in the DNA. Determine how much irradiation should be used to generate, on average, 4 kb fragments and use this dose. Ligate linkers onto the ends of the irradiated DNA, digest the linkers with a restriction enzyme to leave sticky overhanging ends, ligate the DNA into a similarly digested plasmid vector, and then transform the ligation products into bacterial cells.

Which student's strategy will insure that the inserts are representative of *all* of the genomic sequences? Why are the other student's strategies flawed?

*8.16 Genomic libraries are important resources for isolating genes and for studying the functional organization of chromosomes. List the steps you would use to make a genomic library of yeast in a plasmid vector. In what fundamental way would you modify this procedure if you were making the library in a BAC vector?

*8.17 The human genome contains about 3×10^9 bp of DNA. How many 200-kb fragments would you have to clone into a BAC library to have a 90 percent probability of including a particular sequence?

8.18 Some restriction enzymes leave sticky ends, while others leave blunt ends. It is more efficient to clone DNA fragments with sticky ends than DNA fragments with blunt ends. What is the best way to efficiently clone a set of DNA fragments having blunt ends?

8.19 A molecular genetics research laboratory is working to develop a mouse model for bovine spongiform encephalopathy (BSE) ("mad cow") disease, which is caused by misfolding of the prion protein. As part of their investigation, they want to investigate the structure of the gene for the prion protein in mice. They have a mouse genomic DNA library made in a BAC vector and a 2.1-kb long cDNA for the gene. List the steps they should take to screen the BAC library with the cDNA probe.

*8.20 Suppose a researcher wants to clone the genomic sequences that include a human gene for which a cDNA has already been obtained. She has available a variety of genomic libraries that can be screened with a probe made from the cDNA using the method described in Figure 8.11.
a. Assuming that each library has an equally good representation of the 3×10^9 base pairs in a haploid human genome, about how many clones should be screened if the researcher wants to be 95 percent sure of obtaining at least one hybridizing clone and
   i. the library is a plasmid library with inserts that are, on average, 7 kb?
   ii. the library is a YAC library with inserts that are, on average, 1 Mb?
b. What advantages and disadvantages are there to screening these different libraries?
c. What kinds of information might be gathered from the analysis of genomic DNA clones that could not be gathered from the analysis of cDNA clones?

8.21 A scientist has carried out extensive studies on the mouse enzyme phosphofructokinase. He has purified the enzyme and studied its biochemical and physical properties. As part of these studies, he raised antibodies against the purified enzyme. What steps should he take to clone a cDNA for this enzyme?

*8.22 A researcher interested in the control of the cell cycle identifies three different yeast mutants whose rate of cell division is temperature-sensitive. At low, permissive temperatures, the mutant strains grow normally and produce yeast colonies having a normal size. However, at elevated, restrictive temperatures, the mutant strains are unable to divide and produce no colonies. She has a yeast genomic library made in a plasmid shuttle vector, and wants to clone the genes affected by the mutants. What steps should she take to accomplish this objective?

8.23 The amino acid sequence of the actin protein is conserved among eukaryotes. Outline how you would use a genomic library of yeast prepared in a bacterial plasmid vector and a cloned cDNA for human actin to identify the yeast actin gene.

*8.24 It's 3 am. Your best friend has awakened you with yet another grandiose scheme. He has spent the last two years purifying a tiny amount of a potent modulator of the immune response. He believes that this protein, by stimulating the immune system, could be the ultimate cure for the common cold. Tonight, he has finally been able to obtain the sequence of the first seven amino acids at the N-terminus of the protein: Met-Phe-Tyr-Trp-Met-Ile-Gly-Tyr. He wants your help in cloning a cDNA for the gene so that he can express large amounts of the protein and undertake further testing of its properties. After you drag yourself out of bed and ponder the sequence for a while, what steps do you propose to take to obtain a cDNA for this gene?

8.25 Explain how gel electrophoresis can be used to determine the size of a PCR product.

8.26 Restriction endonucleases are used to construct restriction maps of linear or circular pieces of DNA. The DNA usually is produced in large amounts by recombinant DNA techniques. Generating restriction maps is like putting the pieces of a jigsaw puzzle together. Suppose we have a circular piece of double-stranded DNA that is 5,000 base pairs long. If this DNA is digested completely with restriction enzyme I, four DNA fragments are generated: fragment *a* is 2,000 base pairs long, *b* is 1,400 base pairs long, *c* is 900 base pairs long, and *d* is 700 base pairs long. If, instead, the DNA is incubated with the enzyme for a short time, the result is incomplete digestion of the DNA: Not every restriction enzyme site in every DNA molecule will be cut by the enzyme, and all possible combinations of adjacent fragments can be produced. From an incomplete digestion experiment of this type, fragments of DNA were produced from the circular piece of DNA that contained the following combinations of the above fragments: *a-d-b*, *d-a-c*, *c-b-d*, *a-c*, *d-a*, *d-b*, and *b-c*. Lastly, after digesting the original circular DNA to completion with restriction enzyme I, the DNA fragments are treated with restriction enzyme II under conditions conducive to complete digestion. The resulting fragments are 1,400, 1,200, 900, 800, 400, and 300 bp. Analyze all the data to locate the restriction enzyme sites as accurately as possible.

*8.27 A piece of DNA 5,000 bp long is digested with restriction enzymes A and B, singly and together. The DNA fragments produced are separated by DNA electrophoresis and their sizes are calculated, with the following results:

Digestion with		
A	**B**	**A + B**
2,100 bp	2,500 bp	1,900 bp
1,400 bp	1,300 bp	1,000 bp
1,000 bp	1,200 bp	800 bp
500 bp		600 bp
		500 bp
		200 bp

Each A fragment is extracted from the gel and digested with enzyme B, and each B fragment is extracted from the gel and digested with enzyme A. The sizes of the resulting DNA fragments are determined by gel electrophoresis, with the following results:

A Fragment	Fragments Produced by Digestion with B	B Fragment	Fragment Produced by Digestion with A
2,100 bp	→ 1,900, 200 bp	2,500 bp	→ 1,900, 600 bp
1,400 bp	→ 800, 600 bp	1,300 bp	→ 800, 500 bp
1,000 bp	→ 1,000 bp	1,200 bp	→ 1,000, 200 bp
500 bp	→ 500 bp		

Construct a restriction map of the 5,000-bp DNA fragment.

8.28 A colleague has sent you a 4.5-kb DNA fragment excised from a plasmid cloning vector with the enzymes *Pst*I and *Bgl*II (see Table 8.1 for a description of these enzymes and the sites they recognize). Your colleague tells you that within the fragment there is an *Eco*RI site that lies 0.49 kb from the *Pst*I site.

a. List the steps you would take to clone the *Pst*I-*Bgl*II DNA fragment into the plasmid vector pUC19 (described in Figure 8.4).

b. How would you verify that you have cloned the correct fragment and determine its orientation within the pUC19 cloning vector?

*8.29 A 10-kb genomic DNA *Eco*RI fragment from a newly discovered insect is ligated into the *Eco*RI site of the pUC19 plasmid vector and transformed into *E. coli*. Plasmid DNA and genomic DNA from the insect are prepared and each DNA sample is digested completely with the restriction enzyme *Eco*RI. The two digests are loaded into separate wells of an agarose gel, and electrophoresis is used to separate the products by size.

a. What will be seen in the lanes of the gel after it is stained to visualize the size-separated DNAs?

b. What will be seen if the gel is transferred to a membrane to make a Southern blot, and the blot is probed with the 10-kb *Eco*RI fragment? (Assume the fragment does not contain any repetitive DNA sequence.)

*8.30 During Southern blot analysis, DNA is separated by size using gel electrophoresis, and then transferred to a membrane filter. Before it is transferred, the gel is soaked in an alkaline solution to denature the double-stranded DNA, and then neutralized. Why is it important to denature the double-stranded DNA? (Hint: Consider how the membrane will be probed.)

*8.31 A researcher digests genomic DNA with the restriction enzyme *Eco*RI, separates it by size on an agarose gel, and transfers the DNA fragments in the gel to a membrane filter using the Southern blot procedure. What result would she expect to see if the source of the DNA and the probe for the blot is as described as follows?

a. The genomic DNA is from a normal human. The probe is a 2.0-kb DNA fragment obtained by excision with the enzyme EcoRI from a plasmid containing single-copy genomic DNA.

b. The genomic DNA is from a normal human. The probe is a 5.0-kb DNA fragment that is a copy of a LINE sequence (see Chapter 2, p. 36 and Chapter 7, p. 165) with an internal EcoRI site.

c. The genomic DNA is from a normal human. The probe is a 5.0-kb DNA fragment that is a copy of a LINE sequence that lacks an internal EcoRI site.

d. The genomic DNA is from a human heterozygous for a translocation (exchange of chromosome parts) between chromosomes 14 and 21. The probe is a 3.0-kb DNA fragment that is obtained by excision with the enzyme EcoRI from a plasmid containing single-copy genomic DNA from a normal chromosome 14. The translocation breakpoint on chromosome 14 lies within the 3.0-kb genomic DNA fragment.

e. The genomic DNA is from a normal female. The probe is a 5.0-kb DNA fragment containing part of the *testis determining factor* gene, a gene located on the Y chromosome.

*8.32 The investigators described in Question 8.19 were successful in purifying a BAC-DNA clone containing the gene for the mouse prion protein. To narrow down which region of the BAC DNA contains the prion-protein gene, they purified the BAC DNA, digested it with the restriction enzyme NotI, and separated the products of the enzymatic digestion by size using gel electrophoresis. Then, they purified each of the relatively large NotI DNA fragments from the gel, digested each individually with the restriction enzyme BamHI, and separated the products of each enzymatic digestion by size using gel electrophoresis. Finally, they transferred the size-separated DNA fragments from the agarose gel onto a membrane filter using the Southern blot technique, and allowed the DNA fragments on the filter to hybridize with a labeled cDNA probe. Figure 8.A shows the results that were obtained: The pattern of DNA bands seen after the BAC DNA is digested with NotI is shown in Panel A, the pattern of DNA bands seen after each NotI fragment is digested with BamHI is shown in Panel B, and the pattern of hybridizing DNA fragments visible after probing the Southern blot is shown in Panel C.

a. Note the scales (in kb) on the left of each figure. Why are relatively larger DNA fragments obtained with NotI than with BamHI?

b. An alternative approach to identify the BamHI fragments containing the prion-protein gene would be to digest the BAC DNA directly with BamHI, separate the products by size using gel electrophoresis, make a Southern blot, and probe it with the labeled cDNA clone. Why might the researchers have added the additional step of first purifying individual large NotI fragments, and then separately digesting each with BamHI before making the Southern blot?

c. Which NotI DNA fragment contains the gene for the mouse prion protein?

Figure 8.A

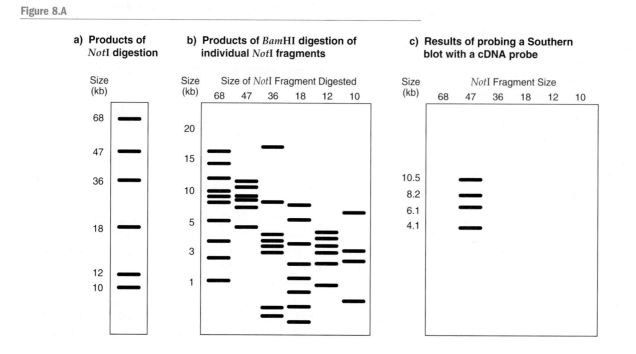

a) **Products of NotI digestion**

b) **Products of BamHI digestion of individual NotI fragments**

c) **Results of probing a Southern blot with a cDNA probe**

Figure 8.B

NotI	BamHI	BamHI		BamHI	BamHI	BamHI	NotI
7.8	6.1	10.5	4.1	8.2		10.1	

d. Which *Bam*HI fragments contain the gene for the mouse prion protein?

e. About what size is the RNA-coding region of the gene for the mouse prion protein? Why is it so much larger than the cDNA?

8.33 Sara is an undergraduate student who is doing an internship in the research laboratory described in Questions 8.19 and 8.32. Just before Sara started working in the lab, the restriction map in Figure 8.B was made of the 47-kb *Not*I restriction fragment containing the prion-protein gene (distances between restriction sites are in kb).

Since smaller DNA fragments cloned into plasmids are more easily analyzed than large DNA fragments cloned into BACs, Sara has been asked to "subclone" the 6.1-, 10.5-, 4.1-, and 8.2-kb *Bam*HI DNA fragments containing the prion-protein gene into the pUC19 plasmid vector. (See Figure 8.4 for a description of the pUC19 vector.) Her mentor gives her some intact pUC19 plasmid DNA, some of the purified 47-kb *Not*I fragment, and shows her where the lab's stocks of DNA ligase and *Bam*HI are stored. Describe the steps Sara should take to complete her task. In your answer, address how she will identify plasmids that contain genomic DNA inserts, and how she will verify that she has identified clones containing each of the desired genomic *Bam*HI fragments.

*8.34 Imagine that you have been able to clone the structural gene for an enzyme in a catecholamine biosynthetic pathway from the adrenal gland of rats. How could you use this cloned DNA as a probe to determine whether this same gene functions in the rat brain?

*8.35 A cDNA library is made with mRNA isolated from liver tissue. When a cloned cDNA from that library is digested with the enzymes *Eco*RI (E), *Hind*III (H), and *Bam*HI (B), the restriction map shown in the following figure, part (**a**) is obtained. When this cDNA is used to screen a cDNA library made with mRNA from brain tissue, three identical cDNAs with the restriction map shown in the following figure, part (**b**) are obtained. When either cDNA is used to synthesize a uniformly labeled ^{32}P-labeled probe and the probe is allowed to hybridize to a Southern blot prepared from genomic DNA digested singly with the enzymes *Eco*RI, *Hind*III, and *Bam*HI, an autoradiograph shows the pattern of bands in the following figure, part (**c**). When either cDNA is used to synthesize a uniformly labeled ^{32}P-labeled probe and used to probe a northern blot prepared with poly(A) RNA isolated

from liver and brain tissues, the pattern of bands in part (**d**) of the figure is seen. Fully analyze these data and then answer the following questions.

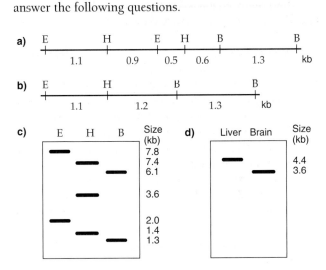

a. Do these cDNAs derive from the same gene?

b. Why are different-sized bands seen on the northern blot?

c. Why do the cDNAs have different restriction maps?

d. Why are some of the bands seen on the whole-genome Southern blot different sizes than some of the restriction fragments in the cDNAs?

*8.36 Draw the pattern of bands you would expect to see on a DNA sequencing gel if you annealed the primer 5′-CTAGG-3′ to the following single-stranded DNA fragment and carried out a dideoxy sequencing experiment. Assume the dNTP precursors were all labeled.

3′-GATCCAAGTCTACGTATAGGCC-5′

8.37 What information and materials are needed to amplify a segment of DNA using PCR?

8.38 In most PCR reactions, a DNA polymerase that can withstand short periods at very high (near boiling) temperatures is used. Why?

8.39 Both PCR and cloning allow for the production of many copies of a DNA sequence. What are the advantages of using PCR instead of cloning to amplify a DNA template?

8.40 PCR and RT-PCR can be used to quantify DNA and RNA levels. If you assume that each step of the PCR process is 100% efficient, how many copies of a template would be amplified after 30 cycles of a PCR reaction if the number of starting template molecules were

a. 10

b. 1,000

c. 10,000

8.41 *Taq* DNA polymerase, which is commonly used for PCR, is a thermostable DNA polymerase that lacks proofreading activity. Other DNA polymerases, such as *Vent*, have proofreading activity.

a. What advantages are there to using a DNA polymerase for PCR that has proofreading activity?

b. Although some DNA polymerases are more accurate than others, all DNA polymerases used in PCR introduce errors at a low rate. Why are errors introduced in the first few cycles of a PCR amplification more problematic than errors introduced in the last few cycles of PCR amplification?

*8.42 Katrina purified a clone from a plasmid library made using genomic DNA and sequenced a 500-bp long segment using the dideoxy sequencing method. Her twin-sister Marina used PCR with *Taq* DNA polymerase to amplify the same 500 bp fragment from genomic DNA. Marina sequenced the fragment using the dideoxy sequencing method, and obtained the same sequence as Katrina did. She then cloned the fragment into a plasmid vector, and, following ligation and transformation into *E. coli*, sequenced several, independently isolated plasmids to verify that she cloned the correct sequence. Most of them have the same sequence as Katrina's clone, but Marina finds that about 1/3 of them have a sequence that differs in one or two base pairs. None of the clones that differ from Katrina's clone are identical. Fearing she has done something wrong, Marina repeats her work, only to obtain the same results: about 1/3 of the fragments cloned from the PCR product have single base differences. Explain this discrepancy.

9

Applications of Recombinant DNA Technology

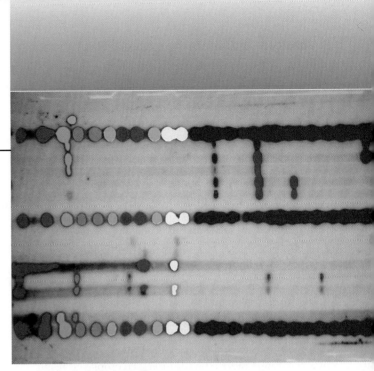

DNA Typing (color enhanced).

PRINCIPAL POINTS

- When a gene has been cloned, specific mutations can be made in that gene in vitro and then studied in vivo. The mutations may be site-specific changes in the protein-coding region, in order to affect protein function, or in the regulatory region, to affect gene expression. Techniques can also be used to delete a gene from a genome—knock out its function—to determine the phenotypic consequences of the lack of a particular protein.

- A DNA polymorphism is one of two or more alternative forms of a locus that differ in nucleotide sequence. Polymorphic loci are DNA markers that like genes, can be used in mapping experiments, as well as for other applications. The phenotypes of polymorphic loci are the DNA variations that are analyzed molecularly. Examples of DNA polymorphisms are single-nucleotide polymorphisms (SNPs), short tandem repeats (STRs), and a variable number of tandem repeats (VNTRs).

- Recombinant DNA and PCR techniques are used in DNA molecular testing for human genetic disease mutations. In general, human genetic testing is done for prenatal diagnosis, newborn screening, or carrier

detection. Many DNA molecular tests are based on restriction fragment length polymorphisms (RFLPs) or on PCR amplification followed by allele-specific oligonucleotide (ASO) hybridization.

- DNA typing, or DNA fingerprinting, is done to distinguish individuals on the basis of the concept that no two individuals, save for identical twins, have the same genome sequence. The variations are manifested in, for example, RFLPs and length variations resulting from different numbers of short tandemly repeated sequences. DNA typing has many applications, including basic biological studies, forensics, detecting infectious species of bacteria, and analyzing old or ancient DNA.

- Recombinant DNA techniques and PCR are widely used in analyzing basic biological processes such as the arrangement of restriction sites in DNA, the sizes of RNA transcripts, the amount of RNA transcription, the steps in RNA processing, and protein–protein interactions in the cell.

- Gene therapy is the treatment of a genetic disorder by introducing a normal gene into the patient to replace or overcome the effects of a mutant gene. For ethical reasons, only somatic gene therapy is being developed for humans. There are only a few examples of successful somatic gene therapy in humans, but great hope is held out for treating many genetic diseases in this way in the future.

- Biotechnology and pharmaceutical companies develop products for the market by using the same kinds of recombinant DNA and PCR techniques in basic biological analysis, DNA molecular testing, gene cloning, DNA typing, and gene therapy.

- Genetic engineering of animals and plants is possible with recombinant DNA and PCR techniques. For example, a number of genetically modified crops already have been developed, and a lot of the processed food we buy contains modified genes. It is expected that many more types of improved crops will result from continued applications of this new technology.

i RECOMBINANT DNA TECHNOLOGY HAS BECOME SO prevalent in our society that on any given day it is likely that you will hear or read a news article about a new application. The most common by far are stories about the use of recombinant DNA in the fields of medicine and agriculture; however, biotechnology has also revolutionized such fields as anthropology, conservation, industry, and forensics. In this chapter, you will learn about some of the specific uses of recombinant DNA technology. After you have read and studied the chapter, you can apply what you've learned by trying the iActivity, in which you'll work with nonhuman DNA to help solve a murder.

In Chapter 8, we discussed a number of the techniques for constructing, cloning, and analyzing recombinant DNA molecules. Since the development of recombinant DNA technology and PCR, it has been possible to investigate many new questions in all areas of biology, and this has resulted in numerous exciting advances. Among the advances of just the past few years is the ability to determine the sequences of a number of genomes, including the human genome. We discuss genome analysis in Chapter 18. In the current chapter, we examine some basic applications of recombinant DNA technologies, going from DNA manipulation and analysis, to gene expression, then to protein analysis, and on to more specialized applications such as gene therapy. The applications are so broad ranging that we can only scratch the surface. The examples have been chosen to describe some of the applications as case studies so that you can learn about the specific example while looking beyond it to see more generally the types of questions and hypotheses that can be investigated.

Site-Specific Mutagenesis of DNA

The study of mutants is a cornerstone of genetics research. In Chapter 7, we learned that mutations can be induced in experimental organisms by treatment with mutagens. In making mutations this way, the whole genome is the target for the mutagen. Thus, each survivor of the mutagenesis likely has many mutations, and the challenge is to find the mutants of interest by an appropriate screen or selection. Further, while mutations of a particular gene might well produce an altered phenotype that can be used in a screen or selection, the precise mutation in the gene is undirected, because mutagenesis is random. However, if a researcher is studying the function of a particular gene, for example, and that gene has been cloned, then specific mutations can be targeted to any part of the gene in vitro. This approach is called **site-specific mutagenesis.**

There are many procedures for site-specific mutagenesis, a number of them using the polymerase chain reaction (PCR; see Chapter 8, Figure 8.21, and pp. 200–202). Figure 9.1 shows one way in which a point mutation or small addition or deletion can be made in cloned DNA (such as a cloned gene) with a PCR-based mutagenesis approach. Four primers are used. Primer 1 is at the left end of the sequence to be amplified, and primer 2 is at the right end. Two other primers, 1M and 2M, match the target DNA sequence within its length, except where the mutation (M) is desired; 1M and 2M are complementary to each other. The mutation is symbolized in the figure as a "blip" in the primers. First, a PCR is done with primers 1 and 1M, and a second PCR is done with primers 2 and 2M. Then the primers are removed, the two products A and B are mixed, and the DNAs are denatured and allowed to reanneal. In some cases, this results in the pairing of a molecule of single-stranded A with a molecule of single-stranded B. DNA polymerase can then extend the 3' ends of the strands in the central paired region, giving a full-length double-stranded DNA. This full-length molecule with the introduced mutation in the central region is then amplified using primers 1 and 2 and transformed into a cell to replace the wild-type sequence.

One application of site-specific mutagenesis is the creation of mutant mice. Since we cannot perform mutational studies with humans, researchers often attempt to mimic human mutations in mice. Such mouse models of human mutations are valuable for furthering our understanding of the gene involved and, in the case of disease genes, may move us toward diagnosis and a cure.

If we have a cloned human gene, we can easily clone the equivalent mouse gene because the two genes likely have a high degree of similarity. The cloned mouse gene can then be mutagenized in vitro—for example, by deleting part of it—to render its product nonfunctional, after which it may be inserted into an appropriate vector. Then,

Figure 9.1

An example of site-specific mutagenesis using PCR.

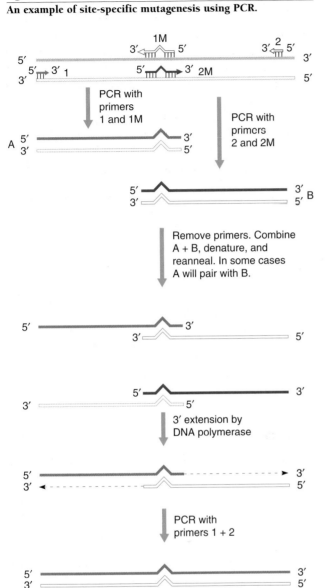

using an extremely fine glass pipette, we can inject the mutated gene insert into mouse embryos, where, in some cells, it will replace one of the resident wild-type genes. The embryos are implanted into a female mouse, and progeny mice are screened molecularly (through such techniques as PCR, restriction analysis, or Southern blot analysis) to identify those carrying the mutation; such mice are heterozygous +/– for the mutation. Mice carrying the mutation are called **knockout mice,** because one of the alleles of the gene has been knocked out by the engineered gene. By interbreeding +/– mice, we can produce knock-out –/– progeny with both alleles mutated. Molecular screens are used to identify these progeny (assuming that they are viable), which can then be studied. Since the mutated genes are completely nonfunctional, the mice that now harbor them show a null phenotype for the gene—that is, a phenotype resulting from a complete absence of gene product.

Many types of knockout mice, both heterozygotes and homozygotes, have been produced to date. One example is knockout mice for the gene *TP53*. (See Chapter 22, pp. 622–623). Mutations of the *TP53* gene are involved in a large proportion of human cancers. *TP53* (−/−) knock-out mice show rapid development of spontaneous tumors and a number of other mutant phenotypes. Knockout mice have also been made for the cystic fibrosis (CF) gene. (CF was described in Chapter 4, p. 79, and the cloning of the human CF gene is described later in the current chapter.) The symptoms in mice are similar to those in humans with CF—notably, major mucous membrane defects in the pharynx, lungs, intestine, and colon caused by defects in chloride ion transport. CF is essentially lethal in humans, although new advances have extended the life expectancy of a CF sufferer to about 30 years. In mice, 40 percent of CF gene knockouts die from intestinal obstruction within 1 week of birth.

The knockout approach for generating specific gene mutations is applicable to any organism that can be transformed by cloned DNA and whose resident genes can be replaced by the introduced DNA.

KEYNOTE

When a gene has been cloned, specific mutations can be made in that gene in vitro and then studied in vivo. The mutations may be site-specific changes in the protein-coding region that affect protein function. Techniques can also be used to delete a gene from a genome—knock out its function—to determine the phenotypic consequences of the lack of a particular protein.

Analysis of DNA Polymorphisms in Genomes

DNA Polymorphism Defined

To this point, we have focused on **genes** as markers for genetic analysis. Genes have different **alleles** which produce different phenotypes that can be followed in crosses. For example, to build a picture of the arrangement of genes in the genome—a **genetic map**—crosses are made between parents differing in alleles of two or more genes and the fraction of recombinant phenotypes among the progeny is determined. Gene mapping was discussed in detail in Chapters 15 and 16. The maps resulting from genetic mapping crosses indicate the location—**locus**—for each gene that was mapped.

Genetic maps have been a highly useful part of efforts to sequence genomes, because the genes are defined landmarks along the chromosomes. That is, the maps provide a skeleton on which to hang the "meat" from the analysis of the DNA organizations of genomes and of sequences of genomes. In the efforts to sequence the human genome, for example, a goal was to construct a high-density genetic map with at least one marker per 1 Mb (million base pairs). However, only a fraction of all the genes have allelic forms that can be seen easily, and in addition, approximately 90% of the genome is noncoding—that is, does not consist of genes. As a result, it is not possible to use only gene markers to develop a high-density map.

Consequently, researchers turned to DNA polymorphisms. A **DNA polymorphism** is one of two or more alternative forms (alleles) of a chromosomal locus that either differ in nucleotide sequence or have variable numbers of tandemly repeated nucleotide units. This definition introduces the concept of an allele being something other than a form of a gene, because a DNA polymorphism can be anywhere in the genome, not necessarily in a gene. In addition, in order that it include the sites of genes and of DNA polymorphisms, we must broaden the concept of a locus to that of any chromosomal location showing variation. Many DNA polymorphisms are useful for genetic mapping studies and hence are called **DNA markers.** (Note that DNA markers are useful for various studies in addition to genetic mapping.) Since there are no products that interact to give a phenotype, the alleles of DNA markers are codominant; that is, they do not show dominance or recessiveness, as is seen with the alleles of most genes. DNA markers are detected by means of molecular tools that focus on the DNA itself, rather than on the gene product or associated phenotype. With genes and DNA markers, map distances can be calculated between genes, between DNA markers, or between a gene and a DNA marker. DNA polymorphisms have a number of other useful applications apart from mapping, as we shall see later in the chapter.

DNA markers solved the problem of making a high-density genetic map of the human genome because the frequency of DNA polymorphisms in the genome is high, about 1 per 350 bp. That is, since the human genome is 3×10^9 bp, there are about 9 million polymorphisms between two haploid genomes (such as in a diploid cell, with one genome copy from the mother and the other from the father). In other words, genome sequences of humans show a great deal of variation. Interestingly, other multicellular organisms have a DNA polymorphism frequency that is close to that of humans.

Classes of DNA Polymorphisms

In this section, we consider three major classes of DNA polymorphisms—single-nucleotide polymorphisms (SNPs), short tandem repeats (STRs), and a variable number of tandem repeats (VNTRs)—and describe ways in which they may be analyzed. Our focus is on the human genome, but these polymorphisms also occur in the genomes of other organisms.

Single-nucleotide polymorphisms (SNPs, or "snips"). A **single-nucleotide polymorphism** is single base-pair change—a point mutation—at a site (the SNP locus). SNPs are the most common type of DNA polymorphism, occurring at a frequency of about 1 per 350 bp and accounting for 90–95 percent of DNA sequence variation. New SNPs arise by spontaneous mutation, such as by replication errors. (See Chapter 3, p. 50.) Since replication errors are rare, the occurrence of new SNPs also is rare.

The vast majority of SNPs occur in noncoding regions of the genome, and these are called *non-coding SNPs.* SNPs in coding regions of the genome (i.e., in genes) are called *coding SNPs* (cSNPs). Systematic studies of cSNPs in humans have shown that each gene has about four cSNPs, half of which cause missense mutations in the encoded protein and half of which cause silent mutations. (See Chapter 7, p. 137). Whether a cSNP affects the phenotype, then, depends on what amino-acid change in the protein the polymorphism causes. It is estimated that one-half of missense mutations, which are cSNPs, as we have just learned, cause genetic disease in humans. Gene function can also be affected by noncoding SNPs when they occur at key locations in promoter regions or other gene regulatory regions.

In the remainder of this section, we discuss how to detect SNPs in genomes. We divide the discussion into methods for detecting SNPs that alter restriction sites and methods for studying all SNPs.

Detection of SNPs that alter restriction sites. A small fraction of SNPs affect restriction sites, either creating them or eliminating them. Such SNPs can be detected by using the restriction enzyme for the site and either Southern blot analysis or, more typically these days, PCR. The different patterns of restriction sites in different genomes result in **restriction fragment length polymorphisms** (RFLPs, "riff-lips"), which are restriction enzyme-generated fragments of different lengths. These will be apparent in the examples that follow.

Figure 9.2 illustrates the Southern blot analysis approach to study SNPs that affect restriction sites. The figure shows a theoretical 7-kb segment in the genome with a pair of SNP alleles, one of which (SNP allele 1) is a TA base pair in a *Bam*HI restriction site and the other of which (SNP allele 2) is a CG base pair which eliminates that site. The site is 2 kb from the left-hand *Bam*HI site. Determining which SNP alleles are present involves the Southern blot analysis steps shown in Figure 8.16, p. 196. That is, genomic DNA is isolated and digested (here) with *Bam*HI, and the fragments are separated by agarose gel electrophoresis. After the fragments are transferred to a membrane filter, DNA fragments of interest are visualized by hybridization with a labeled probe (which, here, spans a large part of the DNA shown in Figure 9.2), followed by autoradiography. The results for possible genotypes are

Figure 9.2

Southern blot analysis method for studying SNPs that affect restriction sites. A 7-kb section of the chromosome has *Bam*HI sites at each end. SNP allele 1 (top) has a *Bam*HI site 2 kb from the left end, whereas SNP allele 2 (bottom) has a CG base pair in place of a TA base pair, so that the *Bam*HI site has been lost. *Bam*HI digestion with DNA samples from individuals with different SNP genotypes, followed by Southern blot analysis using the probe shown, gives the DNA banding patterns at the bottom.

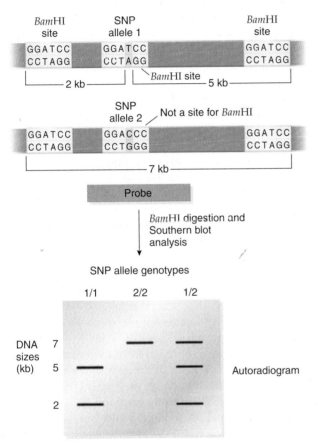

Figure 9.3

PCR method for studying SNPs that affect restriction sites. A 2,000-bp section of the chromosome has SNP alleles 500 bp from the left end. The TA-to-CG change from SNP allele 1 (top) to SNP allele 2 (bottom) alters a *Bam*HI site to a sequence that is not recognized by a restriction enzyme. PCR of DNA samples from individuals with different SNP allele genotypes using the left and right primers shown, followed by *Bam*HI digestion, gives the DNA banding pattern at the bottom.

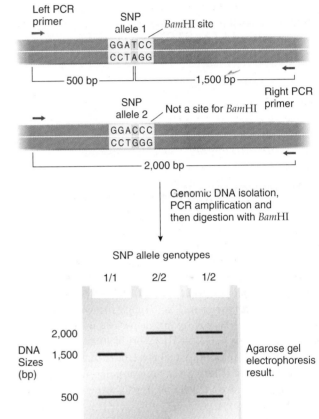

shown at the bottom of the figure. A homozygote for the SNP allele 1 (1,1), which has the intact *Bam*HI site, will show two bands, of 5 kb and 2 kb. A homozygote for the SNP allele 2 (2,2), which has lost the *Bam*HI site, will show one band of 7 kb. A heterozygote for the two SNP alleles (1,2) will show three bands, of 7 kb (from the homolog with allele 2), 5 kb, and 2 kb (the latter two from the homolog with allele 1).

Figure 9.3 illustrates the *PCR–RFLP analysis method.* We consider a 2,000-bp segment of the genome with a pair of SNP alleles similar to the pair in the previous paragraph affecting a *Bam*HI site that is 500 bp from the left end. Primers for PCR are available that recognize the DNA at the left and right ends. PCR analysis of SNP alleles affecting restriction sites involves isolating genomic DNA, amplifying the DNA segment of interest using the left and right primers, digesting the amplified fragment with the restriction enzyme (*Bam*HI, here), and using agarose gel

electrophoresis to examine the sizes of the fragments produced. For our example, the results for possible genotypes are shown at the bottom of the figure. A homozygote for the SNP allele 1 (1,1) will give an amplified DNA fragment that can be digested with *Bam*HI to produce 1,500- and 500-bp fragments. A homozygote for the SNP allele 2 (2,2) will give a 2,000-bp fragment, and a heterozygote for the two alleles (1,2) will give 2,000-, 1,500-, and 500-bp fragments.

Detection of all SNPs. Since most SNPs do not affect restriction sites, other methods of analysis were needed to analyze SNPs generally. You can imagine that analyzing one particular SNP locus is a challenge because, in humans, this is one base pair that is polymorphic out of the 3 billion base pairs in the genome.

Individual SNPs can be analyzed by **allele-specific oligonucleotide (ASO) hybridization analysis** (Figure 9.4). In this procedure, a short oligonucleotide

Figure 9.4

Typing of an SNP by oligonucleotide hybridization analysis.
An oligonucleotide that is completely complementary to the normal allele is hybridized to the target DNA under conditions that favor a perfect match between probe and target. If hybridization occurs, the target DNA has the normal allele, but if hybridization does not occur, the target DNA has a base mismatch—that is, an SNP polymorphism.

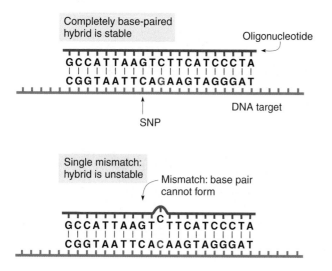

that is complementary to one SNP allele is synthesized. The oligonucleotide is mixed with the target DNA, and hybridization is performed under *high stringency,* meaning that the conditions favor only a perfect match between probe and target DNAs. If hybridization occurs, the target DNA has the common allele. Under these same high-stringency conditions, the oligonucleotide will not hybridize with target DNA that has any other SNP allele at that locus.

Simultaneous typing of hundreds to thousands of SNPs can be done by using **DNA microarrays,** also known as **DNA chips, GeneChip® arrays** (trademark of Affymetrix, Inc.), and **oligonucleotide arrays.** We will now take a small "excursion" to learn about this exciting new technology that is reshaping molecular biology.

First developed in the early 1990s, a DNA microarray is an ordered grid of DNA molecules of known sequence, fixed at known positions on a solid substrate, either a silicon chip, glass, or, less commonly, a nylon membrane. The DNA molecules are placed—arrayed—on the substrate in various ways—for example, by a robotic machine that deposits microspots in known locations or by the synthesis of oligonucleotides in situ at defined locations. The latter method, pioneered by Affymetrix, Inc., generates oligonucleotide arrays—GeneChips®—that are about the size of a small postage stamp, with a density of about 1 million oligonucleotides per square centimeter.

Figure 9.5 illustrates an experiment that uses a DNA microarray. As with Southern and northern

hybridization, experiments involving DNA microarrays involve probes and target nucleic acids. With microarrays, though, the probes are the *unlabeled* DNA molecules fixed to the glass or silicon chip, and the targets are *labeled* free DNA molecules whose identities or quantities are being analyzed. For this reason, a useful generic term for a microarray or a chip is a **probe array.** The labels usually are fluorescent tags such as fluorescein or cyanine dyes, commonly a green Cy3 dye and a red Cy5 dye. In the experiment shown in the figure, the chip has an array of oligonucleotide probes, and the target DNAs are two populations of cDNAs, one labeled with the green Cy3 and the other labeled with the red Cy5 dye. This experimental setup would be used to compare two types of cDNAs, for example, from different types of tissues, from cells or tissues with and without a particular experimental treatment, or from tissues at different developmental stages. After hybridization, the fluorescence pattern is recorded by laser scanning, and the data are analyzed according to the goal of the experiment. For instance, the experimental goal might be to determine the differences in which genes are expressed and the extent to which they are expressed in normal cells versus tumor cells. As can be seen, the hybridization pattern across the array can be quite complicated. A spot that is green indicates that the green-labeled target DNA hybridized to that oligonucleotide spot, and a spot that is red indicates that the red-labeled target DNA hybridized to *that* oligonucleotide spot. Spots that are between green and red (shades of yellow) indicate that both labeled target segments of DNA hybridized to those spots. The particular color of the spot shows the relative amount of hybridization of the two target DNAs.

In analyzing SNPs using a probe array, oligonucleotides that match the common allele and all possible variant alleles of each SNP locus to be analyzed are synthesized on the chip, and the target DNA (fragments of genomic DNA from an individual) is labeled and hybridized. Because the oligonucleotide locations of the probes for each allele are known on the probe array, the pattern of fluorescence observed indicates which SNP alleles the individual has, including whether each allele is homozygous or heterozygous. Also, since a probe array can hold a very large number of oligonucleotide probes, the SNP alleles present at thousands of loci can be determined in a single hybridization experiment.

Short Tandem Repeats (STRs). STRs, also called *microsatellites* and *simple sequence repeats* (SSRs), are 2–6-bp DNA sequences tandemly repeated a few times up to about 100 times. Examples are the dinucleotide repeat, $(GT)_n$, and the trinucleotide repeat, $(CAG)_n$. A recent count for STRs in the human genome is 128,000 two-nucleotide, 8,740 three-nucleotide, 23,680 four-nucleotide, 4,300 five-nucleotide, and 230 six-nucleotide

Figure 9.5

Illustration of an experiment using a DNA microarray.

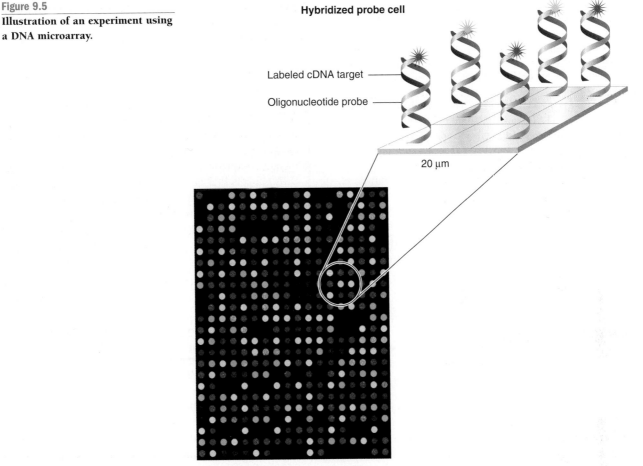

Hybridized probe cell

Labeled cDNA target

Oligonucleotide probe

20 µm

Image of hybridized probe array

repeats. The six-nucleotide repeats include the repeated sequences found at the telomeres.

Many STRs are polymorphic in a population, so they have become valuable in several types of study, including genetic mapping and forensics. Because the overall length of an STR is relatively short, PCR is the preferred method for analyzing STR polymorphic loci (Figure 9.6). Two alleles of an STR locus are shown, one with 6 copies of the GATA repeat, the other with 10 copies. In a population, there will be many alleles of different lengths at an STR locus. One particular human STR locus with the GATA repeat has alleles from 6 to 15 copies, for example. The analysis uses primers that flank the locus. PCR will produce different-length DNA fragments consisting of the STR span plus the DNA from the STR to the leftmost end of the left primer and to the rightmost end of the right primer. For these two alleles, the DNA fragments will differ by 16 bp, due to the four-repeat difference in the repeat length. In analyzing genomic DNA from different individuals, this PCR approach can distinguish homozygotes and heterozygotes, as well as defining the

actual copy number of each repeat—with both results obtained from the lengths of the DNAs amplified.

Variable-Number Tandem Repeats (VNTRs). VNTRs, also called *minisatellites*, are similar to STRs, but the repeating unit is larger than that for STRs, from 7 to a few tens of base pairs long. VNTRs were first discovered by Alec J. Jeffreys in 1985. The discovery was the first demonstration of DNA sequence polymorphism in the human genome. There are far fewer VNTR loci than STR loci in the human genome.

VNTR loci also show polymorphisms. VNTR repeat lengths are longer than those in STRs, so PCR is usually not a convenient way to analyze VNTRs, because of the overall length of DNA that would have to be amplified for the VNTR locus. Instead, restriction digestion and Southern blot analysis is more typically used to study VNTRs. That is, genomic DNA is isolated and cut with a restriction enzyme that cuts on either side of the VNTR locus. The restriction fragments are separated by gel electrophoresis and transferred to a membrane filter by Southern blotting. The length of the VNTR allele is then

Figure 9.6

Using PCR to determine which STR (microsatellite) alleles are present. Genomic DNA is isolated, and PCR primers flanking an STR locus are used to amplify the repeats. The sizes of the DNA fragments produced are determined by agarose gel electrophoresis. In the figure, STR allele 1 has 6 repeats of GATA, and STR allele 2 has 10 repeats of GATA. The gel shows the three possible genotypes for these two alleles: (6,6) [i.e., both homologues have the six-repeat allele], (10,10), and (6,10). In reality, there is likely to be a lot of variation in repeat number at an STR locus.

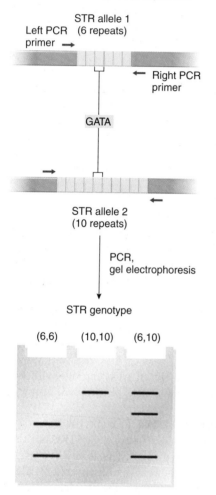

determined by using a probe for the particular repeat sequence of the VNTR locus. As with STR analysis, the results indicate the allele(s) present in the genome being studied. For example, an individual could be homozygous or heterozygous for alleles at a locus. In a population study, the range of alleles for a locus can be determined.

There are two types of VNTR loci: unique loci and multicopy loci. In other words, there may be only one copy of a VNTR locus in an organism's genome, or there may be

a number of copies scattered around the genome. If a probe detects only one VNTR locus, it is called a *monolocus, or single-locus, probe.* Probes that detect VNTR loci at a number of sites in the genome are known as *multilocus probes.*

KEYNOTE

A DNA polymorphism is one of two or more alternative forms of a locus that either differ in nucleotide sequence or have variable numbers of tandemly repeated sequences. Like genes, polymorphic loci are DNA markers that can be used in mapping experiments, as well as for other applications. The phenotypes of polymorphic loci are the DNA variations that are analyzed molecularly. Examples of DNA polymorphisms are single-nucleotide polymorphisms (SNPs), short tandem repeats (STRs), and a variable number of tandem repeats (VNTRs).

DNA Molecular Testing for Human Genetic Disease Mutations

Throughout this text, there are many examples of human genetic diseases. These diseases are caused by enzyme or other protein defects that are the result of mutations at the DNA level. For an increasingly large number of genetic diseases, including Huntington disease (OMIM 143100), hemophilia (OMIM 306700), cystic fibrosis (OMIM 219700), Tay–Sachs disease (OMIM 272800), and sickle-cell anemia (SCA; OMIM 141900), we can perform DNA molecular tests for the presence of mutations associated with the disease. In this section, practical issues of DNA molecular testing are discussed, along with some examples of the testing approaches used. The mutations involved fall into the classes of DNA polymorphisms we have just discussed, so we will be able to see some practical applications of the methods that use those polymorphisms.

Concept of DNA Molecular Testing

Genetic testing determines whether an individual who has symptoms of, or is at a high risk of developing, a genetic disease because of a family history of a heritable disease actually has a particular gene mutation. **DNA molecular testing** is a type of genetic testing that focuses on the molecular nature of mutations associated with disease. Designing DNA molecular tests, then, depends on having knowledge about gene mutations that cause the disease of interest. That knowledge comes about from sequencing the gene involved.

A complication of genetic testing is that many different mutations of a gene can cause a loss of function and therefore lead to the development of the disease. Often, no single molecular test can detect all mutations of the disease gene in question. For example, two genes, *breast cancer one* and (*BRCA1* [OMIM 113705]) and *breast cancer two* (*BRCA2* [OMIM 600185]) are implicated in the development of breast and ovarian cancer. When functioning normally, the *BRCA1* and *BRCA2* gene products help control cell growth in breast and ovarian tissue. However, mutations that cause a loss of function or an abnormal function of the genes' products can lead to the development of cancer. (See Chapter 22, p. 623, for more discussion of the *BRCA* genes and cancer.) Hundreds of mutations have been identified in *BRCA1* and *BRCA2*, but the risk of developing breast cancer varies widely among the mutations. Obviously, this wide variation makes it impossible to develop a single DNA molecular test for *BRCA* gene mutations.

It is important to recognize that a genetic test primarily tells an investigator whether an individual has a mutation known to be associated with a genetic disease. However, genetic testing is distinct from screening for a disease. That is, screening usually is done on people without symptoms or a family history of the disease, whereas genetic testing is done on a targeted population of people with symptoms or a significant family history of the disease. For example, mammograms are clinical screening tests that detect breast lesions that might lead to cancer before clinical symptoms appear. Genetic testing for breast cancer, by contrast, reveals the presence or absence of mutations associated with the development of breast cancer, although it cannot predict whether or when breast cancer will actually develop.

In the same vein, genetic tests are different from diagnostic tests for a disease. Diagnostic tests reveal whether a disease is present and to what extent the disease has developed. For example, a biopsy of a lump in the breast is a diagnostic test to determine whether the lesion is benign or cancerous.

Purposes of Human Genetic Testing

Genetic testing is done for three main purposes: *prenatal diagnosis*, *newborn screening*, and *carrier (heterozygote) detection*.

Prenatal diagnosis is done to assess whether a fetus is at risk for a genetic disorder. Amniocentesis or chorionic villus samples can be taken and analyzed for a specific gene mutation or for biochemical or chromosomal abnormalities. (See Chapter 4, pp. 80–81.) For example, if both parents are asymptomatic carriers (heterozygotes) for a genetic disease, there is a $\frac{1}{4}$ chance of the fetus being homozygous for the mutant allele, and the risk of developing the disease is likely to be high. More recently, techniques have been developed to test embryos for genetic disorders before using them for in vitro fertiliza-

tion. Embryos containing mutated genes that could lead to serious genetic disease can then be removed before implantation.

Newborns can also be tested for specific mutations. For example, we mentioned in Chapter 4 (pp. 73–75) that all newborns in the United States are tested for PKU (phenylketonuria: OMIM 261600) by using the Guthrie test with blood taken from the newborn. Other tests are available for groups at high risk for other genetic disorders, such as sickle-cell anemia (OMIM 141900) in African-Americans and Tay–Sachs disease (OMIM 272800) in Ashkenazi Jews. These genetic tests, including the DNA molecular tests, typically are done on blood samples.

Testing individuals to see whether they are carriers (heterozygotes) for a recessive genetic disease identifies those who may pass on a deleterious gene to their offspring. Carriers can now be detected for a large number of genetic diseases, including Tay–Sachs disease, Duchenne muscular dystrophy (a progressive disease resulting in muscle atrophy and muscle dysfunction; OMIM 310200), and cystic fibrosis (OMIM 602421). DNA molecular tests can readily be done on blood samples.

Examples of DNA Molecular Testing

For DNA molecular testing, DNA samples typically are analyzed by restriction enzyme digestion and Southern blotting or by procedures involving PCR. In this section, we discuss some examples of these testing approaches.

Testing by Restriction Fragment Length Polymorphism Analysis. A mutation associated with a genetic disease may cause the loss or addition of a restriction site either within the gene or in a flanking region. As we learned in the previous section, the chromosomal site where the mutation occurs is an SNP locus, and the different patterns of restriction sites result in restriction fragment length polymorphisms (RFLPs). A restriction map is independent of gene function, so an RFLP is detected whether the DNA sequence change responsible does or does not affect a detectable phenotype. Also, because we are looking directly at DNA, both parental types are seen in heterozygotes, so carriers can be identified easily.

A number of RFLPs are associated with a gene known to cause a disease. In sickle-cell anemia, a single base-pair change in the gene for hemoglobin's β-globin polypeptide results in an abnormal form of hemoglobin, Hb-S, instead of the normal Hb-A. (See Chapter 4, pp. 77–78.) Hb-S molecules associate abnormally, leading to sickling of the red blood cells, tissue damage, and, possibly, death.

The sickle-cell mutation changes an AT base pair to a TA base pair, so the sixth codon for β-globin is changed from GAG to GUG. As a result of this SNP allele, valine instead of glutamic acid is inserted into the polypeptide (Figure 9.7). The mutational change also generates an RFLP for the restriction enzyme *Dde*I ("D-D-E-one"). The *Dde*I restriction site is

$$5'-\text{CTNAG}-3'$$
$$3'-\text{GANTC}-5'$$

where the central base pair can be any of the four possible base pairs. The AT-to-TA mutation changes the fourth base pair in the restriction site. Thus, in the normal β-globin gene, β^A, there are three *Dde*I sites, one upstream of the start of the gene and the other two within the coding sequence (Figure 9.8a). In the sickle-cell mutant β-globin gene, β^S, the mutation has removed the middle *Dde*I site (see Figure 9.7), leaving only two *Dde*I sites (Figure 9.8a). When DNA from normal individuals is cut with *Dde*I and the fragments separated by gel electrophoresis are transferred to a membrane filter by the Southern blot technique and then probed with the 5' end of a cloned β-globin gene, two fragments of 175 bp and 201 bp are seen (Figure 9.8b). DNA from individuals with SCA analyzed in the same way gives one fragment of 376 bp because of the loss of the *Dde*I site. Heterozygotes are detected by the presence of three bands, of 376 bp, 201 bp, and 175 bp.

Not all RFLPs result from changes in restriction sites directly related to the gene mutations; many result from changes to the DNA flanking the gene, sometimes a fair distance away. This is the case for an RFLP that is related to the genetic disease PKU (OMIM 261600; see Chapter 4). Recall that PKU results from a deficiency in the activity of the enzyme phenylalanine hydroxylase. After digestion of genomic DNA with *Hpa*I ("hepaone"), Southern blotting, and probing with a cDNA probe derived from phenylalanine hydroxylase mRNA, different-sized restriction fragments are produced from DNA isolated from individuals with PKU and from DNA isolated from homozygous normal individuals. The RFLP in question results from a difference outside the coding region of the gene, in this case to the 3' side of the gene. The RFLP can be used to test for the PKU mutant gene in fetuses after amniocentesis or chorionic villus sampling. In these cases, detection of the mutation relies on the flanking RFLP segregating most of the time with the gene mutation. Recombination between

Figure 9.8

Detection of sickle-cell gene by the *Dde*I restriction fragment length polymorphism. (a) DNA segments showing the *Dde*I restriction sites. (b) Results of analysis of DNA cut with *Dde*I, subjected to gel electrophoresis, blotted, and probed with a β-globin probe.

a) *Dde*I restriction sites

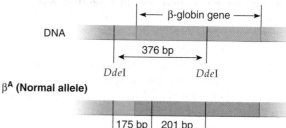

b) *Dde*I fragments detected on a Southern blot by probing with beginning of β-globin gene

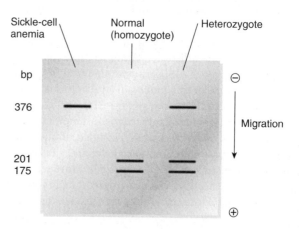

Figure 9.7

The beginning of the β-globin gene, mRNA, and polypeptide showing the normal Hb-A sequences and the mutant Hb-S sequences. The sequence differences between Hb-A and Hb-S are shown in bold. The mutation alters a *Dde*I site (boxed in the Hb-A DNA).

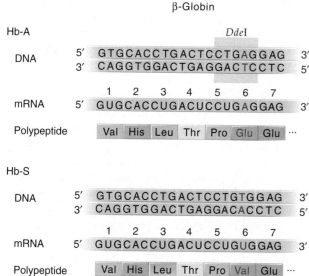

the RFLP and the gene of interest can occur, causing difficulty in interpreting the results.

Testing Using PCR Approaches.

DNA molecular tests using PCR can be developed only when sequence information is available; otherwise the PCR primers cannot be designed. One common test using PCR is allele-specific oligonucleotide (ASO) hybridization. (See Figure 9.4.) Let us illustrate the use of ASO hybridization in testing for mutations of the *GLC1A* (OMIM 137750), one of several genes that, when mutated, cause open-angle glaucoma. Glaucoma generally is caused by increasing pressure in the eye. Open-angle glaucoma is by far the most common form of glaucoma. The condition has no symptoms initially, but as the pressure in the eye builds, at some point peripheral vision is lost, and if it is not diagnosed and treated, total blindness can occur.

The *GLC1A* gene has been sequenced, and a number of glaucoma-causing mutations have been identified. One of the mutations involves a change from CG to TA in the DNA, resulting in a codon change from CCG (Pro) to CUG (Leu). (These two alleles define an SNP locus.) To show the mutation, Figure 9.9a presents the sequence of part of the *GLC1A* gene. The DNA from a heterozygote was sequenced, so both the wild-type C and the mutant T are seen at the location of the mutation.

Based on the *GLC1A* gene sequence, primers were designed for PCR amplification of the region of the gene containing the mutation. The PCR product was separated by agarose gel electrophoresis and then extracted from the gel and dotted onto two membrane filters under conditions that denatured the DNA to single strands. Two ASOs were made, one for the wild-type allele and one for the mutant allele (Figure 9.9b). In this case, each ASO was 19 nt long, with the mutation position approximately in the middle. The ASOs were labeled radioactively, and each was hybridized with the DNA immobilized on one of the filters. The resulting autoradiograms indicated, for each DNA sample, whether the individual from whom the DNA was taken was homozygous normal, heterozygous, or homozygous mutant. As Figure 9.9c shows, for a homozygous normal individual, a signal is seen only for the wild-type ASO; for the heterozygous individual, a signal is seen for both ASOs; and for the homozygous mutant individual, a signal is seen only for the mutant ASO. This method has been used to analyze affected members of glaucoma families for the presence of particular mutations.

As illustrated here, ASO hybridization used one radioactively labeled ASO as a probe for hybridization with a PCR product immobilized on a membrane filter. This approach allows one allele to be probed on each filter and therefore is used to test individuals for the presence of a single particular mutation. A related procedure, called *reverse ASO hybridization*, uses radioactively labeled PCR product as a probe for hybridization with many different

Figure 9.9

DNA molecular testing for mutations of the open-angle glaucoma gene *GLC1A*, using PCR and allele-specific oligonucleotide (ASO) hybridization. (a) Sequence of part of the *GLC1A* gene from a heterozygote showing a mutation from C to T, causing a Pro-to-Leu change in the polypeptide at amino acid 370. (b) Sequences of the two allele-specific oligonucleotides (ASOs), one for the wild-type allele and one for the mutant allele. (c) Results (theoretical) of hybridization with radioactive ASOs for homozygous normal, homozygous mutant, and heterozygous individuals.

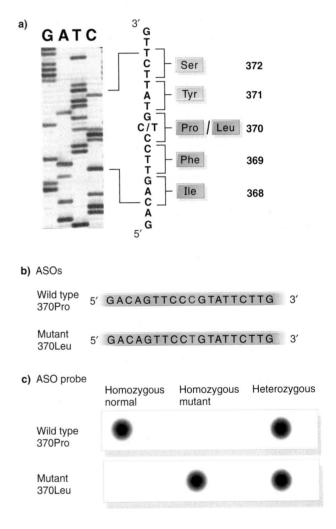

ASOs bound to a membrane filter. This approach is useful for testing DNA samples for the presence of any one of several mutations simultaneously. For example, there are hundreds of mutations in the gene for cystic fibrosis. *Multiplex PCR* can be used to amplify several regions of the gene in DNA samples from patients. The resulting PCR products are labeled radioactively and hybridized with wild-type or mutant oligonucleotides bound to membrane filters. On the autoradiograms, the dot to which the PCR product binds indicates the allele that the individual has.

This method, then, tells us whether an individual has any of the mutant alleles used in the test and, if so, whether the individual is homozygous or heterozygous for that allele. The method, however, cannot rule out an individual having a mutation in the gene that is not covered by the array of ASOs being used.

Availability of DNA Molecular Testing

Genetic testing for a disease is not always available, for one or more reasons, including the following:

1. The gene may not yet have been found in the genome; or the gene has been cloned, but not yet sequenced, so that molecular testing tools have not been developed. For obvious reasons, the more common genetic diseases tend to be the first for which genes have been cloned and molecular tests developed.

2. The gene has been cloned and sequenced, but is subject to many different mutations, making a single molecular test impossible to develop. In this case, there may be tests for a subset of the known mutations, so that a positive test result confirms the presence of a disease gene mutation, but a negative test result does not rule out the presence of such a mutation.

3. For some diseases, mutations in the gene involved do not necessarily cause the disease to develop in every individual. A prime example concerns gene mutations that predispose individuals to the development of cancer (discussed in more detail in Chapter 22, pp. 610–624 and p. 625). In such cases, testing might be limited to high-risk families.

4. Many diseases are caused by multiple gene interactions.

K E Y N O T E

Recombinant DNA and PCR techniques are used in DNA molecular testing for human genetic disease mutations. These tests have become possible as knowledge about the molecular nature of mutations associated with human genetic diseases has increased. In general, human genetic testing is done for prenatal diagnosis, to screen newborns, or to detect carriers of a mutated allele. Many DNA molecular tests are based on restriction fragment length polymorphisms (RFLPs) or on PCR amplification followed by allele-specific oligonucleotide (ASO) hybridization.

Isolation of Human Genes

With a defined gene product isolated by biochemical procedures, a gene can be cloned—for example, by using antibodies against the gene product to screen a cDNA library made in an expression vector. (See Chapter 8, pp. 188–189, and Figure 8.10). However, cloning a gene is

difficult if the gene product that is altered is unknown. Fortunately, a number of approaches are available to solve this problem. One approach is to identify an RFLP marker that is genetically linked to the disease phenotype and then to home in on the gene, starting from the location of the marker on the chromosome. The isolation of a gene associated with a genetic disease on the basis of its approximate chromosomal position is called **positional cloning.** Some of the techniques used in positional cloning are illustrated in the story of how the cystic fibrosis gene was cloned. This cloning was lengthy and laborious. Wherever possible, other approaches are used to clone genes, so positional cloning is not used much these days.

Cloning the Cystic Fibrosis Gene

Cystic fibrosis (CF; OMIM 219700) is the most common lethal genetic disease in the United States today. CF is a disease inherited in an autosomal recessive fashion. Symptoms and genetic properties of the disease are described in Chapter 4, p. 79. The CF gene was the first human disease gene to be cloned solely by positional cloning. The effort took 4 years and the involvement of numerous researchers in many laboratories.

Identifying RFLP Markers Linked to the CF Gene. Many individuals in CF pedigrees were screened with a large number of RFLPs to determine whether any RFLPs were linked genetically to the CF gene. This was done by tracking the inheritance of the CF gene in the families and simultaneously analyzing their DNA by Southern blot analysis and hybridizing with probes to identify any RFLP marker (detected as characteristic DNA fragment sizes) that showed a genetic linkage to the CF locus. One RFLP showed weak linkage to the CF locus.

Identifying the Chromosome on Which the CF Gene Is Located. The RFLP marker was used to identify the chromosome on which the CF gene is located. This was done by *in situ hybridization*, a technique in which chromosomes are spread on a microscope slide and hybridized with a labeled probe, after which the results analyzed by autoradiography. By using a ^{3}H-labeled RFLP probe, the CF gene was shown to be on chromosome 7.

Identifying the Chromosome Region Where the CF Gene Is Located. Other known RFLPs on chromosome 7 were used to find those most closely linked to the CF gene. Two closely linked flanking markers (one marker on each side of the CF gene) were found that are 0.5 map units apart, which corresponds to about 500,000 bp in the human genome. The two markers were known to be located at region 7q31–q32 (7 = chromosome 7, q = the long arm [p is the short arm], 31–32 = subregions 31 and 32), so this result localized the CF gene to that section of chromosome 7.

Cloning the CF Gene between the Flanking Markers. An approach that can be used to find a gene between flanking markers is **chromosome walking,** a process used to identify adjacent clones in a genomic library (Figure 9.10). In chromosome walking, an initial cloned DNA fragment—for example, one of the flanking markers—is used to begin the walk. A labeled end piece of the initial clone (the right end in the figure) is used to screen a genomic library for clones that hybridize with it—in particular, the clones that overlap the original clone. These clones can be analyzed by restriction mapping to determine the extent of overlap. Then, a new labeled probe can be made from the right end of a clone with minimal overlap, and the library is screened again. By repeating this process over and over, we can walk along the chromosome clone by clone.

A procedure related to chromosome walking is also used to move along chromosomes. Called *chromosome jumping*, it is a technique designed to span large amounts of DNA. Whereas in chromosome walking each step is an overlapping DNA clone, in chromosome jumping each jump is from one chromosome location to another without "touching down" on the intervening DNA.

To clone the CF gene, seven chromosome jumps were made toward the locus, and chromosome walks were made from each jump site to identify overlapping clones. In the end, a large number of clones were isolated that spanned over 500 kb of DNA of the CF region.

Figure 9.10

Chromosome walking. (From *Biochemistry,* by Donald Voet and Judith G. Voet. Copyright © 1990 Donald Voet and Judith G. Voet. Reprinted by permission of John Wiley & Sons, Inc.)

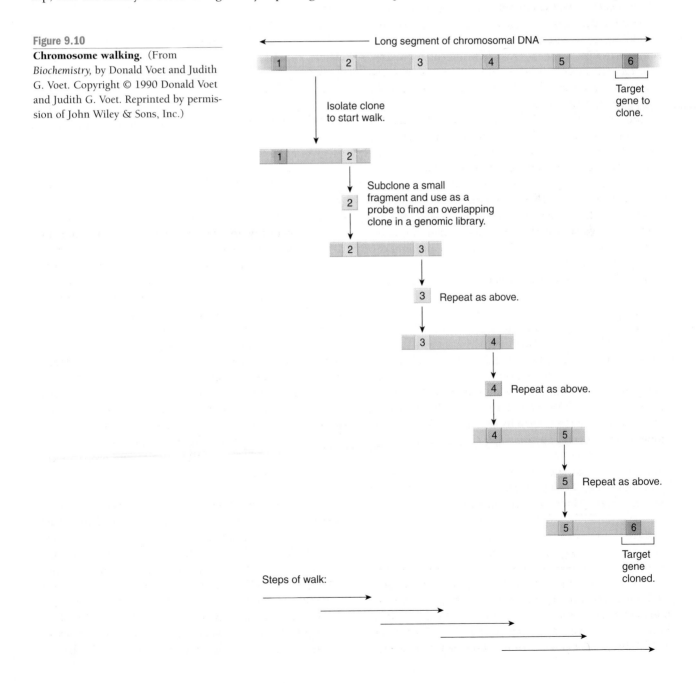

Identifying the CF Gene in the Cloned DNA. The clones spanning the region between the RFLP markers were known to include the CF gene itself, but how does one identify the particular gene of interest in the set of clones?

One way to home in on genes in such clones is to use cloned DNA as probes to see whether they can hybridize with sequences in other species. The reasoning is that genes are conserved in sequence among related species, whereas nongene sequences are less likely to be conserved. The procedure is to digest genomic DNA from other organisms (such as a mouse, a hamster, or a chicken) with restriction enzymes and analyze the fragments by Southern blotting and hybridization with a labeled probe. Since the blot contains DNA from a variety of organisms, it is often called a *zoo blot*. (By the way, a blot with DNA from males and females of a variety of organisms is called a *Noah's ark blot*.) In the CF project, five subcloned segments of the CF region used as probes cross-hybridized with DNA sequences from other organisms, identifying them as candidates for harboring the CF gene. Three of the probes were ruled out either because of linkage data or because no functional gene was involved.

Another property of protein-coding genes is that they produce mRNAs when they are expressed. Thus, a DNA probe made from a protein-coding gene or part of such a gene should hybridize with mRNAs on a northern blot. (See Chapter 8, p. 197.) Northern blotting ruled out a fourth probe, leaving only one. This fifth probe was sequenced and was found to contain a cluster of C and G nucleotides called *CpG islands*. Because the promoters of many protein-coding genes are known to contain CpG islands, the discovery was an encouraging sign that the CF gene was nearby.

A probe made from the fifth sequence was used to screen a cDNA library prepared from cultured normal sweat gland cells. This tissue type was used as the source of mRNA for the library because the symptoms of CF were believed to result from a defect in sodium and chloride transport and such transport is an important function of sweat gland cells. The probe identified a single, positive cDNA clone, which was then used to analyze the genomic clones in more detail: The *candidate CF gene* was shown to span approximately 250 kb of DNA and have 24 exons. Confirmation that the CF gene had been cloned came from comparing the DNA sequences of the candidate CF gene in a normal individual and in an individual with CF. The expectation was that mutational changes would be obvious if the correct gene had been identified. This proved to be the case: A 3-base-pair deletion was detected in the gene from the patient with CF.

The Gene Defects in Cystic Fibrosis. With the CF gene identified, researchers then investigated more fully the nature of the mutations responsible for CF. Sixty-eight percent of Caucasian patients with CF had the 3-bp deletion mentioned

earlier, which results in the loss of the amino acid phenylalanine in the protein encoded by the gene. The remaining patients with CF showed more than 60 different mutations.

KEYNOTE

The isolation of human genes, particularly those associated with disease, is possible through an array of molecular techniques. Where the gene product is not known, the starting point for cloning is knowledge of the genetic linkage between the disease locus and a DNA marker or markers. The isolation of a gene associated with a genetic disease on the basis of its approximate chromosomal position is called positional cloning.

DNA Typing

No two human individuals (except identical twins) have exactly the same genome, base pair for base pair, and this fact has led to the development of **DNA typing** (also called **DNA fingerprinting** or **DNA profiling**) techniques for use in forensic science, in paternity and maternity testing, and for other purposes. DNA typing relies on DNA analysis of DNA polymorphisms (molecular markers) described earlier in the chapter.

DNA Typing in a Paternity Case

Let us consider an example of using DNA typing in a paternity case. In this fictional scenario, a mother of a new baby has accused a particular man of being the father of her child, and the man denies it. The court will decide the case on the basis of evidence from DNA typing, which proceeds as follows (Figure 9.11): DNA samples are obtained from all three individuals involved (Figure 9.11, part 1). In a paternity case, the usual source of DNA is from a blood sample. The DNA is cut with the restriction enzyme for the marker to be analyzed, and the resulting fragments are separated by electrophoresis (Figure 9.11, part 2), transferred to a membrane filter by Southern blotting (Figure 9.11, part 3), and probed with a labeled monolocus STR or VNTR probe (Figure 9.11, parts 4 and 5). After autoradiography or chemiluminescence detection, the DNA banding pattern—the DNA fingerprint, or DNA profile—is analyzed to compare the samples (Figure 9.11, part 6).

The data can be interpreted as follows: Two DNA fragments are detected for the mother, so she is heterozygous for one particular pair of alleles at the STR or VNTR locus under study. Likewise, two DNA fragments are detected for the baby, so the baby is also heterozygous. One of the fragments for the baby matches the larger of the fragments for the mother, and the other fragment for the baby is

Figure 9.11

DNA typing to determine paternity.

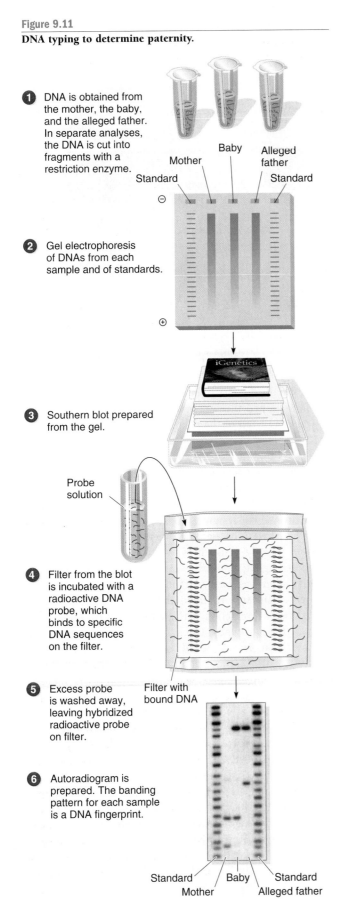

1. DNA is obtained from the mother, the baby, and the alleged father. In separate analyses, the DNA is cut into fragments with a restriction enzyme.

2. Gel electrophoresis of DNAs from each sample and of standards.

3. Southern blot prepared from the gel.

4. Filter from the blot is incubated with a radioactive DNA probe, which binds to specific DNA sequences on the filter.

5. Excess probe is washed away, leaving hybridized radioactive probe on filter.

6. Autoradiogram is prepared. The banding pattern for each sample is a DNA fingerprint.

much larger, indicating many more repeats in that allele. The baby receives one allele from its mother and one from its father. The important question is whether the paternal allele of the baby matches an allele from the alleged father. Inspection of that lane in the autoradiogram leads us to conclude that the answer is "yes," for one fragment is in common with the larger fragment of the baby.

The data indicate that the man shares an allele with the baby, but they do not prove that he contributed that allele to the genome of the baby, although he obviously could have. If the man had no alleles in common with the baby, then the DNA typing data would have proved that he is not the father; this is the *exclusion* result. To establish positive identity—the *inclusion* result—through DNA typing is more difficult; it requires calculating the relative odds that the allele came from the accused or from another person. The calculation depends on knowing the frequencies of STR or VNTR alleles identified by the probe in the ethnic population from which the man comes. Most legal arguments focus on this matter, because good estimates of STR or VNTR allele frequencies are known for only a limited array of ethnic groups, so that calculations of probability of paternity give numbers of questionable accuracy in many cases. To minimize inaccuracy, investigators use a number of different probes (often five or more), so that the combined probabilities calculated for the set of STRs or VNTRs can be high enough to convince a court that the accused is actually the parent (or is guilty in a criminal case), even allowing for problems with knowing true STR or VNTR allele frequencies for the population in question.

It is these combined probabilities that you hear or read about in the media with respect to DNA typing in court cases. (See next section.) In court, the scientific basis for the method is usually not in question; rather, DNA evidence is most commonly rejected for reasons such as possible errors committed in collecting or processing the evidence or weak population statistics. In our paternity case, we would probably be more persuaded that the accused was the father of the child if the data for each of five different monolocus probes indicated that he contributed a particular allele to the child.

Recently, PCR testing to determine paternity has become the chosen method of commercial laboratories performing such tests.

You are a forensic scientist using STR analysis to solve a murder case in the iActivity *Combing Through "Fur"ensic Evidence* on your CD-ROM.

Crime Scene Investigation: DNA Forensics

The reason that DNA typing can be used in paternity testing is that, with the exception of identical twins, no two individuals in the human population have identical genomes. The DNA polymorphisms we have at a very large number of loci make each of our genomes almost unique. On these principles, it is possible to compare two DNA samples to determine the likelihood that they are from the same individual. In crime investigations these days, it is routine to seek out and analyze DNA samples as a means of building a case against, or of exonerating, a suspect. If DNA samples match, probability calculations are made as described in the previous section. Of course, in court cases, DNA evidence is only one type of evidence that is considered.

The methods used in DNA forensics are the ones we have already discussed. The usefulness of DNA typing in forensics is illustrated in the selected case studies that follow. The examples include cases in which the DNA evidence established the guilt of a suspect and cases in which it proved a suspect or already convicted individual innocent.

The Narborough murders: the first murder exoneration and conviction due to DNA evidence.

In 1983 and 1986, two schoolgirls were murdered in the small town of Narborough, Leicestershire ("less-ter-shear"), England. Both girls had been sexually assaulted, and samples of semen recovered from the bodies indicated that the murderer or murderers had the same blood type. The prime suspect in the second murder had that blood type and eventually confessed to the killing, but denied involvement in the first murder. The police were convinced that he had committed both murders, so they contacted Alec Jeffreys (Figure 9.12) at nearby Leicester University to perform DNA typing on samples they had taken. As mentioned earlier, Alec Jeffreys—now Sir Alec Jeffreys—had discovered VNTRs. He had also just demonstrated that DNA could be extracted from stains at crime scenes and could be typed for particular VNTR loci. Using Southern blot analysis with multilocus probes, Dr. Jeffreys showed that the DNA in the semen samples from both murders did not match the police's suspect, whereupon the individual was released, the first person in the world to be exonerated of murder through the use of DNA fingerprinting. In the absence of the DNA evidence, it was almost certain that a court would have convicted him.

But what of the real murderer? The chief superintendent of police overseeing the case decided to embark on the world's first mass screening of DNA in a population. Five thousand adult males in nearby towns were asked to provide blood or saliva samples for forensic analysis. About 10 percent of the samples showed the same blood type as the killer's and hence were followed up with DNA typing. No DNA profiles matched the crime scene profiles, a frustrating result for the police. In a strange twist,

Figure 9.12

Sir Alec Jeffreys, the discoverer of VNTRs. He is holding examples of DNA fingerprints.

though, a woman overheard her work colleague saying that he had given his sample in the name of his friend, Colin Pitchfork. The police arrested Pitchfork. His DNA profile matched the semen samples' profile, and in 1988 he was convicted of the murders and sentenced to life in prison.

The Green River murders: conviction.

On July 8, 1982, Wendy Lee Coffield, age 16, disappeared in Tacoma, Washington. Her body was found in the Green River, in King County, Washington, on July 15, 1982; she had been strangled. Over the next few years, many other young women, usually prostitutes, also disappeared and were found strangled, a number of them in the Green River. A serial killer was loose. An interview of many prostitutes in the Seattle area revealed that some had been raped or had been threatened with being killed by a man driving a blue-and-white truck. The evidence made Gary Ridgway a suspect. When King County sheriffs searched his home in 1987, they had him chew a piece of gauze. At the time, DNA forensics was in its infancy, but increasingly crime investigators collected samples in anticipation of future applications of DNA fingerprinting in forensics. Fortunately, the sample was handled and stored properly, so the DNA in it did not degrade. Ridgway was the prime suspect, but material evidence was not sufficient to arrest him. However, in September 2001, PCR-based STR analysis used on the collected evidence, with the result that Ridgway's DNA profile matched that in sperm samples taken from Carol Christensen, one of the Green River victims. In November 2003, Ridgway admitted in court to killing 48 women, pleading guilty to 48 counts of murder in the first degree. Apparently, he "hated prostitutes" and said that strangling young women was his career.

The Central Park Jogger case: exoneration. In April 1989, a 28-year-old female investment banker was violently raped and beaten while jogging in New York City's Central Park. She was left tied up, bleeding and unconscious with severe injuries. Eventually, she regained consciousness and began a slow recovery. The public was outraged about the savagery of the crime. Police investigators discovered that, at the time of the crime, a group of teenage men had been "wilding"—attacking people at random. Five suspects were arrested in connection with the woman's rape and beating. Four of them confessed, and all five were convicted and imprisoned in 1990. However, supporters of the men argued that the confessions had been coerced, and in addition, there was no other physical evidence to connect any of the men to the crime. Then, in 2002, Matías Reyes, a convict serving time for another rape and murder, confessed to the Central Park Jogger rape. His DNA, and not that of any of the convicted five, was shown to match that of the sample of semen taken from the victim. On the basis of Reyes's confession, the convictions of the five men were overturned.

DNA typing, then, clearly is a powerful tool in criminal investigation. Used properly, it can convict or exonerate an individual of a crime or free a wrongly convicted person. To the latter end, Barry Scheck and Peter Neufeld in 1992 set up The Innocence Project, a nonprofit legal organization that takes cases in which postconviction DNA typing of evidence can result in proof of innocence. Up to April 2004, 143 convicted people have been exonerated by the efforts of The Innocence Project (http://www.innocenceproject.org/).

Other Applications of DNA Typing

There are many uses of DNA typing with present-day samples. The following is a list of a few examples to illustrate the scope of the usefulness of DNA typing for human testing and for tests involving other organisms:

1. Population genetics studies to establish variability in populations or ethnic groups.
2. Proving pedigree status in certain breeds of horses for breed registration purposes.
3. Conservation biology studies of endangered species to determine genetic variability.
4. Forensic analysis in wildlife crimes. Wild animals sometimes are killed illegally, and DNA typing is increasingly helping solve the crimes. For example, a set of six STR markers was used in a poaching investigation in Wyoming involving pronghorn antelope. Six headless pronghorn antelope carcasses were discovered and reported to authorities. An investigation turned up a suspect who had a skull with horns.

DNA samples were taken from the skull and compared with DNA samples from carcasses, and a match was found. At the trial, the suspect was convicted on six counts of wanton destruction of big or trophy game. He received 30 days in jail, was fined $1,300 and ordered to pay $12,000 in restitution, and had his hunting license suspended for 36 years.

5. PCR using strain-specific primers to test for the presence of pathogenic *E. coli* strains in food sources such as hamburger meat.
6. Detecting genetically modified organisms (GMOs), which have been widely introduced into agriculture in the United States. Genetically modified crops typically contain genes that were introduced in the development of the new crop. Often, these genes are expressed using a particular promoter and a particular transcription terminator, enabling PCR primers designed on the basis of these sequences to be used to test for their presence. We can perform these tests with plants themselves or with processed foods. A positive PCR result indicates that the plant is genetically modified or that the food contains one or more GMOs. However, a negative result does not rule out the presence of a GMO. That is, the plant may have been genetically modified using genes that have a different promoter or terminator, and the food may have been made from such organisms. Alternatively, the DNA may have been destroyed in processing. Between 50 and 75 percent of produce and processed foods in a supermarket may be genetically modified or may contain GMOs.

There is also an increasing number of interesting applications of DNA typing with non-present-day samples:

1. Analyzing the DNA extracted from ancient organisms, such as a 40 million-year-old insect in amber, a 17 million-year-old fossil leaf, and a 40,000-year-old mammoth to compare them molecularly with present-day descendants.
2. Some historical controversies and mysteries have been resolved by DNA typing. For example, in 1795, a 10-year-old boy died of tuberculosis in the tower of the Temple Prison in France. The great mystery was whether the boy was the dauphin—the sole surviving son of Louis XVI and Marie Antoinette, who were executed by republicans on the guillotine—or whether he was a stand-in while the true heir to the throne escaped. The dead boy's heart was saved after the autopsy, and despite some very rough handling and storage conditions since his death, two small tissue samples were taken from the heart in December 1999; remarkably, DNA could be extracted from them. This DNA was typed against DNA extracted from locks of the dauphin's hair kept by Marie Antoinette, DNA taken from two of the queen's

sisters, and DNA from present-day descendants. The results showed that the dead boy was the dauphin.

KEYNOTE

DNA typing, or DNA fingerprinting, is done to distinguish individuals on the basis of the concept that no two individuals of a species, save for identical twins, have the same genome sequence. The variations are manifested in restriction fragment length polymorphisms and length variations resulting from different numbers of short tandemly repeated sequences. DNA typing has many applications, including basic biological studies, forensics, detecting infectious species of bacteria, and analyzing old or ancient DNA.

Analysis of Expression of Individual Genes

This section presents two examples illustrating the use of recombinant DNA and PCR techniques to study gene expression.

Regulation of Transcription: Glucose Repression of the Yeast *GAL1* Gene

In Chapter 20, we will discuss in detail the regulation of gene expression in eukaryotes. The following example illustrates how recombinant DNA technology can be used to study gene transcription:

In the yeast *Saccharomyces cerevisiae*, the expression of *GAL* (galactose) genes is induced by galactose, the carbon source in the growth medium. The products of the *GAL* genes are enzymes that catalyze the breakdown of galactose. However, when yeast is grown on glucose, the preferred carbon source, the *GAL* genes are not transcribed. (The genetics of transcriptional regulation of the *GAL* genes is described in Chapter 20, pp. 551–553.) What happens if glucose is added to a culture of yeast already growing in a medium containing galactose? In that case, the *GAL* genes are turned off, and not only is transcription of those genes stopped, but the *GAL* mRNAs in the cell are rapidly degraded. The latter was demonstrated in an experiment, the results of which are illustrated in Figure 9.13. In this experiment, yeast cells were grown so that their *GAL* genes were turned on. Then, at time zero, glucose was added, and samples were taken at various times thereafter. RNA was extracted from each of the samples, and the RNAs were separated by agarose gel electrophoresis. Northern blotting was then performed, and the blot was probed for the mRNA of *GAL1*, one of the *GAL* genes, by using a radioactive probe, such as a riboprobe made as described in Chapter 8 (p. 182). It is easy to see in the figure that the amount of hybridization decreased

Figure 9.13

Regulation of transcription of the yeast *GAL1* gene by glucose. Glucose was added at time zero, and the amount of *GAL1* transcribed was analyzed at various times thereafter by blotting and probing, as described in the text. (From Figure 5, Johnston, M., Flick, J. S., and Pexton, T., 1994. Multiple mechanisms provide rapid and stringent glucose repression of *GAL* gene expression in *Saccharomyces cerevisiae*. *Mol Cell Biol* 14:3834–3841.)

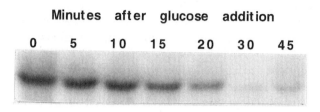

Minutes after glucose addition

| 0 | 5 | 10 | 15 | 20 | 30 | 45 |

rapidly in the 45-minute span of the sampling period. When these results were quantified and plotted on a graph, it became clear that there was a very rapid loss of mRNA in the first 10 minutes and a more gradual loss thereafter.

Alternative Pre-mRNA Splicing: A Role in Sexual Behavior in *Drosophila*

In Chapter 20, we will discuss in detail alternative splicing—the removal of different amounts of pre-mRNA as a result of the use of different splice sites—as one of the levels of regulation of gene expression in eukaryotes. Alternative splicing results in different mRNA molecules that encode proteins with different functions. In this section, we discuss the expression of a gene that is alternatively spliced. The gene controls male sexual behavior in the fruit fly, *Drosophila melanogaster*.

In the fruit fly, as in many species (including humans), sexual reproduction follows elaborate courtship behavior. Courtship in *Drosophila* involves a sequence of behaviors by the male fly, among which are orienting himself toward the female, following the female, tapping the female with his forelegs, singing a species-specific courtship song, and curling the abdomen into a copulation position. This sexual behavior is under the control of a number of sex-determining regulatory genes that are involved in sex determination pathways. Ultimately, the fly's sexual behavior depends on nervous system function, and in fact, regions of the central nervous system (CNS) have been shown to be important for specific steps of male courtship behavior.

By genetic screens, researchers have identified a number of genes involved in the control of sexual behavior in fruit flies. One such gene is *fruitless* (*fru*). Mutants homozygous for *fru* are defective in male courtship behavior. Specifically, the later steps of courtship, from singing to copulation, are abnormal or absent. Since *fru* mutant males are incapable of "persuading" a female to allow copulation, they are functionally sterile, even though they have normal sexual organs and produce normal sperm.

The *fru* gene has been molecularly cloned; it spans at least 140 kb. A complex set of transcripts with different starting and ending points, some of which are sex specific, has been identified through northern blot analysis of poly(A)+ RNAs isolated from flies and the cloned gene as a probe. DNA sequencing of cDNAs prepared from the mRNAs by means of reverse transcriptase (see Chapter 8, pp. 186–187) indicates that some of the sex-specific transcripts are produced by alternative splicing from different 5' splice sites 1,590 nucleotides (nt) apart used with the same 3' splice site (Figure 9.14); the smaller mRNA produced is the male-specific one. The mRNAs were also studied by taking RNA samples from male and female flies and using RT-PCR (reverse transcriptase-PCR; PCR cannot be used directly with RNA because it is a DNA-based procedure). However, cDNA can be made from RNA by using a primer and reverse transcriptase. The DNA produced is then amplified with PCR. By using carefully chosen primers and analyzing the RT-PCR products, researchers found evidence of sex-specific transcripts of different sizes.

It is thought that the *fru* transcripts are translated to produce sex-specific proteins that, in turn, regulate the transcription of other specific genes. In other words, *fru* probably is a regulatory gene. The sex-specific *fru* products are synthesized in only a few neurons in the CNS (about 500 out of 100,000), so perhaps the *fru* protein determines the fates or activities of neurons that are involved in controlling the complex male courtship behaviors.

Analysis of Protein–Protein Interactions

We study genes and their products because we want to understand the structure and function of cells and organisms. As we have been learning about proteins and their roles in the cell, we have discovered that many cellular functions are carried out by proteins that contact one another. We have already seen some examples of this, such as the α-globin and β-globin polypeptides in hemoglobin and the transcription factors interacting with one another and with RNA polymerase to form a complex that initiates transcription. (See Chapter 5, pp. 94–95.)

animation

a The Yeast Two–Hybrid System

One experimental procedure for finding genes which encode proteins that interact with a known protein is the **yeast two-hybrid system** (also called the *interaction trap assay*), developed by Stanley Fields and his coworkers (Figure 9.15). Here is how it works: For the yeast galactose-metabolizing gene *GAL1* to be transcribed, a regulatory protein called Gal4p (encoded by the *GAL4* gene) binds to a promoter element called the upstream activator sequence G, or UAS_G. (See Figure 9.15.) Gal4p has two domains: a DNA binding domain (BD) that binds directly to UAS_G and an activation domain (AD) that facilitates the binding of RNA polymerase to the promoter and the initiation of transcription.

In the two-hybrid system, two types of yeast expression plasmids are used. One type contains the sequence for the Gal4p BD, fused to the sequence for the known protein (X). The other type contains the Gal4p AD sequence, fused to protein-coding sequences encoded by a library of cDNAs (Y). A yeast strain is cotransformed with the BD plasmid and the AD plasmid library so that each transformant has the BD plasmid and one of the plasmids from the AD library. In the chromosome of the yeast strain into which the plasmids are transformed is a *reporter gene*—a gene that encodes a readily assayable product—with a UAS_G. In Figure 9.15, the reporter gene is the *lac Z* gene from *E. coli* that encodes β-galactosidase. Yeast colonies expressing this enzyme turn blue in the presence of the colorless substrate X-gal. (See Chapter 8, p. 181.) The reporter gene is expressed only if the unknown protein (Y) of the AD

Figure 9.14

Alternative sex-specific pre-mRNA splicing in the *fru* (*fruitless*) gene of *Drosophila melanogaster*. A section of the *fru* pre-mRNA transcript is shown. In males and females, different 5' splice sites 1,590 nt apart are used with the same 3' splice site; the result is a smaller mRNA in the male.

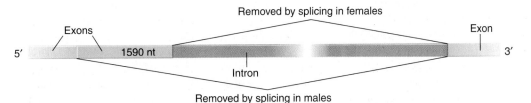

Figure 9.15

The yeast two-hybrid system for detecting protein–protein interactions.

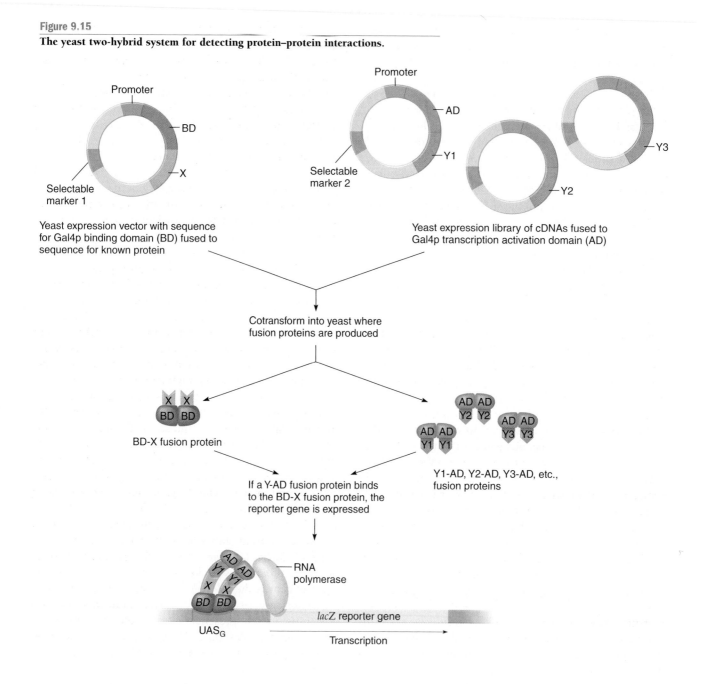

fusion protein interacts with the known protein (X) of the BD fusion protein, thereby bringing the AD and BD domains close together and activating transcription of the reporter gene. If X and Y do not interact, the AD and BD parts of Gal4p stay separate, and transcription of the reporter gene is not activated. In other words, the BD fusion protein acts as *bait* for the interacting protein or proteins. When an interaction is seen, as evidenced by expression of the reporter gene, the AD fusion plasmid from that yeast transformant can be isolated and the cDNA sequence used to find the genomic gene for study.

One example of the use of the two-hybrid system involves studies of interactions between human proteins called peroxins—encoded by *PEX* genes—that are required for peroxisome biogenesis. (The peroxisome is a single-membrane organelle present in nearly all eukaryotic cells; one of the most important metabolic processes of the peroxisome is the β-oxidation of long-chain fatty acids.) The two-hybrid system has shown that the PEX1 and PEX6 proteins interact in normal individuals, but disruption of that interaction is the most common cause of a variety of neurological disorders, such as Zellweger syndrome (OMIM 214100; http://www3.ncbi.nlm.nih.gov/Omim/). Individuals with Zellweger syndrome have lost many peroxisome enzyme functions, have severe neurological, liver, and renal abnormalities and mental retardation, and die in early infancy.

K E Y N O T E

> Recombinant DNA techniques and PCR are widely used in the analysis of basic biological processes. For example, DNA can be analyzed as in the construction of restriction maps (see Chapter 16), RNA transcripts can be sized and quantified, RNA processing events can be monitored, and protein–protein interactions can be studied.

Gene Therapy

Is it possible to treat genetic diseases? Theoretically, two types of gene therapy are possible: *somatic cell therapy*, in which somatic cells are modified genetically to prevent a genetic defect in the offspring; and *germ-line cell therapy*, in which germ-line cells are modified to correct a genetic defect. Somatic cell therapy results in a treatment for the genetic disease in the individual, but his or her progeny could still inherit the mutant gene. Germ-line cell therapy, however, could prevent the disease, because the mutant gene would be replaced by the normal gene and that normal gene would be inherited by the offspring. Both somatic cell therapy and germ-line cell therapy have been used successfully in nonhuman organisms, but only somatic cell therapy has been used in humans, because of ethical issues raised by germ-line cell therapy.

The most promising candidates for somatic cell therapy are genetic disorders that result from a simple defect of a single gene and for which the cloned normal gene is available. Gene therapy involving somatic cells proceeds as follows: A sample of the individual's cells carrying the defective gene is taken. Then, normal, wild-type copies of the mutant gene are introduced into the cells, and the cells are reintroduced into the individual. There, it is hoped, the cells will produce a normal gene product, and the symptoms of the genetic disease will be completely or partially reversed.

The source of the cells varies with the genetic disease. For example, blood disorders, such as thalassemia or sickle-cell anemia, require modification of blood-line cells isolated from the bone marrow. For genetic diseases affecting circulating proteins, a promising approach is the gene therapy of skin fibroblasts—cells that are constituents of the dermis (the lower layer of the skin). Modified fibroblasts can easily be implanted back into the dermis, where blood vessels invade the tissue, allowing gene products to be distributed.

A cell that has had a gene introduced into it by artificial means is called a **transgenic cell,** and the gene involved is called a **transgene.** The introduction of normal genes into a mutant cell poses several problems.

First, procedures for introducing DNA into cells (transformation, although actually called *transfection* for eukaryotic cells) typically are inefficient; perhaps only 1 in 1,000 or 100,000 cells will receive the gene of interest. Thus, a large population of cells is needed to attempt gene therapy. Present procedures use special virus-related vectors to introduce the transgene. Second, in cells that take up the cloned gene, the fate of the foreign DNA cannot be predicted. In some cases the mutant gene is replaced by the normal gene, and in others the normal gene integrates into the genome elsewhere. In the first case, the gene therapy is successful, provided that the gene is expressed. In the second case, successful treatment of the disease results only if the introduced gene is expressed and the resident mutant gene is recessive, so that it does not interfere with the normal gene.

Successful somatic gene therapy has been demonstrated repeatedly in experimental animals such as mice, rats, and rabbits. However, in humans, there have been more failures than successes. In addition, a recent concern is the development of leukemias in therapy patients as a result of the viral vectors used to introduce the transgene.

One successful human somatic gene therapy treatment was done in 1990 with a 4-year-old girl suffering from severe combined immunodeficiency (SCID; OMIM 102700) caused by a deficiency in adenosine deaminase (ADA), an enzyme needed for normal function of the immune system. T cells (cells involved in the immune system) were isolated from the girl and grown in the laboratory, and the normal ADA gene was introduced using a viral vector. The "engineered" cells were then reintroduced into the patient. Since T cells have a finite life in the body, continued infusions of engineered cells have been necessary. The introduced ADA gene is expressed, probably throughout the life of the T cell. As a result, the patient's immune system is functioning more normally, and she now gets no more than the average number of infections. The gene therapy treatment has enabled her to live a more normal life. Recently, some patients who received gene therapy for ADA have developed leukemia for reasons unknown.

Successful somatic gene therapy has also been achieved for sickle-cell anemia (OMIM 141900). In December 1998, 13-year-old Keone Penn's bone marrow cells were replaced with stem cells from the umbilical cord of an unrelated infant, with the hope that the new cells would produce healthy bone marrow, the source of blood cells. After 1 year, there were no signs of sickle cells, so the patient was declared cured of the disease.

With time, many other genetic diseases, including thalassemias, phenylketonuria, Lesch–Nyhan syndrome, cancer, Duchenne muscular dystrophy, and cystic fibrosis, are expected to be treatable with somatic gene therapy. For example, after successful experiments with rats, human clinical trials are under way for transferring the normal

CF gene to patients with cystic fibrosis. As methods are learned for targeting genes to replace their mutant counterparts and for regulating the expression of the introduced genes, increasing success in treating genetic diseases is expected. However, many scientific, ethical, and legal questions must be addressed before gene therapy is implemented on a routine basis.

KEYNOTE

> Gene therapy is the curing of a genetic disorder by introducing into the individual a normal gene to replace or overcome the effects of a mutant gene. For ethical reasons, only somatic gene therapy is being developed for humans. There are few examples of successful somatic gene therapy in humans, but great hope is held out for treating many genetic diseases in this way in the future.

Biotechnology: Commercial Products

The development of cloning and other DNA manipulation techniques has spawned the formation of many *biotechnology* companies, some of which focus on using DNA manipulations to make a wide array of commercial products. Although the details vary, the general approach to making a product is to express a cloned gene or cDNA in an organism that will transcribe the cloned sequence and translate the mRNA. The gene or cDNA is placed into an expression vector (see Chapter 8, p. 182) appropriate for the organism into which it will be transformed. Many different organisms are used, from *E. coli* to mammals, so expression vectors differ in the promoters used for transcription, in their translation start signals, and in their selectable markers. For expression in *E. coli*, for example, the promoter must be recognized by that bacterium's RNA polymerase, and there must be a Shine–Dalgarno sequence so that ribosomes will read the mRNA from the correct AUG. In mammals such as goats or sheep, the simplest way to isolate the product is to have it secreted into the milk, which is, of course, easy to collect. The protein product can then be extracted. The production of recombinant protein products in transgenic mammals (in this case, sheep) is illustrated in Figure 9.16. Here, the gene of interest (*GOI* in the figure) has been manipulated so that it is adjacent to a promoter that is active only in mammary tissue, such as the β-lactoglobulin promoter. The recombinant DNA molecules are microinjected into sheep ova, and each ovum is then implanted into a foster mother. Transgenic offspring are identified by the use of PCR to detect the recombinant DNA sequences. When these transgenic animals mature, the β-lactoglobulin promoter begins to express the associated gene in the mammary tissue, the milk is collected, and the protein of

Figure 9.16

Production of a recombinant protein product (here, the protein encoded by the gene of interest, *GOI*) in a transgenic mammal—in this case, a sheep.

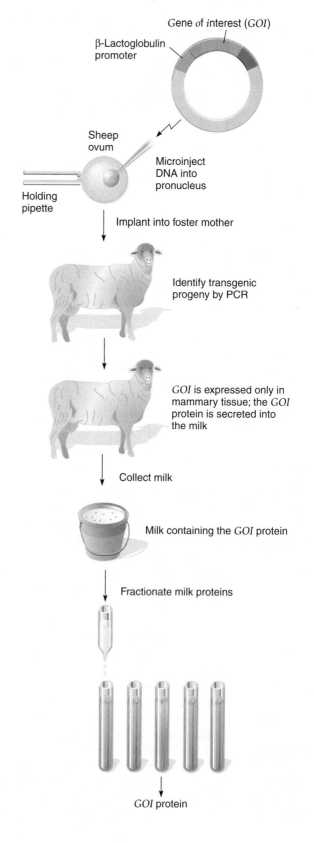

interest is obtained by biochemical separation techniques.

Following are a few examples of the many products produced by biotechnology companies:

1. Tissue plasminogen activator (TPA), used to prevent or dissolve blood clots, thereby preventing strokes, heart attacks, or pulmonary embolisms
2. Human growth hormone, used to treat pituitary dwarfism
3. Tissue growth factor-beta (TGF-β), which promotes new blood vessel and epidermal growth and thus is useful for wound and burn healing
4. Human blood clotting factor VIII, used to treat hemophilia
5. Human insulin ("humulin"), used to treat insulin-dependent diabetes
6. DNase, used to treat cystic fibrosis
7. Recombinant vaccines, used to treat human and animal viral diseases (such as hepatitis B in humans)
8. Bovine growth hormone, used to increase cattle and dairy yields
9. Platelet-derived growth factor (PDGF), used to treat chronic skin ulcers in patients with diabetes
10. Genetically engineered bacteria and other microorganisms, used to improve the production of, for example, industrial enzymes (such as amylases to break down starch to glucose), citric acid (flavoring), and ethanol
11. Genetically engineered bacteria that can accelerate the degradation of oil pollutants or certain chemicals (such as dioxin) in toxic wastes

K E Y N O T E

With the same kinds of recombinant DNA and PCR techniques that are used in basic biological analysis, DNA molecular testing, gene cloning, DNA typing, and gene therapy, biotechnology and pharmaceutical companies develop useful products. Many types of products are now available or are in development, including pharmaceuticals and vaccines for humans and for animals and genetically engineered organisms for improved production of important compounds in the food industry or for cleaning up toxic wastes.

Genetic Engineering of Plants

For many centuries, the traditional genetic engineering of plants involved selective-breeding experiments in which plants with desirable traits were chosen and allowed to produce offspring. As a result, humans have produced hardy varieties of plants (for example, corn, wheat, and oats) and increased yields, all with the use of long-established plant-breeding techniques. (Similar techniques have also been used with animals, such as dogs, cattle, and horses, to produce desired breeds.) Now, vectors developed by recombinant DNA technology are available to transform cells of crop plants, making possible the genetic engineering of plants for agricultural use.

Transformation of Plant Cells

Introducing genes into plant cells is more difficult in some respects than introducing genes into bacteria, yeast, and animals, and this difficulty has slowed plant genetic engineering's rate of progress. Typical plant transformation approaches exploit features of a soil bacterium, *Agrobacterium tumefaciens*, which infects many kinds of plants. Specifically, these approaches take advantage of a natural mechanism in the bacterium for transferring a defined segment of DNA into the chromosome of the plant.

Agrobacterium tumefaciens causes crown gall disease, characterized by tumors (the gall) at wound sites. Most dicotyledonous plants (called *dicots*) are susceptible to crown gall disease, but monocotyledonous plants are not. *Agrobacterium tumefaciens* transforms plant cells at the wound site, causing the cells to grow and divide autonomously and therefore to produce the tumor.

The transformation of plant cells is mediated by a natural plasmid in the *Agrobacterium* called the *Ti plasmid* (the *Ti* stands for *tumor-inducing*; Figure 9.17). Ti plasmids are circular DNA plasmids somewhat analogous to pUC19, but, in comparison, Ti plasmids are huge (about 200 kb versus 2.69 kb for pUC19).

The interaction between the infecting bacterium and the plant cell of the host stimulates the bacterium to excise a 30-kb region of the Ti plasmid called T-DNA (so called because it is *transforming DNA*). T-DNA is flanked by two repeated 25-bp sequences called borders that are involved in T-DNA excision. Excision is initiated by a nick in one strand of the right-hand border sequence. A second nick in the left-hand border sequence releases a single-stranded T-DNA molecule, which is then transferred from the bacterium to the nucleus of the plant cell by a process analogous to bacterial conjugation. Once in the plant cell nucleus, the T-DNA integrates into the nuclear genome. As a result, the plant cell acquires the genes found on the T-DNA, including the genes for plant cell transformation. However, the genes needed for the excision, transfer, and integration of the T-DNA into the host plant cell are not part of the T-DNA. Instead, they are found elsewhere on the Ti plasmid, in a region called the *vir* (for *virulence*) region.

Using recombinant DNA approaches, researchers have found that excision, transfer, and integration of the T-DNA require only the 25-bp terminal repeat sequences. As a result, the Ti plasmid and the T-DNA constitute a

Figure 9.17

Formation of tumors (crown galls) in plants by infection with certain species of *Agrobacterium*. Tumors are induced by the Ti plasmid, which is carried by the bacterium and integrates some of its DNA (the T, or transforming, DNA) into the plant cell's chromosome.

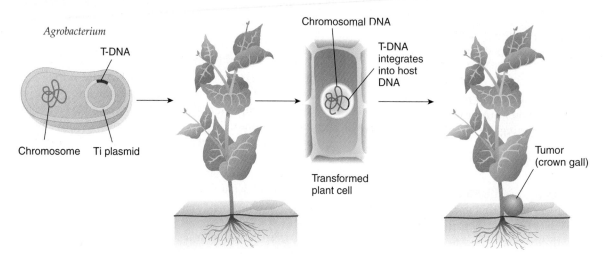

Applications to Plant Genetic Engineering

useful vector for introducing new DNA sequences into the nuclear genome of somatic cells from susceptible plant species. Since any genes placed between the 25-bp borders will integrate into the host genome, a variety of transformation vectors have been derived from the Ti plasmid and T-DNA.

Although the T-DNA-based transformation system is highly effective for dicotyledonous plants, it is not effective for monocotyledonous plants, because they are not part of the normal host range of *Agrobacterium tumefaciens*. This is a serious limitation, since most crop plants are monocotyledonous. Fortunately, alternative transformation procedures have been developed in which the DNA is delivered into the cell physically rather than by a plasmid vector. In the *electroporation* method, DNA is added to plant cell protoplasts and the mixture is "shocked" with high voltage to introduce the DNA into the cell. After the cells are grown in tissue culture to allow them to regenerate their cell walls and begin growing again, appropriate procedures can be applied to select for the cells that were successfully transformed. Another method involves the *gene gun* (the Biolistic Gun, Bio-Rad, Hercules, California). In this method, DNA is coated onto the surface of tiny tungsten beads, which are placed on the end of a plastic bullet. The bullet is fired by a special particle gun. The bullet hits a plate, and the tungsten beads are propelled through a small hole in the plate into a chamber in which target cells have been placed. The force of the "shot" is sufficient to introduce the DNA-carrying beads into the cells. Selection techniques can then be applied to isolate successfully transformed cells, and these can be used to regenerate whole plants.

We mentioned earlier that a very large number of genetically modified crops already have been developed, and a lot of the processed food we buy contains them. Let us briefly consider approaches to generating transgenic plants that are tolerant to the broad-spectrum herbicide Roundup™, to illustrate the types of approaches that are possible. Roundup contains the active ingredient glyphosate, which kills plants by inhibiting EPSPS, a chloroplast enzyme required for the biosynthesis of essential aromatic amino acids. Roundup is used widely to kill weeds because it is active in low doses and is degraded rapidly in the environment by microbes in the soil. If a crop plant is resistant to Roundup, a field can be sprayed with the herbicide to kill weeds without affecting the plant. Approaches to making transgenic, Roundup-tolerant plants include (1) introducing a modified bacterial form of EPSPS that is resistant to the herbicide, so that the aromatic amino acids can still be synthesized even when the chloroplast enzyme is inhibited (Figure 9.18), and (2) introducing genes that encode enzymes for converting the herbicide to an inactive form. Monsanto brought Roundup Ready soybeans to market in 1996, although their use has been controversial because of opposition by groups questioning the safety of genetically engineered plants for human consumption.

With more sophisticated approaches, it will be possible to make transgenic plants that control the expression of genes in different tissues. Examples include controlling the rate at which cut flowers die or the time at which fruit ripens. With regard to the latter, commercially produced, genetically unaltered tomatoes are picked while unripe so

Figure 9.18

Making a transgenic, Roundup™-tolerant tobacco plant by introducing a modified form of the bacterial gene for the enzyme EPSPS that is resistant to the herbicide. The gene encoding the bacterial EPSPS was spliced to a petunia sequence encoding a transit peptide for directing polypeptides into the chloroplast, and the modified gene was inserted into a T-DNA vector and introduced into tobacco by *Agrobacterium*-based transformation. Both the native and the modified bacterial EPSPS are transported into the chloroplast. When plants are sprayed with Roundup, wild-type plants die because only the native chloroplast EPSPS is present, and it is sensitive to the herbicide, but the transgenic plants live because they contain the bacterial EPSPS that is resistant to the herbicide.

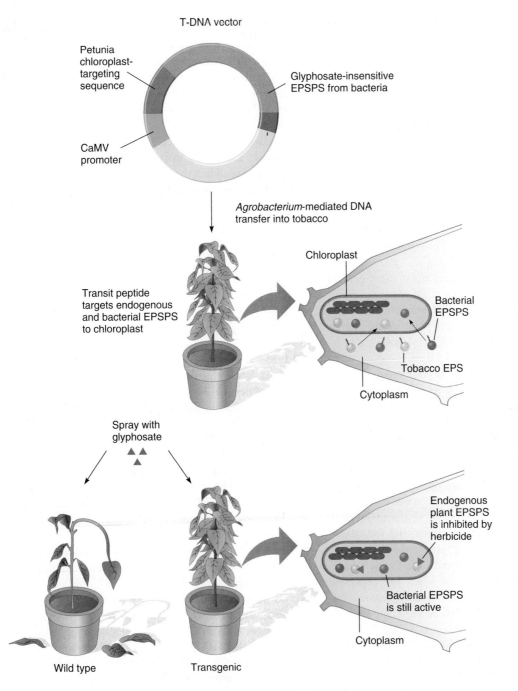

that they can be shipped without bruising. Prior to shipping, they are exposed to ethylene gas, which initiates the ripening process. The tomatoes arrive in the ripened state at the store. Such prematurely picked, artificially ripened tomatoes, however, do not have the flavor of tomatoes picked when they are ripe. Accordingly, the Flavr Savr tomato, genetically engineered by Calgene, Inc., in collaboration with the Campbell Soup Company, was approved for market in 1994. In pursuit of the Flavr Savr, Calgene scientists devised a way to block an ordinary tomato from making the normal amount of polygalacturonase (PG), a fruit-softening enzyme. They introduced into the plant a copy of the PG gene that was backward in its orientation with respect to the promoter. When this gene is transcribed, the mRNA is complementary to the mRNA produced by the normal gene; it is called an **antisense mRNA.**[1] In the cell, the antisense mRNA binds to the normal "sense" mRNA, preventing much of it from being translated. As a result, much less PG enzyme is produced, and ripening is slowed, allowing the tomato to remain longer on the vine without getting too soft for handling and shipping. Once picked, the Flavr Savr tomato was also less susceptible to bruising in shipping or to overripening in the store. The Flavr Savr tomato was advertised as tasting better than store-ripened tomatoes and more like home-grown tomatoes. However, it was expensive and did not achieve commercial success. For economic reasons, it is no longer on the market.

In the near future, we can expect many more genetically engineered plants to be developed. Of particular value will be crop plants with increased yield, resistance to insect pests, and tolerance to herbicides. Such plants could help in alleviating world hunger; however, there is significant public resistance to genetically modified plants in many countries, including a growing resistance in the United States.

Transgenic plants may also be useful for delivering vaccines. The cost of an injected vaccine is relatively high, making it a significant issue in inoculating people in developing countries. However, it could cost just pennies to deliver vaccines in a plant. Such vaccines have been termed *edible vaccines*, and the area of biotechnology dealing with pharmaceuticals in plants or animals has been whimsically termed *pharming*. Basically, transgenic plants are made that express antigens for infections or diseases of interest, so that, when the plant is eaten, the individual could develop antibodies. Indeed, after successful trials with animals, early stage clinical trials with humans have shown that eating raw potatoes can elicit the expected immune responses when those potatoes are expressing, for example, the hepatitis B virus surface antigen, the toxin B subunit of enterotoxigenic *E. coli* (responsible for diarrhea), or the capsid protein of the Norwalk virus. Further research is needed to obtain high levels of antigen produc-

tion in the plants so that sufficient antigen is available after eating to mount a protective immune response.

K E Y N O T E

> Genetic engineering of plants is possible using recombinant DNA or PCR techniques. It is expected that many more types of improved crops will result from future applications of this new technology.

Summary

Recombinant DNA technology and PCR are being widely applied in both basic research and commerce. All areas of modern biology have been revolutionized by these new molecular techniques. Perhaps the most ambitious project that would not have been possible without such techniques is the Human Genome Project (HGP), which has the mandate to generate a complete map of all the genes in the human genome and to obtain the complete sequence of the human genome. We discuss the HGP in detail in Chapter 10.

Recombinant DNA and PCR techniques are also being used to develop new pharmaceuticals (including drugs, other therapeutics, and vaccines), to develop new tools for diagnosing infectious and genetic diseases, for human gene therapy, in forensic analysis, and in agriculture.

Analytical Approaches to Solving Genetics Problems

Q9.1 M. K. Halushka and colleagues used specially designed DNA chips to search for SNPs in 75 protein-coding genes in 74 individuals. They scanned about 189 kb of genomic sequence consisting of 87 kb of coding, 25 kb of introns, and 77 kb of untranslated, but transcribed (i.e., 5′-UTR and 3′-UTR), sequences. They identified a total of 874 possible SNPs, of which 387 were within coding sequences; these are designated cSNPs. Of the cSNPs, 209 would change the amino-acid sequence in one of 62 predicted proteins.

a. In their sample, what is the frequency of SNPs (# bp/SNP)?

b. Are the SNPs evenly distributed in coding and noncoding sequences? Is this an expected result? What implications does the result have?

c. At least 40,000 human ESTs have already been identified, and a reasonable estimate of the human gene number is about 75,000. Use the data given to estimate the number and distribution of SNPs in human genes.

  i. About how many SNPs exist in human genes?

  ii. How many are estimated to be in noncoding regions?

  iii. How many are in coding regions, but do not affect protein structure?

---

[1]The use of antisense mRNA to prevent or inhibit the translation of a natural mRNA is called *antisense technology*. This technology is being tested in a number of systems as a way to control genetic diseases.

iv. How many are in coding regions and could affect protein structure?

d. Many biological traits, including some diseases, are complex in that they are affected by alleles at many different genes. Based on your answers to parts (a)–(c), why is it thought that screens of SNPs using gene chips will allow the identification of genes associated with such complex traits?

A9.1 SNPs are *single-nucleotide polymorphisms*—differences of just one base pair in the DNA of different individuals. These alterations in DNA sequence are not necessarily detrimental to the organism. Rather, they are initially identified simply as differences, or polymorphisms, in DNA sequence. This problem asks you to analyze their frequency and distribution in humans and consider the implications of your analysis.

a. In 189,000 bp, there are 874 SNPs, so, on average, there are 189,000/874 = 216 bp of DNA sequence per SNP. Note that this sampling assesses the number of SNPs *in genes* and does not estimate the number of SNPs in genomic regions in between genes.

b. 387/874 = 44 percent of the SNPs lie in coding sequences, and 487/874 = 56 percent of the SNPs lie in noncoding sequences. The observation that there is a smaller percentage of SNPs in coding sequences suggests that there is less sequence variation in those sequences. This is expected, because coding sequences specify amino acids that confer a function on a protein. An SNP within a coding sequence might result in the insertion of an amino acid that alters the normal function of the protein. This alteration could be disadvantageous and be selected against. Indeed, only 209/874 = 24 percent of the SNPs alter amino-acid sequences, and SNPs that do so are not found in all 75 genes examined. This indicates that, although some sequence constraints may be present in noncoding sequences (for example, if they bind a regulatory protein), more sequence variation is tolerated in noncoding regions.

c.

i. If there are 75,000 genes, one expects to find about 8.74×10^5 SNPs in the human genome: (874 SNP/75 genes) × 75,000 genes = 874,000.

ii. About 487/874 = 56 percent, or 4.9×10^5, will be in noncoding regions.

iii. About 387/874 = 44 percent, or 3.8×10^5, will be in coding regions. About (387 − 209)/874 = 20 percent, or 1.7×10^5, will not affect protein structure because they do not change the amino-acid sequence in a protein.

iv. About 209/874 = 24 percent, or 2.1×10^5, could affect protein structure, because they change the amino-acid sequence in a protein. However, not all of these genes affect protein structure significantly. If an SNP results in the substitution of a similar (conserved) amino acid, it may not significantly alter the structure (or function) of the protein. For example, an SNP might result in aspartate being replaced by glutamate. Both are acidic amino acids, so this substitution may not significantly alter the protein's structure.

d. These data suggest that, even in a relatively small population of individuals (*n* = 74), there will be multiple SNPs for every gene. The data also suggest that SNPs can be identified for most, if not all, genes and much more often than other types of DNA markers. Since gene chip technology can be used to assess a large number of SNP alleles in one genomic DNA sample simultaneously, it should be feasible to obtain comprehensive genotypic information. That is, it is possible to identify the alleles an individual has at many different genes. This possibility has two implications for identifying the genetic contribution to complex traits and diseases, where the aim is to identify the set of alleles at genes that contribute to those traits or diseases. First, SNPs can serve as a very dense set of markers to more easily map genes contributing to complex traits and diseases. Second, SNP analyses allow for a systematic identification of alleles shared by individuals with the traits or diseases.

Q9.2 ROC is a hypothetical polymorphic STR (microsatellite) locus in humans with a repeating unit of CAGA. The locus is shown in Figure 9.A as a box with 25 bp of flanking DNA sequences.

a. You plan to use PCR to type individuals for the ROC locus. If PCR primers must be 18 nucleotides long, what are the sequences of the pair of primers required to amplify the ROC locus?

b. Consider ROC alleles with 10 and 7 copies of the repeating unit. Using the primers you have designed, what will be the sizes of the amplified PCR products for each allele?

c. There are four known alleles of ROC with 15, 12, 10, and 7 copies of the repeating unit. How many possible human genotypes are there for these alleles, and what are they?

d. If one parent is heterozygous for the 15 and 10 alleles of the ROC locus and the other parent is heterozygous for the 10 and 7 alleles, what are the

Figure 9.A

```
5'-CTGATTCTTGATCTCCTTTAGCTTC ┌─────┐ GTATAATTCATTATGTGATAATGCC-3'
3'-GACTAAGAACTAGAGGAAATCGAAG │ ROC │ CATATTAAGTAATACACTATTACGG-5'
                             └─────┘
```

possible genotypes of their offspring for this locus, and in what proportion will they be found?

e. Growing up in the house with the two parents mentioned in (d) are three children. When you type them for the ROC locus, you find that their genotypes are (10,10), (15,10), and (12,7). What can you conclude?

A9.2

a. You need PCR primers that immediately flank the ROC locus. The primers must be of the correct polarity to amplify the DNA between them. Thus, the left primer is 5'-TTGATCTCCTTTAGCTTC-3' (the rightmost 18 nucleotides of the flanking sequence to the left of ROC, reading from left to right on the top strand), and the right primer is 5'-TCACATAATGAATTATAC-3' (the leftmost 18 nucleotides of the flanking sequence to the right of ROC, reading from right to left on the bottom strand).

b. PCR amplifies the DNA between the two primers used in the reaction. The size of a PCR product is the length of the DNA between the primers, plus the lengths of the two primers. So, for a 10-copy allele of the ROC locus, with a repeating unit length of 4 nucleotides, the PCR product is $18 + (10 \times 4) + 10 = 76$ bp. For a 7-copy allele of the ROC locus, the PCR product is $18 + (7 \times 4) + 18 = 64$ bp.

c. Humans are diploid, so there are two copies of each locus in the genome. Individuals can be homozygous or heterozygous for each locus. Figuring out the genotypes involves determining all possible pairwise combinations of alleles. For four STR alleles, there are 10 genotypes, 4 of which are homozygous and 6 of which are heterozygous. The genotypes are (15,15), (12,12), (10,10), (7,7), (15,12), (15,10), (15,7), (12,10), (12,7), and (10,7).

d. This question concerns the segregation of alleles (see, for example Chapter 1, pp. 3–4.) Each diploid parent produces haploid gametes, and the gametes from each parent pair randomly to produce the diploid progeny. Thus, a (15,10) parent produces equal numbers of 15 and 10 gametes, and a (10,7) parent produces equal numbers of 10 and 7 gametes. They will fuse randomly, as in the following figure:

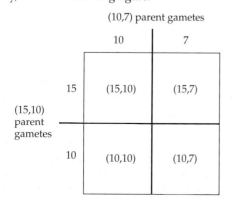

The progeny phenotypes are ¼ (15,10), ¼ (15,7), ¼ (10,10), and ¼ (10,7).

e. In part (d), the possible offspring genotypes for pairings of (15,10) and (10,7) parents were determined. The genotypes of two of the three children match expectations for offspring of the two parents, namely, the (10,10) and (15,10) children. However, the (12,7) child cannot be produced from the two parental genotypes given. Certainly, the (10,7) parent could have contributed the 7 allele, but the 12 allele does not derive from either parent. There is no way to explain the situation here without further information. Hypotheses to explain the (12,7) child include (1) the child is adopted, (2) the child comes from a previous marriage of the (10,7) parent with an individual who had a 12 allele, and (3) the child was somehow switched at birth at the hospital.

Questions and Problems

9.1 What modifications are made to the polymerase chain reaction (PCR) to use this method for site-specific mutagenesis?

9.2 Metalloproteases are enzymes that require a metal ion as a cofactor when they cleave peptide bonds. Members of one family of metalloproteases share the following consensus amino acid sequence in their catalytic site: His-Glu-X-Gly-His-Asp-X-Gly-X-X-His-Asp (X is any amino acid). Structural models of the catalytic site developed from X-ray crystallographic data suggest that the second amino acid, glutamate, is essential for proteolytic activity. Outline the experimental steps you would take to test this hypothesis. Assume you possess a cDNA encoding a metalloprotease having the consensus sequence, and can measure metalloprotease activity in a biochemical assay.

*9.3 Positional cloning is used to identify the gene for an autosomal dominant human disease. Sequence analysis of a mutant allele reveals it to be a missense mutation. Two alternate hypotheses are proposed for how the mutant allele could cause disease. In one hypothesis, the missense mutation alters a critical amino acid in the protein so that the protein is no longer able to function: heterozygotes with just one copy of the normal allele develop the disease because they have half of the normal dose of this protein's function. In the second hypothesis, the missense mutation alters the protein so that it interferes with a normal process: heterozygotes develop the disease because the mutant allele actively disrupts a required function. How could you gather evidence to support one of these alternate hypotheses using knockout mice?

9.4 What different types of DNA polymorphisms exist and what different methods can be used to detect them?

9.5 Do all DNA polymorphisms lead to an alteration in phenotype? Explain why or why not.

*9.6 Abbreviations used in genomics typically facilitate the quick and easy representation of longer tongue-twisting terms. Explore the nuances associated with some abbreviations by stating whether an RFLP, VNTR, or STR could be identified as an SNP? Explain your answers.

*9.7 The frequency of individuals in a population with two different alleles at a DNA marker is called the marker's heterozygosity. Why would an STR DNA marker with nine known alleles and a heterozygosity of 0.79 be more useful for mapping and DNA fingerprinting studies than a nearby STR having three alleles and a heterozygosity of 0.20?

*9.8 DNA was prepared from small samples of white blood cells from a large number of people. These DNAs were individually digested with *Eco*RI, subjected to electrophoresis and Southern blotting, and the blot was probed with a radioactively labeled cloned human sequence. Ten different patterns were seen among all of the samples. The following figure shows the results seen in ten individuals, each of whom is representative of a different pattern.

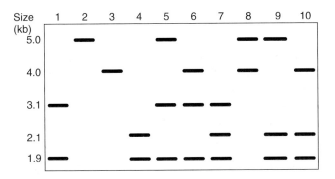

a. Explain the hybridization patterns seen in the 10 representative individuals in terms of variation in *Eco*RI sites.

b. If the individuals whose DNA samples are in lanes 1 and 6 on the blot were to produce offspring together, what bands would you expect to see in DNA samples from these offspring?

*9.9 The maps of the sites for restriction enzyme R in the wild type and the mutated cystic fibrosis genes are shown schematically in the following figure:

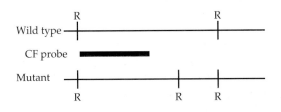

Samples of DNA obtained from a fetus (F) and her parents (M and P) were analyzed by gel electrophoresis fol-

lowed by the Southern blot technique and hybridization with the radioactively labeled probe designated "CF probe" in the previous figure. The autoradiographic results are shown in the following figure:

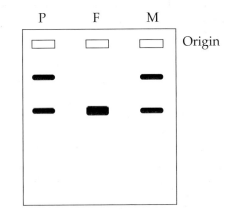

Given that cystic fibrosis is a recessive mutation, will the fetus be affected? Explain.

*9.10 The enzyme *Tsp*45I recognizes the 5-bp site 5′-G-T-(either C or G)-A-C-3′. This site appears in exon 4 of the human gene for α-synuclein, where, in a rare form of Parkinson's disease, it is altered by a single G-to-A mutation. (Note: Not all forms of Parkinson's disease are caused by genetic mutations.)

a. Suppose you have primers that can be used in PCR to amplify a 200 bp segment of exon 4 containing the *Tsp*45I site, and that the *Tsp*45I site is 80 bp from the right primer. Describe the steps you would take to determine if a parkinsonian patient has this α-synuclein mutation.

b. What different results would you see in homozygotes for the normal allele, homozygotes for the mutant allele, and in heterozygotes?

c. How would you determine, in heterozygotes, if the mutant allele is transcribed in a particular tissue?

9.11 Pathologists categorize different types of leukemia, a cancer that affects cells of the blood, using a set of laboratory tests that assess the different types and numbers of cells present in blood. Patients classified into one category using this method had very different responses to the same therapy: some showed dramatic improvement while others showed no change or worsened. This finding raised the hypothesis that two (or more) different types of leukemia were present in this set of patients, but that these types were indistinguishable using existing laboratory tests. How would you test this hypothesis using DNA microarrays and mRNA isolated from blood cells of these leukemia patients?

9.12 Some features that we commonly associate with racial identity, such as skin pigmentation, hair shape, and facial morphology, have a complex genetic basis. However, it turns out that these features are not representative of the genetic differences between racial groups—individuals assigned to different racial categories share many more DNA polymorphisms than not—supporting the contention that race is a social, and not a biological construct. How could you use DNA chips to quantify the percentage of SNPs that are shared between individuals assigned to different racial groups?

9.13 What is DNA fingerprinting and what different types of DNA markers are used in DNA fingerprinting? How could this method be used to establish parentage? How is it used in forensic science laboratories?

*9.14 One application of DNA fingerprinting technology has been to identify stolen children and return them to their parents. Bobby Larson was taken from a supermarket parking lot in New Jersey in 1978, when he was 4 years old. In 1990, a 16-year-old boy called Ronald Scott was found in California, living with a couple named Susan and James Scott, who claimed to be his parents. Authorities suspected that Susan and James might be the kidnappers and that Ronald Scott might be Bobby Larson. DNA samples were obtained from Mr. and Mrs. Larson and from Ronald, Susan, and James Scott. Then DNA fingerprinting was done, using a polymorphic probe for a particular VNTR family, with the results shown in the following figure. From the information in the figure, what can you say about the parentage of Ronald Scott? Explain.

Mrs. Larson | Mr. Larson | Ronald Scott | James Scott | Susan Scott

*9.15 As described in the text and demonstrated in Question 9.14, VNTRs can robustly distinguish between different individuals. Five well-chosen, single-locus VNTR probes used together can almost uniquely identify one individual, as, statistically, they are able to discriminate 1 in 10^9 individuals. However, the use of VNTR markers has largely been supplanted by the use of STR markers. For example, the FBI uses a set of 13 STR markers in forensic analyses. Different fluorescently labeled primers and reaction conditions have been developed so that this marker set can be multiplexed—all of the markers can be amplified in one PCR reaction. The marker set used by the FBI, the number of alleles at each marker, and the probability of obtaining a

random match of a marker in Caucasians is listed in the following table:

STR Marker	Number of Alleles	Probability of a Random Match (Based on an Analysis of Caucasians)
CSF1PO	11	0.112
FGA	19	0.036
TH01	7	0.081
TPOX	7	0.195
VWA	10	0.062
D3S1358	10	0.075
D5S818	10	0.158
D7S820	11	0.065
D8S1179	10	0.067
D13S317	8	0.085
D16S539	8	0.089
D18S51	15	0.028
D21S11	20	0.039

a. Consider the types of DNA samples that the FBI analyzes and the requirements concerning DNA samples in the methods used to analyze STR and VNTR markers. Why is the use of STR markers preferable to the use of VNTR markers?

b. Why is it advantageous to be able to multiplex the PCR reactions used in forensic STR analyses?

c. Suppose the first four STR markers listed in the table are used to characterize an individual's genotype, and the genotype is an exact match with results obtained from a hair sample found at a crime scene. What is the probability that the individual has been misidentified, that is, what is the chance of a random match when just these four markers are used? About how often do you expect an individual to be misidentified if only these four markers are used?

d. Answer the questions posed in (c) if all 13 STR markers are used.

9.16 About midnight on Saturday, the strangled body of a regular patron of the Seedy Lounge is found in an alleyway near the bar. The police interview the workers and patrons remaining in the bar. A few of the patrons indicate that several individuals, including the bartender, owed money to the deceased. The police notice that the bartender and patrons A, C, D, F, K, L, O and R all have recent cuts and scratches on their face and back of their necks, but are told that these happened during mud-wrestling matches earlier in the evening. DNA samples are obtained from the bartender and the bar's patrons, from the deceased and from scrapings of her fingernails. STR analyses are performed on the DNA samples using three of the markers described in Question 9.15: THO1, D18S51, and D21S11. The sizes of the PCR products obtained in each DNA sample for each marker are shown in the following table.

	STR		
DNA Sample	**THO1**	**D21S11**	**D18S51**
Victim	162, 170	221, 239	292, 304
Victim's fingernail scraping	162, 170, 174	221, 225, 233, 239	280, 292, 300, 304
Patron A	159, 174	221, 225	292, 316
Patron B	162	221, 235	296, 304
Patron C	162, 174	225, 233	280, 300
Patron D	170, 174	229, 231	300, 304
Patron E	170, 174	225, 233	288, 292
Patron F	162, 166	229, 243	284, 288
Patron G	174	225, 235	292, 308
Patron H	159, 174	221, 233	296
Patron I	159, 174	233, 235	300, 308
Patron J	170, 174	225	284, 296
Patron K	170, 174	231, 235	288, 292
Patron L	159	237, 239	276, 304
Patron M	159, 170	221, 229	304, 308
Patron N	166, 174	229, 239	292, 304
Patron O	170	221, 225	288, 308
Patron P	162, 170	221	296, 300
Patron Q	159, 174	235, 239	284, 304
Patron R	170, 174	225, 233	288, 292
Bartender	170, 174	221, 231	300, 308

a. How many different alleles are present at each marker in these samples and how does this compare to the total number of alleles that exist? How do you explain the appearance of only one marker allele in some individuals? How do you explain the appearance of three and four marker alleles in the DNA sample obtained from the victim's fingernails?

b. Who should the police investigate further if they consider the results obtained using only the D21S11 marker? Explain your reasoning.

c. Who should the police investigate further if they consider the results obtained using all three STR markers? Explain your reasoning.

*9.17 For rare genetic disorders that have only one mutant allele, genetic tests can be tailored to specifically detect the mutant and normal alleles. However, for more prevalent genetic disorders, such as anemia caused by mutations in α- and β-globin, Duchene muscular dystrophy caused by mutations in the dystrophin gene, and cystic fibrosis caused by mutations in CFTR, there are many different alleles at one gene that can lead to different disease phenotypes. These diseases present a challenge to genetic testing, as for these diseases, a genetic test that identifies only a single type of DNA change is inadequate. How can this challenge be overcome?

9.18 A research team interested in social behavior has been studying different populations of laboratory rats.

By using a selective breeding strategy, they have developed two populations of rats that differ markedly in their behavior: one population is abnormally calm and placid, while the second population is hyperactive, nervous, and easily startled. Biochemical analyses of brains from each population reveal different levels of a catacholamine neurotransmitter, a molecule used by neurons to communicate with each other. Relative to normal rats, the hyperactive population has increased levels while the calm population has decreased levels. Based on these results, the researchers have hypothesized that the behavioral and biochemical differences in the two populations are caused by variations in a gene that encodes an enzyme used in the synthesis of the catacholamine. Suppose you have a set of SNPs that are distributed throughout the rat genomic region containing this gene, including its promoter, coding region, enhancers, and silencers. How could you use these SNPs to test this hypothesis?

9.19 When geneticists sought to identify the gene responsible for cystic fibrosis, they did not know the identity of its protein product. Even though they knew nothing about the nature of the defective protein, they did know that something about salt transport was disrupted. The geneticists used a powerful combination of classic and molecular approaches and in 1989 located the cystic fibrosis gene. What steps did they take to identify the CFTR gene?

*9.20 Chromosome walking and chromosome jumping can both be used to find a gene between flanking markers. What is the difference between chromosome walking and chromosome jumping? Given that these can proceed in either of two opposition directions, what experimental method could you use to verify that a chromosomal walk or jump is going toward the gene you are seeking?

9.21 A positional cloning approach has been used to identify a chromosomal region for a disease gene, and the region is found to contain several different candidate genes. What is meant by a candidate gene? What types of criteria can be used to prioritize the evaluation of different candidate genes?

9.22 Sexual behavior in *Drosophila* (fruit fly) is under the control of several regulatory genes. Recently the *fru* gene has been molecularly cloned. Explain why male mutants homozygous for *fru* are considered sterile even though they have normal sexual organs and produce normal sperm.

*9.23 The *fru* gene has been cloned, and both genomic and cDNA clones for *fru* are available. One means to more fully understand its function in male sexual behavior is to identify genes for proteins that interact with the *fru* gene's protein product. Describe the steps you would take to accomplish this goal.

9.24 Genetic variability is important for maintaining the ability of a species to adapt to different environments. Therefore, it is important to understand how much genetic variation there is in an endangered species, as this type of information can be used to design better strategies to help the species from becoming extinct. Listed below are four strategies that have been proposed for detecting a SNP in a known DNA sequence in several hundred individuals from an endangered species. Critically evaluate them, and explain why each is, or is not, a good strategy for this purpose.

a. Sacrifice each of the animals or plants in the name of science. Isolate their genomic DNA, prepare libraries from each, and screen for clones containing the sequence. Sequence each clone individually. Then compare the sequences of the different clones.

b. Isolate a few cells (e.g., by using a cheek scraping or leaf sampling) from each of the individuals. Prepare DNA from the samples and use the ASO hybridization method.

c. Isolate a few cells from each of the individuals. Prepare DNA from each of the samples and then use the yeast two-hybrid system.

d. Search the literature to find a restriction enzyme that cleaves the sequence containing the SNP and that cleaves the site when only one SNP allele is present. Use the restriction enzyme to measure the site as an RFLP marker. After isolating a few cells from the individuals, prepare DNA from the cells, digest it with the restriction enzyme, separate it by size using electrophoresis, make a Southern blot, and perform an RFLP analysis.

*9.25 A scientist is interested in understanding the physiological basis of alcoholism. She hypothesizes that the levels of the enzyme alcohol dehydrogenase, which is involved in the degradation of ethanol, are increased in individuals who routinely consume alcohol. She develops a rat model system to test this hypothesis. What steps should she take to determine if the transcription of the gene for alcohol dehydrogenase is increased in the livers of rats who are chronically fed alcohol compared to a control, abstinent population?

9.26 In 1990, the first human gene therapy experiment on a patient with adenosine deaminase deficiency was done. Patients who are homozygous for a defective gene for this enzyme have defective immune systems and risk death from diseases as simple as a common cold. Which cells were involved, and how were they engineered?

9.27 What methods are used to introduce genes into plant cells, and how are these methods different than those used to introduce genes into animal cells?

9.28 The ability to place cloned genes into plants raises the possibility of engineering new, better strains of crops such as wheat, maize, and squash. It is possible to identify useful genes, isolate them by cloning, and insert them directly into a plant host. Usually these genes bring out desired traits that allow the crops in question to flourish. Why, then, is there such concern by consumers about this process? Do you feel that the concern is justified? Defend your answer.

10

Genomics

Logo for the Department of Energy Human Genome Project.

PRINCIPAL POINTS

- Genomics is the development and application of new mapping, sequencing, and computational procedures for the analysis of the entire genome of organisms.

- Genomics has three distinct subfields. Structural genomics is the genetic mapping, physical mapping, and sequencing of entire genomes; functional genomics is the comprehensive analysis of the functions of gene and of nongene sequences in entire genomes; and comparative genomics is the comparison of entire genomes of different species, with the goal of enhancing our understanding of the functions of each genome, including its evolutionary relationships.

- One approach to genome sequencing is first to generate high-resolution genetic and physical maps of the genome, in order to define segments of increasing resolution, and then to sequence the segments in an orderly manner. Another approach, the whole-genome shotgun approach, is to break up the genome into random, overlapping fragments and then to sequence

the fragments and assemble the sequences by means of computer algorithms.

- The genomes of many viruses and living organisms have been completely sequenced. Analysis of the sequences has affirmed the division of living organisms into the Bacteria, Archaea, and Eukarya.

- Genomes show a trend of increasing amounts of DNA with increasing complexity of the organism, although the relationship is not perfect. In Bacteria and Archaea, genes make up most of their genomes; that is, gene density is very high. In Eukarya, there is a wide range of gene densities, showing a trend of decreasing gene density with increasing complexity.

- In functional genomics, genomes are studied to define the functions of all the genes, including the patterns and control of gene expression. Gene expression is analyzed at two levels: mRNA transcripts, called the transcriptome, and proteins, called the proteome.

i IF YOU'RE LIKE MOST PEOPLE IN THE UNITED STATES, at some point in your life you have taken a prescription drug. Although your doctor may have considered your medical history when he or she selected the drug, it is very unlikely that the doctor could fully predict how you would react to the medication before you took it. In fact, because of inherited variations in your genes, your ability to metabolize any given drug and the side effects you may experience from that drug differ greatly from those of other people. But in the near future, doctors may be able to prescribe medications, adjust dosage, and select treatments on the basis of the patient's genetic information. The technology that makes this possible is called a DNA microarray. In this chapter, you will learn more about DNA microarrays and other tools and techniques used to analyze the entire genomes of organisms. Then, in the iActivity, you will discover how DNA microarrays can be used to create a personalized drug therapy regimen for a patient with cancer.

The development of molecular techniques for analyzing genes and gene expression has revolutionized experimental biology. Once DNA sequencing techniques were developed (see Chapter 8, pp. 197–200), scientists realized that determining the sequences of whole genomes was possible, although not necessarily easy. The first complete genome sequenced was the 16,159-bp circular genome of the human mitochondrion in 1981. Given the state of the art of sequencing at the time, this was an amazing accomplishment. But the nuclear genome is 200,000 times larger, making the determination of its sequence daunting. However, major advances in automating DNA sequencing and developing computer programs to analyze large amounts of sequence data made the sequencing of large genomes a real possibility by the mid-1980s. The field of genomics was born! The complete sequencing of the human genome was proposed in 1986, and the sequencing initiatives called the **Human Genome Project (HGP)** were started as an international effort in October 1990, with an estimated time span of 15 years. Details of the HGP may be found at http://www.ornl.gov/TechResources/Human_Genome/home.html. The U.S. Human Genome Project consisted of the Department of Energy's (DOE's) Human Genome Program and the National Institutes of Health's (NIH's) National Human Genome Research Institute (NHGRI: http://www.genome.gov). As part of the HGP, for purposes of comparison, parallel sequencing studies were carried out on the genomes of several other important model organisms commonly used in genetic studies: *E. coli* (representing prokaryotes), the yeast *Saccharomyces cerevisiae* (representing single-celled eukaryotes), *Drosophila melanogaster* and *Caenorhabditis elegans* (the fruit fly and nematode worm, respectively, representing multicellular animals of moderate genome complexity),

and *Mus musculus* (the mouse, representing a mammal of genome complexity comparable to that of humans). The HGP goals were stated as follows:
- Identify all the genes in human DNA,
- Determine the sequences of the 3 billion chemical base pairs that make up human DNA,
- Store this information in public databases,
- Develop tools for data analysis,
- Transfer related technologies to the private sector, and
- Address the ethical, legal, and social issues (ELSIs) that might arise from the project.

Importantly, the U.S. funding agents have actively promoted the transfer of technology and information to all researchers, including those in the private sector, so the HGP is also leading to the development of new medical applications and new biotechnology products. Subsequently, the National Science Foundation (NSF) funded the U.S. part of the sequencing of *Arabidopsis thaliana* (representing plants). Outside the United States, a number of other countries, including Great Britain, France, and Japan, initiated similar genome projects coordinated by an international organization called the Human Genome Organisation (HUGO). More recently, private companies became involved in genome sequencing projects. Celera Genomics, for example, was formed in 1998 and set a goal to sequence the genome in three years, using HGP-generated resources.

Mapping and sequencing genomes is a complex subject. This chapter is an overview of the subject and an introduction to the information that is being obtained from genome sequence analysis. Our focus is the sequencing of the human genome, but we will also learn about the successes in sequencing the genomes of other organisms. The sequences of five individual human chromosomes were reported in 1999 and 2000, and in June 2000 there was a join announcement by Francis Collins, representing the HGP, and J. Craig Venter of Celera Genomics that a "working draft" of 3.1 billion nucleotides of the human genome had been completed. Then, on April 14, 2003, the international consortium of the Human Genome Project announced the successful completion of all of the original goals of the HGP, more than two years ahead of schedule. Most importantly, the draft sequence of the human genome has been converted to a finished sequence, meaning that it is both highly accurate and highly contiguous, with gaps only in places where sequences cannot be determined with current technology. The finished sequence produced by the HGP covers about 99 percent of the gene-containing regions of the genome, with an accuracy of 99.99 percent.

As you read through this chapter, recognize that sequencing an organism's genome is *descriptive science* rather than *hypothesis-driven science*. Clearly, no hypotheses play a role in collecting the primary data of an organism's genome. But hypothesis-driven experiments are a major part of researchers' efforts to understand the genome data being generated, especially which genes are present and how they direct the structure and function of the organism.

Structural Genomics

Genome projects are enormous undertakings, and their success depends on significant technical advances. Moreover, they have led to the development of the field of **genomics**—the development and application of new mapping, sequencing, and computational procedures to analyze the entire genome of organisms. There are different subfields of genomics: **structural genomics**, involving the genetic mapping, physical mapping, and sequencing of entire genomes; **functional genomics**, dealing with the comprehensive analysis of the functions of genes and of nongene sequences in entire genomes; and **comparative genomics**, treating the comparison of entire genomes of different species, with the goal of enhancing our understanding of the functions of each genome, including its evolutionary relationships. In this section, you will learn about structural genomics—how a genome sequence is obtained—with an emphasis on sequencing the human genome.

Sequencing Genomes

Two general approaches are used in genome projects. The *mapping approach* described in this section involves dividing the genome into segments by constructing genetic maps and physical maps of increasing resolution and then sequencing the segments. This technique is analogous to ordering the headings in a book first and then finding the words between the headings. The other approach is the *whole-genome shotgun (WGS) approach*, in which the entire genome is broken up into random, overlapping fragments that are subsequently sequenced. The genome sequence is then assembled by computer on the basis of the sequence overlaps between fragments. In this approach, there is no prior knowledge of the location of the fragment, and the technique is analogous to taking 10 books that have been torn randomly into smaller leaflets of a few pages each and, by matching overlapping pages of the leaflets, assembling a complete copy of the book with the pages in the correct order. We will now consider these two approaches briefly.

Genome Sequencing Using a Mapping Approach. In this approach, genetic maps and physical maps are made with the initial goal of having markers mapped relatively closely throughout the genome. In this section, selected examples of genetic-mapping and physical-mapping experiments are described to illustrate the

types of thinking and logic that were involved. In the HGP itself, genetic mapping and physical mapping data from more types of experiments were collected and analyzed.

Genetic mapping of a genome. One of the initial goals of the HGP was a high-density genetic map with at least one genetic marker per 1 Mb (about 1 map unit) of the genome. Genetic maps of genomes are constructed using genetic crosses and, for humans, pedigree analysis. The principles used in genetic mapping are those described in detail in Chapters 15 and 16. The mapping of human genes was the focus of a discussion in Chapter 16 on pp. 443–446. Genetic crosses are used to establish the locations of genetic markers (alleles that can be used to mark a location on a chromosome) on chromosomes and to determine the genetic distance between them. Genetic distance is given by the frequency of crossing-over between the markers. Both genes and DNA markers (see Chapter 9, pp. 214–218) are used for genetic-mapping experiments. For the human genome, there are 24 maps, corresponding to the 22 autosomes, the X chromosome, and the Y chromosome.

Creating a genetic map for two genes is the simplest genetic-mapping analysis one can do. Building a genetic map with mapping information for two genes at a time is not very efficient for a very large genome, however. Even using a handful of gene loci in genetic-mapping experiments makes the assembly of gene maps quite slow. For humans, there are simply not enough known genes to give chromosome maps with closely spaced genes. Fortunately, with the discovery of a large number of polymorphic DNA markers and the development of molecular tools for the rapid typing of those markers, genetic mapping was elevated to a whole new level. In outline, the methods of mapping are the ones described earlier for human pedigree analysis, except that hundreds of loci are involved; all those loci in all individuals in the cross must be typed by molecular protocols and the resulting data analyzed by sophisticated computer algorithms to determine linkage relationships.

In the human genome project, one DNA marker called a **sequence-tagged site (STS)** was used for genetic mapping. An STS is a DNA sequence that is unique in the genome. Typically, it is a couple of hundred base pairs or less in length and is generated by PCR, using primers based on determined DNA sequences and designed to allow the specific amplification of some or all of the sequence. The genomic site for the sequence in question is considered to be *tagged* by its ability to be assayed for that sequence. Polymorphic STRs are the best kind of DNA markers for generating genetic maps of STSs. These were the markers used in making the high-density genetic map of the human genome described in Chapter 16 (p. 444, and Figure 16.12).

KEYNOTE

Genetic maps of genomes are constructed with the use of recombination data from genetic crosses in the case of experimental organisms or from pedigree analysis in the case of humans. Both gene markers and DNA markers are used in genetic-mapping analysis.

Physical mapping of a genome. For organisms with small genomes, genetic maps have sufficient resolution to provide landmarks for a genome sequencing effort. When the *E. coli* genome sequencing project began, for example, there were 1,400 genetic markers with an average of one per 3.3 kb. In humans, however, the genetic map did not have sufficient resolution for beginning sequencing. Researchers had to develop a detailed **physical map**—a map of genetic markers made by analyzing genomic DNA directly, rather than by analyzing recombinants from pedigree analysis (in humans) or genetic crosses (with experimental organisms) as is the case with genetic maps—to supplement the genetic map. As with the genetic map of the human genome, there are 24 maps for the complete physical map, corresponding to the 22 autosomes, the X chromosome, and the Y chromosome. This section describes briefly some of the types of physical maps, presented in order of increasing resolution.

Among the various types of physical maps are, in order of increasing resolution, cytogenetic maps of chromosomal banding patterns (see Chapter 12, pp. 301–302), FISH maps (see Chapter 16, pp. 445–446), restriction maps, radiation hybrid maps (see Chapter 16, p. 446), and clone contig maps. We consider restriction mapping and clone contig maps briefly in what follows.

In Chapter 8 (pp. 192–195), we discussed the principles of **restriction mapping.** At the chromosome scale or the genome scale, restriction mapping with commonly used enzymes such as *Eco*RI, *Hind*III, and *Bam*HI is seriously limited because of the very large number of restriction sites and, therefore, the difficulties of analyzing the fragments produced for their positions in the genome, even with computer assistance. Experimentally, the number of restriction fragments produced may be so large that a continuous smear, rather than discrete bands, is seen on an agarose gel after electrophoresis.

However, restriction mapping of large genomes is possible with restriction enzymes that cut rarely. There are two types of such enzymes. The first includes enzymes that have 7- or 8-bp recognition sequences. An example of the latter is *Not*I (see Table 8.1), which has the recognition sequence 5'-GCGGCCGC-3' 3'-CGCCGGCG-5'. For a genome with 50 percent GC content, this enzyme would cut, on average, once every $4^8 = 65{,}536$ bp (0.065536 Mb; i.e., there is a 1 in 4 chance of a G in the first position of the recognition sequence, a 1 in 4 chance of a C in the second position, a 1 in 4 chance of G

in the third position, and so on, with each probability multiplied to give the total probability of finding the given recognition sequence), compared with $4^6 = 4{,}096$ bp for an enzyme with a 6-bp recognition sequence, such as *Eco*RI. In fact, *Not*I cuts human DNA once every 10 Mb, on average. The second type includes enzymes which have recognition sequences that are uncommon in the DNA being mapped. That is, the genome does not consist of randomly arranged base pairs, and certain sequences are found at low frequencies. For example, the sequence 5'-GC-3' 3'-GC-5' is rare in the human genome. Therefore, enzymes that have 5'-CG-3' 3'-GC-5' in their recognition sequences cut human DNA less frequently than one might expect if that sequence were not rare. The enzyme *Bss*HII (5'-GCGCGC-3' 3'-CGCGCG-5'), for example, cuts only once every 390 kb, on average. The presence of two 5'-CG-3' 3'-GC-5' sections in the *Not*I recognition sequence helps explain this enzyme's very rare cutting of human DNA. Restriction maps made with these types of restriction enzymes have a resolution of several hundred kilobase pairs.

Even with rare cutters such as *Not*I, constructing a restriction map of the entire genome is very difficult. Completed in 1993, the *Not*I map of human chromosome 21 is the only whole chromosome restriction map of a human chromosome. The map spanned 37 Mb, and the *Not*I restriction sites were valuable markers used when the sequence of chromosome 21 was assembled in 2000.

Higher resolution is possible with a **clone contig map**—a set of ordered, partially overlapping clones comprising all the DNA of an entire chromosome or part of a chromosome, without any gaps. (The word *contig* is a shortened form of *contiguous*.) Clone contig maps provide the framework for sequencing the entire genome.

For the entire human genome, 24 clone contig maps are needed to cover the 24 different chromosomes. Because of the size of the human genome, the clones must be made by using vectors that have a large capacity for inserted DNA, so that the number of clones is manageable. Researchers set out to make a first-generation clone contig map of the human genome with YAC (yeast artificial chromosome) vectors (see Chapter 8, pp. 182–183), because they have a cloning capacity of several hundred kilobase pairs of DNA.

Constructing a clone contig map starts by making a library of partially overlapping DNA fragments similar to that discussed in Chapter 8 (p. 184). In brief, high-molecular-weight DNA is isolated and then sheared mechanically—for example, by passing it through a syringe needle to produce a complete set of random, partially overlapping fragments of a genome. The fragments produced by mechanical shearing have blunt ends. For cloning, they are inserted into a vector—a YAC, for example—cut with a restriction enzyme such as *Sma*I that generates blunt ends. (DNAs with blunt ends can be ligated together in the normal reaction catalyzed by DNA ligase.) The YAC clone library produced may be of the whole genome if genomic DNA is used as the source or of a particular chromosome if chromosomes are first separated by flow cytometry before DNA is isolated. (See Chapter 8, p. 185. Flow cytometry is a procedure in which fluorescently labeled cells or cell components such as chromosomes are passed by a sensing point through which laser light is directed. Detectors measure the emitted fluorescent light and the data are entered into a computer. Flow cytometers can measure physical characteristics such as chromosome size and shape, facilitating the separation—sorting—of these characteristics into the different ones of the organism.)

How do we make a YAC clone contig map? YAC clones from the library can be roughly located on a chromosome map by using FISH (see Chapter 16, pp. 445–446), but the best way to assemble clone contigs over large regions of the genome is to use DNA fingerprinting techniques to type clones at random and then to assemble the clones in order on the basis of the overlaps determined between them. We discussed some examples of DNA fingerprinting in Chapter 9 (pp. 224–228). The most useful protocol in this case involves STS mapping. An STS (*sequence-tagged site*) is a DNA sequence marker that is unique in the genome. It is a short sequence that can be amplified by PCR. The genomic site for the sequence is said to be *tagged* by its ability to be arrayed for that sequence. Several types of STSs are known. For genetic mapping, STSs were used that were highly polymorphic because of the need to follow genetic recombination. For constructing contig maps, nonpolymorphic STSs are used, because unique markers, and not variable ones, are needed at the chromosomal loci being mapped. This type of STS includes parts of genes and random nongene sequences.

Figure 10.1 shows a YAC contig map assembled by STS mapping. First, the locations of STSs on YAC clones are mapped by means of PCR-derived probes. Then, the STS maps on each YAC clone are compared to find clones that share some STSs, indicating partial overlap of the cloned DNA.

For example, through STS mapping, we may find that YAC clone 1 has STSs A, D, F, and M; clone 2 has D, M, P, T, and V; and clone 3 has P, W, and Z. We can see that clones 1 and 2 share STSs D and M, and clones 2 and 3 share STS P. Therefore, clones 1 and 2 must overlap with a DNA segment containing STSs D and M, and clones 2 and 3 must overlap with a DNA segment containing STS P. We cannot be sure of the exact order of the STSs (therefore they are shown in parentheses), but that does not affect our ability to demonstrate the overlaps between certain clones, as follows:

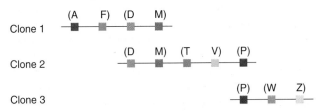

If the library is complete, with no gaps, the result will be a complete contig map for the chromosome or region being analyzed, with the individual STSs ordered and located on the genomic map on the basis of their locations on the cloned DNA.

YAC contig maps for human chromosomes 21 and Y were reported in 1992, and a YAC contig map encompassing 75 percent of the human genome was reported in the same year. Unfortunately, constructing YAC contig maps ran into trouble when it was discovered that, in many cases, the cloned DNA in YAC vectors contained DNA from more than one location. Using such YAC clones in contig mapping could place DNA segments incorrectly on the physical map.

Because of these problems, human genome researchers turned to radiation hybrid mapping (see Chapter 16, p. 446) of STS markers, resulting in the publication in 1995 of a map with 15,806 STS markers with one marker per 199 kb, on average. Later, the map was enhanced by an additional 20,104 STS markers, most of them **ESTs (expressed sequence tags)**. An EST is a marker produced by PCR, using oligonucleotide primers based on the sequence of a cDNA. Because a cDNA is a DNA copy of an mRNA, an EST marker corresponds to a functional protein-coding gene. If that gene is a unique one, then the EST derived from it is a unique STS. The enhanced map had the additional benefit of locating a large number of protein-coding genes on the physical map. The marker density on this new map averaged almost one per 100 kb. Combining the genetic and physical maps gives a map with markers at the targeted density desired for sequencing to be done.

Returning to clone contig maps, more accurate, second-generation clone contig maps of the human genome were made using BAC (bacterial artificial chromosome) cloning vectors. (See Chapter 8, p. 183.) Recall that a BAC vector contains part of the *E. coli* F factor, a circular extrachromosomal element that facilitates mating between *E. coli* strains. (See Chapter 18, pp. 485–487.) BAC vectors have a capacity of 300 kb of DNA or more, and the copy number is one per *E. coli* cell. Recombinant BAC molecules are introduced into *E. coli* by electroporation, the process of using high voltage to make transient pores in the cell membrane through which DNA can pass. (See Chapter 8, p. 181.)

Generating the Sequence of a Genome.   After a high-resolution map has been developed, the next step is the actual sequencing. The principles of automated DNA sequencing were outlined in Chapter 8 (pp. 199–200). In brief, with dideoxy sequencing methods, DNA is synthesized against a template strand and terminates with a fluorescently labeled dideoxy nucleotide. All four reactions, each with a different labeled ddNTP, are done in the same tube, and the products are separated by

Figure 10.1

A representative YAC contig map assembled by STS mapping. The YAC contig map shown is of the telomere proximal region of the small arm of chromosome 7 (7pter). The contig is oriented such that the 7p telomere is leftward and the chromosome 7 centromere is rightward. Maps such as this are difficult to read because of the amount of information they contain. On the map, each STS is positioned along the top (with its name given vertically, starting with "y"), and each YAC is indicated as a horizontal line below, with its name given to the left and its measured YAC size (in kb) in parentheses. The presence of an STS in a YAC is indicated by a closed blue circle at the appropriate position. STSs in red correspond to YAC insert ends, with a square placed around the corresponding circle both at the top (under the STS name) and at the end of the YAC from which the STS was derived. All other STSs are shown in blue.

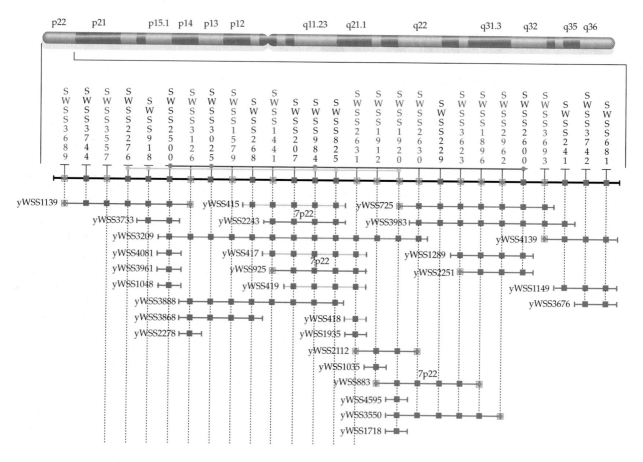

electrophoresis using an automated sequencer. The colored DNA bands are detected by lasers, and the sequencer automatically converts the data into a computer sequence file. Typically, the DNA synthesis reaction is done by using PCR with a single oligonucleotide primer and a thermostable DNA polymerase. The single primer ensures that only one of the two strands is copied onto the newly synthesized DNA molecule. The two advantages of PCR-based DNA sequencing are that (1) double-stranded DNA can be sequenced directly, because the two strands are separated during the melting step of PCR, allowing the single primer to anneal, and (2) only a small amount of DNA is needed for the reaction.

Current technology is limited to sequencing about 500 nucleotides in a single reaction. To obtain an accurate sequence, both strands must be sequenced, and the sequencing must be repeated several times—hence the enormous task of sequencing the 3 billion-bp human genome. Fortunately, improving sequencing and sequence analysis technologies is an important part of genome projects. In recent years, automated sequencing has become much quicker through the development of more efficient machines and the increasing use of robotics to prepare samples for sequencing.

For the human genome, BAC clone contig maps were used for sequencing by those genome centers which followed the mapping approach to genome sequencing. However, the BAC inserts are too large to be sequenced in a single sequencing reaction. Instead, each insert is sequenced using a *shotgun approach*. That is, each insert is cut out, sheared mechanically into a partially overlapping set of fragments of sizes amenable to sequencing, and cloned into a plasmid vector. Each of these subclones is then sequenced, and the overlapping sequences are assembled into a contiguous sequence by computer. Because each BAC clone is mapped onto the chromosome, a complete sequence for a chromosome is assembled by integrating the sequences for the individual clone inserts into one contiguous sequence. Doing this for each chromosome gives the sequence for the complete genome. Theoretically, if the genomic inserts are sequenced so that the total length of sequence produced is 6.5 to 8 times the length of the genome, the sequence contigs generated will span more than 99.8 percent of the genome sequence. The HGP did its original sequencing seven times over and, in its June 2000 announcement, reported that it had decoded 97 percent of the genome. However, the sequence was only partially assembled then; that is, the actual order of only 53 percent of the sequences was known at that time.

In sum, genome sequencing using a mapping approach involves generating genetic and physical maps of increasing resolution, integrating the two maps, and then sequencing a set of contiguous clones that contain partially overlapping DNA fragments.

KEYNOTE

> Except for organisms with small genomes, a genetic map's resolution is insufficient to provide the landmarks necessary for genome sequencing. So a high-resolution physical map—a map of genetic markers made by analyzing genomic DNA directly by molecular means—is made to supplement the genetic map. A variety of physical maps can be constructed, including restriction maps and clone contig maps. The genome sequencing itself is done with automated sequencers, using the dideoxy sequencing method with laser detection of fluorescently labeled dideoxy terminal nucleotides.

Whole-Genome Shotgun Sequencing of Genomes

The first genomic sequence of a living cell, that of the bacterium *Haemophilus influenzae*, was achieved using a *whole-genome shotgun* (WGS) approach to sequencing. In this approach, the whole genome is broken into partially overlapping fragments, each fragment is cloned and sequenced, and the genome sequence is assembled using a computer. Once thought to be of limited usefulness

animation

Genome Sequencing Using a Whole-Genome Shotgun Approach

for sequencing whole genomes greater than 100 kb, sophisticated computer algorithms for assembling sequences from hundreds to thousands of 300- to 500-bp sequences and robotic procedures for preparing DNA for sequencing opened the door for sequencing large genomes via this shotgun approach. The human genome was sequenced in this way by J. Craig Venter's Celera Genomics company. By the time of the announcement in the year 2000, Venter's group had done its sequencing five times over. Celera's human genome sequence was also about 97 percent of the genome, but, in contrast to the HGP's, it had been completely assembled with the exception of gaps resulting from the missing 3 percent of the sequence, much of which was considered to be too difficult or too costly to sequence.

Figure 10.2 outlines the whole-genome shotgun approach to sequencing. First, random, partially overlapping fragments of genomic DNA are generated by mechanical shearing, and the fragments are cloned to form a library. In contrast to the mapping approach, the insert size for each clone is small—about 2 kb—enabling the clones to be made using simple plasmid vectors. Five hundred base pairs are sequenced from each end of each insert, and the sequence data are entered into the computer. Because of the partial overlapping of the clones, the sequence of the central approximately 1 kb of DNA is obtained when the overlapping clones are sequenced. For example, if a second clone overlapped the first clone by 500 bp, then sequencing the second clone would generate 500 bp of sequence from the middle unsequenced section of the first clone. The result of sequencing this library is a number of contig sequences covering most of the genome. There are gaps in the contig sequences because some sequences are missing in the library.

A second library used in the shotgun approach consists of a random, partially overlapping library of genomic DNA fragments of about 10 kb in size in a simple plasmid vector. One important purpose for this library is to sequence regions of the genome containing repeated sequences. Many repeated sequences are around 5 kb, so they can be found in 10-kb clones, but not in 2-kb clones. Here is the dilemma with the 2-kb clone library: In assembling a genomic sequence from the 2-kb clones, a clone with the first part consisting of a unique sequence DNA and the second part consisting of part of a copy of a repeated sequence causes a dead stop in the assembly of the sequence. The reason is that many clones in the library contain parts of the repeated sequence family and they come from all over the genome. As a result, the flanking unique sequence DNA is different in each case, so we cannot determine which clone is the true overlapping one from the genome. The 10-kb clone library allows us to get around this problem because some clones have unique sequence DNA flanking a repeated DNA sequence, and

Figure 10.2

The whole-genome shotgun approach to obtaining the genomic DNA sequence of an organism.

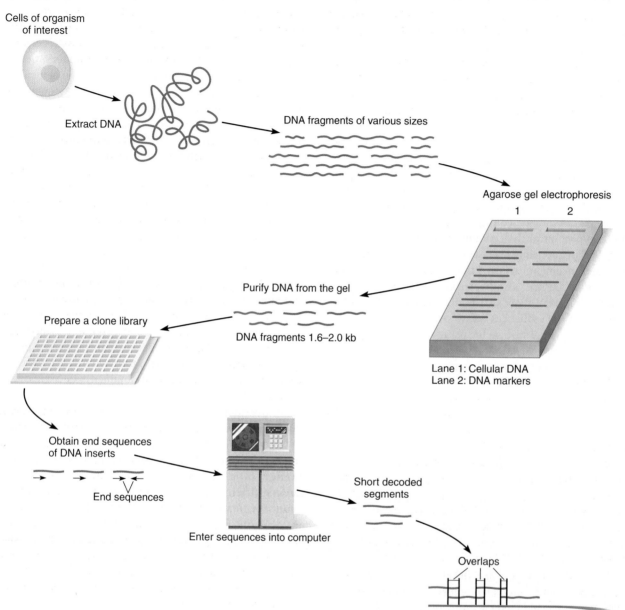

this allows us to proceed with the genome sequence assembly. Another purpose of the library is to obtain sequence information to provide independent confirmation of contig sequence structure.

Advances continue to be made in DNA sequencing automation and in computer algorithms for analyzing the sequences obtained. The whole-genome shotgun approach is now used extensively in genome sequencing projects, even for large genomes. In many cases, BAC maps are also created as part of the projects.

KEYNOTE

Sequencing a genome by the shotgun approach involves constructing a partially overlapping library of genomic DNA fragments, sequencing each clone, and assembling the genomic sequence by computer on the basis of the sequence overlaps.

Assembling and finishing genome sequences. The raw sequences obtained from genome sequencing projects must be *assembled* into sequence contigs; that is, the bases must be pieced together in their correct order as they are found in the genome. Once assembly is complete, that is often the point when "working drafts" of genome sequences are announced. The work actually is not completed at that point, because there are still many gaps in the sequences to fill in, as well as errors from the sequencing. *Finishing* the genome sequence is the next step, producing a highly accurate sequence with fewer than one error per 10,000 bases and as many gaps as possible filled in.

Selected Examples of Genomes Sequenced

We now discuss some of the genomes that have been sequenced and why the particular organisms were chosen or what the sequences are likely to contribute to our knowledge about those organisms. Genome sequences are becoming available at an increasing rate, so for your favorite organism, check the Internet sites for the National Center for Biotechnology Information (http://www.ncbi.nlm.nih.gov/Genomes/index.html) and the Institute for Genomic Research (http://www.tigr.org/) for available sequencing information.

Bacterial Genomes

Haemophilus influenzae. The first cellular organism to have its genome sequenced was the eubacterium *H. influenzae.* This organism was chosen because its genome size is typical among bacteria and the GC content of the genome is close to that of humans. The task was completed by the Institute for Genomic Research in 1995. The only natural host for *H. influenzae* is the human; in some cases, it causes ear and respiratory tract infections. The 1.83-Mb (1,830,137-bp) genome of this bacterium was sequenced by the whole-genome shotgun approach, to test the feasibility of the method, because no genetic or physical map existed at the beginning of the project.

With all genome projects, the next step after obtaining the sequence is *annotation*: the identification and description of putative genes and other important sequences. The annotated genome of *H. influenzae* is shown in Figure 10.3. There are usually no introns in protein-coding genes of bacteria and archaeons, so putative protein-coding genes are identified by using a computer to search both DNA strands for open reading frames (ORFs, or potential amino-acid coding regions—i.e., potential protein-coding genes). An ORF begins with a start codon (AUG; ATG in the DNA sequence obtained) and ends with a stop codon (TAG, TAA, or TGA in the DNA sequence) a multiple of three nucleotides downstream. However, not all ORFs (particularly the short ones) correspond to genes. Thus, somewhat arbitrarily, ORFs larger than 100 codons are considered statistically likely to be protein-coding genes. The identified ORFs for

the entire genome are searched against nucleotide and amino-acid databases to attempt to identify genes specifically. With the current state of computer searching algorithms and the amount of defined information in sequence databases, a complete microbial genome sequence can be annotated for essentially all coding regions and other elements, such as repeated sequences, operons, and transposable elements.

For *H. influenzae*, genome analysis predicted 1,743 protein-coding genes making up 85 percent of the genome. Of these predicted genes, 736 did not match any protein in the databases or matched only proteins designated hypothetical. The remaining 1,007 predicted ORFs matched genes that have known functions in the databases. This sort of result is typical of genome projects. Many genes have known functions, while a significant fraction has unknown functions, requiring much hypothesis-driven science to determine those functions.

Escherichia coli. In 1997, the annotated genome sequence of one of the model organisms targeted by the HGP, the bacterium *E. coli K12*, was reported by researchers at the *E. coli* Genome Center at the University of Wisconsin, Madison. An unannotated sequence of the *E. coli* genome made up of segments from more than one strain was reported at the same time by Takashi Horiuchi of Japan.

E. coli (see Figure 1.10) is an extremely important organism. It is found in the lower intestines of animals, including humans, and survives well when introduced into the environment. Pathogenic *E. coli* strains make the news all too frequently, as humans develop sometimes deadly enteric and other infections after contacting the bacterium at restaurants (e.g., in tainted meat) or in the environment (e.g., in contaminated lakes). In the laboratory, *E. coli* has been an extremely important model system for molecular biology, genetics, and biotechnology. Thus, the complete genome sequence of this bacterium was eagerly awaited. It was the sixth genome of a cellular organism to be reported.

The circular genome was sequenced by means of the whole-genome shotgun approach. The genome of *E. coli* is 4.64 Mb (4,639,221 bp). The 4,288 ORFs make up 87.8 percent of the genome. Thirty-eight percent of the ORFs had unknown functions.

Archaeon Genome. The *Methanococcus jannaschii* genome was the first genome of an archaeon to be completely sequenced. *M. jannaschii* is a hyperthermophilic methanogen that grows optimally at 85°C and at pressures up to 200 atmospheres. It is a strict anaerobe, and it derives its energy from the reduction of carbon dioxide to methane. Sequencing was by the whole-genome shotgun approach. The sequence was reported in 1996. The complete genome

Figure 10.3

The annotated genome of *H. influenzae*. The figure shows the location of each predicted ORF containing a database match, as well as selected global features of the genome. Outer perimeter: Key restriction sites. Outer concentric circle: Coding regions for which a gene was identified. Each location of a coding region is color coded with respect to its function. Second concentric circle: Regions of high GC content are shown in red (>42 percent) and blue (>40 percent) and regions of high AT content are shown in black (>66 percent) and green (>64 percent). Third concentric circle: The locations of the six ribosomal RNA gene clusters (green), the tRNAs (black), and the cryptic mu-like prophage (blue). Fourth concentric circle: Simple tandem repeats. The origin of replication is illustrated by the outward-pointing arrows (green) originating near base 603,000. Two possible replication termination sequences are shown near the opposite midpoint of the circle (red).

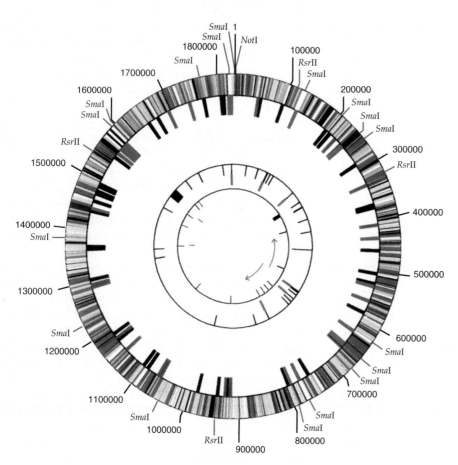

has a large, main circular chromosome of 1,664,976 bp; a circular, extrachromosomal element (ECE) of 58,407 bp; and a smaller, circular ECE of 16,550 bp. The main chromosome has 1,682 ORFs, the larger ECE has 44, and the smaller ECE has 12. Most of the genes involved in energy production, cell division, and metabolism are similar to their counterparts in the Bacteria, whereas most of the genes involved in DNA replication, transcription, and translation are similar to their counterparts in the Eukarya. Clearly, this organism was neither a bacterium nor a eukaryote. Its genome sequence therefore affirmed the existence of a third major branch of life on Earth.

Eukaryotic Genomes

Yeast. For decades, the budding yeast *Saccharomyces cerevisiae* (Figure 10.4) has been a model eukaryote for many kinds of research. Some of the reasons for its utility are that it can be cultured on simpler media, it is highly amenable to genetic analysis, and it is highly tractable for sophisticated molecular manipulations. Moreover, functionally, it resembles mammals in many ways. Therefore, it was logical that its genome would be a target for early genome sequencing efforts. In fact, the *S. cerevisiae* (yeast) genome was the first eukaryotic genome to be sequenced completely, in 1996. The 16-chromosome genome was

Figure 10.4

Scanning electron micrograph of the yeast *Saccharomyces cerevisiae*.

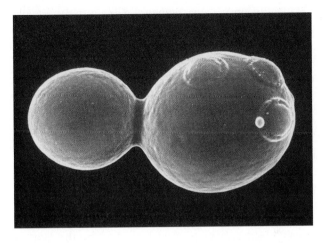

reported to be 12,067,280 bp. Approximately 969,000 bp of repeated sequences were estimated not to be included in the published sequence, which revealed 6,183 ORFs, only 233 of which had introns. At the outset of the yeast genome project, only about 1,000 genes had been defined by genetic analysis. About a third of the protein-coding genes have no known function.

The Nematode Worm *Caenorhabditis elegans*. The genome of the nematode *C. elegans* (Figure 10.5), also called the "worm," was the first multicellular eukaryotic genome to be sequenced. Nematodes are smooth, nonsegmented worms with long, cylindrical bodies. *C. elegans* is about 1 mm long; it lives in the soil, where it feeds on microbes. There are two sexes: a self-fertilizing XX hermaphrodite and an XO male. The former has 959 somatic cells, the latter 1,031. The lineage of each adult cell through development is well understood. The worm has a simple nervous system, exhibits a number of behaviors, and is

Figure 10.5

The nematode worm *Caenorhabditis elegans*.

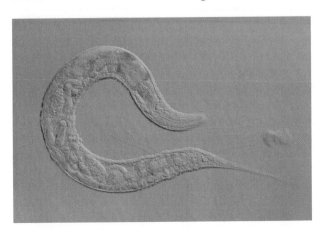

capable of simple learning tasks. *C. elegans* has become an important model organism for studying the genetic and molecular aspects of embryogenesis, morphogenesis, development, nerve development and function, aging, and behavior.

The *C. elegans* genome project was initiated by Sydney Brenner and carried out by an international consortium. A nearly complete physical map was constructed in the 1980s; in the 1990s, sequencing began, and a draft was reported in December 1998. The genome is 100.3 Mb, with 20,443 genes, 1,270 of which do not code for protein.

The Fruit Fly *Drosophila melanogaster*. The genome sequence of an organism of particular historical importance in genetics, the fruit fly *D. melanogaster* (see Figure 1.7b, p. 10), was reported in March 2000. The fruit fly has been the subject of much genetics research and has contributed to our understanding of the molecular genetics of development. Its genome sequence was as eagerly awaited as that of yeast. The fruitfly's genome was sequenced with the whole-genome shotgun approach, supported by a BAC-derived physical map.

The sequence of the euchromatic part of the *Drosophila* genome is 118.4 Mb in size. Approximately another 60 Mb of the genome consists of highly repetitive DNA that essentially cannot be cloned, making the associated sequences unobtainable. There are 14,015 genes, fewer than that found in the worm, but with a similar diversity of functions. Surprisingly, the number of fruit fly genes is just over twice that found in yeast, yet the fruit fly seems to be a much more complex organism. We must conclude that higher complexity in animals such as flies and humans does not require a correspondingly larger repertoire of gene products. The fruit fly's value as a model system for studying human biology and disease was affirmed by the finding that *D. melanogaster* has homologues for 177 of 289 genes known to be involved in human disease, including cancer.

The Flowering Plant *Arabidopsis thaliana*. The genome of *A. thaliana* (see Figure 1.7d, p. 10) was the first flowering plant genome to be sequenced. *Arabidopsis* has been an important model organism for studying the genetic and molecular aspects of plant development. The 120-Mb genome contains about 25,900 genes, almost twice that found in the fruit fly *Drosophila melanogaster* and close to the lower estimates for the number of genes in the human genome. Interestingly, about a hundred *Arabidopsis* genes are similar to disease-causing genes in humans, including the genes for breast cancer and cystic fibrosis. The next step is to fill in the gaps in the sequence and explore the structure and function of the genome in detail. Toward that end, an initiative called the **Arabidopsis 2010** Project has been set up. It has an ambitious set of goals, including defining the function of every gene in the plant, determining where and when every gene is expressed, showing where the encoded protein ends up in the plant, and defining any proteins with which the protein interacts.

Homo sapiens. As mentioned earlier, a working draft of the genome of most interest to us, the genome of *Homo sapiens,* was announced in a joint press conference in June 2000 involving Francis Collins of the National Human Genome Research Institute (NHGRI, representing the HGP) and J. Craig Venter of Celera Genomics. Whose DNA was sequenced? The HGP researchers collected female blood and male sperm from a large number of donors, but used only some of the samples to extract DNA for sequencing. Hence, neither the scientists nor the donors know whose DNA is being sequenced. Celera recruited up to 30 donors via self-referral, newspaper ads, and outreach activities and chose five males and five females from a variety of ethnic backgrounds from whom they obtained DNA for sequencing. (Later, Craig Venter admitted that he was one of the donors.) In both cases, the human genome sequence being generated is an amalgamation of sequences and will not be an exact match for the genome of any one person in the human population.

The sequencing part of the HGP was carried out by an international Human Genome Sequencing Project Consortium of scientists at 16 institutions in the United States, Great Britain, France, Germany, and China. This consortium set out to determine the sequence of the euchromatic part of the human genome. Their sequencing approach was built on a foundation of genetic and physical maps as described earlier in the chapter. At the time of the press conference, the consortium centers had generated more than 22.1 billion nucleotides of raw sequence data, representing sevenfold sequence coverage of the genome. Approximately 50 percent of this genome sequence was considered to be in a near-finished form, and 24 percent was completely finished. The achievement of the consortium's final goal of a finished sequence of the euchromatic part of the genome was announced in April 2003.

Celera Genomics used the whole-genome shotgun approach for sequencing and assembled its sequence data with the help of human genome physical map data in public databases that were generated by the Human Genome Sequencing Project Consortium. The Consortium had assembled only part of the genome, and Celera claimed that it alone had achieved the first complete assembly of the human genome sequence. Celera used two independent methods to assemble the sequence, which consisted of 3.12 billion bp (3.12 Gb, where Gb = 1 gigabase = 1 billion base pairs; current figures indicate that the genome is 2.9 Gb). One method used 26.4 million sequences of 550 bp, for a total of 14.5 billion bp, with 4.6-fold sequence coverage. More than 99 percent of the genome was covered by this

assembled sequence. According to Celera, the calculation to perform the sequence assembly involved 500 million trillion base-to-base comparisons taking more than 20,000 CPU hours on a supercomputer. The second assembly method was used to validate the results from the whole-genome direct shotgun sequence assembly and involved relating the Celera sequence data to BAC clone sequence data in the GenBank database. Recall that BAC clones contain large inserts, so this method helps resolve any ambiguities arising from assembling the sequences of the short fragments involved in the direct shotgun sequencing approach. Unquestionably, Celera's accomplishment is a highly significant event in science. With an assembled sequence in hand, the next step is annotation: determining the genes it contains and analyzing other features of the genome.

The draft genome sequences and initial interpretations of assembled sequences were published by the Human Genome Project Sequencing Consortium in the February 15, 2001, issue of *Nature* and by Celera Genomics in the February 16, 2001, issue of *Science.* In the next two years, the human genome sequence was finished and, as mentioned at the beginning of this chapter, was announced to the public on April 14, 2003. Analysis of the sequence continues with the goal of determining the number of genes and what each gene encodes. So, how many genes make a human? As of October 2004, the best estimate is 20,000–25,000 protein-coding genes, far fewer than the 50,000 to 100,000 often predicted before sequencing began. This low number is making scientists drastically change their thinking about the complexity and development of organisms. Interestingly, the human genome shares 223 genes with bacteria, but those genes are not found in yeast, the worm, or the fruit fly. All in all, the two human genome sequences are proving a great resource for scientists to learn about our species, and *data mining*—searching through genome sequences for information—will continue for many years. There will be a strong focus on human disease genes, with an eye toward treatment and therapy.

Three mammalian genomes have been sequenced completely, from humans, the mouse (*Mus musculus*), and, most recently, the rat (*Rattus norvegicus*) (Figure 10.6). The human genome is the largest, followed by that of the rat, and then that of the mouse (Table 10.1). All three mammals have approximately the same number of genes. Importantly, both the mouse and the rat have been model organisms for studies of mammalian physiology, including those dealing with diseases. The mouse, in particular, has been a model for mammalian genetics due to its genetic tractability, including its susceptibility to gene knockout mutations. (See later in the chapter.) Sequence analysis reveals that approximately 99 percent of the genes of the mouse and the rat have direct counterparts in the human, including, therefore, genes associated with disease.

Figure 10.6

The three mammals whose genomes have been sequenced.
(a) Human (*Homo sapiens*). (b) Mouse (*Mus musculus*). (c) Rat (*Rattus norvegicus*).

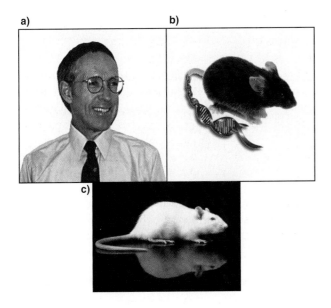

Studies of the mouse and rat genomes will undoubtedly provide valuable knowledge about human diseases and other areas of human biology, as well as about mammalian evolution.

KEYNOTE

Many genomes both of viruses and of living organisms, have now been sequenced. Analysis of the sequences has affirmed their divergence during evolution, to give rise to the present-day division of living organisms into the Bacteria, Archaea, and Eukarya. In general, completed genome sequences are analyzed by computer to identify ORFs and the array of gene functions they encode.

Insights from Genome Analysis: Genome Sizes and Gene Densities

Having a number of genomes sequenced makes it possible to compare genome organizations, particularly with respect to the arrangement of genes and intergenic regions.

In Chapter 2, we discussed the C value paradox, according to which there is no direct relationship between the C value—the amount of DNA in the haploid genome—and the structural or organizational complexity of the organism. This is an old concept based on measuring the amount of DNA in the nuclei of haploid cells. With genome sequences, we can see that there are differences in genome organization that are responsible for this paradox, including the gene density—the number of genes for a given length of DNA.

The genome sizes, estimated number of genes, and gene densities for selected Bacteria, Archaea, and Eukarya are shown in Table 10.1. What follows is an overview of the organizations of the genomes of each of these kingdoms.

Genomes of Bacteria. Bacterial genomes vary over quite a large range. Of the completely sequenced bacterial genomes, that of *Mycoplasma genitalium* is the smallest, with a size of 0.58 Mb. *Bradyrhizobium japonicum* has the largest, with a size of 9.11 Mb. (See Table 10.1.)

Bacterial genomes have similar gene densities of 1 gene per 1–2 kb. For example, *Mycoplasma genitalium*'s 0.58-Mb genome has 523 genes, for a density of 1 gene per 1.15 kb, and *E. coli*'s 4.6-Mb genome has 4,397 genes, for a density of 1 gene per 1.05 kb. The spaces between genes are relatively small (110–125 bp for *Mycoplasma genitalium*), meaning that the genes are very densely packed in the genome. In fact, it is typical of Bacteria and of Archaea that approximately 85–90 percent of their genomes consist of coding DNA.

Genomes of Archaea. Most Archaea are extremophiles, meaning that they thrive in extreme environments. Archaea are found in such conditions as very high temperature, high pressure, extreme pH, high metal ion concentration, and high salt. Members of the Archaea resemble Bacteria morphologically, occurring with shapes such as spheres, rods, and spirals. However, physiological and molecular studies suggested that they resembled Eukarya in a number of respects. Indeed, genes for DNA replication, RNA transcription, and protein synthesis machinery in Archaea more closely resemble those of Eukarya than those of Bacteria. However, there are no introns in Archaea's protein-coding genes, as there are in eukaryotic genes, but there are introns in tRNA genes, as has been found in Eukarya.

As regards the genomes as a whole, Archaeal genomes also show a wide range of sizes, from 1.56 Mb for *Thermoplasma acidophilum* to 5.75 Mb for *Methanosarcina acetivorans*. (See Table 10.1.) As with Bacteria, genes are densely packed in the genome; the two examples just given have one gene per 1.03 and 1.23 kb, respectively.

Genomes of Eukarya. The Eukarya vary enormously in form and complexity, from single-celled organisms such as yeast to multicellular organisms such as humans. There is a trend of increasing genomic DNA content with increasing complexity, although, as already mentioned, there is by no means a direct relationship. For example, the two insects *Drosophila melanogaster* (fruit fly) and *Locusta migratoria* (locust) have similar complexity, yet the 5,000-Mb locust genome is 50 times larger than that of the fruit fly and twice that of the mouse. (See Table 10.1.) Differences in gene density are involved. In

Table 10.1	Genome sizes, estimated number of genes, and gene densities for selected Bacteria, Archaea, and Eukarya		
Organism	**Genome Size (Mb)**	**Number of genes**	**Gene density (kb per gene)**
Bacteria			
Mycoplasma genitalium	0.58	523	1.11
Escherichia coli K-12	4.6	4,481	1.03
Agrobacterium tumefaciens	5.7	5,482	1.04
Bradyrhizobium japonicum	9.1	8,322	1.10
Archaea			
Thermoplasma acidophilum	1.56	1,509	1.03
Methanosarcina acetivoran	5.75	4,662	1.23
Eukarya			
Fungi			
Saccharomyces cerevisae (yeast)	12	~6,000	2.0
Neurospora crassa (orange bread mold)	40	~10,100	3.8
Protozoa			
Tetrahymena thermophila	220	>20,000	11
Invertebrates			
Caenorhabditis elegans (nematode)	97	19,000	5
Drosophila melanogaster (fruit fly)	180	13,700	13
Vertebrates			
Fugu rubripes (pufferfish)	400	>31,000	13
Mus musculus (mouse)	2,600	~29,000	90
Rattus norvegicus (rat)	2,750	~30,200	91
Homo sapiens (human)	2,900	~25,000	116
Plants			
Arabidopsis thaliana	125	25,500	4.9
Oryza sativa (rice)	430	>45,000	9.6

this particular example, there is one gene every 13 kb in the fruit fly genome and, if there are a similar number of genes in the locust genome (the number is not known at present), there is one gene every 365 kb in the locust, a substantial difference in gene density.

In general, gene density in the Eukarya is lower and shows more variability than in Bacteria and Archaea. (See Table 10.1.) The Eukarya show a great range in gene density, although with a definite trend of increasing gene density with increasing complexity. Figure 10.7 illustrates the differences in gene density in yeast, fruit flies, and humans and compares them with that of *E. coli*. Yeast has a gene density which is closest to that of prokaryotes: 1 gene per 2 kb versus 1 gene per 1.03 kb for *E. coli*. Compared with yeast, the fruit fly has a 7-fold, and humans have a 56-fold, lower gene density. Organisms with genomes larger than that of humans are assumed to have lower gene densities than humans.

Of course, the gene density values given are averages. In any particular organism, there will be stretches of chromosomes with significantly more genes than average—*gene-rich regions*—and stretches with significantly fewer genes than average—*gene deserts*. In humans, for example, the most gene-rich region of the genome has 60 genes in a 700-kb segment, and the largest gene desert is 4.1 Mb long. Defining a gene desert as any 1 Mb or more without any genes, we find that there are 82 gene deserts in the human genome.

In short, humans and other complex organisms have a minority of their genomes dedicated to genes, the remainder being intergenic regions. In humans at least, the majority of the intergenic sequences consist of repetitive DNA. (See Chapter 2, pp. 36–37.) With such a gene-sparse genome, it is difficult, and sometimes impossible, to find genes of interest. Another vertebrate with high gene density may help with this problem. That

Figure 10.7

Regions of the chromosomes of *E. coli*, yeast, fruit fly, and human, showing the differences in gene density.

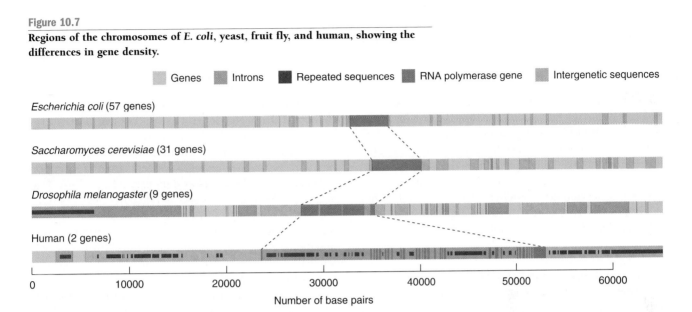

| | Genes | Introns | Repeated sequences | RNA polymerase gene | Intergenetic sequences |

Escherichia coli (57 genes)

Saccharomyces cerevisiae (31 genes)

Drosophila melanogaster (9 genes)

Human (2 genes)

Number of base pairs

vertebrate is *Fugu rubripes*, the pufferfish (Figure 10.8), the genome of which has recently been sequenced. *Fugu* is a spotted fish that puffs up into a ball when threatened. Particularly in Japan, this fish is a delicacy. It has a tangy taste, but brings with it risk: Not prepared properly, it can paralyze and kill. As Table 10.1 shows, *Fugu* has a genome size of 365 Mb, about eightfold smaller than that of humans, but with an estimated gene number similar to that of humans. In other words, the gene density of *Fugu* is eight times that of humans. In part, this is because of smaller and fewer introns in genes, so homologous genes in humans tend to be much larger. It also is because there is very little repetitive DNA and hence much less intergenic DNA present. The smaller gene density makes *Fugu* DNA much easier to work with than human DNA. Also, because many of the *Fugu* genes are homologous to human genes, once genes are identified in *Fugu*, the homologous genes in humans can be identified and studied. Scientists are hopeful that decoding the func-

tions of pufferfish genes will aid in the identification of human genes that may cause disease.

KEYNOTE

Genome sequences are resources that inform us about the number and organization of genes in different organisms. Genomes show a trend of increasing DNA amount with increasing complexity of the organism, although the relationship is not perfect. Genes make up most of the genomes in Bacteria and Archaea; that is, gene density is very high in these organisms. In Eukarya, there is a wide range of gene densities, with a trend of decreasing gene density with increasing complexity.

Functional Genomics

The successes of the HGP have empowered researchers working with a wide range of organisms to obtain genome sequences for those organisms. Research questions can now be asked at the genomic level about gene expression, physiology, development, and so on. In other words, the ability to sequence genomes efficiently and quickly has changed how research in biology, and in genetics in particular, is being done.

Of course, the complete genome sequence for an organism is just a very long string of letters. The sequence must be analyzed in detail. One important research direction is to describe the functions of all the genes in the genomes, including gene expression and control. This task defines the field of functional genomics. The difficulty in assigning gene function is that going from gene sequence to function is the direction opposite of that classically taken in genetic analysis, in

Figure 10.8

The pufferfish, *Fugu rubripes*.

which researchers start with a phenotype and set out to identify and study the genes responsible. Present-day functional genomics relies on laboratory experiments by molecular biologists and sophisticated computer analyses by researchers in the rapidly growing field of **bioinformatics,** which fuses biology with mathematics and computer science. Bioinformatics is used for many things, including finding genes within a genomic sequence; aligning sequences in databases to determine the degree of matching; predicting the structure and function of gene products; describing the interactions between genes and gene products at a global level within the cell, between cells, and between organisms; and postulating phylogenetic relationships for sequences.

Identifying Genes in DNA Sequences

The next step after obtaining the complete sequence of a genome is *annotation*—the identification and description of putative genes and other important sequences. Annotation begins the process of describing the functions of all genes of an organism. Of particular interest are the protein-coding genes, and we focus our attention on them here.

Procedurally, annotation involves using computer algorithms to search both DNA strands of the sequence for protein-coding genes. Putative protein-coding genes are found by searching for ORFs—that is, start codons (AUG) in frame (separated by a multiple of three nucleotides) with a stop codon (UAG, UAA, or UGA). This process is straightforward with prokaryotic genomes, because they have no introns. However, the presence of introns in many eukaryotic protein-coding genes necessitates the use of more sophisticated algorithms designed to include the identification of junctions between exons and introns in scanning for ORFs.

ORFs of all sizes are found in the computer scan, so a size must be set below which it is deemed unlikely that the ORF encodes a protein in vivo, and it is not analyzed further. For the yeast genome, for instance, the lower limit was set to 100 codons. However, a few genes may be below this limit, and not all ORFs above 100 codons encode proteins. The plasma membrane proteolipid gene, *PMP1*, for instance, encodes a protein of only 40 amino acids. It is estimated that, of the 6,183 ORFs in the yeast genome, 6 to 7 percent do not correspond to real genes, leaving 5,800 actual protein-coding genes.

Sequence Similarity Searches to Assign Gene Function

The function of an ORF identified in genome scans may be assigned by searching databases for a sequence match with a gene whose function has been defined. Such searches are called *sequence similarity searches* and involve computer-based comparisons of an input sequence with all sequences in the database. The searches

can be done using an Internet browser to access the computer programs. For example, the BLAST program at the National Center for Biotechnology Information (http://www.ncbi.nlm.nih.gov/) enables a user to paste the sequence to be studied into a window, either in nucleotide form or in amino-acid form, and to get results indicating the degree to which the sequence of interest is similar to sequences in the database.

Similarity searching is an effective way to assign gene function because homology—descent from a common ancestor—is a reflection of evolutionary relationships. That is, a pair of homologous genes in different organisms has a common evolutionary ancestor, so the two genes' nucleotide sequences are similar; their differences have resulted from mutational changes over evolutionary time. Thus, if a newly sequenced gene (e.g., from a genome sequence project) is similar to a previously sequenced gene, the two genes are related in an evolutionary sense, so the function of the new gene probably is the same as, or at least similar to, the function of the previously sequenced gene.

Sequence similarity searching can be done with either a nucleotide (DNA) sequence or an amino-acid sequence, but the latter is preferred because, with 20 different amino acids and only four different nucleotides, unrelated genes appear more different from one another at the amino-acid level than at the nucleotide level. Given the information in current databases, there is less than a 50 percent chance that a new gene sequence will match a gene sequence in the database with a high enough degree of similarity to infer the new sequence's function.

A sequence similarity search can indicate a match either for the whole protein sequence or for parts of it. In the latter case, this means that a domain of the new gene product matches a domain of a previously identified gene product, so at least part of the new protein's function can be inferred. Evolutionarily speaking, such a result implies that the domains have a common ancestor, but the genes as a whole may not.

Sequence similarity searching plays an important part in assigning gene function. For example, Figure 10.9 shows the distribution of ORFs in the yeast genome. About 30 percent of the genes were known as a result of standard genetic analysis before genome sequencing, including direct assays for function. The remaining 70 percent of the ORFs are genes whose functions have not been directly determined. The latter break down as follows: Thirty percent encode a protein that is related to functionally characterized proteins or has a domain related to domains in functionally characterized proteins. Another 10 percent have homologs in databases, but the functions of those homologs are unknown. Such yeast ORFs are called *FUN* (function *un*known) genes, and those genes and their homologs are called *orphan families*. The remaining 30 percent of ORFs have no homologs in the databases. Within this class are the 6 to 7 percent of

Figure 10.9

The distribution of predicted ORFs in the genome of yeast.
(From B. Dijon. 1996. *Trends Genet* 12:263–270.)

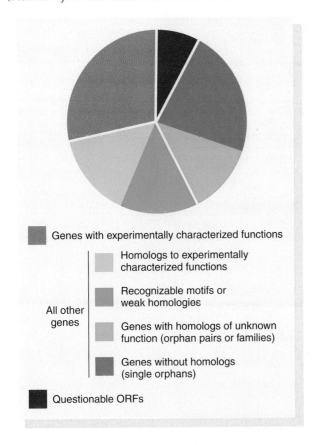

Genes with experimentally characterized functions

Homologs to experimentally characterized functions

Recognizable motifs or weak homologies

Genes with homologs of unknown function (orphan pairs or families)

Genes without homologs (single orphans)

All other genes

Questionable ORFs

ORFs that are questionable in terms of being real genes. The remainder probably are real genes, but at present are unique to yeast; they are called *single orphans*.

The problem of *FUN* genes applies to the genomes of other organisms, both prokaryotic and eukaryotic. As more and more genes with defined functions are added to the databases, the percentage of ORFs with no matches to database sequences is decreasing.

Assigning Gene Function Experimentally

One key approach to assigning gene function experimentally is to knock out the function of a gene and determine what phenotypic changes occur. This is done by deleting the gene, or making a *gene knockout*. Figure 10.10a shows how to do so in yeast, using a PCR-based strategy. Using PCR primers designed on the basis of the known genome sequence, an artificial linear DNA deletion module is constructed and amplified. This module consists of part of the gene sequence upstream of and including the start codon and part of the gene sequence downstream of and including the stop codon, flanking a DNA fragment containing the *kan*R (kanamycin) selectable marker that confers resistance to the inhibitory chemical G418. This linear DNA is trans-

formed into yeast, and G418-resistant colonies are selected. The desired transformants are those generated when the fragment replaces the target ORF by homologous recombination. The strategy completely inactivates—knocks out—the gene, because most of it is replaced. In genetic terms, a *loss-of-function mutation* or a *null allele* is produced.

A molecular screen must be used to confirm that the transformant has resulted in deletion of the ORF of interest. PCR is used for this screen, as illustrated in Figure 10.10b. First, let us consider the condition of an unsuccessful deletion in which the ORF is still present (Figure 10.10b1). Four different PCR primers, A–D, are used. Primers A and D are 200 to 400 bases upstream and downstream, respectively, of the ORF. Primers B and C are from within the ORF itself. DNA is isolated from transformants, and separate PCRs are done with primers A and B, on the one hand, and primers C and D, on the other. If the ORF is still present, these reactions produce DNA fragments of predictable sizes. If the ORF is deleted, no PCR products are seen. However, it is still necessary to show definitively that the deletion has been made, and the scheme for doing so is shown in Figure 10.10b.2. Primers A and D are as in Figure 10.10b.1, and there are two other primers—KanB and KanC—that are specific to the *kan*R DNA fragment. If deletion has been successful, the *kan*R module has replaced the ORF and PCR, using primers A and KanB, and primers KanC and D generate fragments of predictable sizes.

Using the gene deletion approach, a yeast knockout (YKO) project has been completed in which each yeast gene has been systematically deleted. Because some genes have essential functions, deleting them gave a lethal phenotype. However, about 4,200 of the approximately 6,200 genes are nonessential, since knocking each of them out individually results in a viable phenotype. This set of 4,200 strains in the yeast deletion collection is a genomic resource for investigating the functions of nonessential genes in the organism. For example, the deletion strains are being studied under various conditions for changes in phenotype to assign function. The work involved is substantial, because of the many areas of cell function that must be screened for a change in phenotype, including cell cycle events, meiosis, DNA synthesis, RNA synthesis and processing, protein synthesis, DNA repair, energy metabolism, and molecular transport mechanisms. From such work, it has been shown that approximately one-half of the organism's genes show no significant changes in phenotype, and the other one-half do.

Null alleles are important resources in investigations of gene function. Therefore, knockout projects are routinely done on experimental organisms that are amenable to a gene deletion method. Gene knockouts in mice, for example, are being used as models to identify the functions of unknown human genes, because it is unethical to knock out human genes.

Figure 10.10

Creating and verifying a gene knockout in yeast. (a) Schematic of a PCR-based gene deletion strategy involving a DNA fragment constructed by PCR from gene sequences flanking the *kan*^R selectable marker that is transformed into yeast and replaces the chromosomal ORF by homologous recombination. (b) Verification of gene deletion. PCR-based screening method to confirm (1) unsuccessful deletion (ORF still present) and (2) successful deletion (ORF replaced with *kan*^R DNA segment).

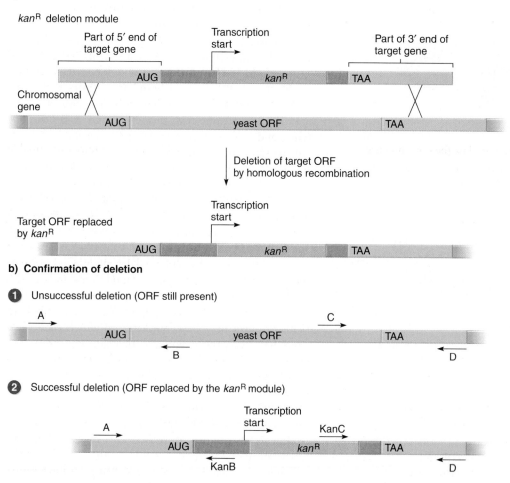

Describing Patterns of Gene Expression

In classical genetic analysis, research begins with a phenotype and leads to the gene or genes responsible. Once the gene is found and isolated, experiments can be done to study the expression of the gene in normal and mutant organisms as a way to understand the role of the gene in determining the phenotype. When the complete genome sequence is obtained for an organism, exciting new lines of research are possible, including the analysis of the expression of *all* genes in a cell at the transcriptional and translational levels and the analysis of all protein–protein interactions. Measuring the levels of RNA transcripts (usually focusing on mRNA transcripts), for example, gives us insight into the global gene expression state of the cell. To go along with this new research, a new term has been coined for the set of mRNA transcripts in a cell: the **transcriptome.** Because the mRNAs specify the proteins that are responsible for cellular function, the transcriptome is a major indicator of cellular phenotype and function. By extension, the complete set of proteins in a cell is called the **proteome.** Studies of the transcriptome and the proteome are described in the remainder of this section.

The Transcriptome. The transcriptome is not stable in a cell. Rather, the types and levels of mRNA transcripts change significantly as a cell responds to its environment and as it proceeds through DNA replication and cell division. By defining exactly which genes are expressed, when they are expressed, and their levels of expression, we can begin to understand cellular function at a global level.

Probe arrays (see Chapter 9, pp. 216–217) are among the most powerful tools for global gene expression experiments. For example, a number of researchers are using probe arrays to study gene expression in yeast. In a collaborative effort, Pat Brown and Ira Herskowitz studied yeast sporulation, the process of producing haploid spores by meiosis (Figure 10.11a). Yeast sporulation involves four major stages: DNA replication and recombination, meiosis I, meiosis II, and spore maturation. The sequential transcription of at least four classes of genes—early, middle, mid–late, and late—correlates with these stages. When Brown and Herskowitz's probe array experiments began, about 150 genes that are differentially expressed during sporulation had been identified. In the new research, the researchers induced diploid yeast cells to sporulate, and at seven timed intervals, they took cell samples and used DNA microarrays containing 97 percent of the known or predicted yeast genes to analyze the temporal program of gene expression during meiosis and spore formation. Both light and electron microscopy were utilized to correlate the sampling time with the exact stage of sporulation. To quantify gene expression, Brown and Herskowitz isolated mRNAs from the cell samples and synthesized fluorescently labeled cDNAs by reverse transcription in the presence of Cy5 (red)-labeled dUTP (Figure 10.11b). For a nonsporulating cell control, they isolated mRNAs from cells at a time point immediately before inducing sporulation, and they synthesized fluorescently labeled cDNAs at this time, using Cy3 (green)-labeled dUTP. For each time point, they hybridized a mixture of reference green-labeled cDNAs and experimental red-labeled cDNAs to DNA microarrays, made by using PCR to amplify each ORF (using primers based on the genome sequence) and printing the sequences onto a glass slide using a robotic printing device. After hybridization, they scanned the microarrays with a laser detector to quantify the red and green fluorescence locations and intensities. Figure 10.11c shows an example of the results obtained. The relative abundance of transcripts from each gene in sporulating versus nonsporulating yeast cells is seen by the ratio of red to green fluorescence. If an mRNA is more abundant in sporulating cells than in nonsporulating cells, as is the case for the *TEP1* gene (see Figure 10.11c), the result is a higher ratio of red-labelled to green-labelled cDNAs prepared from the two types of cells and, therefore, in the same higher ratio of red to green fluorescence detected on the array. In general, a gene whose expression is induced by sporulation is seen as a red spot, and a gene whose expression is repressed by sporulation is seen as a green spot. Genes that are expressed at approximately equal levels in nonsporulating cells and during sporulation are seen as yellow spots.

With this approach, the researchers found that more than 1,000 yeast genes showed significant changes in mRNA levels during sporulation. About one-half of the genes are repressed during sporulation, and one-half are not repressed. At least seven distinct temporal patterns of gene induction are seen, and this observation is providing some insights into the functions of many orphan genes.

Using DNA microarrays is becoming widespread, despite their high cost. For example, DNA microarrays are being used to study part of the transcriptional program for *Drosophila* during metamorphosis and to analyze the expression profiles of 40,000 human genes during cell division. DNA microarrays are also being used to study cancer. For example, by analyzing global gene expression in normal individuals and in individuals with a particular type of cancer, it is theoretically possible to identify characteristic gene expression patterns that could be used to classify the cancer and to screen for its onset. The former has already proved possible for diffuse large B-cell lymphoma, where it has been shown, on the basis of gene expression profiles (*transcriptional "fingerprints"*), that there are previously unknown distinct types of the cancer.

DNA microarrays are also useful for screening for genetic diseases. Of particular interest are genetic diseases that are characterized by a large number of possible mutations, making simple DNA typing methods (see Chapter 9, pp. 224–226) inefficient. For example, the genes *BRCA1* (breast cancer 1) and *BRCA2* cause approximately 60 percent of all cases of hereditary breast and ovarian cancers. However, at least 500 different mutations have been discovered in *BRCA1* that can lead to the development of cancer. Assaying for that many different mutations is well within the scope of DNA chip technology, and such chips are being developed for this and for many other diseases of similar genetic complexity. The principles for their use are similar to those mentioned in the yeast sporulation example. Blood is taken from a patient, and green-labeled DNA is produced by PCR and mixed with red-labeled DNA from a normal individual. The chip in this case consists of oligonucleotides that collectively represent the entirety of the *BRCA1* and *BRCA2* genes. If the patient has a mutation in one or other of the genes, the red (normal) DNA will hybridize to the DNA on the chip, but the green (patient's) DNA will not hybridize in the region where the mutation is located. Normal hybridization is seen as a yellow (red-green) spot, and a mutation is seen as a red spot. Because the position of the spot on the array is known, and because the oligonucleotides for each spot are known, the mutation has been localized within a very narrow region of the *BRCA1* or *BRCA2* gene and can be analyzed in more detail.

In the iActivity *Personalized Prescriptions for Cancer Patients* on your CD-ROM, you are a researcher at the Russellville clinic trying to determine the gene expression profile for a patient with cancer.

Pharmacogenomics. One highly promising area involving genome-based gene expression research is **pharmacogenomics.** The word is a blend of "pharmacology" and

Figure 10.11

Global gene expression analysis of yeast sporulation via a DNA microarray. (a) The stages of sporulation in yeast, correlated with the sequential transcription of at least four classes of genes. (Adapted from Chu et al. 1998. *Science* 282:699–705.) (b) Outline of the DNA microarray experiment. (c) Example of results of a global gene expression analysis in yeast, obtained a DNA microarray. The entire yeast genome is represented on the DNA chip, and the colored dots represent levels of gene expression, as described in the text.

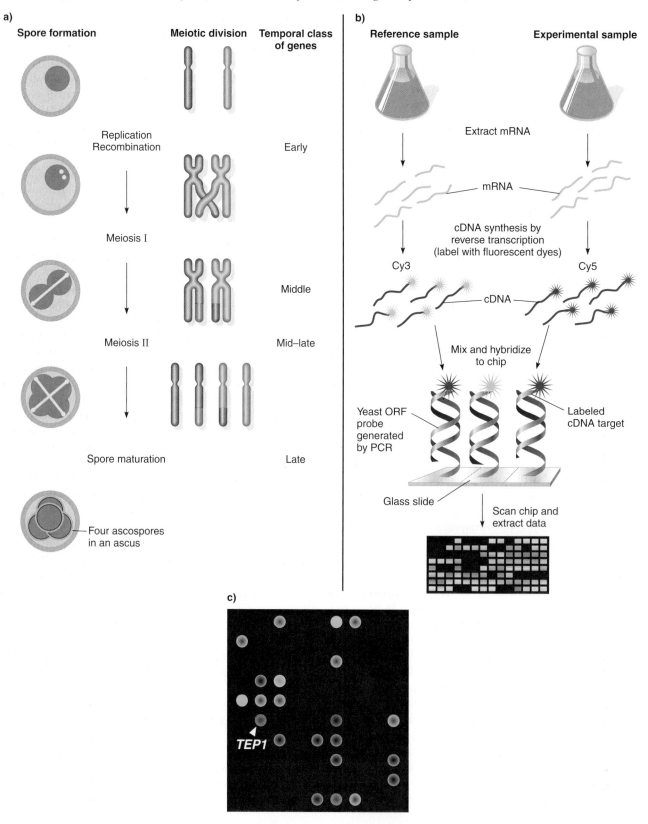

"genomics"; it denotes the study of how an individual's genome affects the body's response. That is, medicine operates mostly on the assumption that all humans are the same, and pharmaceuticals are administered to treat diseases on the basis of that assumption. However, a variety of factors affect a person's response to medicines, notably the genome (actually, the expression of that genome), as well as nongenetic factors such as age, state of health, diet, and the environment. The promise of pharmacogenomics is that drugs may be customized for individuals (i.e., adapted to each person's genome).

Research in pharmacogenomics is grounded in biochemistry (a major component of pharmaceutical science), enhanced with information about genes, proteins, and DNA polymorphisms. The goal is to develop drugs based on the RNA molecules and proteins or enzymes that are associated with genes and diseases. If successful, the drugs would be much more targeted to specific diseases than is presently the case. This would mean that the therapeutic effects of the drugs would be maximized, while, at the same, time the side effects would be minimized. Moreover, drug dosages would be tailored to an individual's genetic makeup; that is, they would take into account how and at what rate a person metabolizes a drug. Presently, dosages are decided upon largely on the basis of weight and age.

Pharmocogenomics is a very young area of research at the moment, so mostly there is a lot of promise, but very few demonstrated successes. One area of application concerns the cytochrome p450 (CYP) family of liver enzymes. These enzymes break down more than 30 different classes of drugs. However, variations in the genes that encode the enzymes result in enzymes with different abilities to metabolize particular drugs. A significant concern here is patients who have inactive or only partially active CYP enzymes, because they are susceptible to a drug overdose. Genetic tests for variations in cytochrome P450 genes are currently in clinical trials. Clearly, such tests would be valuable for adapting drug treatments to patients exhibiting this particular set of genome variations.

The Proteome. The **proteome** is the complete set of expressed proteins in a cell at a particular time. **Proteomics** is the cataloging and analysis of those proteins to determine when a protein is expressed, how much is made, and with what other proteins the protein can interact. The approaches in proteomics are mostly biochemical and molecular.

The goals of proteomics are (1) to identify every protein in the proteome, (2) to determine the sequences of each protein and to enter the data into databases, and (3) to globally analyze protein levels in different cell types and at different stages in development.

Identifying and sequencing all of the proteins from a cell is much more complex than mapping and sequencing a genome. Craig Venter's Celera Genomics is also playing a large role in this area, as it did in genomic sequencing,

working hard to speed up the identification and sequencing of proteins and computer analysis of the data. Coinciding with the publication of the human genome sequences, a global Human Proteome Organisation (HUPO) was launched. HUPO is intended to be the postgenomic analog of HUGO, with a mission to increase awareness of, and support for, proteomics research at scientific, political, and financial levels.

Proteomics is an extremely important field because it focuses on the functional products of genes, which determine the phenotypes of a cell. Of particular human interest are diseases, and proteins and peptides are closer to the actual disease process than are the genes that encode them. However, the challenges for proteomics are much greater than those for genomics. The complexity of the human proteome greatly exceeds that of the genome. Whereas there are an estimated 40,000 genes in the human genome, there may be about 500,000 different proteins. This disparity is the result of variations in gene expression, such as alternative RNA splicing producing different translatable mRNAs, as well as posttranslational modifications of proteins that affect their functions.

Conventional proteome analysis is by two-dimensional acrylamide gel electrophoresis and mass spectometry. These procedures are not well suited to analyzing large numbers of proteins at once, and they are not sensitive enough to detect proteins expressed at low levels. Fortunately, there is a new sensitive tool for analyzing large numbers of proteins at once: **protein arrays.** Similar in concept to DNA microarrays (see Chapter 9, pp. 216–217), protein arrays are rapidly becoming the best way to detect proteins, measure their levels in cells, and characterize their functions and interactions, all on a very large scale. Consequently, they are a central proteomics technology, valuable both for basic research and for biotechnology applications. As with DNA microarrays, the use of protein arrays is becoming highly automated, and this makes it possible to do large numbers of measurements in parallel.

Protein arrays—also called *protein microarrays* and *protein chips*—involve proteins immobilized on solid substrates, such as glass, membranes, or microtiter wells. At the moment, the density of proteins on the arrays is much lower than that for DNA on DNA microarrays. However, with technological advances, we can expect the density of proteins in the arrays to increase. As with DNA microarrays, target proteins are labeled fluorescently (e.g., with Cy3 and Cy5, as is used for DNA), and binding to spots on the arrays is measured by automated laser detection. The resulting complex data are analyzed by computer. Because of the similarities with DNA microarray technology, the same instrumentation used to analyze DNA microarrays can be used to analyze protein arrays.

Let us consider two types of arrays to illustrate how protein arrays can be used. One type is the *capture array*, in which a set of antibodies (usually) bound to the array surface is used to detect target molecules—for example, in cell or tissue extracts. The antibodies are made either by conventional immunization procedures or by using recombinant DNA techniques to make clones from which antibody fragments are made. A capture array can be used as a diagnostic device—for instance, to screen for infections (detecting specific proteins made by the infectious agent) or for the presence of tumors (detecting tumor-specific markers in extracts of biopsied material). In proteomics studies, capture arrays are used for protein expression profiling—that is, defining the proteome qualitatively and quantitatively. For example, one can quantify proteins in different cell types and different tissues, as well as compare proteins under different conditions, such as during differentiation, with and without a drug treatment, and with and without a disease.

Another type of array is the *large-scale protein array*. In these arrays, large numbers of purified proteins are spotted onto an array substrate and are used to assay one of a wide range of biochemical functions, including protein–protein interactions (an alternative to the cell-based yeast two-hybrid system; see Chapter 9, pp. 229–231) and drug–target interactions. The proteins for immobilization on the array substrate are produced from an expression library transformed into a host such as *E. coli* and yeast, from which the expressed proteins are purified. The target proteins are labeled, and binding to the array is as described earlier.

In sum, protein arrays are a promising new technology. They are in their early stages of development at the moment, so there are various limitations and bottlenecks in their use. In the future, we can expect protein arrays to "take off" and become routine for high-throughput analysis of proteins in proteomic studies. Their use will further our understanding of the proteome greatly.

KEYNOTE

> The goal of functional genomics is to define the functions of all the genes in a genome of a particular species, including the patterns and control of gene expression. Gene expression is analyzed at two levels: mRNA transcripts, called the transcriptome, and proteins, called the proteome. Functional genomics relies on laboratory experiments and computer analysis. DNA microarrays are used, for example, to generate a gene expression profile for a cell in which the mRNA transcripts are described qualitatively and quantitatively.

Comparative Genomics

Comparative genomics involves comparing entire genomes of different species, with the goal of enhancing our understanding of the functions and evolutionary relationships of each genome. Comparative genomics is rooted in the tenet that all present-day genomes have evolved from common ancestral genomes. Therefore, studying a gene in one organism can provide meaningful information about the homologous gene in another organism, and, more globally, comparing the overall arrangements of genes and nongene sequences of different organisms can tell us about the evolution of genomes. Since direct experimentation with humans is unethical, comparative genomics provides a valuable way to determine the functions of human genes by studying homologous genes in nonhuman organisms. Identifying and studying homologs to human disease genes in another organism is potentially valuable for developing an understanding of the biochemical function and malfunction of the human gene.

Like functional genomics, comparative genomics focuses on the genome level. Comparative genomics involves analyzing genomes from two or more species, with the goal of defining the extent and specifics of similarities and differences between sequences—either gene sequences or nongene sequences. An obvious issue that comparative genomics can address is the evolutionary relationships between two or more genomes. For example, as we discussed earlier, complete genome sequence analysis affirmed the evolutionary relationships and distinctions among the Eubacteria, Archaea, and Eukarya. Comparative genomics is discussed more in Chapter 25.

KEYNOTE

> Comparative genomics is the comparison of complete genomes of different species, with the goal of increasing our understanding of the gene and nongene sequences of each genome and their evolutionary relationships.

Ethics and the Human Genome Project

The Human Genome Project is raising ethical issues. With the entire human genome sequence in hand, we will be able to identify and isolate all human genes that cause diseases. This will lead to the development of tests for many gene defects that we cannot test for now, including those which may lead to the development of a disease later in life, such as cancer. However, these tests will become available before we have developed a cure for those diseases, as is already the case with most testable genetic diseases. A number of ethical questions arise from this scenario. Should a patient be told if a test for an incurable genetic disease is positive? (This issue applies now for Huntington disease, a dominant lethal disease in which the symptoms typically do not appear until later in life.) Should employers be able to ask for results of a genetic test if the employee does not want to know? Should health insurance companies or employers have access to genetic testing data, and, if so, how can the patient protect his or her insurability and employability? Should states be able to collect genetic data on their populace? The last two questions raise fundamental privacy issues.

Fortunately, these issues are not being ignored. The federal agencies funding the HGP are devoting 3 to 5 percent of their annual budgets to study the ethical, legal, and social issues (ELSIs) related to the availability of genetic information. (See http://www.ornl.gov/sci/techresources/ Human_Genome/elsi/elsi.shtml.) This amounts to the world's largest bioethics program. Four areas are being emphasized by the ELSI program: (1) the privacy of genetic information; (2) the safe and effective introduction of genetic information in the clinical setting; (3) fairness in the use of genetic information; and (4) professional and public education. Appropriate laws and regulations are expected to be developed as a result of the activities of the ELSI program and of continuing dialogues among scientists, physicians, lawmakers, and members of the public.

Summary

With the development of recombinant DNA techniques, it became possible to clone individual genes and therefore to broaden the questions one could ask about phenotypes. With genomic analysis, the scope of questions becomes much broader. That is, one can now ask global questions about gene expression and begin to understand in detail how a cell or an organism functions, rather than how a gene functions.

The analysis of the entire genome of species of organisms defines the field of genomics. Structural genomics, a subfield of genomics, involves the genetic mapping, physical mapping, and sequencing of entire genomes. Two general approaches have been taken to sequence genomes. The mapping approach, conceived originally for large genomes, involves creating genetic maps and physical maps with increasing levels of resolution and then sequencing the segments. The direct shotgun approach, originally shown to be effective with small genomes, involves making a partially overlapping library of genomic DNA fragments, sequencing each clone, and assembling the genomic sequence by computer on the basis of the sequence overlaps. With the increased power of computers to handle large amounts of sequence data, the direct shotgun approach can now be used with large genomes, including that of humans.

Although the original focus of the Human Genome Project was on the genome of humans and those of a few model organisms, it seems that everyone working with a particular organism is interested in sequencing its genome. The genomes of many viruses and living organisms have now been sequenced completely, and many more are in process. Genomic analysis has affirmed the division of living organisms into Bacteria, Archaea, and Eukarya. In April 2003, the finished sequence of the human genome was reported. Through analyses of the sequences involved, we are learning a great deal about the organization of genes in the chromosomes. A general trend is that gene density decreases with the complexity of the organism. That is, in Bacteria and Archaea, gene density is very high, whereas in complex Eukarya, such as humans, it is relatively low.

Through functional genomic analysis, the genes of each species' genome are identified and their patterns of expression described. Functional genomics involves both laboratory analysis and computer analysis (bioinformatics) to identify genes in DNA sequences (annotation), determine function by homology searches and by experimental means such as gene knockouts, and describe the patterns of gene expression at the mRNA and protein levels. The set of mRNA transcripts in a cell is called the transcriptome, and the complete set of proteins in a cell is called the proteome. The transcriptome is specific to the metabolic state of the cell, so by defining the transcriptome qualitatively and quantitatively, we can begin to understand cellular function at a global level. An extremely valuable tool for studying the transcriptome is the DNA microarray.

Since proteins govern the phenotypes of a cell, the study of the proteome—proteomics—will provide much more information about cellular function at a global level. At the moment, though, the tools for analyzing the transcriptome are much more sophisticated than those for analyzing the proteome, although the rapid development of protein arrays (also called protein chips) as a tool for studying proteins and protein interactions is beginning to enrich our information about the proteome significantly.

Comparative genomics, another subfield of genomics, involves the comparison of entire genomes of different species, related or not. The goal of comparative genomics is to enhance our understanding of the functions of each genome for both gene and nongene sequences and to develop an understanding of evolutionary relationships. We can make conclusions about homologous genes in species of different organisms because all present-day genomes have evolved from common ancestral genomes. Comparative genomics will be particularly important for studies of the human genome because direct human experimentation is unethical. Information about a gene in closely related organisms will inform us about the function of that gene in humans.

Finally, the future of human genomics raises ethical issues. Conceivably, we can all have our own genome completely sequenced and analyzed and deposit that information in a database or even on a chip we carry. Our genome sequences will reveal, among other things, whether we have genetic diseases, whether we have the potential to develop a genetic disease or cancer, and whether we have some mental or physical aberration that might influence our life or work. Such information could affect a individual's ability to obtain life insurance or health insurance, or affect his or her job potential. Thus, fundamental privacy issues will have to be addressed as this brave new world of genomics marches forward rapidly.

Analytical Approaches to Solving Genetics Problems

Q10.1 YAC clone contigs can be assembled following STS mapping. First, the locations of STSs on YAC clones are mapped using PCR. Then YAC clones that share STSs are aligned. The first step usually involves repeated screening of a library of YAC clones with different STS markers. In practice, some of the STSs may be well characterized and even localized to specific chromosomal regions or genes, whereas others may be less well characterized. In a typical screen of a YAC library constructed from the genome of a higher eukaryote, most STSs identify only a few YACs. However, a few STSs identify dozens of YACs. What might be the basis of this difference? How does the difference influence the assembly of a contig?

A10.1 An STS marker is a sequence-tagged site—a DNA sequence marker that should identify a unique sequence present at one chromosomal site. For a well-characterized STS, this site may be known and may correspond to a specific chromosomal region or even a specific gene. When a YAC library is screened with an STS marker, all YACs with its sequence are identified. If two STS markers located close together on a single chromosome are each used to screen a YAC library, clones that contain just one or both STS markers can be found. Those clones containing both markers have overlapping, shared DNA sequences, whereas clones that contain just one of the two markers do not overlap in the vicinity of the markers.

This question identifies a practical concern seen when STS markers are used to assemble contigs. Whereas most STSs identify a few YACs, a few identify many YACs. A careful reading of the problem provides two hints as to why this might occur. First, it occurs during the assembly of a YAC clone contig *in a higher eukaryote*. Second, some of the STS markers may not be well characterized. How might each of these facts be related to the observation that

a few STSs identify a large number of YACs? A connection can be made between a larger number of YACs identified with some STSs and the presence of repetitive DNA sequences in higher eukaryotes. Although higher eukaryotes have DNA sequences that occur once and are unique, they also have repetitive DNA sequences. Some are interspersed among the unique sequences, and some are clustered in heterochromatic regions. In this case, if an STS marker were not well characterized, it might be derived from a repetitive sequence instead of a unique sequence. Then it would appear multiple times in the genome. When such an STS is used to screen a YAC library, it would identify all YACs containing the repetitive sequence. Because a repetitive sequence could be distributed over many different chromosomal regions, such an STS would identify many more YACs than could a well-characterized STS that identified a unique sequence. Also, because YACs containing a repetitive STS could derive from different chromosomal regions, their inserts would not necessarily overlap. Consequently, STSs that identify an unusually large number of YAC clones cannot be used to assemble a YAC-clone contig reliably, and data gathered with them must be set aside.

Questions and Problems

10.1 What are STS markers, and in what different ways are they used in structural genomics?

10.2 Discuss the relationship between STSs and ESTs, addressing whether all STSs are capable of being ESTs, and whether all ESTs are capable of being STSs.

*10.3 The average size of fragments, in base pairs, observed after genomic DNA from eight different species was individually cleaved with each of six different restriction enzymes, is shown in Table 10.A:

Table 10.A

Species	Enzyme and Recognition Sequence					
	ApaI GGGCCC	**HindIII** AAGCTT	**SacI** GAGCTC	**SspI** AATATT	**SrfI** GCCCGGGC	**NotI** GCGGCCGC
Escherichia coli	68,000	8,000	31,000	2,000	120,000	200,000
Mycobacterium tuberculosis	2,000	18,000	4,000	32,000	10,000	4,000
Saccharomyces cerevisiae	15,000	3,000	8,000	1,000	570,000	290,000
Arabidopsis thaliana	52,000	2,000	5,000	1,000	No sites	610,000
Caenorhabditis elegans	38,000	3,000	5,000	800	1,110,000	260,000
Drosophila melanogaster	13,000	3,000	6,000	900	170,000	83,000
Mus musculus	5,000	3,000	3,000	3,000	120,000	120,000
Homo sapiens	5,000	4,000	5,000	1,000	120,000	260,000

a. Under the assumption that each genome has equal amounts of A, T, G, and C, and that on average these bases are evenly distributed, what average fragment size is expected following digestion with each enzyme?

b. How might you explain each of the following?

i. There is a large variation in the average fragment sizes when different genomes are cut with the same enzyme.

ii. There is a large variation in the average fragment sizes when the same genome is cut with different enzymes that recognize sites having the same length (e.g., *Apa*I, *Hind*III, *Sac*I, and *Ssp*I).

iii. Both *Srf*I and *Not*I, which each recognize an 8-bp site, cut the *Mycobacterium* genome more frequently than *Ssp*I and *Hind*III, which each recognize a 6-bp site.

c. Based on this data, which enzymes would be good choices for constructing a restriction map of a chromosome (or a large segment of a chromosome) in each organism? Explain your choices.

*10.4 STS mapping has been useful to generate clone contig maps. Perform the following exercise to consider the logistics of locating STSs using PCR.

a. A plasmid library contains 500-bp inserts generated from randomly sheared mouse DNA. How would you identify clones harboring STRs with the dinucleotide repeat $(AT)_N$ or the trinucleotide repeat $(CAG)_N$

b. How would you use these STRs as STSs in a mapping experiment?

c. Nusbaum and colleagues generated a YAC-based physical map of the mouse genome by localizing 8,203 STSs onto 960 YAC clones. If each of the STSs were assayed in each of the 960 YAC clones, how many different PCRs would need to be analyzed?

d. Although PCRs can be performed robotically, each reaction consumes time and material resources, and with each there is a certain chance of a false positive result or other error. It is therefore advantageous to reduce the number of PCR reactions by pooling individual YACs together. First, yeast colonies containing the YACs are grown individually in the wells of ten 96-well plates. Suppose the plates are numbered I, II, . . . X and the wells of each plate are arrayed in an 8-row × 12 grid. The rows are coded A–H and the columns coded 1–12. This allows the position of a single YAC to be specified uniquely by a code (e.g., II-C6 specifies the YAC from plate II, row C, column 6). In one pooling scheme, all YACs from each row of a plate are pooled into one well (e.g., those on plate II, A1–A12 are pooled together into a well designated II-A) and all YACs from each

column of a plate are pooled into one well (those on plate II, A7–H7 are pooled together into a well designated II-7).

i. How many different pools would now have to be screened with each STS?

ii. How many PCRs would have to be performed?

iii. If the II6 and IIF pools had a positive result with STS #6239, what is the code of the YAC containing this STS?

iv. How would you interpret a result where only the IV3 pool was positive for a particular STS?

e. Construct a YAC contig based on the results in the following table.

STS Marker	Positive YAC Pools
63	II-6, II-A
210	II-6, II-A, IV-C, IV-3
522	VII-E, VII-12, X-G, I-C, I-8
713	I-C, I-8
714	VII-E,VII-12
719	X-H, X-9, IV C, IV-3
991	X-H, X-9, VII-E, VII-12
1071	II-6, II-A, IV-C, IV-3, X-H, X-9
2631	II-6, II-A
3097	VII-E, VII-12, I-C, I-8
4630	VII-E, VII-12, I-C, I-8
5192	X-H, X-9, IV-C, IV-3
6193	X-H, X-9, VII-E, VII-12
6892	II-6, II-A, IV-C, IV-3

f. Devise a method to combine the YACs pooled in (c) to further reduce the number of PCRs. In your method, how many pools are there and how many PCRs must be performed?

*10.5 BACs I–VI have been aligned in the contig shown in the following figure after screening a BAC library with a series of numbered STS and EST markers.

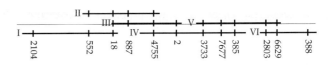

a. During the construction of BAC libraries, genomic DNA segments from different chromosomal regions sometimes are cloned together into a single BAC clone. This generates a chimeric BAC. How would you verify that none of the BACs in this contig was chimeric?

b. How would you identify the distance (in bp) of the markers in each BAC from each other and from the end of the insert?

10.6 Genomes can be sequenced using a whole-genome shotgun approach or a mapping approach.

a. What is the difference between these approaches?

b. Estimate the number of sequencing reactions needed to obtain 97 percent of the sequence of the human genome using the approach taken by the Celera group.

*10.7 When Celera Genomics sequenced the human genome, they obtained 13,543,099 reads of plasmids having an average insert size of 1,951 bp, and 10,894,467 reads of plasmids having an average insert size of 10,800 bp.

a. Dideoxy-chain termination sequencing provides only about 500–550 nucleotides of sequence. About how many nucleotides of sequence did they obtain from sequencing these two plasmid libraries?

b. Why did they sequence plasmids from two libraries with different-sized inserts?

c. They only sequenced the ends of each insert. How did they determine the sequence lying between the sequenced ends?

*10.8 Eukaryotic genomes differ in their repetitive DNA content. For example, consider the typical euchromatic 50-kb segment of human DNA that contains the human β T-cell receptor. About 40 percent of it is composed of various genome-wide repeats, about 10 percent encodes three genes (with introns), and about 8 percent is taken up by a pseudogene. Compare this to the typical 50-kb segment of yeast DNA containing the *HIS4* gene. There, only about 12 percent is composed of a genome-wide repeat, and about 70 percent encodes genes (without introns). The remaining sequences in each case are untranscribed and either contain regulatory signals or have no discernible information. Whereas some repetitive sequences can be interspersed throughout gene-containing euchromatic regions, others are abundant near centromeres. What problems do these repetitive sequences pose for sequencing eukaryotic genomes? When can these problems be overcome, and how?

*10.9 How has genomic analysis provided evidence that Archaea is a branch of life distinct from Bacteria and Eukarya?

10.10 What is bioinformatics, and what is its role in structural, functional, and comparative genomics?

10.11 What is the difference between a gene and an ORF? How might you identify the functions of ORFs whose functions are not yet known?

*10.12 Once a genomic region is sequenced, computerized algorithms can be used to scan the sequence to identify potential ORFs.

a. Devise a strategy to identify potential prokaryotic ORFs by listing features assessable by an algorithm checking for ORFs.

b. Why does the presence of introns within transcribed eukaryotic sequences preclude direct application of this strategy to eukaryotic sequences?

c. The average length of exons in humans is about 100–200 base pairs while the length of introns can range from about 100 to many thousands of base pairs. What challenges do these findings pose for identifying exons in uncharacterized regions of the human genome?

d. How might you modify your strategy to overcome some of the problems posed by the presence of introns in transcribed eukaryotic sequences?

10.13 Annotation of genomic sequences makes them much more useful to researchers. What features should be included in an annotation, and in what different ways can they be depicted? For some examples of current annotations in databases, see the following websites:

http://www.yeastgenome.org/ (*S. cerevisiae*)
http://www.flybase.org and http://flybase.net/annot/ (*Drosophila*)
http://www.tigr.org/tdb/e2k1/ath1/ (*Arabidopsis*)
http://www.ncbi.nlm.nih.gov/genome/guide/human/ (humans)
http://genome.ucsc.edu/cgi-bin/hgGateway (humans)
http://www.h-invitational.jp/

*10.14 One powerful approach to annotate genes is to compare the structures of cDNA copies of mRNAs to the genomic sequences that encode them. However, during the synthesis of cDNA (see Figure 8.8, p. 186), reverse transcriptase may not always copy the entire length of the mRNA and so a cDNA that is not full-length can be generated. This approach to gene annotation often uses ESTs that are not full-length cDNA copies of mRNA. Recently, a large collaboration involving 68 research teams analyzed 41,118 full-length cDNAs to annotate the structure of 21,037 human genes (see http://www.h-invitational.jp/).

a. What types of information can be obtained by comparing the structures of cDNAs with genomic DNA?

b. Why is it desirable, when possible, to use full-length cDNAs in these analyses?

c. The research teams characterized the number of loci per Mb of DNA for each chromosome. Among the autosomes, chromosome 19 had the highest ratio of 19 loci per Mb while chromosome 13 had the lowest ratio of 3.5 loci per Mb. Among the sex chromosomes, the X had 4.2 loci per Mb while the Y had only

0.6 loci per Mb. What does this tell you about the distribution of genes within the human genome? How can these data be reconciled with the idea that chromosomes have gene-rich regions as well as gene deserts?

d. The research teams were able to map 40,140 cDNAs to the current human genome sequence. Of the 978 cDNAs that could not be mapped, 907 could be roughly mapped to the mouse genome. Why might some (human) cDNAs be unable to be mapped to the current human genome sequence while they could be mapped to the mouse genome sequence? (Hint: Consider where errors and limited information might exist.)

*10.15 A central theme in genetics is that an organism's phenotype results from an interaction between its genotype and the environment. Because some diseases have strong environmental components, researchers have begun to assess how disease phenotypes arise from the interactions of genes with their environments, including the genetic background in which the genes are expressed. (See http://pga.tigr.org/desc.shtml for additional discussion.) How might DNA microarrays be useful in a functional genomic approach to understanding human diseases that have environmental components, such as some cancers?

10.16 How does a cell's transcriptome compare with its proteome?

a. For a specific eukaryotic cell, can you predict which has more total members? Can you predict which has more unique members?

b. Suppose you are interested in characterizing changes in the pattern of gene expression in the mouse nervous system during development. Describe how you would efficiently assess changes in the transcriptome from the time the nervous system forms during embryogenesis to its maturation in the adult.

c. How would your analyses differ if you were studying the proteome?

10.17 When cells are exposed to short periods of heat (heat shock), they alter the set of genes they transcribe as part of a protective response.

a. What steps would you take to characterize alterations to the yeast transcriptome following a heat shock?

b. Suppose the transcriptome analyses identify a set of genes whose transcript levels increase following heat shock. How might you experimentally determine which of these genes are required for a protective response following heat shock?

*10.18 Mutations in the dystrophin gene can lead to Duchenne muscular dystrophy. The dystrophin gene is among the largest known: it has a primary transcript that spans 2.5 Mb, and produces a mature mRNA that is about 14 kb. Many different mutations in the dystrophin gene have been identified. What steps would you take if you wanted to use a DNA microarray to identify the specific dystrophin gene mutation present in a patient with Duchenne muscular dystrophy?

*10.19 Distinguish between structural, functional, and comparative genomics by completing the following exercise. The following list describes specific activities and goals associated with genome analysis. Indicate the area associated with the activity or goal by placing a letter (S, structural; F, functional; C, comparative) next to each item. Some items will have more than one letter associated with them.

_____ Aligning DNA sequences within databases to determine the degree of matching

_____ Annotation of sequences within a sequenced genome

_____ Characterizing the transcriptome and proteome present in a cell at a specific developmental stage or in a particular disease state

_____ Comparing the overall arrangements of genes and nongene sequences in different organisms to understand how genomes evolve

_____ Describing the function of all genes in a genome

_____ Determining the functions of human genes by studying their homologs in nonhuman organisms

_____ Developing a comprehensive two-dimensional polyacrylamide gel electrophoresis map of all proteins in a cell

_____ Developing a physical map of a genome

_____ Developing DNA microarrays (DNA chips)

_____ Identifying homologs to human disease genes in organisms suitable for experimentation

_____ Identifying a large collection of simple tandem repeat or microsatellite sequences to use as DNA markers within one organism

_____ Identifying expressed sequence tags

_____ Making gene knockouts and observing the phenotypic changes associated with them

_____ Mapping a gene in one organism using the lod score method

_____ Sequencing individual BAC clones aligned in a contig using a shotgun approach

_____ Using oligonucleotide hybridization analysis to type an SNP

10.20 Genomic analyses of *Mycoplasma genitalium* have not only identified its 523 genes but also allowed identification of the genes that the organism needs for life. The analyses are presented in detail at the website of the

Institute for Genomic Research (http://www.tigr.org/minimal/). Explore this website and then address the following questions.

a. How many protein-coding genes have unknown functions?

b. How do we know which genes are required for life?

c. Are any of the protein-coding genes with unknown functions required for life?

d. What functional genomics approaches might be helpful to discern their function?

e. What practical value might there be in identifying the minimal gene set needed for life?

f. Is it possible to synthesize DNA sequences containing all the genes identified as essential for life and then assemble these on one chromosome? Would this be enough to have a template for a new life form? If not, what additional sequence information would be needed to generate a new life form?

g. What is the difference between the life form generated in (f) and *Mycoplasma genitalium?*

h. What ethical issues arise from these analyses? Who should decide how to address them?

*10.21 Comparative genomics offers insights into the relationship between homologous genes and the organization of genomes. When the genome of *C. elegans* was sequenced, it was striking that some types of sequences were distributed nonrandomly. Consider the data obtained for chromosome V and the X chromosome shown here. The following figure shows the distribution of genes, the distribution of inverted and tandem repeat sequences, and the location of ESTs in *C. elegans* that are highly similar to yeast genes.

a. How do the distributions of genes, inverted and tandem repeat sequences, and conserved genes compare?

b. Based on your analysis in (a), what might you hypothesize about the different rates of DNA evolution (change) on the arms and central regions of autosomes in *C. elegans?*

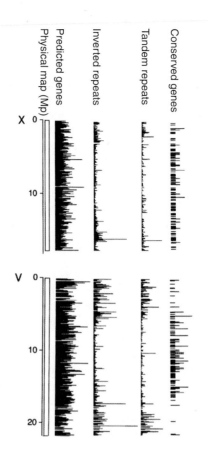

c. Curiously, meiotic recombination (crossing-over, discussed in Chapter 12, pp. 306–309) is higher on the arms of autosomes, with demarcations between regions of high and low crossing-over at the boundaries between conserved and nonconserved genes seen in the physical map. Does this information support your hypothesis in (b)?

10.22 What is the difference between a DNA chip and a protein chip? What different types of protein arrays are there, and how are they used to analyze the proteome?

11

Mendelian Genetics

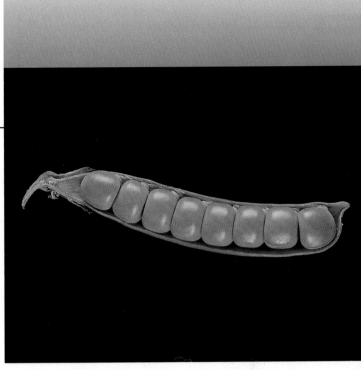

Smooth seeds of the garden pea, *Pisum sativum*.

PRINCIPAL POINTS

- The genotype is the genetic makeup of an organism, whereas the phenotype is the observable characteristic or set of characteristics (structural and functional) of an organism produced by the interaction between its genotype and the environment.

- Genes provide the potential for the development of characteristics; this potential can be affected by interactions with other genes and with the environment.

- Mendel's first law, the principle of segregation, states that the two members of a gene pair segregate from each other in the formation of gametes.

- To determine an unknown genotype (usually in an individual expressing the dominant phenotype), a cross is made between that individual and a homozygous recessive individual. This cross is called a testcross.

- Mendel's second law, the principle of independent assortment, states that members of different gene pairs are transmitted independently of one another during gamete production.

- Mendelian principles apply to all eukaryotes. The study of the inheritance of genetic traits in humans is complicated by the fact that controlled crosses cannot be done within ethical bounds. Instead, human geneticists examine genetic traits by pedigree analysis—that is, by following the occurrence of a trait in family trees in which the trait is segregating.

iActivity

(i) PEOPLE HAVE BRED ANIMALS AND PLANTS FOR specific traits for many centuries. But after Gregor Mendel developed his theory to explain the transmission of hereditary characteristics from generation to generation, breeding became an art form. Now people can use their knowledge of how a characteristic is passed from parent to offspring to produce food crops that are resistant to certain diseases, cows that produce more milk, dogs that are intelligent and gentle enough to make good guide dogs, and even furless cats. What were Mendel's experiments? What is the relationship between genes and traits? How can knowing the way in which characteristics are inherited allow people to breed for specific traits?

Later on, you can try the iActivity for this chapter, which allows you to apply the knowledge you've gained in the effort to breed a very special pet.

The understanding of how genes are transmitted from parent to offspring began with the work of Gregor Johann Mendel (1822–1884), an Augustinian monk. The goal of this chapter is for you to learn the basic principles of the transmission of genes by examining Mendel's work. Be aware that, even though Mendel analyzed the segregation of hereditary traits, he did not know that genes control the traits, that genes are located in chromosomes, or even that chromosomes existed.

Genotype and Phenotype

The characteristics of an individual that are transmitted from one generation to another are sometimes called **hereditary traits** (also called **characters**). These traits are under the control of **genes** (Mendel called them *factors*). The genetic constitution of an organism is called its **genotype,** and the **phenotype** is an observable characteristic or set of characteristics (structural and functional) of an organism produced by the interaction between its genotype and the environment.

Genes provide only the potential for developing a particular phenotypic characteristic. The extent to which that potential is realized depends upon interactions with other genes and their products and, in many cases, upon environmental influences and random developmental events (Figure 11.1). A person's height, for example, is controlled by many genes, the expression of which can be significantly affected by internal and external environmental influences such as the effects of hormones during puberty (an internal environmental influence) and nutrition (an external environmental influence). In other words, genes are a starting point for determining the structure and function of an organism, and the route to the mature phenotypic state is highly complex, involving many interacting biochemical pathways.

Although the phenotype is the product of interaction between genes and environment, the contribution of the environment varies. In some cases, the environmental influence is great, but in others, the environmental contribution is nonexistent. We will develop the relationship between genotype and phenotype in more detail as the text proceeds.

KEYNOTE

The genotype is the genetic constitution of an organism. The phenotype is the observable manifestation of the genetic traits. The genes give the potential for the development of characteristics; this potential often is affected by interactions with other genes and with the environment. Thus, individuals with the same genotype can have different phenotypes, and individuals with the same phenotypes may have different genotypes.

Figure 11.1

Influences on the physical manifestation (phenotype) of the genetic blueprint (genotype): interactions with other genes and their products (such as hormones) and with the environment (such as nutrition).

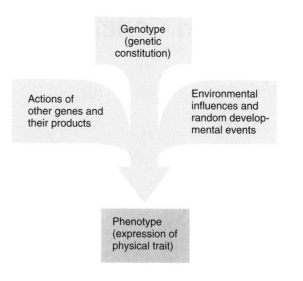

Mendel's Experimental Design

The work of Gregor Johann Mendel (Figure 11.2) is considered the foundation of modern genetics. In 1843, he was admitted to the Augustinian Monastery in Brno (now Brünn, Czech Republic). In 1854, he began a series of breeding experiments with the garden pea *Pisum sativum* to learn something about the mechanisms of heredity. Probably as a result of his creativity, Mendel discovered some fundamental principles of genetics.

From the results of crossbreeding pea plants with different characteristics such as seed shape, seed color, and flower color, Mendel developed a simple theory to explain the transmission of hereditary characteristics or traits from generation to generation. (Mendel had no knowledge of mitosis and meiosis, so he did not know that genes segregate according to chromosome behavior.) Although Mendel reported his conclusions in 1865, their significance was not fully realized until the late 1800s and early 1900s.

Mendel's experimental approach was effective because he made simple interpretations of the ratios of the types of progeny he obtained from his crosses and because he then carried out direct and convincing experiments to test his hypotheses. In his initial breeding experiments, he took the simplest approach of studying the inheritance of one trait at a time. (This is how you should work genetics problems.) He made carefully controlled matings (crosses) between pea strains that had obvious differences in heritable traits and, most importantly, he kept very careful records of the outcome of the crosses. The numerical data he obtained enabled him to

Figure 11.2

Gregor Johann Mendel, founder of the science of genetics.

Figure 11.3

Procedure for crossing pea plants.

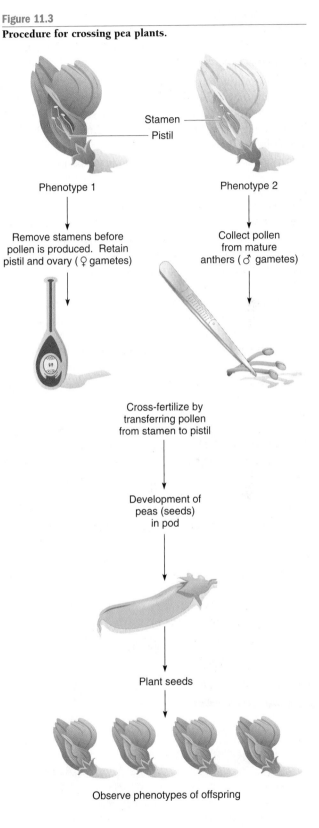

Stamen

Pistil

Phenotype 1   Phenotype 2

Remove stamens before pollen is produced. Retain pistil and ovary (♀ gametes)   Collect pollen from mature anthers (♂ gametes)

Cross-fertilize by transferring pollen from stamen to pistil

Development of peas (seeds) in pod

Plant seeds

Observe phenotypes of offspring

do a rigorous analysis of the hereditary transmission of characteristics.

Generally, genetic crosses are done as follows: Two diploid individuals differing in phenotype are allowed to produce haploid gametes by meiosis. Fusion of male and female gametes produces zygotes from which the diploid progeny individuals are generated. The phenotypes of the offspring are analyzed to provide clues to the heredity of those phenotypes.

Mendel did all his significant genetic experiments with the garden pea. The garden pea was a good choice because it fits many of the criteria that make an organism suitable for use in genetic experiments (see Chapter 1): It is easy to grow, bears flowers and fruit in the same year a seed is planted, and produces a large number of seeds.

Figure 11.3, which presents the procedure for crossing pea plants, begins with a cross section of a flower, showing the stamens (male reproductive organs) and the pistils (female reproductive organs). The pea normally reproduces by **self-fertilization;** that is, the anthers at the ends of the stamen produce pollen (microspore of a flowering plant that germinates to form the male [♂] gametophyte), which lands on the pistil (containing the female [♀] gametophyte) within the same flower and fertilizes the plant. This process is also called **selfing.** Fortunately for the success of his experiment, Mendel was able to prevent self-fertilization of the pea by removing the stamens from a developing flower bud before their anthers produced any mature pollen. Next, he took pollen from the stamens of another flower and dusted them onto the pistil of the emasculated one to pollinate it.

Cross-fertilization, or simply **cross,** is the fusion of male gametes (in this case, pollen) from one individual and female gametes (eggs) from another. Once cross-fertilization has occurred, the zygote develops in the seeds (peas), which are then planted. Finally, the phenotypes of the plants that grow from the seeds are analyzed.

Mendel obtained 34 strains of pea plants that differed in a number of traits. He allowed each strain to self-fertilize for many generations to ensure that the traits he wanted to study were inherited. This preliminary work ensured that Mendel worked only with pea strains in

which the trait under investigation remained unchanged from parent to offspring for many generations. Such strains are called **true-breeding** or **pure-breeding strains.**

Next, Mendel selected seven traits to study in breeding experiments. Each trait (or character) had two easily distinguishable, alternative appearances (phenotypes). Mendel studied the following character pairs (Figure 11.4):

1. Flower and seed coat color (grey versus white seed coats and purple versus white flowers; note that a single gene controls both these color properties of seed coats and flowers)
2. Seed color (yellow versus green)
3. Seed shape (smooth versus wrinkled)
4. Pod color (green versus yellow)
5. Pod shape (inflated versus pinched)
6. Stem height (tall versus short)
7. Flower position (axial versus terminal)

Figure 11.4

Seven character pairs in the garden pea that Mendel studied in his breeding experiments.

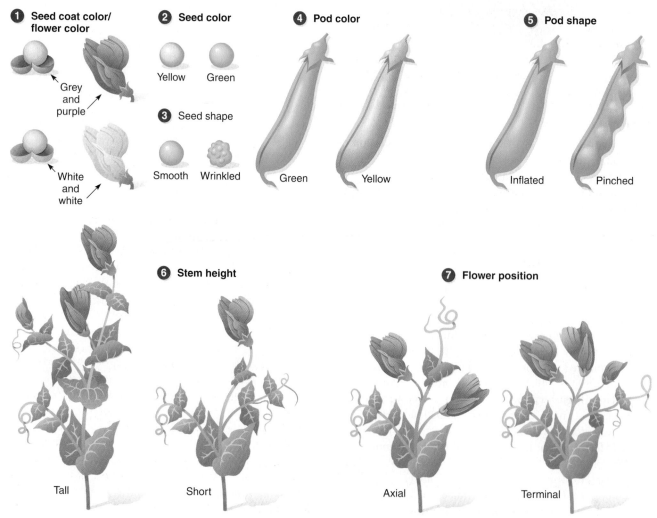

Monohybrid Crosses and Mendel's Principle of Segregation

We must be clear on the terminology used in breeding experiments. The parental generation is called the **P generation.** The progeny of the P mating is called the **first filial generation,** or F_1. The subsequent generation produced by breeding together the F_1 offspring is the F_2 **generation.** Interbreeding the offspring of each generation results in generations F_3, F_4, F_5, and so on.

Mendel first performed crosses between true-breeding strains of peas that differed in a single trait. Such crosses are called **monohybrid crosses.** For example, when he pollinated pea plants that gave rise only to smooth seeds[1] with pollen from a true-breeding variety that produced only wrinkled seeds, the result was all smooth seeds (Figure 11.5). When the parental types were reversed—that is, when the pollen from a smooth-seeded plant was used to pollinate a pea plant that gave wrinkled seeds—the result was the same: all smooth seeds. Matings that are done both ways—smooth female [♀] × wrinkled male [♂] and wrinkled female [♀] × smooth male [♂]—are called **reciprocal crosses.** Conventionally, the female is given first in crosses of plants. If the results of reciprocal crosses are the same, it means that the trait does not depend on the sex of the organism.

The significant point of this cross is that all the F_1 progeny seeds of the smooth × wrinkled reciprocal crosses were smooth: They exactly resembled only one of the parents in this character rather than being a blend of both parental phenotypes. The finding that all offspring

Figure 11.5

Results of one of Mendel's breeding crosses. In the parental generation, he crossed a true-breeding pea strain that produced smooth seeds with one that produced wrinkled seeds. All the F_1 progeny seeds were smooth.

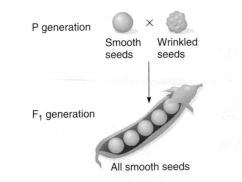

[1]Seeds are the diploid progeny of sexual reproduction. If a phenotype concerns the seed itself, the results of the cross can be seen directly by looking at the seeds. If a phenotype concerns a part of the mature plant, such as flower color, then the seeds must be germinated and grown to maturity before that phenotype can be seen.

Figure 11.6

The F_2 progeny of the cross shown in Figure 11.5. When the plants grown from the F_1 seeds were self-pollinated, both smooth and wrinkled F_2 progeny seeds were produced. Commonly, both seed types were found in the same pod. In his experiments, Mendel counted 5,474 smooth and 1,850 wrinkled F_2 progeny seeds for a ratio of 2.96:1.

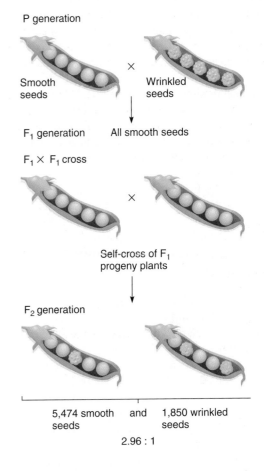

of true-breeding parents are alike is sometimes referred to as the *principle of uniformity in* F_1.

Next, Mendel planted the seeds and allowed the F_1 plants to self-fertilize to produce the F_2 seed. Both smooth and wrinkled seeds appeared in the F_2 generation, and both types could be found within the same pod. Typical of his analytical approach to the experiments, Mendel counted the number of seeds of each type. He found that 5,474 were smooth and 1,850 were wrinkled (Figure 11.6). The calculated ratio of smooth seeds to wrinkled seeds was 2.96:1, which is very close to a 3:1 ratio.

Mendel observed that although the F_1 resembled only one of the parents in their phenotype, they did not breed true, a fact that distinguished the F_1 from the parent they resembled. Moreover, the F_1 could produce some F_2 progeny with the parental phenotype that had disappeared in the F_1. But how can a trait present in the P generation disappear in the F_1 and then reappear in the F_2? Mendel concluded that the alternative traits in the cross—smoothness or wrinkledness of the seeds—were determined by **particulate factors**. He reasoned that these factors, which were transmitted from parents to progeny through the gametes, carried hereditary information. We now know them by another name: *genes*.

Since Mendel was examining a pair of traits (wrinkled and smooth), each factor was considered to exist in alternative forms (which we now call **alleles**), each of which specified one of the traits. For the gene that controls pea seed shape, there is one form (allele) that results in a smooth seed and another, or alternate, allele that results in a wrinkled seed.

Mendel reasoned further that a true-breeding strain of peas must contain a pair of identical factors. Because the F_2 exhibited both traits and the F_1 exhibited only one of those traits, each F_1 individual must have contained both factors, one for each of the alternative traits. In other words, crossing two different true-breeding strains brings together in the F_1 one factor from each strain: The eggs contain one factor from one strain and the pollen grains contain one factor from the other strain. Furthermore, because only one of the traits was seen in the F_1 generation, the expression of the missing trait must somehow have been masked by the visible trait; this masking is called *dominance*. For the smooth × wrinkled cross, the F_1 seeds were all smooth. Thus, the allele for smoothness is masking or **dominant** to the allele for wrinkledness. Conversely, wrinkled is said to be **recessive** to smooth because the factor for wrinkled is masked.

A simple way to visualize the crosses is to use symbols for the alleles, as Mendel did. For the smooth × wrinkled cross we can give the symbol *S* to the allele for smoothness and the symbol *s* to the allele for wrinkledness. The letter used is based on the dominant phenotype, and the convention in this case is that the dominant allele is given the uppercase letter and the recessive allele the lowercase letter. (This convention was used for many years, particularly in plant genetics. Now it is more conventional to base the letter assignment on the recessive phenotype. We will use the newer convention later.)

Using these symbols, we denote the genotype of the parental plant grown from the smooth seeds by *SS* and that of the wrinkled parent by *ss*. True-breeding individuals that contain two copies of the same specific allele of a particular gene are said to be **homozygous** for that gene (Figure 11.7). When plants produce gametes by meiosis (see Chapter 12), each gamete contains only one copy of the gene (one allele); the plants from smooth seeds produce *S*-bearing gametes, and the plants from wrinkled seeds produce *s*-bearing gametes. When the gametes fuse during fertilization, the resulting zygote has one *S* allele and one *s* allele, a genotype of *Ss*. Plants that have two different alleles of a particular gene are said to be **heterozygous.** Because of the dominance of the smooth *S* allele, *Ss* plants produce smooth seeds. (See Figure 11.7.)

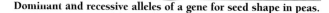

Figure 11.7

Dominant and recessive alleles of a gene for seed shape in peas.

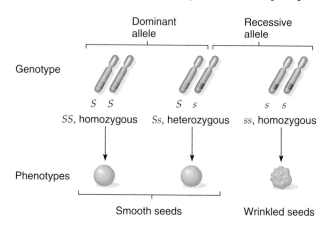

Figure 11.8 diagrams the smooth × wrinkled cross with the use of genetic symbols; the production of the F_1 is shown in Figure 11.8a and that of the F_2, in Figure 11.8b. (In Figures 11.7 and 11.8, the genes are shown on chromosomes. Keep in mind that the segregation of genes from generation to generation follows the behavior of chromosomes.) The true-breeding, smooth-seeded parent has the genotype *SS*, and the true-breeding, wrinkle-seeded parent has the genotype *ss*. Because each parent is true breeding and diploid (that is, has two sets of chromosomes), each must contain two copies of the same allele. All the F_1 plants produce smooth seeds, and all are *Ss* heterozygotes.

The plants grown from the F_1 seeds differ from the smooth parent in that they produce equal numbers of two types of gametes: *S*-bearing gametes and *s*-bearing gametes. All possible fusions of F_1 gametes are shown in the matrix in Figure 11.8b, called a **Punnett square** after its originator, R. Punnett. These fusions give rise to the zygotes that produce the F_2 generation.

In the F_2 generation, three types of genotypes are produced: *SS*, *Ss*, and *ss*. As a result of the random fusing of gametes, the relative proportion of these zygotes is 1:2:1, respectively. However, because the *S* factor is dominant to the *s* factor, both the *SS* and *Ss* seeds are smooth, and the F_2 generation seeds show a phenotypic ratio of 3 smooth : 1 wrinkled.

Mendel also analyzed the behavior of the six other pairs of traits. Qualitatively and quantitatively, the same results were obtained (Table 11.1). From the seven sets of crosses he made the following general conclusions about his data:

1. The results of reciprocal crosses were always the same.
2. All F_1 progeny resembled one of the parental strains, indicating the dominance of one allele over the other.
3. In the F_2 generation, the parental trait that had disappeared in the F_1 generation reappeared. Furthermore, the trait seen in the F_1 was always found in the F_2 at about three times the frequency of the other trait.

Figure 11.8

The same cross as in Figures 11.5 and 11.6, using genetic symbols to illustrate the principle of segregation of Mendelian factors. (a) Production of the F_1 generation. (b) Production of the F_2 generation.

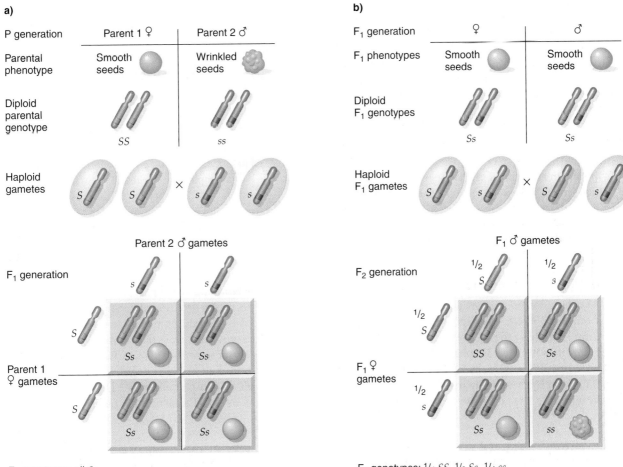

a)

P generation — Parent 1 ♀ | Parent 2 ♂

Parental phenotype: Smooth seeds | Wrinkled seeds

Diploid parental genotype: SS | ss

Haploid gametes: S S × s s

F₁ generation — Parent 2 ♂ gametes

Parent 1 ♀ gametes

Ss Ss
Ss Ss

F₁ genotypes: all Ss

F₁ phenotypes: all smooth (smooth is dominant to wrinkled)

b)

F₁ generation — ♀ | ♂

F₁ phenotypes: Smooth seeds | Smooth seeds

Diploid F₁ genotypes: Ss | Ss

Haploid F₁ gametes: S s × S s

F₂ generation — F₁ ♂ gametes: 1/2 S | 1/2 s

F₁ ♀ gametes: 1/2 S | 1/2 s

SS Ss
Ss ss

F₂ genotypes: 1/4 SS, 1/2 Ss, 1/4 ss

F₂ phenotypes: 3/4 smooth seeds, 1/4 wrinkled seeds

Table 11.1   **Mendel's Results in Crosses between Plants Differing in One of Seven Characters**

Character[a]	F₁	F₂ (Number) Dominant	F₂ (Number) Recessive	Total	F₂ (Ratio) Dominant : Recessive
Seeds: smooth versus wrinkled	All smooth	5,474	1,850	7,324	2.96:1
Seeds: yellow versus green	All yellow	6,022	2,001	8,023	3.01:1
Seed coats: grey versus white[b]	All grey	705	224	929	3.15:1
Flowers: purple versus white	All purple				
Flowers: axial versus terminal	All axial	651	207	858	3.14:1
Pods: inflated versus pinched	All inflated	882	299	1,181	2.95:1
Pods: green versus yellow	All green	428	152	580	2.82:1
Stem: tall versus short	All tall	787	277	1,064	2.84:1
Total or average		14,949	5,010	19,959	2.98:1

[a]The dominant trait is always written first.
[b]A single gene controls both the seed coat and the flower color trait.

The Principle of Segregation

From the sort of data just discussed, Mendel proposed what has become known as his first law, the principle of segregation: *Recessive characters, which are masked in the*

a Mendel's Principle of Segregation

F₁ from a cross between two true-breeding strains, reappear in a specific proportion in the F₂. In modern terms this means that *the two members of a gene pair (alleles) segregate (separate) from each other during the formation of gametes.* As a result, half the gametes carry one allele, and the other half carry the other allele. In other words, each gamete carries only a single allele of each gene. The progeny are produced by the random combination of gametes from the two parents.

In proposing the principle of segregation, Mendel had differentiated between the factors (genes) that determined the traits (the genotype) and the traits themselves (the phenotype). We know now, of course, that genes are on chromosomes. The specific location of a gene on a chromosome is called its **locus** (or **gene locus;** plural *loci*). Furthermore, Mendel's first law means that at the gene level the members of a pair of alleles segregate during meiosis and that each offspring receives only one allele from each parent. Thus, **gene segregation** parallels the separation of homologous pairs of chromosomes at anaphase I in meiosis. (See Chapter 12, pp. 308–309.)

Box 11.1 presents a summary of the genetics concepts and terms we have discussed so far in this chapter.

A thorough familiarity with these terms is essential to your study of genetics.

KEYNOTE

Mendel's first law, the principle of segregation, states that the two members of a gene pair (alleles) segregate (separate) from each other in the formation of gametes; half the gametes carry one allele, and the other half carry the other allele.

Representing Crosses with a Branch Diagram

The use of a Punnett square to represent the pairing of all possible gamete types from two parents in a cross (see Figure 11.8) is a simple way to predict the relative frequencies of genotypes and phenotypes in the next generation. There is an alternative method, one you are encouraged to master: the branch or fork diagram. (Box 11.2 discusses some elementary principles of probability that will help you understand this approach.) To use the branch diagram approach, it is necessary to know the dominance/recessiveness relationship of the allele pair so that the progeny phenotypic classes can be determined. Figure 11.9 illustrates the application of branch diagram analysis of the F₁ selfing of the smooth × wrinkled cross diagrammed in Figure 11.8.

The F₁ seeds from the cross in Figure 11.8 have the genotype *Ss.* In meiosis we expect half of the gametes to be *S* and half to be *s.* (See Figure 11.9.) Thus, ½ is the predicted frequency of each of these two types. But just as

---

Box 11.1   Genetic Terminology

Alleles: Alternative forms of a gene. For example, *S* and *s* alleles represent the smoothness and wrinkledness of the pea seed. (Like gene symbols, allele symbols are italicized.)

Cross: A mating between two individuals, leading to the fusion of gametes.

Diploid: A eukaryotic cell or organism with two homologous sets of chromosomes.

Gamete: A mature reproductive cell that is specialized for sexual fusion. Each gamete is haploid and fuses with a cell of similar origin, but of opposite sex, to produce a diploid zygote.

Gene (Mendelian factor): The determinant of a characteristic of an organism. (Gene symbols are italicized.) A gene's nucleotide sequence specifies a polypeptide or an RNA.

Genotype: The genetic constitution of an organism. A diploid organism in which both alleles are the same at a given gene locus is said to be **homozygous** for that allele. Homozygotes produce only one gametic type with respect to that locus. For example, true-breeding smooth-seeded peas have the genotype *SS,* and true-breeding wrinkle-seeded peas have the genotype *ss;* both are homozygous. The smooth parent is

homozygous dominant; the wrinkled parent is **homozygous recessive.**

Diploid organisms that have two different alleles at a specific gene locus are said to be **heterozygous.** Thus, F₁ hybrid plants from the cross of *SS* and *ss* parents have one *S* allele and one *s* allele. Individuals heterozygous for two allelic forms of a gene produce two kinds of gametes (*S* and *s*).

Haploid: A cell or an individual with one copy of each chromosome.

Locus (gene locus; plural = *loci*): The specific place on a chromosome where a gene is located.

Phenotype: The physical manifestation of a genetic trait that results from a specific genotype and its interaction with the environment. In our example, the *S* allele was dominant to the *s* allele, so in the heterozygous condition the seed is smooth. Therefore, both the homozygous dominant *SS* and the heterozygous *Ss* seeds have the same phenotype (smooth), even though they differ in genotype.

Zygote: The cell produced by the fusion of male and female gametes.

Figure 11.9

Using the branch diagram approach to calculate the ratios of phenotypes in the F$_2$ generation of the cross in Figure 11.8.

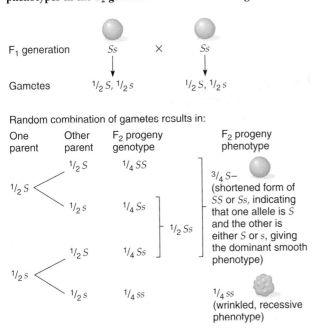

the F$_2$ generation. To produce an *SS* plant, an *S* egg must pair with an *S* pollen grain. The frequency of *S* eggs in the population of eggs is $\frac{1}{2}$, and the frequency of *S* pollen grains in the pollen population is $\frac{1}{2}$. Therefore, the expected proportion of *SS* smooth plants in the F$_2$ is $\frac{1}{2} \times \frac{1}{2} = \frac{1}{4}$. Similarly, the expected proportion of *ss* wrinkled progeny in the F$_2$ is $\frac{1}{2} \times \frac{1}{2} = \frac{1}{4}$.

What about the *Ss* progeny? Again, the frequency of *S* in one gametic type is $\frac{1}{2}$, and the frequency of *s* in the other gametic type is also $\frac{1}{2}$. However, there are two ways in which *Ss* progeny can be obtained. The first involves the fusion of an *S* egg with *s* pollen, and the second is a fusion of an *s* egg with *S* pollen. Using the product rule (see Box 11.2), we find that the probability of each of these events occurring is $\frac{1}{2} \times \frac{1}{2} = \frac{1}{4}$. Using the sum rule (see Box 11.2), we see that the probability of one or the other occurring is the sum of the individual probabilities, or $\frac{1}{4} + \frac{1}{4} = \frac{1}{2}$.

The prediction, then, is that one-fourth of the F$_2$ progeny will be *SS*, half will be *Ss*, and one-fourth will be *ss*, exactly as we found with the Punnett square method shown in Figure 11.8. Either method—the Punnett square or the branch diagram—may be used with any cross, but as crosses become more complicated, the Punnett square method becomes cumbersome.

tossing a coin many times does not always give exactly half heads and half tails, the two gametes may not be produced in an exact 1:1 ratio. However, the more chances (tosses), the closer the observed frequency will come to the predicted frequency.

From the rules of probability, we can predict the expected frequencies of the three possible genotypes in

Confirming the Principle of Segregation: The Use of Testcrosses

When formulating his principle of segregation, Mendel did a number of tests to ensure the correctness of his results. He continued the self-fertilizations at each generation up

---

Box 11.2   Elementary Principles of Probability

A **probability** is the ratio of the number of times a particular event is expected to occur to the number of trials during which the event could have happened. For example, the probability of picking a heart from a deck of 52 cards, 13 of which are hearts, is $P(\text{heart}) = \frac{13}{52} = \frac{1}{4}$. That is, we would expect, on the average, to pick a heart from a deck of cards once in every four trials.

Probabilities and the *laws of chance* are involved in the transmission of genes. As a simple example, consider a couple and the chance that their child will be a boy or a girl. Assume that an exactly equal number of boys and girls are born (which is not precisely true, but we can assume it to be so for the sake of discussion). The probability that the child will be a boy is $\frac{1}{2}$, or 0.5. Similarly, the probability that the child will be a girl is also $\frac{1}{2}$.

Now a rule of probability can be introduced: the **product rule.** The product rule states that the probability of two independent events occurring simultaneously is the product of each of their individual probabilities. Thus, the probability that both

children in a family with two children will be girls is $\frac{1}{4}$. That is, the probability of the first child being a girl is $\frac{1}{2}$, the probability of the second being a girl is also $\frac{1}{2}$, and, by the product rule, the probability of the first and second being girls is $\frac{1}{2} \times \frac{1}{2} = \frac{1}{4}$. Similarly, the probability of having three boys in a row is $\frac{1}{2} \times \frac{1}{2} \times \frac{1}{2} = \frac{1}{8}$.

Another rule of probability, the **sum rule,** states that the probability of occurrence of any of several mutually exclusive events is the sum of the probabilities of the individual events. For example, if one die is thrown, what is the probability of getting a one or a six? The individual probabilities are calculated as follows: The probability of rolling a one, $P(\text{one})$, is $\frac{1}{6}$, because there are six faces to a die. For the same reason, the probability of rolling a six, $P(\text{six})$, is also $\frac{1}{6}$. To roll a one or a six with a single throw of the die involves two mutually exclusive events, so the sum rule is used. The sum of the individual probabilities is $\frac{1}{6} + \frac{1}{6} = \frac{2}{6} = \frac{1}{3}$. To return to our family example, the probability of having two boys or two girls is $\frac{1}{4} + \frac{1}{4} = \frac{1}{2}$.

to the F_6 and found that, in every generation, both the dominant and recessive characters were found. He concluded that the principle of segregation was valid no matter how many generations were involved.

Another important test concerned the F_2 plants. As shown in Figure 11.8, a ratio of 1:2:1 occurs for the genotypes *SS, Ss,* and *ss* for the smooth × wrinkled example. Phenotypically, the ratio of smooth to wrinkled is 3:1. At the time of Mendel's experiments, the presence of segregating factors that were responsible for the smooth and wrinkled phenotypes was only a hypothesis. To test his factor hypothesis, Mendel allowed the F_2 plants to self-pollinate. As he expected, the plants produced from wrinkled seeds bred true, supporting his conclusion that they were pure for the *s* factor (gene).

Selfing the plants derived from the F_2 smooth seeds produced two different types of progeny: One-third of the smooth F_2 seeds produced all smooth-seeded progeny, whereas the other two-thirds produced both smooth and wrinkled seeds in each pod in a ratio of 3 smooth : 1 wrinkled (Figure 11.10). For the plants that produced both seed types in the progeny, the actual ratio of smooth : wrinkled seeds was 3:1, the same ratio as seen for the F_2 progeny. These results support the principle of gene segregation. The random combination of gametes that form the zygotes of the original F_2 produces two genotypes that give rise to the smooth phenotype (see Figures 11.8 and 11.9); the relative proportion of the two genotypes *SS* and *Ss* is 1:2. The *SS* seeds give rise to true-breeding plants, whereas the *Ss* seeds give rise to plants that behave exactly like the F_1 plants when they are self-pollinated in that they produce a 3:1 ratio of smooth : wrinkled progeny. *Mendel explained these results by proposing that each plant had two factors, whereas each gamete had only one. He also proposed that the random combination of the gametes generated the progeny in the proportions he found. Mendel obtained the same results in all seven sets of crosses.*

The self-fertilization test of the F_2 progeny proved a useful way to confirm the genotype of a plant with a given phenotype. A more common test to ascertain the genotype of an organism is to perform a **testcross**, a cross of an individual of unknown genotype (usually expressing the dominant phenotype) with a homozygous recessive individual to determine the unknown genotype.

Consider again the cross shown in Figure 11.8. We can predict the outcome of a testcross of the F_2 progeny showing the dominant, smooth-seed phenotype. If the F_2 individuals are homozygous *SS*, then the result of a testcross with an *ss* plant will be all smooth seeds. As Figure 11.11a shows, the Parent 1 smooth *SS* plants produce only *S* gametes. Parent 2 is homozygous recessive wrinkled, *ss*, so it produces only *s* gametes. Therefore, all zygotes are *Ss*, and all the resulting seeds have the smooth phenotype. In actual practice, then, if a plant with a dominant trait is testcrossed and only the dominant phenotype is seen among the progeny, then the plant must have been homozygous for the dominant allele. In contrast, heterozygous *Ss* F_2 plants testcrossed with a homozygous *ss* plant give a 1:1 ratio of dominant to recessive phenotypes. As Figure 11.11b shows, the Parent 1 smooth *Ss* produces both *S* and *s* gametes in equal proportion, and the homozygous *ss* Parent 2 produces only *s* gametes. As a result, half the progeny of the testcross are *Ss* heterozygotes and have a smooth phenotype because of the dominance of the *S* allele, and the other half are *ss* homozygotes and have a wrinkled phenotype. In actual practice, then, if a plant with a dominant trait is testcrossed and the progeny exhibit a 1:1 ratio of dominant to recessive phenotypes, then the plant must have been heterozygous.

In sum, testcrosses of the F_2 progeny from Mendel's crosses that showed the dominant phenotype resulted in a 1:2 ratio of homozygous dominant : heterozygous genotypes in the F_2 progeny. That is, when crossed with the homozygous recessive, one-third of the F_2 progeny with the dominant phenotype gave rise only to progeny with the dominant phenotype and were therefore homozygous for the dominant allele. The other two-thirds of the F_2 progeny with the dominant phenotype produced progeny with a 1:1 ratio of dominant phenotype to recessive phenotype and therefore were heterozygous.

Figure 11.10

Determining the genotypes of the F_2 smooth progeny of Figure 11.8 by selfing the plants grown from the smooth seeds.

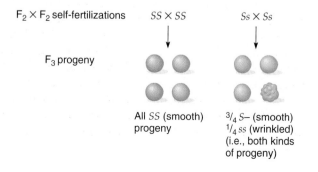

The Wrinkled-Pea Phenotype

Why is the wrinkled phenotype recessive? To answer this question, we must think about genes at the molecular level. The functional allele of a gene that predominates (is present in the highest frequency) in the population of an organism found in the "wild" is called the **wild-type allele.** Wild-type alleles typically encode a product for a particular biological function. Therefore, if a mutation in the gene causes the protein product of a gene to be absent, partially functional, or nonfunctional, then the associated biological function is likely to be lost

Figure 11.11

Determining the genotypes of the F₂ generation smooth seeds (Parent 1) of Figure 11.8 by testcrossing plants grown from the seed with a homozygous recessive wrinkled (ss) strain (Parent 2). (a) If Parent 1 is *SS*, then all progeny seeds are smooth. (b) If Parent 1 is *Ss*, then one-half of the progeny seeds are smooth and one-half are wrinkled.

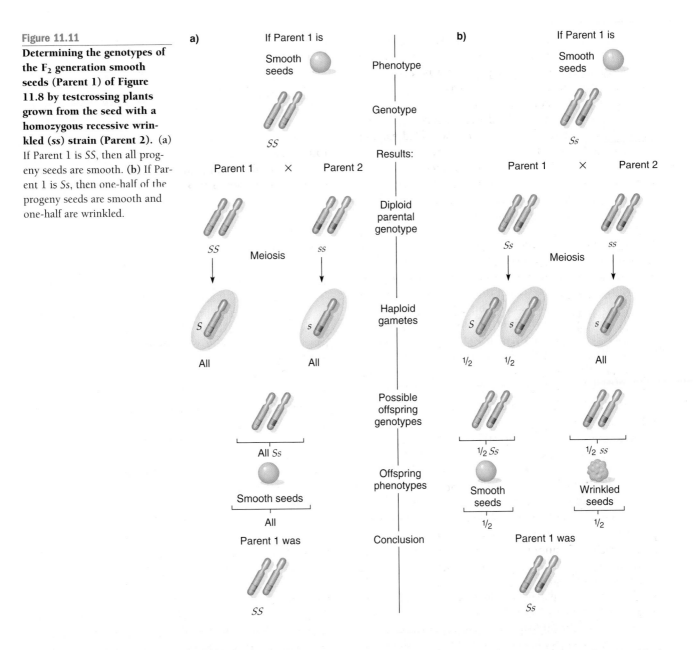

or decreased significantly. Such mutations are called **loss-of-function mutations** and are usually recessive because the function of a wild-type gene in a heterozygote is often sufficient to produce the normal phenotype. Loss-of-function mutations may be caused in various ways but, most commonly, the base-pair sequence of the gene is altered, resulting in a changed amino acid sequence of the encoded protein.

Mendel's wrinkled peas result from a loss-of-function mutation. In *SS* (smooth or wild-type) peas, starch grains are large and simple, while in *ss* (wrinkled) peas they are small and deeply fissured. *SS* seeds contain larger amounts of starch and lower levels of sucrose than *ss* seeds. The sucrose difference leads to a higher water content and larger size of developing *ss* seeds. When the seeds mature, the *ss* seeds lose a larger

proportion of their volume, leading to the wrinkled phenotype. At the molecular level, the seed-shape gene encodes one form of starch-branching enzyme (SBEI) in developing embryos. SBEI is important in determining the starch content of embryos so that, in *ss* plants, starch content is reduced. The wrinkled peas in Mendel's experiments did not have a simple base-pair change in the seed-shape gene that inactivated SBEI, however. Rather, molecular analysis of *ss* plant lines directly descended from those that Mendel used in his experiments shows that they have an 800-bp extra piece of DNA inserted into the *S* gene, resulting in the *s* allele. This inserted piece of DNA is a *transposon*, a piece of DNA that can move ("transpose") to different locations in the genome. We learn more about transposons in Chapter 7.

Dihybrid Crosses and Mendel's Principle of Independent Assortment

The Principle of Independent Assortment

Mendel also analyzed a number of crosses in which two pairs of traits were simultaneously involved. In each case, he obtained the same results. From these experiments, he proposed his **second law,** the **principle of independent assortment,** which states that *the factors for different traits assort independently of one another.* In modern terms, this means that *genes on different chromosomes behave independently in gamete production.*

animation

a Mendel's Principle of Independent Assortment

Consider an example involving smooth (*S*), wrinkled (*s*), yellow (*Y*), and green (*y*) seed traits (yellow is dominant to green). When Mendel made crosses between true-breeding smooth, yellow plants (*SS YY*) and wrinkled, green plants (*ss yy*), he got the results shown in Figure 11.12. All the F$_1$ seeds from this cross were smooth and yellow, as the results of the monohybrid crosses predicted. As Figure 11.12a shows, the smooth, yellow parent produces only *SY* gametes, which give rise to *Ss Yy* zygotes upon fusion with the *s y* gametes from the wrinkled, green parent. Because of the dominance of the smooth and the yellow traits, all F$_1$ seeds are smooth and yellow.

The F$_1$ are heterozygous for two pairs of alleles at two different loci. Such individuals are called dihybrids, and a cross between two of these dihybrids of the same type is called a **dihybrid cross.**

When Mendel self-pollinated the dihybrid F$_1$ plants to give rise to the F$_2$ generation (Figure 11.12b), there were two possible outcomes. One was that the genes for the traits from the original parents would be transmitted together to the progeny. In this case, a phenotypic ratio of 3:1 smooth, yellow : wrinkled, green would be predicted. The other possibility was that the traits would be inherited independently of one another. In this case, the dihybrid F$_1$ would produce four types of gametes: *S Y, S y, s Y,* and *s y.* Given the independence of the two pairs of genes, each gametic type is predicted to occur with equal frequency. In F$_1$ × F$_1$ crosses, the four types of gametes would be expected to fuse randomly in all possible combinations to give rise to the zygotes and hence the progeny seeds. All

Figure 11.12a

The principle of independent assortment in a dihybrid cross. This cross, actually done by Mendel, involves the smooth, wrinkled and yellow, green character pairs of the garden pea. (**a**) Production of the F$_1$ generation. (**b**) The F$_2$ genotypes and 9:3:3:1 phenotypic ratio of smooth, yellow : smooth, green : wrinkled, yellow : wrinkled, green, derived by using the Punnett square. (Note that, compared with previous figures of this kind, only one box is shown in the F$_1$ instead of four. This is because only one class of gametes exists for Parent 2 and only one class for Parent 1. Previously, we showed two gametes from each parent, even though those gametes were identical.)

a)

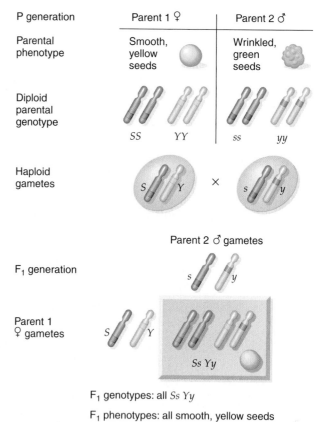

F$_1$ genotypes: all *Ss Yy*

F$_1$ phenotypes: all smooth, yellow seeds

the possible gametic fusions are represented in the Punnett square in Figure 11.12b. In a dihybrid cross, there are 16 possible gametic fusions. The result is nine different genotypes but, because of dominance, only four phenotypes are predicted:

1 *SS YY*, 2 *Ss YY*, 2 *SS Yy*, 4 *Ss Yy*	= 9 smooth, yellow
1 *SS yy*, 2 *Ss yy*	= 3 smooth, green
1 *ss YY*, 2 *ss Yy*	= 3 wrinkled, yellow
1 *ss yy*	= 1 wrinkled, green

According to the rules of probability, if pairs of characters are inherited independently in a dihybrid cross, then the F$_2$ from an F$_1$ × F$_1$ cross will give a 9:3:3:1 ratio of the four possible phenotypic classes. Such a ratio is the

Figure 11.12b

b)

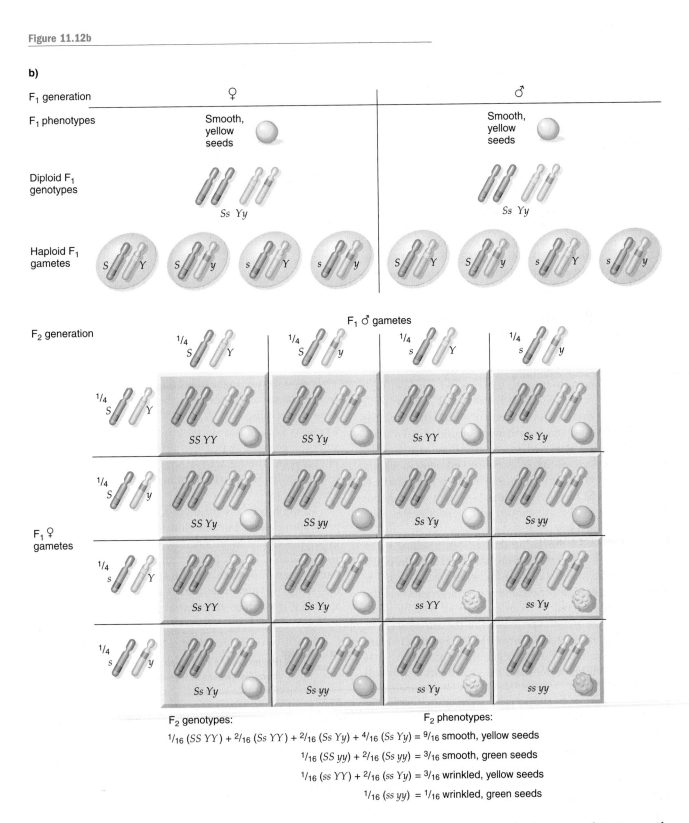

F₂ genotypes:

$^1/_{16}$ (SS YY) + $^2/_{16}$ (Ss YY) + $^2/_{16}$ (Ss Yy) + $^4/_{16}$ (Ss Yy) = $^9/_{16}$ smooth, yellow seeds

$^1/_{16}$ (SS yy) + $^2/_{16}$ (Ss yy) = $^3/_{16}$ smooth, green seeds

$^1/_{16}$ (ss YY) + $^2/_{16}$ (ss Yy) = $^3/_{16}$ wrinkled, yellow seeds

$^1/_{16}$ (ss yy) = $^1/_{16}$ wrinkled, green seeds

result of the independent assortment of the two gene pairs into the gametes and of the random fusion of those gametes.

This prediction was met in all the dihybrid crosses Mendel performed. In every case, the F₂ ratio was close to 9:3:3:1. For our example, he counted 315 smooth, yellow, 108 smooth, green, 101 wrinkled, yellow, and 32 wrinkled, green seeds, very close to the predicted ratio. To Mendel, this result meant that the factors (genes) determining the specific, different character pairs he was analyzing were transmitted independently. Thus, in effect, Mendel rejected the possibility that the two traits were inherited together.

KEYNOTE

> Mendel's second law, the principle of independent assortment, states that genes for different traits assort independently of one another in gamete production.

Branch Diagram of Dihybrid Crosses

Rather than using a Punnett square, it is easier to get into the habit of calculating the expected ratios of phenotypic or genotypic classes by using a branch diagram to apply the laws of probability to the traits one at a time. With practice you should be able to calculate the probabilities of outcomes of various crosses just by using the laws of probability without drawing out the branch diagram. Diligently working problems helps to hone this skill.

Using the same example, in which the two gene pairs assort independently into the gametes, we consider each gene pair in turn. Earlier, we saw that an F_1 self of an Ss heterozygote gave rise to progeny of which three-fourths were smooth and one-fourth were wrinkled. Genotypically, the former class had at least one dominant S allele; that is, they were SS or Ss. A convenient way to signify this situation is to use a dash to indicate an allele that has no effect on the phenotype. Thus, $S-$ means that, phenotypically, the seeds are smooth and, genotypically, they are either SS or Ss.

Now consider the F_2 produced from a selfing of Yy heterozygotes: a 3:1 ratio is seen, with three-fourths of the seeds being yellow and one-fourth being green. Since this segregation occurs independently of the segregation of the smooth, wrinkled pair, we can consider all possible combinations of the phenotypic classes in the dihybrid cross. For example, the expected proportion of F_2 seeds that are smooth and yellow is the product of the probability that an F_2 seed will be smooth and the probability that it will be yellow, or $3/4 \times 3/4 = 9/16$. Similarly, the expected proportion of F_2 progeny that are wrinkled and yellow is $3/4 \times 1/4 = 3/16$. Extending the calculation to all possible phenotypes, as shown in Figure 11.13, we obtain the ratio of 9 $S-Y-$ (smooth, yellow) : 3 $S-yy$ (smooth, green) : 3 $ss Y-$ (wrinkled, yellow) : 1 $ss yy$ (wrinkled, green).

The testcross can be used to check the genotypes of F_1 progeny and F_2 progeny from a dihybrid cross. In our example, the F_1 is a double heterozygote, $Ss Yy$, which produces four types of gametes in equal proportions: $S Y$, $S y$, $s Y$, and $s y$. (See Figure 11.12b.) In a testcross with a doubly homozygous recessive plant—in this case, $ss yy$—the phenotypic ratio of the progeny is a direct reflection of the ratio of gametic types produced by the F_1 parent.

Figure 11.13

Using the branch diagram approach to calculate the F_2 phenotypic ratio of the cross in Figure 11.12.

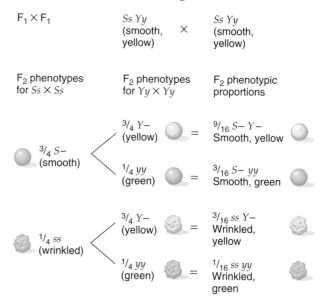

In a testcross such as this one, then, there will be a 1:1:1:1 ratio in the offspring of $Ss Yy : Ss yy : ss Yy : ss yy$ genotypes, which means a ratio of 1 smooth, yellow : 1 smooth, green : 1 wrinkled, yellow : 1 wrinkled, green phenotypes. The 1:1:1:1 phenotypic ratio is diagnostic of testcrosses in which the "unknown" parent is a double heterozygote.

In the F_2 of a dihybrid cross, there are nine different genotypic classes, but only four phenotypic classes. The genotypes can be ascertained by testcrossing, as we have shown. Table 11.2 lists the expected ratios of progeny phenotypes from such testcrosses. No two patterns are the same, so here the testcross is truly a diagnostic approach to confirm genotypes.

Table 11.2 **Proportions of Phenotypic Classes Expected from Testcrosses of Strains with Various Genotypes for Two Gene Pairs**

Testcrosses	Proportion of Phenotypic Classes			
	$A-B-$	$A-bb$	$aa B-$	$aa bb$
$AA\ BB \times aa\ bb$	1	0	0	0
$Aa\ BB \times aa\ bb$	$1/2$	0	$1/2$	0
$AA\ Bb \times aa\ bb$	$1/2$	$1/2$	0	0
$Aa\ Bb \times aa\ bb$	$1/4$	$1/4$	$1/4$	$1/4$
$aa\ bb \times aa\ bb$	0	1	0	0
$Aa\ bb \times aa\ bb$	0	$1/2$	0	$1/2$
$aa\ BB \times aa\ bb$	0	0	1	0
$aa\ Bb \times aa\ bb$	0	0	$1/2$	$1/2$
$aa\ bb \times aa\ bb$	0	0	0	1

Go to the iActivity, *Tribble Traits*, on your CD-ROM and discover how, as a Tribble breeder, you can choose the right combination of traits to produce the cuddliest creature.

Trihybrid Crosses

Mendel also confirmed his laws for three characters segregating in other crosses. Such crosses are called **trihybrid crosses.** Here, the proportions of F_2 genotypes and phenotypes are predicted with precisely the same logic used before: by considering each character pair independently. Figure 11.14 shows a branch diagram derivation of the F_2 phenotypic classes for a trihybrid cross. The independently assorting character pairs in the cross are smooth versus wrinkled seeds, yellow versus green seeds, and purple versus white flowers. There are 64 combinations of eight maternal and eight paternal gametes. Combination of these gametes gives rise to 27 different genotypes and eight different phenotypes in the F_2 generation. The phenotypic ratio in the F_2 is 27:9:9:9:3:3:3:1.

Figure 11.14

Branch diagram derivation of the relative frequencies of the eight phenotypic classes in the F_2 of a trihybrid cross.

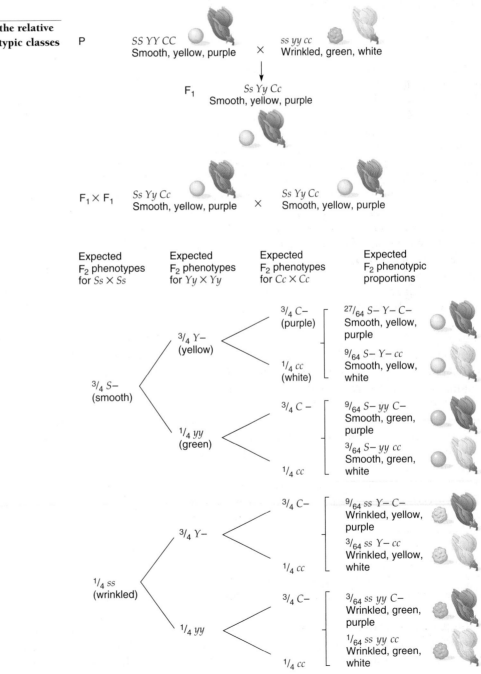

Table 11.3	Number of Phenotypic and Genotypic Classes Expected from Self-Crosses of Heterozygotes in Which All Genes Show Complete Dominance	
Number of Segregating Gene Pairs	Number of Phenotypic Classes	Number of Genotypic Classes
1^a	2	3
2	4	9
3	8	27
4	16	81
n	2^n	3^n

[a]For example from $Aa \times Aa$, two phenotypic classes are expected, with genotypic classes of AA, Aa, and aa.

Now that we have considered enough examples, we can make some generalizations about phenotypic and genotypic classes. In each example discussed, the F_1 is heterozygous for each gene involved in the cross, and the F_2 is generated by selfing (when possible) or by allowing the F_1 progeny to interbreed. In monohybrid crosses, there are two phenotypic classes in the F_2; in dihybrid crosses, there are four; and in trihybrid crosses, there are eight. The general rule is that there are 2^n phenotypic classes in the F_2, where n is the number of independently assorting, heterozygous gene pairs (Table 11.3). (This rule holds *only* when a true dominant–recessive relationship holds for each of the gene pairs.) Furthermore, we saw that there are 3 genotypic classes in the F_2 of monohybrid crosses, 9 in dihybrid crosses, and 27 in trihybrid crosses. A simple rule is that the number of genotypic classes is 3^n, where n is the number of independently assorting, heterozygous gene pairs. (See Table 11.3.)

Incidentally, the phenotypic rule (2^n) can also be used to predict the number of classes that will come from a multiple heterozygous F_1 used in a testcross. Here, the number of genotypes in the next generation will be the same as the number of phenotypes. For example, from $Aa\ Bb \times aa\ bb$, there are four progeny genotypes (2^n, where n is 2)—$Aa\ Bb$, $Aa\ bb$, $aa\ Bb$, and $aa\ bb$—and four phenotypes:

1. Both dominant phenotypes, A and B.
2. The A dominant phenotype and b recessive phenotype.
3. The a recessive phenotype and B dominant phenotype.
4. Both recessive phenotypes, a and b.

The "Rediscovery" of Mendel's Principles

Mendel published his treatise on heredity in 1866 in *Verhandlungen des Naturforschenden Vereines* in Brünn, but it received little attention from the scientific community at the time. In 1985, Iris and Laurence Sandler proposed a possible reason. They contend that it may have been impossible for the scientific community from 1865 to 1900 to understand the significance of Mendel's work because it did not fit into that community's conception of the relationship of heredity to other sciences. To Mendel's contemporaries, heredity included not only those ideas that are today considered as genetic, but also those that are considered developmental. In other words, their concept of heredity included what we now know as genetics and embryology. More pertinently, they also viewed heredity as simply a particular moment in development and not as a distinct process requiring special analysis. By 1900, conceptions had changed enough that the significance of Mendel's work was more apparent.

In 1900, three botanists—Carl Correns, Hugo de Vries, and Erich von Tschermark—independently came to the same conclusions as Mendel. Each was working with different plant hybrids, Correns with maize (corn) and peas, de Vries with several different plant species, and von Tschermark with peas. From their experiments, each deduced the basic laws of genetic inheritance, thinking they were the first to do so. However, in preparing their conclusions for publication, they discovered that those laws had already been published by Mendel several decades earlier. Nonetheless, their work was important in that their rediscovery of Mendelian principles brought to the now more mature scientific world an awareness of the laws of genetic inheritance, and set in motion research on gene structure and function that was so productive in the twentieth century.

That Mendelism applied to animals came in 1902 from the work of William Bateson, who experimented with fowl. Bateson also coined the terms, *character, genetics, zygote,* F_1, F_2, and **allelomorph** (literally, "alternative form," meaning one of an array of different forms of a gene), which other researchers shortened to *allele.* The term *gene* as a replacement for Mendelian *factor* was introduced by W. L. Johannsen in 1909. 'Gene' derives from the Greek word 'genos,' meaning birth.

Statistical Analysis of Genetic Data: The Chi-Square Test

Data from genetic crosses are quantitative. A geneticist typically uses statistical analysis to interpret a set of data from crossing experiments to understand the significance of any deviation of observed results from the results predicted by the hypothesis being tested. The observed phenotypic ratios among progeny rarely exactly match expected ratios due to chance factors inherent in biological phenomena. A hypothesis is developed based on the observations and is presented as a **null hypothesis,** which states that there is no real difference between the observed data and the predicted data. Statistical analysis is used to determine whether the difference is due to chance. If it is not, then the null hypothesis is rejected, and a new hypothesis must be developed to explain the data.

A simple statistical analysis used to test null hypotheses is called the **chi-square** (χ^2) test, which is a type of *goodness-of-fit test*. In the genetic crosses we have examined so far, the progeny seemed to fit particular ratios (such as 1:1, 3:1 and 9:3:3:1), and this is where a null hypothesis can be posed and where the chi-square test can tell us whether the data support that hypothesis.

To illustrate the use of the chi-square test, we will analyze theoretical progeny data from a testcross of a smooth, yellow double heterozygote (*Ss Yy*) with a wrinkled, green homozygote (*ss yy*). (See p. 284 and Table 11.2.) (Additional applications of the chi-square test are given in Chapter 14.) The progeny data are as follows:

154 smooth, yellow

124 smooth, green

144 wrinkled, yellow

146 wrinkled, green

Total 568

We hypothesize that a testcross should give a 1:1:1:1 ratio of the four phenotypic classes if the two genes assort independently, and we use the chi-square test to test the hypothesis, as shown in Table 11.4.

First, in column 1 we list the four classes expected in the progeny of the cross. Then we list the observed (*o*) numbers for each phenotype, using actual numbers, not percentages or proportions (column 2). Next, we calculate the expected (*e*) number for each phenotypic class, given the total number of progeny (568) and the hypothesis under evaluation (in this case, a ratio of 1:1:1:1). Thus, in column 3 we list $^1/_4 \times 568 = 142$. Now we sub-

tract the expected number (*e*) from the observed number (*o*) for each class to find differences, called the deviation value (*d*).

In column 5, the deviation squared (d^2) is computed by multiplying each deviation value in column 4 by itself. In column 6, the deviation squared is then divided by the expected number (*e*). The chi-square value, χ^2 (item 7 in the table), is the total of all the values in column 6. The more the observed data deviate from the data expected on the basis of the hypothesis being tested, the higher chi-square is. In our example, $\chi^2 = 3.43$. The general formula is

$$\chi^2 = \Sigma \frac{d^2}{e}, \text{ where } \Sigma \text{ means "sum" and } d^2 = (o - e)^2.$$

The last value in the table, item 8, is the degrees of freedom (df) for the set of data. The degrees of freedom in a test involving *n* classes are usually equal to $n - 1$. There are four phenotypic classes here, so in this case, df = 3.

The chi-square value and the degrees of freedom are next used to determine the probability (*P*) that the deviation of the observed values from the expected values is due to chance. The *P* value for a set of data is obtained from tables of chi-square values for various degrees of freedom. Table 11.5 is part of a table of chi-square probabilities. For our example—$\chi^2 = 3.43$, with 3 degrees of freedom—the *P* value is between 0.30 and 0.50. This is interpreted to mean that, with the hypothesis being tested, in 30 to 50 out of 100 trials (that is, 30 to 50 percent of the time) we could expect chi-square values of such magnitude or greater due to chance. We can reasonably regard this deviation as simply due to chance. We must be cautious how we use the result obtained, however, because a result like this does not tell us that the hypothesis is *correct*: It indicates only that the experimental data provide no statistically compelling argument against the hypothesis.

As a general rule, if the probability of obtaining the observed chi-square values is greater than 5 in 100 (5 percent of the time, $P > 0.05$), then the deviation of expected from observed is not considered statistically significant, and the data do not indicate that the hypothesis should be rejected.

Suppose that, in another chi-square analysis of a different set of data, we obtained $\chi^2 = 15.85$, with 3 degrees of freedom. By looking up the value in Table 11.5, we see that the *P* value is less than 0.01 and greater than 0.001 ($0.001 < P < 0.01$), which means that, from 0.1 to 1 times out of 100 (0.1 to 1 percent of the time), we could expect chi-square values of this magnitude or greater due to chance with the hypothesis being true. That this *P* value is less than 0.05 indicates that, because of the poor fit, the results are not statistically consistent with the 1:1:1:1 hypothesis being tested.

Table 11.4 Chi-Square Test Example

(1)	(2)	(3)	(4)	(5)	(6)
	Observed	Expected			
	Number	Number	*d*	d^2	d^2/e
Phenotypes	(*o*)	(*e*)	(= *o* : *e*)		
Smooth, yellow	154	142	+12	144	1.01
Smooth, green	124	142	−18	324	2.28
Wrinkled, yellow	144	142	+2	4	0.03
Wrinkled, green	146	142	+4	16	0.11
Total	568	568	0		3.43

(7) $\chi^2 = 3.43$ (8) Degrees of freedom (df) = 3

| Table 11.5 | Chi-Square Probabilities | | | | | | | | | |

| | **Probabilities** | | | | | | | | | |
df	0. 95	0.90	0.70	0.50	0.30	0.20	0.10	0.05	0.01	0.001
1	0.004	0.016	0.15	0.46	1.07	1.64	2.71	3.84	6.64	10.83
2	0.10	0.21	0.71	1.39	2.41	3.22	4.61	5.99	9.21	13.82
3	0.35	0.58	1.42	2.37	3.67	4.64	6.25	7.82	11.35	16.27
4	0.71	1.06	2.20	3.36	4.88	5.99	7.78	9.49	13.28	18.47
5	1.15	1.61	3.00	4.35	6.06	7.29	9.24	11.07	15.09	20.52
6	1.64	2.20	3.83	5.35	7.23	8.56	10.65	12.59	16.81	22.46
7	2.17	2.83	4.67	6.35	8.38	9.80	12.02	14.07	18.48	24.32
8	2.73	3.49	5.53	7.34	9.52	11.03	13.36	15.51	20.09	26.13
9	3.33	4.17	6.39	8.34	10.66	12.24	14.68	16.92	21.67	27.88
10	3.94	4.87	7.27	9.34	11.78	13.44	15.99	18.31	23.21	29.59
11	4.58	5.58	8.15	10.34	12.90	14.63	17.28	19.68	24.73	31.26
12	5.23	6.30	9.03	11.34	14.01	15.81	18.55	21.03	26.22	32.91
13	5.89	7.04	9.93	12.34	15.12	16.99	19.81	22.36	27.69	34.53
14	6.57	7.79	10.82	13.34	16.22	18.15	21.06	23.69	29.14	36.12
15	7.26	8.55	11.72	14.34	17.32	19.31	22.31	25.00	30.58	37.70
20	10.85	12.44	16.27	19.34	22.78	25.04	28.41	31.41	37.57	45.32
25	14.61	16.47	20.87	24.34	28.17	30.68	34.38	37.65	44.31	52.62
30	18.49	20.60	25.51	29.34	33.53	36.25	40.26	43.77	50.89	59.70
50	34.76	37.69	44.31	49.34	54.72	58.16	63.17	67.51	76.15	86.66

$\longleftarrow$ ——— | ——— $\longrightarrow$

Fail to reject | Reject

at 0.05 level

Source: From Table IV in *Statistical Tables for Biological, Agricultural, and Medical Research* by Fisher and Yates, 6th ed., 1974. Reprinted by permission of Addison-Wesley Longman Ltd.

Mendelian Genetics in Humans

After the rediscovery of Mendel's laws, geneticists found that the inheritance of genes follows the same principles in all sexually reproducing eukaryotes, including humans. W. Farabee, in 1905, was the first to document a genetic trait in humans, *brachydactyly* (OMIM 112500 at http://www3.ncbi.nlm.nih.gov/OMIM/), which results in abnormally broad and short fingers. (See Figure 11.15.)

By analyzing the trait in human families, Farabee learned that brachydactyly is inherited. The pattern of transmission of the abnormality over several generations led to the conclusion that the trait is a simple dominant trait. In this section, we explore some of the methods used to determine the mechanism of hereditary transmission in humans and learn about some inherited human traits.

Pedigree Analysis

The study of human genetics is complicated, because controlled matings of humans are not possible for ethical reasons. The inheritance patterns of human traits usually are identified by examining the way the trait occurs in the family trees of individuals who clearly exhibit the trait. Such a study of a family tree, called **pedigree analysis,** involves carefully assembling phenotypic records of the family over several generations. The affected individual through whom the pedigree is discovered is called the **proband** (**propositus** if a male, **proposita** if a female).

Figure 11.15

Photographs of (a) normal hands and (b) hands with brachydactyly.

a)   b)

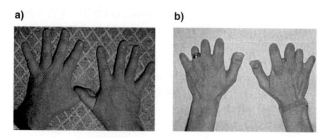

Figure 11.16
Symbols used in human pedigree analysis.

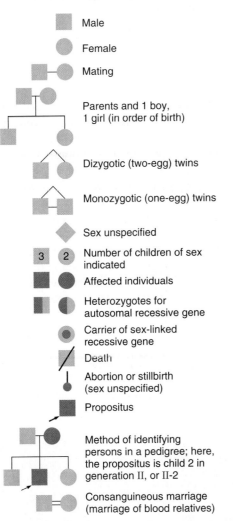

- Male
- Female
- Mating
- Parents and 1 boy, 1 girl (in order of birth)
- Dizygotic (two-egg) twins
- Monozygotic (one-egg) twins
- Sex unspecified
- Number of children of sex indicated
- Affected individuals
- Heterozygotes for autosomal recessive gene
- Carrier of sex-linked recessive gene
- Death
- Abortion or stillbirth (sex unspecified)
- Propositus
- Method of identifying persons in a pedigree; here, the propositus is child 2 in generation II, or II-2
- Consanguineous marriage (marriage of blood relatives)

Figure 11.17
A human pedigree, illustrating the use of pedigree symbols.

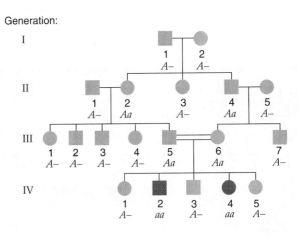

Figure 11.16 summarizes the basic symbols used in pedigree analysis. (The terms *autosomal* and *sex linked* are explained in Chapter 3; they are included here for completeness.) Figure 11.17 presents a hypothetical pedigree to show how the symbols are assigned to the family tree.

The trait presented in Figure 11.17 is determined by a recessive mutant allele *a*. (Note that recessive mutant alleles may be rare or common in a population.) Generations are numbered with Roman numerals, and individuals are numbered with Arabic numerals, which makes it easy to refer to particular people in the pedigree. The trait in the pedigree presented in Figure 11.17 results from homozygosity for the allele, in this case resulting from cousins mating. Since cousins share a fair proportion of their genes, a number of alleles are homozygous in their offspring. Here, one mutant recessive allele became homozygous and resulted in an identifiable genetic trait.

Gene symbols are included in this pedigree to show the deductive reasoning possible with such analysis. The

trait appears first in generation IV. Since neither parent (the two cousins) had the trait, but they produced two children with the trait (IV-2 and IV-4), the simplest hypothesis is that the trait is caused by a recessive allele. Thus, IV-2 and IV-4 would both have the genotype *aa*, and their parents (III-5 and III-6) must have the genotype *Aa*. All other individuals who did not have the trait must have at least one *A* allele—that is, they must be *A–* (either *AA* or *Aa*). Because III-5 and III-6 are both heterozygotes, at least one of each of their parents must have carried an *a* allele. Furthermore, because the trait appeared only after cousins had children, the simplest assumption is that the *a* allele was inherited from individuals with bloodlines shared by III-5 and III-6. This means that II-2 and II-4 probably are both *Aa* and that one of I-1 and I-2 is *Aa* (perhaps both, unless the allele is rare).

Examples of Human Genetic Traits
Recessive Traits. A large number of human traits are known to be caused by homozygosity for mutant alleles that are recessive to the normal allele. Such recessive mutant alleles produce mutant phenotypes because of a *loss of function* or a modified function of the gene product, either of them resulting from the mutation involved.

Many serious abnormalities or diseases result from homozygosity for recessive mutant alleles. Two individuals expressing the recessive trait of *albinism* (deficient pigmentation; OMIM 203100) are shown in Figure 11.18a, and a pedigree for this trait is shown in Figure 11.18b. Individuals with albinism do not produce the pigment melanin, which protects the skin from harmful ultraviolet radiation. As a consequence, their skin and eyes are very sensitive to sunlight. Frequencies of harmful recessive mutant alleles usually are higher than frequencies of harmful dominant mutant alleles because heterozygotes for the recessive

Figure 11.18
Albinism. (a) Two individuals with albinism: blues musicians Johnny (left) and Edgar Winter (right). (b) A pedigree showing the transmission of the autosomal recessive trait of albinism.

a)

b) Pedigree

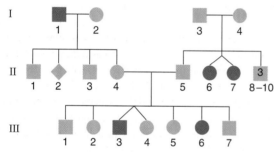

mutant allele are not at a significant selective disadvantage. Nonetheless, individuals homozygous for recessive mutant alleles usually are rare. In the United States, approximately 1 in 17,000 of the white population and 1 in 28,000 of the African-American population have albinism. Among the Irish, about 1 in 10,000 have albinism.

The following are some general characteristics of recessive inheritance for a rare trait:

1. Most affected individuals have two normal parents, both of whom are heterozygous. The trait appears in the F_1 because a quarter of the progeny are expected to be homozygous for the recessive allele. If the trait is rare, an individual expressing the trait is likely to mate with a homozygous normal individual. The next generation from such a mating would be heterozygotes who do not express the trait. In other words, recessive traits often skip generations. In the pedigree in Figure 11.18b, for example, II-6 and II-7 must both be *aa*, and this means both parents (I-3 and I-4) must be *Aa* heterozygotes. I-1 is also *aa*, so II-4 must be *Aa*. Since II-4 and II-5 produce some *aa* children, II-5 also must be *Aa*.

2. Matings between two normal heterozygotes should produce an approximately 3:1 ratio of normal progeny to progeny exhibiting the recessive trait. However, in the analysis of human populations (families), it is difficult to obtain a large enough sample to make the data statistically significant.

3. When both parents are affected, they are homozygous for the recessive trait, and all their progeny usually exhibit the trait.

Dominant Traits. There are many known dominant human traits. Dominant mutant alleles may produce mutant phenotypes because of **gain-of-function mutations** that result in gene products with new functions. In other words, the dominant mutant phenotype is a new property of the mutant gene rather than a decrease in its normal activity. Figure 11.19a illustrates one such trait, called *woolly hair*, in which an individual's hair is very tightly kinked, is very brittle, and breaks before it can grow very long. The best examples of pedigrees for this trait come from Norwegian families; one of these pedigrees is presented in Figure 11.19b. Since it is a fairly rare trait and since not all children of an affected parent show the trait, most woolly haired individuals probably are heterozygous for the dominant allele involved rather than homozygous.

Figure 11.19
Woolly hair. (a) Members of a Norwegian family, some of whom exhibit the trait of woolly hair. (b) Part of a pedigree showing the transmission of the autosomal dominant trait of woolly hair.

a)

b) Generation:

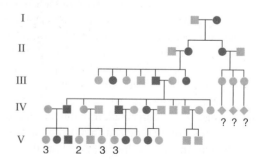

Dominant mutant alleles are expressed in a heterozygote when they are in combination with the wild-type allele. Because many dominant mutant alleles that give rise to recognizable traits are rare, it is extremely unusual to find individuals homozygous for the dominant allele. An affected person in a pedigree is likely to be a heterozygote, and most pairings that involve the mutant allele are between a heterozygote and a homozygous recessive (wild type). Most dominant mutant genes that are clinically significant (that is, cause medical problems) fall into this category.

The following are some general characteristics of dominant inheritance for a rare trait (refer to Figure 11.18b):

1. Every affected person in the pedigree must have at least one affected parent.
2. The trait usually does not skip generations.
3. On average, an affected heterozygous individual will transmit the mutant gene to half of his or her progeny. If the dominant mutant allele is designated A and its wild-type allele is a, then most crosses will be $Aa \times aa$. From basic Mendelian principles, half the progeny will be aa (wild type) and the other half will be Aa (and show the trait).

Other examples of human dominant traits are *achondroplasia* (OMIM 100800; dwarfism resulting from defects in long-bone growth), *brachydactyly* (malformed hands with short fingers), and Marfan syndrome (OMIM 154700; connective tissue defects, potentially causing death by aortic rupture).

KEYNOTE

> Mendelian principles apply to humans and all other eukaryotes. The study of the human inheritance of genetic traits is complicated by the fact that no controlled crosses can be done. Instead, human geneticists analyze genetic traits by pedigree analysis.

Summary

In this chapter, we discussed fundamental principles of gene segregation and gene assortment. Genes are DNA segments that control the biological characteristics transmitted from one generation to another—that is, the hereditary traits. An organism's genetic constitution is its genotype, and the physical manifestation of a genetic trait is its phenotype. An organism's genes provide only the potential for the development of that organism's characteristics. That potential is influenced during development by interactions with other genes and with the environment. Individuals with the same genotype can have different phenotypes, and individuals with the same phenotype may have different genotypes.

The first person to obtain some understanding of the principles of heredity—that is, the inheritance of certain traits—was Gregor Mendel. From his breeding experiments with garden peas, Mendel proposed two basic principles of genetics. In modern terms, the principle of segregation states that the two members of a single gene pair (the alleles) segregate from each other in the formation of gametes. For each gene with two alleles, half the gametes carry one allele, and the other half carry the other allele. The principle of independent assortment, proposed on the basis of experiments involving more than one gene, states that genes for different traits behave independently in the production of gametes. Both principles are recognized by characteristic phenotypic ratios—called Mendelian ratios—in particular crosses. For the principle of segregation, in a monohybrid cross between two true-breeding parents, one exhibiting a dominant phenotype and the other a recessive phenotype, the F_2 phenotypic ratio is 3:1 for the dominant : recessive phenotypes. For the principle of independent assortment, in a dihybrid cross, the F_2 phenotypic ratio is 9:3:3:1 for the four phenotypic classes.

The gene segregation patterns can be studied more definitively by determining the genotypes for each phenotypic class. This is done by using a testcross, in which an individual of unknown genotype is crossed with a homozygous recessive individual to determine the unknown genotype. For example, in a monohybrid cross resulting in an F_2 3:1 phenotypic ratio, the dominant class can be shown to consist of 1 homozygous dominant : 2 heterozygotes by using a testcross.

Geneticists have found that Mendelian principles of gene segregation apply to all eukaryotes, including humans. The study of the inheritance of genetic traits in humans is complicated because no controlled crosses can be done. Instead, human geneticists analyze genetic traits by pedigree analysis—that is, by examining the occurrences of the trait in family trees of individuals who clearly exhibit the trait. Many recessively inherited and dominantly inherited genetic traits have been identified by pedigree analysis.

Analytical Approaches to Solving Genetics Problems

The most practical way to reinforce genetics principles is to solve genetics problems. In this and all following chapters, we discuss how to approach genetics problems by presenting examples of such problems and by discussing the answers to those problems. The following problems use familiar and unfamiliar examples and pose questions designed to get you to think analytically:

Q11.1 A purple-flowered pea plant is crossed with a white-flowered pea plant. All the F_1 plants produce purple flowers.

When the F_1 plants are allowed to self-pollinate, 401 of the F_2s have purple flowers and 131 have white flowers. What are the genotypes of the parental and F_1 generation plants?

A11.1 The ratio of plant phenotypes in the F_2 is 3.06:1, which is very close to the 3:1 ratio expected of a monohybrid cross. More specifically, this ratio is expected to result from an $F_1 \times F_1$ cross in which both are heterozygous for a specific gene pair. In addition, because the two parents differed in phenotype and only one phenotypic class appeared in the F_1, it is likely that both parental plants were true breeding. Furthermore, because the F_1 phenotype exactly resembled one of the parental phenotypes, we can say that purple is dominant to white flowers. Assigning the symbol P to the gene that determines purpleness of flowers and the symbol p to the alternative form of the gene that determines whiteness, we can write the genotypes:

P generation: *PP*, for the purple-flowered plant;
                pp, for the white-flowered plant

F_1 generation: *Pp*, which, because of dominance, is purple flowered

We could further deduce that the F_2 plants have an approximately 1:2:1 ratio of *PP* : *Pp* : *pp* by performing testcrosses.

Q11.2 Consider three gene pairs *Aa*, *Bb*, and *Cc*, each of which affects a different character. In each case, the uppercase letter signifies the dominant allele and the lowercase letter the recessive allele. These three gene pairs assort independently of each other. Calculate the probability of obtaining the following:
a. an *Aa BB Cc* zygote from a cross of individuals that are *Aa Bb Cc* × *Aa Bb Cc*
b. an *Aa BB cc* zygote from a cross of individuals that are *aa BB cc* × *AA bb CC*
c. an *A B C* phenotype (that is, having the dominant phenotypes for each of the three genes) from a cross of individuals that are *Aa Bb CC* × *Aa Bb cc*
d. an *a b c* phenotype (that is, having the recessive phenotypes for each of the three genes) from a cross of individuals that are *Aa Bb Cc* × *aa Bb cc*

A11.2 We must break down the question into simple parts in order to apply basic Mendelian principles. The key is that the genes assort independently, so we must multiply the probabilities of the individual occurrences to obtain the answers.
a. First, we must consider the *Aa* gene pair. The cross is *Aa* × *Aa*, so the probability of the zygote being *Aa* is $2/4$, because the expected distribution of genotypes is 1 *AA* : 2 *Aa* : 1 *aa*. Then, following the same logic, the probability of *BB* from *Bb* × *Bb* is $1/4$, and that of *Cc* from *Cc* × *Cc* is $2/4$. Using the product rule (see Box 2.2), we find that the probability of an *Aa BB Cc* zygote is $1/2 \times 1/4 \times 1/2 = 1/16$.

b. Similar logic is needed here, although, because they differ from one gene pair to another, we must be sure of the genotypes of the parental types. For the *Aa* pair, the probability of getting *Aa* from *AA* × *aa* has to be 1. Next, the probability of getting *BB* from *BB* × *bb* is 0, so on these grounds alone, we cannot get the zygote asked for from the cross given.
c. This question and the next ask for the probability of getting a particular phenotype, so we must start thinking about dominance. Again, we consider each character pair in turn. From basic Mendelian principles, the probability of an *A* phenotype from *Aa* × *Aa* is $3/4$. Similarly, the probability of a *B* phenotype from *Bb* × *Bb* is $3/4$. Lastly, the probability of a *C* phenotype from *CC* × *cc* is 1. Overall, the probability of an *A B C* phenotype is $3/4 \times 3/4 \times 1 = 9/16$.
d. The probability of an *a b c* phenotype from *Aa Bb Cc* × *aa Bb cc* is $1/2 \times 1/4 \times 1/2 = 1/16$.

Q11.3 In chickens, the white plumage of the leghorn breed is dominant over colored plumage, feathered shanks are dominant over clean shanks, and pea comb is dominant over single comb. Each of the gene pairs segregates independently. If a homozygous white, feathered, pea-combed chicken is crossed with a homozygous colored, clean, single-comb chicken and the F_1s are allowed to interbreed, what proportion of the birds in the F_2 will produce only white, feathered, pea-combed progeny if mated to colored, clean-shanked, single-combed birds?

A11.3 This example is typical of a question that presents the unfamiliar in an attempt to get at the familiar. The best approach to such questions is to reduce them to their simplest parts and, whenever possible, to assign gene symbols for each character. We are told which character is dominant for each of the three gene pairs, so we can use *W* for white and *w* for colored, *F* for feathered and *f* for clean shanks, and *P* for pea comb and *p* for single comb. The cross involves true-breeding strains and can be written as follows:

P generation: *WW FF PP* × *ww ff pp*
F_1 generation: *Ww Ff Pp*

Now, the question asks for the proportion of the birds in the F_2 that will produce only white, feathered, peacombed progeny if mated to colored, clean-shanked, single-combed birds. The latter are homozygous recessive for all three genes—that is, *ww ff pp*, as in the parental generation. For the requested result, the F_2 birds must be white, feathered, and pea-combed, and they must be homozygous for the dominant alleles of the respective genes to produce only progeny with the dominant phenotype. What we are seeking, then, is the proportion of the F_2 chickens that are *WW FF PP* in genotype. We know that each gene pair segregates independently; thus, the answer can be calculated by using simple probability rules. We consider each gene pair in turn. For the white versus colored case, the $F_1 \times F_1$ is $Ww \times Ww$, and we know from Mendelian principles that the relative proportion of F_2 genotypes is $1\ WW$: $2\ Ww$: $1\ ww$. Therefore, the proportion of the F_2s that will be WW is $\frac{1}{4}$. The same relationship holds for the other two pairs of genes. Because the segregation of the three gene pairs is independent, we must multiply the probabilities of each occurrence to calculate the probability for $WW\ FF\ PP$ individuals. The answer is $\frac{1}{4} \times \frac{1}{4} \times \frac{1}{4} = \frac{1}{64}$.

Questions and Problems

*11.1 In tomatoes, red fruit color is dominant to yellow. Suppose a tomato plant homozygous for red is crossed with one homozygous for yellow. Determine the appearance of
a. the F_1 tomatoes,
b. the F_2 tomatoes,
c. the offspring of a cross of the F_1 tomatoes back to the red parent,
d. the offspring of a cross of the F_1 tomatoes back to the yellow parent.

11.2 In maize, a dominant allele A is necessary for seed color, as opposed to colorless (a). Another gene has a recessive allele wx that results in waxy starch, as opposed to normal starch (Wx). The two genes segregate independently. An $Aa\ WxWx$ plant is testcrossed. What are the phenotypes and relative frequencies of offspring?

*11.3 F_2 plants segregate $\frac{3}{4}$ colored : $\frac{1}{4}$ colorless. If a colored plant is picked at random and selfed, what is the probability that both colored and colorless plants will be seen among a large number of its progeny?

*11.4 In guinea pigs, rough coat (R) is dominant over smooth coat (r). A rough-coated guinea pig is bred to a smooth one, giving eight rough and seven smooth progeny in the F_1 generation.
a. What are the genotypes of the parents and their offspring?
b. If one of the rough F_1 animals is mated to its rough parent, what progeny would you expect?

11.5 In cattle, the polled (hornless) condition (P) is dominant over the horned (p) phenotype. A particular polled bull is bred to three cows. Cow A, which is horned, produces a horned calf; polled cow B produces a horned calf; and horned cow C produces a polled calf. What are the genotypes of the bull and the three cows,

and what phenotypic ratios do you expect in the offspring of these three matings?

*11.6 In jimsonweed, purple flowers are dominant to white. Self-fertilization of a particular purple-flowered jimsonweed produces 28 purple-flowered and 10 white-flowered progeny. What proportion of the purple-flowered progeny will breed true?

*11.7 Two black female mice are crossed with the same brown male. In a number of litters, female X produced 9 blacks and 7 browns, and female Y produced 14 blacks. What is the mechanism of inheritance of black and brown coat color in mice? What are the genotypes of the parents?

11.8 Bean plants may have different symptoms when infected with a virus. Some show local lesions that do not seriously harm the plant; others show general systemic infection. The following genetic analysis was made:
P   local lesions × systemic infection
F_1 all local lesions
F_2 785 local lesions : 269 systemic infection
What is the likely genetic basis of this difference in beans? Assign gene symbols to the genotypes occurring in the genetic analysis. Design a testcross to verify your assumptions.

11.9 A normal *Drosophila* (fruit fly) has both brown and scarlet pigment granules in its eyes, which appear red as a result. Brown (bw) is a recessive allele on chromosome 2 that, when homozygous, results in brown eyes because of the absence of scarlet pigment granules. Scarlet (st) is a recessive allele on chromosome 3 that, when homozygous, results in scarlet eyes because of the absence of brown pigment granules. Any fly homozygous for recessive alleles at both genes produces no eye pigment and has white eyes. The following results were obtained from crosses:
P   brown-eyed fly × scarlet-eyed fly
F_1 red eyes (both brown and scarlet pigment present)
F_2 $\frac{9}{16}$ red : $\frac{3}{16}$ scarlet : $\frac{3}{16}$ brown : $\frac{1}{16}$ white
a. Assign genotypes to the P and F_1 generations.
b. Design a testcross to verify the F_1 genotype, and predict the results.

*11.10 Grey seed color (G) in garden peas is dominant to white seed color (g). In the following crosses, the indicated parents with known phenotypes, but unknown genotypes, produced the progeny listed:

| Parents | Progeny | | Female Parent |
Female × Male	Grey	White	Genotype
grey × white	81	82	?
grey × grey	118	39	?
grey × white	74	0	?
grey × grey	90	0	?

On the basis of the segregation data, give the possible genotypes of each female parent.

*11.11 Fur color in the babbit, a furry little animal and popular pet, is determined by a pair of alleles, B and b. BB and Bb babbits are black, and bb babbits are white. A farmer wants to breed babbits for sale. True-breeding white (bb) female babbits breed poorly. The farmer purchases a pair of black babbits, and these mate and produce six black and two white offspring. The farmer immediately sells his white babbits, and then he comes to consult you for a breeding strategy to produce more white babbits.

a. If he performed random crosses between pairs of F_1 black babbits, what proportion of the F_2 progeny would be white?
b. If he crossed an F_1 male to the parental female, what is the probability that this cross would produce white progeny?
c. What would be the farmer's best strategy to maximize the production of white babbits?

11.12 In jimsonweed, purple flower (P) is dominant to white (p) and spiny pods (S) are dominant to smooth (s). A true-breeding plant with white flowers and spiny pods is crossed to a true-breeding plant with purple flowers and smooth pods. Determine the phenotype of

a. the F_1 generation;
b. the F_2 generation;
c. the progeny of a cross of the F_1s back to the white, spiny parent; and
d. the progeny of a cross of the F_1 back to the purple, smooth parent.

11.13 Use the information in Problem 11.12 to determine what progeny you would expect from the following jimsonweed crosses (you are encouraged to use the branch diagram approach):

a. $PP\ ss \times pp\ SS$
b. $Pp\ SS \times pp\ ss$
c. $Pp\ Ss \times Pp\ SS$
d. $Pp\ Ss \times Pp\ Ss$
e. $Pp\ Ss \times Pp\ ss$
f. $Pp\ Ss \times pp\ ss$

*11.14 Cleopatra normally is a very refined cat. When she finds even a small amount of catnip, however, she purrs madly, rolls around in the catnip, becomes exceedingly playful, and appears intoxicated. Cleopatra and Antony, who walks past catnip with an air of indifference, have produced five kittens who respond to catnip just as Cleopatra does. When the kittens mature, two of them mate and produce four kittens that respond to catnip and one that does not. When another of Cleopatra's daughters mates with Augustus (a nonrelative), who behaves just like Antony, three catnip-sensitive and two catnip-insensitive kittens are produced. Propose a hypothesis for the inheritance of catnip sensitivity that explains these data.

*11.15 In summer squash, white fruit (W) is dominant over yellow (w), and disk-shaped fruit (D) is dominant over sphere-shaped fruit (d). Determine the genotypes of the parents in each of the following crosses:

a. White, disk × yellow, sphere gives $\frac{1}{2}$ white, disk and $\frac{1}{2}$ white, sphere.
b. White, sphere × white, sphere gives $\frac{3}{4}$ white, sphere and $\frac{1}{4}$ yellow, sphere.
c. Yellow, disk × white, sphere gives all white, disk progeny.
d. White, disk × yellow, sphere gives $\frac{1}{4}$ white, disk; $\frac{1}{4}$ white, sphere; $\frac{1}{4}$ yellow, disk; and $\frac{1}{4}$ yellow, sphere.
e. White, disk × white, sphere gives $\frac{3}{8}$ white, disk; $\frac{3}{8}$ white, sphere; $\frac{1}{8}$ yellow, disk; and $\frac{1}{8}$ yellow, sphere.

*11.16 Genes a, b, and c assort independently and are recessive to their respective alleles A, B, and C. Two triply heterozygous (Aa Bb Cc) individuals are crossed.

a. What is the probability that a given offspring will be phenotypically A B C—that is, will exhibit all three dominant traits?
b. What is the probability that a given offspring will be homozygous for all three dominant alleles?

11.17 In garden peas, tall stem (T) is dominant over short stem (t), green pods (G) are dominant over yellow pods (g), and smooth seeds (S) are dominant over wrinkled seeds (s). Suppose a homozygous short, green, wrinkled pea plant is crossed with a homozygous tall, yellow, smooth one.

a. What will be the appearance of the F_1 generation?
b. If the F_1 plants are interbred, what will be the appearance of the F_2 generation?
c. What will be the appearance of the offspring of a cross of the F_1 back to the short, green, wrinkled parent?
d. What will be the appearance of the offspring of a cross of the F_1 back to the tall, yellow, smooth parent?

11.18 C and c, O and o, and I and i are three independently segregating pairs of alleles in chickens. C and O are dominant alleles, both of which are necessary for pigmentation. I is a dominant inhibitor of pigmentation. Individuals of genotype cc, oo, Ii, or II are white, regardless of what other genes they possess.

Assume that White Leghorns are CC OO II, White Wyandottes are cc OO ii, and White Silkies are CC oo ii. What types of offspring (white or pigmented) are possible, and what is the probability of each, from the following crosses?

a. White Silkie × White Wyandotte
b. White Leghorn × White Wyandotte
c. (Wyandotte–Silkie F_1) × White Silkie

11.19 Two homozygous strains of corn are hybridized. They are distinguished by six different pairs of genes, all of which assort independently and produce an independent phenotypic effect. The F_1 hybrid is selfed to give an F_2 generation.
a. What is the number of possible genotypes in the F_2 plants?
b. How many of these genotypes will be homozygous at all six gene loci?
c. If all gene pairs act in a dominant–recessive fashion, what proportion of the F_2 plants will be homozygous for all dominants?
d. What proportion of the F_2 will show all dominant phenotypes?

*11.20 The coat color of mice is controlled by several genes. The agouti pattern, characterized by a yellow band of pigment near the tip of the hairs, is produced by the dominant allele A; homozygous aa mice do not have the band and are nonagouti. The dominant allele B determines black hairs, and the recessive allele b determines brown. Homozygous $c^h c^h$ individuals allow pigments to be deposited only at the extremities (e.g., feet, nose, and ears) in a pattern called Himalayan. The genotype C– allows pigment to be distributed over the entire body.
a. If a true-breeding black mouse is crossed with a true-breeding brown, agouti, Himalayan mouse, what will be the phenotypes of the F_1 and F_2 generation?
b. What proportion of the non-Himalayan black agouti F_2 animals will be $Aa\ BB\ Cc^h$?
c. What proportion of the Himalayan mice in the F_2 generation is expected to show brown pigment?
d. What proportion of all agoutis in the F_2 generation is expected to show black pigment?

11.21 In cocker spaniels, solid coat color is dominant over spotted coat. Suppose a true-breeding, solid-colored dog is crossed with a spotted dog, and the F_1 dogs are interbred.
a. What is the probability that the first puppy born will have a spotted coat?
b. What is the probability that, if four puppies are born, all of them will have solid coats?

11.22 In the F_2 of his cross of red-flowered × white-flowered *Pisum* (pea plant), Mendel obtained 705 plants with red flowers and 224 with white.
a. Is this result consistent with his hypothesis of factor segregation, which predicts a 3:1 ratio?
b. In how many similar experiments would a deviation as great as or greater than this one be expected? (Calculate χ^2 and obtain the approximate value of P from Table 11.5.)

11.23 In tomatoes, cut leaf and potato leaf are alternative characters, with cut (C) dominant to potato (c). Purple stem and green stem are another pair of alternative characters, with purple (P) dominant to green (p). A true-breeding cut, green tomato plant is crossed with a true-breeding potato, purple plant, and the F_1 plants are allowed to interbreed. The 320 F_2 plants were phenotypically 189 cut, purple; 67 cut, green; 50 potato, purple; and 14 potato, green. Propose a hypothesis to explain the data, and use the chi-square test to test the hypothesis.

*11.24 The simple case of two mating types (male and female) is by no means the only sexual system known. The ciliated protozoan *Paramecium bursaria* has a system of four mating types, controlled by two genes (A and B). Each gene has a dominant and a recessive allele. The four mating types are expressed according to the following scheme:

Genotype	Mating Type
AA BB	A
Aa BB	A
AA Bb	A
Aa Bb	A
AA bb	D
Aa bb	D
aa BB	B
aa Bb	B
aa bb	C

It is clear, therefore, that some of the mating types result from more than one possible genotype. We have four strains of known mating type—"A", "B", "C", and "D"— but unknown genotype. The following crosses were made, with the indicated results:

Cross	Mating Type of Progeny			
	A	**B**	**C**	**D**
"A" × "B"	24	21	14	18
"A" × "C"	56	76	55	41
"A" × "D"	44	11	19	33
"B" × "C"	0	40	38	0
"B" × "D"	6	8	14	10
"C" × "D"	0	0	45	45

Assign genotypes to "A", "B", "C", and "D".

*11.25 In bees, males (drones) develop from unfertilized eggs and are haploid. Females (workers and queens) are diploid and come from fertilized eggs. W (black eyes) is dominant over w (white eyes). Workers of genotype RR or Rr use wax to seal crevices in the hive; rr workers use resin instead. A $Ww\ Rr$ queen founds a colony after being fertilized by a black-eyed drone bearing the r allele.

a. What will be the appearance and behavior of workers in the new hive, and what are their relative frequencies?

b. Give the genotypes of male offspring, with relative frequencies.

c. Fertilization normally takes place in the air during a nuptial flight, and any bee unable to fly would effectively be rendered sterile. Suppose a recessive mutation, *c*, occurs spontaneously in a sperm that fertilizes a normal egg, and suppose also that the effect of the mutant gene is to cripple the wings of any adult not bearing the normal allele *C*. The fertilized egg develops into a normal queen named Madonna. What is the probability that wingless males will be *found in a hive* founded two generations later by one of Madonna's granddaughters?

d. By one of Madonna's great-great-granddaughters?

*11.26 Consider the following pedigree, in which the allele responsible for the trait (*a*) is recessive to the normal allele (*A*):

Generation

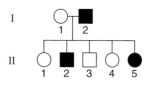

a. What is the genotype of the mother?

b. What is the genotype of the father?

c. What are the genotypes of the children?

d. Given the mechanism of inheritance involved, does the ratio of children with the trait to children without the trait match what would be expected?

11.27 For the pedigrees A and B (see top of next column), indicate whether the trait involved in each case could be recessive or dominant, and explain your answers.

11.28 After a few years of marriage, a woman comes to believe that, among all of the reasonable relatives in her and her husband's families, her husband, her mother-in-law, and her father have so many similarities in their unreasonableness that they must share a mutation. A friend taking a course in genetics assures her that it is unlikely that this trait has a genetic basis and that, even if it did, all of her children would be reasonable. Diagram and analyze the relevant pedigree to evaluate whether the friend's advice is accurate.

*11.29 Gaucher disease is caused by a chronic enzyme deficiency that is more common among Ashkenazi Jews than in the general population. A Jewish man has a sister afflicted with the disease. His parents, grandparents, and

Pedigree A

Generation

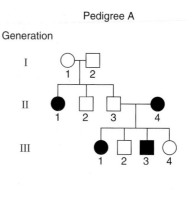

Pedigree B

Generation

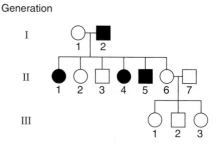

three siblings are not affected. Discussions with relatives in his wife's family reveal that the disease is not likely to be present in her family, although some relatives recall that the brother of his wife's paternal grandmother suffered from a similar disease. Diagram and analyze the relevant pedigree to determine (a) the genetic basis for inheriting the trait for Gaucher disease and (b) the highest probability that, if this couple has a child, the child will be affected (i.e., what is the chance of the worst-case scenario occurring?).

*11.30 a^+, b^+, c^+, and d^+ are independently assorting Mendelian genes controlling the production of a black pigment. The alternate alleles that give abnormal functioning of these genes are *a*, *b*, *c*, and *d*. A black individual of genotype a^+/a^+ b^+/b^+ c^+/c^+ d^+/d^+ is crossed with a colorless individual of genotype *a/a b/b c/c d/d* to produce a black F_1. $F_1 \times F_1$ crosses are then done. Assume that a^+, b^+, c^+, and d^+ act in a pathway as follows:

$$\text{colorless} \xrightarrow{a^+} \text{colorless} \xrightarrow{b^+} \text{colorless} \xrightarrow{c^+} \text{brown} \xrightarrow{d^+} \text{black}$$

a. What proportion of the F_2 progeny is colorless?

b. What proportion of the F_2 progeny is brown?

11.31 Using the genetic information given in Problem 11.30, now assume that a^+, b^+, and c^+ act in a pathway as follows:

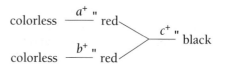

Black can be produced only if both red pigments are present; that is, c^+ converts the two red pigments together into a black pigment.

a. What proportion of the F_2 progeny is colorless?

b. What proportion of the F_2 progeny is red?

c. What proportion of the F_2 progeny is black?

*11.32 Three genes on different chromosomes are responsible for three enzymes that catalyze the same reaction in corn:

$$\text{colorless compound} \xrightarrow{a^+, b^+, c^+} \text{red compound}$$

The normal functioning of any one of these genes is sufficient to convert the colorless compound to the red compound. The abnormal functioning of these genes is designated by a, b, and c, respectively.

a. A red a^+/b^+ b^+/b^+ c^+/c^+ is crossed with a colorless a/a b/b c/c to give a red F_1, a^+/a b^+/b c^+/c. The F_1 is selfed. What proportion of the F_2 progeny is colorless?

b. It turns out that another step is involved in the pathway—one that is controlled by gene d^+, which assorts independently of a^+, b^+, and c^+:

$$\text{colorless} \xrightarrow{d^+} \text{colorless} \xrightarrow{a^+, b^+, c^+} \text{red}$$
$$\text{compound 1} \qquad\qquad \text{compound 2} \qquad\qquad \text{compound}$$

The inability to convert colorless 1 to colorless 2 is designated d. A red a^+/a^+ b^+/b^+ c^+/c^+ d^+/d^+ is crossed with a colorless a/a b/b c/c d/d. The F_1 corn are all red. The red F_1 corn are now selfed. What proportion of the F_2 corn is colorless?

*11.33 In J. R. R. Tolkien's *The Lord of the Rings*, the Black Riders of Mordor ride steeds with eyes of fire. As a geneticist, you are very interested in the inheritance of the fire-red eye color. You discover that the eyes contain two types of pigments—brown and red—that are usually bound to core granules in the eye. In wild-type steeds, precursors are converted by these granules to the afore said pigments, but in steeds homozygous for the recessive X-linked gene w (white eye), the granules remain unconverted and a white eye results. The metabolic pathways for the synthesis of the two pigments are shown in Figure 11.A below.

Each step of the pathway is controlled by a gene: Mutation v results in vermilion eyes, cn results in cinnabar eyes, st results in scarlet eyes, bw results in brown eyes, and se results in black eyes. All these mutations are recessive to their wild-type alleles, and all are unlinked. For the following genotypes, show the proportions of steed eye phenotypes that would be obtained in the F_1 of the given matings:

a. w/w bw^+/bw^+ $st/st \times w^+/Y$ bw/bw st^+/st^+

b. w^+/w^+ se/se $bw/bw \times w/Y$ se^+/se^+ bw^+/bw^+

c. w^+/w^+ v^+/v^+ $bw/bw \times w/Y$ v/v bw/bw

d. w^+/w^+ bw^+/bw $st^+/st \times w/Y$ bw/bw st/st

*11.34 Two different true-breeding strains of corn with colorless kernels A and B are crossed with each of two different true-breeding strains with purple kernels C and D. The F_1s from each cross are selfed, with the following results:

P:	A × C	A × D
F_1:	all purple	all purple
F_2:	3 purple : 1 colorless	3 purple : 1 colorless
P:	B × C	B × D
F_1:	all purple	all purple
F_2:	3 purple : 1 spotted*	3 purple : 1 colorless

*spotted = kernels with purple spots in a colorless background

Propose an explanation for these results based on your understanding of the molecular structures of *Ac* and *Ds* elements, and their transposition, in corn (See Chapter 7, pp. 159–162).

Figure 11.A

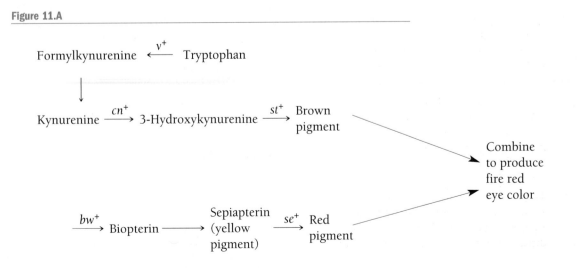

12

Chromosomal Basis of Inheritance

Human X chromosome (left) and Y chromosome (right).

PRINCIPAL POINTS

- Diploid eukaryotic cells have two haploid sets of chromosomes, one set coming from each parent. The members of a pair of chromosomes, one from each parent, are called homologous chromosomes. The complete set of chromosomes in a eukaryotic cell is called its karyotype

- Mitosis is the process of nuclear division in eukaryotic cells represented by M in the cell cycle (that is, G_1, S, G_2, and M). Mitosis results in the production of daughter nuclei that contain identical chromosome numbers and that are genetically identical to one another and to the parent nucleus from which they arose.

- Meiosis occurs in all sexually reproducing eukaryotes. A specialized diploid cell (or cell nucleus) with two haploid sets of chromosomes is transformed through one round of DNA replication and two rounds of nuclear division into four haploid cells (or four nuclei), each with one set of chromosomes.

- Meiosis generates genetic variability through the processes by which maternal and paternal chromosomes are reassorted in progeny nuclei and through

crossing-over between members of a homologous pair of chromosomes.

- The chromosome theory of inheritance states that genes are located on chromosomes.

- In eukaryotes with separate sexes, a sex chromosome is a chromosome or a group of chromosomes that is represented differently in the two sexes. In organisms with sex chromosomes, one sex is homogametic and the other is heterogametic.

- Sex linkage is the physical association of genes with the sex chromosomes of eukaryotes. Such genes are called sex-linked genes.

- The correlation between gene segregation patterns and the patterns of chromosome behavior in meiosis supports the chromosome theory of inheritance.

- In many eukaryotic organisms, sex determination is related to the sex chromosomes. In humans and other mammals, for example, the presence of a Y chromosome specifies maleness, and its absence results in femaleness. Several other sex determination mechanisms are known in eukaryotes.

- In humans, the allele responsible for a trait can be inherited in one of five main ways: autosomal recessive, autosomal dominant, X-linked recessive, X-linked dominant, or Y-linked.

- Since Mendel's time, many exceptions to his rules have been discovered.

i WHEN A CHILD IS BORN, THE FIRST QUESTION MOST people ask is "Is it a boy or a girl?" The answer, at the chromosomal level, depends on the sex chromosomes: Two X chromosomes produce a girl, while an X and a Y chromosome produce a boy. But the genes contained on these chromosomes determine more than just the sex of an individual; they also are responsible for the inheritance of a number of other traits.

In this chapter, you will learn about the behavior of chromosomes during nuclear division in eukaryotes, the ways in which sex is determined in humans and other organisms, and sex-linked traits in humans. After you have read and studied this chapter, you can apply what you've learned by trying the iActivity, in which you will investigate the inheritance of deafness within a family.

On Mendel's foundation, early geneticists began to build genetic hypotheses that could be tested by appropriate crosses, and they began to investigate the nature of Mendelian factors. We now know that Mendelian factors are genes and that genes are located on chromosomes. In this chapter, we focus on the behavior of genes and chromosomes. We start by learning about the transmission of chromosomes from cell division to cell division and from generation to generation by the processes of mitosis and meiosis, respectively. We then consider the evidence for the association of genes and chromosomes. In so doing, we will learn about the segregation of genes located on the sex chromosomes. Next, we learn about various mechanisms of sex determination, and finally, we discuss sex-linked traits in humans. The goal of the chapter is for you to learn how to think about gene segregation in terms of chromosome inheritance patterns.

Chromosomes and Cellular Reproduction

The association between chromosomes and genes was determined as a result of the efforts of cytologists, who examined the behavior of chromosomes, and geneticists, who examined the behavior of genes. In this section, we discuss the general structure of eukaryotic chromosomes and the transmission of chromosomes from cell division

to cell division and from generation to generation by the processes of mitosis and meiosis, respectively.

Eukaryotic Chromosomes

The genetic material of eukaryotes is distributed among multiple, linear chromosomes (see Chapter 11); the number of chromosomes, with rare exceptions, is characteristic of the species. Many eukaryotes have two copies of each type of chromosome in their nuclei, so their chromosome complement is said to be **diploid**, or 2N. Diploid eukaryotes are produced by the fusion of two haploid gametes (mature reproductive cells that are specialized for sexual fusion), one from the female parent and one from the male parent. The fusion produces a diploid **zygote**, which then undergoes embryological development. Each gamete has only one set of chromosomes and is said to be **haploid** (N). The complete compendium of genetic information in a haploid chromosome set is called the **genome.** Two examples of diploid organisms are humans, with 46 chromosomes (23 pairs), and *Drosophila melanogaster*, with 8 chromosomes (4 pairs). By contrast, laboratory strains of the yeast *S. cerevisiae*, with 16 chromosomes, are haploid.

In diploid organisms, the members of a chromosome pair that contain the same genes and that pair during meiosis are called **homologous chromosomes;** each member of a pair is called a **homolog,** and one homolog is inherited from each parent. Chromosomes that contain different genes and that do not pair during meiosis are called **nonhomologous chromosomes.** Figure 12.1 illustrates the chromosomal organization of haploid and diploid organisms.

In animals and in some plants, male and female cells are distinct with respect to their complement of **sex chromosomes**—the chromosomes that are represented differently in the two sexes in many eukaryotic organisms. One sex has a matched pair of sex chromosomes, and the other sex has an unmatched pair of sex chromosomes or

Figure 12.1

Chromosomal organization of haploid and diploid organisms.

Haploid (N)	**Diploid (2N)**
One copy of genetic material subdivided into chromosomes	Two copies of genetic material subdivided into chromosomes
Three nonhomologous chromosomes	Three pairs of homologous chromosomes

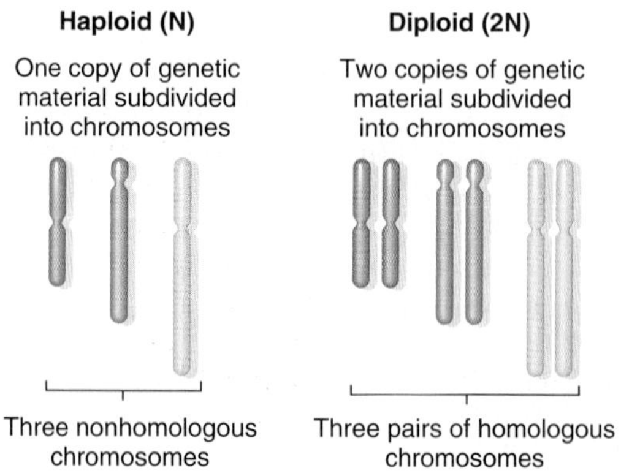

Figure 12.2

General classification of eukaryotic chromosomes as metacentric, submetacentric, acrocentric, and telocentric types, based on the position of the centromere.

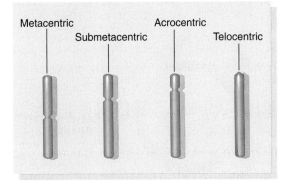

Figure 12.3

G banding in a karyotype of human male metaphase chromosomes.

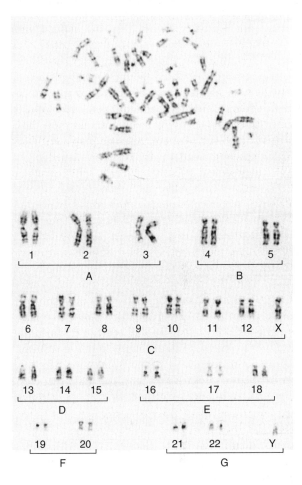

a single sex chromosome. For example, human females have two X chromosomes (XX), whereas human males have one X and one Y (XY). Chromosomes other than sex chromosomes are called **autosomes.** We discuss X chromosomes in more detail later in the chapter.

Under the microscope, we can see that chromosomes differ in size and morphology (appearance) within and between species. Each chromosome often has a constriction along its length called a **centromere,** which is important for the behavior of the chromosomes during cellular division. The location of the centromere in one of four general positions in the chromosome is useful in classifying eukaryotic chromosomes (Figure 12.2). A **metacentric chromosome** has the centromere at about the center, so the chromosome appears to have two approximately equal arms. **Submetacentric chromosomes** have one arm longer than the other, **acrocentric chromosomes** have one arm with a stalk and often with a "bulb" (called a *satellite*) on it, and **telocentric chromosomes** have only one arm, because the centromere is at the end. Chromosomes also vary in relative size. Chromosomes of mice, for example, are all similar in length, whereas those of humans have a wide range of relative lengths. Chromosome length and centromere position are constant for each chromosome and help in identifying individual chromosomes.

A complete set of all the metaphase chromosomes in a cell—consisting of two chromatids per chromosome—is called the cell's **karyotype** ("carry-o-type"; literally, "nucleus type"). Chromosomes at their most condensed in mitosis (metaphase chromosomes; see p. 304) are used for identification, because they are the most compact form during the cell division cycle and therefore are easy to see under the microscope after staining. The karyotype is species specific, so a wide range of numbers, sizes, and shapes of chromosomes is seen among eukaryotic organisms. Even closely related organisms may have quite different karyotypes.

Figure 12.3 shows the karyotype for the cell of a normal human male. It is customary, particularly with human chromosomes, to arrange chromosomes in order according to size and position of the centromere. This karyotype shows 46 chromosomes: two pairs of each of the 22 autosomes ("other chromosomes"—that is, chromosomes other than the sex chromosomes) and one of each of the X and Y sex chromosomes (which differ greatly in size). In a human karyotype, the chromosomes are numbered for easy identification. Conventionally, the largest pair of homologous chromosomes is designated 1, the next largest 2, and so on. Although chromosome 21 is smaller than chromosome 22, it is called 21 for historical reasons. As shown in Figure 12.3, chromosomes with similar morphologies are arranged under the letter designations A through G.

Based on size and morphology alone, the different chromosomes are hard to distinguish unambiguously when they are stained evenly. Fortunately, a number of procedures stain certain regions or *bands* of the chromosomes more intensely than other regions. Banding patterns are specific to each chromosome, enabling us to distinguish each clearly in the karyotype. One of these staining techniques is called G banding. In this procedure, chromosomes are treated with mild heat or proteolytic enzymes (enzymes that digest proteins) to digest the chromosomal proteins partially and then are stained with Giemsa stain to produce dark bands called G bands. (See Figure 12.3.) In humans, approximately 300 G bands can be distinguished in metaphase chromosomes, and approximately 2,000 G bands can be distinguished in chromosomes from the prophase stage of mitosis. Conventionally, drawings (*ideograms*) of human chromosomes show the G banding pattern. Furthermore, a standard nomenclature based on the banding patterns has been established for the chromosomes so that scientists can talk about gene and marker locations with reference to specific regions and subregions. Each chromosome has two arms separated by the centromere. The smaller arm is designated p and the larger arm is designated q. Numbered regions and numbered subregions are then assigned from the centromere outward; that is, region 1 is closest to the centromere. For example, the breast cancer susceptibility gene *BRCA1* is at location 17q21, meaning that it is on the long arm of chromosome 17 in region 21. Subregions are indicated by decimal numerals after the region number. For instance, the cystic fibrosis gene spans subregions 7q31.2–q31.3; that is, it spans both subregions 2 and 3 of region 31 of the long arm of chromosome 7.

KEYNOTE

> Diploid eukaryotic cells have two haploid sets of chromosomes, one set from each parent. The members of a pair of chromosomes, one from each parent, are called homologous chromosomes. Haploid eukaryotic cells have only one set of chromosomes. The complete set of chromosomes in a cell is called its karyotype. The karyotype is species specific. Staining with particular dyes results in characteristic banding patterns, thereby defining chromosome regions and subregions by number.

Mitosis

In both unicellular and multicellular eukaryotes, cellular reproduction is a cyclical process of growth, **mitosis** (*nuclear division* or *karyokinesis*), and (usually) **cell division** (*cytokinesis*). The cycle of growth, mitosis, and cell division is called the **cell cycle.** In proliferating somatic cells, the cell cycle consists of two phases: the mitotic (or division) phase (M) and an interphase between divisions (Figure 12.4). Interphase consists of three stages: G_1 (gap 1), S, and G_2 (gap 2). During G_1 (the presynthesis stage), the cell prepares for DNA and chromosome replication, which take

animation
a Mitosis

Figure 12.4

Eukaryotic cell cycle. This cycle assumes a period of 24 hours, although great variation exists between cell types and organisms.

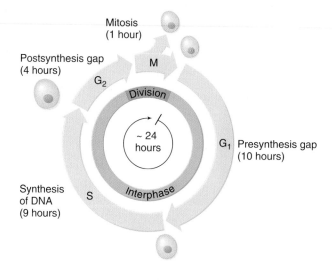

place in the S stage. In G_2 (the postsynthesis stage), the cell prepares for cell division, or the M phase. Put another way, chromosome replication takes place in interphase and then mitosis occurs, resulting in the distribution of a complete chromosome set to each of two progeny nuclei.

The relative time spent in each of the four stages of mitosis varies greatly among cell types. In a given organism, variation in the length of the cell cycle depends primarily on the duration of G_1; the duration of S plus G_2 plus M is approximately the same in all cell types. For example, some cancer cells and early fetal cells of humans spend minutes in G_1, whereas some differentiated adult cells (such as nerve cells) spend years in G_1. Finally, some cells exit the cell cycle from G_1 and enter a quiescent, nondividing state called G_0.

During interphase, the individual chromosomes are elongated and are difficult to see under the light microscope. The DNA of each chromosome is replicated in the S phase, giving two exact copies, called **sister chromatids,** which are held together by the replicated but unseparated centromeres. (Because the centromeres have not separated, only one centromere structure is visible under the microscope.) More precisely, a **chromatid** is one of the two distinct longitudinal subunits of all replicated chromosomes that becomes visible between early prophase and metaphase of mitosis. Later, when the centromeres separate, the sister chromatids become known as **daughter chromosomes.**

Mitosis occurs in both haploid and diploid cells. It is a continuous process, but for purposes of discussion, it is usually divided into four cytologically distinguishable stages called *prophase, metaphase, anaphase,* and *telophase.* Figure 12.5 shows the four stages in simplified diagrams.

Figure 12.5

Interphase and mitosis in an animal cell.

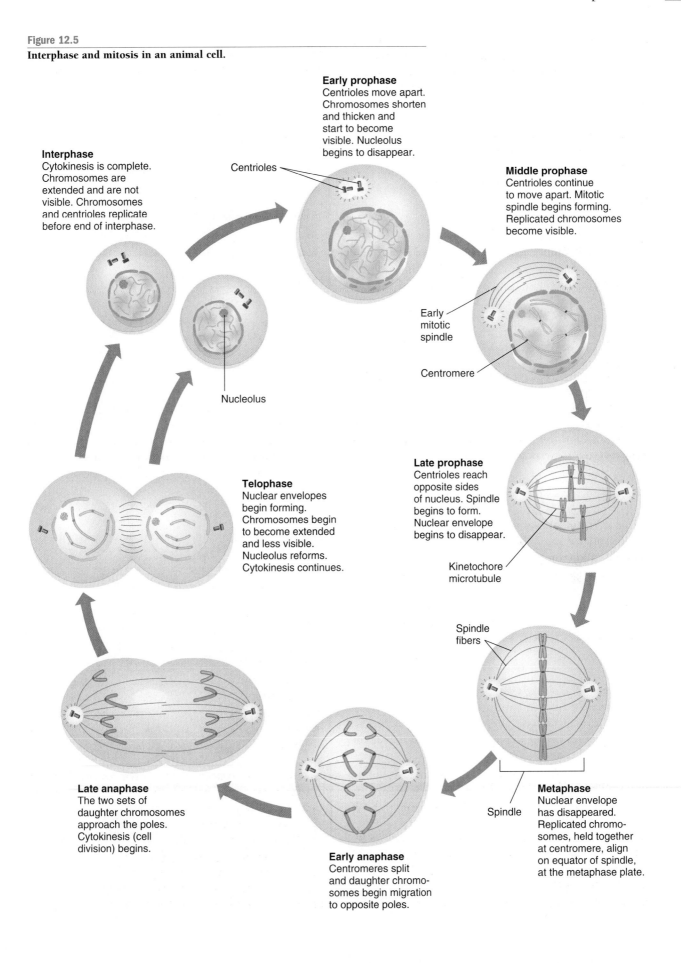

Interphase
Cytokinesis is complete. Chromosomes are extended and are not visible. Chromosomes and centrioles replicate before end of interphase.

Early prophase
Centrioles move apart. Chromosomes shorten and thicken and start to become visible. Nucleolus begins to disappear.

Centrioles

Middle prophase
Centrioles continue to move apart. Mitotic spindle begins forming. Replicated chromosomes become visible.

Early mitotic spindle

Centromere

Nucleolus

Telophase
Nuclear envelopes begin forming. Chromosomes begin to become extended and less visible. Nucleolus reforms. Cytokinesis continues.

Late prophase
Centrioles reach opposite sides of nucleus. Spindle begins to form. Nuclear envelope begins to disappear.

Kinetochore microtubule

Spindle fibers

Late anaphase
The two sets of daughter chromosomes approach the poles. Cytokinesis (cell division) begins.

Early anaphase
Centromeres split and daughter chromosomes begin migration to opposite poles.

Spindle

Metaphase
Nuclear envelope has disappeared. Replicated chromosomes, held together at centromere, align on equator of spindle, at the metaphase plate.

The photographs in Figure 12.6 show the typical chromosome morphology in interphase and in the four stages of mitosis in animal (whitefish early embryo) cells.

Prophase. At the beginning of **prophase** (see Figures 12.5 and 12.6b), the chromatids are very elongated. In preparation for mitosis, they begin to coil tightly, so they appear shorter and fatter under the microscope. By late prophase, each chromosome, which was duplicated during the preceding S phase of interphase, can be seen to consist of two sister chromatids.

Many mitotic events depend on the *mitotic spindle* (spindle apparatus), a structure consisting of fibers composed of microtubules made of special proteins called *tubulins*. The mitotic spindle assembles outside the nucleus during prophase. In most animal cells, the centrioles (see Figure 1.8b) are the focal points for spindle assembly; higher plant cells usually lack centrioles, but they do have a mitotic spindle. Centrioles are arranged in pairs, and, before the S phase, the cell's centriole pair replicates. Then, during mitosis, each new centriole pair becomes the focus of a radial array of microtubules called the *aster*. Early in prophase, the two asters are next to one another and close to the nuclear envelope; by late prophase, the two asters have moved far apart along the outside of the nucleus and are spanned by the microtubular spindle fibers.

Near the end of prophase, a substage called *prometaphase* occurs in which the nucleolus or nucleoli cease to be discrete areas within the nucleus and the nuclear envelope breaks down, allowing the spindle to enter the nuclear area. A specialized multiprotein complex called a **kinetochore** binds to each centromere. The kinetochores are the sites for the attachment of the chromosomes to spindle microtubules. For a pair of sister chromatids, one to many microtubules from one pole attach to one kinetochore, and an equivalent number of microtubules from the other pole attach to the other kinetochore.

Metaphase. Metaphase (see Figures 12.5 and 12.6c) begins when the nuclear envelope has completely disappeared. During metaphase, the microtubules attached to the kinetochores orient the chromosomes so that their centromeres become aligned in one plane halfway between the two spindle poles, with the long axes of the chromosomes at 90 degrees to the spindle axis. The plane where the chromosomes become aligned is called the **metaphase plate.**

Figures 12.7a and b are electron micrographs of human chromosomes in metaphase. Note the highly condensed state of the sister chromatids. The micrograph in Figure 12.7c shows a human chromosome from which much of the protein has been removed. In the center of the chromosome is a dense framework of protein called a

Figure 12.6

Interphase and the stages of mitosis in whitefish early embryo cells. (a) Interphase. (b) Late prophase. (c) Metaphase. (d) Early anaphase. (e) Telophase.

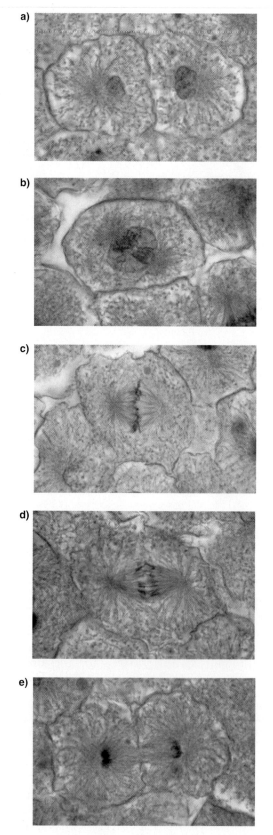

Figure 12.7

Human metaphase chromosome. (a) Transmission electron micrograph of an intact chromosome. (b) Scanning electron micrograph of an intact chromosome. (c) Transmission electron micrograph of an intact chromosome from which much of the protein has been removed. In the center is the proteinaceous chromosome scaffold from which the DNA filaments have uncoiled and spread outward.

a)   b)   c)

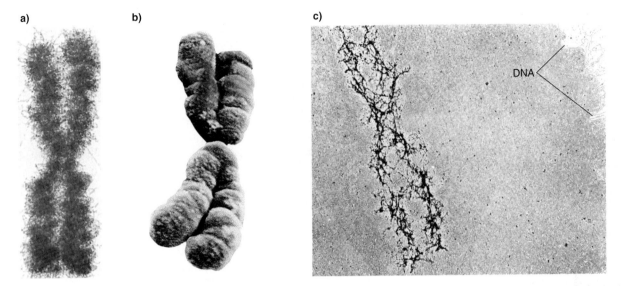

DNA

scaffold, which still has the form of the chromosome. The scaffold is surrounded by a halo of DNA that has uncoiled and spread outward. Such pictures suggest that each chromatid consists of one long, uninterrupted molecule of double-stranded DNA.

Anaphase. Anaphase (see Figures 12.5 and 1.16d) begins when the joined centromeres of sister chromatids separate, giving rise to two daughter chromosomes. Once the paired kinetochores on each chromosome separate, the sister chromatid pairs undergo disjunction (separation), and the daughter chromosomes (the former sister chromatids) move toward the poles. In anaphase, the daughter chromosomes are pulled toward the opposite poles of the cell by the shortening microtubules attached to the kinetochores. As they are pulled, the chromosomes have characteristic shapes related to the location of the centromere along the chromosome's length. For example, a metacentric chromosome is V shaped as the two roughly equal-length chromosome arms trail the centromere in its migration toward the pole. Similarly, a submetacentric chromosome has a J shape with a long and short arm. Cytokinesis (cell division) usually begins in the latter stages of anaphase.

Telophase. During telophase (see Figures 12.5 and 12.6e), the migration of daughter chromosomes to the two poles is completed, and the two sets of daughter chromosomes are assembled into two groups at opposite ends of the cell. The chromosomes begin to uncoil and

assume the elongated state characteristic of interphase. A nuclear envelope forms around each group of chromosomes, the spindle microtubules disappear, and the nucleolus or nucleoli re-form. At this point, nuclear division is complete: The cell has two nuclei.

Cytokinesis. **Cytokinesis** is division of the cytoplasm; usually, it follows the nuclear division stage of mitosis and is completed by the end of telophase. Cytokinesis compartmentalizes the two new nuclei into separate daughter cells, completing mitosis and cell division (Figure 12.8). In animal cells, cytokinesis proceeds with the formation of a constriction in the middle of the cell; the constriction continues until two daughter cells are produced. (See Figure 12.8a.) By contrast, most plant cells do not divide by the formation of a constriction. Instead, a new cell membrane and cell wall are assembled between the two new nuclei to form a *cell plate*. (See Figure 12.8b.) Cell-wall material coats each side of the plate, and the result is two progeny cells.

Gene Segregation in Mitosis. Mitosis maintains a constant amount of genetic material and a constant set of genes from cell generation to cell generation. Mitosis is a highly ordered process in which one copy of each duplicated chromosome segregates into both daughter cells. Thus, for a haploid (N) cell, chromosome duplication produces a cell in which each chromosome has doubled its content. Mitosis then results in two progeny haploid cells, each with one complete set of chromosomes (a genome). For

Figure 12.8

Cytokinesis (cell division). (**a**) Diagram of cytokinesis in an animal cell. (**b**) Diagram of cytokinesis in a plant cell.

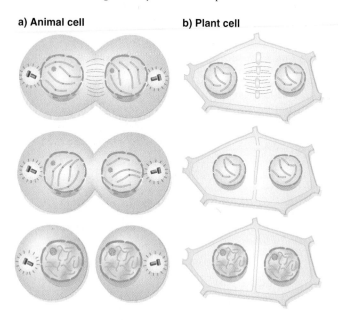

a) Animal cell   **b) Plant cell**

a diploid (2N) cell, which has two sets of chromosomes (two genomes), chromosome duplication produces a cell in which each chromosome set has doubled its content. Mitosis then results in two genetically identical progeny diploid cells, each with two sets of chromosomes (two genomes). As a result, an equal distribution of genetic material occurs, and no genetic material is lost.

K E Y N O T E

Mitosis is the process of nuclear division in eukaryotes. It is one part of the cell cycle (G_1, S, G_2, and M), and it results in the production of daughter nuclei that contain identical chromosome numbers and that are genetically identical to one another and to the parent nucleus from which they arose. Before mitosis, the chromosomes duplicate. Mitosis usually is followed by cytokinesis. Both haploid and diploid cells proliferate by mitosis.

Meiosis

Meiosis is the two successive divisions of a *diploid* ucleus after only one DNA replication (chromosome duplication) cycle. The original diploid nucleus contains one haploid set of chromosomes from the mother and one set from the father (with the exception of self-fertilizing organisms—for example, many plants—in which both sets of chromosomes come from the same parent). Meiosis occurs only at a special point in an organism's life

animation

a Meiosis

cycle. In animals, it results in the formation of haploid gametes (eggs and sperm by **gametogenesis**); in plants, it results in the formation of haploid meiospores (in *sporogenesis*). (A meiospore undergoes mitosis to produce a gamete-bearing, multicellular stage called the gametophyte.) Before meiosis, the DNA that makes up homologous chromosomes replicates, and during meiosis these chromosomes pair and then undergo two divisions—meiosis I and meiosis II—each consisting of a series of stages (Figure 12.9). Meiosis I results in a reduction in the number of chromosomes in each cell from diploid to haploid (reductional division), and meiosis II results in the separation of the sister chromatids. As a consequence, each of the four nuclei that come about from the two meiotic divisions receives one chromosome of each chromosome set (that is, one complete haploid genome). In most cases, the divisions are accompanied by cytokinesis, so the meiosis of a single diploid cell produces four haploid cells.

Meiosis I: The First Meiotic Division. Meiosis I, in which the chromosome number is reduced from diploid to haploid, consists of four stages: *prophase I, metaphase I, anaphase I,* and *telophase I.*

Prophase I. As **prophase I** begins, the chromosomes have already duplicated (producing a temporary 4N stage), with each consisting of two sister chromatids attached at a centromere. (See Figure 12.9.) Prophase I is divided into a number of substages. Except for the behavior of homologous pairs of chromosomes and crossing-over, prophase I of meiosis is similar to prophase of mitosis.

In **leptonema** (early prophase I, the leptotene stage), the extended chromosomes begin to coil and become visible as long, thin threads. Once a cell enters leptonema, it is committed to the meiotic process.

In **zygonema** (early to midprophase I, the zygotene stage), the chromosomes continue to shorten. The homologous pairs of chromosomes actively find each other and align roughly along their lengths. Each pair of homologs then undergoes **synapsis**—the formation along the length of the chromatids of a zipperlike structure called the **synaptonemal complex**, which aligns the two homologs precisely, base pair for base pair. The chromosomes are maximally condensed before the synaptonemal complex forms.

The telomeres of chromosomes play an important role in the initiation of synapsis. That is, during meiosis I, telomeres are clustered on the nuclear envelope to produce an arrangement called a *bouquet* because of its resemblance to the stems from a bouquet of cut flowers. In some way, the telomeres move the chromosomes around so that homologous chromosomes align and undergo synapsis.

Figure 12.9

The stages of meiosis in an animal cell.

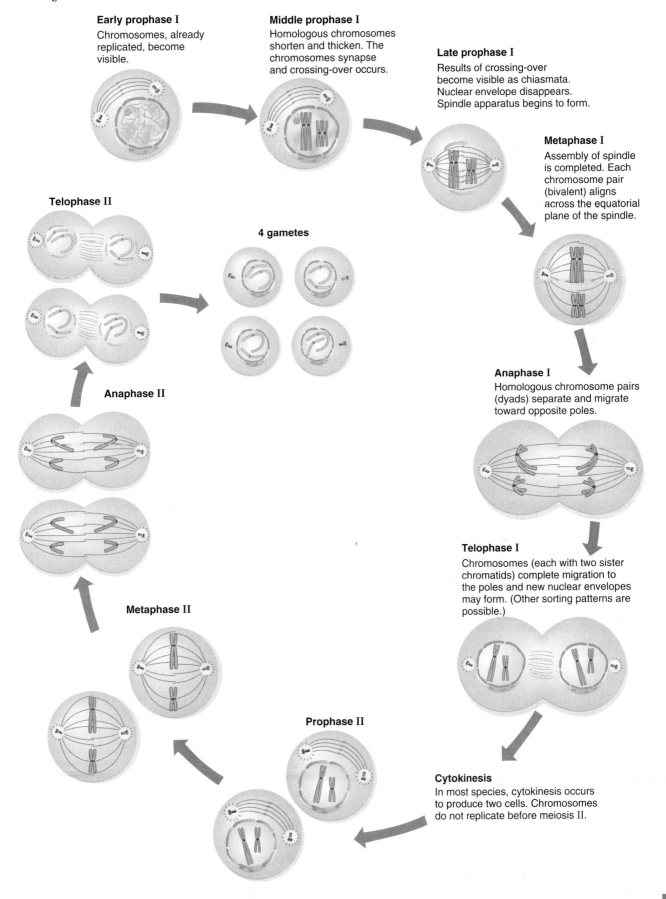

Early prophase I
Chromosomes, already replicated, become visible.

Middle prophase I
Homologous chromosomes shorten and thicken. The chromosomes synapse and crossing-over occurs.

Late prophase I
Results of crossing-over become visible as chiasmata. Nuclear envelope disappears. Spindle apparatus begins to form.

Metaphase I
Assembly of spindle is completed. Each chromosome pair (bivalent) aligns across the equatorial plane of the spindle.

Telophase II

4 gametes

Anaphase I
Homologous chromosome pairs (dyads) separate and migrate toward opposite poles.

Anaphase II

Telophase I
Chromosomes (each with two sister chromatids) complete migration to the poles and new nuclear envelopes may form. (Other sorting patterns are possible.)

Metaphase II

Prophase II

Cytokinesis
In most species, cytokinesis occurs to produce two cells. Chromosomes do not replicate before meiosis II.

Pachynema (midprophase I, the pachytene stage) starts when synapsis is completed. Because of the replication that occurred earlier, each synapsed set of homologous chromosomes consists of four chromatids and is called a **bivalent** or a *tetrad*. During pachynema, a most significant event for genetics occurs: **crossing-over—** the reciprocal physical exchange of chromosome segments at corresponding positions along pairs of homologous chromosomes. This exchange is facilitated by the alignment of the homologous chromosomes brought about by the synaptonemal complex. If there are genetic differences between the homologs, crossing-over can produce new gene combinations in a chromatid. There is usually no loss or addition of genetic material to either chromosome, since crossing-over involves reciprocal exchanges. A chromosome that emerges from meiosis with a combination of genes that differs from the combination with which it started is called a **recombinant chromosome.** Therefore, crossing-over is a mechanism that can give rise to **genetic recombination.** At the end of pachynema, the synaptonemal complex is disassembled and the chromosomes have started to elongate.

In **diplonema** (mid- to late prophase I, the diplotene stage), the synaptonemal complex disassembles and the homologous chromosomes begin to move apart. The result of crossing-over becomes visible during diplonema as a cross-shaped structure called a **chiasma** (plural, *chiasmata*; Figure 12.10). At each chiasma, the homologous chromosomes are very tightly associated. Because all four chromatids may be involved in crossing-over events along the length of the homologs, the chiasma pattern at this stage may be quite complex.

In most organisms, diplonema is followed rapidly by the remaining stages of meiosis. However, in many

Figure 12.10

Appearance of chiasmata, the visible evidence of crossing-over, in diplonema.

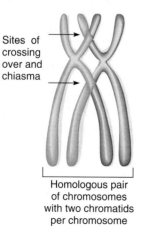

Sites of crossing over and chiasma

Homologous pair of chromosomes with two chromatids per chromosome

animals, the oocytes (egg cells) can remain in diplonema for very long periods. In human females, for example, oocytes go through meiosis I up to diplonema by the seventh month of fetal development and then remain arrested in this stage for many years. At the onset of puberty and until menopause, one oocyte per menstrual cycle completes meiosis I and is ovulated. If the oocyte is fertilized by a sperm as it passes down the fallopian tube, it quickly completes meiosis II, and, by fusion with a haploid sperm, a functional zygote is produced.

In **diakinesis** (late prophase), the nucleolus and nuclear envelope break down. Simultaneously, the spindle is assembled. The chromosomes can be counted most easily at this stage of meiosis.

The synapsis and crossing-over phenomena that take place in prophase I apply to homologous chromosomes—namely, the autosomes. Even though the sex chromosomes are not homologous, the Y chromosome of eutherian (placental) mammals has small regions at each end that are homologous to regions on the X chromosome. These *pseudoautosomal regions* (PARs) pair in male meiosis, and crossing-over occurs between them. When the PAR is deleted from the short arm of the Y chromosome, pairing between the X and Y chromosomes does not occur, and the male is sterile. Thus, pairing and crossing-over of the PAR regions have been considered necessary for the correct segregation of X and Y chromosomes as meiosis proceeds. Interestingly, the genes found in the PARs are variable, even among primates. Even the mouse and human PARs are completely different. PARs are not found in all mammals, however: PARs are absent from some rodents and from all marsupial chromosomes, and the X and Y chromosomes of these animals do not pair or show crossing-over in meiosis. Still, the X and Y chromosomes segregate normally in marsupial meiosis, indicating that a PAR is not essential for mammalian sex chromosome pairing and male fertility.

Metaphase I. By the beginning of **metaphase I** (see Figure 12.9), the nuclear envelope has completely broken down and the bivalents become aligned on the equatorial plane of the cell. The spindle is completely formed now, and the microtubules are attached to the kinetochores of the homologs. Note particularly that the *pairs* of homologs (the bivalents) are found at the metaphase plate. In contrast, in most organisms, replicated homologous chromosomes (sister chromatid pairs) align *independently* at the metaphase plate during mitosis.

Anaphase I. In **anaphase I** (see Figure 12.9), the chromosomes in each bivalent separate, so the chromosomes of each homologous pair disjoin and migrate toward opposite poles, the areas in which new nuclei will form. (At this stage, each of the separated chromosomes is called a *dyad*.) This migration assumes that maternally derived and paternally derived centromeres segregate randomly to each pole (except for the parts of chromosomes exchanged during the crossing-over process) and that, at each pole, there is a haploid complement of replicated centromeres with associated chromosomes. *At this time, the segregated sister chromatid pairs remain attached at their respective centromeres. In other words, a key difference between mitosis and meiosis I is that sister chromatids remain joined after metaphase in meiosis I, whereas they separate in mitosis.*

Telophase I. In **telophase I** (see Figure 12.9), the dyads complete their migration to opposite poles of the cell, and (in most cases) new nuclear envelopes form around each haploid grouping. In most species, cytokinesis follows, producing two haploid cells. Thus, meiosis I, which begins with a diploid cell that contains one maternally derived and one paternally derived set of chromosomes, ends with two nuclei, each of which is haploid and contains one mixed-parental set of dyads. After cytokinesis, each of the two progeny cells has a nucleus with a haploid set of dyads.

Meiosis II: The Second Meiotic Division. The second meiotic division is similar to a mitotic division. (See Figure 12.9.)

In **prophase II**, the chromosomes condense.

In **metaphase II**, each of the two daughter cells organizes a spindle apparatus which attaches to the centromeres that still connect the sister chromatids. The centromeres line up on the equator of the second-division spindles.

During **anaphase II**, the centromeres split, and the chromatids are pulled to the opposite poles of the spindle. One sister chromatid of each pair goes to one pole, and the other goes to the opposite pole. The separated chromatids are considered chromosomes in their own right.

In the last stage, **telophase II**, a nuclear envelope forms around each set of chromosomes, and cytokinesis takes place. After telophase II, the chromosomes become more elongated and are no longer visible under the light microscope.

The end products of the two meiotic divisions are four haploid cells (gametes in animals) from one original diploid cell. (See Figure 12.9.) Each of the four progeny cells has one chromosome from each homologous pair of chromosomes. Moreover, these chromosomes are not exact copies of the original chromosomes because of crossing-over. Figure 12.11 compares mitosis and meiosis.

Gene Segregation in Meiosis. Meiosis has three significant results:

1. Meiosis generates haploid cells with half the number of chromosomes found in the diploid cell that entered the process, because two division cycles follow only one cycle of DNA replication (the S period). The fusion of haploid nuclei restores the diploid number. Therefore, through a cycle of meiosis and fusion, the chromosome number is maintained in sexually reproducing organisms.

2. In metaphase I, each maternally derived chromosome and each paternally derived chromosome has an equal chance of aligning on one or the other side of the equatorial metaphase plate. As a result, each nucleus generated by meiosis has some combination of maternal and paternal chromosomes.

   The number of possible chromosome combinations in the haploid nuclei resulting from meiosis is large, especially when the number of chromosomes in an organism is large. Consider a hypothetical organism with two pairs of chromosomes in a diploid cell entering meiosis. Figure 12.12 shows the two combinations of maternal and paternal chromosomes that can occur at the metaphase plate. Understanding this concept is useful in a consideration of gene segregation. (See Chapter 11.)

   The general formula for the number of possible chromosome arrangements at the metaphase plate in meiosis is 2^{n-1}, where n is the number of chromosome pairs. Similarly, the general formula for the number of possible chromosome combinations in the nuclei resulting from meiosis is 2^n. In *Drosophila*, which has four pairs of chromosomes, the number of possible combinations in nuclei resulting from meiosis is 2^3, or 8; in humans, which have 23 chromosome pairs, more than 4 million combinations are possible. Therefore, because there are many gene differences between the maternally derived and paternally derived chromosomes, the nuclei produced by meiosis are genetically quite different from the parental cell and from one another.

3. The crossing-over between maternal and paternal chromatid pairs during meiosis I generates still more variation in the final combinations. Crossing-over occurs during every meiosis, and because the sites of crossing-over vary from one meiosis to another, the number of different kinds of progeny nuclei produced by the process is extremely large. Given its genetic features an understanding of meiosis is of critical importance for understanding the behavior of genes.

   The events that occur in meiosis are the bases for the segregation and independent assortment of genes according to Mendel's laws, discussed in Chapter 2.

Figure 12.11

Comparison of mitosis and meiosis in a diploid cell.

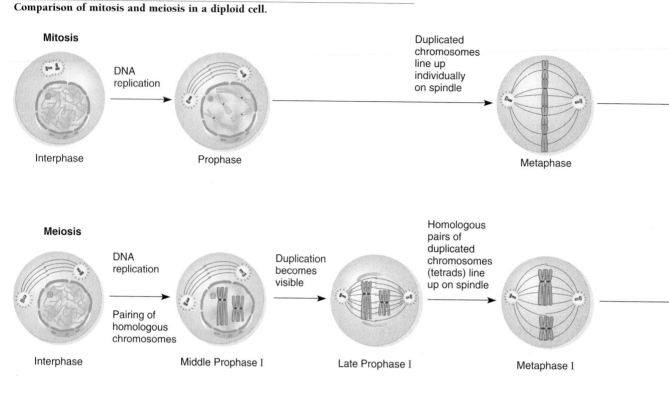

Mitosis

Interphase — DNA replication → Prophase → Duplicated chromosomes line up individually on spindle → Metaphase

Meiosis

Interphase — DNA replication / Pairing of homologous chromosomes → Middle Prophase I — Duplication becomes visible → Late Prophase I — Homologous pairs of duplicated chromosomes (tetrads) line up on spindle → Metaphase I

Figure 12.12

The two possible arrangements of two pairs of homologous chromosomes on the metaphase plate of the first meiotic division. Paternal chromosomes are shown in yellow and green, maternal chromosomes in purple and tan.

Two pairs of homologous chromosomes

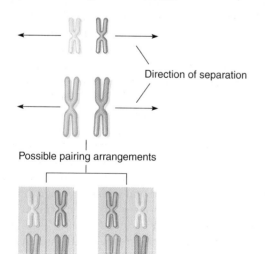

Direction of separation

Possible pairing arrangements

KEYNOTE

Meiosis occurs in all sexually reproducing eukaryotes. It is a process by which a specialized diploid (2N) cell or cell nucleus with two sets of chromosomes is transformed, through one round of chromosome replication and two rounds of nuclear division, into four haploid (N) cells or nuclei, each with one set of chromosomes. In the first of two divisions, pairing, synapsis, and the crossing-over of homologous chromosomes occur. The meiotic process, in combination with fertilization, conserves the number of chromosomes from generation to generation. It also generates genetic variability through the various ways in which maternal and paternal chromosomes are combined in the progeny nuclei and by crossing-over (the physical exchange of chromosome segments at corresponding positions along pairs of homologous chromosomes).

Meiosis in Animals and Plants. Lastly, we discuss briefly the role of meiosis in animals and plants.

Meiosis in Animals. Most multicellular animals are diploid through most of their life cycles. In such animals, meiosis produces haploid gametes, the fusion of two haploid gametes produces a diploid zygote when their nuclei fuse in fertilization, and the zygote then divides by mitosis to produce the new diploid organism; this series of events, involving an alternation of diploid and haploid

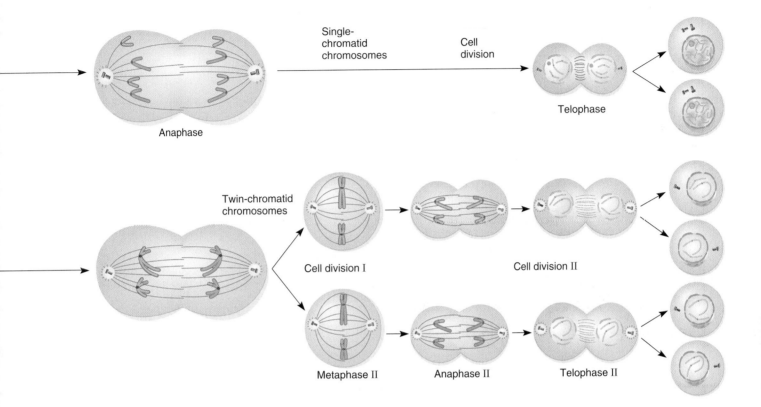

Single-chromatid chromosomes Cell division Telophase

Twin-chromatid chromosomes Cell division I Cell division II

Metaphase II Anaphase II Telophase II

phases, is **sexual reproduction.** Thus, the gametes are the only haploid stages of the life cycle. Gametes are formed only in specialized cells. In males, the gamete is the sperm, produced through a process called **spermatogenesis;** in females, the gamete is the egg, produced by **oogenesis.** Spermatogenesis and oogenesis are illustrated in Figure 12.13.

In male animals, the **sperm cells (spermatozoa)** are produced within the testes, which contain the primordial germ cells (*primary spermatogonia*). Via mitosis, the primordial germ cells produce *secondary spermatogonia*, which then transform into *primary spermatocytes* (*meiocytes*), each of which undergoes meiosis I and gives rise to two *secondary spermatocytes*. Each secondary spermatocyte in turn undergoes meiosis II. The results of these two divisions are four haploid *spermatids* that eventually differentiate into the male gametes: the spermatozoa.

In female animals, the ovary contains the primordial germ cells (*primary oogonia*), which, by mitosis, give rise to *secondary oogonia*. These cells transform into **primary oocytes,** which grow until the end of oogenesis. The diploid primary oocyte goes through meiosis I and unequal cytokinesis to give two cells: a large one called the **secondary oocyte** and a very small one called the *first polar body*. In meiosis II, the secondary oocyte produces

two haploid cells. One is a very small cell called a *second polar body*; the other is a large cell that rapidly matures into the mature egg cell, or **ovum.** The first polar body may or may not divide during meiosis II. The polar bodies have no function in most species and degenerate; only the ovum is a viable gamete. (In many animals, the cell that is actually fertilized is the secondary oocyte; however, nuclear fusion must await completion of meiosis by that oocyte.) Thus, in the female animal, only one mature gamete (the ovum) is produced by meiosis of a diploid cell. In humans, all oocytes are formed in the fetus, and one oocyte completes meiosis I each month in the adult female, but does not progress further unless stimulated to do so through fertilization by a sperm.

Meiosis in Plants. The life cycle of sexually reproducing plants typically has two phases: the **gametophyte** or haploid stage, in which gametes are produced, and the **sporophyte** or diploid stage, in which haploid spores are produced by meiosis.

In angiosperms (the flowering plants), the flower is the structure in which sexual reproduction occurs. Figure 12.14 shows a generalized flower containing both male and female reproductive organs—the **stamens** and **pistils,** respectively. Each stamen consists of a single stalk—the

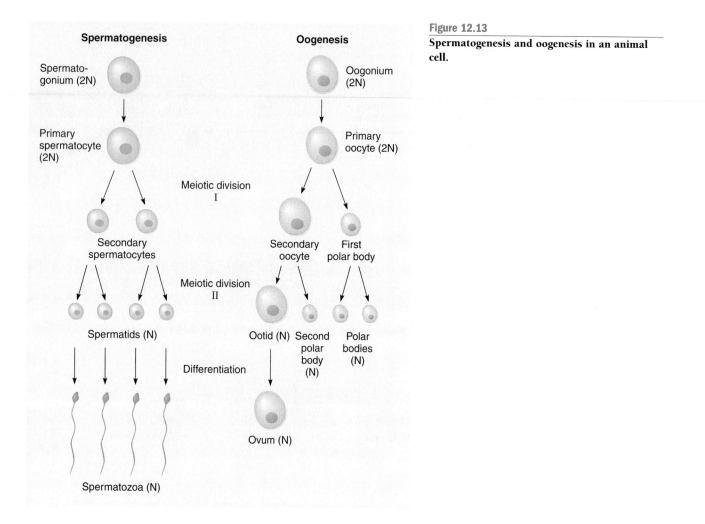

Figure 12.13
Spermatogenesis and oogenesis in an animal cell.

filament—on the top of which is an anther. Pollen grains, which are immature male gametophytes (formed in the gamete-producing phase), are released from the anther. The pistil, which contains the female gametophytes, typically consists of the stigma, a sticky surface specialized to receive the pollen; the style, a thin stalk down which a pollen tube grows from a pollen grain that adheres to the stigma; and, at the base of the structure, the ovary, within which are the ovules. Each ovule encloses a female gametophyte (the embryo sac) containing a single egg cell. When the egg cell is fertilized, the ovule develops into a seed.

Among living organisms, only plants produce gametes from special bodies called gametophytes. Thus plant life cycles have two distinct reproductive phases, called the **alternation of generations** (Figure 12.15). Meiosis and fertilization are the transitions between these stages. The haploid *gametophyte generation* begins with spores that are produced by meiosis. In flowering plants, the spores are the cells that ultimately become pollen and the embryo sac. Fertilization initiates the diploid *sporophyte generation,* which produces the specialized haploid cells called spores, completing the cycle.

Chromosome Theory of Inheritance

Around the turn of the twentieth century, cytologists had established that, within a given species, the total number of chromosomes is constant in all cells, whereas the chromosome number varies widely among species (Table 12.1). In 1902, Walter Sutton and Theodor Boveri independently recognized that the transmission of chromosomes from one generation to the next closely paralleled the pattern of inheritance of Mendelian factors from one generation to the next. This correlation has become known as the **chromosome theory of inheritance.** The theory states that the Mendelian factors—which we now know as genes—are located on chromosomes. In this section, we consider some of the evidence cytologists and geneticists obtained to support that theory.

Sex Chromosomes

The support for the chromosome theory of inheritance came from experiments that related the hereditary behavior of particular genes to the transmission of the sex

Figure 12.14
Generalized structure of a flower.

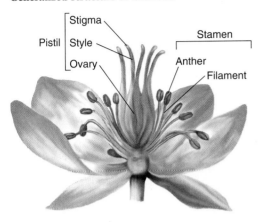

chromosomes—in eukaryotes with different sexes, the chromosomes or groups of chromosomes that are represented differently in the two sexes. The other chromosomes in eukaryotes, which are not represented differently in the two sexes, are the autosomes.

Sex chromosomes were discovered in the early 1900s when Clarence E. McClung, Nettie Stevens, and Edmund B. Wilson, all experimenting with insects, independently obtained evidence that particular chromosomes determined the sex of an organism. Working with grasshoppers and

Figure 12.15
Alternation of gametophyte and sporophyte generations in flowering plants.

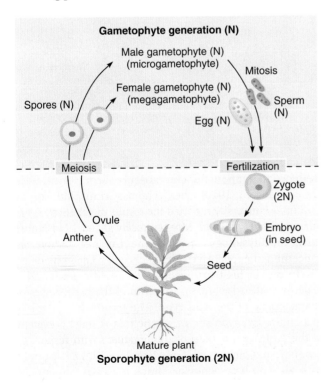

Table 12.1 **Chromosome Number in Various Organisms**[a]

Organism	Total Chromosome Number
Human	46
Chimpanzee	48
Dog	78
Cat	72
Mouse	40
Horse	64
Chicken	78
Toad	36
Goldfish	94
Starfish	36
Fruit fly (*Drosophila melanogaster*)	8
Mosquito	6
Australian ant (*Myrecia pilosula*)	♂1, ♀2
Nematode	♂11, ♀12
Neurospora (haploid)	7
Sphagnum moss (haploid)	23
Field horsetail	216
Giant sequoia	22
Tobacco	48
Cotton	52
Potato	48
Tomato	24
Bread wheat	42
Yeast (*Saccharomyces cerevisiae*) (haploid)	16

[a]Except as noted, all chromosome numbers are for diploid cells.

other members of the order Orthoptera, McClung proposed in 1901 that a particular chromosome, which he called the *accessory chromosome,* was involved in determining each insect's sex during fertilization. In 1905, Stevens and Wilson provided more information concerning McClung's proposal. They found that, in some Orthoptera, the female has an even number of chromosomes while the male has an odd number. There are two copies of one of the chromosomes in the female but only one copy in the male. It was this extra chromosome in the female that McClung had hypothesized as the accessory chromosome. Stevens called the extra chromosome an **X chromosome.** Since this chromosome is directly related to the sex of the organism, the X chromosome is an example of a sex chromosome.

When female grasshoppers form eggs by meiosis, each egg receives one of each chromosome, including an X chromosome. Half the male's sperm cells produced by meiosis receive an X chromosome, and the other half do not. Consequently, the sex of the progeny grasshoppers is determined by the chromosomal composition of the sperm that fertilizes the egg. If the sperm carries an X chromosome, then the resulting fertilized egg will have a pair of X's, and the individual that develops from the zygote will be female. If the sperm does not have an X, the fertilized egg will have an unpaired X, and the individual will be a male.

Unlike grasshoppers, some insects have an unpaired sex chromosome. For example, Stevens found that in the common mealworm, *Tenebrio molitor,* the male has a partner chromosome for the X chromosome. That partner is much smaller and clearly distinguishable from the X chromosome. Stevens called the partner chromosome the **Y chromosome,** and like the X chromosome, it is a sex chromosome. The sperm cells of the mealworm contain either an X or a Y chromosome, and the sex of the offspring is determined by the type of sperm that fertilizes the X-chromosome-bearing egg: XX mealworms are female, and XY mealworms are male.

Similar X–Y sex chromosome complements are found in other organisms, including humans and the fruit fly, *Drosophila melanogaster.* In most cases, the female has two X chromosomes (she is XX with respect to the sex chromosomes), and the male has one X chromosome and one Y chromosomes (he is XY). Figure 12.16a shows male and female *Drosophila,* and Figure 12.16b shows the chromosome sets of the two sexes. Because the male produces two kinds of gametes with respect to sex chromosomes (X or Y) and because the female produces only one type of gamete (X), the male is called the **heterogametic sex** and the female is called the **homogametic sex.** In *Drosophila,* the X and Y chromosomes are similar in size, but their shapes are different. (Note that in some organisms the male is homogametic and the female is heterogametic.)

The pattern of transmission of X and Y chromosomes from generation to generation is straightforward. (See Figure 12.17.) In this figure, the X is represented by a straight structure much like a slash mark, and the Y is represented by a similar structure topped by a hook to the right. The female produces only X-bearing gametes, and the male produces both X-bearing and Y-bearing gametes. Random fusion of male and female gametes produces an F_1 generation with ½ XX (female) and ½ XY (male) flies.

KEYNOTE

In eukaryotes with separate sexes, a sex chromosome is a chromosome or a group of chromosomes that are represented differently in the two sexes. In many of the organisms encountered in genetic studies, one sex possesses a pair of identical chromosomes (the X chromosomes). The opposite sex possesses a pair of visibly different chromosomes: One is an X chromosome, and the other, structurally and functionally different, is called the Y chromosome. Commonly, the XX sex is female, and the XY sex is male. The XX and XY sexes are called the homogametic and heterogametic sexes, respectively.

Figure 12.16

***Drosophila melanogaster* (fruit fly), an organism used extensively in genetics experiments.** (a) Female (left) and male (right). Top, adult flies; bottom, drawings of ventral abdominal surface to show differences in genitalia. (b) Chromosomes of *Drosophila melanogaster* diagrammed to show their morphological differences. A female (left) has four pairs of chromosomes in her somatic cells, including a pair of X chromosomes. The only difference in the male is an XY pair of sex chromosomes instead of two X's.

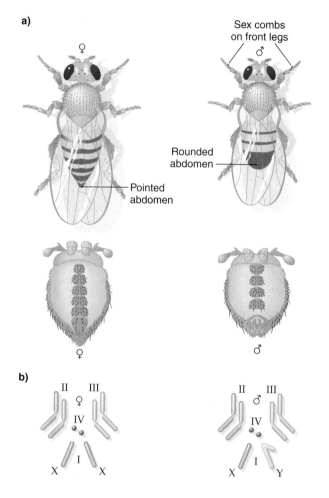

Sex Linkage

Evidence to support the chromosome theory of heredity came about in 1910 when Thomas Hunt Morgan of Columbia University reported the results of genetics experiments with *Drosophila.* Morgan received the 1933 Nobel Prize in Physiology or Medicine for his discoveries concerning the role played by the chromosome in heredity.

In one of his true-breeding stocks, Morgan found a male fly that had white eyes instead of the brick-red eyes characteristic of the **wild type.** The term *wild type* refers to a strain, an organism, or a gene that is most prevalent in the "wild" population of the organism with respect to genotype and phenotype. For example, a *Drosophila* strain with all wild-type genes has brick-red eyes. Variants of

Figure 12.17

Inheritance pattern of X and Y chromosomes in organisms where the female is XX and the male is XY. (a) Production of the F₁ generation. (b) Production of the F₂ generation.

a)

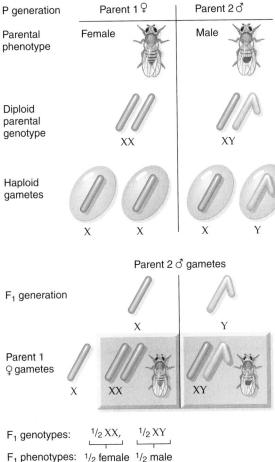

b)

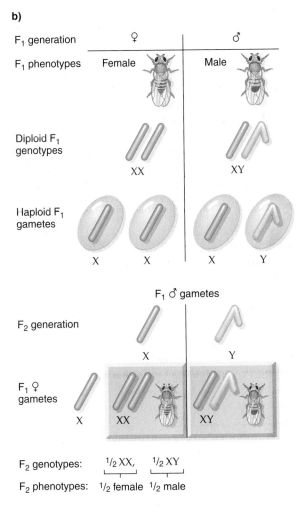

a wild-type strain arise from mutational changes of the wild-type alleles that produce **mutant alleles;** the result is strains with mutant characteristics. Mutant alleles may be recessive or dominant to the wild-type allele; for example, the mutant allele that causes white eyes in *Drosophila* is recessive to the wild-type (red-eye) allele.

Morgan crossed the white-eyed male with a red-eyed female from the same stock and found that all the F₁ flies were red-eyed. He concluded that the white-eyed trait was recessive. Next, he allowed the F₁ progeny to interbreed and counted the phenotypic classes in the F₂ generation; there were 3,470 red-eyed and 782 white-eyed flies. The number of individuals with the recessive phenotype was too small to fit the Mendelian 3:1 ratio. (Later, he determined that the lower-than-expected number of flies with the recessive phenotype was the result of white-eyed flies' lower viability) In addition, *Morgan noticed that all the white-eyed flies were male.*

Figure 12.18 diagrams the crosses. The *Drosophila* gene symbolism used here is different from the symbolism we adopted for Mendel's crosses and is described in Box 12.1. You should understand the *Drosophila* gene symbolism before proceeding with this discussion. As we continue, note that the mother–son inheritance pattern presented in Figure 12.18 is the result of the segregation of genes located on a sex chromosome.

Morgan proposed that the gene for the eye color variant is located on the X chromosome. The condition of X-linked genes in males is said to be **hemizygous,** because the gene is present only once in the organism, since there is no homologous gene on the Y. For example, the white-eyed *Drosophila* males have an X chromosome with a white allele and no other allele of that gene in their genomes; these males are hemizygous for the white allele. Since the white allele of the gene is recessive, the original white-eyed male must have had the recessive allele for

Figure 12.18

The X-linked inheritance of red eyes and white eyes in *Drosophila melanogaster*.
The symbols *w* and *w* indicate the white- and red-eyed alleles, respectively. (**a**) A red-eyed female is crossed with a white-eyed male. (**b**) The F₁ flies are interbred to produce the F₂s.

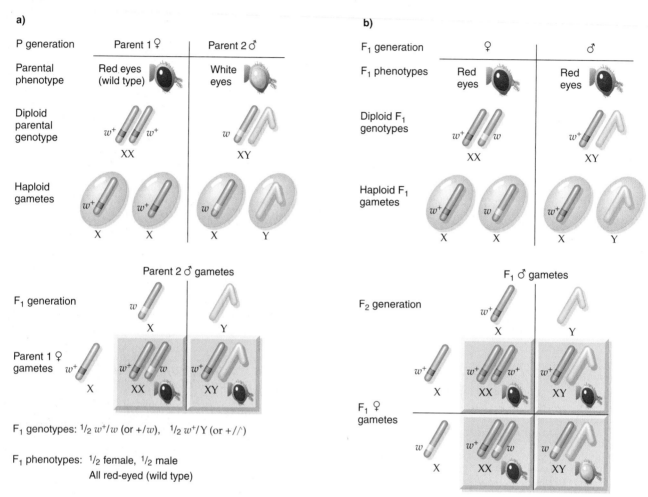

a)

F₁ genotypes: ½ w^+/w (or +/*w*), ½ w^+/Y (or +//∧)

F₁ phenotypes: ½ female, ½ male
All red-eyed (wild type)

b)

F₂ genotypes: 1 w^+/w, 1 w^+/w^+, 1 w^+/Y,   1 w/Y

F₂ phenotypes: ¾ red eyes (2♀, 1♂) ¼ white eyes (1♂)

white eyes (designated *w*; see Box 12.1 on page 317) on his X chromosome. The red-eyed female came from a true-breeding strain, so both of her X chromosomes must have carried the dominant allele for red eyes, w^+ ("w plus").

The F₁ flies are produced in the following way (see Figure 12.18a): The males receive their only X chromosome from their mother and hence have the w^+ allele and are red-eyed. The F₁ females receive a dominant w^+ allele from their mother and a recessive *w* allele from their father, so they are also red-eyed.

In the F₂ produced by interbreeding the F₁ flies, the males that received an X chromosome with the *w* allele from their mother are white eyed; those which received an X chromosome with the w^+ allele are red-eyed. (See

Figure 12.18b.) The gene transmission shown in this cross—from a male parent, to a female offspring ("child"), to a male "grandchild"—is called **crisscross inheritance**.

Morgan also crossed a true-breeding white-eyed female (homozygous for the *w* allele) with a red-eyed male (hemizygous for the w^+ allele; Figure 12.19). This cross is the *reciprocal cross* of Morgan's first cross—white male × red female—shown in Figure 12.18. All the F₁ females receive a w^+-bearing X from their father and a *w*-bearing X from their mother. (See Figure 12.19a.) Consequently, they are heterozygous w^+/w and have red eyes. All the F₁ males receive a *w*-bearing X from their mother and a Y from their father, so they have white eyes. (see Figure 12.19a.) This result is distinct from that of the cross

---

Box 12.1 | **Genetic Symbols Revisited**

Unfortunately, no single system of gene symbols has been adopted by geneticists; the gene symbols used for *Drosophila* are different from those used for peas in Chapter 11. The *Drosophila* symbolism is commonly, but not exclusively, employed in genetics today. In this system, the symbol [+] indicates a wild-type allele of a gene. A lowercase letter designates mutant alleles of a gene that are recessive to the wild-type allele, and an uppercase letter is used for alleles that are dominant to the wild-type allele. *The letters are chosen on the basis of the phenotype of the organism expressing the mutant allele.* For example, a variant strain of *Drosophila* has bright orange eyes instead of the usual brick red. The mutant allele involved is recessive to the wild-type brick-red allele, and because the bright orange eye color is close to vermilion in tint, the allele is designated *v* and is called the vermilion allele. The wild-type allele of *v* is v^+, but when there is no chance of confusing it with other genes in the cross, it is often shortened to +. In the "Mendelian" terminology used up to now, the recessive mutant allele would be *v*, and its wild-type allele would be *V*.

A conventional way to represent the chromosomes (instead of the way we have been doing so in the figures) is to use a slash (/). Thus v^+/v or +/*v* indicates that there are two homologous chromosomes, one with the wild-type allele

(v^+ or +) and the other with the recessive allele (*v*). The Y chromosome usually is symbolized as a Y or a bent slash ∧. Morgan's cross of a true-breeding red-eyed female fly with a white-eyed male could be written $w^+/w^+ \times w/Y$ or +/+ × *w*/∧.

The same rules apply when the alleles involved are dominant to the wild-type allele. For instance, some *Drosophila* mutants, called *Curly*, have wings that curl up at the end rather than the normal straight wings. The symbol for this mutant allele is *Cy*, and the wild-type allele is Cy^+, or + in the shorthand version. Thus, a heterozygote would be Cy^+/Cy or +/*Cy*.

In the rest of the book, the *A/a* ("Mendelian"), a^+/a (*Drosophila*), and other symbols will be used. Because it is easier to verbalize the "Mendelian" symbols (e.g., big *A*, little *a*), many of our examples follow that symbolism, even though the *Drosophila* symbolism in many ways is more informative. That is, with the *Drosophila* system, the wild-type and mutant alleles are readily apparent because the wild-type allele is indicated by a^+. The "Mendelian" system is commonly used in animal and plant breeding. A good reason for this is that, after many years (sometimes centuries) of breeding, it is no longer apparent what the "normal" (wild-type) gene is.

---

in Figure 12.18. Furthermore, all the results obtained are different from the normal results of a reciprocal cross because of the inheritance pattern of the X chromosome.

Interbreeding of the F_1 flies (see Figure 12.19b) involves a *w*/Y male and a w^+/w female, giving approximately equal numbers of male and female red- and white-eyed flies in the F_2. This ratio differs from the approximately 3:1 ratio of red-eyed : white-eyed flies obtained in the first cross, after which none of the females and approximately half the males exhibited the white-eyed phenotype. The difference in phenotypic ratios in the two sets of crosses reflects the transmission patterns of sex chromosomes and the genes they contain.

Morgan's crosses of *Drosophila* involved eye color characteristics that we now know are coded for by a gene found on the X chromosome. These characteristics and the genes that give rise to them are called **sex-linked**—or, more correctly, **X-linked**—because the gene locus is part of the X chromosome. *X-linked inheritance* is the term used for the pattern of hereditary transmission of X-linked genes. When the results of reciprocal crosses are not the same, and different ratios are seen for the two sexes of the offspring, sex-linked characteristics may well be involved. By comparison, the results of reciprocal crosses are always the same when they involve genes located on the autosomes. Most significantly, Morgan's results strongly

animation

a X-Linked Inheritance

supported the hypothesis that genes were located on chromosomes. Morgan found many other examples of genes on the X chromosome in *Drosophila* and in other organisms, thereby showing that his observations were not confined to a single species. Later in this chapter, we discuss the analysis of X-linked traits in humans.

KEYNOTE

Sex linkage is the linkage of genes with the sex chromosomes of eukaryotes. Such genes, as well as the phenotypic characteristics these genes control, are called sex-linked. Genes that are only on the X chromosome are called X-linked. Morgan's pioneering work with the inheritance of sex-linked genes of *Drosophila* strongly supported, but did not prove, the chromosome theory of inheritance.

Nondisjunction of X Chromosomes

Proof of the chromosome theory of inheritance came from the work of Morgan's student Calvin Bridges. Morgan's work showed that, from a cross of a white-eyed female (*w*/*w*) with a red-eyed male (w^+/Y), all the F_1 males should be white-eyed and all the females should be red-eyed. Bridges found rare exceptions to this result: About 1 in 2,000 of the F_1 flies from such a cross are either white-eyed *females* or red-eyed *males*.

Figure 12.19

Reciprocal cross of that shown in Figure 12.18. (a) A homozygous white-eyed female is crossed with a red-eyed (wild-type) male. (b) The F₁ flies are interbred to produce the F₂s. The results of this cross differ from those in Figure 12.18 because of the way sex chromosomes segregate in crosses.

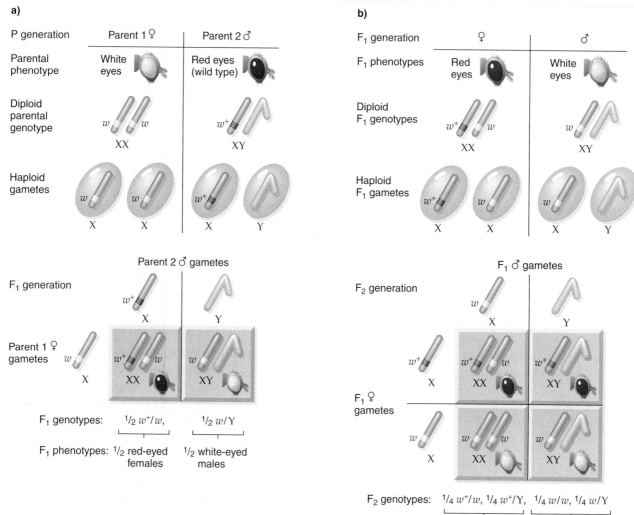

To explain the exceptional flies, Bridges hypothesized that a problem had occurred with chromosome segregation in meiosis. (See pp. 306–309.) Normally, homologous chromosomes (in meiosis I) or sister chromatids (in meiosis II or mitosis) move to opposite poles at anaphase (see p. 308); when this movement fails to take place, chromosome **nondisjunction** results. Nondisjunction can involve either autosomes or the sex chromosomes. For the crosses Bridges analyzed, occasionally the two X chromosomes failed to separate, so eggs were produced either with two X chromosomes or with no X chromosomes instead of the usual single X chromosome. This particular type of

animation

ⓐ **Nondisjunction**

nondisjunction is called **X chromosome nondisjunction** (Figure 12.20). When it occurs in an individual with a normal set of chromosomes, it is called **primary nondisjunction.** Normal disjunction of the X chromosomes is illustrated in Figure 12.20a, and nondisjunction of the X chromosomes in meiosis I and meiosis II is shown in Figures 12.20b and 12.20c, respectively.

How can nondisjunction of the X chromosomes explain the exceptional flies in Bridges's cross? When nondisjunction occurs in the w/w female (Figure 12.21), two classes of exceptional eggs result with equal (and low) frequency: those with two X chromosomes and those with no X chromosomes. The XY male is w^+/Y and produces equal numbers of w^+- and Y-bearing sperm. When these

Figure 12.22

Results of a cross between the exceptional white-eyed XXY female of Figure 12.21 with a normal red-eyed XY male. Again, XXX and YY progeny usually die. **(a)** Normal disjunction of the X chromosomes in the XXY female. **(b)** Secondary nondisjunction of the homologous X chromosomes in meiosis I of the XXY female.

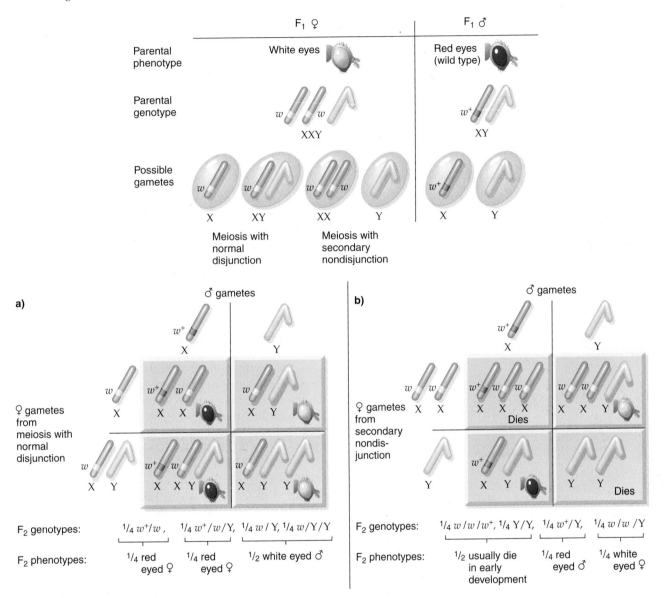

F₂ genotypes (a): ¹/₄ w⁺/w , ¹/₄ w⁺/w/Y, ¹/₄ w / Y, ¹/₄ w/Y/Y

F₂ phenotypes (a): ¹/₄ red eyed ♀   ¹/₄ red eyed ♀   ¹/₂ white eyed ♂

F₂ genotypes (b): ¹/₄ w /w /w⁺, ¹/₄ Y/Y,   ¹/₄ w⁺/Y,   ¹/₄ w /w /Y

F₂ phenotypes (b): ¹/₂ usually die in early development   ¹/₄ red eyed ♂   ¹/₄ white eyed ♀

determines the sex of an individual. Individuals with a Y chromosome are genetically male, and individuals without a Y chromosome are genetically female. This dichotomy occurs because the Y chromosome uniquely carries an important gene (or perhaps genes) that sets the switch toward male sexual differentiation. The gene product is called **testis-determining factor,** and the corresponding gene is the *testis-determining factor gene.* Testis-determining factor causes the tissue that will become gonads to differentiate into testes instead of ovaries. In the absence of a Y chromosome, the gonads

develop as ovaries. The testis-determining factor gene and the way a Y chromosome determines sex in mammals are examined further in Chapter 21.

Evidence for the Y Chromosome Mechanism of Sex Determination. Early evidence for the Y chromosome mechanism of sex determination in mammals came from studies in which nondisjunction in meiosis produced an abnormal sex chromosome complement. Nondisjunction, for example, can lead to XO individuals. In humans, XO individuals with the normal two sets of autosomes are

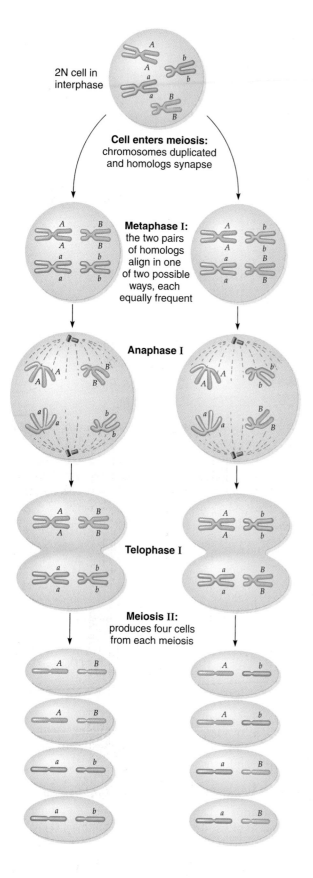

2N cell in interphase

Cell enters meiosis: chromosomes duplicated and homologs synapse

Metaphase I: the two pairs of homologs align in one of two possible ways, each equally frequent

Anaphase I

Telophase I

Meiosis II: produces four cells from each meiosis

Figure 12.23

The parallel behavior between Mendelian genes and chromosomes in meiosis. This hypothetical *Aa Bb* diploid cell contains a homologous pair of metacentric chromosomes, which carry the *A/a* gene pair, and a homologous pair of telocentric chromosomes, which carry the *B/b* gene pair. The independent alignment of the two homologous pairs of chromosomes at metaphase I results in equal frequencies of the four meiotic products *A B*, *a b*, *A b*, and *a B*, illustrating Mendel's principle of independent assortment.

female and sterile, and they exhibit **Turner syndrome.** One such individual is shown in Figure 12.24a, and her karyotype is shown in Figure 12.24b. Turner syndrome individuals have only one sex chromosome: an X chromosome. These aneuploid females have a genomic complement designated as 45,X, indicating that they have a total of 45 chromosomes (one sex chromosome plus 22 pairs of autosomes), in contrast to the normal 46, and that the sex chromosome complement consists of one X chromosome.

Turner syndrome individuals occur with a frequency of 1 in every 10,000 females born. Up to 99 percent of all 45,X embryos die before birth. Surviving Turner syndrome individuals have few noticeable major defects until puberty, when they fail to develop secondary sexual characteristics. They tend to be shorter than average, and they have weblike necks, poorly developed breasts, and immature internal sexual organs. They have a reduced ability to interpret spatial relationships, and they are usually infertile. All of these defects in XO individuals indicate that two X chromosomes are needed during normal development if a female is to develop normally.

Nondisjunction can also result in the generation of XXY humans, who are male and have **Klinefelter syndrome** (Figure 12.25). About 1 in 1,000 males born have the syndrome. These 47,XXY males have underdeveloped testes and often are taller than the average male. Some degree of breast development is seen in about 50 percent of affected individuals, and some show subnormal intelligence. Individuals with similar phenotypes are also found with higher numbers of X or Y chromosomes (or both)—for example, 48,XXXY, and 48,XXYY. The defects in Klinefelter individuals indicate that one X and one Y chromosome are needed for normal development in males.

Some individuals have one X and two Y chromosomes; they have *XYY syndrome*. These 47,XYY individuals are male because of the Y. The XYY karyotype results from nondisjunction of the Y chromosome in meiosis. About 1 in 1,000 males born have XYY syndrome. They tend to be taller than average, and occasionally there are adverse effects on fertility.

About 1 in 1,000 females born have three X chromosomes instead of the normal two. These 47,XXX (triplo-X)

Figure 12.24
Turner syndrome (XO). Individual (left) and karyotype (right).

a)

b)

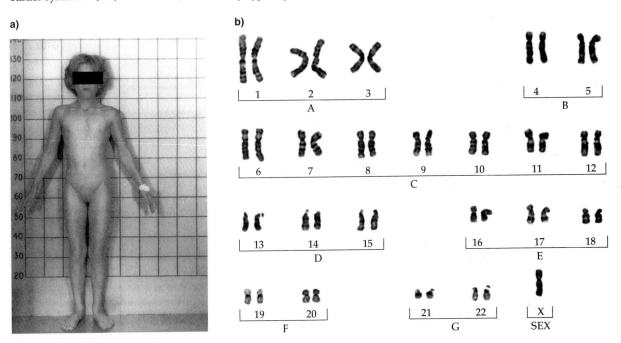

females are mostly normal, although they are slightly less fertile and a small number have lower-than-average intelligence.

Table 12.2 summarizes the consequences of exceptional X and Y chromosomes in humans. In every case, the normal two sets of autosomes are associated with the sex chromosomes. The Barr bodies mentioned in the table are discussed next.

Dosage Compensation Mechanism for X-Linked Genes in Mammals. Organisms with sex chromosomes have an inequality in gene dosage (the number of gene

Figure 12.25
Klinefelter syndrome (XXY). Individual (left) and karyotype (right).

a)

b)

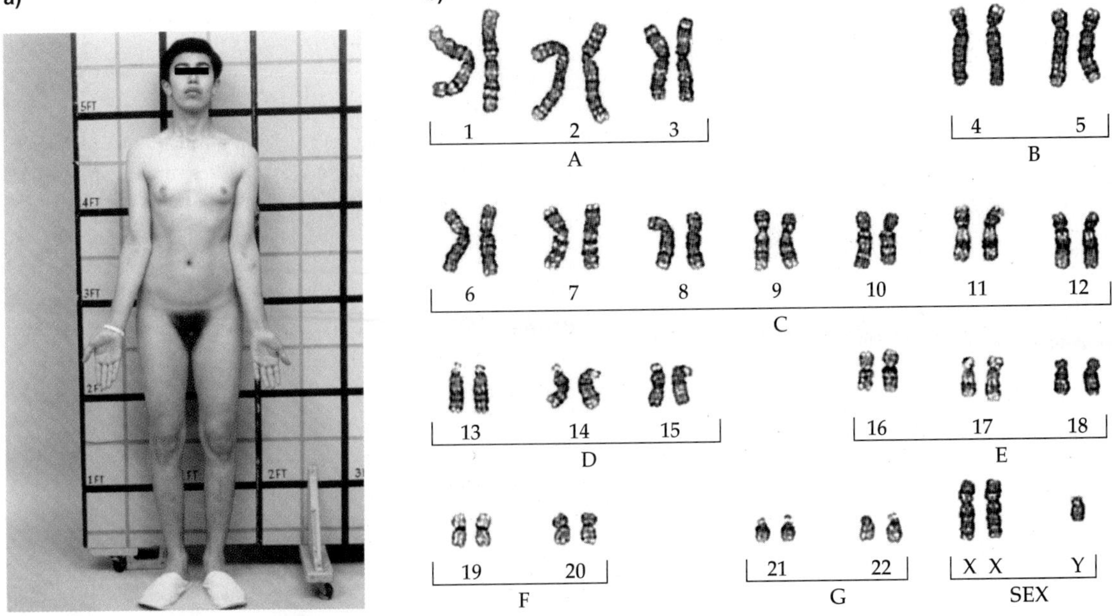

Table 12.2	Consequences of Various Numbers of X- and Y-Chromosome Abnormalities in Humans, Showing Role of the Y in Sex Determination	
Chromosome Constitution[a]	**Designation of Individual**	**Expected Number of Barr Bodies**
46,XX	Normal ♀	1
46,XY	Normal ♂	0
45,X	Turner syndrome ♀	0
47,XXX	Triplo-X ♀	2
47,XXY	Klinefelter syndrome ♂	1
48,XXXY	Klinefelter syndrome ♂	2
48,XXYY	Klinefelter syndrome ♂	1
47,XYY	XYY syndrome ♂	0

[a]The first number indicates the total number of chromosomes in the nucleus, and the Xs and Ys indicate the sex chromosome complement.

copies) between the sexes; that is, there are two copies of X-linked genes in females and one copy in males. In many such organisms, if gene expression on the X chromosome is not equalized, the condition is lethal early in development. A number of different systems for **dosage compensation** have evolved. In mammals, the somatic cell nuclei of normal XX females contain a highly condensed mass of chromatin—named the **Barr body** after its discoverer, Murray Barr—not found in the nuclei of normal XY male cells. That is, somatic cells of XX individuals have one Barr body, and somatic cells of XY individuals have no Barr bodies (Figure 12.26 and Table 12.2). In 1961, Mary Lyon and Lillian Russell expanded this concept into what is now called the **Lyon hypothesis,** which proposed the following:

1. The Barr body is a highly condensed and (mostly) genetically inactive X chromosome. (It has become "lyonized" in a process called **lyonization.**) This leaves a single X chromosome that is transcriptionally equivalent to the single X chromosome of the male.

2. The X chromosome that is inactivated is randomly chosen from the maternally derived and paternally derived X chromosomes in a process that is independent from cell to cell. (Once a maternal or paternal X chromosome is inactivated in a cell, all descendants of that cell inherit the inactivation pattern.)

X inactivation is an example of an **epigenetic** phenomenon—that is, a heritable change in gene expression which occurs without a change in DNA sequence. In other words, X inactivation is an epigenetic silencing of one X chromosome. X inactivation occurs at about the 16th day after fertilization in humans (at about the 500- to 1,000-cell stage), and between 3.5 and 6.5 days after fertilization in mice. Because of X inactivation, mammalian females which are heterozygous for X-linked traits are effectively *genetic mosaics*; that is, some cells show the phenotypes of one X chromosome, and the other cells show the phenotypes of the other X chromosome. This mosaicism is readily visible in, for example, the orange and black patches on calico cats (Figure 12.27). A calico cat is a female cat with the genotype *Oo B–*. That is, a calico is homozygous or heterozygous for the dominant *B* allele of an autosomal gene for black hair. A calico is also heterozygous for an X-linked gene for orange hair. If the dominant *O* allele of that gene is expressed, orange hair results no matter what other genes for coat color the cat has. So orange and black patches are produced because of random X inactivation as the female develops. The orange patches are where the chromosome with the *O* allele was *not* inactivated, so that the active *O* allele masks the *B* alleles, and the black patches are where the chromosome with the *O* allele *was* inactivated, allowing the *B* alleles to be expressed. (The white areas on calico cats are the

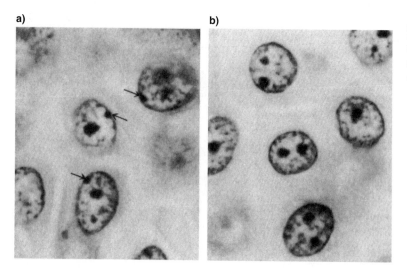

a)

b)

Figure 12.26

Barr bodies. (a) Nuclei of normal human female cells (XX), showing Barr bodies (indicated by arrows). (b) Nuclei of normal human male cells (XY), showing no Barr bodies.

Figure 12.27
A calico kitten.

result of the activity of yet another coat color gene that, when expressed, masks the expression of any other color gene, leaving white hairs. Very rarely is a calico cat male; it is an XXY cat with the appropriate coat color gene genotype.)

A similar, but less visible, phenotype is seen in human females who are heterozygous for an X-linked mutation that causes the absence of sweat glands (anhidrotic ectodermal dysplasia; OMIM 305100). In this condition, there is a mosaic of skin patches lacking sweat glands.

The X inactivation process explains how mammals tolerate abnormalities in the number of sex chromosomes quite well, whereas, with rare exceptions, mammals with an unusual number of autosomes usually die. When lyonization operates in cells with extra X chromosomes, all but one of the X chromosomes typically become inactivated to produce Barr bodies; no such mechanism exists for extra autosomes. A general formula for the number of Barr bodies is the number of X chromosomes minus one. Table 12.2 lists the number of Barr bodies associated with abnormal human X chromosome numbers we have discussed.

The molecular events involved in X inactivation are discussed in Chapter 21.

Sex Determination in *Drosophila* and *Caenorhabditis*. In *Drosophila melanogaster* and the nematode *Caenorhabditis elegans* (*C. elegans*), sex is determined by the *ratio of the number of X chromosomes to the number of sets of autosomes*. In this **X chromosome–autosome balance system** of sex determination, the Y chromosome has no

effect on sex determination, but it is required for male fertility.

In *Drosophila*, the homogametic sex is the female (XX) and the heterogametic sex is the male (XY). That the Y chromosome is not sex determining is seen by the fact that an XXY fly is female and an XO fly is male. Table 12.3 presents some chromosome complements and the sex of the resulting flies, to illustrate the relationship between sex and the ratio of X chromosomes to sets of autosomes. A normal female has two X's and two sets of autosomes; the X:A ratio is 1.00. A normal male has a ratio of 0.50. If the X:A ratio is greater than or equal to 1.00, the fly is female; if the X:A ratio is less than or equal to 0.50, the fly is male. If the ratio is between 0.50 and 1.00, the fly is neither male nor female; it is an intersex. Intersex flies are variable in appearance, generally having complex mixtures of male and female attributes for the internal sex organs and external genitalia. Such flies are sterile. Some molecular details of the complicated regulatory cascade for sex determination in *Drosophila* are presented in Chapter 21, pp. 586–590.

Dosage compensation of X-linked genes also occurs in *Drosophila*, but in a different way than in mammals. That is, the transcription of X-linked genes in males is higher than in females, so as to equal the sum of the expression levels of the two X chromosomes in females.

In *C. elegans*, there are two sexual types: hermaphrodites and males. Most individuals are **hermaphroditic**; that is, they have both sex organs: an ovary and two testes. They make sperm when they are larvae and store

Table 12.3	**Sex Balance Theory of Sex Determination in *Drosophila melanogaster***			

Sex Chromosome Complement	Autosome Complement (A)	X:A Ratio[a]	Sex of Flies
XX	AA	1.00	♀
XY	AA	0.50	♂
XXX	AA	1.50	Metafemale (sterile)
XXY	AA	1.00	♀
XXX	AAAA	0.75	Intersex (sterile)
XX	AAA	0.67	Intersex (sterile)
X	AA	0.50	♂ (sterile)

[a] If the X chromosome autosome ratio is greater than, or equal to, 1.00 (X:A ≥ 1.00), the fly is a female. If the X chromosome:autosome ratio is less than, or equal to, 0.50 (X:A ≤ 0.50), the fly is mate Between these two ratios, the fly is an intersex.

those sperm as development continues. In adults, the ovary produces eggs that are fertilized by the stored sperm as the eggs migrate to the uterus. Self-fertilization in this way almost always produces more hermaphrodites. However, 0.2 percent of the time, males are produced from self-fertilization. These males can fertilize hermaphrodites if the two mate, and such matings result in about equal numbers of hermaphrodite and male progeny, because the sperm from males has a competitive advantage over the sperm stored in the hermaphrodite. Genetically, hermaphrodites are XX and males are XO with respect to sex chromosomes; both have five pairs of autosomes. That is, an X chromosome : autosome ratio of 1.00 results in hermaphrodites, and a ratio of 0.50 results in males.

Dosage compensation of X-linked genes in *C. elegans* occurs by yet another mechanism. In this case, genes on both X chromosomes in an XX hermaphrodite are transcribed at half the rate of the same gene on the single X chromosome in the XO male.

Sex Chromosomes in Other Organisms. In birds, butterflies, moths, and some fish, the sex chromosome composition is the opposite of that in mammals. The male is the homogametic sex and the female is the heterogametic sex. To prevent confusion with the X and Y chromosome convention, we designate the sex chromosomes in these organisms as Z and W: The males are ZZ, and the females are ZW. Genes on the Z chromosome behave just like X-linked genes, except that hemizygosity is found only in females. All the daughters of a male homozygous for a Z-linked recessive gene express the recessive trait. Interestingly, examination of the locations of genes on the sex chromosomes has revealed that the W and Z chromosomes of birds are quite different from the X and Y chromosomes of mammals. That is, mammalian X- and Y-chromosome genes typically are on bird chromosomes 1 and 4, while bird W- and Z-chromosome genes are on chromosomes 5 and 9 of mammals. The interpretation is that mammalian and bird sex chromosomes have evolved from different autosomal pairs.

Plants exhibit a variety of arrangements of sex organs. Some species (the ginkgo, for example) have plants of separate sexes, with male plants producing flowers that contain only stamens and female plants producing flowers that contain only pistils. These species are called **dioecious** ("two houses"). Other species have both male and female sex organs on the same plant; such plants are said to be **monoecious** ("one house"). If both sex organs are in the same flower, as in the rose and the buttercup, the flower is said to be a *perfect flower*. If the male and female sex organs are in different flowers on the same plant, as in corn, the flower is said to be an *imperfect flower*.

Some dioecious plants have sex chromosomes that differ between the sexes, and a large proportion of these plants have an X–Y system. Such plants typically have an X-chromosome–autosome balance system of sex determination like that in *Drosophila*. However, we see many other sex determination systems in dioecious plants.

Genic Sex Determination

Many other eukaryotic species, particularly eukaryotic microorganisms, do not have sex chromosomes, but instead rely on a *genic system* for sex determination. In this system, the sexes are specified by simple allelic differences at a small number of gene loci. For example, the yeast *Saccharomyces cerevisiae* is a haploid eukaryote that has two "sexes"—a and α—called **mating types**. The mating types have the same morphologies, but crosses can occur only between individuals of opposite type. These mating types are determined by the *MATa* and *MATα* alleles, respectively, of a single gene.

KEYNOTE

Many eukaryotic organisms have sex chromosomes that are represented differentially in the two sexes; in humans and many other mammals, the male is XY and the female is XX. In other eukaryotes with sex chromosomes, the male is ZZ and the female is ZW. In many cases, sex determination is related to the sex chromosomes. For humans and many other mammals, for instance, the presence of the Y chromosome confers maleness, and its absence results in femaleness. *Drosophila* and *Caenorhabditis* have an X-chromosome–autosome balance system of sex determination: The sex of the individual is related to the ratio of the number of X chromosomes to the number of sets of autosomes. Several other sex-determining systems are known in the eukaryotes, including genic systems, found particularly in the lower eukaryotes, and environmental systems.

Analysis of Sex-Linked Traits in Humans

In Chapter 11, we introduced the analysis of recessive and dominant traits in humans; those traits were not sex-linked, but instead were the result of alleles carried on autosomes. In this section, we discuss examples of the analysis of X-linked and Y-linked traits in humans.

For the analysis of all pedigrees, whether the trait is autosomal or X-linked, collecting reliable human pedigree data is a difficult task. For example, one often has to rely on a family's recollections. Also, there may not be enough affected people to enable a clear determination of the mechanism of inheritance involved, especially when the trait is rare and the family is small. Furthermore, the expression of a trait may vary, resulting in some individuals erroneously being classified as normal. Finally, because the same mutant phenotype could result from mutations in more than one gene, it is possible that different pedigrees will indicate, correctly, that different mechanisms of inheritance are involved in the "same" trait.

> **iActivity**
>
> Go to the iActivity *It Runs in the Family* on your CD-ROM, and assume the role of a genetic counselor helping a couple determine whether deafness could be passed on to their children.

X-Linked Recessive Inheritance

A trait resulting from a recessive mutant allele carried on the X chromosome is called an **X-linked recessive trait.** At least 100 human traits are known for which the gene has been traced to the X chromosome. Most of the traits involve X-linked recessive alleles. The best known X-linked recessive trait is hemophilia A (OMIM 306700), in Queen Victoria's family (Figure 12.28). Hemophilia is a serious ailment in which the blood lacks a clotting factor, so a cut or even a bruise can be fatal to a hemophiliac. In Queen Victoria's pedigree, the first instance of hemophilia was in one of her sons. Since she passed the mutant allele on to some of her other children (carrier daughters), she must have been a carrier (heterozygous) herself. It is thought that the mutation occurred on an X chromosome in the germ cells of one of her parents.

In X-linked recessive traits, females usually must be homozygous for the recessive allele in order to express the mutant trait. The trait is expressed in males who possess only one copy of the mutant allele on the X chromosome. Therefore, affected males normally transmit the mutant gene to all their daughters, but to none of their sons. The instance of father-to-son inheritance of a rare trait in a pedigree tends to rule out X-linked recessive inheritance.

Other characteristics of X-linked recessive inheritance are the following (see Figure 12.28):

1. For X-linked recessive mutant alleles, many more males than females should exhibit the trait because of the different number of X chromosomes in the two sexes.

2. All sons of an affected (homozygous mutant) mother should show the trait, since males receive their only X chromosome from their mothers.

3. The sons of heterozygous (carrier) mothers should show an approximately 1:1 ratio of normal individuals to individuals expressing the trait; that is, $a^+/a \times a^+/Y$ gives half a^+/Y and half a/Y sons.

4. From a mating of a carrier female with a normal male, all daughters will be normal, but half will be carriers; that is, $a^+/a \times a^+/Y$ gives half a^+/a^+ and half a/a females. In turn, half the sons of these carrier females will exhibit the trait.

5. A male expressing the trait, when mated with a homozygous normal female, will produce all normal children, but all the female progeny will be carriers; that is, $a^+/a^+ \times a/Y$ gives a^+/a females and a^+/Y (normal) males.

Other examples of human X-linked recessive traits are Duchenne muscular dystrophy (progressive muscle degeneration that shortens the person's life) and two forms of color blindness.

X-Linked Dominant Inheritance

A trait resulting from a dominant mutant allele carried on the X chromosome is called an **X-linked dominant trait.** Only a few X-linked dominant traits have been identified.

One example of an X-linked dominant trait—faulty tooth enamel and dental discoloration (hereditary enamel hypoplasia, OMIM 130900)—is shown in Figure 12.29a; a pedigree for this trait is shown in Figure 12.29b. Note that all the daughters and none of the sons of an affected father (III-1) are affected and that heterozygous mothers (IV-3) transmit the trait to half of their sons and half of their daughters. Other X-linked dominant mutant traits are webbing to the tips of the toes in a family in South Dakota (studied in the 1930s) and a severe bleeding anomaly called constitutional thrombopathy. In the latter (also studied in the 1930s), bleeding is caused, not by the absence of a clotting factor (as in hemophilia), but instead by interference with the formation of blood platelets, which are needed for clotting.

X-linked dominant traits follow the same sort of inheritance rules as the X-linked recessives, except that heterozygous females express the trait. In general, X-linked dominant traits tend to be milder in females than in males. Also, because females have twice the number of X chromosomes as males, X-linked dominant traits are more frequent in females than in males. If the trait is rare, females with the trait are likely to be heterozygous. These females pass the trait onto ½ of their male progeny and to ½ of their female progeny. Males with an X-linked dominant trait pass the trait onto all of their daughters and to none of their sons.

Y-Linked Inheritance

A trait resulting from a mutant gene that is carried on the Y chromosome but has no counterpart on the X is called a **Y-linked,** or **holandric** ("wholly male"), trait. Such traits should be easily recognizable because every son of an affected male should have the trait, and no females should ever express it. Several traits with Y-linked inheritance have been suggested. In most cases, the genetic evidence for such inheritance is poor or nonexistent. A number of genes on the Y chromosome have been identified, however, including the *SRY* gene for testis-determining factor mentioned earlier and other testis-specific genes present in multiple copies.

Figure 12.28

X-linked recessive inheritance. (a) Painting of Queen Victoria as a young woman. (b) Pedigree of Queen Victoria (III-2) and her descendants, showing the inheritance of hemophilia. (See Figure 11.16, p. 289, for an explanation of symbols used in pedigrees. In the pedigree shown here, marriage partners who were normal with respect to the trait may have been omitted to save space.) Since Queen Victoria was heterozygous for the sex-linked recessive hemophilia allele, but no cases occurred in her ancestors, the trait may have arisen as a mutation in one of her parents' germ cells (the cells that give rise to the gametes).

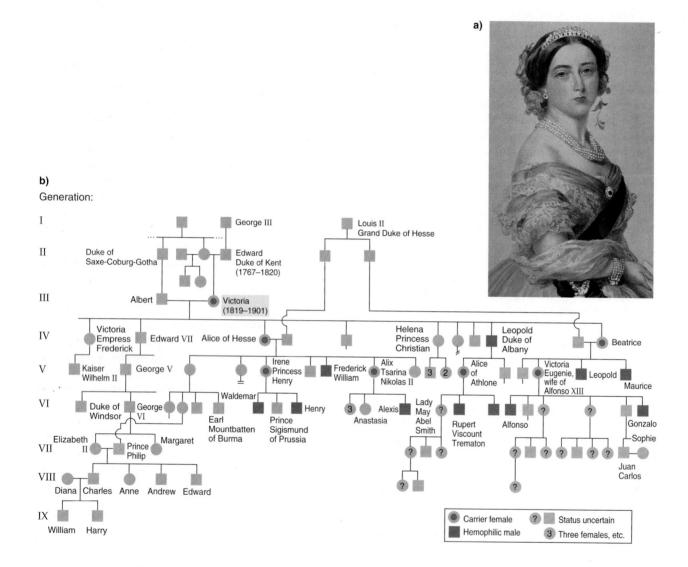

A possible example of Y-linked inheritance is the hairy ears trait (OMIM 425500), in which bristly hairs of atypical length grow from the ears. This trait is common in parts of India, although some other populations also exhibit it. Although the trait shows father-to-son inheritance, there is no doubt that it is a complex phenotype, and many of the collected pedigrees can be interpreted in other ways, such as autosomal inheritance. The trait could also be the result of the interaction of a gene with the male hormone testosterone, similar to the appearance of hair on the face and chest.

K E Y N O T E

Analysis of the inheritance of genes in humans typically relies on pedigree analysis: the careful study of the phenotypic records of the family extending over several generations. The data obtained from pedigree analysis enable geneticists to make judgments, with varying degrees of confidence, about whether a mutant gene is inherited as an autosomal recessive, an autosomal dominant, an X-linked recessive, an X-linked dominant, or a Y-linked allele.

Figure 12.29

X-linked dominant inheritance. (a) The teeth of a person with the X-linked dominant trait of faulty enamel. (b) A pedigree showing the transmission of the faulty enamel trait. This pedigree illustrates a shorthand convention that omits parents who do not exhibit the trait. Thus, it is a given that the female in generation I paired with a male who did not exhibit the trait.

a)

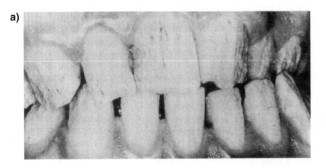

b) Pedigree

Generation:

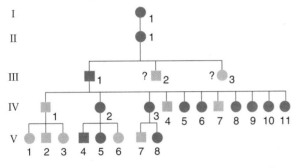

Summary

Chromosomes and Cellular Reproduction

Eukaryotic cells contain either one or two sets of chromosomes; the former are haploid cells and the latter are diploid cells. Both haploid and diploid cells proliferate by mitosis, which involves one round of DNA replication followed by one round of nuclear division, often accompanied by cell division. Thus, mitosis results in the production of daughter nuclei that contain identical chromosome numbers and that are genetically identical to one another and to the parent nucleus from which they arose.

In all sexually reproducing organisms, meiosis occurs at a particular stage in the life cycle. Meiosis is the process by which a diploid cell (never a haploid cell) or cell nucleus undergoes one round of DNA replication and two rounds of nuclear division to produce four specialized haploid cells or nuclei. The products of meiosis are gametes or meiospores. Unlike mitosis, meiosis generates genetic variability in two main ways: through the various ways in which maternal and paternal chromosomes are combined in progeny nuclei and through crossing-over between maternally derived and paternally derived homologs to produce recombinant chromosomes with some maternal and some paternal genes.

Chromosome Theory of Inheritance

In Chapter 11, genes—the modern term for Mendelian factors—were considered abstract entities that control hereditary characteristics. Cytologists working in the nineteenth century accumulated information about cell structure and cell division. The fields of genetics and cytology came together in 1902 when Sutton and Boveri independently hypothesized that Mendelian factors are on chromosomes. Their hypothesis was called the chromosome theory of heredity. Evidence for the theory came from the observation that the patterns of chromosome transmission from one generation to the next closely parallel the patterns of transmission of Mendelian factors (i.e., genes) from one generation to the next.

Support for the chromosome theory of inheritance came from experiments that related the hereditary behavior of particular genes to the transmission of the sex chromosome. The sex chromosome in eukaryotic organisms is the chromosome that is represented differently in the two sexes. In most organisms with sex chromosomes, the female has two X chromosomes, whereas the male has one X and one Y chromosome. The Y chromosome is structurally and genetically different from the X chromosome. The association of genes with the sex chromosomes of eukaryotes is called sex linkage. Such genes and the phenotypes they control are called sex-linked.

Sex Determination Mechanisms

In many cases, sex determination is related to the sex chromosomes. In humans, for example, the presence of a Y chromosome specifies maleness and its absence results in femaleness. Several other sex determination systems are known in eukaryotes, including X-chromosome–autosome balance systems (in which sex is determined by the ratio of the number of X chromosomes to the number of sets of autosomes), and genic systems (in which a simple allelic difference determines the sex of an individual).

Analysis of Sex-Linked Traits in Humans

In Chapter 11, we discussed the fact that the inheritance patterns of traits in humans usually are studied by charting the family trees of individuals exhibiting the traits—a method called pedigree analysis. In this chapter, we considered examples of X- and Y-linked human traits to illustrate the features of those mechanisms of inheritance in pedigrees. Collecting reliable human pedigree data is a difficult task. In many cases, the accuracy of recordkeeping within the families involved is open to question. Also, particularly with small families, there may not be enough affected people to allow an unambiguous determination of the inheritance mechanism. Moreover, the degree to which a trait is expressed may vary, so some individuals may be erroneously classified as normal. It is also possible for the same mutant phenotype to be produced by mutations in different genes; therefore, different pedigrees may correctly indicate different mechanisms of inheritance of the "same" trait.

Analytical Approaches to Solving Genetics Problems

The concepts introduced in this chapter may be reinforced by solving genetics problems similar to those introduced in Chapter 11. Remember that, when sex linkage is involved, one sex has two kinds of sex chromosomes, whereas the other sex has only one; this feature alters the inheritance patterns slightly. Most of the problems presented in this section center on interpreting data and predicting the outcomes of particular crosses.

Q12.1 A female from a true-breeding strain of *Drosophila* with vermilion-colored eyes is crossed with a male from a true-breeding wild-type, red-eyed strain. All the F_1 males have vermilion-colored eyes, and all the females have wild-type red eyes. What conclusions can you draw about the mechanism of inheritance of the vermilion trait, and how could you test them?

A12.1 The observation is the classic one which suggests that a sex-linked trait is involved. Since none of the F_1 daughters have the trait and all the F_1 males do, the trait is presumably X-linked recessive. The results fit this hypothesis because the F_1 males receive the X chromosome with the *v* gene from their homozygous *v/v* mother. Furthermore, the F_1 females are v^+/v, because they receive a v^+-bearing X chromosome from the wild-type male parent and a *v*-bearing X chromosome from the female parent. If the trait were autosomal recessive, all the F_1 flies would have had wild-type eyes. If it were autosomal dominant, both the F_1 males and females would have had vermilion-colored eyes. If the trait were X-linked dominant, all the F_1 flies would have had vermilion eyes.

The easiest way to verify the hypothesis is to let the F_1 flies interbreed. This cross is v^+/v ♀ × v/Y ♂, and the expectation is that there will be a 1:1 ratio of wild-type : vermilion eyes in both sexes in the F_2s. That is, half the females are v^+/v and half are v/v; half the males are v^+/Y and half are v/Y. This ratio is certainly not the 3:1 ratio that would result from an $F_1 \times F_1$ cross for an autosomal gene.

Q12.2 In humans, hemophilia is caused by an X-linked recessive gene. A woman who is a nonbleeder had a father who was a hemophiliac. She marries a nonbleeder, and they plan to have children. Calculate the probability of hemophilia in the female and male offspring.

A12.2 Since hemophilia is an X-linked trait, and because her father was a hemophiliac, the woman must be heterozygous for this recessive gene. If we assign the symbol *h* to this recessive mutation and h^+ to the wild-type (nonbleeder) allele, she must be h^+/h. The man she marries is normal with regard to blood clotting and hence must be hemizygous for h^+—that is, h^+/Y. All their daughters receive an X chromosome from the father, so each must have an h^+ gene. In fact, half the daughters are h^+/h^+ and the other half are h^+/h. Because the wild-type allele is dominant, none of the daughters are hemophiliacs. However, all the sons of the marriage receive their X chromosome from their mother. Therefore, they have a probability of $\frac{1}{2}$ that they will receive the chromosome carrying the *h* allele, in which case they will be hemophiliacs. Thus, the probability of hemophilia among daughters of this marriage is 0; among sons, it is $\frac{1}{2}$.

Q12.3 Tribbles are hypothetical animals that have an X–Y sex determination mechanism like that of humans. The trait blotchy (*b*), with pigment in spots, is X-linked and recessive to solid color (b^+), and the trait light color (*l*) is autosomal and recessive to dark color (l^+). If you make reciprocal crosses between true-breeding blotchy, light-colored tribbles and true-breeding solid, dark-colored tribbles, do you expect a 9:3:3:1 ratio in the F_2 of either or both of these crosses? Explain your answer.

A12.3 This question focuses on the fundamentals of X-chromosome and autosome segregation during a genetic cross, and it tests whether you have grasped the principles involved in gene segregation. Figure 12.A diagrams the two crosses involved, and we can discuss the answer by referring to the figure.

First, consider the cross of a wild-type female tribble (b^+/b^+, l^+/l^+) with a male double-mutant tribble (b/Y, l/l). Part (a) of the figure diagrams this cross. These F_1s are all normal—that is, they are solid and light colored—because, for the autosomal character, both sexes are heterozygous, and for the X-linked character, the female is heterozygous and the male is hemizygous for the b^+ allele donated by the normal mother. For the production of the F_2 progeny, the best approach is to treat the X-linked and autosomal traits separately. For the X-linked trait, a random combination of the gametes produced gives a genotypic ratio of 1 b^+/b^+ (solid female) : 1 b^+/b (solid female) : 1 b^+/Y (solid male) : 1 b/Y (blotchy male) progeny. Categorizing by phenotypes, $\frac{1}{2}$ of the progeny are solid females, $\frac{1}{4}$ are solid males, and $\frac{1}{4}$ are blotchy males. For the autosomal leg trait, the $F_1 \times F_1$ is a cross of two heterozygotes, so we expect a 3:1 phenotypic ratio of dark : light tribbles in the F_2s. Since autosome segregation is independent of the inheritance of the X chromosome, we can multiply the probabilities of occurrence of the X-linked and autosomal traits to calculate their relative frequencies. The calculations are presented at the bottom of part (a) of the figure.

Figure 12.A

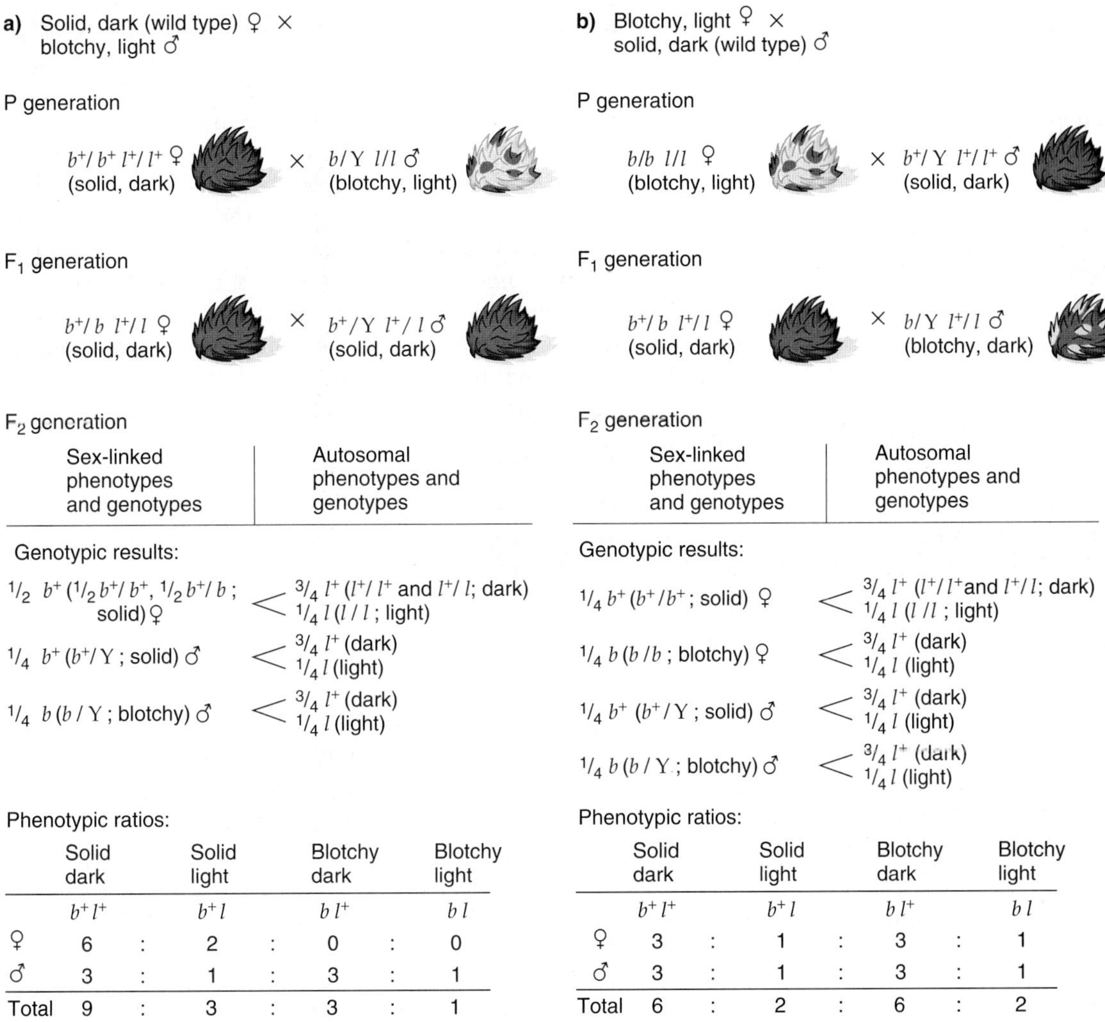

a) Solid, dark (wild type) ♀ ×
blotchy, light ♂

P generation

b^+/b^+ l^+/l^+ ♀   × b/Y l/l ♂
(solid, dark)   (blotchy, light)

F_1 generation

b^+/b l^+/l ♀   × b^+/Y l^+/l ♂
(solid, dark)   (solid, dark)

F_2 generation

Sex-linked phenotypes and genotypes	Autosomal phenotypes and genotypes

Genotypic results:

$1/2$ b^+ ($1/2$ b^+/b^+, $1/2$ b^+/b; solid) ♀   < $3/4$ l^+ (l^+/l^+ and l^+/l; dark) / $1/4$ l (l/l; light)

$1/4$ b^+ (b^+/Y; solid) ♂   < $3/4$ l^+ (dark) / $1/4$ l (light)

$1/4$ b (b/Y; blotchy) ♂   < $3/4$ l^+ (dark) / $1/4$ l (light)

Phenotypic ratios:

	Solid dark		Solid light		Blotchy dark		Blotchy light
	$b^+ l^+$		$b^+ l$		$b l^+$		$b l$
♀	6	:	2	:	0	:	0
♂	3	:	1	:	3	:	1
Total	9	:	3	:	3	:	1

b) Blotchy, light ♀ ×
solid, dark (wild type) ♂

P generation

b/b l/l ♀   × b^+/Y l^+/l^+ ♂
(blotchy, light)   (solid, dark)

F_1 generation

b^+/b l^+/l ♀   × b/Y l^+/l ♂
(solid, dark)   (blotchy, dark)

F_2 generation

Sex-linked phenotypes and genotypes	Autosomal phenotypes and genotypes

Genotypic results:

$1/4$ b^+ (b^+/b^+; solid) ♀   < $3/4$ l^+ (l^+/l^+ and l^+/l; dark) / $1/4$ l (l/l; light)

$1/4$ b (b/b; blotchy) ♀   < $3/4$ l^+ (dark) / $1/4$ l (light)

$1/4$ b^+ (b^+/Y; solid) ♂   < $3/4$ l^+ (dark) / $1/4$ l (light)

$1/4$ b (b/Y; blotchy) ♂   < $3/4$ l^+ (dark) / $1/4$ l (light)

Phenotypic ratios:

	Solid dark		Solid light		Blotchy dark		Blotchy light
	$b^+ l^+$		$b^+ l$		$b l^+$		$b l$
♀	3	:	1	:	3	:	1
♂	3	:	1	:	3	:	1
Total	6	:	2	:	6	:	2

The first cross, then, has a 9:3:3:1 ratio of the four possible phenotypes in the F_2. However, note that the ratio in each sex is not 9:3:3:1, because of the inheritance pattern of the X chromosome. This result contrasts markedly with the pattern of two autosomal genes segregating independently, in which the 9:3:3:1 ratio is found for both sexes.

The second cross (a reciprocal cross) is diagrammed in part **(b)** of the figure. Because the parental female in this cross is homozygous for the sex-linked trait, all the F_1 males are blotchy. Genotypically, the F_1 males and females differ from those in the first cross with respect to the sex chromosome, but are just the same with respect to the autosome. Again, considering the X chromosome first as we go to the F_2s, we find a genotypic ratio of 1 solid females : blotchy females : solid males : blotchy males. In this case, then, half of both males and females are solid, and half are blotchy, in contrast to the results of the first cross, in which no blotchy females were produced in the F_2s. For the autosomal trait, we expect a 3:1 ratio of dark : light in the F_2s, as before. Putting the two traits together, we get the calculations presented in part **(b)** of the figure. (*Note*: We use the total

6:2:6:2 here rather than 3:1:3:1 because the numbers add to 16, as does 9:3:3:1.) Hence, in this case, we do not get a 9:3:3:1 ratio; moreover, the ratio is the same in both sexes.

This question has forced us to think through the segregation of two types of chromosomes and has shown that we must be careful about predicting the outcomes of crosses in which sex chromosomes are involved. Nonetheless, the basic principles for the analysis are the same as those used before: Reduce the questions to their basic parts, and then put the puzzle together step by step.

Questions and Problems

*12.1 Interphase is a period corresponding to the cell cycle phases of
a. mitosis
b. S
c. $G_1 + S + G_2$
d. $G_1 + S + G_2 + M$

12.2 Chromatids joined together by a centromere are called
a. sister chromatids
b. homologs
c. alleles
d. bivalents (tetrads)

*12.3 Mitosis and meiosis always differ in regard to the presence of
a. chromatids
b. homologs
c. bivalents
d. centromeres
e. spindles

12.4 State whether each of the following statements is true or false, and explain your choice:
a. The chromosomes in a somatic cell of any organism are all morphologically alike.
b. During mitosis, the chromosomes divide and the resulting sister chromatids separate at anaphase, ending up in two nuclei, each of which has the same number of chromosomes as the parental cell.
c. At zygonema, a chromosome can synapse with any other chromosome in the same cell.

12.5 For each mitotic event described in the table that follows, write the name of the event in the blank provided in front of the description. Then put the events in the correct order (sequence). Start by placing a 1 next to the description of interphase, and continue through 6, which should correspond to the last event in the sequence.

Name of event		Order of event
	The cytoplasm divides and the cell contents are separated into two separate cells.	
	Chromosomes become aligned along the equatorial plane of the cell.	
	Chromosome replication occurs.	
	The migration of the daughter chromosomes to the two poles is complete.	
	Replicated chromosomes begin to condense and become visible under the microscope.	
	Sister chromatids begin to separate and migrate toward opposite poles of the cell.	

*12.6 Answer these questions with a "yes" or "no," and then explain the reasons for your answer:
a. Can meiosis occur in haploid species?
b. Can meiosis occur in a haploid individual?

12.7 Which of the following sequences describes the general life cycle of a eukaryotic organism?
a. 1N → meiosis → 2N → fertilization → 1N
b. 2N → meiosis → 1N → fertilization → 2N
c. 1N → mitosis → 2N → fertilization → 1N
d. 2N → mitosis → 1N → fertilization → 2N

*12.8 Which statement is true?
a. Gametes are 2N; zygotes are 1N.
b. Gametes and zygotes are 2N.
c. The number of chromosomes can be the same in gamete cells and in somatic cells.
d. The zygotic and the somatic chromosome numbers cannot be the same.
e. Haploid organisms have haploid zygotes.

12.9 All of the following happen in prophase I of meiosis, except
a. chromosome condensation
b. pairing of homologues
c. chiasma formation
d. terminalization
e. segregation

*12.10 Give the name of each stage of mitosis and meiosis at which each of the following events occurs:
a. Chromosomes are located in a plane at the center of the spindle.
b. The chromosomes move away from the spindle equator to the poles.

12.11 Consider the diploid, meiotic mother cell shown here:
Diagram the chromosomes as they would appear

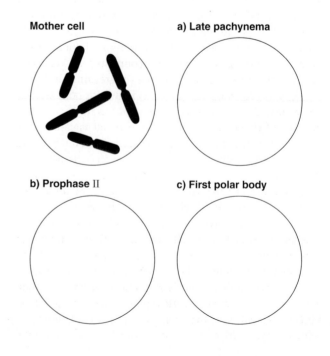

Mother cell

a) Late pachynema

b) Prophase II

c) First polar body

a. in late pachynema
b. in a nucleus at prophase of the second meiotic division
c. in the first polar body resulting from oogenesis in an animal

12.12 The cells in the following figure were all taken from the same individual (a mammal):

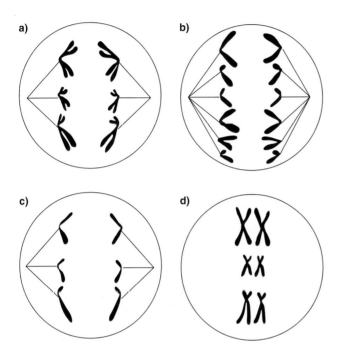

a)   b)
c)   d)

Identify the cell division events occurring in each cell and explain your reasoning. What is the sex of the individual? What is the diploid chromosome number?

12.13 Does mitosis or meiosis have greater significance in the study of heredity? Explain your answer.

*12.14 Consider a diploid organism that has three pairs of chromosomes. Assume that the organism receives chromosomes A, B, and C from the female parent and A′, B′, and C′ from the male parent. Answer the following questions, assuming that crossing-over does not occur:
a. What proportion of the gametes of this organism would be expected to contain all the chromosomes of maternal origin?
b. What proportion of the gametes would be expected to contain some chromosomes of both maternal and paternal origin?

*12.15 Normal diploid cells of a theoretical mammal are examined cytologically at the mitotic metaphase stage for their chromosome complement. One short chromosome, two medium-length chromosomes, and three long chromosomes are present. Explain how the cells might have such a set of chromosomes.

12.16 Explain whether the following statement is true or false: "Meiotic chromosomes can be seen after appropriate staining in nuclei from rapidly dividing skin cells."

*12.17 Explain whether the following statement is true or false: "All the sperm from one human male are genetically identical."

12.18 The horse has a diploid set of 64 chromosomes, and the donkey has a diploid set of 62 chromosomes. Mules are the viable, but usually sterile, progeny of a mating between a male donkey and a female horse. How many chromosomes will a mule cell contain?

*12.19 The red fox has 17 pairs of large, long chromosomes. The arctic fox has 26 pairs of shorter, smaller chromosomes.
a. What do you expect to be the chromosome number in somatic tissues of a hybrid between these two foxes?
b. The first meiotic division in the hybrid fox shows a mixture of paired and single chromosomes. Why do you suppose this occurs? Can you suggest a possible relationship between the mixed chromosomes and the observed sterility of the hybrid?

*12.20 At the time of synapsis preceding the reduction division in meiosis, the homologous chromosomes align in pairs, and one member of each pair passes to each of the daughter nuclei. In an animal with five pairs of chromosomes, assume that chromosomes 1, 2, 3, 4, and 5 have come from the father, and 1′, 2′, 3′, 4′, and 5′ have come from the mother. Assuming no crossing-over, in what proportion of the gametes of this animal will all the paternal chromosomes be present together?

12.21 Depict each of the crosses that follow, first using Mendelian and then using *Drosophila* notation (Box 12.1, p. 317). Give the genotype and phenotype of the F_1 progeny that can be produced.
a. In humans, from a mating between two individuals, each heterozygous for the recessive trait phenylketonuria, whose locus is on chromosome 12.
b. In humans, from a mating between a female heterozygous for both phenylketonuria and X-linked color blindness and a male with normal color vision and heterozygous for phenylketonuria.
c. In *Drosophila*, from a mating between a female with white eyes, curled wings, and normal long bristles and a male that has normal red eyes, normal straight wings, and short, stubble bristles. In these individuals, curled wings result from a heterozygous condition at a gene whose locus is on chromosome 2, whereas the short, stubble bristles result from a heterozygous condition at a gene whose locus is on chromosome 3.

d. In *Drosophila*, from a mating between a female from a true-breeding line that has eyes of normal size that are white, black bodies (a recessive trait on chromosome 2), and tiny bristles (a recessive trait called *spineless* on chromosome 3) and a male from a true-breeding line that has normal red eyes, normal grey bodies, normal long bristles, and a reduced eye size (a dominant trait called *eyeless* on chromosome 4).

12.22 In *Drosophila*, white eyes are a sex-linked character. The mutant allele for white eyes (w) is recessive to the wild-type allele for brick-red eye color (w^+). A white-eyed female is crossed with a red-eyed male. An F_1 female from this cross is mated with her father, and an F_1 male is mated with his mother. What will be the eye color of the offspring of these last two crosses?

***12.23** One form of color blindness (c) in humans is caused by a sex-linked recessive mutant gene. A woman with normal color vision (c^+) and whose father was color blind marries a man of normal vision whose father was also color blind. What proportion of their offspring will be color blind? (Give your answer separately for males and females.)

12.24 In humans, red–green color blindness is recessive and X linked, whereas albinism is recessive and autosomal. What types of children can be produced as the result of marriages between two homozygous parents—a normal-visioned albino woman and a color-blind, normally pigmented man?

***12.25** In *Drosophila*, vestigial (partially formed) wings (vg) are recessive to normal long wings (vg^+), and the gene for this trait is autosomal. The gene for the white eye trait is on the X chromosome. Suppose a homozygous white-eyed, long-winged female fly is crossed with a homozygous red-eyed, vestigial-winged male.
a. What will be the genotypes and phenotypes of the F_1 flies?
b. What will be the genotypes and phenotypes of the F_2 flies?
c. What will be the genotypes and phenotypes of the offspring of a cross of the F_1 flies back to each parent?

12.26 In *Drosophila*, two red-eyed, long-winged flies are bred together and produce the offspring listed in the following table:

	Females	Males
red eyed, long winged	3/4	3/8
red eyed, vestigial winged	1/4	1/2
white eyed, long winged	—	3/8
white eyed, vestigial winged	—	1/8

What are the genotypes of the parents?

12.27 In chickens, a dominant sex-linked gene (B) produces barred feathers, and the recessive allele (b), when homozygous, produces nonbarred (solid-color) feathers. Suppose a nonbarred cock is crossed with a barred hen.
a. What will be the appearance of the F_1 birds?
b. If an F_1 female is mated with her father, what will be the appearance of the offspring?
c. If an F_1 male is mated with his mother, what will be the appearance of the offspring?

***12.28** A man (A) suffering from defective tooth enamel, which results in brown-colored teeth, marries a normal woman. All their daughters have brown teeth, but the sons are normal. The sons of man A marry normal women, and all their children are normal. The daughters of man A marry normal men, and 50 percent of their children have brown teeth. Explain these facts genetically.

12.29 In humans, differences in the ability to taste phenylthiourea result from a pair of autosomal alleles. Inability to taste is recessive to ability to taste. A child who is a nontaster is born to a couple who can both taste the substance. What is the probability that their next child will be a taster?

***12.30** Cystic fibrosis is inherited as an autosomal recessive. Two parents without cystic fibrosis have two children with cystic fibrosis and three children without. The parents come to you for genetic counseling.
a. What is the probability that their next child will have cystic fibrosis?
b. Their unaffected children are concerned about being heterozygous. What is the probability that a given unaffected child in the family is heterozygous?

12.31 Huntington disease is a human disease inherited as a Mendelian autosomal dominant. The disease results in choreic (uncontrolled) movements, progressive mental deterioration, and, eventually, death. In carriers of the trait, the disease appears between 15 and 65 years of age. The American folk singer Woody Guthrie died of Huntington's disease, as did just one of his parents. Marjorie Mazia, Woody's wife, had no history of this disease in her family. The Guthries had three children. What is the probability that a particular Guthrie child will die of Huntington's disease?

***12.32** Suppose gene *A* is on the X chromosome, and genes *B*, *C*, and *D* are on three different autosomes. Thus, *A*– signifies the dominant phenotype in the male or female. An equivalent situation holds for *B*–, *C*–, and *D*–. The cross *AA BB CC DD* ♀ × *aY bb cc dd* ♂ is made.

a. What is the probability of obtaining an $A-$ individual in the F_1 progeny?

b. What is the probability of obtaining an a male in the F_1 progeny?

c. What is the probability of obtaining an $A- B- C- D-$ female in the F_1 progeny?

d. How many different F_1 genotypes will there be?

e. What proportion of F_2 individuals will be heterozygous for the four genes?

f. Determine the probabilities of obtaining each of the following types in the F_2 individuals (1) $A- bb\ CC\ dd$ (female), (2) $aY\ BB\ Cc\ Dd$ (male), (3) $AY\ bb\ CC\ dd$ (male), (4) and $aa\ bb\ Cc\ Dd$ (female).

*12.33 As a famous mad scientist, you have cleverly devised a method to isolate *Drosophila* ova that have undergone primary nondisjunction of the sex chromosomes. In one experiment, you used females homozygous for the sex-linked recessive mutation causing white eyes (w) as your source of nondisjunction ova. The ova were collected and fertilized with sperm from red-eyed males. The progeny of this "engineered" cross were then backcrossed separately to the two parental strains. What classes of progeny (genotype and phenotype) would you expect to result from these backcrosses? (The genotype of the original parents may be denoted as ww for the females and w^+/Y for the males.)

12.34 In *Drosophila*, the bobbed gene (bb^+) is located on the X chromosome: bb mutants have shorter, thicker bristles than wild-type flies. Unlike most X-linked genes, however, a bobbed gene is also present on the Y chromosome. The mutant allele bb is recessive to bb^+. If a wild-type F_1 female that resulted from primary nondisjunction in oogenesis in a cross of bobbed female with a wild-type male is mated to a bobbed male, what will be the phenotypes and their frequencies in the offspring? List males and females separately in your answer. (*Hint*: Refer to the chapter for information about the frequency of nondisjunction in *Drosophila*; see p. 318.)

12.35 An individual with Turner syndrome would be expected to have how many Barr bodies in the majority of cells?

12.36 An XXY individual with Klinefelter syndrome would be expected to have how many Barr bodies in the majority of cells?

*12.37 In human genetics, pedigrees are used to analyze inheritance patterns. Females are represented by a circle, males by a square. The figure that follows presents three 2-generation family pedigrees for a trait in humans. Normal individuals are represented by unshaded symbols, people with the trait by shaded symbols. For each pedigree (A, B, and C), state (by

answering "yes" or "no" in the appropriate blank space) whether transmission of the trait can be accounted for on the basis of each of the listed simple modes of inheritance:

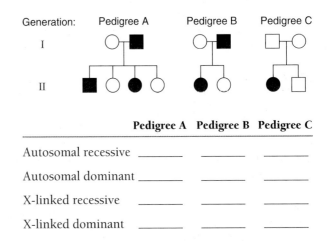

	Pedigree A	Pedigree B	Pedigree C
Autosomal recessive	_____	_____	_____
Autosomal dominant	_____	_____	_____
X-linked recessive	_____	_____	_____
X-linked dominant	_____	_____	_____

12.38 Shaded symbols in the following pedigree represent a trait:

Which of the progeny eliminate X-linked recessiveness as a mode of inheritance for the trait?

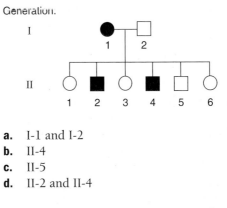

a. I-1 and I-2
b. II-4
c. II-5
d. II-2 and II-4

*12.39 When constructing human pedigrees, geneticists often refer to particular individuals by a number. The generations are labeled with Roman numerals, the individuals in each generation with Arabic numerals. For example, in the pedigree in the following figure, the female with the asterisk is I-2:

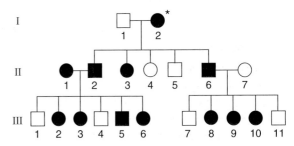

Figure 12.B

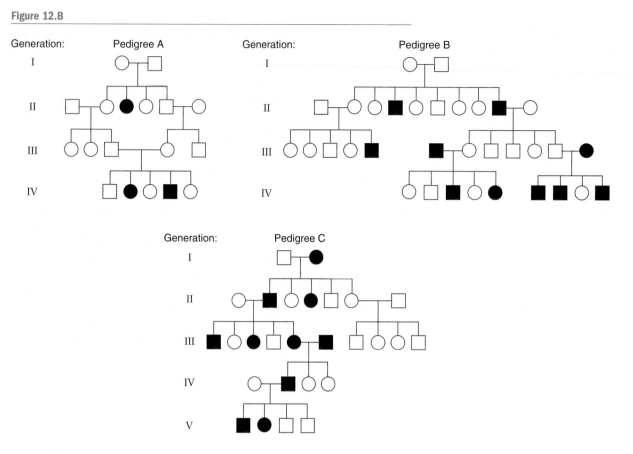

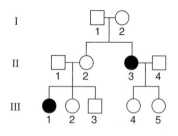

Use this method to designate specific individuals in the pedigree. Determine the probable inheritance mode for the trait shown in the affected individuals (the shaded symbols) by answering the following questions (assume that the condition is caused by a single gene):

a. Y-linked inheritance can be excluded at a glance. What two other mechanisms of inheritance can be definitely excluded? Why can these be excluded?

b. Of the remaining mechanisms of inheritance, which is the most likely? Why?

12.40 A three-generation pedigree for a particular human trait is shown in the following figure:

a. What is the mechanism of inheritance for the trait?

b. Which persons in the pedigree are known to be heterozygous for the trait?

c. What is the probability that III-2 is a carrier (heterozygous)?

d. If III-3 and III-4 marry, what is the probability that their first child will have the trait?

12.41 For each of the more complex pedigrees shown in Figure 12.B, determine the probable mechanism of inheritance: autosomal recessive, autosomal dominant, X-linked recessive, X-linked dominant, or Y linked.

In 12.42 through 12.44, select the correct answer.

*12.42 A genetic disease is inherited on the basis of an autosomal dominant gene. Which of the statements that follow are true, and which are false? Why?

a. Affected fathers have only affected children.

b. Affected mothers never have affected sons.

c. If both parents are affected, all of their offspring have the disease.

d. If a child has the disease, one of his or her grandparents also had the disease.

*12.43 A genetic disease is inherited as an autosomal recessive. Which of the following statements are true, and which are false? Why?

a. Two affected individuals never have an unaffected child.

b. Two affected individuals have affected male offspring, but no affected female children.

c. If a child has the disease, one of his or her grand-parents had it.

d. In a marriage between an affected individual and an unaffected one, all the children are unaffected.

12.44 Which of the following statements is not true for a disease that is inherited as a rare X-linked dominant trait?

a. All daughters of an affected male will inherit the disease.

b. Sons will inherit the disease only if their mothers have it.

c. Both affected males and affected females will pass the trait to half the children.

d. Daughters will inherit the disease only if their fathers have it.

*12.45 Women who were known to be carriers of the X-linked, recessive hemophilia gene were studied to determine the amount of time required for blood to clot. It was found that the time required for clotting was extremely variable from individual to individual. The values obtained ranged from normal clotting time at one extreme to clinical hemophilia at the other. What is the most probable explanation for these findings?

12.46 Hurler syndrome is a genetically transmitted disorder of mucopolysaccharide metabolism resulting in short stature, mental retardation, and various bony malformations. Two specific types are described with extensive pedigrees in the medical genetics literature:

Type I: recessive autosomal
Type II: recessive X linked

You are a consultant in a hospital ward with several patients with Hurler syndrome who have asked you for advice about their relatives' offspring. Being aware that both types are extremely rare and that afflicted individuals almost never reproduce, what counsel would you give to a woman with Type I Hurler syndrome (whose normal brother's daughter is planning marriage) about the offspring of the proposed marriage? In your answer, state the probabilities that the offspring will be affected and whether male and female offspring have an equal probability of being affected.

13

Extensions of Mendelian Genetic Principles

Palomino horse.

PRINCIPAL POINTS

- More than two allelic forms of a gene can exist. However, any given diploid individual can possess only two different alleles of a given gene.

- With complete dominance, the same phenotype results whether the dominant allele is heterozygous or homozygous. In incomplete dominance, the phenotype of the heterozygote is intermediate between those of the two homozygotes. In codominance, the heterozygote exhibits the phenotypes of both homozygotes.

- In many cases, different genes interact to determine phenotypic characteristics. In epistasis, for example, modified Mendelian ratios occur because of gene interactions: The phenotypic expression of one gene depends on the genotype of one or more gene loci.

- Alleles of certain genes may be fatal to the individual. The existence of such lethal alleles indicates that the product usually produced by the nonlethal allele is essential to the functioning of the organism; the gene is called an essential gene.

- Penetrance is the frequency (in percent) at which an allele manifests itself phenotypically within a population. Expressivity is the kind or degree of phenotypic manifestation of a gene or genotype in a particular individual.

- The zygote's genetic constitution specifies only the organism's potential to develop and function. As the organism develops and cells differentiate, many things can influence gene expression. One such influence is the organism's environment, both internal and external. Examples of the internal environment include age and sex; examples of the external environment include nutrition, light, chemicals, temperature, and infectious agents.

- Variation in most of the genetic traits considered in the earlier discussion of Mendelian principles is determined predominantly by differences in genotype; that is, phenotypic differences result from genotypic differences. For many traits, however, the phenotypes are influenced by both genes and the environment.

i A HALF-CENTURY BEFORE WATSON AND CRICK determined the structure of DNA, Karl Landsteiner discovered that different individuals have different blood types and that these blood types are inherited. However, the inheritance of blood type does not always follow the inheritance patterns predicted by Mendel's principles. As it turns out, blood types are one example of a trait that has an inheritance pattern more complex than Mendel described. In this chapter, you will learn about the inheritance of blood types and other traits that are exceptions to, and extensions of, Mendel's principles. Then, in the iActivity, you can use your understanding of these inheritance patterns to help solve a paternity suit involving the actor Charlie Chaplin.

Mendel's principles apply to all eukaryotic organisms and form the foundation for predicting the outcome of crosses in which segregation and independent assortment occur. As more and more geneticists did experiments, though, they found exceptions to, and extensions of, Mendel's principles. Several of these cases are discussed in this chapter, with the goal of gaining a broader knowledge of genetic analysis, particularly in terms of how genes relate to the phenotypes of an organism.

Determining the Number of Genes for Mutations with the Same Phenotype

Up to this point in the book, each mutation we have discussed has affected a different gene. In this chapter, we will encounter cases where that is not so. To help us analyze and understand those cases, we need to understand the relationship between the phenotype and the gene in more detail.

We have learned that the general genetic approach to studying a biological phenomenon is to isolate mutants which affect that phenomenon. Those mutants are identified by their phenotype—the mutant phenotype—which is distinct from the wild-type phenotype. Consider a genetic study in which a large number of mutants is isolated, with each mutant having the same altered phenotype. Our aim is to understand the structure and function of the genes controlling the phenotype. Does each mutant define a different gene, or not? We can answer that question with the **complementation test,** which determines whether two independently isolated mutations with the same phenotype have affected the same or different genes. Also called the *cis–trans* test, the complementation test was developed by Edward Lewis to study genes in *Drosophila.*

In a complementation test, two mutations resulting in the same phenotype are crossed and the phenotype of the progeny is observed. If the two mutations affect different genes, then the progeny will be wild-type/mutant heterozygotes for each of the two genes involved. Because there is a wild-type copy of each gene, the phenotype will be wild type, not mutant (Figure 13.1a). We say that the two mutants complement each other. However, if the two mutations affect the same gene, then the progeny will have a different mutant version of the gene on each of the two homologues,

Figure 13.1

Complementation test to determine whether two mutations resulting in the same phenotype are in the same or different genes. (a) Complementation occurs when mutations are in different genes. (b) Complementation does not occur when mutations are in the same gene.

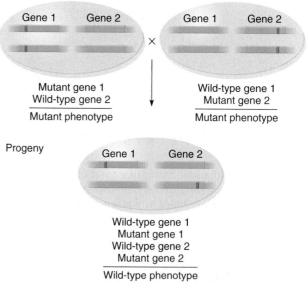

a) Mutations in different genes: complementation

Parents

Mutant gene 1
Wild-type gene 2
Mutant phenotype

Wild-type gene 1
Mutant gene 2
Mutant phenotype

Progeny

Wild-type gene 1
Mutant gene 1
Wild-type gene 2
Mutant gene 2
Wild-type phenotype

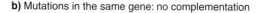

b) Mutations in the same gene: no complementation

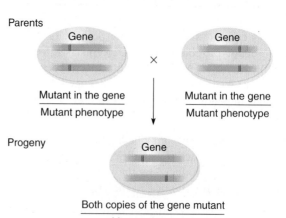

Parents

Mutant in the gene
Mutant phenotype

Mutant in the gene
Mutant phenotype

Progeny

Both copies of the gene mutant
Mutant phenotype

Figure 13.2

Complementation between two black-body mutations of *Drosophila melanogaster*.

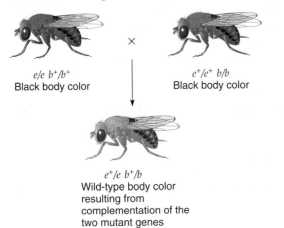

$e/e\ b^+/b^+$
Black body color

×

$e^+/e^+\ b/b$
Black body color

$e^+/e\ b^+/b$
Wild-type body color resulting from complementation of the two mutant genes

and the phenotype will be mutant (Figure 13.1b). In this case, we say that the two mutants do *not* complement each other. Of course, the test is done on unknowns, so the interpretation is the other way around. That is, if two mutations complement each other, they must be in different genes, and if two mutations do not complement each other, they must be in the same gene. How many genes are defined by a set of mutations depends on the number of genes involved in the biological process under genetic study.

Let us consider an actual example from *Drosophila*. Two true-breeding mutant strains have black body color instead of the wild-type grey yellow. When the two strains are crossed, all the F_1 flies have wild-type body color (Figure 13.2). How can these data be interpreted? The simplest explanation is that complementation has occurred between mutations in two genes, each of which is involved in the body color phenotype. That is, a recessive autosomal gene, *ebony* (*e*), when homozygous mutant, results in a black body color. On another autosome, a different recessive gene, *black* (*b*), also results a black body color when homozygous mutant. Because the two parents are homozygotes, they are genotypically $e/e\ b^+/b^+$ and $e^+/e^+\ b/b$, and each is phenotypically black. The F_1 genotype is $e^+/e\ b^+/b$. The F_1 flies have wild-type body color because there is now one wild-type allele of each gene—complementation has occurred. Note that no recombination is involved in the complementation of body color genes: The double heterozygote was produced simply by the fusion of gametes produced by the two true-breeding parents.

What if the F_1 flies from the cross between two independently isolated, true-breeding recessive black-bodied mutant strains were all phenotypically black? We would conclude that, because the mutant phenotype occurred, the two mutations involved did not show complementation and therefore the mutations are in the same gene.

Multiple Alleles

So far in our genetic analyses, we have considered genes that have only two alleles, such as smooth versus wrinkled seeds in peas, red versus white eyes in *Drosophila*, and unattached versus attached earlobes in humans. The allele that predominates in populations of the organism in the wild is the wild-type allele, and the alternative allele is the mutant allele. In a population of individuals, however, a given gene may have several alleles (one wild type and the rest mutant), not just two. Such genes are said to have **multiple alleles,** and the alleles are said to constitute a *multiple allelic series* (Figure 13.3). Although a gene may have multiple alleles in a given population of individuals, *a single diploid individual can have only a maximum of two of these alleles, one on each of the two homologous chromosomes carrying the gene locus.* With experimental organisms, it is straightforward to determine whether mutants are alleles by performing the complementation test described in the previous section.

ABO Blood Groups

An example of multiple alleles of a gene is found in the human ABO blood group series, which was discovered by Karl Landsteiner in the early 1900s. He received the 1930 Nobel Prize in Physiology or Medicine for this discovery. Since certain ABO blood groups are incompatible, these alleles are of particular importance when blood transfusions are done. (There are many blood group series other than ABO, and they also can cause problems in blood transfusions. Hence, they need to be checked through the process of cross-matching.)

There are four blood group phenotypes in the ABO system: O, A, B, and AB. Different combinations of three ABO blood group alleles—I^A, I^B, and *i*—give rise to the four

Figure 13.3

Allelic forms of a gene.

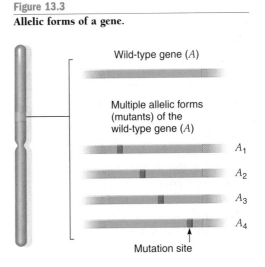

Wild-type gene (*A*)

Multiple allelic forms (mutants) of the wild-type gene (*A*)

A_1

A_2

A_3

A_4

Mutation site

Table 13.1	ABO Blood Groups in Humans, Determined by the Alleles I^A, I^B, and i	
Phenotype (Blood Group)		**Genotype**
O		i/i
A		I^A/I^A or I^A/i
B		I^B/I^B or I^B/i
AB		I^A/I^B

phenotypes (Table 13.1). People who are homozygous for the recessive i allele are of blood group O. Both I^A and I^B are dominant to i. Individuals of blood group A are either I^A/I^A or I^A/i, and those of blood group B are either I^B/I^B or I^B/i. Heterozygous I^A/I^B individuals are of blood group AB—that is, essentially of both blood groups A and B. (See the discussion of codominance in this chapter, pp. 346–347.)

The genetics of this system follows Mendelian principles. An individual who expresses blood group O, for example, must be i/i in genotype. Both parents of this person could be O ($i/i \times i/i$), both could be A ($I^A/i \times I^A/i$, to produce one-fourth i/i progeny), both could be B ($I^B/i \times I^B/i$), or one could be A and one could be B ($I^A/i \times I^B/i$). Each parent must be either homozygous i or heterozygous, with i as one of the two alleles.

Blood typing (determining an individual's blood group) and analyzing blood group inheritance sometimes are used in cases of disputed paternity or maternity or when babies are inadvertently switched in a hospital. In such cases, genetic data cannot prove the identity of the parent. Genetic analysis on the basis of blood group can be used only to show that an individual is *not* the parent of a particular child. For example, a child of phenotype AB (genotype I^A/I^B) could not be the child of a parent of phenotype O (genotype i/i). (*Note*: In most states, blood-type data alone usually are not sufficient for a legal decision about paternity or maternity. In addition, DNA fingerprinting typically is used; see Chapter 9, pp. 224–225.)

With blood transfusions, the blood types of donors and recipients must be carefully matched, because the blood group alleles specify molecular groups, called *cellular antigens*, that are attached to the outsides of the red blood cells. An **antigen** (*antibody generating substance*) is any molecule that is recognized as foreign by an organism and that therefore stimulates the production of specific protein molecules called antibodies, which bind to the antigen. An **antibody** is a protein molecule that recognizes and binds to the foreign substance (antigen) introduced into the organism. A given individual has a large number of antigens on cells and tissues, many of which are foreign to another individual—hence the concern over blood type in

blood transfusions and tissue type in organ transplants. Antigens usually are not recognized as foreign by the organism expressing them. (Autoimmune diseases are an exception.)

The I^A, I^B, and i alleles of the ABO locus specify the four blood types—O, A, B, and AB—as just described. The I^A allele specifies the A antigen; people of blood type A (genotype I^A/I^A or I^A/i) have the A antigen on their red blood cells; and blood serum prepared from them contains naturally occurring antibodies against the B antigen (called anti-B antibodies), but none against the A antigen. Antibodies against the B antigen agglutinate, or clump, any red blood cells that have the B antigen on them. Since clumped cells cannot move through the fine capillaries, agglutination may lead to organ failure and, possibly, death. Conversely, people of blood type B (genotype I^B/I^B or I^B/i) have the B antigen on their red blood cells, and blood serum prepared from them contains naturally occurring anti-A antibodies, but no anti-B antibodies. People of AB blood type (genotype I^A/I^B) have both A and B antigens on the blood cells and neither anti-A nor anti-B antibodies in their blood serum. Lastly, in people with blood type O (i/i), the red blood cells have neither A nor B antigen, and their blood serum contains both anti-A and anti-B antibodies. The antigen–antibody relationships are summarized in Figure 13.4. Agglutination (clumping) of the red blood cells is seen in each case where an antibody interacts with the antigen for which it is specific.

What transfusions are safe between people with different blood groups in the ABO system?

1. People with blood type A produce the A antigen, so their blood can be transfused only into recipients who do not have the anti-A antibody—that is, people of blood type A or AB.

2. Those with blood type B produce the B antigen, so their blood can be transfused only into recipients who do not have the anti-B antibody—that is, people of blood type B or AB.

3. People with blood type AB produce both the A and B antigens, so their blood can be transfused only into recipients who do not have either the anti-A antibody or the anti-B antibody—that is, people of blood type AB.

4. People with blood type O produce neither A nor B antigens, so their blood can be transfused into any recipient—that is, people of blood type A, B, AB, or O.

The preceding discussion implies that people of blood type AB can receive transfusions of blood from people of any of the four blood types. Thus, people with blood type AB are called *universal recipients*. Similarly, type O blood can be used as donor blood for any recipient, because it elicits no reaction. Thus, people with blood type O are called *universal donors*.

Figure 13.4

Antigenic reactions that characterize the human ABO blood types. Blood serum from each of the four blood types was mixed with blood cells from the four types in all possible combinations. In some cases, such as a mix of B serum with A cells, the cells become clumped.

Serum from blood type	Antibodies present in serum	Cells from blood type			
		O	A	B	AB
O	Anti-A Anti-B				
A	Anti-B				
B	Anti-A				
AB	—				

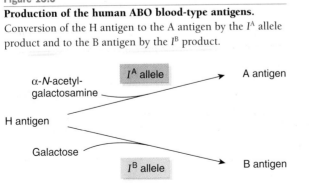

The relationship between the ABO alleles and the antigens on the red blood cells is as follows: The ABO locus encodes *glycosyltransferases,* enzymes that add sugar groups to a preexisting polysaccharide (Figure 13.5). The polysaccharides in question are those which have combined with lipids to form glycolipids. The glycolipids then associate with red blood cell membranes to form the blood-group antigens. Most people produce a glycolipid called the H antigen. The I^A allele produces a glycosyltransferase enzyme that adds the sugar α-N-acetylgalactosamine to the end of the polysaccharide to produce the A antigen. The I^B allele produces a different glycosyltransferase, which also recognizes the H antigen, but adds galactose to its polysaccharide to produce the B antigen. (Note that this small difference in the structure of the A and B antigens is sufficient to induce an antibody response.) In both cases, some H antigen remains unconverted.

In the I^A/I^B heterozygote, both enzymes are produced, so some H antigen is converted to the A antigen and some

to the B antigen. The red blood cell has both antigens on the surface, so the person is of blood group AB.

People who are homozygous for the *i* allele produce no enzymes to convert the H antigen glycolipid. Therefore, their red blood cells carry only the H antigen. This antigen does not elicit an antibody response in people of other blood groups because its polysaccharide component is the basic component of the A and the B antigens as well, so it is not detected as a foreign substance. People who are heterozygous for the *i* allele have the blood type of the other allele. For example, in I^B/i people, the I^B allele results in the conversion of some of the H antigen to the B antigen, determining the person's blood type.

The H antigen is produced by the action of the dominant *H* allele at a locus distinct from the ABO locus. People who are homozygous for the recessive mutant allele, *h*, do not make the H antigen; therefore, regardless of the presence of I^A or I^B alleles at the ABO locus, no A or B antigens can be produced. These very rare *h/h* people are like blood-group O people in the sense that they lack A and B antigens; they are said to have the Bombay blood type. However, people in the Bombay blood group produce anti-O antibodies (antibodies against the H antigen), whereas people in blood group O do not.

Figure 13.5

Production of the human ABO blood-type antigens. Conversion of the H antigen to the A antigen by the I^A allele product and to the B antigen by the I^B product.

α-N-acetyl-galactosamine — I^A allele → A antigen
H antigen
Galactose — I^B allele → B antigen

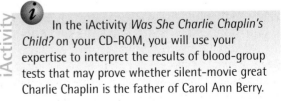

iActivity In the iActivity *Was She Charlie Chaplin's Child?* on your CD-ROM, you will use your expertise to interpret the results of blood-group tests that may prove whether silent-movie great Charlie Chaplin is the father of Carol Ann Berry.

Drosophila Eye Color

Another example of multiple alleles concerns the white (*w*) locus of *Drosophila*. Recall from Chapter 12 (pp. 314–317) that the w^+ allele results in wild-type brick-red eyes and that the recessive *w* allele, when homozygous or hemizygous, results in white eyes. There are actually more than 100 mutant alleles at the white locus. With w^+ symbolizing the wild-type (brick-red) allele of the white-eye gene, *w* is the recessive mutant white-eye allele, and w^e is the recessive mutant eosin (reddish-orange) allele.

It was Thomas Hunt Morgan's work with a white-eyed variant of *Drosophila* that indicated the presence of genes on the X chromosome. Soon after that discovery, he found evidence of other distinct genes on the *Drosophila* X chromosome (such as the bar-eye shape and the vermilion eye color genes). Morgan experimented with strains that had different eye color genes. When he crossed a white-eyed female with a vermilion-eyed male, the results were unexpected to him: The F_1 females were all red eyed. His explanation was that the white and vermilion eye color traits are specified by *two different genes*. Using genetic symbols, we can write his cross as $w\ v^+/w\ v^+ \female \times w^+\ v/Y \male$.[1] The F_1 females from this cross would be doubly heterozygous $w\ v^+/w^+\ v$ and therefore would show the dominant effect resulting from the presence of the wild-type allele for each allelic pair, namely, the brick-red eyes characteristic of wild-type flies.

In 1912, Morgan obtained data for X-linked eye color genes that could not be explained so conveniently. The eye color variants were *white* and *eosin* (reddish orange). Like *white*, *eosin* is X-linked recessive to the wild-type eye color. However, when a female from an eosin-eyed strain is crossed with a male from a white-eyed strain, all F_1 females have eosin eyes. In 1913, Alfred Sturtevant observed that (1) red (wild-type) eye color is dominant to *eosin* and to *white* and (2) *eosin* is recessive to the wild type, but dominant to *white*. He concluded that *eosin* and *white* are both mutant alleles of a single gene. In other words, there are multiple alleles of the white-eyed gene.

Sturtevant's concept becomes clearer if we assign symbols to the alleles. A lowercase letter (or letters) designates the gene, and superscripts designate the different alleles, as in the following example: w^+ is the wild-type allele of the white-eye gene, *w* is the recessive white-eye allele, and w^e is the eosin allele. Figure 13.6a uses this notation to track the cross of an eosin-eyed female with a white-eyed male. The F_1 females are w^e/w and have eosin eyes, because w^e is dominant over *w*. When these F_1 females are crossed with red-eyed males (Figure 13.6b), all the female progeny are heterozygous and red-eyed, because they contain the w^+ allele; they

are either w^+/w^e or w^+/w. Half the male progeny are eosin-eyed (w^e/Y), and the other half are white-eyed (w/Y).

Many known alleles of the white-eyed gene are distinguishable because they produce different eye colors, ranging from white to near-wild type when the alleles are homozygous or hemizygous. The eye color phenotype depends on the amount of pigment deposited in the eye cells. Table 13.2 lists the relative amounts of pigment present in wild-type females and in females that are homozygous for different mutant alleles of the white locus. As expected, the original mutant allele *w* has the least amount of pigment. Between white and wild, there is a wide range of pigment amounts. The phenotypic expressions of different alleles of the same gene reflect how the biological activity of the protein encoded by the white gene (its *gene product*) has been altered to different extents.

As Table 13.3 shows, the number of allelic forms of a gene is not limited to the three we have discussed; indeed, many hundreds of alleles are known for some genes. The number of possible genotypes in a multiple allelic series depends on the number of alleles involved. With one allele, only one genotype is possible. With two alleles A^1 and A^2, three genotypes are possible: A^1/A^1 and A^2/A^2 homozygotes and the A^1/A^2 heterozygote. The general formula for *n* alleles is $n(n-1)/2$ possible genotypes, of which *n* are homozygotes and $n(n-1)/2$ are heterozygotes.

Relating Multiple Alleles to Molecular Genetics

The base-pair sequence of a gene specifies the amino-acid sequence of a protein, and the function of a protein depends on its amino-acid sequence. From this modern

Table 13.2	Eye Pigment Quantification for *Drosophila* White Alleles	
Genotypes	**Relative Amount of Total Pigment**
w^+/w^+ (wild type)	1.0000
w/w (white)	0.0044
w^t/w^t (tinged)	0.0062
w^a/w^a (apricot)	0.0197
w^{bl}/w^{bl} (blood)	0.0310
w^e/w^e (eosin)	0.0324
w^{ch}/w^{ch} (cherry)	0.0410
w^{a3}/w^{a3} (apricot-3)	0.0632
w^w/w^w (wine)	0.0650
w^{co}/w^{co} (coral)	0.0798
w^{sat}/w^{sat} (satsuma)	0.1404
w^{col}/w^{col} (colored)	0.1636

---

[1]As a reminder, the slash represents the homologous chromosomes on which the alleles are found.

Figure 13.6
Results of crosses of *Drosophila melanogaster* involving two mutant alleles of the same locus: white (w) and white-eosin (w^e). (a) white-eosin-eyed (w^e/w^e) ♀ × white eyed (w/Y) ♂. (b) F$_1$ (w^e/w) ♀ × red-eyed (wild type) (w^+/Y) ♂.

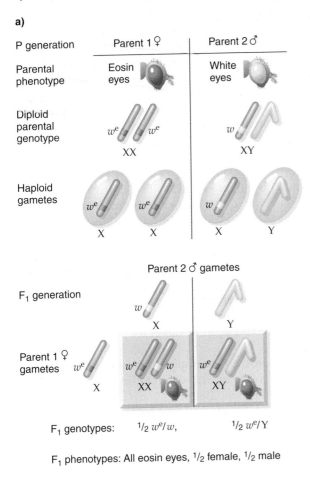

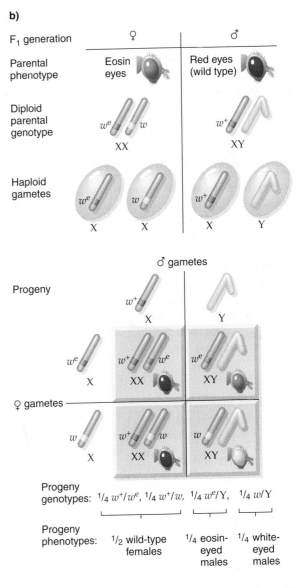

perspective, we should not be surprised to find multiple alleles of a gene. That is, an amino acid change at one of many places in the protein affects its function adversely, and the position and type of the change determine the extent of loss of function of the protein. The *Drosophila* eye color illustrates this concept (see Table 13.2), and if you look at the entries for many human genetic diseases in OMIM (http://www3.ncbi.nlm.nih.gov/Omim/), such as the breast cancer susceptibility gene *BRCA1* (OMIM 113705), you will find that several to many alleles often have been identified. Two practical consequences of multiple alleles in the case of human genetic diseases are that the symptoms of a disease may vary with the allele and that devising a single DNA-based test for

individuals with a disease or carriers of a disease is difficult or impossible.

KEYNOTE

Many allelic forms of a gene can exist in a population. When they do, the gene is said to show multiple allelism, and the alleles involved constitute a multiple allelic series. However, any given diploid individual can possess only two different alleles of a given gene. Multiple alleles obey the same rule of transmission as do alleles of which there are only two types, although the dominance relationships among multiple alleles vary from one group to another.

Table 13.3	Genotype Number of Multiple Alleles		
Number of Alleles	Kinds of Genotypes	Kinds of Homozygotes	Kinds of Heterozygotes
1	1	1	0
2	3	2	1
3	6	3	3
4	10	4	6
5	15	5	10
n	$n(n+1)/2$	n	$n(n-1)/2$

Modifications of Dominance Relationships

Complete dominance is the phenomenon in which one allele is dominant to another, so that the phenotype of the heterozygote is the same as that of the homozygous dominant. With **complete recessiveness,** the recessive allele is phenotypically expressed only when the organism is homozygous. Complete dominance and complete recessiveness are the two extremes of a range of dominance relationships. Whereas all the allelic pairs Mendel studied showed complete dominance–complete recessiveness relationships, many allelic pairs do not.

Incomplete Dominance

When one allele is not completely dominant to another allele, it is said to show **incomplete,** or **partial dominance.** With incomplete dominance, the heterozygote's phenotype is intermediate to those of individuals that are homozygous for either allele involved.

animation

a Incomplete Dominance and Codominance

Plumage color in chickens is an example of incomplete dominance. Crosses between a true-breeding black strain (homozygous C^BC^B) and a true-breeding white strain (homozygous C^WC^W) give F_1 birds with bluish-grey plumage (Figure 13.7a), called Andalusian blues by chicken breeders. (Here, the symbols are designed to give equal weight to the two alleles, because neither dominates the phenotype. In this particular example, the C signifies color, while "B" and "W" indicate black and white, respectively.) An Andalusian blue does not breed true because it is heterozygous; hence, in Andalusian × Andalusian crosses (Figure 13.7b), the two alleles segregate in the offspring and produce 1 black : 2 Andalusian blue : 1 white fowl. The most efficient way to produce Andalusian blues is to cross black × white because all progeny of this mating are Andalusian blues. In very loose molecular terms, then, two doses of C^B are needed to give black, whereas in the heterozygote the white dilutes the black to grey.

Another example of incomplete dominance is the palomino horse, which has a golden-yellow body color and a mane and tail that are almost white. (See chapter opener figure.) Palominos do not breed true. When they are interbred, the progeny are ¼ cremellos (extremely light colored) ½ palominos, and ¼ light chestnuts. This 1:2:1 ratio resulting from interbreeding is characteristic of incomplete dominance. The interpretation is that there are two alleles, C and C^{cr}, involved in producing palominos. That is, the genotype C/C, in combination with other genes, results in horses with a light chestnut (also called sorrel) color. The genotype C/C^{cr}, in combination with the same genes as above, results in a dilution of the normal reddish brown of light chestnuts to yellow or cream, giving the palomino color. The genotype C^{cr}/C^{cr} brings about an extreme dilution of the phenotype, giving the cremello color. In other words, the C^{cr} allele is a modifier allele that affects the expression of other coat-color genes in the horse. The C allele at that locus allows full expression of the other coat-color genes in the animal.

There are many examples of incomplete dominance in plants, such as flower color in the snapdragon, involving the two alleles C^R and C^W. Crosses of red-flowered (C^RC^R) snapdragons have red flowers, C^RC^W plants have pink flowers, and C^WC^W plants have white flowers.

Codominance

A modification of the dominance relationship, **codominance** is related to incomplete dominance. In codominance, the heterozygote exhibits the phenotypes of *both* homozygotes. By contrast in incomplete dominance, the heterozygote exhibits a phenotype intermediate between the two homozygotes.

The ABO blood series discussed earlier in this chapter provides a good example of codominance. Heterozygous I^A/I^B individuals are of blood group AB because both the A antigen (product of the I^A allele) and the B antigen (product of the I^B allele) are produced. Thus, the I^A and I^B alleles are codominant.

The human M–N blood group system is another example of codominance. For transfusion compatibility, this system is of less clinical importance than the ABO system. In the M–N system, three blood types occur: M, MN, and N, respectively determined by the genotypes L^M/L^M, L^M/L^N, and L^N/L^N. As in the ABO system, the M–N

Figure 13.7

Incomplete dominance in chickens. (a) A cross between a white bird and a black bird produces F_1 birds of intermediate grey color, called Andalusian blues. (b) The F_2 generation shows the 1:2:1 phenotype ratio characteristic of incomplete dominance.

a)

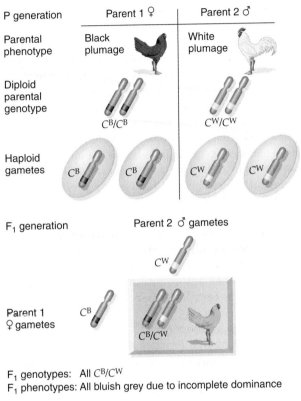

F_1 genotypes: All C^B/C^W
F_1 phenotypes: All bluish grey due to incomplete dominance

b)

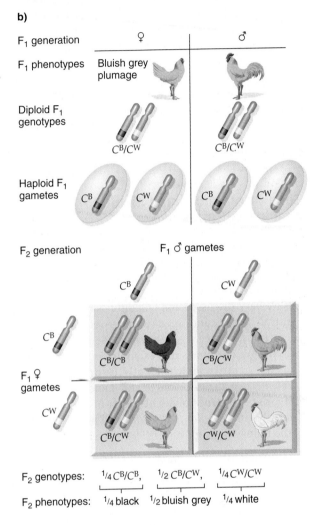

F_2 genotypes: $1/4\,C^B/C^B$, $1/2\,C^B/C^W$, $1/4\,C^W/C^W$
F_2 phenotypes: $1/4$ black $1/2$ bluish grey $1/4$ white

---

alleles result in the formation of antigens on the surface of the red blood cell. The heterozygote in this case has both the M and the N antigens and shows the phenotypes of both homozygotes.

Molecular Explanations of Incomplete Dominance and Codominance

What explains incomplete dominance and codominance at the molecular level? A general interpretation is that, in codominance, products result from both alleles in a heterozygote, so that often (but not always) both homozygote phenotypes are observed. (For example, L^M/L^N individuals express both M and N blood-group antigens). In incomplete dominance, only one allele in a heterozygote is expressed to produce a product. A homozygote for the expressed allele, then, has two doses of the gene product, and full phenotypic expression results (for example, black chickens). In a homozygote for the allele that is not expressed, a phenotype characteristic of no gene expression results (white chickens or white snapdragons). In a heterozygote, the single allele expressed results in only enough product for an intermediate phenotype (grey chickens or pink snapdragons). By contrast, in a heterozygote for a gene showing normal dominance, half the amount of protein produced by the homozygote may be sufficient for normal cell function. In this case, the gene is said to be **haplosufficient.** Alternatively, the expression of the one normal allele in the heterozygote may be increased to produce protein levels that give normal cell function.

K E Y N O T E

With complete dominance, the same phenotype results whether the dominant allele is heterozygous or homozygous. With complete recessiveness, the allele is phenotypically expressed only when the genotype is homozygous recessive; the recessive allele has no effect on the phenotype of the heterozygote. Complete dominance and complete recessiveness are two extremes between which all transitional degrees of dominance are possible. In incomplete dominance, the phenotype of the heterozygote is intermediate between those of the two homozygotes, whereas in codominance the heterozygote exhibits the phenotypes of both homozygotes.

Gene Interactions and Modified Mendelian Ratios

No gene acts by itself in determining an individual's phenotype; instead, the phenotype is the result of highly complex and integrated patterns of molecular reactions that are under direct gene control. All the genetic examples we have discussed and will discuss have discrete biochemical bases, and in a number of cases complex interactions between genes can be detected by genetic analysis. We examine some examples in this section.

Consider two independently assorting gene pairs, each with two alleles: A and a; and B and b. The outcome of a cross between individuals, each of which is doubly heterozygous ($A/a\ B/b \times A/a\ B/b$), will be nine genotypes in the following proportions:

$1/16\ A/A\ B/B$
$2/16\ A/A\ B/b$
$1/16\ A/A\ b/b$
$2/16\ A/a\ B/B$
$4/16\ A/a\ B/b$
$2/16\ A/a\ b/b$
$1/16\ a/a\ B/B$
$2/16\ a/a\ B/b$
$1/16\ a/a\ b/b$.

If the phenotypes determined by the two allelic pairs are distinct—for example, smooth versus wrinkled peas, or long versus short stems—and there is complete dominance, then we get the familiar dihybrid phenotypic ratio of 9:3:3:1. (See Figure 11.12b.) Any deviation from this standard 9:3:3:1 ratio indicates that the phenotype is the product of the interaction of two or more genes.

As we shall show in Chapter 3, the 9:3:3:1 phenotypic ratio can be represented genotypically in a shorthand way as $A/–\ B/–$, $A/–\ b/b$, $a/a\ B/–$, $a/a\ b/b$, respectively. The dash indicates that the phenotype is the same whether the gene is homozygous dominant or heterozygous. (For example, $A/–$ means either A/A or A/a.) This

system cannot be used when incomplete dominance or codominance is involved, because the A/A and A/a genotypes have different phenotypes.

The next several sections discuss the main processes that result in modified Mendelian ratios. We begin with examples of interactions between different genes that control the same general phenotypic attribute. Then we discuss examples of interactions of different genes in which an allele of one gene masks or modifies the expression of alleles of another gene. This second type of interaction is called *epistasis* (literally, "to stand upon"). In both cases, the discussions are confined to dihybrid crosses in which the two pairs of alleles assort independently. In the real world, there are many more complex examples of gene interactions involving more than two pairs of alleles or genes that do not assort independently.

For many of the examples we will discuss, *hypothetical* molecular explanations are presented. *It is important to note that these presentations are theoretical.* They are included because the processes of experimental science typically involve proposing hypotheses based on theories or models and doing experiments designed to test the hypotheses. Therefore, it is appropriate to consider models that are compatible with the modified Mendelian ratios being discussed.

Gene Interactions That Produce New Phenotypes

If the two allelic pairs in a dihybrid cross affect the same phenotypic characteristic, there is a chance that the interaction of their gene products will give novel phenotypes, and the result may or may not be modified phenotypic ratios, depending on the particular interaction between the products of the nonallelic genes.

Comb Shape in Chickens. In chickens, different comb-shape phenotypes (Figure 13.8) result from interactions between the alleles of two gene loci. Each of the four comb types will breed true if the alleles involved are homozygous.

Crosses made between true-breeding rose-combed and single-combed varieties show that rose is completely dominant over single. When the F_1 rose-combed birds are interbred, a ratio of 3 rose : 1 single results in the F_2. Similarly, pea comb is completely dominant over single, with a 3 pea : 1 single ratio in the F_2. When true-breeding rose and pea varieties are crossed, however, the result is different and interesting (Figure 13.9a). Instead of showing either rose or pea combs, all birds in the F_1 show yet another comb form, which is called walnut comb because it resembles half a walnut meat.

When the F_1 walnut-combed birds are interbred, not only do walnut-, rose-, and pea-combed birds appear, but so do single-combed birds (Figure 13.9b). These four comb types occur in a ratio of 9 walnut : 3 rose : 3 pea : 1 single.

Figure 13.8

Four distinct comb-shape phenotypes in chickens, resulting from all possible combinations of a dominant and a recessive allele at each of two gene loci. (a) Rose comb (*R/– p/p*). (b) Walnut comb (*R/– P/–*). (c) Pea comb (*r/r P/–*). (d) Single comb (*r/r p/p*).

a) Rose comb

b) Walnut comb

c) Pea comb

d) Single comb

Such a ratio is characteristic in F₂ progeny from two parents, each of which is heterozygous for two genes. The class in the F₂ with at least one copy of each dominant allele is walnut, and the proportion of the single-combed birds indicates that this class contains both recessive alleles.

The explanation of the preceding results is as follows (see Figure 13.9b): The walnut comb depends on the presence of two dominant alleles—*R* and *P*—both located at two independently assorting gene loci. In *R/– p/p* birds, a rose comb results; in *r/r P/–* birds, a pea comb results; and in *r/r p/p* birds, a single comb results.

Thus, it is the interaction of two dominant alleles, each of which individually produces a different phenotype, that produces a different phenotype. No modification of typical Mendelian ratios is involved. The molecular basis for the four comb types is not known. At a very general level, we can propose that the single-comb phenotype results from the activities of a number of genes other than the *R* and *P* genes. In other words, *r/r p/p* birds do not produce any functional gene product that influences the comb phenotype beyond the basic single appearance. The dominant *R* allele might produce a gene product that interacts with the products of genes controlling the single-comb phenotype to produce a rose-shaped comb. Similarly, the dominant *P* allele might produce a gene product that interacts with the products of the

single-comb genes to produce a pea-shaped comb. When the products of both the *R* and *P* alleles are present, they interact to produce still another comb variation: the walnut comb.

Fruit Shape in Summer Squash. Two of the many varieties of summer squash have long fruit and sphere-shaped fruit, respectively. This trait also shows complete dominance at two gene pairs, and interaction between both dominants results in a new phenotype.

The long-fruit varieties are always true breeding. However, in some crosses between different varieties of true-breeding sphere-shaped plants, the F₁ fruit is disk-shaped (Figure 4.10). In such instances, the F₂ fruit are approximately ⁹⁄₁₆ disk-shaped, ⁶⁄₁₆ sphere-shaped, and ¹⁄₁₆ long-shaped; this is a modification of the typical Mendelian ratio, so two genes are involved. The explanation is as follows: A dominant allele of either gene and homozygosity for the recessive allele of the other gene (*A/– b/b* or *a/a B/–*) results in the same phenotype: spherical fruit. Two dominant alleles, one of each gene (*A/– B/–*), interact together to produce a new phenotype: disk-shaped fruit. The doubly homozygous recessive (*a/a b/b*) gives a long fruit shape. Thus, the cross of two sphere-shaped squashes here is *A/A b/b* and *a/a B/B*. The F₁s are disk-shaped and genotypically *A/a B/b*. The F₂ disk-shaped fruits are *A/– B/–*, the spherical fruits are *A/– b/b* or *a/a B/–*, and the long-shaped fruits are *a/a b/b*, giving the 9:6:1 ratio.

The precise molecular bases for the different shapes of squash fruit are not known. In the *a/a b/b* squash, in the absence of *A* and *B* allele products, the functions of other genes determine the long fruit shape. If either the *A* or the *B* allele product, but not both, is present, the basic long fruit shape is modified into a sphere shape. The disk fruit shape presumably occurs through a modification of the sphere shape caused by the interaction of *A* and *B* gene products.

Epistasis

Epistasis is the interaction between two or more genes to control a single phenotype. The interaction involves one gene masking or modifying the phenotypic expression of another gene. No new phenotypes are produced by this type of gene interaction. A gene that masks another gene's expression is said to be *epistatic,* and a gene whose expression is masked is said to be *hypostatic.* If we think about the F₂ genotypes *A/– B/–, A/– b/b, a/a B/–,* and *a/a b/b,* epistasis may be caused by the presence of homozygous recessives of one gene pair, so that *a/a* masks the effect of the *B* allele. Or epistasis may result from the presence of one dominant allele in a gene pair. For example, the *A* allele might mask the effect of the *B* allele. Epistasis can also occur in both directions between

Figure 13.9

Complete dominance in chickens. The genetic crosses show the interaction of genes for comb shape. (a) The cross of a true-breeding rose-combed bird with a true-breeding pea-combed bird gives all walnut-combed offspring in the F_1. (b) When the F_1 birds are interbred, a 9:3:3:1 ratio of walnut : rose : pea : single occurs in the F_2.

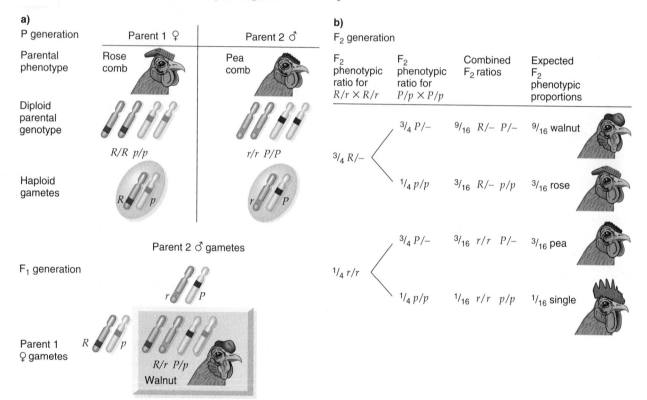

F$_1$ genotypes: All R/r P/p

F$_1$ phenotypes: All walnut comb

two gene pairs. All these possibilities produce quite a number of modifications of the 9:3:3:1 ratio in a dihybrid cross. Some examples of epistasis follow.

Recessive Epistasis. In *recessive epistasis*, $A/-$ b/b and a/a b/b individuals have the same phenotype, so the phenotypic ratio in the F_2 is 9:3:4 rather than 9:3:3:1. An example is coat color in rodents. Wild mice have a greyish color, because the hairs in the fur have alternating bands of black and yellow. This coloration—the agouti pattern—aids in camouflage and is found in many wild rodents, including guinea pigs and grey squirrels, besides wild mice. Several other coat colors are seen in domesticated rodents. Albinos, for example, have no pigment in the fur or in the irises of the eyes, so they have a white coat and pink eyes. Albinos are true breeding, and this variation behaves as a complete recessive to any other color. Another variant has black coat color as the result of the absence of the yellow pigment found in the agouti pattern. Black is recessive to agouti.

When true-breeding agouti mice are crossed with albinos, the F_1 progeny are all agouti, and when these F_1 agoutis are interbred, the F_2 progeny consist of approximately $9/16$ agouti animals, $3/16$ black, and $4/16$ albino (Figure 13.11). This pattern occurs because the parents differ in whether they have a dominant allele C of a gene for the development of any color (black mice are $C/-$ and albinos are c/c) and in whether they have a dominant allele A of a gene for the agouti pattern, which is a yellow banding of the black hairs ($A/-$ are agouti and a/a are nonagouti). (*Note:* The symbols here are the actual ones used for the genes involved in rodent coat color. Do not confuse the a and c loci here with the a and b loci referred to in our continuing discussion of modified ratios that began on p. 348.) Phenotypically, $A/-$ $C/-$ are agouti, a/a $C/-$ are black, and $A/-$ c/c and a/a c/c are albino, giving a 9:3:4 phenotypic ratio of agouti : black : albino. Thus, this example demonstrates epistasis of c/c over $A/-$. In other words, white hairs are produced in c/c mice, regardless of the genotype at the other locus.

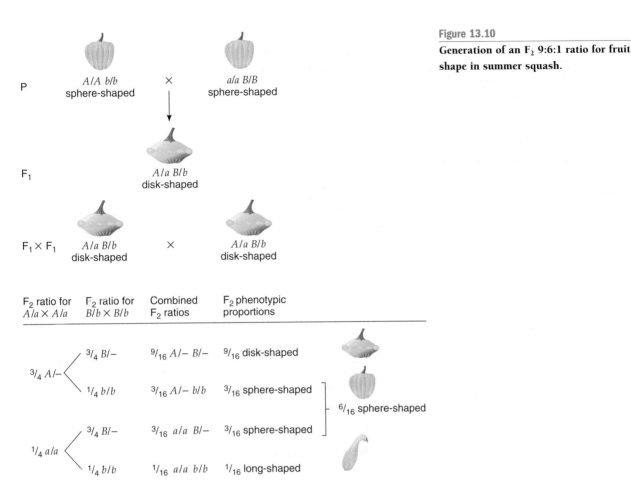

Figure 13.10

Generation of an F$_2$ 9:6:1 ratio for fruit shape in summer squash.

In fact, three gene loci are involved in the phenotypes of rodent coat color. At one locus, the dominant *C* allele allows pigment formation; the recessive *c* allele, when homozygous, prevents pigment formation, regardless of the genotypes of other coat-color genes. At a second locus, the dominant allele *A* determines agouti, and its recessive allele *a* in the homozygous state produces nonagouti mice. The dominant allele *B* of the third locus specifies a black pigment, and its recessive allele *b*, when homozygous, results in brown pigment. All the mice involved in Figure 13.11 must have had at least one *B* allele; otherwise some brown mice would have been seen. (For example, *A/– C/– b/b* are brown.)

Another example of recessive epistasis involves coat color in labrador retrievers (labs). At one gene, *B/–* specifies the formation of a black pigment, while *b/b* specifies the formation of a brown pigment. At an independent gene, *E/–* allows the expression of the *B* gene, while *e/e* does not allow the expression of the *B* gene, and this situation gives yellow. Therefore, genotype *B/– E/–* produces a black lab, *b/b E/–* produces a chocolate (brown) lab, and *–/– e/e* (*–/–* means either *B* or *b* on each chromosome) produces a yellow lab (Figure 13.12). There is a slight variation among the yellow labs in that dogs with the *B/– e/e* genotype have dark noses and lips, whereas dogs with

the *b/b e/e* genotype have pale noses and lips. So, if we cross a true-breeding black lab of genotype *B/B E/E* with a true-breeding yellow lab of genotype *b/b e/e*, the F$_1$s will all be black labs of genotype *B/b E/e*. Intercrossing dogs with this genotype will produce 9 *B/– E/–* black : 3 *B/– e/e* yellow : 3 *b/b E/–* chocolate : 1 *b/b e/e* yellow, for a phenotypic ratio of 9 black : 3 chocolate : 4 yellow.

Dominant Epistasis. In dominant epistasis, *A/– B/–* and *A/– b/b* individuals have the same phenotype, so the phenotypic ratio in the F$_2$ is 12:3:1 rather than 9:3:3:1. In other words, in dominant epistasis, one gene, when dominant, is epistatic to the other gene.

An example of dominant epistasis may be seen in the fruit color of summer squash, which has three common fruit colors: white, yellow, and green. In crosses of white and yellow and of white and green, white is always expressed. In crosses of yellow and green, yellow is expressed. Yellow thus is recessive to white, but dominant to green.

Now, consider two gene pairs: *W/w* and *Y/y*. In squashes that are *W/–* in genotype, the fruit is white no matter what genotype is at the other locus. In *w/w* plants, the fruit will be (1) yellow if a dominant allele of the other locus is present and (2) green if it is absent. In other words,

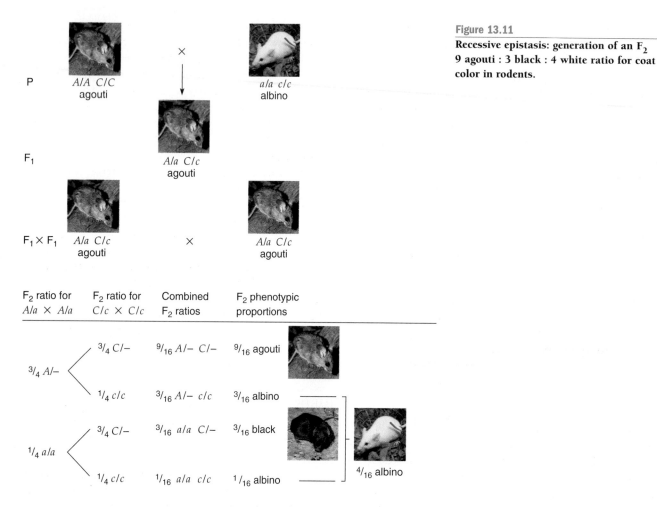

Figure 13.11
Recessive epistasis: generation of an F$_2$ 9 agouti : 3 black : 4 white ratio for coat color in rodents.

W/– Y/– and W/– y/y plants have white fruits, w/w Y/– plants have yellow fruits, and w/w y/y plants have green fruits. The F$_2$ progeny of an F$_1$ self of doubly heterozygous individuals shows a ratio of 12 white : 3 yellow : 1 green fruits in the plants (Figure 13.13).

A theoretical biochemical pathway to explain the 12:3:1 ratio of squash color is shown in Figure 13.14. The hypothesis is that a white substance is converted to a yellow end product via a green intermediate. The dominant Y allele is needed for the conversion of the green substance to yellow, and the dominant W allele specifies a product that inhibits the white-to-green conversion step. Thus, all plants that have at least one W allele will be white-fruited, no matter which alleles are present at the Y locus, since the W gene product prevents the making of green substance (Figure 13.14a). So, $^{12}/_{16}$ of the F$_2$ squashes are white-fruited and have the genotypes W/– Y/– and W/– y/y. F$_2$ plants with the w/w genotype are colored: $^{3}/_{16}$ are yellow-fruited with the genotype w/w Y/– (Figure 13.14b), and $^{1}/_{16}$ are green-fruited with the genotype w/w y/y (Figure 13.14c). The yellow-fruited plants occur because the green substance is made (there is no inhibition of the white-to-green step) and functional Y allele product catalyzes the conversion of green substance to yellow substance. The

white-fruited plants occur because the green substance is produced, but in the absence of a dominant Y allele, the green substance cannot be converted to yellow substance.

Figure 13.12

Recessive epistasis in labrador retrievers. Left: Black lab, genotype B/– E/–. Middle: Yellow lab, genotype –/– e/e. Right: Chocolate lab, genotype b/b E/–.

Another example of dominant epistasis causes greying in horses (Figure 13.15). If a horse is genotypically *GG* or *Gg*, it will show a progressive silvering of the coat color with which it is born until the coat becomes grey (really, essentially white) as a mature animal. The *G* allele does not affect skin or eye pigmentation, however. Horses with the genotype *gg* do not go grey as they mature; instead, they remain the color they exhibited at the time they were born. In other words, if one or more copies of the dominant *G* allele is present, the expression of other coat-color genes becomes masked over a period of several years.

Epistasis Involving Duplicate Genes. A gene or genotype at one locus may produce a phenotype identical to that produced by a gene or genotype at a second locus. In that case, we say that *duplicate genes* are involved. As with the other types of epistasis we have discussed, epistasis involving duplicate genes results in a modification of the 9:3:3:1 ratio for progeny of crosses of doubly heterozygous parents. How the ratio is changed depends on the dominance–recessiveness states of the two loci that give the same phenotype. Two examples will make this clear.

In the sweet pea, purple flower color is dominant to white and gives a typical 3:1 ratio in the F_2. White-flowered varieties of sweet peas breed true, and crosses between different white varieties usually produce white-flowered progeny. In some cases, however, crosses of two true-breeding white varieties give only purple-flowered F_1 plants. When these F_1 hybrids are self-fertilized, they produce an F_2 consisting of about 9/16 purple-flowered sweet peas and 7/16 white-flowered. The 9:7 ratio is a modification of the 9:3:3:1 ratio. Even though they are not all homozygous for the alleles in question, all the F_2 white-flowered plants breed true when self-fertilized.

One-ninth of the purple-flowered F_2 plants—the *C/C P/P* genotypes—breed true.

These results can be explained by the interaction of two genes. The 9/16 purple-flowered F_2 plants suggest that colored flowers appear only when two independent dominant alleles are present together and that the color purple results from some interaction between them. White flower color would then result from homozygosity for the recessive allele of one or both genes. Thus, gene pair *C/c* specifies whether the flower can be colored, and gene pair *P/p* specifies whether the purple flower color will result. An interaction of two genes in this way to give rise to a specific product is a form of epistasis called *duplicate recessive epistasis* or *complementary gene action*. In other words, here we get the same phenotype when one or other of two loci is homozygous for a recessive allele.

The following purely theoretical pathway—probably too simplistic for the actual phenotypes—can be envisioned for the production of the purple pigment:

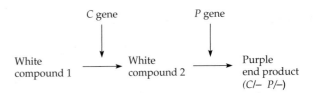

In this pathway, a colorless precursor compound (not shown) is converted to a purple end product in a series of steps. Each step is controlled by the product of a gene. To explain the F_2 ratio, we can propose that gene *C* controls the conversion of white compound 1 to white compound 2 and that gene *P* controls the conversion of white compound 2 to the purple end product. Therefore, homozygosity for the recessive allele of either or both of

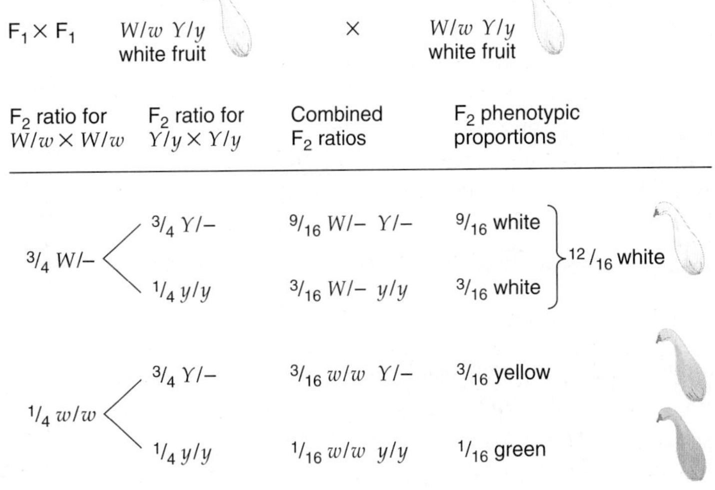

Figure 13.13

Dominant epistasis: generation of an F_2 12 white : 3 yellow : 1 green ratio for fruit color in summer squash.

	$F_1 \times F_1$	*W/w Y/y* white fruit	$\times$	*W/w Y/y* white fruit

F_2 ratio for *W/w × W/w*	F_2 ratio for *Y/y × Y/y*	Combined F_2 ratios	F_2 phenotypic proportions
3/4 *W/–*	3/4 *Y/–*	9/16 *W/– Y/–*	9/16 white
	1/4 *y/y*	3/16 *W/– y/y*	3/16 white	12/16 white
1/4 *w/w*	3/4 *Y/–*	3/16 *w/w Y/–*	3/16 yellow
	1/4 *y/y*	1/16 *w/w y/y*	1/16 green

Figure 13.14

Dominant epistasis: hypothetical pathway to explain the F$_2$ ratio of 12 white : 3 yellow : 1 green color in summer squash. (a) Production of white color. (b) Production of yellow color. (c) Production of green color.

a) White fruit-producing pathway

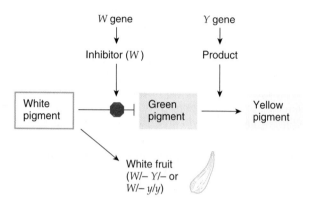

b) Yellow fruit-producing pathway

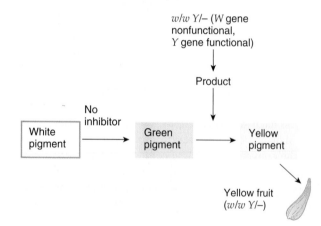

c) Green fruit-producing pathway

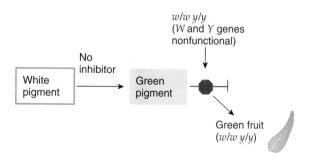

Figure 13.15

Dominant epistasis: Dominant greying allele in the horse causes the coat to turn grey as the horse matures. A horse (my Lipizzaner) is shown at age 4 (top) and age 7 (bottom).

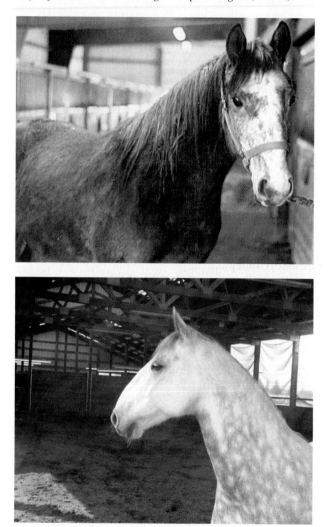

the C and P genes will result in a block in the pathway, and only white pigment will accumulate. That is, the $C/– \, p/p$, $c/c \, P/–$, and $c/c \, p/p$ genotypes will all be white. The only plants that will produce purple flowers are those in which both steps of the pathway are completed, so that the colored pigment is produced. This situation occurs only in $C/– \, P/–$ plants.

In the cross of two true-breeding white sweet peas that we introduced this example with, the two white parentals were $C/C \, p/p$ and $c/c \, P/P$, and the F$_1$ plants were purple and doubly heterozygous: $C/c \, P/p$ (Figure 13.16). Interbreeding the F$_1$ gives a 9 purple : 7 white in the F$_2$. Recessive epistasis occurs in both directions between two gene pairs. The consequence of this gene interaction is that the same phenotype (white) is exhibited whenever one or the other gene pair is homozygous recessive.

Fruit shape in the shepherd's purse plant provides an example of *duplicate dominant epistasis*. Two different fruit shapes are seen in the shepherd's purse: heart shaped and narrow. When a true-breeding plant that produces heart-shaped fruit is crossed with a narrow fruit plant that produces narrow fruit, the F$_1$ plants all produce heart-shaped fruit. When the heart-shaped F_1 plants are crossed, the F_2 show a ratio of 15 heart-shaped fruit plants : 1 narrow fruit plant. This is a modification of the 9:3:3:1 ratio, with the genotypes *A/– B/–*, *A/– b/b*, and *a/a B/–* all producing the phenotype—heart-shaped fruit—and the genotype *a/a b/b* producing the other phenotype of narrow fruit. In other words, there are duplicate genes involved in the fruit-shape phenotype, and one dominant allele of either gene is sufficient to give the heart-shaped fruit phenotype.

In sum, many types of phenotypic modifications are possible as a result of interactions between the products of different gene pairs. Geneticists detect such interactions when they observe deviations from the expected phenotypic ratios in crosses. We have discussed some examples in which two genes assort independently and in which complete dominance is exhibited in each allelic pair. The ratios we found would necessarily be modified further if the genes did not assort independently or if incomplete dominance or codominance prevailed. Table 13.4 shows examples of epistatic F_2 phenotypic ratios from an *A/a B/b* × *A/a B/b* cross with both allelic pairs showing complete dominance.

Epistasis plays a role in many human genetic diseases, further complicating their analysis. In these cases, complex interrelationships exist. For example, the majority of cases of bipolar disorder (also called manic–depressive illness), a complex human genetic disorder involving pathological mood disturbances, involves epistasis between multiple genes and may include other, more complex genetic mechanisms.

KEYNOTE

In many instances, alleles of different genes interact to determine phenotypic characteristics. Sometimes the interaction between genes results in new phenotypes without modification of typical Mendelian ratios. In epistasis, interaction between genes causes modifications of Mendelian ratios because one gene interferes with the phenotypic expression of another gene (or genes). The phenotype is controlled largely by the former gene, and not the latter, when both genes occur together in the genotype. The analysis of epistasis is complicated further when one or both gene pairs involve incomplete dominance or codominance or when gene pairs do not assort independently.

Table 13.4 Examples of Epistatic F_2 Phenotypic Ratios from an *A/a B/b* × *A/a B/b* in Which Complete Dominance is Shown for Each Gene Pair

		A/A B/B	A/A B/b	A/a B/B	A/a B/b	A/A b/b	A/a b/b	a/a B/B	a/a B/b	a/a b/b
More than four phenotypic classes	A and B both incompletely dominant	1	2	2	4	1	2	1	2	1
	A incompletely and B completely dominant	3		6		1	2	3		1
Four phenotypic classes	A and B both completely dominant (classic ratio)	9				3		3		1
Fewer than four phenotypic classes	a/a epistatic to B and b; recessive epistasis	9				3		4		
	A epistatic to B and b; dominant epistasis	12						3		1
	A epistatic to B and b; b/b epistatic to A and a; dominant and recessive epistasis	13[a]						3		
	a/a epistatic to B and b; b/b epistatic to A and a; duplicate recessive epistasis	9						7		
	A epistatic to B and b; B epistatic to A and a; duplicate dominant epistasis	15								1
	Duplicate interaction	9				6				1

[a] The 13 is composed of the 12 classes immediately above plus the one *a/a b/b* from the last column.

Source: Science of Genetics, 6th ed., by George W. Burns and Paul J. Bottino. Copyright © 1989. Reprinted by permission of Prentice Hall, Inc., Upper Saddle River, NJ.

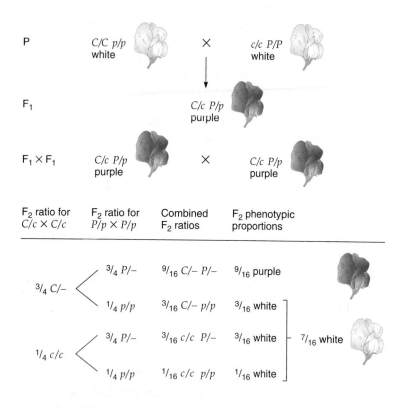

Figure 13.16
Duplicate recessive epistasis: generation of an F$_2$ 9 purple : 7 white ratio for flower color in sweet peas.

Essential Genes and Lethal Alleles

For a few years after the rediscovery of Mendel's principles, geneticists believed that mutations only changed the appearance of a living organism, but then they discovered that a mutant allele could cause death. In a sense, this mutation is still a change in phenotype, with the new phenotype being lethality. An allele that results in the death of an organism is called a **lethal allele,** and the gene involved is called an essential gene. **Essential genes** are genes that, when mutated, can result in a lethal phenotype. If the mutation is caused by a **dominant lethal allele,** both homozygotes and heterozygotes for that allele show the lethal phenotype. If the mutation is caused by a **recessive lethal allele,** only homozygotes for that allele have the lethal phenotype.

An example of a recessive lethal gene is the allele for yellow body color in mice, first studied by Lucien Cuenot in 1905. He reported that he had not been able to produce homozygous yellow mice. When yellows were bred with yellows, he obtained 283 yellow : 100 nonyellow, or approximately a 3:1 ratio. (The nonyellow color depends on other coat-color genes.) However, when he tested 81 of the yellows, he found that all were heterozygous. If normal dominance–recessiveness was involved, you would expect that 1/3 of the dominant phenotype class would be homozygous and 2/3 heterozygous, so the all-heterozygous result was unexpected. In 1910, William Ernest Castle and Clarence Cook Little repeated the yellow × yellow cross and got 800 yellows and 435 nonyellows, for a ratio of 2:1. They concluded that the yellow homozygotes are aborted in utero; in other words, the yellow allele has a *dominant* effect with regard to coat color, but acts as a *recessive* allele with respect to the lethality phenotype because only homozygotes die. We now know that homozygotes for the yellow allele die at the embryo stage.

The yellow allele is an allele of the agouti locus (a) and has been given the symbol A^Y. The yellow × yellow cross is shown in Figure 4.17. Genotypically, the cross is $A^Y/A \times A^Y/A$. We expect a genotypic ratio of 1/4 A^Y/A^Y : 2/4 A^Y/A : 1/4 A/A among the progeny. The 1/4 A^Y/A^Y mice die before birth, giving a birth ratio of 2/3 A^Y/A (yellow) : 1/3 A/A (nonyellow). Because the A^Y allele causes lethality in the homozygous state, it is called a recessive lethal. Characteristically, when two heterozygotes are crossed, recessive lethal alleles are recognized by a 2:1 ratio of progeny types.

The agouti locus of mouse has been molecularly cloned, permitting analysis of the lethal yellow allele. In wild-type agouti mice, the agouti gene is expressed in skin samples taken a few days after birth, at the time when the yellow band in the hair is being produced, in skin during regeneration of hair after plucking, and in no other tissues and at no other time. In heterozygous lethal yellow (A^Y/A) mice, the agouti-yellow allele is expressed at high levels in all tissues and at all developmental stages, indicating that tissue-specific regulation of expression has been lost. The explanation is that the A^Y allele has resulted from the deletion of a large DNA segment between the agouti locus and an upstream gene called *Raly,* such that the *Raly* promoter and the first part of that gene are now fused to the agouti gene. The *Raly* promoter thus controls the expression of the attached agouti gene. The expression in all tissues is

Figure 13.17

Inheritance of a lethal gene A^Y in mice. A mating of two yellow mice gives ¼ nonyellow (black) mice, ½ yellow mice, and ¼ dead embryos. The viable yellow mice are heterozygous A^Y/A, and the dead individuals are homozygous A^Y/A^Y.

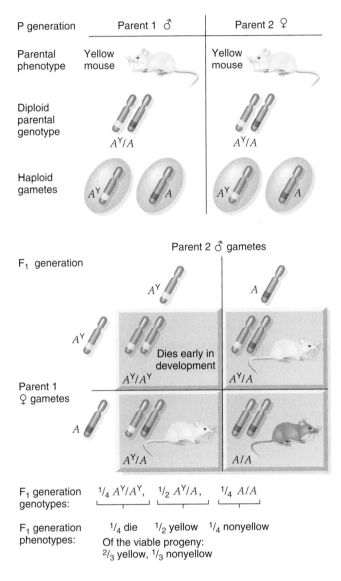

P generation — Parent 1 ♂ — Parent 2 ♀

Parental phenotype: Yellow mouse / Yellow mouse

Diploid parental genotype: A^Y/A — A^Y/A

Haploid gametes: A^Y A — A^Y A

F₁ generation

Parent 2 ♂ gametes: A^Y A

Parent 1 ♀ gametes: A^Y / A

Dies early in development A^Y/A^Y — A^Y/A

A^Y/A — A/A

F₁ generation genotypes: ¼ A^Y/A^Y, ½ A^Y/A, ¼ A/A

F₁ generation phenotypes: ¼ die ½ yellow ¼ nonyellow

Of the viable progeny: ⅔ yellow, ⅓ nonyellow

caused by regulatory signals in the *Raly* promoter. The embryonic lethality of yellow homozygotes probably results from the absence of *Raly* gene activity rather than a defective agouti gene.

Essential genes are almost certainly found in all diploid organisms. In 1907, soon after Cuenot reported his work with yellow mice, Erwin Baur showed that *aurea*, a form of snapdragon with yellowish leaves, gave a leaf color ratio of 2 yellowish : 1 green when selfed. He reasoned that the homozygous yellowish embryos died, and in 1908 he showed that they germinated, but produced seedlings that were almost white, so that they died due to the inability to photosynthesize. In 1912, Morgan reported the first sex-linked recessive lethal gene in

Drosophila. Unlike the other examples just described, this gene had no dominant effect in heterozygotes, but males carrying it died, resulting in a 2:1 sex ratio. There are many known recessive lethal alleles in humans. One example is Tay–Sachs disease (OMIM 272800; see pp. 73–74). The gene involved, *HEXA*, encodes the enzyme hexosaminidase A. Homozygotes appear normal at birth, but before about 1 year of age they show symptoms of central nervous system deterioration. Progressive mental retardation, blindness, and loss of neuromuscular control follow. Afflicted children usually die at 3 to 4 years of age. The genetic defect in Tay–Sachs results in an enzyme deficiency that prevents proper nerve function. Most disease-causing mutations of the *HEXA* gene are single base-pair substitutions, either causing amino-acid changes in the protein or altering the splicing of the gene's pre-mRNA.

There are X-linked lethal mutations as well as autosomal lethal mutations, and there are also dominant lethal mutations as well as recessive lethals. In humans, for example, the genetic disease hemophilia (OMIM 306700) is caused by an X-linked recessive allele. Untreated hemophilia is lethal. Dominant lethals exert their effect in heterozygotes, resulting in the death of the organism, usually at conception or at a fairly young age. Dominant lethals cannot be studied genetically unless death occurs after the organism has reached reproductive age. For example, the symptoms of the autosomal dominant trait Huntington disease (OMIM 143100)—involuntary movements and progressive central nervous system degeneration—may not begin until affected individuals reach their early thirties; as a result, parents may unknowingly pass on the gene to their offspring. Death usually occurs when the afflicted persons are in their forties or fifties. The well-known American folk singer Woody Guthrie died from Huntington disease.

KEYNOTE

A lethal allele is fatal to the individual. There are recessive lethal and dominant lethal alleles, and they can be X-linked or autosomal. The existence of lethal alleles of a gene indicates that the gene's normal product is essential to the functioning of the organism; therefore, the gene is an essential gene.

Gene Expression and the Environment

The *development* of a multicellular organism from a zygote is a process of *regulated growth and differentiation* that results from the interaction of the organism's genome with both the internal cellular environment and the external environment. Development is a tightly controlled, programmed series of phenotypic changes that, under normal environmental conditions, is essentially irreversible. Four major processes interact to constitute the complex process of development: replication of the genetic material, growth, differentiation of the various cell types, and the arrangement of differentiated cells into defined tissues and organs.

Think of development as a series of intertwined, complex biochemical pathways. The internal or external environment may influence any of these pathways by affecting the products of the genes controlling the pathways. This phenomenon is most readily studied in experimental organisms whose genotypes are unequivocally known. The extent to which the gene manifests its effects under varying environmental conditions can then be seen. We consider some examples in the next section.

Penetrance and Expressivity

In some cases, not all individuals with a particular genotype show the expected phenotype. The frequency with which a dominant or homozygous recessive gene manifests itself in individuals in a population is called the **penetrance** of the gene. Penetrance depends on both the genotype (for example, the presence of epistatic or other genes) and the environment. Figure 13.18a illustrates the concept of penetrance. Penetrance is complete (100 percent) when all the homozygous recessives show one phenotype, when all the homozygous dominants show another phenotype, and when all the heterozygotes are alike. For example, if all individuals carrying a dominant mutant allele show the mutant phenotype, the allele is completely penetrant. Many genes show complete penetrance; the seven gene pairs in Mendel's experiments and the alleles in the human ABO blood group system are examples.

If less than 100 percent of the individuals with a particular genotype exhibit the phenotype expected, penetrance is incomplete. For example, an organism may be genotypically $A/-$ or a/a, but may not display the phenotype typically associated with that genotype. If, for instance, 80 percent of the individuals carrying a particular gene show the corresponding phenotype, we say that there is 80 percent penetrance.

Figure 13.18

Illustrations of the concepts of penetrance and expressivity in the phenotypic expression of a genotype. (a) Incomplete penetrance compared with complete penetrance. (b) Variable expressivity compared with constant expressivity. (c) Incomplete penetrance and variable expressivity.

a)

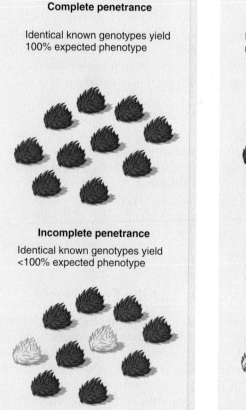

Complete penetrance

Identical known genotypes yield 100% expected phenotype

Incomplete penetrance

Identical known genotypes yield <100% expected phenotype

b)

Constant expressivity

Identical known genotypes with no expressivity effect yield 100% expected phenotype

Variable expressivity

Identical known genotypes with an expressivity effect yield a range of phenotypes

c)

Incomplete penetrance with variable expressivity

Identical known genotypes produce a broad range of phenotypes, due to varying degrees of gene activation and expression

In humans, many genes show reduced penetrance. For example, brachydactyly (OMIM 112500), an autosomal dominant trait that causes shortened and malformed fingers, shows 50 to 80 percent penetrance. It is also thought that a number of genes that confer a predisposition to cancer exhibit low to moderate penetrance, making it more difficult to identify and characterize those genes.

Genes may influence a phenotype to different degrees. **Expressivity** is the degree to which a penetrant gene or genotype is phenotypically expressed in an individual. Figure 13.18b illustrates the concept of expressivity. Like penetrance, expressivity depends on both the genotype and the environment, and it may be constant or variable. Molecularly, we can think of expressivity at a simple level as the result of different degrees of function of the protein encoded by the gene.

An example of variation in expressivity is found in the human condition called osteogenesis imperfecta (OMIM 166200). The three main features of this disease are blueness of the sclerae (the whites of the eyes), very fragile bones, and deafness. Osteogenesis imperfecta is inherited as an autosomal dominant with almost 100 percent penetrance. However, the trait shows variable expressivity: A person with the gene may have any one or any combination of the three traits. Moreover, the fragility of the bones for those who exhibit the condition is also highly variable.

Lastly, some genes exhibit both incomplete penetrance and variable expressivity. Figure 13.18 illustrates this concept. For example, individuals with neurofibromatosis (OMIM 162200), an autosomal dominant trait, develop tumorlike growths (neurofibromas) over the body. This genetic disease shows 50 to 80 percent penetrance and also shows variable expressivity (Figure 13.19). In its mildest form, the disease causes individuals to have only a few pigmented areas on the skin (called *café-au-lait spots* because they are the color of coffee with milk). In more severe cases, one or more other symptoms may be seen, including neurofibromas of various sizes; high blood pressure; speech impediments; headaches; a large head; short stature; tumors of the eye, brain, or spinal cord; and curvature of the spine. Therefore, in medical genetics it is important to recognize that a gene may vary widely in its expression, a qualification that makes the task of genetic counseling that much more difficult.

KEYNOTE

Penetrance is the frequency in the population with which a dominant or homozygous recessive allele manifests itself in the phenotype of an individual. Expressivity is the type or degree of phenotypic manifestation of a penetrant allele or genotype in a particular individual.

Effects of the Environment

The phenotypic expression of a gene depends on a number of factors, including the influences of the environment. Next, we consider some examples of environmental influences on gene expression.

Age of Onset. The age of the organism creates internal environmental changes that can affect gene function. All genes are not active all the time instead, over time, pro-

Figure 13.19

Variable expressivity in individuals with neurofibromatosis. Top: Café-au-lait spot. Middle: Café-au-lait spot and freckling. Bottom: Large number of cutaneous neurofibromas (tumorlike growths).

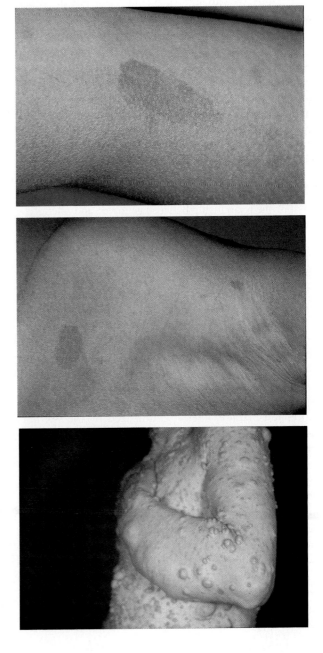

grammed activation and deactivation of genes occurs as the organism develops and functions. Numerous age-dependent genetic traits occur in humans; pattern baldness (OMIM 109200) appears in males between 20 and 30 years of age, and Duchenne muscular dystrophy (DMD, OMIM 310200) appears in children between 2 and 5 years of age. In most cases, the nature of the age dependency is not understood.

Sex. The expression of particular genes may be influenced by the sex of the individual. In the case of sex-linked genes, as mentioned earlier, differences in the phenotypes of the two sexes are related to different complements of genes on the sex chromosomes. However, in some cases, genes that are on autosomes affect a particular character that appears in one sex, but not the other. Traits of this kind are called **sex-limited traits.**

Examples of sex-limited traits in animals are milk production in dairy cattle (the genes involved obviously operate in females, but not in males), the appearance of horns in certain species of sheep (males with genes for horns have horns, and females with genes for horns do not have horns), and the ability to produce eggs or sperm. An example in humans is the distribution of facial hair.

A slightly different situation is found in **sex-influenced traits,** which, like sex-limited traits, often are controlled by autosomal genes. Such traits appear in both sexes, but either the frequency of occurrence in the two sexes is different or the relationship between genotype and phenotype is different.

Pattern baldness is an example of a sex-influenced trait in humans (Figure 13.20). Pattern baldness is controlled by an autosomal gene that acts as a dominant in males and as a recessive (or at least it is expressed at lower levels) in females. That is, the b/b genotype specifies pattern baldness in both males and females, and the b^+/b^+ genotype gives a nonbald phenotype in both sexes. The difference lies in the heterozygote: In males b^+/b leads to the bald phenotype, and in females it leads to the nonbald phenotype. In other words, the b allele acts as a dominant in males, but as a recessive in females. The expression of the b allele is influenced by the sex hormones of the individual; that is, the male hormone testosterone is responsible for the expression of the pattern baldness allele b when it is present in one dose. The sex-influenced pattern of inheritance and gene expression explains why pattern baldness is far more common among men than among women. From a large sample of the progeny of matings between two heterozygotes, $3/4$ of the daughters are nonbald and $1/4$ are bald, and $3/4$ of the sons are bald and $1/4$ are nonbald. Finally, baldness is not a straightforward trait to study. One reason is that there is variable expressivity in the baldness phenotype: As is apparent in the adult population, baldness may occur early in life or late, it may appear first on the crown (for

example, Prince Charles) or on the forehead, and the degree of baldness varies from minimal to total. In fact, a number of genes can affect the presence of hair on the head, including the pattern baldness gene. The final phenotype, then, is mostly the result of the interaction between the environment and the particular set of those genes present. For example, although b/b females show pattern baldness, the onset of baldness in these women occurs much later in life than is the case in men, because of the influence of hormones in the female internal environment.

Other human examples of sex-influenced traits are cleft lip and palate (incomplete fusion of the upper lip and palate; OMIM 119530), in which there is a 2:1 ratio of the trait in males:females; clubfoot (OMIM 119800; 2:1 ratio); gout (OMIM 138900; 8:1 ratio); rheumatoid arthritis (OMIM 180300; 1:3 ratio); osteoporosis (OMIM 166710; 1:3 ratio); and systemic lupus erythematosus (OMIM 152700, an autoimmune disease; 1:9 ratio).

Temperature. Biochemical reactions in the cell are catalyzed by enzymes. Normally, enzymes are unaffected by temperature changes within a reasonable range. However, some alleles of an enzyme-coding gene may give rise to an enzyme that is temperature sensitive; that is, it may function normally at one temperature, but be nonfunctional at another temperature. An example of a temperature effect on gene expression is fur color in Himalayan rabbits. Certain genotypes of this white rabbit cause dark fur to develop at the ears, nose, and paws, where the local surface temperature is lower (Figure 13.21a). (A similar situation applies to Siamese cats.) Since all body cells develop from a single zygote, this distinct fur pattern cannot be the result of a genotypic difference of the cells in those areas. Rather, it can be hypothesized that the fur pattern results from environmental influences. This hypothesis can be tested by rearing Himalayan rabbits under different temperatures. When a rabbit is reared at a temperature above 30°C, all its fur, including that of the ears, nose, and paws, is white (Figure 13.21b). If a rabbit is raised at 25°C, the typical Himalayan phenotype results (Figure 13.21c). Finally, if a rabbit is raised at 25°C while part of its body is artificially cooled to a temperature below 25°C, the rabbit develops the Himalayan coat phenotype and exhibits an additional patch of dark fur on the cooled area (Figure 13.21d).

Chemicals. Certain chemicals can have significant effects on an organism, as the next two examples show.

Phenylketonuria. The human disease phenylketonuria (PKU; OMIM 261600) is an autosomal recessive, with a defect in the biochemical pathway for the metabolism of the amino acid phenylalanine. In individuals who are homozygous for the recessive allele, various symptoms

Figure 13.20

Sex-influenced inheritance of pattern baldness in humans. The *b* allele is recessive in one sex and dominant in the other. (**a**) Cross of a nonbald female with a bald male who is homozygous *b/b*. (**b**) A cross between two F$_1$ heterozygotes produces a 3:1 ratio of bald : nonbald in males and a 1:3 ratio of bald : nonbald in females.

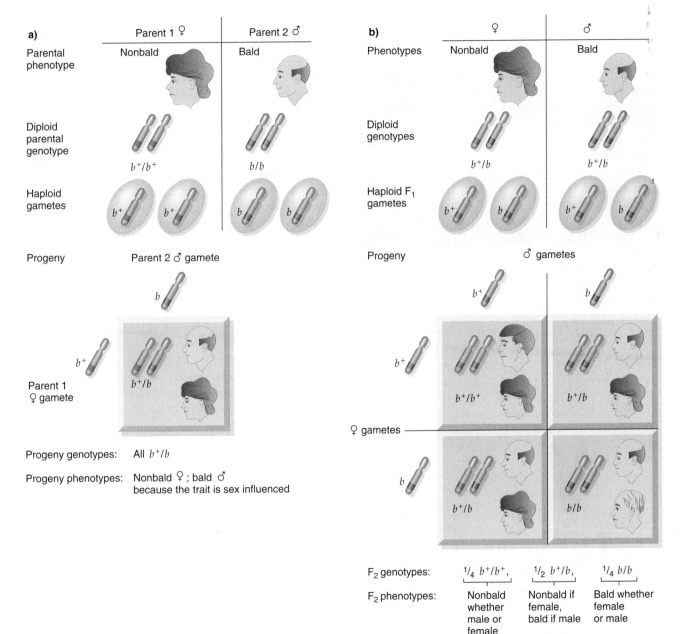

appear, most notably mental retardation at an early age. The diet determines how severe the symptoms of PKU will be. Problem foods include protein containing phenylalanine, such as the protein of mother's milk. PKU can be treated by restricting the amount of phenylalanine in the diet. (PKU is discussed further in Chapter 4, pp. 73–76.)

Phenocopies Induced by Chemicals. Changes in the chemical composition of the environment can also influ-

ence the expression of one or more genes. The most sensitive period is during early development, because small changes at that time can result in great changes later. When a developing embryo is exposed to certain drugs, chemicals, and viruses, these agents may produce a **phenocopy** (phenotypic copy)—a nonhereditary phenotypic modification, caused by special environmental conditions, that mimics a similar phenotype caused by a known gene mutation. In other words, although the individual expresses a

Figure 13.21

Effect of temperature on gene expression. (a) Himalayan rabbit. (b) White extremities result when Himalayan rabbits are reared at above 30°C. (c) Normal Himalayan pattern when rabbits are reared at 25°C. (d) Normal Himalayan pattern when rabbits are reared at 25°C, with a dark patch on the side where the flank was cooled to below 25°C.

a)

b)

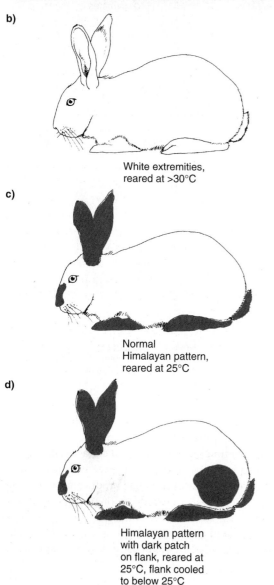

White extremities,
reared at >30°C

c)

Normal
Himalayan pattern,
reared at 25°C

d)

Himalayan pattern
with dark patch
on flank, reared at
25°C, flank cooled
to below 25°C

mutant phenotype, *the genotype is normal*. The agent that produces a phenocopy is called a *phenocopying agent*. There are many examples of phenocopies, and, in some instances, studies of phenocopies have provided useful information about the actual molecular defects caused by the phenocopies' mutant counterparts.

In humans, cataracts, deafness, and heart defects sometimes are produced when an individual is homozygous for rare recessive alleles; these disorders may also result if the mother is infected with rubella (German measles) virus during the first 12 weeks of pregnancy. Another human trait for which there is a phenocopy is *phocomelia*, a suppression of the development of the long bones of the limbs, which is caused by a rare dominant allele with variable expressivity. Between 1959 and 1961, similar phenotypes were produced by the sedative thalidomide when it was taken by expectant mothers between the 35th and 50th days of gestation. The drug was removed from the market when its devastating effects were discovered.

KEYNOTE

The phenotypic expression of a gene depends on several factors, including its dominance relationships, the genetic constitution of the rest of the genome (for example, the presence of epistatic or modifier genes), and the influences of the internal and external environments. In some cases, special environmental conditions can cause a phenocopy—a nonhereditary and phenotypic modification that mimics a similar phenotype caused by a gene mutation.

Nature versus Nurture

We are left with the nature–nurture question: What are the relative contributions of genes and the environment to the phenotype? (The nature–nurture question is discussed more fully in Chapter 14.) Up to this point, variation in most of the traits we have examined has been determined largely by differences in genotype; that is, phenotypic differences have reflected genetic differences. However, we have already seen that the phenotypes of many traits are influenced by both genes and the environment. Let us consider the nature–nurture issue in the context of some human examples.

Human height, or stature, is definitely influenced by genes. On the average, tall parents tend to have tall offspring, and short parents tend to have short offspring. There are also a number of genetic forms of dwarfism in humans. Achondroplasia is a type of dwarfism in which the bones of the arms and legs are shortened, but the trunk and head are of normal size; achondroplasia results from a single dominant allele. But the environment also plays a role in determining height. For example, human height has increased about 1 inch per generation over the past 100 years as a result of better diets and improved health care. Genes and the environment have interacted in determining human height.

For a trait such as height, genes set certain limits (or specify a potential) for the phenotype. Within these limits, the phenotype an individual develops depends on the environment. The range of potential phenotypes that a single genotype could develop if exposed to a range of environmental conditions is called the **norm of reaction.** For some genotypes, the norm of reaction is small; that is, the phenotype produced by a genotype is nearly the same in different environments. For other genotypes, the norm of reaction is large, and the phenotype produced by the genotype varies greatly in different environments.

Many human *behavioral* traits are the result of interaction between genes and the external environment. One example is alcoholism, which is a major health problem in the United States: About 14 million Americans have alcohol use disorders. Numerous studies have shown that alcoholism is influenced by genes. For instance, sons of alcoholic fathers who are separated from their biological parents at birth and adopted into a family with nonalcoholic parents are four times more likely to become alcoholic than sons adopted at birth whose biological fathers were not alcoholic. However, no gene *forces* a person to drink alcohol. That is, one cannot become an alcoholic unless one is exposed to an environment in which alcohol is available and drinking is encouraged. What genes do is make certain people more or less susceptible to alcohol abuse; they increase or decrease the risk of developing alcoholism. How genes influence our susceptibility to alcohol abuse is not yet clear. They may affect the way we metabolize alcohol, which in turn might affect how much we drink. Or genes may influence certain of our personality traits, making us more or less likely to drink heavily. The important point is that a behavioral trait such as alcoholism may be influenced by genes, but the genes alone do not produce the phenotype.

Nowhere has the role of genes and environment been more controversial than in the study of human intelligence. In the past, people tended to think of human intelligence as either genetically preprogrammed or produced entirely by the environment. The clash of these opposing views was called the nature–nurture controversy. Today, geneticists recognize that neither of these extreme views is correct; human intelligence is the product of both genes and the environment.

That genes influence human intelligence is clearly evidenced by genetic conditions that produce mental retardation, such as PKU (OMIM 261600; see p. 100) and Down syndrome (OMIM 190685; see Chapter 17, pp. 469–470). Numerous studies also indicate that genes influence differences in IQ among nonretarded people. (IQ, or intelligence quotient, is a standardized measure of mental age compared with chronological age; it is fairly stable over time.) For example, adoption studies show that the IQ of adopted children are closer to that of their biological parents than to the IQ of their adoptive parents.

However, IQ is also influenced by environment. Identical twins frequently differ in IQ, a fact that can be explained only by environmental differences. Family size, diet, and culture are environmental factors that are known to affect IQ. Thus, IQ results from the interaction of genes and the environment. Consequently, if two people (other than identical twins) differ in IQ, it is impossible to attribute that difference solely to either genes or the environment because both interact in determining the phenotype. So, although we cannot change our genes, we can alter the environment and thus affect a phenotypic trait such as intelligence.

KEYNOTE

Variation in most of the genetic traits considered in the discussion of Mendelian principles is determined predominantly by differences in genotype; that is, phenotypic differences result from genotypic differences. For many traits, however, the phenotypes are influenced by both genes and the environment. The debate over the relative contribution of genes and the environment to the phenotype has been called the nature–nurture controversy.

Summary

A variety of exceptions to, and extensions of, Mendel's principles were discussed in this chapter, including the following:

1. *Multiple alleles:* A gene may have many allelic forms in a population, and these alleles are called multiple alleles. A diploid individual can have only two different alleles of a given set of multiple alleles.

2. *Modified dominance relationships:* In complete dominance, the same phenotype results whether an allele is heterozygous or homozygous. In incomplete dominance, the phenotype of the heterozygote is intermediate between those of the two homozygotes. In codominance, the heterozygote shows the phenotypes of both homozygotes.

3. *Gene interactions and modified Mendelian ratios:* In many cases, genes do not function independently in determining phenotypic characteristics. In epistasis, modified Mendelian ratios occur because of interactions of nonallelic genes: The phenotypic expression of one gene depends on the genotype of another gene locus. In other interactions, a new phenotype is produced.

4. *Essential genes and lethal alleles:* Alleles of certain genes result in the failure to produce a necessary

functional gene product, and this deficiency gives rise to a lethal phenotype. Such lethal alleles may be recessive or dominant. The existence of lethal alleles of a gene indicates that the normal product of the gene is essential to the viability the organism.

5. *Penetrance and expressivity:* Penetrance is the condition in which not all individuals who are known to have a particular allele show the phenotype specified by that allele. That is, penetrance is the frequency with which a dominant or homozygous recessive allele manifests itself in individuals in the population. The related phenomenon of expressivity describes the degree to which a penetrant gene or genotype is phenotypically expressed in an individual. Both penetrance and expressivity depend on the genotype and the external environment.

6. *Dual influence of genes and the environment on phenotype:* An organism's potential to develop and function is specified by the zygote's genetic constitution. As an organism develops and differentiates, gene expression is influenced by a number of factors, including dominance relationships, the genetic constitution of the rest of the genome, and the influences of the internal and external environments. That is, the phenotypes of many traits are influenced by both genes and the environment. The debate over the relative contributions of each to the phenotype has been called the nature–nurture controversy.

Analytical Approaches to Solving Genetics Problems

Q13.1 In snapdragons, red flower color (C^R) is incompletely dominant to white flower color (C^W); the heterozygote has pink flowers. Also, normal broad leaves (L^B) are incompletely dominant to narrow, grasslike leaves (L^N); the heterozygote has an intermediate leaf breadth. If a red-flowered, narrow-leaved snapdragon is crossed with a white-flowered, broad-leaved one, what will be the phenotypes of the F_1 and F_2 generations, and what will be the frequencies of the different classes?

A13.1 This basic question on gene segregation involves the issue of incomplete dominance. In the case of incomplete dominance, remember that the genotype can be determined directly from the phenotype. Therefore, we do not need to ask whether a strain is true breeding, because all phenotypes have a different (and therefore known) genotype.

The best approach here is to assign genotypes to the parental snapdragons. Let $C^R/C^R L^N/L^N$ represent the red, narrow plant and $C^W/C^W L^B/L^B$ represent the white, broad plant. The F_1 plants from this cross will all be double het-

erozygotes, $C^R/C^W L^B/L^N$. Because of the incomplete dominance, these plants are pink flowered and have leaves of intermediate breadth. Interbreeding the F_1 plants give the F_2 generation, but it does not have the usual 9:3:3:1 ratio. Instead, there is a different phenotype for each genotype. These genotypes and phenotypes and their relative frequencies are shown in Figure 13.A.

Q13.2 In snapdragons, red flower color is incompletely dominant to white, with the heterozygote being pink; normal flowers are completely dominant to peloric-shaped ones; and tallness is completely dominant to dwarfness. The three gene pairs segregate independently. If a homozygous red, tall, normal-flowered plant is crossed with a homozygous white, dwarf, peloric-flowered one, what proportion of the F_2 plants will resemble the F_1 plants in appearance?

A13.2 Let us assign symbols: C^R = red and C^W = white; N = normal flowers and n = peloric; T = tall and t = dwarf. The initial cross, then, becomes $C^R/C^R T/T N/N \times C^W/C^W t/t n/n$. From this cross, we see that all the F_1 plants are triple heterozygotes with the genotype $C^R/C^W T/t N/n$ and with the phenotype pink, tall, normal flowered. Interbreeding the F_1 generation will produce 27 different genotypes among the F_2 plants; this answer follows from the rule that the number of genotypes is 3^n, where n is the number of heterozygous gene pairs involved in the $F_1 \times F_1$ cross. (See Chapter 11.)

Here, we are asked specifically for the proportion of F_2 progeny that resemble the F_1s in appearance. We can calculate this proportion directly without needing to display all the possible genotypes and then grouping the progeny in classes according to phenotype. First, we calculate the frequency of pink-flowered plants in the F_2; then we determine the proportion of these plants that have the other two attributes. From a $C^R/C^W \times C^R/C^W$ cross, we calculate that half of the progeny will be heterozygous C^R/C^W and therefore pink. Next, we determine the proportion of F_2 plants that are phenotypically like the F_1 with respect to height (tall). Either T/T or T/t plants will be tall, so ¾ of the F_2 will be tall. Similarly, ¾ of the F_2 plants will be normal flowered like the F_1s. To obtain the probability of all three of these phenotypes occurring together (pink, tall, normal), we must multiply the individual probabilities, because the gene pairs segregate independently. The answer is ½ × ¾ × ¾, or %32.

Q13.3
a. An $F_1 \times F_1$ self gives a 9:7 phenotypic ratio in the F_2. What phenotypic ratio would you expect if you test-crossed the F_1s?
b. Answer the same question for an $F_1 \times F_1$ cross that gives a 9:3:4 ratio.
c. Answer the same question for a 15:1 ratio.

Figure 13.A

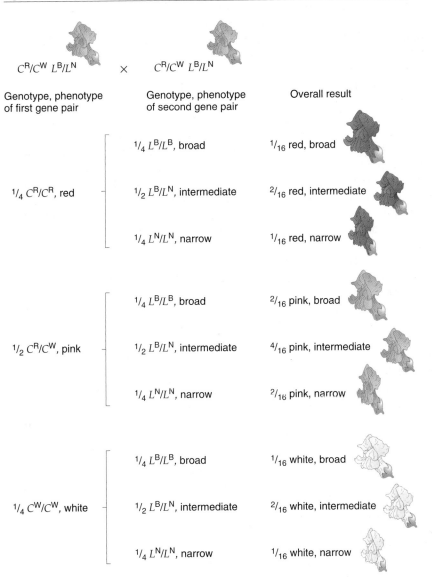

A13.3 This question deals with epistatic effects. In answering the question, we must consider the interaction between the different genotypes in order to proceed with the testcross. Let us set up the general genotypes that we will deal with throughout. The simplest are allelic pairs a^+ and a and b^+ and b, where the wild-type alleles are completely dominant to the other member of the pair.

a. A 9:7 ratio in the F_2 implies that both members of the F_1s are double heterozygotes and that epistasis is involved. Essentially, any genotype with a homozygous recessive condition has the same phenotype, so the 3, 3, and 1 parts of a 9:3:3:1 ratio are phenotypically combined into one class. In terms of genotype, $9/16$ are $a^+/-\ b^+/-$ types and the other $7/16$ are $a^+/-\ b/b$, $a/a\ b^+/-$, and $a/a\ b/b$. (As always, the

use of the dash after a wild-type allele signifies that the same phenotype results, whether the missing allele is a wild type or a mutant.) The testcross asked for is $a^+/a\ b^+/b \times a/a\ b/b$, and we can predict a 1:1:1:1 ratio of $a^+/a\ b^+/b : a^+/a\ b/b : a/a\ b^+/b : a/a\ b/b$. The first genotype will have the same phenotype as the $9/16$ class of the F_2 but because of epistasis, the other three genotypes will have the same phenotype as the $7/16$ class of the F_2. In sum, the answer is a phenotypic ratio of 1:3 in the progeny of a testcross of the F_1s.

b. We are asked to answer the same question for a 9:3:4 ratio in the F_2. Again, this question involves a modified dihybrid ratio where two classes of the 9:3:3:1 have the same phenotype. Complete dominance for each of the two gene pairs occurs here also, so the F_1 individuals are $a^+/a\ b^+/b$. Perhaps both the $a^+/-\ b^+/b$ and $a/a\ b/b$ classes in the F_2 will have the same phenotype, whereas the $a^+/-\ b^+/-$ and $a/a\ b^+/-$ classes will have phenotypes distinct from each other and from the interaction class. The genotypic ratio of a testcross of the F_1s is the same as in part (a) of this question. Considering them in the same order as we did there, we find that the second and fourth classes would have the same phenotype because of epistasis. So there are only three possible phenotypic classes, instead of the four found in the testcross of a dihybrid F_1, where there is complete dominance and no interaction. The phenotypic ratio here is 1:1:2.

c. This question is yet another example of epistasis. Since $15 + 1 = 16$, this number gives the outcome of an F_1 self of a dihybrid where there is complete dominance for each gene pair and interaction between the dominant alleles. In this case, the $a^+/-\ b^+/-$, $a^+/-\ b/b$, and $a/a\ b^+/-$ classes have one phenotype and include $^{15}/_{16}$ of the F_2 progeny, and the $a/a\ b/b$ class has the other phenotype and $^1/_{16}$ of the F_2. The genotypic results of a testcross of the F_1s are the same as in parts (a) and (b) of this question; that is, the F_2 progeny exhibit a 1:1:1:1 ratio of $a^+/a\ b^+/b : a^+/a\ b/b : a/a\ b^+/b : a/a\ b/b$. The first three classes have the same phenotype, which is the same as that of the $^{15}/_{16}$ of the F_2, and the last class has the other phenotype. The answer, then, is a 3:1 phenotypic ratio.

Questions and Problems

13.1 In rabbits, C = agouti coat color, c^{ch} = chinchilla, c^h = Himalayan, and c = albino. The four alleles constitute a multiple allelic series. The agouti C is dominant to the other three alleles, c is recessive to the other three alleles, and chinchilla is dominant to Himalayan. Determine the phenotypes of progeny from the following crosses:
a. $C/C \times c/c$
b. $C/c^{ch} \times C/c$
c. $C/c \times C/c$
d. $C/c^h \times c^h/c$
e. $C/c^h \times c/c$

*13.2 If a given population of diploid organisms contains only three alleles of a particular gene (say, w^1, w^2, and w^3), how many different diploid genotypes are possible in the populations? List all possible genotypes of diploids. (Consider only the three given alleles.)

13.3 The genetic basis of the ABO blood types is
a. multiple alleles
b. polyexpressive hemizygotes

c. allelically excluded alternatives
d. three independently assorting genes

13.4 In humans, the three alleles I^A, I^B, and i constitute a multiple allelic series that determines the ABO blood group system, as described in this chapter. For the following problems, state whether the child mentioned can be produced, and explain your answer:
a. An O child by a mating of two A individuals
b. An O child by an A and a B mating
c. An AB child by an A and an O mating
d. An O child by an AB and an A mating
e. An A child by an AB and a B mating

13.5 A man is blood type O, M. A woman is blood type A, M, and her child is type A, MN. The man cannot be the father of this child because
a. O men cannot have type-A children.
b. O men cannot have MN children.
c. An O man and an A woman cannot have an A child.
d. An M man and an M woman cannot have an MN child.

*13.6 A woman of blood group AB marries a man of blood group A whose father was of group O. What is the probability that
a. their two children will both be of group A?
b. one child will be of group B, the other of group O?
c. the first child will be a son of group AB and the second child a son of group B?

13.7 If a mother and her child belong to blood group O, what blood group could the father not belong to?

*13.8 A man of what blood group could not be the father of a child of blood type AB?

13.9 In snapdragons, red flower color (C^R) is incompletely dominant to white (C^W); the C^R/C^W heterozygotes are pink. A red-flowered snapdragon is crossed with a white-flowered one. Determine the flower color of
a. the F_1 snapdragons,
b. the F_2 snapdragons,
c. the progeny of a cross of the F_1 snapdragons to the red parent, and
d. the progeny of a cross of the F_1 snapdragons to the white parent.

*13.10 In shorthorn cattle, the heterozygous condition of the alleles for red coat color (C^R) and white coat color (C^W) is roan coat color. If two roan cattle are mated, what proportion of the progeny will resemble their parents in coat color?

13.11 What progeny will a roan shorthorn have if bred to
a. a red shorthorn,

b. a roan shorthorn, or

c. a white shorthorn?

*13.12 In peaches, fuzzy skin (*F*) is completely dominant to smooth (nectarine) skin (*f*), and the heterozygous condition of oval glands at the base of the leaves (G^O) and no glands (G^N) gives round glands. A homozygous fuzzy, no-gland peach variety is bred to a smooth, oval-gland variety.

a. What will be the appearance of the F_1 peaches?

b. What will be the appearance of the F_2 peaches?

c. What will be the appearance of the offspring of a cross of the F_1 peaches back to that smooth, oval-glanded parent?

13.13 In guinea pigs, short hair (*L*) is dominant to long hair (*l*), and the heterozygous conditions of yellow coat (C^Y) and white coat (C^W) give cream coat. A short-haired, cream guinea pig is bred to a long-haired, white guinea pig, and a long-haired, cream baby guinea pig is produced. When the baby grows up, it is bred back to the short-haired, cream parent. What phenotypic classes, and in what proportions, are expected among the offspring?

13.14 The shape of radishes may be long (S^L/S^L), oval (S^L/S^S), or round (S^S/S^S), and the color of radishes may be red (C^R/C^R), purple (C^R/C^W), or white (C^W/C^W). If a long, red radish plant is crossed with a round, white plant, what will be the appearance of the F_1 and the F_2 plants?

13.15 In poultry, the dominant alleles for rose comb (*R*) and pea comb (*P*), if present together, give a walnut comb. The recessive alleles of each gene, when present together in a homozygous state, give single comb. What will be the comb characters of the offspring of the following crosses?

a. *R/R P/p × r/r P/p*

b. *r/r P/P × R/r P/p*

c. *R/r p/p × r/r P/p*

13.16 For the following crosses involving the comb character in poultry, determine the genotypes of the two parents:

a. A walnut crossed with a single produces offspring that are ¼ walnut, ¼ rose, ¼ pea, and ¼ single.

b. A rose crossed with a walnut produces offspring that are ⅜ walnut, ⅜ rose, ⅛ pea, and ⅛ single.

c. A rose crossed with a pea produces five walnut and six rose offspring.

d. A walnut crossed with a walnut produces one rose, two walnut, and one single offspring.

13.17 In poultry, feathered (*F*) shanks (part of the legs) are dominant to clean (*f*), and white plumage (*I*) of white leghorns is dominant to black (*i*). Comb phenotypes and genotypes are given in Figure 13.8.

a. A feathered-shanked, white, rose-combed bird crossed with a clean-shanked, white, walnut-combed bird produces these offspring: 2 feathered, white, rose; 4 clean, white, walnut; 3 feathered, black, pea; 1 clean, black, single; 1 feathered, white, single; and 2 clean, white, rose. What are the genotypes of the parents?

b. A feathered-shanked, white, walnut-combed bird crossed with a clean-shanked, white, pea-combed bird produces a single offspring that is clean shanked, black, and single combed. In additional offspring from this cross, what proportion can be expected to resemble each parent?

*13.18 F_2 plants segregate ⁹⁄₁₆ colored : ⁷⁄₁₆ colorless. If just one colored plant from the F_2 generation is chosen at random and selfed, what is the probability that there will be *no* segregation of the two phenotypes among its progeny?

*13.19 In peanuts, a plant may be either "bunch" or "runner." Two different strains of peanut, V4 and G2, in which "bunch" occurred were crossed, with the following results:

V4 bunch × V4 bunch
↓
all bunch

G2 bunch × G2 bunch
↓
all bunch

The two true-breeding strains of bunch were crossed in the following way:

V4 bunch × G2 bunch
↓
F_1 runner

F_1 × F_1
↓
F_2 9 runner : 7 bunch

What is the genetic basis of the inheritance pattern of runner and bunch in the F_2 peanuts?

*13.20 In rabbits, one enzyme (the product of a functional gene *A*) is needed to produce a substance required for hearing. Another enzyme (the product of a functional gene *B*) is needed to produce another substance required for hearing. The genes responsible for the two enzymes are not linked. Individuals homozygous for either one or both of the nonfunctional recessive alleles, *a* or *b*, are deaf.

a. If a large number of matings were made between two double heterozygotes, what phenotypic ratio would be expected in the progeny?

b. The phenotypic ratio found in part (a) is a result of what well-known phenomenon?

c. What phenotypic ratio would be expected if rabbits homozygous recessive for trait A and heterozygous for trait B were mated to rabbits heterozygous for both traits?

13.21 In doodlewags (hypothetical creatures), the dominant allele S causes solid coat color; the recessive allele s results in white spots on a colored background. The black coat color allele B is dominant to the brown allele b, but these genes are expressed only in the genotype a/a. Individuals that are $A/-$ are yellow regardless of B alleles. Six pups are produced in a mating between a solid yellow male and a solid brown female. Their phenotypes are 2 solid black, 1 spotted yellow, 1 spotted black, and 2 solid brown.

a. What are the genotypes of the male and female parents?

b. What is the probability that the next pup will be spotted brown?

13.22 The allele l in *Drosophila* is recessive, sex linked, and lethal when homozygous or hemizygous (the condition in the male). If a female of genotype L/l is crossed with a normal male, what is the probability that the first two surviving progeny will be males?

*13.23 A locus in mice is involved in pigment production; when parents heterozygous at this locus are mated, $3/4$ of the progeny are colored and $1/4$ are albino. Another phenotype concerns coat color; when two yellow mice are mated, $2/3$ of the progeny are yellow and $1/3$ are agouti. The albino mice cannot express whatever alleles they have at the independently assorting agouti locus.

a. When yellow mice are crossed with albino, they produce F_1 mice consisting of $1/2$ albino, $1/3$ yellow, and $1/6$ agouti. What are the probable genotypes of the parents?

b. If yellow F_1 mice are crossed among themselves, what phenotypic ratio would you expect among the progeny? What proportion of the yellow progeny produced here would be expected to be true breeding?

13.24 In *Drosophila melanogaster*, a recessive autosomal allele, ebony (e), produces a black body color when homozygous, and an independently assorting autosomal allele, black (b), also produces a black body color when homozygous. Flies with genotypes $e/e\ b^+/-$, $e^+/-\ b/b$, and $e/e\ b/b$ are phenotypically identical with respect to body color. Flies with genotype $e^+/-\ b^+/-$ have a grey body color. True-breeding $e/e\ b^+/b^+$ ebony flies are crossed with true-breeding $e^+/e^+\ b/b$ black flies.

a. What will be the phenotype of the F_1 flies?

b. What phenotypes and what proportions would occur in the F_2 generation?

c. What phenotypic ratios would you expect to find in the progeny of these backcrosses?

i. $F_1 \times$ true-breeding ebony
ii. $F_1 \times$ true-breeding black

*13.25 In *Drosophila*, mutants A, B, C, D, E, F, and G all have the same phenotype: the absence of red pigment in the eyes. In pairwise combinations in complementation tests, the following results were produced, where $+$ = complementation and $-$ = no complementation:

	A	B	C	D	E	F	G
G	+	−	+	+	+	+	−
F	−	+	+	−	+	−	
E	+	+	−	+	−		
D	−	+	+	−			
C	+	+	−				
B	+	−					
A	−						

a. How many genes are present?

b. Which mutants have defects in the same gene?

*13.26 In four-o'clock plants, two genes, Y and R, affect flower color. Neither is completely dominant, and the two interact with each other to produce seven different flower colors:

$Y/Y\ R/R$ = crimson $Y/y\ R/R$ = magenta
$Y/Y\ R/r$ = orange-red $Y/y\ R/r$ = magenta-rose
$Y/Y\ r/r$ = yellow $Y/y\ r/r$ = pale yellow
$y/y\ R/R$, $y/y\ R/r$, and $y/y\ r/r$ = white

a. In a cross of a crimson-flowered plant with a white one ($y/y\ r/r$), what will be the appearances of the F_1 plants, the F_2 plants, and the offspring of the F_1 plants backcrossed to their crimson parent?

b. What will be the flower colors in the offspring of a cross of orange-red × pale yellow?

c. What will be the flower colors in the offspring of a cross of a yellow with a $y/y\ R/r$ white?

13.27 Two four-o'clock plants were crossed and gave the following offspring: $1/8$ crimson, $1/8$ orange-red, $1/4$ magenta, $1/4$ magenta-rose, and $1/4$ white. Unfortunately, the person who made the crosses was color blind and could not record the flower colors of the parents. From the results of the cross, deduce the genotypes and flower colors of the two parents.

13.28 In *Drosophila*, the recessive allele bw causes a brown eye and the (unlinked) recessive allele st causes a scarlet eye. Flies homozygous for both recessives have white eyes. The genotypes and corresponding phenotypes, then, are as follows:

$bw^+/-\ st^+/-$	red eye
$bw/bw\ st^+/-$	brown
$bw^+/-\ st/st$	scarlet
$bw/bw\ st/st$	white

Outline a hypothetical biochemical pathway that would produce this type of gene interaction. Demonstrate why each genotype shows its specific phenotype.

*13.29 Genes *A*, *B*, and *C* are independently assorting and control the production of a black pigment.

a. Suppose that *A*, *B*, and *C* act in the following pathway:

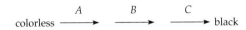

The alternative alleles that give abnormal functioning of these genes are designated *a*, *b*, and *c*, respectively. A black *A/A B/B C/C* is crossed with a colorless *a/a b/b c/c* to give a black F_1. The F_1 is selfed. What proportion of the F_2 individuals is colorless? (Assume that the products of each step except the last are colorless, so only colorless and black phenotypes are observed.)

b. Suppose instead that a different pathway is utilized. In it, the *C* allele produces an inhibitor that prevents the formation of black by destroying the ability of *B* to carry out its function. The mechanism is as follows:

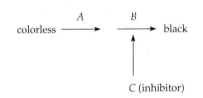

A colorless *A/A B/B C/C* individual is crossed with a colorless *a/a b/b c/c*, giving a colorless F_1. The F_1 is selfed to give an F_2. What is the ratio of colorless to black in the F_2 individuals? (Only colorless and black phenotypes are observed, as in part (a).)

c. How would you evaluate which of the biochemical pathways hypothesized in parts (a) and (b) is more likely?

*13.30 Alleles of a wild-type gene can be thought of as giving a normal phenotype because they confer a particular, normal amount of gene function. Mutant alleles can alter the level of function in a variety of ways. Loss-of-function alleles can be thought of as eliminating or decreasing gene function. Among the loss-of-function alleles are amorphic alleles (which eliminate gene function) and hypomorphic alleles (which decrease gene function). A hypermorphic allele is an overproducer that makes more of the wild-type product. Gain-of-function alleles have novel functions. Among the gain-of-function alleles are antimorphic alleles (which are antagonistic to the function of a wild-type allele) and neomorphic alleles (which provide a new, substantially altered function to the gene).

Consider hypothetical situation: In the production of purple pigment in sweet peas (see p. 353), the wild-type (+) product of the p^+ gene is an enzyme P that converts white pigment to purple pigment. This enzyme activity can be measured in tissue extracts and mixtures of tissue extracts. The table below describes the results obtained when enzyme activity is measured in extracts and mixtures of extracts from strains with different p^+ genotypes (all are c^+/c^+).

Complete the columns of the table, filling in (a) the phenotype expected in a homozygote; (b) the phenotype expected in a heterozygote (allele /+); (c) the phenotype expected in a hemizygote (an individual that is heterozygous for the allele and a deletion for the locus); (d) the classification of the allele, using the definitions set forth at the beginning of the problem.

13.31 In cats, two alleles (*B*, *O*) at an X-linked gene control whether black or orange pigment is deposited. A dominant allele at an autosomal gene *I/i* partially inhibits the deposition of pigment, lightening the coat color from black to grey or from orange to pale orange. A dominant allele at the autosomal gene *T/t* determines whether a tabby, or vertically striped, pattern is present. The tabby pattern depends on a dominant agouti (*A*) allele for its expression,

Genotype	Percent of +/+ Activity	Percent of +/+ Activity When Mixed 50:50 with +/+ Extract	(A) Homozygote Phenotype	(B) Heterozygote Phenotype	(C) Hemizygote Phenotype	(D) Allele Classification
p^+/p^+	100	100	_____	_____	_____	_____
p^1/p^1	20	60	_____	_____	_____	_____
p^2/p^2	0	50	_____	_____	_____	_____
p^3/p^3	300	200	_____	_____	_____	_____
p^4/p^4	0	5	_____	_____	_____	_____
p^5/p^5	0^a	50^b	_____	_____	_____	_____

[a] Produces red, not purple, pigment.
[b] Produces red and purple pigments.

with nonagouti (*a*) epistatic to tabby. The agouti allele also causes a speckled, rather than solid, coat color. Judy, a stray cat with a speckled, tortoiseshell pattern with grey and pale orange spots and no trace of a tabby pattern, gives birth to four kittens. Of the three female offspring, two are solid grey and the third is speckled grey and light orange like her mother, but also shows traces of a tabby pattern. The single male offspring is solid grey.

a. Explain how the tortoiseshell pattern arises in cats. That is, how can a cat have distinct patches of fur with different deposits of pigment?

b. Cats with a tortoiseshell pattern usually are female. Explain why this is the case and also why, when an unusual male tortoiseshell male cat is found, he is atypically large and typically not very swift.

c. What genotype(s) might Judy and her kittens have?

d. Assuming that there is just one father, what phenotype(s) should be considered in assessing the neighborhood males for paternity?

13.32 In *Drosophila*, a mutant strain has plum-colored eyes. A cross between a plum-eyed male and a plum-eyed female gives $2/3$ plum-eyed and $1/3$ red-eyed (wild-type) progeny flies. A second mutant strain of *Drosophila*, called stubble, has short bristles instead of the normal long bristles. A cross between a stubble female and a stubble male gives $2/3$ stubble and $1/3$ normal-bristled flies in the offspring. Assuming that the plum gene assorts independently from the stubble gene, what will be the phenotypes and their relative proportions in the progeny of a cross between two plum-eyed, stubble-bristled flies? (Both genes are autosomal.)

13.33 In *Drosophila*, a recessive, temperature-sensitive mutation in the *transformer-2* (*tra-2*) gene on chromosome 2 causes XX individuals raised at 29°C to be transformed into phenotypic males. At 16°C, these individuals develop as normal females. The sex type of XY individuals is unaffected by the *tra-2* mutation. Suppose you are given three true-breeding, unlabeled vials containing different strains of *Drosophila*, all raised at 16°C. Two of the strains have white eyes, and one has red eyes. You are told that one of the white-eyed strains also carries the *tra-2* mutation. Devise two different methods to determine which white-eyed strain has the *tra-2* mutation. Is there a reason to prefer one method over the other?

13.34 Normal *Drosophila* have straight wings and smooth, well-ordered compound eyes. A strain with curly wings and rough eyes has the following properties: Interbreeding its males and females always gives progeny identical to the parents. An outcross of a male from this strain to a normal female gives 45 curly and 49 rough progeny. An outcross of a female from the same strain to a normal male gives 53 curly and 47 rough progeny.

Crossing a curly F_1 male and female from the first outcross gives 81 curly and 53 straight progeny. The same curly F_1 male mated to a normal female gives 57 curly and 61 normal progeny. Crossing a rough F_1 male and female from the first outcross gives 78 rough and 42 smooth progeny. When the same rough F_1 male is mated to a normal female, 46 rough and 48 normal progeny are recovered. Develop hypotheses to explain these data, and test them using chi-square tests.

*13.35 In sheep, white fleece (*W*) is dominant over black (*w*), and horned (*H*) is dominant over hornless (*h*) in males, but recessive in females. If a homozygous horned white ram is bred to a homozygous hornless black ewe, what will be the appearances of the F_1 and the F_2 sheep?

13.36 A horned black ram bred to a hornless white ewe produces the following offspring: Of the males, $1/4$ are horned, white; $1/4$ are horned, black; $1/4$ are hornless, white; and $1/4$ are hornless, black. Of the females, $1/2$ are hornless and black, and $1/2$ are hornless and white. What are the genotypes of the parents?

*13.37 A horned white ram is bred to the following four ewes and has one offspring by the first three and two by the fourth: Ewe A is hornless and black; the offspring is a horned white female. Ewe B is hornless and white; the offspring is a hornless black female. Ewe C is horned and black; the offspring is a horned white female. Ewe D is hornless and white; the offspring are one hornless black male and one horned white female. What are the genotypes of the five parents?

*13.38 Pattern baldness is more frequent in males than in females. This appreciable difference in frequency is assumed to result from

a. Y linkage of this trait

b. X-linked recessive mode of inheritance

c. sex-influenced autosomal inheritance

d. excessive beer drinking in males, with the consumption of gin being approximately equal between the sexes

13.39 King George III, who ruled England during the period of the Revolutionary War in the United States, is an ancestor of Queen Elizabeth II. (See Figure 12.28 on p. 328.) Near the end of his life, he exhibited sporadic periods of "madness." In retrospect, it appears that he showed symptoms of porphyria, an autosomal dominant disorder of heme metabolism. In addition to "madness," the symptoms of porphyria, which include a variety of physical ailments that King George III exhibited, are sporadic, are variable in severity, can be affected by diet, and, currently, can be treated with medication.

a. How would you describe this disease in terms of penetrance and expressivity?

b. If, indeed, King George III had porphyria, what is the chance that the current Prince of Wales (Charles) carries a disease allele? State all of your assumptions.

13.40 Jasper Rine and his colleagues at the University of California at Berkeley have launched the Dog Genome Initiative to study canine genes and behavior. They mated Pepper, a vocal, highly affectionate, very social Newfoundland female that is not good at fetching tennis balls, but loves water, to Gregor, a quiet, less affectionate, less social border collie that is exceptionally good at fetching tennis balls, but avoids water. They obtained 7 F_1 and 23 F_2 progeny. When the aforesaid behavioral traits were analyzed, it was found that all 7 F_1 dogs were similar, each showing a mixture of the parents' behavioral traits. When the behaviors of the F_2 dogs were analyzed, differences were more evident. In particular, two of the F_2 dogs (Lucy and Saki) shared Pepper's love of water. (For more information, see "California Geneticists Are Going to the Dogs," by Donald McCaig, in *Smithsonian*, Vol. 27, 1996, pp. 126–141.)

a. Develop hypotheses to explain the various observations and, when appropriate, test them using a chi-square test.

b. What practical value might there be in studying the genes of canines?

13.41 Parkinson disease is a progressive neurological disease that causes slowness of movement, stiffness, and shaking, and eventually leads to disability. Actor Michael J. Fox has this disease. Parkinson disease affects about 2 percent of the U.S. adult population over 50 years of age and appears most often in individuals who are between their fifth and seventh decades. There has been much discussion among scientists as to whether the disease is caused by environmental factors, genetic factors, or both. Support for the environmental hypothesis stems from the observation that the disease seems not to have been reported until after the Industrial Revolution and from the discovery that some chemicals can cause symptoms. Support for the genetic hypothesis stems from pedigree analysis.

Consider the pedigree in Figure 13.B (modified to protect patient confidentiality), which shows the incidence of parkinsonism in a family of European descent. The shaded portions of the pedigree indicate family members who reside in the United States. The remaining portions of the pedigree reside in Europe. Members of the U.S. branches of the family have not visited Europe for any extensive period since the initial emigration from Europe.

a. If the disease in this family has a genetic basis, what is its basis? Explain your answer.

b. Why might this pedigree be particularly helpful in distinguishing between an environmental and a genetic cause of Parkinson disease?

c. What reservations, if any, do you have about concluding that the disease has a genetic basis in some individuals?

Figure 13.B

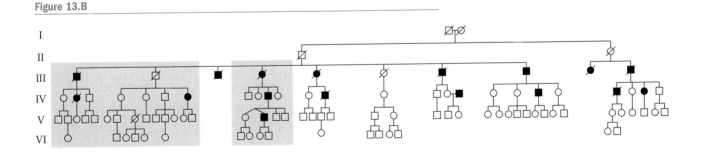

14

Quantitative Genetics

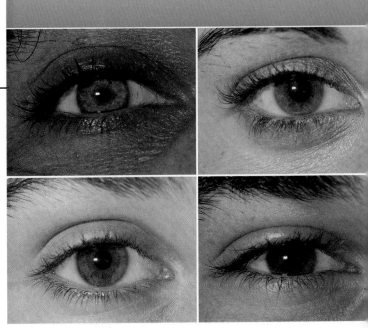

Various human eye colors.

PRINCIPAL POINTS

- Discontinuous traits exhibit only a few distinct phenotypes. Continuous, or quantitative, traits display a range of phenotypes.

- Continuous traits have a range of phenotypes because many loci contribute to the phenotype (polygenic inheritance) and because environmental factors influence the phenotype produced by a genotype.

- Continuous traits can be studied by using samples of populations and statistics such as the mean, variance, and correlations between characters, combined with statistical techniques such as analysis of variance and regression analysis.

- Variation among individuals can be partitioned into genetic and environmental components. However, genotypes may respond differently in distinct environments, so caution must be exercised when designing and interpreting experiments that measure genetic and environmental contributions to phenotypic variation.

- The broad-sense heritability of a trait is the proportion of the phenotypic variance that results from genetic differences among individuals. The narrow-sense heritability is the proportion of the phenotypic variance due only to additive genetic variance. Both measures depend on a particular population in a particular environment.

- The amount that a trait changes in one generation as a result of selection on the trait is called the response to selection. The magnitude of the response to selection depends on the selection differential and the narrow-sense heritability.

- Genetic correlations arise when two traits are influenced by the same genes or linked genes. When a trait is selected, genetically correlated traits also exhibit a response to selection.

- Quantitative trait loci (QTLs) that determine continuous traits can be identified through marker-based mapping. QTL mapping provides an estimate of the number and relative importance of genes influencing quantitative genetic variation.

i JUST LIKE SNOWFLAKES, NO TWO FINGERPRINTS are alike, even those of identical twins. Yet research shows that the patterns of ridges on our fingers and palms are inherited. What factors would produce such a variety of phenotypes? Is the trait encoded by more than one locus? Are there environmental factors involved? Is there a relationship between a person's fingerprints and another trait, such as hair color or blood type? In this chapter, you will learn the answers to questions such as these. Then, in the iActivity, you can apply what you've learned as you investigate whether a relationship exists between fingerprint patterns and high blood pressure.

The Nature of Continuous Traits

By isolating phenotypic mutants, then crossing and comparing the mutants with the wild type, Mendel was able to originally describe the basic laws of heredity, and later geneticists extended our understanding down to the molecular basis of mutant phenotypes. The mutations used in these studies, and in fact most of the traits we have studied up to this point, have been characterized by the presence of only a few distinct phenotypes. The seed coats of pea plants, for example, were either grey or white, the seedpods were green or yellow, and the plants were tall or short. In each trait, the phenotypes were markedly different, and each phenotype was easily separated from all other phenotypes. Traits such as these, with only a few distinct phenotypes (Figure 14.1), are called **discontinuous traits.**

For discontinuous traits, a simple relationship usually exists between the genotype and the phenotype. In most cases, the effects of variant alleles at the single locus are observable at the level of the organism, so the phenotype can be used as a quick assay for the genotype. When dominance occurs, the same phenotype may be produced by two different genotypes, but the relationship between the genes and the trait remains simple. Chapter 4 introduced situations where the relationship between genotype and phenotype is not so simple: variable **penetrance** and **expressivity,** as well as **pleiotropy** and **epistasis.** In addition, single genotypes can give rise to a range of phenotypes as the genotype interacts with variable environments during development to give rise to a **norm of reaction.** As a result of these and other factors, there are not many traits with phenotypes that fall into a few distinct categories. Many traits (probably most), such as human birth weight (illustrated in Figure 14.2) and adult height, protein content in corn, and number of eggs laid by *Drosophila*, exhibit a wide range of possible phenotypes. Traits such as these, with a continuous distribution of phenotypes, are called **continuous traits.** Since the

Figure 14.1

Discontinuous distribution of shell color in the snail *Cepaea nemoralus* from a population in England.

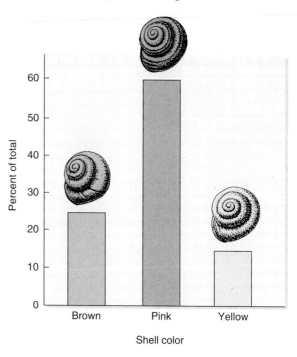

phenotypes of continuous traits must be described by quantitative measures, such traits are also known as **quantitative traits,** and the field of **quantitative genetics** studies the inheritance of these traits.

Figure 14.2

Distribution of birth weight of babies (males + females) born to teenagers in Portland, Oregon, in 1992.

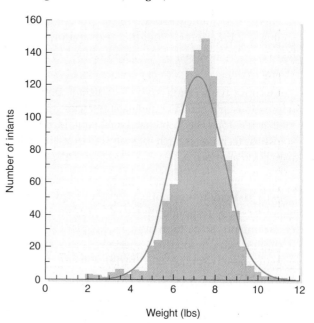

Questions Studied in Quantitative Genetics

There is a great deal of genetic variation among individuals. The amount of variation and how it is distributed determines the population's genetic structure. In this chapter, we shift attention from a purely genetic perspective to consider the phenotypic structure of a population, as well as the relationship between the genetic structure and the phenotypic structure. As we will see, quantitative genetics plays an important role in our understanding of evolution, conservation, and complex human traits. Quantitative genetics is especially important in agricultural genetics, where traits such as crop yield, rate of weight gain, milk production, and fat content are all studied by quantitative genetics. In psychology, methods of quantitative genetics are used to study IQ, learning ability, and personality. Human geneticists also use these methods to study traits such as blood pressure, antibody titer, fingerprint pattern, and birth weight.

In transmission genetics, we frequently determine the probability of inheriting a particular phenotype. With quantitative traits, however, individuals differ in the quantity of a trait, so it makes no sense to ask about the probability of inheriting a continuous trait, as we did for simple discontinuous traits. Instead, the following are examples of questions frequently studied by quantitative geneticists:

1. To what degree does the observed variation in phenotype result from differences in genotype and to what degree does this variation reflect the influence of different environments? In our study of discontinuous traits, this question assumed little importance because the differences in phenotype were assumed to reflect only genotypic differences.

2. How many genes determine the phenotype? When only a few loci are involved and the trait is discontinuous, the number of loci can often be determined by examining the phenotypic ratios in genetic crosses. With complex, continuous traits, however, determining the number of loci involved is more difficult.

3. Are the contributions of the determining genes equal? Or do a few genes have major effects on the trait and other genes modify the phenotype only slightly?

4. Are the effects of alleles additive? To what degree do alleles at the different loci interact with one another?

5. When selection occurs for a particular phenotype, how rapidly can the trait change? Do other traits change at the same time?

6. What is the best method for selecting and mating individuals to produce desired phenotypes in the progeny?

The Inheritance of Continuous Traits

Biologists began developing techniques for the study of continuous traits during the late nineteenth century, even before they were aware of Mendel's principles of heredity. Francis Galton and his associate Karl Pearson demonstrated that for many traits in humans, such as height, weight, and mental traits, the phenotypes of parents and their offspring are statistically associated. From this result, they were able to infer that these traits are inherited, but they were not successful in determining how genetic transmission occurs. Even after the rediscovery of Mendel's work, considerable controversy arose over whether continuous traits also follow Mendel's principles or whether they are inherited in some different fashion.

Polygene Hypothesis for Quantitative Inheritance

A trait may have a range of phenotypes because environmental factors affect the trait. When environmental factors exert an influence, the same genotype may produce a range of phenotypes (the norm of reaction) or multiple genotypes may produce the same phenotype. Which phenotype is expressed depends both on the genotype and on the specific environment in which the genotype is found. In 1903, Wilhelm Johansen published a study demonstrating that quantitative variation in seed weight in bean had both environmental and genetic determinants. His study was a crucial step in recognizing that both environment and genotype influence some quantitative traits; such traits are referred to as **multifactorial traits.** Because inheritance of quantitative traits cannot be explained by a single locus, the simplest alternative explanation is that they are controlled by many genes. This explanation, called the **polygene** or **multiple-gene hypothesis for quantitative inheritance,** is one of the landmarks of genetic thought.

Polygene Hypothesis for Wheat Kernel Color

The polygene hypothesis can be traced back to 1909 and the classic work of Hermann Nilsson-Ehle, who studied the color of wheat kernels. Like Mendel, Nilsson-Ehle started by crossing true-breeding lines of red kernel plants and white kernel lines. The F_1 had grains that were all the same shade of an intermediate color between red and white. When he intercrossed the F_1s, the F_2 prog-

animation

a Polygene Hypothesis for Wheat Kernel Color

eny displayed kernels that were white and many shades of red, in a ratio of approximately 15 red (all shades): 1 white kernels. While this was clearly a deviation from a 3:1 ratio expected for a monohybrid cross, he could recognize four discrete shades of red among the progeny. When he counted the relative number of each class, he found a 1:4:6:4:1 phenotypic ratio of plants with dark red, medium red, intermediate red, light red, and white kernels.

How can the data be interpreted in genetic terms? Recall from Chapter 13 (Analytical Approaches to Solving Genetics Problems, Question 13.3c, p. 366) that a 15:1 ratio of two alternative characteristics resulted from the interaction of the products of two genes that affect the same trait. Let us hypothesize that there are two independently segregating loci that control the production of pigment: the *red* locus with alleles *R* and *r*, and the *crimson* locus with alleles *C* and *c*. Nilsson-Ehle's parental cross and the F_1 genotypes can then be shown as follows:

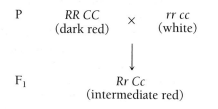

P   *RR CC*   ×   *rr cc*
        (dark red)   (white)

F_1 *Rr Cc*
         (intermediate red)

When the F_1 is interbred, the distribution of genotypes in the F_2 is typical of dihybrid inheritance, that is, $^1/_{16}$ *RR CC* + $^2/_{16}$ *Rr CC* + $^1/_{16}$ *rr CC* + $^2/_{16}$ *RR Cc* + $^4/_{16}$ *Rr Cc* + $^2/_{16}$ *rr Cc* + $^1/_{16}$ *RR cc* + $^2/_{16}$ *Rr cc* + $^1/_{16}$ *rr cc*. If *R* and *C* have a simple dominant relationship to *r* and *c*, the 9:3:3:1 phenotypic ratio characteristic of dihybrid inheritance should result. From the kernel phenotypic ratio, then, dominance is not the simple answer because the observed phenotypes fell into five classes with a ratio that approximates 1:4:6:4:1. An alternative explanation is that alleles can be classified as either functional (**contributing alleles**) or not (**noncontributing alleles**) in pigment production, and that each contributing allele allows for the synthesis of a certain amount of pigment. Under this hypothesis, the intensity of kernel coloration is a function of the number of *R* or *C* alleles in the genotype: *RR CC* would be dark red and *rr cc* would be white. Table 14.1 shows this would explain the five phenotypic classes observed. Nilsson-Ehle concluded that the inheritance of red kernel color in wheat is an example of a polygene series of two loci with as many as four contributing alleles.

We must be cautious in extending this explanation of the genetic basis of this trait to other situations. Some F_2 populations show only three phenotypic classes with a 3:1 ratio of red to white, whereas other F_2 populations show a 63:1 ratio of red to white, with other discrete classes of color falling between the dark red and the white. These results suggest that there is quite a bit of variability segregating among strains at the loci that contribute to kernel color, so that a locus does not always contribute to quantitative variation in a cross. As a general rule, in a hybrid cross the number of possible genotypes is 3^n, where *n* equals the number of independent loci with two alleles. If more than two alleles are present at a locus, the number of genotypes is even greater.

The multiple-gene hypothesis that fits the wheat kernel color example so well has been applied to other examples of quantitative inheritance. In its basic form, the multiple-gene hypothesis proposes that quantitative inheritance can be explained by the action and segregation of allelic pairs at a number of loci, called **polygenes**, each with a small effect on the overall phenotype. For the most part, the multiple-gene hypothesis is satisfactory as a working hypothesis for interpreting many quantitative traits. The whole picture of quantitative traits is, of course, more complicated. For example, the proposal that a number of alleles each function to produce a particular amount of pigment is an attractive hypothesis, but what does that hypothesis mean at the molecular level? Is some regulation of product output exerted at the translational level or in a biochemical pathway? In a large polygenic series, how many biochemical pathways are controlled? Polygenic inheritance provides an explanation for the inheritance of continuous traits that is compatible with Mendel's laws and, as we shall see later in the chapter, with the application of molecular techniques, we are slowly developing a better understanding of the nature of the genes involved.

K E Y N O T E

> Discontinuous traits exhibit only a few distinct phenotypes and can be described in qualitative terms. Continuous traits, on the other hand, display a range of phenotypes and must be described in quantitative terms. The multiple-gene hypothesis assumes that multiple loci contribute to quantitative phenotypes, and as the number of contributing alleles increases, there is an additive (or occasionally multiplicative) effect on the phenotype. The relationship between the genotype and the phenotype for quantitative traits may be complex because multiple alleles at multiple loci allow many genotypes, and the genotypic response to environmental factors can modulate the range of phenotypes.

Table 14.1	Genetic Explanation for the Number and Proportions of F_2 Phenotypes for the Quantitative Trait Red Kernel Color in Wheat		
Genotype	**Number of Contributing Alleles for Red**	**Phenotype**	**Fraction of F_2**
RR CC	4	Dark red	$^1/_{16}$
RR Cc or *Rr CC*	3	Medium red	$^4/_{16}$
RR cc or *rr CC* or *Rr Cc*	2	Intermediate red	$^6/_{16}$
rr Cc or *Rr cc*	1	Light red	$^4/_{16}$
rr cc	0	White	$^1/_{16}$

Statistical Tools

When multiple genes and environmental factors influence a trait, the relationship between individual loci and their contribution to the phenotype may be obscured. The same rules of transmission genetics and gene function still apply, but defining the effect of a single locus requires the ability to determine the genotypes of many individuals at loci across the genome. Prior to the advent of modern genotyping methods, and even today in systems where such tools are lacking, it was impractical to try and understand the action and role of each individual gene, so quantitative geneticists applied statistical and analytical procedures to understand the overall influence of genes on continuous traits.

As indicated earlier, one of the fundamental questions addressed in the study of quantitative traits is how much of the variation that exists among individuals in populations is genetically determined and how much is environmentally induced. This is an important question because in many situations an understanding of the contribution of both factors is needed to make informed decisions. For example, in agricultural yield trials, we may be interested in identifying superior lines of wheat (genotypes) from data gathered across many field sites (environments). Thus, at the heart of the field of quantitative genetics (the only field of science that addresses this question explicitly) is the perennial question of *nature versus nurture,* or the relative roles of genes versus environment. Notice that the traditional phrasing of the problem pits one against the other, implying that they are mutually exclusive. This oversimplified phrasing has resulted in much bad science and needless debate. Nevertheless, in quantitative genetic terms, we phrase the problem in terms of variation from these two sources: How much of the variation in some aspect of the phenotype (V_P) results from genetic variation (V_G) and how much from environmental variation (V_E)? This relationship is expressed as

$$V_P = V_G + V_E$$

To work this equation we must learn how to measure variation in phenotype and how to partition the variation into genetic and environmental components. To do this, we need to understand some statistical methodology, much of which was developed specifically to deal with quantitative genetics.

Samples and Populations

Suppose we want to describe some aspect of a trait for a large group of individuals. For example, we might be interested in the average birth weight of infants born in New York City during 1987. One way to answer this question is to collect the weight of each of the thousands of babies born in New York City in 1987. An alternative method, which is less laborious, would be to collect these data from a subset of the group, say birth weights of 100 infants born in New York City during 1987, and then use the average obtained from this subset as an estimate of the average for all the infants from the entire city. Scientists commonly use this sampling procedure in data collection. The group of ultimate interest (in our example, all infants born in New York City during 1987) is called the **population,** and the subset (our set of 100 babies) used to give us an estimate of the population is called a **sample.** For a sample to give us confidence in our estimates for the population, it must be large enough that chance differences between the sample and the population are not misleading. If our sample consisted of only a few babies, and these infants were unusually large, then our estimate of the average birth weight of all babies would not be very accurate. The sample must also be a random subset of the population. If all the babies in our sample came from Hope Hospital for Premature Infants, then we would grossly underestimate the true average birth weight of the population. Although this might seem obvious, a great many errors are made because data are not collected randomly.

KEYNOTE

To describe and study a large group of individuals, scientists frequently examine a subset of the group. This subset is called a sample, and the sample provides estimates for the larger group, which is called the population. The sample must be of reasonable size and it must be a random subset of the larger group for it to provide accurate information about the population.

Distributions

When we studied discontinuous traits in Chapters 2 to 4, we defined alternate phenotypes found within a group of individuals and described the group by stating the proportion of individuals falling into each phenotypic class. Because continuous traits exhibit a range of phenotypes, describing a group of individuals is more complicated. One means of summarizing the phenotypes of a continuous trait is with a **frequency distribution,** which is a summary of a group in terms of the proportion of individuals that fall within a certain phenotypic range (see Figure 14.2).

To make a frequency distribution, first classes are constructed that consist of a specified range of the phenotypic measure, and the number of individuals in each class is counted. Table 14.2 presents the data from Johannsen's study of the inheritance of seed weight in the bean *Phaseolus vulgaris.* As shown in the table, Johannsen weighed 5,494 beans from the F_2 progeny of a cross and

Table 14.2 Weight of 5,494 F$_2$ Beans (Seeds of *Phaseolus vulgaris*) Observed by Johannsen in 1903

Weight (mg)	50–150	150–250	250–350	350–450	450–550	550–650	650–750	750–850	850–950
(Midpoint of range)	(100)	(200)	(300)	(400)	(500)	(600)	(700)	(800)	(900)
Number of beans	5	38	370	1,676	2,255	928	187	33	2

classified them into nine classes, each of which covered a 100-mg range of weight. Frequency data such as these can be displayed graphically in a frequency histogram, as shown in Figure 14.3. In the histogram, the phenotypic classes are indicated along the horizontal axis, and the number present in each class is plotted on the vertical axis.

There are certain shapes of frequency distributions that correspond to known, mathematically described probability distributions. For example, many continuous phenotypes exhibit a symmetrical, bell-shaped distribution similar to the curve overlaid on the data in Figure 14.3 (see also Figure 14.2). This type of distribution is called a **normal distribution.** Data that conform to a normal distribution can be accurately described by a few statistics, namely the mean and variance, which are described below. In addition, data that are normally distributed allow us to make simplifying assumptions that facilitate complicated analyses.

The Mean

A frequency distribution of a normally distributed phenotypic trait can be summarized by two statistics, the mean and the variance. The sample **mean** ($\bar{x}$), also known as the average, tells us where the center of the distribution of the phenotypes from a sample is located. The mean of a sample is calculated by simply adding up all the individual measurements ($x_1, x_2, x_3, \ldots, x_n$) and dividing by the number of measurements we added (n).

The mean is one statistic used frequently in quantitative genetics to characterize the phenotypes of a group of individuals. For example, in an early study of continuous variation, Edward M. East examined the inheritance of flower length using a cross between a long-flowered and a short-flowered strain of tobacco. Within each strain, flower length varied, so East reported that the mean phenotype of the short strain was 40.4 mm and the mean phenotype of the long strain was 93.1 mm. The F$_1$ progeny, which consisted of 173 plants, had a mean flower length of 63.5 mm. In this situation, the mean provides a convenient way for quickly characterizing and comparing the phenotypes of parents and offspring.

The Variance and the Standard Deviation

A second statistic that provides key information about a distribution is the **variance.** The variance is a measure of how much the individual observations spread out around the mean—how variable the individuals and their measurements are. Two distributions may have the same mean, but when they have different variances, as shown in Figure 14.4, the distributions differ markedly. A broad curve implies high variability in the quantity measured and a correspondingly large variance. A narrow curve, in contrast, indicates little variability in the quantity

Figure 14.4

Graphs showing three distributions with the same mean but with different variances.

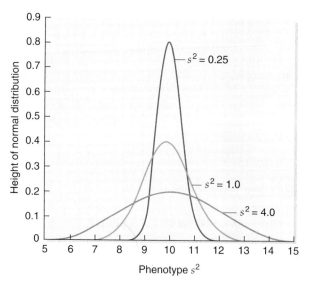

Figure 14.3

Frequency histogram for bean weight in *Phaseolus vulgaris* plotted from data in Table 5.2. A normal curve has been fitted to the data and is superimposed on the frequency histogram.

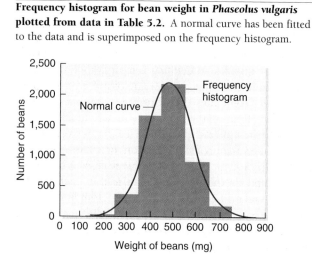

measured and a correspondingly small variance. The sample variance, symbolized as s^2, is defined as the average squared deviation from the mean.

$$\text{Variance} = s^2 = \frac{\sum (x_i - \bar{x})^2}{n - 1}$$

The sample variance is calculated by first subtracting the sample mean from each individual measurement. This difference is then squared (so that the variance describes distance from the mean without regard to direction) and all the squared values are added up. The sum of these squared values is then divided by the number of original measurements minus 1 (for mathematical reasons that we will not discuss here, the sample variance is obtained by dividing by $n - 1$ instead of by n.)

The **standard deviation** is often preferred over the variance because the standard deviation shares the same units as the original measurements (whereas the variance is in the units squared). The standard deviation for a sample is simply the square root of the sample variance:

$$\text{Standard deviation} = s = \sqrt{s^2}$$

A theoretical normal distribution is completely specified by the mean and standard deviation. It always has the shape indicated in Figure 14.5, where 66 percent of the individual observations have values within one standard deviation above or below ($\pm 1s$) the mean of the distribution, about 95 percent of the values fall within two standard deviations ($\pm 2s$) of the mean, and more than 99 percent fall within three standard deviations

Figure 14.5

Normal distribution curve showing the proportions of the data in the distribution that are included within certain multiples of the standard deviation.

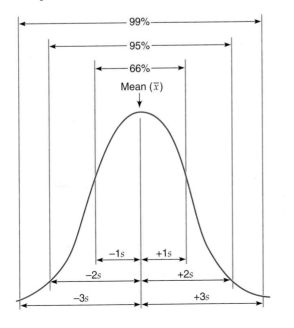

Table 14.3	Sample Calculations of the Mean, Variance, and Standard Deviation for Body Length of 10 Spotted Salamanders from Penobscot County, Maine

Body Length (x_i) (mm)	$(x_i - \bar{x})$	$(x_i - \bar{x})^2$
65	$(65 - 57.1) =$ 7.9	$7.9^2 =$ 62.41
54	$(54 - 57.1) =$ -3.1	$-3.1^2 =$ 9.61
56	$(56 - 57.1) =$ -1.1	$-1.1^2 =$ 1.2
60	$(60 - 57.1) =$ 2.9	$2.9^2 =$ 8.41
56	$(56 - 57.1) =$ -1.1	$1.1^2 =$ 1.21
55	$(55 - 57.1) =$ -2.1	$2.1^2 =$ 4.41
53	$(53 - 57.1) =$ -4.1	$-4.1^2 =$ 16.81
55	$(55 - 57.1) =$ -2.1	$-2.1^2 =$ 4.41
58	$(58 - 57.1) =$ 0.9	$0.9^2 =$ 0.81
59	$(59 - 57.1) =$ 1.9	$1.9^2 =$ 3.61
$\Sigma x_i = 571$		$\sum (x_i - \bar{x})^2 = 112.9$

$$\text{Mean} = \bar{x} = \frac{\sum x_i}{n} = \frac{571}{10} = 57.1$$

$$\text{Variance} = s_x^2 = \frac{\sum (x_i - \bar{x})^2}{n-1} = \frac{112.9}{9} = 12.54$$

$$\text{Standard deviation} = s_x = \sqrt{12.54} = 3.54$$

($\pm 3s$). Table 14.3 presents the body lengths of 10 spotted salamanders from Penobscot County, Maine, along with calculations of the sample mean, variance, and standard deviation for these data. We can infer many things about our sample data and experiments using these objectively determined percentages. More important for genetic analysis, we can use the statistical technique analysis of variance, developed by Sir Ronald Fisher (see Figure 24.1a), to help partition the observed variation into components, an important part of quantitative genetic analyses.

The variance and the standard deviation provide valuable information about the phenotypes of a group of individuals. In our discussion of the mean, we saw how East used the mean to describe flower lengths of parents and offspring in crosses between tobacco lines. When East crossed a strain of tobacco with short flowers ($\bar{x} = 40.4$ mm) to a strain with long flowers ($\bar{x} = 93.1$ mm), the F_1 offspring had a mean flower length of 63.5 mm, which was intermediate to the phenotypes of the parents. When he intercrossed the F_1, the mean flower length of the F_2 offspring was 68.8 mm, approximately the same as the mean phenotype of the F_1. However, the F_2 progeny differed from the F_1 in an important attribute not apparent from the means: The F_2 were more variable in phenotype than the F_1. The variance in the flower length of the F_2 was 42.4 mm^2, whereas the variance in the F_1 was only 8.6 mm^2. Thus, the mean and the variance are both necessary to fully describe the distribution of phenotypes among a group of individuals.

Correlation

A difficulty encountered when thinking about the overall phenotype of an organism is that it is somewhat artificial to pick out traits and study them in isolation. Organisms are composites of many traits. Some of these traits, such as height and weight, may actually be two members of a more general trait called size. It isn't unreasonable to think that genes and environmental factors that affect development may have pleiotropic effects that affect both height and weight. In other words, the values of two or more traits are often *correlated*, meaning that if one variable changes, the other is also likely to change. For example, arm and leg length are correlated in humans and most other animals: Individuals with long arms usually have correspondingly long legs, and vice versa.

The **correlation coefficient** is a statistic that measures the strength of the association between two variables in the same experimental unit, which in genetics is usually an individual. Suppose we have two variables, arm length and leg length, where x equals arm length and y equals leg length. To calculate the correlation between these variables, we begin by obtaining the **covariance** of x and y, which is a measure of how much variation is shared by an individual for both traits. The covariance is computed by taking the same deviations from the mean used in calculating the variance for each trait x and y, but instead of squaring these values as for the variance, the product of the two is taken for each pair of x and y values, and the products are added together. The sum is then divided by $n - 1$, where n is the number of xy pairs, to give the covariance of x and y:

$$\text{cov}_{xy} = \frac{\sum (x_i - \bar{x})(y_i - \bar{y})}{n - 1}$$

An algebraically equivalent equation, which is easier to compute, is

$$\text{cov}_{xy} = \frac{\sum x_i y_i - n\overline{xy}}{n - 1}$$

where $\sum x_i y_i$ is the sum of each value of x multiplied by each corresponding value of y, $\sum x_i$ is the sum of all x values, and $\sum y_i$ is the sum of all y values.

The correlation coefficient r can then be obtained by dividing the covariance by the product of the standard deviations of x and y,

$$\text{Correlation coefficient} = r = \frac{\text{cov}_{xy}}{s_x s_y}$$

where s_x equals the standard deviation of x, and s_y equals the standard deviation of y. Table 14.4 gives a sample calculation of the correlation coefficient between two variables.

When the covariance is divided by the two standard deviations, the resulting correlation coefficient becomes a unitless, standardized measure that can range from -1 to $+1$. The sign of the correlation coefficient indicates the direction of the correlation. If the correlation coefficient is positive, then an increase in one variable tends to be associated with an increase in the other variable. If the number of flowering heads and seed number are positively correlated in a species of flower, for example, plants with a greater number of flowering heads will also tend to produce more seeds. Positive correlations are illustrated in Figure 14.6b, c, d, and f. A negative correlation coefficient indicates that an increase in one variable is associated with a decrease in the other. If seed size and seed number are negatively correlated, for example, plants with large seeds tend to produce fewer seeds on average than do plants with smaller seeds. Figure 14.6e presents a negative correlation. The absolute value of the correlation coefficient provides information about the strength of the association. When the correlation coefficient is close to -1 or $+1$, the correlation is strong, meaning that a change in one variable is almost always associated with a corresponding change in the other variable. For example, the x and y variables in Figure 14.6f are strongly associated and have a correlation coefficient of 0.9. On the other hand, a correlation coefficient near 0 indicates only a weak relationship, if any, exists between the variables, as is illustrated in Figure 14.6b.

Several important points about correlation coefficients warrant emphasis. First, a correlation between variables means only that the variables are associated: *Correlation does not imply that a cause-effect relationship exists.* The classic example of a noncausal correlation between two variables is the positive correlation between the number of ministers and liquor consumption in cities with population size over 10,000. One should not conclude from this correlation that ministers are the direct or indirect cause of increasing alcohol consumption. Alcohol consumption and the number of ministers are associated because both are positively correlated with a third factor, population size: Larger cities contain more ministers and also have higher alcohol consumption due to their larger populations. Assuming that two factors are causally related because they are correlated may lead to erroneous conclusions.

Another important point is that because the correlation coefficient is unitless, correlation means only that a change in one variable is associated with a corresponding change in the other variable: *Two variables can be highly correlated and yet have different values.* For example, the overall height and knee height of elderly Mexican females are highly correlated; however, the knee height is always much less than the overall height of a person. Thus, it is important to remember that correlation demonstrates only the trend between two variables.

Table 14.4 Sample Calculation of the Correlation Coefficient for Body Length and Head Width of Tiger Salamanders

Body Length (mm)			Head Width (mm)			
x_i	$x_i - \bar{x}$	$(x_i - \bar{x})^2$	y_i	$y_i - \bar{y}$	$(y_i - \bar{y})^2$	$x_i y_i$
72.00	−7.92	62.67	17.00	−0.75	0.56	1224
62.00	−17.92	321.01	14.00	−3.75	14.06	868
86.00	6.08	37.01	20.00	2.25	5.06	1720
76.00	−3.92	15.34	14.00	−3.75	14.06	1064
64.00	−15.92	253.34	15.00	2.75	7.56	960
82.00	2.08	4.34	20.00	2.25	5.06	1640
71.00	−8.92	79.51	15.00	−2.75	7.56	1065
96.00	16.08	258.67	21.00	3.25	10.56	2016
87.00	7.08	50.17	19.00	1.25	1.56	1653
103.00	23.08	532.84	23.00	5.25	27.56	2369
86.00	6.08	37.01	18.00	0.25	0.06	1548
74.00	−5.92	35.01	17.00	−0.75	0.56	1258
$\Sigma x_i =$ 959.00		$\Sigma(x_i - x)^2 =$ 1,686.92	$\Sigma y_i =$ 213.00		$\Sigma(y_i - y)^2 =$ 94.25	$\Sigma x_i y_i =$ 17,385

$\bar{x} = \Sigma x_i/n = 959/12 = 79.92$
$\bar{y} = \Sigma y_i/n = 213/12 = 17.75$
Variance of $x = s_x^2 = \Sigma(x_i - x)^2/n - 1 = 1,686.92/11 = 153.35$
Standard deviation of $\bar{x} = s_x = \sqrt{s_x^2} = \sqrt{153.35} = 12.38$
Variance of $y = s_y^2 = \Sigma(y_i - \bar{x})^2/n - 1 = 94.25/11 = 8.57$
Standard deviation of $y = s_y = \sqrt{s^2} = \sqrt{8.57} = 2.93$
Covariance $= \text{cov}_{xy} = (\Sigma x_i y_i - 1/n(\Sigma x_i \, \Sigma y_i))/n - 1$
$\text{cov}_{xy} = \dfrac{(17,385 - 1/12(959 \times 213))}{12 - 1}$
$\text{cov}_{xy} = 32.97$
Correlation coefficient $= r = \text{cov}_{xy}/(s_x s_y) = 32.97/(12.38 \times 2.93)$
$r = 0.91$

KEYNOTE

> The correlation coefficient is a measure of how strongly two variables are associated. A positive correlation coefficient indicates that the two variables change in the same direction: An increase in one variable usually is associated with a corresponding increase in the other variable. When the correlation coefficient is negative, an increase in one variable is most often associated with a decrease in the other. The absolute value of the correlation coefficient provides information about the strength of the association. Strong correlation does not imply that a cause-effect relationship exists between the two variables.

Regression

The correlation coefficient tells us about the strength of association between variables and indicates whether the relationship is positive or negative, but it provides little information about the precise quantitative relationship between the variables. For example, if we know there is a correlation between heights of father and son, we might ask, "If a father is six feet tall, what is the most likely height of his son?" To answer this question, **regression** analysis is used.

The relationship between two variables can be expressed in the form of a **regression line,** as shown in Figure 14.7 for the relationship between the heights of fathers and sons. Each point on the graph represents the actual height of a father (value on the x axis) and the height of his son (value on the y axis). Regression finds the line that best fits the data, found by minimizing the squared vertical distances from the points to the regression line. The regression line can be represented with the equation

$$y = a + bx$$

where x and y represent the values of the two variables (in Figure 14.7, the heights of father and son, respectively),

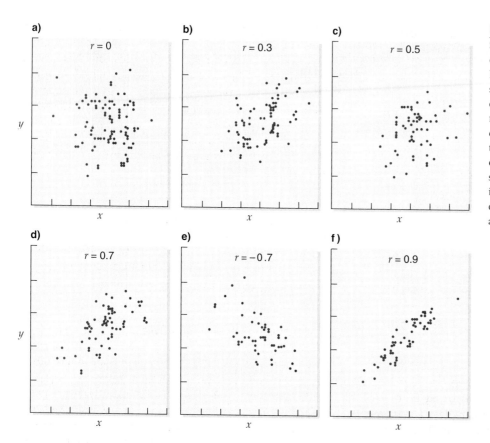

a) $r = 0$

b) $r = 0.3$

c) $r = 0.5$

d) $r = 0.7$

e) $r = -0.7$

f) $r = 0.9$

Figure 14.6

Scatter diagrams showing the correlation of x and y variables. Diagrams (b), (c), (d), and (f) show positive correlations, whereas diagram (e) shows a negative correlation. The absolute value of the correlation coefficient (r) indicates the strength of the association. For example, diagram (f) illustrates a strong correlation and diagram (b) illustrates a weak correlation. In diagram (a), the x and y variables are not correlated.

b represents the **slope of the line**, also called the **regression coefficient**, and a is the y intercept. The slope can be calculated from the covariance of x and y and the variance of x in the following manner:

$$\text{slope} = b = \frac{\text{cov}_{xy}}{s_x^2}$$

Figure 14.7

Regression of sons' height on fathers' height. Each point represents a pair of data for the height of a father and his son. The regression equation is $y = 36.05 + 0.49x$.

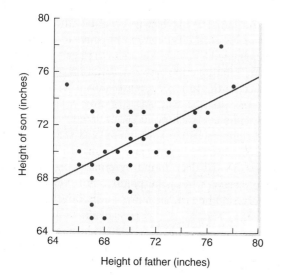

Height of son (inches) vs. Height of father (inches)

The slope indicates how much of an increase in the variable on the y axis is associated with a unit increase in the variable on the x axis. For example, a slope of 0.5 for the regression of father and son height would mean that for each 1-inch increase in height of a father, the expected height of the son would increase 0.5 inches. The y intercept is the expected value of y when x is zero (the point at which the regression line crosses the y axis). Examples of regression lines with different slopes are presented in Figure 14.8. Regression analysis is one method that is commonly used for measuring the extent to which variation in a trait is genetically determined, as described later in the section on heritability.

Analysis of Variance

One last statistical technique that we will mention briefly is **analysis of variance (ANOVA).** Analysis of variance is a powerful statistical procedure for determining whether differences in means are significant (larger than we would expect from chance alone) and for dividing the variance into components. For example, we might be interested in knowing whether males with the XYY karyotype differ in height from males with a normal XY karyotype. (See Chapter 12.) We would proceed by first calculating the mean height of a sample of XYY males and the mean height of a sample of XY males. Suppose we found that the mean height of our sample of XYY males was 74 inches, and the

Figure 14.8

Regression lines with different slopes. The slope indicates how much of a change in the y variable is associated with a change in the x variable.

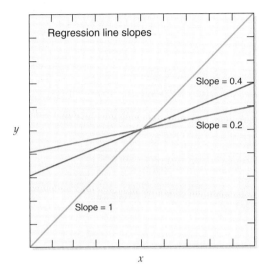

mean height of our sample of XY males was 70 inches. The means are different, but our result could be caused by chance differences between the samples rather than by some factor associated with the extra Y chromosome. Analysis of variance can provide us with the probability that the difference in means of the two samples results from chance. For example, it might indicate that there is less than a 1 percent probability (often expressed as $p > 0.01$) that the difference we observed in the mean heights resulted from chance. We would probably conclude, then, that the difference in mean heights of XYY males and XY males is not caused by chance differences in our samples but results from some significant factor associated with the difference in chromosomes (i.e., karyotype).

More important to the understanding of genetic influence on traits such as height, ANOVA can also be used to determine how much of the variation in height is associated with the difference in karyotype. We might find that the difference in the karyotypes is associated with 40 percent of the overall variation in height among the individuals in our samples. Factors other than difference in the number of Y chromosomes (such as genes on other chromosomes, diet, and health care) would be responsible for the other 60 percent of the variation.

The calculations involved in analysis of variance are beyond the scope of this book, but the concept of breaking down the variance into components—called *partitioning the variance*—is fundamental to quantitative genetics. For example, frequently we are interested in knowing how much of the variation in a trait is associated with genetic differences among individuals and how much of the variation is associated with environmental factors. Suppose that we wanted to increase milk production in a herd of dairy

cattle. We might use analysis of variance to determine how much of the variation in milk production among the cows results from environmental differences and how much arises because of genetic differences. If much of the variation is genetic, we could increase milk production by selective breeding. On the other hand, if most of the variation is environmental, selective breeding will do little to increase milk production, and our efforts would be better directed toward providing the optimum environment for high production.

Quantitative Genetic Analysis

Quantitative traits require a different kind of analysis from discontinuous traits. The genetic analysis of quantitative traits is illustrated in this section.

Inheritance of Ear Length in Corn

With a statistical background, we can see how quantitative geneticists apply these methods to understand multifactorial traits. An organism that has been the subject of genetic and cytological studies for many years is corn, *Zea mays*. Ear length is one trait that was examined in a classic study using some of these techniques to demonstrate a genetic basis for a quantitative trait. In this study, reported in 1913, Rollins Emerson and Edward East started their experiments with two pure-breeding strains of corn, each of which displayed little variation in ear length. The two varieties were black Mexican sweet corn (which had short ears of mean length 6.63 cm) and Tom Thumb popcorn (which had long ears of mean length 16.80 cm).

Emerson and East crossed the two strains and then interbred the F_1 plants. The parental plants were inbred lines, so we can assume that each was homozygous for the genes controlling ear length. The F_1 plants were heterozygous for all genes, and all plants should have the same genotype. The top two panels of Figure 14.9 present the phenotypes of the parents and F_1 plants in photographs and histograms. The range of ear length phenotypes seen in the parents and F_1 plants must, therefore, result from factors other than genetic differences. These factors are most likely environmental because it is almost impossible to grow plants in exactly identical conditions.

In the F_2, the mean ear length of 12.89 cm is about the same as the mean for the F_1 population, but the F_2 population has a much larger variation around the mean than does the F_1 population. This variation is intuitively easy to see in Figure 14.9b, and it can also be shown by calculating the standard deviation (s). The standard deviation of the long-eared parent is 1.887, and that of the short-eared parent is 0.816. In the F_1, $s = 1.519$, and in the F_2, $s = 2.252$, confirming that the F_2 has greater variability.

Figure 14.9

Inheritance of ear length in corn. (a) Representative corn ears from the parental, F_1, and F_2 generations from an experiment in which two pure-breeding corn strains that differ in ear length were crossed and then the F_1s interbred. (b) Histograms of the distributions of ear length (in centimeters) of ears of corn from the experiment represented in (a); the vertical axes represent the percentages of the different populations found at ear length. (From E. W. Sinnott and L. C. Dunn, 1925. "Principles of Genetics" [New York McGraw-Hill].)

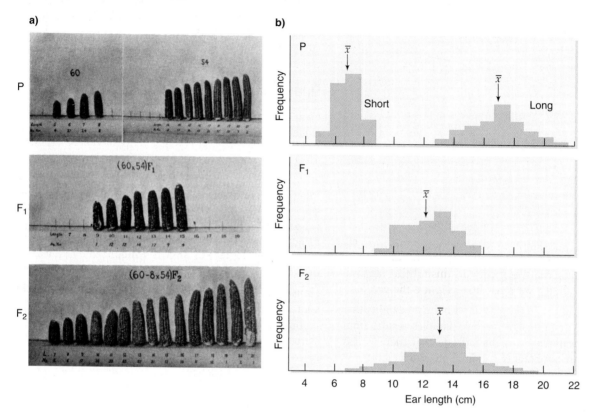

The key to this experiment is to examine the patterns of variation. Is the variation seen in the F_2 solely the result of environmental factors? If we assume the environment was responsible for variation in the parental and the F_1 generations, then we have every reason to believe that it would have a similar effect on the F_2. However, we have no reason to suppose that the environment would have a greater influence on the F_2 than on the other two generations, so there must be another explanation for the greater variation in ear length in the F_2 generation. A more reasonable hypothesis is that the increased variability of the F_2 results from the presence of greater genetic variation in the F_2, which must have been inherited from the parents.

Setting aside the environmental influence for the moment, the data reveal four observations that apply generally to similar quantitative inheritance studies:

1. The mean value of the quantitative trait in the F_1 is usually intermediate between the means of the two true-breeding parental lines.

2. The mean value for the trait in the F_2 is usually approximately equal to the mean for the F_1 population.
3. The F_2 almost always shows more variability around the mean than the F_1 does.
4. The extreme values for the quantitative trait in the F_2 extend closer to the two parental values than do the extreme values of the F_1, and may sometimes even surpass the parental values.

K E Y N O T E

For a quantitative trait, the F_1 progeny of a cross between two phenotypically distinct, pure-breeding parents usually has a phenotype intermediate between the parental phenotypes. The F_2 shows more variability than the F_1, with a mean phenotype close to that of the F_1. The extreme phenotypes of the F_2 extend well beyond the range of the F_1 and into the ranges of the two parental values.

Heritability

Heritability is the proportion of a population's phenotypic variation that is attributable to genetic factors. As we have seen, continuous traits are influenced by multiple genes and by environmental factors. To make informed management decisions, plant and animal breeders, for example, need to know the genetic contribution to traits such as weight gain in cattle, number of eggs laid by chickens, and amount of fleece produced by sheep. Moreover, many ecologically important traits, such as variation in body size, fecundity, and developmental rate, are also polygenic, and the genetic contribution to this variation is important for understanding how natural populations evolve. Heritability is nonetheless often misunderstood and the term is frequently misused or tossed about without firm scientific basis. For example, in humans, when individuals in a family resemble each other in some aspect of the phenotype, be it stature or intelligence, a genetic basis often is assumed to be responsible for the similarity. But all the resemblance among family members could just as easily result from their shared environment as opposed to their shared genes. Carefully planned quantitative genetic experiments are the only way to distinguish between these alternatives in any organism being studied. To assess heritability, we must first measure the variation in the trait, and then we must partition that variance into components attributable to different causes.

Components of the Phenotypic Variance

As an analogy, we can consider the variance among individuals as a stick that can be divided into various pieces. The **phenotypic variance**, represented by V_P, is a measure of all variability for a trait (i.e., the whole stick). Recall that it is the sum of squared deviations from the mean for all individuals, as outlined in this chapter's section "Statistical Tools". Some of the differences from the mean for each individual may arise because of genetic differences between individuals (different genotypes within the group); alternatively, some of the differences from the mean may be due to different environments experienced by individuals. The genetic contribution to the phenotypic variation is called **genetic variance,** represented by the symbol V_G. Figure 14.10a shows a situation in which all the variation is due to genotype.

As noted, additional variation often results from environmental differences experienced by the individuals. The **environmental variance** is symbolized by V_E, and by definition includes any nongenetic source of variation. Temperature, nutrition, and parental care are examples of

Figure 14.10

Hypothetical example of genetic and environmental effects on plant height. In each plot, the ovals represent the groups of data points for each genotype, illustrating a small amount of variability. The blue ovals represent the points for genotype 1 (G1), red ovals represent the points for genotype 2 (G2), and green ovals denote an overlap. (a) Plant height variation is influenced predominantly by genotype, with G1 individuals having on average a greater height than G2 individuals. Plant height is independent of the temperature in which the plants are raised. (b) In this case, variation is predominantly due to the environment, but the two genotypes are indistinguishable across the range of temperatures tested. (c) Both genotype and the environment exert an additive influence on height. (d) Both genotype and the environment exert an influence, but the response of each genotype is dependent on the environment. This is an example of genotype-by-environment interaction.

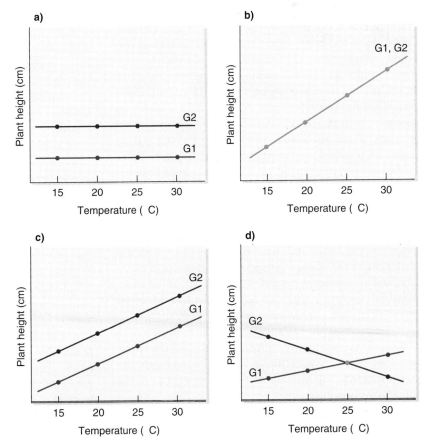

obvious environmental factors that may cause differences among individuals during development. Figure 14.10b shows a situation in which all the variation results from the environment, and Figure 14.10c shows a situation in which both genes and environment contribute to the variation. Thus, we have two pieces of our stick that correspond to the basic nature-nurture issue discussed earlier:

$$V_P = V_G + V_E$$

One hundred percent of the variation among individuals is accounted for by genetic and environmental influences; however, the partitioning of phenotypic variance can often be complicated. The sum of the genetically caused variance and environmentally caused variance may not add up to the total phenotypic variance, because the genetically caused variance and environmentally caused variance covary. For example, let's say milk production in cows is influenced by genes, but it is also influenced by the amount of feed a farmer provides. The farmer knows his cows and provides the offspring of good milking cows more feed and poor milking cows less feed. Because individuals of above-average genetic quality therefore receive above-average resources, a covariance between genotype and environment is produced. In this way, the variance in milk production is increased beyond that expected on the basis of genes and environment operating independently. To account for situations such as these, another term ($COV_{G,E}$) is needed.

There is another source of phenotypic variance, genotype-by-environment interaction, or G×E. Variance caused by G×E exists when the relative effects of the genotypes differ among environments. For example, suppose that in a cold environment, *AA* plants are on average 40 cm tall and *aa* plants are on average 35 cm tall. However, when the genotypes are moved to a warm climate, *aa* plants are now on average 60 cm tall, and *AA* plants are on average 50 cm tall. This relationship is illustrated in Figure 14.10d. In this example, both genotypes grow taller in the warm environment, so there is an environmental effect on variance. There is also a genetic effect, but the genetic effect depends on the environment. The relative performance of the genotypes switches in the two environments. Therefore, while both environmental differences (temperature) and genetic differences (genotypes) contribute to the phenotypic variance, the effects of genotype and environment cannot simply be added together. An additional component that accounts for how genotype and environment interact must be considered, $V_{G×E}$.

The phenotypic variance, composed of differences arising from genetic variation, environmental variation, genetic-environmental covariation, and genetic-environmental interaction, can be represented by the following equation:

$$V_P = V_G + V_E + 2COV_{G,E} + V_{G×E}$$

It is important to note that while this is the full equation, there may be situations in which individual components equal zero, depending on the genetic composition of the population, the specific environment, and the manner in which the genes interact with the environment.

KEYNOTE

Variation among individuals can be partitioned into genetic and environmental components, and the interaction between the two. The fact that genotypes might not be distributed randomly across environments and that genotypes may behave differently in different environments means that care must be taken to correctly identify the contributions of these factors.

The pieces of the stick corresponding to genetic and environmental variation can be further divided to reveal more precise components of causal influence. Genetic variance, V_G, can be subdivided into components arising from different types of gene action and interactions between genes. Some of the genetic variance occurs as a result of the additive effects of the different alleles on the phenotype. For example, an allele g may, on average, contribute 2 cm in height to a plant, while the allele G contributes on average 4 cm. In this case, the gg homozygote would contribute 2 + 2 = 4 cm in height, the Gg heterozygote would contribute 2 + 4 = 6 cm in height, and the GG homozygote would contribute 4 + 4 = 8 cm in height. To determine the genetic contribution to height, we would then add the effects of alleles at this locus to the effects of alleles at other loci that might influence the phenotype. Genes such as these are said to have additive effects, and variation resulting from this sort of gene action is called *additive genetic variance*, symbolized by V_A. You may recall that the genes studied by Nilsson-Ehle that determine kernel color in wheat are strictly additive in this way. Some alleles contribute to the pigment of the kernel and others do not; the added effects of all the individual contributing alleles determines the phenotype of the kernel. Thus, the genotypes *AA bb*, *aa BB*, and *Aa Bb* all produce the same phenotype because each genotype has two contributing alleles. The phenotypic variance arising from the additive effects of genes is the additive genetic variance.

Additivity among alleles at a locus is not always the case. Other genes may exhibit dominance, and this is the source of the *dominance variance* (V_D). When dominance is present, the individual effects of the alleles are not strictly additive, so we must also factor in how genotypes contribute to the phenotypic variation. A locus exhibiting dominance will contribute to V_P only when both the recessive homozygote and either the heterozygote or dominant homozygote are present in the population. Under these conditions, for example, an F_2 or a backcross to the recessive parent, loci showing dominance increase variability; in the case of a backcross to a homozygous dominant parent, loci with dominant alleles wouldn't produce phenotypic variation. Finally, epistatic interactions may occur among alleles at different loci. Recall that when epistasis exists, alleles at different loci interact in determining the phenotype. Thus, we might have three genotypes at one locus, but their penetrance could be affected by variation at other loci. The presence of epistasis adds another source of genetic variation, called epistatic or **interaction variance** (V_I). Thus, we can partition the genetic variance as follows:

$$V_G = V_A + V_D + V_I$$

and the total phenotypic variance up to this point can then be summarized as

$$V_P = V_A + V_D + V_I + V_E + 2COV_{G,E} + V_{G\times E}$$

Just as we partitioned the genetic component of variance into three parts, the environmental component of variance can also be partitioned. Individuals in a population may be exposed to varying temperature or nutritional environments during development, resulting in somewhat irreversible differences called *general environmental effects* (V_{Eg}). For example, an individual that is raised in a nutritionally deprived region may have a smaller body size. Other environmental variation results in immediate, transient changes in the phenotype, such as skin pigment differences upon exposure to sun, called *special environmental effects* (V_{Es}). Finally, environmental effects may be shared by all members of a family. These common *family environmental effects* (V_{Ecf}) are especially important because they contribute to differences among families and can be confounded with genetic influences. For example, many insects deposit eggs on specific host plants, and their larvae develop by feeding on these plants. Individual plants vary with respect to nutritional quality and levels of toxins present. As a result, insects that sequester compounds from host plants for their own defense, such as monarch butterflies and the moth *Utetheisa*, can exhibit family differences simply because of the plant they feed on as larvae. Such differences often are misinterpreted as evidence of genetic variation.

A special category of common family environmental effects, **maternal effects** (V_{Em}), are prevalent enough to deserve special mention. For example, variation in the size of mammals at birth has both genetic and environmental components. The genetic component results from the specific genotypic differences between offspring. Their environment up to birth is their mother's uterus, and because there is variation among mothers for variables such as litter size, gestation period, and so on, these constitute maternal effects. Maternal effects can continue after birth as well. In mammals, much of early growth is affected by the amount and nutritional makeup of milk available to offspring; therefore, milk quantity and quality are further maternal influences on offspring that can increase phenotypic variation.

At this point, our variation stick comprises many small segments, and the nature versus nurture equation is partitioned as follows:

$$V_P = V_A + V_D + V_I + V_{Eg} + V_{Es} + V_{Ecf} + V_{Em} + 2COV_{G,E} + V_{G\times E}$$

Partitioning the phenotypic variance into these components is useful for thinking about the contribution of different factors to the variation in phenotype. It is very difficult to design experiments that can analyze all these components simultaneously, so assumptions about some components usually are made. For example, it is often assumed that there is no covariance between genotype and environment ($2COV_{G,E}$) or G×E variance, but the well-trained geneticist will always remember that the results of such an experiment must be presented with appropriate caution.

Broad-Sense and Narrow-Sense Heritability

One of the most important questions for quantitative geneticists is the extent to which variation between individuals results from genetic differences. Thus, they are interested in how much of the phenotypic variance, V_P, can be attributed to genetic variance, V_G. This quantity is called the **broad-sense heritability** and can be thought of as how much of the stick of variation is made up of genetic variance. Broad-sense heritability is calculated as a proportion:

$$\text{Broad-sense heritability} = H_B^2 = \frac{V_G}{V_P}$$

Heritability of a trait can range from 0 to 1. A broad-sense heritability of 0 indicates that none of the variation in phenotype among individuals results from genetic differences, while a broad-sense heritability of 1 suggests that all the phenotypic variance is genetically determined. Broad-sense heritability ignores partitioning the genetic variance into additive, dominance, or interactive components, and assumes that genotype by environment interaction ($V_{G\times E}$) is not important.

More frequently, we are interested in the proportion of the phenotypic variation that results only from additive genetic effects. This is because of the important relationship between additive genetic variation and both artificial and natural selection. To understand the reason for this, recall that only in the case of additive interactions between alleles can we unambiguously determine an individual's genotype from their phenotype. With either dominance or epistasis, we must know either the genotypes of the parents of the individual or conduct testcrosses in hope of knowing an individual's genotype. Only the additive portion of genetic variation allows accurate predictions of the average phenotype of the offspring from the phenotype of an individual. Dominance and epistasis thus represent added complications and expense to plant and animal breeders, and a hindrance to evolution by natural selection.

Since the additive genetic variance allows one to make accurate predictions about the resemblance between parents and offspring, quantitative geneticists frequently determine the **narrow-sense heritability,** which is the proportion of the phenotypic variance that results only from additive genetic variance:

$$\text{Narrow-sense heritability} = H_N^2 = \frac{V_A}{V_P}$$

Because additive genetic variance determines resemblance across generations in a predictable way, it is also the variation that responds to selection in a predictable way. For this reason the narrow-sense heritability provides valuable information about how a trait will evolve under natural selection for the trait or how it can be modified through artificial selection.

Understanding Heritability

Despite their utility, heritability estimates have several significant limitations. Unfortunately, these limitations are often ignored, making heritability one of the most misunderstood and widely abused concepts in genetics. Before we discuss how heritability is determined and used, it is important to list some of the important qualifications and limitations of heritability:

1. **Broad-sense heritability does not define the complete genetic basis of a trait.** What broad-sense heritability does measure is the *proportion of the phenotypic variance* that results from genetic differences among individuals in a specific population. Estimates of broad-sense heritability for a population depend on genetic variation, which may or may not be present. When a trait is not variable, such as the number of eyes or ears in humans, we cannot estimate the heritability, but it is determined by genes. If all individuals in a population have identical genes at the loci that control the trait, then the genetic variance is zero ($V_G = 0$). Although the heritability in this case is zero, it would be incorrect to assume that genes play no role in the development of the trait. Similarly, a high heritability does not negate the importance of environmental factors influencing a trait; a high heritability might simply mean that the environmental factors that influence the trait are uniform among the individuals studied.

2. **Heritability does not indicate what proportion of an individual's phenotype is genetic.** Since it is based on the variance, which can be calculated only for a group of individuals, heritability is a characteristic of a population. An individual does not have heritability, a population does.

3. **Heritability is not fixed for a trait.** The heritability value for a trait depends on the genetic makeup and the specific environment of the population.

   To illustrate this point, suppose that we calculated broad-sense heritability for adult height in individuals living in a small New England town. A value of 0.7 would indicate that 70 percent of the variation in adult height among these individuals results from genetic variation. Heritability for other populations might not be the same, even assuming that all populations are large enough to avoid sample size effects. Residents of San Francisco, for example, might be more heterogeneous than the inhabitants of a small New England town; therefore, the San Francisco population would have more genetic variation for height. If we assume the environmental variances of the two populations are similar, but the genetic variance is greater in San Francisco, then heritability calculated for the height of San Francisco residents also would be greater.

   Genes are not the only factors that influence height in humans. Diet, an environmental effect, is also a major determinant of height. Since most individuals in our small New England town probably receive an adequate diet, at least in terms of calories, this part of the environmental variance for height would not be large. In a developing nation, however, some individuals might receive adequate nutrition, whereas the diet of others might be severely deficient. Since greater differences in diet exist, the environmental variance for height would be larger, and as a result, the heritability of height would be less. Thus, heritability calculated for human height might differ substantially for residents of the small New England town, residents of San Francisco, and residents of a developing country.

   These examples illustrate that heritability can be applied only to a specific group of individuals in a specific environment. If the genetic composition of the group is different, or the environment is different, heritability estimates cannot be transferred. Changing groups or environments does not alter the way in which genes affect the trait, but it may change the amount of genetic and environmental variance for the trait, which would then alter the heritability.

4. **High heritability for a trait does not imply that differences between populations for the same trait are genetically determined.** For example, suppose that you obtain some genetically variable

mice and divide them into two groups. You feed one group a nutritionally rich diet, and you are careful to provide each mouse with exactly the same amount of food, living space, water, and other environmental necessities. The mice grow to a large size because of the rich diet. When you measure heritability for adult body weight, you obtain a high value of 0.93. The high heritability is not surprising because the mice were genetically variable and environmental differences were kept at a minimum. The second group of mice comes from the same genetic stock, but you feed them an impoverished diet, lacking in calories and essential nutrients; again, each mouse gets exactly the same amount of food, living space, water, and necessities. Because of the poor diet, the mice of this second group are all smaller than those in the first group. When you calculate heritability for adult weight in the small mice, you again obtain a high value of 0.93 because the mice were genetically variable and the environmental differences were kept to a minimum.

Because the heritability of body weight is high in both groups and the mice of the two groups differ in adult weight, some people might suggest that the two groups of mice are genetically different with respect to body size. Yet any claim that the mice of the two groups differ genetically is clearly wrong: Both groups came from the same stock. The important point is that heritability cannot be used to draw conclusions about the nature of differences between populations. If we draw an analogy between this example of body size in mice and book-reading ability in humans, we see how easy it is to misapply quantitative genetic approaches to socially loaded human issues. Let's say we had two groups of humans and determined that variation in book-reading ability within each group had a high heritability. One group was raised in a book-rich environment, and most individuals could read well. The other group was raised in a book-poor environment, and individuals read poorly. What conclusions would you draw about the genetic differences between the two populations? Can social intervention programs enhance overall book-reading levels, or is it hopeless because book-reading ability is "genetic"?

5. **Traits shared by members of the same family do not necessarily have high heritability.** A characteristic that is shared by members of a family is referred to as a **familial trait.** As mentioned earlier, familial traits may arise because family members share genes or because they are exposed to the same environmental factors. Thus, familiality is not the same as heritability.

K E Y N O T E

The broad-sense heritability of a trait represents the proportion of the phenotypic variance in a group that results from genetic differences between individuals. Narrow-sense heritability measures only the proportion of the phenotypic variance that results from additive genetic variance. Narrow-sense heritability is the part of the phenotypic variance that responds to natural or artificial selection in a predictable manner, allowing quantitative geneticists to make predictions about the resemblance between parents and offspring.

How Heritability Is Calculated

Several different methods are available for calculating heritability that involve comparing individuals with different degrees of relatedness. All of these comparisons are based on the premise that if we control the environment, and genes are important in determining the phenotypic variance, then closely related individuals should be more similar in phenotype because they have more genes in common. Alternatively, if environmental factors are responsible for determining differences in the trait, then related individuals should be no more similar in phenotype than unrelated individuals. An important point to remember is that the related individuals studied must not share a more similar environment than unrelated individuals do, as this would increase the covariance. Manipulation of environmental factors can often be achieved in domestic plants and animals, and environments may vary among family members in the wild, providing the requisite conditions for heritability experiments. These conditions are very difficult to obtain in humans, however, where family structure and extended parental care create common environments for many related individuals. If related individuals share a more common environment than unrelated individuals, separating the effects of genes and environment is practically impossible.

Heritability from Parent-Offspring Regression. If the additive genetic component of variation is important in determining the differences among individuals, then we expect that offspring should resemble their parents. To quantify the degree to which genes influence a trait, we can measure the phenotypes of parents and offspring in a series of families and then statistically analyze the relationship between their phenotypes using correlation and regression. An important quantity in this and other analyses is called the midparent value, which is the mean of the two parents' phenotypic values. If the variation between parents is due completely to additive genetic variation, then the midparent value predicts the mean phenotype of the offspring.

Figure 14.11

Three hypothetical regressions of mean parental wing length on mean offspring wing length in *Drosophila*. In each case, the slope of the regression line (*b*) equals the narrow-sense heritability (H_N^2). (See text for explanation.)

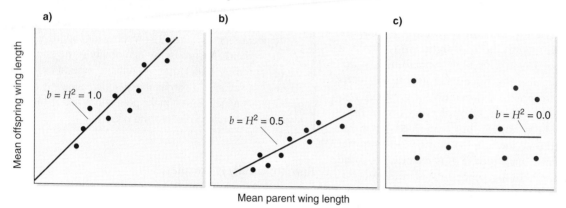

Response to Selection

We can represent the relationship between offspring phenotype and parental phenotype by plotting the midparent value against the mean phenotype of the offspring, as shown in Figure 14.11. In this graph, each point represents one family. The slope of the regression line of midparent values and mean offspring phenotype provides information about the magnitude of the narrow-sense heritability. When the slope of the parent-offspring regression is 1, as in Figure 14.11a, the mean offspring phenotype is exactly intermediate to the phenotype of the two parents, and genes with additive effects determine all the phenotypic differences (assuming that no common environmental effects between parents and offspring have influenced the trait). If the slope is less than 1 but greater than 0, as in Figure 14.11b, additive genes, genes with dominance or epistasis, and environmental factors all likely affect phenotypic variation. If the points are randomly scattered across the plot and the slope is 0, as in Figure 14.11c, then the narrow-sense heritability H_N^2 is 0.

When the mean phenotype of the offspring is regressed against the phenotype of only one parent, the narrow-sense heritability is twice the slope because an offspring shares only half its genes with one of its parents. Similarly, other combinations of relatives can be used (full and half sibs, identical and fraternal twins, etc.); in these cases, the factor by which the slope must be multiplied to obtain heritability increases with increased distance between relatives.

Heritability values for several traits in different species are given in Table 14.5. Most estimates of heritability have large standard errors, and heritabilities calculated for the same traits in the same organism often vary widely. Thus, heritability values calculated for human traits must be viewed with special caution, given the difficulties of separating genetic and environmental influences.

Quantitative genetics has played a particularly important role in evolutionary biology and plant and animal breeding. Both fields are concerned with genetic change within groups of organisms, the definition of **evolution.** Evolution by **natural selection** occurs because individuals with certain traits leave more offspring than others. Humans bring about evolution in domestic plants and animals through the similar process of **artificial selection,** where only superior individuals are used to create agricultural varieties. If the selected traits have a genetic basis in either situation, then the genetic structure of the selected population will change over time and evolve. Artificial selection can be a powerful tool in bringing about rapid evolutionary change, as evidenced by the extensive variation observed in domesticated plants and animals. For example, all breeds of domestic dogs are derived from wolves. The large number of breeds that exist today, encompassing a tremendous variety of sizes, shapes, colors, and even behaviors, has been produced by artificial selection and breeding during the past 10,000 years.

Both natural selection and artificial selection depend on the presence of genetic variation. Furthermore, the amount and the type of genetic variation present are crucial in determining how rapidly change can occur. Therefore, both evolutionary biologists and plant and animal breeders use quantitative genetics to estimate the amount of genetic variation and to predict the rate and magnitude of genetic change.

Estimating the Response to Selection

If genetic variation underlying a trait is present in a population, and natural or artificial selection is imposed on a phenotype, then the mean value of the phenotype in the population changes from one generation to the next.

Table 14.5 Heritability Values for Some Traits in Humans, Domesticated Animals, and Natural Populations [a]

Organism	Trait	Heritability
Humans	Stature	0.65
	Serum immunoglobulin (IgG) level	0.45
Cattle	Milk yield	0.35
	Butterfat content	0.40
	Body weight	0.65
Pigs	Back-fat thickness	0.70
	Litter size	0.05
Poultry	Egg weight	0.50
	Egg production (to 72 weeks)	0.10
	Body weight (at 32 weeks)	0.55
Mice	Body weight	0.35
Drosophila	Abdominal bristle number	0.50
Jewelweed	Germination time	0.29
Milkweed bugs	Wing length (females)	0.87
	Fecundity (females)	0.50
Spring peeper (frog)	Size at metamorphosis	0.69
Wood frog	Development rate (mountain population)	0.31
	Size at metamorphosis (mountain population)	0.62

[a]The estimates given in this table apply to particular populations in particular environments; heritability values for other individuals may differ.

The amount that the phenotype changes in one generation is called the response to selection or **selection response, R.**

To illustrate the concept of selection response, imagine a geneticist working to produce a strain of *Drosophila melanogaster* with large body size. The geneticist starts by examining flies from a genetically diverse population and, measuring the body sizes, finds the mean body weight in the population to be 1.3 mg. Suppose that at this point, the geneticist divides the population in half at random with respect to body size. One half of the population is allowed to interbreed normally. In the other half, the geneticist selects only the flies with large bodies (assume the mean body weight of the selected flies is 3.0 mg), and places them in a separate culture vial to interbreed. After both F_1 offspring (selected and unselected) emerge, the geneticist measures the body weights of both F_1 populations. If genetic variation underlies the variation in body size in the original population, the offspring of the selected flies should resemble their parents, and the mean body size of the selected F_1 population will be greater than the mean body size of the unselected F_1 population. The selection response for body size can then be calculated as the difference between the mean body size in the selected F_1 population minus the mean body size of the unselected F_1 population. If the selected F_1 flies have a mean body weight of 2.0 mg and the unselected F_1 flies again have a mean body weight of 1.3 mg, as was observed in the original population, a response to selection of 0.7 mg has occurred.

The selection response depends on two factors: the narrow-sense heritability and the **selection differential, s.** The selection differential is defined as the difference between the mean phenotype of the selected parents and the mean phenotype of the population before selection. In our example of body size in fruit flies, the original population had a mean weight of 1.3 mg, and the mean weight of the selected parents was 3.0 mg, so the selection differential is 3.0 mg − 1.3 mg = 1.7 mg. The selection response is related to the selection differential and the narrow-sense heritability by the following formula, known as the breeder's equation:

$$R = H_N^2 s$$

With values for two of the three parameters in the preceding equation, the selection response (0.7 mg) and the selection differential (1.7 mg), we can solve for the narrow-sense heritability:

$$\text{Narrow-sense heritability} = H_N^2 = \text{Selection response/selection differential}$$
$$H_N^2 = 0.7 \text{ mg}/1.7 \text{ mg} = 0.41$$

Selection experiments such as this provide another means for estimating the narrow-sense heritability.

Table 14.6	Approximate Heritabilities of Some Important Morphological and Behavioral Traits in Domestic Dogs[a]	
Phenotype		H^2
Litter size		0.1–0.2
Chest depth		0.5
Chest width		0.8
Muzzle length		0.5
Hip dysplasia		0.2–0.5
Nervousness		0.5
Hunting traits		0.1–0.3
Success as guide dog		0.5

[a]Heritabilities of some traits depend on breed.

A trait will continue to respond to selection, generation after generation, as long as heritable variation for the trait exists within the population. Recalling the example of dog evolution under domestication, Table 14.6 shows some of the heritabilities for a variety of morphological and behavioral traits in dogs. The results from an actual, long-term selection experiment on phototaxis in *Drosophila pseudoobscura* are presented in Figure 14.12. Phototaxis is a behavioral response to light. In this study, flies were scored for the number of times each moved toward light in a total of 15 light-dark choices. Two different experiments were carried out. In one, attraction to light was selected, and in the other, avoidance of light was selected. As can be seen in Figure 14.12, the fruit flies

Figure 14.12

Selection for phototaxis in *Drosophila pseudoobscura.* The upper graph is the line selected for avoidance of light. The lower graph is the line selected for attraction to light. The phototactic score is the number of times the fly moved toward the light out of a total of 15 light-dark choices.

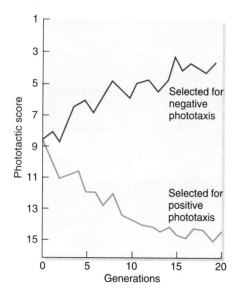

responded to selection for positive and negative phototactic behavior for a number of generations. Eventually, however, the response to selection tapered off, and finally no further directional change in phototactic behavior occurred. One possible reason for this lack of response in later generations is that no more genetic variation for phototactic behavior existed within the population. In other words, all flies at this point were homozygous for all the alleles affecting the behavior, and phototactic behavior could not undergo further evolution in this population unless input of additional genetic variation occurred. More often, some variation still exists for the trait, even after the selection response levels off, but the population fails to respond to selection because the genes for the selected trait have detrimental effects on other traits. These detrimental effects occur because of genetic correlations, which are discussed in the next section.

KEYNOTE

The amount that a trait changes in one generation as a result of selection on the trait is called the selection response. The magnitude of the selection response depends on both the intensity of selection, called the selection differential, and the narrow-sense heritability.

Genetic Correlations

When two or more phenotypes are correlated, the traits do not vary independently. For example, fair skin, blond hair, and blue eyes often are found together in the same individual. The association is not perfect—we sometimes see individuals with dark hair, fair skin, and blue eyes—but the traits are found together with enough regularity for us to say that they are correlated. The **phenotypic correlation** between two quantitative traits can be computed by measuring the two phenotypes on a number of individuals and then calculating a correlation coefficient for the two traits.

One reason for a phenotypic correlation among traits is pleiotropy (multiple phenotypic effects resulting from a single mutant gene among the loci determining the traits. (See Chapter 13.) Indeed, this is the most likely reason for the association among hair color, eye color, and skin color in humans. Genes rarely affect only a single trait, and this is particularly true for the polygenes that influence continuous traits. For example, the genes that affect growth rates in humans also influence both weight and height, so these two phenotypes tend to be correlated. Pleiotropy is one of the main causes of **genetic correlations** for quantitative traits.

Another significant cause of genetic correlations is genetic linkage. Recall that linkage is one violation of Mendel's law of independent assortment, and that the closer loci are on a chromosome, the greater the frequency that their alleles will be inherited together. When new alleles are first produced by mutation, they are associated with the other alleles that exist on that particular chromosome. These new alleles will be inherited with the other closely linked alleles, causing genetic correlations, and the persistence of these correlations over time depends in part on the amount of recombination between the loci. An important distinction between linkage and pleiotropy as causes of genetic correlations is that over evolutionary time, even the tightest linkages can be broken, allowing new associations between alleles; with pleiotropy, however, functional constraints of an individual protein might not be able to be disociated, causing a correlation to persist. Table 14.7 presents some genetic correlations that have been detected in studies of quantitative genetics.

Care must be taken when making a correlation, because beyond pleiotropy and linkage, environmental factors may also influence several traits simultaneously to cause nonrandom associations between phenotypes. For example, adding fertilizer to soil often causes plants both to grow taller and to produce more flowers. If we measured plant height and counted the number of flowers on a group of responsive plants, some of which received fertilizer and some of which did not, we would find that the two traits are correlated; plants receiving fertilizer would be tall and would have many flowers, and those without fertilizer would be short and have few flowers. This phenotype is due to gene action, but the correlation is a result of the common effect of an environmental factor, the fertilizer, on both traits.

Genetic correlations may be positive or negative. A positive correlation means that genes causing an increase in the magnitude of one trait bring about a simultaneous increase in the magnitude of the other. In chickens, body weight and egg weight have a positive genetic correlation. If breeders select for heavier chickens, both the size of the chickens and the mean weight of the eggs produced by these chickens will increase. This increase in egg weight occurs because the genes that produce heavier chickens presumably have a pleiotropic effect on egg weight. In the case of negative genetic correlations, genes that cause an increase in one trait tend to produce a corresponding decrease in another trait. For example, when breeders select for chickens that produce larger eggs, the average egg size increases, but the number of eggs laid by each chicken decreases.

Negative correlations between traits represent trade-offs or genetic constraints that must be balanced when under selection pressures. For example, faster speed is important to garter snakes for both hunting and escaping predators. In the western United States, some garter snakes prey on toxic newts that produce the neurotoxin tetrodotoxin in their skin. The negative genetic correlation between speed and neurotoxin resistance in garter snake populations shown in Figure 14.13 is apparently an evolutionary constraint. Although tetrodotoxin resistance had evolved independently at least twice, it seems there have been no mutations that have increased resistance without decreasing speed. Negative genetic correlations often place practical constraints on the ability of plant and animal breeders to make progress from selection. As an example, milk yield and butterfat content have a negative genetic correlation in cattle. The same genes that cause an increase in milk production bring about a decrease in butterfat content of the milk. Thus,

Table 14.7 **Genetic Correlations Between Traits in Humans, Domesticated Animals, and Natural Populations[a]**

Organism	Traits	Genetic Correlation
Humans	IgG, IgM	0.07
Cattle	Butterfat content, milk yield	−0.38
Pigs	Weight gain, back-fat thickness	0.13
	Weight gain, efficiency	0.69
Chickens	Egg weight, egg production	−0.31
	Body weight, egg weight	0.42
	Body weight, egg production	−0.17
Mice	Body weight, tail length	0.29
Jewelweed	Seed weight, germination time	−0.81
Milkweed bugs	Wing length, fecundity	−0.57
Wood frogs	Developmental rate, size at metamorphosis	−0.86
Drosophila	Early life fecundity, resistance to starvation	−0.91

[a]The estimates given in this table apply to particular populations in particular environments; genetic correlations for other individuals may differ.

Negative genetic correlation ($r = -0.45$) between speed and resistance to tetrodotoxin in garter snakes illustrated by family means. The correlation of family means approximates the genetic correlation when families are large.

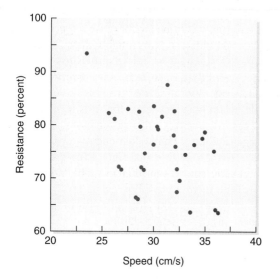

when breeders select for increased milk yield, the amount of milk produced by the cows may go up, but the butterfat content decreases. Knowing the amount and type of genetic correlations before undertaking a breeding program is essential to ensure success.

An organism's ability to adapt to a particular environment is strongly influenced by genetic correlations among traits, as we saw in the garter snake example; therefore, genetic correlations are of great interest to evolutionary biologists. As another illustration, consider two traits in tadpoles: developmental rate and size at metamorphosis. Most tadpoles are found in small ponds and pools, where fish (potential predators) are absent and food is abundant. A major liability in using this type of aquatic habitat is that ponds often dry up, frequently before the tadpoles have developed sufficiently to metamorphose into frogs and leave the water. One might expect, then, that natural selection would favor a maximum rate of development in tadpoles, so that the tadpoles could quickly metamorphose into frogs. However, many species of tadpoles fail to develop at maximum rates, contrary to this prediction. One reason for a slower rate of development is a negative genetic correlation between developmental rate and body size at metamorphosis. Genes that accelerate development also tend to cause metamorphosis at a smaller size, at least in some populations. Thus, selection for fast metamorphosis also produces smaller frogs, and size is extremely important in determining the survival of young frogs. Small frogs tend to lose water more rapidly in the terrestrial environment, are more likely to be eaten by predators, and have more difficulty finding sufficient food. The negative genetic correlation between developmental rate and body size at metamorphosis places constraints on the frogs' ability to develop rapidly and to attain a large body size at metamorphosis. Knowing about such genetic correlations is important for understanding how animals adapt or fail to adapt to a particular environment.

KEYNOTE

Genetic correlations arise from pleiotropy or linkage. When a trait is selected, any genetically correlated traits also exhibit a selection response. Thus, the evolution of a population from artificial breeding or in response to natural selection depends on the simultaneous integration of many aspects of the phenotype.

You are a researcher trying to determine whether fingerprint patterns are correlated with high blood pressure in the iActivity *Your Fate in Your Hands?* on your CD-ROM.

Quantitative Trait Loci

The statistical approach to understanding quantitative inheritance has been useful in analyzing components of variation and response to selection. However, to understand quantitative traits fully, we need to characterize the individual quantitative trait loci (QTLs) that affect them. Recent advances combining molecular genotyping and statistical tools have enabled geneticists to begin to make the connection between quantitative phenotypes and the specific QTLs that control them.

QTL identification is an exercise in finding segments of the genome associated with phenotypic differences between individuals. As such, it is most powerful when analyzing a population with a detailed linkage map, substantial phenotypic variation, and a large numbers of individuals. Typically, inbred lines that have been selected for differing phenotypes are crossed and then either backcrossed, intercrossed to generate an F_2, or intercrossed and selfed to create a series of recombinant inbred strains. The population is then grown, measured, and genotyped. While the analytical methods used to determine which genomic regions are correlated with phenotypic variation are increasingly sophisticated, the essence of finding QTL is to split the individuals into groups on the basis of a marker genotype, then test to see whether the groups have similar or different means. In fact, the earliest QTL identification methods took the genotypic and phenotypic data from a population and conducted an ANOVA using each marker as a factor and the phenotype as the response. If a marker locus is unlinked to a QTL, the average phenotype is the same for all genotype classes. If the marker locus is linked to a QTL, then genotypes should differ in their mean value of the trait examined. The difference in phenotypic means for the marker genotype classes depends on the size of the effect of a QTL and also on how tightly linked the QTL is with the marker.

Because of the many QTL identification studies conducted to date, we can now begin to understand not only how many QTLs underlie these traits but also the magnitude of their effects and distribution in the genome. Some of the most significant applications of this work have led to the identification of QTLs responsible for important agronomic traits and QTLs responsible for adaptive differences between closely related species. For example, some of the most important differences between closely related plant

species are the suites of floral traits that attract pollinators, including color, shape, and nectar rewards. Monkeyflowers have diverged into hummingbird-pollinated species, such as *Mimulus cardinalis*, exhibiting red coloration, deep tubular flowers with lots of nectar at the base, and flared back (reflexed) petals, and species such as *M. lewisii* with little nectar reward, broad petal landing pads, and pink flowers characteristic of bee-pollinated flowers (Figure 14.14). Crosses between these species have revealed that differences in each of the pollinator attraction and efficiency traits appear to be controlled by at least one QTL that influences 25 percent or more of the phenotypic variation (see Figure 14.15). This finding, along with data showing substantial adaptive differences between these QTL alleles, suggests that, in at least some cases, important phenotypic shifts between species may have occurred by mutations in

Figure 14.14

***Mimulus lewisii* (A, C) and *M. cardinalis* (B, D) flowers.** Flowers are shown from the front (A, B) as an approaching pollinator views them. In side views (C, D), the relative positions of the stigma and anthers are shown.

Figure 14.15

QTL maps for 12 floral traits in monkeyflowers. Boxes show the position of markers correlated to phenotypic traits. Taller boxes indicate QTLs that explain ≥25 percent of the variance in a trait.

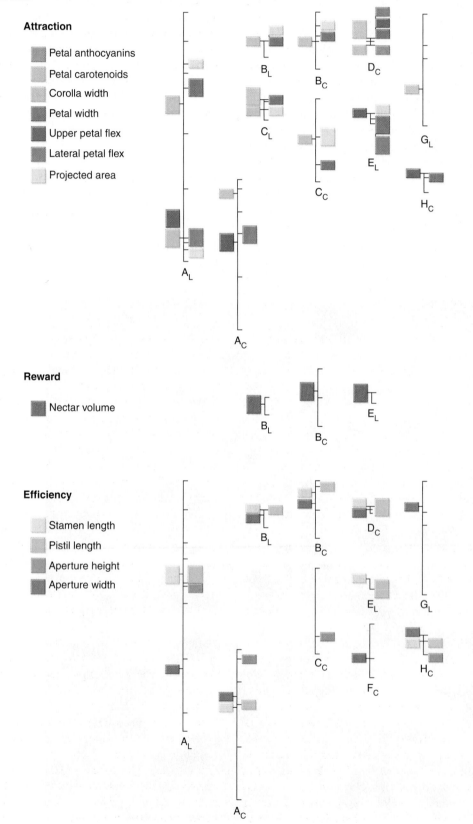

single loci. Interestingly, many of the QTLs for different traits are physically close, suggesting that floral characters are genetically correlated.

Moving from associated genomic regions to cloning the actual QTL depends on the availability of a suite of molecular tools. Some of the better examples of QTL that have been characterized to date come from intensively studied model species, such as corn, tomato (*Solanum lycopersicum*), and *Drosophila melanogaster*. In corn, the *teosinte branched 1 (tb1)* QTL controlling the number of axillary branches was cloned in 1997. Branching is a key difference between cultivated corn and its wild relative teosinte, with both evolutionary and commercial importance. *tb1* has been shown to be a member of a DNA-binding transcriptional regulator gene family that acts in corn to suppress growth in specific tissues. The important allelic differences between corn and teosinte that appear to have been selected during domestication of maize are in a region of the gene that is not transcribed. The tomato QTL *fw2.2*, which controls up to 30 percent of the difference in fruit weight seen in crosses between cultivated and wild tomato species, was cloned in 2000. The FW2.2 protein is expressed early in fruit development, and variation between alleles has been found in the timing and levels of gene expression, but there are not major differences in the FW2.2 protein sequence. In addition, there appears to be a genetic correlation caused by a pleiotropic effect of the small fruit allele: Isogenic lines that are homozygous for this allele also produce more fruits than plants that are homozygous for the large fruit allele.

Another approach to finding QTLs involves testing for associations between phenotypic differences and allelic variation at candidate loci. Candidate loci can be defined on the basis of known or suggested function, proximity to genomic regions implicated in QTL studies, or both. In *Drosophila*, for example, mapping experiments defined a QTL for sternopleural bristle number in a region of the genome that also contained a key developmental gene, *hairy (h)*, whose classically defined mutants have extra bristles. Surveys of natural variation in bristle number and molecular variation at the *h* locus have shown that *h* is indeed a QTL for bristle number. Analysis of candidate loci does not, however, always yield similar results. An experiment to determine whether variation in the structural genes of the carotenoid biosynthetic pathway affected mature red fruit color in tomato showed that most of the QTL did not correspond to these genes. Continued refining of phenotypes and understanding of metabolic and genetic pathways, combined with new analytical techniques, will be crucial for identifying and analyzing QTLs for years to come.

The genes underlying quantitative traits in humans cannot be determined through traditional pedigree analysis as they are with discontinuous traits, because environmental differences through time and the action of other segregating genes tend to obscure the effects of single loci. However,

important segments of the genome that play a part in phenotypic variation can still be identified through association studies. Association studies utilize widely distributed DNA markers (sequences at particular locations in the genome that vary in the population) in the human genome and populations showing a trait and random control populations to look for QTLs. The most useful DNA markers here are SNPs. (See Chapter 9, pp. 216–218.) The methodology for these studies is roughly similar to experimental populations, in that individuals are phenotyped and genotyped, then statistical methods are used to associate QTL and SNPs.

KEYNOTE

Marker-based mapping approaches can be used to correlate segments of the genome with phenotypic variation in quantitative traits in natural and experimental populations. Studies identifying quantitative trait loci (QTLs) provide estimates of the number of genes and size and mechanism of effects influencing variation in continuous traits.

Summary

The field of quantitative genetics studies the inheritance of continuous, or quantitative, traits, and strives to clarify the relationship between the environment, complex phenotypes, and complex genetic architecture. Continuous phenotypes usually result from the influence of multiple genes and environmental factors. Statistics such as the mean, variance, and standard deviation can be used to describe continuous traits; correlation, regression, and analysis of variance are statistical techniques that are used to study continuous traits.

The multiple loci that affect a quantitative trait all follow the principles of Mendelian inheritance. The multiple-gene hypothesis assumes that the effects of the alleles across loci are small and additive, but specific quantitative trait loci (QTLs) that determine variation in continuous traits identified through marker-based mapping have shown that QTL with large effects are quite common.

The broad-sense heritability is the proportion of the phenotypic variance in a population due to genetic differences. The narrow-sense heritability includes only the proportion of the phenotypic variance that results from additive genetic variance. The response to selection is the amount a trait changes in one generation as a result of the narrow-sense heritability and the selection differential. When a trait responds to selection, traits genetically correlated due to linkage or pleiotropy will also exhibit a selection response. These factors greatly influence the direction and speed of both natural and artificial selection. Responses from the action of artificial or natural selection depend on narrow-sense heritability and integration of the phenotype.

Analytical Approaches to Solving Genetics Problems

Q14.1 Assume that loci *A, B, C,* and *D* are members of a multiple-gene series that controls a quantitative trait, and that each gene assorts independently. Each *A, B, C,* and *D* allele has a cumulative effect that contributes 3 cm of height when present, and alleles *a, b, c,* and *d* do not contribute anything to the height of the organism. In addition, gene *L* is always present in the homozygous state, and the *LL* genotype contributes a constant 40 cm of height. If we ignore height variation caused by environmental factors, an organism with genotype *AA BB CC DD LL* would be 64 cm high, and one with genotype *aa bb cc dd LL* would be 40 cm. A cross is made of *AA bb CC DD LL* × *aa BB cc DD LL* and is carried into the F_2 by selfing of the F_1.

a. How does the size of the F_1 individuals compare with the size of each of the parents?

b. Compare the mean of the F_1 with the mean of the F_2, and comment on your findings.

c. What proportion of the F_2 population would show the same height as the *AA bb CC DD LL* parent?

d. What proportion of the F_2 population would show the same height as the *aa BB cc DD LL* parent?

e. What proportion of the F_2 population would breed true for the height shown by the *aa BB cc DD LL* parent?

f. What proportion of the F_2 population would breed true for the height characteristic of F_1 individuals?

A14.1 This question explores our understanding of the basic genetics involved in a multiple-gene series that controls a quantitative trait. The approach we will take is essentially the same as the approach used with a series of independently assorting genes that controls different traits. That is, we make predictions on the basis of genotypes and relate the results to phenotypes, or we make predictions on the basis of phenotypes and relate the results to genotypes.

a. Each allele represented by a capital letter contributes 3 cm of height to the base height of 40 cm provided by *LL* homozygosity. Therefore, the *AA bb CC DD LL* parent, which has six capital-letter alleles from the *A–D* multiple-gene series, is $40 + (6 \times 3) = 58$ cm high. Similarly, the *aa BB cc DD LL* parent has four capital-letter alleles and therefore is $40 + 12 = 52$ cm high. The F_1 from a cross between these two individuals would be heterozygous for the *A, B,* and *C* loci and homozygous for *D* and *L*, that is, *Aa Bb Cc DD LL*. This progeny has five capital-letter alleles apart from *LL* and therefore is $40 + 15 = 55$ cm high.

b. The F_2 is derived from a self of the *Aa Bb Cc DD LL* F_1. All the F_2 individuals will be *DD LL*, making

them at least $40 + 6 = 46$ cm high. Now we must deal with the heterozygosity at the other three loci. What we need to calculate is the relative frequencies of all possible genotypes for the three loci and collect those with zero, one, two, three, four, five, and six capital-letter alleles. The probability of getting an individual with two capital-letter alleles for a particular locus is $\frac{1}{4}$, the probability of getting an individual with one capital-letter allele for the locus is $\frac{1}{2}$, and the probability of getting an individual with no capital-letter alleles for the locus is $\frac{1}{4}$. Therefore, the probability of getting an F_2 individual with six capital-letter alleles for the *A, B,* and *C* loci is $(\frac{1}{4})^3 = \frac{1}{64}$, and the same probability is obtained for an individual with no capital-letter alleles. The simplest approach to find the expected numbers for genotypes with one to five capital-letter alleles is to compute the coefficients in the binomial expansion of $(a + b)^6$. Recall that in the wheat kernel color example, a phenotypic ratio of 1:4:6:4:1 was observed. These numbers are the coefficients in the binomial expansion of $(a + b)^4$, and correspond to the numbers of different genotypes with zero, one, two, three, and four contributing alleles. Similarly, the expansion of $(a + b)^6$ gives a 1:6:15:20:15:6:1 distribution of zero, one, two, three, four, five, and six capital-letter alleles, respectively. Since each capital-letter allele in the *A, B,* and *C* set contributes 3 cm of height over the 46-cm height given by the *DD LL* genotype common to all, the F_2 individuals would fall into the following distribution:

Number of Capital-Letter Alleles	Height Added to 46 cm *DD LL* Genotype (cm)	Height of Individuals (cm)	Number Expected per 64 F_2
6	18	64	1
5	15	61	6
4	12	58	15
3	9	55	20
2	6	52	15
1	3	49	6
0	0	46	1

The distribution is clearly symmetrical, giving an average of 55 cm, the same height shown in F_1 individuals.

c. The *AA bb CC DD LL* parent was 58 cm, so we can read the proportion of F_2 individuals that show this same height directly from the table in part (b). The answer is $\frac{15}{64}$.

d. The *aa BB cc DD LL* parent was 52 cm, and from the table in part (b) the proportion of F_2 individuals that show this same height is $\frac{15}{64}$.

e. We are asked to determine the proportion of the F_2 population that would breed true for the height shown by the *aa BB cc DD LL* parent, which was 52 cm. To breed true, the organism must be homozygous. We have also established that *DD LL* is a constant genotype for the F_2 individuals, giving a basic height of 46 cm. Therefore, for a height of 52 cm, two additional, active, capital-letter alleles must be present apart from those at the *D* and *L* loci. With the requirement for homozygosity there are only three genotypes that give a 52-cm height; they are *AA bb cc DD LL*, *aa BB cc DD LL*, and *aa bb CC DD LL*. The probability of each combination occurring in the F_2 is $\frac{1}{64}$, so the answer to the problem is $\frac{1}{64} + \frac{1}{64} + \frac{1}{64} = \frac{3}{64}$. (Note that the individual probability for each genotype can be calculated. That is, probability of $AA = \frac{1}{4}$, probability of $bb = \frac{1}{4}$, probability of $cc = \frac{1}{4}$, and probability of $DD\ LL = 1$, giving an overall probability for *AA bb cc DD LL* of $\frac{1}{64}$.)

f. We are asked to determine the proportion of the F_2 population that would breed true for the height characteristic of F_1 individuals. Again, the basic height given by *DD LL* is 46 cm. The F_1 height is 55 cm, so three capital-letter alleles must be present in addition to *DD LL* to give that height because (3×3) cm $= 9$ cm, and 9 cm $+$ 46 cm $=$ 55 cm. However, because an individual must be homozygous to be true-breeding, the answer to this question is none, because 3 is an odd number, meaning that at least one locus must be heterozygous to get the 55-cm height.

Q14.2 Five field mice collected in Texas had weights of 15.5 g, 10.3 g, 11.7 g, 17.9 g, and 14.1 g. Five mice collected in Michigan had weights of 20.2 g, 21.2 g, 20.4 g, 22.0 g, and 19.7 g. Calculate the mean weight and the variance in weight for mice from Texas and for mice from Michigan.

A14.2 To answer this question, we use the formula given in the section "Statistical Tools." The formula for the mean is

$$\bar{x} = \frac{\sum x_i}{n}$$

The symbol Σ means to add, and the x_i represents all the individual values. We begin by summing up all the weights of the mice from Texas:

$$\sum x_i = 15.5 + 10.3 + 11.7 + 17.9 + 14.1 = 69.5$$

Next, we divide this summation by n, which represents the number of values added together. In this case, we added together five weights, so $n = 5$. The mean for the Texas mice is therefore

$$\frac{\sum x_i}{n} = \frac{69.5}{5} = 13.9$$

To calculate the variance in weight among the Texas mice, we use the formula

$$s^2 = \frac{\sum (x_i - \bar{x})^2}{n - 1}$$

We must take each individual weight and subtract it from the mean weight of the group. Each value obtained from this subtraction is then squared, and all squared values are added up, as shown below.

$15.1 - 13.9 =$	1.6	$(1.6)^2 = 2.56$
$10.3 - 13.9 =$	-3.6	$(-3.6)^2 = 12.96$
$11.7 - 13.9 =$	-2.2	$(-2.2)^2 = 3.84$
$17.9 - 13.9 =$	4.0	$(4.0)^2 = 16.00$
$14.1 - 13.9 =$	0.2	$(10.2)^2 = 0.04$
		36.4

The sum of all the squared values is 36.4. All that remains for us to do is to divide this sum by $n - 1$, which is $5 - 1 = 4$.

$$s^2 = \frac{\sum (x_i - \bar{x})^2}{n - 1} = \frac{36.4}{4} = 9.1$$

The mean and the variance for the Texas mice are 13.9 and 9.1.

We now repeat these steps for the mice from Michigan.

$$\sum x_i = 20.2 + 21.2 + 20.4 + 22.0 + 19.7 = 103.5$$

$$\frac{\sum x_i}{n} = \frac{103.5}{4} = 20.7$$

$$s^2 = \frac{\sum (x_i - \bar{x})^2}{n - 1}$$

$20.2 - 20.7 =$	-0.5	$(-0.5)^2 = 0.25$
$21.2 - 20.7 =$	0.5	$(0.5)^2 = 0.25$
$20.4 - 20.7 =$	-0.3	$(-0.3)^2 = 0.09$
$22.0 - 20.7 =$	1.3	$(1.3)^2 = 1.69$
$19.7 - 20.7 =$	-1.0	$(-1.0)^2 = 1.0$
		3.28

$$s^2 = \frac{\sum (x_i - \bar{x})^2}{n - 1} = \frac{3.28}{4} = 0.82$$

The mean and the variance for the Michigan mice are 20.7 and 1.023.

We conclude that the Michigan mice are much heavier than the Texas mice, and the Michigan mice also exhibit less variance in weight.

Questions and Problems

*14.1 The following measurements of head width and wing length were made on a series of steamer ducks:

Specimen	Head Width (cm)	Wing Length (cm)
1	2.75	30.3
2	3.20	36.2
3	2.86	31.4
4	3.24	35.7
5	3.16	33.4
6	3.32	34.8
7	2.52	27.2
8	4.16	52.7

a. Calculate the mean and the standard deviation of head width and of wing length for these eight birds.
b. Calculate the correlation coefficient for the relationship between head width and wing length in this series of ducks.
c. What conclusions can you make about the association between head width and wing length in steamer ducks?

*14.2 Given the following sets of 30 phenotypic measurements for different traits, decide whether each trait is qualitative or quantitative and explain your answer.
a. Trait 1: 38.9, 47.0, 53.1, 39.1, 62.8, 46.8, 57.5, 54.9, 48.9, 56.3, 52.5, 60.8, 46.7, 48.0, 52.3, 40.7, 50.4, 51.0, 46.5, 47.9, 55.4, 53.1, 58.5, 51.1, 60.2, 50.6, 48.6, 52.5, 54.5, 51.4, 48.1, 49.5, 55.8, 52.9, 42.9, 44.4, 56.4, 38.9, 42.2, 42.2
b. Trait 2: 25.7, 8.8, 11.2, 5.7, 20.6, 34.3, 13.0, 28.8, 20.5, 24.1, 21.2, 14.3, 17.7, 18.7, 24.3, 30.2, 20.2, 25.1, 30.6, 21.2, 31.2, 23.0, 16.9, 10.5, 14.1, 10.2, 30.5, 22.5, 34.1, 10.6, 19.5, 21.0, 20.9, 27.7, 33.0, 7.7, 20.1, 16.9, 18.8, 15.7
c. Trait 3: 31.1, 22.0, 28.1, 14.1, 43.4, 52.8, 32.5, 39.0, 43.1, 52.2, 45.1, 35.8, 36.4, 38.7, 52.8, 42.6, 42.6, 54.8, 43.4, 45.1, 45.1, 49.5, 34.2, 26.1, 35.2, 25.6, 43.1, 48.3, 52.2, 26.4, 40.9, 44.5, 44.3, 36.4, 49.5, 19.4, 42.4, 34.2, 39.0, 31.1

*14.3 The F_1 generation from a cross of two pure-breeding parents that differ in a size character usually is no more variable than the parents. Explain.

*14.4 Two pairs of independently segregating genes with two alleles each, A/a and B/b, determine plant height additively in a population. The homozygote $AA\ BB$ is 50 cm tall, and the homozygote $aa\ bb$ is 30 cm tall.
a. What is your prediction of the F_1 height in a cross between the two homozygous stocks?
b. What genotypes in the F_2 will show a height of 40 cm after an $F_1 \times F_1$ cross?

c. What will be the F_2 frequency of the 40-cm plants?
d. What assumptions have you made in answering this question?

14.5 Assume that in squashes the difference in fruit weight between a 3-lb type and a 6-lb type results from three independently segregating allelic pairs, A/a, B/b, and C/c. Each capital-letter allele contributes a half pound to the weight of the squash. From a cross of a 3-lb plant ($aa\ bb\ cc$) with a 6-lb plant ($AA\ BB\ CC$), what will be the phenotypes (weights) of the F_1 and the F_2? What will be their distribution?

14.6 Refer to the assumptions stated in Problem 14.5. Determine the range in fruit weight of the offspring in the following squash crosses:
a. $AA\ Bb\ CC \times aa\ Bb\ Cc$
b. $AA\ bb\ Cc \times Aa\ BB\ cc$
c. $aa\ BB\ cc \times AA\ BB\ cc$.

14.7 Three independently segregating genes (A, B, C), each with two alleles, determine height in a plant. Each capital-letter allele adds 2 cm to a base height of 2 cm.
a. What are the heights expected in the F_1 progeny of a cross between homozygous strains $AA\ BB\ CC \times aa\ bb\ cc$?
b. What is the distribution of heights (frequency and phenotype) expected in an $F_1 \times F_1$ cross?
c. What proportion of F_2 plants will have heights equal to the heights of the original two parental strains?
d. What proportion of the F_2 will breed true for height?

14.8 Repeat Problem 14.7, but assume that one of the loci shows dominance instead of additivity.

14.9 Assume that three independently segregating, equally and additively contributing pairs of alleles control flower length in nasturtiums. A completely homozygous plant with 10-mm flowers is crossed to a completely homozygous plant with 30-mm flowers. The F_1 plants all have flowers about 20 mm long. The F_2 plants show a range of lengths from 10 to 30 mm, with about $\frac{1}{64}$ of the F_2 having 10-mm flowers and $\frac{1}{64}$ having 30-mm flowers. What distribution of flower length would you expect to see in the offspring of a cross between an F_1 plant and the 30-mm parent?

*14.10 An experiment found that the mean internode length in spikes (the floral structures) of the barley variety *asplund* to be 2.12 mm. In the variety *abed binder*, the mean internode length was found to be 3.17 mm. The mean of the F_1 of a cross between the two varieties was approximately 2.7 mm. The F_2 population included individuals similar to both parents, as well as intermediate types. Analysis of the F_3 generation showed that 8 out of the total 125 F_2 individuals of the *asplund* type were true breeding, giving a mean of 2.19 mm. Nine other F_2 individuals were similar to *abed binder*, and they bred true to type, with a mean internode length of 3.24 mm. Is the internode length in spikes of barley a discontinuous or a quantitative trait? Why?

14.11 Assume that the difference between a corn plant 10 dm (decimeters) high and one 26 dm high results from four pairs of equal and cumulative multiple alleles, with the 26-dm plants being *AA BB CC DD* and the 10-dm plants being *aa bb cc dd*. Make and detail your assumptions, then predict the following:
a. What will be the size and genotype of an F_1 from a cross between these two true-breeding types?
b. Determine the limits of height variation in the offspring from the following crosses:
  i. *Aa bb cc dd × Aa bb cc dd*
  ii. *aa BB cc dd × Aa Bb Cc dd*
  iii. *AA BB Cc DD × aa BB cc Dd*
  iv. *Aa bb cc dd × Aa bb cc dd*

14.12 Refer to the assumptions given in Problem 14.11. For this problem, two 14-dm corn plants, when crossed, give nothing but 14-dm offspring (case A). Two other 14-dm plants give one 18-dm, four 16-dm, six 14-dm, four 12-dm, and one 10-dm offspring (case B). Two other 14-dm plants, when crossed, give one 16-dm, two 14-dm, and one 12-dm offspring (case C). What genotypes for each of these 14-dm parents (cases A, B, and C) would explain these results? Would it be possible to get a plant that is taller than 18 dm by selection in any of these families?

*14.13 Transgressive segregation is the phenomenon in which two pure-breeding strains, differing in a trait, are crossed and produce F_2 individuals with phenotypes that are more extreme than either grandparent (i.e., that are larger than the largest or smaller than the smallest in the original generation). Even if two pure-breeding strains are the same for a quantitative trait, it is possible to see transgressive segregation in an F_2. Propose scenarios with specific assumptions for each of these examples of transgressive segregation.

*14.14 Pigmentation in the imaginary river-bottom dweller *Mucus yuccas* is a quantitative character controlled by a set of five independently segregating polygenes with two alleles each: *A/a, B/b, C/c, D/d,* and *E/e*. Pigment is deposited at three different levels, depending on the threshold of gene products produced by the capital-letter alleles. Greyish-brown pigmentation is seen if at least four capital-letter alleles are present, light-tan pigmentation is seen if two or three capital-letter alleles are present, and whitish-blue pigmentation is seen if these thresholds are not met. If an *AA BB CC DD EE* animal is crossed to an *aa bb cc dd ee* animal and the progeny are intercrossed, what kinds of phenotypes are expected in the F_1 and F_2?

*14.15 Alzheimer disease (AD) is the leading cause of dementia in older adults. Evidence that genetic alterations are involved in AD comes from three sources; the incidence of AD in first-degree relatives, the incidence in pairs of twins, and pedigree analysis. There is a 24–50 percent risk of AD by age 90 in first-degree relatives of individuals with AD, a 40–50 percent risk of AD in the identical (monozygotic) twin of an individual with AD, and a 10–50 percent risk of AD in the fraternal (dizygotic) twin of an individual with AD. Individuals with AD in a subset of families showing AD have an alteration in the *APP* (amyloid protein) gene on chromosome 21. Individuals with AD in another subset of AD families have a particular allele (*E4*) at the *APOE* (apolipoprotein E) gene on chromosome 19. Individuals homozygous for the *E4* allele have increased risk of AD and earlier disease onset than heterozygotes. Population studies have shown that 40–50 percent of AD cases are associated with alterations in the *APOE* gene, but less than 1 percent of AD cases are associated with mutations in the *APP* gene.
a. In what sense might AD be considered a polygenic trait?
b. If AD has a genetic basis, why are identical twins not equally affected?

14.16 Since monozygotic twins share all their genetic material and dizygotic twins share, on average, half of their genetic material, twin studies sometimes can be useful for evaluating the genetic contribution to a trait. Consider the following two instances:

An intelligence quotient (IQ) assesses intellectual performance on a standardized test that involves reasoning, ability, memory, and knowledge of an individual's language and culture. IQ scores are transformed so that the population mean score is 100 and 95 percent of the individuals have scores in the range between 70 and 130. Observations in the United States and England found that monozygotic twins had an average difference of 6 IQ points, dizygotic twins had an average difference of 11 points, and random pairs of individuals had an average difference of 21 points.

In a large sample of pairs of twins in the United States where one twin was a smoker, 83 percent of monozygotic twins both smoked, whereas 62 percent of dizygotic twins both smoked.

From these data, can you infer the genetic determination of IQ or smoking?

14.17 A quantitative geneticist determines the following variance components for leaf width in a population of wildflowers growing along a roadside in Kentucky:

Additive genetic variance (V_A) = 4.2
Dominance genetic variance (V_D) = 1.6
Interaction genetic variance (V_I) = 0.3
Environmental variance (V_E) = 2.7
Genetic-environmental variance $(V_{G \times E})$ = 0.0

a. Calculate the broad-sense heritability and the narrow-sense heritability for leaf width in this population of wildflowers.

b. What do the heritabilities obtained in (a) indicate about the genetic nature of leaf width variation in this plant?

*14.18 Members of the inbred rat strain SHR are salt sensitive: They respond to a high-salt environment by developing hypertension. Members of a different inbred rat strain, TIS, are not salt sensitive. Imagine that you placed a population consisting only of SHR rats in an environment that was variable in regard to distribution of salt, so that some rats would be exposed to more salt than others. What would be the heritability of blood pressure in this population?

14.19 In Kansas, a farmer is growing a variety of wheat called TK138. He calculates the narrow-sense heritability for yield (the amount of wheat produced per acre) and finds that the heritability of yield for TK138 is 0.95. The next year, he visits a farm in Poland and observes that another variety of wheat, UG334, growing there has only about 40 percent as much yield as the TK138 grown on his farm in Kansas. Since he found the heritability of yield in his wheat to be very high, he concludes that the TK138 wheat is genetically superior to the UG334 wheat, and he tells the Polish farmers that they can increase their yield by using TK138. Is his conclusion correct? Why or why not?

*14.20 Dermatoglyphics are the patterns of the ridged skin found on the fingertips, toes, palms, and soles of the feet. (Fingerprints are dermatoglyphics.) Classification of dermatoglyphics frequently is based on the number of triradii: A triradius is a point from which three ridge systems separate at angles of 120°. The number of triradii on all 10 fingers was counted for each member of several families, and the results are tabulated here.

Family	Mean Number of Triradii in the Parents	Mean Number of Triradii in the Offspring
I	14.5	12.5
II	8.5	10.0
III	13.5	12.5
IV	9.0	7.0
V	10.0	9.0
VI	9.5	9.5
VII	11.5	11.0
VIII	9.5	9.5
IX	15.0	17.5
X	10.0	10.0

a. Calculate the narrow-sense heritability for the number of triradii by the regression of the mean phenotype of the parents against the mean phenotype of the offspring.

b. What does your calculated heritability value indicate about the relative contributions of genetic variation and environmental variation to the differences observed in number of triradii?

14.21 The heights of nine college-age males and the heights of their fathers are presented here.

Height of Son (inches)	Height of Father (inches)
70	70
72	76
71	72
64	70
66	70
70	68
74	78
70	74
73	69

a. Calculate the mean and the variance of height for the sons and for the fathers.

b. Calculate the correlation coefficient for the relationship between the height of father and height of son.

c. Determine the narrow-sense heritability of height in this group by regression of the son's height on the height of father.

*14.22 A scientist wants to determine the narrow-sense heritability of tail length in mice. He measures tail length among the mice of a population and finds a mean tail length of 9.7 cm. He then selects the 10 mice in the population with the longest tails: Mean tail length in these selected mice is 14.3 cm. He interbreeds the mice with the long tails and examines tail length in their progeny. The mean tail length in the F_1 progeny of the selected mice is 13 cm.

Calculate the selection differential, the response to selection, and the narrow-sense heritability for tail length in these mice.

14.23 Assume that all phenotypic variance in seed weight in beans is genetically determined and is additive. From a population in which the mean seed weight was 0.88 g, a farmer selected two seeds, each weighing 1.02 g. He planted these and crossed the resulting plants to each other, then collected and weighed their seeds. The mean weight of their seeds was 0.96 g. What is the narrow-sense heritability of seed weight?

14.24 The narrow-sense heritability of egg weight in a particular flock of chickens is 0.60. A farmer selects for

Figure 14.A

Parental White Leghorn Strain (196 eggs, 51 g)

→ Select Set A (279 eggs, 57 g)   → Select Set B (310 eggs, 53 g)
  ↓   ↓
  F_1 (208 eggs, 54 g) F_1 (217 eggs, 52 gm)
    → Select F_1 (271 eggs, 60 g)   → Select F_1 (310 eggs, 55 g)
      ↓   ↓
      F_2 (214 eggs, 55 g) F_2 (224 eggs, 53 g)
        → Select F_2 (292 eggs, 61 g)   → Select F_2 (315 eggs, 57 g)
          ↓   ↓
          Line A (218 eggs, 56 g)   Line B (230 eggs, 54 g)

Line A × Line B → Hybrid used for commercial production (262 eggs, 57 g)

increased egg weight in this flock. The difference in the mean egg weight of the unselected chickens and the selected chickens is 10 g. How much should egg weight increase in the offspring of the selected chickens?

14.25 Members of a strain of white leghorn chickens are selectively crossed to produce two lines, A and B, that show improved egg production. The progeny from a cross of lines A and B are used for commercial egg production. The selection strategy is shown in Figure 14.A. The mean number of eggs produced in the first egg production year and the mean egg weight (in grams) from hens at an age of 240 days is given for animals at each step of the selection procedure.
  a. What is the narrow-sense heritability for the traits at each selection step?
  b. Why does the response of the traits to selection change during the selection process?
  c. What percentage increase in numbers of eggs produced is obtained when lines A and B are crossed?
  d. With the possible exception of dairy cattle, commercial livestock are hybrids produced by crossing breeds, lines, or strains already selected for a set of desirable traits. Why?

14.26 The following variances were determined for measurements of body length, antenna bristle number, and egg production in a species of moth. Which of these characters would be most rapidly changed by natural selection? Which character would be most slowly affected by natural selection?

Variance	Body Length	Antenna Bristle Number	Egg Production
Phenotypic (V_P)	798	342	145
Additive (V_A)	132	21	21
Dominance (V_D)	122	126	24
Interaction (V_I)	118	136	34
Genetic-environmental ($V_{G \times E}$)	81	23	21
Maternal effects (V_{Em})	345	36	45

14.27 Imagine that you have made the following initial crosses to start your tomato breeding program. You need to cross a cultivated tomato with a small fruited, late flowering wild tomato because the wild tomato has a disease-resistance gene critical for agriculture. The data for each cross below include the lines used as parents (C = cultivated, W = wild), followed by its average fruit weight and days to first flower, along with averages from the progeny (P = F_1 progeny).

Which crosses would be your first choices for starting your improvement program? Explain your reasoning.

Cross Number	Cultivated Parent	Wild Parent	F_1
1	C1 (68 g, 32 d) ×	W1 (6 g, 42 d)	P1 (30 g, 40 d)
2	C1 (68 g, 32 d) ×	W2 (6 g, 41 d)	P2 (38 g, 36 d)
3	C1 (68 g, 32 d) ×	W3 (8 g, 44 d)	P3 (40 g, 41 d)
4	C2 (72 g, 31 d) ×	W1 (6 g, 42 d)	P4 (38 g, 36 d)
5	C2 (72 g, 31 d) ×	W2 (6 g, 41 d)	P5 (34 g, 32 d)
6	C2 (72 g, 31 d) ×	W3 (8 g, 44 d)	P6 (42 g, 42 d)

14.28 Suppose that the narrow-sense heritability of wool length in a breed of sheep is 0.92, and the narrow-sense heritability of body size is 0.87. The genetic correlation between wool length and body size is −0.84. If a breeder selects for sheep with longer wool, what will be the most likely effects on wool length and body size?

15

Gene Mapping in Eukaryotes

The fruit fly, *Drosophila melanogaster*, with the *vestigial* wing mutation.

PRINCIPAL POINTS

- Genetic recombinants result from physical exchanges between homologous chromosomes in meiosis. A chiasma is the site of *crossing-over*: the reciprocal exchange of chromosome parts at corresponding positions along homologous chromosomes by symmetrical breakage and rejoining.

- Crossing-over is a reciprocal event that, in eukaryotes, occurs at the four-chromatid stage in prophase I of meiosis.

- The map distance between genes is measured in map units (mu) (also called centiMorgans, cM); 1 mu is defined as the interval in which 1 percent of crossing-over takes place. However, gene-mapping crosses produce data in the form of recombination frequencies, which are used to estimate the map distance, where, in this case, 1 mu is equivalent to a recombination frequency of 1 percent.

- As the distance between genes increases, the incidence of multiple crossovers causes the recombination frequency to be an underestimate of the crossover frequency and hence of the true map distance. Mapping functions can be used to correct for this problem and thereby give a more accurate estimate of map distance.

i THE HUMAN GENOME PROJECT MAY BE THE BEST-known gene-mapping project in the world. The aim of the project is to determine both the locations of all the genes in the human genome and the exact nucleotide sequence that comprises the 3 billion nucleotide pairs that make up the genome. But long before the development of the recombinant technologies that allowed the Human Genome Project to come into being, scientists were creating genetic maps of eukaryotic organisms. What do these maps tell us? How are they constructed? How can they be used? After you have read and studied this chapter, you can further explore the answers to these and other questions by trying the iActivity.

iActivity

Genes on nonhomologous chromosomes assort independently during meiosis. In many instances, however, certain genes (and hence the phenotypes they control) are inherited together because they are located on the same chromosome. Genes that are on the same chromosome are said to be *syntenic*. Genes that do not appear to assort independently because they are located on the same chromosome exhibit **linkage** and are called **linked genes.** These genes belong to a *linkage group*.

Genetic analysis is the dissection of the structure and function of the genetic material. In classic genetic analysis, progeny from crosses between parents with different genetic characters are analyzed to determine the frequency with which differing parental alleles are associated in new combinations. Progeny showing the parental combinations of alleles are called *parentals*, and progeny showing nonparental combinations of alleles are called *recombinants*. The process by which the recombinants are produced is called **genetic recombination.** Through testcrosses, we can determine which genes are linked to each other and can then construct a *linkage map*, or *genetic map*, of each chromosome.

Classic genetic mapping has provided information that is useful in many aspects of genetic analysis. For example, knowing the locations of genes on chromosomes has been useful in recombinant DNA research and in experiments directed toward understanding the DNA sequences in and around genes. These days, the focus of mapping studies is on constructing genetic maps of genomes with the use of both gene markers and DNA markers. A *marker*, or **genetic marker,** is another name for a mutation that gives a distinguishable phenotype. In other words, it is an allele that marks a chromosome or a gene. **Gene markers** are alleles of the kind we have discussed to this point in the text. **DNA markers** are molecular markers—that is, DNA regions in the genome that differ sufficiently between individuals and thus can be detected by the molecular analysis of DNA. The goal of genome-mapping studies is to generate high-resolution maps of the chromosomes. Such maps are useful for investigating genes and their functions. The ultimate genetic maps will be of the base-pair sequences of organisms' genomes. (See Chapter 10.)

The goal of this chapter is to learn how genetic linkage affects Mendelian gene segregation patterns and how genes are mapped classically in eukaryotes.

Early Studies of Genetic Linkage: Morgan's Experiments with *Drosophila*

By 1911, Thomas Hunt Morgan had identified a number of X-linked genes, including *w* (white eye) and *m* (miniature wing; the wing is smaller than normal). Morgan crossed a white miniature (*w m/w m*) female fly with a wild-type male (*w⁺ m⁺/Y*; Figure 15.1). For the former genotype, the slash signifies the pair of homologous chromosomes and indicates that the genes on either side of the slash are linked. For the latter genotype, because the genes are X linked, a slash indicates the X chromosome and Y indicates a Y chromosome. We will also use another special genetic symbol for genes on the same chromosome: $\frac{a\ b}{a\ b}$, which signifies that genes *a* and *b* are on the same chromosome, with the chromosome represented by the horizontal line. In this system, X-linked genes in a female are indicated by allele symbols separated by one or two continuous lines to denote the homologous chromosomes, as in

$$\frac{w\ m}{w\ m} \text{ or } \frac{w\ m}{w\ m},$$

and X-linked genes in a male are shown as, for example,

$$\underrightarrow{w\ m}$$

where the straight line designates the X chromosome and the bent line the Y chromosome. This genotype representation is the same as *w m/^* (where the bent slash is the Y chromosome) or *w m*/Y. (*Note*: If a discontinuous line is used between a series of allele pairs, the extent of each segment signifies a different chromosome.)

In the cross, the F_1 males were white eyed and had miniature wings (genotype *w m*/Y), whereas all females were heterozygous and wild type for both eye color and wing size (genotype *w⁺ m⁺/w m*). The F_1 flies were interbred, and 2,441 F_2 flies were analyzed. In crosses of X-linked genes set up as in Figure 15.1, the $F_1 \times F_1$ is equivalent to doing a testcross, because the F_1 males produce X-bearing gametes with recessive alleles of both genes and Y-bearing gametes that have no alleles for the genes being studied. In the F_2, the most common phenotypic classes in both sexes were the *grandparental phenotypes* of white eyes plus miniature wings or normal red eyes plus large wings. Conventionally, we call the original genotypes of the two chromosomes **parental genotypes, parental classes,** or, more simply, **parentals.** The term is also used to describe phenotypes, so the original white miniature females and wild-type males in these particular crosses are defined as the parentals.

Morgan observed that 900 of the 2,441 F_2 flies, or 36.9 percent, had nonparental phenotypic combinations of white eyes plus normal wings and red eyes plus miniature wings. Nonparental combinations of linked genes are called **recombinants.** Fifty percent recombinant phenotypes is expected if independent assortment is the case; thus, the lower percentage observed is evidence of linkage of the two genes. To explain the recombinants, Morgan proposed that, in meiosis, exchanges of genes had occurred between the two X chromosomes of the F_1 females.

Morgan's group analyzed a large number of other crosses of this type. *In each case, the parental phenotypic classes were the most frequent and the recombinant classes occurred less frequently*. Approximately equal numbers of

Figure 15.1

Morgan's experimental crosses of white-eye and miniature-wing variants of
***Drosophila melanogaster,* showing evidence of linkage and recombination**
in the X chromosome.

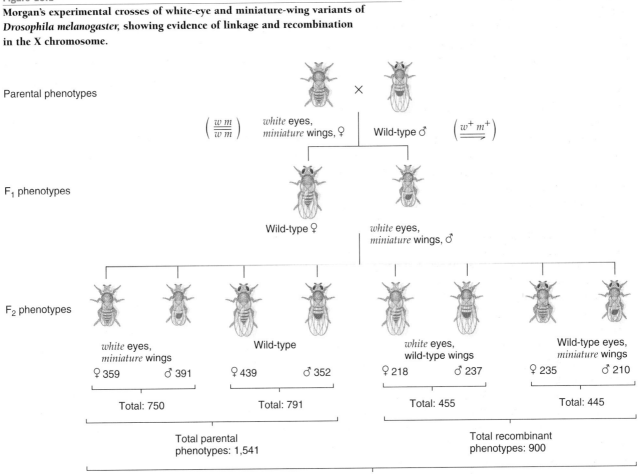

each of the two parental classes were obtained, and similar results were seen for the recombinant classes. Morgan's general conclusion was that, *during meiosis, alleles of some genes assort together because they lie near each other on the same chromosome.* To turn this statement around, the closer two genes are on the chromosome, the more likely they are to remain together during meiosis. The reason is that the recombinants are produced as a result of crossing-over between homologous chromosomes during meiosis, and the closer two genes are together, the less likely there will be a recombination event between them.

The terminology related to the physical exchange of homologous chromosome parts can be confusing. To clarify,

1. A chiasma (plural, *chiasmata;* see Figure 12.10) is the place on a homologous pair of chromosomes at which a physical exchange is occurring; it is the site of crossing-over.

2. Crossing-over is the reciprocal exchange of chromatid segments at corresponding positions along homologous chromosomes; the process involves the breaking and rejoining of two chromatids.

3. Crossing-over is also defined as the events leading to genetic recombination between linked genes in both prokaryotes and eukaryotes.

Crossing-over occurs at the four-chromatid stage in prophase I of meiosis. Each crossover involves two of the four chromatids, and, along the length of a chromosome, all chromatids can be involved in crossing-over.

KEYNOTE

The production of genetic recombinants results from physical exchanges between homologous chromosomes during meiotic prophase I. A chiasma is the site of crossing-over. Crossing-over is the reciprocal exchange of chromosome parts at corresponding positions along homologous chromosomes by the breaking and rejoining of two chromatids. *Crossing-over* is also used to describe the events leading to genetic recombination between linked genes. Crossing-over in eukaryotes takes place at the four-chromatid stage in prophase I of meiosis.

Gene Recombination and the Role of Chromosomal Exchange

Two key experiments in the 1930s, one using corn and the other using *Drosophila*, established that the appearance of genetic recombinants is associated with crossing-over.

Corn Experiments

In 1931, Harriet B. Creighton and Barbara McClintock reported the results of crosses made with a strain of corn (*Zea mays*) that was heterozygous for two genes on chromosome 9 (Figure 15.2a). One of the genes determines colored (*C*) or colorless (*c*) seeds. The other gene determines the forms of starch synthesized by the plants; standard-type plants (*Wx*) produce two forms of starch—amylose and amylopectin—and waxy plants (*wx*) produce only amylopectin. One of the chromosomes had a normal appearance and had the genotype *c Wx*. Its homolog had the genotype *C wx*, had a large, darkly staining knob at the end nearer to *C*, and was longer than the *c Wx* chromosome because a piece of chromosome 8 was attached to the end nearer to *wx*. (The process of a chromosome segment breaking off from one chromosome and reattaching to another is called *translocation*; see Chapter 17, pp. 460–462.) Cytologically distinguishable features such as these are called **cytological markers.**

During meiosis, crossing-over occurs between the two gene loci (Figure 15.2b). When the two classes of recombinants—*c wx* and *C Wx*—that occurred in the progeny were examined, Creighton and McClintock found that whenever the genes had recombined, the cyto-

logical features (the knob and the extra piece) had also recombined (Figure 16.2c). No such physical exchange of cytological markers was evident in the parental (nonrecombinant) classes of progeny. These data provided strong evidence that genetic recombination is associated with the physical exchange of parts between homologous chromosomes.

Drosophila Experiments

Within a few weeks of the publication of Creighton and McClintock's results, Curt Stern reported identical conclusions for experiments done with *Drosophila melanogaster*. In these experiments, the approach was identical: Strains containing appropriate genetic markers and cytological markers were crossed, and the two types of markers were analyzed in the next generation (Figure 15.3).

In Stern's work, two X-linked gene loci were involved: the *car* (carnation) gene and the *B* (bar-eye) gene. Mutants of the *car* gene are recessive and, when homozygous, result in a carnation-colored eye instead of the wild-type red. Mutants of the *B* gene are incompletely dominant, resulting in a bar-shaped eye instead of the round eye of the wild type. In Stern's crosses, the male parent was *car B*[+]/*Y*. These flies had carnation-colored and nonbar (wild-type-shaped) eyes. The female parent had two abnormal, cytologically distinct X chromosomes. One X chromosome, genotype *car*[+] *B*[+], had a portion of the Y chromosome attached to it. The other X chromosome, genotype *car B*, was distinctly shorter than normal X, because part of it had broken off and was attached to the small chromosome 4. (See the unattached piece in Figure 15.3.) In females, the shape of the eye depends on the number of copies of the mutant *B* allele. In *B/B* homozygotes, the eye is very narrow, whereas in *B/B*[+] heterozygotes, the eye is kidney shaped. Thus, phenotypically, the

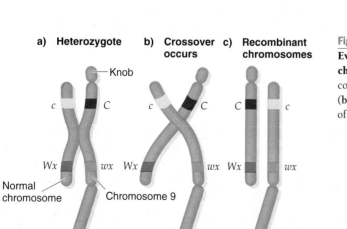

a) **Heterozygote** b) **Crossover occurs** c) **Recombinant chromosomes**

Knob

c   *C*   *c*   *C*   *C*   *c*

Wx   *wx* *Wx*   *wx* *Wx*   *wx*

Normal chromosome

Chromosome 9

Part of chromosome 8

Figure 15.2

Evidence of the association of gene recombination with chromosomal exchange in corn. (a) Physical and genetic constitutions of the two chromosomes in the heterozygote. (b) Crossover occurs. (c) Physical and genetic constitutions of the recombinant chromosomes.

Figure 15.3

Stern's experiment to demonstrate the relationship between genetic recombination and chromosomal exchange in *Drosophila melanogaster*.

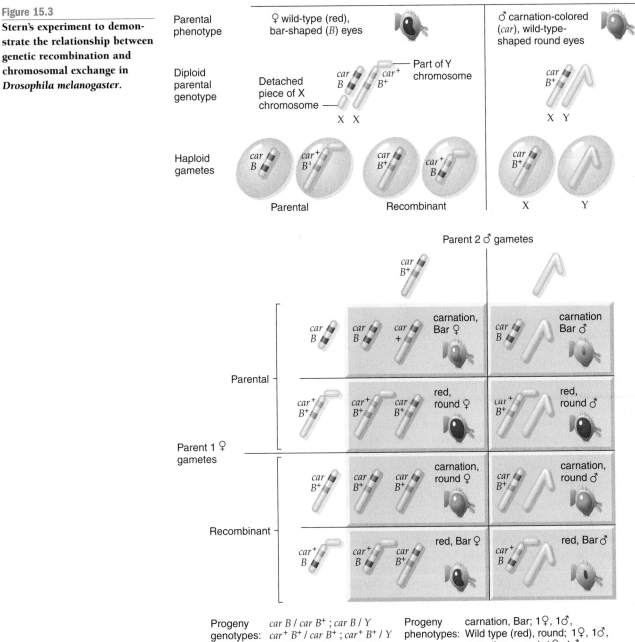

Progeny genotypes: *car B / car B⁺* ; *car B / Y*
car⁺ B⁺ / car B⁺ ; *car⁺ B⁺ / Y*
car B⁺ / car B⁺ ; *car B⁺ / Y*
car⁺ B / car B⁺ ; *car⁺ B / Y*

Progeny phenotypes: carnation, Bar; 1♀, 1♂,
Wild type (red), round; 1♀, 1♂,
carnation, round; 1♀, 1♂,
Wild type (red), Bar; 1♀, 1♂,

parental females had a red-colored, kidney-shaped eye because they were heterozygous *car⁺/car* and *B/B⁺*. (In the figure, the kidney shape is referred to as bar for simplicity.)

As with the corn experiments, analysis of the progeny showed that every case in which genetic recombination occurred was accompanied by an exchange of identifiable chromosome segments. That is, if no recombination occurred, the two phenotypic classes of progeny were (1) carnation-colored eye with a kidney shape in females and bar shape in males, genotypically *car B/car B⁺* in females and *car B/Y* in males; and (2) red eye with round shape (i.e., wild type for both genes), genotypically *car⁺ B⁺/car B⁺* in females and *car⁺ B⁺/Y* in males. No exchanges of chromosome parts were seen among these nonrecombinants. There were also two classes of recombinants: (1) carnation-colored, round eyes, genotypically *car B⁺/car B⁺* females and *car B⁺/Y* males; and (2) red-colored, bar-shaped eyes, genotypically *car⁺ B/car B⁺* females and *car⁺ B/Y* males. The carnation flies had a complete X chromosome, and the bar flies had a shorter-than-normal X chromosome to which a piece of the Y chromosome was attached, whereas the rest of the X chromosome was attached to chromosome 4. This chromosomal makeup could have resulted only from physical exchanges of homologous chromosome parts.

There is no doubt, therefore, that genetic recombination results from physical crossing-over between chromosomes.

KEYNOTE

The proof that genetic recombination occurs when crossing-over takes place during meiosis came from breeding experiments in which the parental chromosomes differed with respect to both genetic and cytological markers. These experiments showed that whenever recombinant phenotypes occurred, the cytological markers indicated that crossing-over had also occurred.

Box 15.1   DNA Recombination

Crossing-over involves breaking and rejoining DNA at the same position in two homologous DNA molecules with no loss or addition of base pairs. Thus a very early step in recombination is the generation of a break in each double helix involved. In this box we discuss some molecular aspects of DNA recombination.

In the mid-1960s Robin Holliday proposed a model for reciprocal recombination. Since then, the **Holliday model** has been refined and embellished by other geneticists, notably Matthew Meselson and Charles Radding, T. Orr-Weaver and Jack Szostak. To give just a flavor for the recombination process at the molecular level, we present the Holliday model here.

The Holliday model is diagrammed in Box Figure 15.1 for genetically distinguishable homologous chromosomes, one with alleles a^+ and b^+ at opposite ends and the other with alleles a and b. The two DNA double helices in the figure participate in the recombination event. The first stage of the recombination process in recognition and alignment (Box Figure 15.1, part 1), in which two homologous DNA double helices become aligned precisely. In the second stage, one strand of each double helix breaks; each broken strand invades the opposite double helix and base pairs with the complementary nucleotides of the invaded helix (Box Figure 15.1, part 2). Enzymes are responsible for each of these steps. DNA polymerase and DNA ligase seal the gaps that are left, producing what is called a *Holliday intermediate*, with an internal branch point (Box Figure 15.1, part 3). The hybrid DNA molecules evident at this stage are called *heteroduplexes*; that is, the two strands of the double-stranded DNA molecules do not have completely complementary sequences. The two DNA double helices in the Holliday intermediate can rotate, causing the branch point to move to the right or the left. Box Figure 15.1, part 4, shows a branch migration event that has occurred to the right. The four-armed structure for the DNA strands is produced simply by pulling the four chromosome ends apart. Branch migration generates complementary regions of hybrid DNA in both double helices (diagrammed as stretches of DNA helices with two different colors in Box Figure 15.1, parts 5 through 8).

The cleavage and ligation phase of recombination is best visualized if the Holliday intermediate is redrawn so that no DNA strand passes over or under another DNA strand. Thus, if the four-armed Holliday intermediate after Box Figure 15.1,

part 4, is taken as a starting point and the lower two arms are rotated 180 degrees relative to the upper arms, the structure shown in Box Figure 15.1, part 5, is produced.

Next, enzymes cut the Holliday intermediate at two points in the single-stranded DNA region of the branch point (Box Figure 15.1, part 6). The cuts can be in either the horizontal or vertical plane; both kinds of cuts occur with equal probability. Endonuclease cleavage in the horizontal plane (Box Figure 15.1, part 6, left) produces the two double helices shown in Box Figure 15.1, part 7, left. In each helix is a single-stranded gap. DNA ligase seals the gaps to produce the double helices shown in Box Figure 15.1, part 8, left. Since each of the resulting helices contains a segment of single-stranded DNA from the other helix, flanked by nonrecombinant DNA, these double helices are called *patched duplexes*.

If the endonuclease cleavage in Box Figure 15.1, part 5, is in the vertical plane (Box Figure 15.1, part 6, right), the gapped double helices of Box Figure 15.1, part 7, right, are produced. In this case, there are segments of hybrid DNA in each duplex, but they are formed by what looks like a splicing together of two helices. The result is the double helices in Box Figure 15.1, part 8, right, which are called *spliced duplexes*.

In the example diagrammed in Box Figure 15.1, the parental duplexes contain different genetic markers at the ends of the molecules. One parent is $a^+ b^+$ and the other is $a\ b$ as would be the case for a doubly heterozygous parent. However, the reciprocal recombination events shown in Box Figure 15.1 result in different products. In the patched duplexes (Box Figure 15.1, part 8, left) the markers are $a^+ b^+$ and $a\ b$, which is the parental configuration. However, in the spliced duplexes (Box Figure 15.1, part 8, right), the markers are recombinant, that is $a^+ b$ and $a\ b^+$. Since the enzymes that cut in the branch region during the final cleavage and ligation phase (Box Figure 15.1, part 6, left and right) cut randomly with respect to the plane of the cut (that is, horizontal or vertical), the Holliday model predicts that a physical exchange between two gene loci on homologous chromosomes should result in the genetic exchange of the outside chromosome markers about half of the time.

Although the basics features of the Holliday model are generally accepted, a number of other models attempt to explain recombination at a more detailed level or in special systems. Discussion of these other models in beyond the scope of this text.

Box 15.1   Continued

Box Figure 15.1

Holliday model for reciprocal genetic recombination. Shown are two homologous DNA double helices that participate in the recombination process. Inset: Electron micrograph of a Holliday intermediate with some single-stranded DNA in the branch point region. (David Dressler, Oxford University, UK; from *Proc. National Academy of Sciences USA* 75:605, 1978.)

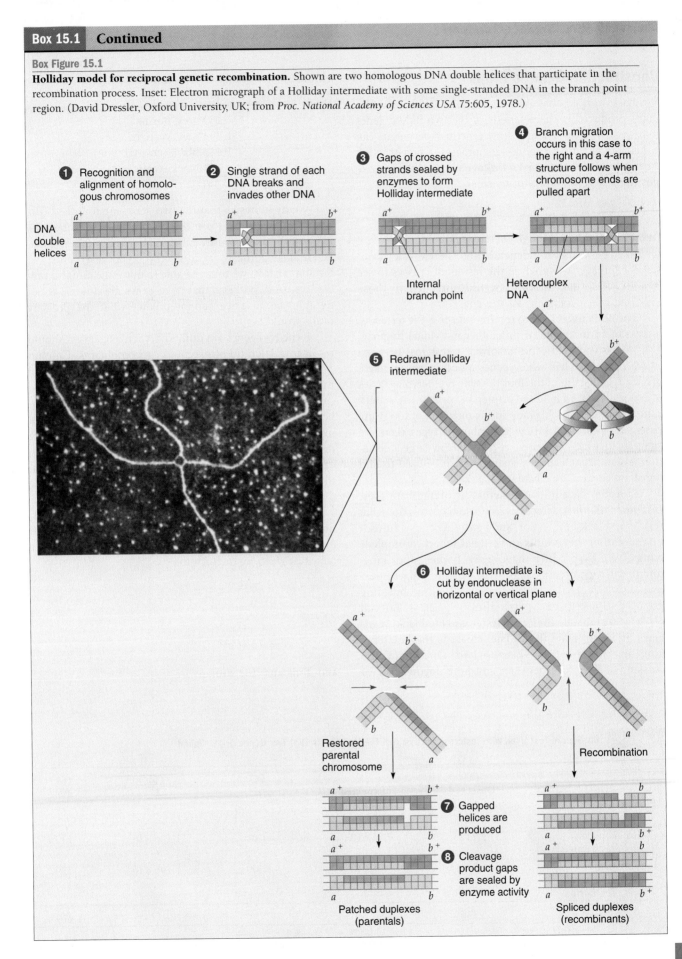

Constructing Genetic Maps

We have learned that the number of genetic recombinants produced is characteristic of the two linked genes involved. We now examine how genetic experiments can be used in genetic mapping—the process of constructing a **genetic map** (also called a **linkage map**) of the relative positions of genes on a chromosome.

Detecting Linkage through Testcrosses

Unlinked genes assort independently. Therefore, a way to test for linkage is to analyze the results of crosses to see whether the data deviate significantly from those expected by independent assortment.

The best cross to use to test for linkage is the testcross, a cross of an individual with another individual homozygous recessive for all genes involved. A testcross between $a^+/a\ b^+/b$ and $a/a\ b/b$, where genes a and b are unlinked, gives a 1:1:1:1 ratio of the four possible phenotypic classes $a^+\ b^+ : a^+\ b : a\ b^+ : a\ b$. (See Chapter 11, p. 284.) A significant deviation from this ratio in the direction of too many parental types and too few recombinant types therefore suggests that the two genes are linked. How large a deviation is significant? The *chi-square test* can be used to find the significance. (See Chapter 11, p. 287.)

Consider data from a testcross involving fruit flies. In *Drosophila*, b is a recessive autosomal mutation that results in black body color, and vg is a recessive autosomal mutation that results in vestigial (short, crumpled) wings. Wild-type flies have grey bodies and long, uncrumpled (normal) wings. True-breeding black, normal ($b/b\ vg^+/vg^+$) flies were crossed with true-breeding grey, vestigial ($b^+/b^+\ vg/vg$) flies. F_1 grey, normal ($b^+/b\ vg^+/vg$) female flies were testcrossed to black, vestigial ($b/b\ vg/vg$) male flies. (The female is the heterozygote in this testcross because, in *Drosophila*, no crossing-over occurs between any homologous pair of chromosomes in males.) The testcross progeny data were as follows:

	283 grey, normal
	1,294 grey, vestigial
	1,418 black, normal
	241 black, vestigial
Total	3,236 flies

We hypothesize that the two genes are unlinked (the null hypothesis) and use the chi-square test to test the hypothesis, as shown in Table 15.1. We use this particular null hypothesis because the hypothesis must be testable; that is, we must be able to make meaningful predictions. A hypothesis that "two genes are linked" is not testable, because we cannot predict what the progeny ratios would be.

If the two genes are unlinked, then a testcross should result in a 1:1 ratio of parentals : recombinants. Column 1 lists the parental and recombinant phenotypes expected in the progeny of the cross, column 2 lists the observed (o) numbers, and column 3 lists the expected (e) numbers for the parentals and recombinants, given the total number of progeny (3,236) and the hypothesis being tested (1:1 in this case). Column 4 lists the deviation (d), calculated by subtracting the expected number (e) from the observed number (o) for each class. The sum of the d values is always zero.

Column 5 lists the deviation squared (d^2), and column 6 lists the deviation squared divided by the expected number (d^2/e). The chi-square value, χ^2 (item 7 in the table), is given by the formula

$$\chi^2 = \Sigma\ \frac{d^2}{e},$$

where

$$d^2 = (o - e)^2$$

and Σ means "the sum of."

Table 15.1 Chi-Square Test Used with Testcross Data to Test the Hypothesis That Two Genes Are Unlinked

(1) Phenotypes	(2) Observed Number (o)	(3) Expected Number (e)	(4) d ($= o - e$)	(5) d^2	(6) d^2/e
Parentals: (black, normal and grey, vestigial)	2,712	1,618	1,094	1,196,836	739.7
Recombinants (black, vestigial and grey, normal)	524	1,618	−1,094	1,196,836	739.7
Total	3,236	3,236			1,479.4
	(7) $\chi^2 = 1{,}479.4$	(8) df 1			

In the table, chi-square is the sum of the two values in column 6. In our example, $\chi^2 = 1,479.4$. The last value in the table, item 8, is the degrees of freedom (df) for the set of data; there is $n - 1 = 1$ degree of freedom in this case.

The chi-square value and the degrees of freedom are used with a table of chi-square probabilities (see Table 11.5, p. 288) to determine the probability (P) that the deviation of the observed values from the expected values is due to chance. For $\chi^2 = 1,479.4$ with 1 degree of freedom, the P value is much lower than 0.001; in fact, it is not in the table. This means that independent repetitions of the experiment would produce chance deviations from what was expected as large as those observed in many fewer than 1 out of 1,000 trials. As a reminder, if the probability of obtaining the observed chi-square values is greater than 5 in 100 ($P > 0.05$), the deviation is considered not statistically significant and could have occurred by chance alone. If $P \leq 0.05$, the deviation from the expected values is statistically significant and not due to chance alone; then the hypothesis may well be invalid. If $P \leq 0.01$, the deviation is highly statistically significant, and the data are not consistent with the null hypothesis. In that case, we would reject the independent assortment hypothesis, and, genetically, the only alternative hypothesis that could logically apply is that the genes are linked.

The Concept of a Genetic Map. In an individual that is doubly heterozygous for the w and m alleles, for example, the alleles can be arranged in two ways:

$$\frac{w^+ \; m^+}{w \; m} \quad \text{or} \quad \frac{w^+ \; m}{w \; m^+}$$

In the arrangement on the left, the two wild-type alleles are on one homolog and the two recessive mutant alleles are on the other homolog, an arrangement called **coupling** (or the *cis* configuration). Crossing-over between the two loci produces $w^+ m$ and $w m^+$ recombinants. In the arrangement on the right, each homolog carries the wild-type allele of one gene and the mutant allele of the other gene, an arrangement called **repulsion** (or the *trans* configuration). Crossing-over between the two genes produces $w^+ m^+$ and $w m$ recombinants.

The data obtained by Morgan from *Drosophila* crosses indicated that the frequency of crossing-over (and hence of recombinants) for linked genes is characteristic of the gene pairs involved: For the X-linked genes white (w) and miniature (m), the recombination frequency is 36.9 percent. Moreover, the recombination frequency for two linked genes is the same, regardless of whether the alleles of the two genes involved are in coupling or in repulsion. *Although the actual phenotypes of the recombinant classes are different for the two*

arrangements, the percentage of recombinants among the total progeny will be the same in each case (within experimental error).

In 1913, a student of Morgan's, Alfred Sturtevant, suggested that recombination frequencies could be used as a quantitative measure of the genetic distance between two genes on a genetic map. The genetic distance between genes is measured in **map units (mu)**, where 1 map unit is defined as the interval in which 1 percent crossing-over takes place. The map unit is sometimes called a **centimorgan** (cM) in honor of Morgan. It is important to note that, for a pair of linked genes, the crossover frequency is *not* the same as the recombination frequency. The former refers to the frequency of physical exchanges between chromosomes in meiosis for the region between the genes, and the latter refers to the frequency of recombination of genetic markers in a cross, as determined by analyzing the phenotypes of the progeny. Geneticists follow genetic markers in crosses, so the data obtained are in the form of recombination frequencies. In our discussions, we will use recombination frequencies as geneticists often do: as working estimates of map distances between genes, where a map unit is equivalent to a recombination frequency of 1 percent. Later we will discuss how such data relate to crossover frequencies and therefore to true map units.

The genes on a chromosome, then, can be represented by a one-dimensional genetic map that shows, in linear order, the genes belonging to the chromosome. Crossover and recombination frequencies give the linear order of the genes on a chromosome and provide information about the genetic distance between any two genes. The farther apart two genes are, the greater is the *crossover frequency*.

The first genetic map ever constructed was based on *recombination frequencies* from *Drosophila* crosses involving the sex-linked genes w, m, and y, where w gives white eyes, m gives miniature wings, and y gives yellow body. From these mapping experiments, the recombination frequencies for the $w \times m$, $w \times y$, and $m \times y$ crosses were established as 32.6, 1.3, and 33.9 percent, respectively. (In this independent experiment, the recombination frequency for w and m is a little lower than in the experiment discussed previously on pp. 406–407.) The percentages are quantitative measures of the distances between the genes involved.

We can construct a genetic map on the basis of the recombination frequency data. The recombination frequencies show that w and y are closely linked and that m is quite far from the other two genes. Since the w–m genetic distance is less than the y–m distance (as shown by the smaller recombination frequency in the $w \times m$ cross), the order of genes must be $y \; w \; m$ (or $m \; w \; y$); thus, the three genes are ordered and spaced with 1.3 mu between y and w and 32.6 mu between w and m:

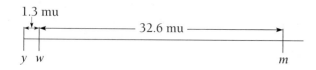

Gene Mapping with Two-Point Testcrosses

We have seen that the recombination frequency may be used to obtain an estimate of the genetic distance between two linked genes. By carrying out two-point testcrosses such as those shown in Figure 15.4, we can determine the relative numbers of parental and recombinant classes in the progeny. For autosomal recessives (as in the figure), a double heterozygote is crossed with a doubly homozygous recessive mutant strain. When the double heterozygous $a^+ b^+/a\ b$ F_1 progeny from a cross of $a^+ b^+/a^+ b^+$ with $a\ b/a\ b$ are testcrossed with $a\ b/a\ b$, four phenotypic classes are found among the F_2 progeny. Two of these classes have the parental phenotypes $a^+ b^+$ and $a\ b$, and the other two have the recombinant phenotypes $a^+ b$ and $a\ b^+$.

Two-point testcrosses for the purpose of mapping are set up in similar ways for X-linked recessive genes. That is, a double heterozygous female is crossed with a hemizygous male carrying the recessive alleles:

$$\frac{a^+\ b^+}{a\ b} \times \frac{a\ b}{\longrightarrow}$$

In all cases, a two-point testcross should yield a pair of parental types that occur with about equal frequency and a pair of recombinant types that also occur with about equal frequency. Of course, the actual phenotypes depend on the relative arrangement of the two allelic pairs in the homologous chromosomes—that is, whether they are in coupling (*cis*) or in repulsion (*trans*). The following formula is used to calculate the recombination frequency:

$$\frac{\text{number of recombinants}}{\text{number of testcross progeny}} \times 100 = \frac{\text{recombination}}{\text{frequency}}$$

The recombination frequency is used directly as an estimate of map units.

The two-point method of mapping is most accurate when the two genes examined are close together; when genes are far apart, there are inaccuracies, as we will see later. Large numbers of progeny must also be counted (scored) to ensure a high degree of accuracy. From mapping experiments carried out in all types of organisms, we know that genes are linearly arranged in linkage groups. There is a one-to-one correspondence between linkage

Figure 15.4

Testcross to show that two genes are linked. Genes a and b are recessive mutant alleles linked on the same autosome. A homozygous $a^+ b^+/a^+ b^+$ individual is crossed with a homozygous recessive $a\ b/a\ b$ individual, and the doubly heterozygous F_1 progeny $(a^+ b^+/a\ b)$ are testcrossed with homozygous $a\ b/a\ b$ individuals.

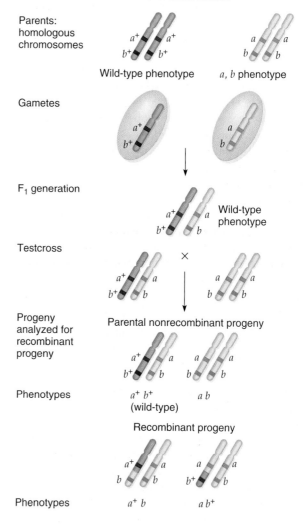

groups and chromosomes, so the sequence of genes on the linkage group reflects the sequence of genes on the chromosome.

Generating a Genetic Map

We can now discover how a genetic map is generated from an estimation of the number of times crossing-over occurred in a particular segment of the chromosome out of all meioses examined. In many cases, the probability of a crossing-over event is not uniform along a chromosome, so we must be cautious about how far we extrapolate the genetic map (derived from data produced by genetic crosses) to the physical map of the chromosome (derived from determinations of the locations of genes along the chromosome itself—for example, from sequencing the DNA).

The recombination frequencies observed between genes may also be used to predict the outcome of genetic crosses. For example, a recombination frequency of 20 percent between genes indicates that, for a doubly heterozygous genotype (such as $a^+ b^+/a b$), 20 percent of the gametes produced, on average, will be recombinants ($a^+ b$ and $a b^+$ in the example, with 10 percent of each expected).

For any testcross, the recombination frequency in the progeny cannot exceed 50 percent. That is, if the genes are assorting independently, an equal number of recombinants and parentals are *expected* in the progeny, so the recombination frequency is 50 percent. If we get a recombination frequency of 50 percent from a cross, then we state that the two genes are unlinked. Genes may be unlinked (that is, show 50 percent recombination) either when the genes are on different chromosomes (a case we discussed before), or when *the genes are far apart on the same chromosome.*

The second case can be illustrated by referring to Figure 15.5, which shows the effects of single crossovers and double crossovers on the production of parental and recombinant chromosomes for two loci that are far apart on the same chromosome. (In reality, in such a situation, there would be multiple crossovers between the two loci in each meiosis.) Single crossovers between any pair of non-sister chromatids result in two parental and two recombinant chromosomes; that is, for two loci, 50 percent of the products are recombinant. (See Figure 15.5a.)

Double crossovers can involve two, three, or all four of the chromatids. (See Figure 15.5b.) For double crossovers involving the same two nonsister chromatids (a *two-strand double crossover*), all four resulting chromosomes are parental for the two loci of interest. For *three-strand double crossovers* (double crossovers involving three of the four chromatids), two parental and two recombinant chromosomes result. For a *four-strand double crossover*, all four resulting chromosomes are recombinant. Considering all possible double crossover patterns together, we find that 50 percent of the products are recombinant for the two loci. Similarly, for any multiple number of crossovers between loci that are far apart, examination of a large number of meioses will show that 50 percent of the resulting chromosomes are recombinant. This is the reason for the recombination frequency limit of 50 percent exhibited by unlinked genes on the same chromosome.

The point is, if two genes show 50 percent recombination in a cross, they might not be on different chromosomes. More data would be needed to determine whether the genes are on the same chromosome or on different chromosomes. One way to find out is to map

a number of other genes in the linkage group. For example, if *a* and *m* show 50 percent recombination, perhaps we will find that *a* shows 27 percent recombination with *e* and *e* shows 36 percent recombination with *m*. This result would indicate that *a* and *m* are in the same linkage group approximately 63 mu apart, as shown here:

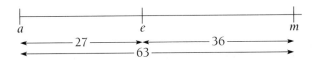

Gene Mapping with Three-Point Testcrosses

Although genetic maps can be built by using a series of two-point testcrosses, geneticists typically have mapped several linked genes at a time in single testcrosses. Here, we illustrate this more complex type of mapping analysis for three linked genes with the use of a **three-point testcross.** In diploid organisms, the three-point testcross is a cross of a triple heterozygote with a triply homozygous recessive. If the mutant genes in the cross are all recessive, a typical three-point testcross might be

$$\frac{a^+ b^+ c^+}{a \ b \ c} \times \frac{a \ b \ c}{a \ b \ c}$$

In a testcross involving sex-linked genes, the female is the heterozygous strain (assuming that the female is the homogametic sex) and the male is hemizygous for the recessive alleles.

Suppose we have a hypothetical flowering plant in which there are three linked genes, all of which control fruit phenotypes. A recessive allele *p* of the first gene determines purple fruit color, versus yellow color of the wild type. A recessive allele *r* of the second gene results in a round fruit shape, versus elongated fruit in the wild type. A recessive allele *j* of the third gene gives a juicy fruit, versus the dry fruit of the wild type. The task before us is to determine the order of the genes on the chromosome and the map distances between the genes. To do so, we perform a testcross of a triple heterozygote ($p^+ r^+ j^+/p r j$) with a triply homozygous recessive ($p \ r \ j/p \ r \ j$) and then count the different phenotypic classes in the progeny (Figure 15.6).

For each gene in the cross, two different phenotypes occur in the progeny; therefore, for the three genes, $(2)^3 = 8$ phenotypic classes appear in the progeny, representing all possible combinations of phenotypes. In an actual experiment, not all the phenotypic classes may be generated. The absence of a phenotypic class is also important information, and the experimenter should enter a 0 in the class for which no progeny are found.

Figure 15.5

Demonstration that the recombination frequency between two genes located far apart on the same chromosome cannot exceed 50 percent. (a) Single crossovers produce one-half parental and one-half recombinant chromatids. (b) Two-strand, three-strand, and four-strand double crossovers collectively produce one-half parental and one-half recombinant chromatids.

Parental genotypes

a) Single crossover   **Products**   **Resulting genotypes**   **Sum**

a^+ b^+	Parental	
a^+ b	Recombinant	Recombinants = 2
a b^+	Recombinant	Total = 4
a b	Parental	Therefore, 2/4 recombinants

b) Double crossovers

Two-strand double crossover

a^+ b^+	Parental
a^+ b^+	Parental
a b	Parental
a b	Parental

Total: 0/4 recombinants

Three-strand double crossover (two ways)

One way

a^+ b^+	Parental
a^+ b	Recombinant
a b	Parental
a b^+	Recombinant

Total: 2/4 recombinants

Second way

a^+ b	Recombinant
a^+ b^+	Parental
a b^+	Recombinant
a b	Parental

Total: 2/4 recombinants

Four-strand double crossover

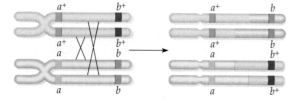

a^+ b	Recombinant
a^+ b	Recombinant
a b^+	Recombinant
a b^+	Recombinant

Total: 4/4 recombinants

Sum: Recombinants = 0 + 2 + 2 + 4 = 8
Total = 4 + 4 + 4 + 4 = 16
Therefore, recombinants = 50%

Figure 15.6

Three-point mapping, showing the testcross used and the resultant progeny.

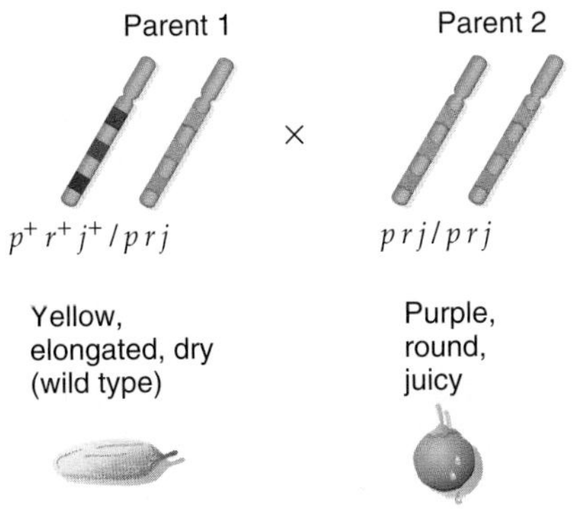

Testcross progeny

	Class	Phenotype	Number	Genotype of gamete from heterozygous parent
Parental phenotypes	1	Wild type (yellow, elongated, dry)	179	p^+ r^+ j^+
	2	purple, round, juicy	173	p r j
Recombinant phenotypes	3	yellow, round, juicy	46	p^+ r j
	4	purple, elongated, dry	52	p r^+ j^+
	5	yellow, round, dry	22	p^+ r j^+
	6	purple, elongated, juicy	22	p r^+ j
	7	yellow, elongated, juicy	4	p^+ r^+ j
	8	purple, round, dry	2	p r j^+
			Total = 500	

iActivity

You've discovered new genes that determine different traits in tomatoes. Now you can construct a genetic map showing the locations of those genes in the iActivity *Crossovers and Tomato Chromosomes* on your CD-ROM.

Establishing the Order of Genes. The first step in mapping the three genes is to determine the order of the genes on the chromosome. One parent carries the recessive alleles for all three genes; the other is heterozygous for all three. Therefore, the phenotype of each of the progeny is determined by the alleles in the gamete from the triply heterozygous parent; the gamete from the other parent carries only recessive alleles. We know from the genotypes of the original parents that all three genes are in coupling. Since the heterozygous parent in the testcross was $p^+ r^+ j^+/p\ r\ j$, classes 1 and 2 in Figure 15.6 are parental progeny: Class 1 is produced by the fusion of a $p^+ r^+ j^+$ gamete with a $p\ r\ j$ gamete from the triply homozygous recessive parent, and class 2 is produced by the fusion of a $p\ r\ j$ gamete from the heterozygous parent and a $p\ r\ j$ gamete. These classes are generated from meioses in which no crossing-over occurs in the region of the chromosome in which the three genes are located.

The other six progeny classes result from crossovers within the region spanned by the three genes that gave rise to recombinant gametes. There may have been a single

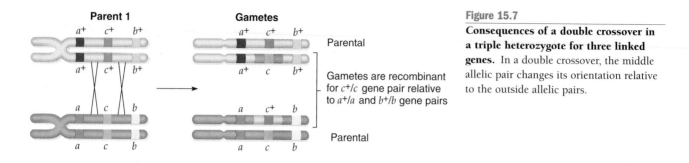

Figure 15.7

Consequences of a double crossover in a triple heterozygote for three linked genes. In a double crossover, the middle allelic pair changes its orientation relative to the outside allelic pairs.

crossover between a pair of linked genes, or there may have been a double crossover—that is, two crossovers, one between *each* pair of linked genes. Statistically, the frequency of double crossovers in the region is less than the frequency of either single crossover, so *double-crossover gametes are the least frequent pair found.* Therefore, to identify the double-crossover progeny, we examine the progeny to find the pair of classes that has the lowest number of representatives. In Figure 15.6, classes 7 and 8 are such a pair. The genotypes of the gametes from the heterozygous parent that give rise to these phenotypes are $p^+ r^+ j$ and $p r j^+$.

Figure 15.7 illustrates the consequences of a double crossover in a triple heterozygote for three linked genes *a*, *b*, and *c*, where the alleles are in coupling and the *c* gene is in the middle. A double crossover changes the orientation of the allelic pair in the middle of the three genes (here, c^+/c) with respect to the two flanking allelic pairs. That is, after the double crossover, the *c* allele is now on the chromatid with the a^+ and b^+ alleles, and the c^+ allele is on the chromosome with the *a* and *b* alleles. Therefore, genes *p*, *r*, and *j* must be arranged in such a way that the center gene switches from the parental arrangement to give classes 7 and 8. To determine the arrangement, we first check the relative organization of the genes in the parental heterozygote to be sure which alleles are in coupling and which are in repulsion. In this example, the parental (noncrossover) gametes are $p^+ r^+ j^+$ and $p r j$, so all are in coupling. The double-crossover gametes are $p^+ r^+ j$ and $p r j^+$, so the only possible gene order compatible with the data is *p j r*, with the genotype of the heterozygous parent being $p^+ j^+ r^+/p j r$.

Calculating the Recombination Frequencies for Genes. The cross data can be rewritten as shown in Figure 15.8 to reflect the newly determined gene order. For convenience in the analysis, the region between genes *p* and *j* is called region I, and that between genes *j* and *r* is called region II.

The recombination frequency can now be calculated for two genes at a time. For the *p – j* distance, all the crossovers that occurred in region I are added together. Thus, we must add the recombinant progeny resulting from a single crossover in that region (classes 3 and 4) and the recombinant progeny produced by a double crossover in which one crossover is between

p and *j* and the other is between *j* and *r* (classes 7 and 8). The double crossovers must be included because each double crossover includes a single crossover in region I and therefore involves recombination between genes *p* and *j*. From Figure 15.8, there are 98 recombinant progeny in classes 3 and 4 and 6 in classes 7 and 8, giving a total of 104 progeny that result from recombination in region I. There are 500 progeny in all, so the percentage of progeny generated by crossing over in region I is 20.8 percent, determined as follows (sco = single crossovers; dco = double crossovers):

$$\frac{\text{sco in region I } (p - j) + \text{dco}}{\text{total progeny}} \times 100\%$$

Figure 15.8

Rewritten form of the testcross and testcross progeny in Figure 15.6, based on the actual gene order *p j r*.

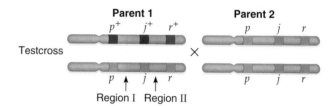

Testcross progeny

Class	Genotype of gamete from heterozygous parent			Number	Origin
1	p^+	j^+	r^+	179	Parentals, no crossover
2	p	j	r	173	
3	p^+	j	r	52	Recombinants, single crossover region I
4	p	j^+	r^+	46	
5	p^+	j^+	r	22	Recombinants, single crossover region II
6	p	j	r^+	22	
7	p^+	j	r^+	4	Recombinants, double crossover
8	p	j^+	r	2	

Total = 500

$$= \frac{(52 + 46) + (4 + 2)}{500} \times 100\%$$

$$= \frac{98 + 6}{500} \times 100\%$$

$$= \frac{104}{500} \times 100\%$$

$$= 20.8\%$$

In other words, the recombination frequency for genes *p* and *j* is 20.8, which gives us an estimated map distance of 20.8 mu. This map distance, which is quite large, is chosen mainly for illustration. We will see later that a recombination frequency of 20.8 in an actual cross would underestimate the true map distance.

The same method is used to calculate the recombination frequency for genes *j* and *r*. That is, we calculate the frequency of crossovers in the cross that gave rise to progeny recombinant for genes *j* and *r* and directly relate that frequency to map distance. In this case, all the crossovers that occurred in region II (see Figure 15.8) must be added (classes 5, 6, 7, and 8). The percentage of crossovers is calculated in the following manner.

$$\frac{\text{sco in region II}(j - r) + \text{dco}}{\text{total progeny}} \times 100\%$$

$$= \frac{(22 + 22) + (4 + 2)}{500} \times 100\%$$

$$= \frac{44 + 6}{500} \times 100\%$$

$$= \frac{50}{500} \times 100\%$$

$$= 10.0\%$$

Thus, the recombination frequency for genes *j* and *r* is 10.0, which gives us an estimated map distance of 10.0 map units.

To summarize, we have generated a genetic map of the three genes in the example (Figure 15.9). The example has illustrated that the three-point testcross is an effective way to establish the order of genes and calculate map distances.

To compute the map distance between the two outside genes, we simply add the two map distances. In the example, the *p – r* distance is 20.8 + 10.0 = 30.8 mu. This map distance also can be computed directly from the data by combining the two formulas discussed previously:

$$\text{distance} = \frac{(\text{sco in region I}) + (\text{dco}) + (\text{sco in region II}) + (\text{dco})}{\text{total progeny}} \times 100\%$$

$$= \frac{(\text{sco in region I}) + (\text{sco in region II}) + (2 \times \text{dco})}{\text{total progeny}} \times 100\%$$

$$= \frac{52 + 46 + 22 + 22 + 2(4 + 2)}{500} \times 100\%$$

$$= \frac{98 + 44 + 2(6)}{500} \times 100\%$$

$$= 30.8 \text{ map units}$$

KEYNOTE

The map distance between genes can be calculated from the results of testcrosses between strains carrying appropriate genetic markers. The unit of genetic distance is the map unit (mu), where 1 mu is defined as the interval in which 1 percent crossing-over takes place. Gene mapping crosses produce data in the form of recombination frequencies, which are used to estimate map distance, where 1 mu is equivalent to a recombination frequency of 1 percent. Recombination frequencies are not identical to crossover frequencies and typically underestimate the true map distance.

Interference and Coincidence. The recombination frequencies determined by three-point mapping are useful in elaborating the overall organization of genes on a chromosome and in telling us a little about the recombination mechanisms themselves. For example, we computed a recombination frequency of 20.8 between genes *p* and *j* and a recombination frequency of 10.0 between genes *j* and *r* in the previous three-point testcross example. If crossing-over in region I is independent of crossing-over in region II, then the probability of a double crossover in the two regions is equal to the product of the probabilities of the two events occurring separately; that is,

$$\frac{\substack{\text{recombination frequency,} \\ \text{region I}}}{100} \times \frac{\substack{\text{recombination frequency,} \\ \text{region II}}}{100}$$

$$= 0.208 \times 0.100 = 0.0208$$

or 2.08 percent double crossovers are expected to occur. However, only $^6\!/_{500} = 1.2$ percent double crossovers occurred in this cross (classes 7 and 8).

It is characteristic of mapping crosses that double-crossover progeny typically do not appear as often as the map distances between the genes lead us to expect. That is, in some way, the presence of one crossover interferes with the formation of another crossover

Figure 15.9

Genetic map of the *p–j–r* region of the chromosome computed from the recombination data in Figure 15.8.

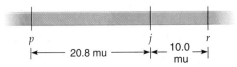

nearby, a phenomenon called **interference**. The extent of interference is expressed as a **coefficient of coincidence**; that is,

$$\text{coefficient of coincidence} = \frac{\text{observed double crossover frequency}}{\text{expected double crossover frequency}}$$

and

interference = 1 − coefficient of coincidence

For the portion of the map in our example, the coefficient of coincidence is

0.012/0.0208 = 0.577

A coefficient of coincidence of 1 means that, in a given region, all double crossovers occurred that were expected on the basis of two independent events; there is no interference, so the interference value is zero. If the coefficient of coincidence is zero, none of the expected double crossovers occurred. Here, there is total interference, with one crossover completely preventing a second crossover in the region under examination; the interference value is 1. These examples show that coincidence values and interference values are inversely related. In our mapping example, the coefficient of coincidence of 0.577 means that the interference value is 0.423. Only 57.7 percent of the expected double crossovers took place in the cross.

KEYNOTE

The occurrence of a crossover may interfere with the occurrence of a second crossover nearby. The extent of interference is expressed by the coefficient of coincidence, which is calculated by dividing the number of observed double crossovers by the number of expected double crossovers. The coefficient of coincidence ranges from zero to unity, and the extent of interference is measured as 1 minus the coefficient of coincidence.

Calculating Accurate Map Distances

Map units between linked genes, strictly speaking, are defined in terms of the crossover frequency, whereas, operationally, geneticists quantify the frequency of recombinants in genetic crosses. The crossover frequency and the recombination frequency are not identical, so the latter often leads to an underestimation of the true map distance. How, then, do we obtain accurate map distances for linked genes?

To answer this question, we need to focus on the consequences of crossovers between linked genes. Consider a hypothetical case of two allelic pairs (a^+/a and

Figure 15.10

Progeny of single and double crossovers. (a) A single crossover between linked genes generates recombinant gametes. (b) A double crossover between linked genes gives parental gametes.

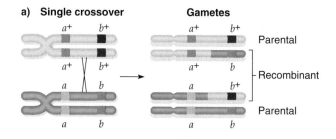

a) Single crossover

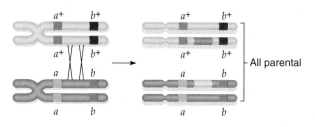

b) Double crossover (two-strand)

b^+/b) linked in coupling and separated by quite a distance on the same chromosome. Figure 15.10a shows that a single crossover results in recombination of the two allelic pairs, producing two parental and two recombinant gametes. The same result will occur for any odd number of crossovers in the region between the genes. Figure 15.10b shows that a double crossover involving two of the four chromatids does not result in recombination of the allelic pairs, so only parental gametes result. Parental gametes also result for any even number of crossovers between the two linked genes. However, the crossover frequency between genes is a measure of the distance between them. Therefore, because the double crossover in Figure 15.10b did not generate recombinant gametes, two crossover events will go uncounted, and the map distance based on recombination frequency between genes a and b will be underestimated.

In gene mapping, if no more than a single crossover occurs between linked genes, there is a direct linear relationship between the genetic map distance and the observed recombination frequency, because the recombination frequency then equals the crossover frequency. In practice, we see this relationship only when genetic map distances are small—that is, when genes are between 0 and approximately 7 mu apart. To turn this statement around, map distances based on recombination frequencies of 7 percent or less are highly accurate. As the distance between genes increases beyond this point, the chance of multiple crossovers increases, and there is no longer an exact linear relationship between map distance and recombination frequency because some crossovers go uncounted. As a result, it is difficult to obtain an accurate measure of map distance when multiple crossovers are involved.

Fortunately, mathematical formulas, called **mapping functions,** have been derived and define the relationship between map distance and recombination frequency. A particular mapping function that assumes no interference between crossovers is shown in Figure 15.11. You can see the direct relationship between map distance and recombination frequency at 7 mu or less, and the curve slowly approaches the limit recombination frequency of 50 percent. Just to pick a couple of points, when the recombination frequency is 20 percent, the true map distance is almost 30 mu, and when the recombination frequency is 30 percent, the true map distance is almost 50 mu. In general, mapping functions require some basic assumptions about the frequency of crossovers compared with distance between genes. Therefore, the usefulness of applying the mapping functions depends on the validity of the assumptions.

KEYNOTE

At genetic distances greater than about 7 mu, the incidence of multiple crossovers causes the recombination frequency to be an underestimate of the crossover frequency and hence of the true map distance. Mapping functions can be used to correct for the effects of multiple crossovers and thereby give a more accurate map distance.

Figure 15.11

A mapping function for relating map distance and recombination frequency. This particular mapping function was developed by J. B. S. Haldane and assumes no interference between crossovers. The variable d is the crossover frequency and e is the base of natural logarithms.

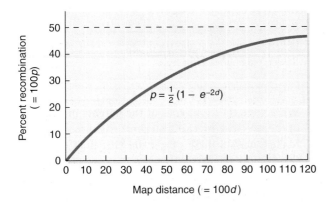

$$p = \tfrac{1}{2}(1 - e^{-2d})$$

Summary

In this chapter, we discussed linkage, crossing-over, and gene mapping in eukaryotes. The production of genetic recombinants results from physical exchanges between homologous chromosomes in meiosis. The exchange of parts of chromatids is called crossing-over, and the site of crossing-over is called a chiasma. Crossing-over is a reciprocal event that, in eukaryotes, occurs at the four-strand stage in prophase I of meiosis.

Gene mapping is the process of locating the position of genes in relation to one another on the chromosome. The first step is to show that genes are linked (located on the same chromosome), which is indicated by the fact that they do not assort independently in crosses. Then crosses are done to determine the map distance between the linked genes. The unit of genetic distance is the map unit (mu), where 1 mu is the interval in which 1 percent crossing-over takes place. However, gene-mapping crosses generate recombination frequency data; when the genes are not closely linked, multiple crossovers often occur between them, leading to underestimates of map distances. Nonetheless, traditionally, recombination frequencies have been used by geneticists to give an approximation of map distances. Mathematically derived mapping functions can be used to calculate more accurate map distances based on recombination frequencies.

Analytical Approaches to Solving Genetics Problems

Q15.1 In corn, the gene for colored (C) seeds is completely dominant to the gene for colorless (c) seeds. Similarly, a single gene pair controls whether the endosperm (the part of the seed that contains the food stored for the embryo) is full or shrunken. Full (S) is dominant to shrunken (s). A true-breeding colored, full-seeded plant was crossed with a colorless, shrunken-seeded one. The F_1 colored, full plants were testcrossed to the doubly recessive type—that is, colorless and shrunken. The result was as follows:

colored, full	4,032
colored, shrunken	149
colorless, full	152
colorless, shrunken	4,035
Total	8,368

Is there evidence that the gene for color and the gene for endosperm shape are linked? If so, what is the distance between the two loci?

A15.1 The best approach is to begin by diagramming the cross, using gene symbols:

P: colored and full × colorless and shrunken
      CC SS   *cc ss*
                    ↓
F$_1$:   colored and full
                         Cc Ss

Testcross: colored and full × colorless and shrunken
              Cc Ss   *cc ss*

If the genes were unlinked, a 1:1:1:1 ratio of colored and full : colored and shrunken : colorless and full : colorless and shrunken would be the progeny of this testcross. By inspection, we can see that the actual progeny deviate a great deal from this ratio, showing a 27:1:1:27 ratio instead. If we did a chi-square test (using the actual numbers, not the percentages or ratios), we would see immediately that the hypothesis that the genes are unlinked is invalid, and we must consider the two genes to be linked in coupling. More specifically, the parental combinations (colored, full and colorless, shrunken) are more numerous than expected, whereas the recombinant types (colorless, full and colored, shrunken) are correspondingly less numerous than expected. This result comes directly from the inequality of the four gamete types produced by meiosis in the colored and full F$_1$ parent.

Given that the two genes are linked, the crosses can be diagrammed to reflect their linkage as follows:

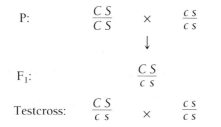

To calculate the map distance between the two genes, we need to compute the frequency of crossovers in that region of the chromosome during meiosis. We cannot do that directly, but we can compute the percentage of recombinant progeny:

ental types: colored, full   4,032
                     colorless, shrunken   4,035
                                            ―――――
                                            8,067

nt types: colored, shrunken   149
                     colorless, full   152
                                            ―――――
                                             301

…es about 3.6 percent recombinant types
…nd about 96.4 percent parental types

(8,067/8,368 × 100). Since the recombination frequency can be used directly as an indication of map distance, especially when the distance is small, we can conclude that the distance between the two genes is 3.6 mu (3.6 cM).

We would get approximately the same result if the two genes were in repulsion rather than in coupling. That is, the crossovers are occurring between homologous chromosomes, regardless of whether there are genetic differences in the two homologues that we, as experimenters, use as markers in genetic crosses. This same cross in repulsion would be as follows:

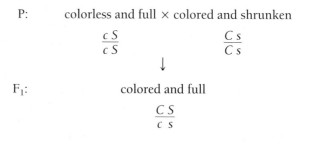

Data from an actual testcross of the F$_1$ with colorless and shrunken (*cc ss*) gave 638 colored and full (recombinant) : 21,379 colored and shrunken (parental) : 21,906 colorless and full (parental) : 672 colorless and shrunken (recombinant), with a total of 44,595 progeny. Thus, 2.94 percent were recombinants, for a map distance between the two genes of 2.94 mu, a figure reasonably close to the results of the cross made in coupling.

Q15.2 In the Chinese primrose, slate-colored flower (*s*) is recessive to blue flower (*S*), red stigma (*r*) is recessive to green stigma (*R*), and long style (*l*) is recessive to short style (*L*). All three genes involved are on the same chromosome. The F$_1$ of a cross between two true-breeding strains, when testcrossed, gave the following progeny:

Phenotype	Number of Progeny
slate flower, green stigma, short style	27
slate flower, red stigma, short style	85
blue flower, red stigma, short style	402
slate flower, red stigma, long style	977
slate flower, green stigma, long style	427
blue flower, green stigma, long style	95
blue flower, green stigma, short style	960
blue flower, red stigma, long style	27
Total	3,000

a. What were the genotypes of the parents in the cross of the two true-breeding strains?

b. Make a map of these genes, showing their order and the distances between them.

c. Derive the coefficient of coincidence for interference between the given genes.

A15.2

a. With three gene pairs, eight phenotypic classes are expected, and eight are observed. The reciprocal pairs of classes with the most representatives are those resulting from no crossovers, and these pairs can tell us the genotypes of the original parents. The two classes are slate, red, long and blue, green, short. Thus, the F_1 triply heterozygous parent of this generation must have been *S R L/s r l*, so the true-breeding parents were *S R L/S R L* (blue, green, short) and *s r l/s r l* (slate, red, long).

b. The order of the genes can be determined by inspecting the reciprocal pairs of phenotypic classes that represent the results of double crossing-over. These classes have the least numerous representatives, so the double-crossover classes are slate, green, short (*s R L*) and blue, red, long (*S r l*). The gene pair that has changed its position relative to the other two pairs of alleles is the central gene, *S/s* in this case. Therefore, the order of genes is *R S L* (or *L S R*). We can diagram the F_1 testcross as follows:

$$\frac{R\ S\ L}{r\ s\ l} \times \frac{r\ s\ l}{r\ s\ l}$$

A single crossover between the *R* and *S* genes gives the green, slate, long (*R s l*) and red, blue, short (*r S L*) classes, which have 427 and 402 members, respectively, for a total of 829. The double-crossover classes have already been defined, and they yield 54 progeny. The map distance between *R* and *S* is given by the crossover frequency in that region, which is the sum of the single crossovers and double crossovers, divided by the total number of progeny and then multiplied by 100 percent. Thus,

$$\frac{829 + 54}{3{,}000} \times 100\% = \frac{883}{3{,}000} \times 100\%$$

$$= 29.43\% \text{ or } 29.43 \text{ map units}$$

With similar logic, the distance between *S* and *L* is given by the crossover frequency in that region, which is the sum of the single-crossover and double-crossover progeny classes, divided by the total number of progeny. The single-crossover progeny classes are green, blue, long (*R S l*) and red, slate, short (*r s L*), which have 95 and 85 members, respectively, for a total of 180. The map distance is given by

$$\frac{180 + 54}{3{,}000} \times 100\% = \frac{234}{3{,}000} \times 100\%$$

$$= 7.8\% \text{ or } 7.8 \text{ map units}$$

The data we have derived give us the following map:

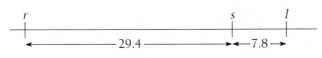

c. The coefficient of coincidence is given by

$$\frac{\text{frequency of observed double crossovers}}{\text{frequency of expected double crossovers}}$$

The frequency of observed double crossovers is $54/3{,}000 = 0.018$. The frequency of expected double crossovers is the product of the map distances between *r* and *s* and between *s* and *l*—that is, $0.294 \times 0.078 = 0.023$. The coefficient of coincidence, therefore, is $0.018/0.023 = 0.78$ In other words, 78 percent of the expected double crossovers did indeed take place; thus, there was 22 percent interference.

Questions and Problems

15.1 The cross $a^+a^+\ b^+b^+ \times aa\ bb$ produces an F_1 that is phenotypically $a^+\ b^+$. Its F_2 phenotypes appear in the following numbers:

$a^+\ b^+$	110
$a^+\ b$	16
$a\ b^+$	19
$a\ b$	15
Total	160

What F_2 numbers would be expected if the *a* and *b* loci assort independently? Use a chi-square test, to evaluate whether the two loci are linked or assort independently.

*****15.2** In corn, a dihybrid for the recessives *a* and *b* is testcrossed. The distribution of the phenotypes is as follows:

A B	122
A b	118
a B	81
a b	79

Test the hypothesis that these genes are assorting independently using a chi-square test. Explain tentatively any deviation from expected values, and tell how you would test your explanation.

15.3 The F_1 from a cross of *A B/A B* × *a b/a b* is testcrossed, resulting in the following phenotypic ratios:

A B	308
A b	190
a b	292
a B	210

What is the frequency of recombination between gen*A* and *b*?

15.4 In *Drosophila*, the mutant black (*b*) has a black body, and the wild type has a grey body; the mutant vestigial (*vg*) has wings that are much shorter and crumpled compared with the long wings of the wild type. In the following cross, the true-breeding parents are listed together. In the counts of offspring of F_1 females × black and vestigial males:

P   black and normal × grey and vestigial
F$_1$ females × black and vestigial males

Progeny: grey, normal   283
            grey, vestigial   1,294
            black, normal   1,418
            black, vestigial   241

Use these data to calculate the map distance between the black and vestigial genes.

*15.5 In *Drosophila*, the vestigial (*vg*) gene is located on chromosome 2. Homozygous *vg/vg* animals have incompletely formed vestigial wings; *vg*$^+$/− animals have wild-type long wings. A new eye mutation called maroonlike (*m*) is isolated. Homozygous *m/m* animals have maroon-colored eyes; *m*$^+$/− animals have wild-type bright-red eyes. The location of the *m* gene is unknown, and you are asked to design an experiment to determine whether it is on chromosome 2.

You cross virgin maroon-eyed females to vestigial males and obtain all wild-type F$_1$ progeny. Then you allow the F$_1$ offspring to interbreed. As soon as the F$_2$ offspring start to hatch, you begin to classify the flies. Among the first six newly hatched flies, you find four wild type, one vestigial-winged red-eyed fly, and one vestigial-winged maroon-eyed fly. You immediately conclude that (1) *m* is not X linked and (2) *m* is not linked to *vg*. How could you tell on the basis of this small sample? On what chromosomes might *m* be located? (*Hint:* There is no crossing-over in male *Drosophila* flies.)

*15.6 Use the following two-point recombination data to map the genes concerned, and show the order and the length of the shortest intervals:

Gene Loci	% Recombination	Gene Loci	% Recombination
a,b	50	b,d	13
a,c	15	b,e	50
a,d	38	c,d	50
a,e	8	c,e	7
b,c	50	d,e	45

15.7 Use the following two-point recombination data to map the genes concerned, and show the order and the length of the shortest intervals:

Gene Loci	% Recombination	Gene Loci	% Recombination
a,	50	c,d	50
a,d	17	c,e	50
a,e	50	c,f	7
a,f	50	c,g	19
a,g	12	d,e	7
b,c	3	d,f	50
b,d	50	d,g	50
b,e	2	e,f	50
b,f		e,g	50
b,g		f,g	15

15.8 The following data are from Bridges and Morgan's work on recombination between the genes black (black body color), curved (curved wings), purple (purple eyes), speck (black specks on wings), and vestigial (crumpled wings) in chromosome 2 of *Drosophila*:

Genes in cross	Total progeny	Number of recombinants
black, curved	62,679	14,237
black, purple	48,931	3,026
black, speck	685	326
black, vestigial	20,153	3,578
curved, purple	51,136	10,205
curved, speck	10,042	3,037
curved, vestigial	1,720	141
purple, speck	11,985	5,474
purple, vestigial	13,601	1,609
speck, vestigial	2,054	738

On the basis of the data, map the chromosome for these five genes as accurately as possible. (Remember that determinations for short distances are more accurate than those for long ones.)

15.9 A corn plant known to be heterozygous at three loci is testcrossed. The progeny phenotypes and numbers are as follows:

+	+	+	455
a	b	c	470
+	b	c	35
a	+	+	33
+	+	c	37
a	b	+	35
+	b	+	460
a	+	c	475
		Total	2,000

Give the gene arrangement, linkage relationships, and map distances.

*15.10 Genes *a* and *b* are linked, with 10 percent recombination. What would be the phenotypes, and the probability of each, among progeny of the following cross?

$$\frac{a\ b^+}{a^+\ b} \times \frac{a\ b}{a\ b}$$

*15.11 Genes *a* and *b* are sex linked and are located 7 mu apart on the X chromosome of *Drosophila*. A female of genotype *a*$^+$ *b*/*a* *b*$^+$ is mated with a wild-type male (*a*$^+$ *b*$^+$/Y).

a. What is the probability that one of her sons will be either *a*$^+$ *b*$^+$ or *a* *b*$^+$ in phenotype?

b. What is the probability that one of her daughters will be *a*$^+$ *b*$^+$ in phenotype?

15.12 In maize, the dominant genes A and C are both necessary for colored seeds. Homozygous recessive plants give colorless seed, regardless of the genes at the second locus. Genes A and C show independent segregation, and the recessive mutant gene waxy endosperm (wx) is linked with C (20 percent recombination). The dominant Wx allele results in starchy endosperm.

a. What phenotypic ratios would be expected when a plant of constitution $c\ Wx/C\ wx\ A/A$ is testcrossed?

b. What phenotypic ratios would be expected when a plant of constitution $c\ Wx/C\ wx\ A/a$ is testcrossed?

15.13 In tomatoes, tall vine is dominant over dwarf, and spherical fruit shape is dominant over pear shape. Vine height and fruit shape are linked, showing 20 percent recombination. A certain tall, spherical-fruited tomato plant is crossed with a dwarf, pear-fruited plant. The progeny are 81 tall, spherical; 79 dwarf, pear; 22 tall, pear; and 17 dwarf, spherical. Another tall and spherical plant crossed with a dwarf and pear plant produces 21 tall, pear; 18 dwarf, spherical; 5 tall, spherical; and 4 dwarf, pear. What are the genotypes of the two tall and spherical plants? It they were crossed, what types and frequencies of offspring would they produce?

*15.14 Genes a and b are on one chromosome, 20 mu apart; c and d are on another chromosome, 10 mu apart. Genes e and f are on yet another chromosome and are 30 mu apart. A homozygous $A\ B\ C\ D\ E\ F$ individual is crossed to an $a\ b\ c\ d\ e\ f$ individual, and the resulting F_1 is crossed back to an $a\ b\ c\ d\ e\ f$ individual. What are the chances of getting individuals of the following phenotypes in the progeny?

a. $A\ B\ C\ D\ E\ F$
b. $A\ B\ C\ d\ e\ f$
c. $A\ b\ c\ D\ E\ f$
d. $a\ B\ C\ d\ e\ f$
e. $a\ b\ c\ D\ e\ F$

*15.15 Genes d and p occupy loci 5 mu apart in the same autosomal linkage group. Gene h is in a different autosomal linkage group. What types of offspring are expected, and what is the probability of each, when individuals of the following genotypes are testcrossed?

a. $\dfrac{D\ P}{d\ p}\ \dfrac{h}{h}$ **b.** $\dfrac{d\ P}{D\ p}\ \dfrac{H}{h}$

15.16 A hairy-winged (h) *Drosophila* female is mated with a yellow-bodied (y), white-eyed (w) male. The F_1 are all wild type. The F_1 progeny are then crossed, and the F_2 that emerge are as follows:

Females:	wild type	757
	hairy	243
Males:	wild type	390
	hairy	130
	yellow	4
	white	3
	hairy, yellow	1
	hairy, white	2
	yellow, white	360
	hairy, yellow, white	110

Give genotypes of the parents and the F_1 and show the linkage relations and distances where possible.

15.17 In the Maltese bippy, amiable (A) is dominant to nasty (a), benign (B) is dominant to active (b), and crazy (C) is dominant to sane (c). A true-breeding amiable, active, crazy bippy was mated, with some difficulty, to a true-breeding nasty, benign, sane bippy. An F_1 individual from this cross was then used in a testcross (to a nasty, active, sane bippy) and produced, in typical prolific bippy fashion, 4,000 offspring. From an ancient manuscript titled *The Genetics of the Bippy, Maltese and Other*, you discover that all three genes are autosomal, that a is linked to b, but not to c, and that the map distance between a and b is 20 mu.

a. Predict all the expected phenotypes and the numbers of each type from this cross.

b. Which phenotypic classes would be missing had a and b shown complete linkage?

c. Which phenotypic classes would be missing if a and b were unlinked?

d. Again, assuming a and b to be unlinked, predict all the expected phenotypes of nasty bippies and the frequencies of each type resulting from a self-cross of the F_1.

15.18 In the following table, continuous bars indicate linkage and the order of linked genes is as shown:

Parent Genotypes	Number of Different Possible Gametes	Least-Frequent Classes	
$\dfrac{A\ b\ C}{a\ B\ c}$	_____	____	____
$\dfrac{A\ b\ C}{a\ B\ c}$	_____	____	____
$\dfrac{A\ b\ C\ D}{a\ B\ c\ d}$	_____	____	____
$\dfrac{A\ b\ C\ D\ E\ f}{a\ B\ C\ d\ e\ f}$	_____	____	____
$\dfrac{b\ D}{B\ d}$	_____	____	____

Fill in the blanks in the table. In the rightmost column, show two gamete genotypes, unless all types are equally frequent, in which case write "none."

*15.19 Genes at loci f, m, and w are linked, but their order is unknown. The F_1 heterozygotes from a cross of $FF\ MM\ WW \times ff\ mm\ ww$ are testcrossed. The most frequent phenotypes in testcross progeny will be $F\ M\ W$ and $f\ m\ w$, regardless of what the gene order turns out to be.

a. What classes of testcross progeny (phenotypes) will be least frequent if locus m is in the middle?

b. What classes will be least frequent if locus f is in the middle?

c. What classes will be least frequent if locus w is in the middle?

15.20 The following numbers were obtained for testcross progeny in *Drosophila* (phenotypes):

$+\ m\ +$	218
$w\ +\ f$	236
$+\ +\ f$	168
$w\ m\ +$	178
$+\ m\ f$	95
$w\ +\ +$	101
$+\ +\ +$	3
$w\ m\ f$	1
Total	1,000

Construct a genetic map.

*15.21 Three of the many recessive mutations in *Drosophila melanogaster* that affect body color, wing shape, or bristle morphology are black (b) body, versus grey in the wild type; dumpy (dp), obliquely truncated wings, versus long wings in the wild type; and hooked (hk) bristles at the tip, versus nonhooked bristles in the wild type. From a cross of a dumpy female with a black, hooked male, all the F_1 were wild type for all three characters. The testcross of an F_1 female with a dumpy, black, hooked male gave the following results:

wild type	169
black	19
black, hooked	301
dumpy, hooked	21
hooked	8
hooked, dumpy, black	172
dumpy, black	6
dumpy	304
Total	1,000

a. Construct a genetic map of the linkage group (or groups) these genes occupy. If applicable, show the order and give the map distances between the genes.

b. Determine the coefficient of coincidence for the portion of the chromosome involved in the cross. How much interference is there?

15.22 The frequencies of gametes of different genotypes, determined by testcrossing a triple heterozygote, are as shown in the following table:

Gamete genotype	%
$+\ +\ +$	12.9
$a\ b\ c$	13.5
$+\ +\ c$	6.9
$a\ b\ +$	6.5
$+\ b\ c$	26.4
$a\ +\ +$	27.2
$a\ +\ c$	3.1
$+\ b\ +$	3.5
Total	100.0

a. Which gametes are known to have been involved in double crossovers?

b. Which gamete types have not been involved in any exchanges?

c. The order shown is not necessarily correct. Which gene locus is in the middle?

15.23 Two normal-looking *Drosophila* are crossed and yield the following phenotypes among their progeny:

Females:	$+\ +\ +$	2,000
Sons:	$+\ +\ +$	3
	$a\ b\ c$	1
	$+\ b\ c$	839
	$a\ +\ +$	825
	$a\ b\ +$	86
	$+\ +\ c$	90
	$a\ +\ c$	81
	$+\ b\ +$	75
	Total	4,000

Show the parental genotypes, the gene arrangement in the female parent, the map distances, and the coefficient of coincidence.

15.24 The questions that follow make use of this genetic map:

Calculate

a. the frequency of $j\ b$ gametes from a $J\ B/j\ b$ genotype

b. the frequency of $A\ M$ gametes from an $a\ M/A\ m$ genotype

c. the frequency of $J\ B\ D$ gametes from a $j\ B\ d/J\ b\ D$ genotype

d. the frequency of $J\ B\ d$ gametes from a $j\ B\ d/J\ b\ D$ genotype

e. the frequency of $j\ b\ d/j\ b\ d$ genotypes in a $j\ B\ d/J\ b\ D \times j\ B\ d/J\ b\ D$ mating

f. the frequency of $A\ k\ F$ gametes from an $A\ K\ F/a\ k\ f$ genotype

*15.25 A female *Drosophila* carries the recessive mutations a and b in repulsion on the X chromosome. (She is heterozygous for both.) She is also heterozygous for an X-linked recessive lethal allele, *l*. When she is mated to a true-breeding, normal male, she yields the following progeny:

Females: 1,000 + +

Males: 405 *a* +
        44 + *b*
        48 + +
         2 *a b*

Draw a chromosome map of the three genes, in the proper order and with map distances as nearly as you can calculate them.

*15.26 The following *Drosophila* cross is done:

$$\frac{a + b}{+ c +} \times \xrightarrow{a \, c \, b}$$

Predict the numbers of phenotypes of male and female progeny that will emerge if the gene arrangement is as shown, the distance between *a* and *c* is 14 mu, the distance between *c* and *b* is 12 mu, the coefficient of coincidence is 0.3, and the number of progeny is 2,000.

15.27 A farmer who raises rabbits wants to break into the Easter market. He has stocks of two true-breeding lines. One is hollow and long eared, but not chocolate, and the second is solid, short eared, and chocolate. Hollow (*h*), long ears (*le*), and chocolate (*ch*) are all recessive and autosomal and are linked as shown in the following map:

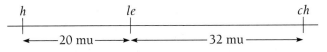

The farmer can generate a trihybrid by crossing his two lines, and at great expense he is able to obtain the services of a male who is homozygous recessive at all three loci to cross with his F₁ females. The farmer has buyers for both solid and hollow bunnies; however, all must be chocolate and long eared. Assuming that interference is zero, if he needs 25 percent of the progeny of the desired phenotypes to be profitable, should he continue with his breeding? Calculate the percentage of the total progeny that will be the desired phenotypes.

15.28 Three different semidominant mutations affect the tails of mice. These mutations are alleles of linked genes, and all three are lethal in the embryo when homozygous. Fused-tail (*Fu*) and kinky-tail (*Ki*) mice have kinky-appearing tails, whereas brachyury (*T*) mice have short tails. A fourth gene, histocompatibility-2 (*H-2*), is linked to the three tail genes and is concerned with tissue transplantation. Mice that are *H-2/+* will accept tissue grafts, whereas *+/+* mice will not. The phenotypes of the progeny are as follows for four crosses, with the normal allele represented by *+*:

(1) $\dfrac{Fu+}{+\ Ki} \times \dfrac{++}{++}$
$\begin{cases} \text{Fused tail} & 106 \\ \text{Kinky tail} & 92 \\ \text{Normal tail} & 1 \\ \text{Fused–kinky tail} & 1 \end{cases}$

(2) $\dfrac{Fu\ H\text{-}2}{+\ \ +} \times \dfrac{++}{++}$
$\begin{cases} \text{Fused tail, accepts grafts} & 88 \\ \text{Normal tail, rejects graft} & 104 \\ \text{Normal tail, accepts graft} & 5 \\ \text{Fused tail, rejects graft} & 3 \end{cases}$

(3) $\dfrac{T\ H\text{-}2}{+\ +} \times \dfrac{++}{++}$
$\begin{cases} \text{Brachy tail, accepts graft} & 1,048 \\ \text{Normal tail, rejects graft} & 1,152 \\ \text{Brachy tail, rejects graft} & 138 \\ \text{Normal tail, accepts graft} & 162 \end{cases}$

(4) $\dfrac{Fu+}{+\ T} \times \dfrac{++}{++}$
$\begin{cases} \text{Fused tail} & 146 \\ \text{Brachy tail} & 130 \\ \text{Normal tail} & 14 \\ \text{Fused–brachy tail} & 10 \end{cases}$

Make a map of the four genes involved in these crosses, giving gene order and map distances between the genes. If more than one map is possible, draw all possible maps.

*15.29 In *Drosophila*, a cross of

$$\frac{a^+ \ b^+ \ c \ \ d \ \ e}{a \ \ b \ \ c^+ d^+ e^+} \times \frac{a \, b \, c \, d \, e}{a \, b \, c \, d \, e}$$

gave 1,000 progeny of the following 16 phenotypes:

Genotype					Number
(1)	a^+	b^+	c	d	e	220
(2)	a^+	b^+	c	d	e^+	230
(3)	a	b	c^+	d^+	e	210
(4)	a	b	c^+	d^+	e^+	215
(5)	a	b^+	c^+	d^+	e	12
(6)	a	b^+	c^+	d^+	e^+	13
(7)	a^+	b	c	d	e^+	16
(8)	a^+	b	c	d	e	14
(9)	a	b^+	c^+	d	e^+	14
(10)	a	b^+	c^+	d	e	13
(11)	a^+	b	c	d^+	e^+	8
(12)	a^+	b	c	d^+	e	8
(13)	a^+	b^+	c^+	d	e^+	7
(14)	a^+	b^+	c^+	d	e	7
(15)	a	b	c	d^+	e^+	6
(16)	a	b	c	d^+	e	7

a. Draw a genetic map of the chromosome, indicating the linkage of the five genes and the number of map units separating each.

b. From the single-crossover frequencies, what would be the expected frequency of $a^+\ b^+\ c^+\ d^+\ e^+$ flies?

15.30 In *Drosophila*, many different mutations have been isolated that affect a normally brick-red eye color caused by the deposition of brown and bright-red pigments. Two X-linked recessive mutations are *w* (white eyes, map position 1.5) and *cho* (chocolate-brown eyes, map position 13.0), with *w* epistatic to *cho*.

a. A white-eyed female is crossed to a chocolate-eyed male, and the normal, red-eyed F_1 females are crossed to either wild-type or white-eyed males. Determine the frequency of the progeny types produced in each cross.

b. The recessive mutation *st* causes scarlet (bright-red) eyes and maps to the third chromosome at position 44. Mutant flies with only *st* and *cho* alleles have white eyes, and *w* is epistatic to *st*. Suppose a true-breeding *w* male is crossed to a true-breeding *cho, st* female. Determine the frequency of the progeny types you would expect if the F_1 females are crossed to true-breeding scarlet-eyed males.

*15.31 Breeders of thoroughbred horses used for racing keep extensive information on pedigrees. Such information can be useful in determining simple inheritance patterns (e.g., chestnut coat color has been determined to be recessive to bay coat color) and in speculating whether racehorses that win competitive races ("classy" horses) share genetic traits. Sharpen Up was a chestnut stallion that was only somewhat successful as a racehorse. At age 4, he was retired from horse racing and put out to stud.

His progeny were very successful: Of 367 foals fathered in the United States, 43 were prizewinners in highly competitive races, and of 200 foals fathered in England, 40 were prizewinners in highly competitive races. A commentator who analyzed Sharpen Up's progeny (and that of other chestnut prize-winners) has suggested that whatever gene combinations produced class (winning horses) were tied to the horses' chestnut coat. Indeed, of the 83 progeny that have shown class (won highly competitive races), about 45 were also chestnut in color. Use a chi-square test to assess whether there is any reason to believe that if there is a gene (or genes) for class, it is linked to the gene for chestnut coat color. Examine this issue using two different assumptions: (1) Sharpen Up was mated equally frequently to homozygous bay, heterozygous bay/chestnut, and homozygous chestnut mares, and (2) Sharpen Up was mated equally frequently to heterozygous bay/chestnut and homozygous chestnut horses. State carefully any additional assumptions and your hypothesis.

*15.32 In an $a^+ b^+/a\, b$ individual, a physical exchange between the *a* and *b* loci occurs in 14 percent of meioses. What percentage of meiosis are expected to produce $a^+ b$ or $a\, b^+$ chromosomes? Explain your answer.

16

Advanced Gene Mapping in Eukaryotes

Ordered tetrads of the fungus, *Neurospora Crassa*.

PRINCIPAL POINTS

- Tetrad analysis is a mapping technique that can be used to map the genes of certain haploid eukaryotic organisms in which the products of a single meiosis—the meiotic tetrad—are contained within a single structure. In these situations, the map distance between genes is computed by analyzing the relative proportion of tetrad types, rather than by analyzing individual progeny.

- In organisms in which the meiotic tetrads are arranged linearly, it is easy to map the distance of a gene from its centromere.

- Crossing-over can also occur during mitosis, although at a frequency much lower than during meiosis. As in meiosis, mitotic crossing-over occurs at a four-strand stage.

- For organisms amenable to such analysis, such as the fungus *Aspergillus nidulans*, mitotic recombination can be used to determine gene order and map distances between genes.

- Due to both ethical and practical issues, human genes cannot be mapped by making crosses and analyzing progeny. A number of approaches are used to determine the linkage relationships between human genes, including analyzing pedigree data recombinationally and physically locating genes on chromosomes by molecularly aided methods.

i SPECIAL TECHNIQUES ARE AVAILABLE FOR constructing linkage maps for eukaryotic organisms beyond the standard genetic mapping crosses. In this chapter you will learn about these techniques. Then, in the iActivity, you can apply what you've learned and further explore one of the advanced mapping techniques.

iActivity

In the previous chapter, we considered the classical principles for mapping genes in eukaryotes by means of recombination analysis. We saw that the outcome of crosses can be used to construct genetic maps, with distances between genes given in map units (centimorgans), and we also saw that map distances are useful in predicting the outcome of other crosses. In the current chapter, we discuss the determination of map distance by tetrad analysis in certain appropriate haploid organisms and the rare incidence of crossing-over in mitosis, and we give an overview of some of the traditional methods used in the construction of genetic linkage maps of the human genome.

Tetrad Analysis in Certain Haploid Eukaryotes

Tetrad analysis is a special mapping technique that can be used to map the genes of haploid eukaryotic organisms in which the products of a single meiosis—the **meiotic tetrad**—are contained within a single structure. The eukaryotic organisms in which this phenomenon occurs are either fungi or single-celled algae, all of which are found predominantly in the haploid state. During sexual reproduction in these organisms, a diploid is formed (transiently in some) which then undergoes meiosis to produce spores that germinate to produce the next haploid generation. In particular, tetrad analysis is used with the yeast

 Go to the iActivity *Mapping Genes by Tetrad Analysis* on your CD-ROM and assume the role of researcher to construct a detailed genetic map using tetrad analysis.

Saccharomyces cerevisiae, the orange bread mold *Neurospora crassa,* and the single-celled alga *Chlamydomonas reinhardtii.* Tetrad analysis allows the researcher to study of the details of meiotic events that would be impossible to study in any other system.

By analyzing the phenotypes of the meiotic tetrads, geneticists can infer the genotypes of each member of the tetrad directly. That is, because haploid organisms have only one copy of each gene, the phenotype is the direct result of the allele that is present. In other words, dominance and recessiveness issues do not come into play as they do in diploid organisms. We will outline this technique shortly.

Before we discuss the principles of tetrad analysis, let us learn a little about how meiotic tetrads arise in the life cycles of organisms upon which tetrad analysis can be performed. The life cycle of baker's yeast (also called budding yeast), *Saccharomyces cerevisiae,* is diagrammed in Figure 16.1. Yeast has two mating types: *MAT*a and *MAT*α (*MAT = mating type*). The haploid cells of this organism reproduce mitotically (the vegetative life cycle), with each new cell arising from the parental cell by budding. Fusion of haploid *MAT*a and *MAT*α cells produces a diploid cell that is stable

Figure 16.1

Life cycle of the yeast *Saccharomyces cerevisiae*.

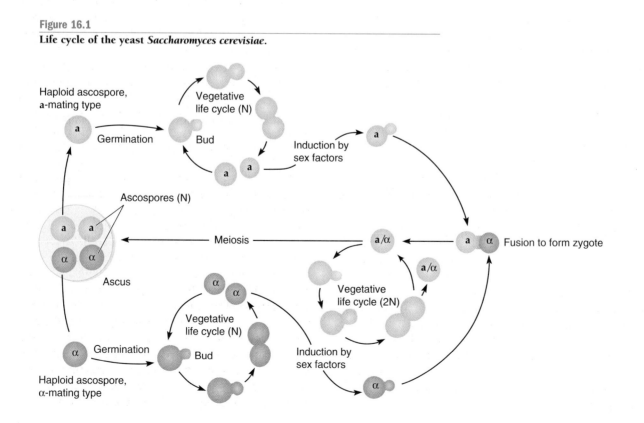

and that also reproduces by budding. Diploid *MATa/MATα* cells sporulate; that is, they go through meiosis. The four haploid meiotic products of a diploid cell (the meiotic tetrad) are called *ascospores* and are contained within a roughly spherical *ascus*. Two of the ascospores are of mating type *MATa*, and two are of mating type *MATα*. When the ascus is ripe, the ascospores are released, and they germinate to produce haploid cells that, on a solid medium, grow and divide to produce a colony. In yeast, the four ascospores are arranged randomly within the ascus in what is called an *unordered tetrad*.

Like yeast, *Chlamydomonas reinhardtii* has haploid vegetative cells. Each individual is a single green algal cell that can swim freely as a result of the motion of its two flagella. (See Figure 1.7j, p. 10). When nitrogen is limited, the cells change morphologically to become gametes so that mating is possible. There are two mating types, designated plus (*mt*⁺) and minus (*mt*⁻). Only gametes of opposite mating types can fuse to produce diploid zygotes. After a maturation process, the zygote enters meiosis. The four haploid meiotic products are contained within a sac as an unordered tetrad, and there are two *mt*⁺ and two *mt*⁻ cells. When these cells are released, they are free swimming, and, by mitosis, they give rise to clones of those meiotic products

The haploid fungus *Neurospora crassa* has a somewhat similar life cycle (see Figure 4.2, p. 69), but in this organism the ascospores are arranged in a linear ascus, an *ordered tetrad*. There are actually eight ascospores in the ascus of this fungus because each of the four meiotic products divides again by mitosis. The order of the four spore pairs within an ascus reflects exactly the orientation of the four chromatids of each tetrad at the metaphase plate in meiosis I. The spores can be isolated in the same order as they are in the ascus, or they can be isolated randomly from the ascus.

Using Random-Spore Analysis to Map Genes in Haploid Eukaryotes

In the three haploid eukaryotes we have discussed, the meiotic products can be collected after they have been released from the ascus (in yeast and *Neurospora*) or sac (in *Chlamydomonas*). In the case of the fungi (yeast and *Neurospora*), spores are germinated and the resulting cultures are analyzed. The free-swimming meiotic products of *Chlamydomonas* are analyzed directly. The haploid nature of the mature stage of all three organisms simplifies the analysis, because this stage is exactly equivalent to that of the gametes produced after meiosis in a diploid eukaryote, so genotypes can be determined directly from the phenotypes. In short, we can make two-point and three-point crosses to map genes on a chromosome, using the same approach as the one we used for a diploid eukaryote.

Calculating Gene–Centromere Distance in Organisms by Using Ordered Tetrads

As mentioned earlier, *Neurospora* produces ordered tetrads. In actuality, there are eight spores in each linear ascus, produced when the four meiotic products undergo one more mitotic division before the ascus is mature. The mitotic division represents the replication of the DNA molecules that have passed through the same meiosis. For the purposes of our genetic discussions, the eight spores can be considered as four pairs. The most interesting thing about ordered tetrads is that their genetic content directly reflects the orientation of the four chromatids of each chromosome pair in the diploid zygote nucleus at metaphase I. *This fact allows us to map the distance between genes and their centromeres.* Locating the centromeres on the genetic maps of chromosomes makes the maps more complete. In more complex organisms, centromeres are located primarily through cytological studies (which often are not possible in haploid eukaryotes because their chromosomes are too small).

In this example, we will map the position of the mating-type locus of *Neurospora* in relation to its centromere. Mating type is a function of whichever allele, *A* or *a,* is present at a locus in linkage group I. If a *Neurospora* strain of mating type *A* is crossed with one of mating type *a*, a diploid zygote of genotype *A/a* results. Figure 16.2 shows the various ways in which this zygote can give rise to the four meiotic products—the four pairs of ascospores in an ascus. For the purposes of illustration, the symbol ● is used to indicate the centromere of the *A* parent and the symbol ○ is used to indicate the centromere of the *a* parent. (In reality, there is no difference between the two.)

If no crossing-over occurs between the mating-type locus and the centromere, the resulting ascospores have the genotypes shown in Figure 16.2a. Centromeres do not separate until just before the second meiotic division, so the spores in the top half of the ascus always have the centromere from one parent (the ● centromere in this case), and the spores in the bottom half of the ascus always have the centromere from the other parent (○ here).

Since the two types of centromeres segregate to different nuclear areas after the first meiotic division, we say that they show *first-division segregation*. Also, because no crossing-over occurs between the mating-type locus and the centromere in Figure 16.2a, *each allelic pair also shows first-division segregation.* That is, after meiosis I, both copies of the *A* allele are at one pole, and both copies of the *a* allele are at the other pole; the *A* and *a* alleles have segregated into different nuclear areas. So after completion of meiosis II and the ensuing mitosis, there is a 4:4 segregation of the *A* and *a* alleles in terms of their positions in the ascus. More specifically, all eight spores in an ascus that show first-division segregation of alleles are parental types: The *A* allele is on the chromosome with a ● centromere, and the *a* allele is on the

Figure 16.2

Determination of gene–centromere distance of the mating-type locus in *Neurospora*. (a) Production of an ascus from a diploid zygote in which no crossing-over occurred between the centromere and the mating-type locus; first-division segregation for the mating-type alleles. (b) Production of asci after a single crossover occurs between the mating-type locus and its centromere. Chromosomes are shown after the crossover is complete. The asci show second-division segregation for the mating-type locus, and the four types of asci are produced in equal proportions.

a) No crossover

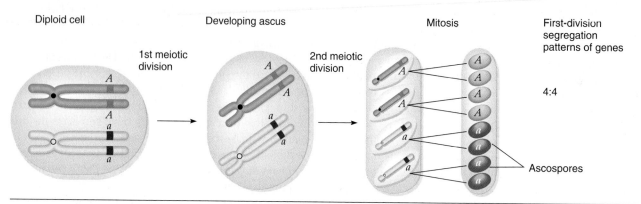

b) Crossover between gene and centromere (four ways)

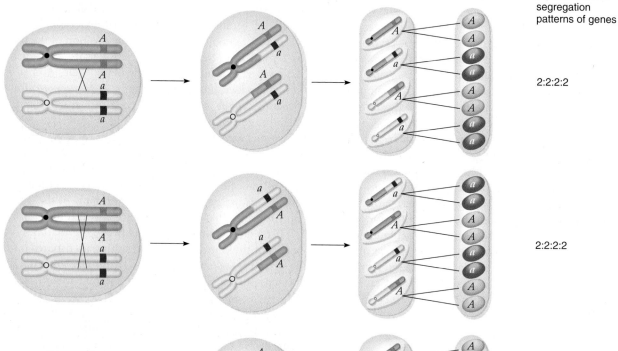

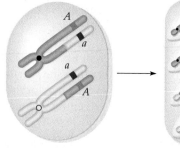

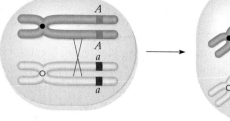

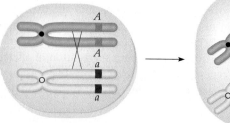

2N zygote nucleus

chromosome with the ○ centromere. Furthermore, because it is equally likely that the four chromatids in the diploid zygote will be rotated 180 degrees, we expect equal numbers of first-division segregation asci in which the four ● *A* spores are in the bottom half and the four ○ *a* spores are in the top half.

To determine the map distance between a gene and its centromere, we measure the crossover frequency between the two chromosomal sites. When a single crossover between the mating-type locus and the centromere occurs, one of four possible ascus types is produced (Figure 16.2b). These four types are generated in equal frequencies because they reflect the four possible orientations of the four chromatids in the diploid zygote at metaphase I. Each has first-division segregation of the centromere; that is, segregation occurred during meiosis I. By contrast, *A* and *a* in Figure 16.2b are both present in each of the two nuclear areas after the first division; they do not segregate into separate nuclei until the second division. This situation is called *second-division segregation* of the alleles; that is, segregation of *A* from *a* occurs during meiosis II. Here, the pattern of gene segregation depends on which chromatids are involved in the crossover event. The 2:2:2:2 (*AAaaAAaa* and *aaAAaaAA*) and 2:4:2 (*AAaaaaAA* and *aaAAAAaa*) second-division segregation patterns are readily distinguishable from the 4:4 (*AAAAaaaa* and *aaaaAAAA*) first-division segregation pattern.

By analyzing ordered tetrads, we can count the number of asci that show second-division segregation for a particular gene marker. For the mating-type locus, about 14 percent of the asci show second-division segregation. This value must then be converted to a map distance via the formula

$$\text{gene–centromere distance} = \frac{\text{percent of second-division tetrads}}{2}$$

That is, if we consider the centromere to be a gene marker, then the nonrecombinant parental types are ● *A* and ○ *a*, and the recombinant types are ● *a* and ○ *A*. In a first-division segregation ascus (Figure 16.2a) all the spores are parentals (i.e., nonrecombinants), whereas in a second-division segregation ascus (Figure 16.2b) half are parentals and half are recombinants. Therefore, to convert the tetrad data to crossover or recombination data, we divide the percentage of second-division segregation asci (14 percent) by 2. Thus, the mating-type gene is 7 mu from the centromere of linkage group I.

In essence, the computation of gene–centromere distance is a special case of mapping the distance between two genes. If ordered tetrads are isolated, then not only can genes be mapped one to another, but each can be mapped to its centromere.

Using Tetrad Analysis to Map Two Linked Genes

Let us see how we can analyze gene linkage relationships by analyzing unordered tetrads.

By making an appropriate cross, a diploid is constructed that is heterozygous for both genes. Then, after meiosis, the resulting tetrads are analyzed. Consider the cross $a^+ b^+ \times a\,b$. Figure 16.3 shows the three different tetrad types that result. A **parental ditype** (PD) tetrad contains only two types of meiotic products, both of which are of the parental type (hence the name). A **tetratype** (T) tetrad contains two parentals (one of each type, here $a^+ b^+$ and $a\,b$) and two recombinants (one of each type, here $a^+ b$ and $a\,b^+$). A **nonparental ditype** (NPD) tetrad contains two types of meiotic products, both of which are of the nonparental (recombinant) types, here $a^+ b$ and $a\,b^+$. (Notice that, in each tetrad type, there is a 2:2 segregation of alleles. Occasionally 3:1 or 1:3 ratios of alleles are seen as a result of a phenomenon called *gene conversion*, which is discussed in Box 16.1.)

Tetrad analysis determines whether two genes are linked. This determination is based on the ways each tetrad type is produced when genes are unlinked and when genes are linked.

Figure 16.4 shows how the three tetrad types are produced when two genes are on different chromosomes. The PD and NPD tetrads result from events in which no crossovers are involved; the metaphase plate orientation of the four chromatids for the two chromosomes determines whether a PD or an NPD tetrad results. Since the two sets of four chromatids align at the metaphase plate independently, the PD and NPD orientations occur with approximately equal frequency. Thus, *if two genes are unlinked, the*

animation

Mapping Linked Genes by Tetrad Analysis

Figure 16.3

Three types of tetrads produced from a cross of $a^+ b^+ \times a\ b$: parental ditype (PD), tetratype (T), and nonparental ditype (NPD).

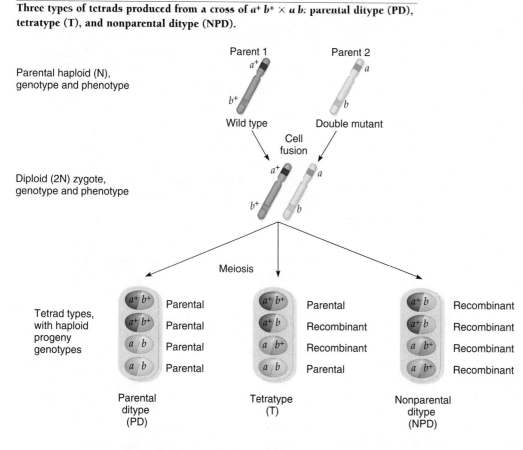

frequency of PD tetrads equals the frequency of NPD tetrads. T tetrads are produced when there is, for example, a single crossover between one of the genes and the centromere on that chromosome. The sum of all possible progeny types equals 50 percent parental types and 50 percent recombinant types when the two genes are unlinked.

Figure 16.5 shows the origins of each tetrad type when two genes are linked on the same chromosome. A PD tetrad results if no crossing-over occurs between the genes (Figure 16.5a). A single crossover (Figure 16.5b) produces two parental and two recombinant chromatids and, hence, a T tetrad. For double crossovers, the chromatid strands involved must be considered. In a two-strand double crossover (Figure 16.5c) the two crossover events involve the same two chromatids, resulting in a PD tetrad. Three-strand double crossovers (Figure 16.5d) involve three of the four chromatids. In either of the two ways in which this event can occur, two recombinant and two parental progeny types are produced in each tetrad, which is a T tetrad. Finally, in four-strand double crossovers (Figure 16.5e), each crossover event involves two distinct chromatids, so all four chromatids of the tetrad are involved. *This rare event is the only way in which NPD tetrads are produced.* In sum, PD tetrads are produced either when there is no crossing-over—the most frequent event—or when there is a two-strand double

crossover, whereas NPD tetrads are produced only by four-strand double crossovers (which are ¼ of all the possible double crossovers and therefore are rare). Accordingly, two genes are considered to be linked if the frequency of PD tetrads is far greater than the frequency of NPD tetrads (i.e., PD ≫ NPD).

Once we know that two genes are linked, and once we have data on the relative numbers of each type of meiotic tetrad, the distance between the two genes can be computed by using a modification of the basic mapping formula:

$$\frac{\text{number of recombinants}}{\text{total number of progeny}} \times 100$$

In tetrad analysis, we analyze types of tetrads rather than individual progeny. To convert the basic mapping formula into tetrad terms, the recombination frequency between genes *a* and *b* becomes

$$\frac{\tfrac{1}{2}\text{T} + \text{NPD}}{\text{total tetrads}} \times 100$$

In essence, we are looking at tetrads with recombinants and determining the proportion of spores in those tetrads that are recombinant. So, in the formula, the ½ T and the NPD represent the recombinants from the cross; the other

Box 16.1 Gene Conversion and Mismatch Repair

The molecular processes of DNA recombination (see Box 15.1, pp. 410–411) can be used to explain some unusual gene segregation patterns. As the main text indicates, each tetrad type—PD, NPD, and T—from a cross of $a\ b \times a^+\ b^+$ in which a and b are linked genes, there is a 2:2 segregation of the alleles.

Occasionally, however, 3:1 and 1:3 rations are seen. Tetrads showing these unusual ratios are proposed to derive from a process called *gene conversion*. **Gene conversion** is a meiotic process of directed change in which one allele directs the conversion of a partner allele to its own form. Any pair of alleles is subject to gene conversion. Consider, for example, the cross $a^+\ m\ b^+ \times a\ m^+\ b$, where gene m is located between genes a and b. The generation by *mismatch repair* of a tetrad showing gene conversion is diagrammed in Box Figure 16.1. The starting point is parental homologous chromosomes synapsed in meiotic prophase (Box Figure 16.1, part 1). If a recombination event occurs between the two inner chromatids (Box Figure 16.1, part 2), a patched duplex can be produced with two mismatches. That is, in these mismatches (heteroduplexes), one of the two DNA strands has the m sequence from one parent and other DNA strand in the helix has the m sequence of the other parent. Both mismatches can be excised and repaired as shown in Box Figure 16.1, part 3. That is, an exonuclease removes a segment of one DNA strand of the molecule, DNA polymerase catalyzes new DNA synthesis, and DNA ligase seals the gap between the new DNA and the preexisting DNA. After the subsequent two meiotic divisions, a tetrad is produced, showing a 3:1 gene conversion for the + allele of m while the outside genetic markers a^+/a and b^+/b retain their parental configuration Box Figure 16.1, part 4.

Finally, if two markers are very close together, they undergo co-conversion so that both exhibit a 1:3 or 3:1 segregation in a tetrad, whereas outside markers segregate in a normal 2:2 ratio.

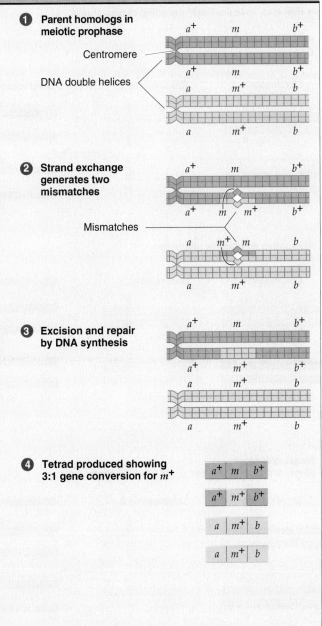

① Parent homologs in meiotic prophase

Centromere

DNA double helices

② Strand exchange generates two mismatches

Mismatches

③ Excision and repair by DNA synthesis

④ Tetrad produced showing 3:1 gene conversion for m^+

Box Figure 16.1

Gene conversion by mismatch repair at two sites. (1) Parent homologs. (2) Recombination between inner two chromatids produces a patched duplex with two mismatches. (3) Both mismatches are repaired by excision and DNA synthesis. (4) Two meiotic divisions produce a tetrad with a 3:1 conversion for the m^+ allele.

$\frac{1}{2}$T and the PD represent the nonrecombinants (the parentals). Thus, the formula does indeed compute the percentage of recombinants. For instance, if there are 200 tetrads with 140 PD, 48 T, and 12 NPD, then the recombination frequency between the genes is

$$\frac{\frac{1}{2}(48) + 12}{200} \times 100 = 18\%$$

If more than two genes are linked in a cross, the data can best be analyzed by considering two genes at a time and by classifying each tetrad into PD, NPD, and T for each pair.

KEYNOTE

In organisms in which all products of meiosis are contained within a single structure, the analysis of the relative proportion of tetrad types provides another way to compute the map distance between genes. If PD = NPD, the two genes are unlinked, whereas if PD > NPD, the two genes are linked. The general formula when two linked genes are being mapped is

$$\frac{\frac{1}{2}T + NPD}{\text{total tetrads}} \times 100$$

Figure 16.4

Origin of tetrad types for a cross *a b* × *a⁺ b⁺* in which the two genes are located on different, independently assorting chromosomes.

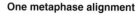

**No crossover:
One metaphase alignment**

Products

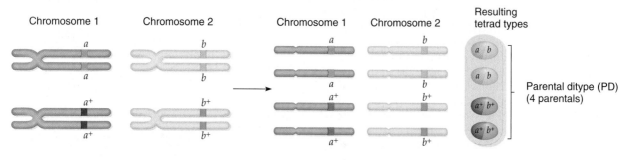

Alternative metaphase alignment

Products

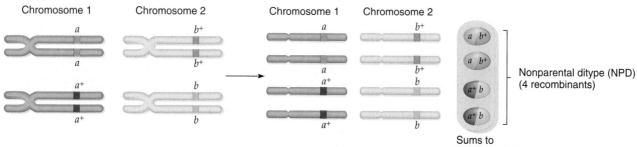

**Single crossover:
Single crossover between *a* gene and its centromere**

Products

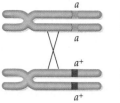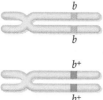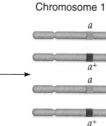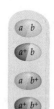

Mitotic Recombination

Discovery of Mitotic Recombination

Crossing-over occurs during mitosis as well as during meiosis. **Mitotic crossing-over** (*mitotic recombination*) produces a progeny cell with a combination of genes which differs from that of the diploid parental cell that entered the mitotic cycle. *Mitotic crossing-over occurs at a stage similar to the four-strand stage of meiosis.* This stage forms only rarely, so the incidence of mitotic crossing-

over between linked genes is much lower than the incidence of meiotic crossing-over.

Mitotic crossing-over was first observed by Curt Stern in 1936 in crosses involving *Drosophila* strains carrying recessive sex-linked mutations that cause short, twisty bristles (singed, *sn*), instead of the normal long, curved bristles, and yellow body color (*y*), instead of the normal grey body color. In flies with a wild-type grey body color, all bristles are black; in yellow-bodied (mutant) flies, the bristles are yellow. From a cross of homozygous *sn y⁺/sn y⁺*

Figure 16.5

Origin of tetrad types for a cross *a b* × *a⁺ b⁺* in which both genes are located on the same chromosome. (a) No crossover. (b) Single crossover. (c–e) Three types of double crossovers.

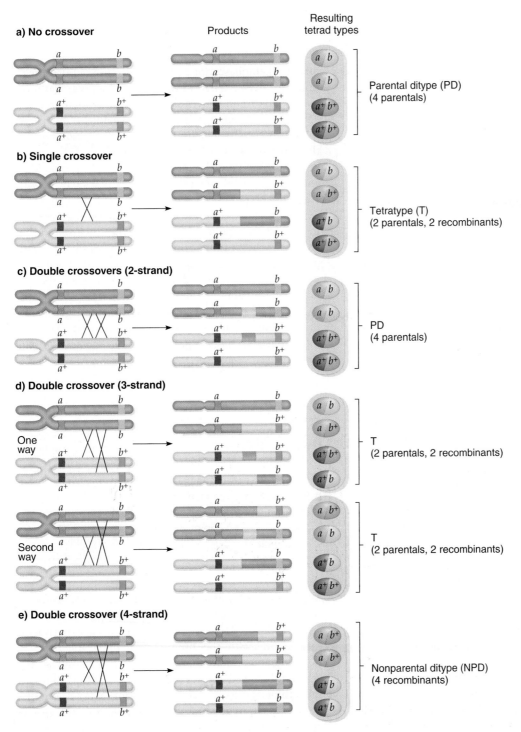

females (grey bodies, singed bristles) with *sn⁺ y*/Y (yellow bodies, normal bristles), Stern found, as expected, that the female F₁ progeny were mostly wild type in appearance: They had grey bodies and normal bristles (genotype *sn y⁺/sn⁺ y*). Some females, however, had patches of yellow

or singed bristles that were not explained by regular gene segregation (Figure 16.6a,b). The origin of these flies could have been explained by chromosome nondisjunction or by chromosomal loss. Other females had **twin spots**—two adjacent regions of mutant bristles, one showing the

Figure 16.6

**Body surface phenotype segregation in a *Drosophila* strain
y⁺ sn/y sn⁺.** The *sn* allele causes short, twisted (singed) bristles,
and the *y* allele results in a yellow body coloration. (a) Single
yellow spot in normal body color background. (b) Single
singed-bristle spot in normal-bristle phenotype background.
(c) Twin spot of yellow color and singed bristles. (From
Principles of Genetics by Robert H. Tamarin, Copyright © 1996.
Reproduced with permission of the McGraw-Hill Companies.)

**a) Single
yellow spot** **b) Single
singed spot** **c) Twin spot**

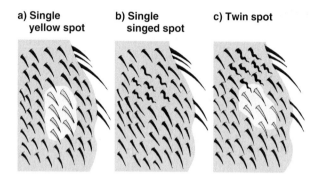

yellow phenotype and the other showing the singed phenotype—a *mosaic* phenotype (Figure 16.6c). For the rest of
the animal, the phenotype was wild type. Stern reasoned
that, because the two parts of a twin spot were always adjacent, the spots must be the reciprocal products of the same
genetic event. The best explanation was that they were
generated by a *mitotic crossing-over* event, an event that
occurs rarely.

The production of twin spots by mitotic crossing-
over is shown in Figure 16.7. This figure also serves to
illustrate the principles underlying the production of
genetic recombinants by mitotic crossing-over. The genotype of the F₁ flies was *sn y⁺/sn⁺ y*, with the mutant alleles in repulsion. As the flies develop, in some cells mitotic
tetrads rarely form after chromosome duplication. In
these tetrads, mitotic crossing-over can occur. For our
example, consider mitotic crossing-over either between
the centromere and the *sn* locus (Figure 16.7, left) or
between the *sn* and the *y* locus (Figure 16.7, right). The
chromatids are numbered, so we can track them to the
progeny cells. After crossing-over has occurred, the chromatid pairs separate and become oriented on the *mitotic*
metaphase plate in one of two possible ways, each of
which is equally likely.

For the crossover between the centromere and the *sn*
locus (see Figure 16.7, left), in one orientation chromatids 1 and 3 will segregate to one daughter nucleus
and chromatids 2 and 4 will segregate to the other
daughter nucleus. The former nucleus is genotypically
homozygous *sn⁺ y/sn⁺ y*, and the latter is homozygous
sn y⁺/sn y⁺. When these cells divide, they produce a yellow patch of tissue and a singed patch of tissue—a twin
spot has been produced. The surrounding tissue, not

involved in any mitotic crossing-over, will be wild type
in phenotype because it is *sn⁺ y/sn⁺ y*. In the other orientation, chromatids 1 and 4 segregate to one nucleus and
chromatids 2 and 3 to the other nucleus, to give genotypes *sn⁺ y/sn y⁺* and *sn y⁺/sn⁺ y*, respectively. Both have
a wild-type phenotype.

In the case of crossing-over between *sn* and *y* (see
Figure 16.7, right), we can consider the two possible
mitotic metaphase orientations and chromatid segregation patterns similarly. In one orientation, segregation of
chromatids 1 and 3 gives *sn⁺ y/sn y* cells, which produces
a yellow spot, and segregation of chromatids 2 and 4 gives
sn⁺ y⁺/sn y⁺ cells, which are wild type in phenotype. In
the other orientation, segregation of chromatids 1 and 4
gives *sn⁺ y/sn y⁺* cells and segregation of chromatids 2
and 3 gives *sn⁺ y⁺/sn y* cells; both cell types have a wild-
type phenotype.

In general, mitotic crossing-over makes all genes distal
to the crossover point (i.e., between the crossover and the
end of the chromosome arm) homozygous *if* the chromatid
pairs align appropriately at the metaphase plate. Such alignment occurs one-half of the time. This generation of
homozygosity applies only to the genes on the *same* chromosome arm as the crossover—that is, to those genes from
the centromere outward. A mitotic crossover on one arm of
the chromosome *has no effect on genes on the other arm of the
chromosome*.

KEYNOTE

Crossing-over can occur during mitosis as well as
during meiosis, although it occurs much more rarely
during mitosis. As in meiosis, mitotic crossing-over
occurs at a four-strand stage. Single crossovers during
mitosis can be detected in a heterozygote because loci
distal to the crossover and on the same chromosome
arm may become homozygous when passed to the
same daughter cell.

Mitotic Recombination in the Fungus *Aspergillus nidulans*
Mitotic recombination has been studied most extensively in fungi. Here we will look at some experiments
carried out with the fungus *Aspergillus nidulans*; the
experiments showed that mitotic recombination analysis can be used to construct genetic maps. Meiotic
recombination analysis is not feasible with *Aspergillus*
because the organism selfs; hence, controlled crosses
cannot be made.

The fungus *Aspergillus nidulans* is a mycelial-form fungus like *Neurospora*, and its colonies are greenish because
of the color of the asexual spores. *Aspergillus* is a suitable organism for genetic studies that use mitotic recombination because (1) the asexual spores of this fungus are

Figure 16.7

Production of the twin spot and the single yellow spot shown in Figure 16.6 by mitotic crossing-over.

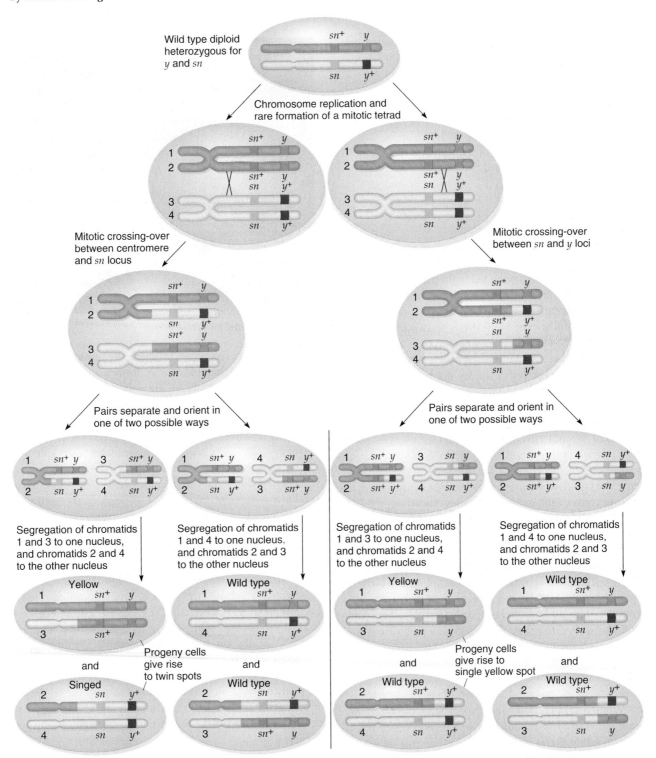

uninucleate—they have a single nucleus; (2) the phenotype of each asexual spore is controlled by the genotype of the nucleus it carries; and (3) two haploid strains can be fused by mixing them together.

As in meiotic recombination analysis, in mitotic recombination analysis a strain must be constructed that is heterozygous for the genes to be studied. In *Aspergillus*, this is done by fusing two haploid strains that differ in nuclear genotypes. The result of such fusion is a mycelium in which the two nuclear types coexist and divide mitotically within the same cytoplasm. Cells of this type are called **heterokaryons** (literally, "different nuclei").

For example, consider the following two haploid strains:

> Strain 1: *w ad⁺ pro paba⁺ y⁺ bi*
> Strain 2: *w⁺ ad pro⁺ paba y bi⁺*

The alleles *ad*, *pro*, *paba*, and *bi* are recessive, and they specify that adenine, proline, para-aminobenzoic acid, and biotin, respectively, must be added to the growth medium in order for the strain carrying those mutant alleles to survive. Thus, either parent strain alone cannot grow without the appropriate supplements. However, a heterokaryon resulting from the fusion of the two strains requires no growth supplements, since all four genes are then heterozygous.

The recessive *w* and *y* alleles control the color of the asexual spores and hence the overall color of the colony. A strain with genotype *w⁺ y⁺* is green, a *w y⁺* strain is white, a *w⁺ y* strain is yellow, and a *w y* strain is white because of epistatic effects. (See Chapter 13, pp. 349–355.) The heterokaryon of strain 1 and strain 2, however, is *not* green, because the color of the uninucleate spores (and hence the colony coloration) is controlled by the genotype of the nucleus contained in each spore. Therefore, the heterokaryon has a mixture of mostly yellow and white spores and has a mottled appearance.

Rarely, two haploid nuclei in the heterokaryon will fuse to produce a diploid nucleus, in a process called diploidization. The spores from these diploid cells will be diploid also and will be green because of their *w⁺/w y⁺/y* genotype. The diploid spores (easily distinguished from haploid cells because they are larger) may then be isolated and cultured for study. The diploid cultures derived from these spores require no growth supplements, because wild-type alleles for all the nutritional requirement genes are present: *w ad⁺ pro paba⁺ y⁺ bi/w⁺ ad pro⁺ paba y bi⁺*.

When a diploid *Aspergillus* spore with the foregoing genotype is used to inoculate a solid growth medium, a predominantly green colony will be produced, with haploid or diploid, and white or yellow (or both), sectors occurring rarely (Figure 16.8). (Green haploid sectors are also produced, but they cannot be distinguished easily from the parental diploid colony.) Whether the sector is haploid or diploid is shown by the spore size.

Figure 16.8

Photo of a green diploid *Aspergillus* colony, genotype *w ad⁺ pro paba⁺ y⁺ bi/w⁺ ad pro⁺ paba y bi⁺* with a yellow and a white sector.

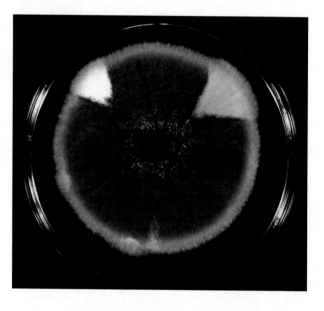

The haploid sectors are produced by **haploidization**— the formation of haploid nuclei from a diploid nucleus. That is, the diploid nuclei are unstable and eventually divide by mitosis (without any chromosome duplication) to produce haploid progeny nuclei, called *haploid segregants*. Let us consider the haploid white sectors in order to exemplify the information these haploids can give us. About half the white haploid sectors have the genotype *w ad⁺ pro paba⁺ y⁺ bi*, and half have the genotype *w ad pro⁺ paba y bi⁺*. With the exception of the common white allele, these two genotypes are reciprocals.

The 50:50 segregation of the two sets of five alleles indicates that they are located on a different chromosome from that carrying the *w* gene. Thus, in some haploid white sectors a chromosome with *ad pro⁺ paba y bi⁺* has segregated, while in others its homolog with *ad⁺ pro paba⁺ y⁺ bi* alleles has segregated. The interpretation of the haploid white-sector data is that the six gene loci are located on two nonhomologous chromosomes; the white gene is on one chromosome, and the other five genes are on the other. The yellow haploid sectors could be analyzed in a similar way. It is not possible to determine gene order by analyzing haploid segregants, although the correct gene order is given in the following diagram:

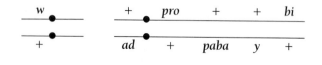

Once the information about which genes are linked on which chromosomes is known, the next stage of mitotic analysis is to establish gene order and determine map distances between genes on the same chromosome. To do this, we must study the diploid white and yellow segregants.

The diploid segregants are produced by mitotic crossing-over. Since this event is very rare, only single crossovers need to be considered. As was discussed earlier, a crossing-over event in mitosis makes all those genes distal to the crossing-over point (on the same chromosome arm) homozygous. Thus, any recessive alleles that are heterozygous in the diploid cell may become homozygous recessive as a result of the crossover, and the recessive phenotype will be seen.

One way in which a diploid yellow sector can arise is diagrammed in Figure 16.9. A mitotic crossover between the *pro* and *paba* genes has produced a segregant that is homozygous for the *y* allele and hence is yellow. The

Figure 16.9

Possible mitotic crossing-over event between the *pro* and *paba* loci that can give rise to a diploid yellow sector in the green diploid *Aspergillus* strain of Figure 16.8.

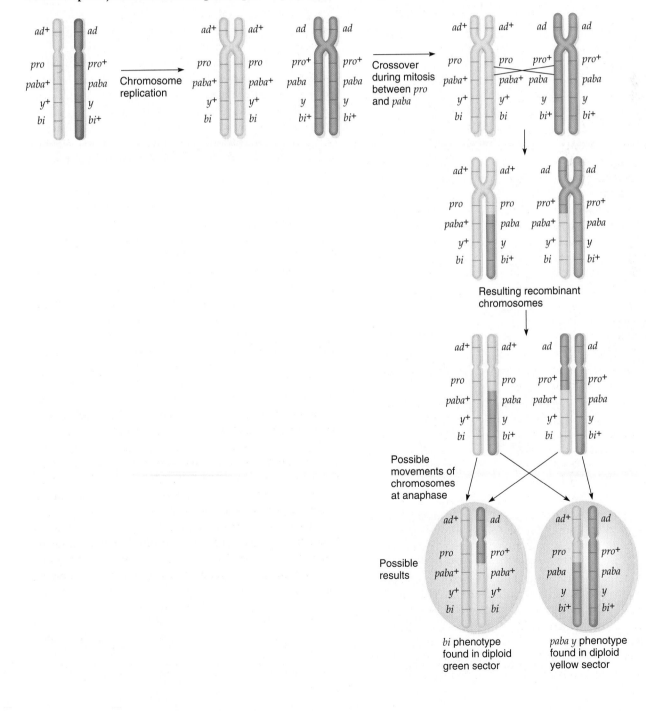

same crossover produces a twin spot that is homozygous y^+/y^+, but since it is green, it is not detected in the overall green color of the colony. The crossover diagrammed has made all genes distal to that point homozygous. Therefore, the yellow segregant is also *paba/paba* and requires para-aminobenzoic acid in order to grow. The homozygosity for the bi^+ allele goes undetected.

One other crossover that could produce a diploid yellow sector is a crossover between the *paba* and *y* loci (Figure 16.10). In this case, the yellow sector is still

Figure 16.10

Production of diploid yellow sector in a green diploid *Aspergillus* strain by mitotic crossing-over between the *paba* and *y* loci.

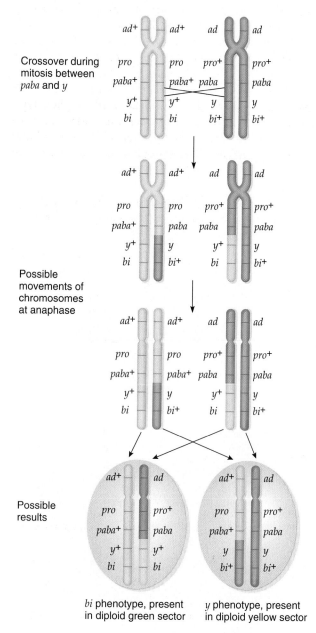

Crossover during mitosis between *paba* and *y*

Possible movements of chromosomes at anaphase

Possible results

bi phenotype, present in diploid green sector

y phenotype, present in diploid yellow sector

heterozygous for *paba*, since mitotic crossing-over produces homozygotes only for genes distal to the crossover point. These facts give us a way to determine the gene order for each chromosome arm. A gene marker that is far away from the centromere is chosen, and then diploid segregants homozygous for that gene as a result of mitotic recombination are isolated. With a number of genes marking the chromosome from the centromere out to the distal marker, it is then a simple matter to see which sets of genes become homozygous in the various segregants. In the example, *y* and *paba y* genotypes were found in the various yellow sectors. From the mechanics of mitotic recombination, the order of genes is centromere–*paba*–*y*. We cannot assign a position to those genes that become homozygous wild type, so their relative positions must be determined from other mitotic recombination experiments.

After diploid segregants, which are produced by mitotic recombination, are obtained, they can be counted. Such quantitative data can be used to compute map distance between the genes; that is, the mitotic recombination frequency between genes can be computed from the formula used in meiotic recombination studies. The map distance between the *paba* and *y* genes, for example, is given by the percentage of yellow segregants that result from crossing-over in the *paba*–*y* region. Those particular segregants are still heterozygous for *paba* and *pro*.

Systems that achieve genetic recombination by means other than the regular alternation of meiosis and fertilization are called **parasexual systems.** In fungi (such as *Aspergillus*), the parasexual cycle consists of the following sequence of events: the formation of a heterokaryon; the rare fusion of haploid nuclei with different genotypes within the heterokaryon, to produce a heterozygous diploid nucleus; mitotic crossing-over within that diploid nucleus; and the subsequent haploidization of the diploid nucleus.

Retinoblastoma, a Human Tumor That Can Be Caused by Mitotic Recombination

Retinoblastoma is a childhood cancer of the eye (OMIM 18020); a patient with retinoblastoma is shown in Figure 22.9, and the cancer is described in more detail in Chapter 22, pp. 619–621. There are two forms of retinoblastoma. In *sporadic* (or *nonhereditary*) *retinoblastoma*, the development of an eye tumor is a spontaneous event in a patient from a family with no history of the disease. In these cases, a *unilateral tumor* will develop; that is, the tumor is in one eye only. In *hereditary retinoblastoma*, the susceptibility to develop the eye tumors is inherited. Patients with this form of retinoblastoma typically develop multiple eye tumors involving both eyes (*bilateral tumors*), usually at an earlier age than is the case for unilateral tumor formation in sporadic retinoblastoma patients.

Mutations in a gene called *RB* are responsible for retinoblastoma. In patients with hereditary retinoblastoma, the tumor cells always have mutations in both *RB* genes, while in all normal cells from the *same* individuals, one of the two *RB* genes has the mutation. Thus, a second mutation is all that is needed in the normal cells to change it to a tumor cell. Using the molecularly cloned *RB* gene for molecular analysis of hereditary retinoblastoma patients, researchers have found that, in a significant fraction of cases, the second mutation produces a mutated allele that is *identical* to the inherited mutated allele. This means that the normal wild-type copy of the retinoblastoma gene is somehow replaced by a duplicated copy of the homologous chromosome region carrying the mutant allele. While there are several mechanisms that could account for this phenomenon, one possibility is that mitotic recombination is the cause.

K E Y N O T E

The parasexual cycle describes genetic systems that achieve genetic recombination by means other than the regular alternation of meiosis and fertilization. The parasexual cycle in fungi such as *Aspergillus* consists of (1) the formation of a heterokaryon by mycelial fusion and then fusion of the two haploid nuclei to give a diploid nucleus; (2) mitotic crossing-over within the diploid nucleus; and (3) haploidization of the diploid nuclei without meiosis, a process that produces haploid nuclei into which one or the other parental chromosome has segregated randomly. Using the parasexual cycle, gene order and map distances between genes can be calculated.

Mapping Human Genes

For practical and ethical reasons, with humans it is not possible to do genetic-mapping experiments of the kind performed on other organisms. Nonetheless, we have had a strong interest in mapping genes in human chromosomes, since there are so many known diseases and traits that have a genetic basis. (Of course, with the human genome completely sequenced, we now have the data in hand to identify every human gene and to learn the location of each gene with precision on the chromosomes.) In Chapters 11 and 12, we saw that pedigree analysis could be used to determine the mode by which a particular genetic trait is inherited. In this way, many genes have been localized to the X chromosome. However, pedigree analysis cannot show on which chromosome a particular autosomal gene is located. In this section, we discuss some of the methods used to map human genes.

Mapping Human Genes by Recombination Analysis

It is not possible to set up appropriate testcrosses for human genetic mapping by recombination analysis. Only in a very few cases have multigenerational pedigrees included individuals with segregating genotypes that were appropriate to permit any analysis of linkage between autosomal genes. Recombination analysis in humans is simpler for X-linked genes, however, because the hemizygosity of the X chromosome in males provides a rich source of useful genotypic pairings in pedigrees.

Consider the following theoretical example (Figure 16.11): A male with two rare X-linked recessive alleles *a* and *b* marries a woman who expresses neither of the traits involved. Since the traits are rare, it is likely that the woman is homozygous for the wild-type allele of each gene; that is, she is $a^+ b^+/a^+ b^+$. A female offspring from these parents would be doubly heterozygous $a^+ b^+/a\,b$. Recombinant gametes from this female would be produced by crossing-over between the two genes at a frequency related to the genetic distance that separates them.

If this female offspring pairs with a normal $a^+ b^+/Y$ male, all female progenies will be $a^+ b^+$ in phenotype, because of the $a^+ b^+$ chromosome transmitted from the father. The male progenies, however, will express all four possible phenotype classes—that is, the male parental

Figure 16.11

Calculation of recombination frequency for two X-linked human genes by analyzing the male progenies of a woman doubly heterozygous for the two genes.

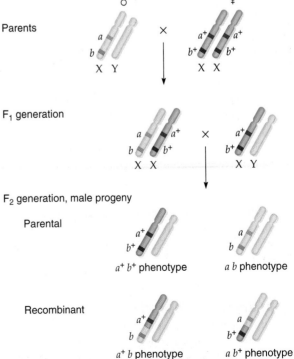

$a^+ b^+$ and $a\ b$, and the recombinant $a^+ b$ and $a\ b^+$—because of the hemizygosity of the X chromosome. Thus, analysis of the male progenies from pairings such as this ($a^+ b^+/a\ b \times a^+ b^+/Y$) in *a large number of pedigrees* will produce a value for the frequency of recombination between the two loci involved, and an estimate of genetic map distance can be obtained.

Using this approach, researchers mapped a number of genes along the human X chromosome. For example, it was found that the distance between the green weakness gene, *g* (the recessive allele responsible for a form of color blindness), and the hemophilia A gene, *h*, was 8 map units. Clearly, though, this approach has a limited application.

lod Score Method for Analyzing Linkage of Human Genes

The rarity of suitable pedigrees also makes it hard to test for linkage and to calculate map distance between genetic markers. So, in most cases, a statistical test known as the **lod** (logarithm of odds) **score method**, invented by mathematical geneticist Newton Morton in 1955, is used to test for possible linkage between two loci. The lod score method is usually done by computer programs that use pooled data from a number of pedigrees. A full discussion of the method is beyond the scope of this text, so only a brief presentation is given here.

The lod score method compares the probability of obtaining the pedigree results if two markers are linked with a certain amount of recombination between them to the probability that the results would have been obtained if there was no linkage (i.e., 50% recombination) between the markers. The results are expressed as the $\log_{10}$ of the ratio of the two probabilities. By convention, a hypothesis of linkage between two genes is accepted if the lod score at a particular recombination frequency is +3 or more, because a score of +3 means that the odds are 10^3 to 1 (1,000:1) in favor of linkage between two genes or markers (the $\log_{10}$ of 1,000 is +3). Similarly, a hypothesis of linkage between two genes is rejected when the lod score reaches −2 or less because a score of −2 means that the odds are 10^2 to 1 (100:1) against the two genes or markers being linked.

Once linkage is established between genetic markers, the map distance is computed from the recombination frequency giving the highest lod score. (The higher the lod score, the closer to each other are the two genes.) This is done by solving lod scores for a range of proposed map units. For the human genome, 1 mu corresponds, on average, to approximately 1 million base pairs (1 megabase, or 1 Mb).

High-Density Genetic Maps of the Human Genome

Creating a genetic map for two genes is the simplest genetic mapping analysis one can do. Building a genetic map with mapping information for two genes at a time

is not very efficient for a very large genome, however. Even using a handful of gene loci in genetic-mapping experiments makes the assembly of gene maps quite slow—and, for humans, there are simply not enough known genes to give chromosome maps with closely spaced genes. Fortunately, another type of genetic marker was discovered that elevated genetic mapping in humans to a new level. **DNA markers** are molecular markers in which DNA regions in the genome differ sufficiently between individuals so that those regions can be detected easily and rapidly by molecular analysis of DNA. (See Chapter 9, p. 214.) Briefly, the methods of mapping are the ones outlined earlier for human pedigree analysis, except that hundreds of loci are involved. Each DNA marker corresponds to a particular sequence at a site in the genome. If more than one type of sequence is found in the population at the site, then the DNA marker is polymorphic; in essence, we have alleles of a locus that differ in a *molecular* phenotype, rather than a phenotype such as eye color or plant height. Therefore, to analyze a pedigree for the segregation of DNA markers, all individuals in the pedigree must be analyzed—typed—for the particular DNA alleles present at each locus and the results analyzed by sophisticated computer algorithms to determine linkage relationships.

Making genetic maps of a large number of DNA markers is too much for a single research lab, because of the combinatorics of manipulations needed. For example, to type 5,000 DNA marker loci in 500 individuals would require performing 2,500,000 typing tests, each involving molecular techniques, and then entering 2,500,000 results into a database. So, typically for such analyses, geneticists set up a collaboration among many laboratories to do the work and, most importantly, *to have the consortium work on the same set of DNA samples from the same set of individuals.* The set of DNA samples used in this type of analysis is called a *mapping panel.*

One of the initial goals of the Human Genome Project (HGP; described in more detail in Chapter 10) was a genetic map with a density of at least one genetic marker per million base pairs of the genome by 1998. For this key genetic-mapping study, a mapping panel was used from a human DNA collection held at the Centre d'Étude du Polymorphisme Humain (CEPH), a research center in Paris, France. This panel is from 517 individuals representing 40 three-generation families. Eight of the families were used for most of the mapping analysis in the HGP study. The result was a high-density genetic map completed in 1994, with 5,264 of a particular type of DNA marker localized to 2,335 chromosomal loci (Figure 16.12). (The reason the two numbers do not match is because some markers are too close together to be separated definitively.) In this map, the average density was one marker per 599 kb (kilobase pairs = 1,000 base pairs) for all chromosomes considered together, with a range from one marker per 495 kb to one marker per 767 kb.

Figure 16.12

A high-density genetic map with 5,264 microsatellites localized to 2,335 chromosomal loci. (From Dib et al. 1996. *Nature* 380:152–154.)

1
2
3
4
5
6
7
8
9
10
11
12
13
14
15
16
17
18
19
20
21
22
X

KEYNOTE

Genetic maps of genomes are constructed with the use of recombination data from genetic crosses in the case of experimental organisms or from pedigree analysis in the case of humans. Both gene markers and DNA markers are used in genetic-mapping analysis.

Physical Mapping of Human Genes

An alternative to mapping genes via crosses and the analysis of recombinants is to use physical methods to locate genes on chromosomes. This is physical mapping

of genes in the genome, and a variety of approaches may be employed. We consider two examples here.

In **fluorescent in situ hybridization (FISH)** mapping, individual eukaryotic chromosomes are colored fluorescently at the locations of specific genes or DNA sequences. Human metaphase chromosomes on a microscope slide are treated to cause the two DNA strands of each DNA molecule in the chromosomes to separate, but stay in the same physical location. Specific DNA sequences corresponding to a gene or a DNA marker are molecularly cloned and tagged with fluorescent chemicals. The tagged single-stranded DNA sequences—the DNA probes—are added to the chromosomes, where they pair to the single-stranded chromosomal DNA sequence that they match. This pairing process is called *hybridization* (described in more detail in Chapter 8, pp. 187–189). In this way, the chromosome sites corresponding to the probe are identified by the fluorescent emissions of the tag. By utilizing chemicals that fluoresce at different wavelengths, it is possible to use a number of different probes in the same experiment. Computer imaging analysis of the sample examined under a fluorescence microscope then identifies the locations where the probes have bound.

Figure 16.13 shows the results of FISH with six different DNA probes. The probe colors are not the true colors from the fluorescence, but are generated by the computer. The complete chromosomes are visualized by staining them with a chemical that colors all the DNA (blue in this case). Each chromosome to which a probe has hybridized has two dots, because the DNA in metaphase chromosomes is already duplicated in preparation for cell division. For example, the yellow dots identify an uncharacterized DNA sequence on chromosome 5, and the red dots identify the Duchenne muscular dystrophy gene on the X chromosome.

Figure 16.13

An example of the results of fluorescent in situ hybridization (FISH) in which fluorescently tagged DNA probes were hybridized to human metaphase chromosomes.

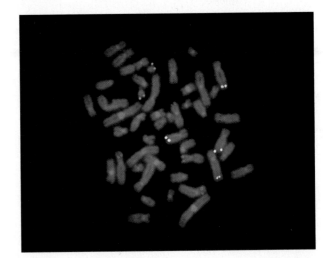

Another physical mapping approach involves determining gene linkage through **radiation hybrids (RH)**. A **radiation hybrid (RH)** is a rodent (hamster, rat, mouse) cell line that carries a small fragment of the genome of another organism, such as a human. The genome fragments are produced by irradiating human cells with X rays to cause random breakage of the DNA (Figure 16.14). The size of the fragments decreases as the dosage of X rays increases. The irradiation kills the human cells, but the chromosome fragments can be "rescued" by fusing the irradiated cells with rodent cells.

For RHs irradiated with human fragments the human fragment typically is a few megabase pairs long. The human DNA in the RH is then analyzed for the genetic markers it carries. The principle of RH mapping is straightforward: The closer two markers are, the greater is the probability that those markers will be on the same DNA fragment and therefore end up in the same RH. Both gene markers and DNA markers can be used in RH mapping. A detailed RH map of the human genome was published in 1997.

Figure 16.14

Making a radiation hybrid. Human cells are irradiated by X rays to fragment the chromosomes. The cells are killed, but the chromosome fragments are "rescued" by fusing the irradiated cells with rodent cells. The chromosome fragments become integrated into the rodent chromosomes.

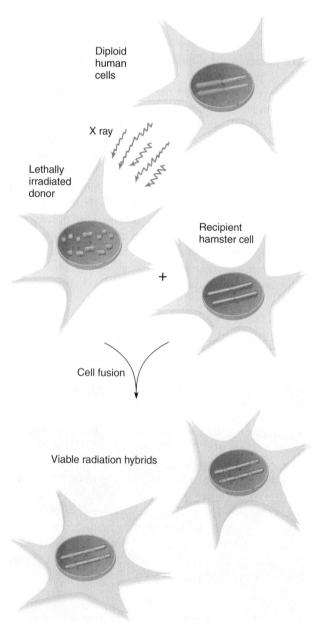

Diploid human cells

X ray

Lethally irradiated donor

Recipient hamster cell

+

Cell fusion

Viable radiation hybrids

KEYNOTE

Gene markers and DNA markers can be mapped on human chromosomes by physical mapping approaches. A variety of such approaches is available, including FISH and radiation hybrid mapping.

Summary

In this chapter, we learned how to map genes in certain haploid microorganisms by using tetrad analysis. In some of those microorganisms, the meiotic tetrads are ordered in a way that reflects the orientation of the four chromatids of each homologous pair of chromosomes at metaphase I. Ordered tetrads make it possible to map a gene's location relative to its centromere.

Next, we learned about the rare incidence of crossing-over in mitosis and how mitotic recombination may be used in certain organism to map genes. Finally, we learned about gene mapping in humans, in which controlled crosses obviously cannot be conducted. Instead, recombination analysis is done by means of multigenerational pedigrees, and genes may be pinpointed to chromosomal locations via physical mapping approaches.

Analytical Approaches to Solving Genetics Problems

Q16.1 A *Neurospora* strain that required both adenine (*ad*) and tryptophan (*trp*) for growth was mated to a wild-type strain (*ad⁺ trp⁺*), and this cross produced seven types of ordered tetrads in the following frequencies:

Spore Pair 1:	*ad trp*	*ad +*	*ad trp*	*ad trp*
Spore Pair 2:	*ad trp*	*ad +*	*ad +*	*+ trp*
Spore Pair 3:	*+ +*	*+ trp*	*+ trp*	*ad +*
Spore Pair 4:	*+ +*	*+ trp*	*+ +*	*+ +*
Type	(1) 63	(2) 15	(3) 3	(4) 9

Spore Pair 1:	*ad trp*	*ad +*	*+ trp*
Spore Pair 2:	*+ +*	*+ trp*	*+ +*
Spore Pair 3:	*ad trp*	*ad +*	*ad trp*
Spore Pair 4:	*+ +*	*+ trp*	*ad +*
Type	(5) 3	(6) 1	(7) 6

a. Determine the gene–centromere distance for the two genes.

b. From the data given, calculate the map distance between the two genes.

A16.1

a. The gene–centromere distance is given by the formula

$$\frac{\text{percent second-division tetrads}}{2} = x \text{ mu from centromere}$$

For the *ad* gene, tetrads 4, 5, and 6 show second-division segregation; the total number of such tetrads is 13. There are 100 tetrads, so the *ad* gene is ($13/2$)% = 6.5 mu from its centromere. For the *trp* gene tetrads, 3, 5, 6, and 7 show second-division segregation, and the total number of such tetrads is 25, indicating that the *trp* gene is 12.5 mu from its centromere.

b. The linkage relationship between the genes can be determined by analyzing the relative number of parental ditype (PD), nonparental ditype (NPD), and tetratype (T) tetrads. Tetrads 1 and 5 are PD, 2 and 6 are NPD, and 3, 4, and 7 are T. If two genes are unlinked, the frequency of the PD tetrads will approximately equal the frequency of the NPD tetrads. If two genes are linked, the frequency of the PD tetrads will greatly exceed that of the NPD tetrads. Here, the latter case prevails, so the two genes must be linked. The map distance between two genes is given by the general formula

$$\frac{\frac{1}{2}\text{T} + \text{NPD}}{\text{total}} \times 100$$

For this example, the number of T tetrads is 30 and the number of NPD tetrads is 4. Thus, the map distance between *ad* and *trp* is

$$\frac{\frac{1}{2}(30) + 4}{100} \times 100 = 19 \text{ mu}$$

Hence, we have the following map, with c indicating the centromere:

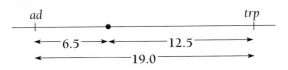

Q16.2 In *Aspergillus*, forced diploids were constructed between a wild-type strain and a strain containing the

mutant genes *y* (yellow), *w* (white), *pro* (proline requirement), *met* (methionine requirement), and *ad* (adenine requirement). All these genes are known to be on a single chromosome.

Homozygous yellow and homozygous white segregants were isolated and analyzed for the presence of the other gene markers. The following phenotypic results were obtained:

y/y segregants:	$w^+ \, pro^+ \, met^+ \, ad^+$	15
	$w^+ \, pro \;\; met^+ \, ad^+$	28
w/w segregants:	$y^+ \, pro^+ \, met \;\; ad$	6
w/w segregants:	$y^+ \, pro^+ \, met^+ \, ad$	12

Draw a map of the chromosome, giving the order of the genes and the position of the centromere.

A16.2 The segregants are all diploid. In mitotic recombination, a single crossover renders all gene loci distal to that point homozygous. In this regard, the crossover events in one chromosome arm are independent of those in the other chromosome arm. Therefore, we must inspect the data with these concepts in mind.

There are two classes of *y/y* segregants: wild-type segregants and proline-requiring segregants, which are homozygous for the *pro* gene. Thus, of the four loci other than yellow, only *pro* is in the same chromosome arm as *y*. Furthermore, since not all the *y/y* segregants are *pro* in phenotype, the *pro* locus must be closer to the centromere than the *y* locus is, as shown in the following map:

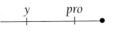

A single crossover between *pro* and *y* will give *y/y* segregants that are wild type for all other genes, whereas a single crossover between the centromere and *pro* will give homozygosity for both *pro* and *y*—hence the *pro* requirement.

Similar logic can be applied to the *w/w* segregants. Again, there are two classes. Both are also phenotypically *ad*, indicating that the *ad* locus is further from the centromere than the *w* locus. Consequently, every time *w* becomes homozygous, so does *ad*. The remaining gene to be located is *met*. Some of the *w/w* segregants are *met*$^+$ and some are *met*, so the *met* gene is closer to the centromere than the *w* gene is. The reasoning here is analogous to that for the placement of the *pro* gene in the other arm. Taking all the conclusions together, we have the following gene order:

Questions and Problems

16.1 A cross was made between a pantothenate-requiring (*pan*) strain and a lysine-requiring (*lys*) strain of *Neurospora crassa*, and 750 random ascospores were analyzed for their ability to grow on a minimal medium (a medium lacking pantothenate and lysine). Thirty colonies subsequently grew. Map the *pan* and *lys* loci.

16.2 Four different albino strains of *Neurospora* were each crossed to the wild type. All crosses resulted in half wild-type and half albino progeny. Crosses were made between the first strain and the other three, with the following results:

1 × 2:	975 albino, 25 wild type
1 × 3:	1,000 albino
1 × 4:	750 albino, 250 wild type

Which mutations represent different genes, and which genes are linked? How did you arrive at your conclusions?

*16.3 Genes *met* and *thi* are linked in *Neurospora crassa*; we want to locate *arg* with respect to *met* and *thi*. From the cross *arg* + + × + *thi met*, the following random ascospore isolates were obtained:

arg thi met	26	*arg* + +	51
arg thi +	17	+ *thi* +	4
arg + *met*	3	+ + *met*	14
+ *thi met*	56	+ + +	29

Map the three genes.

16.4 Double exchanges between two loci can be of several types, called two-strand, three-strand, and four-strand doubles.

a. Four recombinant gametes would be produced from a tetrad in which the first of two exchanges is depicted in the following figure:

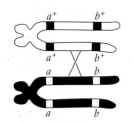

Draw in the second exchange.

b. In the following figure, draw in the second exchange so that four nonrecombinant gametes would result:

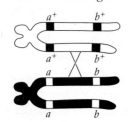

*16.5 A cross between a pink (*p⁻*) yeast strain of mating type **a** and a cream strain (*p⁺*) of mating type α produced the following tetrads:

18	*p⁺* **a**	*p⁺* **a**	*p⁻* α	*p⁻* α
8	*p⁺* **a**	*p⁻* **a**	*p⁺* α	*p⁻* α
20	*p⁺* α	*p⁺* α	*p⁻* **a**	*p⁻* α

On the basis of these results, are the *p* and mating-type genes on separate chromosomes?

16.6 The following asci were obtained from the cross *leu* + × + *rib* in yeast:

110	45	6	39
leu +	*leu rib*	+ +	*leu* +
+ *rib*	*leu* +	*leu rib*	+ *rib*
leu +	+ +	*leu rib*	+ +
+ *rib*	+ *rib*	+ +	*leu rib*

Draw the linkage map and determine the map distance.

*16.7 The genes *a*, *b*, and *c* are on the same chromosome arm in *Neurospora crassa*. The following ordered asci were obtained from the cross *a b* + × + + *c*:

45	5	146	1
a b +	*a b* +	*a b* +	*a b* +
+ *b c*	*a* + +	*a b* +	+ + +
a + +	+ *b c*	+ + *c*	*a b c*
+ + *c*	+ + *c*	+ + *c*	+ + *c*

10	20	15	58
a b +	*a b* +	*a b* +	*a b* +
a + *c*	+ + *c*	*a b c*	+ *b* +
+ *b* +	*a b* +	+ + +	*a* + *c*
+ + *c*	+ + *c*	+ + *c*	+ + *c*

Determine the correct gene order and calculate all gene–gene and gene–centromere distances.

*16.8 Two mutant strains of *Neurospora* lack the ability to make compound Z. When crossed, the strains usually yield asci of two types: (1) those with spores that are all mutant and (2) those with four wild-type and four mutant spores. The two types occur in a 1:1 ratio.

a. Let *c* represent one mutant and let *d* represent the other. What are the genotypes of the two mutant strains?

b. Are *c* and *d* linked?

c. Wild-type strains can make compound Z from the constituents of the minimal medium. Mutant *c* can make Z if supplied with X, but not if supplied with Y, while mutant *d* can make Z from either X or Y. Construct the simplest linear pathway of the synthesis of Z from the precursors X and Y, and show where the pathway is blocked by mutations *c* and *d*.

16.9 Under transmitted light, spores of wild-type (+) *Neurospora* appear black, while spores of an albino mutant (*al*) appear white.

a. Assume that there is no chromatid interference—that is, that crossing-over occurs equally frequently between any of the four chromatids during meiosis. What patterns of ordered asci do you expect to see, and in what frequencies, if there is exactly one crossover between *al* and its centromere in every meiosis?

b. Under the preceding conditions, what is the map distance between *al* and its centromere? Are *al* and its centromere linked or unlinked?

*16.10 The frequency of mitotic recombination in experimental organisms can be increased by exposing them to low levels of ionizing radiation (such as X rays) during development. Hans Becker used this method to examine the patterns of clones produced by mitotic recombination in the *Drosophila* retina. (*Drosophila* has a compound eye consisting of many repetitive units called ommatidia.) What type of spots would be produced in the *Drosophila* retina if you irradiated a developing *Drosophila* female obtained from crossing a white-eyed male with a cherry-eyed female? (See Table 4.2 [p. 74] for a description of the *w* and *w*^ch^ alleles.)

16.11 A diploid strain of *Aspergillus nidulans* (forced between wild type and a multiple mutant) that was heterozygous for the recessive mutations *y* (yellow), *w* (white), *ad* (adenine), *sm* (small), *phe* (phenylalanine), and *pu* (putrescine) produced haploid segregants. Forty-one haploid white and yellow segregants were tested and were found to have the following genotypes and numbers:

white
$$\begin{bmatrix} y\ w\ pu\ ad\ sm\ phe & 7 \\ y\ w\ pu\ ad\ +\ + & 11 \end{bmatrix}$$

yellow
$$\begin{bmatrix} y\ +\ +\ +\ sm\ phe & 16 \\ y\ +\ +\ +\ +\ + & 7 \end{bmatrix}$$

What are the linkage relationships of these genes?

16.12 A heterokaryon was established in the fungus *Aspergillus nidulans* between a *met⁻ trp⁻* auxotroph and a *leu⁻ nic⁻* auxotroph. A diploid strain was selected from this heterokaryon. From the diploid strain, the following eight haploid strains were obtained from conidial isolates, via the parasexual cycle, in approximately equal frequencies:

1. *nic⁺ leu⁺ met⁻ trp⁻*
2. *met⁺ leu⁺ nic⁺ trp⁺*
3. *trp⁻ met⁻ leu⁺ nic⁻*
4. *leu⁻ nic⁻ trp⁺ met⁺*
5. *leu⁺ nic⁻ met⁺ trp⁺*
6. *met⁻ nic⁻ leu⁻ trp⁻*
7. *trp⁺ leu⁻ met⁺ nic⁺*
8. *nic⁺ met⁺ trp⁻ leu⁻*

Which, if any, of these four marker genes are linked, and which are unlinked?

*16.13 A (green) diploid of *Aspergillus nidulans* is heterozygous for *each* of the following recessive mutant genes: *sm*, *pu*, *phe*, *bi*, *w* (white), *y* (yellow), and *ad*. Analysis of white and yellow haploid segregants from this diploid indicated several classes with the following genotypes:

Genotype						
sm	**pu**	**phe**	**bi**	**w**	**y**	**ad**
sm	pu	phe	+	w	y	ad
+	pu	+	+	w	y	ad
+	pu	+	bi	w	+	ad
sm	+	phe	+	+	y	+
+	+	+	+	+	y	+
sm	pu	phe	bi	w	+	ad

How many linkage groups are involved, and which genes are on which linkage group?

16.14 A (green) diploid of *Aspergillus nidulans* is homozygous for the recessive mutant gene *ad* and heterozygous for the following recessive mutant genes: *paba*, *ribo*, *y* (yellow), *an*, *bi*, *pro*, and *su-ad*. Those recessive alleles which are on the same chromosome are in coupling. The *su-ad* allele is a recessive suppressor of the *ad* allele: The *+/su-ad* genotype does not suppress the adenine requirement of the *ad/ad* diploid, whereas the *su-ad/su-ad* genotype does suppress that requirement. Therefore, the parental diploid requires adenine for growth. From this diploid, two classes of segregants were selected: yellow and adenine independent. The following table lists the types of segregants obtained:

Segregant Type Selected	Phenotype
Adenine-independent	+
	ribo
	ribo an
	ribo an pro paba y bi
Yellow	y ad bi
	paba y ad bi
	pro paba y ad bi
	ribo an pro paba y bi

Analyze these results as completely as possible to determine the location of the centromere and the relative locations of the genes.

16.15 High-density genetic maps can be generated through the use of mapping panels with a set of DNA markers and lod score methods. The same DNA markers can be mapped by means of radiation hybrid methods.

a. In what ways will maps generated by these two methods be identical, and in what ways will they differ?

b. Much or most of the entire genomic sequence has been obtained for a number of complex eukaryotes, including humans, mice, and the plant *Arabidopsis thaliana.* What is the value of high-density genetic maps in the genetic analysis of organisms whose gerome has been sequenced?

*__16.16__ Two panels of radiation hybrids were produced by irradiating human tissue culture cells and then fusing them with hamster tissue culture cells. The differing properties of the two panels are shown in the following table (1 Mb = 10^6 bp of DNA):

	Panel GB4	Panel G3
X-ray dosage used to generate cell hybrids	3,000 rad	10,000 rad
Number of cell lines established	93	83
Average retention of human genome per hybrid	32%	16%
Average human DNA fragment size	25 Mb	2.4 Mb
Effective map resolution	1 Mb	0.25 Mb

a. A haploid human genome has about 3×10^9 bp of DNA. About how many different human DNA segments are present, on average, in the hybrid cells of each panel?

b. Two human markers are found together in some, but not all, cell hybrids. Are they necessarily linked?

c. How do these panels differ in their advantages with respect to mapping genes and markers?

d. DNA markers *A, B,* and *C* derive from a single chromosomal region. Their presence or absence is assessed in DNA isolated from the hybrids of each panel, with the following results:

	Number of Hybrids Testing Positive For Markers	
Marker(s) Present	**Panel GB4**	**Panel G3**
A only	0	4
B only	1	6
C only	2	15
A and *B* only	2	11
A and *C* only	1	2
B and *C* only	0	0
A, B and *C*	27	0

Why do the two panels give such different results? What reasonable hypothesis can you generate concerning the arrangement of these three markers?

__16.17__ As discussed in Chapter 12, XO individuals have Turner syndrome. Some individuals who display a Turner phenotype are mosaic individuals with 45,X/46,XX or 46,XY/45,X karyotypes. It is clinically important to address mosaicism in Turner individuals, as some types of mosaics have an increased risk of gonadal cancer.

a. What chromosomal events could lead to a mosaic Turner individual? When do such events occur?

b. How might physical mapping methods be adapted to determine, reliably and readily, whether a Turner individual is mosaic?

*__16.18__ Some dogs love water, while others avoid it. A dog that loved water was mated to a dog that avoided it, and their F_1 progeny were interbred to give an F_2. The parental, F_1, and F_2 generations were evaluated by DNA typing, and the lod-score method was used to assess linkage between DNA markers and genes for water affection (*waf* genes). Suppose that the following data were obtained for one marker, where θ gives the value of the recombination frequency between the marker and a *waf* gene used in calculating the lod score:

θ	lod Score
0	$-\infty$
0.05	-12.51
0.10	-2.34
0.15	-1.32
0.20	2.66
0.25	4.01
0.30	3.21
0.35	2.14
0.40	1.56
0.50	0

Graph these lod scores and evaluate whether the marker is linked to a *waf* gene. If it is, estimate the physical distance between the marker and the gene.

__16.19__ What is gene conversion? How does the Holliday model for genetic recombination (See Box 15.1) allow for gene conversion?

*__16.20__ Crosses were made between strains, each of which carried one of three different alleles of the same gene, *a*, in yeast. For each cross, some unusual tetrads resulted at low frequencies. Explain the origin of each of these tetrads:

Cross:	$a1$ $a2^+$	$a1$ $a3^+$	$a2$ $a3^+$
	$\times$	$\times$	$\times$
	$a1^+$ $a2$	$a1^+$ $a3$	$a2^+$ $a3$
Tetrads:	$a1^+$ $a2$	$a1^+$ $a3$	$a2^+$ $a3$
	$a1^+$ $a2^+$	$a1^+$ $a3$	$a2^+$ $a3^+$
	$a1$ $a2$	$a1^+$ $a3^+$	$a2$ $a3^+$
	$a1$ $a2^+$	$a1$ $a3^+$	$a2$ $a3^+$

__16.21__ From a cross of $y1$ $y2^+ \times y1^+$ $y2$, where $y1$ and $y2$ are both alleles of the same gene in yeast, the following tetrad type occurs at very low frequencies:

$$y1^+ \ y2$$
$$y1 \ y2$$
$$y1 \ y2$$
$$y1 \ y2^+$$

Explain the origin of this tetrad at the molecular level.

16.22 In *Neurospora*, the *a, b,* and *c* loci are situated in the same arm of a particular chromosome. The location of *a* is near the centromere; *b* is near the middle, and *c* is near the telomere of the arm. Among the asci resulting from a cross of *ABC* × *abc*, the following ascus was found (the eight spores are indicated in the order in which they were arranged in the ascus): *ABC, ABC, ABc, ABc, aBC, aBC, abc, abc.* How might this ascus have arisen?

*16.23 Mutants at the autosomal *rosy* (*ry*) locus in *Drosophila* have rosy eyes instead of the normal deep-red color. Two mutations, ry^{206} and ry^{209}, are point mutations at the *ry* gene. The *kar* and *ace* loci are each about 0.2 map units from *ry*, with the order *kar–ry–ace*. Mutants at *kar* have karmoisin (bright-red) eyes, while mutants at *ace* lack the enzyme acetylcholinesterase and are recessive lethal. A rosy-eyed *kar* ry^{206} *ace/kar⁺* ry^{209} *ace⁺* female was crossed to a rosy-eyed *kar⁺* ry^{209} *ace⁺/kar⁺* ry^{209} *ace⁺* male. The vast majority of progeny had rosy eyes, but normal-eyed males and females were produced at a very low frequency. Testcrossing the normal-eyed male progeny (recall that recombination does not occur in *Drosophila* males) revealed that they received one of four types of chromosomes from their mother: *kar⁺ ry⁺ ace, kar ry⁺ ace⁺, kar ry⁺ ace,* or *kar⁺ ry⁺ ace⁺*. Explain the origin of each of these chromosomes at the molecular level.

17

Variations in Chromosome Structure and Number

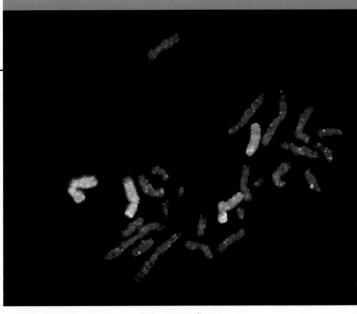

Human chromosomes "painted" by FISH (fluorescent in situ hybridization).

PRINCIPAL POINTS

• Chromosomal mutations are variations from the normal condition in chromosome number or chromosome structure. Chromosomal mutations can occur spontaneously, or they can be induced by chemicals or radiation.

• Deletion is the loss of a DNA segment, duplication is the addition of one or more extra copies of a DNA segment, inversion is a reversal of orientation of a DNA segment in a chromosome, and translocation is the movement of a DNA segment to another chromosomal location in the genome.

• Variations in the chromosome number of a cell or an organism include aneuploidy, monoploidy, and polyploidy. In aneuploidy there are one, two, or more whole chromosomes greater or fewer than the diploid number. In monoploidy, each body cell of the organism has only one set of chromosomes, and in polyploidy more than two sets of chromosomes are present.

• A change in chromosome number or chromosome structure can have serious, and even lethal, consequences for the organism. In eukaryotes, abnormal phenotypes typically result from abnormal chromosome segregation during meiosis, from gene disruptions where chromosomes break, or from altered gene expression

levels when the number of copies of a gene or genes (gene dosage) is altered or when rearrangement separates a gene from its regulatory sequence. Some human tumors, for example, have chromosomal mutations associated with them—either a change in the number of chromosomes or a change in chromosome structure.

ⓘ *iActivity*

YOU LIE ON A TABLE IN A SOFTLY LIT ROOM watching a black-and-white monitor. You see the image of a long needle being inserted into your uterus as you simultaneously feel the pressure of the needle against your abdomen. The doctor collects some of the amniotic fluid that surrounds your 16-week-old fetus. When the procedure is done, you get up, get dressed, and go home. Six weeks later, you go back to the clinic, where a counselor gently informs you that your unborn child has an extra chromosome 21, which causes Down syndrome.

Down syndrome is just one example of a number of human disorders that are the result of variations in the normal set of chromosomes. In this chapter, you will learn about the causes and effects of different chromosomal mutations. After you have read and studied the chapter, you can try the iActivity, in which you use your understanding of chromosomal mutations to help a couple who are trying to conceive a child.

In previous chapters, we learned many of the fundamental principles of transmission genetics, as applied to eukaryotes. With our understanding of the relationship between genes and chromosomes, we now consider chromosomal mutations—changes in normal chromosome structure or chromosome number. Chromosomal mutations affect both prokaryotes and eukaryotes, as well as viruses. The association of genetic defects with changes in chromosome structure or chromosome number indicates that not all genetic defects result from simple mutations of single genes. The study of normal and mutated chromosomes and their behavior is called *cytogenetics*. Your goal in this chapter is to learn about the various types of chromosomal mutations in eukaryotes and about some of the human disease syndromes that result from chromosomal mutations.

Types of Chromosomal Mutations

Chromosomal mutations (or **chromosomal aberrations**) are *variations from the normal (wild-type) condition in chromosome structure or chromosome number*. In Bacteria, Archaea, and Eukarya, chromosomal mutations arise spontaneously or can be induced experimentally by certain chemicals or radiation. Chromosomal mutations are detected by genetic analysis—that is, by observing changes in the linkage arrangements of genes. In eukaryotes, chromosomal mutations can be detected under the microscope during mitosis and meiosis. In this chapter we limit our discussion to chromosomal changes in eukaryotes.

We often have the impression that reproduction in humans usually occurs without significant problems affecting chromosome structure or number. After all, most babies appear normal, as does the majority of the adult population. However, chromosomal mutations are more common than we once thought, and they contribute significantly to spontaneously aborted pregnancies and stillbirths, as well as to some forms of cancer. For example, major chromosomal mutations are present in approximately half of spontaneous abortions, and a visible chromosomal mutation is present in about 6 out of 1,000 live births. Other studies have shown that some 11 percent of men with serious fertility problems and about 6 percent of people institutionalized with mental deficiencies have chromosomal mutations. Chromosomal mutations are significant causes of developmental disorders.

KEYNOTE

Chromosomal mutations are variations from the wild-type condition in chromosome number or chromosome structure. Chromosomal mutations can occur spontaneously, or they can be induced by treatment with chemicals or radiation.

Variations in Chromosome Structure

There are four common types of chromosomal mutations involving changes in chromosome structure: deletions and duplications (both of which involve a change in the amount of DNA on a chromosome), inversions (which involve a change in the orientation of a chromosomal segment), and translocations (which involve a change in the location of a chromosomal segment).

All four classes of chromosomal structure mutations begin with one or more breaks in the chromosome. If a break occurs within a gene, then the function of that gene may be lost. Wherever the break occurs, broken ends remain without the specialized sequences found at the ends of chromosomes (the telomeres) that prevent their degradation. The broken end of a chromosome is "sticky" and can adhere to other broken chromosome ends. This property of stickiness can help us understand the formation of the types of chromosomal structure mutations we will discuss.

We have learned a lot about changes in chromosome structure from the study of **polytene chromosomes** (Figure 17.1)—special kinds of chromosomes found in certain tissues (such as the salivary glands in the larval stages) of insects of the order *Diptera* (e.g., *Drosophila*). Polytene chromosomes consist of chromatid bundles resulting from repeated cycles of chromosome duplication without nuclear or cell division, a process called *endoreduplication*. Polytene chromosomes can be a thousand times the size of corresponding chromosomes at meiosis or in the nuclei of ordinary somatic cells and are easily detectable under the microscope. In each polytene chromosome, the homologous chromosomes are tightly paired; therefore, the observed number of polytene chromosomes per cell is reduced to half the diploid number of chromosomes. Polytene chromosomes are joined together at their centromeres by a proteinaceous structure called the *chromocenter*.

As a result of the intimate pairing of multiple copies of chromatids, characteristic banding patterns are easily seen when the chromosomes are stained, enabling cytogeneticists to identify any segment of a polytene chromosome. In *Drosophila melanogaster*, for example, more than 5,000 bands and interbands can be counted in the four polytene chromosomes. Each band contains an average of 30,000 base pairs (30 kb) of DNA, enough to encode several average-sized proteins. DNA cloning and sequencing studies have shown that many bands contain up to seven genes. Genes are also found in the interbands. Polytene chromosomes are mentioned throughout this chapter because it is easy to see the different types of chromosomal mutations in *Drosophila* salivary gland polytene chromosomes.

Figure 17.1

Diagram of the complete set of *Drosophila* polytene chromosomes in a single salivary gland cell. There are four chromosome pairs, but each pair is tightly synapsed, so only a single chromosome is seen for each pair. The four chromosome pairs are linked together by regions near their centromeres to produce a large chromocenter.

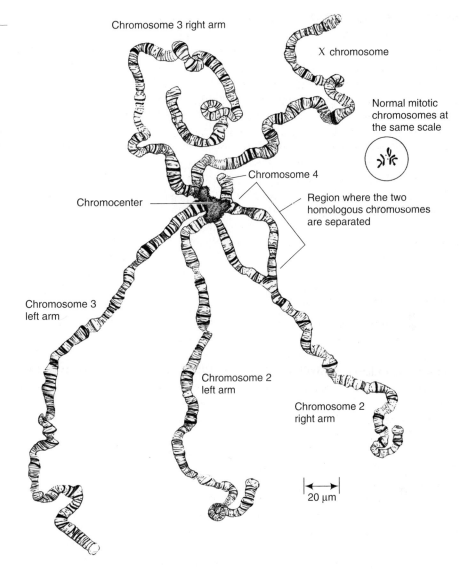

Chromosome 3 right arm

X chromosome

Normal mitotic chromosomes at the same scale

Chromosome 4

Chromocenter

Region where the two homologous chromosomes are separated

Chromosome 3 left arm

Chromosome 2 left arm

Chromosome 2 right arm

20 μm

Deletion

A **deletion** is a chromosomal mutation in which part of a chromosome is missing (Figure 17.2). A deletion starts where breaks occur in chromosomes. Breaks can be induced by agents such as heat, radiation (especially ionizing radiation; see Chapter 7), viruses, chemicals, and transposable elements (see Chapter 7), or by errors in recombination. Because a segment of chromosome is missing, deletion mutations cannot revert to the wild-type state.

The consequences of a deletion depend on the genes or parts of genes that have been removed. In diploid organisms, an individual heterozygous for a deletion may be normal. However, if the homolog contains recessive genes with deleterious effects, the consequences can be severe. If the deletion involves the loss of a centromere, the result is an acentric chromosome, which is usually lost during meiosis. This deletion of an entire chromosome from the genome may have very serious or even

Figure 17.2

A deletion of a chromosome segment (here, *D*).

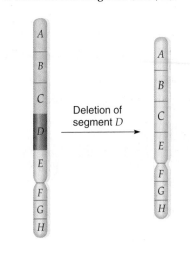

Deletion of segment *D*

lethal consequences, depending on the particular chromosome deleted and the organism. For example, no known living humans have one whole chromosome of a homologous pair of autosomes deleted from the genome. (Recall from Chapter 12 that the human XO female can be viable despite a sex chromosome deletion. Survival here is a consequence of the sex chromosome dosage compensation mechanism.)

In organisms in which karyotype analysis (analysis of the chromosome complement; see Chapter 12) is practical, deletions can be detected by that procedure if the losses are large enough. In that case, a mismatched pair of homologous chromosomes is seen, one shorter than the other. In individuals heterozygous for a deletion, unpaired loops are seen when the two homologous chromosomes pair at meiosis.

Deletions can be used to determine the physical location of a gene on a chromosome. In *Drosophila*, for example, the banding patterns of polytene chromosomes are useful visible landmarks for *deletion mapping* of genes. The principle behind the method is that the deletion of the dominant allele of a heterozygote results in the appearance of the phenotype of the recessive allele. This unexpected expression of a recessive trait, caused by the absence of a dominant allele, is called **pseudodominance.** Figure 17.3 shows how Demerec and Hoover used deletion mapping in 1936 to localize genes to specific physi-

cal sites on *Drosophila* polytene chromosomes. The fly strain studied was heterozygous for the X-linked recessive mutations *y*, *ac*, and *sc*. Genetic analysis had shown that the three loci were linked at the left end of the X chromosome. The banding pattern of that end of the chromosome is shown in Figure 17.3a. The regions labeled A, B, and C are major cytological subdivisions of the X chromosome, and the numbers within each region refer to the chromosome bands. Recall that a single polytene chromosome is actually a tightly fused pair of homologous chromosomes. Deletions of this region of the chromosome were used to localize the gene loci. In strain 260-1, bands A1–7 and B1–4 are deleted, so that pseudodominance is observed for *y*, *ac*, and *sc*. The extent of the deletion is shown in Figure 17.3a, and the appearance of the polytene chromosomes in the deletion heterozygote is shown in Figure 17.3b. In strain 260-2, bands A1–7 and B1 are deleted from the chromosome bearing the wild-type alleles, so that pseudodominance is observed for *y* and *ac*. The extent of this deletion is shown in Figure 17.3a also, and the appearance of the polytene chromosomes in the deletion heterozygote is shown in Figure 17.3c. Since the wild-type *sc* locus was lost in deletion strain 260-1, but was not lost in deletion strain 260-2, *sc* must be located in the region of the X chromosome that distinguishes the two deletions, namely, bands B2–B4. (See Figure 17.3a.) This method of analysis was used to construct the detailed

Figure 17.3

Use of deletions to determine the physical locations of genes on *Drosophila* polytene chromosomes. (a) Cytological appearance of the left end of the X chromosome heterozygous for the recessive mutations *y*, *ac*, and *sc*, showing major regions A, B, and C and the chromosome bands they contained. Region 260-1 shows the extent of a deletion that produced pseudodominance for *y*, *ac*, and *sc*, and region 260-2 shows the extent of a deletion that produced pseudodominance for *y* and *ac*. (b) Cytological appearance of the polytene X chromosome in flies heterozygous for the 260-1 deletion. These flies show pseudodominance for *y*, *ac*, and *sc*, (c) Cytological appearance of the polytene X chromosome in flies heterozygous for the 260-2 deletion. These flies show pseudodominance for *y* and *ac*.

a) Wild type

260-1 (*y ac sc*)⁻

260-2 (*y ac*)⁻

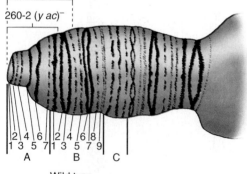

2 4 6 | 2 4 6 8
1 3 5 7 | 1 3 5 7 9
A   B   C

Wild type

b) Polytene X chromosome in flies heterozygous for the 260-1 deletion

260-1

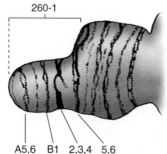

A5,6   B1   2,3,4   5,6

Pseudodominant for
y, *ac*, and *sc*

c) Polytene X chromosome in flies heterozygous for the 260-2 deletion

260-2

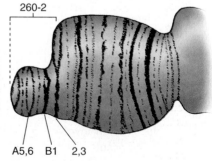

A5,6   B1   2,3

Pseudodominant for
y and *ac*

Figure 17.4

Cri-du-chat syndrome results from the deletion of part of one of the copies of human chromosome 5. (a) Karyotype of individual with cri-du-chat syndrome. (b) A child with cri-du-chat syndrome.

a)

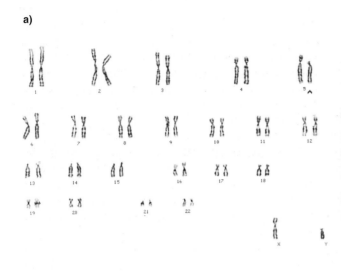

b)

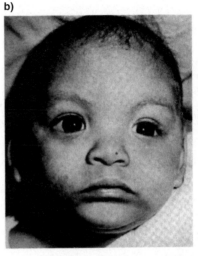

physical map of *Drosophila* polytene chromosomes that has been so valuable to geneticists.

Some human disorders are caused by deletions of chromosome segments. In many cases, the abnormalities are found in heterozygous individuals; homozygotes for deletions usually die if the deletion is large. This distinction tells us that, in humans at least, the number of copies of genes is important for normal development and function. Typically, several to many genes are lost in a deletion, so the syndrome that results is because of the loss of the combined functionality of those genes, rather than the loss of just one gene.

One human disorder caused by a heterozygous deletion is *cri-du-chat syndrome* (OMIM 123450 at http://www3.ncbi.nlm.nih.gov/OMIM), which results from an observable deletion of part of the short arm of chromosome 5, one of the larger human chromosomes (Figure 17.4). Children with cri-du-chat syndrome are severely mentally retarded, have a number of physical abnormalities, and cry with a sound like the mew of a cat (hence the name, which is French for "cry of the cat"). About 1 infant in 50,000 live births has cri-du-chat syndrome.

Another example is *Prader–Willi syndrome* (OMIM 176270), which results from heterozygosity for a deletion of part of the long arm of chromosome 15. Many individuals with the syndrome go undiagnosed, so its frequency of occurrence is not known accurately, although it is estimated to affect between 1 in 10,000 and 1 in 25,000 people, predominantly males. Infants with this syndrome are weak because their sucking reflex is poor, making feeding difficult. As a result, growth is poor. By age 5 to 6, for reasons not yet understood, children with Prader–Willi syn-

drome become compulsive eaters, producing obesity and related health problems. Left untreated, afflicted individuals may feed themselves to death. Other phenotypes associated with the syndrome include poor sexual development in males, behavioral problems, and mental retardation. (Molecular information about Prader–Willi syndrome is presented in Chapter 20, p. 561.)

Duplication

A **duplication** is a chromosomal mutation that results in the doubling of a segment of a chromosome (Figure 17.5). The size of the duplicated segment varies widely, and

Figure 17.5

Duplication, with a chromosome segment (here, *BC*) repeated.

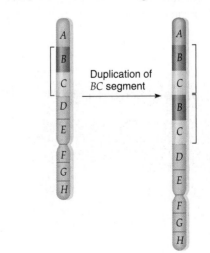

Figure 17.6

Forms of chromosome duplications are tandem, reverse tandem, and terminal tandem.

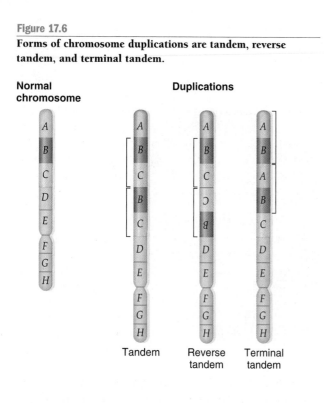

duplicated segments may occur at different locations in the genome or in a *tandem* configuration (that is, adjacent to each other). When the order of genes in the duplicated segment is the opposite of the order of the original, the mutation is a *reverse tandem duplication*; when the duplicated segments are arranged in tandem at the end of a chromosome, the mutation is a *terminal tandem duplication* (Figure 17.6). Heterozygous duplications result in unpaired loops similar to those described for chromosome deletions and therefore may be detected cytologically.

Duplications of particular genetic regions can have unique phenotypic effects, as in the *Bar* mutant on the X chromosome of *Drosophila melanogaster*, first studied by Alfred Sturtevant and Thomas H. Morgan in the 1920s. In strains homozygous for the *Bar* mutation (not to be confused with the Barr body), the number of facets of the compound eye is less than that of the normal eye (shown in Figure 17.7a), giving the eye a bar-shaped (slitlike), rather than an oval, appearance (shown in Figure 17.7b). *Bar* resembles an incompletely dominant mutation, because females heterozygous for *Bar* have more facets, and hence a somewhat larger bar-shaped eye, than do females homozygous for *Bar*. Males hemizygous for *Bar* have very small eyes like those of homozygous *Bar* females. The *Bar* trait is the result of a duplication of a small segment (16A) of the X chromosome (Figure 17.7b).

Duplications have played an important role in the evolution of multiple genes with related functions (a **multigene family**). For example, hemoglobin molecules contain two copies each of two different subunits: the α-globin polypeptide and the β-globin polypeptide. At different developmental stages, from the embryo to the adult, a human individual has different hemoglobin molecules assembled from different types of α-globin and β-globin polypeptides. The genes for each of the α-globin type of polypeptides are clustered together on one chromosome, while the genes for each of the β-globin type of polypeptides are clustered together on another chromosome. The sequences of the α-globin genes are all similar, as are the sequences of the β-globin genes. It is thought that each assembly of genes evolved from a different ancestral gene by duplication and subsequent divergence in the sequences of the duplicated genes.

Figure 17.7

Chromosome constitutions of *Drosophila* strains, showing the relationship between duplications of region 16A of the X chromosome and the production of reduced-eye size phenotypes. (a) Wild type. (b) Homozygous *Bar* mutant.

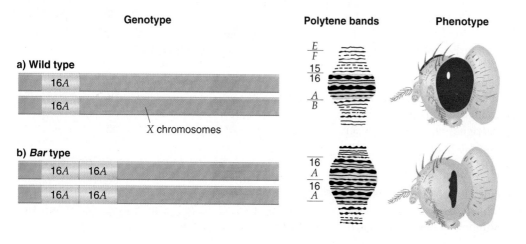

Inversion

An **inversion** is a chromosomal mutation that results when a segment of a chromosome is excised and then reintegrated at an orientation 180 degrees from the original orientation (Figure 17.8). There are two types of inversions: A **paracentric inversion** does not include the centromere (Figure 17.8a), and a **pericentric inversion** includes the centromere (Figure 17.8b).

Typically, genetic material is not lost when an inversion takes place, although there can be phenotypic consequences when the breakpoints (inversion ends) occur within genes or within regions that control gene expression. Homozygous inversions can be identified through the non-wild-type linkage relationships that result between the genes within the inverted segment and the genes that flank the inverted segment. For example, if the order of genes on the normal chromosome is *ABCDEFGH* and the *BCD* segment is inverted (shown next in bold), the gene order will be *ADCBEFGH*, with *D* now more closely linked to *A* than to *E* and *B* now more closely linked to *E* than to *A*. (See Figure 17.8a.)

The meiotic consequences of a chromosome inversion depend on whether the inversion occurs in a homozygote or a heterozygote. If the inversion is homozygous, then meiosis is normal and there are no problems related to gene duplications or deletions. There are also no meiotic problems for an inversion heterozygote if crossing-over is absent, but serious genetic consequences ensue if crossing-over occurs in the inversion, as we will now see.

Let us consider a paracentric inversion heterozygote, genotype ○*ABCDEFGH*/○*ADCBEFGH*, with the centromere (○) to the left of gene *A*. In meiosis, the homologous chromosomes attempt to pair such that the best possible base pairing occurs. Because of the inverted segment on one homolog, pairing of homologous chromo-

somes requires the formation of loops containing the inverted segments, called *inversion loops*. Inversion heterozygotes, then, may be identified by looking for those loops. If no crossovers occur in the inversion loop of a paracentric inversion heterozygote, then all resulting gametes receive a complete set of genes (two gametes with a normal gene order, ○*ABCDEFGH*, and two gametes with the inverted segment, ○*ADCBEFGH*), and they are all viable. Figure 17.9 shows the effects of a single crossover in the inversion loop, here between genes *B* and *C*. During the first meiotic anaphase, the two centromeres migrate to opposite poles of the cell. Because of the crossover, one recombinant chromatid becomes stretched across the cell as the two centromeres begin to migrate in anaphase, forming a **dicentric bridge**—that is, a chromosome with two centromeres (a **dicentric chromosome**). With continued migration, the dicentric bridge breaks due to tension. The other recombinant product of the crossover event is a chromosome without a centromere (an acentric fragment). This acentric fragment is unable to continue through meiosis and is usually lost. (It is not found in the gametes.)

In the second meiotic division, each daughter cell receives a copy of each chromosome. Two of the gametes—the gamete with the normal order of genes (○*ABCDEFGH*) and the gamete with the inverted segment of genes (○*ADCBEFGH*)—have complete sets of genes and are viable. The other two gametes are inviable because they are unbalanced: Many genes are deleted. Thus, *the only gametes that can give rise to viable progeny are those containing the chromosomes that did not involve crossing-over.* However, in many cases in female animals, the dicentric chromosomes or acentric fragments arising as a result of inversion are shunted to the polar bodies, so the reduction in fertility may not be so great. In short, for paracentric inversion heterozygotes, viable recombinants are reduced significantly or suppressed altogether. That is, the frequency of crossing-over is not lower in the heterozygotes than in normal cells, but gametes or zygotes derived from recombined chromatids are inviable.

The consequences of a single crossover in the inversion loop of an individual heterozygous for a pericentric inversion are shown in Figure 17.10. The normal chromosome is *ABC*○*DEFGH* and the inversion chromosome is *AD*○*CBEFGH*; the centromere is between *C* and *D*. The crossover event and the ensuing meiotic divisions result in two viable gametes with the nonrecombinant chromosomes *ABC*○*DEFGH* (normal) and *AD*○*CBEFGH* (inversion) and in two recombinant gametes that are inviable, each as a result of the deletion of some genes and the duplication of other genes.

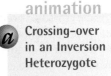

animation

a Crossing-over in an Inversion Heterozygote

Figure 17.8

Inversions. (a) Paracentric inversion. (b) Pericentric inversion.

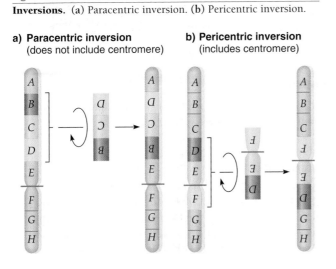

a) Paracentric inversion
(does not include centromere)

b) Pericentric inversion
(includes centromere)

Figure 17.9

Consequences of a paracentric inversion. Meiotic products resulting from a single crossover within a heterozygous, paracentric inversion loop Crossing-over occurs at the four-strand stage involving two nonsister homologous chromatids.

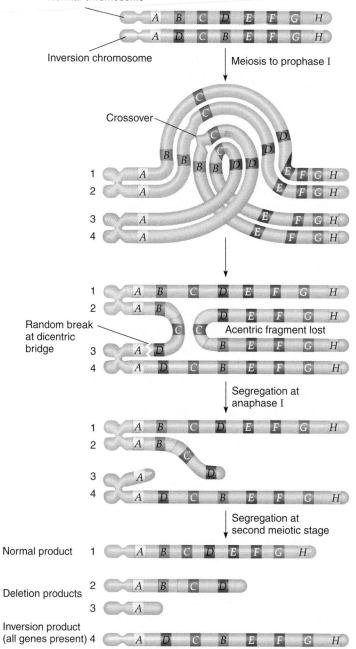

Some crossover events within an inversion loop do not affect gamete viability. For example, a double crossover close together and involving the same two chromatids (a two-strand double crossover; see Chapter 16) produces four viable gametes. A second exception occurs when the duplicated and deleted segments of the recombinant chromatids do not affect gene expression, and hence viability, to a significant degree, as when the chromosome segments involved are very small. Also, recent studies with mammals show that inverted segments may remain unpaired. Since crossing-over cannot occur between unpaired segments, no inviable gametes are generated.

Translocation

A **translocation** is a chromosomal mutation in which there is a change in position of chromosome segments and the gene sequences they contain to a different location in the genome (Figure 17.11). No gain or loss of genetic material is involved in a translocation. If a chromosome segment

Figure 17.10

Meiotic products resulting from a single crossover within a heterozygous, pericentric inversion loop. Crossing-over occurs at the four-strand stage involving two nonsister homologous chromatids.

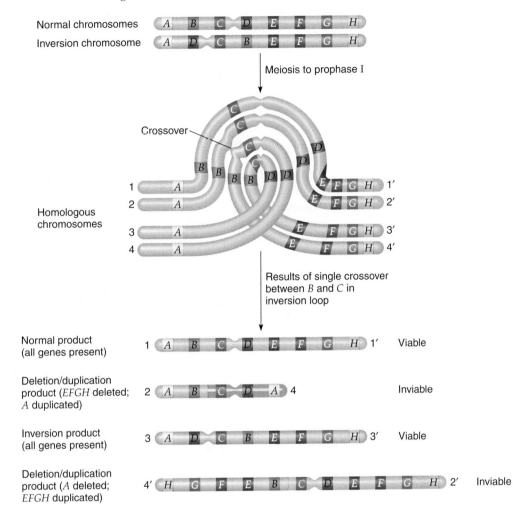

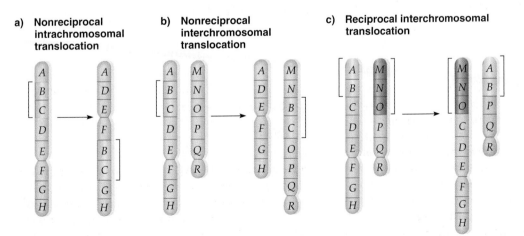

Figure 17.11

Translocations. (a) Nonreciprocal intrachromosomal. (b) Nonreciprocal interchromosomal. (c) Reciprocal interchromosomal.

changes position within the same chromosome, the translocation is a *nonreciprocal intrachromosomal* (within a chromosome) *translocation* (Figure 17.11a). If a chromosome segment is transferred from one chromosome to another, the translocation is a *nonreciprocal interchromosomal* (between chromosomes) *translocation* if a one-way transfer is involved (Figure 17.11b) and a *reciprocal interchromosomal translocation* if an exchange of segments between the two chromosomes is involved (Figure 17.11c).

In organisms homozygous for the translocations (i.e., when both copies of the genome in the diploid have the translocation), the genetic consequence is an alteration in the linkage relationships of genes. For example, in the nonreciprocal intrachromosomal translocation shown in Figure 17.11a, the *BC* segment has moved to the other chromosome arm and has become inserted between the *F* and *G* segments. As a result, genes in the *F* and *G* segments are now farther apart than they are in the normal strain, and genes in the *A* and *D* segments are now more closely linked. Similarly, in reciprocal translocations, new linkage relationships are produced.

Translocations typically affect the products of meiosis. In many cases, some of the gametes produced are unbalanced, in that they have duplications or deletions, and consequently are inviable. In other cases, such as familial Down syndrome resulting from a duplication stemming from a translocation, the gametes are viable. (See later in the chapter). We focus here on reciprocal translocations.

In strains *homozygous* for a reciprocal translocation, meiosis takes place normally because all chromosome pairs can pair properly and crossing-over does not produce any abnormal chromatids. In strains heterozygous for a reciprocal translocation, however, all homologous chromosome parts pair as best they can. Since one set of normal chromosomes (N) and one set of translocated chromosomes (T) are involved, the result is a crosslike configuration in meiotic prophase I (Figure 17.12). These crosslike figures consist of four associated chromosomes, each partially homologous to two other chromosomes in the group.

animation

a Meiosis in a Translocation Heterozygote

Segregation at anaphase I may occur in three different ways. (We ignore the complication of crossing-over in this discussion.) In one way, called *alternate segregation*, alternate centromeres migrate to the same pole (Figure 17.12, left: N_1 and N_2 migrate to one pole, T_1 and T_2 to the other pole). This produces two gametes, each of which is viable because it contains a complete set of genes—no more, no less. One of these gametes has two normal chromosomes, and the other has two translocated chromosomes. In the second way, called *adjacent 1 segregation*, adjacent

nonhomologous centromeres migrate to the same pole (Figure 17.12, middle: N_1 and T_2 migrate to one pole, N_2 and T_1 to the other pole). Both gametes produced contain gene duplications and deletions and are often inviable. Adjacent 1 segregation occurs about as frequently as alternate segregation. In the third way, called *adjacent 2 segregation*, different pairs of adjacent *homologous* centromeres migrate to the same pole (Figure 17.12, right: N_1 and T_1 migrate to one pole, N_2 and T_2 to the other pole). Both products have gene duplications and deletions and are always inviable. Adjacent 2 segregation seldom occurs.

In sum, of the six theoretically possible gametes, the two from alternate segregation are functional, the two from adjacent 1 segregation usually are inviable (because of gene duplications and deficiencies), and the two from adjacent 2 seldom occur and are inviable if they do. Moreover, because alternate segregation and adjacent 1 segregation occur with about equal frequency, the term *semisterility* is applied to this condition. (The term is also used for inversion heterozygotes.)

In practice, animal gametes that have large duplicated or deleted chromosome segments may function, but the zygotes formed by such gametes typically die. In contrast, if the duplicated and deleted chromosome segments are small, the gametes may function normally and viable offspring may result. In plants, pollen grains with duplicated or deleted chromosome segments typically do not develop completely and hence are nonfunctional.

Chromosomal Mutations and Human Tumors

Most human malignant tumors have chromosomal mutations. In fact, the most common class of mutation associated with cancer is a translocation. The exact chromosomal abnormality actually varies quite a bit among tumors, ranging from simple rearrangements to complex changes in chromosome structure and number. In many tumors, there is no specific associated chromosomal mutation. Rather, a variety of different chromosomal mutations is seen. This is the case with most solid tumors, for instance, which have complex patterns of chromosomal mutations. Examples are epithelial tumors of the ovary, lung, and pancreas and many sarcomas (connective-tissue tumors), such as osteosarcoma. By contrast, certain tumors are associated with specific chromosomal anomalies. For example chronic myelogenous leukemia (CML; OMIM 151410; involves chromosomes 9 and 22) and Burkitt lymphoma (BL; OMIM 113970; involves chromosomes 8 and 14) are associated with reciprocal translocations. Untreated, CML is an invariably fatal cancer involving the uncontrolled replication of myeloblasts (stem cells of white blood cells). A new targeted drug developed recently, STI571 (Gleevec®), is showing promise in treating the disease.

Figure 17.12

Meiosis in a translocation heterozygote in which no crossover occurs.

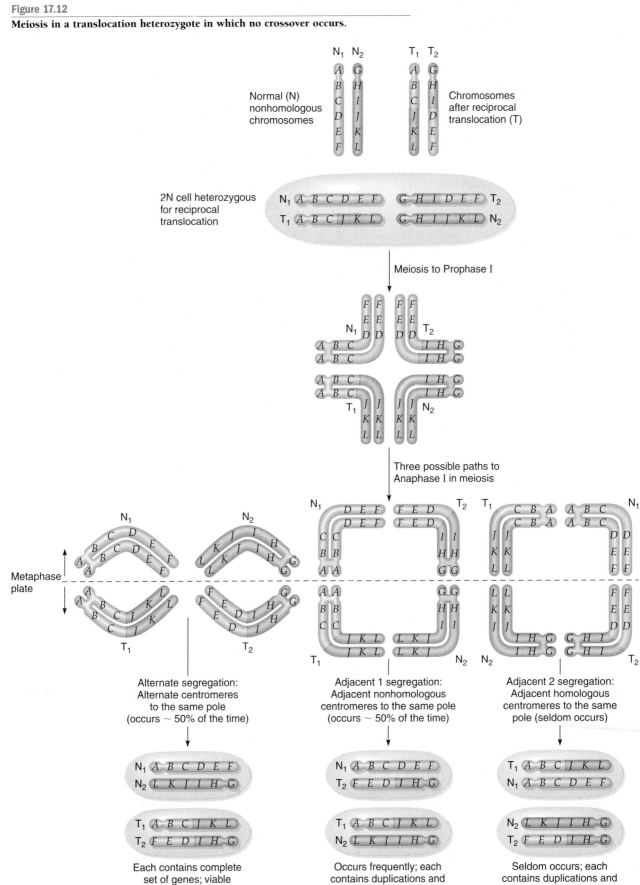

Ninety percent of patients with CML have a chromosomal mutation called the *Philadelphia chromosome* (*Ph*[1]) in the leukemic cells. The mutation was so named because it was discovered in Philadelphia. The Philadelphia chromosome results from a reciprocal translocation involving the movement of part of the long arm of chromosome 22 (the second-smallest human chromosome) to chromosome 9 and the movement of a small part from the tip of the long arm of chromosome 9 to chromosome 22 (Figure 17.13). This reciprocal translocation apparently converts a **proto-oncogene**—a gene that, in normal cells, controls the normal proliferation of cells—to an **oncogene** (see Chapter 22)—a gene that encodes a protein which plays a role in the transition from a differentiated cell to a tumor cell with an uncontrolled pattern of growth. Specifically, the *ABL* ("able"; named for *Abelson*) proto-oncogene, normally located on chromosome 9, is translocated to chromosome 22 in patients with CML. (See Figure 17.13.) The translocation positions the *ABL* gene within the *BCR* (breakpoint cluster region) gene. The hybrid *BCR–ABL* gene is the oncogene responsible for CML. The drug STI571 targets the Bcr–Abl fusion protein transcribed from the hybrid gene and thus is a potentially a highly effective treatment for this cancer because the fusion protein is found only in cells in which the translocation has occurred.

Burkitt lymphoma, a particularly common disease in Africa, is a viral-induced tumor that affects cells of the immune system called B cells. Characteristically, the tumorous B cells secrete antibodies. Ninety percent of the tumors in Burkitt lymphoma patients are associated with a reciprocal translocation involving chromosomes 8 and 14. As with CML, a proto-oncogene becomes activated as a result of the translocation event: The distal end of chromosome 8, starting with the *MYC* proto-oncogene, exchanges with the distal end of chromosome 14. The *MYC* gene becomes positioned next to a transcriptionally active immunoglobulin gene, resulting in overexpression of the *MYC* gene. The overexpressed *MYC* gene is the oncogene involved in the development of Burkitt lymphoma, and the activity of the immunoglobulin gene explains the secretion of antibodies (immunoglobulin molecules) associated with the disease.

Figure 17.13

Origin of the Philadelphia chromosome in chronic myelogenous leukemia (CML) by a reciprocal translocation involving chromosomes 9 and 22. The arrows show the sites of the breakage points.

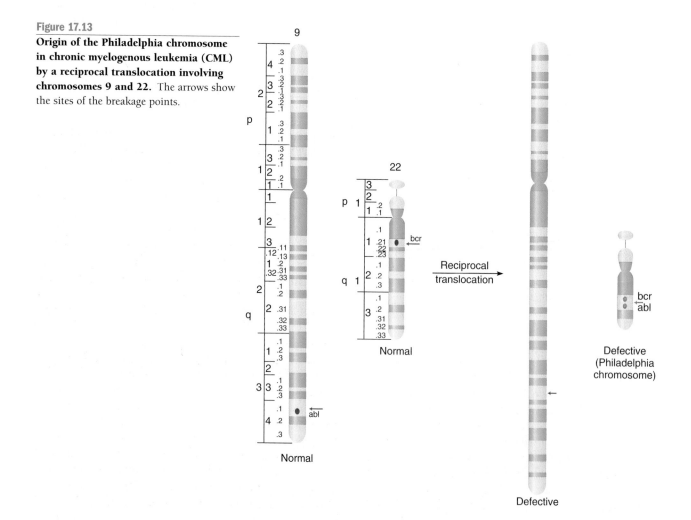

In the iActivity *Deciphering Karyotypes* on your CD-ROM, you are a genetic counselor who must determine whether there are any chromosomal abnormalities that could be affecting a couple's ability to have children.

KEYNOTE

Chromosomal mutations may involve parts of individual chromosomes, rather than whole chromosomes or sets of chromosomes. The four major types of structural alterations are deletions and duplications (both of which involve a change in the amount of DNA on a chromosome), inversions (which involve no change in the amount of DNA on a chromosome, but rather a change in the arrangement of a chromosomal segment), and translocations (which also involve no change in the amount of DNA, but instead a change in the chromosomal location of one or more DNA segments). Problems associated with inversions and translocations are often manifested only during meiotic crossing-over.

Position Effect

Unless inversions or translocations involve breaks within a gene, those chromosomal mutations do not produce mutant phenotypes. Rather, as we have seen, they have significant consequences in meiosis when they are heterozygous with normal sequences. In some cases, however, phenotypic effects of inversions or translocations occur because of a different phenomenon called **position effect**—a change in the phenotypic expression of one or more genes as a result of a change in position in the genome. This is another example of an **epigenetic** phenomenon—a heritable change in gene expression that does not involve a change in the DNA sequence of the affected gene.

For example, position effect may be exhibited if a gene that is normally located in euchromatin (chromosomal regions, representing most of the genome, that are condensed during division, but become uncoiled during interphase) is brought near heterochromatin (chromosomal regions that remain condensed throughout the cell cycle and are genetically inactive) by a chromosomal rearrangement. (Gene transcription typically occurs in euchromatin, but not in heterochromatin; the difference in chromosome condensation is responsible for the distinction. (See Chapter 20.)) An example of this kind of position effect involves the X-linked white-eye (*w*) locus in *Drosophila*. One inversion moves the w^+ gene from a euchromatic region near the end of the X chromosome to a position next to the heterochromatin at the centromere of the X. In a w^+ male, or in a w^+/w female in which the w^+ is involved in the inversion, the eye exhibits a mottled pattern of red and white rather than being completely red as expected. The explanation is that, in flies with the inversion, some eye cell clones have the w^+ allele inactivated because of the position effect of w^+ near heterochromatin. Those clones produce white spots in the eyes. Clones of cells in which the w^+ allele is not inactivated produce red spots in the eye. Since the inactivation event is variable, the eye exhibits a mottled pattern of red and white spots.

Some human genetic diseases are associated with position effects. An example is aniridia (literally, "without iris"; OMIM 106210), a congenital eye condition characterized by severe hypoplasia (underdevelopment) of the iris, typically associated with cataracts and clouding of the cornea. Aniridia is caused by loss of function of the *PAX6* gene, which is involved in eye development. In individuals with a nonfunctional *PAX6* gene, eye development stops too early, and, at the time of birth, most of the eye is underdeveloped. The loss of function may be due to a deletion of the gene or simple mutations within the gene. In addition, some affected individuals have translocations with chromosomal breakpoints somewhat distant from the *PAX6* gene. It appears that, in this case, the expression of *PAX6* is suppressed by a position effect brought about by the new chromosomal environment surrounding the gene generated by the translocation.

Fragile Sites and Fragile X Syndrome

When human cells are grown in culture, some of the chromosomes develop narrowings or unstained areas (gaps) called *fragile sites*. The chromosome may break spontaneously at a fragile site, resulting in deletion of the chromosome distal to the site. More than 40 fragile sites have been identified since the first one was discovered in 1965. One particular fragile site on the long arm of the X chromosome at position Xq27.3 (Figure 17.14) is associated with *fragile X syndrome* (also called *fragile site mental retardation*). After Down syndrome, fragile X syndrome is the leading genetic cause of mental retardation in the United States, with an incidence of about 1 in 1,250 males and 1 in 2,500 females (heterozygotes). As with all recessive X-linked traits, males predominantly exhibit this type of mental retardation. An individual with fragile X syndrome is shown in Figure 17.15.

The fragile X chromosome is inherited as a typical Mendelian gene. Male offspring of carrier females have a 50 percent chance of receiving a fragile X chromosome. However, only 80 percent of males with a fragile X chromosome are mentally retarded; the rest are normal. These phenotypically normal males are called *normal transmitting males* and carry a *premutation* because they can pass on the fragile X chromosome to their daughters. (A premutation could be considered a silent mutation.) The sons of those daughters frequently show symptoms of mental retardation. Female offspring of carrier (heterozygous) females also have a 50 percent chance of

Figure 17.14

Fragile site on the X chromosome. (a) Scanning electron micrograph and (b) diagram of a human X chromosome showing the location of the fragile site responsible for fragile X syndrome. (From Gerald Stine, *The New Human Genetics*. Copyright © 1989. Reproduced by permission of The McGraw-Hill Companies.)

a)

b)

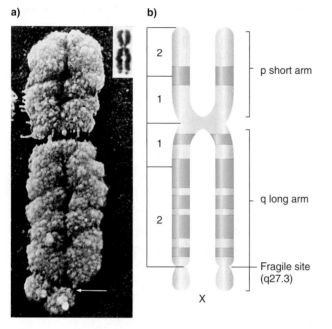

p short arm

q long arm

Fragile site (q27.3)

X

Figure 17.15

Individual with fragile X syndrome.

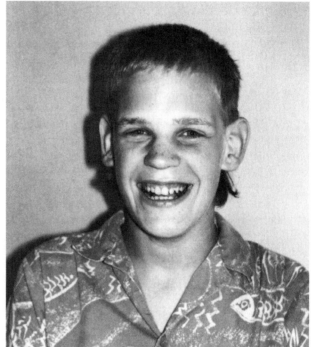

inheriting a fragile X chromosome. Up to 33 percent of the carrier females show mild mental retardation.

Modern molecular techniques brought to bear on this disease have resulted in an understanding of the disease at the DNA level. There is a repeated 3-base pair sequence, CGG, in a gene called *FMR-1* (*fragile X mental retardation-1*; OMIM 309550) located at the fragile X site. Normal individuals have an average of 29 CGG repeats (the range is from 6 to 54) in the coding region of the *FMR-1* gene. Phenotypically normal transmitting males and their daughters, as well as some carrier females, have a significantly larger number of CGG repeats, ranging from 55 to 200 copies. These individuals do not show symptoms of fragile X syndrome, and the increased number of repeats they have is the aforementioned premutation. Males and females with fragile X syndrome have even larger numbers of the CGG repeats, ranging from 200 to 1,300 copies; these are considered to be the full mutations. In other words, the triplet repeat, CGG, in the *FMR-1* gene becomes duplicated (amplified) in a tandem manner. The process has been termed *triplet repeat amplification*. Below a certain threshold number of copies (about 200 or fewer) there are no clinical symptoms, and above that threshold number (greater than 200) clinical symptoms are seen. Interestingly, amplification of the CGG repeats does not occur in males, but only in females. Therefore, a phenotypically normal transmitting male (who has the premutation) transmits his X chromosome to his daughter. In a slipped

mispairing process during DNA replication in his daughter, perhaps, the triplets may amplify, and she can transmit the amplified X to her offspring. Thus, affected males inherit the mutation from their grandfather.

The *FMR-1* gene encodes an RNA-binding protein that apparently binds to some of the mRNAs in the cell. However, the triplet repeat expansion in *FMR-1* is not in the protein-coding region. Instead, it affects the expression of *FMR-1*. In other words, the fragile X syndrome symptoms result from loss of gene activity, rather than altered protein function. We do not understand how the triplet repeat amplification that occurs within this gene produces mental retardation.

Triplet repeat amplification has also been shown to cause other human diseases, such as myotonic dystrophy (MD; OMIM 160900), spinobulbar muscular atrophy (also called Kennedy disease; OMIM 313200), and Huntington disease (HD; OMIM 143100; see Chapter 13). In each of these cases, no fragility of the associated chromosome is seen. Also, they differ from fragile X syndrome in that the amplification can occur in both sexes at each generation. For each, there is a threshold number of triplet repeat copies above which symptoms of the disease are produced.

Variations in Chromosome Number

When an organism or a cell has one complete set of chromosomes or an exact multiple of complete sets, that organism or cell is said to be **euploid.** Thus, eukaryotic organisms that are normally diploid (such as humans and fruit flies) and eukaryotic organisms that are normally haploid (such as yeast) are euploids. Chromosome mutations that result in variations in the number of chromosome sets occur in nature, and the resulting organism or cells are also euploid. Chromosome mutations resulting in variations in the number of individual chromosomes are examples of **aneuploidy.** An aneuploid organism or cell has a chromosome number that is not an exact multiple of the haploid set of chromosomes. Both euploid and aneuploid variations affecting whole chromosomes are discussed in this section.

Changes in One or a Few Chromosomes

Generation of Aneuploidy. Changes in chromosome number can occur in both diploid and haploid organisms. The nondisjunction of one or more chromosomes during meiosis I or meiosis II typically is responsible for generating gametes with abnormal numbers of chromosomes. Nondisjunction was discussed in Chapter 12 in the context of unusual complements of X chromosomes, with Figure 12.20 (p. 319) illustrating the consequences of nondisjunction at the first and second meiotic divisions. Referring to that figure and considering just one particular chromosome, one can see that nondisjunction at meiosis I produces four abnormal gametes: two with a chromosome duplicated and two with the corresponding chromosome missing. In a male, nondisjunction at meiosis I can produce a gamete with both the X and the Y chromosome; in a female, it produces a gamete with both sets of homologs (and thus possible heterozygotes). Fusion of the former gamete type with a normal gamete produces a zygote with three copies of the particular chromosome instead of the normal two and, unless nondisjunction also involves other chromosomes, there will be two copies of all other chromosomes. Similarly, fusion of the latter gamete type with a normal gamete produces a zygote with only one copy of the particular chromosome instead of the normal two, and two copies of all other chromosomes. Nondisjunction in meiosis II (see Figure 12.20) is different from nondisjunction in meiosis I in that some normal gametes are produced. As Figure 12.21 shows, nondisjunction in meiosis II results in two normal gametes and two abnormal gametes—that is, a single gamete with two sister chromosomes and one gamete with that same chromosome missing. Fusion of these with normal gametes gives the zygote types just discussed. Nondisjunction can occur in mitosis, giving rise to somatic cells with unusual chromosome complements.

Types of Aneuploidy. In aneuploidy, one or more chromosomes are lost from or added to the normal set of chromosomes (Figure 17.16). Aneuploidy can occur, for example, from the loss of individual chromosomes in meiosis or (rarely) in mitosis by nondisjunction. In ani-

Figure 17.16

Normal (theoretical) set of metaphase chromosomes in a diploid (2N) organism *(top)* and examples of aneuploidy *(bottom).*

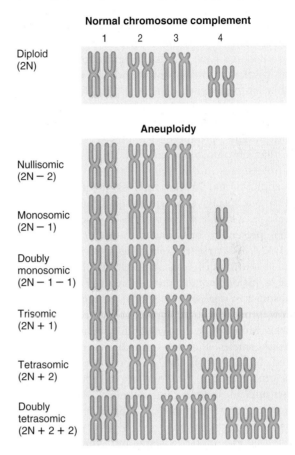

mals, autosomal aneuploidy is lethal in most cases, so in mammals it is detected mainly in aborted fetuses. Aneuploidy is tolerated more by plants, especially in species that are considered polyploid.

In diploid organisms, there are four main types of aneuploidy (see Figure 17.16):

1. **Nullisomy** (a nullisomic cell) involves a loss of one homologous chromosome pair—the cell is 2N − 2. (Nullisomy can arise, for example, if nondisjunction occurs for the same chromosome in meiosis in both parents, producing gametes with no copies of that chromosome and one copy of all other chromosomes in the set.)

2. **Monosomy** (a monosomic cell) involves a loss of a single chromosome—the cell is 2N − 1. (Monosomy can arise, for example, if nondisjunction in meiosis in a parent produces a gamete with no copies of a particular chromosome and one copy of all other chromosomes in the set.)

3. **Trisomy** (a trisomic cell) involves a single extra chromosome—the cell has three copies of a particular chromosome and two copies of all other chromo-

somes. A trisomic cell is 2N + 1. (Trisomy can arise, for example, if nondisjunction in meiosis in a parent produces a gamete with two copies of a particular chromosome and one copy of all other chromosomes in the set.)

4. **Tetrasomy** (a tetrasomic cell) involves an extra chromosome pair; that is, there are four copies of one particular chromosome and two copies of all other chromosomes—the cell is 2N + 2. (Tetrasomy can arise, for example, if nondisjunction occurs for the same chromosome in meiosis in both parents, producing gametes with two copies of that chromosome and one copy of all other chromosomes in the set.)

Aneuploidy may involve the loss or the addition of more than one specific chromosome or chromosome pair. For example, a *double monosomic* has two separate chromosomes present in only one copy each; that is, it is 2N − 1 − 1. A *double tetrasomic* has two chromosomes present in four copies each; that is, it is 2N + 2 + 2. In both cases, meiotic nondisjunction involved two different chromosomes in one parent's gamete production.

Most forms of aneuploidy have serious consequences in meiosis. Monosomics, for example, produce two kinds of haploid gametes: N and N − 1. Alternatively, the odd, unpaired chromosome in the 2N − 1 cell may be lost during meiotic anaphase and not be included in either daughter nucleus, thereby producing two N − 1 gametes. For trisomics, there are more segregation possibilities in meiosis. Consider a trisomic of genotype +/+/a in an organism that can tolerate trisomy, and assume no crossing-over between the a locus and its centromere. Then, as shown in Figure 17.17, random segregation of the three types of chromosomes produces four genotypic classes of gametes: 2 (+ a) : 2 (+) : 1 (+ +) : 1 (a). In a cross of a +/+/a trisomic with an a/a individual, the predicted phenotypic ratio among the progeny is 5 wild type (+) : 1 mutant (a). This ratio is seen in many actual crosses of this kind.

In the sections that follow, we examine some examples of aneuploidy as they are found in the human population. Table 17.1 summarizes various aneuploid abnormalities for autosomes and for sex chromosomes in the human population. Examples of aneuploidy of the X and Y chromosomes are discussed in Chapter 12. Recall that, in mammals, aneuploidy of the sex chromosomes is more often found in adults than is aneuploidy of the autosomes, because of a dosage compensation mechanism (lyonization) by which excess X chromosomes are inactivated.

In humans, autosomal monosomy is rare. Presumably, monosomic embryos do not develop significantly and are lost early in pregnancy. In contrast, autosomal trisomy accounts for about one-half of chromosomal abnormalities producing fetal deaths. In fact, only a few autosomal

Figure 17.17

Meiotic segregation possibilities in a trisomic individual. Shown is segregation in an individual of genotype +/+/a when two chromosomes migrate to one pole and one goes to the other pole, and assuming no crossing-over between the a locus and its centromere.

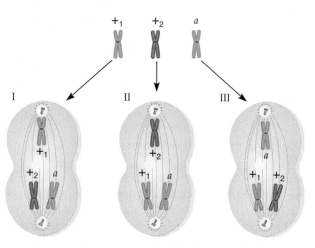

Gametes produced after 2nd meiotic division

	haploid	disomic
I	+$_1$	+$_2$/a
II	+$_2$	+$_1$/a
III	a	+$_1$/+$_2$

In sum: 2 +/a : 2 + : 1 +/+ : 1 a

Table 17.1 **Aneuploid Abnormalities in the Human Population**

Chromosomes	Syndrome	Frequency at Birth
Autosomes		
Trisomic 21	Down	14.3/10,000
Trisomic 13	Patau	2/10,000
Trisomic 18	Edwards	2.5/10,000
Sex chromosomes, females		
XO, monosomic	Turner	4/10,000 females
XXX, trisomic	Viable; most	
XXXX, tetrasomic	are fertile	14.3/10,000 females
XXXXX, pentasomic		
Sex chromosomes, males		
XYY, trisomic	Normal	25/10,000 males
XXY, trisomic		
XXYY tetrasomic	Klinefelter	40/10,000
XXXY, tetrasomic		

trisomies are seen in live births. Most of these (trisomy-8, -13, and -18) result in early death. Only in trisomy-21 (Down syndrome) does survival to adulthood occur.

Trisomy-21. Trisomy-21 (OMIM 190685) occurs when there are three copies of chromosome 21 (Figure 17.18a) and with a frequency of about 3,510 per 1 million conceptions and about 1,430 per 1 million live births. Individuals with trisomy-21 have Down syndrome (Figure 17.18b), characterized by such abnormalities as low IQ, epicanthal folds over the eyes, short and broad hands, and below-average height. Down syndrome is named for the late-nineteenth-century English physician John Langdon Down, who was the first to publish (in 1866) an accurate description of a person with the condition.

There is a relationship between the age of the mother and the probability of her having a trisomy-21 individual (Table 17.2). (For many years, it was thought that there was no correlation with age of the father. Recent evidence, however, indicates that paternal age has an effect on Down syndrome if the mother is 35 years old or older; in younger women, there is no paternal effect.) During the development of a female fetus before birth, the primary oocytes in the ovary undergo meiosis, but stop at prophase I. In a fertile female, each month at ovulation the nucleus of a secondary oocyte (see Chapter 12) begins the second meiotic division, but progresses only to metaphase, when division again stops. If a sperm penetrates the secondary oocyte, the second meiotic division is completed. The

Table 17.2	Relationship Between Age of Mother and Risk of Trisomy-21
Age of Mother	**Risk of Trisomy-21 in Child**
16–26	7.7/10,000
27–34	4/10,000
35–39	29/10,000
40–44	100/10,000
45–47	333/10,000
All mothers combined	14.3/10,000

probability of nondisjunction increases with the length of time the primary oocyte is in the ovary. It is important, then, that older mothers-to-be consider testing—for example, by undergoing amniocentesis or chorionic villus sampling (see Chapter 4, pp. 80–81)—to determine whether the fetus has a normal complement of chromosomes.

Are there other risk factors for having a Down syndrome baby? Where a person lives, social class, and race have no influence on the chance of having a baby with Down syndrome. However, mothers under 35 years of age who smoke and have an error in meiosis II are at an increased risk of having children with the syndrome. If mothers with these characteristics use cigarettes and oral contraceptives, the risk is increased over using cigarettes alone. Oral contraceptive use alone for this class of mothers has no effect on the incidence of Down syndrome.

Figure 17.18

Trisomy-21 (Down syndrome). (a) Karyotype. (b) Individual.

a)   b)

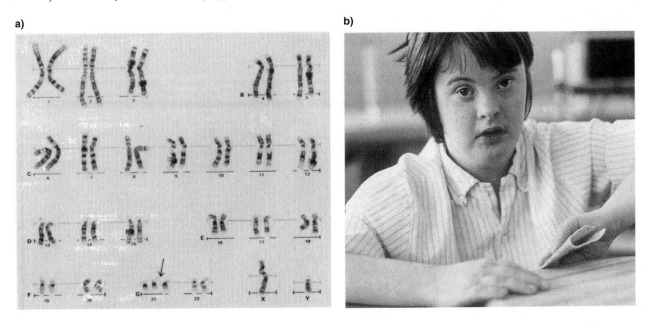

Down syndrome can also result from a different sort of chromosomal mutation called centric fusion or **Robertsonian translocation,** which produces three

animation

@ **Down Syndrome Caused by a Robertsonian Translocation**

copies of the long arm of chromosome 21. This form of Down syndrome is called familial Down syndrome. A Robertsonian translocation is a type of nonreciprocal translocation in which two nonhomologous

acrocentric chromosomes (chromosomes with centromeres near their ends) break at their centromeres and then the long arms become attached to a single centromere (Figure 17.19). The short arms also join to form the reciprocal product, which typically contains nonessential genes and usually is lost within a few cell divisions. In humans, when a Robertsonian translocation joins the long arm of chromosome 21 with the long arm of chromosome 14 (or 15), the heterozygous carrier is phenotypically normal, because there are two copies of all major chromosome arms and hence two copies of all essential genes.

There is a high risk of Down syndrome among the offspring of pairings between heterozygous carriers and normal individuals (Figure 17.20). The normal parent produces gametes with one copy each of chromosomes 14 and 21. The heterozygous carrier parent produces three reciprocal pairs of gametes, each as a result of different segregation of the three chromosomes involved. As the figure shows, the zygotes produced by pairing these gametes with gametes of normal chromosomal constitution are theoretically as follows: One-sixth have normal chromosomes 14 and 21, one-sixth are heterozygous carriers like the parent and are phenotypically normal, one-sixth are inviable

Figure 17.19

Robertsonian translocation. Production of a Robertsonian translocation (centric fusion) by breakage of two acrocentric chromosomes at their centromeres (indicated by arrows) and fusion of the two large chromosome arms and of the two small chromosome arms.

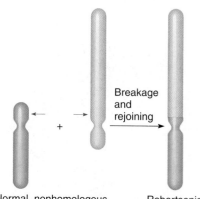

Normal, nonhomologous, acrocentric chromosomes

Robertsonian translocation

Figure 17.20

The three segregation patterns of a heterozygous Robertsonian translocation involving human chromosomes 14 and 21. Fusion of the resulting gametes with gametes from a normal parent produces zygotes with various combinations of normal and translocated chromosomes.

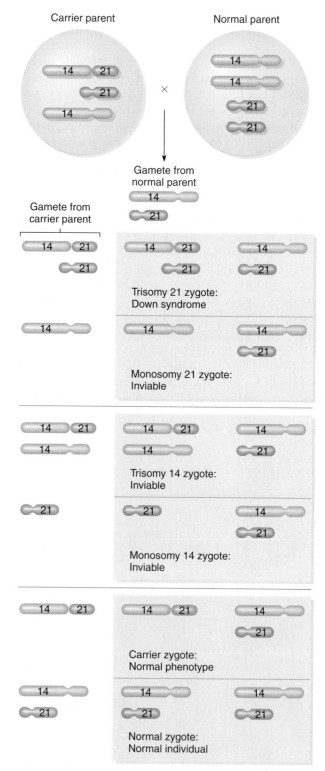

Figure 17.21

Trisomy-13 (Patau syndrome). (a) Karyotype. (b) Individual.

a)

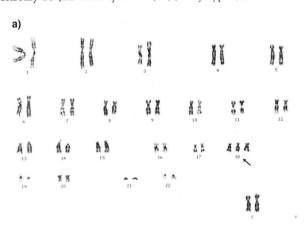

b)

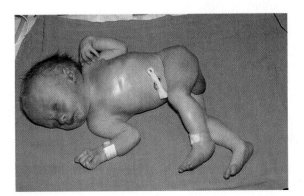

because of monosomy for chromosome 14, one-sixth are inviable because of monosomy for chromosome 21, one-sixth are inviable because of trisomy for chromosome 14, and one-sixth are trisomy-21 and therefore produce a Down syndrome individual. (The latter individuals actually have the normal diploid number of 46 chromosomes, but, because of the Robertsonian translocation, they have three copies of the long arm of chromosome 21, which is sufficient to produce Down syndrome symptoms. Similarly, the trisomy-14 zygotes shown in Figure 17.20 have 46 chromosomes, but they have three copies of the long arm of chromosome 14. The dosage of the genes involved on this larger chromosome is more critical, so trisomy-14 individuals are inviable.) In sum, one-half of the zygotes produced are inviable, and, theoretically, one-third of the viable zygotes give rise to an individual with familial Down syndrome, a much higher risk than that for nonfamilial Down syndrome associated with the mother's age. The observed risk is lower.

Trisomy-13. **Trisomy-13** produces Patau syndrome (Figure 17.21). About 2 in 10,000 live births produce individuals with trisomy 13. Characteristics of individuals with trisomy-13 include cleft lip and palate, small eyes, polydactyly (extra fingers and toes), mental and developmental retardation, and cardiac anomalies, among many other abnormalities. Most die before the age of 3 months.

Trisomy-18. **Trisomy-18** produces Edwards syndrome (Figure 17.22), which occurs in about 2.5 in 10,000 live births. For reasons that are not known, about 80 percent of infants with Edwards syndrome are female. Individuals with trisomy-18 are small at birth and have multiple congenital malformations affecting almost every organ in the body. Clenched fists, an elongated skull, low-set malformed ears, mental and developmental retardation, and many other abnormalities are associated with the syndrome. Ninety percent of infants with trisomy-18 die within 6 months, often from cardiac problems.

Figure 17.22

Trisomy-18 (Edwards syndrome). (a) Karyotype. (b) Individual.

a)

b)

Changes in Complete Sets of Chromosomes

Monoploidy and **polyploidy** involve variations from the normal state in the number of complete sets of chromosomes. Because the number of complete sets of chromosomes is involved in each case, monoploids and polyploids are euploids. Monoploidy and polyploidy are lethal in most animal species, but are less consequential in plants. Both have played significant roles in plant speciation and diversification.

Changes in complete sets of chromosomes can result when the first or second meiotic division is abortive (lack of cytokinesis) or when meiotic nondisjunction occurs for all chromosomes, for example. If such nondisjunction occurs at meiosis I, half of the gametes have no chromosome sets, and half have two chromosome sets. (See Figure 12.20b, p. 319.) If such nondisjunction occurs at meiosis II, half of the gametes have the normal one set of chromosomes, one-quarter have two sets of chromosomes, and one-quarter have no chromosome sets (See Figure 12.20b.) Fusion of a gamete with two chromosome sets with a normal gamete produces a polyploid zygote—in this case, one with three sets of chromosomes, which is a *triploid* (3N). Similarly, fusion of two gametes, each with two chromosome sets, produces a *tetraploid* (4N) zygote. Polyploidy of somatic cells can also occur following the mitotic nondisjunction of complete chromosome sets. Monoploid (haploid) individuals, by contrast, typically develop from unfertilized eggs.

Monoploidy. A monoploid individual has only one set of chromosomes instead of the usual two sets (Figure 17.23a). Monoploidy is sometimes called haploidy, although the term *haploidy* typically is used to describe the chromosome complement of gametes. Some fungi and males of haploid/diploid species (ants, bees, wasps) are haploid, for example.

Monoploidy is seen only rarely in normally adult diploid organisms. Because of the presence of recessive lethal mutations (which are usually counteracted by dominant alleles in heterozygous individuals) in the chromosomes of many diploid eukaryotic organisms, many monoploids probably do not survive. Certain species produce monoploid organisms as a normal part of their life cycle. Some male wasps, ants, and bees, for example, are monoploid because they develop from unfertilized eggs.

Cells of a monoploid individual are very useful for producing mutants, because there is only one dose of each of the genes. Thus, mutants can be isolated directly.

Polyploidy. Polyploidy is the chromosomal constitution of a cell or an organism that has three or more sets of chromosomes (Figure 17.23b). Polyploids may arise spontaneously or be induced experimentally. They

Figure 17.23

Variations in number of complete chromosome sets.
(a) Monoploidy (only one set of chromosomes instead of two).
(b) Polyploidy (three or more sets of chromosomes).

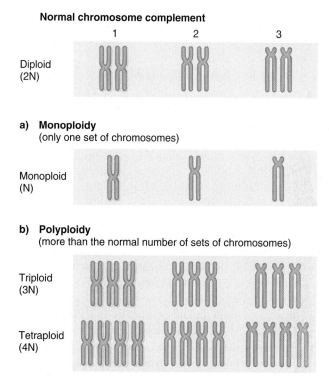

often occur as a result of a breakdown of the spindle apparatus in one or more meiotic divisions or in mitotic divisions. Almost all plants and animals probably have some polyploid tissues. For example, the endosperm of plants is triploid, the liver of mammals and perhaps other vertebrates is polyploid, and the giant abdominal neuron of the sea hare *Aplysia* has about 75,000 copies of the genome. Plants that are completely polyploid include wheat, which is hexaploid (6N), and the strawberry, which is octaploid (8N). Some animal species, such as the North American sucker (a freshwater fish), salmon, and some salamanders, are polyploid.

There are two general classes of polyploids: those with an *even* number of chromosome sets and those with an *odd* number of sets. Polyploids with an even number of chromosome sets have a better chance of being at least partially fertile, because there is the potential for homologs to be segregated equally during meiosis. Polyploids with an odd number of chromosome sets always have an unpaired chromosome for each chromosome type, so the probability of producing a balanced gamete is extremely low; such organisms usually are sterile or have an increased incidence of abortion of zygotes.

In triploids, the nucleus of a cell has three sets of chromosomes. As a result, triploids are highly unstable in meiosis because, as in trisomics, two of the three homologous chromosomes go to one pole and the other goes to the other pole. The segregation of each chromosome from its homologs in the triploid is random, so the probability of producing balanced gametes that contain either a haploid or a diploid set of chromosomes is small; many of the gametes are unbalanced, with one copy of one chromosome, two copies of another, and so on. In general, the probability of a triploid producing a haploid gamete is $(1/2)^n$, where n is the number of chromosomes.

In humans, the most common type of polyploidy is triploidy, and it is always lethal. Triploidy is seen in 15 to 20 percent of spontaneous abortions and about 1 in 10,000 live births, but most affected infants die within 1 month. Triploid infants have many abnormalities, including a characteristically enlarged head. Tetraploidy in humans is also always lethal, usually before birth. It is seen in about 5 percent of spontaneous abortions. Very rarely is a tetraploid human born, but such an individual does not survive long.

Polyploidy is less consequential to plants. One reason is that many plants undergo self-fertilization, so if a plant is produced with an even polyploid number of chromosome sets (for example, 4N) it can still produce fertile gametes and reproduce.

Two types of polyploidy are encountered in plants. In **autopolyploidy,** all the sets of chromosomes originate in the same species. The condition probably results from a defect in meiosis that leads to diploid or triploid gametes. If a diploid gamete fuses with a normal haploid gamete, the zygote and the organism that develops from it will have three sets of chromosomes; in other words, it will be triploid. The cultivated banana is an example of a triploid autopolyploid plant. Because it has an odd number of chromosome sets, the gametes have a variable number of chromosomes, and few fertile seeds are set, thereby making most bananas seedless and highly palatable. Because of the triploid state, cultivated bananas are propagated vegetatively (by cuttings). In general, the development of "seedless" fruits such as grapes and watermelons relies on odd-number polyploidy. Triploidy has also been found in grasses, garden flowers, crop plants, and forest trees.

In **allopolyploidy,** the sets of chromosomes involved come from different, though usually related, species. This situation can arise if two different species interbreed to produce an organism with one haploid set of each parent's chromosomes (one set from each species) and then both chromosome sets double. For example, the fusion of haploid gametes of two diploid plants that can cross may produce an $N_1 + N_2$ hybrid plant that has a haploid set of chromosomes from plant species 1 and a haploid set from plant species 2. However, because of the differences between the two chromosome sets, no chromosomes pair at meiosis, and no viable gametes are produced. As a result, the hybrid plants are sterile. Rarely, through a division error, the two sets of chromosomes double, producing tissues of $2N_1 + 2N_2$ genotype. (That is, the cells in the tissue have a diploid set of chromosomes from plant species 1 and a diploid set from plant species 2.) Each diploid set can function normally in meiosis, so that gametes produced from the $2N_1 + 2N_2$ plant are $N_1 + N_2$. Such fusion of two gametes can produce fully fertile, allotetraploid, $2N_1 + 2N_2$ plants.

A classic example of allopolyploidy resulted from crosses made between cabbages (*Brassica oleracea*) and radishes (*Raphanus sativus*) by Karpechenko in 1928. Both parents have a chromosome number of 18, and the F_1 hybrids also have 18 chromosomes, 9 from each parent. The hybrids produced are morphologically intermediate between cabbages and radishes. The F_1 plants are mostly sterile because of the failure of chromosomes to pair at meiosis. However, a few seeds are produced through meiotic errors, and some of those seeds are fertile. The somatic cells of the plants produced from those seeds have 36 chromosomes—that is, full diploid sets of chromosomes from both the cabbage and the radish. These plants are completely fertile and belong to a breeding species named *Raphanobrassica*, a fusion of the two parental genus names. Morphologically, the plants look a lot like the F_1 hybrids.

Finally, many commercial grains, most crops, and many common commercial flowers are polyploid. In fact, polyploidy is the rule rather than the exception in agriculture and horticulture. For example, the cultivated bread wheat, *Triticum aestivum*, is an allohexaploid with 42 chromosomes. This plant species is descended from three distinct species, each with a diploid set of 14 chromosomes. Meiosis is normal because only homologous chromosomes pair, so the plant is fertile.

KEYNOTE

Variations in the chromosome number of a cell or an organism give rise to aneuploidy, monoploidy, and polyploidy. In aneuploidy, a cell or organism has one, two, or a few whole chromosomes more or less than the basic number of the species under study. In monoploidy, an organism that is usually diploid has only one set of chromosomes. In polyploidy, an organism has more than the normal number of complete sets of chromosomes. Any or all of these abnormal conditions may have serious consequences for the organism.

Summary

In this chapter, we have considered several kinds of chromosomal mutations—variations from the normal (wild-type) condition in chromosome structure or chromosome number. Chromosomal mutations can occur spontaneously, or their frequency can be increased by exposure to radiation or chemical mutagens. There are four major types of chromosomal structural mutations: (1) deletion, in which a DNA segment is lost; (2) duplication, in which there are one or more extra copies of a DNA segment; (3) inversion, in which a DNA segment in a chromosome has reversed its orientation; and (4) translocation, in which a DNA segment has moved to a new location in the genome.

The consequences of these kinds of structural mutations depend on the specific mutation involved. First, each kind of mutation involves one or more breaks in a chromosome. If a break occurs within a gene, then a gene mutation has been produced. In deletions, genes may be lost and multiple mutant phenotypes may result. In some cases, deletions and duplications result in lethality or severe defects as a consequence of a deviation from the normal gene dosage. Second, chromosomal mutations in the heterozygous condition can result in the production of some gametes that are inviable because of duplications or deficiencies. This effect is common after meiotic crossovers that produce inversions and translocations in heterozygotes.

Variations in chromosome number involve a departure from the normal diploid (or haploid) state of the organism. For diploids, the three classes of such mutations are (1) aneuploidy, in which one to a few whole chromosomes are lost from or added to the normal chromosome set; (2) monoploidy, in which only one set of chromosomes is present in a usually diploid organism; and (3) polyploidy, in which a cell or organism has three or more sets of chromosomes. The consequences of these chromosomal mutations depend on the organism. In general, plants are more tolerant than animals of variations in the number of chromosome sets; for example, wheat is hexaploid. Although some animal species are naturally polyploid, in the majority of instances monoploidy and polyploidy are lethal, probably because gene expression problems occur when abnormal numbers of gene copies are present. Even in viable individuals, viable gametes may not result because of segregation problems during meiosis.

Analytical Approaches to Solving Genetics Problems

Q17.1 Diagram the meiotic pairing behavior of the four chromatids in an inversion heterozygote $a\ b\ c\ d\ e\ f\ g/$ $a'\ b'\ f'\ e'\ d'\ c'\ g'$. Assume that the centromere is to the left of gene a. Next, diagram the early anaphase configuration if a crossover occurred between genes d and e.

A17.1 Answering this question requires a knowledge of meiosis (see Figure 12.9, p. 307) and the ability to draw and manipulate an appropriate inversion loop. Part (a) of the following figure shows the diagram for the meiotic pairing:

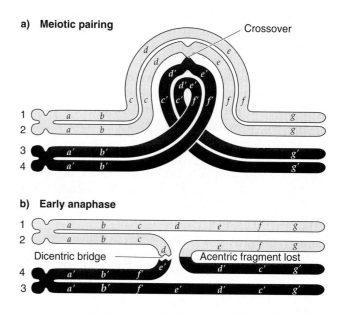

Note that the lower pair of chromatids (a', b', etc.) must loop over in order for all the genes to align; this looping is characteristic of the pairing behavior expected for an inversion heterozygote.

Once the first diagram has been constructed, answering the second part of the question is straightforward. We diagram the crossover and then trace each chromatid from the centromere end to the other end. It is convenient to distinguish maternal and paternal genes, perhaps by a' versus a, and so on, as we did in part (a) of the figure. The result of the crossover between d and e is shown in part (b).

In anaphase I of meiosis, the two centromeres, each with two chromatids attached, migrate toward the opposite poles of the cell. At anaphase, the noncrossover chromatids (top and bottom chromatids in the figure) segregate to the poles normally. As a result of the single crossover between the other two chromatids, however, unusual chromatid configurations are produced, and these configurations are found by tracing the chromatids from left to right. If we begin by tracing the second chromatid from the top, we get

$$\underset{\circ}{a\ b\ c\ d\ e\ f'\ b'\ a'}\circ$$

which is a dicentric chromatid (where ○ is a centromere); in other words, we have a single chromatid attached to two centromeres. This chromatid also has duplications and deletions for some of the genes. Thus, during anaphase, this so-called dicentric chromosome becomes stretched between the two poles of the cells as the centromeres separate, and the chromosome eventually breaks at a random location. The other product of the single crossover is an acentric fragment (a fragment without a centromere) that can be traced starting from the right with the second chromatid from the top. This chromatid,

$$g\ f\ e\ d'\ c'\ g'$$

contains neither a complete set of genes nor a centromere; it is an acentric fragment that will be lost as meiosis continues.

Thus, the consequence of a crossover within the inversion in an inversion heterozygote is the production of gametes with duplicated or deleted genes. These gametes often are inviable. However, viable gametes are produced from the noncrossover chromatids: One of these chromatids (1 in part [b] of the figure) has the normal gene sequence, and the other (3 in part [b] of the figure) has the inverted gene sequence.

Q17.2 *Eyeless* is a recessive gene (*ey*) on chromosome 4 of *Drosophila melanogaster*. Flies homozygous for *ey* have tiny eyes or no eyes at all. A male fly trisomic for chromosome 4 with the genotype +/+/*ey* is crossed with a normal diploid, *eyeless* female of genotype *ey/ey*. What expected genotypic and phenotypic ratios would result from random assortment of the chromosomes to the gametes?

A17.2 To answer this question, we must apply our understanding of meiosis to the unusual situation of a trisomic cell. Regarding the *ey/ey* female, only one gamete class can be produced, namely, eggs of genotype *ey*. Gamete production with respect to the trisomy for chromosome 4 occurs by a random segregation pattern in which, during meiosis I, two chromosomes migrate to one pole and the other chromosome migrates to the other pole. (This pattern is similar to the meiotic segregation pattern shown in the secondary nondisjunction of XXY cells; see Chapter 12.) Three types of segregation are possible in the formation of gametes in the trisomy, as shown in part (**a**) of the following figure: The random union of these sperm with eggs of genotype *ey* occurs as shown in part (**b**), and the resulting genotypic and phenotypic ratios are listed in part (**c**).

a) Segregation

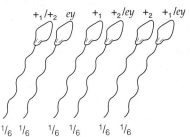

b) Union

		Eggs *ey*	Phenotype
	+/+	+/+/*ey*	+
	ey	*ey/ey*	*ey*
Sperm	+	+/*ey*	+
	+/*ey*	+/*ey/ey*	+
	+	+/*ey*	+
	+/*ey*	+/*ey/ey*	+

c) Summary of genotypes and phenotypes

Ratios:

Genotypes	Phenotypes
$1/6$ +/+/*ey*	$5/6$ wild type
$1/3$ +/*ey/ey*	$1/6$ eyeless
$1/3$ +/*ey*	
$1/6$ *ey/ey*	

Questions and Problems

*17.1 A normal chromosome has the following gene sequence:

$$A\ B\ C\ D\ \circ\ E\ F\ G\ H$$

Determine the chromosomal mutation illustrated by each of the following chromosomes:

a. $A\ B\ C\ F\ E\ \circ\ D\ G\ H$
b. $A\ D\ \circ\ E\ F\ B\ C\ G\ H$
c. $A\ B\ C\ D\ \circ\ E\ F\ E\ F\ G\ H$
d. $A\ B\ C\ D\ \circ\ E\ F\ F\ E\ G\ H$
e. $A\ B\ D\ \circ\ E\ F\ G\ H$

*17.2 Distinguish between pericentric and paracentric inversions.

17.3 In some instances, very small deletions behave like recessive mutations. Why are some recessive mutations known not to be deletions?

*17.4 Inversions are known to affect crossing-over. The following homologs have the indicated gene order (the filled and open circles are homologous centromeres):

$$\overset{\bullet}{\underline{A\ B\ C\ D\ E}}$$

$$\overset{\circ}{\underline{A\ D\ C\ B\ E}}$$

a. Considering the position of the centromere, what is this sort of inversion called?

b. Diagram the alignment of these chromosomes during meiosis.

c. Diagram the results of a single crossover between homologous genes *B* and *C* in the inversion.

17.5 Single crossovers within the inversion loop of inversion heterozygotes give rise to chromatids with duplications and deletions. What happens if, within the loop, there is a two-strand double crossover in such an inversion heterozygote when the centromere is outside the loop?

17.6 An inversion heterozygote possesses one chromosome with genes in the normal order:

$$\overset{}{\underline{a\ b\ c\ d\ e\ f\ g\ h}}$$

It also contains one chromosome with genes in the inverted order:

$$\overset{}{\underline{a\ b\ f\ e\ d\ c\ g\ h}}$$

A four-strand double crossover occurs in the areas *e–f* and *c–d*. Diagram and label the four strands at synapsis (showing the crossovers) and at the first meiotic anaphase.

*17.7 The following gene arrangements in a particular chromosome are found in *Drosophila* populations in different geographic regions:

a. *A B C D E F G H I*
b. *H E F B A G C D I*
c. *A B F E D C G H I*
d. *A B F C G H E D I*
e. *A B F E H G C D I*

Assuming that the arrangement in part (**a**) is the original arrangement, in what sequence did the various inversion types probably arise?

17.8 The following figure shows map distances observed for genes in one chromosomal region (the open circle represents a centromere):

A new paracentric inversion bears a recessive mutation *d* at the *D* locus. Its proximal breakpoint lies between *B* and *C*, 1 mu from *B*, and its distal breakpoint lies between *G* and

H, 1 mu from *G*. A heterozygote for this inversion and the wild-type arrangement mates with a *dd* individual homozygous for the inversion.

a. In the absence of multiple crossovers, what is the chance that a *dd* offspring will have a homolog with a wild-type arrangement?

b. What type of event is required to produce a *dd* offspring having a homolog with a wild-type arrangement? What is the likelihood of such an event?

c. On the basis of your answers to (**a**) and (**b**), how might spontaneously arising inversions contribute to the maintenance of genetic differences between subpopulations of a species?

*17.9 The deep-red eye color of normal *Drosophila* results from pigment deposition controlled by the *white* gene, which lies on the X chromosome at map position 1.5, far from centromeric heterochromatin (which starts at about map position 66). Hermann Müller screened for new *white* mutants by irradiating wild-type *Drosophila* males (w^+/Y) and mating them to white-eyed (*w/w*) females. He isolated several mutant females bearing mottled red eyes—red eyes with varying amounts of white spotting. One mutant, w^{M5}, was associated with a reciprocal translocation with breakpoints near the *white* locus and centromeric heterochromatin of chromosome 4. A different mutant, w^{M4}, was associated with an X-chromosome inversion with breakpoints near the *white* locus and centromeric heterochromatin. Kenneth Tartof screened for revertants of the mottled-eye phenotype of w^{M4} by crossing irradiated w^{M4}/Y males with *w/w* females. He recovered three different normal-eyed female revertants, each associated with a new X-chromosome inversion. In addition to having the original w^{M4} breakpoints near the *white* locus and in centromeric heterochromatin, each had a third euchromatic breakpoint.

a. On the basis of these results, is the mottled-eye phenotype in Müller's mutants due to a mutation within the *white* gene? If not, what is its most likely cause?

b. How can the w^{M4} mutation be reverted by an additional inversion with a euchromatic breakpoint?

*17.10 Human abnormalities associated with chromosomal mutations often exhibit a range of symptoms, of which only some subsets appear in a particular individual. Recombinant 8 [Rec(8)] syndrome is an inherited chromosomal abnormality found primarily in individuals of Hispanic origin. Phenotypic characteristics associated with the syndrome include congenital heart disease, urinary system abnormalities, eye abnormalities, hearing loss, and abnormal muscle tone. Most reported cases of Rec(8) have been found in the offspring of phenotypically normal parents who are heterozygous for an inversion of chromosome 8 with breakpoints at p23.1 and q22.1. Individuals with Rec(8) syndrome typically have a duplication of part of 8q (from q22.1 to the terminus of the q arm) and a deletion of 8p (from p23.1 to the terminus of the p arm).

a. Using diagrams, explain why individuals with Rec(8) syndrome typically have a duplication and a deletion for part of chromosome 8.

b. An individual is heterozygous for an inversion on chromosome 8 with breakpoints at p23.1 and q22.1. If a crossover occurs within the inverted region during a particular meiosis, what is the chance that the resulting offspring will have Rec(8) syndrome?

c. Why might the phenotypes of Rec(8) individuals vary?

d. A child with some of the symptoms of Rec(8) syndrome is referred to a human geneticist. The karyotype of the child reveals heterozygosity for a large pericentric inversion in chromosome 8 with breakpoints at p23.1 and q22.1. Cytogenetic analysis of her phenotypically normal mother and phenotypically normal maternal grandmother reveals a similar karyotype. According to the child's mother, the father has a normal phenotype, but he is unavailable for examination. Propose at least two explanations for why the child, but not her mother or maternal grandmother, is affected with some of the symptoms of Rec(8) syndrome. (Hint: Consider the limitations of karyotype analysis using G-banding methods [see Chapter 12, p. 302], and also consider what is unknown about the father.)

*17.11 A particular plant species that had been subjected to radiation for a long time in order to produce chromosomal mutations was then inbred for many generations until it was homozygous for all of these mutations. It was then crossed to the original unirradiated plant, and the meiotic process of the F_1 hybrids was examined. It was noticed that a cell with a dicentric chromosome (bridge) and a fragment occurred at low frequency in anaphase I of the hybrid.

a. What kind of chromosomal mutation occurred in the irradiated plant? In your answer, indicate where the centromeres are.

b. Explain, in words and with a clear diagram, where crossover(s) occurred and how the bridge chromosome of the cell arose.

17.12 On a normal-ordered chromosome, two loci, a and b, lie 15 map units apart on the left arm of a metacentric chromosome. A third locus, c, lies 10 map units to the right of b on the right arm of the chromosome. What frequency of progeny phenotypes do you expect to see in a testcross of an $a\ b\ c/a^+\ b^+\ c^+$ individual if the $a^+\ b^+\ c^+$ chromosome

a. has a normal order?

b. has an inversion with breakpoints just proximal (toward the centromere) to a and just distal (away from the centromere) to b?

c. has an inversion with breakpoints just proximal to a and just proximal to c?

d. has an inversion with breakpoints just distal to a and just distal to c?

*17.13 Mr. and Mrs. Lambert have not yet been able to produce a viable child. They have had two miscarriages and one severely defective child who died soon after birth. Studies of banded chromosomes of father, mother, and child showed that all chromosomes were normal except for pair number-6. The number 6 chromosomes of mother, father, and child are shown in the following figure:

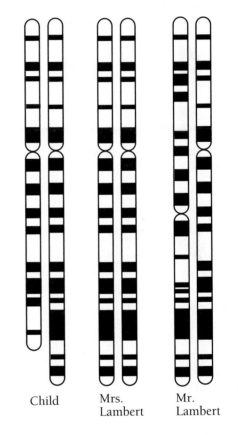

Child   Mrs.   Mr.
             Lambert   Lambert

a. Does either parent have an abnormal chromosome? If so, what is the abnormality?

b. How did the chromosomes of the child arise? Be specific as to what events in the parents gave rise to those chromosomes.

c. Why is the child not phenotypically normal?

d. What can be predicted about future conceptions by this couple?

17.14 Mr. and Mrs. Simpson have been trying for years to have a child, but have been unable to conceive. They consulted a physician, and tests revealed that Mr. Simpson had a markedly low sperm count. His chromosomes were studied, and a testicular biopsy was done. His chromosomes proved to be normal, except for pair 12. The following figure shows Mrs. Simpson's normal pair of number-12 chromosomes and Mr. Simpson's number-12 chromosomes.

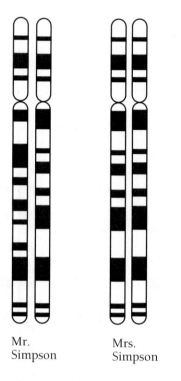

Mr. Simpson

Mrs. Simpson

a. What is the nature of the abnormality of pair number 12 in Mr. Simpson's chromosomes?

b. What abnormal feature would you expect to see in the testicular biopsy? (Cells in various stages of meiosis can be seen.)

c. Why is Mr. Simpson's sperm count low?

d. What can be done about Mr. Simpson's low sperm count?

*17.15 Chromosome I in maize has the gene sequence *ABCDEF*, whereas chromosome II has the sequence *MNOPQR*. A reciprocal translocation resulted in *ABCPQR* and *MNODEF*. Diagram the expected pachytene (see Chapter 12, p. 308) configuration in the F_1 of a cross of homozygotes of these two arrangements.

17.16 Diagram the pairing behavior at prophase of meiosis I (see Chapter 12, p. 306) of a translocation heterozygote that has normal chromosomes of gene order *abcdefg* and *tuvwxyz* and has the translocated chromosomes *abcdvwxyz* and *tuefg*. Assume that the centromere is at the left end of all chromosomes.

*17.17 Mr. and Mrs. Denton have been trying for several years to have a child. They have experienced a series of miscarriages, and last year they had a child with multiple congenital defects. The child died within days of birth.

The birth of this child prompted the Dentons' physician to order a chromosome study of parents and child. The results of the study are shown in the accompanying figure. Chromosome banding was done, and all chromosomes were normal in these individuals, except some copies of number 6 and number 12. The number-6 and number-12 chromosomes of mother, father, and child are shown in the figure (the number 6 chromosomes are the larger pair):

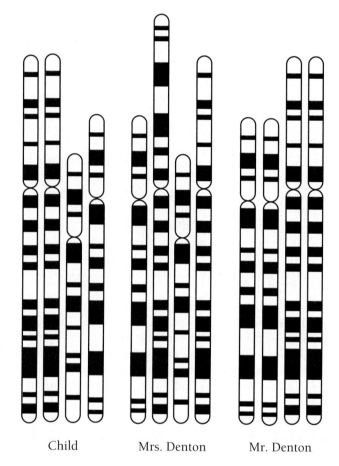

Child   Mrs. Denton   Mr. Denton

a. Does either parent have an abnormal karyotype? If so, which parent has it, and what is the nature of the abnormality?

b. How did the child's karyotype arise? (What pairing and segregation events took place in the parents?)

c. Why is the child phenotypically defective?

d. What can this couple expect to happen in subsequent conceptions?

e. What medical help, if any, can be offered to the couple?

17.18 Irradiation of *Drosophila* sperm produces translocations between the X chromosome and autosomes, between the Y chromosome and autosomes, and between different autosomes. Translocations between the X and Y chromosomes are not produced. Explain the absence of X–Y translocations.

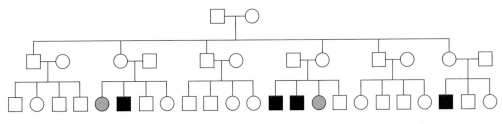

*17.19 In the above pedigree, mental retardation is indicated by shaded symbols, with individuals shaded black being more severely affected than individuals shaded gray.

a. What features of this pedigree suggest that the mental retardation phenotype may be associated with fragile X syndrome?

b. What molecular and cytological means would you employ to evaluate if mental retardation in this pedigree is due to fragile X syndrome, and what would you expect to find if it is?

c. If the mental retardation in this pedigree is due to fragile X syndrome, which individual(s)
i. must carry a permutation?
ii. must be normal transmitting males?
iii. must carry a fragile X chromosome?
iv. in generation III are phenotypically normal but may still carry a fragile X chromosome?

d. Why might females in this pedigree be less severely affected?

17.20 Define the terms *aneuploidy, monoploidy*, and *polyploidy*.

17.21 If a normal diploid cell is 2N, what is the chromosome content of the following?
a. a nullisomic
b. a monosomic
c. a double monosomic
d. a tetrasomic
e. a double trisomic
f. a tetraploid
g. a hexaploid

*17.22 In humans, how many chromosomes would be typical of nuclei of cells that are
a. monosomic?
b. trisomic?
c. monoploid?
d. triploid?
e. tetrasomic?

*17.23 An individual with 47 chromosomes, including an additional chromosome 15, is said to be
a. triplet
b. trisomic
c. triploid
d. tricycle

*17.24 A color-blind man marries a homozygous normal woman, and after four joyful years of marriage they have two children. Unfortunately, both children have Turner syndrome, although one has normal vision and one is color blind. The type of color blindness involved is a sex-linked recessive trait.

a. For the color-blind child, did nondisjunction occur in the mother or the father? Explain your answer.

b. For the child with normal vision, in which parent did nondisjunction occur? Explain your answer.

*17.25 The frequency of chromosome loss in *Drosophila* can be increased by a recessive chromosome 2 mutation called *pal*. The mutation causes the preferential loss of chromosomes contributed to a zygote by *pal/pal* fathers. The paternally contributed chromosomes are lost during the first few mitotic divisions after fertilization. What phenotypic consequences do you expect in offspring of the following crosses (keep in mind how sex is determined in *Drosophila*, that the loss of an entire chromosome 2 or chromosome 3 is lethal, and that the loss of one copy of the small chromosome 4 is tolerated)?

a. X-chromosome loss at the first mitotic division in a cross between a true-breeding *yellow* (recessive, X-linked mutation causing yellow body color) female and a *pal/pal* father.

b. Chromosome 4 loss at the first mitotic division in a cross between a true-breeding *eyeless* (recessive mutation on chromosome 4 causing reduced eye size) female and a *pal/pal* father.

c. Chromosome 3 loss at the first mitotic division in a cross between a true-breeding *ebony* (recessive mutation on chromosome 3 causing black body color) female and a *pal/pal* father.

17.26 Assume that *x* is a new mutant gene in corn. A female *x/x* plant is crossed with a triplo-10 individual (trisomic for chromosome 10) carrying only dominant alleles at the *x* locus. Trisomic progeny are recovered and crossed back to the *x/x* female plant.

a. What ratio of dominant to recessive phenotypes is expected if the *x* locus is not on chromosome 10?

b. What ratio of dominant to recessive phenotypes is expected if the *x* locus is on chromosome 10?

17.27 Why are polyploids with even multiples of the chromosome set generally more fertile than polyploids with odd multiples of the chromosome set?

17.28 One plant species (N = 11) and another (N = 19) produced an allotetraploid. For the following statements, select the correct answer from the key:

I. The chromosome number of this allotetraploid is 30.
II. The number of linkage groups of this allotetraploid is 30.
Key:
a. Statement I is true and Statement II is true.
b. Statement I is true but Statement II is false.
c. Statement I is false but Statement II is true.
d. Statement I is false and Statement II is false.

*****17.29** According to Mendel's first law, genes *A* and *a* segregate from each other and appear in equal numbers among the gametes. But Mendel did not know that his plants were diploid. In fact, because plants are frequently tetraploid, he could have been unlucky enough to have started with peas that were 4N rather than 2N. Let us assume that Mendel's peas were tetraploid, that every gamete contains two alleles, and that the distribution of alleles to the gamete is random. Suppose we have a cross of *AAAA* × *aaaa*, where *A* is dominant, regardless of the number of *a* alleles present in an individual.
a. What will be the genotype of the F_1 peas?
b. If the F_1 peas is selfed, what will be the phenotypic ratios in the F_2 peas?

17.30 What phenotypic ratio of *A* to *a* is expected if *AAaa* plants are testcrossed against *aaaa* individuals? (Assume that the dominant phenotype is expressed whenever at least one *A* is present, no crossing-over occurs, and each gamete receives two chromosomes.)

17.31 The root-tip cells of an autotetraploid plant contain 48 chromosomes. How many chromosomes were contained by the gametes of the diploid from which this plant was derived?

17.32 A number of species of the birch genus have a somatic chromosome number of 28. The paper birch is reported as occurring with several different chromosome numbers; *fertile* individuals with the somatic numbers 56, 70, and 84 are known. How should the 28-chromosome individuals be designated with regard to chromosome number?

*****17.33** How many chromosomes would be found in somatic cells of an allotetraploid derived from two plants, one with N = 7 and the other with N = 10?

17.34 Plant species A has a haploid complement of four chromosomes. A related species, B, has five. In a geographic region where A and B are both present, C plants are found that have some characters of both species and somatic cells with 18 chromosomes. What is the chromosome constitution of the C plants likely to be? With what plants would they have to be crossed to produce fertile seed?

18

Genetics of Bacteria and Bacteriophages

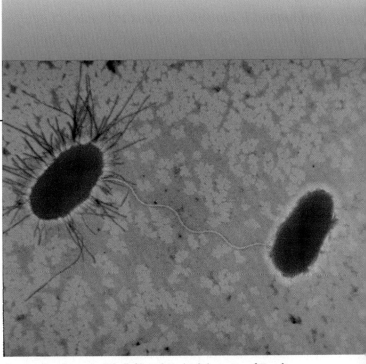

Conjugation between *Hfr* (left) and *F⁻* (right) strains of *E. coli*.

PRINCIPAL POINTS

- Conjugation is a process in which there is a unidirectional transfer of genetic information through direct cellular contact between a donor and a recipient bacterial cell. The donor state is conferred on that cell by the presence of a plasmid called an *F* factor. Conjugation results in the unidirectional transfer of a copy of the *F* factor from donor to recipient.

- The *F* factor can integrate into the bacterial chromosome. Strains in which this has occurred—*Hfr* strains—can conjugate with recipient strains and transfer the bacterial chromosome. The sequence and distances between genes can be determined by the order and time of acquisition of the genes by the recipient from the donor during conjugation.

- Transformation is the transfer of genetic material between organisms by small extracellular pieces of DNA. Through genetic recombination, part of the transforming DNA molecule can exchange with a portion of the recipient's chromosomal DNA. Transformation can be used experimentally to determine gene order and map distances between genes.

- Transduction is a process whereby bacteriophages (phages) mediate the transfer of bacterial DNA from one bacterium (the donor) to another (the recipient). Transduction can be used experimentally to map bacterial genes.

- The same principles used to map eukaryotic genes are used to map phage genes. That is, a bacterial host is simultaneously infected with two strains of phages differing from each other in one or more gene loci. The percentages of recombinants are determined, and the sequence and distances between genes are then inferred.

- The same principles of recombinational mapping in eukaryotes can be applied to mapping the distance between mutational sites in different genes (intergenic mapping) and to mapping mutational sites within the same gene (intragenic mapping)

- From fine-structure analysis of the *rII* region of bacteriophage T4, it was determined that the unit of mutation and recombination is the DNA base pair.

• The number of genes that cause a particular mutant phenotype is determined by the complementation, or *cis-trans*, test. If two viral mutants, each carrying a mutation in a different gene, are combined in a single host cell, the mutations make up for each other's defect (that is, they complement) and a wild-type phenotype results. If two mutants, each carrying a mutation in the same gene, are combined, the mutations do not complement and the mutant phenotype is still expressed.

i BACTERIA AND BACTERIOPHAGES HAVE LONG played a key role in genetics. Griffith's and Avery's experiments with *Streptococcus* were key in the discovery of DNA as the genetic material. Using bacteria, Herbert Boyer and Stanley Cohen first cloned a recombinant DNA molecule. F. Peyton Rous helped determine the nature of certain inherited cancers with his investigations of a virus that infects chickens.

In this chapter, you will learn about the genetics of bacteria and bacteriophages: how these organisms reproduce, how new strains are produced, and the experimental techniques that geneticists use to map bacterial and viral genes. After you have read and studied the chapter, you can apply what you've learned by trying the iActivity, in which you will create a genetic map of the *E. coli* chromosome.

In Chapters 15 and 16, we considered the principles of gene mapping in eukaryotic organisms. To map genes in bacteria and bacteriophages, geneticists use essentially the same experimental strategies. Crosses are made between strains that differ in genetic markers, and recombinants—the products of the exchange of genetic material—are detected and counted. The analysis of data obtained from such crosses is the same as for eukaryotes: The frequency with which crossing-over occurs between two sets of genes relates to the map distance between the two gene loci. The major difference lies in the experimental techniques involved.

Recently, emphasis in genetic research has shifted from localizing individual genes on chromosomes by making crosses to determining the complete sequence of nucleotides in the genome. With such DNA sequence information, scientists can identify genes directly, so that the ultimate genetic map of a species can be constructed. When all the genes of an organism are identified, at least at the nucleotide level, we can study the function of each gene. In the case of pathogenic microorganisms, the genomic sequence information is an extremely valuable resource in efforts to identify and understand the genes responsible for pathogenesis. Complete genomic sequences have already been determined for many species of Bacteria and Archaea. (See Chapter 10.) Bacterial genomes sequenced include the 4.6-million-base-pair (4.6-megabase) genome of *E. coli*,

the 1.44-megabase genome of the Lyme disease causative agent *Borrelia burgdorferi*, the 1.66-megabase genome of the stomach-ulcer-causing *Helicobacter pylori*, and the 1.14-megabase genome of the syphilis bacterium *Treponema pallidum*. Among the archaeon genomes sequenced is the 1.66-megabase genome of *Methanococcus jannaschii*, a hyperthermophilic methanogen that grows optimally at 85°C and at pressures up to 200 atmospheres.

In this chapter, you will learn about the classic genetic studies of bacteria and bacteriophages. You will also learn about a series of classic genetic experiments that investigated the fine structure of the gene—that is, the detailed molecular organization of the gene as it relates to the mutational, recombinational, and functional events in which the gene is involved. A bacteriophage gene was the subject of these experiments.

Genetic Analysis of Bacteria

Genetic material can be transferred between bacteria by three main processes: conjugation, transformation, and transduction. It is possible to map bacterial genes using methods involving any one of these processes. In each case, (1) the transfer is unidirectional and (2) no complete diploid stage is formed (unlike the situation in eukaryotes). However, not all methods of genetic analysis can be used for all bacterial species, and the size of the region that can be mapped varies according to the method.

Among bacteria, *E. coli* has been used extensively for genetic analysis. It is found commonly in the large intestines of most animals (including humans). This bacterium is a good subject for study because it can be grown on a simple, defined medium and can be handled with simple microbiological techniques.

E. coli is a cylindrical organism about 1–3 μm long and 0.5 μm in diameter (see chapter opening photo, p. 481); like most bacteria, it is small compared with eukaryotic cells. *E. coli* has a single circular DNA chromosome, with no membrane between the chromosome and the rest of the cell.

Like other bacteria, *E. coli* can be grown both in a liquid culture medium and on the surface of a growth medium solidified with agar. Genetic analysis of bacteria typically is done by spreading (plating) cells on the surface of an agar medium. Wherever a single bacterium lands on the agar surface, it will grow and divide repeatedly, ultimately forming a visible cluster of genetically identical cells called a *colony* (Figure 18.1). Each colony consists of a clone of cells each of which are genetically identical to the parental cell that initiated the colony. The concentration of bacterial cells in a liquid culture—the *titer*—can be determined by spreading known volumes of the culture or of a known dilution of

Figure 18.1

Bacterial colonies growing on a nutrient medium in a Petri dish.

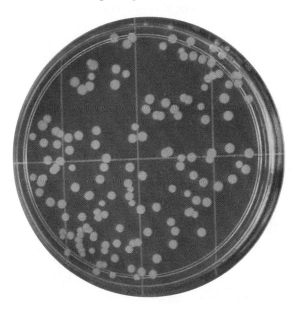

the culture on the agar surface, incubating the plates at a temperature of 37°C, and then counting the number of resulting colonies. The number obtained is converted to colony-forming units (cfu) per milliliter (mL). For example, if 100 μL of a thousandfold dilution of a culture is spread on a plate and 165 colonies are produced, then there were 165 bacteria in 100 μL of the thousandfold dilution. Thus, in the original culture, there were 165 (colonies) × 1,000 (dilution factor) × 10 (because 0.1 mL was plated) = 1,650,000 cfu/mL = 1.65×10^6 cfu/mL.

The composition of the culture medium used depends on the experiment and the genotypes of the strains being examined. Each bacterial species (or any other microorganism, such as yeast) has a characteristic **minimal medium** on which it will grow. A minimal medium contains only the nutrients required for the growth of wild-type cells. The minimal medium for wild-type *E. coli*, for example, consists of a sugar (a carbon source) and some salts and trace elements. From the minimal medium, the organism can synthesize all the other components it needs for growth and reproduction, including amino acids, vitamins, and the precursors for DNA and RNA. By contrast, the **complete medium** for a microorganism supplies vitamins and amino acids and all kinds of substances that might be expected to be essential metabolites and whose biosynthesis might be interfered with by mutation.

The genetic analysis of bacteria (and other microorganisms) typically involves studying mutants that are defective in their abilities to make one or more molecules essential for growth and that are perhaps also defective in genes affecting other metabolic processes.

Strains unable to synthesize essential nutrients are called **auxotrophs** (also called **auxotrophic mutants** or **nutritional mutants**). A strain that is wild type and thus that can synthesize all essential nutrients is a **prototroph.** Prototrophs (also called **prototrophic strains**) need no nutritional supplements in the growth medium. By definition, the wild type or prototroph grows on the minimal medium for that organism, whereas an auxotroph grows on a complete medium or on a minimal medium plus the appropriate nutritional supplement or supplements.

For example, an *E. coli* strain with the genotype *trp ade thi*$^+$ will not grow on a minimal medium, because it has mutations for tryptophan and adenine biosynthesis. It will grow either on a complete medium or on a minimal medium supplemented with the amino acid tryptophan (because of the *trp* mutation) and the purine adenine (because of the *ade* mutation). It does not need the vitamin thiamine to grow, because it carries the wild-type *thi* allele, as signified by the superscript +.

Some genes are involved, not in biosynthetic pathways, but in utilization pathways. For example, a number of different genes use various carbon sources, such as lactose, arabinose, and maltose. In this case, the superscript + next to the gene symbol means that the gene is wild-type and therefore that the bacterium can metabolize the substance. For instance, a *lac*$^+$ strain can metabolize lactose, whereas a *lac* mutant strain cannot.

In genetic experiments with microorganisms such as *E. coli*, crosses are made between strains differing in genotype (and, therefore, phenotype), and progeny are analyzed for parental and recombinant phenotypes. When auxotrophic mutations are involved, the determination of parental and progeny phenotypes (and, therefore, genotypes, because bacteria are haploid) involves testing colonies for their growth requirements. One convenient procedure for such testing is *replica plating*, invented by Joshua and Esther Lederberg. (See Figure 7.15.) In replica plating, some of the bacteria of colonies on a plate of complete medium (the master plate) are transferred onto a sterile velveteen cloth mounted on a replica plater. Replicas of the original colony pattern on the cloth are then made by gently pressing new plates onto the velveteen. If the new plate contains minimal medium, only prototrophic colonies can grow. Then researchers can readily identify auxotrophic colonies because they will be on the master plate, but not on the minimal medium plate. Using other plates containing minimal medium plus combinations of nutritional supplements appropriate for the strain or strains involved, one can determine the phenotypes and genotypes of all auxotrophic colonies.

Gene Mapping in Bacteria by Conjugation

Discovery of Conjugation in *E. coli*

Conjugation is a process in which there is a unidirectional transfer of genetic information through direct cellular contact between a donor bacterial cell and a recipient bacterial cell. The contact is followed by the formation of a physical bridge between the cells. Then a segment (rarely all) of the donor's chromosome may be transferred into the recipient cell and may undergo genetic recombination with a homologous chromosome segment of that cell. Recipients that have incorporated a piece of donor DNA into their chromosomes are called **transconjugants.**

animation

ⓐ Mapping Bacterial Genes by Conjugation

Conjugation was discovered in 1946 by Joshua Lederberg and Edward Tatum. They studied two *E. coli* strains that differed in their nutritional requirements. Strain A had the genotype *met bio thr⁺ leu⁺ thi⁺*, meaning that it could grow only on a medium supplemented with the amino acid methionine (*met*) and the vitamin biotin (*bio*), but it did not need the amino acids threonine (*thr*) or leucine (*leu*) or the vitamin thiamine (*thi*). Strain B had the genotype *met⁺ bio⁺ thr leu thi*, meaning that it could grow only on a medium supplemented with threonine, leucine, and thiamine, but it did not require methionine or biotin.

Lederberg and Tatum mixed *E. coli* strains A and B together and plated them onto minimal medium (Figure 18.2). The mixed culture gave rise to some prototrophic colonies (*met⁺ bio⁺ thr⁺ leu⁺ thi⁺*) at a frequency of about 1 in 10 million cells. Since no colonies appeared when each strain was plated separately on minimal medium, mutation was ruled out as the cause of the prototrophic colonies. The mixing, then, was a genetic cross that produced recombinants.

In a separate experiment, Bernard Davis placed strains A and B in a liquid medium on either side of a U-tube apparatus (Figure 18.3) separated by a filter with pores too small to allow bacteria to move through. The medium was moved between compartments by alternating suction and pressure, and then the cells were plated on minimal medium to check for the appearance of prototrophic colonies. No prototrophic colonies appeared, meaning that cell-to-cell contact was required for the genetic exchange to occur. These experiments indicated that *E. coli* has the type of mating system called *conjugation*.

The Sex Factor *F*

In 1953, William Hayes showed that genetic exchange in *E. coli* occurs in only one direction, with one cell acting as a donor and the other cell acting as a recipient. Hayes proposed that the transfer of genetic material between the

Figure 18.2

Lederberg and Tatum experiment showing that sexual recombination occurs between cells of *E. coli*. After the cells from strain A and strain B have been mixed and the mixture plated, a few colonies grow on the minimal medium, indicating that they can now make the essential constituents. These colonies are recombinants produced by an exchange of genetic material between the strains.

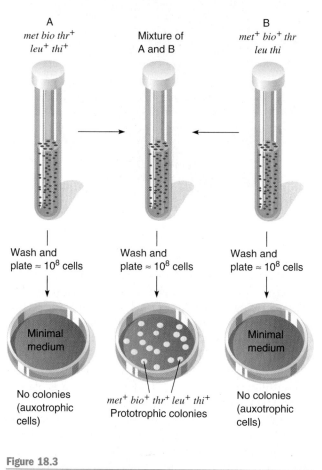

Figure 18.3

Davis's U-tube experiment showing that physical contact between the two bacterial strains of the Lederberg and Tatum experiment was needed for genetic exchange to occur.

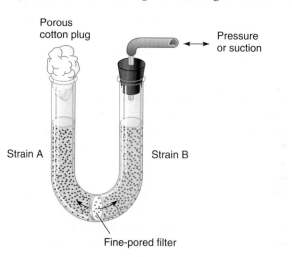

strains is mediated by a *sex factor* named *F* that the donor cell possesses (*F*⁺) and the recipient cell lacks (*F*⁻). The *F* factor found in *E. coli* is an example of a *plasmid*—a self-replicating, circular DNA distinct from the main bacterial chromosome. About ¹⁄₄₀ of the size of the host chromosome, the *F* factor contains a region of DNA called the **origin** (or O)—the point where DNA transfer to the recipient begins—as well as a number of genes, including those which specify hairlike host cell surface components called **F-pili** (singular, F-pilus), or *sex-pili*, which permit the physical union of *F*⁺ and *F*⁻ cells.

When *F*⁺ and *F*⁻ cells are mixed, they may conjugate ("mate") (Figure 18.4 and Figure 18.5a1). No conjugation can occur between two cells of the same mating type (that is, two *F*⁺ bacteria or two *F*⁻ bacteria). During conjugation, genetic material is transferred from donor to recipient when one DNA strand of the *F* factor is nicked at the origin; DNA replication then proceeds from that point (Figure 18.5a2). Beginning at the origin, a single strand of DNA is transferred to the *F* as replication maintains the remaining circular *F* factor in a double-stranded form (Figure 18.5a3; see also Figure 3.10 and Chapter 3, pp. 55). Think of the process loosely like a roll of paper towels unraveling. Once the *F* factor single-stranded DNA enters the *F*⁻ recipient, the complementary strand is synthesized (Figure 18.5a4). When the complete *F* factor has been transferred, the *F*⁻ cell becomes an *F*⁺ cell (Figure 18.5a5). In *F*⁺ × *F*⁻ crosses, none of the bacterial chromosome is transferred; only the *F* factor is.

KEYNOTE

> Some *E. coli* bacteria possess a plasmid, called the *F* factor, that is required for mating. *E. coli* cells containing the *F* factor are designated *F*⁺ and those without it are *F*⁻. The *F*⁺ cells (donors) can mate with *F*⁻ cells (recipients) in a process called conjugation, which leads to the one-way transfer of a copy of the *F* factor from donor to recipient during replication of the *F* factor. As a result, both donor and recipient are *F*⁺. None of the bacterial chromosome is transferred during *F*⁺ × *F*⁻ conjugation.

High-Frequency Recombination Strains of *E. coli*

Producing recombinants for chromosomal genes by conjugation involves special derivatives of *F*⁺ strains, called **Hfr (high-frequency recombination)** strains. Discovered separately by William Hayes and Luca Cavalli-Sforza, *Hfr* strains originate by a rare crossover event in which the *F* factor integrates into the bacterial chromosome (Figure 18.5b1–2). Plasmids such as *F* that are also capable of integrating into the bacterial chromosome are called **episomes.** When the *F* factor is integrated, it no longer

Figure 18.4

Electron micrograph of conjugation between an *F*⁺ donor and *F*⁻ recipient *E. coli* bacterium. (The spherical structure on the upper F-pilus bridging the two bacteria is a donor-specific RNA phage MS-2.) Magnification 30,000 ×.

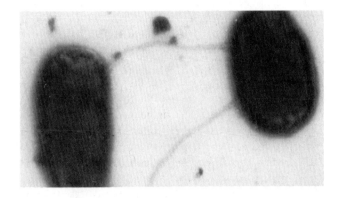

replicates independently, but is replicated as part of the host chromosome.

Because of the *F* factor genes, *Hfr* cells can conjugate with *F*⁻ cells (Figure 18.5b3). When mating happens, events similar to those in the *F*⁺ × *F*⁻ mating occur. The integrated *F* factor becomes nicked at the origin, and replication begins (Figure 18.5b4). During replication, part of the *F* factor starting with the origin moves into the recipient cell, where the transferred strand is copied. In a short time, the donor bacterial chromosome begins to transfer into the recipient. If there are allelic differences between donor genes and recipient genes, recombinants can be isolated (Figure 18.5b5). The recombinants are produced by double crossovers between the linear donor DNA and the circular recipient chromosome, which switch a segment of donor DNA for the homologous segment of recipient DNA.

In *Hfr* × *F*⁻ matings, the *F*⁻ cell almost never acquires the *Hfr* phenotype. To become *Hfr*, the recipient cell must receive a complete copy of the *F* factor. However, only part of the *F* factor is transferred at the beginning of conjugation; the rest lies at the end of the donor chromosome. All of the donor chromosome would have to be transferred for a complete functional *F* factor to be found in the recipient, and that would take about 100 minutes at 37°C. This is an extremely rare event, because mating pairs typically break apart long before the second part of the *F* factor is transferred.

The low-frequency recombination of chromosomal gene markers in *F*⁺ × *F*⁻ crosses can be understood when we consider that only about 1 in 10,000 *F*⁺ cells in a population become *Hfr* cells by *F* factor integration. The reverse process, excision of the *F* factor, also occurs spontaneously and at low frequency, producing an *F*⁺ cell from an *Hfr* cell. In excision, the *F* factor loops out of the *Hfr* chromosome, and by a single crossing-over event (just

Figure 18.5

Transfer of genetic material during conjugation in *E. coli*. (a) Transfer of the *F* factor from donor to recipient cell during $F^+ \times F^-$ matings. (b) Production of *Hfr* strain by integration of *F* factor and transfer of bacterial genes from donor to recipient cell during *Hfr* $\times F^-$ matings.

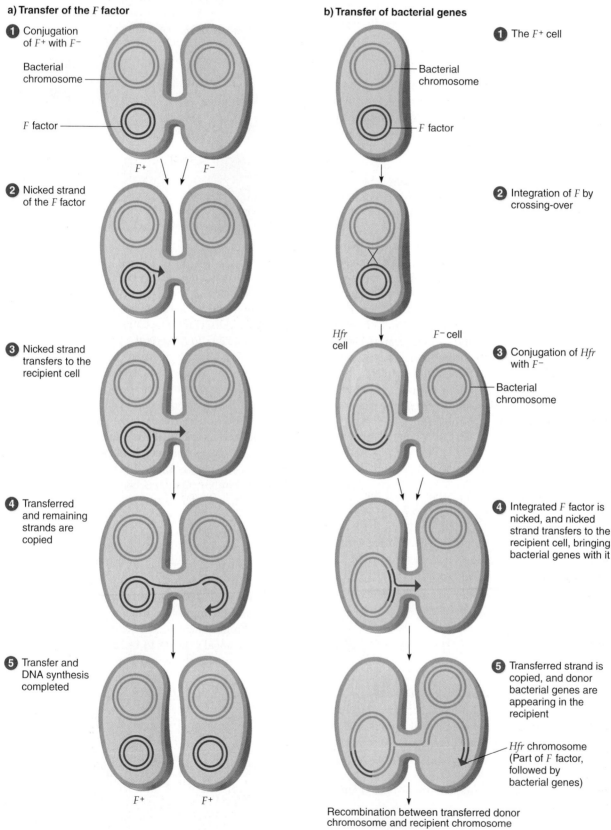

a) Transfer of the *F* factor

1 Conjugation of F^+ with F^-

Bacterial chromosome

F factor

F^+ F^-

2 Nicked strand of the *F* factor

3 Nicked strand transfers to the recipient cell

4 Transferred and remaining strands are copied

5 Transfer and DNA synthesis completed

F^+ F^+

b) Transfer of bacterial genes

1 The F^+ cell

Bacterial chromosome

F factor

2 Integration of *F* by crossing-over

Hfr cell F^- cell

3 Conjugation of *Hfr* with F^-

Bacterial chromosome

4 Integrated *F* factor is nicked, and nicked strand transfers to the recipient cell, bringing bacterial genes with it

5 Transferred strand is copied, and donor bacterial genes are appearing in the recipient

Hfr chromosome (Part of *F* factor, followed by bacterial genes)

Recombination between transferred donor chromosome and recipient chromosome like the integration event), a circular host chromosome and a circular extrachromosomal F factor are generated.

F′ Factors

Occasionally, excision of the F factor from the chromosome of an *Hfr* cell is not precise, and an F factor is produced with a small section of the host chromosome that was adjacent to the integrated F factor. Since the F factor integrates at one of many sites on the chromosome, many different host chromosome segments can be picked up in this way. Consider an *E. coli* strain in which the F factor has integrated next to the *lac*⁺ region, a set of genes required for the breakdown of lactose (Figure 18.6a). If the looping out is not precise, then the adjacent *lac*⁺ host chromosomal genes may be included in the loop (Figure 18.6b). Then, by a single crossover, the looped-out DNA is separated from the host chromosome (Figure 18.6c) to produce an F factor carrying the host's *lac*⁺ genes. F factors containing bacterial genes are called F′ (F prime) factors, and they are named for the genes they have picked up. An F′ with the *lac* genes is called F′ (*lac*).

Cells with F′ factors can conjugate with F⁻ cells. As in F⁺ × F⁻ conjugation, a copy of the F′ factor is transferred to the F⁻ cell, which then becomes F′. The recipient also receives a copy of the bacterial gene(s) on the F factor (*lac* in our example). Since the recipient has its own copy of that DNA, the resulting cell line is partially diploid (*merodiploid*), having two copies of one or a few genes and only one copy of all the others. This particular type of conjugation is called **F-duction,** or *sexduction,* and it provides a way to study particular genes in a diploid state in *E. coli.*

Using Conjugation to Map Bacterial Genes

In the late 1950s, François Jacob and Elie Wollman studied the transfer of chromosomal genes from *Hfr* strains to F⁻ cells that had allelic differences for a number of genes. Their experimental design involved making an *Hfr* × F⁻ mating and, at various times after conjugation began, using a kitchen blender to break apart the conjugating pairs and then analyzing the transconjugants for which donor genes they had received. This approach is called an *interrupted-mating experiment.*

The use of interrupted mating to map bacterial genes is illustrated by the following cross (Figure 18.7a):

Donor:
   HfrH thr⁺ *leu*⁺ *azi*ᴿ *ton*ᴿ *lac*⁺ *gal*⁺ *str*ˢ

Recipient:
   F⁻ *thr leu azi*ˢ *ton*ˢ *lac gal str*ᴿ

(The superscript S means "sensitive" and the superscript R means "resistant.")

The *HfrH* strain is prototrophic and is resistant to growth inhibition by the chemical sodium azide (*azi*ᴿ),

Figure 18.6

Production of an F′ factor. (a) Region of bacterial chromosome into which the F factor has integrated. (b) The F factor looping out incorrectly, so it includes a piece of bacterial chromosome, the *lac* genes. (c) Excision, in which a single crossover between the looped-out DNA segment and the rest of the bacterial chromosome results in an F′ factor, called F′(*lac*).

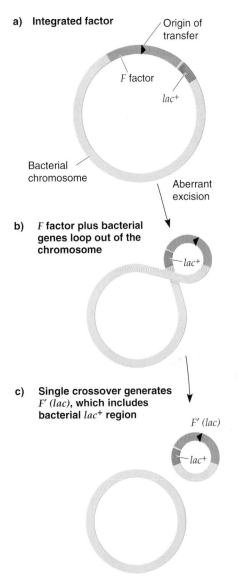

a) Integrated factor

Origin of transfer

F factor

lac⁺

Bacterial chromosome

Aberrant excision

b) F factor plus bacterial genes loop out of the chromosome

lac⁺

c) Single crossover generates F′ (lac), which includes bacterial lac⁺ region

F′ (*lac*)

lac⁺

resistant to infection by the bacteriophage T1 (*ton*ᴿ), and sensitive to the antibiotic streptomycin (*str*ˢ). The F⁻ strain is auxotrophic for threonine (*thr*) and leucine (*leu*), sensitive to growth inhibition by the chemical sodium azide (*azi*ˢ), sensitive to infection by the bacteriophage T1 (*ton*ˢ), unable to ferment lactose (*lac*) or galactose (*gal*), and resistant to growth inhibition by the antibiotic streptomycin (*str*ᴿ).

In such a conjugation experiment, the two cell types are mixed together in a liquid medium at 37°C. Samples

Figure 18.7

Interrupted-mating experiment involving the cross *HfrH thr⁺ leu⁺ aziᴿ tonᴿ lac⁺ gal⁺ strˢ × F⁻ thr leu aziˢ tonˢ lac gal strᴿ*. The progressive transfer of donor genes with time is illustrated. Recombinants are generated by an exchange of a donor fragment with the homologous recipient fragment resulting from a double crossover event. (a) At various times after mating commences, the conjugating pairs are broken apart and the transconjugant cells are plated on selective agar media to determine which genes have been transferred from the *Hfr* to the *F⁻*. (b) The graph shows the frequency (percentage) of *Hfr* genetic markers among *thr⁺, leu⁺* recombinants and their time of appearance in the recipient. (c) Genetic map of the genes. The marker positions represent the time of entry of the genes into the recipient during the experiment.

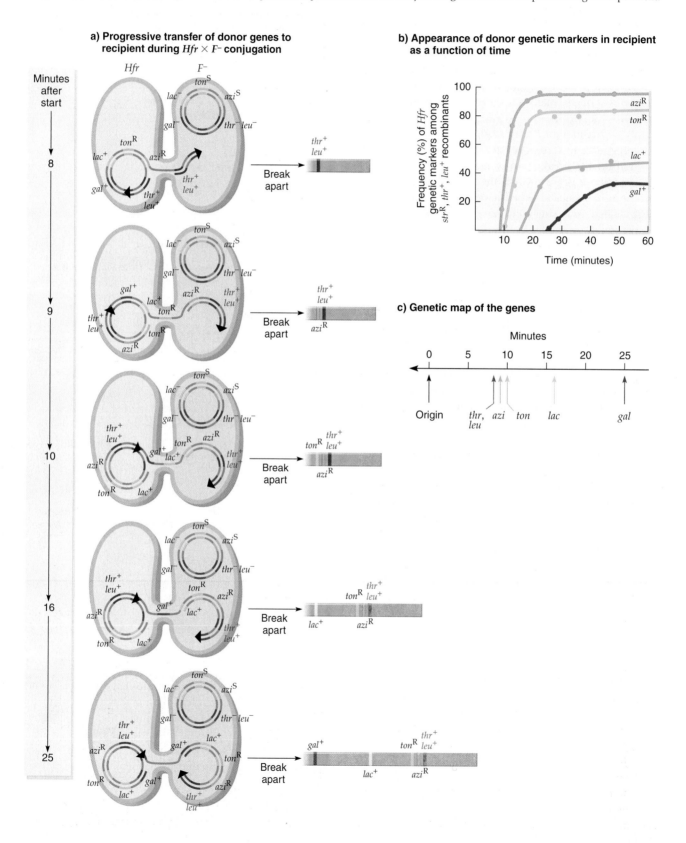

are removed from the mating mixture at various times and are then agitated to break the pairs apart. Through plating on selective agar media, recombinant recipients (the transconjugants) are then searched for and analyzed with respect to the time at which the first donor genes entered the recipient and produced recombinants.

For this particular cross, the medium contains streptomycin to kill the *HfrH* parent and lacks threonine and leucine, so that the *F⁻* parent cannot grow. The threonine (*thr⁺*) and leucine (*leu⁺*) genes are the first donor genes to be transferred to the *F⁻* to produce a merodiploid, so recombinants formed by the exchange of those genes with the *thr leu* genes of the *F⁻* recipient grow on the selective medium. Appropriate media can be used to test for the appearance of other donor genes (*azi^R*, *ton^R*, *lac⁺*, and *gal⁺*) among the selected *thr⁺ leu⁺ str^R* transconjugants. For example, a medium with sodium azide added can test for the presence of *azi^R* from the donor.

Figure 18.7b shows the results. The threonine (*thr⁺*) and leucine (*leu⁺*) genes are the first donor genes to be transferred to the *F⁻*, at 8 minutes. (The two genes are inseparable timewise in a conjugation experiment because they are physically very close to one another.) The next gene to be transferred is *azi^R*, and recombinants for this gene are seen at 9 minutes after the start of conjugation—that is, 1 minute after the *thr⁺* and *leu⁺* genes entered. Then *ton^R* recombinants are seen at 10 minutes, followed by *lac⁺* recombinants at about 16 minutes and *gal⁺* recombinants at about 25 minutes. Clearly, the maximum frequency of recombinants becomes smaller the later the gene enters the recipient, because, with time, there is an increasing chance that mating pairs will break apart.

In this experiment, each gene from the *Hfr* bacterium appears in recombinants at a different, but reproducible, time after mating begins. Thus, from the time intervals for the experiment described, the genetic map in Figure 18.7c may be constructed, with map units in minutes; the entire *E. coli* chromosome takes about 100 minutes to transfer.

Circularity of the *E. coli* Map

Only one F factor is integrated into each *Hfr* strain. Different *Hfr* strains have the F factor integrated into the chromosome at different locations and in different orientations. Therefore, *Hfr* strains differ with respect to where the transfer of donor genes begins and what the order of transfer of donor genesis. Figure 18.8a shows the order of chromosomal gene transfer for four different *Hfr* strains: *H, 1, 2,* and *3*. In each case, only one *Hfr* strain was used to cross with the recipient, and the order of gene transfer and the time between the appearance of each gene in the recipient were determined. The genetic distance in time units between a particular pair of genes

Figure 18.8

Interrupted-mating experiments with a variety of *Hfr* strains, showing that the *E. coli* linkage map is circular. (a) Orders of gene transfer for the *Hfr* strains *H, 1, 2,* and *3*; (b) Alignment of gene transfer for the *Hfr* strains. (c) Circular *E. coli* chromosome map derived from the *Hfr* gene transfer data. The map is a composite showing various locations of integrated F factors. A given *Hfr* strain has only one integrated F factor.

a) Orders of gene transfer

Hfr strains:

H	*origin–thr–pro–lac–pur–gal*
1	*origin–thr–thi–gly–his*
2	*origin–his–gly–thi–thr–pro–lac*
3	*origin–gly–his–gal–pur–lac–pro*

b) Alignment of gene transfer for the *Hfr* strains

H	*thr–pro–lac–pur–gal*
1	*his–gly–thi–thr*
2	*his–gly–thi–thr–pro–lac*
3	*pro–lac–pur–gal–his–gly*

c) Circular *E. coli* chromosome map derived from *Hfr* gene transfer data

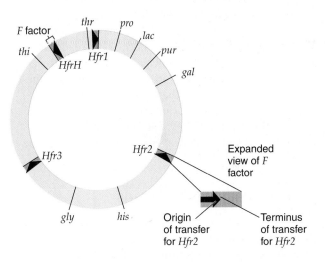

is constant, no matter which *Hfr* strain is used as donor; for example, the genetic distance between *thr* and *pro* is the same in *H, 1, 2,* and *3*. This sameness validates the use of time units as a measure of genetic distance in *E. coli*.

From the preceding sort of data, a genetic map of the chromosome is constructed by aligning the genes transferred by each *Hfr* as shown in Figure 18.8b. In view of the overlap of the genes, the simplest map that can be drawn from these data is a circular one, as shown in Figure 18.8c. The map is a composite of the results of the individual matings. The circularity of the map was itself a significant finding, because all previous genetic maps of eukaryotic chromosomes were linear.

A complete genetic map of the *E. coli* chromosome eventually was constructed from conjugation experiments;

the map is 100 minutes long. Like genetic maps of other organisms, this one provides information about the relative locations of *E. coli* genes on the circular chromosome. In 1997, the ultimate genetic map of *E. coli* was completed—that of the 4.6×10^6 (4.6-megabase) base-pair sequence of the bacterium's genome.

You are assisting Elie Wollman and François Jacob as they construct a genetic map of *E. coli*, using the newly discovered interrupted-mating procedure in the iActivity *Conjugation in E. coli* on your CD-ROM.

KEYNOTE

The circular *F* factor can integrate into the circular bacterial chromosome by a single crossover event. Strains in which this integration has happened can conjugate with F^- strains, and transfer of the bacterial chromosome occurs. The strains containing the integrated *F* factor are called *Hfr* (high-frequency recombination) strains. In $Hfr \times F^-$ matings, the chromosome is transferred in a one-way fashion from the *Hfr* cell to the F^- cell, beginning at a specific site called the origin (O). The farther a gene is from O, the later it is transferred to the F^-, and this temporal difference is the basis for mapping genes by their times of entry into the F^- cell. Conjugation and interrupted mating allow mapping of the chromosome.

Genetic Mapping in Bacteria by Transformation

Transformation is the unidirectional transfer of extracellular DNA into cells, resulting in a phenotypic change in the recipient. Bacterial transformation was first observed by Frederick Griffith in 1928, and in 1944 Oswald Avery and his colleagues showed that DNA was responsible for the genetic change that was observed. (See Chapter 2.) Bacterial transformation has been used to map the genes of certain bacterial species in which mapping by other methods (conjugation or transduction) was not possible. In mapping experiments using transformation, DNA from a donor bacterial strain is extracted, purified, and broken into small fragments. This DNA is then added to recipient bacteria with a different genotype. If the donor DNA is taken up by a recipient cell and recombines with the homologous parts of the recipient's chromosome, a recombinant chromosome is produced. Recipients whose phenotypes are changed by transformation are called **transformants.**

Bacterial species vary in their ability to take up DNA. To enhance the efficiency of transformation, cells typically

are treated chemically or are exposed to a strong electric field in a process called *electroporation,* making the cell membrane more permeable to DNA. Cells prepared to take up DNA by transformation are called *competent cells.*

There are two types of bacterial transformation: *natural transformation,* in which bacteria are naturally able to take up DNA and be genetically transformed by it; and *engineered transformation,* in which bacteria have been altered to enable them to take up and be genetically transformed by added DNA. *Bacillus subtilis* exemplifies bacteria amenable to natural transformation; *E. coli* exemplifies bacteria responsive to engineered transformation. *B. subtilis* is a cylindrical, spore-forming bacterium, about 3–8 μm long and 1–1.5 μm wide.

Only a small proportion of the cells involved in transformation actually take up DNA. Consider an example of the transformation of *B. subtilis* (Figure 18.9). (Other systems may differ in the details of the process.) The donor double-stranded DNA fragment is wild type (a^+) for a mutant allele a in the recipient cell (Figure 18.9a). During DNA uptake, one of the two DNA strands is degraded, so only one intact linear DNA strand is left inside the cell (Figure 18.9b). This single linear strand pairs with the homologous DNA of the recipient cell's circular chromosome to form a triple-stranded region (Figure 18.9c). Recombination then occurs by a double crossover event involving the single-stranded DNA strand of the donor and the double-stranded DNA of the recipient (Figure 18.9d). The result is a recombinant recipient chromosome: In the region between the two crossovers, one DNA strand has the donor a^+ DNA segment, and the other strand has the recipient a DNA segment. In other words, in that region, *the two DNA strands are part donor, part recipient, for the genetic information.* A region of DNA with different sequence information on the two strands is called **heteroduplex DNA.** (The other product of the double crossover event, a single-stranded piece of DNA carrying an a DNA segment, is degraded.)

After replication of the recipient chromosome, one progeny chromosome has donor genetic information on both DNA strands and is an a^+ transformant. The other progeny chromosome has recipient genetic information on both DNA strands and is an a nontransformant. Equal numbers of a^+ transformants and a nontransformants are produced. Given highly competent recipient cells, the transformation of most genes occurs at a frequency of about 1 cell in every 10^3 cells.

Transformation can be used to determine whether genes are linked (in this case, meaning physically close to one another on the single bacterial chromosome), to determine the order of genes on the genetic map, and to determine map distance between genes. The principles of determining whether two genes are linked are as follows: The efficient transformation of DNA involves fragments with a size sufficient to include only a few genes. If two

Figure 18.9

Transformation in *Bacillus subtilis*. (a) The linear donor double-stranded bacterial DNA fragment carries the a^+ allele, and the recipient bacterium carries the a allele. (b) One donor DNA strand enters the recipient. (c) The single linear DNA strand pairs with the homologous region of the recipient's chromosome, forming a triple-stranded structure. (d) A double crossover produces a recombinant a^+/a recipient chromosome and a linear a DNA fragment. The linear fragment is degraded, and by replication, one-half of the progeny are a^+ transformants and one-half are a nontransformants.

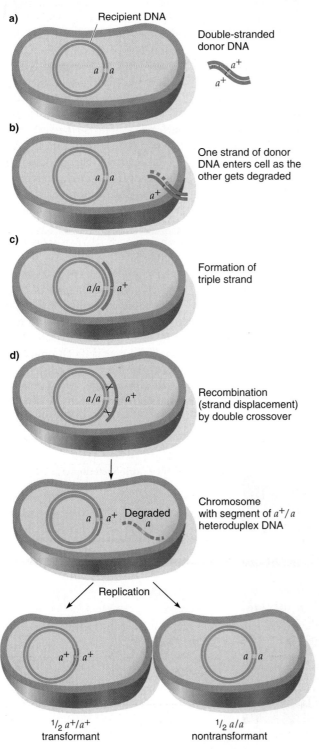

a)

Recipient DNA

Double-stranded donor DNA

a a

a^+

a^+

b)

a a

a^+

One strand of donor DNA enters cell as the other gets degraded

c)

a/a a^+

Formation of triple strand

d)

a/a a^+

Recombination (strand displacement) by double crossover

a a^+ Degraded

a

Chromosome with segment of a^+/a heteroduplex DNA

Replication

a^+ a^+

$\frac{1}{2}$ a^+/a^+ transformant

a a

$\frac{1}{2}$ a/a nontransformant

genes, x^+ and y^+, are far apart on the donor chromosome, they will always be found on different DNA fragments. Thus, given an $x^+ y^+$ donor and an $x y$ recipient, the probability of simultaneous transformation (cotransformation) of the recipient to $x^+ y^+$ (from the product rule) is the product of the probability of transformation with each gene alone. If transformation occurred at a frequency of 1 in 10^3 cells per gene, $x^+ y^+$ transformants would be expected to appear at a frequency of 1 in 10^6 recipient cells ($10^{-3} \times 10^{-3}$). So if two genes are close enough that they often are carried on the same DNA fragment, the cotransformation frequency would be close to the frequency of transformation of a single gene. As determined experimentally, if the frequency of cotransformation of two genes is substantially higher than the products of the two individual transformation frequencies, the two genes must be close together.

Gene order can be determined from cotransformation data as shown in Figure 18.10. If genes p and q are often transmitted to the recipient together, then these two genes must be closely linked. Similarly, if genes q and o are often transmitted together, those two genes must be close to one another. To determine gene order, we now need information about genes p and o. Theoretically, there are two possible orders: p-o-q and p-q-o. If the order is p-o-q, then p and o should be cotransformed, because they are more closely linked than p and q, whereas if the order is p-q-o, then p and o should be cotransformed rarely or not at all, because they are far apart. The data show no cotransformants for p and o, indicating that the gene order must be p-q-o.

Figure 18.10

Demonstration of the determination of gene order by cotransformation.

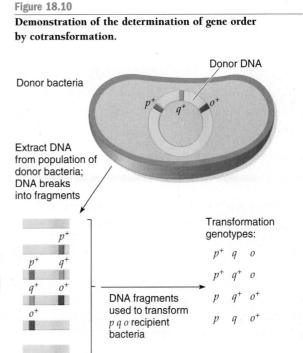

Donor bacteria

Donor DNA

p^+ q^+ o^+

Extract DNA from population of donor bacteria; DNA breaks into fragments

p^+

p^+ q^+

q^+ o^+

o^+

DNA fragments used to transform $p\,q\,o$ recipient bacteria

Transformation genotypes:

p^+ q o

p^+ q^+ o

p q^+ o^+

p q o^+

K E Y N O T E

Transformation is the transfer of small extracellular pieces of DNA between organisms. In transformation, DNA is extracted from a donor strain and added to recipient cells. A DNA fragment taken up by the recipient cell may associate with the homologous region of the recipient's chromosome. Part of the transforming DNA molecule can exchange with part of the recipient's chromosomal DNA. Frequent cotransformation of donor genes indicates close physical linkage of those genes. Cotransformants can be analyzed to determine gene order and map distance between genes. Transformation has been used to construct genetic maps of bacterial species for which conjugation or transduction analyses are not possible.

Genetic Mapping in Bacteria by Transduction

Transduction (literally, "leading across") is a process by which bacteriophages (bacterial viruses; phages, for short) transfer genes from one bacterium (the donor) to another (the recipient); such phages are called **phage vectors.** Since the amount of DNA a phage can carry is limited, the amount of genetic material that can be transferred usually is less than 1 percent of that in the bacterial chromosome. Once the donor genetic material has been introduced into the recipient, it may undergo genetic recombination with a homologous region of the recipient chromosome. The recombinant recipients are called **transductants.**

Bacteriophages

Most bacterial strains can be infected by specific phages. A phage has a relatively simple structure consisting of a single chromosome of DNA or RNA surrounded by a coat of protein molecules. Variation in the number and organization of the proteins gives the phages their characteristic appearances. Phages T2, T4, and λ (lambda) were introduced in Chapter 2 (pp. 16–17 and 28). Phages T2 (Figure 2.4, p. 17) and T4 are *virulent phages*, meaning that they follow the *lytic cycle* when they infect *E. coli* (see Figure 2.5, p. 17). That is, the phage injects its chromosome into the cell, and phage genes take over the function of the cell, leading to the production of progeny phages that are released from the bacterium as the cell is broken open (lysed). The suspension of released progeny phages is called a **phage lysate** and phage genes take over the function of the cell.

We can follow the phage lytic cycle visually. A mixture of phages and bacteria is plated on a solid medium. The concentration of bacteria is chosen so that an entire "lawn" of bacteria grows. Phages are present in much lower concentrations. Each phage infects a bacterium on the plate surface. Progeny phages released from the first infected bacterium infect neighboring bacteria, and the lytic cycle is repeated. The result is a cleared patch in the lawn of bacteria. The clearing is called a **plaque,** and each plaque derives from one of the original bacteriophages that was plated (Figure 18.11).

The λ life cycle, shown in Figure 18.12, is more complex than that of a T2 phage. When phage λ DNA is injected into *E. coli*, the phage follows one of two alternative paths. One is a lytic cycle, exactly like that of the T phages. The other is the **lysogenic pathway** (or *lysogenic cycle*). In the lysogenic pathway, the λ chromosome does not replicate; instead, it inserts (integrates) itself physically into a specific region of the host cell's chromosome, much like *F* factor integration. In this integrated state, the phage chromosome is called a **prophage.** Every time the host cell chromosome replicates, the integrated λ chromosome replicates as part of it. A bacterium that contains

Figure 18.11

Plaques of the *E. coli* bacteriophage T2.

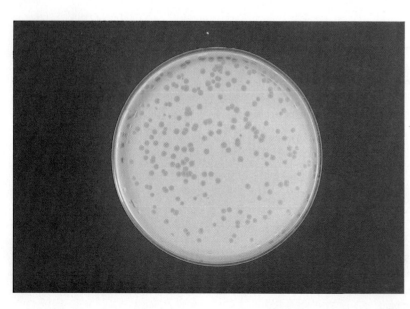

Figure 18.12

Life cycle of the temperate phage λ. When a temperate phage infects a cell, the phage may go through the lytic or lysogenic cycle.

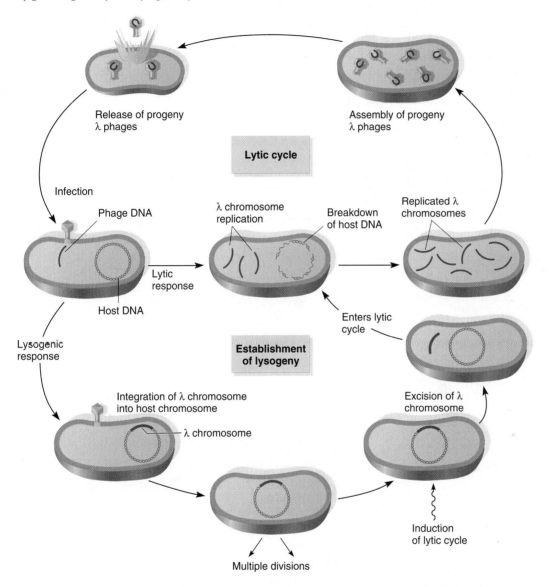

a phage in the prophage state is said to be **lysogenic** for that phage; the phenomenon of the insertion of a phage chromosome into a bacterial chromosome is called **lysogeny.** Phages that have a choice between lytic and lysogenic pathways are called **temperate phages.**

The prophage state is maintained by the action of a specific phage gene product (a repressor protein) that prevents the expression of λ genes essential to the lytic cycle. When the repressor that maintains the prophage state is destroyed—for example, by environmental factors such as ultraviolet light irradiation—the lytic cycle is induced. Upon induction, the integrated λ chromosome is excised from the bacterial chromosome and the lytic cycle begins, resulting in the production and release of progeny λ phages from the cell.

Transduction Mapping of Bacterial Chromosomes

Transduction may be used to map bacterial genes. Two types of transduction occur: In **generalized transduction,** any gene can be transferred between bacteria; in **specialized transduction,** only specific genes are transferred.

Generalized Transduction. Joshua Lederberg and Norton Zinder discovered transduction in 1952. These researchers tested whether conjugation occurred in the bacterial species *Salmonella typhimurium.* Their experiment was similar to the one showing that conjugation existed in *E. coli.* They mixed together two multiple auxotrophic strains—*phe⁺ trp⁺ tyr⁺ met his* (required methionine and histidine) and *phe trp tyr met⁺ his⁺* (required phenylalanine, tryptophan, and tyrosine)—and found

Figure 18.13

Generalized transduction between strains of *E. coli*. (1) Wild-type donor cell of *E. coli* infected with the temperate bacteriophage P1. (2) The host cell DNA is broken up during the lytic cycle. (3) During the assembly of progeny phages, some pieces of the bacterial chromosome are incorporated into some of the progeny phages to produce transducing phages. (4) After cell lysis, a low frequency of transducing phages is found in the phage lysate. (5) The transducing phage infects an auxotrophic recipient bacterium. (6) A double crossover results in the exchange of the donor a^+ gene with the recipient mutant a gene. (7) The result is a stable a^+ transductant, with all descendants of that cell having the same genotype.

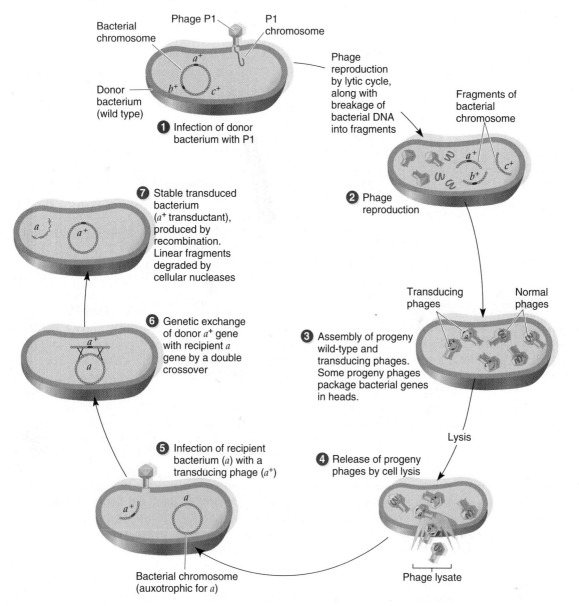

prototrophic recombinants—*phe⁺ trp⁺ tyr⁺ met⁺ his⁺*—at a low frequency. However, unlike what transpired in the conjugation experiment, when they used the U-tube apparatus (see Figure 18.3), they still found prototrophs. This result indicated that recombinants were being produced by a mechanism that did not require cell-to-cell contact. The interpretation was that the agent responsible for the formation of recombinants was a *filterable agent*, because it could pass through a filter with pores small enough to block bacteria. In this particular case, the filterable agent was identified as the temperate phage P22.

As an example, Figure 18.13 shows the mechanism for the generalized transduction of *E. coli* by temperate phage P1. Normally, the P1 phage enters the lysogenic state when it infects *E. coli* (Figure 18.13, part 1). If the lysogenic state is not maintained, the phage goes through the lytic cycle and produces progeny phages (Figure 18.13, part 2). During the lytic cycle, the bacterial DNA is degraded and, rarely, a piece of bacterial DNA is packaged into a phage head instead of phage DNA (Figure 18.13, part 3). These phages are called **transducing phages,** because they are the vehicles by which genetic material is carried between bacteria. In the example shown in the figure, the transducing phages are those carrying the donor bacterial genes a^+, b^+, or c^+. The population of phages in the phage lysate (Figure 18.13, part 4), consisting mostly of normal phages, but with about 1 in 10^5 transducing phages present, can now be used to infect a new population of bacteria (Figure 18.13, part 5). The recipient bacteria are *a* in genotype. If a transducing phage carrying the a^+ gene infects the recipient, genetic exchange of the donor a^+ gene with the recipient *a* gene can occur by double crossing-over (Figure 18.13, part 6). The result is a stable transduced bacterium called a *transductant*—in this case, an a^+ transductant.

Typically, a transduction experiment is designed so that the donor cell type and the recipient cell type have different genetic markers; then the transduction events can be followed. For instance, if the donor cell is thr^+ and the recipient cell is *thr*, prototrophic transductants can be detected because the cell no longer requires threonine to grow. In this way, researchers can pick out the extremely low number of transductants from among all the cells present by selecting for those cells which are able to do something the nontransduced cells cannot, namely, grow on a minimal medium. In such case, thr^+ is called a *selected marker*. Other markers in the experiment are termed *unselected markers*.

The process just described is called *generalized transduction*; the piece of bacterial DNA that the phage erroneously picks up is a *random* piece of the fragmented bacterial chromosome. Thus, any genes can be transduced; only an appropriate phage and bacterial strains carrying different genetic markers are needed. Gene order and map distances between cotransduced genes can be determined by generalized transduction, and it is also by this procedure that fine-structure (i.e., detailed) linkage maps of bacterial chromosomes have been constructed. The logic is identical as that for mapping genes by transformation. For example, consider the mapping of some *E. coli* genes by using transduction with the temperate phage P1. The donor *E. coli* strain is leu^+ thr^+ azi^R (is able to grow on a minimal medium and is resistant to the metabolic poison sodium azide). The recipient cell is leu thr azi^S (requires leucine and threonine as supplements in the medium and is sensitive to sodium azide). The P1 phages are grown on the bacterial donor cells, and the phage lysate is used to infect the recipient bacterial cells. Transductants are selected for any one of the donor

Table 18.1 **Transduction Data for Deducing Gene Order**

Selected Marker	Unselected Markers
leu^+	50% = azi^R
	2% = thr^+
thr^+	3% = leu^+
	0% = azi^R

markers and are then analyzed for the presence of the other unselected markers (in this case, two). Typical data from such an experiment are shown in Table 18.1.

Consider the leu^+ selected transductants. We look to see if other donor markers are also present—that is, whether they have been *cotransduced* with the selected marker. Such *cotransductants* can occur in one of two possible ways: (1) If two genes are close enough so that they can be packaged physically into a phage head and be injected into a cell by a single phage; and (2) If two genes are not closely linked and are introduced into the same bacterium by simultaneous infection with two different phages. The transduction of two genes into a single bacterium by two phages is rare. Therefore, if two genes are close enough that they often are packaged into the phage head on the same DNA fragment, the **cotransduction** frequency would be close to the frequency of transduction of a single gene; cotransduction of two or more genes is a good indication that the genes are closely linked. Of the leu^+ transductants, 50 percent were azi^R and 2 percent were thr^+. This means that the *leu* and *azi* genes often are cotransduced on the same DNA molecule; much less frequently, the *thr* gene is on the same transducing DNA with the *leu* gene. For the thr^+ transductants, 3 percent are leu^+ and 0 percent are azi^R. This distribution confirms that the *thr* and *leu* genes can be on the same transducing DNA and also indicates that the *azi* gene is distant enough never to be included on the same DNA. Taken together, these two sets of results tell us that the *leu* gene is closer to the *thr* gene than is the *azi* gene and that the *leu* and *azi* genes are closer together than are the *leu* and *thr* genes. The order of genes and rough map must then be as follows:

The transductants are produced by crossing-over between the piece of donor bacterial chromosome brought in by the infecting phage and the homologous region on the recipient bacterial chromosome. The infected donor DNA finds the region of the recipient chromosome to which it is homologous, and the exchange of parts is accomplished by double (or some other even-numbered) crossovers. (See Figure 18.13.)

Map distance can be obtained from transduction experiments involving two or more genes. As before, transductants for one of two or more donor markers are selected and are then analyzed for the presence or absence of other donor markers. For example, transduction from an $a^+ b^+$ donor to an $a\ b$ recipient produces various transductants for a^+ and b^+, namely, $a^+ b$, $a\ b^+$, and $a^+ b^+$. If we select for one or more donor markers, we can determine linkage information for the two genes. If we select for a^+ transductants, map distance between genes a and b is given by

$$\frac{\text{number of single-gene transductants}}{\text{number of total transductants}} \times 100\%$$

$$= \frac{(a^+ b)}{(a^+ b) + (a^+ b^+)} \times 100\%$$

If we select for b^+ transductants, map distance between a and b is given by

$$\frac{(a\ b^+)}{(a\ b^+) + (a^+ b^+)} \times 100\%$$

This method of gene mapping produces map distances only if the genes involved are close enough on the chromosome so that they can be cotransduced.

Specialized Transduction. Some temperate bacteriophages can transduce only certain sections of the bacterial chromosome, in contrast to generalized transducing phages, which can carry any part of the bacterial chromosome. An example of such a **specialized transducing phage** is λ, which infects *E. coli*.

The life cycle of λ was described earlier. (See Figure 18.12.) In the bacterial cell, the λ genome integrates into the bacterial chromosome at a specific site between the *gal* region and the *bio* region, producing a *lysogen* (Figure 18.14a). That site on the *E. coli* chromosome is called *att* λ (attachment site for lambda) and is homologous with a site called *att* in the λ DNA. By a single crossover, the λ chromosome integrates. In the integrated state, the phage, now called a *prophage*, is maintained by the action of a phage-encoded repressor protein.

The particular *E. coli* strain that λ lysogenizes is *E. coli K12*, and when it contains the λ prophage, it is called *E. coli K12*(λ). Let us focus just on the *gal* gene and assume that the particular *K12* strain that lysogenized is *gal*+; that is, it can ferment galactose as a carbon source. This phenotype is readily detectable by plating the cells on a solid medium containing galactose as a carbon source, together with a dye that changes color in response to the products of galactose fermentation. On this medium, the *gal*+ colonies are pink, the *gal* colonies white. If we induce the prophage (see p. 492)—that is, reverse the inhibition of phage functions—the lytic cycle is initiated.

When the lytic cycle is initiated, the phage chromosome loops out, generating a separate circular λ chromosome by a single crossover at the *att* λ/*att* sites (Figure 18.14b). In most cases, the excision of the phage chromosome is precise and the complete λ chromosome is produced (Figure 18.14b1). In rare cases, crossing-over occurs at sites other than the homologous recognition sites, giving rise to an abnormal circular DNA product (Figure 18.14b2). In the case diagrammed, a piece of λ chromosome has been left in the bacterial chromosome, and a piece of bacterial chromosome, including the *gal*+ gene, has been added to the rest of the λ chromosome. Because a bacterial gene (or genes) is included in a progeny phage, we have a transducing phage—here, λ*d gal*+. The *d* stands for "defective," because not all phage genes are present, and the *gal* indicates that the bacterial host cell *gal* gene has been acquired. This outcome of transduction is similar to *F'* production by defective excision of the F factor. The λ*d gal*+ can replicate and lyse the host cell in which it is produced, however, because all λ genes are still present: Some are on the phage chromosome, and the others are in the bacterial chromosome.

The abnormal looping-out phenomenon is a rare event, so the phage lysate produced from the initial infection diagrammed contains mostly normal phages and a few *gal*+ transducing phages ($1/10^5$). Because of the small proportion of transducing phages, the lysate is called a *low-frequency transducing (LFT)* lysate. Infection of *gal* bacterial cells with the LFT lysate produces two types of transductants (Figure 18.14c). In one type, the wild-type λ integrates at its normal *att* λ site, and then the λ*d gal*+ phage integrates by crossing over within the common λ sequences to produce a double lysogen (Figure 18.14c1). In this case, both types of phages are integrated into the bacterial chromosome and the bacterium is heterozygous *gal*+/*gal* and therefore can ferment galactose.

This type of transductant is unstable, because the lytic cycle can be initiated by induction. The wild-type λ has a complete set of genes for virus replication, so it controls the outlooping and replication of itself and the λ*d gal*+. In this capacity, the wild-type λ phage acts as a *helper phage*. Since as many as one-half of the progeny phages could be λ*d gal*+, this new lysate is called a *high-frequency transducing (HFT)* lysate.

The second type of transductant produced by the initial lysate is stable: It is produced when only a λ*d gal*+ phage infects a cell (Figure 18.14c2). The *gal*+ gene carried by the phage may be exchanged for the bacterial *gal* gene by a double crossover. Such a transductant is stable because the bacterial chromosome contains only one type of *gal* gene, and no phage genes are integrated.

Because of the mechanisms involved, specialized transduction can transduce only small segments of the bacterial chromosome that are on either side of the prophage. Specialized transduction is used to move

Figure 18.14

Specialized transduction by bacteriophage λ (for detail, see text). (a) Production of a lysogenic bacterial strain by crossing-over in the region of homology between the circular bacterial chromosome (*att*λ) and the circularized phage chromosome (*att*). (b) Production of initial low-frequency transducing (LFT) lysate: Induction of the lysogenic bacterium causes outlooping. Normal outlooping (1) produces normal λ phages, and rare abnormal outlooping (2) produces transducing λd *gal*⁺ phages. (c) Transduction of *gal* bacteria by the initial lysate produces either (1) unstable transductants by the integration of both λ and λd *gal*⁺ (resulting in a double lysogen) or (2) stable transductants (single lysogens) by crossing-over around the *gal* region. Induction of the unstable double lysogen produces about equal numbers of λd *gal*⁺ and wild-type λ phages; the result is a high-frequency transducing (HFT) lysate.

a) Production of lysogen

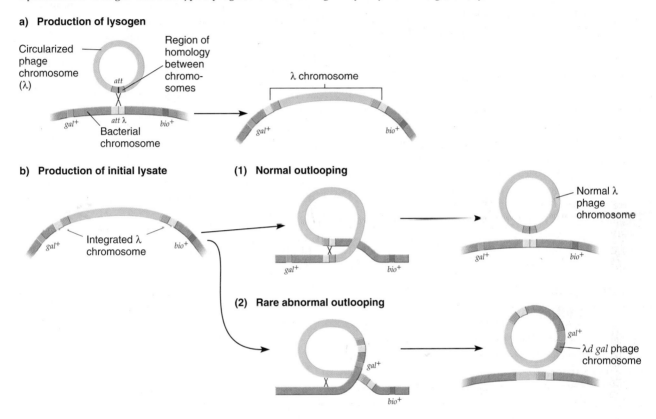

b) Production of initial lysate

(1) Normal outlooping

(2) Rare abnormal outlooping

c) Transduction of *gal* bacteria by initial lysate, consisting of λ and λd *gal* phage

1) Lysogenic transductant

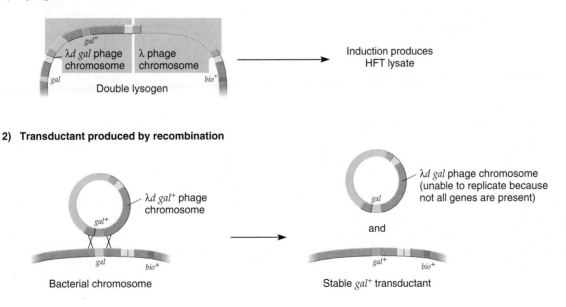

2) Transductant produced by recombination

specific genes between bacteria—for example, for constructing strains with particular genotypes. (A discussion of the process is beyond the scope of the text.)

KEYNOTE

> Transduction is the process by which bacteriophages mediate the transfer of genetic information from one bacterium (the donor) to another (the recipient). The capacity of the phage particle is limited, so the amount of DNA transferred usually is less than 1 percent of that in the bacterial chromosome. In generalized transduction, any bacterial gene can be incorporated accidentally into the transducing phage during the phage's life cycle and subsequently transferred to a recipient bacterium. Specialized transduction is mediated by temperate phages (such as λ) in which prophages associate with only one site of the bacterial chromosome. In this case, the transducing phage is generated by abnormal excision of the prophage from the host chromosome, so the prophage includes both bacterial and phage genes. Transduction allows the fine-structure mapping of small chromosome segments.

Mapping Bacteriophage Genes

The same principles used to map eukaryotic genes are used to map phage genes. Crosses are made between phage strains that differ in genetic markers, and the proportion of recombinants among the progeny is determined. The basic procedure for mapping phage genes in two-, three-, or four-gene crosses involves the mixed infection of bacteria with phages of different genotypes and analyzing the progeny—that is, the plaques.

To perform a genetic analysis of bacteriophages, we must have phage phenotypes to study. Several mutations affect the phage life cycle, giving rise to differences in the appearance of plaques on a bacterial lawn. For example, there are strains of T2 that differ in either plaque morphology (the size and shape of the edge of the plaque) or host range (which bacterial strain the phage can lyse). Consider two phage strains. One has the genotype $h^+ r$, meaning that it is wild type for the host range gene (h^+—able to lyse the *B* strain, but not the *B/2* strain, of *E. coli*; that is, strain *B* is the *permissive host* and strain *B/2* is the *nonpermissive host* for h^+ phages) and mutant for the plaque morphology gene (*r*—producing large plaques with distinct borders). The other phage strain has the genotype $h r^+$, meaning that it is mutant for the host range gene (able to lyse both the *B* and *B/2* strains of *E. coli*) and wild type for the plaque morphology gene (r^+—producing small plaques with fuzzy borders). When plated on a lawn containing both the *B* and *B/2* strains, any phage carrying the mutant host range allele *h*

(infects both *B* and *B/2*) produces clear plaques, whereas phages carrying the wild-type h^+ allele produce cloudy plaques. The latter characteristic arises because phages bearing the h^+ allele can infect only the *B* bacteria, leaving a background cloudiness of uninfected *B/2* bacteria.

To map these two genes, we make a genetic cross by infecting *E. coli* strain *B* with the two (parental) phages $h^+ r$ and $h r^+$ (Figure 18.15a). Once the two genomes are within the bacterial cell, each replicates (Figure 18.15b). If an $h^+ r$ and an $h r^+$ chromosome come together, a

Figure 18.15

The principles of performing a genetic cross with bacteriophages. (a) Bacteria of *E. coli* strain *B* are coinfected with the two parental bacteriophages, $h^+ r$ and $h r^+$. (b) Replication of both parental chromosomes. (c) Pairing of some chromosomes of each parental type occurs, and crossing-over takes place between the two gene loci to produce $h^+ r^+$ and $h r$ recombinants. (d) Progeny phages are assembled and are released into the medium when the bacteria lyse; both parental and recombinant phages are found among the progeny.

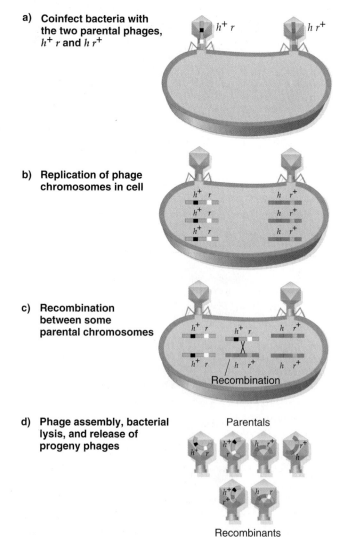

a) **Coinfect bacteria with the two parental phages, $h^+ r$ and $h r^+$**

b) **Replication of phage chromosomes in cell**

c) **Recombination between some parental chromosomes**

Recombination

d) **Phage assembly, bacterial lysis, and release of progeny phages**

Parentals

Recombinants

Figure 18.16

Plaques produced by progeny of a cross of T2 strains $h\ r^+ \times$ $h^+\ r$. Four plaque phenotypes, representing both parental types and the two recombinants, can be discerned. The parental $h\ r^+$ phage produces a small clear plaque with a fuzzy border; the other parental $h^+\ r$ phage produces a large cloudy plaque with a distinct border. The recombinant $h^+\ r^+$ phage produces a small cloudy plaque with a fuzzy border, and the recombinant $h\ r$ phage produces a large clear plaque with a distinct border.

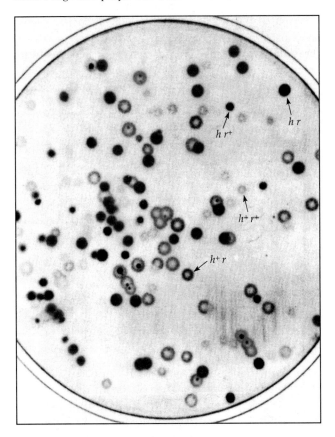

crossover can occur between the two gene loci to produce $h^+\ r^+$ and $h\ r$ recombinant chromosomes (Figure 18.15c), which are assembled into progeny phages. When the bacterium lyses, the recombinant progeny are released into the medium, along with nonrecombinant (parental) phages (Figure 18.15d).

After the life cycle is completed, the progeny phages are plated onto a bacterial lawn containing a mixture of *E. coli* strains B and B/2. Four plaque phenotypes—two parental types and two recombinant types—are found from the experiment. The parental type $h\ r^+$ gives a small clear plaque with a fuzzy border; the other parental $h^+\ r$ gives a large cloudy plaque with a distinct border (Figure 18.16). The reciprocal recombinant types give recombined phenotypes: The $h^+\ r^+$ plaques are cloudy and small with a fuzzy border, and the $h\ r$ plaques are clear and large with a distinct border. (See Figure 18.16.)

Once the progeny plaques are counted, we can find the recombination frequency between h and r from the formula

$$\frac{(h^+\ r^+) + (h\ r)\ \text{plaques}}{\text{total plaques}} \times 100$$

As with eukaryotes, this recombination frequency reflects the relative genetic distance between the phage genes. When the genes are close enough together so that multiple crossovers are not likely to occur, the recombination frequency equals the crossover frequency, and in that case the recombination frequency can be converted directly to map units.

KEYNOTE

The same principles used to map eukaryotic genes are used to map phage genes. That is, genetic material is exchanged between strains differing in genetic markers, and recombinants are detected and counted.

Fine-Structure Analysis of a Bacteriophage Gene

The recombinational mapping of the distance between genes, called *intergenic mapping*, can be used to construct chromosome maps for both eukaryotic and prokaryotic organisms. Historically, the early picture of a gene was that it was like a bead on a string with mutation changing the bead from wild type to mutant or vice versa and with recombination occurring between the beads. We now know, of course, that the gene is subdivisible by mutation and recombination and that the same general principles of recombinational mapping can be applied to mapping the distance between mutational sites within the same gene, a process called *intragenic mapping*.

The first evidence that the gene was subdivisible by mutation and recombination came from the work of C. P. Oliver in 1940. Oliver studied two mutations that were considered to be alleles of the X-linked *lozenge* (*lz*) locus of *Drosophila*; that is, females heterozygous for the two mutations showed the mutant lozenge-shaped eye phenotype. When female flies heterozygous for these two alleles were crossed with male flies hemizygous for either allele, progeny flies with wild-type eyes were seen with a frequency of about 0.2 percent. Oliver showed that these wild-type offspring had resulted from recombination between the alleles. In other words, he had shown that the gene was divisible by recombination, rather than being an indivisible "bead on a string." Using genetic symbols, we can represent the last cross as

$$\frac{lz^A\ +}{+\ lz^B} \times \xrightarrow{lz^A\ +}$$

where lz^A and lz^B are the two lozenge alleles. Recombination in the female between the two alleles produces $+\ +$ gametes and hence wild-type progeny.

Oliver's discovery spawned investigations of the detailed organization of alleles within a gene. As we now know, such intragenic mapping is possible because each gene consists of many nucleotide pairs of DNA, linearly arranged along the chromosome. The impetus to analyze the fine details of gene structure came largely from the elegantly detailed work by Seymour Benzer in the 1950s and 1960s with bacteriophage T4. His genetic experiments revealed much about the relationship between mapping and gene structure. His initial experiments involved **fine-structure mapping:** the detailed genetic mapping of sites within a gene.

Benzer used strains of phage T4 carrying mutations of the *rII* region. *rII* mutants have both a distinct *plaque morphology* phenotype and distinct *host range properties*. Specifically, when cells of *E. coli* growing on a solid medium are infected with wild-type (r^+) T4, small turbid plaques with fuzzy edges are produced, whereas plaques produced by *rII* mutants are large and clear (Figure 18.17). Regarding host range properties, wild-type T4 can grow in and lyse cells of either *E. coli* strain B or K12(λ), whereas *rII* mutants can grow in B, but not K12(λ). That is, strain B is the permissive host for *rII* mutants, and strain K12(λ) is the nonpermissive host.

Recombination Analysis of *rII* Mutants

Benzer realized that the growth defect of *rII* mutants on *E. coli* K12(λ) could serve as a powerful selective tool for detecting the presence of a very small proportion of r^+ phages within a large population of *rII* mutants. Initially, he set out to construct a fine-structure genetic map of the *rII* region. Using *E. coli* B as the permissive host, he crossed 60 independently isolated *rII* mutants in all possible combinations and then collected the progeny phages once the cells had lysed. For each cross

Figure 18.17

The r^+ and mutant *rII* plaques on a lawn of *E. coli* B. The r^+ plaque is turbid, with a fuzzy edge; the *rII* plaque is larger and clear and has a distinct boundary.

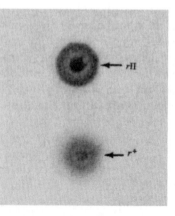

*rII*x × *rII*y, where x and y are different mutations, there can be four types of progeny: two parental classes, *rII*x and *rII*y; and two recombinant classes, the double mutant *rII*x,y and the r^+ wild-type. Roughly equal numbers of the two parental classes will be produced and roughly equal numbers of the two recombinant classes will be produced. The relative frequencies of the parentals and recombinants will depend on how far apart the two alleles are. For his analysis, Benzer plated a sample of the phage progeny on *E. coli* B, the permissive host. Then, from the number of plaques produced, the total number of progeny phage per milliliter were determined. He plated another sample on *E. coli* K12(λ), the nonpermissive host, to find the frequency of r^+ recombinants. In this way, Benzer calculated the percentage of very rare r^+ recombinants produced by crossing-over between closely linked alleles. The recombination frequency for the two alleles is given by the formula

$$\frac{2 \times \text{number of } r^+ \text{ recombinants}}{\text{total number of progeny}} \times 100\%$$

We multiply the number of r recombinants by 2 to account for the other class of recombinants—the double mutants—that we cannot detect phenotypically from single mutants.

An important control was set up for each cross. Each *rII* parent alone was used to infect the permissive *E. coli* B host, and the progenies were tested on plates of B and of K12(λ). Just as a mutation can generate an *rII* mutant from the r^+, a mutation can occur whereby an *rII* mutant changes back (reverts) to the r^+. Thus, it is extremely important to calculate the reversion frequencies for the two *rII* mutations in a cross and subtract the combined value from the computed recombination frequency. Fortunately, the reversion frequency for an *rII* mutation is at least an order of magnitude lower than the smallest recombination frequency that was found.

Benzer constructed a linear genetic map from the recombination data obtained from all possible pairwise crosses of the 60 *rII* mutants (Figure 18.18). Some pairs produced no r^+ recombinants when they were crossed, meaning that those pairs carried mutations at exactly the same site. Mutations that change the same nucleotide pair within a gene are called *homoallelic*. However, most pairs of *rII* mutants did produce r^+ recombinants when crossed, indicating that they carried different altered nucleotide pairs in the DNA. Mutations that change different nucleotide pairs within a gene are called *heteroallelic*. The map showed that the lowest frequency with which r^+ recombinants were formed in any pairwise crosses of *rII* mutants carrying heteroallelic mutations was 0.01 percent.

Figure 18.18

Preliminary fine-structure genetic map of the *rII* region of phage T4; map derived by Benzer from crosses of an initial set of 60 *rII* mutants. Lower levels in the figure show finer detail of the map. In the lowest level, the numbered vertical lines indicate individual *rII* point mutants; the blue rectangles indicate the individual *rII* deletion mutants 47, 312, 295, 164, 196, 187, and 102; and the decimals indicate the percentage of *r*⁺ recombinants found in crosses between the two *rII* mutants connected by an arrow.

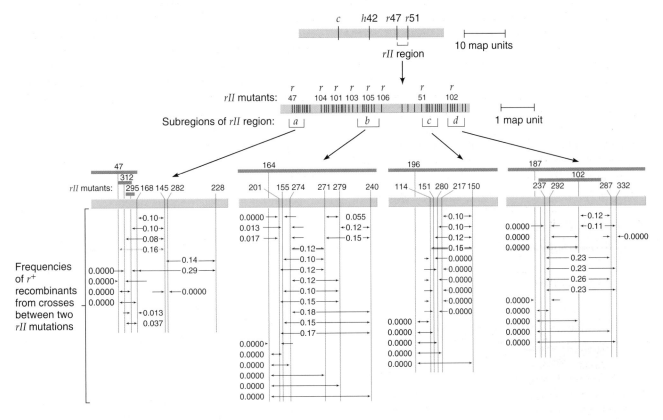

The minimum map distance of 0.01 percent can be used to make a rough calculation of the molecular distance—the distance in nucleotide pairs—between mutant markers. The genetic map of phage T4 is about 1,500 map units. If two *rII* mutants produce 0.01 percent *r*⁺ recombinants, then the mutations are separated by 0.02 map unit, or by about $0.02/1,500 = 1.3 \times 10^{-5}$ of the total T4 genome. Since the total T4 genome contains about 2×10^5 nucleotide pairs (base pairs), the smallest recombination distance that was observed was $(1.3 \times 10^{-5}) \times (2 \times 10^5)$, or about 3 base pairs. This means that Benzer's data showed that genetic recombination can occur within distances on the order of 3 base pairs. Later experiments by others demonstrated that recombination can occur between mutations that affect adjacent base pairs in the DNA. That is, genetic experiments have shown that the *base pair* is both the *unit of mutation* and the *unit of recombination*. These definitions replaced the classic definitions that the *gene* was the unit of mutation and the unit of recombination—that is, that the gene was indivisible by the processes of mutation and recombination.

KEYNOTE

The same general principles of recombinational mapping can be applied to mapping the distance between mutational sites in different genes (intergenic mapping) and to mapping mutational sites within the same gene (intragenic mapping).

Deletion Mapping

After his initial series of crossing experiments, Benzer continued to map more than 3,000 *rII* mutants to complete his fine-structure map. Mapping that number of mutants would have required approximately 5 million crosses, an overwhelming task even in phages, with which up to 50 crosses can be done per day. Therefore, Benzer developed some genetic tricks to simplify his mapping studies. These tricks involved using *deletion mapping* to localize unknown mutations, as we will now see.

Most of the *rII* mutants Benzer isolated were **point mutants;** their phenotype resulted from an alteration of a single nucleotide pair. A point mutant can revert to the wild-type state spontaneously or after treatment with an appropriate mutagen. However, some of Benzer's *rII* mutants did not revert, nor did they produce *r*⁺ recombinants in crosses with a number of *rII* point mutants that were known to be located at different places on the *rII* map. These mutants were *deletion mutants*—mutants that had lost a segment of DNA. Benzer found a wide range in the extent and location of deleted genetic material among the *rII* deletion mutants he studied. Some deletion mutants are shown in Figure 18.19.

In actual practice, an unknown *rII* point mutant was first crossed with each of the seven standard deletion mutants that defined seven main segments of the *rII* region (segments *A1–A6* and *B* in Figure 18.19). For example, if an *rII* point mutant produced *r*⁺ recombinants when crossed with deletion mutants *rA105* (deficient in *A6* and *B*) and *r638* (deficient in *B*), but did not

produce *r*⁺ recombinants when crossed with deletion mutants *r1272* (deficient in all segments), *r1241* (deficient in *A2–A6* and *B*), *rJ3* (deficient in *A3–A6* and *B*), *rPT1* (deficient in *A4–A6* and *B*), and *rPB242* (deficient in *A5–A6* and *B*), then the point mutation had to be in the segment of DNA that the five nonrecombinant deletion mutants lacked. *r*⁺ recombinants cannot be produced in crosses with deletion mutants if the deleted segment overlaps the region of DNA containing the point mutation. In Benzer's experiment, all five nonrecombinant deletion mutants lacked the segment *A5*, and both recombinant deletion mutants contained that segment, so the point mutation had to be in the *A5* region.

Once the main segment in which the mutation occurred was known, the point mutant was crossed with each of the relevant secondary set of reference deletions: *r1605*, *r1589*, and *rPB230*. (See Figure 18.19.) With

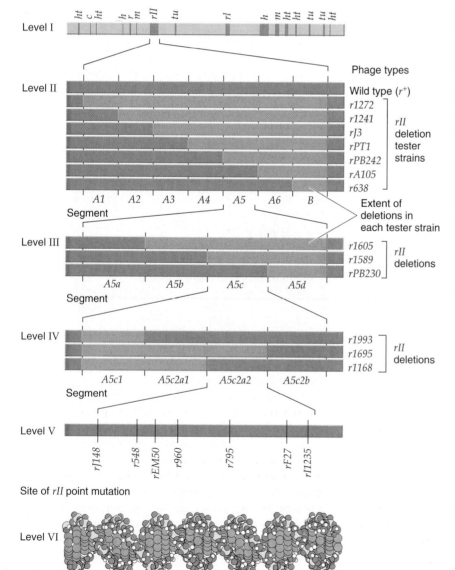

Figure 18.19

Segmental subdivision of the *rII* region of phage T4 by means of deletion. Level I shows the whole T4 genetic map. In Level II, seven deletions define seven segments of the *rII* region. In Level III, three deletions define four subsegments of the *A5* segments. In Level IV, three deletions define four subsegments of the *A5c* subsegment. Level V shows the order and spacing of the sites of the *rII* mutations in the *A5c2a2* subsegment, as established by pairwise crosses of seven point mutants. Level VI is a model of the DNA double helix, indicating the approximate scale of the level-V map.

Figure 18.20

Map of deletions used to divide the *rII* region into 47 small segments. Here, the segments are shown as small boxes at the bottom of the figure.

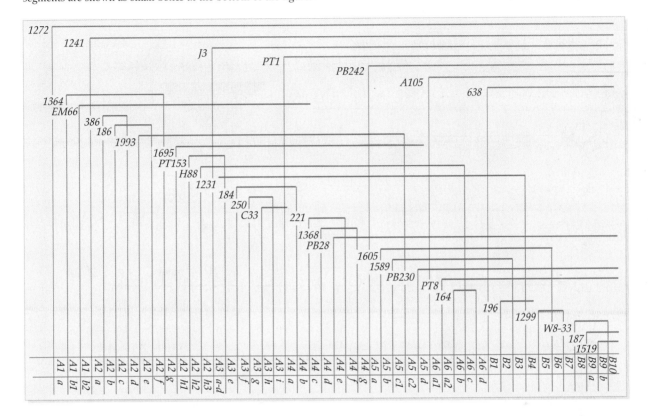

segment *A5*, for example, three deletions divide *A5* into the four subsegments *A5a* through *A5d*. The presence or absence of *r⁺* recombinants in the progeny of the crosses of the *A5 rII* mutant with the secondary set of deletions enabled Benzer to localize the mutation more precisely to a smaller region of the DNA. For example, if the mutation was in segment *A5c*, then *r⁺* recombinants were produced with deletion *rPB230*, but not with either of the other two deletions. Other deletion mutants defined even smaller regions of each of the four subsegments *A5a* through *A5d*; for instance, *A5c* was divided into *A5c1, A5c2a1, A5C2a2,* and *A5c2b* by deletions *r1993, r1695,* and *r1168.*

As shown in Figure 18.20, deletions divide the *rII* region into 47 segments. Any given *rII* point mutant can be localized to one of these segments in three sequential sets of crosses of point mutants with deletion mutants. Then, all those point mutants within a given segment can be crossed in all possible pairwise crosses to construct a detailed genetic map. In this way, Benzer used the more than 3,000 *rII* mutants to prove that the *rII* region is subdivisible into more than 300 mutable sites that were separable by recombination (Figure 18.21). The distribution of mutants is not random: Certain sites, called *hot spots*, are represented by a large number of independently isolated point mutants.

KEYNOTE

Through fine-structure analysis of the *rII* region of bacteriophage T4 and other experiments, it was determined that the unit of mutation and recombination is the base pair in DNA.

Defining Genes by Complementation (*Cis–Trans*) Tests

From the classic point of view, the gene is a unit of function; that is, each gene specifies one function. Benzer designed genetic experiments to determine whether this classic view was true of the *rII* region. To find out whether two different *rII* mutants belonged to the same gene (unit of function), Benzer adapted the ***cis–trans* test**, or **complementation test**, developed by Edward Lewis to study the nature of the functional unit of the gene in *Drosophila*. We introduced the complementation test in Chapter 13, pp. 340–341. As you read the discussion that follows, it will help you to know that the complementation tests showed that the *rII* region actually consists of two genes (units of function): *rIIA*

animation

a Defining Genes by Complementation Tests

Figure 18.21

Fine-structure map of the rII region derived from Benzer's experiments. The number of independently isolated mutations that mapped to a given site is indicated by the number of blocks at the site. Hot spots are represented by a large number of blocks.

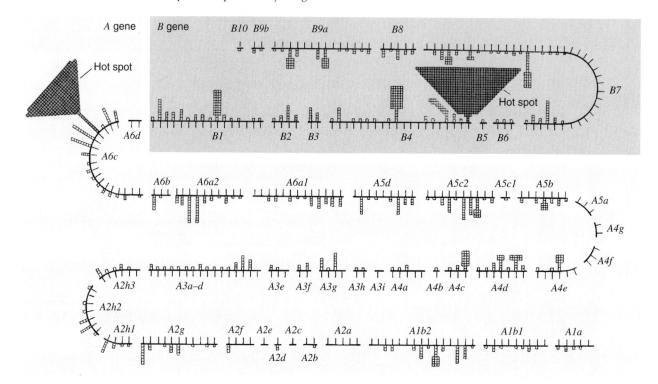

and *rIIB*. A mutation anywhere in either gene produces the *rII* plaque morphology phenotype and host range property. In other words, *rIIA* and *rIIB* each specify a different product needed for growth in *E. coli K12(λ)*.

The complementation test is used to establish how many units of function (genes) are defined by a given set of mutations that express the same mutant phenotypes. In Benzer's work with the *rII* mutants, the nonpermissive strain *K12(λ)* was infected with a pair of *rII* mutant phages to see whether the two mutants, each unable by itself to grow in strain *K12(λ)*, could work together to produce progeny phages. If the phages do produce progeny, the two mutants are said to complement each other, meaning that the two mutations must be in different genes (units of function) that encode different products. That is, those two products work together to allow progeny to be produced. If no progeny phages are produced, the mutants are not complementary, indicating that the mutations are in the same functional unit. In this case, both mutants produce the same defective product, so the phage life cycle cannot proceed and no progeny phages result. (Note that genetic recombination is not necessary for complementation to occur; if genetic recombination does take place, a few plaques may occur on the lawn, but if

complementation occurs, the entire lawn of bacteria will be lysed.)

The two situations are diagrammed in Figure 18.22. In the first case, the bacterium is infected with two phage genomes, one with a mutation in the *rIIA* gene and the other with a mutation in the *rIIB* gene (Figure 18.22a). The *rIIA* mutant makes a nonfunctional *A* product and a functional *B* product, whereas the *rIIB* mutant makes a functional *A* product and a nonfunctional *B* product. Complementation occurs because the *rIIA* mutant still makes a functional *B* product and the *rIIB* mutant makes a functional *A* product, so phage propagation in *E. coli K12(λ)* can occur. In the second case, the bacterium is infected with two phage genomes, each with a different mutation in the same gene, *rIIA* (Figure 18.22b). No complementation occurs in this case, because, although both mutants produce a functional *B* product, neither makes a functional *A* product, so the *A* function cannot take place and phage propagation in *E. coli K12(λ)* cannot occur.

On the basis of the results of such complementation tests, Benzer found that each *rII* mutant falls into one of two units of function: *rIIA* and *rIIB* (also called *complementation groups*, which directly correspond to genes). That is, all *rIIA* mutants complement all *rIIB* mutants, but *rIIA* mutants fail to complement other *rIIA*

Figure 18.22

Complementation tests for determining the units of function in the *rII* region of phage T4; the nonpermissive host *E. coli* K12(λ) is infected with two different *rII* mutants. (a) Complementation occurs. (b) Complementation does not occur.

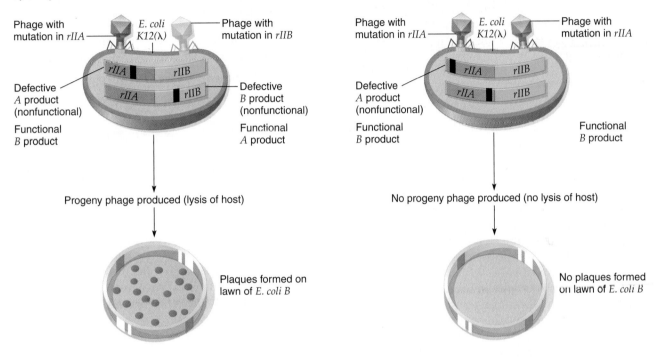

mutants and *rIIB* mutants fail to complement other *rIIB* mutants. The dividing line between the *rIIA* and *rIIB* units of function is indicated in the fine-structure map in Figure 18.21. Point mutants and deletion mutants in the *rII* region obey the same rules in the complementation tests. The only exceptions are deletions that span parts of both the *A* and the *B* functional units. Such deletion mutants do not complement either *A* or *B* mutants.

In the complementation test examples shown in Figure 18.22, each phage that coinfects the nonpermissive *E. coli* strain K12(λ) carries one *rII* mutation, which is actually a configuration of mutations called the *trans* configuration. In this configuration, the two mutations are carried by different phages. As a control, it is usual to coinfect *E. coli* K12(λ) with an *r*+ (wild-type) phage and an *rII* mutant phage carrying both mutations to see whether the expected wild-type function results. When both mutations under investigation are carried on the same chromosome, the configuration is called the *cis* configuration of mutations. (It is because of the *cis* and *trans* configurations of mutations used in the complementation test that it is also called the *cis–trans* test.) In the *cis* test, the *r*+ is expected to be dominant over the two mutations carried by the *rII* mutant phage, so progeny phages are

produced. Therefore, the failure to produce progeny would not prove that the mutations are in different functional genes.

Benzer called the genetic unit of function revealed by the *cis–trans* test the cistron. A cistron may be defined as the smallest segment of DNA that encodes a piece of RNA. At present, *gene* is commonly used and *cistron* is used less often. Genetically, the *rIIA* cistron is about 6 mu and 800 base pairs long, and the *rIIB* cistron is about 4 mu and 500 base pairs long. Presumably, their two products act in common processes necessary for T4 propagation in strain K12(λ).

The principles underlying a complementation test are the same in other organisms; only the practical details of performing the test are organism specific. For example, in yeast, one could select two haploid cells that are of different mating types (*MAT*a and *MAT*α) and that carry different mutations conferring the same mutant phenotype. Mating these two types would produce a diploid, which would then be analyzed for complementation of the two mutations. In animal cells, two cells, each exhibiting the same mutant phenotype, can be fused and analyzed; a wild-type phenotype indicates that complementation has occurred. Again, in neither of these cases is recombination necessary for complementation to occur.

KEYNOTE

The complementation, or *cis–trans*, test is used to determine how many units of function (genes) define a given set of mutations expressing the same mutant phenotypes. If two mutants, each carrying a mutation in a different gene, are combined, the mutations complement and a wild-type function results. If two mutants, each carrying a mutation in the same gene, are combined, the mutations do not complement and the mutant phenotype is exhibited.

Summary

In this chapter, we have seen how genes are mapped in prokaryotes such as bacteria and in bacteriophages. The same experimental strategy is used for all gene mapping; that is, genetic material is exchanged between strains differing in genetic markers, and recombinants are detected and counted. In bacteria, the mechanism of gene transfer may be transformation, conjugation, or transduction. In each process, there is a donor strain and a recipient strain.

Conjugation is a plasmid-mediated process in which there is a unidirectional transfer of genetic information through direct cellular contact between a donor and a recipient. Transformation is the unidirectional transfer of extracellular DNA into cells whereby a phenotypic change is produced in the recipient. Transformation is the transfer of genetic material as small extracellular pieces of DNA between organisms. Transduction is a process in which bacteriophages mediate the transfer of bacterial DNA from the donor bacterium to the recipient. Bacteriophage DNA can be mapped by infecting bacteria simultaneously with two phage strains and analyzing the resulting phage progeny for parental and recombinant phenotypes. Formally, this method of mapping is the same as that used to map genes in haploid eukaryotes.

Insights into the relationships between mapping and gene structure were obtained from a fine-structure analysis of the bacteriophage T4 *rII* region. The mutational sites within a gene were mapped through intragenic mapping. The resulting map indicated that the unit of mutation and the unit of recombination are the same: the base pair in DNA. These definitions replaced the classic definition of a gene as a unit of mutation, recombination, and function.

The number of units of function (genes) is determined by complementation tests. Given a set of mutations expressing the same mutant phenotype, two mutants are combined and the phenotype is determined. If the phenotype is wild-type, the two mutations have complemented and must be in different units of function. If the phenotype is mutant, the two mutations have not complemented and must be in the same unit of function.

Analytical Approaches to Solving Genetics Problems

Q18.1 In *E. coli*, the following *Hfr* strains donate the genes shown in the order given:

Hfr Strain	Order of Gene Transfer
1	G E B D N A
2	P Y L G E B
3	X T J F P Y
4	B E G L Y P

All the *Hfr* strains were derived from the same F⁺ strain. What is the order of genes in the original F⁺ chromosome?

A18.1 This question is an exercise in piecing together various segments of the circumference of a circle. The best approach is to draw a circle and label it with the genes transferred from one *Hfr* and then see which of the other *Hfrs* transfers an overlapping set. For example, *Hfr 1* transfers E, then B, then D, and so on, and *Hfr 4* transfers B, then E, and so forth. Now we can juxtapose the two sets of genes transferred by the two *Hfrs* and deduce the fact that the polarities of transfer are opposite:

Hfr 1	G E B D N A
Hfr 4	P Y L G E B

Extending this reasoning to the other *Hfrs*, we can draw an unambiguous map (see the following figure), with the arrowheads indicating the order of transfer:

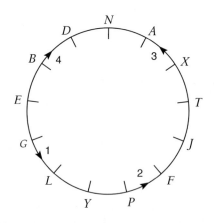

The same logic would be used if the question gave the relative time units of entry of each of the genes. In that case, we would expect that the time distance between any two genes would be approximately the same, regardless of the order of transfer or how far the genes were from the origin.

Q18.2 In a transformation experiment, donor DNA from an $a^+ b^+$ strain was used to transform a recipient strain of genotype $a\ b$. The transformed classes were isolated and their frequencies determined to be

$$
\begin{array}{ll}
a^+ b^+ & 307 \\
a^+ b & 215 \\
a\ b^+ & 278
\end{array}
$$

The total number of transformants was 800. What is the frequency with which the b locus is cotransformed with the a locus?

A18.2 The frequency with which b^+ is cotransformed with the a^+ gene is calculated with the use of values for the total number of a^+ transformants and the number of transformants for both a^+ and b^+. The formula is

$$
\frac{\text{number of } a^+ b^+ \text{ cotransformants}}{\text{total number of } a^+ \text{ transformants}} \times 100\%
$$

The $a^+ b^+$ cotransformants number 307. The a^+ transformants are represented by two classes: $a^+ b^+$ (307) and $a^+ b$ (215), for a total of 522. The $a\ b^+$ class is irrelevant to the question because its members are not transformants for a^+. Thus, the cotransformation frequency for a^+ and b^+ is $307/522 \times 100 = 58.8\%$.

Q18.3 In a transduction experiment, the donor was $c^+ d^+ e^+$ and the recipient was $c\ d\ e$. Selection was for c^+. The four classes of transductants from this experiment are shown in the following table:

Class	Genetic Composition	Number of Individuals
1	$c^+ d^+ e^+$	57
2	$c^+ d^+ e$	76
3	$c^+ d\ e$	365
4	$c^+ d\ e^+$	2
	Total	500

a. Determine the cotransduction frequency for c^+ and d^+.
b. Determine the cotransduction frequency for c^+ and e^+.
c. Which of the cotransduction frequencies calculated in (a) and (b) represents the greater actual distance between genes? Why?

A18.3
a. The analysis is similar to the cotransformation frequency analysis described in Q9.2. The formula for the cotransduction frequency for c^+ and d^+ is

$$
\frac{\text{number of } c^+ d^+ \text{ contransductants}}{\text{total number of } c^+ \text{ transductants}} \times 100\%
$$

From the data presented, classes 1 and 2 are the $c^+ d^+$ cotransductants, and the total number of c^+ trans-

ductants is the sum of classes 1 through 4. Thus, the number of c^+ and d^+ transductants is $57 + 76 = 133$, and the cotransductant frequency is $133/500 \times 100 = 26.6\%$.

b. The analysis is identical in approach to (a). The formula for the cotransduction frequency for c^+ and e^+ is

$$
\frac{\text{number of } c^+ e^+ \text{ cotransductants}}{\text{total number of } c^+ \text{ transductants}} \times 100\%
$$

From the data presented, classes 1 and 4 are the $c^+ e^+$ cotransductants, and the total number of c^+ transductants is the sum of classes 1 through 4. Thus, the number of c^+ and e^+ transductants is $57 + 2 = 59$, and the cotransductant frequency is $59/500 \times 100 = 11.8\%$.

c. The greater actual distance is between the c^+ and e^+ genes. The principle involved is that the closer two genes are on the chromosome, the greater is the likelihood that they will be cotransduced. Thus, as the distance between genes increases, the cotransduction frequency decreases. Since the $c^+ e^+$ cotransduction frequency is 11.8 percent and the $c^+ d^+$ cotransduction frequency is 26.6 percent, genes c^+ and e^+ are farther apart than genes c^+ and d^+.

Q18.4 Five different *rII* deletion strains of phage T4 were tested for recombination by pairwise crossing in *E. coli* B. The following results were obtained, where $+ = r^+$ recombinants produced and $0 =$ no r^+ recombinants produced:

	A	B	C	D	E
E	0	+	0	+	0
D	0	0	0	0	
C	0	0	0		
B	+	0			
A	0				

Draw a deletion map compatible with these data.

A18.4 The principle here is that if two deletion mutations overlap, then no r^+ recombinants can be produced. Conversely, if two deletion mutations do not overlap, then r^+ recombinants can be produced. To approach a question of this kind, we must draw overlapping and nonoverlapping lines from the given data.

Starting with A and B, we see that these two deletions do not overlap, because r^+ recombinants are produced. Therefore, these two mutations can be represented as follows:

$$
\underline{\qquad A \qquad} \qquad \underline{\qquad B \qquad}
$$

The next deletion, *C*, does not produce *r*⁺ recombinants with any of the other four deletions. We must conclude, therefore, that *C* is an extensive deletion that overlaps the other four, with endpoints that cannot be determined from the data given. One possibility is as follows:

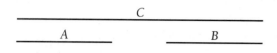

Deletion *D* does not produce *r*⁺ recombinants with *A*, *B*, or *C*, but it does with *E*. In turn, *E* produces *r*⁺ recombinants with *B* and *D*, but not with *A* or *C*. Thus, *D* must overlap both *A* and *B*, but not *E*, and *E* must overlap *A* and *C*, but not *B*. A compatible map for this situation is as follows:

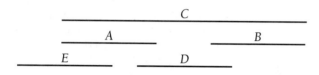

Other maps can be drawn in terms of the endpoints of the deletions.

Q18.5 Seven different *rII* point mutants (1 to 7) of phage T4 were tested for recombination crosses in *E. coli* B with the five deletion strains described in Question 18.4. The following results were obtained, where + = *r*⁺ recombinants produced and 0 = no *r*⁺ recombinants produced:

	A	B	C	D	E
1	0	+	0	+	+
2	+	0	0	+	+
3	0	+	0	+	0
4	+	+	0	+	0
5	+	0	0	0	+
6	0	+	0	0	+
7	+	+	0	0	+

In which regions of the map can you place the seven point mutations?

A18.5 If an *r*⁺ recombinant is produced, the *rII* point mutation cannot overlap the region missing in the deletion mutation with which it was crossed. Thus, the table of results enables us to localize the point mutations to the regions defined by the deletion mutants. Potentially, the results define the relative extent of deletion overlap. For example, point mutation 7 produces *r*⁺ recombinants with *A*, *B*, and *E*, but not with *D*. Logically, then, 7 is located in the region defined by the part of deletion *D* that is not

involved in the overlap with *A* and *B*. Similarly, point mutation 4 gives *r*⁺ recombinants with *A*, *D*, and *B*, but not with *E*. Hence, 4 must be in a region defined by a segment of deletion *E* that does not overlap deletion *A*. Furthermore, because 4 does not produce *r*⁺ recombinants with *C* either, deletion *C* must overlap the site defined by point mutation 4. This result, then, refines the deletion map with regard to the *E*, *C*, and *A* endpoints. The map we can draw from the matrix of results is as follows:

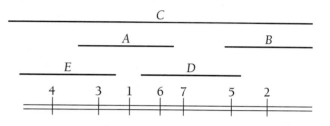

Questions and Problems

*18.1 In *F*⁺ × *F*⁻ crosses, the *F*⁻ recipient is converted to a donor with very high frequency. However, it is rare for a recipient to become a donor in *Hfr* × *F*⁻ crosses. Explain why.

*18.2 With the technique of interrupted mating, four *Hfr* strains were tested for the sequence in which they transmitted a number of different genes to an *F*⁻ strain. Each *Hfr* strain was found to transmit its genes in a unique order, as shown in the accompanying table. (Only the first six genes transmitted were scored for each strain.)

Order of Transmission	Hfr Strain			
	1	2	3	4
First	O	R	E	O
	F	H	M	G
	B	M	H	X
	A	E	R	C
	E	A	C	R
Last	M	B	X	H

What is the gene sequence in the original strain from which these *Hfr* strains derive? On a diagram, indicate the origin and polarity of each of the four *Hfr* strains.

18.3 At time zero, an *Hfr* strain (*Hfr* 1) was mixed with an *F*⁻ strain, and at various times after mixing, samples were removed and agitated to separate conjugating cells. The cross may be written as

$$Hfr\ 1: \quad a^+\ b^+\ c^+\ d^+\ e^+\ f^+\ g^+\ h^+\ str^S$$
$$F^-: \quad a\ b\ c\ d\ e\ f\ g\ h\ str^R$$

(No order is implied in listing the markers.) The samples were then plated onto selective media to measure the frequency of *h*⁺ *str*ᴿ recombinants that had received certain genes from the *Hfr* cell. A graph of the number of recombinants against time is shown in the accompanying figure.

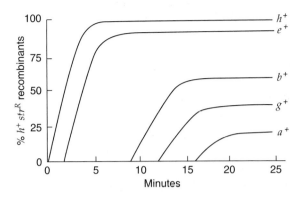

a. Indicate whether each of the following statements is true or false:

   i. All *F⁺* cells that received *a⁺* from the *Hfr* in the chromosome transfer process must also have received *b⁺*.

   ii. The order of gene transfer from *Hfr* to *F⁻* was *a⁺* (first), then *g⁺*, then *b⁺*, then *e⁺*, and, finally, *h⁺*.

   iii. Most *e⁺ strᴿ* recombinants are likely to be *Hfr* cells.

   iv. None of the *b⁺ strᴿ* recombinants plated at 15 minutes are also *a⁺*.

b. Draw a linear map of the *Hfr* chromosome, indicating

   i. the point of nicking (the origin) and the direction of DNA transfer

   ii. the order of the genes *a⁺*, *b⁺*, *e⁺*, *g⁺*, and *h⁺*

   iii. the shortest distance between consecutive genes on the chromosomes

18.4 What steps would you take to selectively grow each of the bacterial cell types found in the following mixtures A through D?

Mixture	Genotypes Present	Phenotypes
A	*his*, *his⁺*	*his* cells require supplemental histidine; *his⁺* cells are able to grow without supplemental histidine.
B	*aziᴿ*, *aziˢ*	*aziᴿ* cells are able to grow even in the presence of the poison sodium azide; *aziˢ* cells die in the presence of sodium azide.
C	*lac*, *lac⁺*	*lac⁺* cells can grow even if lactose is the only sugar present; *lac ⁻*cells cannot utilize lactose for growth, they require a sugar other than lactose for growth whether or not lactose is present
D	*pcsA⁺*, *pcsA*	*pcsA* cells are cold sensitive and grow at 37°C, but not at 30°C; *pcsA⁺* cells can grow at both 37°C and 30°C.

*18.5 If an *E. coli* auxotroph *A* could grow only on a medium containing thymine, and an auxotroph *B* could grow only on a medium containing leucine, how would you test whether DNA from *A* could transform *B*?

*18.6 Three different prototrophic strains (1, 2, and 3) that are all sensitive to the antibiotic streptomycin are isolated. Each is individually mixed with an auxotrophic *F⁻* strain that is *a b c d e f g h* (and therefore requires compounds A, B, C, D, E, F, G, and H to grow) and that is also resistant to the antibiotic streptomycin. At 1-minute intervals after the initial mixing, a sample of the mixture is removed, shaken violently, and plated on media to select for *c⁺ strᴿ* recombinants. Recombinants are then tested for the presence of other genes. The following results are obtained:

   Strain 1 × *F⁻*: No *c⁺* recombinants are ever obtained, even after 25 minutes.

   Strain 2 × *F⁻*: *c⁺* recombinants are obtained at 6 minutes, *g⁺* at 8 minutes, *h⁺* at 11 minutes, *a⁺* at 14 minutes, *b⁺* at 16 minutes. No *d⁺*, *e⁺*, or *f⁺* recombinants are obtained.

   Strain 3 × *F⁻*: *c⁺* recombinants are obtained at 1 minute, and *c⁺ g⁺* recombinants are obtained at or after 3 minutes. No *a⁺*, *b⁺*, *d⁺*, *e⁺*, *f⁺*, or *h⁺* recombinants are obtained.

If *c⁺* recombinants obtained at 16 minutes from the cross involving strain 2 are mixed with an *ampᴿ* (ampicillin-resistant) *F⁻* strain, no *c⁺ ampᴿ* recombinants are ever recovered. However, if *c⁺* recombinants obtained at 16 minutes from the cross involving strain 3 are mixed with an *ampᴿ F⁻* strain, *ampᴿ c⁺* recombinants can be recovered after 1 minute of mating.

a. How was the initial selection for *c⁺ strᴿ* recombinants done? How were the subsequent selections done?

b. Use the given data to ascertain, as best you can, whether each strain is *F⁻*, *Hfr*, *F⁺*, or *F′*. If these data do not allow you to make an unambiguous determination, indicate the possibilities.

c. To the extent you can, draw a map of the chromosomes that might be present in each of strains 1, 2, and 3. Indicate the location and distance between genes *a⁺*, *b⁺*, *d⁺*, *e⁺*, *f⁺*, and *h⁺* as best you can.

18.7 You are given a prototrophic *strᴿ* (streptomycin-resistant) *Hfr* strain and an *ampᴿ* (ampicillin-resistant) *F⁻* auxotrophic strain that requires leucine (*leu*), arginine (*arg*), lysine (*lys*), purine (*pur*), and biotin (*bio*).

a. Devise a strategy to determine quickly which gene (*leu⁺*, *arg⁺*, *lys⁺*, *pur⁺*, or *bio⁺*) lies closest to the F-factor origin of replication.

b. Even when the prototrophic *Hfr* strain is mixed with the *F⁻* strain for very long periods, streptomycin resistance is not transferred. State two hypotheses that explain this finding.

*18.8 Indicate whether each of the following lettered items
Occurs or is a characteristic of generalized transduction (GT)
Occurs or is a characteristic of specialized transduction (ST)
Occurs in both (B)
Occurs in neither (N)

a. Phage carries DNA of bacterial or viral DNA origin, never both.

b. Phage carries viral DNA covalently linked to bacterial DNA.

c. Phage integrates into a specific site on the host chromosome.

d. Phage integrates at a random site on the host chromosome.

e. "Headful" of bacterial DNA is packaged into phage.

f. Host is lysogenized.

g. Prophage state exists.

h. Temperate phage is involved.

i. Virulent phage is involved.

18.9 Consider the following transduction data:

Donor	Recipient	Selected Marker	Unselected Marker	%
$aceF^+$ dhl	$aceF$ dhl^+	$aceF^+$	dhl	88
$aceF^+$ leu	$aceF$ leu^+	$aceF^+$	leu	34

Is dhl or leu closer to $aceF$?

*18.10 Consider the following data pertaining to P1 transduction:

Donor	Recipient	Selected Marker	Unselected Marker	%
$aroA$ $pyrD^+$	$aroA^+$ $pyrD$	$pyrD^+$	$aroA$	5
$aroA^+$ $cmlB$	$aroA$ $cmlB^+$	$aroA^+$	$cmlB$	26
$cmlB$ $pyrD^+$	$cmlB^+$ $pyrD$	$pyrD^+$	$cmlB$	54

Choose the correct order:

a. $aroA$ $cmlB$ $pyrD$
b. $aroA$ $pyrD$ $cmlB$
c. $cmlB$ $aroA$ $pyrD$

18.11 Order the mutants trp, $pyrF$, and qts on the basis of the following three-factor transduction cross:

Donor trp^+ pyr^+ qts
Recipient trp pyr qts^+
Selected Marker trp^+

Unselected Markers	Number
pyr^+ qts^+	22
pyr^+ qts	10
pyr qts^+	68
pyr qts	0

*18.12 Order $cheA$, $cheB$, eda, and $supD$ from the following data:

Markers	% Cotransduction
$cheA$-eda	15
$cheA$-$supD$	5
$cheB$-eda	28
$cheB$-$supD$	2.7
eda-$supD$	0

*18.13 Wild-type phage T4 grows on both *E. coli* B and *E. coli* K12(λ), producing turbid plaques. The rII mutants of T4 grow on *E. coli* B, producing clear plaques, but do not grow on *E. coli* K12(λ). This host range property permits the detection of a very low number of r^+ phages among a large number of rII phages. With such a sensitive system, it is possible to determine the genetic distance between two mutations within the same gene—in this case, the rII locus. Suppose *E. coli* B is mixedly infected with $rIIx$ and $rIIy$, two separate mutants in the rII locus. Suitable dilutions of progeny phages are plated on *E. coli* B and *E. coli* K12(λ). A 0.1-mL sample of a thousandfold dilution plated on *E. coli* B produces 672 plaques. A 0.2-mL sample of undiluted phage plated on *E. coli* K12(λ) produces 470 turbid plaques. What is the genetic distance between the two rII mutations?

18.14 Construct a map from the following two-factor phage cross data (show the map distance):

Cross	% Recombination
$r1 \times r2$	0.10
$r1 \times r3$	0.05
$r1 \times r4$	0.19
$r2 \times r3$	0.15
$r2 \times r4$	0.10
$r3 \times r4$	0.23

*18.15 The following two-factor crosses were made to analyze the genetic linkage between four genes in phage λ: c, mi, s, and co.

Parents	Progeny
$c +$ × $+ mi$	1,213 $c +$, 1,205 $+mi$, 84 $++$, 75 c mi
$c +$ × $+ s$	566 $c +$, 808 $+s$, 19 $++$, 20 cs
$co +$ × $+ mi$	5,162 $co +$, 6,510 $+mi$, 311 $++$, 341 co mi
$mi +$ × $+ s$	502 $mi +$, 647 $+s$, 65 $++$, 56 mi s

Construct a genetic map of the four genes.

18.16 Wild-type (r^+) strains of T4 produce turbid plaques, whereas rII mutant strains produce larger, clearer plaques. Five rII mutations (a–e) in the A cistron of the rII region of T4 give the following percentages of wild-type recombinants in two-point crosses:

Cross	% of Wild-Type Recombinants	Cross	% of Wild-Type Recombinants
$a \times b$	0.2	$e \times d$	0.7
$a \times c$	0.9	$e \times c$	1.2
$a \times d$	0.4	$e \times b$	0.5
$b \times c$	0.7	$b \times d$	0.2
$e \times a$	0.3	$d \times c$	0.5

What is the order of the mutational sites, and what are the map distances between the sites?

*18.17 Given the following map with point mutants and given the data in the following table, draw a topological representation of deletion mutants $r21$, $r22$, $r23$, $r24$, and $r25$ (be sure to show the endpoints of the deletions. $+ = r^+$ recombinants are obtained, $0 = r^+$ recombinants are not obtained):

Map
```
       r12   r16   r11   r15   r13    r14    r17
```

Deletion Mutants	Point Mutants						
	r11	r12	r13	r14	r15	r16	r17
r21	0	+	0	+	0	+	+
r22	+	+	0	0	+	+	0
r23	0	0	0	+	0	0	+
r24	+	+	0	0	+	+	+
r25	+	+	0	0	0	+	+

*18.18 Given the following deletion map with deletions $r31$, $r32$, $r33$, $r34$, $r35$, and $r36$, place the point mutants $r41$, $r42$, etc., on the map (be sure to show where they lie with respect to end points of the deletions):

```
                    r31
                  r32
            r33
               r34
         r35              r36
```

Point Mutants	Deletion Mutants (+ = recombinants produced; 0 = no r^+ recombinants produced)					
	r31	r32	r33	r34	r35	r36
r41	0	0	0	0	+	0
r42	0	0	0	+	0	+
r43	0	0	+	+	+	0
r44	0	0	0	0	+	+
r45	0	+	0	+	+	+
r46	0	0	+	0	+	0

Show the dividing line between the A cistron and the B cistron on your map from the following data

$[+ =$ growth on strain $K12(\lambda)$, $0 =$ no growth on strain $K12(\lambda)]$:

Mutant	Complementation with	
	rIIA	rIIB
r31	0	0
r32	0	0
r33	0	+
r34	0	0
r35	0	+
r36	0	0
r41	0	+
r42	0	+
r43	+	0
r44	0	+
r45	0	+
r46	0	+

18.19 Some adenine-requiring mutants of yeast are pink because of the intracellular accumulation of a red pigment. Diploid strains were made by mating haploid mutant strains. The diploids exhibited the following phenotypes:

Cross	Diploid Phenotypes
1×2	pink, adenine requiring
1×3	white, prototrophic
1×4	white, prototrophic
3×4	pink, adenine requiring

How many genes are defined by the four different mutants? Explain.

18.20 Specialized transduction can be used to develop fine-structure maps. Five different λd *gal* phage were isolated (1, 2, 3, 4, 5). Each was infected into five different *gal* point mutants (a, b, c, d, e), and gal^+ recombinants were selected by plating the cells on media containing galactose as the sole carbon source. The results are shown in the following table (+ indicates that gal^+ recombinants were obtained; – indicates that no gal^+ recombinants were obtained):

E. coli gal mutant	λd gal phage				
	1	2	3	4	5
a	–	+	–	–	–
b	–	+	–	+	–
c	+	+	+	+	+
d	+	+	+	+	–
e	+	+	–	+	–

Each λd *gal* phage was then coinfected into *E. coli* with each of five lambda point mutants (j, k, l, m, n), and a selection was performed for wild-type lambda progeny. The following table shows the results (+ indicates that wild-type λ recombinants were obtained; – indicates that no wild-type λ recombinants were obtained):

λ mutant	λd gal phage				
	1	*2*	*3*	*4*	*5*
j	+	−	+	+	+
k	+	−	−	−	+
l	+	−	+	−	+
m	+	−	−	−	−
n	−	−	−	−	−

Draw the *gal–bio* region of an *E. coli* λ lysogen, and label the location of the *att* site, the *gal* and *bio* genes, and λ. Below your drawing, indicate the relative map positions of the five *gal* mutations, the relative map positions of the five λ mutations, and the regions of the *gal* gene and λ genome that are present in each λd *gal* phage.

*18.21 In *E. coli*, eight spontaneously and independently arising *leu* mutants were isolated from a parental F^- str^R strain. Each mutant requires supplemental leucine to grow, but is resistant to streptomycin. Interrupted-mating experiments were performed with each of the eight *leu* mutants and a prototrophic *E. coli* Hfr strain sensitive to streptomycin. In each cross, str^R leu^+ recombinants were recovered just after 4 minutes of mating. The fine structure of the *leu* region was then evaluated with the use of generalized transduction. Each of the eight mutants was individually infected with a generalized transducing phage. The resulting lysate was used to infect the other mutants, and *leu^+* recombinants were selected. The following table shows the results (+ indicates that *leu^+* recombinants were recovered, − indicates that *leu^+* recombinants were not recovered):

leu mutant	*1*	*2*	*3*	*4*	*5*	*6*	*7*	*8*
1	−	−	+	−	+	+	+	+
2		−	−	+	−	−	+
3			−	+	+	+	−	+
4				−	−	+	+	+
5					−	+	+	+
6						−	+	+
7							−	−
8								−

a. Draw a map showing the relative order and locations of the mutant sites in this region. (*Hint:* First identify the deletions.)
b. Can you infer whether any of these mutations are point mutants? If not, how would you address this issue?
c. Explain whether you can infer how many cistrons involved in leucine biosynthesis are present in the region of interest.

18.22 A homozygous white-eyed Martian fly (w_1/w_1) is crossed with a homozygous white-eyed fly from a differ-

ent stock (w_2/w_2). It is well known that wild-type Martian flies have red eyes. The cross produces all white-eyed progeny. State whether each of the following is true or false, and explain your answer:
a. w_1 and w_2 are allelic genes.
b. w_1 and w_2 are nonallelic.
c. w_1 and w_2 affect the same function.
d. The cross was a complementation test.
e. The cross was a *cis–trans* test.
f. w_1 and w_2 are allelic by the terms of the functional test.

The F_1 white-eyed flies are allowed to interbreed, and when you classify the F_1, you find 20,000 white-eyed flies and 10 red-eyed progeny. Concerned about contamination, you repeat the experiment and get exactly the same results. How can you best account for the presence of the red-eyed progeny? As part of your explanation, give the genotypes of the F_1 and F_2 generation flies.

18.23 Propose a genetic explanation for the ugly-duckling phenomenon: Two white parents have a rare black offspring amid a prolific number of white offspring.

*18.24 Both *trpA* and *trpB* mutants of *E. coli* lack tryptophan synthetase activity. All *trpA* mutants complement all *trpB* mutants. Explain how two different complementing mutants (*trpA* and *trpB*) can affect the activity of the same enzyme.

18.25 Four strains of *Neurospora*, all of which require arginine, but have an unknown genetic constitution, have the following nutrition and accumulation characteristics:

	Growth on				
Strain	Minimal Medium	Ornithine	Citrulline	Arginine	Accumulates
1	−	−	+	+	Ornithine
2	−	−	−	+	Citrulline
3	−	−	−	+	Citrulline
4	−	−	−	+	Ornithine

Pairwise complementation tests of the four strains gave the following results (+ = growth on minimal medium and 0 = no growth on minimal medium):

	4	3	2	1
1	0	+	+	0
2	0	0	0	
3	0	0		
4	0			

Crosses among mutants yielded prototrophs in the following percentages:

1 × 2:	25 percent
1 × 3:	25 percent
1 × 4:	none detected among 1 million ascospores
2 × 3:	0.002 percent
2 × 4:	0.001 percent
3 × 4:	none detected among 1 million ascospores

Analyze the data and answer the following questions:
a. How many distinct mutational sites are represented among these four strains?
b. In this collection of strains, how many types of polypeptide chains (normally found in the wild type) are affected by mutations?
c. Write the genotypes of the four strains, using a consistent and informative set of symbols.
d. Determine the map distances between all pairs of linked mutations.
e. Determine the percentage of prototrophs expected among ascospores of the following types: (1) strain 1 × wild type; (2) strain 2 × wild type; (3) strain 3 × wild type; (4) strain 4 × wild type

*18.26 Herpes simplex virus type 1 (HSV-1) is a large eukaryotic virus whose growth proceeds sequentially. Progression from one stage to the next requires completion of the earlier stage. Understanding how different viral genes are used at each stage should aid in the development of therapies for viral infection. Nine different mutations (B2, B21, B27, B28, B32, 901, LB2, D, c75) block viral growth at a very early stage. Each mutant grows at the permissive temperature of 34°C and fails to grow at the restrictive temperature of 39°C. All of the mutants except for c75 spontaneously revert to wild-type at about the same low frequency; c75 reverts to wild-type, but much less frequently the others. Complementation analysis of these mutants and a tenth temperature-sensitive mutation that blocks growth at a later stage, J12g, was performed by coinfecting pairs of mutants into cells at 39°C, collecting the cell culture media, and assaying their virus content by infecting cells at 34°C. Virus production was quantified with an index I. For two mutants A and B, I = [yield (coinfection of A and B)]/[yield(infection of A) + yield (infection of B)]. I must be over 2 to be considered positive. The following table gives the values of I for pairs of coinfected mutants:

Virus	B21	B27	B28	B32	901	LB2	D	c75	J12g
B2	1.5	0.70	0.81	0.37	0.40	0.55	0.48	0.70	170
B21		0.31	0.27	0.86	0.76	0.88	0.33	0.32	4.9
B27			0.28	0.18	1.9	1.5	0.61	0.68	19
B28				0.20	0.42	0.68	0.13	0.50	72
B32					1.4	0.84	0.28	0.38	580
901						0.45	0.10	0.20	570
LB2							0.44	0.91	22
D								0.35	444
c75									30

Pairwise recombination frequencies of the eight mutants that block very early growth were determined. The recombination frequency for two mutants A and B was calculated as RF = [yield (coinfection of A and B) at 39°C]/[yield (coinfection of A and B) at 34°C] × 2 × 100. The following table gives RF values for pairs of mutants:

Mutant	B21	B27	B28	B32	901	LB2	D	c75
B2	0.36	0.51	2.6	2.5	6.0	1.7	4.0	0.87
B21		0.91	3.0	2.8	6.4	2.3	4.5	1.6
B27			2.1	2.2	5.5	1.4	3.6	1.8
B28				0.11	3.8	0.71	1.55	0.31
B32					3.4	0.73	1.45	0.55
901						4.1	1.9	0.89
LB2							2.2	0.40
D								0.0

Analyze these results and answer the following questions about the nine mutants that block HSV-1 growth at a very early stage:
a. Are the mutants point mutations or deletions?
b. How many functions are affected by the mutants?
c. Do any of the mutants affect the same site?
d. Do any of the mutants affect multiple sites?
e. What is the rationale behind the calculations of I and RF?
f. What are the relative map positions of B2, B21, B27, B28, B32, 901, LB2, and D?
g. RF values for the c75 mutant are inconsistent with those of the other mutants. Assuming that there are no technical errors, what might explain this inconsistency?

*18.27 A large number of mutations in *Drosophila* alter the normal deep-red eye color. As we discussed in Chapter 13, even alleles at one gene (*white*, *w*) can display a variety of phenotypes.

You are given a wild-type, deep-red strain and six independently isolated, true-breeding mutant strains that have varying shades of brown eyes, with the assurance that each mutant strain has only a single mutation. How would you determine
a. whether the mutation in each strain is dominant or recessive?
b. how many different genes are affected in the six mutant strains?
c. which mutants, if any, are allelic?
d. whether any of these mutants are alleles of genes already known to affect eye color?

18.28 In *Drosophila*, the *kar*, *ry*, and *l(3)26* loci are located on chromosome 3 at map positions 51.7, 52.0, and 52.2, respectively. Mutants at *kar* have karmoisin (bright-red) eyes. Mutants at $l(3)26$ are recessive lethal. Mutants at ry lack the enzyme xanthine dehydrogenase. They survive and have rosy eyes if their dietary purine is limited, but die if it is not. Wild-type ry^+ animals have deep-red eyes and survive if fed a diet rich in purine. You want to test whether Benzer's findings at the rII locus in the T4 phage can be replicated in eukaryotes, so embark on a fine-structure analysis of the ry locus. Over the years, hundreds of mutants with rosy eyes have been identified by different researchers, and you have obtained many of them. Describe your experimental design, and address each of the following concerns:

a. There are many loci that affect eye color in *Drosophila*. What methods will you use to ensure that a rosy-eyed mutant is caused by a mutation at the ry locus?

b. What sets of crosses would you perform? In general terms, what progeny and frequencies do you expect to see in each cross?

c. How will you efficiently select for intragenic recombinants at the ry locus?

d. If you undertake both fine-structure recombination and a complementation analyses, what results do you expect to see if Benzer's findings are replicated?

19

Regulation of Gene Expression in Bacteria and Bacteriophages

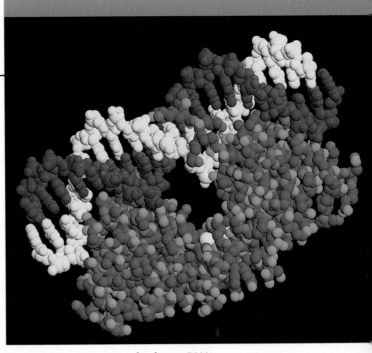

lac operon repressor protein binding to DNA.

PRINCIPAL POINTS

- In the lactose system of *E. coli,* the addition of lactose to cells brings about a rapid synthesis of three enzymes. In the absence of lactose, the synthesis of the three enzymes is turned off. The genes for the enzymes are contiguous on the *E. coli* chromosome and are adjacent to two regulatory sequences: a promoter and an operator. The promoter, operator, and genes constitute an operon. Transcription of the genes results in a single polycistronic mRNA. A regulatory gene is associated with an operon. To turn on gene expression in the lactose system, a lactose metabolite binds with a repressor protein (the product of the regulatory gene), inactivating it and preventing it from binding to the operator. As a result, RNA polymerase can bind to the promoter and transcribe the three genes as a single polycistronic mRNA. Operons are commonly involved in the regulation of gene expression in a large number of prokaryotic and bacteriophage systems.

- The expression of a number of bacterial amino-acid synthesis operons is controlled by a repressor–operator system and through attenuation at a second regulatory sequence, called an attenuator. The repressor–operator

system functions essentially like that for the *lac* operon, except that the addition of amino acid to the cell activates the repressor, thereby turning the operon off. An attenuator located between the operator region and the first structural gene is a transcription termination site that allows only a fraction of RNA polymerases to transcribe the rest of the operon. Attenuation requires, a coupling between transcription and translation, and the formation of particular RNA secondary structures that signal whether transcription can continue.

- Bacteriophages such as lambda are especially adapted for undergoing reproduction within a bacterial host. Many genes related to the production of progeny phages or to the establishment or reversal of lysogeny in temperate phages are organized into operons. Like bacterial operons, these operons are controlled through the interaction of regulatory proteins with operators that are adjacent to clusters of structural genes. Phage lambda has been an excellent model for studying the genetic switch that controls the choice between lytic and lysogenic pathways in a lysogenic phage.

(i) ONE OF THE BEST STRATEGIES FOR AN ORGANISM'S survival is to be able to adapt quickly to changes in its environment. In fact, a fundamental property of living cells is their ability to turn their genes on and off in response to extracellular signals. This control of gene expression makes it possible for cells to produce specific kinds of proteins when and where they are needed. In this chapter, you will learn some of the ways in which gene expression in microorganisms is regulated. Then, in the iActivity, you can investigate how mutations affect the process of regulation in *E. coli*.

Bacteria are free-living organisms that grow by increasing in mass and then divide by binary fission. Growth and division are controlled by genes, the expression of which must be regulated appropriately. Genes whose activity is controlled in response to the needs of a cell or organism are called **regulated genes.** An organism also has a large number of genes whose products are essential to the normal functioning of a growing and dividing cell, no matter what the conditions are. These genes are always active in growing cells and are known as **constitutive genes** or *housekeeping genes;* examples include genes that code for the enzymes needed for protein synthesis and glucose metabolism. Note that all genes are regulated on some level. If normal cell function is impaired for some reason, the expression of all genes, including constitutive genes, is reduced by regulatory mechanisms. Thus, the distinction between regulated and constitutive genes is somewhat arbitrary.

The goal of this chapter is to learn about some of the mechanisms by which gene expression is regulated in bacteria and bacteriophages. Significantly, genes which encode proteins that work together in the cell typically are organized into operons; that is, the genes are adjacent to each other and are transcribed together onto a **polycistronic mRNA,** so called because it contains the information from more than one gene. (Here, the word "cistron" is used synonymously with "gene".) Regulation of the synthesis of this mRNA depends on interactions between a regulatory protein and a regulatory sequence called an operator that is next to the gene array. Such studies of bacterial and bacteriophage gene regulation have provided important insights into how genes are regulated in higher organisms also, including humans. Of course, much remains to be done to understand completely the regulation of gene expression in bacteria. The 4.6-megabase (4.6×10^6 base pair) genome of *E. coli,* for example, has 4,288 protein-coding genes according to the genomic sequence, but researchers have been unable to attribute a function to more than 35 percent of them.

The *lac* Operon of *E. coli*

When gene expression is turned on in a bacterium by adding a substance (such as lactose) to the medium, the genes involved are said to be *inducible.* The regulatory substance that brings about this gene induction is called an **inducer,** and the phenomenon of producing a gene product in response to an inducer is called **induction.** The inducer is an example of a class of small molecules, called **effectors** or **effector molecules,** that help control the expression of many regulated genes. An inducible gene is transcribed in response to a regulatory event occurring at a specific regulatory DNA sequence adjacent to or near the protein-coding sequence (Figure 19.1). The regulatory event typically involves an inducer and a regulatory protein, and when it occurs, RNA polymerase initiates transcription at the promoter (usually upstream of the regulatory sequence). The gene is turned on, mRNA is made, and the protein coded for by the gene is produced. The regulatory sequence itself does not code for any product. As an example of such gene regulation, let us examine regulation of the genes of the *E. coli lac* operon, an **inducible operon.**

Lactose as a Carbon Source for *E. coli*

E. coli can grow in a simple medium containing salts (including a nitrogen source) and a carbon source such as glucose. The energy for biochemical reactions in the cell comes from glucose metabolism. The enzymes required for glucose metabolism are coded for by constitutive genes. If lactose is provided to *E. coli* as a carbon source instead of glucose, a number of enzymes that are required to metabolize lactose are rapidly synthesized. (The same series of events, each involving a sugar-specific set of enzymes, is triggered by other sugars as well.) The enzymes are synthesized because the genes that code for them become actively transcribed in the presence of the sugar; the same genes are inactive if the sugar is absent. In other words, the genes are regulated genes whose products are needed only at certain times.

Figure 19.1

General organization of an inducible gene.

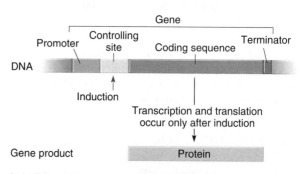

Inducible genes are expressed only in the absence of a repressor and/or presence of an effector/inducer molecule.

Lactose is a disaccharide consisting of the monosaccharides glucose and galactose. When lactose is present as the sole carbon source in the growth medium, three proteins are synthesized:

1. **β-Galactosidase.** This enzyme breaks down lactose into glucose and galactose, as well as catalyzing the isomerization (conversion to a different form) of lactose to *allolactose,* a compound that is important in regulating expression of the *lac* operon (Figure 19.2). (In the cell, the galactose is converted to glucose by enzymes encoded by a gene system specific to galactose catabolism. The glucose is then utilized by constitutively produced enzymes.)

2. *Lactose permease* (also called *M protein*). This protein, found in the *E. coli* cytoplasmic membrane, actively transports lactose into the cell.

3. *Transacetylase.* The function of this enzyme in the cell is poorly understood.

In wild-type *E. coli* growing in a medium containing glucose, only a low concentration of each of these three proteins is produced. For example, only an average of three molecules of β-galactosidase is present in the cell under these conditions. In the presence of lactose, but the absence of glucose, the amount of each enzyme increases coordinately (simultaneously) about a thousand-fold (e.g., to about 3,000 molecules of β-galactosidase), because the three essentially inactive genes are now being actively transcribed. The process is called **coordinate induction.** Allolactose, not lactose, is the inducer molecule directly responsible for the increased production of the three enzymes. (See Figure 19.2.) Furthermore, the mRNAs for the enzymes have a short half-life, so the transcripts must be made continually in order for the enzymes to be produced. When lactose is no longer present, transcription of the three genes is stopped and any mRNAs already present are broken down, so no more of these proteins are made. Existing proteins are diluted out by cell growth and division.

Experimental Evidence for the Regulation of *lac* Genes

Our basic understanding of the organization of the genes, the regulatory sequences involved in lactose utilization, and the control of expression of the *lac* genes of *E. coli* came largely from the genetic experiments of

animation

a **Regulation of Expression of the *lac* Operon Genes**

Figure 19.2

Reactions catalyzed by the enzyme β-galactosidase. Lactose brought into the cell by the permease is converted to glucose and galactose *(top)* or to allolactose *(bottom),* the true inducer for the lactose operon of *E. coli.*

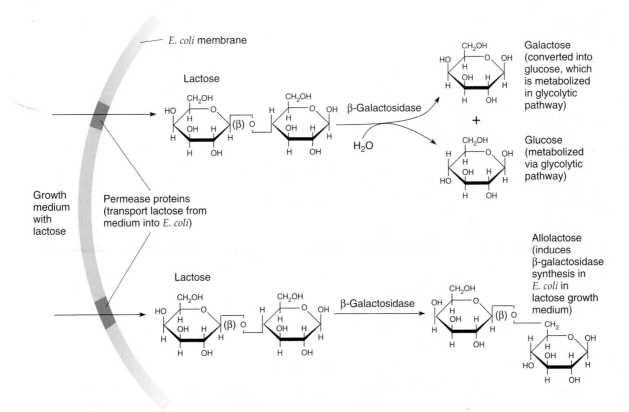

François Jacob and Jacques Monod, for which they shared (along with André Lwoff, for his work on the genetic control of virus synthesis) the 1965 Nobel Prize in Physiology or Medicine. Let us summarize their experiments.

Mutations in the Protein-Coding Genes. Mutations in the protein-coding (structural) genes were obtained after treatment of cells with mutagens (chemicals that induce mutations). The β-galactosidase gene was named *lacZ*, the permease gene *lacY*, and the transacetylase gene *lacA*. The *lacZ⁻*, *lacY⁻*, and *lacA⁻* mutations obtained were used to map the locations of the three genes via standard mapping experiments. The experiments showed that the three genes are tightly linked in the order *lacZ-lacY-lacA*.

For the *lac* genes, a **missense mutation**—a base-pair change in the DNA that causes a change in an mRNA codon, resulting in the substitution of one amino acid for another in a polypeptide (see Chapter 7, p. 136)—affects only the function of the product of the gene with the mutations. It was expected that a nonsense mutation—a base-pair change in DNA that changes an amino-acid coding codon in mRNA to a stop codon—would act similarly. However, nonsense mutations in the *lacZ* gene not only knocked out the function of β-galactosidase, but also knocked out the functions of permease and transacetylase. The *lacY* nonsense mutations resulted in nonfunctional permease and transacetylase, but β-galactosidase activity was unaffected. Finally, *lacA* nonsense mutants lack

transacetylase activity, but have normal β-galactosidase and permease activities. In sum, nonsense mutations in the cluster of three genes involved in lactose utilization have different effects, depending on where they are located within the cluster: The nonsense mutations are said to exhibit *polar effects*, and the phenomenon is called **polarity.** (Nonsense mutations that show polar effects are therefore called *polar mutations.*)

The interpretation of the polar effects of nonsense mutations in the *lac* structural genes is that all three genes are clustered in the genome and are transcribed onto a single mRNA molecule—called a polycistronic mRNA—rather than onto three separate mRNAs. That is, RNA polymerase initiates transcription at a single promoter, and a polycistronic mRNA is synthesized with the gene transcripts in the order 5'-*lacZ⁺*-*lacY⁺*-*lacA⁺*-3'. In translation, a ribosome loads onto the polycistronic mRNA at the 5' end, synthesizes β-galactosidase, then reinitiates translation at the permease sequence, synthesizes permease, then reinitiates at the transacetylase sequence, synthesizes transacetylase, and finally dissociates from the mRNA (Figure 19.3a).

A nonsense mutation in the *lacZ* gene exerts its effect on the translation of a polycistronic mRNA as shown in Figure 19.3b. The ribosome begins to translate the *lacZ* sequence and stops at the premature nonsense codon; a partially completed and therefore nonfunctional β-galactosidase is released. The ribosome continues to

Figure 19.3

Translation of the polycistronic mRNA encoded by *lac* utilization genes in (a) wild-type *E. coli* and (b) a mutant strain with a nonsense mutation in the β-galactosidase (*lacZ*) gene.

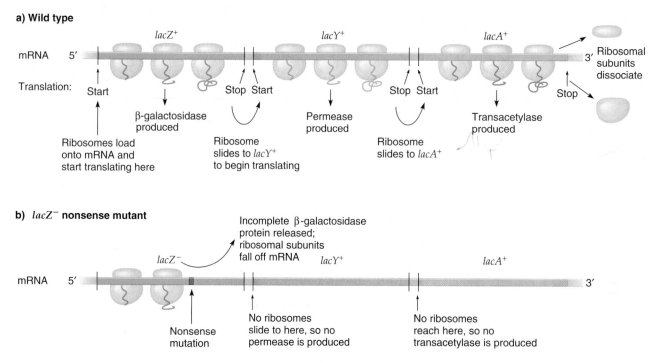

slide along the polycistronic mRNA, but typically dissociates before the start codon for permease because of the greater distance between the premature stop codon and the start codon. Thus, very few, if any, ribosomes translate the downstream permease and transacetylase sequences, so no enzymes or, at most, very few enzymes can be produced. By extension, you can see how a *lacY* (permease) nonsense mutation will affect downstream transacetylase production, but not upstream β-galactosidase translation.

Mutations Affecting the Regulation of Gene Expression. In wild-type *E. coli*, the three gene products are induced coordinately when lactose is present. Jacob and Monod isolated mutants in which all gene products of the operon were synthesized *constitutively*; that is, they were synthesized regardless of whether the inducer was present. The researchers hypothesized that the mutations were regulatory mutations that affected the normal mechanisms controlling the expression of the structural genes for the enzymes. They then identified two classes of constitutive mutations: One class mapped to a small region next to the *lacZ* gene they called the **operator** (*lacO*), and the other class mapped to a gene upstream of the operator that they called the *lacI* gene or Lac **repressor gene**. Figure 19.4 depicts the organization of the *lac* structural gene cluster and the associated regulatory sequences. This complex is the *lac* operon.

Operator Mutations The mutations of the operator were called operator-constitutive, or *lacO^c*, mutations. Through the use of partial diploid strains (*F'* strains in which a few chromosomal genes on an extrachromosomal genetic element called the *F* factor are introduced into a bacterial cell; see Figure 18.6, p. 487), Jacob and Monod were able to define better the role of the operator in regulating expression of the lac genes. One such partial diploid was $\dfrac{F'\ lacO^+\ lacZ^-\ lacY^+}{lacO^c\ lacZ^+\ lacY^-}$. (Both gene sets have a normal promoter, and the *lacA* gene is omitted because it is not important to our discussion.)

One *lac* region in the partial diploid has a normal operator (*lacO^+*), a mutated β-galactosidase gene (*lacZ^-*), and a normal permease gene (*lacY^+*). The other *lac* region has a constitutive operator mutation (*lacO^c*), a normal β-galactosidase gene (*lacZ^+*), and a mutated permease gene (*lacY^-*). This partial diploid was tested for the production of β-galactosidase (from the *lacZ^+* gene) and of permease (from the *lacY^+* gene), both in the presence and in the absence of the inducer.

Jacob and Monod found that active β-galactosidase is synthesized in the absence of the inducer and that permease is synthesized, but is inactive because of the mutation. Only when lactose is added to the culture and the allolactose inducer is produced is active permease synthesized. That is, the *lacZ^+* gene (which is on the same DNA molecule as *lacO^c*) is constitutively *expressed* (meaning that the gene is active in the presence or absence of the inducer), whereas the *lacY^+* gene is under normal inducible control: It is inactive in the absence of the inducer and active in the presence of the inducer. In other words, a *lacO^c* mutation affects only the genes downstream from it on the same DNA molecule. Similarly, the *lacO^+* region controls only *lac* structural genes adjacent to it and has no effect on the genes on the other DNA molecule. A gene or DNA sequence that controls only genes located on the same, contiguous piece of DNA is said to be **cis-dominant.** The *lacO^c* mutation is cis-dominant because the defect affects the adjacent genes only and cannot be overcome by a normal *lacO^+* region elsewhere in the cell. That is, the operator must not encode a diffusible product. If it did, then, in the *lacO^+*/*lacO^c* diploid state, one or the other of the regions would have controlled all the lactose utilization genes, regardless of their location.

lacI Gene Regulatory Mutations. The second class of *lac* constitutive mutants defined the *lacI* gene. Again, the use of partial diploid strains illuminated the normal function of the gene.

Figure 19.4

Organization of the *lac* genes of E. coli and the associated regulatory elements: the operator, promoter, and regulatory gene. The promoter, operator, and adjacent *lac* genes are together called the *lac* operon.

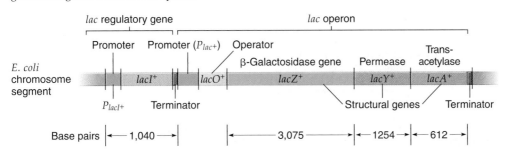

The partial diploid here is $\dfrac{lacI^+\ lacO^+\ lacZ^-\ lacY^+}{lacI^-\ lacO^+\ lacZ^+\ lacY^-}$; both gene sets have normal operators and normal promoters. In the absence of the inducer, no β-galactosidase or permease was produced; both were synthesized in the presence of the inducer. In other words, the expression of both operons was inducible. This means that the $lacI^+$ gene in the cell can overcome the defect of the $lacI^-$ mutation. Since the two $lacI$ genes are located on different DNA molecules (i.e., they are in a trans configuration), the $lacI^+$ gene is said to be **trans-dominant** to the $lacI^-$ gene.

Because the $lacI^+$ gene controls the genes on the other DNA molecule, Jacob and Monod proposed that the $lacI^+$ gene is a repressor gene that produces a **repressor** molecule. No functional repressor molecules are produced in $lacI^-$ mutants. Thus, in a haploid $lacI^-$ bacterial strain, the lac operon is constitutive. In a partial diploid with both a $lacI^+$ and a $lacI^-$, however, the functional Lac repressor molecules produced by the $lacI^+$ gene control the expression of both lac operons in the cell, making both operons inducible.

Promoter Mutations. The promoter for the structural genes (located at the $lacZ$ end of the cluster of lac genes; see Figure 19.4) is also affected by mutations. Most of the known promoter mutants (P_{lac-}) affect all three structural genes. Even in the presence of inducer, the lactose utilization enzymes are not made or are made only at very low rates. Since the promoter is the recognition sequence for RNA polymerase and does not code for any product, the effect of a P mutation is confined to the genes that it controls on the same DNA strand. The P_{lac-} mutations are another example of cis-dominant mutations.

Jacob and Monod's Operon Model for the Regulation of *lac* Genes

On the basis of the results of their studies, Jacob and Monod proposed their now-classic *operon model*. By definition, an **operon** is a *cluster of genes, the expressions of which are regulated together by operator–repressor protein interactions, plus the operator region itself and the promoter.* The promoter was not part of Jacob and Monod's original model; its existence was demonstrated in later studies. The order of the controlling elements and genes in the *lac* operon is promoter-operator-*lacZ-lacY-lacA*, and the regulatory gene *lacI* is located close to the structural genes, just upstream of the promoter. (See Figure 19.4.) The *lacI* gene has its own constitutive promoter and terminator. The promoter for *lacI* is a weak promoter, so few repressor molecules are present in the cell. The Lac repressor protein encoded by *lacI* is made constitutively, but its ability to bind to the operator is affected by the presence of the inducer.

The description that follows of the Jacob–Monod model for regulation of the *lac* operon has been embellished with up-to-date molecular information. Figure 19.5 depicts the state of the *lac* operon in wild-type *E. coli* growing in the absence of lactose. The repressor gene ($lacI^+$) is transcribed constitutively, and the translation of its mRNA produces a 360-amino acid polypeptide. Four of these polypeptides associate together to form a tetramer, the functional Lac repressor protein (Figure 19.6).

The Lac repressor binds to the operator ($lacO^+$). The DNA sequence covered by the repressor overlaps the DNA sequence recognized by RNA polymerase. Therefore, when the repressor is bound to the operator, RNA polymerase cannot bind to the promoter and transcription can occur, the *lac* operon is said to be under

Figure 19.5

Functional state of the *lac* operon in wild-type *E. coli* growing in the absence of lactose.

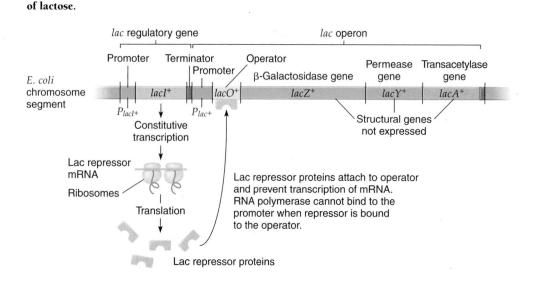

Figure 19.6

Molecular model of the *lac* repressor tetramer. The four monomers are colored green, violet, red, and yellow.

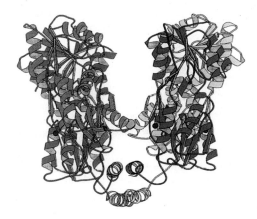

negative control. The low level of gene transcription that produces a few molecules of the enzymes, even in the absence of the inducer, occurs because repressors do not just bind and stay; they bind and dissociate. In the split second after one repressor unbinds and before another binds, an RNA polymerase could initiate transcription of the operon, even in the absence of the inducer.

When wild-type *E. coli* grows in the presence of lactose as the sole carbon source (Figure 19.7), some lactose is converted by β-galactosidase into allolactose (see Figure 19.2). Allolactose binds to the Lac repressor and changes its shape; this is called an *allosteric shift.* As a result, the repressor loses its affinity for the *lac* operator and dissociates from the site. Free repressor proteins are also altered so that they cannot bind to the operator. Thus, allolactose induces production of the *lac* operon enzymes.

With no Lac repressor bound to the operator, RNA polymerase initiates synthesis of a single polycistronic mRNA molecule for the *lacZ⁺*, *lacY⁺*, and *lacA⁺* genes. The polycistronic mRNA for the *lac* operon is translated by a string of ribosomes to produce the three proteins specified by the operon. This efficient mechanism ensures the coordinate production of proteins of related function.

Effect of *lacO^c* Mutations. The *lacO^c* mutations lead to constitutive expression of the *lac* operon genes and are cis-dominant to *lacO⁺* (Figure 19.8). This is because base-pair alterations of the operator DNA sequence make it unrecognizable to the repressor protein. Since the repressor cannot bind, the structural genes physically linked to the *lacO^c* mutation become constitutively expressed.

Effects of *lacI* gene Mutations. The *lacI* mutations map within the repressor structural gene and result in changes to amino acids in the repressor. The repressor's shape is changed, and it can neither recognize nor bind to the operator. As a consequence, in a haploid strain, transcription cannot be prevented, even in the absence of lactose, and constitutive expression of the *lac* operon results (Figure 19.9a).

Figure 19.7

Functional state of the *lac* operon in wild-type *E. coli* growing in the presence of lactose as the sole carbon source.

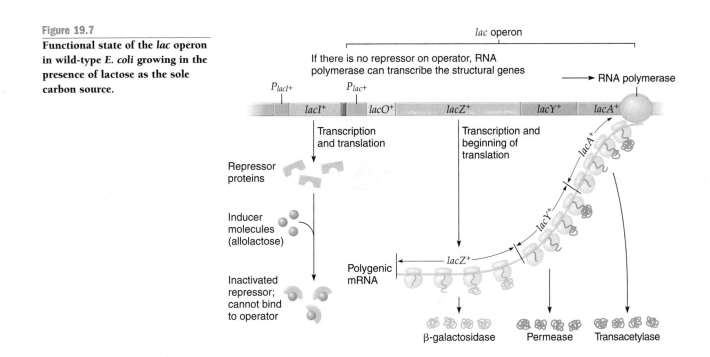

Figure 19.8

Cis-dominant effect of *lacO*ᶜ mutation in a partial diploid strain of *E. coli*. (a) In the absence of the inducer, the *lacO*⁺ operon is turned off, whereas the *lacO*ᶜ operon produces functional β-galactosidase from the *lacZ*⁺ gene and nonfunctional permease molecules from the *lacY*⁻ gene with a missense mutation. (b) In the presence of the inducer, the functional β-galactosidase and defective permease are produced from the *lacO*ᶜ operon, whereas the *lacO*⁺ operon produces nonfunctional β-galactosidase from the *lacZ*⁻ gene (a missense mutation) and functional permease from the *lacY*⁺ gene. Between the two operons in the cell, functional β-galactosidase and permease are produced.

a) Partial diploid in the absence of inducer

b) Partial diploid in the presence of inducer

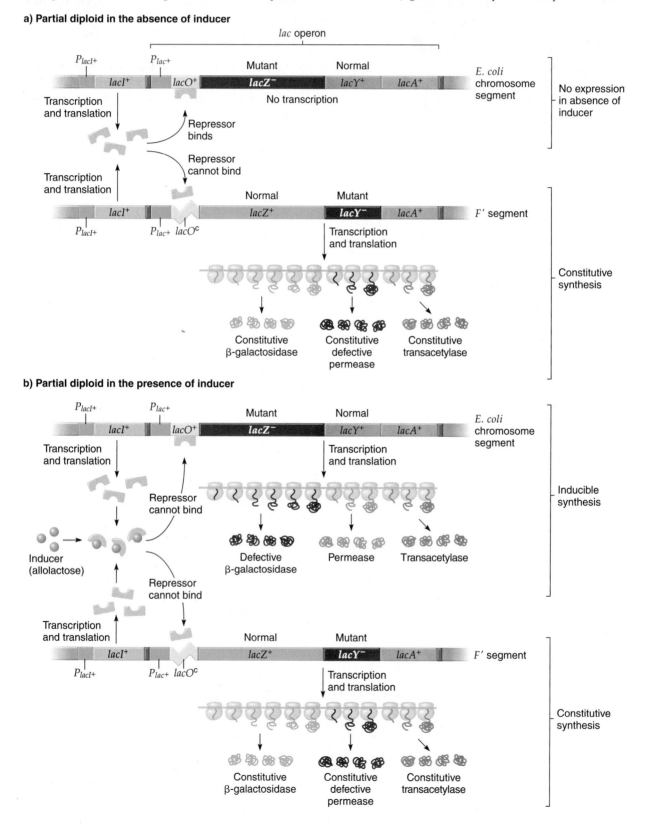

The dominance of the *lacI⁺* (wild-type) gene over *lacI⁻* mutants is illustrated for the partial diploid $\frac{lacI^+\ lacO^+\ lacZ^-\ lacY^+}{lacI^-\ lacO^+\ lacZ^+\ lacY^-}$ described earlier. In the absence of the inducer (Figure 19.9b), the defective *lacI⁻* repressor cannot bind to either normal operator (*lacO⁺*) in the cell, but sufficient normal Lac repressors, produced from the *lacI⁺* gene, are present, and they bind to the two operators and block the transcription of both operons. When the inducer is present (Figure 19.9c), the wild-type repressors are inactivated, so both operons are transcribed. One produces a defective β-galactosidase and a normal permease, and the other produces a normal β-galactosidase and a defective permease; between them,

active β-galactosidase and permease are produced. Thus, in *lacI⁺/lacI⁻* partial diploids, both operons present in the cell are under inducible control.

Other classes of *lacI* gene mutants have been identified since Jacob and Monod studied the *lacI⁻* class of mutants. One of these classes, the *lacIˢ* (*superrepressor*) mutants, shows no production of *lac* enzymes in the presence or absence of lactose. In partial diploids with a *lacI⁺/lacIˢ* genotype, the *lacIˢ* allele is trans-dominant, affecting both operon copies (Figure 19.10). In this situation, the mutant repressor gene produces a superrepressor protein that can bind to the operator, but cannot recognize the inducer allolactose. Therefore, the mutant superrepressors bind to the operators even in the presence of the

Figure 19.9

Effects of a *lacI⁻* mutation. (a) Effect on expression of *lac* operon in a haploid cell, where mutant, inactive Lac repressor molecules that cannot bind to the operator *lacO⁺* are produced; the structural genes are transcribed constitutively. Effect in a partial diploid strain *lacI⁺ lacO⁺ lacZ⁻ lacY⁺ /lacI⁻ lacO⁺ lacZ⁺ lacY⁻* in (b) the absence or (c) the presence of inducer. (The *lacZ* and *lacY* mutations are missense mutations.)

a) Haploid strain (in presence or absence of inducer)

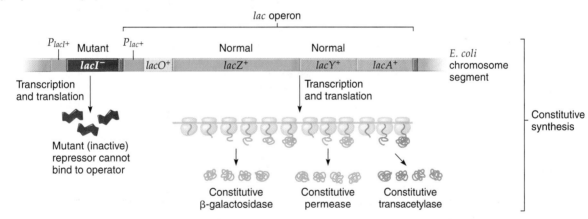

b) Partial diploid in the absence of inducer

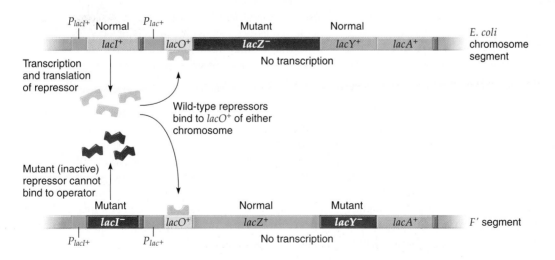

Figure 19.9 (continued)

c) Partial diploid in the presence of inducer

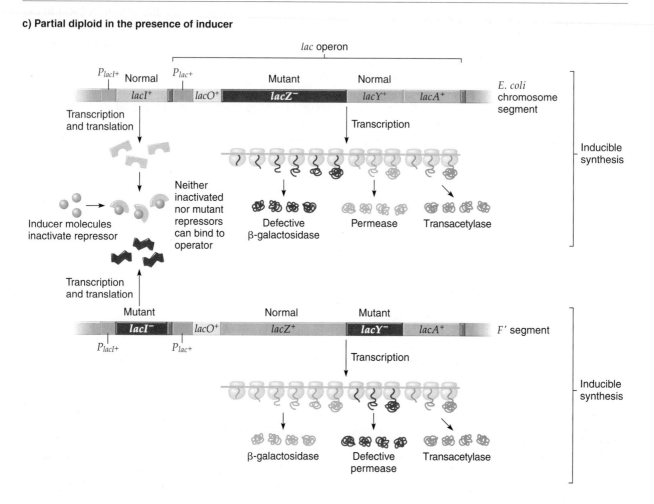

inducer, and the operons can never be transcribed. The presence of normal repressors in the cell has no effect, because, once a $lacI^S$ repressor is on the operator, the repressor cannot be induced to fall off. Low levels of transcription occur because the superrepressor is not permanently (covalently) bound to the operator. Cells with a $lacI^S$ mutation cannot use lactose as a carbon source.

A third type of repressor gene mutation is the $lacI^{-d}$ (dominance) class. These missense mutations cluster toward the 5′ end of the $lacI$ gene. In haploid cells, the $lacI^{-d}$ mutants have a constitutive phenotype like the other $lacI^-$ mutants; the lac enzymes are made in the presence or absence of lactose. Unlike the lac mutations, the $lacI^{-d}$ mutations are trans-dominant to $lacI^+$ in $lacI^{-d}/lacI^+$ partial diploids, so lac enzymes are produced constitutively even in the presence of the normal repressor.

The dominance of $lacI^{-d}$ mutants is explained as follows: The Lac repressor protein is a tetramer consisting of four identical polypeptides. In $lacI^{-d}$ mutants, the repressor subunits do not combine normally, so no complete repressor is formed and no operator-specific binding is possible. The $lacI^{-d}/lacI^+$ diploids have a mixture of nor-

mal and mutant polypeptides, which combine randomly to form repressor tetramers. There are only about a dozen repressor molecules in the cell. The presence of one or more defective polypeptide subunits in the repressor is enough to block normal binding to the operator, so there is a good chance that no normal repressor proteins will be produced, because there are so few molecules per cell. As a result of the absence or near absence of complete, functional repressors, a constitutive enzyme phenotype results.

Finally, some mutations in the repressor gene promoter affect the expression of the repressor gene. Earlier, we mentioned that the extent of transcription of a gene is a function of the affinity of that gene's promoter for RNA polymerase molecules. Since few repressor molecules are synthesized in wild-type E. coli cells, the repressor gene promoter must be of low affinity (i.e., it is a weak promoter). Base-pair mutations have been found that decrease and that increase transcription rates. For example, $lacI^Q$ and $lacI^{SQ}$ mutants (where Q stands for "quantity" and SQ stands for "super quantity") result in an increase in the rate of transcription of the repressor gene, with the $lacI^{SQ}$ giving the greater increase. These

Figure 19.10

Dominant effect of *lacI*[S] **mutation over wild-type** *lacI*[+] **in a**
lacI[+] *lacO*[+] *lacZ*[+] *lacY*[+] *lacA*[+]/*lacI*[S] *lacO*[+] *lacZ*[+] *lacY*[+] *lacA*[+] **partial diploid**
cell growing in the presence of lactose.

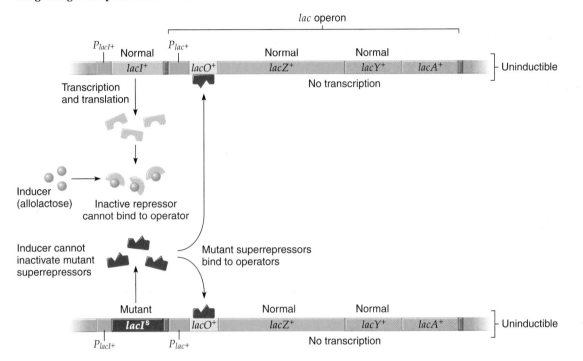

mutants were useful historically because they produce large numbers of repressor molecules, which facilitated their isolation and purification and, consequently, the determination of the amino-acid sequence of the repressor polypeptide. Since *lacI*[Q] and *lacI*[SQ] mutants produce more Lac repressor molecules than the wild type produces, these mutants reduce the efficiency of induction of the *lac* operon. Note that *lacI*[Q] and *lacI*[SQ] can be induced at high lactose concentrations.

The mutants of the *lacI* gene point out the different functions of the Lac repressor. That is, the repressor is involved in three recognition interactions: (1) binding of the repressor to the operator region; (2) binding of the inducer to the repressor; and (3) binding of individual repressor polypeptides to each other to form the active repressor tetramer.

Positive Control of the *lac* Operon

The Lac repressor protein exerts a negative effect on the expression of the *lac* operon by blocking RNA polymerase's binding to the promoter if the inducer is absent. Several years after Jacob and Monod proposed their operon model, researchers found a positive control system that also regulates the *lac* operon—a system that functions to turn on the expression of the

animation

a **Positive Control of the *lac* Operon**

operon. This system ensures that the *lac* operon will be expressed at high levels only if lactose is the sole carbon source *and not if glucose is present as well.* Glucose is a preferred carbon source because it can be used directly by the glycolytic pathway to produce "energy" for the cell. First, lactose and other sugars are converted to glucose in reactions consuming energy; then the glucose enters the glycolytic pathway. Therefore, more energy can be obtained for the cell from glucose than from other sugar sources.

Figure 19.11 shows the positive regulation of the *lac* operon if lactose is present and glucose is absent. First, a protein called **catabolite activator protein** (CAP) binds with **cAMP** (**cyclic AMP,** or cyclic adenosine 3′,5′-monophosphate; see Figure 19.12) to form a CAP-cAMP complex. This complex is the positive-regulator molecule. The CAP protein is a dimer of two identical polypeptides. Next, the CAP-cAMP complex binds to the *CAP site,* which is upstream of the site at which RNA polymerase binds to the promoter. CAP then recruits RNA polymerase to the promoter, and transcription is initiated.

When glucose is in the medium along with lactose, the glucose is used preferentially because **catabolite repression** (also called the **glucose effect**) occurs. In catabolite repression, the *lac* operon is expressed at only very low levels even though lactose is present in the medium. This occurs because glucose causes the amount of cAMP in the cell to be greatly reduced. As a result,

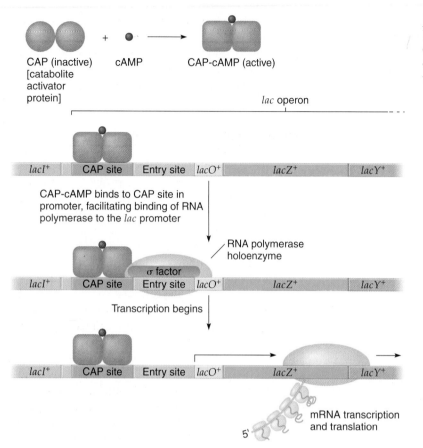

Figure 19.11

Role of cyclic AMP (cAMP) in the functioning of glucose-sensitive operons such as the *lac* operon of *E. coli*. Shown is the condition in which lactose is present and glucose is absent.

insufficient CAP-cAMP complex is available to recruit RNA polymerase to the *lac* promoter, and transcription is lowered significantly, even though repressors are removed from the operator by the presence of allolactose. In other words, RNA polymerase cannot bind to the promoter without the aid of the CAP-cAMP complex. That cAMP plays a crucial role in catabolite repression was shown by a number of experiments, including one in which transcription of the *lac* operon was restored by the addition of cAMP to the cell, even though glucose was still present.

The model is that catabolite repression acts on adenylate cyclase, the enzyme that makes cAMP. (See Figure 19.12.) In *E. coli*, adenylate cyclase is activated by the phosphorylated form of an enzyme called III^{Glc}. When glucose is transported across the cell membrane into the cell, it triggers a series of events that includes the dephosphorylation of III^{Glc}. As a result, adenylate cyclase is inactivated and no new cAMP is produced. This, along with the breakdown of cAMP by phosphodiesterase, reduces the level of cAMP in the cell.

Catabolite repression occurs in the same way in a number of other bacterial operons related to the catabolism of sugars other than glucose. These operons all have in common a CAP site in their promoters to which a specific CAP-cAMP complex binds to facilitate RNA polymerase binding.

Molecular Details of *lac* Operon Regulation

From DNA- and RNA-sequencing experiments, we know the nucleotide sequences of the significant *lac* operon regulatory sequences. One general approach to obtaining this information has been to purify the protein known to bind to a regulatory sequence and to let it bind to isolated *lac* operon DNA in vitro. For example, if the repressor is bound to the *lac* operator, it will protect that region of the operon from deoxyribonuclease digestion. If DNase is allowed to digest the rest of the DNA, the operator sequence can be isolated, cloned by using recombinant DNA technology, and sequenced.

Promoter Region of the *lac* Repressor Gene (*lacI*).

Figure 19.13 shows the nucleotide pair sequence for the *lacI* gene promoter region, the sequence for the 5′ end of the repressor mRNA, and the first few amino acids of the repressor protein itself. The nucleotide sequence of the repressor mRNA can be aligned with this promoter sequence, with its start approximately in the middle. As with all gene transcripts, translation does not start right at the end of the mRNA molecule. The translation start sequence (ribosome binding site) is a Shine–Dalgarno sequence (AGGG) at 12 to 9 bases upstream from the start codon (Chapter 16). In this

Figure 19.12

Structure of cyclic AMP (cAMP, or cyclic adenosine 3′,5′-monophosphate). cAMP is synthesized from ATP in a reaction catalyzed by adenylate cyclase and is broken down in a reaction catalyzed by phosphodiesterase.

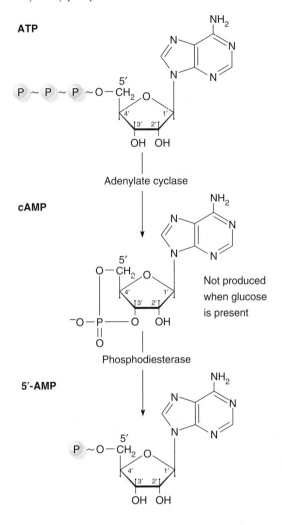

unusual case, the start codon is GUG rather than AUG, at nucleotides 27 to 29 from the 5′ end of the messenger. The figure also shows the single base-pair change found for a particular *lacI*^Q mutant; this change, from CG to TA, brings about a tenfold increase in repressor production.

lac Operon Regulatory Sequences. Figure 19.14 shows the nucleotide pair sequence of the *lac* operon regulatory sequences. The orientation of this sequence was put together from several different pieces of information. First, the amino-acid sequences of the repressor protein and of β-galactosidase were completely known, and that information made it possible to identify the coding regions of the *lacI* gene and of the *lacZ*⁺ gene. Then the other regions were identified on the basis of "protection" experiments of the kind described previously. Here, CAP-cAMP complex, RNA polymerase, and repressor protein were used separately to bind to the DNA, and DNase-resistant regions were then sequenced.

The beginning of the promoter region is defined as position −84 in the figure (i.e., 84 base pairs upstream from the mRNA initiation site), immediately next to the stop codon for the *lacI* gene. The consensus sequence matches for the CAP–cAMP binding site are nucleotide pairs −54 to −58 and −65 to −69, and the DNA covered by RNA polymerase spans nucleotide pairs −44 to −8, including −10 and −35 consensus sequence matches. (See Figure 19.11.) Together, the region from −84 to −8, which includes the CAP protein and the RNA polymerase interaction sites (including a Pribnow box), essentially defines the *lac* operon promoter region.

Adjacent to the promoter region is the operator. The region protected by the Lac repressor protein is the area containing nucleotide pairs −3 to +21. When the Lac repressor is bound to the operator, RNA polymerase cannot bind to the promoter.

Figure 19.13

Base-pair sequences of the lac operon lacI⁺ gene promoter (P_{lac}⁺) and of the 5′ end of the repressor mRNA. Also shown is the amino-acid sequence of the first part of the repressor protein itself. Note that GUG is the initiation codon for methionine in this case.

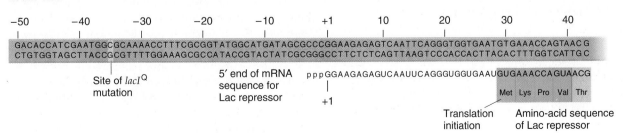

Figure 19.14

Base-pair sequence of the promoter and operator for the lactose operon of _E. coli._
Also shown are locations of some known _lacO^c_ mutations (indicated by arrows).

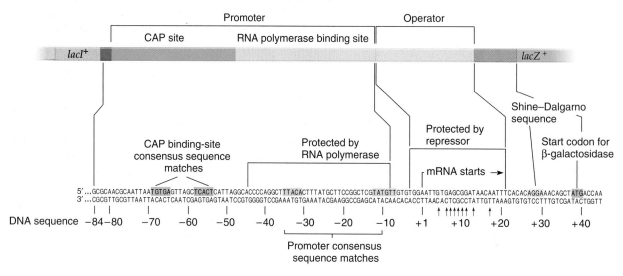

The β-galactosidase mRNA has a _leader region_ before the start codon is encountered. The actual start of the mRNA here is nucleotide pair +1 in Figure 19.14, which is very close to the beginning of the repressor binding site. Transcription of the _lac_ operon includes a large proportion of the operator region, in addition to the protein-coding genes themselves. The AUG start codon for β-galactosidase, which defines the beginning of the _lacZ_ coding sequence, is at nucleotide pairs +39 to +41. Thus, the first 38 bases of the _lac_ mRNA are not translated.

Figure 19.14 also shows the sites of base-pair substitutions that have been identified for some of the _lacO^c_ mutations studied. In each case, a single base-pair change is responsible for the altered regulatory control of the _lac_ operon.

In conclusion, the _lac_ operon has proved to be a model system for understanding gene regulation in prokaryotic organisms. Jacob and Monod's original work on this system had a great impact on further studies. As the first molecular model for the regulation of gene expression in any organism, it sparked numerous studies in both prokaryotes and eukaryotes to see whether operons were generally present. We now know that operons are prevalent in bacteria and bacteriophages, but they are very rarely encountered in eukaryotes.

 Your job is to determine the location and effect of a mutation in strains of _E. coli_ in the iActivity _Mutations and Lactose Metabolism_ on your CD-ROM.

KEYNOTE

Studies of the synthesis of the lactose-utilizing enzymes of _E. coli_ generated a model that is the basis for the regulation of gene expression in a large number of bacterial and bacteriophage systems. In the lactose system, the addition of lactose to the cells brings about a rapid synthesis of three enzymes. The genes for these enzymes are contiguous on the _E. coli_ chromosome and are adjacent to two regulatory sequences: a promoter and an operator. The promoter, operator, and genes constitute an operon, which is transcribed as a single unit. In the absence of lactose, the operon is turned off by a repressor.

A positive control system also regulates the _lac_ operon. That is, CAP-cAMP binds to the promoter, and this facilitates the binding of RNA polymerase to the promoter. If glucose is present, however, no CAP-cAMP is produced, so RNA polymerase cannot bind and the _lac_ genes are not transcribed.

The _trp_ Operon of _E. coli_

E. coli has certain operons and other gene systems that enable it to manufacture any amino acid that is lacking in the medium in which it is placed, so that it can grow and reproduce. When an amino acid is present in the growth medium, though, the genes encoding the enzymes for that amino acid's biosynthetic pathway are turned off. Unlike the _lac_ operon, wherein gene activity is induced when a chemical (lactose) is added to the medium, in this case gene activity is repressed when a chemical (an amino acid) is added. We call amino-acid biosynthesis operons controlled in this way **repressible operons.** In general, operons for anabolic (biosynthetic) pathways are repressed (turned off) when the end product is readily available. One repressible operon in *E. coli* that has been extensively studied is the operon for the biosynthesis of the amino acid tryptophan (Trp).

Gene Organization of the Tryptophan Biosynthesis Genes

Figure 19.15 shows the organization of the regulatory sequences and of the genes that code for the tryptophan biosynthetic enzymes and how they relate to the biosynthetic steps. Much of the work we will discuss is that of Charles Yanofsky and his collaborators.

Five structural genes (*A–E*) occur in the *trp* operon. The promoter and operator regions are upstream from the *trpE* gene. Between the promoter–operator region and

trpE is a short region called *trpL*, the leader region. Within *trpL*, close to *trpE*, is an *attenuator site* (*att*) that plays an important role in the regulation of the *trp* operon.

The entire *trp* operon is approximately 7,000 base pairs long. Transcription of the operon results in the production of a polycistronic mRNA for the five structural genes.

Regulation of the *trp* Operon

Two regulatory mechanisms are involved in controlling the expression of the *trp* operon. One mechanism uses a repressor–operator interaction, and the other determines whether initiated transcripts include the structural genes or are terminated before those genes are reached.

Expression of the *trp* Operon in the Presence of Tryptophan. The regulatory gene for the *trp* operon is *trpR*, located some distance from the operon (and therefore not

Figure 19.15

Organization of regulatory sequences and the structural genes of the *E. coli trp* operon. Also shown are the steps catalyzed by the products of the structural genes *trpA, trpB, trpC, trpD,* and *trpE*.

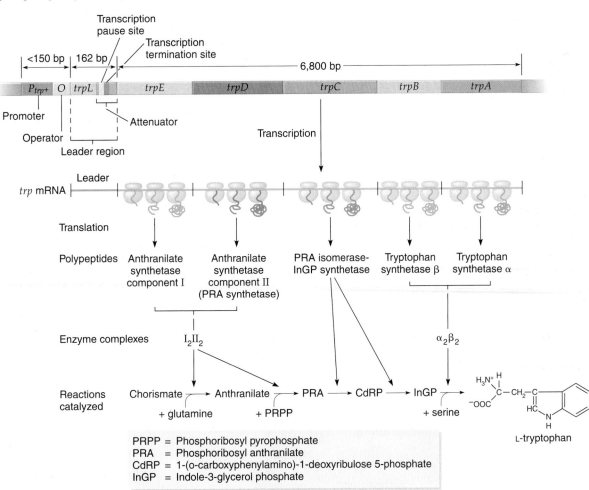

PRPP = Phosphoribosyl pyrophosphate
PRA = Phosphoribosyl anthranilate
CdRP = 1-(o-carboxyphenylamino)-1-deoxyribulose 5-phosphate
InGP = Indole-3-glycerol phosphate

shown in Figure 19.15). The product of *trpR* is an **aporepressor protein,** which is basically an inactive repressor that alone cannot bind to the operator. When tryptophan is abundant within the cell, it interacts with the aporepressor and converts it to an active Trp repressor. (Tryptophan is an example of an effector molecule, just as allolactose is the effector molecule for the *lac* operon.) The active Trp repressor binds to the operator and prevents the initiation of transcription of the *trp* operon protein-coding genes by RNA polymerase. As a result, the tryptophan biosynthesis enzymes are not produced. By repression, transcription of the *trp* operon can be reduced about seventy-fold.

Expression of the *trp* Operon in the Presence of Low Concentrations of Tryptophan. The second regulatory mechanism is involved in the expression of the *trp* operon under conditions of tryptophan starvation or tryptophan limitation. Under severe tryptophan starvation, the *trp* genes are expressed maximally; under less severe starvation conditions, the *trp* genes are expressed at less than maximal levels. This is accomplished by a mechanism that controls the ratio of full-length transcripts that include the five *trp* structural genes to short, 140-bp transcripts that have terminated at the attenuator site within the *trpL* region. (See Figure 19.15.) The short transcripts are terminated by a process called **attenuation.** The proportion of the transcripts that include the structural genes is inversely related to the amount of tryptophan in the cell; the more tryptophan there is, the greater is the proportion of short transcripts. Attenuation can reduce transcription of the *trp* operon by a factor of 8 to 10. Thus, repression and attenuation together can regulate the transcription of the *trp* operon by a factor of about 560 to 700.

Molecular Model for Attenuation. The mRNA transcript of the leader region includes a sequence that can be translated to produce a short polypeptide. Just before the stop codon of the transcript are two adjacent codons for tryptophan that play an important role in attenuation.

animation

a **Attenuation in the *trp* Operon of E. coli**

There are four regions of the leader peptide mRNA that can fold and form secondary structures by complementary base pairing (Figure 19.16). The pairing of regions 1 and 2 results in a transcription *pause signal,* that of 3 and 4 is a *termination of transcription signal,* and the pairing of 2 and 3 is an *antitermination signal* for transcription to continue.

Crucial to the attenuation model is the fact that transcription and translation are tightly coupled in prokaryotes, made possible by the absence of a nuclear envelope and the lack of processing of mRNA transcripts. In the *trp* regulatory system, this coupling of transcription and translation is brought about by a pause of the RNA polymerase caused by the pairing of RNA regions 1 and 2 just after they have been synthesized. (See Figure 19.16.) The pause lasts long enough for the ribosome to load onto the mRNA and to begin translating the leader peptide so that translation of the leader mRNA transcript occurs just behind the RNA polymerase.

As coupled transcription and translation continues, the position of the ribosome on the leader transcript plays an important role in the regulation of transcription termination at the attenuator. If the cells are starved for tryptophan, the amount of Trp-tRNA molecules (charged tryptophanyl-tRNA) drops dramatically, since very few tryptophan molecules are available for the aminoacylation of the tRNA. A ribosome translating

Organization of region:

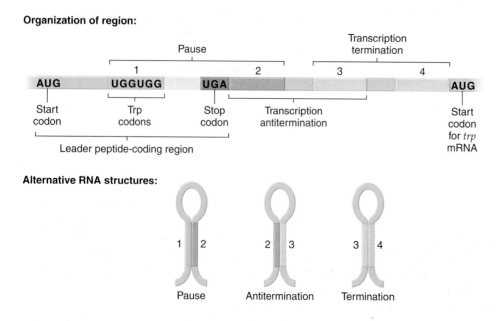

Figure 19.16

Four regions of the *trp* operon leader mRNA and the alternative secondary structures they can form by complementary base pairing.

Figure 19.17

Models for attenuation in the *trp* operon of E. coli. The light-blue structures are ribosomes that are translating the leader transcript. (a) Tryptophan-starved cells. (b) Non-tryptophan-starved cells.

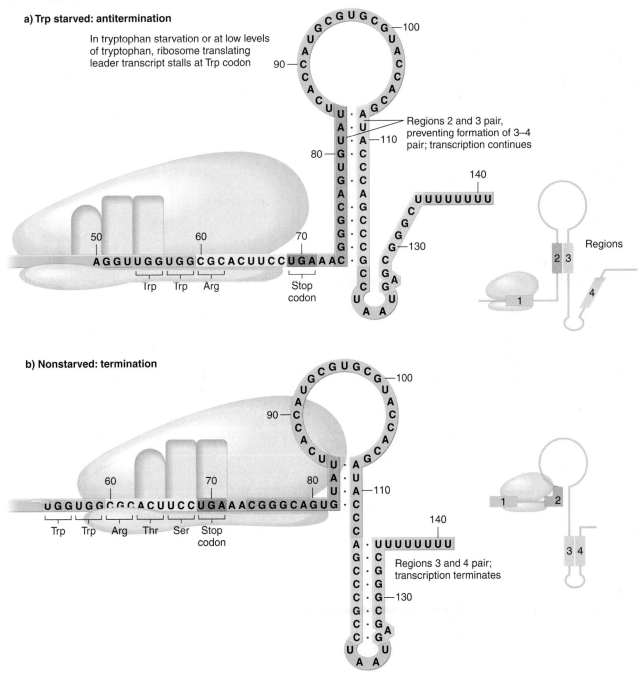

a) Trp starved: antitermination

In tryptophan starvation or at low levels of tryptophan, ribosome translating leader transcript stalls at Trp codon

Regions 2 and 3 pair, preventing formation of 3–4 pair; transcription continues

Trp   Trp   Arg

Stop codon

Regions

b) Nonstarved: termination

Trp   Trp   Arg   Thr   Ser   Stop codon

Regions 3 and 4 pair; transcription terminates

the leader transcript stalls at the tandem Trp codons in region 1 because the next specified amino acid in the peptide is in short supply; the leader peptide cannot be completed (Figure 19.17a). Since the ribosome now "covers" region 1 of the attenuator region, the 1:2 pairing cannot happen, as region 1 is no longer available. However, RNA region 2 will pair with RNA region 3 once region 3 is synthesized. Because region 3 is paired with region 2, region 3 cannot pair with region 4 when it is synthesized. The 2:3 pairing is an antitermination signal, since the termination signal of 3 paired with 4 does not form, thereby allowing RNA polymerase to continue past the attenuator and transcribe the structural genes.

If, instead, enough tryptophan is present that the ribosome can translate the Trp codons (Figure 19.17b), then the ribosome continues to the stop codon for the leader peptide. Since the ribosome is then covering part of RNA region 2, that region is unable to pair with region 3, and region 3 is then able to pair with region 4 when it is transcribed. The bonding of region 3 with region 4 is a termination signal. The 3:4 structure is called the *attenuator*. The key signal for attenuation is the concentration of Trp-tRNA in the cell because that determines how far the ribosome gets on the leader transcript, either to the Trp codons or to the stop codon.

Genetic evidence for the attenuation model has been obtained through the study of mutants. One type of mutant shows less efficient transcription termination at the attenuator, increasing structural gene expression. The mutations involved are single base-pair changes leading to the base changes in the leader transcript shown in Figure 19.18. In each case, the change is in the regions of 3:4 pairing; each causes a disruption of the pairing so that the structure is less stable. In the less stable state, the structure is less able to prevent transcription from proceeding into the structural genes.

Further direct evidence for the attenuation model came from DNA manipulations in which the DNA sequences for the two Trp codons were changed to encode another amino acid. In those mutant strains, attenuation was not seen in response to changing levels of tryptophan, but it was seen in response to changing levels of the amino acid now specified by the codons.

Attenuation is involved in the genetic regulation of a number of other amino-acid biosynthetic operons of *E. coli* and *Salmonella typhimurium*. In every case, there is a leader sequence with two or more codons for the particular amino acid, the synthesis of which is controlled by the enzymes encoded by the operon (Figure 19.19). For example, the *his* operon of *E. coli* has a string of seven histidines in the leader peptide, and 7 of the 15 amino

Figure 19.18

In the *trpL* region, mutation sites that show less efficient transcription at the attenuator site. The mutations map to DNA regions that correspond to regions 3 and 4 in the RNA.

Part of leader transcript

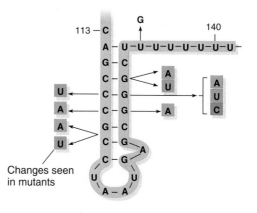

acids in the leader for the *pheA* operon of *E. coli* are phenylalanine. Attenuation has also been shown to regulate a number of genes not involved with amino-acid biosynthesis, such as the rRNA operons (*rrn*) and the *ampC* gene of *E. coli* (for resistance to ampicillin).

KEYNOTE

Regulation of the tryptophan (*trp*) operon of *E. coli* is at the level of initiating or completing a transcript of the operon. This is accomplished through a repressor–operator system, which responds to free tryptophan levels, and through attenuation at a second regulatory sequence called an attenuator, which responds to Trp-tRNA levels. The attenuator is located in the leader region between the operator region and the first *trp* structural gene. The attenuator acts to

Figure 19.19

Predicted amino-acid sequences of the leader peptides of a number of attenuator-controlled bacterial operons. Shown are the peptides for the *pheA*, *his*, *leu*, *thr*, and *ilv* (isoleucine and valine) operons of *E. coli* or *Salmonella typhimurium*. The amino acids that regulate the respective operons are highlighted in yellow.

pheA: Met – Lys – His – Ile – Pro – Phe – Phe – Phe – Ala – Phe – Phe – Phe – Thr – Phe – Pro – –

his: Met – Thr – Arg – Val – Gln – Phe – Lys – His – His – His – His – His – His – His – Pro – Asp – –

leu: Met – Ser – His – Ile – Val – Arg – Phe – Thr – Gly – Leu – Leu – Leu – Leu – Asn – Ala – Phe –
          Ile – Val – Arg – Gly – Arg – Pro – Val – Gly – Ile – Gln – His – –

thr: Met – Lys – Arg – Ile – Ser – Thr – Thr – Ile – Thr – Thr – Thr – Ile – Thr – Ile – Thr – Thr –
          Gly – Asn – Gly – Ala – Gly – –

ilv: Met – Thr – Ala – Leu – Leu – Arg – Val – Ile – Ser – Leu – Val – Val – Ile – Ser – Val – Val –
          Val – Ile – Ile – Ile – Pro – Pro – Cys – Gly – Ala – Ala – Leu – Gly – Arg – Gly – Lys – Ala – –

terminate transcription, depending on the concentration of tryptophan. In the presence of large amounts of tryptophan, attenuation is highly effective; that is, enough Trp-tRNA is present so that the ribosome can move past the attenuator and allow the leader transcript to form a secondary structure that causes transcription to be blocked. In the absence of tryptophan or at low amounts of the amino acid, the ribosomes stall at the attenuator and the leader transcript forms a secondary structure that permits transcription to continue.

Regulation of Gene Expression in Phage Lambda

Bacteriophages exist by invading and manipulating bacterial cells. Many or all of the essential components for phage reproduction are provided by the bacterial host cell, and the use of those components is controlled by the products of phage genes. Most genes of a phage, then, code for products that control the life cycle and the production of progeny phage particles. Much is known about gene regulation in a number of bacteriophages. In this section, we discuss the regulation of gene expression as it relates to the lytic cycle and lysogeny in bacteriophage lambda (λ). (Recall from Chapter 18, p. 492, that, in the lytic cycle, the phage takes over the bacterium and directs its growth and reproductive abilities so that it expresses the phage's genes and produces progeny phages. Lysogeny involves the insertion of a temperate phage chromosome into a bacterial chromosome, with the former replicating whenever the latter does. In this state, the phage genome is repressed and is said to be in the *prophage* state.)

Early Transcription Events

Figure 19.20 shows the genetic map of λ. The mature λ chromosome is linear and has complementary "sticky"

Figure 19.20

A map of phage λ, showing the major genes. (Promoters discussed in text: P_L = promoter for leftward transcription of the left early operon, P_R = promoter for rightward transcription of the right early operon, P_{RE} = promoter for repressor establishment, and P_{RM} = promoter for repressor maintenance.)

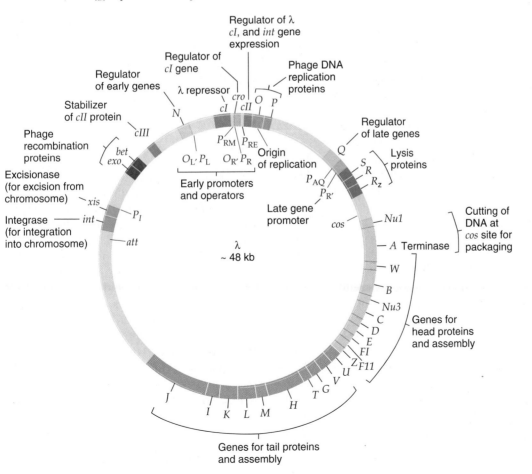

ends. Once free in the host cell, the λ chromosome circularizes, so we show the genetic map in a circular form. Recall that λ is a temperate phage (see Figure 18.12, p. 493), so when it infects a bacterial cell, the phage has a choice of whether to enter the lytic pathway (when progeny phages are assembled and released from the cell) or the lysogenic pathway (when the λ chromosome integrates into the chromosome and no progeny phages are produced). The regulatory system involved in this choice is an excellent model for a genetic switch and, as such, has contributed to our thinking about how genetic switches might operate in eukaryotic systems.

The choice between the lytic and lysogenic pathways occurs soon after λ infects the cell and its genome circularizes. The choice involves a sophisticated *genetic switch*. First, transcription begins at promoters P_L and P_R (Figure 19.21, part 1). Promoter P_L is for leftward transcription of the left early operon, and promoter P_R is for rightward transcription of the right early operon.

The first gene to be transcribed from P_R is *cro* (control of repressor and other), the product of which is the Cro protein. This protein plays an important role in setting the genetic switch to the lytic pathway. The first gene to be transcribed from P_L is N. The resulting N protein is a transcription *antiterminator* that allows RNA synthesis to proceed past certain transcription terminators, in this case leftward of N and rightward of *cro*, thereby including all the early genes (Figure 19.21, part 2). Genes transcribed as a result of the action of the N protein are *cII, O, P,* and *Q*. Gene *cII* encodes protein cII, which can turn on gene *cI* (which encodes the λ repressor) and gene *int* (which encodes the integrase required for integrating the lambda chromosome into the host chromosome during the lysogenic pathway). However, cII protein performs this function only when the phage follows the lysogenic pathway. Genes *O* and *P* encode two DNA replication proteins, and gene *Q* encodes a protein needed to turn on late genes for lysis and phage particle proteins. The Q protein is another antiterminator, permitting transcription to continue into the late genes involved in the lytic pathway. However, only when the switch is set to the lytic pathway and transcription continues from P_R for a sufficient time does enough Q protein accumulate to function effectively.

The Lysogenic Pathway

After the early transcription events, either the lysogenic or lytic pathway is followed (Figure 19.21, part 3). The switch is set for the lysogenic pathway as follows:

The establishment of lysogeny requires the protein products of the *cII* (right early operon) and *cIII* (left

early operon; see Figure 19.20) genes. The cII protein (stabilized by cIII protein) activates transcription of the *cI* gene (located between the P_L and P_R promoters; see Figure 19.21, part 4a) leftward from a promoter called P_{RE} (promoter for repressor establishment). The product of the *cI* gene, the λ repressor, binds to two operator regions, O_L and O_R (see Figure 19.20 and Figure 19.21, part 5a), whose sequences overlap the P_L and P_R promoters, respectively. The binding of the λ repressor prevents the further transcription by RNA polymerase of the early operons controlled by P_L and P_R. As a result, transcription of the N and *cro* genes is blocked, and because the two proteins specified by these genes are unstable, the concentrations of those two proteins in the cell drop dramatically. Furthermore, a repressor bound to O_R stimulates the synthesis of more repressor mRNA from a different promoter, P_{RM} (promoter for repressor maintenance), thereby maintaining repressor concentrations in the cell. (See Figure 19.21, part 5a.) Thus, if enough λ repressors are present, lysogeny is established by the binding of the repressor to operators O_L and O_R, followed by the integration of λ DNA catalyzed by integrase, which is the product of the cII-regulated promoter P_I. As the concentration of cII drops, P_I transcription shuts off, leaving P_{RM} as the only active promoter.

In sum, the lysogenic pathway is favored when enough λ repressor is made so that early promoters are turned off, thereby repressing all the genes needed for the lytic pathway. One important lytic pathway gene that is repressed is *Q*; the Q protein is a positive regulatory protein required for the production of phage coat proteins and lysis proteins. (See the next section.)

The Lytic Pathway

Let us consider the induction of the lytic pathway caused by ultraviolet light irradiation. Inducers such as ultraviolet light typically damage DNA, and this somehow causes a change in the function of the bacterial protein RecA (the product of the *recA* gene). Normally RecA functions in DNA recombination, but when DNA is damaged, RecA stimulates the λ repressor polypeptides to cleave themselves in two and therefore become inactivated. The resulting absence of repressor at O_R allows RNA polymerase to bind at P_R, and the *cro* gene is then further transcribed. The Cro protein that is produced then acts to decrease RNA synthesis from P_L and P_R, and this reduces the synthesis of the cII protein, the regulator of λ repressor synthesis, and blocks the synthesis of λ repressor mRNA from P_{RM} (Figure 19.21, part 4b). At the same time, transcription of the right early operon genes from P_R is decreased, but enough Q proteins are accumulated to set the genetic switch for

Figure 19.21

Expression of λ genes after infection of *E. coli*, and the transcriptional events that occur when either the lysogenic or lytic pathway is followed.

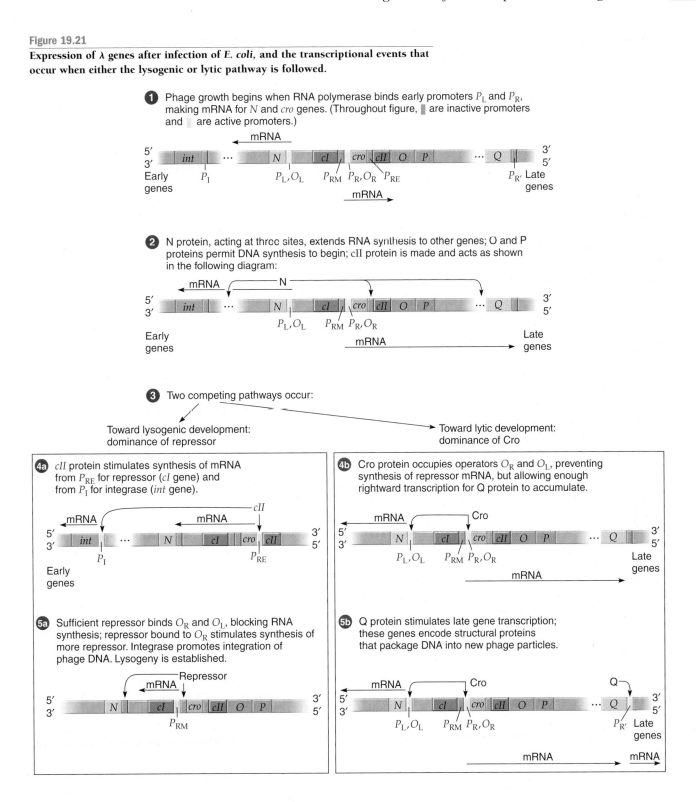

1 Phage growth begins when RNA polymerase binds early promoters P_L and P_R, making mRNA for *N* and *cro* genes. (Throughout figure, ▌ are inactive promoters and ▌ are active promoters.)

2 N protein, acting at three sites, extends RNA synthesis to other genes; O and P proteins permit DNA synthesis to begin; cII protein is made and acts as shown in the following diagram:

3 Two competing pathways occur:

Toward lysogenic development: dominance of repressor

Toward lytic development: dominance of Cro

4a cII protein stimulates synthesis of mRNA from P_{RE} for repressor (*cI* gene) and from P_I for integrase (*int* gene).

4b Cro protein occupies operators O_R and O_L, preventing synthesis of repressor mRNA, but allowing enough rightward transcription for Q protein to accumulate.

5a Sufficient repressor binds O_R and O_L, blocking RNA synthesis; repressor bound to O_R stimulates synthesis of more repressor. Integrase promotes integration of phage DNA. Lysogeny is established.

5b Q protein stimulates late gene transcription; these genes encode structural proteins that package DNA into new phage particles.

transcription of the late genes for starting the lytic pathway (Figure 19.21, part 5b).

In sum, lambda uses complex regulatory systems to choose either the lytic or the lysogenic pathway. The decision depends on a sophisticated genetic switch that involves competition between the products of the *cI* ("c-one") gene (the repressor) and the *cro* gene (the gene for the Cro protein). If the repressor dominates, the lysogenic pathway is followed; if the Cro protein dominates, the lytic pathway is followed.

Summary

From studies of the regulation of expression of the three lactose-utilization genes in *E. coli*, Jacob and Monod developed a model that today is still the basis for the regulation of gene expression in a large number of bacterial and bacteriophage systems. The genes for the enzymes are contiguous in the chromosome and are adjacent to a regulatory sequence (an operator) and a single promoter. This complex constitutes a transcriptional regulatory unit called an operon. A regulator gene, which may or may not be nearby, is associated with the operon. The addition of an appropriate substrate to the cell results in the coordinate induction of the operon's structural genes. Induction of the *lac* operon occurs as follows: Lactose binds with a repressor protein that is encoded by the regulator gene, inactivating the repressor protein and preventing it from binding to the operator. As a result, RNA polymerase can transcribe the three genes from a single promoter onto a single polycistronic mRNA. As long as lactose is present, mRNA continues to be produced and the enzymes are made. When lactose is no longer present, the Lac repressor protein is no longer inactivated, and it binds to the operator, thereby preventing RNA polymerase from transcribing the *lac* genes.

If both glucose and lactose are present in the medium, the lactose operon is not induced, because glucose (which requires less energy to metabolize than does lactose) is the preferred energy source. This phenomenon is called catabolite repression and involves cellular levels of cyclic AMP. That is, in the presence of lactose and in the absence of glucose, cAMP complexes with CAP to form a positive regulator needed for RNA polymerase to bind to the promoter. The addition of glucose results in a lowering of cAMP concentration, so no CAP-cAMP complex is produced, and therefore RNA polymerase cannot bind efficiently to the promoter and transcribe the *lac* genes.

The genes for a number of bacterial amino-acid biosynthesis pathways are also arranged in operons. Expression of these operons is accomplished by a repressor–operator system or, in some cases, through attenuation at a second regulatory sequence called an attenuator (or by both mechanisms). The *trp* operon is an example of an operon with both types of transcription regulation systems. The *trp* repressor–operator system functions essentially like that of the *lac* operon, except that the addition of tryptophan to the cell activates the repressor, turning the operon off. The attenuator, located downstream from the operator in a leader region that is translated, is a transcription termination site which allows only a fraction of the RNA polymerases that initiate at the *trp* promoter to transcribe the rest of the operon. Attenuation requires a coupling between transcription and translation and the formation of particular RNA secondary structures that signal whether transcription can continue.

Key to attenuation is the presence of multiple copies of codons for the amino acid synthesized by the enzymes encoded by the operon. When enough of the amino acid is present in the cell, enough charged tRNAs are produced so that the ribosome can quickly translate the key codons in the leader region, and this causes the RNA being made by the RNA polymerase ahead of the ribosome to assume a secondary structure that signals transcription to stop. However, when the cell is starved for that amino acid, there are insufficient charged tRNAs to be used at the key codons, so the ribosome stalls at that point. As a result, the RNA ahead of that point assumes a secondary structure which permits continued transcription into the structural genes. The combination of repressor–operator regulation and attenuation control permits a fine degree of control of operon transcription.

Operons are also used extensively by bacteriophages to coordinate the synthesis of proteins with related functions. Many genes related to the production of progeny phages or to the establishment or reversal of lysogeny in temperate phages are organized into operons. Like bacterial operons, these operons are controlled through the interaction of regulatory proteins with operators that are adjacent to clusters of structural genes. Bacteriophage lambda has been an excellent model for studying the elaborate genetic switch that controls the choice between lytic and lysogenic pathways in a lysogenic phage. The switch involves two regulatory proteins—the lambda repressor and the Cro protein—which have opposite affinities for three binding sites within each of the two major operators that control the lambda life cycle.

In sum, operons are commonly encountered in bacteria and their viruses. They provide a simple way to coordinate the expression of genes with related functions. The following generalizations can be made about the regulation of operons: First, for most operons, a regulator protein (e.g., the Lac repressor) plays a key role in the regulation process, because it is able to bind to a regulatory sequence in the operon, namely, the operator. Second, the transcription of a set of clustered structural genes is controlled by an adjacent operator through interaction with a regulator protein, which can exert positive or negative control, depending upon the operon. Third, the trigger for changing the state of an operon from off to on and vice versa is an effector molecule (e.g., the inducer allolactose in the case of the *lac* operon), which controls the conditions under which the regulator protein will or will not bind to the operator.

Analytical Approaches to Solving Genetics Problems

Q19.1 In the laboratory, you are given 10 strains of *E. coli* with the following *lac* operon genotypes, where I = *lacI* (the Lac repressor gene), $P = P_{lac}$ (the promoter), O = the *lacO* operator), and Z = *lacZ* (the β-galactosidase gene):

1. $I^+ P^+ O^+ Z^+$
2. $I^- P^+ O^+ Z^+$
3. $I^+ P^+ O^c Z^+$
4. $I^- P^+ O^c Z^+$
5. $I^+ P^+ O^c Z^-$
6. $\dfrac{F'\ I^+ P^+ O^c\ Z^-}{I^+ P^+ O^+\ Z^+}$
7. $\dfrac{F'\ I^+ P^+ O^+\ Z^-}{I^+ P^+ O^c\ Z^+}$
8. $\dfrac{F'\ I^+ P^+ O^+\ Z^+}{I^- P^+ O^+\ Z^-}$
9. $\dfrac{F'\ I^+ P^+ O^c\ Z^-}{I^- P^+ O^+\ Z^+}$
10. $\dfrac{F'\ I^- P^+ O^+\ Z^-}{I^- P^+ O^c\ Z^+}$

For each strain, predict whether β-galactosidase will be produced (a) if lactose is absent from the growth medium and (b) if lactose is present in the growth medium. Glucose is absent from the medium in every case. (*Note*: In the partial diploid strains (6–10), one copy of the *lac* operon is in the host chromosome and the other copy is in the extrachromosomal *F* factor.)

A19.1 The answers are as follows, where "+" = β-galactosidase is produced and "−" = β-galactosidase is not produced:

Genotype	Noninduced: Lactose Absent	Induced: Lactose Present
(1)	−	+
(2)	+	+
(3)	+	+
(4)	+	+
(5)	−	−
(6)	−	+
(7)	+	+
(8)	−	+
(9)	−	+
(10)	+	+

To answer this question completely, we need a good understanding of how the *lac* operon is regulated in the wild type and of the consequences of particular mutations on the regulation of the operon.

Strain (1) is the standard wild-type operon. No enzyme is produced in the absence of lactose because the Lac repressor produced by the I^+ gene binds to the operator (O^+) and blocks the initiation of transcription. When lactose is added, it binds to the repressor, changing its conformation so that it no longer can bind to the O^+ region, thereby facilitating transcription of the structural genes for RNA polymerase.

Strain (2) is a haploid strain with a mutation in the *lacI* gene (I^-). The consequence is that the Lac repressor protein cannot bind to the (normal) operator region, so there is no inhibition of transcription, even in the absence of lactose. This strain, then, is constitutive, meaning that β-galactosidase is produced by the *lacZ*$^+$ gene in the presence or absence of lactose.

Strain (3) is another constitutive mutant. In this case, the repressor gene is a wild type and the β-galactosidase gene *lacZ*$^+$ is a wild type, but there is a mutation in the operator region (O^c). Therefore, the Lac repressor protein cannot bind to the operator, and transcription occurs in the presence or absence of lactose.

Strain (4) carries both regulatory mutations of the previous two strains. Functional Lac repressor is not produced, but even if it were, the operator is changed so that it cannot bind. The consequence is the same: constitutive enzyme production.

Strain (5) produces functional Lac repressor, but the operator (O^c) is mutated. Therefore, transcription cannot be blocked, and *lac* polycistronic mRNA is produced in the presence or absence of lactose. However, because there is also a mutation in the β-galactosidase gene (Z^-), no functional enzyme is generated.

In the partial diploid strain (6), one *lac* operon is completely wild type and the other carries a constitutive operator mutation and a mutant β-galactosidase gene. In the absence of lactose, no functional enzyme is produced. For the wild-type operon, the Lac repressor binds to the operator and blocks transcription. For the operon with the two mutations, the operator region is mutated and cannot bind repressor, so the mRNA for the mutated operon is produced; however, the *lacZ* gene is also mutated, so that functional enzyme cannot be produced. In the presence of lactose, functional enzyme *is* produced, because repression of the wild-type operon is relieved, so that the Z^+ gene can be transcribed. This type of strain provided one of the pieces of evidence that the operator region does not produce a diffusible substance.

In partial diploid (7), functional enzyme is produced in the presence or absence of lactose because one of the operons has an O^c mutation that does not respond to a repressor and that is linked to a wild-type Z^+ gene. That operon is transcribed constitutively. The other operon is inducible, but because there is a Z^- mutation, no functional enzyme is produced.

Partial diploid (8) has a wild-type operon and an operon with an I^- regulatory mutation and a Z^- mutation. The I^+ gene product is diffusible, so that it can bind to the O^+ region of both operons, thereby putting both operons under inducer control. This strain demonstrates that the I^+ gene is trans-dominant to an I^- mutation. In this case, the particular location of the one Z^- mutation is irrelevant: The same result would have been obtained had the Z^+ and Z^- been switched between the two operons. In this partial diploid, β-galactosidase is not produced unless lactose is present.

In strain (9), β-galactosidase is produced only when lactose is present, because the O^c region controls only the genes that are adjacent to it on the same chromosome (cis dominance) and in this case one of the adjacent genes is Z^-, which codes for a nonfunctional enzyme. The partial diploid is heterozygous I^+/I^-, but I^+ is trans-dominant, as discussed in regard to strain (8). Thus, the only normal Z^+ gene is under inducer control.

Partial diploid (10) has a defective repressor protein as well as an O^c mutation adjacent to a Z^+ gene. On the latter ground alone, this partial diploid is constitutive. The other operon is also constitutively transcribed, but because there is a Z^- mutation, no functional enzyme is generated from it.

Questions and Problems

19.1 What is meant by constitutive gene expression? How is constitutive gene expression unlike regulated gene expression?

*19.2 Give two examples of effector molecules, and discuss how effector molecules function to regulate gene expression.

19.3 Operons produce polygenic mRNA when they are active. What is a polygenic mRNA? What advantages, if any, does it confer in terms of the cell's function?

19.4 How does lactose trigger the coordinate induction of the synthesis of β-galactosidase, permease, and transacetylase? Why does the synthesis of these enzymes not occur when glucose is also in the medium?

*19.5 An E. coli mutant strain synthesizes β-galactosidase whether or not the inducer is present. What genetic defect(s) might be responsible for this phenotype?

19.6 How did the discovery of polarity and polar mutations contribute to the formulation of the operon hypoth-

esis? How do polar mutations prevent the initiation of translation at downstream genes in a polycistronic mRNA?

*19.7 Distinguish the effects you would expect from (a) a missense mutation and (b) a nonsense mutation in the *lacZ* (β-galactosidase) gene of the *lac* operon.

19.8 Elucidation of the regulatory mechanisms associated with the enzymes of lactose utilization in E. coli was a landmark in our understanding of regulatory processes in microorganisms. In formulating the operon hypothesis as applied to the lactose system, Jacob and Monod found that results from particular partial diploid strains were invaluable. In terms of their operon hypothesis, what specific information did analyses of partial diploids provide that analyses of haploids could not?

*19.9 For the E. coli *lac* operon, write the partial diploid genotype for a strain that will produce β-galactosidase constitutively and permease by induction.

19.10 Mutants were instrumental in elaborating the model for regulation of the *lac* operon.
a. Discuss why P_{lac-} and $lacO^c$ mutants are cis-dominant but not trans-dominant.
b. Explain why $lacI^S$ and $lacI^{-d}$ mutants are trans-dominant to the wild-type $lacI^+$ allele but $lacI^-$ mutants are recessive.
c. Discuss the consequences of mutations in the repressor gene promoter as compared with mutations in the structural gene promoter.

*19.11 This question involves the *lac* operon of E. coli, where $I = lacI$ (the repressor gene), $P = P_{lac}$ (the promoter), $O = lacO$ (the operator), $Z = lacZ$ (the β-galactosidase gene), and $Y = lacY$ (the permease gene). Complete Table 19.A, using + to indicate that the enzyme in question will be synthesized and − to indicate that the enzyme will not be synthesized.

19.12 A new sugar, sugarose, induces synthesis of two enzymes from the *sug* operon of E. coli. Some properties of deletion mutations affecting the appearance of these enzymes are as follows (here, + = enzyme induced normally, i.e., synthesized only in the presence of the inducer; C = enzyme synthesized constitutively; 0 = enzyme cannot be detected):

Mutation of	Enzyme 1	Enzyme 2
Gene A	+	0
Gene B	0	+
Gene C	0	0
Gene D	C	C

Table 19.A

Genotype	Inducer Absent		Inducer Present	
	β-Galactosidase	Permease	β-Galactosidase	Permease
a. $I^+\ P^+\ O^+\ Z^+\ Y^+$				
b. $I^+\ P^+\ O^+\ Z^-\ Y^+$				
c. $I^+\ P^+\ O^+\ Z^+\ Y^-$				
d. $I^-\ P^+\ O^+\ Z^+\ Y^+$				
e. $I^S\ P^+\ O^+\ Z^+\ Y^+$				
f. $I^+\ P^+\ O^c\ Z^+\ Y^+$				
g. $I^S\ P^+\ O^c\ Z^+\ Y^+$				
h. $I^+\ P^+\ O^c\ Z^+\ Y^-$				
i. $I^{-d}\ P^+\ O^+\ Z^+\ Y^+$				
j. $\dfrac{I^-\ P^+\ O^+\ Z^+\ Y^+}{I^+\ P^+\ O^+\ Z^-\ Y^-}$				
k. $\dfrac{I^-\ P^+\ O^+\ Z^+\ Y^+}{I^+\ P^+\ O^+\ Z^-\ Y^-}$				
l. $\dfrac{I^S\ P^+\ O^+\ Z^+\ Y^-}{I^+\ P^+\ O^+\ Z^-\ Y^+}$				
m. $\dfrac{I^+\ P^+\ O^c\ Z^-\ Y^+}{I^+\ P^+\ O^+\ Z^+\ Y^-}$				
n. $\dfrac{I^-\ P^+\ O^c\ Z^+\ Y^-}{I^+\ P^+\ O^+\ Z^-\ Y^+}$				
o. $\dfrac{I^S\ P^+\ O^+\ Z^+\ Y^+}{I^+\ P^+\ O^c\ Z^+\ Y^+}$				
p. $\dfrac{I^{-d}\ P^+\ O^+\ Z^+\ Y^-}{I^+\ P^+\ O^+\ Z^-\ Y^+}$				
q. $\dfrac{I^+\ P^-\ O^c\ Z^+\ Y^-}{I^+\ P^+\ O^+\ Z^-\ Y^+}$				
r. $\dfrac{I^+\ P^-\ O^+\ Z^+\ Y^-}{I^+\ P^+\ O^c\ Z^-\ Y^+}$				
s. $\dfrac{I^-\ P^-\ O^+\ Z^+\ Y^+}{I^+\ P^+\ O^+\ Z^-\ Y^-}$				
t. $\dfrac{I^-\ P^+\ O^+\ Z^+\ Y^-}{I^+\ P^-\ O^+\ Z^-\ Y^+}$				

a. The genes are adjacent, in the order $A\ B\ C\ D$. Which gene is most likely to be the structural gene for enzyme 1?

b. Complementation studies using partial diploid (F') strains were made. The extrachromosomal element (F') and chromosome each carried one set of *sug* genes. The results were as follows:

Genotype of F'	Chromosome	Enzyme 1	Enzyme 2
$A^+\ B^-\ C^+\ D^+$	$A^-\ B^+\ C^+\ D^+$	+	+
$A^+\ B^-\ C^-\ D^+$	$A^-\ B^+\ C^+\ D^+$	+	0
$A^-\ B^+\ C^-\ D^+$	$A^+\ B^-\ C^+\ D^+$	0	+
$A^-\ B^+\ C^+\ D^+$	$A^+\ B^-\ C^+\ D^-$	+	+

From all the evidence given, determine whether the following statements are true or false:

i. It is possible that gene D is a structural gene for one of the two enzymes.

ii. It is possible that gene D produces a repressor.

iii. It is possible that gene D produces a cytoplasmic product required to induce genes A and B.

iv. It is possible that gene D is an operator locus for the *sug* operon.

v. The evidence is also consistent with the possibility that gene C could be a gene that produces a cytoplasmic product required to induce genes A and B.

vi. The evidence is also consistent with the possibility that gene *C* could be the controlling end of the *sug* operon (the end from which mRNA synthesis presumably commences).

19.13 Four different polar mutations, *1, 2, 3,* and *4,* in the *lacZ* gene of the *lac* operon were isolated after mutagenesis of *E. coli.* Each caused total loss of β-galactosidase activity. Two revertant mutants, due to suppressor mutations in genes unlinked to the *lac* operon, were isolated from each of the four strains: Suppressor mutations of polar mutation *1* are *1A* and *1B*, those of polar mutation *2* are *2A* and *2B*, and so on. Each of the eight suppressor mutations was then tested, by appropriate crosses, for its ability to suppress each of the four polar mutations; the test involved examining the ability of a strain carrying the polar mutation and the suppressor mutation to grow with lactose as the sole carbon source. The results follow (+ = growth on lactose and − = no growth):

Polar Mutation	Suppressor Mutation							
	1A	1B	2A	2B	3A	3B	4A	4B
1	+	+	+	+	+	+	+	+
2	+	−	+	+	+	+	−	−
3	+	−	+	−	+	+	−	−
4	+	+	+	+	+	+	+	+

A mutation to a UAG codon is called an amber nonsense mutation, and a mutation to a UAA codon is called an ochre nonsense mutation. Suppressor mutations allowing reading of UAG and UAA are called amber and ochre suppressors, respectively.

a. Which of the polar mutations are probably amber? Which are probably ochre?

b. Which of the suppressor mutations are probably amber suppressors? Which are probably ochre suppressors?

c. How would you explain the anomalous failure of suppressor *2B* to permit growth with polar mutation *3*? How could you test your explanation most easily?

d. Explain precisely why ochre suppressors suppress amber mutants but amber suppressors do not suppress ochre mutants.

*19.14 What consequences would a mutation in the catabolite activator protein (CAP) gene of *E. coli* have for the expression of a wild-type *lac* operon?

19.15 The presence of glucose in the medium along with lactose leads to catabolite repression. Explain why catabolite repression is considered to be a form of positive control, while repression by the *lac* repressor is considered to be a form of negative control.

*19.16 DNase protection experiments were helpful to elucidate the functions of different DNA sequences in the *lac* promoter.

a. What is a DNase protection experiment and how does it provide this information?

b. How are the binding sites for the *lac* repressor, RNA polymerase, and CAP-cAMP arranged at the 5′-end of the *lac* operon?

c. What effects would you expect each of the following mutations to have on the coordinate induction of the *lac* operon by lactose (in the absence of glucose)? Explain your reasoning. [The base-pair coordinates used here are those specified in Figure 19.14.]
  i. a deletion of base pairs from +3 to +18
  ii. a TA-to-GC transversion at −12
  iii. a TA-to-GC transversion at −69
  iv. a GC-to-AT transition at +28
  v. a GC-to-AT transition at +9
  vi. Would any of the mutations listed in (c) affect catabolite repression of the *lac* operon?

19.17 The *lac* operon is an inducible operon, whereas the *trp* operon is a repressible operon. Discuss the differences between these two types of operons.

19.18 Transcription of the *trp* operon can be reduced through a combination of repression using an aporepressor and attenuation.

a. How much of a reduction in transcription can be achieved using the aporepressor, and how much of a reduction in transcription can be achieved using attenuation? Speculate why the aporepressor might be unable to completely silence expression of the *trp* operon, and why this might be advantageous to *E. coli.*

b. Explain how the mechanism of attenuation is dependent on translation of transcripts at the *trp* operon.

*19.19 In the presence of high intracellular concentrations of tryptophan, only short transcripts of the *trp* operon are synthesized because of attenuation of transcription 5′ to the structural genes. This is mediated by the recognition of two Trp codons in the leader sequence. What affect would mutating these two codons to UAG stop codons have on the regulation of the operon in the presence or absence of tryptophan? Explain.

*19.20 The mutant *E. coli* strains described in the following table are individually inoculated into two different media, one with supplemental tryptophan and one without:

Mutant	Phenotype
1	Aporepressor is unable to bind to tryptophan.
2	A point mutation in the *trp* operator prevents binding by an active Trp repressor.
3	The *trpE* gene has a nonsense mutation.
4	The levels of Trp-tRNA. Trp are decreased due to a mutation in a gene for Trp-aminoacyl synthetase.
5	Three adjacent G-C base pairs in region 4 (see Figures 19.16 and19.17) are mutated to A-T base pairs.

For each mutant and medium, state whether the level of tryptophan synthetase will be increased or decreased relative to the level found in a wild-type strain and, where possible, by how much, and why.

19.21 In the bacterium *Salmonella typhimurium*, seven of the genes coding for histidine biosynthetic enzymes are located adjacent to one another in the chromosome. If excess histidine is present in the medium, the synthesis in all seven enzymes is coordinately repressed, whereas in the absence of histidine all seven genes are coordinately expressed. Most mutations in this region of the chromosome result in the loss of activity of only one of the enzymes. However, mutations mapping to one end of the gene cluster result in the loss of all seven enzymes, even though none of the structural genes have been lost. What is the counterpart of these mutations in the *lac* operon system?

19.22 On infecting an *E. coli* cell, bacteriophage λ has a choice between the lytic and lysogenic pathways. Discuss the molecular events that determine which pathway is taken.

19.23 How do the lambda repressor protein and the Cro protein regulate their own synthesis?

*19.24 A mutation in the phage lambda *cI* gene results in a nonfunctional *cI* gene product. What phenotype would you expect the phage to exhibit?

19.25 Bacteriophage λ can form a stable association with the bacterial chromosome because the virus manufactures a repressor. This repressor prevents the virus from replicating its DNA and making lysozyme and all the other tools used to destroy the bacterium. When you induce the virus with UV light, you destroy the repressor, and the virus undergoes its normal lytic cycle. The repressor is the product of a gene called the *cI* gene and is a part of the wild-type viral genome. A bacterium that is lysogenic for λ$^+$ is full of repressor protein, which confers immunity against any λ virus added to these bacteria. Added viruses can inject their DNA, but the repressor from the resident virus prevents

replication, presumably by binding to an operator on the incoming virus. Thus, this system has many elements analogous to the *lac* operon. We could diagram a virus as shown in the following figure. Several mutations of the *cI* gene are known. The c_i mutation results in an inactive repressor.

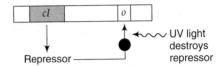

a. If you infect *E. coli* with λ containing a c_i mutation, can it lysogenize (form a stable association with the bacterial chromosome)? Why or why not?

b. If you infect a bacterium simultaneously with a wild-type c^+ and a c_i mutant of λ, can you obtain stable lysogeny? Why or why not?

c. Another class of mutants called c^{IN} makes a repressor that is insensitive to UV destruction. Will you be able to induce a bacterium lysogenic for c^{IN} with UV light? Why or why not?

*19.26 Five λ mutants have the molecular phenotypes shown in the left column of the following table:

Mutant	Molecular Phenotype	Lytic Growth	Lysogenic Growth	Inducible by UV Light
1	The Cro protein is unable to bind DNA.			
2	The N protein does not function.			
3	The cII protein does not function.			
4	The Q protein does not function.			
5	P_{RM} is unable to bind RNA polymerase.			

Fill in the table to indicate whether each mutant will be able to undergo lytic or lysogenic growth. For mutants able to follow a lysogenic pathway, state whether they can be induced by UV-light.

20

Regulation of Gene Expression in Eukaryotes

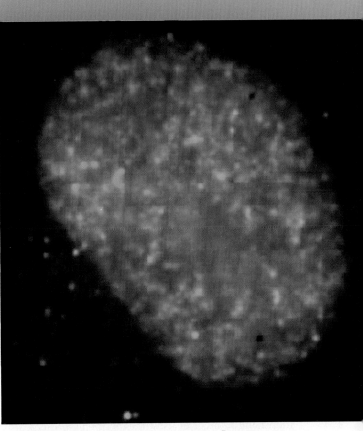

A human interphase cell nucleus stained with a fluorescent antibody for histone deacetylase (HDAC) showing the punctate (spotted) distribution of the enzyme.

PRINCIPAL POINTS

- With very rare exceptions, there are no operons in eukaryotes. Even in the absence of operons, genes for related functions often are regulated coordinately.

- In eukaryotes, gene expression is regulated at a number of distinct levels. That is, there are regulatory systems for the control of transcription, precursor-RNA processing, transport of the mature RNA out of the nucleus, translation of the mRNAs, degradation of the mature RNAs, and degradation of the protein products.

- Chromatin configuration poses a barrier to transcription. Transcriptionally active regions have looser chromatin structure than transcriptionally inactive regions. It is the chromatin structure at the core promoter of a nonexpressed gene that is repressive to transcription. Remodeling of this chromatin region is necessary for activation of transcription. Chromatin remodeling occurs when activators bind to enhancers and recruit large, multiprotein remodeling complexes that either acetylate nucleosomes, loosening their association with the DNA, or move or restructure nucleosomes, allowing the transcription machinery to access the promoter.

- Eukaryotic protein-coding genes contain both promoter and enhancer elements which interact with a number of regulatory proteins (transcription factors) that activate or repress transcription.

- In complex, multicellular eukaryotes, steroid hormones regulate the expression of particular sets of genes. To function in this short-term regulatory system, a steroid hormone enters a cell. If that cell contains a receptor molecule specific for the hormone, the hormone binds to the receptor and activates it, and the hormone–receptor complex binds to hormone regulatory elements next to genes in the nucleus, thereby regulating the expression of those genes. Other, polypeptide hormones act at the cell surface, activating a system to produce cyclic AMP, a substance that acts as a second messenger to control gene activity. The specificity of hormone action results from the presence of hormone receptors in only certain cell types and from interactions of steroid–receptor complexes with cell-type-specific regulatory proteins.

- Gene silencing is the phenomenon of turning off the transcription of a gene as a result of its position in the chromosome. Gene silencing involves changes in chromatin structure to produce heterochromatin, which usually affects a zone of genes. A gene may also be silenced through the methylation of cytosines in the promoter upstream of the gene. Sometimes the methylation pattern is associated with genomic imprinting, a phenomenon in which the expression of a gene is determined by whether the gene is inherited from the female or male parent.

- Regulation at the level of RNA includes processing, nuclear transport, translation, and stability of the mRNA. Regulation at the level of proteins involves a mechanism that specifies the lifetime and rate of degradation of a protein.

- Individual genes can be silenced by a mechanism called RNA interference (RNAi) whereby double-stranded RNA is cut into small fragments of short interfering RNA (siRNA), which, in a complex with proteins, binds to complementary sequences in mRNAs and triggers the cleavage of those mRNAs. The result is a knockout or knockdown of gene expression. Among the natural functions of RNAi are protection against viral infections and the regulation of developmental processes. In addition, RNAi has an increasing number of uses in experimental research.

i **TO RESPOND TO A CHANGING ENVIRONMENT OR** allow for differentiation, cells must often turn several genes on or off in a coordinated manner. In this chapter you will learn about eukaryotic gene regulation. Then, in the iActivity, you can see how researchers work out the details of eukaryotic gene regulation mechanisms.

In the previous chapter, we learned about the regulation of gene expression in bacteria and bacteriophages. Most of this regulation occurs at the level of transcription. We discovered that specific DNA sequences are needed for this regulation—specifically, the promoter, where RNA polymerase binds, and regulatory sequences, where regulatory proteins bind. In the *lac* operon, for instance, the regulatory sequence is *lacO*, the operator, where the Lac repressor regulatory protein binds. More generally, the regulatory protein may be a repressor (inhibiting transcription—e.g., Lac) or an activator (enhancing transcription—e.g., CAP). We will learn in this chapter that eukaryotic transcription is regulated mostly by activators. While not discussed explicitly in the previous chapter, regulation does occur for some genes at the posttranscriptional level.

Compared with prokaryotes, eukaryotes have many similarities, yet some differences in the regulation of gene expression. Moreover, there is variation in the complexity of the regulatory systems among eukaryotes, with the simplest being in single-celled eukaryotes such as yeast and the most complex in mammals. The key similarities in gene regulation between prokaryotes and eukaryotes include (1) promoter sequences that vary to specify the rate of transcription initiation, (2) regulatory sequences that determine the response of the gene to effector molecules, and (3) Regulatory proteins—both activators and repressors—with specific DNA-binding domains that interact with regulatory sequences to control transcription. There are also key differences, resulting from the greater organizational complexity of the eukaryotic cell, including (1) a role of chromatin structure in regulating gene expression; (2) the necessity to add a 5' cap and a 3' poly(A) tail to a pre-mRNA molecule and then splice the pre-mRNA to remove introns and produce the mature mRNA; (3) in the case of a number of genes for alternative splicing of the pre-mRNA, the possibility of producing different mRNAs; and (4) regulation of the transport of mRNA from the nucleus to the cytoplasm. Finally, we saw in the previous chapter that operons are a common regulatory unit in prokaryotes. As we shall see, operons are rare in eukaryotes.

In this chapter, we discuss the regulation of gene expression in eukaryotes at its various levels. As we do so, we must keep in mind that the regulatory needs of prokaryotes and eukaryotes—particularly multicellular eukaryotes—are different. Prokaryotes grow and divide, whereas multicellular eukaryotes develop and differentiate, requiring a more sophisticated control of gene expression in space and time.

Operons in Eukaryotes

In the previous chapter, much of the discussion was about the regulation of operons. Indeed, the discovery in 1961 of the *lac* operon in *E. coli* was a landmark event. Not only did it show for the first time that there were regulatory sequences and regulatory proteins involved in controlling gene transcription, but it also showed that a cluster of genes could be controlled from a single regulatory sequence. In the *lac* operon, for example, three contiguous genes are transcribed from a promoter onto a single polygenic mRNA. The exciting discovery of operons in bacteria naturally spawned efforts to see whether the operon model was a general feature of gene regulation in all organisms. The answer for prokaryotes and their viruses is that operons are common. For many years, though, there was no evidence of operons in eukaryotes, but then some were found. We start this chapter with a brief overview of eukaryotic operons to compare and contrast them with prokaryotic operons.

About 10 years ago, researchers discovered that the nematode *C. elegans* (see Figure 1.7c, p. 10) has

Figure 20.1

An operon of *C. elegans* and the production of monogenic mRNAs from a polygenic mRNA by trans-splicing and polyadenylation/cleavage.

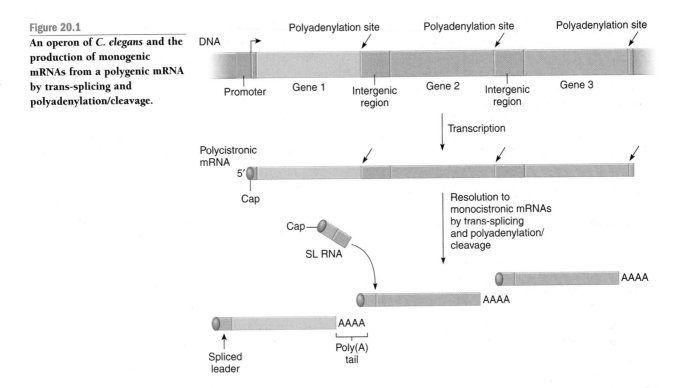

operons. Subsequently, operons were also found in other nematodes. As with prokaryotic operons, a single promoter controls the expression of each gene cluster. Unlike the situation with prokaryotes, however, the polygenic mRNA transcribed from the *C. elegans* operons is not translated sequentially. Such translation is not possible because eukaryotic ribosomes cannot reinitiate translation at a new start codon once translation has been terminated. Rather, for these eukaryotic operons, a pre-mRNA is produced by transcription and is processed to generate monogenic mRNAs that can be translated individually.

The processing events are illustrated in Figure 20.1. Shown is an operon with three genes and a single promoter. Between each gene is an *intergenic region* of about 100 bp. Immediately upstream of each gene is a 3′ splice site like that of introns, and immediately downstream of each gene is a conventional polyadenylation and cleavage site. Transcription by RNA polymerase II produces a single, capped, polygenic pre-mRNA molecule that is processed cotranscriptionally, so the full-length pre-mRNA is not seen. Monogenic mRNAs are resolved from the polygenic pre-mRNA by two events: **trans-splicing** and 3′ end generation by conventional cleavage and polyadenylation. In trans-splicing, an approximately 100-nucleotide, capped *spliced leader (SL)-RNA* that is in an snRNP (small nuclear ribonucleoprotein particle; see p. 101) recognizes each 3′ splice site, and, in a reaction using much the same machinery as intron removal, the pre-mRNA is cleaved adjacent to the 5′ end of each gene and the first 22 nucleotides of the SL-RNA are spliced to

its 5′ end. In essence, we can think of the donated SL-RNA segment as the first exon for every gene in the operon and as providing the same 5′ leader sequence for each gene. (Indeed, it was the observation that many mRNAs had the same 5′ leader sequence and that the genes involved were in clusters that led to the discovery of operons in *C. elegans*.) Simultaneously, cleavage at the polyadenylation site and addition of the poly(A) tail generate the 3′ ends of each gene.

C. elegans genes also contain introns (not shown in Figure 20.1). These introns are removed by the conventional splicing mechanisms described in Chapter 5.

About 15 percent of *C. elegans* genes—approximately 2,600 genes—are organized into operons. An average operon contains 2.6 genes, with the largest containing 8 genes. One of the features of prokaryotic operons is that the genes they contain have related functions. This is not so for *C. elegans* operons: In no case does a single operon contain the genes for all the steps in a pathway or for a multiprotein complex. There are many cases, though, wherein genes which encode proteins that function together are in the same operon, along with other genes of unrelated function.

KEYNOTE

Genes of related function are often organized into operons in bacteria and bacteriophages. With rare exceptions, eukaryotic genes of related function are dispersed rather than being clustered. The exceptions are operons found in nematodes such as *C. elegans*, in which 15 percent of the animal's genes are organized into operons. These operons sometimes contain genes of related function, but that is never exclusively the case. Unlike the situation with prokaryotic operons, the eukaryotic operons are transcribed into polygenic mRNAs, but those molecules are processed by trans-splicing into monogenic mRNAs in order for each gene's transcript to be translated.

Levels of Control of Gene Expression in Eukaryotes

Most prokaryotic organisms are unicellular and respond quickly to the environment by making changes in gene regulation, primarily at the transcriptional level, with some translational control. (See Chapter 19.) Transcriptional response is accomplished through the interaction of regulatory proteins with upstream regulatory DNA sequences. Rapid changes in levels of protein synthesis are achieved by switching off gene transcription and by rapid degradation of the mRNA molecules.

In eukaryotes, both unicellular and multicellular, the control of gene expression is more complicated than in prokaryotes. The reason, in part, stems from the compartmentalization of eukaryotic cells and the demands imposed by the need for multicellular eukaryotes to generate large numbers and types of cells. Notably, the absence of a membrane-bound nucleus in prokaryotes enables translation to proceed on an mRNA that is still being made. The presence of a nucleus in a eukaryotic cell separates the processes of transcription and translation. Consequently, there are more levels at which the expression of protein-coding genes can be regulated in eukaryotes. Figure 20.2 diagrams some of these levels: mRNA *transcription, processing, transport, translation* and *degradation,* and *protein processing* and *degradation.* We consider each of these in turn.

Control of Transcription Initiation

The regulation of protein-coding gene expression in eukaryotes is mostly at the level of transcription initiation. As we learned in Chapter 5 (pp. 89–112), initiation of the transcription of protein-coding genes is under the control of the **promoter** immediately upstream of the gene and of **enhancers** that are distant from the gene. The summary view from our discussion in Chapter 5 is that the general transcription machinery which assembles on the core promoter alone is capable of only a basal level of transcription. Regulated tran-

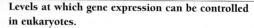

Figure 20.2

Levels at which gene expression can be controlled in eukaryotes.

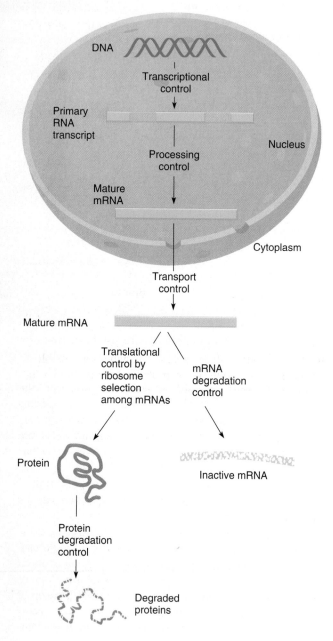

scription up to the maximal level possible for the gene depends upon regulatory proteins (activators) binding to promoter proximal elements and to enhancer elements. Such binding leads to the recruitment of proteins needed to make the chromatin accessible to the transcription machinery and then to recruit the transcription machinery to the promoter and prepare for transcription. On a gene-to-gene basis, there are variations of this general scheme.

Chromatin Remodeling

As we learned in Chapter 2, the eukaryotic chromosome consists of DNA complexed with histones (to form nucleosomes) and nonhistone chromosomal proteins. The nucleosome organization of chromosomes has a generally repressive effect on gene expression, because access to the transcription machinery is physically impeded. Due to this chromosome-level gene repression system, in most cases there is no need for gene repressors. Therefore, cell resources are not wasted making repressors to silence genes specifically. Contrast this situation with the *lac* operon and other bacterial operons, wherein repressor proteins bind to regulatory sequences in naked DNA to silence gene expression.

Repression of Gene Activity by Histones. The effect of chromatin structure on gene expression is seen in several ways. First, transcriptionally active genes in chromatin have increased sensitivity to the DNA-degrading enzyme DNase I, compared with transcriptionally inactive genes. This distinction can be shown experimentally by isolating chromatin from cells, treating the sample with DNase I, digesting with a restriction enzyme that cuts on either side of the gene, and then using Southern-blot analysis with a gene-specific probe to assess the state of the gene. If the gene is accessible to DNase I, either it will be in fragments, or no gene fragments will remain. However, if a gene is inaccessible to DNase I, the gene will be seen to be whole. Note that increased DNase I sensitivity for transcriptionally active genes does not mean that the DNA is not organized into nucleosomes; instead, it means only that the chromosome is less highly coiled in these regions.[1]

Second, the regions of DNA around transcriptionally active genes have certain sites, called **hypersensitive sites** or **hypersensitive regions,** that are even more highly sensitive to digestion by DNase I. These sites or regions typically are the first to be cut with DNase I. Most DNase-hypersensitive sites are in the regions upstream from the start of transcription—including the promoter region—probably corresponding to the DNA sequences where RNA polymerase and other gene regulatory proteins bind.

Third, in vitro experiments have shown directly that histones can repress gene expression:

1. If DNA is mixed simultaneously with histones and promoter-binding proteins, the histones compete more strongly for promoters and form nucleosomes at TATA boxes of core promoters. As a result, the promoter-binding proteins cannot bind and transcription cannot occur.

2. If DNA is mixed first with promoter-binding proteins, the proteins assemble on TATA boxes and other promoter elements and block nucleosome assembly on those sites when histones are added. As a result, transcription occurs.

3. If DNA is mixed *simultaneously* with histones, promoter-binding proteins, and enhancer-binding proteins (activators), the enhancer-binding proteins bind to enhancers and help promoter-binding proteins bind to TATA boxes by blocking access by the histones. As a result, transcription occurs.

These results indicate that histones are effective repressors of transcription, but other proteins can overcome their repression.

Activating Genes by Remodeling Chromatin. For a eukaryotic gene to be activated, the chromatin structure must be altered in the vicinity of the core promoter. This process is called **chromatin remodeling.** Two classes of protein complexes can cause chromatin remodeling.

One class involves enzymes that modify nucleosomes by *acetylating* or *deacetylating* core histones (Figure 20.3a). The enzymes for histone acetylation are the *histone acetyl transferases* (HATs). Found in multiprotein complexes, HATs are recruited to the chromatin when activators—regulatory proteins that stimulate the initiation of transcription of a gene (see Chapter 5, p. 94)—bind to their DNA binding sites and acetylate lysines of the amino-terminal tails of core histones. Acetyl groups are negatively charged. With increasing acetylation, the positively charged histones slowly lose affinity for negatively charged DNA, and the 30-nm chromatin fiber (see Chapter 2, pp. 32–33) loses histone H1 and changes conformation to a 10-nm chromatin fiber. In this form, the promoter is more accessible for activation of transcription. For example, the TFIID complex (see Chapter 5, Figure 5.7, p. 97) binds better to acetylated nucleosomes than to unacetylated nucleosomes. This form of chromatin remodeling can be reversed in response to signals to remove the added acetyl groups. The result is restoration of the 30-nm chromatin fiber conformation. The removal of acetyl groups is catalyzed by *histone deacetylases* (HDACs).

The other class of chromatin remodelers are ATP-dependent **nucleosome remodeling complexes,** large multiprotein complexes that remodel chromatin by using the energy of ATP hydrolysis (Figure 20.3b). The process is similar to that for the HAT complexes. Activators bound to their DNA-binding sites recruit a nucleosome remodeling complex, which alters nucleosome position or structure (see shortly), facilitating binding of the transcription machinery to the core promoter.

A cell contains different types of nucleosome remodeling complexes. Depending on the type, the

---

[1]Protein-coding genes typically fit the pattern described. Highly transcribed genes such as the rRNA genes (transcribed by RNA polymerase I) are devoid of nucleosomes.

Figure 20.3

Chromatin modeling by (a) histone acetylases and (b) nucleosome remodeling complexes. In either case, the result is access to the promoter by the transcription machinery.

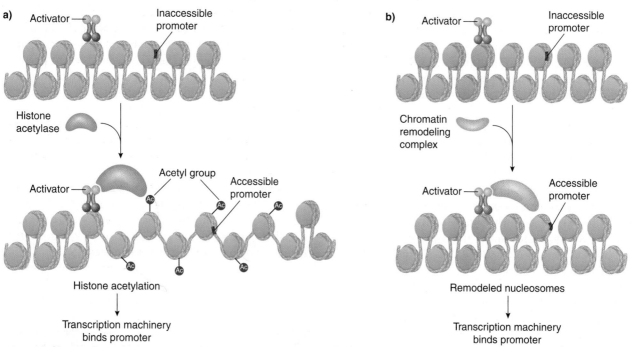

complex may *slide* a nucleosome along the DNA, exposing sites for DNA-binding proteins, *restructure* the nucleosome in place to facilitate the association of a DNA-binding protein with a DNA sequence, or *transfer* the nucleosome from one DNA molecule to another. Some complexes are able to do two or all three of these remodeling events.

One ATP-dependent nucleosome remodeling complex is SWI/SNF. This complex can catalyze all three of the aforementioned remodeling events. SWI/SNF was discovered indirectly as a result of genetic studies in yeast. As was described in Chapter 16, pp. 430–431, yeast has two mating types: *MATa* and *MATα*. Yeast can switch mating types under genetic control. An enzyme called HO endonuclease is needed for this phenomenon. Mutants that are unable to switch mating type fall into three complementation groups that define genes *SWI1*, *SWI2*, and *SWI3* (*SWI* = <u>switch</u>). The HO endonuclease activity in *swi* mutants is a hundredfold lower than in wild-type strains, but, in addition, *swi* mutants cause a large reduction in the expression of many genes unrelated to mating-type switching. In separate studies, yeast mutants were isolated that could not ferment sucrose. These mutants defined the sucrose *non*fermenting genes *SNF1*, *SNF5*, and *SNF6*. The mutants also showed marked decreases in the expression of many genes unrelated to sugar utilization. Then it was discovered that *SWI2* and *SNF2* are the same

gene, suggesting that the *SWI* and *SNF* proteins might be in a complex with a more general function than mating-type switching or sucrose fermentation. That complex, which has now been found in organisms from yeast to mammals, has 8–11 subunits, depending on the organism. SWI/SNF affects the expression of many genes because, as we have discussed, it is one type of chromatin-remodeling complex.

KEYNOTE

Chromosome regions that are transcriptionally active have looser DNA–protein structures than chromosome regions that are transcriptionally inactive, resulting in sensitivity of the DNA to digestion by DNase I. The promoter regions of active genes typically have an even looser DNA–protein structure, resulting in hypersensitivity to DNase I. In other words, the chromatin structure at the core promoter of a nonexpressed gene is repressive to transcription. Remodeling of the chromatin in this region is necessary to activate transcription and is brought about by the binding of activators to enhancers. The activators recruit chromatin remodeling complexes, either a type that acetylates nucleosomes, thereby loosening their association with the DNA, or a type that moves or restructures nucleosomes, allowing the transcription machinery to access the promoter.

Activation of Transcription by Activators and Coactivators

Having learned that chromatin remodeling is an integral part of the activation of gene transcription, let us now discuss the general molecular events of the process.

There are three classes of proteins involved in transcription activation (Figure 20.4). The first class comprises the **general transcription factors** (GTFs) that we discussed earlier (Chapter 5, Figure 5.7, p. 95). Recall that these TFs are required for basal transcription. However, they do not themselves influence the rate of transcription initiation.

The second class constitutes the **activators** (also called *trans-activators* because they influence transcription at a distance). We have already learned that activators play a role in chromatin remodeling. As the name indicates, the function of activators is to activate transcription. Activators have two key domains—a DNA-binding domain and a transcription activation domain—separated by a flexible region. Often, the functional form of an activator is a homodimer—two copies of the same protein. The DNA-binding domain can bind to a particular DNA sequence—its DNA-binding site. Studies of DNA-binding domains have shown that some common structural motifs are involved in the recognition of, and binding, to DNA. Examples are the *helix-turn-helix* (HTH), *zinc finger*, and *leucine zipper* (Figure 20.5). Turning this phenomenon around, if a researcher detects one of these sequences in a protein encoded by a newly discovered gene (e.g., in the computer analysis of a genome sequence), then it may be concluded that that protein is likely a DNA-binding protein.

At the other end of the molecule, the activation domains vary considerably and do not have readily classifiable motifs. Activation domains stimulate transcription initiation up to about a hundred-fold.

Activators activate transcription by recruiting a third class of protein: the **coactivator.** (See Figure 20.4.) A coactivator is a large multiprotein complex that does not bind directly to DNA, but participates in the activation of transcription by interacting both with activators and with

Figure 20.4

Activation of transcription by general transcription factors, activators, and a coactivator ("Mediator"). TBP = TATA binding protein. CTD = C-terminal domain (tail) of RNA polymerase II.

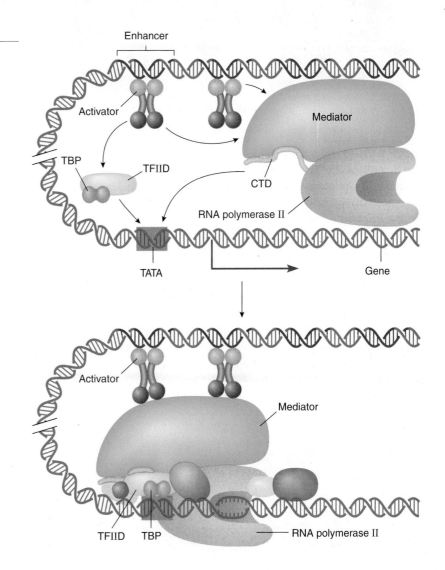

Figure 20.5

Examples of the structural motifs (DNA-binding domains) found in DNA-binding proteins such as transcription factors and transcription regulator proteins. (a) Helix-turn-helix motif: Shown is a computer-generated model of the helix-turn-helix-containing yeast telomere-binding protein, Rap1, bound to DNA. (b) Zinc finger motif: Shown is pyrimidine pathway regulator 1 protein bound to DNA and to zinc (light blue). Zinc fingers constitute the DNA recognition domains of many DNA regulatory proteins and are so named for their resemblance to fingers projecting from the protein. Characteristically, two cysteine amino acids and two histidine amino acids are positioned to bind a zinc molecule. The region containing the histidines is in the form of an α-helix; it is this coiled coil that binds in the major groove of DNA. (c) Leucine zipper motif: Shown is the yeast Gcn4 protein bound to DNA. Leucine zipper proteins are dimers, with each leucine zipper domain consisting of two helical regions. The name derives from the presence of leucines (L) at every seventh position in the region at the carboxy end of the protein. This positioning puts the leucines all on the same face of the amino-acid helix and facilitates the binding together of the two proteins to form a coiled coil because of the hydrophobicity of the leucines. The amino terminal recognition helices of the proteins have a high level of positively charged amino acids (+); this end of the dimer binds to the DNA.

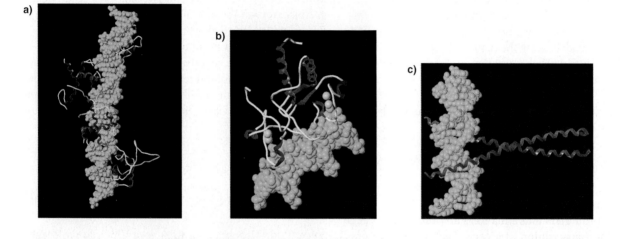

transcription factors. Specifically, the recruitment of a coactivator leads to the recruitment of RNA polymerase II, which contacts the TFs in the appropriate orientation for initiating transcription.[2] The interactions between activators and coactivator, between coactivator and RNA polymerase II, and between RNA polymerase II and the TFs cause the DNA to loop back onto itself. (See Figure 20.4).

There are several types of coactivators in a cell; the details of which are used with each regulated gene and how they are used remain to be worked out. The task is difficult because of the large numbers of proteins in the complexes. One coactivator is the Mediator Complex. Found originally in yeast, this complex consists of at least 20 polypeptides. One surface of the complex binds to the carboxy-terminal domain of RNA polymerase II, and other surfaces bind to activators. Homologs of Mediator have been found in other eukaryotes, including mammals.

Blocking Transcription with Repressors

For a few genes, there are **repressors** that counteract the action of activators, thereby blocking transcription. Like

activators, repressors have two domains: a DNA-binding domain and a repressing domain.

Recall that bacterial transcription repressors often work by binding to DNA sequences that overlap the promoter and, hence, prevent RNA polymerase from binding. This mechanism of repression is not seen in eukaryotes. Instead, eukaryotic repressors work in various other ways. In one way, a repressor protein binds to a binding site near an activator's binding site, and, through interaction of the repressor domain of the repressor with the activation domain of the activator, the activator's action is blocked. In another way, an activator binding site and a repressor binding site overlap, and binding of the repressor prevents the activator from binding. Finally, as discussed previously, chromatin remodeling can block transcription if a repressor binds to its binding site and recruits a histone deacetylase (HDAC) complex, bringing about chromatin compaction. A case study that is described later (pp. 550–552) presents a specific example of repression.

Combinatorial Gene Regulation

Protein-coding eukaryotic genes contain both promoter elements and enhancers. (See Chapter 5.) The promoter elements are located just upstream of the site at which transcription begins. The enhancers usually are some dis-

---

[2]In an alternative model, the Mediator, RNA polymerase II, and some TFs are recruited as one very large complex.

tance away, either upstream or downstream. We can think of the different promoter elements as modules that function in the regulation of expression of the gene. Certain promoter elements, such as the TATA element in the core promoter, are required to specify where transcription is to begin. Other promoter elements, in the promoter proximal region, control whether transcription of the gene occurs; specific regulatory proteins bind to these elements.

A regulatory promoter element is specialized with respect to the gene (or genes) it controls because it binds a signaling molecule—activator or repressor—that is involved in the regulation of that gene's expression. Depending on the particular gene, there can be one, a few, or many regulatory promoter elements, because, under various conditions, there may be one, a few, or many regulatory proteins that control the gene's expression. The remarkable specificity of regulatory proteins in binding to their specific regulatory element in the DNA and to no others ensures careful control of which genes are turned on and which are turned off.

Whereas promoter elements are crucial for determining whether transcription can occur, enhancer elements ensure maximal transcription of the gene. At an enhancer element, depending on its sequence, an activator or a repressor will bind.

Both promoters and enhancers are important in regulating the transcription of a gene. Each regulatory promoter element and enhancer element binds a special regulatory protein. Some regulatory proteins are found in most or all cell types, whereas others are found in only a limited number of cell types. Because some of the regulatory proteins activate transcription when they bind to the enhancer or promoter element, whereas others repress transcription, the net effect of a regulatory element on transcription depends on the combination of different proteins bound. If activators are bound at both the enhancer and promoter elements, the result is activation of transcription. However, if a repressor binds to the enhancer and an activator binds to the promoter element, the result depends on the interaction between the two regulatory proteins. If the repressor has a strong effect, the gene is repressed. In this case, the enhancer is called a **silencer element.**

Enhancer and promoter elements appear to bind many of the same proteins, implying that both types of regulatory elements affect transcription by a similar mechanism, probably involving interactions of the regulatory proteins, as described earlier. Interestingly, there appear to be a small number of regulatory proteins that control transcription. Therefore, by combining a few regulatory proteins in particular ways, the transcription of different arrays of genes is regulated, and a large number of cell types is specified. The process is called **combinatorial gene regulation.** A theoretical example of

combinatorial gene regulation is shown in Figure 20.6. Maximal transcription of gene *A* involves the binding of activators 1, 2, 3, and 4 to corresponding enhancer sites 1, 2, 3, and 4. Maximal transcription of gene *B* requires the binding of activators 2, 4, and 6 to their corresponding enhancer sequences. That is, each of the two genes requires activators 2 and 4 for full activation, in combination with different additional activators.

KEYNOTE

Eukaryotic protein-coding genes contain both promoter elements and enhancer elements. Promoter elements in the core promoter are required for transcription to begin. Promoter elements in the promoter proximal region have a regulatory function and are specialized for the gene they control, binding specific regulatory proteins that control expression of the gene. Specific regulatory proteins bind also to the enhancer elements and activate transcription through their interaction with proteins bound to the promoter elements. Enhancer elements and promoter elements appear to bind many of the same proteins, implying that both types of regulatory elements affect transcription by a similar mechanism involving interactions of the regulatory proteins.

Case Study: Regulation of Galactose Utilization in Yeast

The regulation of galactose utilization in yeast affords an opportunity to see a number of the principles we have discussed applied in a real regulated system.

Three genes—*GAL1*, *GAL7*, and *GAL10*—encode enzymes needed to utilize the monosaccharide sugar

Figure 20.6

Combinatorial gene regulation. A theoretical example in which (a) the transcription of gene *A* is controlled by activators 1, 2, 3, and 4 and (b) the transcription of gene *B* is controlled by activators 2, 4, and 6.

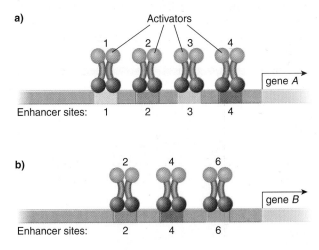

galactose as a source of carbon. In the absence of galactose, the *GAL* genes are not transcribed. When galactose is added, there is a rapid, coordinate induction of transcription of the *GAL* genes and therefore a rapid production of the three galactose-utilizing enzymes if glucose is absent or its concentration is low. In a manner analogous to the repression of the bacterial *lac* operon, glucose in the yeast *GAL* gene system exerts catabolite repression.

Genetic studies have shown that the *GAL1, GAL7,* and *GAL10* genes are located near each other, but do not constitute an operon (Figure 20.7a). Adjacent to each gene are promoter sequences. Genetic studies have also identified an unlinked regulatory gene, *GAL4,* that encodes the protein Gal4p. *GAL4* is expressed only in the absence of glucose. Gal4p is activator protein; its DNA-binding domain is a zinc finger. As a homodimer, Gal4p binds to a regulatory sequence called UAS_G, for "upstream activator sequence for GAL." This sequence is similar to an enhancer sequence, but because UAS

elements function only upstream of the genes they control and not downstream, they do not fit the definition of an enhancer. Within UAS_G are four binding sites for Gal4p: One Gal4p dimer can bind to each site. For simplicity, we will talk about just one Gal4p dimer binding as we discuss the model for regulation of these genes. There is a UAS_G upstream of the *GAL7* gene and a single UAS_G between the *GAL1* and *GAL10* genes. The latter UAS_G controls the expression of both genes. When the genes are activated, transcription proceeds in opposite directions from downstream of the promoters for the genes. For the *GAL1* and *GAL10* genes, the single regulatory sequence controls the transcription of both genes; those genes are *transcribed divergently* (in opposite directions).

In the absence of galactose, a Gal4p dimer binds to each UAS_G (Figure 20.7b). Another protein, Gal80p (encoded by the *GAL80* gene), binds to the Gal4p activation domain that is needed to turn on transcription. (Gal80p is an example of a repressor; its activity to block

a) *GAL* structural genes

b) Absence of galactose

Gal80p binds to Gal4p activation domain, blocking it from activating transcription.

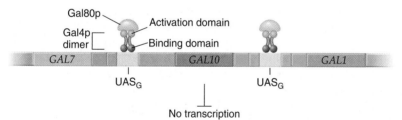

c) Presence of galactose

Gal3p converts galactose to the inducer which binds to Gal80p, causing it to move on Gal4p. The now exposed Gal4p activation domain activates transcription.

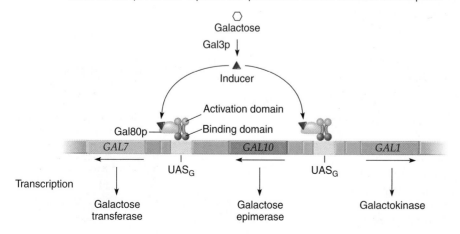

Figure 20.7

Regulation of galactose utilization in yeast. (a) Organization of the *GAL1, GAL7,* and *GAL10* structural genes of yeast on chromosome II. (b) Repression of transcription of the genes in the absence of galactose. (c) Activation of transcription of the genes in the presence of galactose.

activation by Gal4p is called *quenching*.) No transcription occurs under these circumstances.

When yeast is grown in the presence of galactose, the product of the *GAL3* gene, Gal3p, converts galactose into the inducer for the system (Figure 20.7c). The nature of the inducer molecule is not known. The inducer binds to Gal80p and causes it to change position on Gal4p, so no quenching occurs. The Gal4p activation domain then activates the transcription of *GAL1*, *GAL7*, and *GAL10*. In this system, then, Gal4p acts as a positive regulator (activator), Gal80p acts as a repressor, and galactose is an effector molecule.

Recall from our discussion of the *lac* operon that when glucose is present, the *lac* operon is repressed even if lactose is present. This glucose effect occurs because glucose is the preferred carbon source, requiring less energy to metabolize than other sugars, such as lactose. Similarly, glucose is the preferred carbon source for eukaryotic cells, so the *GAL* genes are transcriptionally inactive if glucose and galactose are present. This inactive state of transcription occurs using a repression system. Glucose causes the activation of the repressor Mig1p (product of the *MIG1* gene), which binds to a site with an upstream repressing sequence for galactose (URS$_G$) within the *GAL* gene promoters and blocks Gal4p transcription activation.

Regulation of Gene Expression by Steroid Hormones

Steroid Hormone Regulation of Gene Expression in Animals.

Animals are differentiated into a number of cell types, each of which carries out a specialized function or functions. The cells of animals are not exposed to rapid changes in environment, as are cells of bacteria and of microbial eukaryotes. This is because most cells of animals are exposed to the intercellular fluid, which is nearly constant in the nutrients, ions, and other important molecules it supplies. The constancy of the cell's environment is maintained in part through the action of chemicals called *hormones*, which are secreted by various cells in response to signals and which circulate in the blood until they stimulate their target cells. Elaborate feedback loops control the amount of hormone secreted, as well as the response, so that appropriate levels of chemicals in the blood and tissues are maintained.

A hormone, then, is an effector molecule that is produced by one cell and causes a physiological response in another cell. As shown in Figure 20.8, some hormones—

animation

a Regulation of Gene Expression by Steroid Hormones

Figure 20.8

Mechanisms of action of polypeptide hormones and steroid hormones.

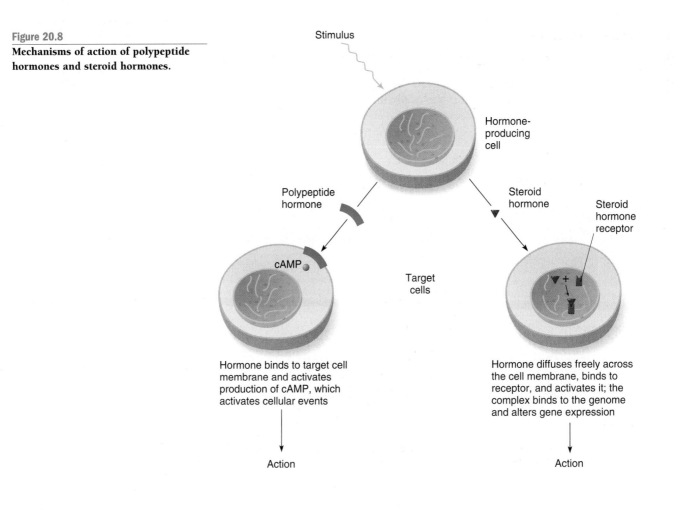

Stimulus

Hormone-producing cell

Polypeptide hormone

Steroid hormone

Steroid hormone receptor

Target cells

cAMP

Hormone binds to target cell membrane and activates production of cAMP, which activates cellular events

Hormone diffuses freely across the cell membrane, binds to receptor, and activates it; the complex binds to the genome and alters gene expression

Action

Action

for example, steroid hormones—act by binding to a specific cytoplasmic receptor called a steroid hormone receptor (SHR), and then the complex binds directly to the cell's genome to regulate gene expression. Other hormones—for example, polypeptide hormones—may act at the cell surface to activate a membrane-bound enzyme, adenyl cyclase, that produces cyclic AMP (cAMP) from ATP. (See Chapter 19, pp. 525–527.) The cAMP acts as an intracellular signaling compound (called a *second messenger*) in a process called *signal transduction* to activate the cellular events associated with the hormone.

A key to hormone action is each hormone acts on specific target cells that have receptors capable of recognizing and binding that particular hormone. For most of the polypeptide hormones (e.g., insulin, ACTH, vasopressin), the receptors are on the cell surface, whereas the receptors for steroid hormones are inside the cell.

We concentrate on steroid hormones here. These hormones have been shown to be important in the development and physiological regulation of organisms ranging from fungi to humans. Figure 20.9 gives the structures of four of the most common mammalian steroid hormones. All have a common four-ring structure; the differences in the side groups are responsible for their different physiological effects.

Table 20.1 presents examples of the induction of specific proteins by selected steroid hormones. As can be seen, there are tissue-specific effects. For example, estrogen induces the synthesis of the protein prolactin in the rat pituitary gland, the protein vitellogenin in frog liver, and the proteins conalbumin, lysozyme, ovalbumin, and ovomucoid in the hen oviduct. Glucocorticoids induce the synthesis of growth hormone in the rat pituitary gland and the enzyme phosphoenolpyruvate carboxykinase in the rat kidney. The specificity of the response to steroid hormones is controlled by the hormone receptors. With the exception of receptors for the steroid hormone glucocorticoid, which are widely distributed among tissue types, steroid receptors are found in a limited number of target tissues. Steroid hormones have well-substantiated effects on transcription, and they also can affect the stability of mRNAs and, possibly, the processing of mRNA precursors.

Mammalian cells contain between 10,000 and 100,000 steroid hormone receptor (SHR) molecules, which are proteins with structures similar to the activators and repressors we have already discussed. That is, they have a DNA-binding domain (BD) and either an activation domain (AD) or a repression domain, depending on the particular SHR. In addition, an SHR has a third domain: the binding domain for the steroid hormone for which it is specific.

All steroid hormones work in the same general way. For example, in the absence of a particular steroid hormone, the appropriate SHR is found in the cytoplasm associated with a large complex of proteins called *chaperones*, one of which is *Hsp90*. In this association, the SHR is inactive. Data from studies of mutant yeast strains suggest that chaperone proteins have an active role in keeping the SHRs functional. When a steroid hormone such as glucocorticoid enters a cell, it binds to the hormone-binding domain (HBD) of its specific SHR molecule, displacing Hsp90 and forming a glucocorticoid-SHR complex (Figure 20.10).

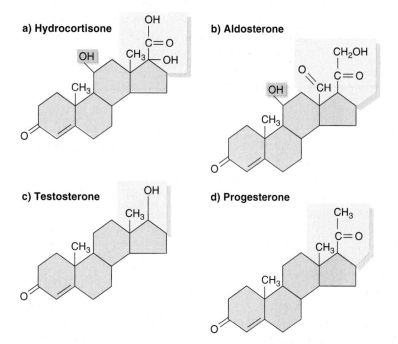

a) Hydrocortisone

b) Aldosterone

c) Testosterone

d) Progesterone

Figure 20.9

Structures of some mammalian steroid hormones.
(a) Hydrocortisone, which helps regulate carbohydrate and protein metabolism. (b) Aldosterone, which regulates salt and water balance. (c) Testosterone, which is used for the production and maintenance of male sexual characteristics. (d) Progesterone, which, with estrogen, prepares and maintains the uterine lining for implantation of an embryo.

Table 20.1	Examples of Proteins Induced by Steroid Hormones		

Hormone	Tissue	Induced Protein
Estrogen	Oviduct (hen)	Conalbumin[a]
		Lysozyme[a]
		Ovalbumin[a]
		Ovomucoid[a]
	Liver (rooster)	Vitellogenin
		Apo, very low density Lipoprotein (apoVLDL)
	Liver (frog)	Vitellogenin
	Pituitary gland (rat)	Prolactin
Glucocorticoids	Liver (rat)	Tyrosine aminotransferase
		Tryptophan oxygenase
	Kidney (rat)	Phosphoenolypyruvate carboxykinase
	Pituitary gland (rat)	Growth hormone
	Oviduct (hen)	Avidin[a]
		Conalbumin[a]
		Lysozyme[a]
		Ovomucoid[a]
	Uterus (rat)	Uteroglobin
Testosterone	Liver (rat)	α-2u Globulin
	Prostate gland (rat)	Aldolase

[a] Egg-white proteins.

This complex now enters the nucleus and binds to specific DNA regulatory sequences, activating or repressing the transcription of the specific genes controlled by the hormone. For genes turned on by the hormone, the new mRNAs appear within minutes after a steroid hormone encounters its target cell, enabling new proteins to be produced rapidly. The DNA-binding domains of many steroid hormone receptor proteins are zinc fingers.

All genes regulated by a specific steroid hormone have in common a DNA sequence to which the steroid–receptor complex binds. The binding regions are called **steroid hormone response elements (HREs).** The "H" in the abbreviation is replaced with another letter to indicate the specific steroid involved. Thus, GRE is the glucocorticoid response element and ERE is the estrogen response element. The HREs are located, often in multiple copies, in the promoters of genes they control. The GRE, for example, is located about 250 bp upstream from the transcription start point. The consensus sequence for GRE is AGAACANNNTGTTCT, where N is any nucleotide. The ERE consensus sequence is AGGTCANNNTGACCT. Note that, for both of these HREs, the sequences on each side of the N's are complementary; that is, the sequences show twofold symmetry.

How the hormone–receptor complexes, once bound to the correct HREs, regulate transcriptional levels is not completely known. Potentially, functional interactions arise among the hormone–receptor complexes and with transcription factors in the transcription initiation complex. To this end, recall that multiple HREs are present for many genes, so multiple hormone–receptor complexes can bind to each gene. These interactions may facilitate the initiation of transcription by RNA polymerase II. It is presumed that each steroid hormone regulates its specific transcriptional activation by the same general mechanism. The unique action of each type of steroid results from the different receptor proteins and HREs involved.

Finally, it is of particular interest that, in different types of cells, the same steroid hormone may activate different sets of genes, even though the various cells have the same SHR. This is because many regulatory proteins bind to both promoter elements and enhancers to regulate gene expression. (See the discussions earlier in this chapter and in Chapter 19). Thus, a steroid–receptor complex can activate a gene only if the correct array of other regulatory proteins is present. Since the other regulatory proteins are specific to the cell type, different patterns of gene expression can result.

In sum, steroid hormones act as effector molecules and SHRs act as regulatory molecules. When the two combine, the resulting complex binds to DNA and regulates gene transcription, producing a large and specific increase

Figure 20.10

Model for the action of the steroid hormone glucocorticoid in mammalian cells.
For the receptor, AD = activation domain, BD = DNA-binding domain,
and HBD = hormone-binding domain.

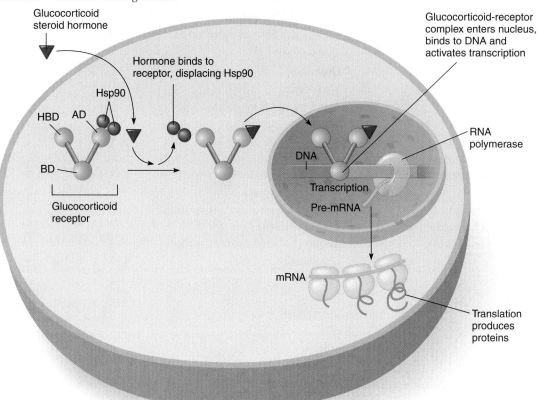

or decrease in cellular mRNA levels. The specific responses characteristic of each steroid hormone result from the fact that receptors are found only in certain cell types, and each of those cell types contains different arrays of other cell-type-specific regulatory proteins that interact with the steroid–receptor complex to activate specific genes.

> ℹ️ Go to the iActivity *Sorting the Signals of Gene Regulation* on your CD-ROM and assume the role of a researcher tracking down some methods used by a eukaryotic cell for the synchronized regulation of genes.

Hormone Control of Gene Expression in Plants. Many chemicals have been identified that play important roles in controlling growth and development in plants. These chemicals, called *plant hormones*, are of five main types—ethylene, abscisic acid, auxins, cytokinins, and gibberellins (Figure 20.11)—and are responsible for many activities. Ethylene, for example, induces ripening in fruits, maturation of flowers, and senescence in plants, among other things.

The *gibberellins* are one of the best-characterized plant hormones. Like many steroid hormones in mammals,

gibberellins stimulate transcription and thereby result in an increase in the production of specific proteins that are responsible for profound changes in plant cell form and cell differentiation.

Gibberellins have been shown to have many effects when applied to plants, such as making certain mutant dwarf plants grow tall or making normal plants grow taller. Gibberellins also are important because they stimulate the germination of some seeds, such as those of barley (Figure 20.12). In this process, the embryo of the seed produces gibberellins that diffuse to the aleurone layer, the outermost layer of the endosperm. As a result of the effect of the gibberellins on gene expression—particularly on the gene for the enzyme α-amylase—the aleurone layer synthesizes and secretes α-amylase, which breaks down the endosperm, releasing nutrients that enable the embryo to grow. By the time the endosperm has been exhausted, the young, developing plant can rely on photosynthesis to obtain nutrients for continued growth.

Even though the action of some gibberellins has been well described, no gibberellin receptor has been detected to date, and the molecular aspects of gibberellins' action are incompletely understood.

Figure 20.11
Chemical structures of the five plant hormones.

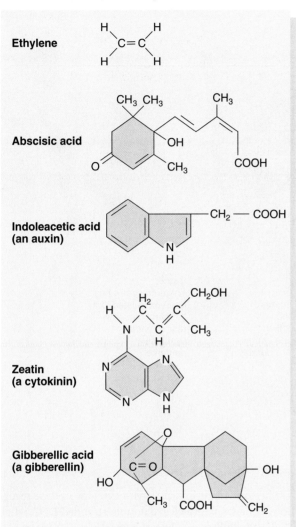

Gene Silencing and Genomic Imprinting

To this point, we have discussed the regulation of transcription at the individual gene level. That is, we saw how specific activators turn on genes and specific repressors turn off genes. By contrast, **gene silencing** is a phenomenon whereby a gene is transcriptionally silent due to its location, not because of the action of a specific repressor. Commonly, gene silencing is a property of heterochromatin (see Chapter 2, p. 34) and, therefore, may involve large sections of DNA and many genes. Heterochromatin is found, for example, at telomeres and the centromere, as well as dispersed throughout the genome.

We have also discussed the role of chromatin structure in regulating the initiation of transcription. Gene silencing in heterochromatin is a related phenomenon. Let us consider gene silencing at the yeast telomere (*TEL*). In yeast, the last 1–1.5 kb of the chromosome consists of telomere repeat sequences folded into a complex hairpin structure (Figure 20.13). Normally, no protein-coding genes are found at telomeres. However, when active genes are moved to a telomere region, those genes are silenced, a phenomenon called *telomere position effect*. This effect is associated with a physical grouping of the telomeres into four or five bouquets physically bound to the nuclear envelope.

A gene moved to a telomere region that becomes silenced can be used to search for mutants that relieve silencing. Such mutants define the silent information regulation genes, *SIR2, SIR3,* and *SIR4,* for the Sir2p, Sir3p, and Sir4p proteins, respectively. Another protein, Rap1p (product of the *repressor–activator protein gene, RAP1*), binds to telomere repeat sequences. Once bound, Rap1p recruits the Sir silencing complex consisting of Sir2p, Sir3p, and Sir4p. The Sir complex also contacts the histones, and Sir2p, a histone deacetylase, catalyzes the local removal of acetyl groups from histone tails. Deacetylated histones are now recognized directly by the silencing complex, causing a wave of binding and deacetylation to spread along the chromosome and generating the heterochromatin structure. The spreading of silencing occurs only for a limited distance; it is stopped by another kind of histone modification: the *methylation* of histone H3 tails. Methylation is catalyzed by *histone methyl transferases* (HMTs).

Transcription can also be silenced by the methylation of particular DNA sequences. This type of silencing is common in many eukaryotes, but is not found in yeast, where DNA methylation occurs at only a very low level. As with chromatin silencing, silencing by DNA methylation can involve large segments of DNA. DNA methylation involves DNA methylase modifying cytosines to produce 5-methylcytosine (5^{m}C) (Figure 20.14). The distribution of 5^{m}C is nonrandom, with most (60–90

KEYNOTE

In multicellular eukaryotes, one of the well-studied systems of short-term gene regulation is the control of protein synthesis by hormones. Polypeptide hormones act at the cell surface, activating a system to produce cAMP, which acts as a second messenger to control gene activity. Steroid hormones exert their action by forming a complex with a specific receptor protein, thereby activating the receptor; the complex then binds directly to particular sequences on the cell's genome to regulate the expression of specific genes. The specificity of hormone action is caused by the presence of hormone receptors only in certain cell types and by interactions of steroid–receptor complexes with cell-type-specific regulatory proteins. Plant gene expression is also, in part, under hormonal control.

Figure 20.12
Effect of gibberellins on the germination of barley seeds.

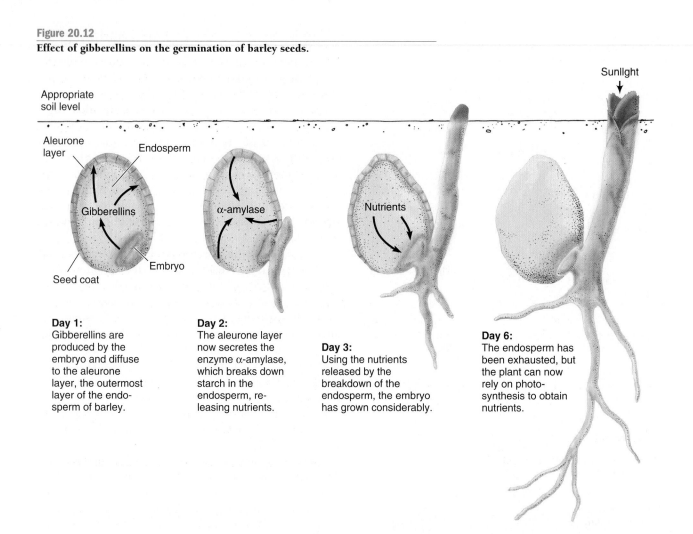

Day 1:
Gibberellins are produced by the embryo and diffuse to the aleurone layer, the outermost layer of the endosperm of barley.

Day 2:
The aleurone layer now secretes the enzyme α-amylase, which breaks down starch in the endosperm, releasing nutrients.

Day 3:
Using the nutrients released by the breakdown of the endosperm, the embryo has grown considerably.

Day 6:
The endosperm has been exhausted, but the plant can now rely on photosynthesis to obtain nutrients.

percent in vertebrate DNA) found in the dinucleotide CpG. This sequence is symmetrical in double-stranded DNA: $\begin{smallmatrix}5'\text{-CG-}3'\\3'\text{-GC-}5'\end{smallmatrix}$. These CG sequences form part of some restriction sites that allow the use of restriction enzymes for the study of the methylation of a segment of

DNA, because many enzymes with cytosine in their recognition sequence fail to cleave double-stranded DNA when cytosine is methylated. The enzyme *Hpa*II ("hepa-two"), for example, cleaves DNA at the sequence 5'-CCGG-3', but not if the internal cytosine of the two is methylated (i.e., if it is 5'-C^{m}CGG-3'). The enzyme *Msp*I ("M-S-P-one") also

Figure 20.13
Gene silencing at a yeast telomere.

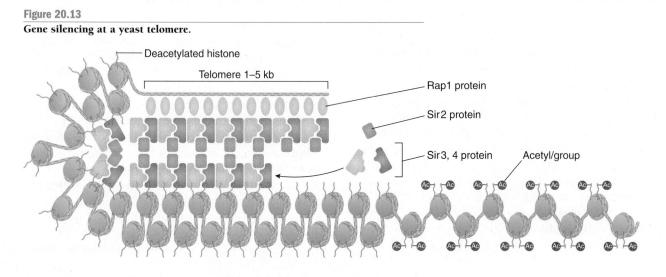

Figure 20.14

Production of 5-methylcytosine from cytosine in DNA by the action of the enzyme DNA methylase.

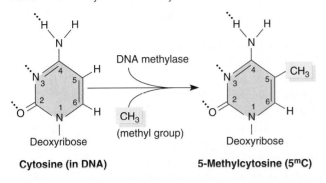

cleaves the same CCGG sequence, but, unlike *Hpa*II, it *will* cleave the methylated sequence C^mCGG.

The use of *Hpa*II and *Msp*I to analyze the extent of methylation of a segment of DNA is illustrated by the following theoretical example: Consider a fragment of genomic DNA that contains three CCGG sequences (Figure 20.15). If the sequences are not methylated (Figure 20.15a), then both *Hpa*II and *Msp*I cleave the DNA at those sequences to produce four DNA fragments of discrete sizes. These DNA fragments can be identified after (1) electrophoresis of the digested DNA on agarose gels, (2) Southern blotting to a membrane filter, and (3) probing of the fragments with a labeled DNA probe that will hybridize with all of the fragments. In this case, DNA fragments of identical sizes are produced from both digests. If one of the CCGG sequences is methylated—for example, the first one from the left in Figure 20.15b— then all three sequences are cleaved by *Msp*I, whereas only the two unmethylated sequences are cleaved by *Hpa*II. As a result, the arrays of DNA fragments produced by digestion of the same DNA piece with the two enzymes are different, as indicated in the stylized blot result. In practice, a piece of DNA is digested separately by the two enzymes *Hpa*II and *Msp*I, and if the results show that the *Msp*I has digested the DNA into more

Figure 20.15

Effect of 5-methylcytosine on cleavage of DNA with *Hpa*II and *Msp*I. (a) Unmethylated (CCGG) sequence. (b) Methylated (C^mCGG) sequence. (The methyl group is on the second C.)

a) Unmethylated

DNA fragment, no methylated CCGG sequences; all sites can be cleaved with *Hpa*II and *Msp*I

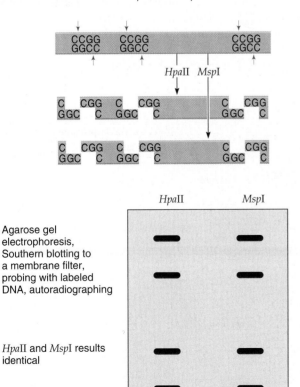

Agarose gel electrophoresis, Southern blotting to a membrane filter, probing with labeled DNA, autoradiographing

*Hpa*II and *Msp*I results identical

b) Methylated

DNA fragment with one methylated C^mCGG site that can be cleaved by *Msp*I, but not by *Hpa*II

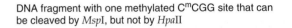
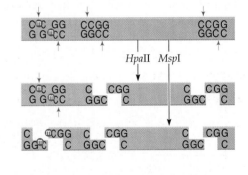

Patterns different for the two enzymes since *Hpa*II was unable to cut the C^mCGG sequence

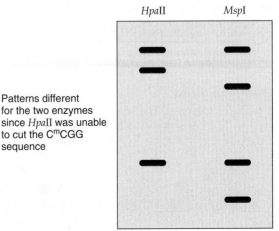

pieces than *Hpa*II, the interpretation is that one of the CCGG sequences is methylated—that is, C^mCGG.

CpG dinucleotides are not distributed randomly throughout vertebrate genomes. That is, some regions of genomes have CpG-rich segments with many copies of the dinucleotides, called **CpG islands.** In the human genome, many protein-coding genes have CpG islands in their promoters. These CpG islands usually are unmethylated, a state that facilitates transcription initiation. However, when CpG dinucleotides become methylated, transcription is repressed. An example of methylation affecting gene expression is found in the development of fragile X syndrome (OMIM 309550; Chapter 17, pp. 466–467), which is the leading cause of inherited mental retardation. The syndrome develops after expansion (a significant increase in the number of copies) of a triplet repeat (a repeated three-base-pair sequence) in the *FMR-1* gene and abnormal methylation of the gene to the point that transcription of the *FMR-1* gene is silenced.

Another example of DNA methylation affecting gene expression also illustrates **genomic imprinting,** a phenomenon in which the expression of certain genes is determined by whether the gene is inherited from the female or male parent. (Imprinting is an example of an epigenetic phenomenon; see Chapter 12, p. 324). This example involves a pair of linked genes—*Igf2* (insulinlike growth factor 2) and *H19* (which encodes an untranslated mRNA of unknown function)—located about 80 kb apart in humans and mice (Figure 20.16). In mice, studies of the inheritance of a deletion that removed gene *Igf2* showed that progeny inheriting the deletion chromosome from the male parent were small, but progeny inheriting the same deletion from the female parent were of normal

size. This indicated that the deletion of *Igf2* gives a mutant phenotype only when the deletion is inherited from the male parent. The interpretation is that *Igf2* is an imprinted gene, being expressed from the paternal chromosome. Similarly, the *H19* gene is an imprinted gene expressed from the maternal chromosome.

The imprinting process for *Igf2* and *H19* is as follows: A single enhancer located downstream of the *H19* gene controls the expression of both genes (Figure 20.16). When activators bind to the enhancer, the transcription machinery could be recruited to both genes. However, another regulatory element located between the genes affects this activation. This regulatory element is an **insulator,** so named because, when functional, it blocks the activation of a promoter to one side of it by activators bound to an enhancer on the other side of it. On the maternal chromosome, the genes and regulatory sequences are not methylated, allowing protein CTCF to bind to the insulator, blocking the activation of *Igf2* transcription by the activator at the enhancer. But the activator can activate *H19* transcription (Figure 20.16a). In other words, on the maternal chromosome, *Igf2* is inactive and *H19* is active. On the paternal chromosome, the DNA is methylated for a segment of the chromosome encompassing the promoter of the *H19* gene and the insulator (Figure 20.16b). Therefore, CTCF cannot bind to the insulator, and the activator is then able to activate the transcription of *Igf2*. However, because of the methylation of its promoter, the activator cannot activate *H19*. In other words, on the paternal chromosome, *Igf2* is active and *H19* is inactive.

Key to imprinting is the methylation of specific DNA sequences and the inheritance of those methylated sequences. For mitotic cell divisions, this inheritance is

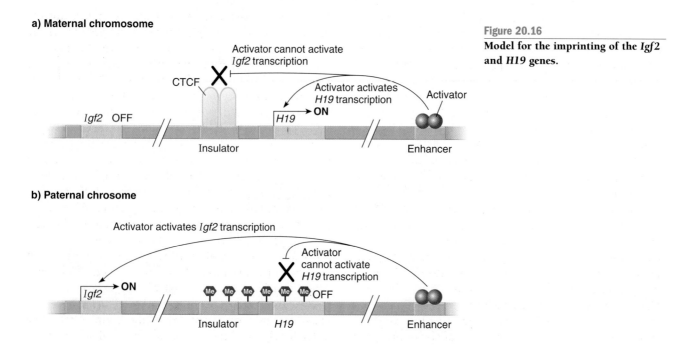

a) **Maternal chromosome**

b) **Paternal chrosome**

Figure 20.16
Model for the imprinting of the *Igf 2* and *H19* genes.

straightforward. After DNA replication, each daughter DNA double helix is hemimethylated; that is, one strand will have the parental methylation pattern and the other strand will be unmethylated. Maintenance methylases recognize the hemimethylation and methylate the new DNA strand to restore the parental methylation pattern. However, the situation is different in passage through meiosis. That is, the methylation imprint is established in the parental germ line and is reset each generation. Consider an allele that is imprinted on the paternal chromosome, such as *H19*. In oogenesis in a female, both maternal and paternal alleles are reactivated, but in spermatogenesis in a male, both maternal and paternal alleles are imprinted (methylated by de novo methylases). Progeny of these two parents inherit a silenced allele from the father and an active allele from the mother.

Some human genetic diseases, such as the Prader–Willi and Angelman syndromes, appear to result from imprinting. Prader–Willi syndrome (PWS; OMIM 176270) occurs in about 1 in 25,000 births. Individuals with PWS typically are small and weak at birth, and their symptoms include retardation and poor feeding caused by diminished swallowing and sucking reflexes. The feeding difficulties improve by the age of 6 months, and from about 12 months onward a pattern of uncontrollable eating develops, leading to obesity and associated psychological problems. Adolescents have poor motor skills and insatiable hunger. Adults are short compared with their family members and often develop a form of diabetes because of the eating disorder. Individuals with PWS rarely live beyond 30 years, unless they maintain strict weight control programs to control the diabetes.

PWS is caused by the deletion or disruption of a gene or several genes in region 15q11–q13 of chromosome 15. Pedigree analysis has shown that, in 70–80% of cases examined, the deletion or disruption occurred in the father and that genomic imprinting plays a role. That is, in a child with PWS, the activities of some genes in region 15q11–q13 on maternal chromosome 15 normally are suppressed as a result of genomic imprinting. The suppression occurs by methylation of the genes. The paternally inherited alleles are necessary for normal development, but because of the gene deletion or disruption event in the father, those genes are also inactive, and the PW phenotype results.

Individuals with Angelman syndrome (AS; OMIM 105830) have symptoms that include severe motor and intellectual retardation, a smaller-than-normal head size, jerky limb movements, hyperactivity, and frequent unprovoked laughter. In about 50% of patients with AS, a deletion of region 15q11–q13 is seen. This is the same region affected in individuals with PWS. Indeed, it seems that AS can be caused in much the same way as PWS, except that in AS maternally inherited alleles of the genes

involved are needed for normal development. That is, the paternally inherited alleles are inactive because of methylation brought about by genomic imprinting, which causes AS to develop if the maternally inherited alleles are deleted or disrupted.

KEYNOTE

Gene silencing is the phenomenon of turning off the transcription of a gene as a result of its position in the chromosome. Gene silencing involves changes in chromatin structure to produce heterochromatin, which usually affects a zone of genes. A gene may also be silenced through the methylation of cytosines in the promoter upstream of the gene. Sometimes the methylation pattern is associated with genomic imprinting, a phenomenon in which the expression of a gene is determined by whether the gene is inherited from the female or male parent.

Posttranscriptional Control

RNA Processing Control

RNA processing control regulates the production of mature RNA molecules from precursor-RNA molecules. As we discussed in Chapter 5, all three major classes of RNA molecules—mRNA, tRNA, and rRNA—are synthesized mostly as larger precursor molecules. In that chapter, we discussed the synthesis of pre-mRNA and its processing to mature mRNA. When we examine living systems, we do not always see the "textbook" processing steps. For example, there are many cases in which **alternative polyadenylation** sites may be used to produce different pre-mRNA molecules and **alternative splicing** (also called *differential splicing*) may be used to produce different functional mRNAs. Which product is generated depends on regulatory signals. The products of alternative polyadenylation or alternative splicing are proteins that are encoded by the same gene, but that differ structurally and functionally. Such proteins are called *protein isoforms,* and their synthesis may be tissue specific. Alternative polyadenylation is independent of alternative splicing.

Figure 20.17 gives an example of how alternative polyadenylation *and* alternative splicing together result in tissue-specific products of the human calcitonin gene (*CALC*). *CALC* consists of five exons and four introns and is transcribed in certain cells of the thyroid gland and in certain neurons of the brain. Alternative polyadenylation occurs with the polyadenylation

Figure 20.17

Alternative polyadenylation and alternative splicing resulting in tissue-specific products of the human calcitonin gene *CALC*. In the thyroid gland, calcitonin is produced, whereas in certain neurons, CGRP (calcitonin-gene-related peptide) is produced.

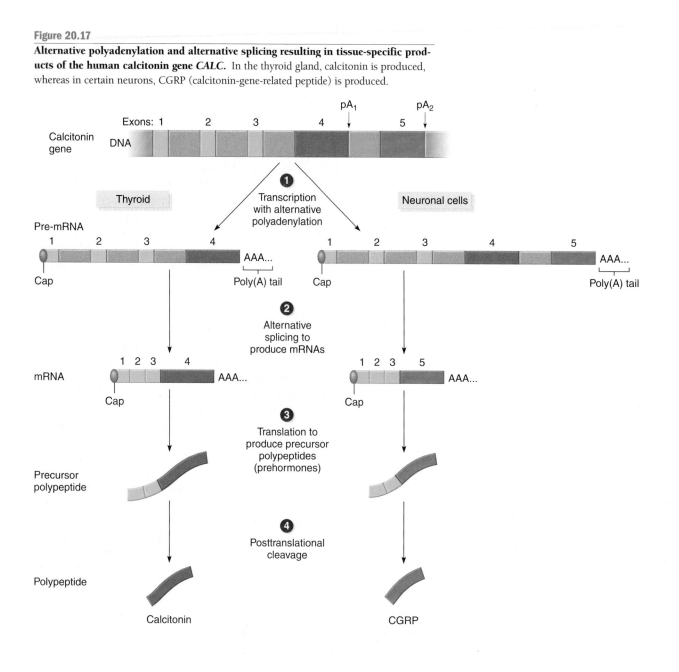

site next to exon 4, pA_1, used in thyroid cells, and the polyadenylation site next to exon 5, pA_2, used in the neuronal cells (Figure 20.17, part 1).

Alternative splicing occurs at the next stage of intron removal (Figure 20.17, part 2). The pre-mRNA in the thyroid is spliced to remove the three introns and bring together exons 1, 2, 3, and 4. The pre-mRNA in the neuronal cells is spliced to remove introns and to bring together introns 1, 2, 3, and 5; exon 4 is excised and discarded. The mRNAs that are produced are translated to yield precursor polypeptides (prehormones; see Figure 20.17, part 3), from which the functional hormones are generated posttranslationally by protease cleavage (Figure 20.17, part 4). The two products are

calcitonin in the thyroid, with its amino-acid sequence encoded by exon 4, and CGRP (calcitonin gene-related peptide), with its amino-acid sequence encoded by part of exon 5. (The remainder of exon 5 is the 3′ trailer part of the mRNA.) The mechanisms by which this alternative polyadenylation and alternative splicing occur are not known. The outcome is two different polypeptides encoded by the same gene and synthesized in two different tissues. The thyroid hormone calcitonin is a circulating calcium ion homeostatic hormone that aids the kidney in retaining calcium. CGRP is found in the hypothalamus and appears to have neuromodulatory and trophic (growth-promoting) activities.

KEYNOTE

Gene expression in eukaryotes can be regulated at the level of RNA processing. This type of regulation operates to direct the production of mature RNA molecules from precursor-RNA molecules. Two regulatory events that exemplify this level of control are the choice of poly(A) site and the choice of splice site. In both cases, different types of mRNAs are produced, depending on the choices made.

mRNA Transport Control

The next level of control is *mRNA transport control,* which is the regulation of the number of transcripts that exit the nucleus to the cytoplasm. A profound difference between prokaryotes and eukaryotes is the presence of a nucleus with a surrounding membrane in eukaryotes. This nuclear envelope can be a control point in gene expression.

We know that primary transcripts are processed extensively in the nucleus. Experiments have also shown that perhaps one-half of the primary transcripts of protein-coding genes (the hnRNA population) never leave the nucleus, but are degraded there. How is the transport of mature mRNA from the nucleus regulated? The mRNA exits through the nuclear pore complexes. The mRNAs must be capped for them to exit. In addition, the export process involves proteins that bind to mRNA molecules and interact with proteins at the nuclear pore complexes to direct the movement of the mRNAs to the cytoplasm. Unspliced pre-mRNAs are not exported. Why not? Recall that the processing of pre-mRNA involves snRNPs (small nuclear ribonucleoprotein particles; see Chapter 5, pp. 98–99). There is some evidence that snRNPs are important for retaining unspliced mRNAs in the nucleus. For example, in mutants of yeast that prevent the assembly of spliceosomes (the splicing complexes formed by the association of several snRNPs bound to the pre-mRNA; see Chapter 5, p. 98), mRNA export from the nucleus is facilitated. This has led to a *spliceosome retention* model in which spliceosome assembly competes with nuclear export. Thus, while pre-mRNAs are in spliceosomes undergoing processing, the RNA is retained in the nucleus, unable to interact with the nuclear pore. However, when processing is complete, the mRNA dissociates from the spliceosome, and the spliceosome remains associated with the intron. The free mRNA (using the proteins mentioned earlier) can interact with the nuclear pore, but the intron cannot.

KEYNOTE

The transport of mRNAs from the nucleus to the cytoplasm is another important control point in eukaryotic gene regulation. Proteins are needed to facilitate the exit of mature mRNAs. For pre-mRNAs, the presence of spliceosomes signals retention of the RNA in the nucleus. That is, during intron removal, the spliceosome keeps the immature RNA in the nucleus, but when all introns have been removed and the spliceosome has dissociated, the free, mature, capped mRNA can interact with the nuclear pore complex and exit.

mRNA Translation Control

Messenger RNA molecules are subject to *translational control* by ribosome selection among mRNAs. Differential translation can greatly affect gene expression. For example, mRNAs are stored in many unfertilized vertebrate and invertebrate eggs. In the unfertilized egg, the rate of protein synthesis is very slow; however, protein synthesis increases significantly after fertilization. Since this increase occurs without new mRNA synthesis, it is likely that translational control is responsible. The mechanisms involved are not completely understood, but typically, stored mRNAs are associated with proteins that both protect the mRNAs and inhibit their translation. Furthermore, the poly(A) tail is known to promote the initiation of translation. It has been found that, in general, stored, inactive mRNAs have shorter poly(A) tails (15–90 As) than active mRNAs have (100–300 As). mRNAs synthesized in growing oocytes that are destined for storage and later translation have short poly(A) tails.

In principle, an mRNA molecule can have a short poly(A) tail either because only a short string of A nucleotides was added at the time of polyadenylation or because a normal-length poly(A) tail was added that was subsequently trimmed. At least for some messages stored in growing oocytes of mouse and frog, the latter mechanism is involved. In one example, the examination of one particular mRNA in this class has shown that the pre-mRNA still in the process of intron removal has a long poly(A) tail (300–400 As), whereas the mature, stored message has a short poly(A) tail (40–60 As). It has been shown that the decrease in length of the poly(A) tail for this message class occurs rapidly in the cytoplasm by a deadenylation enzyme. What pinpoints a particular mRNA for rapid deadenylation, rather than a default, slow decrease in poly(A) length, is a sequence in the 3′ untranslated region (3′ UTR) of the mRNA upstream of the AAUAAA polyadenylation sequence. This signal for deadenylation is called the *adenylate/uridylate (AU)-rich element (ARE)* and has the consensus sequence UUUUUAU. Interestingly, to activate a stored mRNA in this class, a cytoplasmic polyadenylation enzyme recognizes the ARE and adds 150 A nucleotides or so. Thus, the same sequence element is used to control the poly(A) tail length, and therefore mRNA translatability, at different times and in opposite ways.

mRNA Degradation Control

Once in the cytoplasm, all RNA species are subjected to **degradation control,** in which the rate of RNA breakdown (also called RNA turnover) is regulated. Usually, both rRNA (in ribosomes) and tRNA are highly stable species. By contrast, mRNA molecules exhibit a diverse range of stability, with some mRNA types known to be stable for many months whereas others degrade within minutes. The stability of particular mRNA molecules may change in response to regulatory signals. For example, the addition of a regulatory molecule to a cell type can lead to an increase in the synthesis of a particular protein or proteins, accomplished by an increase in the rate of transcription of the genes involved or an increase in the stability of the mRNAs produced. Table 20.2 presents examples of systems in which changes in mRNA stability for a number of cell types occur in the presence and absence of specific effector molecules.

mRNA degradation is an important control point in the regulation of gene expression in eukaryotes. Various sequences or structures have been shown to affect the half-lives of mRNAs, including the AU-rich elements (ARE) discussed earlier and various secondary structures. Two major mRNA decay pathways are the *deadenylation-dependent decay pathway* and the *deadenylation-independent decay pathway*. In the deadenylation-dependent decay pathway, the poly(A) tails are deadenylated until the tails are too short (5–15 As) to bind PAB (poly(A) binding protein). In yeast, the product of the *PAN1* gene, PAB-dependent poly(A) nuclease, catalyzes the deadenylation. Once the tail is almost removed, the 5′ cap structure is removed in a step called *decapping*, an enzyme-catalyzed process. In yeast, the decapping enzyme, or at least an essential part of it, is encoded by the *DCP1* gene. After an mRNA molecule is decapped, it is degraded from the 5′ end by a 5′-to-3′ exonuclease. In yeast, this enzyme—encoded by the *XRN1* gene—is highly aggressive, attesting to the importance of the 5′ cap in protecting active mRNAs in the cell.

Yeast strains with a mutant *DCP1* gene are viable, and mRNA degradation still occurs, providing evidence for the existence of mRNA degradation pathways other than the pathway just described. In these deadenylation-independent decay pathways, mRNAs may be decapped without being deadenylated, thereby exposing them to rapid degradation by 5′-to-3′ exonucleases, or they may be cleaved internally by endonucleases without being deadenylated and then may be broken down further.

Note that, although our understanding of mRNA degradation in yeast is becoming clearer, the details of mRNA degradation in mammalian cells are not as well known. Both deadenylation-dependent and deadenylation-independent decay pathways exist in mammals, and decapping is an important step in at least the former pathway.

Protein Degradation Control

Regulatory mechanisms also exist at the posttranslational level. These mechanisms determine the lifetime of a protein. This topic is peripheral to our discussion of gene expression, so it is discussed only very briefly here.

A wide variety of possibilities exist to regulate the amount of a particular protein in a cell. A constitutively produced mRNA may be translated continuously, with the level of protein product controlled by the degradation rate of that protein, or a short-lived mRNA may encode a protein that is highly stable so that it persists for very long periods in the cell. Proteins in the lens of higher vertebrate eyes, for example, are long lived. Their mRNAs have long since been degraded, but the protein itself persists, usually for the lifetime of the individual. By contrast, steroid receptors and heat-shock proteins have short half-lives.

The degradation of proteins (*proteolysis*) in eukaryotes has been shown to require the protein cofactor *ubiquitin* (a protein found apparently ubiquitously). The binding of ubiquitin to a protein identifies it for degradation by proteolytic enzymes. Ubiquitin is released intact

---

Table 20.2 **Examples of Tissues or Cells in Which Regulation of mRNA Stability Occurs in Response to Specific Effector Molecules**[a]

mRNA	Tissue or Cell	Regulatory Y Signal (= Effector Molecule)	Half-Life of mRNA	
			With Effector	Without Effector
Vitellogenin	Liver (frog)	Estrogen	500 h	16 h
Vitellogenin	Liver (hen)	Estrogen	~24 h	<3 h
Apo-very low density lipoprotein	Liver (hen)	Estrogen	~20–24 h	<3 h
Ovalbumin, conalbumin	Oviduct (hen)	Estrogen, progesterone	>24 h	2–5 h
Casein	Mammary gland (rat)	Prolactin	92 h	5 h
Prostatic steroid-binding protein	Prostate (rat)	Androgen	Increases 30 ×	

[a] Note that the effector molecule in each case results in an increase in transcription, as well as stabilization, of the mRNA.

during degradation, enabling it to be used to tag other proteins for degradation.

The amino acid at the N-terminus of a protein is the key to how the protein initially is targeted for ubiquitin binding. In what has become known as the *N-end rule,* the particular N-terminal amino acid relates directly to the half-life of the protein. In a yeast test system in which the lifetime of the same protein was measured with different N-terminal amino acids, arginine, lysine, phenylalanine, leucine, and tryptophan each specified a half-life of 3 minutes or less, whereas cysteine, alanine, serine, threonine, glycine, valine, proline, and methionine all specified a half-life of more than 20 hours. The same general hierarchy is seen in an *E. coli* system. The N-terminal amino acid directs the rate at which ubiquitin molecules can bind to the protein, which, in turn, determines the half-life of the protein.

In sum, in prokaryotes, gene expression is controlled mainly at the transcriptional level, in association with the rapid turnover of mRNA molecules. In eukaryotes, gene expression is regulated at transcriptional, posttranscriptional, and posttranslational levels. Regulatory systems exist for transcription control, precursor-RNA processing, transport out of the nucleus, degradation of mature RNA species, translation of the mRNA, and degradation of the protein product. The intertwining of the regulatory events at these different levels leads to the fine-tuning of the amount of protein in the cell.

KEYNOTE

> Gene expression is also regulated by the control of mRNA translation and degradation. mRNA degradation is believed to be a major control point in regulating gene expression. Structural features of individual mRNAs have been shown to be responsible for the range of mRNA degradation rates, although the precise roles of cellular factors and enzymes have yet to be determined. Protein degradation is also regulated. Which amino acid is at the protein's terminus correlates with the stability of the protein and directs the rate at which ubiquitin binds to the protein. In turn, that rate determines the rate of protein break down.

RNA Interference: A Mechanism for Silencing Gene Expression

RNA interference (RNAi) is a mechanism by which a small fragment of double-stranded (ds)RNA whose sequence matches part of a gene's sequence, interferes with (i.e., silences) the expression of that gene. RNAi was discovered in research with the nematode *C. elegans.* In 1995, S. Guo and K. Kemphues had shown that antisense RNA (see Chapter 9, p. 236) made in vitro and injected into the worm could interfere with a gene's activity. Surprisingly, the control-sense RNA had a similar effect. Later, in 1998, A. Fire and C. Mello discovered that it was dsRNA made from the sense and antisense RNAs synthesized in the in vitro reactions that was responsible for the interfering activity. Indeed, injecting dsRNA into an adult worm results in the loss of the corresponding mRNA in a highly specific way in that worm and in its progeny. Since then, RNAi by dsRNA has been found to occur in a wide variety of organisms. RNAi has natural functions, including protection against viral infections and the regulation of developmental processes. RNAi is highly specific, and only a few dsRNA molecules per cell are needed to achieve effective interference. Hence, the use of RNAi has become very valuable in research. Some researchers praise RNAi as the most important discovery in molecular biology in the past decade.

How does dsRNA result in silencing of the expression of a specific gene? By using dsRNA corresponding to different parts of a gene, it was shown that the target of RNAi is the mature mRNA. That is, dsRNA matching exons causes silencing, but dsRNA matching promoter or intron sequences does not. Genetic and biochemical investigations involving a number of model organisms have shown that RNAi proceeds by an essentially common mechanism, presented in Figure 20.18. Double-stranded RNA in the cell is cleaved by the enzyme *Dicer* into approximately 21–23-bp dsRNA fragments called **short interfering RNA (siRNA).** Dicer cuts so as to leave 3′ overhangs. The siRNA–Dicer complex recruits other proteins, to which the siRNA is transferred to form the *RNA-induced silencing complex (RISC).* RISC is then activated in an ATP-dependent manner, which leads to unwinding of the double-stranded siRNA. Activated RISC uses the single-stranded siRNA to pair with a complementary mRNA, and then, using an as-yet unidentified endoribonuclease, it cleaves that mRNA. As a result, gene expression is silenced. The cleaved mRNA is subsequently degraded.

The activated RISC may function in other ways. When the incorporated single-stranded siRNA binds to an mRNA, instead of causing cleavage of the mRNA, it can remain bound and therefore block translation by ribosomes. Or the complex can migrate into the nucleus, where the siRNA directs binding to its complementary DNA region. This leads to the recruitment of a chromatin remodeling complex that modifies the chromosome around the gene's promoter and silences transcription of the gene.

The effectiveness of RNAi is the result of another property of activated RISC—that of amplification. That is, when the siRNA in that complex binds to its complementary mRNA sequence, another option is to prime RNA synthesis, resulting in the generation of a dsRNA molecule from the point of binding to the 5′ end of the

Figure 20.18

RNA interference (RNAi) by small interfering RNAs (siRNAs). RISC* = activated RISC.

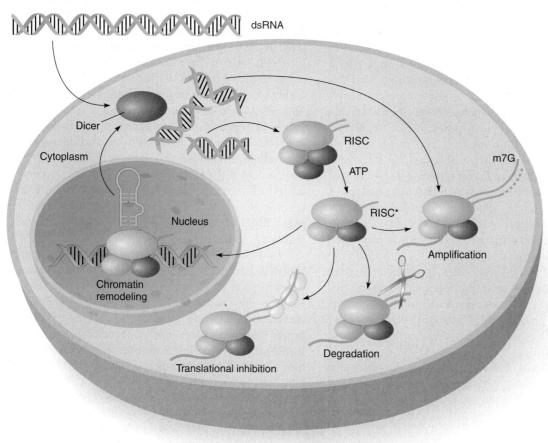

mRNA. This dsRNA becomes a new substrate for Dicer, and the cycle repeats, amplifying the interference signal.

In vivo, RNAi has been shown to have various functions. One function is the regulation of gene expression, and this is exemplified by the role of siRNAs in *C. elegans* development. RNAi also plays a role in blocking the expression of foreign genes, for example, after viral infections, or from transposon activity.

RNAi is rapidly becoming an accepted molecular tool for the manipulation of gene expression in basic and applied research. That is, RNAi is being used as an alternative to gene deletion techniques to "knock out" gene function in model organisms. More accurately, the procedure is a "knock-down," because gene expression may not be abolished completely. For example, in early 2003, researchers at Harvard showed that RNAi had therapeutic possibilities; more specifically, they demonstrated that an siRNA could protect against disease. In their experiments, a gene called *Fas* was targeted. This gene was chosen because its product, a cell surface receptor, Fas, is required for cell death in the liver. A severe hepatitis was induced in mice, and the effects of injecting an siRNA

targeted against the *Fas* gene were analyzed. The scientists used a chemical to induce hepatitis, thereby turning on the Fas-dependent cell death pathway in the liver. Introducing siRNA against *Fas* blocked the development of the hepatitis in a high proportion of the animals and protected most from liver failure and death. All of the mice in the untreated control group died.

K E Y N O T E

Individual genes can be silenced by a mechanism called RNA interference (RNAi). In RNAi, double-stranded RNA is cut into small fragments of short interfering RNA (siRNA) which, in a complex with proteins, binds to complementary sequences in mRNAs and triggers the cleavage of those mRNAs. The result is a knockout or knockdown of gene expression. RNAi has natural functions—for example, protection against viral infections and the regulation of developmental processes—as well as an increasing number of uses in experimental research.

Summary

In this chapter, we have considered a number of examples of gene regulation in eukaryotes. The general picture is one of much greater complexity than in prokaryotes. For most eukaryotes, genes are not organized into operons; nonetheless, genes with related functions typically are regulated coordinately. The exception is found in the nematodes, which have operons. That is, single promoters lead to the synthesis of polygenic mRNAs. However, the individual genes on those mRNAs are not translated sequentially; rather, the polygenic mRNA is processed into monogenic mRNAs in a process called trans-splicing.

In eukaryotes, protein-coding gene expression is regulated mostly at the level of transcription initiation. In contrast to its counterpart in prokaryotes, the chromosome structure of eukaryotes is a significant factor in regulating gene expression. That is, the association of DNA with histones has a repressing action on gene expression. Then, for a gene to be activated, the chromatin structure must be altered in the area of the core promoter. This chromatin remodeling process involves large protein complexes, some of which acetylate histones to loosen their association with DNA and others of which alter nucleosome position or structure, thereby opening the way for binding of the transcription machinery to the core promoter.

Transcription activation requires three classes of proteins: general transcription factors, activators, and coactivators. The general transcription factors bind to the core promoter and are required for basal transcription. The activators play a role in chromatin remodeling as well as activate transcription. Activators bind to enhancers and proximal promoter elements and activate transcription by recruiting a coactivator—a large multiprotein complex that does not by itself bind to DNA, but bridges between activator proteins and general transcription factors. Once bound to activators, coactivators have the important function of recruiting RNA polymerase II, which, through interaction with the transcription factors, is oriented correctly for transcription initiation.

While much of the regulation of gene expression occurs by a positive regulatory system—activation of transcription—for some genes negative regulation occurs using repressors. These repressors bind to the DNA and act in various ways to block or limit transcription initiation.

Another phenomenon affecting gene expression at the transcriptional level is gene silencing. Rather than affecting individual genes, gene silencing affects a zone of genes by changing the chromosome to the condensed heterochromatin state. Thus, a gene is silent in this case, not because of the effect of a specific repressor or the absence of an activator, but because of its position on the chromosome. Transcription of genes can also be silenced by the methylation of particular DNA sequences. One special version of this methylation-based silencing is genomic imprinting, in which the expression of certain genes is determined by whether the gene is inherited from the female or male parent. This epigenetic phenomenon depends upon methylation silencing of either the gene on the maternal chromosome or the gene on the paternal chromosome and the inheritance of that methylation pattern through subsequent cell divisions.

Gene expression can also be regulated at a number of posttranscriptional levels. One such level is RNA processing. This type of regulation operates to determine the production of mature RNA molecules from precursor-RNA molecules. Two regulatory events that exemplify this level of control are the choice of poly(A) site and the choice of splice site. In both cases, different types of mRNAs are produced, depending on the choice made. Examples are known of both types of regulation in developmental systems. For example, different classes of immunoglobulin molecules can result from poly(A) site selection, and the choice of splice site plays a key role in the *Drosophila* sex determination system.

mRNA transport from the nucleus to the cytoplasm is another important control point in the regulation of gene expression in eukaryotes. A spliceosome retention model has been proposed for this regulatory system. In that model, spliceosome assembly on a precursor-mRNA prevents nuclear export of the molecule. It is argued that, during intron removal, the spliceosome that is complexed with the precursor-mRNA prevents the RNA molecule from leaving the nucleus. Then, when all introns have been removed and the spliceosome has dissociated from the now-mature mRNA, the free mRNA molecule can interact with the nuclear pore and move to the cytoplasm.

Gene expression is also regulated by mRNA translation control and by mRNA degradation control. The latter is believed to be a major control point in the regulation of gene expression, as evidenced by the wide range of mRNA stabilities found within organisms. It is clear that nucleases are ultimately responsible for the degradation of the RNAs, and the signals for the differential mRNA stabilities seem to be a property of the structural features of individual mRNAs. For example, a group of AU-rich sequences in the 3′-untranslated regions of some short-lived mRNAs is responsible for their instability, although how the associated signal is read by the cell and transmitted to the cellular factors and enzymes that must be directly involved in the mRNA degradation remains under investigation.

In sum, a great deal of information has been learned in the past decade or so about gene regulation in eukaryotes. We have merely scratched the surface in this chapter. Thousands of researchers are working to elaborate the molecular details of gene regulation in model systems. Much of our advancing knowledge has been made possible by the application of recombinant DNA and related technologies, and we can look forward to sustained and rapid increases in our understanding of eukaryotic gene regulation.

Analytical Approaches to Solving Genetics Problems

Q20.1 A region of the yeast chromosome specifies three enzyme activities in the histidine biosynthesis pathway; these activities are synthesized coordinately. How would you distinguish between the following three models?

a. Three genes are not organized into an operon. They code for three discrete mRNAs that are translated into three different enzymes.

b. Three genes are arranged in an operon. The operon is transcribed to produce a single polygenic mRNA whose translation produces three distinct enzymes.

c. One gene (a supergene) is transcribed to produce a single mRNA whose translation produces a single polypeptide with three different enzyme activities.

A20.1 A key feature of an operon (model b) is that a contiguously arranged set of genes is transcribed onto a single polygenic mRNA. Thus, a nonsense mutation in a structural gene will result in the loss of not only the enzyme activity coded for by that gene, but also the enzyme activities coded for by the structural genes that are more distant from the promoter. (See Chapter 8.) If there is a supergene coding for a single polypeptide with three different enzyme activities (model c), then nonsense mutations will cause effects similar to polar effects in polygenic mRNAs; that is, a nonsense mutation in the coding region for the first enzyme activity will cause a loss of that activity, as well as of the two other enzyme activities. However, if there are three genes that are closely linked, but each with its promoter (model a), then transcription will produce three distinct mRNAs, so a nonsense mutation will affect only the gene in which it is located and no other gene. Therefore, if nonsense mutations are shown to have polar effects, then either model b or model c is correct, but model a cannot be correct. If nonsense mutations do not have polar effects, no matter in which gene they are located, then model a must be correct.

Characterizing the enzyme activities coded for by the three genes would enable us to distinguish between models b and c. That is, in the operon model b, three distinct polypeptides would be produced. These polypeptides could be isolated and purified individually by using standard techniques.

Note that it would be particularly important to make sure that inhibitors of protein-degrading enzymes (i.e., proteases) are present during cell disruption, so that if a

trifunctional polypeptide is present, it is not cleaved by the proteases to produce three separable enzyme activities.

Thus, if the operon model b is correct, we could show that there are three distinct polypeptides, each exhibiting only one of the enzyme activities—that is, three polypeptides and three enzyme activities. By contrast, if the supergene model c is correct, it should be possible only to isolate a large polypeptide with all three enzyme activities (again assuming careful isolation and purification procedures); no polypeptides with only one of the enzyme activities should exist.

Questions and Problems

20.1 How common is it for eukaryotic organisms to have genes organized into operons? How are these operons different from those found in prokaryotic organisms? In what ways are they similar?

*20.2 A nonsense mutation occurs in the first transcribed gene in a eukaryotic operon. If the mutation is close to the 5′ end of the open reading frame of the gene, do you expect it to show polarity? Why or why not?

20.3 Three genes *a*, *b*, and *c* in the nematode *C. elegans* are very tightly linked in the order *a–b–c* and have unrelated functions. An analysis of cDNA and genomic DNA sequences for each gene reveals that their mRNAs share a common 22-basepair 5′ region, but that these 22 base pairs are not present in the genomic sequences of these genes. Furthermore, the *b* and *c* genes lack promoters. Explain the mechanism that results in each mRNA gaining the same 22-bp 5′ sequence, and how *b* and *c* can be transcribed without promoters.

20.4 Critically evaluate the following contention: Prokaryotes and eukaryotes use fundamentally different mechanisms to control gene expression.

*20.5 In *Drosophila*, pulses of the steroid-hormone ecdysone trigger molting between the larval stages and then, at the end of the larval stages, trigger the formation of a pupa, where the larva will metamorphose into an adult fly. Immediately after the ecdysone pulse at the end of the larval stages, transcription of several genes, including *Eip93F*, is dramatically increased. To investigate how ecdysone regulates *Eip93F*, chromatin is isolated from staged wild-type animals just before and just after the ecdysone pulse at the end of the larval stages. The chromatin is distributed to separate test tubes where it is digested for two minutes with different concentrations of DNase I. DNA is then purified from each sample and digested with *Eco*RI. The resulting DNA fragments are then resolved by size using gel electrophoresis and a Southern blot is made. The Southern blot is probed with two *Eco*RI fragments from the *Eip93F* gene: a 4.0-kb fragment from its promoter and a 3.0-kb fragment from its protein-coding region. The following figure shows the results, where the thickness of the band corresponds to the intensity of hybridization signal:

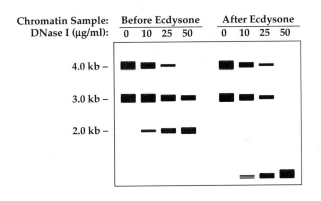

a. Explain why the 4-kb band, but not the 3-kb band, diminishes in intensity when chromatin that was isolated before the pulse of ecdysone is treated with increasing concentrations of DNase I. How do you explain the increasing amounts of the 2-kb band in these samples?

b. Explain why both the 4-kb and 3-kb bands diminish in intensity when chromatin isolated after the pulse of ecdysone is treated with increasing amounts of DNase I. How do you interpret the increasing amounts of low molecular weight digestion products in these samples?

20.6 Chromatin remodeling is essential for gene activation and can be achieved using different mechanisms.

a. What different types of enzymes are used to modify histones, and how do these enzymatic modifications lead to chromatin remodeling?

b. In what other ways can chromatin be remodeled?

c. What phenotype(s) would you expect to see in a mutant where a protein involved in chromatin remodeling failed to function?

*20.7 DNA, histones, promoter-binding proteins, and enhancer-binding proteins are mixed together in the following orders:

a. first histones and DNA, then promoter-binding proteins

b. first histones and promoter-binding proteins, then DNA

c. first DNA and promoter-binding proteins, then histones

d. first histones, promoter-binding proteins and enhancer-binding proteins, then DNA

For each case, state whether transcription can occur. Explain your answers.

20.8 Promoters, enhancers, general transcription factors, activators, coactivators, and repressors that regulate the expression of one gene often have structural features that are similar to those regulating the expression of other genes. Nonetheless, the transcriptional control of a gene can be exquisitely specific: It will be specifically transcribed in some tissues at very defined times. Explore how this specificity arises by addressing the following questions:

a. Distinguish between the functions of promoters and enhancers in transcriptional regulation.

b. Distinguish between the functions of general transcription factors, activators, coactivators, and repressors in transcriptional regulation.

c. What structural features are found in activators, and what role do these play in transcriptional activation?

d. How is the mechanism by which eukaryotic repressors function different from that by which prokaryotic repressors function?

e. How can the same enhancer stimulate as well as quench transcription?

f. Given that several different genes may contain the same types of promoter and enhancer elements, and a number of the proteins that bind these elements contain the same structural features, how is transcriptional specificity generated?

*20.9 Eukaryotic organisms have a large number of copies (usually more than one hundred) of the genes that code for ribosomal RNA, yet they have only two copies (one on each of two homologs) of genes that code for a ribosomal protein. Explain how eukaryotes can produce the same number of ribosomal RNAs and ribosomal proteins, given this disparity in gene copy number.

20.10 Both peptide and steroid hormones can affect gene regulation of a targeted population of cells.

a. What is a hormone?

b. Distinguish between the mechanisms by which a peptide and a steroid hormone affect gene expression.

c. What role does each of the following have in a physiological response to a peptide or a steroid hormone?
   i. steroid hormone receptor (SHR)
   ii. chaperone
   iii. steroid hormone response element (HRE)
   iv. second messenger
   v. cAMP and adenylate cyclase

d. How can the same steroid hormone simultaneously activate distinct patterns of gene expression in two different cell types and have no affect on a third cell type?

*20.11 The following figure shows the effect of the hormone estrogen on ovalbumin synthesis in the oviduct of 4-day-old chicks. Chicks were given daily injections of estrogen ("Primary stimulation") and then after 10 days the injections were stopped. Two weeks after withdrawal (25 days), the injections were resumed ("Secondary stimulation").

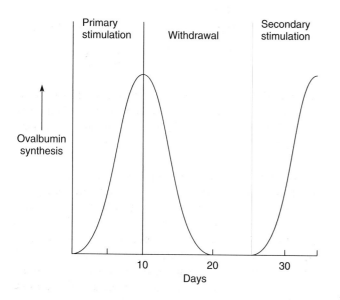

Provide possible explanations of these data.

20.12 In what different ways are the tails of histones chemically modified? What are the biological consequences of each type of modification?

20.13 A cloned DNA sequence was used to probe a Southern blot. There were two DNA samples on the blot, one from white blood cells and the other from a liver biopsy of the same individual. Both samples had been digested with *Hpa*II. The probe bound to a single 2.2-kb band in the white blood cell DNA but bound to two bands (1.5 and 0.7 kb) in the liver DNA.

a. Is this difference likely to result from a somatic mutation in a *Hpa*II site? Explain.

b. How would it affect your answer if you knew that white blood cell and liver DNA from this individual both showed the two-band pattern when digested with *Msp*I?

20.14 In what different ways can DNA methylation affect gene expression?

20.15 When male mice heterozygous for a small deletion on chromosome 2 are mated to normal females, deletion-bearing offspring have thin bodies and are slow moving, while non-deletion-bearing offspring are normal. However, when females heterozygous for the same deletion are mated to normal males, all offspring are normal.

a. How can these findings be explained in terms of imprinting?

b. When, and in what cell types, does imprinting occur?

c. *Neuronatin* is a gene that lies within the deleted region. A DNA polymorphism exists in the 3′-UTR of the *Neuronatin* gene that can be distinguished using PCR-RFLP (see Chapter 9 for a discussion of PCR-RFLP). How would you determine if the *Neuronatin* gene is expressed in a manner consistent with its being imprinted in embryos produced by the cross? Explain what results you would expect if it is imprinted, and what results you would expect if it is not imprinted.

*20.16 Both fragile X syndrome and Huntington disease are caused by trinucleotide repeat expansion. Individuals with fragile X syndrome have at least 200 CGG repeats at the 5′ end of the *FMR-1* gene. In contrast, individuals with Huntington disease have 36 or more in-frame CAG repeats within the protein-coding region of the Huntington gene.

a. Do you expect gene expression at the two genes to be affected in the same way by these repeat expansions? Explain your answer.

b. Based on your answer to **(a)**, why might fragile X syndrome be recessive, whereas Huntington disease is dominant?

c. Generate a hypothesis to explain why the number of trinucleotide repeats needed to cause a disease phenotype is different at each gene.

20.17 Although the primary transcript of a gene may be identical in two different cell types, the translated mRNAs can be quite different. Consequently, in different tissues, distinct protein products can be produced from the same gene. Discuss two different mechanisms by which the production of mature mRNAs can be regulated to this end; give a specific example for each mechanism.

20.18 How are pre-mRNAs prevented from being transported into the cytoplasm until after their introns have been removed?

20.19 Although many mRNAs are present in the cytoplasm of unfertilized vertebrate and invertebrate embryos, the rate of protein synthesis is very low. After fertilization, the rate of protein synthesis increases dramatically without new mRNA transcription.

a. What differences are seen in the length of poly(A) tails between inactive, stored mRNAs and actively translated mRNAs?

b. What role does cytoplasmic polyadenylation have in this process?

c. What signals are present in mRNAs that control polyadenylation and deadenylation?

d. In what way is deadenylation also important for controlling mRNA degradation?

*20.20 Although most eukaryotes lack operons such as those found in prokaryotes, the exceptional conserved organization of the *ChAT/VAChT* locus in *Drosophila* is reminiscent of a prokaryotic operon. *ChAT* is the gene for the enzyme choline acetyltransferase, which synthesizes acetylcholine, a neurotransmitter released by one neuron to signal another neuron. *VAChT* is the gene for the vesicular acetylcholine transporter protein, which packages acetylcholine into vesicles before its release from a neuron. Both *ChAT* and *VAChT* are expressed in the same neuron.

Part of the *VAChT* gene is nested within the first intron of the *ChAT* gene, and the two genes share a common regulatory region and a first exon. The structure of a primary mRNA and two processed mRNA transcripts produced by this locus are diagrammed in the following figure. The common regulatory region important for transcription of the locus in neurons is shown in the DNA, black rectangles in RNA represent exons, lines connecting the exons represent spliced intronic regions, and AUG indicates the translation start codons within the *ChAT* and *VAChT* mRNAs. Polyadenylation sites are not shown.

a. In what ways is the organization of the *VAChT/ChAT* locus reminiscent of a bacterial operon?

b. Why is the organization of this locus not structurally equivalent to a bacterial operon?

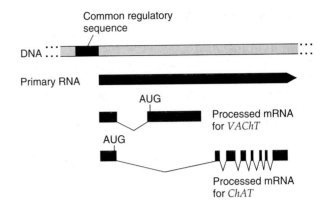

c. Based on the transcript structures shown, what modes of regulation might be used to obtain two different protein products from the single primary mRNA?

*20.21 Four different cDNAs were identified when a cDNA library was screened with a probe from one gene. The locations of introns and exons in the gene were determined by comparing the cDNA and genomic DNA sequences. The results are summarized in the following figure: Exons are represented by filled rectangles with protein-coding regions shaded black and 5'- and 3'-UTR regions shaded grey; introns are represented by thin lines.

a. How many different protein isoforms are encoded by this gene?

b. Carefully inspect these data and generate a specific hypothesis about the type(s) of posttranscriptional control that could generate these different protein isoforms.

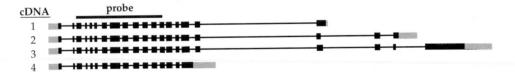

21

Genetic Analysis of Development

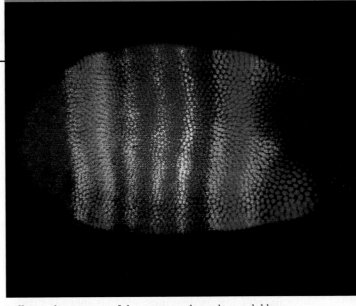

Differential expression of three genes—shown by a red, blue, and yellow immunofluorescence reaction to the proteins they produce—in a developing *Drosophila* embryo.

PRINCIPAL POINTS

- Development is regulated growth resulting from the interaction of the genome with the cytoplasm and with the extracellular environment. Development begins when a zygote is formed. The zygote, and the cells in the subsequent few generations, are totipotent, meaning they can develop into any cell type of the organism. At some point, the genetic program sets the fate of a cell in a process called determination. After determination, differentiation occurs, in which determined cells undergo developmental programs to produce their specific cell types. A process related to differentiation is morphogenesis, in which anatomical structures or cell shape and size are produced by a regulated pattern of cell division and changes in cell shape.

- Development results from differential gene activity of a genome that contains a constant amount of DNA from the zygote stage to the mature organism stage. Nonetheless, the genes are only part of the equation for development; environmental factors can affect the phenotype of an adult organism, as evidenced by phenotypic differences in cloned mammals from the parent that donated the nucleus for cloning.

- Antibodies are specialized proteins called immunoglobulins, which bind specifically to antigens (*antibody generators*: chemicals, recognized as foreign by an organism, that induce an immune response). Antibody molecules consist of two light chains and two heavy chains. In germ-line DNA the coding regions for immunoglobulin chains are scattered in tandem arrays of gene segments. During development, somatic recombination occurs to bring particular gene segments together to form functional antibody chain genes. A large number of different antibody chain genes result from the many possible ways in which the gene segments can recombine.

- In mammals, the sex of the individual is determined by the presence or absence of the Y chromosome. If the Y chromosome is present, the *SRY* gene on that chromosome is transcribed to produce a transcription factor that regulates genes required for directing the gonad to form a testis. In the absence of the *SRY* gene, as in an XX female, the gonad develops into an ovary by default.

- There is a different dosage of genes on the X chromosome in males and females. In mammals, a dosage compensation mechanism equates expression of X-linked genes in males and females by inactivating one of the two X chromosomes in a female early in development, leaving only one X chromosome transcriptionally active, as is the case in males. The inactivation process is complex, involving the transcription of the *XIST* gene on the chromosome to be inactivated; the RNA made then coats that chromosome, triggering chromatin changes that silence the genes.

- In *Drosophila*, the sex of the individual is determined by the ratio of the number of X chromosomes to the number of sets of autosomes. A ratio of 1 results in a female, and a ratio of 0.5 results in a male. The different ratios of the two types of chromosomes results in different levels of proteins encoded by genes on the chromosomes. These proteins influence a master regulatory gene for sex determination that controls a cascade of regulated alternative RNA splicing events, which ultimately leads to differentiation into female-specific or male-specific cells.

- As with mammals, there is a different dosage of genes on the X chromosomes in female and male *Drosophila*. In *Drosophila*, dosage compensation for X-linked genes occurs by increasing the transcriptional level of male X-linked genes twofold to match the gene expression of X-linked genes in the female. The mechanism for dosage compensation relates to the molecular steps for sex determination. That is, the absence of the *Sxl*-encoded protein in males enables a key protein to be translated from its mRNA. That protein associates with other proteins to form a complex that binds to many sites on the X chromosome. The complex triggers chromatin remodeling events spreading in each direction from the binding sites until the whole chromosome is affected. The key chromatin remodeling event involved is acetylation of histone H4; it is this modification that is responsible for the twofold increase in transcriptional activity.

- *Drosophila* has become an important model system in which to study the genetic control of development. *Drosophila* body structures result from specific gradients in the egg and the subsequent determination of embryo segments that directly correspond to adult body segments. Both processes are under genetic control, as shown by mutations that disrupt the development events. Studies of the mutations indicate that *Drosophila* development is directed by a temporal regulatory cascade.

- Once the basic segmentation pattern has been laid down in *Drosophila*, homeotic genes determine the developmental identity of the segments.

Homeotic genes share common DNA sequences called homeoboxes. Homeoboxes have been found in developmental genes in other organisms, and homeodomains—the regions of the proteins the homeoboxes encode—probably play a role in regulating transcription by binding to specific DNA sequences.

HOW DOES THE SINGLE CELL CREATED BY THE FUSION of sperm and egg transform itself into a complex organism? How do the cells of a developing human "know" how to arrange themselves in the shape of a human? What makes dividing cells form a leg rather than an eye, a heart, or a hand? As you will learn in this chapter, the development of humans and other eukaryotic organisms requires the precise regulation of groups of genes. After you have read and studied this chapter, you can apply what you've learned by trying the iActivity, in which you will attempt to identify some of the genes responsible for changing embryonic stem cells into different forms of tissue.

It is natural to follow the chapter on regulation of gene expression in eukaryotes with a chapter on developmental genetics because genes program development, and an understanding of how genes are regulated therefore helps researchers in their genetic analysis of development. While developmental genetics is just a subfield of developmental biology, the amount of important information known is far too much to cover in one chapter for an introductory genetics course. Therefore, we will focus on some key principles and discuss a few examples to illustrate aspects of the genetic analysis of development.

Basic Events of Development

Development is the process of regulated growth that results from the interaction of the genome with cytoplasm and the cellular external environment and that involves a programmed sequence of cellular-level phenotypic events that are typically irreversible. The total of the phenotypic changes constitutes the life cycle of an organism.

For a multicellular organism, development starts when a zygote is formed by fusion of sperm and egg. The zygote is **totipotent,** meaning that that cell has the potential to develop into any cell type of the complete organism. Of course it must be able to do that. Cells later in development may also be totipotent; this is common in plants, but uncommon in animals past the four-cell embryo stage. The different cell types that a cell can become during development is called its *developmental potential*. As development progresses, the developmental potential of most individual cells decreases.

By following a cell through development, researchers can discover what that cell will become. This is called the *fate* of the cell. More specifically, the fates of all the cells in an embryo can be followed, resulting in the construction of a **fate map**, which is a diagram of the fate of each cell of an embryo at a later developmental stage. For instance, in 1983 John Sulston and his coworkers painstakingly observed the development of embryonic cells of *C. elegans* under the microscope and produced a fate map showing the complete lineage of each adult cell.

When the genetic program sets the fate of a cell, the cell is said to be *determined*, and the process is called **determination.** This is still a relatively early stage of development, so a determined cell is not morphologically distinct from its neighbors. The cellular changes that occur during determination are directed, and lead to a stable state. That is, once the fate of the cell is determined, it does not change. Of course, the corollary is that a determined cell now has zero developmental potential: There is no longer a range of cell types that the cell can become.

There are two principal mechanisms for cell determination. In most cases, determination occurs by **induction;** that is, an inductive signal produced by one cell or group of cells affects the development of another cell or group of cells. For example, the signal can move by diffusion through the space between cells and be detected by a surface receptor on the target cell. Or, cells in contact can lead to interaction of transmembrane proteins in the plasma membranes resulting in the production of the signal in one cell type. In some cases, there is an asymmetric distribution of cell-determining molecules when a cell divides. As a result, the two daughter cells differ in the signals they have for future differentiation.

After determination, the most spectacular aspect of development takes place: **differentiation.** Differentiation is the process by which determined cells undergo cell-specific developmental programs to produce cell types with specific identifies, such as nerve cells, antibody-producing cells, skin cells, and so on, in animals, and phloem cells, leaf guard cells, meristematic cells, and so on, in plants. Differentiation in most cases results from differential gene expression, rather than from a differential loss of DNA, to leave different sets of genes in different cell types. That is, expression of different sets of genes in different kinds of determined cells leads to different proteins in the cells, and the proteins guide the progression to the various differentiated states.

Related to differentiation is **morphogenesis,** literally the "generation of form" and, by definition, the developmental process by which anatomical structures or cell shape and size are generated and organized. In both animals and plants, morphogenesis involves regulated patterns of cell division and changes in cell shapes. In animal morphogenesis, cell movement is an important component.

KEYNOTE

Development is regulated growth resulting from the interaction of the genome with the cytoplasm and with the extracellular environment. Development begins when a zygote is formed. The zygote, and cells in the subsequent few generations, are totipotent, meaning they can develop into any cell type of the organism. At some point, the genetic program sets the fate of a cell in a process called determination. After determination, differentiation occurs, in which determined cells undergo developmental programs to produce their specific cell types. A related process to differentiation is morphogenesis, in which anatomical structures or cell shape and size are produced by a regulated pattern of cell division and changes in cell shape.

Model Organisms for the Genetic Analysis of Development

For genetic analysis of development, researchers must have mutants that affect development. These mutants may be naturally occurring or they may be induced but, either way, it must be possible to map the mutations, and then to clone the genes involved so as to study them in detail. Thus, while a wide range of organisms have been the subjects for descriptive studies of development, relatively few organisms qualify as models for the genetic analysis of development. Several of the mutants that have contributed most to our understanding of the genetics of development are introduced here. For many of these organisms the genome has been completely sequenced.

Saccharomyces cerevisiae. The single-celled organism yeast (see Figure 1.7a, p. 10) has a limited developmental repertoire, but notably yeast cells signal each other through secreted extracellular pheromones as an essential part of mating. In this way a *MATa* cell and a *MATα* cell can identify each other; only mating between these two mating types can produce a zygote. Moreover, the actual differentiation of yeast cells into the two mating types has similarities to developmental processes found in multicellular organisms.

Drosophila melanogaster. The fruit fly (see Figure 1.7b, p. 10) has been a model organism for genetics since Morgan's work prior to and following 1910. Among the thousands of mutants isolated

Figure 21.1

Drosophila developmental mutant with four eyes instead of the normal two.

Figure 21.2

Three stages of *C. elegans* development from the two-cell stage to the adult.

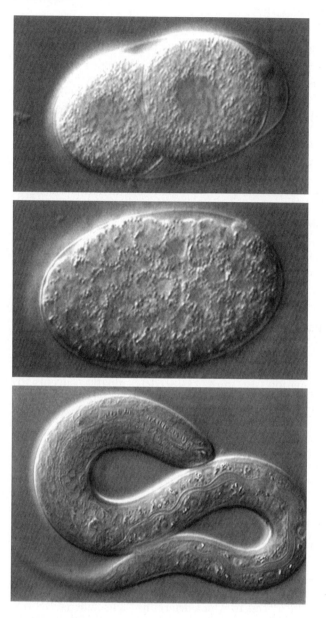

are many that affect development; Figure 21.1 shows a mutant with four eyes instead of the normal two. The study of *Drosophila* developmental mutants is providing a rich array of data about the molecular aspects of development. We will discuss how mutants have helped us understand sex determination, pattern formation in embryogenesis, and how genes program segments in the adult organism later in this chapter.

Caenorhabditis elegans. This nematode worm is transparent, permitting developmental processes to be followed easily under the microscope. As already mentioned, the fate map for every adult cell is known, so it is easy to see where mutations affect developmental processes. Figure 21.2 shows three stages of *C. elegans* development from the two-cell stage to the adult.

Arabidopsis thaliana. This small plant (see Figure 1.7d, p.10) has become popular for genetic and molecular analysis. Its genome has been completely sequenced and genetic analysis is relatively straightforward. As a model for the genetic analysis of the development of plants, *Arabidopsis* has been valuable in particular for a genetic dissection of floral development. Figure 21.3 shows a wild-type flower next to the developmental mutant *agamous (ag)* in which petals have formed instead of stamens, and sepals have formed instead of carpels.

Danio rerio. The zebrafish (Figure 21.4, left) is a model vertebrate for developmental genetics. The embryos are transparent, facilitating observation of developmental stages (Figure 21.4, right). Genetic crosses can be made, large numbers of fish can be bred in the laboratory, and genetic screens have been developed to search for genes

that affect embryogenesis. The genome of the zebrafish currently is being sequenced.

Mus musculus. The mouse (see Figure 1.7e, p10) is a mammal, of course, which makes it a model organism particularly close to humans. The mouse has been used for many years as a subject for genetic analysis, including the genetic analysis of development. The genome sequence of the mouse is known, and making gene knockouts (see Chapter 10, pp. 259–260) is technically straightforward. While many developmental mutants are known, their study in vivo is made difficult by the fact that embryogenesis takes place in utero.

Figure 21.3

Wild-type *Arabidopsis* flower (*left*) and flower of the developmental mutant, *agamous* (*ag*) (*right*). Flowers with a mutation in *ag* have petals replacing stamens in one whorl, and sepals replacing carpels in another whorl.

Development Results from Differential Gene Expression

In this section, we discuss selected experiments showing that, in most cases, development is the result of differential gene expression.

Constancy of DNA in the Genome During Development

In early studies of the genetic control of development, an important area of research focused on whether development involves a *loss* of genetic information (i.e., of DNA), or whether it involves differential gene expression of a constant genome, that is, a genome that is the same in all adult cells as it is in the zygote. More precisely, the loss-of-DNA model proposed that each type of differentiated cell retained only those genes required for that cell type, the remaining DNA having been discarded as part of the differentiation process. Experiments with plants and animals indicate that the DNA remains constant during development.

Regeneration of Carrot Plants from Mature Single Cells. In the 1950s, Frederick Steward dissociated phloem tissue

of a carrot to separate the cells and then attempted to culture new carrot plants from those cells by using plant tissue culture techniques (Figure 21.5). Mature plants with edible carrots were successfully produced (cloned) from the phloem cells. That the mature cells had the potential to act as zygotes and develop into complete plants indicated that mature cells had all the DNA found in zygotes. Steward's findings supported the notion that the DNA content of a cell remains constant during development and provided evidence *against* the model that development involves losses of genetic information.

Cloning Animals. An objection to the general applicability of the results of the carrot experiments is that plants are much more able to propagate themselves vegetatively than are animals; horticulturists have long been able to regenerate plants from cuttings, for example. Researchers also wanted to determine whether the DNA content of animal cells remained constant during development or whether this condition was particular to plants. Some success came in 1975 when John Gurdon and his colleagues showed that a nucleus from a skin cell of an adult frog injected into an enucleated egg could direct development to the tadpole stage. In those experiments, very few adults were produced, and all of them were sterile.

Gurdon's results left unresolved the question of whether a nucleus from adult differentiated tissue is genetically capable of directing development from the egg cell stage to fertile adulthood. That question was answered in the affirmative in 1997, when Ian Wilmut and his colleagues reported the cloning of a mammal (a sheep), starting with an adult cell.

Wilmut's group tested the ability of nuclei from embryonic, fetal, and adult cells to direct the development of sheep. Their experimental approach was as follows (Figure 21.6):

1. Embryonic cells, fetal fibroblast (muscle-forming) cells, and mammary epithelial cells from donor ewes (poll Dorset, black Welsh, and Finn Dorset breeds, respectively) were grown in tissue culture and then induced to enter the quiescent state (the G_0 phase of the cell cycle) by reducing the concentration of the growth serum.

Figure 21.4

Adult (*left*) and embryo (*right*) of the zebrafish, *Danio rerio*.

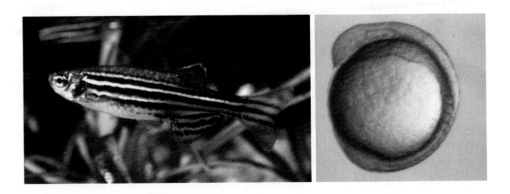

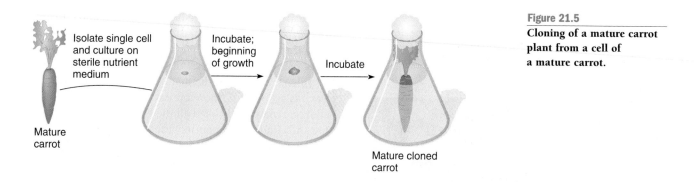

Isolate single cell and culture on sterile nutrient medium

Incubate; beginning of growth

Incubate

Mature carrot

Mature cloned carrot

Figure 21.5

Cloning of a mature carrot plant from a cell of a mature carrot.

2. The cells were fused with enucleated oocytes (egg cells), and the fusion cells were allowed to grow and divide for 6 days to produce embryos.

3. The embryos were implanted into recipient ewes, and the establishment and progression of pregnancy were monitored.

The results were as follows: Four of 385 embryo-derived cells, two of 172 fetal fibroblast-derived cells, and one of 277 adult mammary epithelium-derived cells gave rise to live lambs. The most significant of these results is the last because it demonstrates that the adult nucleus contains all the genetic information required to specify a new organism. That lamb, designated 6LL3 and named Dolly, progressed normally to sexual maturity and became pregnant with offspring Bonnie, born in 1998. In 1999, Dolly delivered a set of triplets. Dolly was euthanized at the age of 6 after being diagnosed with a fatal, virus-induced lung disease that commonly affects sheep of her age. Wilmut's group believes that cloning was not a factor in Dolly becoming infected.

Evidence that Dolly was truly the result of the cell fusion experiment is of two kinds. First, the fusion cell contained the nucleus from a (whiteface) Finn Dorset ewe and was implanted into a Scottish blackface recipient ewe; Dolly is morphologically Finn Dorset. Second, and more definitively, analysis of polymorphic STR (microsatellite) DNA markers (see Chapter 9, pp. 216–217) at four loci showed that the DNA of Dolly matched that of the donor mammary epithelial cells perfectly, but did not match that of the recipient ewe.

In sum, although the success rate for the experiment was not high (for technical reasons), the highly significant accomplishment here was the development of a live lamb directed by an adult nucleus. When the result of the experiment was published, concerns were raised internationally about the possible cloning of humans. Cloning technology is undoubtedly applicable to humans, and the ethical issues it raises will continue to be debated.

Mammal Cloning Problems.   Since the announcement of the cloning of Dolly, several groups have reported the successful cloning of other mammals, including mice, rats, goats, cattle, horses, cats, and monkeys. Biotechnology companies

have a particular interest in cloning mammals because, once they have invested large sums of monies in making transgenic mammals (e.g., for producing pharmaceuticals), it is much more efficient and economical to propagate those animals by cloning than by repeatedly making the transgenic animals afresh.

Unfortunately, the cloning of mammals has not been as straightforward as was hoped. Not only is the process itself very inefficient, but many problems have arisen with the clones produced. Consider, for instance, the cat Cc (carbon copy) that was cloned at Texas A&M University (Figure 21.7a). This calico female has a coat pattern that is not identical to that of Rainbow (Figure 21.7b), the parent that donated the nucleus from which Cc developed. Rainbow has the typical calico pattern of patches of black tabby and orange on white, while her clone Cc has black tabby patches on white with no orange patches. Recall from Chapter 12, pp. 324–325, that a calico results because of the process of X-chromosome inactivation in a female that is heterozygous for an X-linked gene for orange pigment production (O/o) and homozygous or heterozygous for an autosomal gene for black pigment production ($B/-$). (The molecular basis of X-chromosome inactivation is explained later in this chapter.) When the dominant O allele is expressed, orange pigment is produced regardless of other color genes present in the cat. The coat color pattern differences can be explained by the fact that X-chromosome inactivation is a random process in different cells, and that the movements of pigment-producing cells in the skin are mostly environmentally determined rather than genetically determined. However, Cc also differs from Rainbow in body shape and personality, both of which have genetic components. Thus, while Cc and Rainbow are genetically identical, they are not phenotypically identical. This fact argues that the genetic program is not alone in specifying the adult organism. Notably, environmental factors (in the internal environment in particular) play an important role.

More serious problems than variations in coat color and personality have turned up in cloned mammals. As we have already mentioned, mammal cloning is extremely inefficient; usually, most clones die before or soon after

Figure 21.6

Representation of Wilmut's sheep cloning experiment, which showed the totipotency of the nucleus of a differentiated, adult cell.

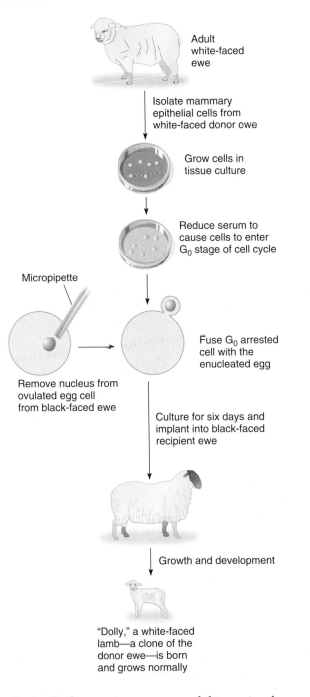

Adult white-faced ewe

Isolate mammary epithelial cells from white-faced donor ewe

Grow cells in tissue culture

Reduce serum to cause cells to enter G_0 stage of cell cycle

Micropipette

Remove nucleus from ovulated egg cell from black-faced ewe

Fuse G_0 arrested cell with the enucleated egg

Culture for six days and implant into black-faced recipient ewe

Growth and development

"Dolly," a white-faced lamb—a clone of the donor ewe—is born and grows normally

birth. The few survivors seem to exhibit varying degrees of developmental abnormalities, suggesting problems at the gene expression level. Rudolph Jaenisch and his colleagues at the Whitehead Institute for Biomedical Research have looked at this by studying the expression of more than 10,000 genes in the livers and placentas of cloned mice using DNA microarray analysis. They found hundreds of genes in the set regulated abnormally; those

Figure 21.7

Problems with cloning mammals. The cloned cat, Cc (**a**) has a different calico pattern from her mother, Rainbow (**b**).

genes represent about 4 percent of the protein-coding genes in the mouse's genome. Notably, the same genes showed abnormal expression whether taken from the cloned mice or from cultured cells containing the donor nuclei (prior to implanting in a surrogate mother). The interpretation is that the transfer of a donor nucleus into an enucleated cell is the cause of many gene expression changes. Practically speaking, it means that the clones that survive are likely not to be normal. The problem is that the donor nucleus is taken from a differentiated cell, and it must become reprogrammed to start the determination/differentiation process anew. This is a major issue, and at present researchers have no tools to apply to this problem. We can expect the production of cloned mammals to continue to be inefficient, and the living clones to show various problems resulting from abnormal gene expression. In view of this, any serious attempts to clone a human should be put out of mind at present.

Examples of Differential Gene Activity During Development

The following classic examples illustrate differential gene activity during development.

iActivity

As part of your job at the Institute of Animal Development, you investigate how gene expression patterns change as mouse stem cells differentiate into specific tissues in the iActivity *The Great Divide* on your CD-ROM.

Hemoglobin Types and Human Development. Human adult hemoglobin, Hb-A, has been examined in this book in many contexts. Hb-A is a tetrameric protein made up of two α and two β polypeptides. Each type of polypeptide is coded by a separate gene, α globin and β globin. The two genes appear to have arisen during evolution by duplication of a single

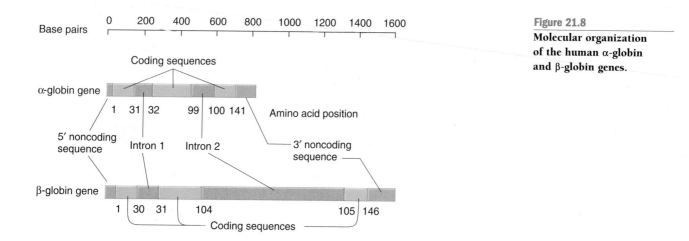

Figure 21.8

Molecular organization of the human α-globin and β-globin genes.

ancestral gene, followed by alteration of the base sequences in each gene. The organization of the genes (Figure 21.8) shows that each contains two introns (intron 1 and intron 2), which are transcribed with the coding sequences but are removed to produce a mature mRNA transcript.

Hb-A is only one type of hemoglobin found in humans. Genetic studies have shown that several distinct genes code for α- and β-like globin polypeptides, which form different types of hemoglobin at different times during human development. Figure 21.9 shows the globin chains synthesized at different stages of human development. In the human embryo, the hemoglobin initially made in the yolk sac consists of two ζ (zeta) polypeptides and two ε (epsilon) polypeptides. From comparisons of the amino acid sequences, ζ is an α-like polypeptide, and ε is a β-like polypeptide. After about three months of development, synthesis of embryonic hemoglobin ceases (i.e., the ζ and ε genes are no longer transcribed), and the site of hemoglobin synthesis shifts to the fetal liver and spleen. Here, *fetal hemoglobin* (Hb-F) is made. Hb-F contains two α polypeptides and two β-like γ

Figure 21.9

Comparison of synthesis of different globin chains at given stages of embryonic, fetal, and postnatal development.

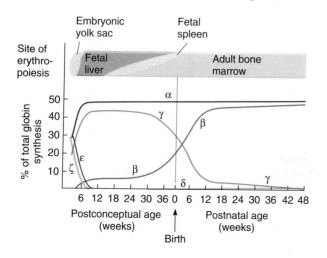

(gamma) polypeptides, either two γA or two γG. γA and γG differ from each other by only one out of 146 amino acids, and each is coded for by distinct genes.

Fetal hemoglobin is made until just before birth, when synthesis of the two types of γ chains stops, and the site of hemoglobin synthesis switches to the bone marrow. In that tissue, α and β polypeptides are made, along with some β-like δ (delta) polypeptides. In the newborn through adult human, most of the hemoglobin is our familiar $\alpha_2\beta_2$ tetramer (Hb-A), with about one in 40 molecules having the constitution $\alpha_2\delta_2$ (Hb-A2). Thus, globin gene expression switches during human development, and this switching involves a sophisticated gene regulatory system that turns appropriate globin genes on and off over a long time period.

In the genome, the α-like genes (two α genes and one ζ gene) are all on chromosome 16, and the β-like genes (ε, γA, γG, δ, and β) are all on chromosome 11. Significantly, the α-like genes and the β-like genes are arranged in the chromosome in an order that exactly parallels the timing in which the genes are transcribed during human development. Recall that embryonic hemoglobin consists of ζ and ε polypeptides; these genes are the first functional genes at the left of the clusters. Next, the α and γ genes are transcribed to produce fetal hemoglobin (Hb-F), and these genes are the next functional genes that can be transcribed from the clusters. Finally, the δ and β polypeptides are produced, and these genes are last in line in the β-like globin gene cluster. Although such an arrangement surely must occur by more than coincidence, there is no insight as yet about how the gene order relates to the regulation of expression of these genes during development.

Polytene Chromosome Puffs During Dipteran (Two-Winged Fly) Development. Recall (Chapter 17, pp. 454–455, and Figure 17.1) that **polytene chromosomes** are a special type of chromosome that consists of a bundle of chromatids produced by repeated cycles of chromosome duplication without nuclear division, and that they are readily visible after staining under the light microscope. After staining, distinct and characteristic bands are visible along the chromosomes. Genes are located both in the bands and in the interband regions. Polytene chromosomes are found, for instance, in *Diptera* in the salivary glands in the larval stages or in the nuclei of ordinary somatic cells.

At characteristic times during development, specific bands unwind locally to form *puffs* (Figure 21.10). The fact that puffs appear and disappear in specific patterns at certain chromosomal loci as development proceeds indicates they are developmentally controlled.

The puffing occurs as a result of very high levels of gene transcription. (Most genes are expressed at low levels and do not puff.) Evidence that the puffs are the visual manifestation of transcriptionally active genes has come from radioactive tracer experiments. If radioactive uridine (an RNA precursor) is added to developing flies, it is incorporated into the RNA being synthesized. If salivary glands are dissected from larvae and their polytene chromosomes are prepared and autoradiographed, radioactivity is localized at the puff sites, indicating that RNA, presumably transcripts, is associated with the puffs. That is, when a polytene chromosome gene is being transcribed during development, the chromosome structure loosens to permit efficient transcription of that region of the DNA When transcription is completed, the puff disappears and the chromosome resumes its compact configuration.

Puffing is under hormonal control in many cases, the key hormone being the steroid ecdysone. (The regulation of gene expression by steroid hormones was discussed in Chapter 20, pp. 553–556.) A model for the control of sequential gene activation by ecdysone is as follows. Ecdysone binds to a receptor protein, and this complex binds to both early (the early-puffing genes) and late genes (the expression of which is seen later in development). The complex turns on the early genes and represses the late

genes. One or more early genes encode a protein, which accumulates during development. When the level of this protein reaches a certain threshold, it displaces the ecdysone-receptor complex from both early and late genes. This turns off the early genes and removes the repression of (i.e., turns on) the late genes. In support of the model, some of the early genes have been shown to encode DNA-binding proteins, products expected of regulatory genes. Furthermore, an ecdysone receptor gene has been cloned and shown to encode a steroidlike receptor protein.

KEYNOTE

> Development results from differential gene activity of a genome that contains a constant amount of DNA from the zygote stage to the mature organism stage. Nonetheless, the genes are only part of the equation for development; environmental factors can affect the phenotype of an adult organism, as evidenced by phenotypic differences in cloned mammals from the parent that donated the nucleus for cloning.

Exception to the Constancy of Genomic DNA During Development: DNA Loss in Antibody-Producing Cells

There are exceptions to the rule that no DNA is lost during development. One such example involves the loss of genetic information during the development of cells that produce antibodies.

Antibody Molecules. The cells responsible for immune specificity are *lymphocytes*, specifically *T cells* and *B cells*. We focus our discussion on B cells. B lymphocytes develop in the adult bone marrow. When activated by an **antigen**, B cells develop into plasma cells that make proteins called **antibodies.** Antibody molecules are inserted in the plasma membrane of the plasma cells, and they are also released into the blood and lymph, where they are responsible for the humoral (*humor*, meaning fluid) immune responses. The antibodies bind specifically to the antigens that stimulated their production.

The establishment of immunity against a particular antigen results from **clonal selection.** This is a process whereby cells that have antibodies displayed on their surfaces that are specific for the antigen are stimulated to proliferate and secrete that antibody. During development, each lymphocyte becomes committed to react with a particular antigen, even though the cell has *never been exposed* to the antigen. For the humoral response system, there is a population of B cells, *each of which can recognize a single antigen*. A particular B cell recognizes an antigen because the B cell has made antibody molecules, which are attached to the outer membrane of the cell and act as receptor molecules. When an antigen encounters a B cell that has the appropriate antibody receptor capable of binding to the antigen, that

Figure 21.10

Light micrograph of a polytene chromosome from *Chironomus* showing two puffs that result from localized uncoiling of the chromosome structure and indicate transcription of those regions. DNA is shown in blue, RNA in red/violet.

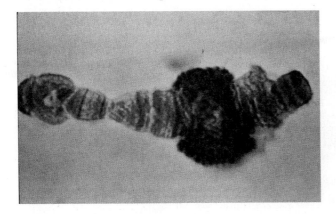

B cell is stimulated selectively to proliferate. This produces a clonal population of plasma cells, each of which makes and secretes the identical antibody. It is important to note that *any given cell makes only one specific kind of antibody toward one specific antigen.* However, the actual immune response may involve the binding of many different antibodies to an array of antigens on invaders such as an infecting virus or a bacterium. This binding mediates a variety of other mechanisms that inactivate the invading antigen.

All antibody molecules made by a given plasma cell are identical; that is, they have the same protein chains and bind the same antigen. There are millions of B cells in the whole organism, and millions of different antibody types can be produced, each with a different amino acid sequence and a different antigen-binding specificity. As a group, antibodies are proteins called **immunoglobulins** (Igs). A stylized antibody (immunoglobulin) molecule of the type IgG is shown in Figure 21.11a, and a model of an antibody molecule based on X ray crystallography is shown in

Figure 21.11b. Both figures show the molecule's two short polypeptide chains, called *light (L) chains*, and two long polypeptide chains, called *heavy (H) chains*. (All antibody molecules also have carbohydrates attached to the regions of H chains not involved in binding with L chains.) The two H chains are held together by disulfide (−S−S−) bonds, and an L chain is bonded to each H chain by disulfide bonds. Other disulfide bonds within each L and H chain cause the chains to fold up into their characteristic shapes.

The overall structure resembles a Y, with the two arms containing the two antigen-binding sites. The two L chains in each Ig molecule are identical, as are the two H chains, so the two antigen-binding sites are identical. The hinge region (see Figure 21.11a) allows the two arms to move in space, making it easier for the antibody to bind an antigen. Also, one arm can bind an antigen on, say, one virus, while the other arm binds the same antigen on a different virus of the same type. Such cross-linking of antibody molecules helps inactivate infecting agents.

Figure 21.11

IgG antibody molecule.
(a) Diagram showing the two heavy and two light chains and the antigen-binding sites. The heavy and light chains are held together by disulfide bonds. (b) Model of IgG antibody molecule.

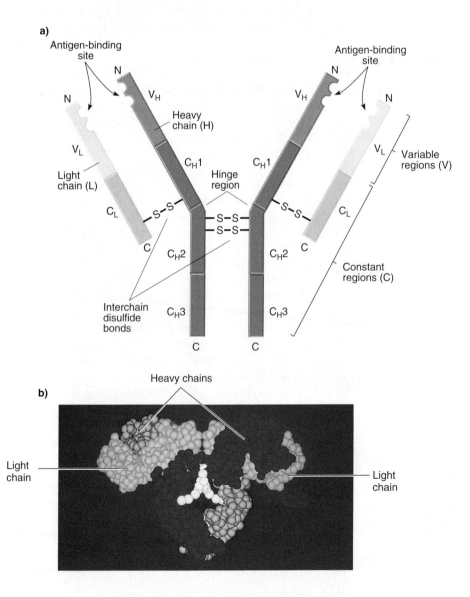

In mammals there are five major classes of antibodies: IgA, IgD, IgE, IgG, and IgM. They have different H chains: α (alpha), δ (delta), ε (epsilon), γ (gamma), and μ (mu), respectively. Two types of L chains are found: κ (kappa) and λ (lambda). Both L chain types are found in all Ig classes, but a given antibody molecule has either two identical κ chains or two identical λ chains. A complete discussion of the functions of the five Ig classes is beyond the scope of this text. For our purposes, we need to be aware that the most abundant class of immunoglobulin in the blood is IgG, and that IgM plays an important role in the early stages of an antibody response to a previously unrecognized antigen. We will focus on these two antibody classes from now on.

Each polypeptide chain in an antibody is organized into domains of about 110 amino acids (see Figure 21.11a). Each L chain (κ or λ) has two domains, and the H chain of IgG (the γ chain) has four domains, whereas the IgM's H chain (the μ chain) has five domains. The N-terminal domains of the H and L chains have highly variable amino acid sequences that constitute the antigen-binding sites. These domains, representing in IgG the N-terminal half of the L chain and the N-terminal quarter of the H chain, are called the *variable*, or V, regions. The V regions are symbolized generically as V_L (for the light chain) and V_H (for the heavy chain). The V_L and V_H regions comprise the antigen-binding sites (see Figure 21.11a). The amino acid sequence of the rest of the L chain is constant for all antibodies (with the same L chain type, i.e., κ or λ) and is called C_L. Similarly, the amino acid sequence of the rest of the H chain is constant and is called C_H. For IgG, there are three approxi-

mately equal domains of C_H called C_H1, C_H2, and C_H3 (see Figure 21.11a). For IgM, there are four approximately equal domains of C_H called C_H1, C_H2, C_H3, and C_H4. Thus, the production of antibody molecules involves synthesizing polypeptide chains, one part of which varies from molecule to molecule and the other part of which is constant. How this occurs is discussed in the following section.

Assembly of Antibody Genes from Gene Segments During B Cell Development.

A mammal may produce 10^6 to 10^8 different antibodies. Since each antibody molecule consists of one kind of L chain and one kind of H chain, these antibodies theoretically would require 10^3 to 10^4 different L chains and 10^3 to 10^4 different H chains, if L and H chains paired randomly. However, there are not nearly enough genes in the human genome to specify that many different molecules. Instead, variability in L and H chains results from particular DNA rearrangements that occur during B cell development. These rearrangements involve the joining of different gene segments to form a gene that is transcribed to produce an Ig chain; the process is called *somatic recombination*. The process is now illustrated for mouse immunoglobulin chains.

Light Chain Gene Recombination.
In mouse germline DNA, there are three types of gene segments, and one of each type is needed to make a complete, functional κ light chain gene (Figure 21.12):

1. Each of the 350 or so L–V_κ segments consists of a leader sequence (L) and a V_κ segment, which varies

Figure 21.12

Production of the kappa (κ) light chain gene in mouse by recombination of V, J, and C gene segments during development. The rearrangement shown is only one of many possible recombinations.

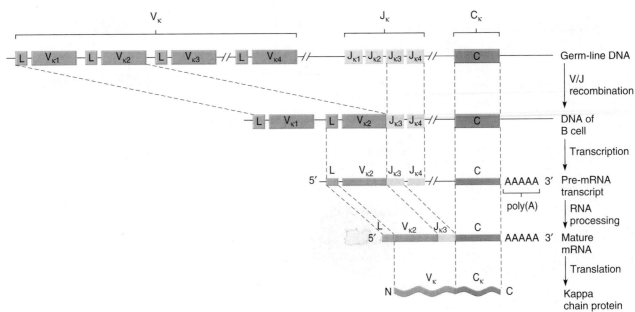

from segment to segment. Each V_κ segment encodes most of the amino acids of the light chain variable domain. The leader sequence also encodes a special sequence called a signal sequence (see Chapter 6) that is required for secretion of the Ig molecule; this signal sequence is subsequently removed and is not part of the functional antibody molecule.

2. A C_κ segment specifies the constant domain of the κ λ light chain.

3. Four J_κ segments (joining segments) are used to join V_κ and C_κ segments to produce a functional κ light chain gene.

In the pre-B cell, the $L–V_\kappa$, J_κ, and C_κ segments, in that order, are widely separated on the chromosome. As the B cell develops, a particular $L–V_\kappa$ segment becomes associated with one of the J_κ segments and with the C_κ segment. In the example in Figure 21.12, $L–V_{\kappa2}$ has recombined next to $J_{\kappa3}$. Transcription of this new DNA arrangement produces the primary RNA transcript. As is typical of eukaryotic mRNAs, a poly(A) tail is added to the end of the transcript posttranscriptionally. Removal of the intron from the primary RNA transcript produces the mature mRNA, which has the organization $L–V_{\kappa2}J_{\kappa3}C_\kappa$; translation and leader removal produces the κ light chain that the B cell is committed to make.

In the mouse, there are about 350 $L–V_\kappa$ gene segments, four functional J_κ segments, and one C_κ gene segment. Thus, the number of possible κ chain variable regions that can be produced by this mechanism is $350 \times 4 = 1{,}400$. Further diversity results from imprecise joining of the V_κ and J_κ gene segments. That is, during the joining process a few nucleotide pairs from V_κ and a few nucleotide pairs from J_κ are lost from the DNA at the $V_\kappa J_\kappa$ joint, generating significant diversity in sequence at that point. Thus, diversity of κ light chains results from (1) variability in the sequences of the multiple V_κ gene segment; (2) variability in the sequences of the four J_κ gene segments; and (3) variability in the number of nucleotide pairs deleted at $V_\kappa J_\kappa$ joints.

A similar mechanism exists for mouse λ light chain gene assembly. In this case, there are only two $L–V_\lambda$ gene segments and four C_λ gene segments, each with its own J_λ gene segment. Thus, fewer λ variable regions can be produced than is the case for κ chains.

Heavy Chain Gene Recombination. The immunoglobulin heavy chain gene is also encoded by V_H, J_H, and C_H segments. In this case, additional diversity is provided by another gene segment, D (diversity), which is located between the V_H segments and the J_H segments (Figure 21.13). For an IgG heavy chain, in the germ line of mouse DNA there is a tandem array of about 500 $L–V_H$ segments, then a spacer, then 12 D segments, then a spacer, and then 4 J_H segments. After another spacer, the

Figure 21.13

Production of heavy chain genes in mouse by recombination of V, D, J, and C gene segments during development. Depending on the C_H segment used, the resulting antibody molecule is IgM, IgD, IgE, or IgA. Shown here is the assembly of an IgG heavy chain. This rearrangement is only one of the many thousands possible.

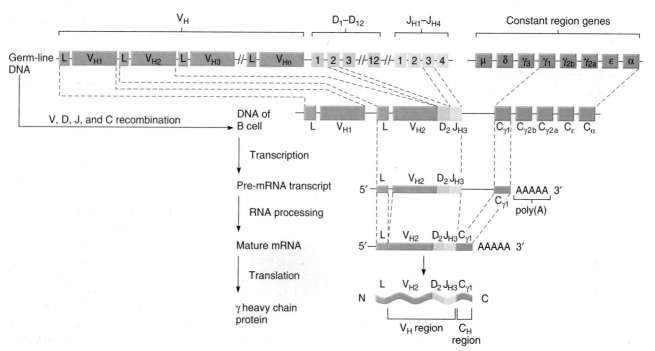

constant region gene segments are arranged in a cluster that, in mouse, has the order μ, δ, γ (four different sequences for four different, but similar, IgG H chain constant domains), ε, and α for the H chain constant domains of IgM, IgD, IgG, IgE, and IgA, respectively. Thus, for the assembly of a heavy chain, there are $500 \times 12 \times 4 = 24,000$ possibilities for each heavy chain type. As in L chain gene rearrangements, further antibody diversity results from imprecise joining of the gene segments that make up the variable region of the chain. Taken together with the light chain variation, an enormous variety of antibody molecules can be produced. Just for antibodies with one of the heavy chain types and a κ light chain, $24,000 \times 1,400 = 33,600,000$ possible antibody molecules can be produced.

KEYNOTE

Antibodies are specialized proteins called immunoglobulins, which bind specifically to antigens. Immunity against a particular antigen results from clonal selection, in which cells already making the required antibody are stimulated to proliferate by the specific antigen. Antibody molecules consist of two light (L) chains and two heavy (H) chains. The amino acid sequence of one domain of each type of chain is variable; this variation is responsible for the different antigen-binding sites on different antibody molecules. The other domains of each chain are constant in amino acid sequence. In germ-line DNA, the coding regions for immunoglobulin chains are scattered in tandem arrays of gene segments. Thus, for light chains, there are many variable region (V) gene segments, a few joining (J) gene segments, and one constant region (C) gene segment. During development, somatic recombination occurs to bring particular gene segments together into a functional L chain gene. A large number of different L chain genes result from the many possible ways in which the gene segments can recombine. Similar rearrangements occur for H chain genes but with the addition of several D (diversity) segments that are between V and J, which increase the possible diversity of H chain genes.

Case Study: Sex Determination and Dosage Compensation in Mammals and *Drosophila*

For the remainder of this chapter, we discuss specific examples of the genetic regulation of developmental processes. In this section we discuss sex determination and dosage compensation in mammals and *Drosophila*. These two topics are related because the sex chromosomes play a role in sex determination in both organisms, yet the different copy number of X chromosomes in the two sexes necessitates regulation of X-linked genes in order to equalize their expression in males and females.

Sex Determination in Mammals

In humans and other placental mammals, sex is determined by the Y chromosome mechanism of sex determination. That is, the presence of the Y chromosome specifies maleness in that gonads develop as testes, while in the absence of a Y chromosome, the gonads develop as ovaries.

The Y chromosome mechanism of sex determination means that the Y uniquely carries an important gene (or perhaps genes) that sets the switch toward male sexual differentiation. The product of this gene is called **testis-determining factor,** and the corresponding hypothesized gene is the *testis-determining factor* gene or *TDF*. Testis-determining factor causes the tissue that will become gonads to differentiate into testes instead of ovaries. This is the central event in sex determination of many mammals; all other differences between the sexes are secondary effects resulting from hormone action or from the action of factors produced by the gonads. Therefore, sex determination is equivalent to testis determination.

The testis-determining factor gene was found by studying rare so-called *sex reversal* individuals, that is, males who are XX (instead of XY) and females who are XY (instead of XX). In the XX males, a small fragment from near the tip of the small arm of the Y chromosome had broken off during the production of gametes and become attached to one of the X chromosomes. The XY females had deletions of the same region of the Y chromosome. These findings suggested that the testis-determining factor gene is in that small segment of the Y chromosome. More careful molecular analysis of the DNA of XX males and XY females showed the presence of a male-specific gene sequence near the end of the small arm. This DNA is present in XY males and in the unusual XX males, and it is absent in XX females and in the unusual XY females. This gene is the *SRY* (sex determining region of the Y) gene. An equivalent gene, *Sry*, has been cloned from mice. That the cloned *SRY* gene is most likely the testis-determining factor gene was shown by the homologs of this gene being found on the Y chromosomes of all eutherian mammals examined, as well as on the Y chromosomes of marsupials and monotremes.

Several lines of evidence indicate that the *SRY* gene is the testis-determining gene. First, the mouse *Sry* gene is expressed only at the time and place expected for the testis-determining factor, that is, in the undifferentiated genital ridges of the embryo just before the formation of testes. Second, when a 14-kb DNA fragment with the mouse *Sry* gene is introduced into XX mouse embryos by microinjection, the transgenic mice produced are males with normal testis differentiation and subsequent normal male secondary sexual development. In other words, *Sry* alone is sufficient to cause a full phenotypic sex reversal in an XX chromosomally female mouse. Third, there are rare XY human females that, instead of having lost a section of the Y chromosome as described earlier, have a simple mutation in the *SRY* gene.

The proteins encoded by *SRY* and *Sry* are transcription factors that specify development of the gonad into a testis. The testes produce the masculinizing steroid hormone testosterone. If the *SRY* gene is absent, the gonad develops into an ovary by default. Ovaries produce the feminizing steroid hormone estrogen. Steroid hormones regulate gene expression, as we discussed earlier in the chapter. In the case of testosterone, this circulating hormone binds to the dihydrotestosterone receptor in target cells and regulates gene expression in those cells.

Dosage Compensation Mechanism for X-Linked Genes in Mammals

Organisms with sex chromosomes have an inequality in gene dosage (number of gene copies) between the sexes; that is, there are two copies of X-linked genes in females and one copy in males. In many such organisms, if gene expression on the X chromosome is not equalized, lethality results early in development. Fortunately, there is a **dosage compensation** mechanism for dealing with this problem. In female mammals, this involves inactivating one of the two X chromosomes in somatic cells at an early stage in development (see Chapter 12, pp. 324–325). The X chromosome inactivated is randomly chosen from the maternally derived and paternally derived X chromosomes in a process that is independent from cell to cell. Once a maternal or paternal X chromosome is inactivated in a cell, all descendants of that cell inherit the inactivation pattern.

Three steps are involved in X inactivation: chromosome counting (determining the number of X chromosomes in the cell), selection of an X for inactivation, and X inactivation itself. Significant information has been obtained at the molecular level for each of these steps. First, there is a key region on each X chromosome called the *X inactivation center* (*XIC* in humans, *Xic* in mice); these regions are involved in the chromosome counting mechanism. Two or more *XIC*s must be present for X inactivation to occur. Some evidence for this has come from an experiment in which a 450-kb (450,000 base pairs) piece of the mouse X chromosome containing *Xic* was introduced into autosomes of male mouse cells in tissue culture. In normal male cells with one X chromosome, X inactivation does not occur. However, in the transgenic mouse cells with a *Xic* added to an autosome, chromosome inactivation was turned on, with either the X or the autosome becoming inactivated in a random fashion. This means that the genetically modified male

cells showed properties typical of X inactivation in normal female cells. The 450 kb of DNA with *Xic* must contain the sequences for chromosome counting and for the initiation of X inactivation.

Female somatic cells have a choice mechanism that determines which X chromosome is inactivated and which X chromosome remains active. The choice is made at the *X-controlling element* (*Xce*), which is in the *XIC/Xic* region. A gene called *XIST/*(humans)/*Xist* (mice), for X inactivation-specific transcripts, is also located in the *XIC/Xic* region. *XIST* is expressed from the *inactive* X rather than from the active X, which is the opposite of the expression pattern of other X-linked genes. The *Xist* gene has been shown to be essential for X inactivation in cultured cells and in mice. *XIST/Xist* is transcribed into an unusually large (17 kb) RNA that is not translated. This RNA coats the X chromosome from which it is transcribed, spreading out in both directions from the *XIC/Xic* and triggering the silencing of genes on that chromosome. That is, immediately after the chromosome is coated with *XIST* RNA, histone H3 is methylated as a first step of chromatin modification. It is the chromatin remodeling that ultimately silences the genes on that X chromosome.

KEYNOTE

In mammals, the sex of the individual is determined by the presence or absence of the Y chromosome. If the Y chromosome is present, the *SRY* gene on that chromosomes is transcribed to produce a transcription factor that regulates genes required for directing the gonad to form a testis. In the absence of the *SRY* gene, as in an XX female, the gonad develops into an ovary by default.

Genes on the X chromosome are unusual in that their dosage is different between males and females. In mammals, a dosage compensation mechanism operates to equate expression of X-linked genes in males and females. This mechanism involves the inactivation of one of the two X chromosomes in a female early in development, leaving only one X chromosome transcriptionally active, as is the case in males. The inactivation process is complex, involving the transcription of the *XIST* gene on the chromosome to be inactivated; the RNA made then coats that chromosome in both directions from *XIC*, triggering chromatin changes that silence the genes.

Sex Determination in *Drosophila*

Recall from Chapter 12 (pp. 327–328) that sex in *Drosophila* is determined by the X chromosome : autosome (X:A) ratio. Our understanding of sex determination in this organism has come from studies of a number of

Figure 21.14

Regulatory cascade for sex determination in Drosophila. For details, see text.

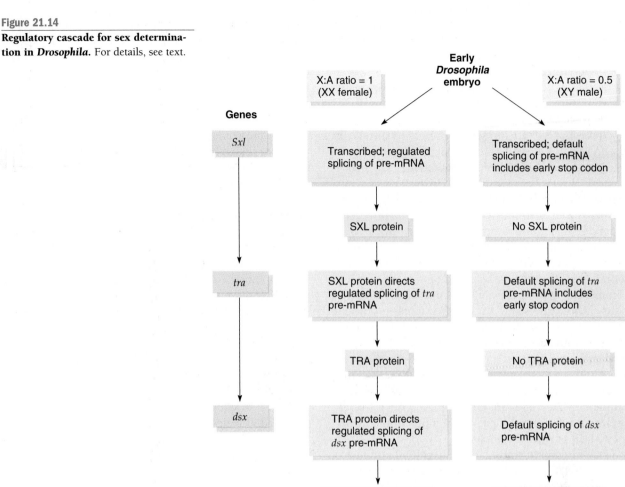

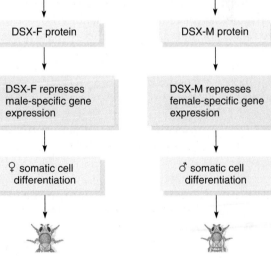

mutations that disrupt normal sex determination. These studies have led to a *regulation cascade model* for sex determination in *Drosophila*, summarized in Figure 21.14. First, the X:A ratio is read during development. For wild-type *Drosophila*, the ratio for females (XX) is 2X : 2 sets of autosomes = 1.0, and the ratio for males (XY) is 1X : 2 sets

of autosomes = 0.5. This information is transmitted to the sex determination genes, which make the choice between the alternative female and male developmental pathways, starting with the master regulatory gene *Sxl* (*sex lethal*). Loss-of-function mutants of *Sxl* are lethal for female embryo development (meaning that *Sxl* needs to be active in females), but they have no effect on male embryo development (meaning that *Sxl* expression is not necessary for male development). However, gain-of-function mutants are lethal for male embryo development, which means that *Sxl* needs to be inactive in males. Alternative splicing of the *Sxl* pre-mRNA in embryos destined to become females or males sets in motion the two different pathways. Steps in each pathway are regulated by alternative splicing of pre-mRNAs, as we will see.

The initial switch for sex determination is set by the X:A ratio. If it is 1.0, development is set toward femaleness, and if it is 0.5, development is set toward maleness. How is the ratio detected? On the X chromosome are the *sisterless* numerator genes *sis-a*, *sis-b*, and *sis-c*, and on an autosome is the *deadpan* (*dpn*) denominator gene. The numerator genes are expressed to produce protein subunits that can form either homodimers or heterodimers with the subunit encoded by the denominator gene (Figure 21.15). In females, an excess of numerator subunits versus denominator subunits results from expression of the two copies of each numerator gene, so there are many numerator homodimers formed. These numerator homodimers are transcription factors that activate *Sxl* expression. In males, there is only one copy of each numerator gene, so most expressed numerator subunits are found in heterodimers with denominator subunits. As a result, there are no (or insufficient) numerator homodimers for activating *Sxl* expression.

Early in embryogenesis in the female, the numerator-numerator dimer transcription factor activates transcription of the *Sxl* gene from P_E (promoter early), one of two promoters for this gene, the other being a more upstream promoter, P_L (promoter late) (Figure 21.16a). The pre-mRNA transcribed from P_E has eight exons; exons 2 and 3 are skipped to produce the mature mRNA consisting of exons E1, 4, 5, 6, 7, and 8. Translation of this mRNA produces the SXL early protein. In males, *Sxl* expression from P_E does not occur because sufficient numerator-numerator transcription factors are absent: No SXL protein is produced in males.

Later in embryogenesis (after gastrulation), *Sxl* is transcribed constitutively from the late promoter, P_L, in all cells, regardless of the X:A ratio (Figure 21.16b). In other words, this transcription does not depend on the numerator transcription factors. The pre-mRNA produced is longer than the transcript from P_E and is subject to alternative splicing (see Chapter 20, pp. 561–562) depending on the presence or absence of SXL early protein. In females, the SXL early protein binds to the *Sxl* pre-mRNA and causes regulated splicing: Exons E1 and 3 are skipped. The result is a mature mRNA with exons L1, 2, 4, 5, 6, 7, and 8. Translation of this mRNA produces the SXL late protein. In males, the absence of SXL early protein results in the default splicing of pre-mRNA. As a result, the mature mRNA includes exon 3. This exon has a stop codon in frame with the start codon at the beginning of exon 2. Therefore, no SXL late protein is produced in males.

The events just described set the switch to either female or male differentiation. A cascade of alternative splicing events follows as outlined in Figure 21.14. In the female embryo, SXL late protein regulates splicing of *transformer* (*tra*) pre-mRNA (Figure 21.17). In this case,

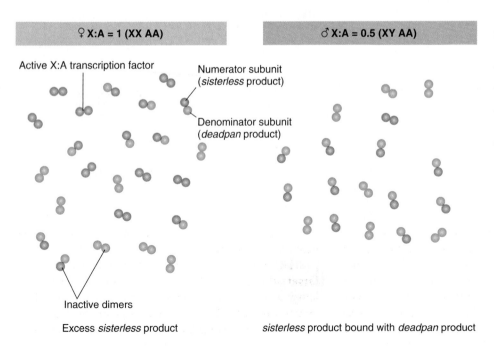

♀ X:A = 1 (XX AA)

♂ X:A = 0.5 (XY AA)

Active X:A transcription factor

Numerator subunit (*sisterless* product)

Denominator subunit (*deadpan* product)

Inactive dimers

Excess *sisterless* product

sisterless product bound with *deadpan* product

Figure 21.15

Manifestation of the X:A ratio in *Drosophila* sex determination by numerator and denominator gene-encoded proteins. In females, the excess of numerator proteins produces numerator-numerator dimers that function as transcription factors to activate *Sxl*.

Figure 21.16

Expression of *Sex-lethal* (*Sxl*) during embryogenesis. (a) In early embryogenesis in females, the numerator-numerator dimer transcription factors activate transcription of *Sxl* from P$_E$. Splicing of the pre-mRNA removes exons 2 and 3 (the angled lines above the pre-mRNA indicate the segments that are removed during splicing); the resulting mRNA is translated to produce SXL early protein. (b) Later in embryogenesis, the *Sxl* gene is transcribed constitutively from PL in both female and male embryos. The pre-mRNA is spliced in a regulated fashion in female embryos owing to the presence of SXL early protein, and in a default fashion in male embryos owing to the absence of SXL early protein. As a result, SXL late protein is produced in female embryos, but not in male embryos.

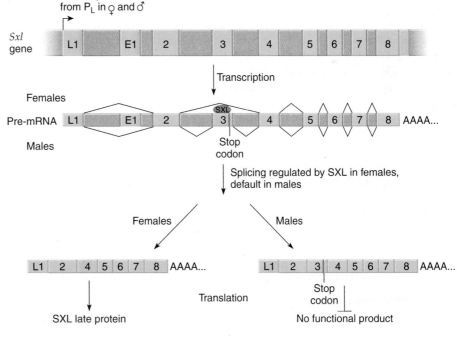

a) Early embryogenesis

b) Later in embryogenesis

a stop codon–containing exon segment upstream of and contiguous with exon 2 is removed, resulting in an mRNA with exons 1, 2, and 3. Translation of this mRNA produces the active TRA protein. In males, default splicing occurs because of the absence of SXL late protein. This means that the stop codon–containing segment is not removed. Translation of the resulting

mRNA halts at the stop codon in that segment; no TRA protein is produced.

TRA protein is also an RNA splicing regulator. The target is the pre-mRNA of the *doublesex* (*dsx*) gene (Figure 21.18). In females, the mRNA produced encodes the DSX-F (F for female) protein, a transcription factor that represses male-specific gene expression

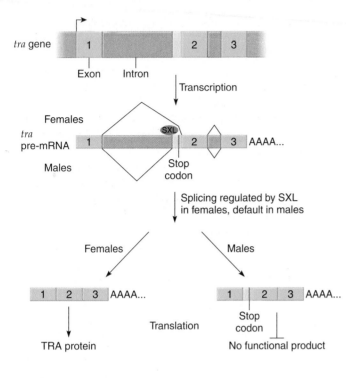

Figure 21.17
Expression of *transformer* (*tra*) during embryogenesis.
SXL late protein in female embryos regulates splicing of
tra pre-mRNA; the resulting mRNA is translated to generate the TRA protein that regulates splicing of the
doublesex transcript (see Figure 21.18). In male embryos,
tra pre-mRNA is spliced in a default fashion because SXL
late protein is absent. The resulting mRNA has a stop
codon prior to exon 2 and so no TRA protein is made.

in all cells. As a result, female-specific somatic cell differentiation occurs. In males, the mRNA produced encodes the DSX-M (M for male) protein, a transcription factor that represses female-specific gene expression in all cells. As a result, male-specific somatic cell differentiation occurs. Knockout mutants of *dsx* have male and female characteristics, which occurs

because of the lack of repression of male- and female-specific genes.

Dosage Compensation in *Drosophila*

In mammals, dosage compensation occurs by decreasing transcriptional activity in females to match that in males.

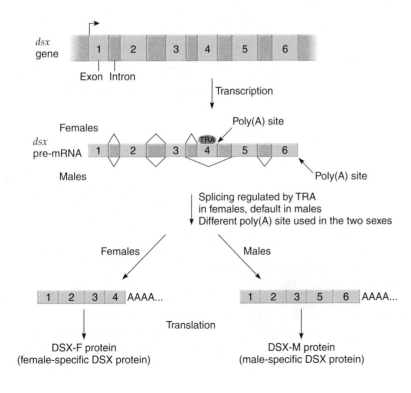

Figure 21.18
Expression of *doublesex* (*dsx*) during embryogenesis. TRA protein in female embryos regulates splicing, so exon 4 is included and cleavage and polyadenylation occurs at the poly(A) site following exon 4. In male embryos, default splicing occurs in the absence of TRA protein, leading to the exclusion of exon 4, but to the inclusion of exons 5 and 6 because of cleavage and polyadenylation at the poly(A) site following exon 6. Translation of the two different mRNAs produces the female-specific DSX protein, DSX-F, in females, and the male-specific DSX protein, DSX-M, in males.

In *Drosophila*, the opposite occurs: Transcriptional activity is increased twofold in males to match that in females, who have twice the number of X chromosomes. Despite the different approaches to dosage compensation, in both cases chromatin remodeling is involved in regulating transcriptional activity.

An understanding of dosage compensation in *Drosophila* has been illuminated by studies of mutants of genes that are essential for male viability. Key male-specific lethal genes are *mle (maleness)*, *msl-1 (male-specific lethal-1)*, *msl-2*, *msl-3*, and *mof (males absent on the first)*. Males with mutations in these genes die at the late larval stage, while females with the same mutations develop normally. The products of these genes are collectively called the male-specific lethal (MSL) proteins.

The SXL late protein (see previous section) plays a key role in dosage compensation. In females, the SXL late protein binds to the transcript of *msl-2*, blocking its translation; no MSL2 protein is produced. In males, the *msl-2* transcript can be translated because SXL late protein is absent. MSL2 forms a complex with the other MSL proteins, MLE, MSL1, MSL3, and MOF. This MSL complex binds to about 35 chromatin entry sites (CES) on the *Drosophila* male X chromosome and then MSL complexes spread from those sites in both directions into the flanking chromatin. The MOF protein of the MSL complex is a histone acetyltransferase and its chromatin remodeling activity (see Chapter 20, pp. 547–548) as it spreads along the X chromosome is responsible for the twofold higher level of transcription of X chromosome genes in males than in females.

In females, the MSL proteins other than MSL2 are produced. However, because MSL2 is essential for the binding of the MSL complex to the X chromosome, no chromatin remodeling can occur in XX females.

K E Y N O T E

In *Drosophila*, the sex of the individual is determined by the ratio of the number of X chromosomes to the number of sets of autosomes. A ratio of 1.0 results in a female, and a ratio of 0.5 results in a male. The ratio of chromosomes results in different amounts of proteins encoded by numerator genes on the X chromosomes versus a denominator gene on an autosome. In female embryos, there is an excess of numerator proteins that form transcription factors that activate a master regulatory gene for sex determination, *Sex-lethal (Sxl)*. In male embryos, numerator proteins are bound to denominator proteins, so no activation of *Sxl* occurs. This key transcription regulatory event sets in motion a cascade of regulated alternative RNA splicing events that ultimately leads to differentiation into female-specific or male-specific cells.

As with mammals, there is a different dosage of genes on the X chromosomes in female and male *Drosophila*. Dosage compensation occurs in fruit flies also, but in this case the transcriptional level of genes on the male's X chromosome is increased twofold to match the gene expression of X-linked genes in the female. The mechanism for dosage compensation relates to the molecular steps for sex determination. That is, the absence of the *Sxl*-encoded protein in males enables a key protein to be translated from its mRNA. That protein associates with other proteins to form a complex that binds to many sites on the X chromosome. The complex triggers chromatin remodeling events spreading in each direction from the binding sites until the whole chromosome is affected. The key chromatin remodeling event involved is acetylation of histone H4; it is this modification that is responsible for the twofold increase in transcriptional activity.

Case Study: Genetic Regulation of the Development of the *Drosophila* Body Plan

Significant progress has been made in understanding the genetic regulation of development in *Drosophila*. Many developmental mutants have been isolated after extensive genetic screens so that almost all areas of *Drosophila* development can be studied in detail at genetic and molecular levels. The discoveries made from such studies have become even more important as discoveries in other systems (e.g., nematode, mouse) indicate that the genes discovered in *Drosophila* have counterparts in all higher organisms, including humans. This implies that the same mechanisms that control development in *Drosophila* are used in higher organisms as well. In this section, we provide a brief overview of what is known.

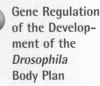

animation

a **Gene Regulation of the Development of the *Drosophila* Body Plan**

Drosophila Developmental Stages

The production of an adult *Drosophila* from a fertilized egg involves a well-ordered sequence of developmentally programmed events under strict genetic control (Figure 21.19). About 24 hours after fertilization, a *Drosophila* egg hatches into a larva, which undergoes three molts, after which it is called a pupa. The pupa metamorphoses into an adult fly. The whole process from egg to adult fly takes about 10 to 12 days at 25°C.

Embryonic Development

Development commences with a single fertilized egg, giving rise to cells that have different developmental fates. What follows is a brief discussion of the information that

Figure 21.19

Development of an adult *Drosophila* from a fertilized egg.

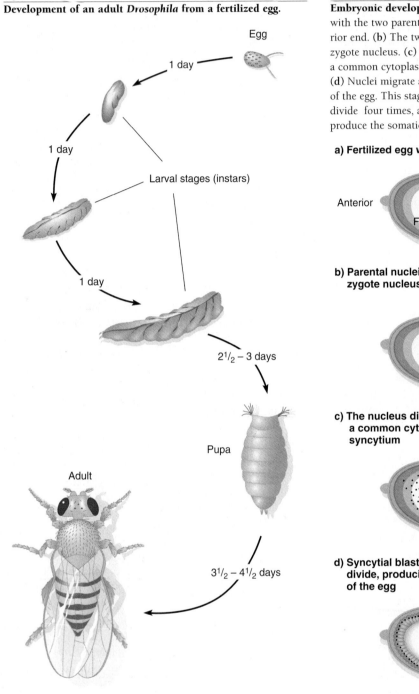

Egg

1 day

1 day

Larval stages (instars)

1 day

2¹/₂ – 3 days

Pupa

Adult

3¹/₂ – 4¹/₂ days

Figure 21.20

Embryonic development in *Drosophila*. (a) The fertilized egg, with the two parental nuclei. Polar cytoplasm indicates the posterior end. (b) The two parental nuclei fuse to produce a diploid zygote nucleus. (c) The nucleus undergoes nine divisions in a common cytoplasm to produce a multinucleate syncytium. (d) Nuclei migrate and divide, producing a layer at the periphery of the egg. This stage is the syncytial blastoderm. (e) Nuclei divide four times, and a membrane forms around each to produce the somatic cells of the cellular blastoderm.

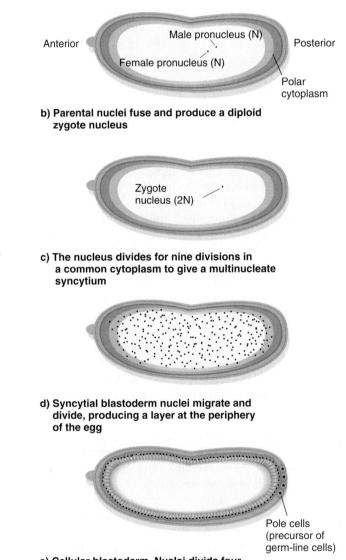

a) Fertilized egg with two parental nuclei

Anterior

Male pronucleus (N)

Female pronucleus (N)

Posterior

Polar cytoplasm

b) Parental nuclei fuse and produce a diploid zygote nucleus

Zygote nucleus (2N)

c) The nucleus divides for nine divisions in a common cytoplasm to give a multinucleate syncytium

d) Syncytial blastoderm nuclei migrate and divide, producing a layer at the periphery of the egg

Pole cells (precursor of germ-line cells)

e) Cellular blastoderm. Nuclei divide four times; membranes form around them and produce somatic cells

has been obtained about the relationship between the developmental events in the egg and the determination of adult body parts.

Before a mature egg is fertilized, particular molecular gradients are established within it, as we will discuss. The posterior end is indicated by the presence of a region called the *polar cytoplasm* (Figure 21.20a). At fertilization, the two parental nuclei are roughly centrally located in the egg. The two nuclei fuse to produce

Figure 21.21

***Drosophila* development results from gradients in the egg that define parasegments in the cellular blastoderm and segments in the embryo and adult.** The adult segment organization directly reflects the segment pattern of the embryo.

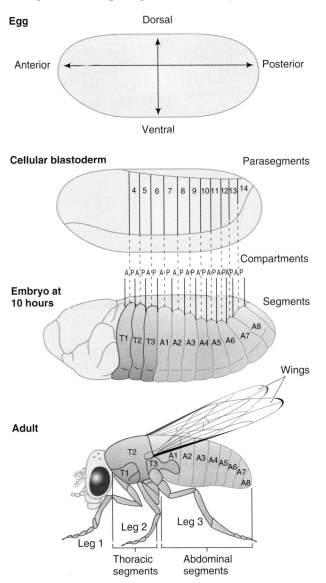

Subsequent development of body structures depends on two processes (Figure 21.21):

1. Gradients of molecules are produced along the anterior-posterior axis and the dorsal-ventral axis of the egg. Gene expression is affected by the position of a nucleus in the molecular concentration in the two intersecting gradients.

2. Regions are determined in the embryo that correspond to adult body segments. In the cellular blastoderm stage, where the regions are not well defined, they are called *parasegments*. In subsequent stages, where they are visible, they are called *segments*, forming a striped pattern along the anterior-posterior axis of the embryo. The embryonic segments give rise to the body segments of the adult fly.

The adult body plan—head, thoracic, and abdominal segments—derives from the larval body plan. In fact, two types of cells are specified by cellular blastoderm cells: those that will produce larval tissues and those that will develop into the adult tissues and organs. For the latter, certain groups of undifferentiated cells form larval structures called **imaginal discs** (*imago* means "imitate"), each of which differentiates into a specific structure of the adult fly. Imaginal discs are characteristic of holometamorphic insects. When an imaginal disc consists of about 20 to 50 cells it is already programmed to specify its given adult structure; its fate is determined. From then on, the number of cells in each disc increases by mitotic division, until by the end of the larval stages there are many thousands of cells per disc. Each imaginal disc differentiates into a specific part of the adult fly, including mouth parts, antennae, eyes, wings, halteres, legs, and the external genitalia. Other structures, such as the nervous system and gut, do not develop from imaginal discs. Figure 21.22 shows the positions of some imaginal discs in a mature larva and the adult structures that develop from them.

Genes involved in regulating *Drosophila* development are revealed by mutations that have a lethal phenotype early in development or that result in the development of abnormal structures (such as embryos with abnormal striping, or two anterior ends). Three major classes of developmental genes are involved: maternal effect genes, segmentation genes, and homeotic genes. We will discuss each of the classes and see how the polarity of the *Drosophila* egg specifies the segments of the adult fly body.

Maternal Effect Genes. **Maternal effect genes** are expressed by the mother during oogenesis; these genes are responsible for the polarity of the egg and, therefore, of the embryo. Through genetic screens a large number of maternal effect genes have been identified that are

a 2N zygote nucleus (Figure 21.20b). For the first nine divisions, only the nuclei divide in a common cytoplasm—cytokinesis does not occur—to produce what is called a multinucleate *syncytium* (Figure 21.20c). After seven divisions, some nuclei migrate into the polar cytoplasm, where they become precursors to germ-line cells. Next, the other nuclei migrate and divide to form a layer at the surface of the egg, producing the *syncytial blastoderm* (Figure 21.20d). After four more divisions, membranes form around the nuclei to produce somatic cells, about 4,000 of which make up the *cellular blastoderm* (Figure 21.20e).

Imaginal discs in larva   **Adult structures**

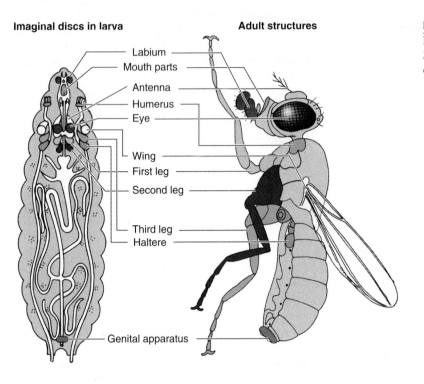

- Labium
- Mouth parts
- Antenna
- Humerus
- Eye
- Wing
- First leg
- Second leg
- Third leg
- Haltere
- Genital apparatus

Figure 21.22

Locations of imaginal discs in a mature *Drosophila* larva and the adult structures derived from each disc.

required for the normal patterning of the fly body. We focus here on two major groups of these genes, one responsible for normal patterning at the anterior of the embryo, and the other responsible mostly for normal patterning at the posterior of the embryo.

The *bicoid* (*bcd*) gene is the key maternal effect gene involved in the formation of the anterior structures of the embryo. In embryos of *bicoid* mutants, head and thoracic structures are converted to that of the abdomen, producing an embryo with posterior structures at each end, a lethal phenotype (Figure 21.23). We now know that the wild-type *bicoid* gene encodes a protein that is a **morphogen;** that is, it participates in controlling development. The *bicoid* gene is transcribed in

Figure 21.23

Scanning electron micrograph of a *bicoid* (*bcd*) mutant embryo.

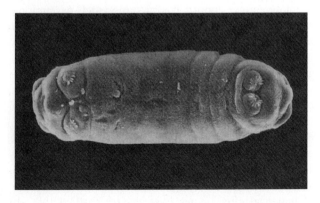

the mother during oogenesis, and the products of other anterior group maternal effect genes function to localize *bicoid* mRNA near the anterior pole (A) of the egg cytoplasm (Figure 21.24). Translation of the mRNA occurs after fertilization and BICOID protein diffuses to form a gradient with its highest concentration at the anterior end of the egg, fading to nothing in the posterior third of the egg (see Figure 21.24). As the phenotype of the *bcd* mutant suggests, the BICOID protein gradient specifies head and thorax development. The BICOID protein acts both as a transcription factor that activates and represses genes along the anterior-posterior axis of the embryo, and as a translational repressor to block translation of the mRNA of the *caudal* (*cad*) gene. The *cad* mRNAs are evenly distributed in the egg before fertilization but, after the *bicoid* mRNA is translated, translation of the *cad* mRNAs produces a gradient of CAUDAL proteins that is lowest at the anterior end and highest at the posterior end. This is opposite to the gradient of the BICOID protein. The CAUDAL protein functions later in the segmentation phase of development to activate genes needed for the formation of posterior structures.

Similarly, the *nanos* (*nos*) gene is the key maternal effect gene involved in the formation of the posterior structures of the embryo. Nullmutations of *nanos* result in a no-abdomen phenotype. The *nanos* gene is also transcribed in the mother during oogenesis, and its mRNA is localized to the posterior pole of the egg cytoplasms by products of other posterior group maternal effect genes. These mRNAs are translated after fertilization to produce

Figure 21.24

Gradients of *bicoid* mRNA and BICOID protein in the egg.
The mRNA gradient is localized near the anterior (A) pole
of the end, and BICOID protein forms a gradient with the highest
concentration at the anterior pole and the lowest concentration
at the posterior (P) pole.

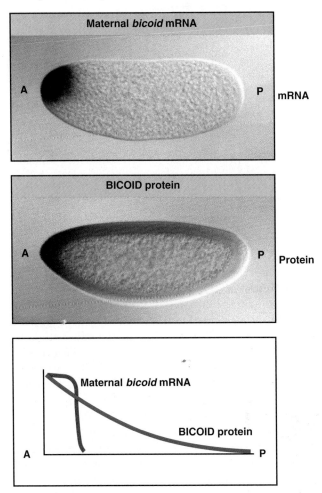

the NANOS protein, which forms a posterior-to-anterior
gradient and acts as a morphogen that directs abdomen
formation. The NANOS protein is a *translational repressor*, repressing the translation particularly of mRNA transcripts of the *hunchback* (*hb*) gene. These transcripts are
deposited in the egg during oogenesis and are distributed
evenly. However, for development to proceed correctly,
HUNCHBACK protein—a transcription factor—must be
present in a gradient that is decreasing in amount from
anterior to posterior. The NANOS protein is present in a
high-posterior-to-low-anterior gradient, and its translational repression activity creates the necessary HUNCH-
BACK protein gradient.

Segmentation Genes.   Next, the embryo is subdivided
into regions through the action of **segmentation genes,**
which determine the segments of the embryo and the
adult. Mutations in segmentation genes alter the number
of segments or their internal organization but do not
affect the overall organizational polarity of the egg. The
segmentation genes are subclassified on the basis of their
mutant phenotypes into gap genes, pair rule genes, and
segment polarity genes (Figure 21.25). Mutations in *gap
genes* (e.g., *Krüppel, hunchback, giant, tailless,* which all
encode transcription factors) result in the deletion of
regions consisting of several adjacent segments; mutations in *pair rule genes* (e.g., *hairy, even-skipped, runt,
fushi tarazu,* which all encode transcription factors) result
in the deletion of the same part of the pattern in every
other segment; and mutations in *segment polarity genes*
(e.g., *engrailed,* which encodes a transcription factor
hedgehog, which encodes a signaling protein; *armadillo,*
which encodes a signal transducing protein; and
gooseberry, which encodes a transcription factor) have
portions of segments replaced by mirror images of adjacent half segments.

Segmentation genes have specific roles in specifying regions of the embryo. Gap genes are activated or
repressed by maternal effect genes. For example, many
gap genes are activated by the BICOID protein. Gap
gene transcription leads to an organization of the
embryo into broad regions, each of which covers areas
that will later develop into several distinct segments.
Critical to this broad definition of regions is expression
of the *hunchback* gene, the control of which was already
described.

Next, through the transcription-regulating action of
the gap genes, the pair rule genes are expressed, leading
to a division of the embryo into a number of regions, each
of which contains a pair of parasegments (see Figure
21.21). The transcription factors encoded by the pair rule
genes regulate the expression of the segment polarity
genes, which determines regions that will become the
segments seen in larvae and adults.

Homeotic Genes.   Once the segmentation pattern has
been determined, a major class of genes called the
homeotic (structure-determining) **genes** (also called
selector genes) specifies the identity of each segment
with respect to the body part that will develop at metamorphosis. The homeotic genes have been defined by
mutations that affect the development of the fly. That
is, **homeotic mutations** alter the identity of particular
segments, transforming them into copies of other segments. The principal pioneer of genetic studies of
homeotic mutants is Edward Lewis, and the more
recent molecular analysis has been done in many
laboratories, including those of Thomas Kaufman,
Walter Gehring, W. McGinnis, Matthew Scott, and
Welcome Bender.

Figure 21.25

Functions of segmentation genes as defined by mutations.

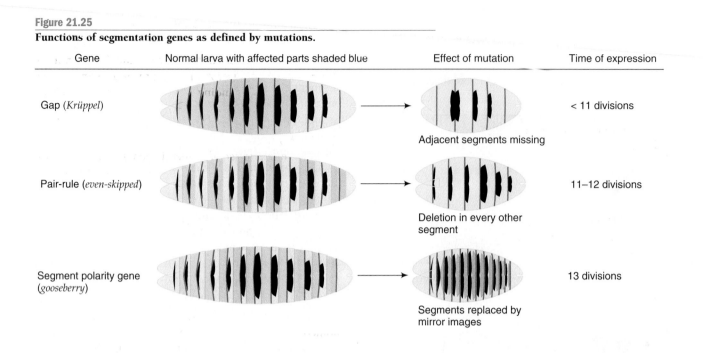

Gene	Normal larva with affected parts shaded blue	Effect of mutation	Time of expression
Gap (*Krüppel*)		Adjacent segments missing	< 11 divisions
Pair-rule (*even-skipped*)		Deletion in every other segment	11–12 divisions
Segment polarity gene (*gooseberry*)		Segments replaced by mirror images	13 divisions

Lewis's pioneering studies were on a cluster of homeotic genes called the *bithorax* complex (*BX-C*). *BX-C* determines the posterior identity of the fly, namely, thoracic segment T3 and abdominal segments A1–A8. *BX-C* contains three complementation groups called *Ultrabithorax* (*Ubx*), *abdominal-A* (*abd-A*), and *Abdominal-B* (*Abd-B*), each of which constitutes one protein-coding transcription unit. Mutations in these homeotic genes often are lethal, and the fly typically does not survive past embryogenesis. Some nonlethal mutations have been characterized, however, that allow an adult fly to develop. Figure 21.26 shows the abnormal adult structures that can result from *bithorax* mutations. A diagram showing the segments of a normal adult fly is in Figure 21.26a; note that the wings are located on segment Thorax 2 (T2), and the pair of halteres (rudimentary wings used as balancers in flight) are on segment T3. A photograph of a normal adult fly clearly showing the wings and halteres is presented in Figure 21.26b. Figure 21.26c shows one type of developmental abnormality that can result from nonlethal homeotic mutations in *BX-C*; shown is a fly that is homozygous for three separate mutations in the *Ubx* gene: *abx*, *bx3*, and *pbx*. Collectively, these mutations transform segment T3 into an adult structure similar to T2. The transformed segment exhibits a fully developed set of wings. The fly lacks halteres, however, because no normal T3 segment is present.

Another well-studied group of mutations defines another large cluster of homeotic genes called the *Antennapedia* complex (*ANT-C*). *ANT-C* determines the anterior identity of the fly, namely, the head and thoracic segments T1 and T2. *ANT-C* contains five genes: *labial* (*lab*), *proboscipedia* (*Pb*), *Deformed* (*Dfd*), *Sex combs reduced* (*Scr*), and *Antennapedia* (*Antp*). Most *ANT-C* mutations are lethal. Among the nonlethal mutations is a group of mutations in *Antp* that result in leg parts instead of an antenna growing out of the cells near the eye during the development of the eye disc (Figure 21.27a and b). Note that the leg has a normal structure, but it is obviously positioned in an abnormal location. A different mutation in *Antp*, called *Aristapedia*, has a different effect: Only the distal part of the antenna, the arista, is transformed into a leg (Figure 21.27c). Therefore, the homeotic genes *ANT-C* and *BX-C* encode products that are involved in controlling the normal development of the relevant adult fly structures.

The *Antennapedia* complex (*ANT-C*) and the *bithorax* complex (*BX-C*) have been cloned. Both complexes are very large. In *ANT-C*, for example, the *Antp* gene is 103 kb long, with many introns; this gene encodes a mature mRNA of only a few kilobases. *BX-C* covers more than 300 kb of DNA and contains only three protein-coding regions amounting to about 50 kb of that DNA: *Ubx*, *abd-A*, and *Abd-B* (Figure 21.28). At least *Ubx* and *abd-A* have introns. Other RNA products are known to be transcribed from *BX-C*, but they do not code for proteins, and they have no known functions. The other 250 kb of DNA in the complex is not transcribed and is believed to consist of regulatory regions of significant size and complexity. The functions of these regulatory regions are to control the expression of the protein-coding genes.

Since the *ANT-C* and *BX-C* protein-coding genes have similar functions, Lewis predicted that the genes

Figure 21.26

Adult structures that result from *bithorax* mutations.
(a) Drawing of a normal fly. T = thoracic segment.
A = abdominal segment. The haltere (rudimentary wing) is
on T3 (see Figure 21.28). (b) Photograph of a normal fly with
a single set of wings. (c) Photograph of a fly homozygous for
three mutant alleles (*bx3, abx, pbx*) that results in the transformation of segment T3 into a structure like T2: a segment with
a pair of wings. These flies therefore have two sets of wings
but no halteres.

a)

Haltere
(rudimentary
wing)

b)

c)

would have related sequences. Analysis of the DNA
sequences for the genes revealed the presence of similar sequences of about 180 bp that has been named
the **homeobox.** The homeobox is part of the protein-
coding sequence of each gene, and the corresponding
60-amino acid part of each protein is called the
homeodomain.

Homeoboxes have been found in more than 20
Drosophila genes, most of which regulate development.
All homeodomain-containing proteins are DNA-binding

proteins. The homeodomain of such proteins binds
strongly to an 8-bp consensus recognition sequence
upstream of all genes controlled as a unit by the homeodomain-containing protein. Helix-turn-helix motifs are
used in the DNA-binding property of homeodomains.
Thus, homeodomain-containing proteins play a role in
transcriptional regulation through interaction with specific DNA sequences.

The complete set of homeotic genes and complexes
in *Drosophila*—generically, the *Hox* genes—consists of
lab, pb, Dfd, Scr, Antp, Ubx, abdA, and *AbdB*. Most interestingly, these complexes are arranged in the same order
along the chromosome as they are expressed along the
anterior-posterior body axis; this is known as the colinearity rule. Homeotic gene complexes are found also in
all major animal phyla with the exception of sponges and
coelenterates. The homeobox sequences in the *Hox* genes
are highly conserved, indicating common function in the
wide range of organisms involved. As in *Drosophila*, the
homeotic genes of vertebrates—the *Hox* genes—follow
the colinearity rule. In mammals, for example, there are
four clusters of homeotic genes designated *HoxA–D*. Each
cluster is thought to have originated by duplication of a
primordial gene cluster followed by evolutionary divergence. The patterns of *Hox* gene expression, the effects of
mutations, and embryological analyses all indicate that
the vertebrate genes have homeotic effects similar to
those of *Drosophila* homeotic genes. Furthermore, the
studies indicate that the *Hox* genes specify the vertebrate
body plan. Figure 21.29 compares the organization
and expression of *Hox* genes in *Drosophila* and in
the mouse.

Homeotic genes are also found in plants. For example,
many homeotic mutations that affect flower development
have been identified in *Arabidopsis*. Studies of homeotic
mutants have led to models of flower development and,
more generally, of plant development. In parallel with
Drosophila homeotic genes, plant homeotic genes appear
to be part of a sequential array of genes that regulates
development.

KEYNOTE

Development of *Drosophila* body structures results
from gradients along the posterior-anterior and
dorsal-ventral axes of the egg and from the subsequent determination of regions in the embryo that
directly correspond to adult body segments. As defined
by mutations, genes control *Drosophila* development
in a temporal regulatory cascade. First, maternal effect
genes specify the gradients in the egg, then segmentation genes (gap genes, pair rule genes, and segment
polarity genes) determine the segments of the embryo
and adult, and homeotic genes next specify the identity of the segments.

Figure 21.27

Effects of some mutations in the *Antennapedia* complex. (a) Scanning electron micrograph (*left*) and drawing (*right*) of the antennal area of a wild-type fly. (b) Scanning electron micrograph (*left*) and drawing (*right*) of the antennal area of the homeotic mutant of *Drosophila*, Antennapedia, in which the antenna is transformed into a leg. (c) Scanning electron micrograph (*left*) and drawing (*right*) of the homeotic mutant of *Drosophila*, Aristapedia, in which the arista is transformed into a leg.

a) Normal

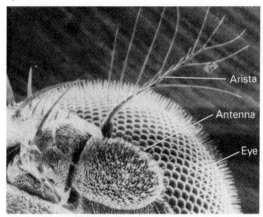 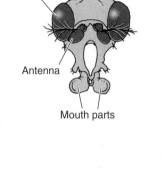

b) Antennapedia

 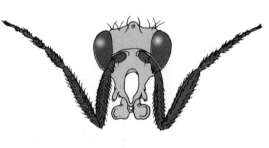

c) Aristapedia

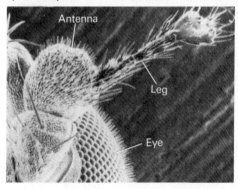 

Microarray Analysis of *Drosophila* Development

With the *Drosophila* genome sequence completed, studies are now underway to find ways of using that information to enrich our molecular understanding of *Drosophila* development. For example, DNA microarrays (see Chapter 9,

pp. 216–217) are being used to study changes in gene expression patterns in various developmental transitions. One such study has examined metamorphosis brought about by the hormone ecdysone from the time frame of 18 hours before pupal formation (BFP) to 12 hours after pupal

Figure 21.28

Organization of the *bithorax* complex (*BX-C*). The DNA spanned by this complex is 300 kb long. T = thoracic segment, A = abdominal segment. The transcription units for *Ubx*, *abdA*, and *abdB* are shown below the DNA; the exons are shown by colored blocks and the introns by bent, dotted lines. All three genes are transcribed from right to left. Shown above the DNA are regulatory mutants that affect the development of different fly segments.

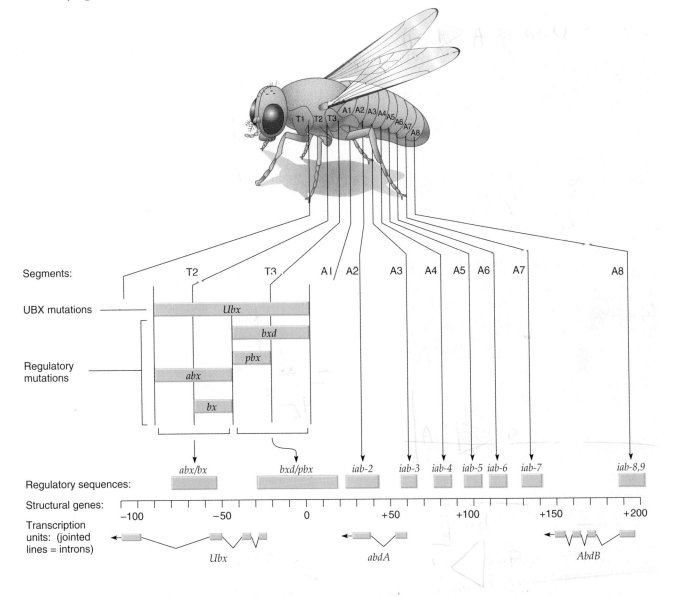

formation (APF). The expression of 6,240 cDNA clones (about 40 percent of the estimated genes of *Drosophila*) was analyzed. The results indicated that 534 genes were differentially expressed during the metamorphosis, some being repressed and some being induced. At a more specific level, the study catalogued the ecdysone-caused induction of a number of genes involved in the dramatic differentiation of the central nervous system during early metamorphosis at four hours BPF. Similarly, the repression of a number of

genes encoding proteins required for muscle formation was shown to occur at four hours BPF; this prepares the metamorphosing *Drosophila* for breakdown of larval muscle tissues beginning at two hours APF.

Undoubtedly, there will be more of these kinds of studies in the future. Thus, we can expect to see continued rapid advances in our knowledge of many aspects of *Drosophila* development, and of the development of other important model organisms.

Figure 21.29

Organization and expression of *Hox* genes in *Drosophila* and in the mouse.

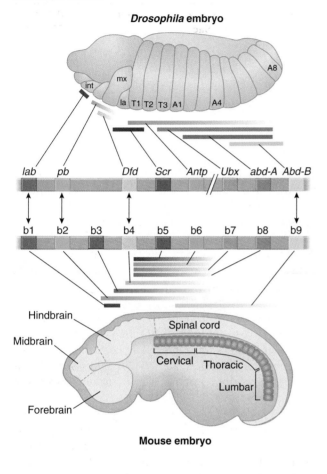

Drosophila embryo

Mouse embryo

heavy polypeptide chains of antibody molecules into the genes that are transcribed. Specifically, each light chain and each heavy chain has a variable (V) domain and a constant (C) domain. For the light chain, in germ-line DNA there are many V domain gene segments tandemly arranged, a few joining (J) gene segments, and one C gene segment. During development, somatic recombination occurs to bring together particular V, J, and C gene segments to form a functional L chain gene. A similar process occurs for assembling a functional H chain gene, in this case with the addition of several D (diversity) segments between V and J. In each cell, a different light chain gene and heavy chain gene are assembled during development, and this is the basis for the huge number of different types of antibodies that can be produced in mammals.

In the remainder of the chapter, we discussed two case studies. The first involved sex determination and dosage compensation in mammals and *Drosophila*. While both mammals and fruit flies have females with two X chromosomes and males with one X chromosome, sex determination is different in the two. In mammals, the presence of the Y chromosome specifies that the individiual will become a genetic male, and its absence means the individual will become a genetic female. That is, the Y chromosome carries the gene for the testis-determining factor that causes the tissue that will become gonads to differentiate into testes. In the absence of that gene, and hence its product, the gonads differentiate into ovaries by default. The gene for the testis-determining factor is *SRY*. The product of this gene is a transcription factor that acts as a testis-determining factor by turning on the genes that specify testis differentiation. In fruit flies, the Y chromosome has no role in sex determination. Instead, the ratio of X chromosomes to number of sets of autosomes determines the sex of the individual, with 1.0 being a female, and 0.5 being a male. Molecularly, the ratio is manifested in terms of numerator proteins encoded by particular X chromosome genes, and denominator proteins encoded by an autosomal gene. In females, who have two X chromosomes, there are excess numerator proteins over denominator proteins. These numerator proteins act as transcription factors for turning on a key regulatory gene in sex determination, *Sex-lethal* (*Sxl*). In males, there are no excess numerator proteins, so the gene remains silent. All subsequent steps toward sex determination in fruit fly embryos stem from this initial event, and involve a cascade of regulated versus default splicing of pre-mRNAs. In the end, the initial expression of *Sxl* in XX embryos leads to the differentiation of cells with female-specific gene expression, whereas the initial absence of expression of *Sxl* in XY embryos leads to the differentiation of cells with male-specific gene expression.

Summary

In this chapter, we discussed selected aspects of the genetic analysis of development. Development is defined as regulated growth resulting from the interaction of the genome with the cytoplasm and with the extracellular environment. Development commences when a zygote is formed, and involves a progression through the key events of determination and differentiation.

The chapter covered some key principles of development, and some case studies to illustrate some of the genetic events involved. With regard to principles, we discussed evidence that development results from differential gene expression, rather than by selective loss of genetic material during differentiation. An exception to the latter may be found in cells that differentiate to produce antibody molecules. To mount an immune response against an antigen, mammals must be capable of generating enormous antibody diversity. Here a particularly interesting process of chromosomal rearrangements is involved during cell development to bring together parts of the coding regions for the light and

In both mammals and *Drosophila*, the inequality of X chromosomes in the two sexes means that there is an issue of the regulation of expression of X-linked genes in the two sexes. In mammals, the expression of X-linked genes is equalized by a process that inactivates one of the two X chromosomes, so an XX female expresses X-linked genes at the same level as the same genes on the male's single X chromosome. In fruit flies, the expression of the genes on the single X chromosome in males is increased twofold to match that of the X-linked genes in females. In both of these dosage compensation processes, significant changes are made in chromatin structure.

The second case study involved the genetic regulation of the development of the *Drosophila* body plan. The presence of an extensive number of mutants affecting fruit fly embryonic development has made it possible to research this process at a detailed molecular level. We learned that development of *Drosophila* body structures results from gradients along the posterior-anterior and dorsal-ventral axes of the egg and from the subsequent determination of regions in the embryo that directly correspond to adult body segments. As defined by mutations, genes control *Drosophila* development in a temporal regulatory cascade. As a first step, maternal effect genes specify the gradients in the egg, then segmentation genes (gap genes, pair rule genes, and segment polarity genes) determine the segments of the embryo and adult, and homeotic genes next specify the identity of the segments.

Analytical Approaches to Solving Genetics Problems

Q21.1 We learned in this chapter that in humans, there are several distinct genes that code for α- and β-like globin polypeptides. These α-, β- γ-, δ-, ε-, and ζ-globin genes are transcriptionally active at specific stages of development, resulting in the synthesis of polypeptides that are assembled in specific combinations to form different types of hemoglobin (see p. 580 and Figure 21.9). Fill in the following table, indicating whether the globin gene in question is sensitive (S) or resistant (R) to DNase I digestion at each of the developmental stages listed.

Globin Gene	Tissue		
	Embryonic Yolk Sac	Fetal Spleen	Adult Bone Marrow
α			
β			
γ			
δ			
ζ			
ε			

A21.1 The correctly filled in table is as follows:

Globin Gene	Tissue		
	Embryonic Yolk Sac	Fetal Spleen	Adult Bone Marrow
α	R	S	S
β	R	R	S
γ	R	S	R
δ	R	R	S
ζ	S	R	R
ε	S	R	R

The explanation for the answers is as follows: DNase I typically digests regions of DNA that are transcriptionally active, while not digesting regions of DNA that are transcriptionally inactive. This is because transcriptionally inactive DNA is more highly coiled than transcriptionally active DNA. R means, then, that the gene was transcriptionally inactive, while S means that the gene was transcriptionally active.

To consider each globin gene in turn, the α gene is transcriptionally inactive in the embryonic yolk sac, but active in fetal spleen and adult bone marrow. That is, in the spleen fetal hemoglobin (Hb-F) is made, Hb-F contains two α polypeptides and two γ polypeptides. In the bone marrow, Hb-A is made, which contains two α and two β polypeptides.

The β-globin gene is inactive in yolk sac and spleen and is active in bone marrow, making one of the two polypeptides found in Hb-A, the main adult form of hemoglobin. The β-like γ polypeptide is found in Hb-F, which is made only in the liver and the spleen; thus, the γ gene is active in spleen and inactive in yolk sac and bone marrow. The β-like δ polypeptide is found in $\alpha_2\delta_2$ hemoglobin, which is a minor class of hemoglobin found in adults; thus, the δ gene is active only in adult bone marrow.

The ζ gene makes an α-like polypeptide found only in the hemoglobin of the embryo, so the ζ gene is active in the yolk sac but inactive in spleen and bone marrow. Finally, the ε gene encodes the β-like polypeptide of the embryo's hemoglobin, so this gene is also active in the yolk sac but inactive in spleen and bone marrow.

Questions and Problems

21.1 Distinguish between the terms *development, determination,* and *differentiation.*

21.2 What is totipotency? Give an example of the evidence for the existence of this phenomenon. What two mechanisms are used to restrict a cell's totipotency during development?

*21.3 It is possible to excise small pieces of early embryos of the frog, transplant them to older embryos, and follow the course of development of the transplanted material as the older embryo develops. A piece of tissue is excised from a region of the late blastula or early gastrula that would later develop into an eye and is transplanted to three different regions of an older embryo host (see part a of Figure 21.A). If the tissue is transplanted to the head region of the host, it will form eye, brain, and other material characteristic of the head region. If the tissue is transplanted to other regions of the host, it will form organs and tissues characteristic of those regions in normal development (e.g., ear, kidney). In contrast, if tissue destined to be an eye is excised from a neurula and transplanted into an older embryo host to exactly the same places as used for the blastula or gastrula transplants, in every case the transplanted tissue differentiates into an eye (see part b of Figure 21.A). Explain these results.

21.4 With respect to the genetic analysis of development, what is meant by a *model* organism? Describe the

features that model organisms possess to make them attractive for the genetic analysis of development, using specific examples.

21.5 In the set of experiments used to clone Dolly, six additional live lambs were obtained. Why is the production of Dolly more significant than the production of the other lambs? What is the evidence that Dolly resulted from the fusion of a nucleus from one cell with the cytoplasm of another?

*21.6 In Woody Allen's 1973 film *Sleeper,* the aging leader of a futuristic totalitarian society has been dismembered in a bomb attack. The government wants to clone the leader from his only remaining intact body part, a nose. The characters Miles and Luna thwart the cloning by abducting the nose and flattening it under a steamroller.

a. In light of the 1996 cloning of the sheep Dolly, how should the cloning have proceeded if Miles and Luna had not intervened?

b. If methods like those used for Dolly had been successful, in what genetic ways would the cloned leader be unlike the original?

c. Suppose that instead of a nose, only mature B cells (B lymphocytes of the immune system) were available. What genetic deficits would you expect in the "new leader"?

d. Based on what has been discovered about cloned cats and mice, in what nongenetic ways might the cloned leader differ from the original? What (constructive) advice would you give the totalitarian government based on these findings?

e. If the cloning of the leader had succeeded, can you make any prediction about whether the "cloned leader" would be interested in perpetuating the totalitarian state?

21.7 Discuss some of the evidence for differential gene activity during development. How have microarray analyses enhanced our understanding of this process?

21.8 Discuss the expression of human hemoglobin genes during development.

21.9 How are the hemoglobin genes organized in the human genome, and how is this organization related to their temporal expression during development?

21.10 In humans, β-thalassemia is a disease caused by failure to produce sufficient β-globin chains. In many cases, the mutation causing the disease is a deletion of all or part of the β-globin structural gene. Individuals homozygous for certain of the β-thalassemia mutations are able to survive because their bone marrow cells produce γ-globin chains. The γ-globin chains combine with

Figure 21.A

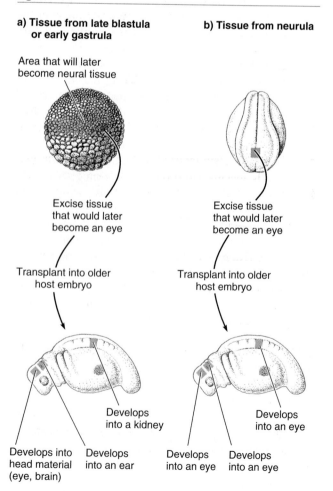

a) **Tissue from late blastula or early gastrula**

Area that will later become neural tissue

Excise tissue that would later become an eye

Transplant into older host embryo

Develops into a kidney

Develops into head material (eye, brain)

Develops into an ear

b) **Tissue from neurula**

Excise tissue that would later become an eye

Transplant into older host embryo

Develops into an eye

Develops into an eye

Develops into an eye

α-globin chains to produce fetal hemoglobin. In these people, fetal hemoglobin is produced by the bone marrow cells throughout life, whereas normally it is produced in the fetal liver. Use your knowledge about gene regulation during development to suggest a mechanism by which this expression of γ-globin might occur in β-thalassemia.

21.11 What are polytene chromosomes? Discuss the molecular nature of the puffs that occur in polytene chromosomes during development.

21.12 Puffs of regions of the polytene chromosomes in salivary glands of *Drosophila* are surrounded by RNA molecules. How would you show that this RNA is single stranded and not double stranded?

*21.13 In experiment A, ^{3}H-thymidine (a radioactive precursor of DNA) is injected into larvae of *Chironomus*, and the polytene chromosomes of the salivary glands are later examined by autoradiography. The radioactivity is seen to be distributed evenly throughout the polytene chromosomes. In experiment B, ^{3}H-uridine (a radioactive precursor of RNA) is injected into the larvae, and the polytene chromosomes are examined. The radioactivity is first found only around puffs; later, radioactivity is also found in the cytoplasm. In experiment C, actinomycin D (an inhibitor of transcription) is injected into larvae and then ^{3}H-uridine is injected. No radioactivity is found associated with the polytene chromosomes, and few puffs are seen. The puffs that are present are much smaller than the puffs found in experiments A and B. Interpret these results.

21.14 Explain how it is possible for both of the following statements to be true: The mammalian genome contains about 10^5 genes. Mammals can produce about 10^6 to 10^8 different antibodies.

21.15 Antibody molecules (Ig) are composed of four polypeptide chains (two of one light-chain type and two of one heavy-chain type) held together by disulfide bonds.
a. If for the light chain there were 300 different V_κ segments and four J_κ segments, how many different light chain combinations would be possible?
b. If for the heavy chain there were 200 V_H segments, 12 D segments, and 4 J_H segments, how many heavy-chain combinations would be possible?
c. Given the information in (a) and (b), what would be the number of possible types of IgG molecules (L + H chain combinations)?

21.16 How was the testis-determining factor gene (*TDF*) identified, and what evidence is there to support the contention that the *SRY* gene is the TDF gene?

*21.17 A male mouse cell line has been generated in which the *gfp* (*green fluorescent protein*) gene has been inserted onto the X chromosome under the control of a constitutively expressed promoter. Cells expressing the *gfp* gene exhibit bright-green fluorescence under UV light.
a. What pattern of green fluorescence do you expect to see in this cell line?
b. The cell line is modified by introducing a segment of DNA containing *Xic* into an autosome. How do you expect the pattern of green fluorescence to change? Why?
c. A cell of the modified cell line described in (b) exhibits green fluorescence. Which copy of *Xist*—the one on the X chromosome or the one on the autosome—is being expressed? On which chromosome does its expression lead to chromatin remodeling, and how?

*21.18 In *Drosophila*, sex type is determined by the X:A ratio.
a. How is this ratio detected early in development, and how does it lead to sex-specific transcription at *Sxl*?
b. A mutation in *Sxl* affects P_E so that early transcription of *Sxl* does not occur. The upstream P_L promoter is unaffected, so constitutive transcription from this late promoter occurs in all cells regardless of their X:A ratio. What phenotype do you expect this mutation to have in animals with an X:A ratio of 1:2? In animals with an X:A ratio of 2:2?
c. A *tra* mutant has a nonsense mutation into exon 2. What phenotype do you expect this mutant to have in animals with an X:A ratio of 1:2? In animals with an X:A ratio of 2:2?
d. The TRA protein targets the *dsx* pre-mRNA for alternatively splicing. However, if TRA is not present, male differentiation ensues. Why then do animals with knockout mutations in *dsx* have both male and female characteristics?

21.19 The SXL protein binds to its own mRNA as well as the mRNAs of *tra* and *msl-2*. How does it regulate its own expression through mRNA binding? Is this mechanism the same as, or different from, the mechanism by which it regulates the expression of *tra* and *msl-2*?

21.20 In *Drosophila*, mutations at five genes (*mle, msl-1, msl-2, msl-3* and *mof*) lead to male-specific lethality during the larval stages due to defective dosage compensation.
a. What common process does each of these genes function in, and how does it lead to dosage compensation in *Drosophila* males?
b. Explain why females with these mutations develop normally.

c. How is the mechanism by which dosage compensation occurs in *Drosophila* related to the molecular steps regulating sex determination?

d. Why do *Sxl* gain-of-function mutations cause male lethality, while *Sxl* loss-of-function mutations have no effect on male development?

*21.21 The following figure shows the percentage of ribosomes found in polysomes in unfertilized sea urchin oocytes (0 hours) and at various times after fertilization:

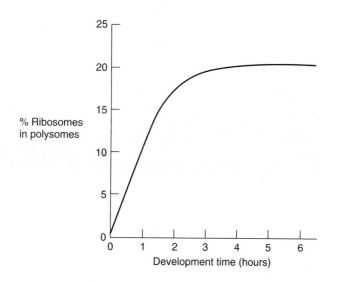

In the unfertilized egg, less than 1 percent of ribosomes are present in polysomes, and at 2 hours postfertilization, about 20 percent of ribosomes are present in polysomes. It is known that no new mRNA is made during the time period shown. How can these data be interpreted?

21.22 Define *imaginal disc* and *homeotic mutant*.

21.23 Both *bicoid* and *nanos* are maternal effect genes whose mRNAs are transcribed in the mother and then localized in the cytoplasm of developing embryos.

a. How does maternal deposition of these mRNAs lead to gradients of morphogens along the anterior-posterior axis of the developing embryo?

b. What would be the effect of loss-of-function mutations in *bicoid* on the distribution of CAUDAL protein, and how does this contribute to the phenotype of these *bicoid* mutations?

c. What would be the affect of loss-of-function mutations in *nanos* on the distribution of HUNCHBACK protein, and how does this contribute to the phenotype of these *nanos* mutations?

d. What is the function of the morphogen gradients established by these two genes?

*21.24 Imagine that you observed the following mutants (*a–e*) in *Drosophila*.

a. Mutant *a:* In homozygotes, phenotype is normal, except wings are oriented backward.

b. Mutant *b:* Homozygous females are normal but produce larvae that have a head at each end and no distal ends. Homozygous males produce normal offspring (assuming the mate is not a homozygous female).

c. Mutant *c:* Homozygotes have very short abdomens, which are missing segments A2 through A4.

d. Mutant *d:* Affected flies have wings growing out of their heads in place of eyes.

e. Mutant *e:* Homozygotes have shortened thoracic regions and lack the second and third pair of legs.

Based on the characteristics given, assign each of the mutants to one of the following categories: maternal effect gene, segmentation gene, or homeotic gene.

21.25 What is the evidence that homeotic genes specify not only the invertebrate body plan, but also the vertebrate body plan?

*21.26 If actinomycin D, an antibiotic that inhibits RNA synthesis, is added to newly fertilized frog eggs, there is no significant effect on protein synthesis in the eggs. Similar experiments have shown that actinomycin D has little effect on protein synthesis in embryos up until the gastrula stage. After the gastrula stage, however, protein synthesis is significantly inhibited by actinomycin D, and the embryo does not develop further. Interpret these results.

22

Genetics of Cancer

p53 protein binding to DNA.

PRINCIPAL POINTS

- Progression through the cell cycle is tightly controlled by the activities of many genes. Checkpoints at key points determine whether a cell has DNA damage or has problems with its cell-cycle machinery and only permits normal cells to continue. The key molecules used at these checkpoints are cyclins and cyclin-dependent kinases (Cdks). In addition, healthy cells grow and divide only when the balance of stimulatory and inhibitory signals received from outside the cell favor cell proliferation. A cancerous cell does not respond to the usual signals and divides without constraints.

- Mutant forms of three classes of genes—proto-oncogenes, tumor suppressor genes, and mutator genes—have the potential to contribute to the transformation of a cell to a cancerous state. The products of proto-oncogenes normally stimulate cell proliferation, the products of tumor suppressor genes normally inhibit cell proliferation, and the products of mutator genes are involved in DNA replication and repair.

- Some DNA viruses and RNA viruses cause cancers. All RNA tumor-causing viruses are retroviruses—viruses that replicate via a DNA intermediate—but not all retro-

viruses cause cancer. When a retrovirus infects a cell, the RNA genome is released from the viral particle, and through the action of reverse transcriptase a cDNA copy of the genome, called the proviral DNA, is synthesized. The proviral DNA integrates into the genome of the host cell. Then, using host transcriptional machinery, viral genes are transcribed, and full-length viral RNAs are produced. Progeny viruses are assembled and exit the cell, where they can infect other cells.

- Tumor induction can occur after retrovirus infection, because of the activity of a viral oncogene (v-*onc*) in that retroviral genome. Retroviruses carrying an oncogene are known as transducing retroviruses.

- Normal animal cells contain genes with DNA sequences that are similar to those of the viral oncogenes. These cellular genes are proto-oncogenes. When a proto-oncogene is mutated to produce a cellular oncogene (c-*onc*), it induces tumor formation.

- The two-hit mutation model for cancer states that two mutational events are necessary for cancer to develop, one in each allele of a cancer-causing gene. In familial (hereditary) cancers, one mutation is inherited, predisposing the person to cancer; the other mutation occurs later in somatic cells. In sporadic (nonhereditary) cancers, both mutations occur in somatic cells. This simple two-hit model applies for very few cancers; other cancers involve mutations in many genes.

- The normal products of tumor suppressor genes have inhibitory roles in cell growth and division. Therefore, when both alleles of a tumor suppressor gene are inactivated or lost, the inhibitory activity is lost, and uncontrolled cell proliferation can occur.

- Mutator genes are genes that, when mutated, increase the spontaneous mutation frequencies of other genes. In the cell, the normal (unmutated) forms of mutator genes are involved in key activities such as DNA replication and DNA repair.

- While telomerase (the enzyme for maintaining the ends of chromosomes) is not active in most normal human cells, the enzyme is active in all types of human cancers. Reactivation of telomerase is not a cause of cancer, but enables cancer cells that have telomeres too short for chromosome replication to divide indefinitely by lengthening the telomeres and stabilizing the chromosomes.

- The development of most cancers involves the accumulation of mutations in a number of genes over a significant period of a person's life. This multistep path typically involves mutational events that change proto-oncogenes to oncogenes and inactivate tumor suppressor genes and mutator genes, thereby breaking down the multiple mechanisms that safeguard growth and differentiation.

- Various types of radiation and many chemicals increase the frequency with which cells become cancerous. These agents are known as carcinogens. Practically all carcinogens act by causing changes in the genome of the cell. In the case of chemical carcinogens, a few act directly on the genome, but most act indirectly by being converted to active derivatives by cellular enzymes. All carcinogenic forms of radiation act directly.

i AT CURRENT RATES, OVER A THIRD OF THE PEOPLE who read this will die of cancer. Cancers are diseases characterized by the uncontrolled and abnormal division of eukaryotic cells. When cells divide unchecked within the body, they can give rise to tissue masses known as tumors. Some of these are not life threatening, but others can invade and disrupt surrounding tissues. What are the mechanisms that regulate cell growth and division? What causes uncontrolled growth in a cell? What genes are involved in the development of cancer? Is cancer inherited? In this chapter, you will learn the answer to these, and other, questions. Then, in the iActivity, you can apply what you learned as you investigate the origins of a form of bladder cancer.

In Chapter 21 we learned about some of the genetically controlled processes involved in development and differentiation. During development, specific tissues and organs arise by genetically programmed cell division and differentiation. In an adult, the many different types of cells of the body proliferate only in a controlled way. For example, for many tissues, programmed cell division occurs only to replace cells lost normally or through injury. Other cells, such as those of the intestinal lining, and those that give rise to blood cells, must divide routinely to replace cells that have died.

Occasionally, dividing and differentiating cells deviate from their normal genetic program and give rise to tissue masses called **tumors,** or *neoplasms* ("new growth"). Figure 22.1 shows a mammogram indicating the presence of a tumor. The process by which a cell loses its ability to remain constrained in its growth properties is called **transformation** (not to be confused with transformation of a cell by uptake of exogenous DNA). If the transformed cells stay together in a single mass, the tumor is said to be *benign*. Benign tumors usually are not life threatening, and their surgical removal generally results in a complete cure. Exceptions include many brain tumors, which are life threatening because they impinge on essential cells.

The unregulated cell division of transformed cells can be seen easily in the culture dish. Normal fibroblast cells (cells that make the structural fibers and ground substance of connective tissue) grown in culture attach to the dish surface and divide until they contact each other. The

Figure 22.1

A mammogram showing a tumor.

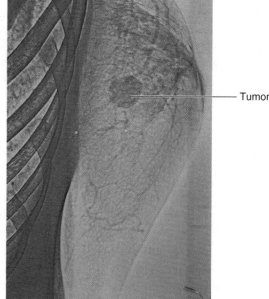

Tumor

result is a *monolayer*—a single layer of cells—covering the dish. This phenomenon is brought about by *contact inhibition*, a process whereby cells in contact communicate with one another and cell division is stopped. Contact inhibition is reflective of the regulated nature of cell division shown by normal cells. By contrast, transformed cells do not show contact inhibition, instead continuing to grow and divide after contacting neighbors and piling up in multiple layers.

If the cells of a tumor can invade and disrupt surrounding tissues, the tumor is said to be *malignant* and is identified as a **cancer.** Sometimes, cells from malignant tumors can also break off and move through the blood system or lymphatic system, forming new tumors at other locations in the body. The spreading of malignant tumor cells throughout the body is called **metastasis** ("change" of "state"). Malignancy can result in death because of damage to critical organs, secondary infection, metabolic problems, second malignancies, or hemorrhage.

The initiation of tumors in an organism is called **oncogenesis** (*onkos*, "mass" or "bulk"; *genesis*, "birth"). In this chapter, we focus on the genetic basis of oncogenesis.

Relationship of the Cell Cycle to Cancer

We discussed development and differentiation in Chapter 21. Recall that, during development, a tissue is produced by cell proliferation. During a series of divisions, progeny cells begin to express genes that are specific for the tissue, a process called cell differentiation. In some tissues, cell differentiation is also associated with the progressive loss of the ability of cells to proliferate: The most highly differentiated cell, the one that is fully functional in the tissue, can no longer divide. Such cells are known as *terminally differentiated cells*. They have a finite life span in the tissue and are replaced with younger cells produced by division of unspecialized cells, called *stem cells*, which are a small fraction of cells in the tissue that are capable of *self-renewal*. To understand both benign and malignant neoplastic diseases, we must realize that the linkage of growth with differentiation of any tissue is not necessary. That is, cells *can* divide without undergoing terminal differentiation. For example, in malignant neoplasms most daughter cells from any replication event fail to express fully the genetic programs that regulate terminal differentiation.

Molecular Control of the Cell Cycle

In every cell cycle, all chromosomes must be duplicated faithfully and a copy of each distributed to both progeny cells. Recall from Chapter 12 that the cell cycle in most somatic cells of higher eukaryotes is divided into four stages: gap 1 (G_1), synthesis (S), gap 2 (G_2), and mitosis (M) (see Figure 12.4).

Progression through the cell cycle is tightly controlled by the activities of many genes in an elaborate system of checks and balances (Figure 22.2). **Checkpoints** at different points in the cell cycle are control points at which the cell cycle is arrested if there is damage to the genome or cell-cycle machinery. This allows the damage to be repaired or, if it is not, the cell is destroyed. These processes are necessary to prevent the possibility of damaged cells dividing in an unprogrammed way, that is, from becoming cancerous.

As a cell proceeds through G_1, it prepares for DNA replication and chromosome duplication in the S phase. A major checkpoint in G_1—called *START* in yeast and G_1-*to-S checkpoint* in mammalian cells—determines whether the cell is able to or should continue into S. The cell will stay in G_1 unless it grows a large enough and the environment is favorable. Another major checkpoint, the G_2-*to-M checkpoint*, occurs at the junction between G_2 and M. Unless all the DNA has replicated, the cell is big enough, and the environment is favorable, the cell cannot enter the mitotic phase of the cell cycle. A third checkpoint occurs during M: The chromosomes must be attached properly to the mitotic spindle to trigger the separation of chromatids and the completion of mitosis.

Proteins known as **cyclins** (named because their concentration increases and decreases in a regular pattern through the cell cycle) and enzymes known as **cyclin-dependent kinases** (Cdks) are the key components in the regulatory events that occur at checkpoints. At the G_1-to-S checkpoint, two different G_1 cyclin/Cdk complexes form, resulting in activation of the kinases. The kinases catalyze a series of phosphorylations (addition of phosphate groups) of cell-cycle control proteins, affecting the functions of those proteins and leading, therefore, to transition into the S phase.

A similar process occurs at the G_2-to-M checkpoint. A G_2 cyclin binds to a Cdk to form a complex. Until the cell is ready to enter mitosis, phosphorylation of the Cdk by another kinase keeps the Cdk inactive. At that time, a phosphatase removes the key phosphate from the Cdk, activating the enzyme. Phosphorylations of proteins by the Cdk move the cell into mitosis.

Regulation of Cell Division in Normal Cells

Control of cell division of a normal cell is handled by both extracellular and cellular molecules that operate in a complicated signaling system. The extracellular molecules are steroids and protein hormones made in other tissues that influence the growth and division of cells in other tissues. For example, a *growth factor* is a molecule that stimulates cell division of a

animation

a **Regulation of Cell Division in Normal Cells**

• Cert
parti
lymp
Chap
mal l
alter
cell c

Genes

Three cla
These are
mutator
mally sti
genes, cal
more acti
The produ
mally inh
genes fou
function.
needed to
of genome
have lost
prone to a

Viral
envelc

Viral
glycoproteins

Host
cell-
rece

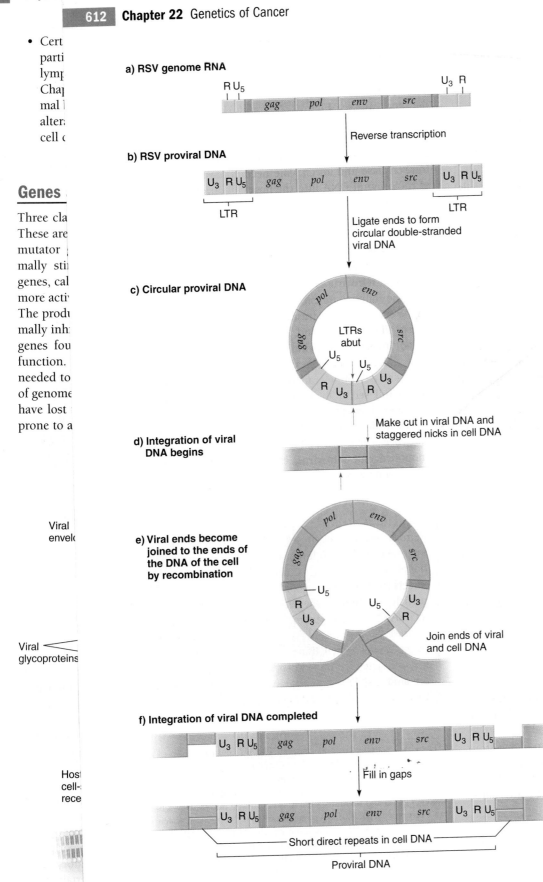

a) RSV genome RNA

Reverse transcription

b) RSV proviral DNA

LTR LTR

Ligate ends to form circular double-stranded viral DNA

c) Circular proviral DNA

LTRs abut

Make cut in viral DNA and staggered nicks in cell DNA

d) Integration of viral DNA begins

e) Viral ends become joined to the ends of the DNA of the cell by recombination

Join ends of viral and cell DNA

f) Integration of viral DNA completed

Fill in gaps

Short direct repeats in cell DNA

Proviral DNA

Figure 22.5

The Rous sarcoma virus (RSV) RNA genome and a suggested mechanism for the integration of the proviral DNA into the host (chicken) chromosome. (a) RSV genome RNA. (b) RSV proviral DNA produced by reverse transcriptase. (c) Circularization of the proviral DNA. (d) Staggered nicks are made in viral and cellular DNAs. (e) By recombination, the viral ends become joined to the ends of the DNA of the cell. (f) The single-stranded gaps are filled in, and a complete, double-stranded, integrated RSV provirus results.

Figure 22.6
Life cycle of a nononcogenic retrovirus.

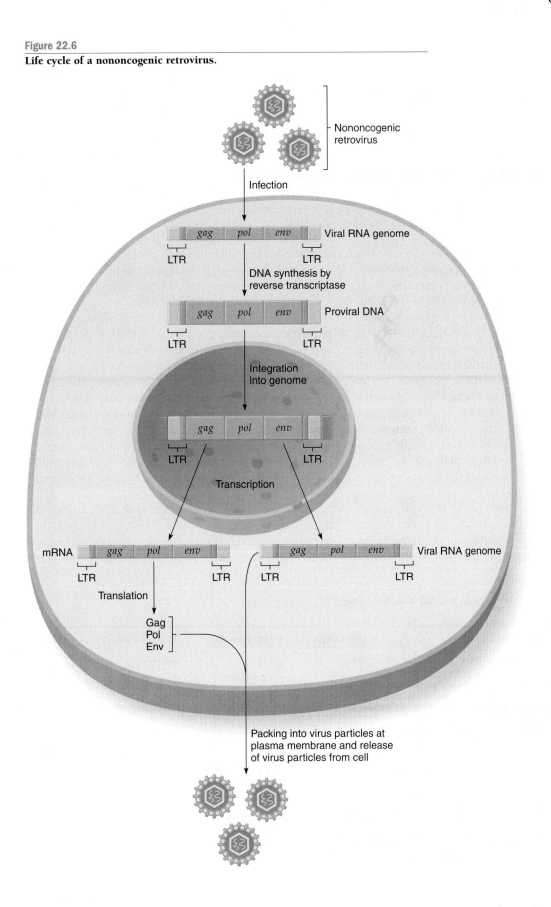

Viral Oncogenes. An oncogenic retrovirus carries a particular viral oncogene in its genome. Viral oncogenes (generically called v-*oncs*) are responsible for many different cancers. The v-*onc* genes are named for the tumor that the virus causes, with the prefix "v" to indicate that the gene is of viral origin. Thus, the v-*onc* gene of RSV is v-*src*. Bacteriophages that have picked up cellular genes are said to transduce the genes to other cells, so such retroviruses are called **transducing retroviruses** because they have picked up an oncogene from the genome of the cell. Table 22.1 lists some transducing retroviruses and their viral oncogenes. Retroviruses that do not carry oncogenes are called *nontransducing retroviruses*.

How is a transducing retrovirus produced? The location at which retroviral DNA (the provirus) integrates into cellular DNA is random. Sometimes there occurs a genetic rearrangement by which the transcriptional unit of the provirus connects to nearby cellular genes, often by a deletion event involving the loss of some or all of the *gag, pol,* and *env* genes. In this way, viral RNA contains all or parts of a cellular gene. All viral progeny then carry the cellular gene and, under the influence of viral promoters in the LTR, express the cellular protein in infected cells. If the cellular gene picked up was an oncogene, the modified retrovirus is oncogenic. If the cellular gene picked up is a proto-oncogene, the modified retrovirus may still be oncogenic if the increased expression of the proto-oncogene causes oncogenesis.

In the case of RSV, cells infected by this virus rapidly transform into the cancerous state because of the activity of the v-*src* gene. Since RSV contains all the genes necessary for viral replication (*gag, env,* and *pol*), an RSV-trans-formed cell produces progeny RSV particles. In this ability, RSV is an exception; all other transducing retroviruses have deletions in their genomes, causing loss of function of one or more of their viral replication genes. They can transform cells, but are unable to produce progeny viruses because they lack one or more genes needed for virus reproduction. These defective retroviruses can produce progeny viral particles if cells containing them are also infected with a normal virus (a *helper virus*) that can supply the missing gene products.

Cellular Proto-Oncogenes. In the mid-1970s, J. Michael Bishop, Harold Varmus, and others demonstrated that normal animal cells contain genes with DNA sequences that are very closely related to the viral oncogenes. These genes are called proto-oncogenes. (Bishop and Varmus received the 1989 Nobel Prize in Physiology or Medicine for their "discovery of the cellular origin of retroviral oncogenes.") In the early 1980s, R. A. Weinberg and M. Wigler showed independently that a variety of human tumor cells contain oncogenes. These genes, when introduced into other cells growing in culture, transformed those cells into cancer cells. The human oncogenes were found to be very similar to viral oncogenes that had been characterized previously, even though viruses did not induce the human cancers involved. These human oncogenes also were shown to be closely related to proto-oncogenes found in normally growing cells.

In short, human and other animal oncogenes are mutant forms of normal cellular genes. Such genes in their normal state are called **proto-oncogenes.** Proto-oncogenes have important roles in regulating the cell cycle. When

Table 22.1	Some Transducing Retroviruses and Their Viral Oncogenes			
Oncogene	**Retrovirus Isolate**	**v-onc Origin**	**v-onc Protein**	**Type of Cancer**
src	Rous sarcoma virus	Chicken	pp60^{src}	Sarcoma
abl	Abelson murine leukemia virus	Mouse	P90–P160$^{gag-abl}$	Pre–B cell leukemia
erbA	Avian erythroblastosis virus	Chicken	P75$^{gag-erbA}$	Erythroblastosis and sarcoma
erbB	Avian erythroblastosis virus	Chicken	gp65^{erbB}	Erythroblastosis and sarcoma
fms	McDonough (SM)-FeSV	Cat	gp180$^{gag-fms}$	Sarcoma
fos	FBJ (Finkel-Biskis-Jinkins)-MSV	Mouse	pp55^{fos}	Osteosarcoma
myc	MC29	Chicken	P100$^{gag-myc}$	Sarcoma, carcinoma, and myelocytoma
myb	Avian myeloblastosis virus (AMV)	Chicken	p45^{myb}	Myeloblastosis
	AMV-E26	Chicken	P135$^{gag-myb-ets}$	Myeloblastosis and erythroblastosis
raf	3611-MSV	Mouse	P75$^{gag-raf}$	Sarcoma
H-*ras*	Harvey MSV	Rat	pp21^{ras}	Sarcoma and erythroleukemia
	RaSV	Rat	P29$^{gag-ras}$	Sarcoma?
K-*ras*	Kirsten MSV	Rat	pp21^{ras}	Sarcoma and erythroleukemia

proto-oncogenes become mutated or translocated to induce tumor formation, they are called oncogenes (oncs). Only one proto-oncogene of a homologous pair needs to be mutated to produce an oncogene and, therefore, loss of cell cycle control. Thus, these mutations are dominant mutations.

If oncogenes are carried by a virus, oncogenes are known as v-*oncs*. If they reside in the host chromosome, oncogenes are called *cellular oncogenes,* or c-*oncs*. A transducing retrovirus, then, carries a significantly altered form of a cellular proto-oncogene (now a v-*onc*). When the transducing retrovirus infects a normal cell, the hitchhiking oncogene transforms the cell into a cancer cell.

One significant difference between a cellular proto-oncogene and its viral oncogene counterpart is that most proto-oncogenes contain introns that are not present in the corresponding v-*onc*. This is the result of splicing that

occurs in the transcription of the genomic viral RNA from proviral DNA. For example, the chicken *src* proto-oncogene is more than 7 kb long, with twelve exons separated by introns (Figure 22.7a). In the RSV RNA genome, the v-*src* oncogene is 1.7 kb long with no introns. Genomic RNA produced by the full-length transcription of the proviral DNA is packaged into virus particles. Three types of mRNA are produced from the RSV proviral DNA: a full-length transcript and two mRNAs generated by differential splicing of the full-length transcript (Figure 22.7b). mRNA transcription starts at a promoter in the left U_3 sequence, and the addition of the poly(A) tail is signaled by a sequence in the right R sequence.

Protein Products of Proto-Oncogenes. About 100 proto-oncogenes have been identified through their oncogene derivatives. Proto-oncogenes fall into several distinct

Figure 22.7

The chicken c-*src* proto-oncogene and its relationship to v-*src* in Rous sarcoma virus. (a) At the top is the molecular organization of the chicken *src* proto-oncogene. The gene contains 12 exons (shown as purple boxes). Below is the molecular organization of the Rous sarcoma virus RNA genome to indicate the relationship of nucleotide sequences in the cellular *src* proto-oncogene and v-*src*; v-*src* was generated mostly by intron removal from the cellular *src* proto-oncogene. (b) mRNAs produced by transcription of RSV proviral DNA genome.

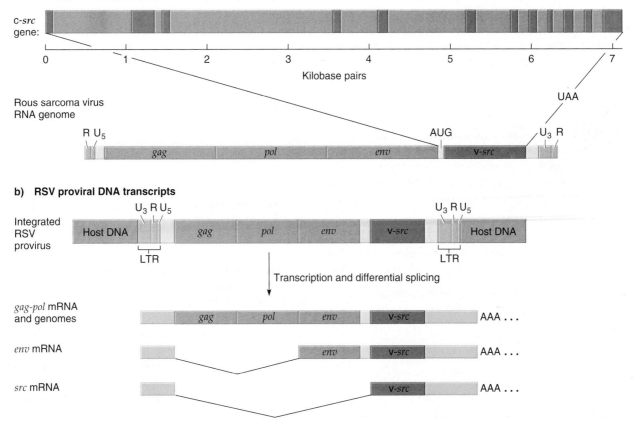

classes, based on DNA sequence similarities and similarities in amino acid sequences of the protein products (Table 22.2).

In each case, the proto-oncogene product is involved in the positive control of cell growth and division; that is, the products stimulate growth or are in a pathway involved in growth stimulation. Three examples, growth factors, protein kinases, and membrane-associated G proteins, are described in the remainder of this section.

Growth Factors. The effect of oncogenes on cell growth and division led to an early hypothesis that proto-oncogenes might be regulatory genes involved with the control of cell multiplication during differentiation. Evidence supporting this hypothesis came when the product of the viral oncogene v-*sis* was shown to be identical to part of platelet-derived growth factor (PDGF, a factor found in blood platelets in mammals), which is released after tissue damage. PDGF affects only one type of cell, fibroblasts, causing them to grow and divide. The fibroblasts are part of the wound-healing system. The causal link between PDGF and tumor induction was demonstrated in an experiment in which the cloned PDGF gene was introduced into a cell that normally does not make PDGF (i.e., a fibroblast); that cell was transformed into a cancer cell.

We can generalize and say that some cancer cells can result from the excessive or untimely synthesis of growth factors in cells that do not normally produce those factors. An altered growth factor gene such as a v-*onc* or mutation of a c-*onc* are examples of changes that can alter growth factor levels.

Protein Kinases. Many proto-oncogenes encode protein kinases, enzymes that add phosphate groups to target proteins, thereby modifying the function of the proteins. Protein kinases are integral members of signal transduction pathways. The *src* gene product, for example, is a nonreceptor protein kinase called pp60src. The viral protein, pp60v-*src*, and the protein encoded by the cellular oncogene, pp60c-*src*, differ in only a few amino acids, and both proteins bind to the inner surface of the plasma membrane.

What is particularly interesting about the *src* protein kinases is that both versions add a phosphate group to the amino acid tyrosine; that is, they are *tyrosine protein kinases*. Before this discovery, the protein kinases that had been characterized had all been shown to add phosphates only to the amino acids serine or threonine. Because protein phosphorylation was known to be important in effecting a multitude of metabolic changes in cells, the *src* discovery was exciting. It suggested a possible explanation of how the *src* and other tyrosine protein kinase-coding oncogenes might transform a normal cell into a metabolically different cancer cell. For example, a large class of proteins, including the receptors for growth factors, uses protein phosphorylation to transmit signals through the membrane. Thus, the action of protein kinases is also to be linked to growth factors and their activities.

Membrane-Associated G Proteins Activated by Surface Receptors. Earlier we discussed how a signal produced by binding of a growth factor to its membrane-embedded receptor is transduced through the cell, activating key nuclear genes that control the cell cycle. The steps from the growth factor receptor to the nucleus are many, forming what is known as a *signaling cascade*. Membrane-associated G proteins are involved in this cascade; they are activated by the binding of the growth factor to the cell-surface receptor.

Table 22.2	Classes of Proto-Oncogene Products

Growth factors

sis	PDGF B-chain growth factor
int-2	FGF-related growth factor

Receptor and nonreceptor protein-tyrosine and protein-serine/threonine kinases

src	Membrane-associated nonreceptor protein-tyrosine kinase
fgr	Membrane-associated nonreceptor protein-tyrosine kinase
fps/fes	Nonreceptor protein-tyrosine kinase
kit	Truncated stem cell receptor protein-tyrosine kinase
pim-1	Cytoplasmic protein-serine kinase
mos	Cytoplasmic protein-serine kinase (cytostatic factor)

Receptors lacking protein kinase activity

mas	Angiotensin receptor

Membrane-associated G proteins activated by surface receptors

H-*ras*	Membrane-associated GTP-binding/GTPase
K-*ras*	Membrane-associated GTP-binding/GTPase
gsp	Mutant-activated form of Gα

Cytoplasmic regulators

crk	SH-2/3 protein that binds to (and regulates?) phosphotyrosine-containing proteins

Nuclear transcription factors (gene regulators)

myc	Sequence-specific DNA-binding protein
fos	Combines with c-*jun* product to form AP-1 transcription factor
jun	Sequence-specific DNA-binding protein; part of AP-1
erbA	Dominant negative mutant thyroxine (T3) receptor
ski	Transcription factor?

An example of such a G protein is the Ras protein, encoded by the *ras* gene. This gene is mutated in many cancers. Figure 22.8 shows part of the signaling cascade involving Ras. Binding of the growth factor EGF to the EGF receptor stimulates autophosphorylation of the receptor. Protein Grb2 can then bind, and the complex recruits SOS protein to the plasma membrane. SOS displaces GDP from Ras (which is anchored to the inner side of the plasma membrane), and allows Ras now to bind GTP. Ras-GTP recruits Raf-1 and activates it. This initiates a cascade of cytoplasm-based phosphorylations of proteins (the MAP kinase cascade), eventually producing phosphorylated ERK. This molecule moves from the cytoplasm into the nucleus, where it phosphorylates several transcription factors, including Elk-1, activating them. The activated transcription factors then turn on the transcription of specific sets of cell-cycle specific target genes.

Turning the Ras signal off in normal cells involves GAP (GTPase activating protein), which makes Ras hydrolyze the GTP bound to it back to GDP. This inactivates Ras, and cancels the cell cycle stimulatory signal.

How does *ras* become an oncogene? One way is by a mutation that abolishes its ability to hydrolyze GTP to

GDP. Therefore, even with stimulation from GAP, the Ras-GTP complex remains and the signal is continuously on.

Changing Cellular Proto-Oncogenes into Oncogenes. In normal cells, expression of proto-oncogenes is tightly controlled so that cell growth and division occur only as appropriate for the cell type involved. However, when proto-oncogenes are changed into oncogenes, the tight control can be lost, and unregulated cell proliferation can take place.

Three examples of the types of changes that have been found follow:

1. *Point mutations* (base pair substitutions). Point mutations in the coding region of a gene or in the controlling sequences (promoter, regulatory elements, enhancers) can change a proto-oncogene into an oncogene by causing an increase in either the activity of the gene product or the expression of the gene, leading in turn to an increase in the amount of gene product. For example, the *ras* mutations described in the previous section typically are point mutations.
2. *Deletions.* Deletions of part of the coding region or of part of the controlling sequences of a proto-oncogene

Figure 22.8

Role of the membrane-associated G protein, Ras, in the activation of transcription of cell-cycle specific target genes. *Source:* http://oregonstate.edu/instruction/bb492/fignames/ras3.html.

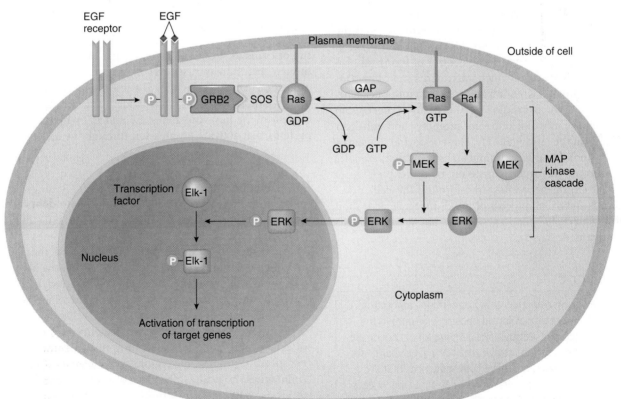

have been found frequently in oncogenes. The deletions cause changes in the amount or activity of the encoded growth stimulatory protein, causing unprogrammed activation of some cell proliferation genes.

For example, the *myc* oncogene can arise from its proto-oncogene by deletion. The normal gene consists of three exons and two introns; in some commonly found *myc* oncogenes, the first exon and most of the first intron are deleted. Transcription is then controlled from sequences in exon 2, which can function as a promoter. The *myc* proto-oncogene encodes a nuclear transcription factor that positively regulates genes involved in cell proliferation. Thus, the deletions in the oncogene forms have brought about a change in the amount or activity of the remaining *myc* protein chain that activates those genes.

3. *Gene amplification* (increased number of copies of the gene). Some tumors have multiple (sometimes hundreds of) copies of proto-oncogenes. These probably result from a random overreplication of small segments of the genomic DNA. In general, extra copies of the proto-oncogene in the cell result in an increased amount of gene product, thereby inducing or contributing to unscheduled cell proliferation. For example, multiple copies of *ras* are found in mouse adrenocortical tumors.

Cancer Induction by Retroviruses. Retroviruses are common causes of cancer in animals, although only one type of cancer thus caused is known for humans. A retrovirus can cause cancer if it is a transducing retrovirus and the v-*onc* it carries is expressed. In this case, transcription of the v-*onc* takes place under the control of retroviral promoters. Another way in which a retrovirus can cause cancer is if the proviral DNA integrates near a proto-oncogene. In this situation, expression of the proto-oncogene can come under control of retroviral promoter and enhancer sequences in the retroviral LTR. These retroviral sequences do not respond to the environmental signals that normally regulate proto-oncogene expression, so overexpression of the proto-oncogene occurs, transforming the cell to the tumorous state. The process of proto-oncogene activation is called *insertional mutagenesis*. It is rare in both animals and humans.

K E Y N O T E

After retrovirus infection, tumor induction can occur as a result of the activity of a viral oncogene (v-*onc*) in the retroviral genome. Retroviruses carrying an oncogene are known as transducing retroviruses. Normal cellular genes, called proto-oncogenes, have DNA sequences that are similar to those of the viral oncogenes. Proto-oncogenes encode proteins that stimulate cell growth and division.

In their mutated state, proto-oncogenes are called cellular oncogenes (c-*oncs*), and they may induce tumors. Retroviral oncogenes are modified copies of the cellular proto-oncogenes that have been picked up by the retrovirus.

DNA Tumor Viruses. DNA tumor viruses are oncogenic—they induce cell proliferation—but they do not carry oncogenes like those in RNA tumor viruses. As mentioned previously (pp. 610–614), their mechanism for transforming cells is completely different. DNA tumor viruses transform cells to the cancerous state through the action of a gene or genes in the viral genome. Examples of DNA tumor viruses are found among five of six major families of DNA viruses: papovaviruses, hepatitis B viruses, herpes viruses, adenoviruses, and pox viruses.

DNA tumor viruses normally progress through their life cycles without transforming the cell to a cancerous state. Typically, the virus produces a viral protein that activates DNA replication in the host cell. Then, through the use of host proteins, the viral genome is replicated and transcribed, ultimately producing a large number of progeny viruses, resulting in lysis and death of the cell. The released viruses can then infect other cells. Rarely, the viral DNA is not replicated and becomes integrated into the host cell genome. If the viral protein that activates DNA replication of the host cell is now synthesized, this protein transforms the cell to the cancerous state by stimulating the quiescent host cell to proliferate; that is, it causes the cell to move from the G_0 phase to the S phase of the cell cycle.

Examples of DNA tumor viruses are found among the papillomaviruses. Some of these viruses cause benign tumors such as skin and venereal warts in humans. Other human papillomaviruses (*HPV-16, HPV-18,* or both) cause cervical cancer, which is a leading cause of cancer deaths among women worldwide. Transformation is caused by the key viral genes, *E6* and *E7,* which encode proteins that activate progression through the cell cycle.

Tumor Suppressor Genes

In the late 1960s, Henry Harris fused normal rodent cells with cancer cells and observed that some of the resultant hybrid cells did not form tumors but instead established a normal growth pattern. Harris hypothesized that the normal cells contained gene products that had the ability to suppress the uncontrolled cell proliferation that is characteristic of cancer cells. The genes involved were called **tumor suppressor genes.**

Further evidence for the existence of tumor suppressor genes came from data indicating that in certain cancers specific chromosome regions were deleted from both homologues. Logically, if the loss of function of particular genes is correlated with tumor development, then the normal alleles of those genes must suppress tumor

Table 22.3	Examples of Tumor Suppressor Genes				
Gene	**Cancer Type**	**Protein Function**	**Hereditary Syndrome**	**Chromosome Location**
APC	Colon carcinoma	Cell adhesion	Familial adenomatous polyposis (FAP)	5q21–q22
BRCA1	Breast cancer	Has possible transcription activation domain; interacts with DNA damage repair machinery	Breast cancer and ovarian cancer	17q21
BRCA2	Breast cancer	Has possible transcription activation domain; interacts with DNA damage repair machinery	Breast cancer	13q12–q13
DCC	Colon carcinoma	Cell adhesion	Involved in colorectal cancer	18q21.3
NF1	Neurofibromas	GTPase activating protein	Neurofibromatosis type I	17q11.2
NF2	Schwannomas and meningiomas	Links cell surface glycoprotein to the actin cytoskeleton?	Neurofibromatosis type II	22q12.2
p16	Melanoma and others	Cdk inhibitor	Familial melanoma	9p21
RB	Retinoblastoma	Cell cycle and transcription regulation	Retinoblastoma	13q14.1–q14.2
TP53	Wide variety	p53 is a transcription factor	Li-Fraumeni syndrome	17p13.1
VHL	Kidney carcinoma, pancreatic tumors, and others	Transcription elongation	von Hippel-Lindau syndrome	3p26–p25
WT1	Nephroblastoma	Transcription activator or repressor depending on cell	Wilms tumor 1	11p13

formation. In other words, the normal products of tumor suppressor genes have an inhibitory role in cell growth and division. Thus, when tumor suppressor genes are inactivated, the inhibitory activity is lost, and unprogrammed cell proliferation can begin. Inactivation of tumor suppressor genes has been linked to the development of a wide variety of human cancers, including breast, colon, and lung cancers.

Finding Tumor Suppressor Genes. In contrast to mutations that change proto-oncogenes to oncogenes, mutations of tumor suppressor genes are recessive; that is, cell proliferation can be affected only if both alleles are inactivated. Thus, oncogenes can be identified in the laboratory because they can stimulate the growth of cells in culture, but that is not the case for tumor suppressor genes. Introducing tumor suppressor genes into cells in culture either results in no change or kills the cell. The isolation of individual tumor suppressor genes was made possible only with the development of positional cloning (Chapter 9, pp. 222–224). In this method, genetic variations are searched for in cancer cells or in cells of patients with inherited cancer predisposition. Such genetic variations indicate the existence of mutant genes in those cells, and that enables researchers to home in on the mutation and therefore to clone and study

the genes. Several tumor suppressor genes were isolated in this way. Table 22.3 lists some of the known tumor suppressor genes in humans. The products of tumor suppressor genes are found throughout the cell.

The Retinoblastoma Tumor Suppressor Gene, *RB*

Retinoblastoma and Knudson's Two-Hit Mutation Model for Cancer. Retinoblastoma (OMIM, http://www3.ncbi.nlm.nih.gov/Omim/, entry 180200) is a childhood cancer of the eye (Figure 22.9). Retinoblastoma develops during the period from birth to age 4 years and is the most common eye tumor in children. If discovered early enough, more than 90 percent of the eye tumors can be permanently destroyed, usually by gamma radiation. There are two forms of retinoblastoma. In *sporadic retinoblastoma* (60 percent of cases), an eye tumor develops spontaneously in a patient from a family with no history of the disease. In these cases, a *unilateral tumor* develops in one eye only. In *hereditary retinoblastoma* (40 percent of cases), the susceptibility to develop the eye tumors is inherited. Patients with this form of retinoblastoma typically develop multiple eye tumors involving both eyes (*bilateral tumors*), usually at an earlier age than is the case for unilateral tumor formation in patients with sporadic retinoblastoma.

Figure 22.9

An eye tumor in a patient with retinoblastoma.

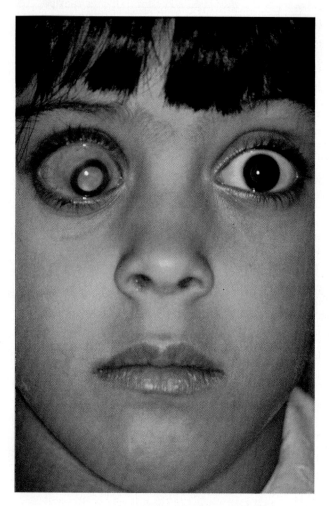

Figure 22.10

Knudson's two-hit mutation model for familial cancers. This model was proposed to explain (a) sporadic retinoblastoma by two independent mutations of the retinoblastoma (*RB*) gene and (b) hereditary retinoblastoma by a single mutation of the wild-type retinoblastoma gene in retinal cells in which a mutant *rb* was inherited through the germ line.

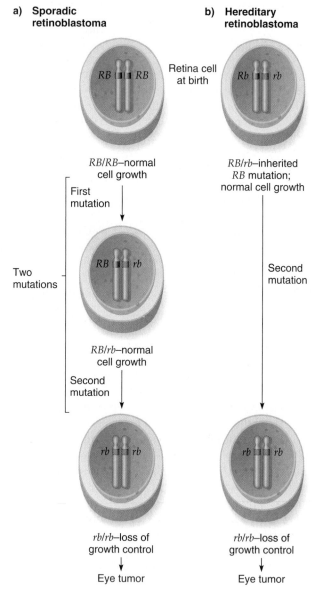

In 1971, Alfred Knudson proposed the following to explain the occurrence of hereditary and sporadic forms of retinoblastoma: "Retinoblastoma is a cancer caused by two mutational events.... One mutation is inherited via the germinal cells and the second occurs in somatic cells. In the nonhereditary form, both mutations occur in somatic cells."[3] This model relates in a general way to all forms of familial cancer.

Knudson's two-hit mutational model, as exemplified by retinoblastoma, is illustrated in Figure 22.10. In sporadic retinoblastoma (Figure 22.10a), a child is born with two wild-type copies of the retinoblastoma gene (genotype *RB/RB*), and mutation of each to a mutant *rb* allele must then occur in the same eye cell.[4] Since the chance of

---

[3]Knudson, A. G., Jr. 1971. Mutation and cancer: statistical study of retinoblastoma. *Proc. Natl. Acad. Sci. USA* 68:820–823.

[4]Retinoblastoma is among a very few cancers for which only one gene is critical for its development, in this case a mutation in a gene for a growth inhibitory factor, that is, a tumor suppressor gene. In most cases, cancer develops as a multistep process involving mutations in several different key genes related to cell growth and division.

having two independent mutational events in the same cell is very low, sporadic retinoblastoma patients would be expected to develop mostly unilateral tumors, as is the case. Furthermore, the rarity of the mutation event means that the two gene copies are mutated at different times, the first mutation producing an *RB/rb* cell, and the second mutation in that cell giving rise to the *rb/rb* genotype that results in eye tumor development. In hereditary retinoblastoma, patients inherit one copy of the mutated retinoblastoma gene through the germ line; that is, they are *RB/rb* heterozygotes (Figure 22.10b). Only a single additional mutation of the retinoblastoma gene in an eye cell is needed to produce an *rb/rb* homozygote that would result in tumor formation. Given the number of cells in a developing retina and the rate of mutation per cell, *loss of heterozygosity* (LOH; here, a mutation in the *RB* allele) is very likely for at least a few cells. Furthermore, because only a single mutation is needed to produce homozygosity for *rb*, hereditary retinoblastoma is characterized on the average by earlier onset than sporadic retinoblastoma and by multiple bilateral tumors.

According to Knudson's model, the *retinoblastoma mutation is recessive* because cancer develops only if both alleles are mutant. However, if one mutation is inherited through the germ line, tumor formation requires only a mutational event in the remaining wild-type allele in any one of the cells in that particular tissue. Because of the high likelihood of such an event, the *disease appears dominant* in pedigrees. Thus, for hereditary retinoblastoma and in hereditary neoplasms in general, we say that inheritance of just one gene mutation predisposes a person to cancer but does not cause it directly; a second mutation is required for loss of heterozygosity. Commonly, therefore, we talk about there being a hereditary disposition for cancer in such families.

Support for Knudson's hypothesis came in the 1980s from the analysis of the chromosomes of tumor cells and normal tissues in patients with retinoblastoma. Many patients carried deletions of a region of chromosome 13, and through genetic analysis the *RB* gene was mapped to chromosome location 13q14.1-q14.2.

KEYNOTE

The two-hit mutation model for cancer explains the difference between familial (hereditary) cancers and sporadic (nonhereditary) cancers. In familial cancers, one

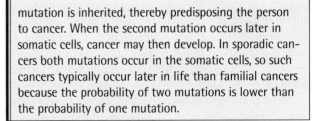

mutation is inherited, thereby predisposing the person to cancer. When the second mutation occurs later in somatic cells, cancer may then develop. In sporadic cancers both mutations occur in the somatic cells, so such cancers typically occur later in life than familial cancers because the probability of two mutations is lower than the probability of one mutation.

Function of the RB Tumor Suppressor Gene. The human *RB* tumor suppressor gene (OMIM 180200) has been mapped to 13q14.1–q14.2. The *RB* gene was cloned in 1986. It spans 180 kb of DNA and encodes 4.7-kb mRNA that is translated to produce the 928 amino acid nuclear phosphoprotein (a protein that can be phosphorylated), pRB.

pRB is involved in regulating the cell cycle at the G_1-to-S checkpoint (see Figure 22.2) as follows. Two Cdk-cyclin complexes are formed during G_1, Cdk4-cyclin D and Cdk2-cyclin E (Figure 22.11). These complexes cause the progression to the S phase by catalyzing a series of phosphorylations of cell-cycle controlling proteins, including pRB. pRB is found in a complex with transcription factor E2F and, when pRB is unphosphorylated, the activity of E2F is inhibited. Then, when pRB becomes phosphorylated, the inhibition of E2F activity is removed and the now-active transcription factor turns on the transcription of genes for DNA synthesis. Then, as the cell begins DNA synthesis, cyclins D and E are degraded, and cyclin A is made. A Cdk2-cyclin A complex forms and activates DNA replication. After the S phase, pRB is dephosphorylated again, rendering E2F inactive.

In a cell with two mutant *rb* alleles, pRB often is truncated and unstable and does not bind to E2F, which is then free to activate DNA synthesis genes. As a result, unprogrammed cell division takes place. Interestingly, several different DNA tumor viruses (e.g., adenovirus, SV40) exert their tumorigenic effects in part by a process in which

Figure 22.11

Role of pRB in regulating the cell cycle at the G_1-to-S checkpoint.

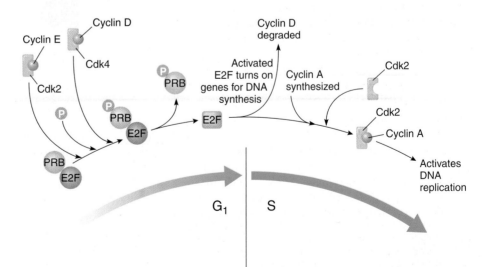

proteins encoded by their oncogenes form complexes with pRB in the cell, thereby blocking its ability to bind to E2F and inactivating the suppressive function of the protein. In other words, these tumor viruses transform cells to the neoplastic state by inactivating a mechanism that inhibits cell cycle progression.

Both point mutations and deletions in the gene have been shown to lead to loss of function of pRB in patients with retinoblastoma. In approximately 5 percent of patients with retinoblastoma, the genetic abnormality can be detected by karyotype analysis. The remainder are more difficult to detect, even with molecular analysis. Interestingly, in hereditary retinoblastoma, the second *rb* mutation often results in an identically mutated allele to the inherited one. Apparently, the chromosomal region with the wild-type *RB* allele is replaced by the homologous chromosome region that carries the mutant *rb* allele. This may occur through mitotic recombination, chromosomal nondisjunction, or gene conversion.

The *TP53* Tumor Suppressor Gene. The tumor suppressor gene *TP53* (*Tumor protein 53*: OMIM 191170) encodes a protein of molecular weight 53 kDa called p53. When

animation

The Tumor Suppressor Gene, *TP53*

both alleles are mutated, *TP53* may be involved in the development of perhaps 50 percent of all human cancers, including breast, brain, liver, lung, colorectal, bladder, and blood cancers. This does not mean that *TP53* causes 50 percent of human cancers, but rather that mutations in *TP53* are among the several genetic changes usually found in those cancers.

Genetics of the **TP53** *Tumor Suppressor Gene.* The human *TP53* gene, isolated by positional cloning (see Chapter 9, pp. 222–224), is at chromosome location 17p13.1. Individuals who inherit one mutant copy of *TP53* develop Li-Fraumeni syndrome, a rare form of cancer that is an autosomal dominant trait because the cancer develops when the second copy of *TP53* becomes mutated. Individuals with this syndrome develop cancers in a number of tissues, including breast and blood.

Function of p53. The 393-amino acid p53 tumor suppressor protein is a transcription factor that is regulated by phosphorylation and by its interaction with another phosphoprotein, the negative regulator Mdm2 (Figure 22.12). In a normal cell, both proteins are unphosphorylated, which allows them to bind together. Mdm2 stimulates degradation of p53, and, as a result, the amount of p53 in the cell is low. When DNA damage occurs, p53 initiates a cascade of events leading to arrest in G_1. DNA damage results in phosphorylation of both p53 and Mdm2 on the domains where they normally

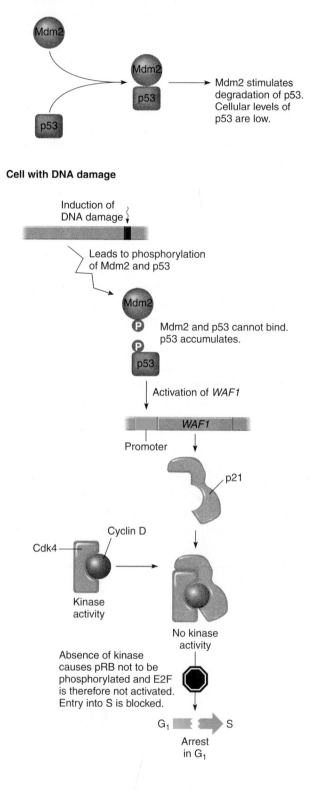

Figure 22.12

Function of p53 in cell cycle control.

Normal cell

Mdm2 stimulates degradation of p53. Cellular levels of p53 are low.

Cell with DNA damage

Induction of DNA damage

Leads to phosphorylation of Mdm2 and p53

Mdm2 and p53 cannot bind. p53 accumulates.

Activation of *WAF1*

WAF1

Promoter

p21

Cyclin D

Cdk4

Kinase activity

No kinase activity

Absence of kinase causes pRB not to be phosphorylated and E2F is therefore not activated. Entry into S is blocked.

G_1 → S

Arrest in G_1

interact. Therefore, a p53-Mdm2 complex cannot form and p53 degradation is not promoted, so p53 accumulates. Functioning as a transcription factor, p53 turns on transcription of DNA repair genes and of *WAF1*, which encodes a 21-kDa protein called p21. The p21 protein binds to the G_1-to-S checkpoint Cdk4-cyclin D complexes and inhibits their activity.[5] As a result, pRB in the pRB-E2F complex does not become phosphorylated, thereby keeping E2F inhibited. Entry into S is blocked (see Figure 22.10) and the cell arrests in G_1.

p53 provides some protection against oncogenes. Expression of viral or cellular oncogenes such as *ras* induces expression of the *ARF* gene. The p14 protein produced binds to Mdm2 in the p53-Mdm2 complex and blocks Mdm2's stimulation of p53 degradation. Somehow, the requirement of phosphorylation of p53 for activation of gene transcription is bypassed.

p53 also plays a role in **programmed cell death (apoptosis)**, a process by which a cell with a high level of DNA damage commits suicide. During apoptosis, DNA is degraded and the nucleus condenses, and the cell may be devoured by phagocytes. In this process, p53 does not induce DNA repair genes or *WAF1*, but activates the *BAX* gene for the apoptosis pathway. The BAX protein blocks the function of the BCL-2 protein, which is a repressor of the apoptosis pathway. Without an active BCL-2 repressor, the apoptotic pathway is activated and the cell commits suicide.

If both alleles of *TP53* carry loss-of-function mutations, no active p53 can be produced. Thus, *WAF1* cannot be activated, and no p21 is available to block Cdk activity, so the cell is unable to arrest in G_1. Therefore, the cell may proceed to S, which is undesirable. Similarly, a cell with a high level of DNA damage will not be able to undergo programmed cell death. Most loss-of-function mutations of *TP53* occur in the part of the gene that encodes the DNA-binding domain of the p53 transcription factor. Logically, the mutant p53 molecules produced are unable to activate transcription of its target genes.

Transgenic mice have been produced with deletions of both *TP53* alleles. That these *TP53⁻/TP53⁻* knockout mice actually developed and were fully viable indicated that the *TP53* gene is not essential for the processes of cell growth, cell division, or cell differentiation, at least in the mouse. The knockout mice showed only one major phenotype, that of a very high frequency of cancers from the sixth month (in 75 percent of the mice) to the tenth month (in 100 percent of the mice). These results support the roles of p53 in tumor suppression and in maintaining the genetic stability of cells.

Breast Cancer Tumor Suppressor Genes. In the United States, more than 185,000 new cases of breast cancer are diagnosed each year, representing more than 31 percent of all new cancers in women, and more than 46,000 women die each year from this cancer. In developed countries there is a 1 in 10 chance that a woman will be diagnosed with breast cancer during her lifetime. The average age of onset is 55. Approximately 5 percent of breast cancers are hereditary. As with hereditary retinoblastoma, this form of the cancer has an earlier age of onset than the sporadic form, and the cancer often is bilateral.

Among the several genes that play a role in familial breast cancer, two genes—*BRCA1* (OMIM 113705) and *BRCA2* (OMIM 600185)—have been hypothesized to be tumor suppressor genes. (Some studies have led to the alternative hypothesis that these two genes are mutator genes; see next section.) It is believed that most hereditary breast cancer in the United States results from mutations in *BRCA1* or *BRCA2*, with most of the mutations occurring in *BRCA1*.

The *breast cancer* susceptibility gene *BRCA1* is at chromosome location 17q21. Mutations of the *BRCA1* gene also lead to susceptibility to ovarian cancer. The *BRCA1* gene encompasses more than 100 kb of DNA; it is transcribed in numerous tissues, including breast and ovary, to produce a 7.8-kb mRNA that is translated to produce a 190-kDa protein of uncertain function with 1,863 amino acids. Various functions have been proposed for the BRCA1 protein, including roles in homologous recombination, cellular responses to DNA damage, and mRNA transcription. The latter function is inferred from evidence that the *BRCA1* protein has been found in the RNA polymerase II holoenzyme, and it also has a possible transcription activation domain.

BRCA2 is at chromosome location 13q12–q13. Unlike *BRCA1*, *BRCA2* does not have an associated high risk of ovarian cancer. The *BRCA2* encompasses approximately 70 kb of DNA and encodes a 3,418-amino acid protein of uncertain function. The various functions proposed for BRCA2 are similar to that of BRCA1.

KEYNOTE

Tumor suppressor genes, like proto-oncogenes, are involved in the regulation of cell growth and division. Whereas the normal products of proto-oncogenes have a stimulatory role in those processes, the normal products of tumor suppressor genes have inhibitory roles. Therefore, when both alleles of a tumor suppressor gene are inactivated or lost, the inhibitory activity is lost, and unprogrammed cell proliferation can occur. Inactivation of tumor suppressor genes is involved in the development of a wide variety of human cancers, including breast, colon, and lung cancers.

[5]The p21 protein can also bind to other checkpoint Cdk-cyclin complexes and inhibit their activity, thereby blocking the cell cycle at any stage. The example here focuses on the retinoblastoma protein and the G_1-to-S checkpoint.

Mutator Genes

A **mutator gene** is any gene that, when mutant, increases the spontaneous mutation frequencies of other genes. In a cell, the normal (unmutated) forms of mutator genes are involved in important activities such as DNA replication and DNA repair. Mutations of these genes can significantly impair those processes and can make the cell error prone so that it accumulates mutations. For an illustration of how a mutation in a mutator gene can result in cancer, we consider hereditary nonpolyposis colon cancer (HNPCC; OMIM 120435).

HNPCC is an autosomal dominant genetic disease in which there is an early onset of colorectal cancer. Unlike hereditary (or familial) adenomatous polyposis (FAP; see next section), no adenomas (benign tumors or polyps) are seen in HNPCC, hence its name. HNPCC accounts for perhaps 5 to 15 percent of colorectal cancers.

Four human genes, *hMSH2*, *hMLH1*, *hPMS1*, and *hPMS2*, have been identified, any one of which gives a phenotype of hereditary predisposition to HNPCC when it is mutated. The first two together are responsible for about 90 percent of all HNPCC cases, with the other two each accounting for 5 percent. Tumor formation requires only one mutational event to inactivate the remaining normal allele. Thus, because of the high probability of such an event, HNPCC appears dominant in pedigrees. All four genes are homologous to *E. coli* and yeast genes known to be involved in DNA repair. For example, *hMSH2* is homologous to *E. coli mutS*, and the other three genes have homologies to *E. coli mutL*. The *E. coli* genes have well-characterized roles in mismatch repair, a process for correcting mismatched base pairs left after DNA replication. (Mismatch repair is described in detail in Chapter 7, pp. 150–152.) The yeast genes have similar functions. In other words, the human, yeast, and *E. coli* genes described here are all mutator genes because they are involved in DNA repair systems. Mutations in these genes make the DNA replication error prone, and mutation rates are significantly higher than they are in normal cells. That the human *hMSH2* gene is indeed a mutator gene was confirmed by an experiment in which an *hMSH2* cDNA was cloned into an *E. coli* plasmid and expressed in *E. coli*. The result was a tenfold increase in the accumulation of mutations. DNA-based assays are available for all four genes, allowing carriers to be detected through analysis of blood samples.

i

iActivity You are a researcher at a cancer clinic investigating the origins of a rare form of bladder cancer in the iActivity *Tracking Down the Causes of Cancer* on your CD-ROM.

Telomere Shortening, Telomerase, and Human Cancer

Telomeres are the ends of eukaryotic chromosomes. (See Chapter 2.) In most eukaryotes, telomeres consist of repeated short sequences; in humans, the sequence is 5′-TTAGGG-3′. During successive cell cycles the telomeres shorten because the primer needed by DNA polymerase for new DNA synthesis is removed. (See Chapter 3). This shortening can be counteracted by the action of telomerase, which adds new telomere repeat sequences to the ends of chromosomes. In recent years, the role of telomere shortening and telomerase activity has been studied with respect to the development of human cancers.

Human cells in culture are limited in their ability to proliferate. For example, fibroblasts in culture are capable of up to about 50 division cycles. After that, the cells stop dividing and normally never divide again—they undergo *replicative senescence*. Replicative senescence is caused by changes in the structure of telomeres. That is, human cells, with the exception of germ line cells and certain stem cells do not have significant telomerase activity, so their telomeres shorten each cell cycle. Eventually, the telomeres become so short that the complex between telomere sequences and the proteins that bind to them is disrupted, and DNA damage occurs. This damage is similar to that which triggers apoptosis involving p53 discussed earlier (p. 622) and, therefore, for a normal senescent cell, further cell division is blocked. However, suppose one of these cells undergoes a mutation in a gene that controls normal cell cycle arrest, such as *TP53*; that cell will divide even with too-short telomeres and will proliferate. The loss of p53 will also result in chromosomal instability and accumulate other mutations. It is unlikely, though, that these cells could divide for many generations because of the too-short telomeres. Such cells can become immortal (that is, proliferate in an unlimited way) though, if telomerase is turned on anew, allowing the ends of the chromosomes to be fixed and the chromosomes to be stabilized. These cells already carry mutations affecting the cell cycle, and typically accumulate other mutations that push the cells toward the cancerous state.

In sum, the primary cause of cancer is mutations in cells. Telomerase is not present in most normal cells and tissues, but the enzyme is reactivated as a secondary event in all major human cancer types. The telomerase enables cancer cells to maintain telomere length, stabilizing the chromosomes, and giving them the ability to proliferate indefinitely. In some tumor animal models, inhibiting telomerase activity in cancer cells has been shown to lead to telomere shortening and, thence, to replicative senescence of death of the cell. There is potential promise here for this strategy in cancer treatment.

The Multistep Nature of Cancer

The development of most cancers is a stepwise process involving an accumulation of mutations in a number of genes. It appears that perhaps six or seven independent mutations are needed over several decades of life for cancer to be induced. The multiple mutational events typically involve both the change of proto-oncogenes to oncogenes and the inactivation of tumor suppressor genes, with a resulting breakdown of the multiple cellular mechanisms that regulate growth and differentiation.

As an example, Figure 22.13 illustrates Bert Vogelstein's molecular model of multiple mutations leading to hereditary FAP, a form of colorectal cancer (OMIM 175100). Patients with FAP inherit the loss of a chromosome 5 tumor suppressor gene called *APC* (adenomatous polyposis coli). The same gene can be lost early in carcinogenesis in sporadic tumors. Once both alleles of *APC* are lost in a colon cell, increased cell growth results. Then, other gene changes are needed in order for the cancer to develop. The following describes just one possible sequence of events; recognize that the molecular mechanism of colorectal cancer development is more complicated and is yet to be completely understood.

If hypomethylation (decreased methylation) of the DNA occurs, a benign tumor called an *adenoma class I* (a small polyp from the colon or rectum epithelium) can develop. Then, if a mutation converts the chromosome 12 *ras* proto-oncogene into an oncogene, the cells can progress to a larger benign tumor known as *adenoma class II* (a larger polyp). Next, if both copies of the chromosome 18 tumor suppressor gene *DCC* (deleted in colon cancer) are lost, the cells progress into *adenoma class III* (large benign polyps). Deletion of both copies of the chromosome 17 tumor suppressor gene *p53* results in the progression to a carcinoma (an epithelial cancer); with other gene losses, the cancer metastasizes. Note that this is only one path whereby adenomatous polyposis can occur; others are possible. However, in all paths observed, deletions of *APC* and mutations of *ras* usually occur earlier in carcinogenesis than do deletions of *DCC* and *TP53*. Progressive changes in function of oncogenes and tumor suppressor genes are also thought to occur for other cancers.

KEYNOTE

The development of most cancers involves an accumulation of mutations in a number of genes over a significant period of life. This multistep nature of cancer typically involves mutational events that convert proto-oncogenes to oncogenes and inactivate tumor suppressor genes, thereby breaking down the multiple mechanisms that regulate growth and differentiation.

Figure 22.13

A multistep molecular event model for the development of hereditary adenomatous polyposis (FAP), a colorectal cancer.

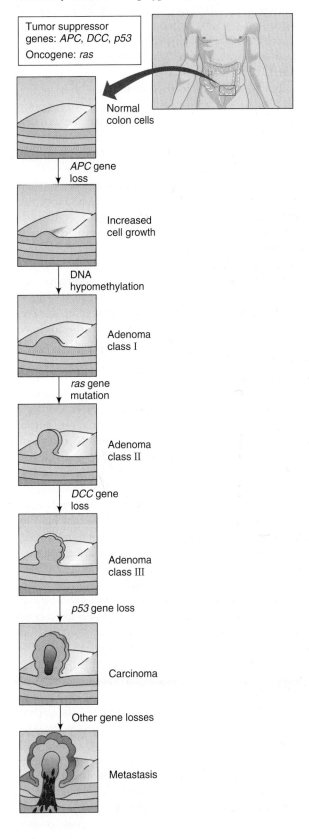

Tumor suppressor genes: *APC*, *DCC*, *p53*
Oncogene: *ras*

Normal colon cells

APC gene loss

Increased cell growth

DNA hypomethylation

Adenoma class I

ras gene mutation

Adenoma class II

DCC gene loss

Adenoma class III

p53 gene loss

Carcinoma

Other gene losses

Metastasis

Chemicals and Radiation as Carcinogens

Several natural and artificial agents increase the frequency with which cells become cancerous. These agents, mostly chemicals and types of radiation, are known as **carcinogens.** Because of the obvious human relevance, there is a vast amount of information about carcinogenesis spanning many areas of biology; only an overview is given here.

Although we have focused much attention in this chapter on viruses as causes of cancer, chemicals are responsible for more human cancers than are viruses. Chemical carcinogenesis was discovered in the eighteenth century by Sir Percival Pott, an English surgeon who correlated the incidence of scrotal skin cancer in some of his patients with occupational exposure to coal soot when they worked as chimney sweeps as children. From the beginning of industrial development in the eighteenth century to the present day, workers in many areas have been exposed to carcinogenic agents and have developed occupationally related cancers. For example, radiologists using X rays and radium (sources of ionizing radiation) and farmers exposed to the sun's ultraviolet (UV) light (nonionizing radiation) have developed skin cancer, asbestos and insulation workers exposed to asbestos have developed bronchial and lung cancers, and workers exposed to vinyl chloride have developed liver cancer.

Chemical Carcinogens

Chemical carcinogens are responsible for most cancer deaths in the United States, with the top two causes of cancer—tobacco smoke and diet—being responsible for 50 to 60 percent of cancer-related deaths. Chemical carcinogens include both natural and synthetic chemicals.

Two major classes of chemical carcinogens are recognized. *Direct-acting carcinogens* are chemicals that bind to DNA and act as mutagens. The second class, *procarcinogens*, must be converted metabolically to become active carcinogens; almost all of these so-called *ultimate carcinogens* also bind to DNA and act as mutagens. In both cases, the mutations typically are point mutations. (The mutagenicity of ultimate carcinogens can be demonstrated in a number of screening tests, including the Ames test described in Chapter 7, pp. 147–148.) Thus, direct-acting and most ultimate carcinogens bring about transformation of cells and the formation of tumors by binding to and causing changes in DNA. Direct-acting carcinogens include alkylating agents. Examples of procarcinogens are polycyclic aromatic hydrocarbons (multiringed organic compounds found in the separate smoke produced by burning wood, coal, and cigarettes, for example), azo dyes and natural metabolites (such as aflatoxin produced from fungal contamination of food), and nitrosamines (produced by nitrites in food). Most chemical carcinogens are procarcinogens.

The metabolic conversion of procarcinogens to ultimate carcinogens is carried out by normal cellular enzymes that function in a variety of pathways that involve hydrolysis, oxidation, and reduction, for example. If a procarcinogen interacts with the active site of one of the enzymes, then it can be modified by the enzyme to give rise to the derivative ultimate carcinogen.

Radiation

We are exposed to radiation from the sun, cellular telephones, radioactive radon gas, electric power lines, and some household appliances, for example. Only about 2 percent of all cancer deaths are caused by radiation, and most of the cancers involved are the highly aggressive melanoma skin cancers that can be induced by exposure to the sun's UV light. Ionizing radiation, such as that emitted by X ray machines, decay of some radioactive materials, and radon gas, for example, can be carcinogenic, although the risk to the public generally is low. Ionizing radiation most commonly causes leukemia and thyroid cancer.

Radiation causes mutations in DNA. The mutagenic effects of ultraviolet light and X rays are discussed in Chapter 7, pp. 141–143. We discuss ultraviolet light as a carcinogen in more detail in the following paragraphs.

UV light is emitted by the sun, along with visible light and infrared radiation. The UV light that reaches Earth is classified into ultraviolet A (UVA, spanning 320–400 nm) and ultraviolet B (UVB, spanning 290–320 nm). The intensity of UVA and UVB reaching an individual on Earth depends on factors such as time of day, altitude, and materials in the atmosphere, including dust and other particles. Generally, the ambient level of UVA is one to three orders of magnitude higher than that of UVB.

UV light causes several forms of skin cancer, the most dangerous of which are directly related to long-term exposure to UV light radiation. Both UVA and UVB play a role in carcinogenesis. Sunburn is caused mainly by UVB, which also induces skin cancer because the radiation in the wavelength range of UVB is mutagenic (see Chapter 7, pp. 141–143, for a discussion). UVA plays a role in skin cancer by increasing the carcinogenic effects of UVB. Fortunately, many skin cancers are easy to detect and can be removed surgically.

KEYNOTE

Various types of radiation and many chemicals increase the frequency with which cells become cancerous. These agents are known as carcinogens. All carcinogens act by causing changes in the genome of the cell. A few chemical carcinogens act directly on the genome; the majority act indirectly. The latter are metabolically converted by cellular enzymes to ultimate carcinogens that bind to DNA and cause mutations. All carcinogenic forms of radiation act directly.

Summary

To understand the development of cancer (neoplasia), it is necessary to understand how normal cell division is controlled. We know that a normal eukaryotic cell moves through the cell cycle in steps that are tightly controlled by a number of molecular factors, notably cyclins and cyclin-dependent kinases. Healthy cells grow and divide only when the balance of stimulatory and inhibitory signals received from outside favors cell proliferation and the signals are appropriately transduced to the cytoplasm and nucleus. A cancerous cell does not respond to the usual signals and reproduces without constraints.

Three classes of genes have been shown to be mutated frequently in cancer: proto-oncogenes (the mutant forms are called oncogenes), tumor suppressor genes, and mutator genes. The products of proto-oncogenes stimulate cell proliferation, the products of wild-type tumor suppressor genes inhibit cell proliferation, and the products of wild-type mutator genes are involved in the replication and repair of DNA. Mutant forms of these three types of genes all have the potential to contribute to the transformation of a cell to a tumorous state.

Some forms of cancer are caused by tumor viruses. Both DNA and RNA tumor viruses are known. DNA tumor viruses transform cells to the cancerous state through the action of a gene or genes that are essential parts of the viral genome and that usually stimulate the transition from G_0 or G_1 to S. All RNA tumor viruses are retroviruses—RNA viruses that replicate via a DNA intermediate—but not all retroviruses cause cancer. When a retrovirus infects a cell, the RNA genome is released from the viral particle, and reverse transcriptase makes a cDNA copy of the genome, called the proviral DNA, that integrates into the host cell's genome. Expression of the proviral genes leads to the production of progeny viruses that exit the cell and infect other cells.

Tumor-causing RNA retroviruses contain cancer-inducing genes called oncogenes. Tumor-causing retroviruses have picked up normal cellular genes, called proto-oncogenes, while simultaneously losing some of their genetic information. Proto-oncogenes in normal cells function in various ways to regulate cell proliferation and differentiation. However, in the retrovirus these genes have been modified or their expression regulated differently so that the oncogene protein product, now synthesized under viral control, is altered. The oncogene products, which include growth factors and protein kinases, are directly responsible for transforming cells to the cancerous state. Cellular proto-oncogenes may also mutate, resulting in a stimulatory effect on cell division. In their mutated state, proto-oncogenes are called cellular oncogenes. Since only one allele of a proto-oncogene must be mutated to cause changes in cell growth and division, the mutations are dominant mutations.

Tumor suppressor genes, like proto-oncogenes, are involved in the regulation of cell growth and division. In this case, the normal products of tumor suppressor genes have inhibitory roles. Therefore, when both alleles of a tumor suppressor gene are inactivated or lost, the inhibitory activity is lost, and unprogrammed cell proliferation can occur. In familial (hereditary) cancers, one mutation is inherited and the other mutation occurs later in somatic cells, leading to loss of heterozygosity. In other words, the inheritance of one gene mutation predisposes a person to cancer. In sporadic (nonhereditary) cancers, both mutations occur in the somatic cells.

All types of human cancers are associated with a reactivation of telomerase, the enzyme that adds telomere repeats to the ends of chromosomes. Telomerase does not itself cause cancer. Rather, cells with the potential to be cancerous have unstable chromosomes because of shortened telomeres, as well as mutations in cell cycle control genes. Such cells can divide only a few times but, when telomerase is reactivated, the chromosomes are stabilized and indefinite proliferation of the damaged cell occurs, producing a cancer.

The development of most cancers involves an accumulation of mutations in a number of genes over a significant period of life. This multistep nature of cancer typically involves mutational events that activate oncogenes and inactivate tumor suppressor genes, thereby breaking down the multiple mechanisms that regulate growth and differentiation. Mutations of mutator genes can also contribute to the development of cancer by adversely affecting the normal maintenance of the integrity of the genome through accurate DNA replication and efficient DNA repair, with the result that the cell accumulates mutations.

Various types of radiation and many chemicals, collectively known as carcinogens, increase the frequency with which cells become cancerous. Practically all carcinogens act by causing changes in the genome of the cell. A few chemical carcinogens act directly on the genome; the majority act indirectly by being converted by cellular enzymes to active derivatives called ultimate carcinogens. By understanding what carcinogens exist in the environment, we can position ourselves to minimize their effects on us and thereby perhaps decrease our risk of cancer.

Analytical Approaches to Solving Genetics Problems

Q22.1 An investigator has found a retrovirus capable of infecting human nerve cells. This is a complete virus, able to reproduce itself in the cell, and it contains no oncogenes. People who are infected suffer a debilitating encephalitis. The investigator has shown that when he infects nerve cells in culture with the complete virus, the nerve cells are killed as the virus reproduces. But if he infects cultured nerve cells with a virus in which he has created deletions in the *env* or *gag* genes, no cell death occurs. The investigator is interested in finding ways to bring about nerve cell growth or regeneration in people who have suffered nerve damage. For example, in a patient with a severed spinal cord, nerve regeneration might relieve paralysis. The investigator has cloned the human nerve growth factor gene and wants to insert it into the genome of the retrovirus from which he has deleted parts of the *env* and *gag* genes. He would then use the engineered retrovirus to infect cultured nerve cells. Adult nerve cells do not normally produce large amounts of nerve growth factor. If he is successful in inducing growth in them without causing any cell death, he would like to move on to clinical trials on injured patients. When the investigator applied for grant support to do this work, his application was denied on the grounds that there were inadequate safeguards in the plan. Why might this work be dangerous?

A22.1 In engineering the retrovirus in the way he plans, the investigator probably would be creating a new cancer virus in which the cloned nerve growth factor gene would be the oncogene. Of course, it is an advantage that the engineered virus would not be able to reproduce itself, but we know that many "wild" cancer viruses are also defective and reproduce with the help of other viruses. If the engineered virus were to infect cells carrying other viruses (for example, wild-type versions of itself) that could supply the *env* and *gag* functions, then the new virus could be reproduced and subsequently spread. Presumably, infection of normal nerve cells in vivo by the engineered retrovirus would sometimes result in abnormally high levels of nerve growth factor and thus perhaps in the production of cancers of the nervous system.

Questions and Problems

22.1 Progression of cells through the cell cycle is tightly regulated, and cancerous cells fail to respond to signals that normally regulate cellular proliferation.
a. Which stages of the cell cycle are subject to regulation?
b. What types of proteins regulate progression through the cell cycle, and what is the role of protein phosphorylation in this process?

22.2 Explain why HIV-1, the causative agent of AIDS, is considered a nononcogenic retrovirus, even though numerous types of cancers are frequently seen in patients with AIDS.

22.3 Distinguish between a transducing retrovirus and a nontransducing retrovirus.

22.4 Cellular proto-oncogenes and viral oncogenes are related in sequence, but they are not identical. What is the fundamental difference between the two?

*22.5 An autopsy of a cat that died from feline sarcoma revealed neoplastic cells in the muscle and bone marrow but not in the brain, liver, or kidneys. To gather evidence for the hypothesis that the virus FeSV contributed to the cancer, Southern blot analysis (see Chapter 8, pp. 195–197) was performed on DNA isolated from these tissues and on a cDNA clone of the FeSV viral genome. The DNA was digested with the enzyme *Hind*III, separated by size on an agarose gel, and transferred to a membrane. The resulting Southern blot was hybridized with a ^{32}P-labeled probe made from a 1.0-kb *Hind*III fragment of the feline *fes* proto-oncogene cDNA. The autoradiogram revealed a 3.4-kb band in each lane, with an additional 1.2-kb band in the lanes with muscle and bone marrow DNA. Only a 1.2-kb band was seen in the lane loaded with *Hind*III-cut FeSV cDNA. Explain these results, including the size of the bands seen. Do these results support the hypothesis?

*22.6 The sequences of proto-oncogenes are highly conserved among a large number of animal species. Based on this fact, what hypothesis can you make about the functions of the proto-oncogenes?

*22.7 Proto-oncogenes produce a diverse set of gene products.
a. What types of gene products are made by proto-oncogenes? Do these gene products share any features?
b. Which of the following mutations might result in an oncogene?
  i. a deletion of the entire coding region of a proto-oncogene
  ii. a deletion of a silencer that lies 5' to the coding region
  iii. a deletion of an enhancer that lies 3' to the coding region
  iv. a deletion of a 3' splice site acceptor region
  v. the introduction of a premature stop codon
  vi. a point mutation (single base-pair change in the DNA)
  vii. a translocation that places the coding region near a constitutively transcribed gene
  viii. a translocation that places the gene near constitutive heterochromatin

*22.8 Explain the likely mechanism underlying the formation of an oncogene in each of the following mutations:
a. a mutation in the promoter for platelet-derived growth factor that leads to an increase in the efficiency of transcription initiation
b. a mutation affecting a regulatory domain of a nonreceptor tyrosine kinase that causes it always to be active
c. a mutation affecting the structure of a membrane-associated G protein that eliminates its ability to hydrolyze GTP.

22.9 Describe a pathway leading to altered nuclear transcription following the binding of a growth factor to a surface-membrane-bound receptor. How can mutations in genes involved in this pathway result in oncogenes?

22.10 What are the three main ways in which a proto-oncogene can be changed into an oncogene?

22.11 List two ways in which cancer can be induced by a retrovirus.

*22.12 After a retrovirus that does not carry an oncogene infects a particular cell, northern blots indicate that the amount of mRNAs transcribed from a particular proto-oncogene became elevated approximately thirteen fold compared with uninfected control cells. Propose a hypothesis to explain this result.

*22.13 In what ways is the mechanism of cell transformation by transducing retroviruses fundamentally different from transformation by DNA tumor viruses? Even though the mechanisms are different, how are both able to cause neoplastic growth?

*22.14 You have a culture of normal cells and a culture of cells dividing uncontrollably (isolated from a tumor). Experimentally, how might you determine whether uncontrolled growth was the result of an oncogene or a mutated pair of tumor suppressor alleles?

*22.15 What is the difference between a hereditary cancer and a sporadic cancer?

22.16 Why are mutations in tumor suppressors, such as mutations in *RB* and *TP53*, said to cause recessive disorders when they appear to be dominant in pedigrees?

22.17 Individuals with hereditary retinoblastoma are heterozygous for a mutation in the *RB* gene; however, their cancerous cells often have two identically mutated *RB* alleles. Describe three different mechanisms by which the normal *RB* allele can be lost. Illustrate your answer with diagrams.

*22.18 Although there has been a substantial increase in our understanding of the genetic basis for cancer, the vast majority of cases of many types of cancer are not hereditary.
a. How might studying a familial form of a cancer provide insight into a similar, more frequent sporadic form?
b. The incidence of cancer in several members of an extended family might reasonably raise concern as to whether there is a genetic predisposition for cancer in the family. What does the term *genetic predisposition* mean? What might be the basis of a genetic predisposition to a cancer that appears as a dominant trait? What issues must be addressed before concluding that

a genetic predisposition for a specific type of cancer exists in a particular family?

22.19 Explain how progression through the cell cycle is regulated by the phosphorylation of the retinoblastoma protein pRB. What phenotypes might you expect in cells where
a. pRB was constitutively phosphorylated?
b. pRB was never phosphorylated?
c. a severely truncated pRB protein was produced that could not be phosphorylated?
d. a normal pRB protein was produced at higher than normal levels?
e. a normal pRB protein was produced at lower than normal levels?

22.20 Mutations in the *TP53* gene appear to be a major factor in the development of human cancer.
a. Discuss the normal cellular functions of the *TP53* gene product and how alterations in these functions can lead to cancer.
b. Explain how mutations at both alleles of *TP53* may be involved in 50 percent of all human cancers when familial cancers caused by mutations in *TP53* are rare and associated with a specific type of cancer, Li-Fraumeni syndrome.
c. Suppose cells in a cancerous growth are shown to have a genetic alteration that results in diminished *TP53* gene function. Why can we not immediately conclude that the mutation *caused* the cancer? How would the effect of the mutation be viewed in light of the current, multistep model of cancer?

22.21 The p53 protein can influence multiple pathways involved in tumor formation.
a. Explain how the functions of p53 are regulated by phosphorylation.
b. Through what pathway does the phosphorylation of p53 influence phosphorylation of pRB to control cell cycle progression?
c. What pathways can be activated by p53 in response to DNA damage? What determines which pathway is activated?

*22.22 What is apoptosis? Why is the cell death associated with apoptosis desirable, and how is it regulated?

22.23 Two alternative hypotheses have been proposed to explain the functions of *BRCA1* and *BRCA2*. One hypothesis proposes that these are tumor suppressor genes, while the other proposes that these are mutator genes.
a. Distinguish between a tumor suppressor gene and a mutator gene.
b. What cellular roles do *BRCA1* and *BRCA2* have that is consistent with their being tumor suppressor genes? Mutator genes?

22.24 Telomerase activity is not normally present in differentiated cells, but is almost always present in cancerous cells. Explain whether telomerase activity alone can lead to cancer. If it cannot, why is it present in most cancerous cells, and what would be the biological consequences of eliminating it from cancerous cells?

*22.25 Material that has been biopsied from tumors is useful for discerning both the type of tumor and the stage to which a tumor has progressed. It has been known for a long time that biopsied tissues with more differentiated cellular phenotypes are associated with less advanced tumors. Explain this finding in terms of the multistep nature of cancer.

*22.26 Some tumor types have been very frequently associated with specific chromosomal translocations. In some cases, these translocations are found as the only cytogenetic abnormality. In each case examined to date, the chromosome breaks that occur result in a chimeric gene (pieces of two genes fused together) that encodes a fusion protein (a protein consisting of parts of two proteins fused together, corresponding to the coding sequences of the chimeric gene). A partial list of tumor-specific chromosomal translocations in bone and soft tissue tumors is given here.

Tumor Type	Translocation	Characteristic	Genes
Ewing sarcoma	t(11;22) (q24;q12)	Malignant	FLI1, EWS,
	t(21;22) (q22;q12)		ERD, EWS,
	t(7;21) (p22;q12)		ATV1, EWS
Soft tissue clear cell carcinoma	t(12;22) (q13;q12)	Malignant	ATF1, EWS
Myxoid chondrosarcoma	t(9;22)(q22–31;q12)	Malignant	CHN, EWS
Synocial sarcoma	t(11;22) (p13;q12)	Malignant	SSX1, SSX2, SYT
Lipoma	t(var;12) (var;q13–15)	Benign	HMGI-C
Leiomyoma	t(12;14) (q13–15;q23-24)	Benign	HMGI-C

a. What conclusions might you draw from the fact that in some cases these translocations are found as the only cytogenetic abnormality?

b. How might the formation of a chimeric protein result in tumor formation?

c. Based on the data presented here, can you infer whether the genes near the breakpoints of these translocations are tumor suppressor genes or proto-oncogenes? If so, which?

d. Can you speculate on how multiple translocations involving the *EWS* gene result in different sarcomas?

e. It is often difficult to diagnose individual sarcoma types based solely on tissue biopsy and clinical symptoms. How might the cloning of the genes involved in translocation breakpoints associated with specific tumors have a practical value in improving tumor diagnosis and management?

22.27 What mechanisms ensure that cells with heavily damaged DNA are unable to replicate?

*22.28 Distinguish between direct-acting carcinogens, procarcinogens, and ultimate carcinogens in the induction of cancer.

22.29 What sources of radiation exist, and how does radiation induce cancer?

23

Non-Mendelian Inheritance

Mitochondria of osteocarcoma cells stained with Mito Tracker Red.

PRINCIPAL POINTS

- Both mitochondria and chloroplasts contain their own DNA genomes. The DNA in most species' mitochondria and in all chloroplasts is circular, double-stranded, and supercoiled. The mitochondrial DNA of some species is linear. Typically, mitochondria and chloroplasts contain several nucleoid regions in which the DNA is located, and each nucleoid contains several copies of the DNA molecule.

- The mitochondrial and chloroplast genomes contain genes for the rRNA components of the ribosomes of these organelles, for many (if not all) of the tRNAs used in organellar protein synthesis, and for a few proteins that remain in the organelles and perform functions specific to the organelles. All other proteins required by these organelles are nuclear-encoded, synthesized on cytoplasmic ribosomes, and transported into them. The mitochondrial genetic code of some organisms is different from the universal nuclear genetic code. Mitochondrial DNA analysis is used to study genetic relationships between individuals because of the maternal inheritance of mitochondria and polymorphisms of mitochondrial DNA.

- The inheritance of mitochondrial and chloroplast genes follows rules different from those for nuclear genes: Meiosis-based Mendelian segregation is not seen, uniparental (usually maternal) inheritance is typically exhibited, extranuclear genes cannot be mapped to the known nuclear linkage groups, and a phenotype resulting from an extranuclear mutation persists after nuclear substitution.

- Examples of gene mutations inherited in a non-Mendelian fashion, involving genes in mitochondrial DNA or chloroplast DNA, include shoot variegation in four o'clock plants, certain slow-growing mutants in fungi, and certain antibiotic resistance or dependence traits in *Chlamydomonas*.

- Not all cases of non-Mendelian inheritance result from genes on mitochondrial DNA or chloroplast DNA. Many other examples in eukaryotes result from infectious heredity, in which symbiotic, cytoplasmically located bacteria or viruses are transmitted when cytoplasms mix.

- Maternal effect is a phenotype in an individual that is established by the maternal nuclear genome as the result of mRNA or proteins that are deposited in the oocyte before fertilization. These inclusions direct early development of the embryo. Maternal effect is different from non-Mendelian inheritance. The maternal inheritance pattern of extranuclear genes occurs because the zygote obtains most of its cellular organelles (containing the extranuclear genes) from the female parent.

iActivity

IN 1909, A PLANT GENETICIST NAMED CARL CORRENS was investigating the inheritance of leaf color in the four o'clock plant. Much to his surprise, Correns found that one trait did not conform to the expected pattern of Mendelian inheritance. Instead, the progeny phenotype was always the same as that of its female parent. In addition, the progeny phenotypes did not appear in Mendelian proportions. Years later, scientists discovered the reason for this unusual inheritance: The genes that controlled the color pattern were inherited via DNA from the plant's chloroplasts rather than from the nucleus! Since that time, scientists have identified other traits that are controlled by genes found in both chloroplasts and mitochondria.

In this chapter, you will learn about the structure and function of these extranuclear genes and the rules that govern non-Mendelian inheritance. Then, in the iActivity, you can use your understanding of both Mendelian and non-Mendelian inheritance patterns to study an unusual human disorder.

In our discussion of eukaryotic genetics up to now, we have analyzed the structure and expression of the genes located on chromosomes in the nucleus and have defined rules for the segregation of nuclear genes. Outside the nucleus, DNA is found in the mitochondrion (found in both animals and plants) and the chloroplast (found only in green plants and certain protists). The genes in these mitochondrial and chloroplast genomes are known as extrachromosomal genes, cytoplasmic genes, non-Mendelian genes, organellar genes, or extranuclear genes.

The term *non-Mendelian* is informative because extranuclear genes do *not* follow the rules of Mendelian inheritance, as do nuclear genes. Cytoplasm is inherited from the mother in many organisms, so the inheritance of such cytoplasmic factors in these organisms is strictly maternal. Recently, the application of molecular biology techniques has led to rapid advances in knowledge about the organization of extranuclear genomes. In this chapter, we first examine some of these advances and then we discuss the inheritance patterns of extranuclear genes.

Origin of Mitochondria and Chloroplasts

First, let us consider how mitochondria and chloroplasts might have arisen. The **endosymbiont theory** is that mitochondria and chloroplasts originated as free-living prokaryotes that invaded primitive eukaryotic cells and established a mutually beneficial (symbiotic) relationship. According to this hypothesis, eukaryotic cells started out as anaerobic organisms that lacked mitochondria and chloroplasts. At some point in their evolution, about a billion years ago, a eukaryotic cell established a symbiotic relationship with a purple nonsulfur photosynthetic bacterium. Over time, the oxidative phosphorylation activities of the bacterium became used for the benefits of the eukaryotic cell, and, in the presence of atmospheric oxygen, photosynthetic activity (which is a generator of oxygen) was no longer needed and so was lost. Eventually, the eukaryotic cell became dependent on the intracellular bacterium for survival, and the mitochondrion was formed. The chloroplasts of plants and algae are hypothesized to have occurred either simultaneously or later by the ingestion of an oxygen-producing photosynthetic bacterium (a cyanobacterium) by a eukaryotic cell.

As we learn in the next section, many proteins found in mitochondria and chloroplasts are encoded by nuclear genes. Thus, further evolution of mitochondria and chloroplasts involved the transfer of genes from the organelles to the nuclear DNA. As a result, mitochondria can no longer exist as independent organisms, and the eukaryotic cell cannot live without mitochondria. Similarly, chloroplasts cannot exist as independent organisms, and the plants and protists that contain them depend on them to survive.

Organization of Extranuclear Genomes

Mitochondrial Genome
Mitochondria (Figure 23.1; see also Figure 1.8, p. 11), organelles found in the cytoplasm of all aerobic eukaryotic

Figure 23.1

Cutaway diagram of a mitochondrion. Energy-processing mechanisms involve interrelationships between the organelle's intermembranal space, the inner membrane, and the matrix.

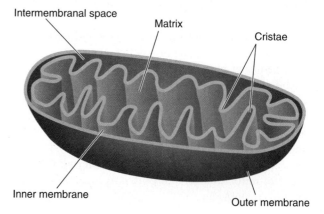

cells, are involved in *cellular respiration*. Cellular respiration is the procecss of oxidizing food molecules, such as glucose, to carbon dioxide and water. The energy released is in the form of ATP, which is used in all the energy-consuming activities of the cell. The first step of cellular respiration is *glycolysis*, the breakdown of glucose to pyruvate. Glycolysis occurs in the cytosol. The second step of cellular respiration, the oxidation of pyruvate to carbon dioxide and water, with the concomitant production of ATP, takes place in mitochondria. A number of mitochondrial enzymes are involved in this step, including pyruvate dehydrogenase and those for electron transport and oxidative phosphorylation, the citric acid cycle, and fatty acid oxidation.

The complete DNA sequences of a number of mitochondrial (mt) DNA genomes are now known. Some of the properties of mitochondrial DNA (mtDNA) are described in this section.

Genome Structure. Many mitochondrial genomes are circular, double-stranded, supercoiled DNA molecules (Figure 23.2). Linear mitochondrial genomes are found in some protozoa and some fungi. In many cases, the GC content of mtDNA differs greatly from that of the nuclear DNA, which allows mtDNA to be separated from nuclear DNA by cesium chloride equilibrium density gradient centrifugation (see Box 3.1, p. 45). No histones or similar proteins are associated with mtDNA. Multiple copies of the genomes are found within mitochondria that are located in multiple nucleoid regions (similar to those of bacterial cells).

The gene content is very similar in both number and function among mitochondrial genomes from different species. However, the size of the genome varies tremendously from organism to organism. In animals, the circular

Figure 23.2

Electron micrograph of mitochondrial DNA.

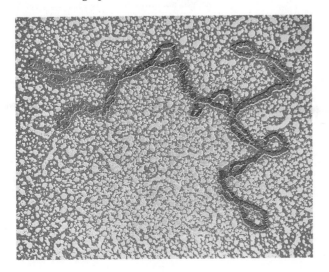

mitochondrial genome is less than 20 kb; for example, human mtDNA is 16,569 bp. In contrast, the mtDNA of yeast is about 80 kb (80,000 bp), and that of plants ranges from 100,000 to 2 million bp. The main difference between animal, plant, and fungal mitochondria is that essentially the entire mitochondrial genomes of animals encode products, whereas the mitochondrial genomes of fungi and plants have extra DNA that does not code for products.

Genome Replication. Replication of mtDNA is semiconservative, and uses DNA polymerases specific to the mitochondrion. The mitochondrial DNA replication process occurs throughout the cell cycle, without any preference for the S phase of the cell cycle, which is when nuclear DNA replicates. The details of mtDNA replication differ among eukaryotes, but the displacement loop (D loop) model (deduced from observing animal mitochondria in vivo) is useful as a general scheme (Figure 23.3). The model is as follows.

In most animals the two strands of mtDNA have different densities because the bases are not equally distributed on each strand, so they are called the H (heavy) and L (light) strands. The synthesis of the new H strand is started at a replication origin for the H strand and forms a D loop structure. As the new H strand extends to about halfway around the molecule, initiation of synthesis of the new L strand at a second replication origin takes place. Both strands are completed by continuous replication. Finally, the circular DNAs are each converted to a supercoiled form.

As for duplication of the mitochondrion itself, experimental evidence obtained by David Luck in 1963 indicates that mitochondria grow and divide.

Genes of mtDNA and Their Transcription. Our present-day understanding of the gene organization of mtDNA derives from mitochondrial genome sequencing experiments, and analysis of the mRNA, rRNA, and tRNA transcripts from mitochondria. Figure 23.4 is a map of the genes of human mtDNA. In general, mtDNA contains information for a number of mitochondrial components such as tRNAs, rRNAs, and *some* of the polypeptide subunits of the proteins cytochrome oxidase, NADH-dehydrogenase, and ATPase. The other components found in the mitochondria—in fact, most of the proteins in the organelles—are encoded by nuclear genes and are imported into the mitochondria. These components include the DNA polymerase and other proteins for mtDNA replication, RNA polymerase and other proteins for transcription, ribosomal proteins for ribosome assembly, protein factors for translation, the aminoacyl tRNA synthetases, and the other polypeptide subunits for cytochrome oxidase, NADH-dehydrogenase, and ATPase.

The mitochondrial protein-coding genes are found in both DNA strands. Their positions were identified in

Figure 23.3

Model for mitochondrial DNA replication that involves the formation of a D loop structure.

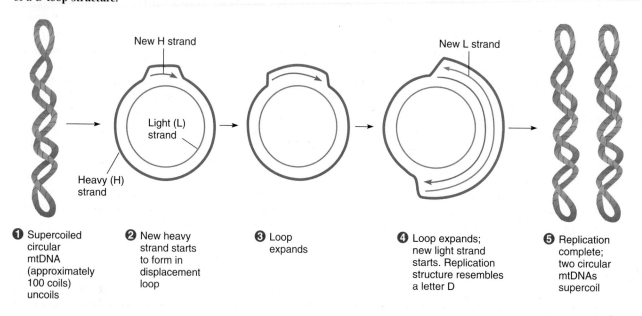

❶ Supercoiled circular mtDNA (approximately 100 coils) uncoils

❷ New heavy strand starts to form in displacement loop

❸ Loop expands

❹ Loop expands; new light strand starts. Replication structure resembles a letter D

❺ Replication complete; two circular mtDNAs supercoil

two ways. One method involved computer-based searching of the DNA sequence for possible start and stop signals in the same reading frame (an open reading frame, or ORF, also called unidentified reading frame, or URF). The second method (used primarily by G. Attardi and his group) was to align the 5′ and 3′ end proximal sequences of mitochondrial mRNAs with the corresponding mitochondrial DNA sequence. All human mitochondrial ORFs have been assigned a function.

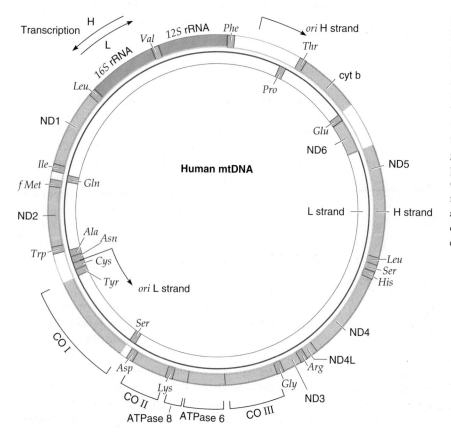

Figure 23.4

Map of the genes of human mitochondrial DNA. The outer circle shows the genes transcribed from the H (heavy) strand, and the inner circle shows the genes transcribed from the L (light) strand. The origins and directions of replication (ori) and the directions of transcription for the H and L strands are indicated. rRNA genes are shown in blue, tRNA genes in purple, and protein-coding genes in yellow. Code: ATPase 6 and 8: components of the mitochondrial ATPase complex. COI, COII, and COIII: cytochrome c oxidase subunits. cyt b: cytochrome b. ND1–6: NADH dehydrogenase components.

Messenger RNAs that are synthesized within the mitochondria remain in the organelle and are translated by mitochondrial ribosomes that are assembled within the organelle. Like their cytoplasmic counterparts, mitochondrial ribosomes consist of two subunits. In humans, the 60S mitochondrial ribosomes consist of 45S and 35S subunits. There are only two rRNAs in the mitochondrial ribosome of most organisms: 16S rRNA in the large subunit and 12S rRNA in the small subunit of animal ribosomes. There is usually one gene for each rRNA in the mitochondrial genome. The ribosomal proteins found in mitochondrial ribosomes are encoded by nuclear genes and transported into the mitochondria from the cytoplasm. In a few organisms, one or more mitochondrial ribosomal proteins are encoded by mitochondrial genes.

Transcription of mammalian mtDNA is unusual in that each strand is transcribed into a *single* RNA molecule that is then cut into smaller pieces. The origins for transcription of the two strands are near the DNA replication origin for the H strand (see Figure 23.4). In the large RNA transcripts that are produced, most of the genes encoding the rRNAs and the mRNAs are separated by tRNA genes. The tRNAs in the transcript are recognized by specific enzymes and are cut out, at the same time liberating the mRNAs and rRNAs. A poly(A) tail is then added to the 3′ end of each mRNA and CCA is added to the 3′ end of each tRNA. Mitochondrial mRNAs have no 5′ caps.

In the much larger mitochondrial genomes of yeast and plants, tRNA genes do not separate the other genes, and the gaps between genes are large. In these systems, transcription termination is signaled by other, non-tRNA sequences in the DNA. Introns are never found in animal mitochondrial genomes, but they are found in yeast, plants, and some other organisms.

Interestingly, the DNA sequences for some animal mitochondrial mRNAs do not encode complete chain-terminating codons. Instead, the processed transcripts end with either U or UA. The subsequent addition of a poly(A) tail completes the missing part(s) of a UAA stop codon.

Translation in Mitochondria. The start codon is very near the 5′ end of the uncapped mRNA so that there is almost no 5′ leader sequence. Therefore, the initiation of translation is different from that for cytoplasmic mRNAs, where both of those features are present. In other words, mitochondrial ribosomes in animals must bind to the mRNAs and orient themselves to start translation in a unique way.

In all mitochondria, a special initiator tRNA—fMet-tRNA—is used in the initiation of protein synthesis, as is the case in *E. coli*. Special mitochondrial initiation factors (IFs), elongation factors (EFs), and release factors (RFs) distinct from cytoplasmic factors are used for translation.

Mitochondrial ribosomes are sensitive to most inhibitors of bacterial ribosome function, including streptomycin, neomycin, and chloramphenicol. Also, mt ribosomes generally are insensitive to antibiotics or other agents to which cytoplasmic ribosomes are sensitive, such as cycloheximide. By the selective use of antibiotics, the site of synthesis of proteins found in mitochondria can be investigated. For example, mt proteins that are synthesized in the presence of cycloheximide, an inhibitor of cytoplasmic ribosomes, must be made on mt ribosomes. Conversely, mt proteins made in the presence of chloramphenicol, an inhibitor of mt ribosomes, must be made on cytoplasmic ribosomes. Using this approach, it was shown that four of the seven subunits of the mitochondrial enzyme cytochrome oxidase in yeast are made in the cytoplasm, and the other three are made in the mitochondria (Figure 23.5).

For protein synthesis, only plant mitochondria use the "universal" nuclear genetic code. Mitochondria of other organisms have differences from that universal code, although there is no one pattern for the differences. Table 23.1 summarizes the points of difference between the nuclear code and the human and yeast mitochondrial genetic codes.

Investigating Genetic Relationships by mtDNA Analysis. In Chapter 9, we discussed some types of DNA analysis involving polymorphisms in DNA sequences. In human mtDNA, one 400-bp region is highly polymorphic. This polymorphism, along with the fact that most mitochondria are maternally inherited, means that maternal lineages are practically unique. Thus, maternal line relationships between individuals can be investigated by using PCR to analyze mtDNA for polymorphisms.

An example of using mtDNA analysis involves the last tsar and tsarina of Russia and their children. During the Bolshevik Revolution of 1917, Tsar Nicholas Romanov II was overthrown and exiled, and in 1918 the tsar and his family were executed by Bolshevik guards. Rumors persisted that one of the tsar's daughters, Princess Anastasia, escaped the execution. In 1922, a woman came forward in Berlin claiming to be Anastasia. In 1928, using the name Anna Anderson, she came to the United States, where she lived until her death in 1984. Although she claimed until she died that she was Anastasia, there was insufficient information available to prove or disprove her claim during her lifetime. In 1993, mtDNA analysis was done on bones, belived to be those of the tsar's family, found two years earlier in a shallow grave in a Russian town. The DNA samples were compared with a blood sample provided by Prince Philip, Duke of Edinburgh, who is the grand nephew of the Tsarina Alexandra. (Prince Philip's grandmother was Princess Victoria, Alexandra's sister.) The mtDNA patterns of the bones matched perfectly the mtDNA of Prince Philip, indicating that they all belonged to the same maternal

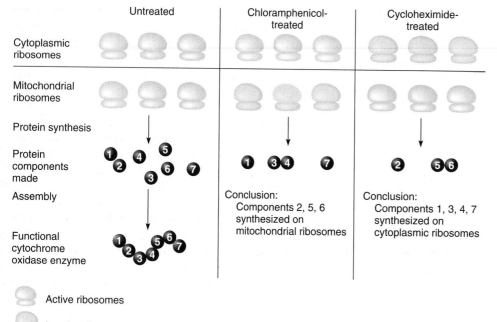

Figure 23.5

Synthesis of the multisubunit protein cytochrome oxidase takes place on both cytoplasmic and mitochondrial ribosomes.

lineage. Further investigation showed unequivocally that the bones were the remains of the tsarina and three of her five children. The bones of the tsar were identified in a similar way by matching mtDNA patterns with those of two living relatives. Soon afterward, mtDNA analysis proved that Anna Anderson was not Anastasia because her mtDNA pattern did not match that of Prince Philip. It is not clear whether any of the three children was Anastasia, although a Russian government commission has stated that there is "definite proof" that one of the skeletons is that of Anastasia.

The case of the Romanovs is an example in which mtDNA analysis has been a powerful tool for analyzing maternal lineages in humans. Mitochondrial DNA analysis is being used to study genetic relationships in many other organisms as well (see Chapter 25, p. 713). Mitochondrial DNA analysis is also being used in conservation biology studies to assess the extent of genetic variability in natural populations. One such study is analyzing the threatened grizzly bear in Yellowstone National Park as a model population for many endangered predator species.

Chloroplast Genome

Chloroplasts (Figure 23.6 and see Figure 1.8, p. 11) are cellular organelles found only in green plants and photosynthetic protists and are the site of *photosynthesis* in the cells containing them. Photosynthesis is carried out in *light reactions* and *dark reactions* (called the *Calvin cycle*). In the light reactions, *chlorophyll* is used to convert light energy into chemical energy, specifically, ATP and NADPH. In the dark reactions, carbon dioxide and water are converted into carbohydrate using chemical energy in the form of ATP and NADPH.

Chloroplasts have a double membrane surrounding an internal, enzyme-rich *stroma*. The stroma is the site of the dark reactions. In the stroma there is a network of stacked, flattened membranous sacs. Each stack is a *granum*, and each of the flattened sacs is a *thylakoid*. The thylakoids are the sites of the *photosystems* and other proteins for the light reactions. The chlorophyll and other pigments are found in the photosystems.

Genome Structure and Replication. The structure of the chloroplast genome is similar to that of mitochondrial genomes. In all cases, the DNA is double-stranded,

Table 23.1	Differences Between Human and Yeast Mitochondrial Genetic Codes		
	Amino Acid		
		Mitochondrial Code	
Codon[a]	**Nuclear Code**	**Mammal**	**Yeast**
UGA	Termination	Tryptophan	Tryptophan
AUA	Isoleucine	Methionine	Isoleucine
CUN[b]	Leucine	Leucine	Threonine
AGG, AGA	Arginine	Termination	Arginine
CGN[b]	Arginine	Arginine	Termination?

[a]All sequences read 5′ to 3′.
[b]N = any one of the four bases A, G, U, and C.

Figure 23.6

Cutaway diagram of a chloroplast. The organelle's energy-harvesting mechanisms involve interrelationships between the intermembranal space, the stroma, the thylakoids stacked in grana, and the area within each thylakoid.

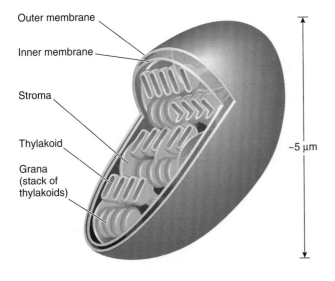

Outer membrane
Inner membrane
Stroma
Thylakoid
Grana (stack of thylakoids)

~5 μm

circular, devoid of structural proteins, and supercoiled. In many cases, the GC content of cpDNA differs greatly from that of the nuclear and mitochondrial DNA, which allows cpDNA to be separated from those two DNAs by CsCl equilibrium density gradient centrifugation.

Chloroplast DNA is much larger than animal mtDNA, with a size between 80 kb and 600 kb. The DNA sequences of the chloroplast genomes of a few organisms have been completely determined. For example, the tobacco genome is 155,844 bp, and the rice genome is 134,525 bp. All chloroplast genomes contain a significant proportion of noncoding DNA sequences.

The number of copies of cpDNA per chloroplast varies between species. In each case, there are multiple copies per chloroplast, typically distributed among a number of nucleoid regions. In the unicellular green alga *Chlamydomonas* ("clammy-da-moan-us"; see Figure 23.12, p. 645), for example, the one chloroplast in the cell contains 500 to 1,500 cpDNA molecules. Chloroplasts themselves grow and divide in essentially the same way mitochondria do.

Gene Organization of cpDNA. The chloroplast genome contains genes for each chloroplast rRNA (16S, 23S, 4.5S, and 5S) and for the tRNAs. It also contains the genes for *some* of the proteins required for transcription and translation of the cp-encoded genes (such as ribosomal proteins, RNA polymerase subunits, and translation factors) and for photosynthesis; most of the proteins found in the chloroplast are encoded by nuclear genes.

The organization of the chloroplast genome in rice is presented in Figure 23.7.

The chloroplast genome contains two copies of each of the rRNA genes. The two sets of chloroplast rRNA genes are located in two identical repeated sequences located in the genome in an inverted orientation. The inverted repeats are designated IR_A and IR_B in Figure 23.7. Other genes are found in the repeated sequence, so they are also duplicated in the chloroplast genome. The location of these repeats defines a short single copy (SSC) region and a long single copy (LSC) region of the chloroplast genome.

Rice has 30 tRNA genes, whereas *Marchantia* ("ma-can-te-ah"), a liverwort, has 32 tRNA genes. Almost 100 ORFs (open reading frames, putative protein-coding genes) have been identified by computer analysis of the cpDNA sequences. Approximately 60 of these ORFs have been correlated with known functional protein-coding genes, and the others remain to be defined.

Translation in Chloroplasts. Chloroplast protein synthesis uses organelle-specific 70S ribosomes that consist of 50S and 30S subunits. The 50S subunit contains one copy each of 23S, 5S, and 4.5S rRNAs, and the 30S subunit contains one copy of a 16S rRNA. Some of the ribosomal proteins in each subunit are nuclear encoded and some are chloroplast encoded.

Protein synthesis in chloroplasts is similar to protein synthesis in prokaryotes. fMet-tRNA is used to initiate all proteins, and the chloroplast uses its own organelle-specific initiation factors (IFs), elongation factors (EFs), and release factors (RFs). The universal genetic code is used in chloroplast protein synthesis.

Like mitochondrial ribosomes, chloroplast ribosomes are resistant to cycloheximide, an inhibitor of cytoplasmic ribosomes, but are sensitive to almost all inhibitors known to block prokaryotic protein synthesis. Using selective antibiotics in essentially the same way we described for the synthesis of multisubunit mitochondrial proteins (see Figure 23.5), the synthesis of chloroplast proteins has been examined. One important chloroplast protein is ribulose bisphosphate carboxylase/oxygenase (rubisco), the first enzyme used in the pathway for fixation of carbon dioxide in the photosynthetic process. This enzyme is a major protein and is found in the chloroplasts of all plant tissues. Since it constitutes about 50 percent of the protein found in green plant tissue, it is the most common protein in the world. Rubisco consists of eight identical large subunits, and eight identical small subunits. The large subunit is encoded by gene *rbcL* on the chloroplast genome (see the 11 o'clock position in Figure 23.7), and is synthesized within the chloroplast. The small subunit is encoded by gene *rbcS* in the nuclear genome, is synthesized on cytosolic ribosomes, and is then imported into the chloroplast.

Figure 23.7

Organizations of the chloroplast genome of rice (*Oryza sativa*). [From "Complete Sequence of the Rice (Oryza sativa) Chloroplast ..." by Hiratsuka et al. in Molecular and General Genetics, Vol. 217, 1989. Reprinted by permission of Springer-Verlag, New York.]

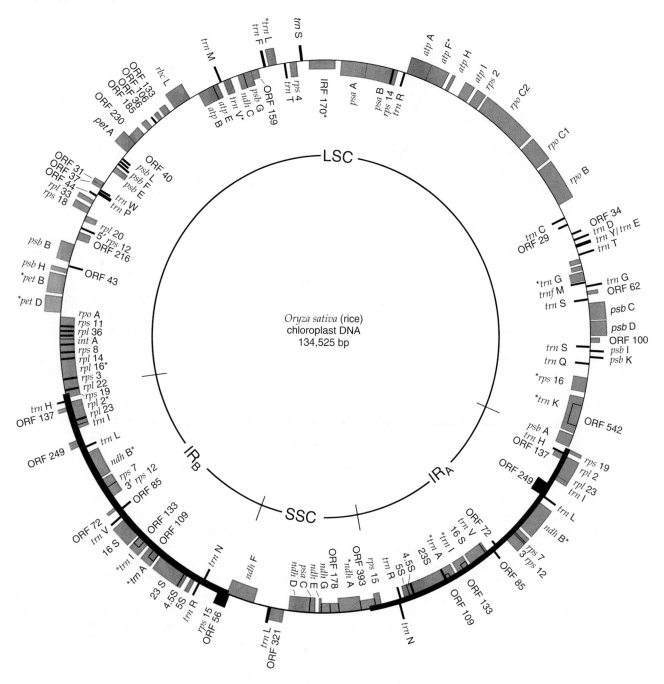

KEYNOTE

> Both mitochondria and chloroplasts contain their own DNA genomes. The DNA in most species' mitochondria and all chloroplasts is circular, double-stranded, and supercoiled. The mitochondrial DNA of some species is linear. The mitochondrial and chloroplast genomes contain genes for the rRNA components of the ribosomes of these organelles, for many (if not all) of the tRNAs used in organellar protein synthesis, and for a few proteins that remain in the organelles and perform functions specific to the organelles. All other proteins are nuclear-encoded, synthesized on cytoplasmic ribosomes, and imported into the organelles. At least in the mitochondria of some organisms, the genetic code is different from that found in nuclear protein-coding genes. Since humans and many other organisms receive most of their mitochondria from their mothers, maternal lineages are practically unique. Thus, maternal line relationships between individuals can be investigated by mtDNA analysis.

Rules of Non-Mendelian Inheritance

Since the pattern of inheritance shown by genes located in organelles differs strikingly from the pattern shown by nuclear genes, the term **non-Mendelian inheritance** is appropriate to use when we discuss extranuclear genes. In fact, if the results obtained from genetic crosses do not conform to predictions based on the inheritance of nuclear genes, there is a good reason to suspect extranuclear inheritance.

Here are the four main characteristics of non-Mendelian inheritance:

1. Ratios typical of Mendelian segregation are not found because meiosis-based Mendelian segregation is not involved.
2. In multicellular eukaryotes, the results of reciprocal crosses involving extranuclear genes are not the same as reciprocal crosses involving nuclear genes because meiosis-based Mendelian segregation is not involved. (Recall from Chapters 2 and 3 that in a reciprocal cross, the sexes of the parents are reversed in each case. For example, if A and B represent contrasting genotypes, A ♀ × B ♂ and B ♀ × A ♂ would be a pair of reciprocal crosses.)

   Mitochondrial and chloroplast genes usually show **uniparental inheritance** from generation to generation. In uniparental inheritance, all progeny (both males and females) have the phenotype of only one parent. Usually for multicellular eukaryotes, the mother's phenotype is expressed exclusively, a phenomenon called **maternal inheritance.** Maternal inheritance occurs because the amount of cytoplasm in the female gamete usually greatly exceeds that in the male gamete. Therefore, the zygote receives most of its cytoplasm (containing the extranuclear genes in the mitochondria and, where applicable, the chloroplasts) from the female parent and a negligible amount from the male parent.

   In contrast, the results of reciprocal crosses between a wild-type and a mutant strain are identical if the genes are located on nuclear chromosomes. One exception occurs when X-linked genes are involved (see Chapter 12), but even then the results are distinct from those for non-Mendelian inheritance.
3. Extranuclear genes cannot be mapped to the chromosomes in the nucleus.
4. Non-Mendelian inheritance is not affected by substituting a nucleus with a different genotype.

KEYNOTE

> The inheritance of extranuclear genes follows rules different from those for nuclear genes. In particular, no meiotic segregation is involved, generally uniparental (and often maternal) inheritance is seen, extranuclear genes are not mappable to the known nuclear linkage groups, and the phenotype persists even after nuclear substitution.

Examples of Non-Mendelian Inheritance

In this section, we discuss the properties of a selected number of mutations in extranuclear chromosomes to illustrate the principles of extranuclear inheritance.

Shoot Variegation in the Four O'Clock

A variegated-shoot phenotype of the albomaculata strain of four o'clocks (*Mirabilis jalapa*, also called the marvel of Peru; Figure 23.8a) involves non-Mendelian inheritance. (A shoot of a plant consists of stem, leaves, and flowers.) This strain has mostly shoots with variegated leaves (leaves with yellowish-white patches) and occasional shoots that are wholly green or wholly yellowish-white (Figure 23.8b).

animation

Shoot Variegation in the Four O'Clock

Table 23.2 summarizes the results of crosses between flowers on variegated, green, and white shoots. The seeds from flowers on green shoots give only green progeny, regardless of whether pollen is from green, white, or variegated shoots. Seeds from flowers on white shoots give only white progeny, regardless of the pollen source. (However, because the whiteness indicates the absence of

Figure 23.8

Variegation in the four o'clock, *Mirabilis jalapa*. (a) Photograph of the four o'clock. (b) Drawing showing variegation. Shoots that are all green, all white, and variegated are found on the same plant, and flowers can form on any of these shoots.

a)

b)

Variegated shoot

All-white shoot

All-green shoot

Variegated main shoot

chlorophyll and hence an inability to carry out photosynthesis, white progeny die soon after seed germination.) Finally, seeds from flowers on variegated shoots all produce three types of progeny—completely green, completely white, and variegated—regardless of the type of pollen. In subsequent generations, maternal inheritance is always seen in these same patterns. In sum, the progeny phenotype in each case was the same as that of the maternal parent (the color of the progeny shoots was the same as the color of the parental flower), which indicates maternal inheritance. Moreover, the results of reciprocal

crosses differed from the expected patterns, and there was a lack of constant proportions of the different phenotypic classes in the segregating progeny. These last two properties are also characteristic of traits showing non-Mendelian inheritance and are not expected for traits showing Mendelian inheritance.

Flowering plants are green because of the presence of the green pigment chlorophyll in large numbers of chloroplasts. Green shoots in four o'clocks have a normal complement of chloroplasts. White shoots have abnormal, colorless chloroplasts called leukoplasts.

Table 23.2 **Results of Crosses of Variegated Plants of *Mirabilis jalapa***

Phenotype of Shoot-Bearing ♀ Parent (Egg)	Phenotype of Shoot-Bearing ♂ Parent (Pollen)	Phenotype of Progeny
White	White	White
	Green	White
	Variegated	White
Green	White	Green
	Green	Green
	Variegated	Green
Variegated	White	Variegated, green, or white
	Green	Variegated, green, or white
	Variegated	Variegated, green, or white

Leukoplasts lack chlorophyll and therefore are incapable of carrying out photosynthesis. The explanation for the inheritance of shoot color in the albomaculata strain of four o'clocks is that the abnormal chloroplasts are defective as a result of a mutant gene in the cpDNA. When a zygote receives both chloroplasts and leukoplasts from the maternal cytoplasm, then during plant growth and development these organelles segregate so that a particular cell and its progeny cells may receive only chloroplasts (leading to green tissues), only leukoplasts (leading to white tissues), or a mixture of chloroplasts and leukoplasts (leading to variegation) (Figure 23.9). Let us take this model one step further. The flowers on a green shoot have only chloroplasts, and so through maternal inheritance these chloroplasts form the basis of the phenotype of the next generation. Similar arguments can be made for the flowers on white or variegated shoots.

This simple model has three assumptions: (1) The pollen contributes essentially no cytoplasmic information—that is, no chloroplasts or leukoplasts—to the egg. This is a reasonable assumption because the egg is much larger than the pollen. Thus, we can assume that in the zygote the extranuclear genetic determinants come from the egg. (2) The chloroplast genome replicates autonomously and, by growth and division of plastids (the general term for photosynthetic organelles such as chloroplasts), the wild-type and mutant cpDNA molecules have the potential to segregate randomly to the new plastids so that pure plastid lines can be generated from an original mixed line. (3) Segregation of plastids to daughter cells is random, so that some daughters receive chloroplasts, some receive leukoplasts, and some receive mixtures.

KEYNOTE

> The leaf color phenotypes in a variegated strain of the four o'clock, *Mirabilis jalapa*, show maternal inheritance. The abnormal chloroplasts in white tissue are the result of a mutant gene in the cpDNA, and the observed inheritance patterns follow the segregation of cpDNA.

The [poky] Mutant of *Neurospora*

The fungus *Neurospora crassa* (see Chapter 16, pp. 430–433) is an obligate aerobe; it requires oxygen to grow and survive, so mitochondrial functions are essential for its growth. The [poky] mutant, which involves a change in the mtDNA, illustrates a number of the classic expectations of non-Mendelian inheritance. Biochemical analysis showed that the [poky] mutant is defective in aerobic respiration as a result of changes in the cytochrome complement of the mitochondria. The change in the cytochrome spectrum affects the ability of the mitochondria to generate sufficient ATP to support rapid growth, so slow growth results.

The *Neurospora* life cycle was presented in Chapter 16 (Figure 16.2, p. 432, and pp. 430–433). The sexual phase of the life cycle is initiated after a fusion of nuclei from mating type *A* and *a* parents. A sexual cross can be made in one of two ways: by putting both parents on the crossing medium simultaneously or by inoculating the medium with one strain and, after three or four days at 25°C, adding the other parent. In the latter case, the first parent on the medium produces all the protoperithecia ("proto-perrytheece-e-ah"), the bodies that will give rise to the true fruiting bodies containing the asci with the sexual ascospores.

Compared with the conidia, the asexual spores, the protoperithecia have a tremendous amount of cytoplasm and therefore can be considered the female parent in much the same way as an egg of a plant or an animal is the female parent. By adding conidia of a strain of the opposite mating type to the crossing medium, we have a *controlled cross* in which one strain acts as the female and the other as the male parent. Using a strain to produce the protoperithecia as the female parent and conidia of another strain as a male parent, geneticists can make reciprocal crosses to determine whether any trait shows extranuclear inheritance. This experiment is now illustrated for the [poky] mutant.

Mary and Herschel Mitchell did reciprocal crosses between [poky] and the wild type, with the following results:

[poky] ♀ × wild type ♂ → all [poky] progeny
wild-type ♀ × [poky] ♂ → all wild-type progeny

In other words, all progeny show the same phenotype as the maternal parent, indicating maternal inheritance as a characteristic for the [poky] mutation.

This analysis can be made more refined by using tetrad analysis (Chapter 16, pp. 443–445) to follow the phenotype more closely. Recall that in *Neurospora* the eight products of a meiosis and subsequent mitosis are retained in linear order within the ascus. The eight ascospores can be removed from the asci, and strains germinated from the spores can be analyzed for the particular phenotype being followed. By doing tetrad analysis for the [poky] ♀ × wild type ♂ cross, we find an 8:0 ratio of [poky] : wild type progeny (Figure 23.10a) and for the wild type ♀ × [poky] ♂ cross, we get a 0:8 ratio of [poky] : wild type progeny with regard to the growth phenotype (Figure 23.10b). At the same time, nuclear genes show the 4:4 segregation expected of Mendelian inheritance. These results show that [poky] is determined by an extranuclear gene (in this case, located in the mitochondrion) that exhibits a pattern of maternal inheritance.

The cytochrome deficiency phenotype in [poky] results from a defect in mitochondrial protein synthesis. The [poky] mutant has been shown to be a 4-bp deletion in the promoter for the gene for the 19S rRNA of the small mitochondrial ribosomal subunit. The mutation causes a

Figure 23.9

Model for the inheritance of shoot color in the four o'clock, *Mirabilis jalapa*.

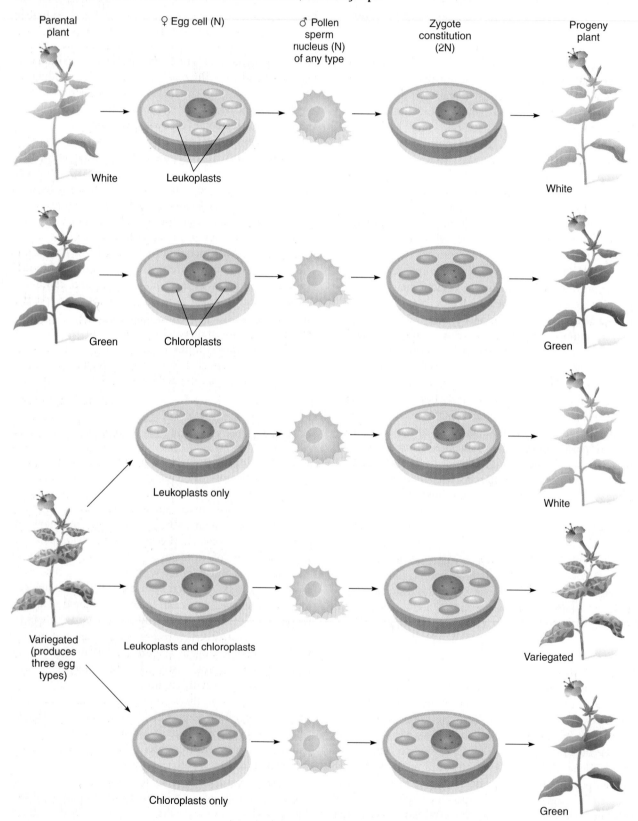

| Parental plant | ♀ Egg cell (N) | ♂ Pollen sperm nucleus (N) of any type | Zygote constitution (2N) | Progeny plant |

Figure 23.10

Results of reciprocal crosses of [*poky*] and normal (wild-type) *Neurospora*. (a) [*poky*] ♀ × normal ♂. (b) normal ♀ × [*poky*] ♂ .

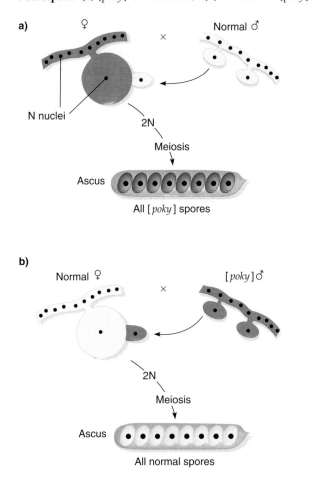

a)

All [*poky*] spores

b)

All normal spores

gross deficiency of 19S rRNA, which leads to a decreased amount of small ribosomal subunits in the organelle and hence to a greatly diminished protein synthesis capability. As a result, synthesis of mitochondrial proteins, including cytochromes, is reduced and the [*poky*] phenotype occurs.

KEYNOTE

> The slow-growing [*poky*] mutant of *Neurospora crassa* shows maternal inheritance and deficiencies for some mitochondrial cytochromes. The molecular defect in [*poky*] is a deletion in the promoter for the rRNA gene of the small mitochondrial ribosomal subunit, which leads to deficiencies for some mitochondrial cytochromes and the slow-growth phenotype.

Yeast *petite* Mutants

Yeast grows as single cells. On solid media, yeast forms discrete colonies consisting of many thousands of individual cells clustered together. Yeast can grow with or without oxygen. In the absence of oxygen, yeast obtains energy for growth and metabolism through fermentation, in which the mitochondria are not involved. In the presence of oxygen, mitochondria carry out aerobic respiration and facilitate a faster growth rate than is the case in the absence of oxygen.

Characteristics of *petite* Mutants. When yeast cells are spread onto a solid medium, allowing growth to occur by aerobic respiration or by fermentation, a very low proportion of the cells give rise to colonies that are much smaller than wild-type colonies. Since the discoverer of this phenomenon, Boris Ephrussi, was French, the small colonies are called *petites* (French for "small"), and the wild-type colonies sometimes are called *grandes* (French for "big"). *Petite* colonies are small because the growth rate of the mutant *petite* strain is significantly slower than that of the wild type, not because the cells are small. That is, there are fewer cells in the *petite* colonies. Biochemical analysis shows that *petites* have cytochrome deficiencies and therefore are essentially incapable of carrying out aerobic respiration; they must obtain their energy primarily from fermentation, which is a less efficient process.

In an unmutagenized population of cells, 0.1 to 1 percent of the cells spontaneously become *petite*. In the presence of an intercalating mutagen (a chemical that can wedge itself between adjacent base pairs in DNA; see Chapter 7, p. 146) such as ethidium bromide, 100 percent of the cells become *petite*. The yeast *petite* system is particularly useful in studies of non-Mendelian inheritance because yeast cells that lack mitochondrial functions can still survive and grow. Such mutations to respiratory deficiencies in yeast are automatically conditional mutants. On a medium that can support fermentation and aerobic growth, the *petites* grow more slowly than the *grandes*, whereas on a medium that supports only aerobic respiration, *petites* are unable to grow.

Different Types of *petite* Mutants. Crosses between *petites* and wild type were made to determine how *petites* are inherited. Yeast are haploid, and crosses involve fusing two cells, one of mating type **a** and one of mating type α, to produce a diploid zygote (see Figure 16.1, p. 430). That zygote can be grown into a colony to check its phenotype. When the zygote goes through meiosis (a process called sporulation), the four haploid meiotic products—the sexual spores (ascospores)—are contained within an ascus. Tetrad analysis can be done, that is, the analysis of all four products of meiosis (here, the four ascospores in an ascus) to determine segregation patterns.

Some *petites*, when crossed with wild type, give a 2:2 segregation of wild type : *petite* in tetrads (Figure 23.11a). This segregation pattern is that found for nuclear gene mutations, so these *petites* are *nuclear petites* (also called *segregational petites*), symbolized *pet⁻*. The existence of

a) Inheritance of *nuclear petites* (*pet⁻*)

b) Inheritance of *neutral petites* ([rho⁻N])

Figure 23.11
**Inheritance of yeast *petite*
mutants.** (a) Mendelian inheritance of *nuclear petites* (*pet⁻.*) (b)
non-Mendelian inheritance
of *neutral petites* ([rho⁻N]).
(c) non-Mendelian inheritance
of *suppressive petites* ([rho⁻S]).

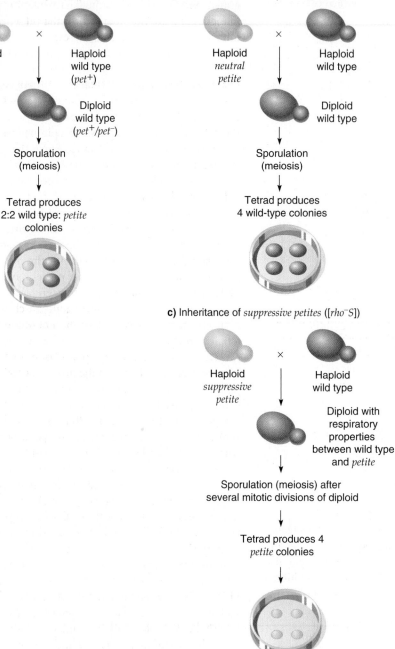

c) Inheritance of *suppressive petites* ([rho⁻S])

nuclear petites is not surprising because some subunits of some mitochondrial proteins are encoded by nuclear genes. Using genetic symbols, when a *pet⁻* mutant is crossed with the wild type (*pet⁺*), the diploid is *pet⁺/pet⁻*, which produces wild-type colonies. When this *pet⁺/pet⁻* cell goes through meiosis, each resulting tetrad shows a 2:2 segregation of wild-type (*pet⁺*) : petite (*pet⁻*) phenotype. *Nuclear petites* occur much less often than extranuclear petites.

Two other classes of petites, the *neutral petites* and the *suppressive petites*, show non-Mendelian inheritance.

Figure 23.11b shows the inheritance pattern of neutral petites (symbolized [rho⁻N]). When a *neutral petite* is crossed with normal wild-type cells ([rho⁺N]), the [rho⁺N]/[rho⁻N] diploids all produce wild-type colonies. When these diploids go through meiosis, all resulting tetrads show a 0:4 ratio of *petite* : wild type while nuclear markers segregate 2:2. The name *neutral*, then, refers to that fact that this class of *petites* does not affect the wild type. This result is a classic example of uniparental inheritance, in which all progeny have the phenotype of only one parent. *This phenomenon is not*

Examples of Non-Mendelian Inheritance

maternal inheritance, however, because the two haploid cells that fuse to produce the diploid are the same size and contribute equal amounts of cytoplasm.

With prolonged treatment with an intercalating agent, the majority of *petites* produced are *neutral petites*. The mitochondrial genome is implicated in these mutants because cytochromes are altered, because there is evidence for extranuclear inheritance, and because the mitochondria are the only other site in the yeast cell in which genetic material is found.

An examination of the mitochondrial genetic material in the *neutral petites* reveals a remarkable characteristic: Essentially 99 to 100 percent of the mtDNA is missing. Not surprisingly, then, the *neutral petites* cannot perform mitochondrial functions. They survive because of nuclear-encoded, cytoplasmically localized fermentation processes. In genetic crosses with the wild type, the normal mitochondria from the wild-type parent form a population from which new, normal mitochondria are produced in all progeny, and therefore the *petite* trait is lost after one generation.

The second class of *petites* that shows non-Mendelian inheritance is the *suppressive petites*, symbolized [*rho⁻S*]. Like the *neutral petites* and the *nuclear petites*, [*rho⁻S*] *petites* are deficient in mitochondrial protein synthesis. The *suppressive petites* are different from the *neutrals* because they do have an effect on the wild type. Most *petite* mutants are of the *suppressive* type.

The inheritance pattern of *suppressive petites* is different from those of *nuclear* and *neutral petites* (Figure 23.11c). When a [*rho⁺*]/[*rho⁻S*] diploid is formed, it has respiratory properties intermediate between those of normal and *petite* strains. If this diploid divides mitotically a number of times, the diploid population produced is up to 99 percent *petites*. The name *suppressive*, then, refers to the fact that this class of *petites* overwhelms the wild type so that a respiratory-deficient phenotype results. Sporulation of any of the *petites* in that population produces tetrads with a 4:0 ratio of *petite* : wild type (see Figure 23.11c). Sporulation of any of the few wild-type diploids in the population produces tetrads with a 0:4 ratio of *petite* : wild type.

The *suppressive petites* have changes in the mtDNA. The changes start out as deletions of part of the mtDNA, and then, by some correction mechanism, sequences that are not deleted become duplicated until the normal amount of mtDNA is restored. During the correction events rearrangements of the mtDNA sometimes occur. These deletions and rearrangements disrupt genes or lead to losses of genes, thereby causing deficiencies in the enzymes involved in aerobic respiration, and so a *petite* colony results. *Suppressive petites* are proposed to have a suppressive effect over normal mitochondria either (1) by replicating faster than normal mitochondria and thereby outcompeting them, or (2) by fusion with normal mitochondria followed by recombination between suppressive mtDNA and normal mtDNA, thereby altering the latter's gene organization.

KEYNOTE

Yeast *petite* mutants grow slowly and have various deficiencies in mitochondrial functions as a result of alterations in mitochondrial DNA. Some *petite* mutants show Mendelian inheritance, while others show non-Mendelian inheritance: Particular patterns of inheritance vary with the latter type of *petite*.

Non-Mendelian Inheritance in *Chlamydomonas*

One of the most thorough analyses of chloroplast genetics was begun by Ruth Sager and her colleagues in their work with the unicellular alga *Chlamydomonas reinhardtii* ("clammy-da-MOAN-us rhine-heart-E-eye"; Figure 23.12).

The motile, haploid unicellular alga *Chlamydomonas reinhardtii* has two flagella and a single chloroplast that contains many copies of cpDNA. There are two mating types, *mt⁺* and *mt⁻*. In a process known as syngamous mating, a zygote is formed by fusion of two equal-sized cells (which therefore contribute an equal amount of cytoplasm), one of each mating type. A thick-walled cyst develops around the zygote. After meiosis, four haploid progeny cells are produced, and because mating type is determined by a nuclear gene, a 2:2 segregation of *mt⁺* : *mt⁻* mating types results.

Erythromycin resistance ([*eryʳ*]) is a chloroplast trait that is inherited in a non-Mendelian manner. Wild-type *Chlamydomonas* cells are erythromycin sensitive ([*eryˢ*]). From a cross of *mt⁺* [*eryʳ*] × *mt⁻* [*eryˢ*], about 95 percent of the offspring are all erythromycin resistant (Figure 23.13a). This result is an example of uniparental inheritance. The reciprocal cross, *mt⁻* [*eryʳ*] × *mt⁺* [*eryˢ*] (Figure 23.13b), also shows uniparental inheritance about 95 percent of the time, although here the

Figure 23.12

The unicellular green alga *Chlamydomonas*. This organism has a single chloroplast and a pair of flagella.

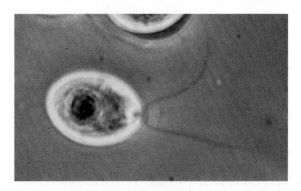

Figure 23.13

Uniparental inheritance in *Chlamydomonas*. (a) From a cross of mt^+ [ery^r] × mt^- [ery^s], 95 percent of the zygotes give tetrads that segregate 2:2 for the nuclear mating-type genes and 4:0 for the extranuclear gene carried by the mt^+ parent (here, [ery^r]). (b) From the reciprocal cross of a mt^- [ery^r] × mt^+ [ery^s], 95 percent of the zygotes give tetrads segregating 2 mt^+ : 2 mt^- and 0 [ery^r] : 4 [ery^s], again showing uniparental inheritance for the extranuclear trait of the mt^+ parent.

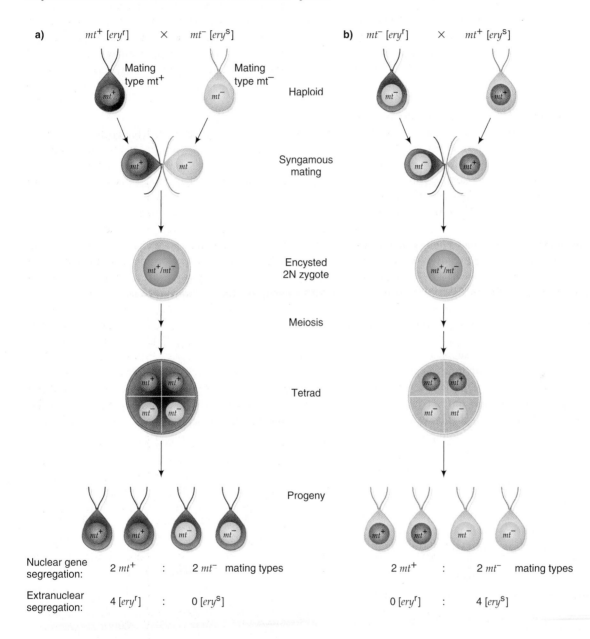

erythromycin-sensitive phenotype is inherited. Thus, even though both parents contribute equal amounts of cytoplasm to the zygote, the progeny resemble the mt^+ parent with regard to chloroplast-controlled phenotypes.

A number of *Chlamydomonas* chloroplast gene mutants in addition to ery^r show uniparental inheritance, with progeny always resembling the phenotype of the mt^+ parent. The explanation for this is as follows: Both mt^+ and mt^-

parents contribute cpDNA equally to the zygote. During zygote maturation and the subsequent zygote germination, the mt^+ cpDNA becomes highly methylated, whereas the mt^- cpDNA becomes only lightly methylated. The highly methylated mt^+ cpDNA replicates much better than the lightly methylated mt^- cpDNA. In addition, within hours after mating, the lightly methylated mt^- cpDNA is destroyed.

As indicated earlier, only about 95 percent of the zygotes in the crosses involving chloroplast genes show uniparental inheritance. The other 5 percent of the zygotes show extranuclear traits from both parents. Such zygotes are said to show **biparental inheritance,** indicating that both types of chloroplast chromosomes are present and active in these zygotes. The genetic condition of these zygotes is defined as a **cytohet** (the term is derived from the term *cyto*plasmically *het*erozygous). In many instances, the non-Mendelian traits of biparental zygotes segregate into pure types (i.e., either erythromycin-sensitive or erythromycin-resistant) on successive mitotic divisions. This phenomenon presumably involves the segregation of the different chloroplast chromosomes, and hence the different chloroplasts, into pure types. This situation parallels that of *Mirabilis* (see pp. 639–641). Since occasionally biparental zygotes are produced, geneticists can make crosses between strains carrying different extranuclear traits that are chloroplast-controlled. Occasionally, a biparental zygote does not segregate the two traits; instead, both traits continue to be expressed in subsequent generations. The explanation is that a recombination event takes place between the two types of cpDNA so that a recombinant chromosome carrying both alleles originally carried by the two parents is produced. From such segregation data, it is possible to construct genetic maps of the chloroplast genome.

K E Y N O T E

> *Chlamydomonas* chloroplast genes are inherited in a non-Mendelian manner, with progeny resembling the phenotype of the mt^+ parent about 95 percent of the time.

Human Genetic Diseases and Mitochondrial DNA Defects

A number of human genetic diseases result from mtDNA gene mutations. These diseases show maternal inheritance. The following are some brief examples:

Leber's hereditary optic neuropathy (LHON; OMIM 535000 at http://www3.ncbi.nlm.nih.gov/Omim/). This disease affects midlife adults and results in complete or partial blindness from optic nerve degeneration. Mutations in the mitochondrial genes for the electron transport chain proteins ND1, ND2, ND4, ND5, ND6, cyt b, COI, COIII, and ATPase 6 (see Figure 23.4) all lead to LHON. The electron transport chain drives cellular ATP production by oxidative phosphorylation. It appears that death of the optic nerve in LHON is a common result of oxidative phosphorylation defects, here brought about by inhibition of the electron transport chain.

Kearns-Sayre syndrome (OMIM 530000). People with this syndrome have three major types of neuromuscular defects: progressive paralysis of certain eye muscles; abnormal accumulation of pigmented material on the retina, leading to chronic inflammation and degeneration of the retina; and heart disease. Large deletions at various positions in the mtDNA cause the syndrome. One model is that each deletion removes one or more tRNA genes, so mitochondrial protein synthesis is disrupted. In some unknown way, this leads to development of the syndrome.

Myoclonic epilepsy and ragged-red fiber (MERRF) disease (OMIM 545000). Individuals with this disease exhibit "ragged-red fibers," an abnormality of tissue when seen under the microscope. The most characteristic symptom of MERRF disease is myoclonic seizures (sudden, short-lived, jerking spasms of limbs or of the whole body). Other symptoms are ataxia (defect in movement coordination) and the accumulation of lactic acid in the blood. There may be additional symptoms associated with the disease, including dementia, loss of hearing, difficulty speaking, optic atrophy, involuntary jerking of the eyes, and short stature. The mitochondria of individuals with MERRF are abnormal in appearance. The disease is caused by a single nucleotide substitution in the gene for a lysine tRNA. The mutated tRNA adversely affects mitochondrial protein synthesis, and somehow this gives rise to the various phenotypes of the disease.

In most diseases resulting from mtDNA defects, the cells of affected individuals have a mixture of mutant and normal mitochondria. This condition is known as *heteroplasmy.* Characteristically, the proportions of the two mitochondrial types vary from tissue to tissue and from individual to individual within a pedigree. The severity of the disease symptoms correlates approximately with the relative amount of mutant mitochondria.

iActivity In the iActivity, *Mitochondrial DNA and Human Disease,* on your CD-ROM, you construct a pedigree to help determine whether a neurological disease has been inherited.

Cytoplasmic Male Sterility and Hybrid Seed Production

Hybrid crops are very important in commercial agriculture. A hybrid is produced by crossing two varieties of the crop plant that are not closely related. The hybrids typically grow more vigorously and produce more seeds than either parent.

This phenomenon is called **heterosis** or **heterozygote superiority,** and is described in more detail in Chapter 24, pp. 693–694. Farmers are sold the hybrid seed —the seed that germinates to produce the hybrids—which means that plant breeders need to make controlled crosses between two parental varieties on a commercial scale.

Corn was the first crop plant used to generate hybrid seed. Corn plants can self-fertilize, but the male (tassel) and female (ear) parts are separate. Therefore, to make a controlled cross manually, you detassel (emasculate) the female plant and fertilize it with pollen from the male plant. (Recall that emasculation was involved in setting up Mendel's controlled crosses.) Emasculation is a relatively easy process in corn, but a laborious one in many other crop plants. Fortunately, plant breeders can exploit mutations that cause male sterility. Male sterility may result from mutations of nuclear genes or of extranuclear genes, producing *genic male sterility* and *cytoplasmic male sterility* (CMS), respectively.

The mutation in CMS is in the mitochondrial genome. Like the chloroplast, the mitochondrion is inherited in plants in a maternal fashion. That is, all the mitochondria in the zygote come from the egg and not from the pollen. The CMS mutation results in a defect in pollen formation, so the plant is male sterile. However, when this CMS plant is used as the female parent in a controlled cross, the hybrid seed will germinate and produce progeny plants. Those plants will be male sterile because they have inherited the CMS mutation, and obviously cannot produce seeds by self-fertilization. Understandably, the farmer would be unhappy with the hybrid plants because male-fertile plants would also have to planted so that the hybrids can be fertilized! The solution to this problem involves a nuclear *restorer of fertility* (*Rf*) gene. The dominant *Rf* allele overrides the CMS mutation, whereas the recessive *rf* allele cannot.

Figure 23.14 shows how hybrid seed can be generated using CMS and the *Rf* gene. The male-sterile female parent is [CMS] *rf/rf* and the fertile male is [CMS] *Rf/rf* (where the [CMS] indicates the cytoplasm for cytoplasmic male sterility). The F$_1$ progeny will show the desired hybrid vigor, and all have the [CMS] cytoplasm. The hybrids will show segregation of 1 *Rf/rf* : 1 *rf/rf*. The *Rf/rf* plants are male fertile because the restorer gene has overridden the CMS mutation. The *rf/rf* plants are male sterile, though, because there is no effect on the [CMS] cytoplasm. However, these latter plants are fertilized readily by pollen from the *Rf/rf* plants in the field.

There is now a genetic engineering approach for making male-sterile plants. This approach does not involve extranuclear genes, but instead involves making transgenic plants using standard plant transformation methods. Two genes are needed, both from the soil bacterium *Bacillus amyloliquefaciens*. One gene encodes *barnase*, an RNase, that is secreted from the bacterium as a defense mechanism against other organisms. The other gene encodes *barnstar*,

Figure 23.14

Production of hybrid seed using cytoplasmic male sterility (CMS) and a *restorer of fertility* gene. This organism has a single chloroplast and a pair of flagella.

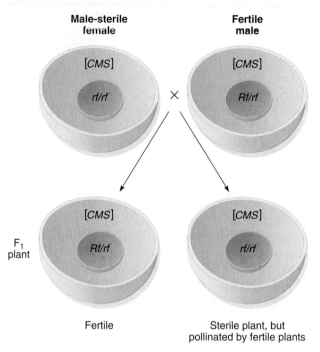

a protein that binds to and inhibits barnase, thereby protecting the bacterium from its own enzyme. A wild-type plant is transformed with the barnase gene fused to the TA29 promoter. The TA29 promoter is from a gene that is expressed only in the tapetum, a tissue that surrounds the pollen sac that is essential for pollen production. In transgenic plants with the TA29-*barnase* gene, barnase is made in the tapetum, and destroys the RNAs in that tissue, making the plant male sterile. This plant is used as the female parent in a cross with a transgenic plant containing a TA29-*barnstar* gene. The seeds from the female produce hybrid plants in which both the barnase and barnstar proteins are produced in the tapetum. The barnstar binds to the barnase, inhibiting the RNase activity, thereby preventing male sterility. The hybrids are male fertile.

Exceptions to Maternal Inheritance

Strict maternal inheritance has been considered to be the case for extranuclear mutations in animals and plants where the female gamete contributes the majority of the cytoplasm to the zygote. However, exceptions are coming to light. The following are some examples:

   1. Exploiting DNA sequence differences between mtDNAs of two inbred lines of mice, researchers have used PCR (polymerase chain reaction; see Chapter 8, pp. 200–202) to demonstrate that paternally inherited

mtDNA molecules are present at a frequency of 10^{-4} relative to maternal mtDNA molecules. This heteroplasmy of paternal and maternal mitochondria has potentially significant evolutionary implications. That is, it has been generally considered that maternal paternal mtDNA remain distinct because of the strict maternal inheritance of mitochondria. However, if heteroplasmy can occur, then there is a likelihood of genetic recombination between maternally derived and paternally derived mtDNA molecules. Such recombination will lead to significant diversity of mtDNA in an individual. The extent to which this phenomenon occurs is unknown, but the fact that it exists at all makes it necessary to be cautious about conclusions made using a purely maternal inheritance of mtDNA.

2. In most angiosperms (flowering plants), the plastids are inherited only from the maternal parent. In some angiosperms, however, plastids are inherited at high frequency from both parents or mostly from the paternal parent. For example, biparental inheritance of plastids is seen in the evening primrose, *Oenothera*. Paternal inheritance of plastids is the rule in conifers, which are gymnosperms.

Infectious Heredity: Killer Yeast

There are examples of eukaryotic extranuclear inheritance that result from the presence of cytoplasmic bacteria or viruses symbiotically coexisting with the eukaryotes. One example is the killer phenomenon in yeast, in which some strains secrete a toxin that kills sensitive yeast strains. (Killer strains are immune to their own toxin.) The killer phenomenon results from the presence in the cell's cytoplasm of two types of viruses, L and M (Figure 23.15a). Neither has adverse effects on the host cell.

The L virus consists of a protein capsid surrounding a 4.6-kb double-stranded (ds) RNA genome, L-dsRNA. L-dsRNA encodes the capsid proteins for both L and M viruses and the viral polymerase required for viral RNA replication. Because all viral particles are encoded by an L-dsRNA, M viruses are found in cells only if L viruses are also present. The M virus consists of a virus particle surrounding two identical copies of a 1.8-kb double-stranded RNA genome, M-dsRNA. M-dsRNA encodes a killer toxin protein, which is secreted from the cell. The same protein confers immunity on the killer cell.

Sensitive yeast cells can be killed by the M-encoded killer toxin. There are two types of sensitive cells (Figure 23.15b): One type has only L viruses, and the other has neither L nor M viruses. In both types, no immunity is produced because killer toxin is not made.

Unlike most viruses, the yeast L and M viruses are not found outside the cell, so sensitive yeast cells cannot be infected by viruses that invade from outside. Rather, virus transmission from yeast to yeast occurs whenever yeast cells mate. All progeny of the mating will inherit

Figure 23.15

The killer phenomenon in yeast. (a) Killer yeast contain two virus types, L and M, each of which contains a double-stranded RNA genome. L-dsRNA encodes both virus particles and the enzyme required for L and M virus replication. M-dsRNA encodes the killer toxin. (b) Sensitive yeast, which can be killed by killer toxin, either have L viruses but no M viruses or have neither virus type.

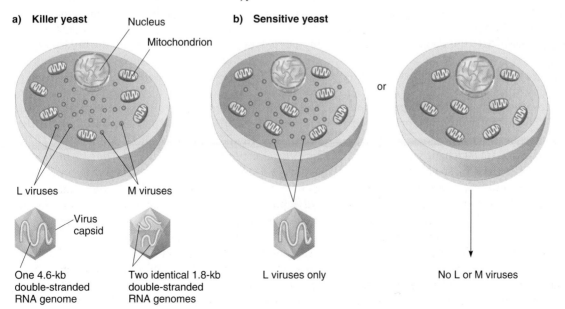

copies of the viruses in the parental cells, illustrating an infectious mechanism of cytoplasmic inheritance.

KEYNOTE

> Not all cases of non-Mendelian inheritance result from genes on mtDNA or cpDNA. Many other examples in eukaryotes result from infectious heredity in which symbiotic, cytoplasmically located bacteria or viruses are transmitted when cytoplasms mix. The killer phenomenon in yeast is an example: It results from the infectious heredity of cytoplasmically located viruses.

Contrasts to Non-Mendelian Inheritance

Maternal Effect

The maternal inheritance pattern of extranuclear genes is distinct from the phenomenon of **maternal effect,** which is defined as the phenotype of an individual that is established by the maternal nuclear genome, as the result of mRNA or proteins that are deposited in the oocyte before fertilization. These inclusions direct early development of the embryo. That is, in maternal inheritance the progeny always have the maternal phenotype, whereas in maternal effect the progeny always have the phenotype specified by the maternal nuclear genotype. Maternal effect does not involve any extranuclear genes and is discussed here to make the distinction from extranuclear inheritance clear.

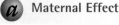
animation
a **Maternal Effect**

Maternal effect is seen in the inheritance of the coiling direction in the shell of the snail *Limnaea peregra*, (Figure 23.16). The shell-coiling trait is determined by a single pair of nuclear alleles: the dominant *D* allele for coiling to the right (dextral coiling) and the recessive *d* allele for coiling to the left (sinistral coiling). The shell-coiling phenotype is always determined by the genotype of the mother. The latter is shown by the results of reciprocal crosses between a true-breeding, dextral-coiling and a sinistral-coiling snail (Figure 23.16). All the F$_1$ snails have the same genotype because a nuclear gene is involved, yet the phenotype is different for the reciprocal crosses.

In the cross of a dextral (*D/D*) female with a sinistral (*d/d*) male (Figure 23.17a), the F$_1$ snails are all *D/d* in genotype and dextral in phenotype. Selfing the F$_1$ produces F$_2$ snails with a 1:2:1 ratio of *D/D*, *D/d*, and *d/d* genotypes. All the F$_2$ snails are dextral, even the *d/d* snails whose genotype seems to indicate sinistral phenotype. Here is our first encounter with maternal effect; the *d/d* snails have a coiling phenotype specified not by the genotype they have but by the genotype of their mother (*D/d*). Selfing the F$_2$ snails gives F$_3$ progeny, ¾ of which are

Figure 23.16

The snail *Limnaea peregra*.

dextral and ¼ of which are sinistral. The latter are the *d/d* progeny of the F$_2$ *d/d* snails; these F$_3$ snails are sinistral because their phenotype reflects their mother's (the F$_2$) genotype.

Similar results are seen in the reciprocal cross of a sinistral (*d/d*) female with a dextral (*D/D*) male (Figure 23.17b). The F$_1$ snails are all *D/d* in genotype, yet they are sinistral in phenotype because the mother is genotypically *d/d*. Selfing the F$_1$ produces F$_2$ snails, all of which are dextral for the same reason as the reciprocal cross just described. The genotypes and phenotypes of the F$_2$ and F$_3$ generations are the same as for the reciprocal cross (see Figure 23.17b), again for the same reasons.

These results do not fit our criteria for non-Mendelian inheritance. That is, if the coil direction phenotype were controlled by an extranuclear gene, the progeny would always exhibit the phenotype of the mother because of maternal inheritance. Here, the coiling phenotype is governed directly by the nuclear genotype of the mother and is an example of maternal effect. But what is the basis for the coiling? The orientation of the mitotic spindle in the first mitotic division after fertilization controls the direction of coiling. The mother encodes products, deposited in the oocyte, that direct the orientation of the mitotic spindle and therefore the direction of cell cleavage. Thus, a mother of genotype *D/–* deposits gene products that specify a dextral (right-handed) coiling, and a mother of genotype *d/d* deposits gene products that specify a sinistral (left-handed) coiling.

Support for this hypothesis has come from the experiments of G. Freeman and J. Lundelius. They injected cytoplasm from dextrally coiling snails into the eggs of *d/d* mothers (who would normally specify sinistrally coiling progeny) and found that the resulting

Figure 23.17

Inheritance of the direction of shell coiling in the snail *Limnaea peregra* is an example of maternal effect. (a) Cross between true-breeding dextral-coiling female (*D/D*) and sinistral-coiling male (*d/d*). (b) Cross between sinistral-coiling female (*d/d*) and true-breeding dextral-coiling male (*D/D*).

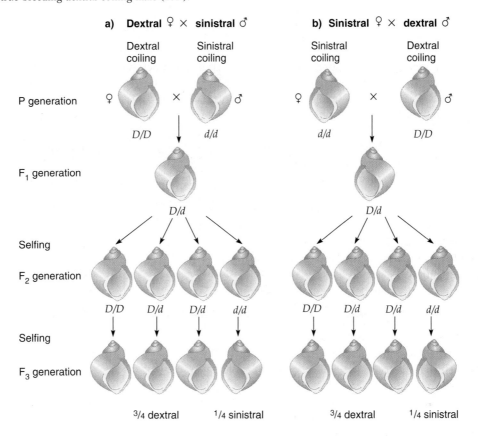

embryos coiled dextrally. In the reciprocal experiment, cytoplasm from sinistrally coiling snails injected into *D/–* mothers had no effect on the resulting embryos; that is, they were still dextrally coiling. The interpretation is that (1) the *D* allele specifies a product that is deposited in the cytoplasm of the egg and causes the next-generation embryos to coil dextrally; and (2) the *d* allele produces either a defective product or none at all, so that *d/d* snails produce embryos that coil in the sinistral direction by default.

KEYNOTE

> Maternal effect is different from non-Mendelian inheritance. The maternal inheritance pattern of extranuclear genes occurs because the zygote receives most of its organelles (containing the extranuclear genes) from the female parent, whereas in the maternal effect the trait inherited is controlled by the maternal *nuclear* genotype before the egg is fertilized and does not involve extranuclear genes.

Summary

Both mitochondria and chloroplasts contain DNA, the length of which varies from organism to organism. The genomes of both organelles usually are circular, double-stranded DNAs without any associated histone-like proteins. The genomes in mitochondria and chloroplasts contain genes that are not duplicated in the nuclear genome; therefore, the organelle genes are contributing different information for the function of the cell. The organellar genes encode the rRNA components of the ribosomes (which are assembled and function in the organelles) and many (if not all) of the tRNAs used in organellar protein synthesis. Some of the organelle proteins are encoded by the organelle; the remainder are nuclear encoded. Nuclear-encoded proteins found in organelles are synthesized on cytoplasmic ribosomes, then imported into the appropriate organelles.

The existence of polymorphic sequences in mtDNA and the fact that humans and many other organisms exhibit maternal inheritance of mitochondria mean that maternal lineages are almost unique. Thus, the genetic relationships between individuals can be investigated readily by mtDNA analysis. This method is used in many areas of biology, including conservation biology and paleoanthropology.

There are many examples of mitochondrial and chloroplast mutants. The inheritance of these extranuclear genes follows rules different from those for nuclear genes, and this is how extranuclear genes were originally identified. For extranuclear genes, meiosis-based Mendelian segregation is not involved, they show uniparental (and usually maternal) inheritance, the trait involved persists after nuclear substitution, and they cannot be mapped to nuclear chromosomes. Not all cases of extranuclear inheritance result from genes on mtDNA or cpDNA. Other examples result from infectious heredity, in which cytoplasmically located bacteria or viruses are transmitted when cytoplasms mix.

Maternal effect is the phenotype in an individual that is established by the maternal nuclear genome, as the result of mRNA or proteins that are deposited in the oocyte before fertilization. These inclusions direct early development of the embryo. Maternal effect is different from extranuclear inheritance. The maternal inheritance pattern of extranuclear genes occurs because the zygote receives most of its organelles (containing the extranuclear genes) from the female parent, whereas in maternal effect the trait inherited is controlled by the maternal nuclear genotype before fertilization of the egg and does not involve extranuclear genes.

Analytical Approaches to Solving Genetics Problems

Q23.1 In *Neurospora*, strains of [*poky*] have been isolated that have reverted to nearly wild-type growth, although they still retain the abnormal metabolism and cytochrome pattern characteristic of [*poky*]. These strains are called *fast*-[*poky*]. A cross of a *fast*-[*poky*] female parent with a wild-type male parent gives a 1:1 segregation of [*poky*] : *fast*-[*poky*] ascospores in all asci. Interpret these results, and predict the results you would expect from the reciprocal of the stated cross.

A23.1 The [*poky*] mutation shows maternal inheritance, so when it is used as a female parent, all the progeny ascospores will carry [*poky*]. We can designate the cross as [*poky*] × [*N*], where *N* signifies normal cytoplasm. All progeny are [*poky*]. The asci show a 1:1 segregation for [*poky*] : *fast*-[*poky*] phenotypes. This ratio is characteristic of a nuclear gene segregating in the cross. Therefore, the simplest explanation is that the factor that causes [*poky*] strains to be *fast*-[*poky*] is a nuclear gene

mutation; we can call it *F* (its actual designated name). The *F* gene segregates in meiosis, as do all nuclear genes. The cross can be rewritten as [*poky*]*F* ♀ × [*N*]+ ♂, which gives a 1:1 segregation of [*poky*]+ : [*poky*]*F*. The former is [*poky*] and the latter is *fast*-[*poky*].

With these results behind us, the reciprocal cross can be diagrammed as [*N*]+ ♀ × [*poky*]*F* ♂. All the progeny spores from this cross are [*N*], half of them being *F* and half of them being +. If *F* has no effect on normal cytoplasm, then these two classes of spores would be phenotypically indistinguishable, which is the case.

Q23.2 Four slow-growing mutant strains of *Neurospora crassa*, coded *a*, *b*, *c*, and *d*, have been isolated. All have an abnormal system of respiratory mitochondrial enzymes. The inheritance patterns of these mutants were tested in controlled crosses with the wild type, with the following results:

Female Parent		Male Parent	Progeny (Ascospores)	
			Wild Type	Slow Growing
Wild type	×	*a*	847	0
a	×	Wild type	0	659
Wild type	×	*b*	1,113	0
b	×	Wild type	0	2,071
Wild type	×	*c*	596	590
Wild type	×	*d*	1,050	1,035

Give a genetic interpretation of these results.

A23.2 This question asks us to consider the expected transmission patterns for nuclear and extranuclear genes. The nuclear genes will have a 1:1 segregation in the offspring because this organism is a haploid organism and therefore should exhibit no differences in the segregation patterns, whichever strain is the maternal parent. On the other hand, a distinguishing characteristic of extranuclear genes is a difference in the results of reciprocal crosses. In *Neurospora*, this characteristic usually is manifested by all progeny having the phenotype of the maternal parent. With these ideas in mind, we can analyze each mutant in turn.

Mutant *a* shows a clear difference in its segregation in reciprocal crosses and is a classic case of maternal inheritance. The interpretation here is that the gene is extranuclear; therefore, the gene must be in the mitochondrion. The [*poky*] mutant described in this chapter shows this type of inheritance pattern.

By the same reasoning, the mutation in strain *b* must also be extranuclear.

Mutants *c* and *d* segregate 1:1, indicating that the mutations involved are in the nuclear genome. In these cases, we need not consider the reciprocal cross because there is no evidence of maternal inheritance. In fact, the actual mutations that are the basis for this question cause sterility, so the reciprocal cross cannot be done. We can confirm that the mutations are in the nuclear genome by doing mapping experiments using known nuclear markers. Evidence of linkage to such markers would confirm that the mutations are not extranuclear.

Questions and Problems

*23.1 The endosymbiont theory provides an explanation for the origin of mitochondria and chloroplasts.
  a. What are the tenets of this theory?
  b. Why are these organelles no longer able to survive as independent organisms?
  c. What role did plant photosynthetic activity and changes in the Earth's atmosphere play in the evolution of mitochondria?
  d. Which attributes of fundamental cellular processes, such as DNA packaging and translation, provide evidence in support of this theory?

23.2 The nuclear, mitochondrial, and chloroplast genomes are different in size, structure, gene content, and gene organization.
  a. Compare the structure of the nuclear genome, the mitochondrial genome, and the chloroplast genome.
  b. How does the gene content and organization of the mitochondrial genome compare to that of the chloroplast genome?
  c. The mitochondrial genome varies widely in size, ranging from less than 20 kb to more than 2,000 kb. What accounts for this variation?

23.3 In what ways is DNA replication in animal mitochondrial genomes different from nuclear DNA replication?

*23.4 Imagine that you have discovered a new genus of yeast. In your studies on this organism, you isolate DNA and subject it to CsCl density gradient centrifugation. In addition to a major peak, you observe a minor peak of lighter density. How could you determine whether the minor peak represents organellar (presumably mitochondrial) DNA as opposed to an AT-rich repeated sequence in the nuclear genome?

23.5 What genes are present in the human mitochondrial genome, and how was this determined?

*23.6 How is transcription and processing of transcripts in animal mitochondria different from that in eukaryotic nuclei or prokaryotes?

23.7 What conclusions can you draw from the fact that most nuclear-encoded mRNAs and all mitochondrial mRNAs have a poly(A) tail at the 3′ end?

*23.8 A substantial body of evidence indicates that defects in mitochondrial energy production may contribute to the neuronal cell death seen in a number of late-onset neurodegenerative diseases, including Alzheimer disease, Parkinson disease, Huntington disease, and amyotrophic lateral sclerosis (ALS, or Lou Gehrig disease). Some of these diseases have been associated with mutations in the nuclear genome. One experimental system that has been developed to evaluate the contributions of the mitochondrial genome to these diseases uses a cytoplasmic hybrid known as a cybid. Cybids are made by repopulating a tissue culture cell line that has been made mitochondria deficient with mitochondria from the cytoplasm of a human platelet cell. The cybids thus have nuclear DNA from the tissue culture cell and mitochondrial DNA from the human platelet cell.

The mitochondrial protein cytochrome oxidase has subunits encoded by both nuclear and mitochondrial genes. Patients with Alzheimer disease have been reported to have lower levels of cytochrome oxidase than do age-matched controls.
  a. What is the experimental evidence that cytochrome oxidase has subunits encoded by both nuclear and mitochondrial genes?
  b. Given the means to assay cytochrome oxidase activity, how would you investigate whether the decreased levels of cytochrome oxidase activity in patients with Alzheimer disease could be ascribed to nuclear or mitochondrial genetic defects? What controls would you create?
  c. If you can demonstrate that the mitochondrial contribution to cytochrome oxidase is responsible for lowered cytochrome oxidase activity, can you conclude that each mitochondrion of an affected individual has an identical defect?

*23.9 Discuss the differences between the universal genetic code of the nuclear genes of most eukaryotes and the code found in human mitochondria. Is there any advantage to the mitochondrial code?

23.10 Compare the cytoplasmic, mitochondrial, and chloroplast protein-synthesizing systems.

*23.11 Rubisco constitutes about half of the protein found in green plant tissue, making it the most common protein on Earth. It consists of eight identical large subunits and eight identical small subunits.
  a. What is the function of rubisco?
  b. Without resorting to DNA sequencing, how would you experimentally determine that the large subunit

gene is in the chloroplast genome and that the small subunit gene is in the nuclear genome?

c. A nonsense mutation occurs near the 5′ end of the protein-coding region in a single copy of the gene for the small subunit. What genetic and phenotypic properties do you expect for this mutation?

d. A nonsense mutation occurs near the 5′ end of the protein-coding region in a single copy of the gene for the large subunit. What genetic and phenotypic properties do you expect for this mutation?

*23.12 What features of non-Mendelian inheritance distinguish it from the inheritance of nuclear genes?

23.13 Distinguish between maternal effect and non-Mendelian inheritance.

23.14 Reciprocal crosses between two types of the evening primrose, *Oenothera hookeri* and *Oenothera muricata*, produce the following effects on the plastids:

O. hookeri female × *O. muricata* male → Yellow plastids
O. muricata female × *O. hookeri* male → Green plastids

Explain the difference between these results, noting that the chromosome constitution is the same in both types.

*23.15 A series of crosses are performed with a recessive mutation in *Drosophila* called *tudor*. Homozygous *tudor* animals appear normal and can be generated from the cross of two heterozygotes, but a true-breeding *tudor* strain cannot be maintained. When homozygous *tudor* males are crossed to homozygous *tudor* females, both of which appear to be phenotypically normal, a normal-appearing F_1 is produced. However, when F_1 males are crossed to wild-type females, or when F_1 females are crossed to wild-type males, no progeny are produced. The same results are seen in the F_1 progeny of homozygous *tudor* females crossed to wild-type males. The F_1 progeny of homozygous *tudor* males crossed to wild-type females appear normal, and they are capable of issuing progeny when mated either with each other or with wild-type animals.

a. How would you classify the *tudor* mutation? Why?
b. What might cause the *tudor* phenotype?

*23.16 A form of male sterility in corn is maternally inherited. Plants of a male-sterile line crossed with normal pollen give male-sterile plants. Some lines of corn carry a dominant, so-called restorer (*Rf*) gene that restores pollen fertility in male-sterile lines.

a. If a male-sterile plant is crossed with pollen from a plant homozygous for gene *Rf*, what will be the genotype and phenotype of the F_1?
b. If the F_1 plants of (a) are used as females in a test-cross with pollen from a normal plant (*rf/rf*), what

would be the result? Give genotypes and phenotypes and designate the type of cytoplasm.

*23.17 In *Neurospora*, a chromosomal gene *F* suppresses the slow-growth characteristic of the [*poky*] phenotype and makes a [*poky*] culture into a *fast*-[*poky*] culture, which still has abnormal cytochromes. Gene *F* in combination with normal cytoplasm has no detectable effect. (Hint: Because both nuclear and extranuclear genes must be considered, it is convenient to use symbols to distinguish the two. Thus, cytoplasmic genes are designated in square brackets, e.g., [N] for normal cytoplasm, [*poky*] for *poky*.)

a. A cross in which *fast*-[*poky*] is used as the female (protoperithecial) parent and a normal wild-type strain is used as the male parent gives half [*poky*] and half *fast*-[*poky*] progeny ascospores. What is the genetic interpretation of these results?
b. What would be the result of the reciprocal cross of the cross described in (a), that is, normal female × *fast*-[*poky*] male?

23.18 Distinguish between *nuclear (segregational)*, *neutral*, and *suppressive petite* mutants of yeast.

*23.19 In yeast, a haploid *nuclear (segregational) petite* is crossed with a *neutral petite*. Assuming that both strains have no other abnormal phenotypes, what proportion of the progeny ascospores are expected to be *petite* in phenotype if the diploid zygote undergoes meiosis?

23.20 When grown on a medium containing acriflavine, a yeast culture produces a large number of very small (*tiny*) cells that grow very slowly. How would you determine whether the slow-growth phenotype was the result of a cytoplasmic factor or a nuclear gene?

23.21 Mating in *Chlamydomonas* is syngamous with cells of each mating type (mt^+, mt^-) contributing an equal amount of cytoplasm. Erythromycin resistance ([ery^r]) is a chloroplast trait. Streptomycin resistance can be a nuclear (str^r) or cytoplasmic trait ([str^r]).

a. For each of the following crosses, provide the expected zygotic phenotypes (erythromycin resistant or sensitive, streptomycin resistant or sensitive) and their approximate proportions:

   i. mt^+ [ery^r] × mt^- [ery^s]
   ii. mt^- [ery^r] × mt^+ [ery^s]
  iii. mt^+ str^r [str^s] × mt^- str^s [str^s]
  iv. mt^+ str^s [str^r] × mt^- str^s [str^s]
   v. mt^+ str^r [str^r] × mt^- str^s [str^s]
  vi. mt^- str^r [str^s] × mt^+ str^s [str^s]
 vii. mt^- str^s [str^r] × mt^+ str^s [str^s]
viii. mt^- str^r [str^r] × mt^+ str^s [str^s]

b. Suppose that in cross (ii), about 5 percent of the zygotes are cytohets. What phenotype will these cells have, and what phenotypes do you expect to observe as these cells undergo successive mitotic divisions?

23.22 Several investigators have demonstrated that chemical and environmental treatments of plants and animals can lead to abnormalities that persist for several generations before disappearing. For example, Hoffman found that treating the bean *Phaseolus vulgaris* with chloral hydrate led to abnormalities in leaf shape that persisted in the female (but not in the male) line for almost six generations before disappearing.
a. In what different ways could you explain the origin of these abnormalities, and their disappearance after several generations?
b. What broader implications might these findings have?

23.23 *Drosophila melanogaster* has a sex-linked, recessive, mutant gene called *maroon-like (ma-l)*. Homozygous *ma-l* females or hemizygous *ma-l* males have light-colored eyes because of the absence of the active enzyme xanthine dehydrogenase, which is involved in the synthesis of eye pigments. When heterozygous *ma-l+/ma-l* females are crossed with *ma-l* males, all the offspring are phenotypically wild type. However, half the female offspring from this cross, when crossed back to *ma-l* males, give all *ma-l* progeny. The other half of the females, when crossed to *ma-l* males, give all phenotypically wild-type progeny. What is the explanation for these results?

*23.24 When females of a particular mutant strain of *Drosophila melanogaster* are crossed to wild-type males, all the viable progeny flies are females. Hypothetically, this result could be the consequence of either a sex-linked, male-specific lethal mutation or a maternally inherited factor that is lethal to males. What crosses would you perform to distinguish between these alternatives?

23.25 Reciprocal crosses between two *Drosophila* species, *D. melanogaster* and *D. simulans*, produce the following results:

melanogaster ♀ × *simulans* ♂ → Females only
simulans ♀ × *melanogaster* ♂ → Males, with few or no females

Propose a possible explanation for these results.

23.26 Some *Drosophila* are very sensitive to carbon dioxide; administering it to them anesthetizes them. The sensitive flies have a cytoplasmic particle called sigma that has many properties of a virus. Resistant flies lack sigma. The sensitivity to carbon dioxide shows strictly maternal inheritance. What would be the outcome of the following two crosses: (a) sensitive ♀ × resistant ♂ and (b) sensitive ♂ × resistant ♀?

*23.27 A few years ago, Chile allowed its government agents to kidnap, torture, and kill many young adults in opposition to the regime in control. The children of abducted young women were often taken and given to government supporters to raise as their own. Now that the political situation has changed, grandparents of these stolen children are trying to locate and reclaim them as their legitimate grandchildren. Imagine that you are the judge in a trial centering on the custody of a child. Mr. and Mrs. Escobar believe Carlos Mendoza is the son of their abducted, murdered daughter. If this is true, then Mr. and Mrs. Sanchez are the paternal grandparents of the child because their son (also abducted and murdered) was the husband of the Escobars' daughter. Mr. and Mrs. Mendoza claim that Carlos is their natural child. The attorney for the Escobar and Sanchez couples informs you that scientists have discovered a series of RFLPs in human mitochondrial DNA. He tells you that his clients are eager to be tested and asks that you order that Mr. and Mrs. Mendoza and Carlos also be tested.
a. Can mitochondrial RFLP data be helpful in this case? In what way?
b. Does the mitochondrial DNA of all 7 parties need to be tested to resolve the case? If not, whose mitochondrial DNA actually needs to be tested in this case? Explain your choices.
c. Assume that the mitochondrial DNA of critical people have been tested and you have received the results. How would the results resolve the question of Carlos's parentage?

23.28 The analysis of mitochondrial DNA has been very useful in assessing the history of specific human populations. For example, a 9-bp deletion in the small intergenic region for cytochrome oxidase subunit II and tRNA.Lys (see Figure 23.3) has been a very informative marker to trace the origins of Polynesians. The deletion is widely distributed across Southeast Asia and the Pacific and is present in 80 to 100 percent of individuals in the different populations within Polynesia. One of the most polymorphic regions of the mitochondrial genome is found in the intergenic region for tRNA.Phe and tRNA.Thr. In Asians with the 9-bp deletion, a specific set of DNA sequence polymorphisms in this region is found. Using the 9-bp deletion and the DNA sequence polymorphisms as markers, comparative analysis of Asian populations has found a genetic trail of mitochondrial DNA variation. The trail begins in Taiwan, winds through the Philippines and Indonesia, proceeds along the coast of New Guinea, and then moves into Polynesia. Based on an estimated rate of mutation in the tRNA.Phe to tRNA.Thr region, this expansion of mitochondrial DNA variants is thought to be about 6,000 years old. This is consistent with linguistic and archeological evidence that associates Polynesian origins with the spread of the Austronesian language family out of Taiwan between 6,000 and 8,000 years ago.

a. Why are these types of mitochondrial DNA polymorphisms such good markers for tracing human migration patterns?

b. Why is it important to correlate findings from mitochondrial DNA polymorphisms with other (non-DNA) assessment methods?

c. Why might sequences in the tRNA.Phe to tRNA.Thr region be more polymorphic than other sequences in the mitochondrial genome?

d. The 9-bp deletion has also been found in human populations in Africa. What different explanations are possible for this, and how might these explanations be evaluated?

23.29 The pedigree in the following figure shows a family in which a rare inherited disease called Leber hereditary optic atrophy is segregating. This condition causes blindness in adulthood. Studies have recently shown that the mutant gene causing Leber hereditary optic atrophy is located in the mitochondrial genome.

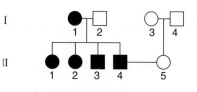

a. What other modes of inheritance (e.g., autosomal dominant, X-linked recessive) are consistent with the inheritance of this rare disease? How could you provide evidence that this disease is not inherited using these modes?

b. Assuming II-5 is normal, what proportion of the offspring of II-4 and II-5 are expected to inherit Leber's hereditary optic atrophy?

c. Assuming that II-2 marries a normal male, what proportion of their sons should be affected? What proportion of their daughters should be affected?

*23.30 The inheritance of shell-coiling direction in the snail *Limnaea peregra* has been studied extensively. A snail produced by a cross between two individuals has a shell with a right-hand twist (dextral coiling). This snail produces only left-hand (sinistral) progeny on selfing. What are the genotypes of the F_1 snail and its parents?

24

Population Genetics

A Darwin's finch.

PRINCIPAL POINTS

- Population genetics seeks to understand the causes of observed patterns of genetic variation within and among populations. In so doing, it seeks to explain the underlying genetic basis for evolutionary change. It includes both empirical studies that measure and quantify genetic variation and theoretical studies that use mathematical models to explain observed patterns of variation in terms of the evolutionary processes that can change gene frequencies.

- The genetic structure of a population is described by the number and frequencies of genotypes at each locus. The total of all alleles at one time (the gene pool) determines the genetic information carried forward from one generation to the next.

- The Hardy-Weinberg law states that in a large, randomly mating population, free from evolutionary forces, the allelic frequencies do not change and the genotypic frequencies stabilize after one generation. In the case of two alleles, A and a, with frequencies p and q, the genotypic proportions at equilibrium are p^2, $2pq$, and q^2.

- The Hardy-Weinberg principle and all other population genetic principles apply to any alleles that are inherited according to Mendelian rules, including segregating factors that influence phenotypes (such as Mendel observed), protein variants, or any of a variety of differences at the DNA level, including single nucleotide polymorphisms (SNPs), insertions, or deletions.

- The genetic structure of a species can vary both geographically and temporally.

- The classical and neutral mutation models generate testable hypotheses and are used to explain how much genetic variation should exist within natural populations and what processes are responsible for the observed variation.

- Mutation, genetic drift, migration, and natural selection are processes that can alter allelic frequencies of a population.

- For most loci, recurrent mutation changes allelic frequencies at such a slow rate that its effects are negligible. Although mutation is the initial source of all variation, other forces predominate in determining its changes in frequency once it has been introduced.

657

- Genetic drift is random change in allelic frequencies that arises from random sampling of gametes that occurs in each generation. Genetic drift produces genetic change within populations, genetic differentiation among populations, and loss of genetic variation within populations.

- Migration, also called gene flow, involves movement of alleles among populations. Migration can alter the allelic frequencies of a population, and it tends to reduce genetic divergence among populations.

- Natural selection is differential reproduction of genotypes. It is measured by Darwinian fitness, which is the relative reproductive success of genotypes. Depending on which genotypes are favored, natural selection can produce a number of different effects on the gene pool of a population.

- Inbreeding and assortative mating affect lead to an increase in homozygosity without affecting the allele frequencies. Negative assortative mating may affect both the allele and genotype frequencies.

- Principles of population genetics can be applied to the management of rare and endangered species. Genetic diversity is best maintained by establishing a population with adequate founders, expanding the population rapidly, avoiding inbreeding, and maintaining an equal sex ratio and equal family size.

iActivity

SOON AFTER MENDEL'S PRINCIPLES WERE rediscovered, geneticists began to look not only at the genetic makeup of individuals but also the genetic makeup of populations. Population genetics is one way that scientists determine whether evolution is occurring in groups of individuals and identify the forces that cause populations to evolve. In this chapter, you will learn about changes in the genetic makeup of populations, how such changes are measured, and the factors that cause these changes. Then, in the iActivity, you will explore the genetics of a type of mussel that is rapidly spreading through North American waterways.

The science of genetics can be broadly divided into four major subdisciplines: transmission genetics, molecular genetics, population genetics, and quantitative genetics. Each of these four areas focuses on a different aspect of heredity. **Transmission genetics** is concerned primarily with genetic processes that occur within individuals and how genes are passed from one individual to another. Thus, the unit of study for transmission genetics is the *individual*. In **molecular genetics,** we are interested largely in the molecular nature of heredity: how genetic information is encoded within the DNA and how biochemical processes of the cell translate the genetic infor-

mation into influencing the phenotype. Consequently, in molecular genetics we focus on the *cell*. **Population genetics,** the subject of this chapter, applies the principles of transmission genetics to large groups of individuals, focusing on the transmission processes at one or a few genetic loci. **Quantitative genetics,** the subject of Chapter 14, also considers the transmission of traits in large groups of individuals, but the traits of concern are simultaneously determined by many genes. Both population and quantitative genetics apply Mendelian principles to groups of organisms, and they are amenable to mathematical treatment. In fact, these areas provide some of the oldest and richest examples of the success of mathematical theory in biology. The impetus for the development of these areas came after the rediscovery of Mendel's work and its great implications for Darwinian theory. In fact, the fusion of Mendelian theory with Darwinian theory is called the neo-Darwinian synthesis and was championed by Sir Ronald Fisher, Sewall Wright, and J. B. S. Haldane (Figure 24.1). The neo-Darwinian synthesis is now the foundation of a large part of modern biology.

Population geneticists investigate the patterns of genetic variation found among individuals within groups (the **genetic structure of populations**) and how these patterns vary geographically and change over time. In this discipline, our perspective shifts away from the individual and the cell and focuses instead on a large group of individuals, a **Mendelian population.** A Mendelian population is a group of *interbreeding* individuals who share a common set of genes. The genes shared by the individuals of a Mendelian population are called the **gene pool.** To understand the genetics of the evolutionary process, we study the gene pool of a Mendelian population rather than the genotypes of its individual members. An understanding of the genetic structure of a population is also a key to our understanding of the importance of genetic resources and the importance of genes for the conservation of species and biodiversity.

The advent of rapid and inexpensive methods for DNA sequencing has resulted in an explosion in the quantity of data showing genetic variation within populations at the DNA sequence level. Since this is the form of genetic variation at its most fundamental level, our abilities to discern the forces that act on this variation have increased dramatically in recent years. These data have also opened up new possibilities for the kinds of questions that population genetics can address. By studying mitochondrial DNA sequence differences, for example, we get a picture of the female lineages of the species, including relative amounts of movement, times of origin of population groups, and periods of population expansion. Similarly, DNA sequence variation in the Y chromosome reveals patterns of past movements of males of a species.

Figure 24.1

(a) Sir Ronald Fisher, (b) Sewall Wright, and (c) J. B. S. Haldane are considered the major architects of neo-Darwinian theory.

a)   b)   c)

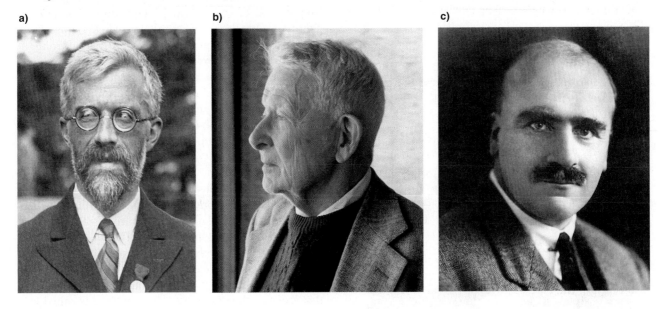

Questions frequently studied by population geneticists include the following:

1. How much genetic variation is found in natural populations, and what processes control the amount of variation observed?
2. What processes are responsible for producing genetic divergence among populations?
3. How do biological characteristics of a population, such as mating system, fecundity, and age structure, influence the genetic structure of the population?

To answer these questions, population geneticists sometimes make direct measurements of genetic variability within and among populations. Often they also develop mathematical models to describe how the gene pool of a population will change under various conditions. An example is the set of equations that describes the influence of random mating on the allele and genotypic frequencies of an infinitely large population, a model called the **Hardy-Weinberg law,** which we discuss later in this chapter. It is important to note that, while the models are simple and require numerous assumptions, many of which seem unrealistic, such models are useful because they strip a process to its essence and allow scientists to test particular attributes of a system in isolation. With such models we can examine what happens to the genetic structure of a population when we deliberately violate one assumption after another and then in combination. Once we understand the results of the simple models, we can incorporate more realistic conditions into the equations, and we can use these more realistic models to help us understand the patterns in data collected from natural populations. In the end, we will see that many attributes of genetic variation in populations can be explained by surprisingly simple models.

KEYNOTE

Population genetics seeks to understand the underlying causes of the observed patterns of genetic variation within and among populations. It uses both empirical tools, measuring variation in natural populations, and theoretical tools, which attempt to explain the observed variation with quantitative modeling.

Genetic Structure of Populations

Genotype Frequencies

To study the genetic structure of a Mendelian population, population geneticists must first describe it quantitatively. They do this by calculating genotype frequencies and allele frequencies within the population. A frequency is a proportion, and it always ranges between 0 and 1. If 43 percent of the people in a group have red hair, the frequency of red hair in the group is 0.43, and the frequency of people who *don't* have red hair is

$$1 - 0.43 = 0.57 \text{ or } 57\%.$$

To calculate the **genotype frequencies** at a specific locus, we count the number of individuals with one

Figure 24.2

Figure 24.2

***Panaxia dominula*, the scarlet tiger moth.** The top two moths are normal homozygotes (*BB*), those in the middle two rows are heterozygotes (*Bb*), and the bottom moth is the rare homozygote (*bb*).

geneticists use frequencies of alleles to describe how the gene pool changes over time. The use of allelic frequencies offers several advantages over the genotypic frequencies. First, in sexually reproducing organisms, genotypes break down to alleles when gametes are formed, and alleles, not genotypes, are passed from one generation to the next. Consequently, only alleles have continuity over time, and the gene pool evolves when allele frequencies change. Furthermore, there are always fewer alleles than genotypes, so the population can be described with fewer parameters when allele frequencies can be used. For example, if there are three alleles segregating at a particular locus, there are six genotypes.

Allele frequencies can be calculated in two equivalent ways: from the observed numbers of different genotypes at a particular locus or from the genotype frequencies. First, we can calculate the allele frequencies directly from the *numbers* of genotypes. In this method, we count the number of alleles of one type at a particular locus and divide it by the total number of alleles at that locus in the population. This method is called *gene counting* and works for a wide variety of cases, including X-linked genes and mitochondrial genes. Expressing the gene counting method as a formula for a nuclear gene with two alleles, we get

$$\text{Allele frequency} = \frac{\text{Number of copies of a given allele}}{\begin{array}{c}\text{Sum of counts of all}\\\text{alleles in the population}\end{array}}$$

For example, imagine a population of 1,000 diploid individuals with 353 *AA*, 494 *Aa,* and 153 *aa* individuals. Each *AA* individual has two *A* alleles, whereas each *Aa* heterozygote possesses only a single *A* allele. Therefore, the number of *A* alleles in the population is (2 × the number of *AA* homozygotes) + (the number of *Aa* heterozygotes), or (2 × 353) + 494 = 1,200. Since every diploid individual has two alleles, the total number of alleles in the population is twice the number of individuals, or 2 × 1,000. Using the equation just given, the allele frequency is 1,200/2,000 = 0.60. When two alleles are present at a locus, we can use the following formula to calculate allele frequencies:

$$p = f(A) = \frac{(2 \times \text{count of } AA) + (\text{count of } Aa)}{2 \times \text{total number of individuals}}$$

particular genotype and divide this number by the total number of individuals in the population. We do this for each of the genotypes at the locus. The sum of the genotype frequencies should be 1. Consider a locus that determines the pattern of spots in the scarlet tiger moth, *Panaxia dominula* (Figure 24.2). Three genotypes are present in most populations, and each genotype produces a different phenotype. E. B. Ford collected moths at one locality in England and found the following numbers of genotypes: 452 *BB*, 43 *Bb,* and 2 *bb,* for a total of 497 moths. The genotype frequencies (where *f* = *frequency* of) are therefore

$$f(BB) = 452/497 = 0.909$$
$$f(Bb) = 43/497 = 0.087$$
$$f(bb) = 2/497 = 0.004$$

Allele Frequencies

Although genotype frequencies at a single locus are useful for examining the effects of certain evolutionary processes on a population, in most cases population

The second method of calculating allelic frequencies goes through the step of first calculating genotype frequencies as demonstrated previously. In this example $f(AA) = 0.353$, $f(Aa) = 0.494$, and $f(aa) = 0.153$. From these genotype frequencies we calculate the allele frequencies as follows:

$$p = f(A) = \text{(frequency of the } AA \text{ homozygote)}$$
$$+ (\tfrac{1}{2} \times \text{frequency of the } Aa \text{ heterozygote)}$$

$$q = f(a) = \text{(frequency of the } aa \text{ homozygote)}$$
$$+ (\tfrac{1}{2} \times \text{frequency of the } Aa \text{ heterozygote)}$$

The frequencies of two alleles, $f(A)$ and $f(a)$, are commonly symbolized as p and q. The allele frequencies for a locus, like the genotype frequencies, should always add up to 1. This is because in a one-locus model that has only two alleles, 100 percent (i.e., the frequency = 1) of the alleles are accounted for by the sum of the percentages of the two alleles. Therefore, once p is calculated, q can be easily obtained by subtraction: $1 - p = q$.

Allele Frequencies with Multiple Alleles. Suppose we have three alleles—A^1, A^2, and A^3—at a locus, and we want to determine the allele frequencies. Here, we use the same rule that we used with two alleles: We add up the number of alleles of each type and divide by the total number of alleles in the population:

$$p = f(A^1) = \frac{(2 \times \text{count of } A^1A^1) + (A^1A^2) + (A^1A^3)}{(2 \times \text{total number of individuals})}$$

$$q = f(A^2) = \frac{(2 \times \text{count of } A^2A^2) + (A^1A^2) + (A^2A^3)}{(2 \times \text{total number of individuals})}$$

$$r = f(A^3) = \frac{(2 \times \text{count of } A^3A^3) + (A^1A^3) + (A^2A^3)}{(2 \times \text{total number of individuals})}$$

To illustrate the calculation of allele frequencies when more than two alleles are present, we will use data from a study on genetic variation in milkweed beetles. Walter Eanes and his coworkers examined allele frequencies at a locus that codes for the enzyme phosphoglucomutase (PGM). Three alleles were found at this locus; each allele codes for a different molecular variant of the enzyme. In one population sample, the following numbers of genotypes were collected:

$$A^1A^1 = 4$$
$$A^1A^2 = 41$$
$$A^2A^2 = 84$$
$$A^1A^3 = 25$$
$$A^2A^3 = 88$$
$$A^3A^3 = 32$$
$$\overline{\text{Total} = 274}$$

The frequencies of the alleles are calculated as follows:

$$p = f(A^1) = \frac{(2 \times 4) + (41) + (25)}{(2 \times 274)} = 0.135$$

$$q = f(A^2) = \frac{(2 \times 84) + (41) + (88)}{(2 \times 274)} = 0.542$$

$$r = f(A^3) = \frac{(2 \times 32) + (88) + (25)}{(2 \times 274)} = 0.323$$

As seen in these calculations, we add twice the number of homozygotes that possess the allele and one times the count of each of the heterozygotes that have the allele. We then divide by twice the number of individuals in the population, which represents the total number of alleles present. In the top part of the equation, notice that for each allelic frequency, we do not add all the heterozygotes because some of the heterozygotes do not have the allele; for example, in calculating the allelic frequency of A^1, we do not add the number of A^2A^3 heterozygotes in the top part of the equation. A^2A^3 individuals do not have an A^1 allele. We can use the same procedure for calculating allelic frequencies when four or more alleles are present.

The second method for calculating allelic frequencies from genotypic frequencies can also be used here. This calculation may be quicker if we have already determined the frequencies of the genotypes. The frequency of the homozygote is added to half of the heterozygote frequency because half of the heterozygote's alleles are A and half are a. If three alleles (A^1, A^2, and A^3) are present in the population, the allelic frequencies are

$$p = f(A^1) = f(A^1A^1) + \frac{f(A^1A^2)}{2} + \frac{f(A^1A^3)}{2}$$

$$q = f(A^2) = f(A^2A^2) + \frac{f(A^1A^2)}{2} + \frac{f(A^2A^3)}{2}$$

$$r = f(A^3) = f(A^3A^3) + \frac{f(A^1A^3)}{2} + \frac{f(A^2A^3)}{2}$$

Although calculating allelic frequencies from genotypic frequencies may be quicker than calculating them directly from the numbers of genotypes, more rounding error will occur. As a result, calculations from direct counts usually are preferred. Calculating genotype frequencies and allelic frequencies is illustrated for a one-locus, three-allele example in Box 24.1.

Allelic Frequencies at an X-Linked Locus. Calculating allelic frequencies at an X-linked locus is slightly more complicated because males have only a single X-linked allele (in mammals and flies, for example). However, we can apply the same principles we used for autosomal loci. Remember that each homozygous female carries two X-linked alleles; heterozygous females have only one of that particular allele, and we will consider the case in which all males have only a single X-linked allele. To determine the number of alleles at an X-linked locus, we multiply the number of homozygous females by 2, then add the number of heterozygous females and the number of hemizygous males. We next

---

Box 24.1 **Sample Calculation of Genotypic and Allelic Frequencies for Hemoglobin Variants Among Nigerians Where Multiple Alleles Are Present**

Hemoglobin Genotypes

AA	AS	SS	AC	SC	CC	Total
2,017	783	4	173	14	11	3,002

Calculation of Genotypic Frequencies

$$\text{Genotypic frequency} = \frac{\text{Number of individuals with the genotype}}{\text{Total number of individuals}}$$

$$f(SS) = \frac{4}{3,002} = 0.0013 \qquad f(AA) = \frac{2,017}{3,002} = 0.672 \qquad f(AC) = \frac{173}{3,002} = 0.058$$

$$f(AS) = \frac{783}{3,002} = 0.261 \qquad f(SC) = \frac{14}{3,002} = 0.0047 \qquad f(CC) = \frac{11}{3,002} = 0.0037$$

Calculation of Allelic Frequencies from the Number of Individuals with a Particular Genotype

$$\text{Allelic frequency} = \frac{\text{Number of copies of a given allele in the population}}{\text{Sum of all alleles in the population}}$$

$$f(S) = \frac{(2 \times \text{number of } SS \text{ individuals}) + (\text{number of } AS \text{ individuals}) + (\text{number of } SC \text{ individuals})}{2 \times \text{total number of individuals}}$$

$$f(S) = \frac{(2 \times 4) + 783 + 14}{(2 \times 3,002)} = \frac{805}{6,004} = 0.134$$

$$f(A) = \frac{(2 \times \text{number of } AA \text{ individuals}) + (\text{number of } AS \text{ individuals}) + (\text{number of } AC \text{ individuals})}{2 \times \text{total number of individuals}}$$

$$f(A) = \frac{(2 \times 2,017) + 783 + 173}{2 \times 3,002} = \frac{4,990}{6,004} = 0.831$$

$$f(C) = \frac{(2 \times \text{number of } CC \text{ individuals}) + (\text{number of } AC \text{ individuals}) + (\text{number of } SC \text{ individuals})}{2 \times \text{total number of individuals}}$$

$$f(C) = \frac{(2 \times 11) + 173 + 14}{(2 \times 3,002)} = \frac{209}{6,004} = 0.035.$$

Calculation of Allelic Frequencies from the Frequencies of Particular Genotypes

$f(S) = f(SS) + \frac{1}{2}f(AS) + \frac{1}{2}f(SC)$

$f(S) = 0.0013 + (\frac{1}{2} \times 0.261) + (\frac{1}{2} \times 0.0047) = 0.134$

$f(A) = f(AA) + \frac{1}{2}f(AS) + \frac{1}{2}f(AC)$

$f(A) = 0.672 + (\frac{1}{2} \times 0.261) + (\frac{1}{2} \times 0.058) = 0.831$

$f(C) = f(CC) + \frac{1}{2}f(SC) + \frac{1}{2}f(AC)$

$f(C) = 0.0037 + (\frac{1}{2} \times 0.0047) + (\frac{1}{2} \times 0.058) = 0.035$

---

divide by the total number of alleles in the population. When determining the total number of alleles, we add twice the number of females (because each female has two X-linked alleles) to the number of males (who have a single allele at X-linked loci). Using this reasoning, the frequencies of two alleles at an X-linked locus (X^A and X^a) are determined with the following equations:

$$p = f(X^A) = \frac{(2 \times X^A X^A \text{ females}) + (X^A X^a \text{ females}) + (X^A Y \text{ males})}{(2 \times \text{number of females}) + (\text{number of males})}$$

$$q = f(X^a) = \frac{(2 \times X^a X^a \text{ females}) + (X^A X^a \text{ females}) + (X^a Y \text{ males})}{(2 \times \text{number of females}) + (\text{number of males})}$$

If the population has the same number of males and females, then the allelic frequencies at an X-linked locus (averaged across sexes) can be determined from the genotypic frequencies as follows:

$$p = f(X^A) = \tfrac{2}{3}\left[f(X^A X^A) + \tfrac{1}{2}f(X^A X^a)\right] + \tfrac{1}{3}f(X^A Y)$$

$$q = f(X^a) = \tfrac{2}{3}\left[f(X^a X^a) + \tfrac{1}{2}f(X^A X^a)\right] + \tfrac{1}{3}f(X^a Y)$$

This formula assumes that the genotypic frequencies were calculated separately for each sex, so that $f(X^A Y) + f(X^a Y) = 1$. Be sure that you understand the logic behind the gene counting method; do not just memorize the formulas. If you understand fully the basis of the calculations, you will not need to remember the exact equations and will be able to determine allelic frequencies for any situation.

KEYNOTE

The genetic structure of a population is determined by the total of all alleles (the gene pool). In the case of diploid, sexually interbreeding individuals, the structure is also characterized by the distribution of alleles into genotypes. The genetic structure can be described in terms of allelic and genotypic frequencies. Except for rare mutations, individuals are born and die with the same set of alleles; what changes genetically over time (evolves) is the hereditary makeup of a group of individuals, reproductively connected in a Mendelian population.

The Hardy-Weinberg Law

The Hardy-Weinberg law serves as a foundation for population genetics because it offers a simple explanation for how the Mendelian principle of segregation influences allelic and genotypic frequencies in a population. The Hardy-Weinberg law is named after the two individuals who independently discovered it in the early 1900s (Box 24.2). We begin our discussion of the Hardy-Weinberg law by simply stating what it tells us about the gene pool of a population. We then explore the implications of this principle and briefly discuss how the Hardy-Weinberg law is derived. Finally, we present some applications of the Hardy-Weinberg law and test a population to determine whether the genotypes are in Hardy-Weinberg proportions.

The Hardy-Weinberg law is divided into three parts: a set of assumptions and two major results. A simple statement of the law follows:

Part 1 (Assumptions): In an infinitely large, randomly mating population, free from mutation, migration, and natural selection (note that there are five assumptions);

Part 2 (Result 1): the frequencies of the alleles do not change over time; and

Part 3 (Result 2): as long as mating is random, the genotypic frequencies remain in the proportions p^2 (frequency of *AA*), $2pq$ (frequency of *Aa*), and q^2 (frequency of *aa*), where p is the allelic frequency of *A* and q is the allelic frequency of *a*. The sum of the genotype frequencies equals 1 (that is, $p^2 + 2pq + q^2 = 1$).

Box 24.2	Hardy, Weinberg, and the History of Their Contribution to Population Genetics

Godfrey H. Hardy (1877–1947), a mathematician at Cambridge University, often met R. C. Punnett, the Mendelian geneticist, at the faculty club. One day in 1908 Punnett told Hardy of a problem in genetics that he attributed to a strong critic of Mendelism, G. U. Yule (Yule later denied having raised the problem). Supposedly, Yule said that if the allele for short fingers (brachydactyly) was dominant (which it is) and its allele for normal-length fingers was recessive, then short fingers ought to become more common with each generation. In time almost everyone in Britain should have short fingers. Punnett believed the argument was incorrect, but he could not prove it.

Hardy was able to write a few equations showing that, given any particular frequency of alleles for short fingers and alleles for normal fingers in a population, the relative number of people with short fingers and people with normal fingers will stay the same generation after generation if no natural selection is involved that favors one phenotype or the other in producing offspring. Hardy published a short paper describing the relationship between genotypes and phenotypes in populations, and within a few weeks a paper was published by Wilhelm Weinberg (1862–1937), a German physician of Stuttgart, that clearly stated the same relationship. The Hardy-Weinberg law signaled the beginning of modern population genetics.

To be complete, we should note that in 1903 American geneticist W. E. Castle of Harvard University was the first to recognize the relationship between allele and genotypic frequencies, but it was Hardy and Weinberg who clearly described the relationship in mathematical terms. Therefore, the law is sometimes called the Castle-Hardy-Weinberg law.

In short, the Hardy-Weinberg law explains what happens to the allelic and genotypic frequencies of a population as the alleles are passed from generation to generation in the absence of evolutionary forces. In other words, if the assumptions listed in part 1 are met, alleles would be expected to combine into genotypes based on simple laws of probability and the population is in Hardy-Weinberg equilibrium. Thus, genotype frequencies can be predicted from allele frequencies.

Assumptions of the Hardy-Weinberg Law

Part 1 of the Hardy-Weinberg law presents certain conditions, or assumptions that must be present for the law to apply. First, the law indicates that the population must be infinitely large. If a population is limited in size, chance deviations from expected ratios can result in changes in allelic frequency, a phenomenon called **genetic drift.** It is true that the assumption of infinite size in part 1 is unrealistic: No population has an infinite number of individuals. But large populations look very similar to populations that are mathematically infinitely large. (We discuss this phenomenon later when we examine genetic drift in more detail.) At this point, it is important to understand that populations need not be infinitely large for the Hardy-Weinberg law to provide a good approximation of genotypic frequencies. In fact, small departures from the assumptions of the law lead only to small departures from the Hardy-Weinberg proportions. Later we will see that in the end one must do a statistical test for goodness of fit of the data to the determine whether the observed proportions are really different from those that are expected. We will see that finite populations with rare mutations, rare migrants, and weak selection depart from Hardy-Weinberg proportions only a little and that only when the deviations from the assumptions become large or when our sample size is very large can we detect departures from the Hardy-Weinberg proportions.

A second condition of the Hardy-Weinberg law is that mating must be random. **Random mating** is mating between genotypes occurring in proportion to the frequencies of the genotypes in the population. More specifically, the probability of a mating between two genotypes is equal to the product of the two genotypic frequencies.

To illustrate random mating, consider the M-N blood types in humans discussed in Chapter 13. The M-N blood type results from an antigen on the surface of a red blood cell, similar to the ABO antigens, except that incompatibility in the M-N system does not cause problems during blood transfusion. The M-N blood type is determined by one locus with two codominant alleles, L^M and L^N. In a population of Eskimos, the frequencies of the three M-N genotypes are $L^M/L^M = 0.835$, $L^M/L^N = 0.156$, and $L^N/L^N = 0.009$. If Eskimos interbreed randomly, the probability of a mating between an L^M/L^M male and an L^M/L^M female is equal to the frequency of L^M/L^M times the frequency of $L^M/L^M = 0.835 \times 0.835 = 0.697$. Similarly, the probabilities of other possible matings are equal to the products of the genotypic frequencies when mating is random.

The requirement of random matings for the Hardy-Weinberg law often is misinterpreted. Many students assume, incorrectly, that the population must be interbreeding randomly for all traits for the Hardy-Weinberg law to hold. If this were true, human populations would never obey the Hardy-Weinberg law because humans do not mate randomly. Humans mate preferentially for height, IQ, skin color, socioeconomic status, and other traits. However, although mating is nonrandom for some traits, most humans still choose mates at random for the M-N blood types; few of us even know what our M-N blood type is. The principles of the Hardy-Weinberg law apply to any locus for which random mating occurs, even if mating is nonrandom for other loci.

Finally, for the Hardy-Weinberg law to work, the population must be free from mutation, migration, and natural selection (described in detail later). In other words, the gene pool must be closed to the addition or subtraction of alleles, and we are interested in how allelic frequencies are related to genotypic frequencies solely on the basis of meiosis and sexual reproduction. Therefore, the influence of other evolutionarily relevant processes must be excluded. Later we discuss these other evolutionary processes and their effect on the gene pool of a population. This condition (that no evolutionary processes act on the population) applies only to the locus in question: A population may be subject to evolutionary processes acting on some genes while still meeting the Hardy-Weinberg assumptions at other loci.

Predictions of the Hardy-Weinberg Law

If the conditions of the Hardy-Weinberg law are met, the population will be in genetic equilibrium, and two results are expected. First, the frequencies of the alleles will not change from one generation to the next. Second, the genotypic frequencies will be in the proportions p^2, $2pq$, and q^2 after one generation of random mating, and they will remain in those proportions in every generation that follows as all the conditions of the Hardy-Weinberg law continue to be met. When the genotypes are in these proportions, the population is said to be in Hardy-Weinberg equilibrium. An important use of the Hardy-Weinberg law is that it allows us to calculate the genotypic frequencies from the allelic frequencies when the population is in equilibrium. If the observed genotype proportions are different from what we expect, we know that one or more of the Hardy-Weinberg assumptions has been violated.

To summarize, the Hardy-Weinberg law makes several predictions about the allelic frequencies and the genotypic frequencies of a population when certain conditions are satisfied. The necessary conditions are that the population is large, randomly mating, and free from mutation, migration, and natural selection. When these conditions are met, the Hardy-Weinberg law indicates that allelic frequencies will not change and genotypic frequencies will be determined by the allelic frequencies, occurring in the proportions p^2, $2pq$, and q^2.

Derivation of the Hardy-Weinberg Law

The Hardy-Weinberg law states that when a population is in equilibrium, the genotypic frequencies will be in the proportions p^2, $2pq$, and q^2. To understand why, consider a hypothetical population in which the frequency of allele A is p and the frequency of allele a is q. In producing gametes, each genotype passes on both alleles that it possesses with equal frequency; therefore, the frequencies of A and a in the gametes are also p and q. If one thinks of the gametes as being in a pool, the random formation of zygotes involves reaching into the pool and drawing two gametes at random. The genotypes that then form after repeatedly drawing two gametes at a time will be in frequencies that are predicted by the probabilities of drawing the particular allele bearing gametes (see product rule, Chapter 11, p. 279). Table 24.1 shows the combinations of gametes when mating is random. This table illustrates the relationship between the allelic fre-

quencies and the genotypic frequencies, which forms the basis of the Hardy-Weinberg law. We see that when gametes pair randomly, the genotypes will occur in the proportions $p^2(AA)$, $2pq(Aa)$, and $q^2(aa)$. These genotypic proportions result from the expansion of the square of the allelic frequencies $(p + q)^2 = p^2 + 2pq + q^2$, and the genotypes reach these proportions after one generation of random mating.

The Hardy-Weinberg law also states that allelic and genotypic frequencies remain constant generation after generation if the population remains large, randomly mating, and free from mutation, migration, and natural selection (i.e., evolutionary forces). This result can also be understood by considering a hypothetical, randomly mating population, as illustrated in Table 24.2. In Table 24.2, all possible matings are given. By definition, random

Table 24.1 Possible Combinations of *A* and *a* Gametes from Gametic Pools for a Population

	Male gametes $A(p)$	$a(q)$
Female gametes $A(p)$	AA (p^2)	Aa (pq)
$a(q)$	Aa (pq)	aa (q^2)

In sum, $p^2\ AA + 2pq\ Aa + q^2\ aa = 1.00$

Table 24.2 Algebraic Proof of Genetic Equilibrium in a Randomly Mating Population for One Gene Locus with Two Alleles

Type of Mating ♀ ♂	Mating Frequency	Offspring Frequencies Contributed to the Next Generation by a Particular Mating AA	Aa	aa
$p^2AA \times p^2AA$	p^4	p^4	—	—
$p^2AA \times 2pq\,Aa$ }[a] $2pq\,Aa \times p^2\,AA$	$4p^3q$	$2p^3q$	$2p^3q$	—
$p^2AA \times q^2\,aa$ } $q^2\,aa \times p^2\,AA$	$2p^2q^2$	—	$2p^2q^2$	—
$2pq\,Aa \times 2pq\,Aa$	$4p^2q^2$	p^2q^2	$2p^2q^2$	p^2q^2
$2pq\,Aa \times q^2\,aa$ } $q^2\,aa \times 2pq\,Aa$	$4pq^3$	—	$2pq^3$	$2pq^3$
$q^2\,aa \times q^2\,aa$	q^4	—	—	q^4
Totals	$(p^2 + 2pq + q^2)^2 = 1$	$p^2(p^2 + 2pq + q^2) = p^2$	$2pq(p^2 + 2pq + q^2) = 2pq$	$q^2(p^2 + 2pq + q^2) = q^2$

Genotype frequencies = $(p + q)^2 = p^2 + 2pq + q^2 = 1$ in each generation afterward.

Gene (allele) frequencies = $p(A) + q(a) = 1$ in each generation afterward.

[a]For example, matings between AA and Aa will occur at $p^2 \times 2pq = 2p^3q$ for $AA \times Aa$ and at $p^2 \times 2pq = 2p^3q$ for $Aa \times AA$ for a total of $4p^3q$. Two progeny types, AA and Aa, result in equal proportions from these matings. Therefore, offspring frequencies are $2p^3q$ (i.e., ½ × $4p^3q$) for AA and for Aa.

mating means that the frequency of mating between two genotypes is equal to the product of the genotypic frequencies. For example, the frequency of an $AA \times AA$ mating is equal to p^2 (the frequency of AA) $\times$ p^2 (the frequency of AA) = p^4. The frequencies of the offspring produced from each mating are also presented in Table 24.2.

We see that the sum of the probabilities of $AA \times Aa$ (or $2p^3q$) and $Aa \times AA$ (or $2p^3q$) matings is $4p^3q$, and we know from Mendelian principles that these crosses produce ½ AA and ½ Aa offspring. Therefore, the probability of obtaining AA offspring from these matings is $4p^3q \times$ ½ $= 2p^3q$. The frequencies of offspring produced by each type of mating are presented in the body of the table. At the bottom of the table, the total frequency for each genotype is obtained by addition. As we can see, after random mating the genotypic frequencies are still p^2, $2pq$, and q^2 and the allelic frequencies remain at p and q. The frequencies of the population can thus be represented in the zygotic and gametic stages as follows:

Zygotes	Gametes
$p^2(AA) + 2pq(Aa) + q^2(aa)$	$p(A) + q(a)$

Each generation of zygotes produces A and a gametes in proportions p and q. The gametes unite to form AA, Aa, and aa zygotes in the proportions p^2, $2pq$, and q^2, and the cycle is repeated indefinitely as long as the assumptions of the Hardy-Weinberg law hold. This short proof gives the theoretical basis for the Hardy-Weinberg law.

The Hardy-Weinberg law indicates that at equilibrium, the genotypic frequencies depend only on the frequencies of the alleles. This relationship between allelic frequencies and genotypic frequencies for a locus with two alleles is represented in Figure 24.3. Several aspects of this relationship should be noted: (1) The maximum frequency of the heterozygote is 0.5, and this maximum value occurs only when the frequencies of A and a are both 0.5; (2) if allelic frequencies are between 0.33 and 0.66, the heterozygote is the most numerous genotype; and (3) when the frequency of one allele is less than 0.33, the homozygote for that allele is the rarest of the three genotypes.

This point is also illustrated by the distribution of genetic diseases in humans, which are frequently rare and recessive. For a rare recessive trait, the frequency of the gene causing the trait is much higher than the frequency of the trait itself because most of the rare alleles are in unaffected heterozygotes (i.e., carriers). Albinism, for example, is a rare recessive condition in humans. In *tyrosinase-negative albinism*, affected individuals have no tyrosinase activity, which is required for normal production of melanin pigment. Among North American whites, the frequency of tyrosinase-negative albinism is roughly

Figure 24.3

Relationship of the frequencies of the genotypes *AA*, *Aa*, and *aa* to the frequencies of alleles *A* and *a* (in values of *p* [top abscissa] and *q* [bottom abscissa], respectively) in populations that meet the assumptions of the Hardy-Weinberg law. Any single population is defined by a single vertical line such as $p = 0.3$ and $q = 0.7$.

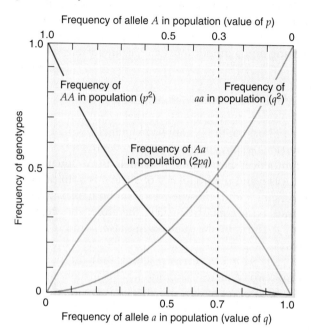

1 in 40,000, or 0.000025. Since albinism is a recessive condition, the genotype of affected individuals is *aa*. If we assume that the population meets the assumptions of the Hardy-Weinberg law, then the frequency of the *aa* genotypes equals q^2. If $q^2 = 0.000025$, then $q = 0.005$ and $p = 1 - q = 0.995$. The heterozygote frequency is therefore $2pq = 2 \times 0.995 \times 0.005 = 0.00995$ (almost 1 percent). Thus, although the frequency of albinism is low (1 in 40,000), individuals heterozygous for albinism are much more common (almost 1 in 100). When an allele is rare, almost all copies of that allele are in heterozygotes, and in this case recessive phenotypes often are very rare.

KEYNOTE

The Hardy-Weinberg law describes what happens to allelic and genotypic frequencies of a large population, when gametes fuse randomly and there is no mutation, migration, or natural selection. If these conditions are met, allelic frequencies do not change from generation to generation, and the genotypic frequencies stabilize after one generation in the proportions p^2, $2pq$, and q^2, where p and q equal the frequencies of the alleles in the population.

Extensions of the Hardy-Weinberg Law to Loci with More Than Two Alleles

When two alleles are present at a locus, the Hardy-Weinberg law tells us that at equilibrium the frequencies of the genotypes is p^2, $2pq$, and q^2, which is the square of the allelic frequencies $(p + q)^2$. This is a simple binomial expansion, and this principle of probability theory can be extended to any number of alleles that are sampled two at a time into a diploid zygote. For example, if three alleles are present (e.g., alleles A, B, and C) with frequencies equal to p, q, and r, the frequencies of the genotypes at equilibrium are also given by the square of the allelic frequencies:

$$(p + q + r)^2 = p^2(AA) + 2pq(AB) + q^2(BB) + 2pr(AC) + 2qr(BC) + r^2(CC)$$

In the blue mussel found along the Atlantic coast of North America, three alleles are common at a locus coding for the enzyme leucine aminopeptidase (LAP). For a population of mussels inhabiting Long Island Sound (discussed later; see Figure 24.6), Richard K. Koehn and colleagues determined that the frequencies of the three alleles were as follows:

Allele	Frequency
LAP^{98}	$p = 0.52$
LAP^{96}	$q = 0.31$
LAP^{94}	$r = 0.17$

If the population were in Hardy-Weinberg equilibrium, the expected genotypic frequencies would be

Genotype	Expected Frequency	
LAP^{98}/LAP^{98}	$p^2 = (0.52)^2$	$= 0.27$
LAP^{98}/LAP^{96}	$2pq = 2(0.52)(0.31)$	$= 0.32$
LAP^{96}/LAP^{96}	$q^2 = (0.31)^2$	$= 0.10$
LAP^{96}/LAP^{94}	$2qr = 2(0.31)(0.17)$	$= 0.11$
LAP^{94}/LAP^{98}	$2pr = 2(0.52)(0.17)$	$= 0.18$
LAP^{94}/LAP^{94}	$r^2 = (0.17)^2$	$= 0.03$

The square of the allelic frequencies can be used in the same way to estimate the expected frequencies of the genotypes when four or more alleles are present at a locus.

Extensions of the Hardy-Weinberg Law to Sex-Linked Alleles

In species like humans or *Drosophila*, in which females are XX and males are XY, the Hardy-Weinberg law must be modified. If alleles are X-linked, females may be homozygous or heterozygous, but males carry only a single allele for each X-linked locus. For X-linked alleles in females, the Hardy-Weinberg frequencies are the same as those for autosomal loci: $p^2(X^AX^A)$, $2pq(X^AX^B)$, and $q^2(X^BX^B)$. In males, however, the frequencies of the genotypes are $p(X^AY)$ and $q(X^BY)$, the same as the frequencies of the alleles in the population. For this reason, recessive

X-linked traits are more frequent among males than among females. To illustrate this concept, consider red-green color blindness, which is an X-linked recessive trait. We actually know that many different defective alleles cause red-green color blindness, but for now let's lump them together. The frequency of the color-blindness allele varies among human ethnic groups; the frequency among African Americans is 0.039. At equilibrium, the expected frequency of color-blind males in this group is $q = 0.039$, but the frequency of color-blind females is only $q^2 = (0.039)^2 = 0.0015$.

When random mating occurs within a population, the equilibrium genotypic frequencies are reached in one generation. However, if the alleles are X-linked and the sexes differ in allelic frequency, the equilibrium frequencies are approached over several generations. This is because males receive their X chromosome from their mother only, whereas females receive an X chromosome from both the mother and the father. Consequently, the frequency of an X-linked allele in males is the same as the frequency of that allele in their mothers, whereas the frequency in females is the average of that in mothers and fathers. With random mating, the allelic frequencies in the two sexes oscillate back and forth each generation, and the difference in allelic frequency between the sexes is reduced by half each generation, as shown in Figure 24.4. Once the allelic frequencies of the males and females are equal, the frequencies of the genotypes are in Hardy-Weinberg proportions after one more generation of random mating.

Testing for Hardy-Weinberg Proportions

When we take a sample from a population and calculate genotypic frequencies, it is rarely the case that the match to Hardy-Weinberg proportions is exact. To test whether the fit to Hardy-Weinberg is acceptable, we ask, "What is the chance that we would get this big a departure by chance alone?" If the observed genetic structure does not

Figure 24.4

Representation of the gradual approach to equilibrium of an X-linked gene with an initial of 1.0 in females and 0 in males.

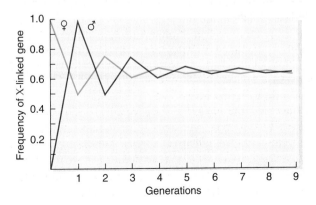

match the expected structure based on the law, we can begin to ask about which of the assumptions are being violated. To determine whether the genotypes of a population are in Hardy-Weinberg proportions, we first compute the allelic frequencies from the observed genotypic frequencies. (Note: It is important not to take the square roots of the homozygote frequencies to obtain allele frequencies because to do that already assumes that the population is in equilibrium. Thus, allele frequencies should be calculated by the gene-counting method; see p. 660.) Once we have obtained the allelic frequencies, we can calculate the expected genotypic frequencies (p^2, $2pq$, and q^2) and compare these frequencies with the actual observed frequencies of the genotypes using a chi-square test (see Chapter 2). The chi-square test gives us the probability that the difference between what we observed and what we expect under the Hardy-Weinberg law is due to chance. To illustrate this procedure, consider a locus that codes for transferrin (a blood protein) in the red-backed vole, *Clethrionomys gapperi*. Three genotypes are found at the transferrin locus: *MM*, *MJ*, and *JJ*. In a population of voles trapped in the Northwest Territories of Canada in 1976, 12 *MM* individuals, 53 *MJ* individuals, and 12 *JJ* individuals were found. To determine whether the genotypes are in Hardy-Weinberg proportions, we first calculate the allelic frequencies for the population using our familiar formula:

$$p = \frac{(2 \times \text{number of homozygotes}) + (\text{number of heterozygotes})}{2 \times \text{total number of individuals}}$$

Therefore,

$$p = f(M) = \frac{(2 \times 12) + (33)}{(2 \times 77)} = 0.50$$

$$q = 1 - p = 0.50$$

Using p and q calculated from the observed genotypes, we can now compute the expected Hardy-Weinberg proportions for the genotypes: $f(MM) = p^2 = (0.50)^2 = 0.25$, $f(MJ) = 2pq = 2(0.50)(0.50) = 0.50$, and $f(JJ) = q^2 = (0.50)^2 = 0.25$. However, for the chi-square test, actual numbers of individuals are needed, not the proportions. To obtain the expected numbers, we simply multiply each expected proportion times the total number of individuals counted (N), as follows:

		Expected	Observed
$f(MM) = p^2 \times N$			
$= 0.25 \times 77 =$		19.3	12
$f(MJ) = 2pq \times N$			
$= 0.50 \times 77 =$		38.5	53
$f(JJ) = q^2 \times N$			
$= 0.25 \times 77 =$		19.3	12

With observed and expected numbers, we can compute a chi-square value to determine the probability that the differences between observed and expected numbers could be the result of chance. The chi-square (χ^2) is computed using the same formula we used for analyzing genetic crosses. That is, d, the deviation, is calculated for each class as (observed − expected); d^2, the deviation squared, is divided by the expected number e for each class; and chi-square (χ^2) is computed as the sum of all d^2/e values. For this example, $\chi^2 = 10.98$. We now need to find this value in the chi-square table (see Table 2.5, p. 30) under the appropriate degrees of freedom. This step is not as straightforward as in our previous chi-square analyses. In those examples, the number of degrees of freedom was the number of classes in the sample minus 1. While there are three classes, there is only one degree of freedom because the frequencies of alleles in a population have no theoretically expected values. Thus, p must be estimated from the observations themselves. So, one degree of freedom is lost for every parameter (p in this case) that must be calculated from the data. Another degree of freedom is lost because, for the fixed number of individuals, once all but one of the classes have been determined, the last class has no degree of freedom and is set automatically. Therefore, with three genotypic classes (*MM*, *MJ*, and *JJ*), two degrees of freedom are lost, leaving one degree of freedom.

In the chi-square table under the column for one degree of freedom, the chi-square value of 10.98 indicates a *P* value less than 0.01. Thus, the probability that the differences between the observed and expected values is due to chance is very low. That is, the observed numbers of genotypes do not fit the expected numbers under Hardy-Weinberg law.

Using the Hardy-Weinberg Law to Estimate Allelic Frequencies

An important application of the Hardy-Weinberg law is the calculation of allelic frequencies when one or more alleles is recessive. For example, we have seen that albinism in humans results from an autosomal recessive gene. Normally, this trait is rare, but among the Hopi Indians of Arizona, albinism is remarkably common. Charles M. Woolf and Frank C. Dukepoo visited the Hopi villages in 1969 and observed 26 cases of albinism in a total population of about 6,000 Hopis (Figure 24.5). This gave a frequency for the trait of 26/6,000, or 0.0043, which is much higher than the frequency of albinism in most populations. Although we have calculated the frequency of the trait, we cannot directly determine the frequency of the gene for albinism because we cannot distinguish between heterozygous individuals and those homozygous for the normal allele. Recall that our computation of the allelic frequency involves counting the number of alleles:

Figure 24.5

Three Hopi girls, photographed about 1900. The middle child has albinism, an autosomal recessive disorder that occurs with high frequency among the Hopi Indians of Arizona.

$$p = \frac{\begin{array}{c}(2 \times \text{number of homozygotes}) \\ + (\text{number of heterozygotes})\end{array}}{2 \times \text{total number of individuals}}$$

But because heterozygotes for a recessive trait such as albinism cannot be identified, this is impossible. Nevertheless, we can determine the allelic frequency from the Hardy-Weinberg law provided that we assume that genotypes in the population are found in Hardy-Weinberg proportions. When genotypes are in Hardy-Weinberg proportions, the frequency of the homozygous recessive genotype is q^2. For albinism among the Hopi, $q^2 = 0.0043$, and q can be obtained by taking the square root of the frequency of the affected genotype. Therefore, $q = \sqrt{0.0043} = 0.065$, and $p = 1 - q = 0.935$. Following the Hardy-Weinberg law, the frequency of heterozygotes in the population is $2pq = 2 \times 0.935 \times 0.065 = 0.122$, or about $\frac{1}{8}$. Thus, one out of eight Hopis, on average, carries an allele for albinism.

We should not forget that this method of calculating allelic frequency rests on the assumption that genotypes are in Hardy-Weinberg proportions. In this case, there are reasons to expect departures from Hardy-Weinberg because strong social factors affect albino Hopi individuals. If the conditions of the Hardy-Weinberg law do not apply, then our estimate of allelic frequency is inaccurate. Also, once we calculate allelic frequencies with these assumptions, we cannot then test the population to determine whether the genotypic frequencies are in the Hardy-Weinberg expected proportions. To do so would involve circular reasoning, for we assumed Hardy-Weinberg proportions in the first place to calculate the allelic frequen-

cies. Before we explore how the model can be used to discern the causes of deviations of observed populations from theoretically expected populations, we will look at the genetic structure of some real populations.

KEYNOTE

The Hardy-Weinberg principle applies to alleles defined in a number of ways, including single nucleotide polymorphisms (SNPs), protein variants, or any other factor that segregates like a Mendelian gene.

Genetic Variation in Space and Time

The genetic structure of populations can vary in space and time. This means that the frequencies and distribution of alleles can vary in samples of the same species in different areas or samples from the same area collected at different times. Figure 24.6 shows how the frequencies of three alleles at the locus for the enzyme leucine aminopeptidase gradually change in a geographic series of samples of blue mussels that inhabit the east coast of the United States. Most populations of plants and animals that have widespread geographic distributions show differences of this sort in the allele frequencies of their component populations. In some cases, the spatial variation shows clear patterns or trends across space. When allele frequencies change in a systematic way across a geographic transect, we call this an allele frequency **cline.** Often clines are associated with changes in a physical attribute in the environment, such as temperature or water availability. In the case of Figure 24.6, there is a clear thermal cline. Although such clines suggest that the geographic pattern is caused by differential selection for the alternate alleles, much additional work is required to exclude other possibilities and to provide evidence for selectively maintained clines. Just as the genetic composition of a species may vary geographically, the genetic structure of individual populations can change over time, as shown in Figure 24.7.

Because of the importance of geographic variation in allele frequencies, population geneticists have devised many statistical tools for quantifying the spatial patterns of genetic variation. The simplest of these quantifies the partitioning of total genetic variance into component parts. At the simplest level we can think of a fraction of the genetic variance that exists within each local population and another fraction of the genetic variance that results from differences between distinct local populations. By this kind of measure, for example, we generally find that only about 12–13 percent of the genetic variance in humans is found between different populations, whereas 87–88 percent of the total genetic variance in humans is found within populations. Geographic patterns of genetic variation may also be of immense importance for conservation. In terms of the

Figure 24.6

Geographic variation in frequencies of the alleles LAP^{94}, LAP^{96}, and LAP^{98} of the locus coding for the enzyme leucine amino peptidase (LAP) in the blue mussel.

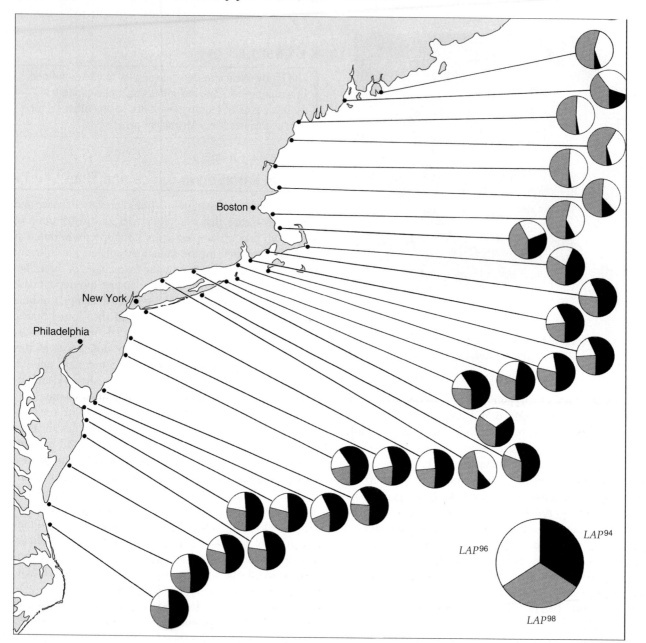

future evolution of a species and conservation of the genetic resources of a species, attention has to be paid to the fact that there is a spatial component to genetic variation. Conservation of genetic diversity demands some quantitative knowledge of the geographic patterns of variability.

KEYNOTE

The genetic structure of a species can vary both geographically and temporally.

Genetic Variation in Natural Populations

One of the most significant questions addressed in population genetics is how much genetic variation exists within natural populations. Genetic variation within populations is important for several reasons. First, it determines the potential for evolutionary change and adaptation. The amount of variation also provides us with important clues about the relative importance of various evolutionary processes because some processes increase variation while others decrease it. The manner in which

Figure 24.7

Temporal variation in the locus coding for the enzyme esterase 4F in the prairie vole, *Microtus ochrogaster*. The four populations are close to each other and near Lawrence, Kansas.

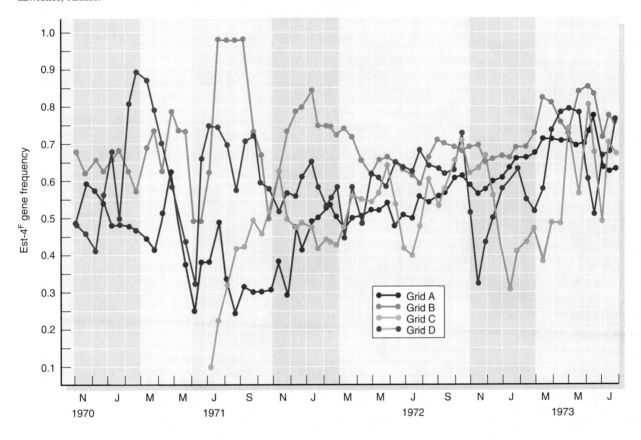

new species arise may depend on the amount of genetic variation harbored within populations. In addition, the ability of a population to persist over time can be influenced by how much genetic variation it has to draw on should environments change. For all these reasons, population geneticists are interested in measuring genetic variation, attempting to understand the evolutionary processes that affect it, and understanding the effects of human environmental disturbance that may alter it.

Measuring Genetic Variation at the Protein Level

For many years, population geneticists were constrained in quantifying how much variation existed within natural populations. Naturalists recognized that plants and animals in nature frequently differ in phenotype (Figure 24.8), but the genetic basis of most traits is too complex to assign specific genotypes to individuals. A few traits and alleles that behaved in a Mendelian fashion, such as spot patterns in butterflies and shell color in snails, provided observable genetic variation, but these isolated cases were too few to provide any general estimate of genetic variation. Then cytological evidence was used such as the morphology of chromosomes and their banding patterns, especially in the

salivary glands of fruit flies. In 1966, Richard Lewontin and John Hubby published a study that applied protein electrophoresis to the study of polymorphism in natural populations. As in the electrophoretic separation of DNA molecules, protein electrophoresis works by separating proteins as they move through a gel matrix. Once they are separated, a specific stain is added to the gel to visualize the protein bands. In one common form of protein electrophoresis, proteins are separated on the basis of a combination of charge, which varies depending on the amino acids, and folding conformation of the protein. The fact that proteins can be separated based on these features implies that as long as alleles of the same gene produce protein products with different charge, they can be separated. For population geneticists, electrophoresis provides a technique for quickly determining the genotypes of many individuals at many loci. This procedure was soon used to examine genetic variation in hundreds of plant and animal species and it paved the way for modern applications of DNA sequencing to answer questions about the forces underlying this genetic variation. The amount of genetic variation within a population was commonly measured with two parameters, the proportion of polymorphic loci and heterozygosity.

Figure 24.8
Extensive phenotypic variation exists in most natural populations, as is apparent in the color patterns of the Cuban tree snail, a species treasured for its varicolored beauty.

A polymorphic locus is any locus that has more than one allele present within a population. The **proportion of polymorphic loci (P)** is calculated by dividing the number of polymorphic loci by the total number of loci examined. For example, suppose we found that of 33 loci in a population of green frogs, 18 were polymorphic. The proportion of polymorphic loci would be $^{18}/_{33} = 0.55$. It is important to realize that the proportion of polymorphic loci depends on the technique used to identify polymorphism and on the sample size. **Heterozygosity (H)** is the proportion of an individual's loci that is heterozygous. Suppose we analyzed the genotypes of green frogs from one population at a locus coding for esterase and found that the frequency of heterozygotes was 0.09. Heterozygosity for this locus would be 0.09. We would average this heterozygosity with those for other loci and obtain an estimate of heterozygosity for the population. Note that protein electrophoresis misses much of the variation that is detected when the DNA sequence of the same gene is determined because silent or synonymous nucleotides may vary at the DNA level but leave no trace of variation at the protein level. This means that estimates of heterozygosity and proportion of polymorphic loci are typically different from estimates based on nucleotide sequence variation.

Table 24.3 presents estimates from protein electrophoresis of the proportion of polymorphic loci and heterozygosity for many species that have been surveyed with electrophoresis. The results of these studies are unambiguous: Most species have large amounts of genetic variation in their proteins. This finding ruled out the classical model for genetic variation that had emerged primarily from the work of laboratory geneticists, which stated that most natural populations have little genetic variation. If the classical model is wrong, then what maintains so much genetic variation within populations? In the late 1960s and early 1970s, observations of amino acid sequences from many species showed that there was considerable variation in amino acid sequence. Motoo Kimura proposed that much of the pattern of evolutionary changes in molecules could be explained by a combination of random mutations and random chance fixation of alleles. This model was also applied to patterns of variation within populations, and initially many predictions from the model seemed to provide good statistical fits to the data. This model, called the **neutral mutation model,** acknowledges the presence of extensive genetic variation in proteins but proposes that this variation is neutral with regard to natural selection. This

Table 24.3 Genic Variation in Some Major Groups of Animals and in Plants

Group	Number of Species or Forms	Mean Number of Loci Examined per Species	Mean Proportion of Loci	
			Polymorphic per Population	Heterozygous per Individual
Insects				
Drosophila	28	24	0.529 ± 0.030^a	0.150 ± 0.010
Others	4	28	0.531	0.151
Haplodiploid wasps	6	15	0.243 ± 0.039	0.062 ± 0.007
Marine invertebrates	9	26	0.587 ± 0.084	0.147 ± 0.019
Snails				
Land	5	18	0.437	0.150
Marine	5	17	0.175	0.083
Fish	14	21	0.306 ± 0.047	0.078 ± 0.012
Amphibians	11	22	0.336 ± 0.034	0.082 ± 0.008
Reptiles	9	21	0.231 ± 0.032	0.047 ± 0.008
Birds	4	19	0.145	0.042
Rodents	26	26	0.202 ± 0.015	0.054 ± 0.005
Large Mammals[b]	4	40	0.233	0.037
Plants[c]	8	8	0.464 ± 0.064	0.170 ± 0.031

[a] Values are mean ± standard error.
[b] Human, chimpanzee, pigtailed macaque, and southern elephant seal.
[c] Predominantly outcrossing species; mean gene diversity is 0.233 ± 0.029.

does not mean that the proteins detected by electrophoresis have no function but rather that the different genotypes are physiologically equivalent. Therefore, natural selection does not act on the neutral alleles, and random processes such as mutation and genetic drift shape the patterns of genetic variation that we see in natural populations. The neutral mutation model proposes that variation at some loci affects fitness, and natural selection eliminates variation at these loci. The neutral mutation model had strong support that was seriously eroded only after full DNA sequences of alleles were available and more complicated models were found to provide better explanations of the data.

KEYNOTE

In population genetics one often encounters competing models that seek to explain the amount of variation in natural populations. In the 1920s and 1930s, the classical model predicted that there was an archetypal "wild type" and that most alleles were of this sort, with a few mutant variants in the population. In the 1960s and 1970s, protein electrophoresis revealed abundant genetic variation in natural populations, thus disproving the classical model. The neutral mutation model proposes that the genetic variation detected by electrophoresis is neutral with regard to natural selection.

iActivity

You use protein electrophoresis to measure genetic variation in mussel populations in the iActivity, *Measuring Genetic Variation*, on your CD-ROM.

Measuring Genetic Variation at the DNA Level

The development of the polymerase chain reaction (PCR; see Chapter 8, pp. 200–202) made it easy for population geneticists to obtain fragments of genes from large numbers of individuals. These fragments could then be separated directly on gels to determine size differences, they could be cut with restriction enzymes to reveal differences, or the fragments could be directly sequenced, either by hand or with modern automated instruments. One such technique for detecting genetic variation uses restriction enzymes. You will recall that restriction enzymes make double-stranded cuts in DNA at specific base sequences (see Chapter 8, pp. 176–177). Most restriction enzymes recognize a sequence of four or six base pairs; see Table 8.1 (p. 178). For example, the restriction enzyme *Bam*HI recognizes the sequence

```
5'-GGATCC-3'
3'-CCTAGG-5'
```
and whenever this sequence appears in DNA, *Bam*HI cuts the DNA. The resulting fragments can be separated by agarose gel electrophoresis and then

observed by staining the DNA or by using probes for specific sequences (see Chapter 8).

Suppose that two individuals differ in one or more nucleotides at a particular DNA sequence and that the differences occur at a site recognized by a restriction enzyme (Figure 24.9). One individual has a DNA molecule with the restriction site, but the other individual does not because the sequences of DNA nucleotides differ. If the DNA from these two individuals is mixed with the restriction enzyme and the resulting fragments are separated on a gel, the two individuals produce different patterns of fragments, as shown in Figure 24.9. The different patterns on the gel are called **restriction fragment length polymorphisms,** or RFLPs ("riff-lips"; see Chapter 9, pp. 214–215). They indicate that the DNA sequences of the two individuals differ. RFLPs are inherited in the same way that alleles coding for other traits are inherited, except that the RFLPs do not produce any outward phenotypes; their

phenotypes are the fragment patterns produced on a gel when the DNA is cut by the restriction enzyme.

RFLPs can be used to provide information about how DNA sequences differ among individuals. Such differences involve only a small part of the DNA, specifically the few nucleotides recognized by the restriction enzyme. However, if we assume that restriction sites occur randomly in the DNA, which is not an unreasonable assumption because the sites are not expressed as traits, then the presence or absence of restriction sites can be used to estimate the overall differences in sequence.

To illustrate the use of RFLPs for estimating genetic variation, suppose we isolate DNA from five wild mice and amplify a polymorphic DNA region we want to test by PCR using oligonucleotide primers complementary to each end of the region. Next, we cut the amplified DNA fragments with the restriction enzyme *Bam*HI and separate the fragments using agarose gel electrophoresis.

Figure 24.9

DNA from individual 1 and individual 2 differ in one nucleotide, found within the sequence recognized by the restriction enzyme *Bam*HI. Individual 1's DNA contains the *Bam*HI restriction sequence and is cleaved by the enzyme. Individual 2's DNA lacks the *Bam*HI restriction sequence and is not cleaved by the enzyme. When placed on an agarose gel and separated by electrophoresis, the DNAs from 1 and 2 produce different patterns on the gel. This variation is called restriction fragment length polymorphism.

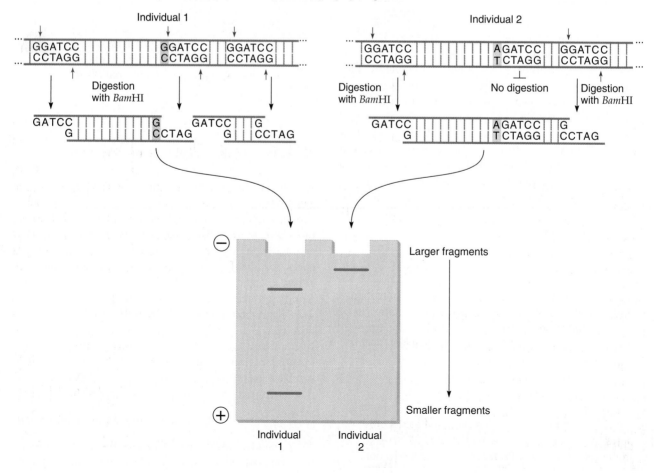

Figure 24.10

Restriction patterns from five mice. The patterns differ in the presence ($+$) or absence ($-$) of a particular restriction site (middle one of the three shown). Each mouse has two homologous chromosomes, each of which potentially carries the restriction site. Thus, a mouse may be $+/+$ (has the restriction site on both chromosomes), $+/-$ (has the restriction site on one chromosome), or $-/-$ (has the restriction site on neither chromosome). When the restriction site is present, the DNA is broken into two fragments after digestion with the restriction enzyme and separation with electrophoresis.

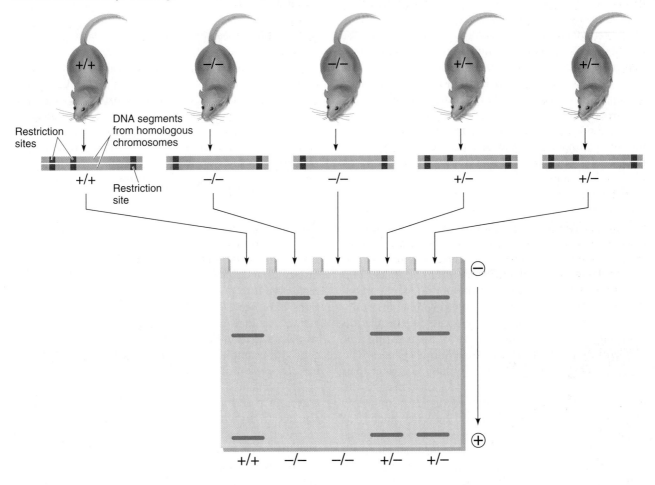

A typical set of restriction patterns that might be obtained is shown in Figure 24.10. Thus, a mouse could be $+/+$ (the polymorphic restriction site is present on both chromosomes), $+/-$ (the restriction site is present on one chromosome and is absent on the other), or $-/-$ (the restriction site is absent on both chromosomes). For the 10 chromosomes present among these particular five mice, four have the restriction site and six do not. Heterozygosity at the nucleotide level can be estimated from restriction site patterns.

Nucleotide heterozygosity has been studied for a number of different organisms. A few examples are shown in Table 24.4. Nucleotide heterozygosity typically varies from 0.001 to 0.02 in eukaryotic organisms. Recent estimates of nucleotide heterozygosity across the entire human genome average around 0.0008, meaning that an individual is heterozygous (contains different nucleotides

on the two homologous chromosomes) at about one in every 50 to 1,000 nucleotides.

One disadvantage of using RFLPs for examining genetic variation is that this method reveals variation at only a small subset of the nucleotides that make up a gene. With RFLP analysis, we are taking a small sample of

Table 24.4	Estimates of Nucleotide Heterozygosity for DNA Sequences		
DNA Sequences		**Organism**	H_{nuc}
β-Globin genes		Humans	0.002
Growth hormone gene		Humans	0.002
Alcohol dehydrogenase gene		Fruit fly	0.006
Mitochondrial DNA		Humans	0.004
H4 gene region		Sea urchin	0.019

the nucleotides (those recognized by the restriction enzyme) and using this sample to estimate the overall level of variation. DNA sequencing, which was described in Chapter 8 (pp. 197–200), provides a method for detecting all nucleotide differences that exist in a set of DNA molecules.

DNA Sequence Variation.   We saw earlier that protein electrophoresis misses genetic differences that do not change the protein charge or conformation. The best method for identifying all genetic variation is to obtain the DNA sequence of the gene from each individual. In the first study to apply this approach, Martin Kreitman sequenced 11 copies (obtained from different fruit flies) of the alcohol dehydrogenase gene in *Drosophila melanogaster*. Among the 11 copies, he found different nucleotides at 43 positions within the 2,659 base-pair segment. Furthermore, only 3 of the 11 copies were identical at all nucleotides examined; thus, there were 8 different alleles (at the nucleotide level) among the 11 copies of this gene. This suggests that populations harbor a tremendous amount of genetic variation in their DNA sequences.

Different regions of a gene apparently are subject to different evolutionary processes, and this is reflected in their different levels of nucleotide diversity. Table 24.5 shows nucleotide diversity estimates for different parts of the *Drosophila Adh* gene. As is observed within most functional genes, the highest diversity occurs at sites that do not change the amino acid sequence of the resulting protein, and these are known as **synonymous** changes. The level of synonymous nucleotide diversity is greater than the observed diversity at nucleotides that change the resulting amino acid sequence of the protein (known as **nonsynonymous** positions). Thus, the large amount of nucleotide diversity seen in synonymous sites is not unexpected because these changes do not affect the functioning of a protein. Roughly ¾ of the positions in a gene are nonsynonymous, so ¾ of all random mutations ought to be nonsynonymous, but in fact the variation one sees is much more likely to be synonymous. For example, in Martin Kreitman's study, there were 13 synonymous sites

that varied. If nonsynonymous mutations were equally likely to be seen, then we ought to see three times this number of nonsynonymous differences, or 39 positions. Instead, Kreitman found only a single nonsynonymous polymorphism. The conclusion is inescapable: Most nonsynonymous mutations are visible to natural selection and have been eliminated from the population, leaving the currently observed excess of synonymous sites that vary.

DNA Length Polymorphisms and Microsatellites.   In addition to evolution of nucleotide sequences through nucleotide substitution, variation frequently occurs in the number of nucleotides found within a gene. These variations are called DNA length polymorphisms, and they arise through deletions and insertions of short stretches of nucleotides. For example, DNA length polymorphisms have been observed in the alcohol dehydrogenase gene of *Drosophila melanogaster*. In addition to extensive variation in nucleotide sequence, Kreitman found six insertions and deletions in the 11 copies of the gene he examined. All of these were confined to introns and flanking regions of the DNA; none was found within exons. Insertions and deletions within exons usually alter the reading frame, so they are selected against. As a result, insertions and deletions are most common in noncoding regions of the DNA. However, some insertions and deletions have been found in the coding regions of certain genes. Another class of DNA length polymorphisms involves variation in the number of copies of a particular gene. For example, among individual fruit flies, the number of copies of ribosomal genes varies widely.

A particularly useful kind of polymorphism that occurs in nearly all organisms is seen in **short tandem repeats** (STRs), also called microsatellites; these were described in Chapter 9, pp. 216–217). More than 8,000 STRs have been mapped in the human genome, and their use in genetic mapping in humans and mice has been critical to the discovery of many genes associated with genetic disorders. In conservation genetics, one is often faced with a need to obtain information on genetic variation in organisms for which there is almost no prior knowledge about gene sequences, and in these cases microsatellites are often used to quantify patterns of genetic variability.

Table 24.5	**Number of Varying Nucleotides and Diversity in Five Different Regions of the Alcohol Dehydrogenase (*Adh*) Gene of *Drosophila melanogaster***		
Gene Region	**Variable Sites**	**Total Sites**	**Diversity**
5′ + 3′ flank	3	335	0.002
Exons			
Synonymous	13	192	0.013
Nonsynonymous	1	576	0.001
Introns	18	789	0.004
3′ nontranscribed	5	767	0.003

Forces That Change Gene Frequencies in Populations

We have looked at the many ways in which genetic variation can be quantified in natural populations and have come to the conclusion that populations harbor enormous amounts of variation. Now it is time to return to the question of what maintains this variation. If the population is large, randomly mating, and free from mutation, migration, and natural selection, then whatever variation there is in the population will remain at the same frequency forever. For many populations, however, the conditions required by the Hardy-Weinberg law do not hold. Populations frequently are small, mating may be nonrandom, and mutation, migration, and natural selection may occur. In these circumstances, allelic frequencies do change, and the gene pool of the population evolves in response to the interplay of these processes. In the following sections we discuss the role of four evolutionary processes—mutation, genetic drift, migration, and natural selection—in changing the allele frequencies of a population. We also discuss the effects of nonrandom mating on genotype frequencies. We first consider violations of the Hardy-Weinberg equilibrium assumptions one by one. We then consider several cases in which two assumptions are violated simultaneously. Be aware, however, that in real populations all of the assumptions can be violated simultaneously, even though our understanding is best enhanced by considering simpler situations.

KEYNOTE

Mutation, random genetic drift, migration, and natural selection are processes that can alter allelic frequencies in a population.

Mutation

One process that can alter the frequencies of alleles within a population is mutation. As we discussed in Chapter 7, **gene mutations** consist of heritable changes in the DNA that occur within a locus. Usually a mutation converts one allelic form of a gene to another. The rate at which mutations arise is generally low but varies between loci and between species (Table 24.6). Certain genes modify overall mutation rates, and many environmental factors, such as chemicals, radiation, and infectious agents, may increase the number of mutations.

Ultimately, mutation is the source of all new genetic variation; new combinations of alleles may arise through recombination, but new alleles occur only as a result of mutation. Thus, mutation provides the raw genetic material on which evolution acts. Some mutations are entirely neutral and have no effect on the reproductive fitness of the organisms. Other mutations are detrimental and are eliminated from the population. However, a few mutations convey some advantage to the individuals that possess them and spread through the population. Whether a mutation is neutral, detrimental, or advantageous depends on the specific environment, and if the environment changes, previously harmful or neutral mutations may become beneficial. For example, after the widespread use of the insecticide DDT, insects with mutations that conferred resistance to DDT were capable of surviving and reproducing; because of this

advantage, the mutations spread and many insect populations quickly evolved resistance to DDT. Mutations are of fundamental importance to the process of evolution because they provide the genetic variation on which other evolutionary processes act.

In general, rates of mutation are so low that once a mutant allele enters a population, the fate of the mutation is determined almost entirely by nonmutational forces. To see why this is so, we can consider a model in which mutation is the only evolutionary force. The mutation of A to a is called a **forward mutation**. For most genes, mutations also occur in the reverse direction; a may mutate to A. These mutations are called **reverse mutations**, and they typically occur at a lower rate than forward mutations. The forward mutation rate—the rate at which A mutates to a ($A \rightarrow a$)—is symbolized with u. The reverse mutation rate—the rate at which a mutates to A ($A \leftarrow a$)—is symbolized with v. Consider a hypothetical population in which the frequency of A is p and the frequency of a is q. We assume that the population is large and that no selection occurs on the alleles. In each generation, a proportion u of all A alleles mutates to a. The actual number mutating depends on both u and the frequency of A alleles. For example, suppose the population consists of 100,000 alleles. If u equals 10^{-4}, one out of every 10,000 A alleles mutates to a. When $p = 1.00$, all 100,000 alleles in the population are A and free to mutate to a, so $10^{-4} \times 100,000 = 10$ A alleles should mutate to a. However, if $p = 0.10$, only 10,000 alleles are A and free to mutate to a. Therefore, with a mutation rate of 10^{-4} only 1 of the A alleles will undergo mutation. The decrease in the frequency of A resulting from mutation of $A \rightarrow a$ is equal to up; the increase in frequency resulting from $A \leftarrow a$ is equal to vq. As a result of mutation, the amount by which A decreases in one generation is equal to the increase in A alleles caused by reverse mutations minus the decrease in A alleles caused by forward mutations. Since we have a forward mutation rate increasing the frequency of a and a reverse mutation rate decreasing the frequency of a, it is intuitively easy to see that eventually the population achieves equilibrium, in which the number of alleles undergoing forward mutation is exactly equal to the number of alleles undergoing reverse mutation. At this point, no further change in allelic frequency occurs, despite the fact that forward and reverse mutations continue to take place. With some simple algebra, population genetics theorists have shown that the equilibrium frequency for a is

$$\hat{q} = \frac{u}{u + v}$$

and the equilibrium value for p is

$$\hat{P} = \frac{u}{u + v}$$

Table 24.6   Spontaneous Mutation Frequencies at Specific Loci for Various Organisms[a]

Organism	Trait	Mutation per 100,000 Gametes[b]
T2 Bacteriophage (virus)	To rapid lysis ($r^+ \rightarrow r$)	7
	To new host range ($h^+ \rightarrow h$)	0.001
E. coli K12 (bacterium)	To streptomycin resistance	0.00004
	To phage T1 resistance	0.003
	To leucine independence	0.00007
	To arginine independence	0.0004
	To tryptophan independence	0.006
	To arabinose dependence	0.2
Salmonella typhimurium (bacterium)	To threonine resistance	0.41
	To histidine dependence	0.2
	To tryptophan independence	0.005
Diplococcus pneumoniae (bacterium)	To penicillin resistance	0.01
Neurospora crassa	To adenine independence	0.0008–0.029
	To inositol independence	0.001–0.010
	(One inos allele, JH5202)	1.5
Drosophila melanogaster males	y^+ to yellow	12
	bw^+ to brown	3
	e^+ to ebony	2
	ey^+ to eyeless	6
Corn	Wx to waxy	0.00
	Sh to shrunken	0.12
	C to colorless	0.23
	Su to sugary	0.24
	Pr to purple	1.10
	I to i	10.60
	R^r to r^r	49.20
Mouse	a^+ to nonagouti	2.97
	b^+ to brown	0.39
	c^+ to albino	1.02
	d^+ to dilute	1.25
	ln^+ to leaden	0.80
	Reverse mutations for above genes	0.27
Chinese hamster somatic cell tissue culture	To azaguanine resistance	0.0015
	To glutamine independence	0.014
Humans	Achondroplasia	0.6–1.3
	Aniridia	0.3–0.5
	Dystrophia myotonica	0.8–1.1
	Epiloia	0.4–1
	Huntington disease	0.5
	Intestinal polyposis	1.3
	Neurofibromatosis	5–10
	Osteogenesis imperfecta	0.7–1.3
	Pelger anomaly	1.7–2.7
	Retinoblastoma	0.5–1.2

[a]Mutations to independence for nutritional substances are from the auxotrophic condition (e.g., leu) to the prototrophic condition (e.g., leu$^+$).
[b]Mutation frequency estimates of viruses, bacteria, Neurospora, and Chinese hamster somatic cells are based on particle or cell counts rather than gametes.
Source: From Genetics, 3rd ed., by Monroe W. Strickberger. Copyright © 1985. Adapted by permission of Prentice Hall, Inc., Upper Saddle River, NJ.

Now consider how slowly this process of pure mutation changes allele frequencies. In a population with the initial allelic frequencies $p = 0.9$ and $q = 0.1$ and mutation rates $u = 5 \times 10^{-5}$ and $v = 2 \times 10^{-5}$, we can calculate the change in allelic frequency in the first generation:

$$\Delta p = vq - up$$
$$= (2 \times 10^{-5} \times 0.1) - (5 \times 10^{-5} \times 0.9)$$
$$\Delta p = -0.000043$$

The frequency of *A* decreases by only four-thousandths of 1 percent. Because mutation rates are so low, the change in allelic frequency due to mutation pressure is exceedingly slow. To change the frequency from 0.50 to 0.49, 2,000 generations are required, and to change it from 0.1 to 0.09, 10,000 generations are necessary. If some reverse mutation occurs, the rate of change is even slower. In practice, mutation by itself changes the allelic frequencies at such a slow rate that populations are rarely in mutational equilibrium. Other processes have more profound effects on allelic frequencies, and mutation alone rarely determines the allelic frequencies of a population. For example, achondroplastic dwarfism is an autosomal dominant trait in humans that arises through recurrent mutation. However, the frequency of this disorder in human populations is determined by an interaction of mutation pressure and natural selection.

KEYNOTE

> When we study what happens when we violate the assumption of the Hardy-Weinberg equilibrium of the absence of mutation, we see that mutation is the only way in which novel genetic material can come to exist within a species. The larger a population, the more potential there is for a novel mutation to arise, but mutation by itself changes allele frequencies only a little.

Random Genetic Drift

Another major assumption of the Hardy-Weinberg law is that the population is infinitely large. Real populations are not infinite in size, but frequently they are large enough that chance factors have small effects on allelic frequencies. Some populations are small, however, and in this case chance factors may produce large changes in allelic frequencies. Random change in allelic frequency due to chance is called random genetic drift, or simply genetic drift. Ronald Fisher and Sewall Wright (see Figure 24.1a and 24.1b), brilliant population geneticists who laid much of the theoretical foundation of the discipline, first described how genetic drift affects the evolution of populations.

Changes in allelic frequency resulting from random events can have important evolutionary implications in small populations. In addition, such changes can have important consequences for the conservation of a rare or endangered species. To see how chance can play a big role in altering the genetic structure of a population, imagine a small group of humans inhabiting a South Pacific island. Suppose that this population consists of only 10 individuals, 5 of whom have green eyes and 5 of whom have brown eyes. For this example, we assume that eye color is determined by a single locus (although actually a number of genes control eye color) and that the allele for green eyes is recessive to brown (*BB* and *Bb* codes for brown eyes and *bb* codes for green). The frequency of the allele for green eyes is 0.6 in the island population. A typhoon strikes the island, killing 50 percent of the population, all of whom have brown eyes. Eye color in no way affects the probability of surviving; the fact that only those with green eyes survive is strictly the result of chance. After the typhoon, the allelic frequency for green eyes is 1.0. Evolution has occurred in this population: The frequency of the green eye allele has changed from 0.6 to 1.0, simply as a result of chance.

Now imagine the same scenario, but this time with a population of 1,000 individuals. As before, 50 percent of the population has green eyes and 50 percent has brown eyes. A typhoon strikes the island and kills half the population. How likely is it that, just by chance, all 500 people who perish will have brown eyes? In a population of 1,000 individuals, the probability of this occurring by chance is extremely remote. This example illustrates an important characteristic of genetic drift: Chance factors are likely to produce rapid changes in allelic frequencies only in small populations.

Random factors producing mortality in natural populations, such as the typhoon in the preceding example, is only one of several ways in which genetic drift arises. Chance deviations from expected ratios of gametes and zygotes also produce genetic drift. We have seen the importance of chance deviations from expected ratios in the genetic crosses we studied in earlier chapters. For example, when we cross a heterozygote with a homozygote (*Aa* × *aa*), we expect 50 percent of the progeny to be heterozygous and 50 percent to be homozygous. We do not expect to get exactly 50 percent every time, however, and if the number of progeny is small, the observed ratio may differ greatly from the expected. Recall that the Hardy-Weinberg law is based on random mating and expected ratios of progeny resulting from each type of mating (see Table 24.2). If the actual number of progeny differs from the expected ratio due to chance, genotypes may not be in Hardy-Weinberg proportions. Simply put, random genetic drift may also result in changes in allelic frequencies.

Chance deviations from expected proportions arise from a general phenomenon called **sampling error.** Imagine that a population produces an infinitely large pool of gametes, with alleles in the proportions p and q. If random mating occurs and all the gametes unite to form viable zygotes, the proportions of the genotypes will be equal to p^2, $2pq$, and q^2, and the frequencies of the alleles in these zygotes will remain p and q. If the number of progeny is limited, however, the gametes that unite to form the progeny constitute a sample from the infinite pool of potential gametes. Just by chance, or by "error," this sample may deviate from the larger pool; the smaller the sample, the larger the potential deviation.

Flipping a coin is analogous to the situation in which sampling error occurs. When we flip a coin, we expect 50 percent heads and 50 percent tails. If we flip the coin 1,000 times, we will get very close to that expected 50:50 ratio. But if we flip the coin only four times, we would not be surprised if by chance we obtain 3 heads and 1 tail, or even all tails. When the sample—in this case, the number of flips—is small, the sampling error can be large. All genetic drift arises from such sampling error.

Mathematicians have worked out the exact probability of getting, for example, 499 heads and 501 tails, and this probability has what is called a binomial distribution. For our purposes, it is important to notice that a population with frequency p of allele A in one generation samples alleles for the next generation, and the expected frequency of that next generation is still p, but there is some variation around p due to drift. In fact, the sampling variance for the allele frequency is simply $pq/2N$, where N is the number of individuals in the population. This is another way to see that small populations have a larger sampling variance, and larger populations have a small sampling variance, so that drift works more slowly in large populations.

Effective Population Size. Genetic drift is random, so we cannot predict what the allelic frequencies will be after drift has occurred. However, since sampling error is related to the size of the population, we can make predictions about the magnitude of genetic drift. Ecologists often measure population size by counting the number of individuals, but not all individuals contribute gametes to the next generation. To determine the magnitude of genetic drift, we must know the **effective population size,** which equals the equivalent number of adults contributing gametes to the next generation. If the sexes are equal in number and all individuals have an equal probability of producing offspring, the effective population size equals the number of breeding adults in the population. However, when males and females are not present in equal numbers, the effective population size (N_e) is

$$N_e = \frac{4 \times N_f \times N_m}{N_f + N_m}$$

where N_f equals the number of breeding females and N_m equals the number of breeding males.

It is not difficult to see why the effective population size is not simply the number of breeding adults. The reason is that males, as a group, contribute half of all genes to the next generation and females, as a group, contribute the other half. Therefore, in a population of 70 females and 2 males, the two males are not genetically equivalent to two females; each male contributes $\frac{1}{2} \times \frac{1}{2} = 0.25$ of the genes to the next generation, whereas each female contributes $\frac{1}{2} \times \frac{1}{70} = 0.007$ of all genes. The small number of males disproportionately influences what alleles are present in the next generation. Using the preceding equation, the effective population size is $N_e = (4 \times 70 \times 2)/(70 + 2) = 7.8$, or approximately 8 breeding adults. This means that in a population of 70 females and 2 males, genetic drift occurs as if the population had only 4 breeding males and 4 breeding females. Therefore, genetic drift has a much greater effect in this population than in one with 72 breeding adults equally divided between males and females.

Other factors, such as differential production of offspring, fluctuating population size, and overlapping generations, can further reduce the effective population size. Most of these factors result in an effective size that is smaller than the count of individual adults in the population.

We can incorporate the effective population size into our formula for the sampling variance of allele frequencies given earlier. This sampling variance, which can also be thought of as the amount of variation in populations resulting from genetic drift, equals

$$s_p^2 = \frac{pq}{2N_e}$$

where N_e is the effective population size and p and q are the allelic frequencies. A more useful measure is the **standard error of allelic frequency,** which is the square root of the variance of allelic frequency:

$$s_p = \sqrt{\frac{pq}{2N_e}}$$

The standard error can be used to calculate the 95 percent confidence limits of allelic frequency, which indicate the expected range of p in 95 percent of such populations. The 95 percent confidence limits are approximately $p \pm 2s_p$. Suppose, for example, that $p = 0.8$ in a population with N_e equal to 50. The standard error in allelic frequency is $s_p = \sqrt{pq/2N_e} = 0.04$. The 95 percent confidence limits for p are therefore $p \pm 2s_p = 0.72 \leq p \leq 0.88$. To interpret the 95 percent confidence limits, imagine that 100 populations with N_e of 50 have p initially equal to 0.8. Genetic drift may cause allelic frequencies in some populations to change; the 95 percent confidence limits tell us that in the next generation, 95 of the original 100 populations should have p within the range of 0.72 to 0.88. Therefore, if we observe a change in p greater than this, say, from 0.8 to 0.68, we know that the probability that this change will occur by genetic drift is less than 0.05. We would probably conclude that some process other than genetic drift contributed to the observed change in allelic frequency.

Bottlenecks and Founder Effects. All genetic drift arises from sampling error, but there are several ways in which sampling error occurs in natural populations. First, as already discussed, genetic drift arises when population size remains continuously small over many generations. Undoubtedly this situation is common, particularly where populations occupy marginal habitats or when competition for resources limits population growth. In such populations, genetic drift plays an important role in the evolution of allelic frequencies. Many species are spread out over a large geographic range. This can result in a species consisting of numerous populations of small size, each undergoing drift independently. In addition, human intervention, such as the clear-cutting of forests, can result in the fragmentation of previously large continuous populations, for example, trees, into small subdivided ones, again, each showing genetic drift.

The effect of genetic drift arising from small population size is seen in a classic laboratory experiment conducted by P. Buri with *Drosophila melanogaster.* Buri examined the frequency of two alleles, bw^{75} and bw, at a locus that determines eye color in the fruit flies. He set up 107 experimental populations, and the initial frequency of bw^{75} was 0.5 in each. The flies in each population interbred randomly, and in each generation Buri randomly selected 8 males and 8 females to be the parents for the next generation. Thus, population size was always 16 individuals. The distribution of allelic frequencies in these 107 populations is presented in Figure 24.11. Notice that the allelic frequencies in the early generations were clumped around 0.5, but genetic drift caused the frequencies in the populations to spread out or diverge over time. By generation 19, the frequency of bw^{75} was 0 or 1 in most populations. What is most elegant about this experiment is that it closely matched the theoretically expected effects of drift demonstrated by Wright and Fisher. The Wright-Fisher model considers the changes in allele frequency each generation as a binomial sampling, but at each successive generation the allele frequency can change, so the binomial parameter (in this case, the allele frequency) changes each generation. The fit of this elegant model to Buri's data was very good, provided that one accounts for the effective population size (which was smaller than 16).

Figure 24.11

Results of Buri's study of genetic drift in 107 populations of *Drosophila melanogaster.* Shown are the distributions of the frequency of the bw^{75} allele among the populations in 19 consecutive generations. Each population consisted of 16 individuals.

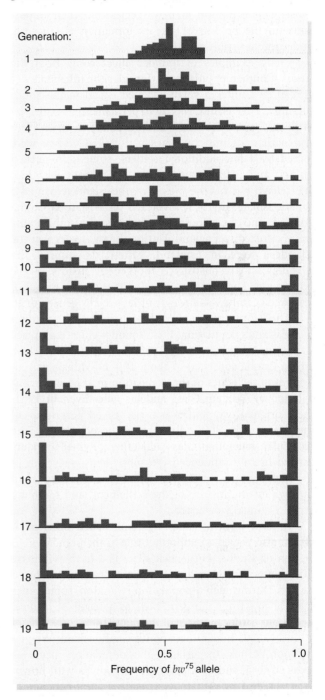

Another way in which genetic drift arises is through **founder effect.** Founder effect occurs when a population is initially established by a small number of breeding individuals. Although the population may subsequently grow in size and later consist of a large number of individuals, the gene pool of the population is derived from the genes present in the original founders. Chance may play a significant role in determining which genes were present among the founders, and this has a profound effect on the gene pool of subsequent generations. Founder effects have frequently been used to explain the subsequent evolution of new species, but their importance to the process of species formation is currently under intense study and debate.

Many examples of founder effect come from the study of human populations. Consider the inhabitants of Tristan da Cunha, a small, isolated island in the South Atlantic. This island was first permanently settled by William Glass, a Scotsman, and his family in 1817. (Several earlier attempts at settlement failed.) They were joined by a few additional settlers, some shipwrecked sailors, and a few women from the distant island of St. Helena, but for the most part the island remained a genetic isolate. In 1961, a volcano on Tristan da Cunha erupted, and the population of almost 300 inhabitants was evacuated to England. During the two years that the islanders were in England, geneticists studied the islanders and reconstructed the genetic history of the population. These studies revealed that the gene pool of Tristan da Cunha was strongly influenced by genetic drift.

Three forms of genetic drift occurred in the evolution of the island's population. First, founder effect took place at the initial settlement. By 1855, the population of Tristan da Cunha consisted of about 100 individuals, but 26 percent of the genes of the population in 1855 were contributed by William Glass and his wife. Even in 1961, these original two settlers contributed 14 percent of all the genes in the 300 individuals of the population. The particular genes that Glass and other original founders carried heavily influenced the subsequent gene pool of the population. Second, population size remained small throughout the history of the settlement, and sampling error continually occurred.

A third form of sampling error, called **bottleneck effect,** also played an important role in the population of Tristan da Cunha. Bottleneck effect is a form of genetic drift that occurs when a population is drastically reduced in size. During such a population reduction, some genes may be lost from the gene pool as a result of chance. Recall our earlier example of the population consisting of 10 individuals inhabiting a South Pacific island. When a typhoon struck the island, the population size was reduced to 5, and by chance all individuals with brown eyes perished in the storm, changing the frequency of green eyes from 0.6 to 1.0. This is an example of bottleneck effect. Bottleneck effect can be viewed as a type of founder effect because the population is refounded by the few individuals that survive the reduction.

Two severe bottlenecks occurred in the history of Tristan da Cunha. The first took place around 1856 and was precipitated by two events: the death of William

Glass and the arrival of a missionary who encouraged the inhabitants to leave the island. At this time many islanders emigrated to America and South Africa, and the population dropped from 103 individuals at the end of 1855 to 33 in 1857. A second bottleneck occurred in 1885. The island of Tristan da Cunha has no natural harbor, and the islanders intercepted passing ships for trade by rowing out in small boats. On November 28, 1885, 15 of the adult males on the island put out in a small boat to make contact with a passing ship. In full view of the entire island community, the boat capsized and all 15 men drowned. Following this disaster, only four adult males were left on the island, one of whom was insane and two of whom were old. Many of the widows and their families left the island during the next few years, and the population size dropped from 106 to 59. Both bottlenecks had a major effect on the gene pool of the population. All the genes contributed by several settlers were lost, and the relative contributions of others were altered by these events. Thus, the gene pool of Tristan da Cunha has been influenced by genetic drift in the form of founder effect, small population size, and bottleneck effect.

As we shall see later, when we discuss migration, gene flow in populations reduces the effects of genetic drift. Small breeding units that lack gene flow are genetically isolated from other groups and often experience considerable genetic drift, even though they are surrounded by much larger populations. A good example is a religious sect known as the Dunkers, found in eastern Pennsylvania. Between 1719 and 1729, 50 Dunker families emigrated from Germany and settled in the United States. Since that time, the Dunkers have remained an isolated group, rarely marrying outside of the sect, and the number of individuals in their communities has always been small.

During the 1950s geneticists studied one of the original Dunker communities in Franklin County, Pennsylvania. At the time of the study, this population had about 300 members, and the population size had remained nearly constant for many generations. The investigators found that some of the allelic frequencies in the Dunkers were very different from the frequencies found among the general population of the United States. The Pennsylvania frequencies were also different from the frequencies of the West German population from which the Dunkers descended. Table 24.7 presents some of the allelic frequencies at the ABO blood group locus. The ABO allele frequencies among the Dunkers are not the same as those in the U.S. population or the West German population. Nor are the Dunker frequencies intermediate between those of the United States and Germany. (Intermediate frequencies might be expected if intermixing of Dunkers and Americans had occurred.) The most likely explanation for the unique Dunker allelic frequencies observed is

Table 24.7	Frequencies of Alleles Controlling the ABO Blood Group System in Three Human Populations						
	Allele Frequencies			**Phenotype (Blood Group) Frequencies**			
Population	I^A	I^B	i	**A**	**B**	**AB**	**O**
Dunker	0.38	0.03	0.59	0.593	0.036	0.023	0.348
United States	0.26	0.04	0.70	0.431	0.058	0.021	0.490
West Germany	0.29	0.07	0.64	0.455	0.095	0.041	0.410

that genetic drift has produced random change in the gene pool. Founder effect probably occurred when the original 50 families emigrated from Germany, and genetic drift probably has continued to influence allelic frequencies in each generation since 1729 because the population size has remained small.

Effects of Genetic Drift. Genetic drift produces changes in allelic frequencies, and these changes have several effects on the genetic structure of populations. First, genetic drift causes the allelic frequencies of a population to change over time. This is illustrated in Figure 24.12. The different lines represent allelic frequencies in several populations over a number of generations. Although all populations begin with an allelic frequency equal to 0.50, the frequencies in each population change over time as a result of sampling error. In each generation, the allelic frequency may increase or decrease, and over time the frequencies wander randomly or drift (hence the name *genetic drift*).

Sometimes, within 30 generations, just by chance, the allelic frequency reaches a value of 0.0 or 1.0. At this point, one allele is lost from the population and the population is said to be *fixed* for the remaining allele in a one locus, two allele example. Once an allele has reached fixation, no further change in allelic frequency can occur unless the other allele is reintroduced through mutation or migration. The probability of fixation in a population increases with time, as shown theoretically by Motoo Kimura in Figure 24.13. If the initial allelic frequencies are equal, which allele becomes fixed is strictly random. On the other hand, if initial allelic frequencies are not equal, the rare allele is more likely to be lost. During this process

Figure 24.13

The average time to fixation or loss of an allele from a population as a function of population size and initial allele frequency as predicted by Kimura. For example, if the initial allele frequency is 0.3 and the population size is 10, it would take just under 2.5×10 generations on average for the allele to be lost or fixed in the population.

Figure 24.12

The effect of genetic drift on the frequency (q) of allele A^2 in four populations. Each population begins with q equal to 0.5, and the effective population size for each is 20. The mean frequency of allele A^2 for the four replicates is indicated by the red line. These results were obtained by a computer simulation.

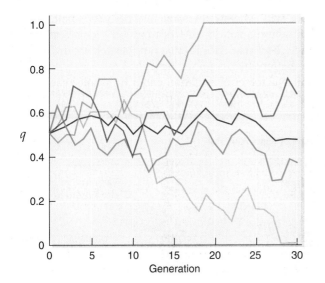

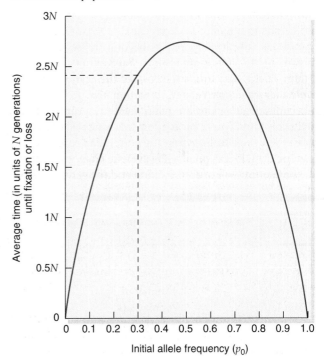

of genetic drift and fixation, the number of heterozygotes in the population also decreases, and after fixation the population heterozygosity is zero. As heterozygosity decreases and alleles become fixed, populations lose genetic variation; thus, the second effect of genetic drift is a reduction in genetic variation within populations.

Since genetic drift causes random change in allelic frequency, the allelic frequencies in separate, individual populations do not change in the same direction. Therefore, populations diverge in their allelic frequencies through genetic drift. This is illustrated in Figures 24.11 and 24.12; all the populations begin with p and q equal to 0.5. After a few generations, the allelic frequencies of the populations diverge, and this divergence increases over generations. The maximum divergence in allelic frequencies is reached when all populations are fixed for one or the other allele. If allelic frequencies are initially 0.5, approximately half of the populations will be fixed for one allele, and half will be fixed for the other.

Since genetic drift is greater in small populations and leads to genetic divergence, we expect more variance in allelic frequency among small populations than among large populations. Such a relationship has been observed in studies of natural populations. Robert K. Selander, for example, studied genetic variation in populations of the house mouse inhabiting barns in Texas. Through systematic trapping, he was able to estimate population size, and using electrophoresis he examined the variance in allelic frequency at two loci, a locus coding for the enzyme esterase (*Est-3*) and a locus coding for hemoglobin (*Hbb*). Selander found that the variance in allelic frequency between small populations was several times larger than that between large populations (Table 24.8), an observation consistent with our understanding of how genetic drift leads to population divergence.

While full development is beyond the scope of this treatment, it is important to note that Kimura's result (see Figure 24.13) also forms the basis of the *neutral theory of molecular evolution*. A new mutation that is not selectively disadvantageous but neutral with respect to natural selection initially will be at a low frequency. Therefore, it most likely will be lost from the population as a result of genetic drift. Occasionally, however, by chance alone, the new mutation will drift to fixation and the gene will have

evolved. Given that mutation is a recurring event, a gene will accumulate differences over time by chance alone. In this way, the genes of two related lineages, such as humans and other primates, can be compared and used to estimate the date at which they last shared a common ancestor. In this way, it is estimated that humans and chimpanzees last shared a common ancestor 6 million years ago. This area is also currently under great debate and study.

KEYNOTE

Genetic drift, or chance changes in allelic frequency caused by sampling error, can have important evolutionary and survival implications for small populations. Genetic drift leads to loss of genetic variation within populations, genetic divergence among populations, and random fluctuation in the allelic frequencies of a population over time. Genetic drift can also explain how molecules in different species accumulate differences on a seemingly regular basis, forming the basis for the neutral theory of molecular evolution.

Balance Between Mutation and Random Genetic Drift. We have seen that mutation continues to introduce new mutation into a population, and the net effect of random genetic drift is to remove that variation from the population. What happens if we combine these two forces, mutation and drift? Will there be a balance achieved such that the rate of gain of variability by mutation is precisely matched by the rate of loss due to chance fixations by drift? Several models have been devoted to this problem, and one of the best known is the *infinitely many alleles model*, often shortened to the *infinite alleles model*. In this model, each mutation that occurs in a gene is assumed to generate a novel allele that has never been seen before. If you imagine a very large gene, consisting of perhaps 10,000 bp, then the chance that two mutations will generate the same allele is very small. So to a first approximation, this assumption seems reasonable. The model also assumes that random genetic drift occurs by the repeated sampling described earlier. In this situation, the forces of mutation and drift balance each other, and we end up with a curious kind of steady state in

Table 24.8 Variance in Allelic Frequency Among Populations as a Function of Population Size

Type of Population	Number of Populations	Mean Allelic Frequency		Variance in Allelic Frequency	
		Est-3[b]	Hbb[8]	Est-3[b]	Hbb[8]
Small ($N < 50$)	29	0.418	0.849	0.051	0.188
Large ($N > 50$)	13	0.372	0.843	0.013	0.008

which new mutations keep on being generated, but alleles are also continuously lost in the population. In this steady state, the number of alleles bounces around a bit but tends not to stray too far from a steady state value. The heterozygosity of the locus, which you can think of as the frequency of heterozygotes in the population, or the chance of drawing two alleles and having them be different, is

$$H = \frac{4N_e\mu}{4N_e\mu + 1}$$

This equation establishes an important point about many models in population genetics: The roles of neutral mutations and population size often are combined into a single term, $4N_e\mu$. This means that if one population is twice as large but has a mutation rate half that of another, the two populations will have the same level of heterozygosity. The reciprocal role is fairly easy to understand: The rate at which a population loses heterozygosity as a result of drift is inversely proportional to its size, while the number of new mutations introduced in each generation is directly proportional to population size. Figure 24.14 shows a plot of the heterozygosity for a range of values of $4N_e\mu$ and shows that plausible population sizes and mutation rates give plausible levels of heterozygosity. One problem with this model was pointed out by John Gillespie, who noted that whereas population sizes vary by several orders of magnitude among organisms, the mutation rates vary only a little. One would expect organisms with higher population sizes to have higher heterozygosity. There are striking exceptions to this prediction, suggesting that a simple mutation-drift balance does not explain everything.

Figure 24.14

Relationship between the neutral parameter $\theta = 4N_e\mu$ and the expected heterozygosity under the infinite alleles models, which gives a balance between mutation and random drift.

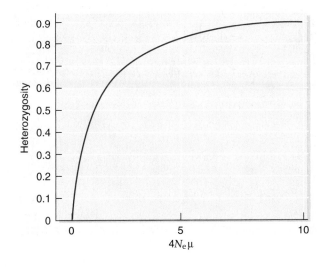

Migration

One of the assumptions of the Hardy-Weinberg law is that the population is isolated and not influenced by other populations. Many populations are not completely isolated, however, and exchange genes with other populations of the same species. Individuals migrating into a population may introduce new alleles to the gene pool and alter the frequencies of existing alleles. Thus, migration has the potential to disrupt Hardy-Weinberg equilibrium and may influence the evolution of allelic frequencies within populations.

The term *migration* usually implies movement of organisms. In population genetics, however, we are interested in the movement of genes, which may or may not occur when organisms move. Movement of genes takes place only when organisms or gametes migrate and contribute their genes to the gene pool of the recipient population. This process is also called **gene flow.**

Gene flow has two major effects on a population. First, it introduces new alleles to the population. Since mutation generally is a rare event, a specific mutant allele may arise in one population and not in another. Gene flow spreads these new alleles to other populations and, like mutation, is a source of genetic variation for the recipient population. Second, when the allelic frequencies of migrants and the recipient population differ, gene flow changes the allelic frequencies within the recipient population. Through exchange of genes, different populations remain similar, and thus migration is a homogenizing force that tends to prevent populations from accumulating genetic differences among them.

To illustrate the effect of migration on allelic frequencies, we consider a simple model in which gene flow occurs in only one direction, from population x to population y. Suppose that the frequency of allele A in population x (p_x) is 0.8 and the frequency of A in population y (p_y) is 0.5. In each generation, some individuals migrate from population x to population y, and these migrants are a random sample of the genotypes in population x. After migration, population y actually consists of two groups of individuals: the migrants, with $p_x = 0.8$, and the residents, with $p_y = 0.5$. The migrants now make up a proportion of population y, which we designate m. The frequency of A in population y after migration (p'_y) is

$$p'_y = mp_x + (1 - m)p_y$$

We see that the frequency of A after migration is determined by the proportion of A alleles in the two groups that now make up population y. The first component, mp_x, represents the A alleles in the migrants; we multiply the proportion of the population that consists of migrants (m) by the allelic frequency of the migrants (p). The second component represents the A alleles in the residents and equals the proportion of the population consisting of residents $(1 - m)$ multiplied by the allelic

Figure 24.15

Theoretical model illustrating the effect of migration on the gene pool of a population. After migration, population y consists of two groups: the migrants with allelic frequency of p_x and the original residents with allelic frequency p_y.

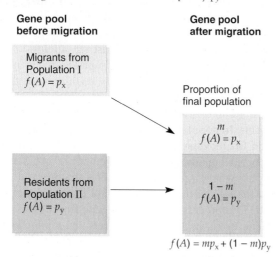

$$f(A) = mp_x + (1 - m)p_y$$

frequency in the residents (p_y). Adding these two components together gives us the allelic frequency of A in population y after migration. This model of gene flow is diagrammed in Figure 24.15.

The change in allelic frequency in population y as a result of migration (Δp) equals the original frequency of A subtracted from the frequency of A after migration:

$$\Delta p = p'_y = p_y$$

In the previous equation, we found that p'_y equaled $mp_x + (1 - m)p_y$, so the change in allelic frequency can be written as

$$\Delta p = mp_x + (1 - m)p_y - p_y$$

Multiplying $(1 - m)$ by p_y in this equation, we obtain

$$\Delta p = mp_x + p_y - mp_y - p_y$$

$$\Delta p = mp_x - mp_y$$

$$\Delta p = m(p_x - p_y)$$

This final equation indicates that the change in allelic frequency from migration depends on two factors: the proportion of the migrants in the final population and the difference in allelic frequency between the migrants and the residents. If no differences exist in the allelic frequency of migrants and residents ($p_x - p_y = 0$), then we can see that the change in allelic frequency is zero. Populations must differ in their allelic frequencies for migration to affect the makeup of the gene pool. With continued migration, p_x and p_y become increasingly similar, and, as a result, the change in allelic frequency due to migration decreases. Eventually, allelic frequencies in the two populations will be equal, and no further change

will occur. However, this is true only when other factors besides migration do not influence allelic frequencies.

In particular, just as a balance between drift and mutation helps to determine the amount of heterozygosity in populations, a balance between drift and migration helps to determine how much populations diverge from one another. When populations are large, drift changes allele frequencies only a little, and a small fraction of migrants each generation will keep populations similar to one another. When populations are small, however, drift changes allele frequencies a lot, and populations will diverge unless the fraction of migrants is large. Extensive gene flow has been shown to occur, for example, among populations of the monarch butterfly (Figure 24.16). Surprisingly, a small number of migrants each generation can reduce the effect of genetic drift, because the roles of migration and drift are combined into the single term $4N_e m$, just as with mutation. Calculations have shown

Figure 24.16

Extensive gene flow occurs among populations of the monarch butterfly. (a) The butterflies overwinter in Mexico (b) and then migrate north during spring and summer to breeding grounds as far away as northern Canada. Extensive gene flow occurs during the migration period, with the result that monarch populations display little genetic divergence.

a)

b)

that a single migrant moving between two populations every other generation will prevent the two populations from becoming fixed for different alleles. In human populations, there is relatively little difference among populations in broad geographical areas, although there are some differences between those areas.

The effects of gene flow have important ramifications, not only for the evolution of species but also for the conservation of species. As discussed earlier, many species that have wide geographic ranges show variation in genetic structure over the species range. Part of the natural genetic structure of a species could include population subdivision in which populations are loosely connected to each other by gene flow. Since gene flow is important in maintaining genetic diversity, this feature of population genetic structure must be taken into account by those interested in conserving genetic structure.

KEYNOTE

Migration of individuals into a population may alter the makeup of the population gene pool if the genes carried by the migrants differ from those of the resident population. Migration, also called gene flow, tends to reduce genetic divergence among populations and increases the effective size of the population. The amount of migration among populations of the same species determines how much genetic substructuring exists and whether different populations of the same species become very different from each other genetically.

Hardy-Weinberg and Natural Selection

We have now examined three major evolutionary processes capable of changing allelic frequencies and producing evolution: mutation, genetic drift, and migration. These processes alter the gene pool of a population, and they certainly influence the evolution of a species. However, mutation, migration, and genetic drift do not result in adaptation. *Adaptation* refers to the tendency for plants and animals to be well suited to the environments in which they are found. **Natural selection** is the process by which traits evolve that make organisms more suited to their immediate environment; these traits increase the organism's chances of surviving and reproducing. Natural selection is responsible for the many extraordinary traits seen in nature: wings that enable a hummingbird to fly backward, leaves of the pitcher plant that capture and devour insects, and brains that allow humans to speak, read, and love. These biological features and countless other exquisite traits are the product of adaptation (Figure 24.17). Genetic drift, mutation, and migration all influence the pattern and process of adaptation, but adaptation arises chiefly from natural

animation

Natural Selection *a*

Figure 24.17

Natural selection produces organisms that are finely adapted to their environment, such as this lizard with cryptic coloration, which allows it to blend in with its natural surroundings.

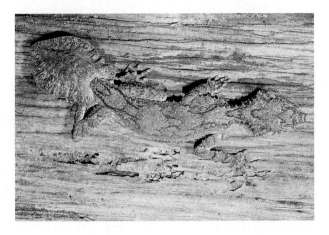

selection. Natural selection is the force responsible for adaptation in plants and animals, and it has shaped much of the phenotypic variation observed in nature.

Charles Darwin and Alfred Russel Wallace (Figure 24.18) independently developed the concept of natural selection in the mid–nineteenth century, although some earlier naturalists had similar ideas. In 1858, Darwin and Wallace's theory was presented to the Linnaean Society of London and was enthusiastically received by other scientists. Darwin pursued the theory of evolution further than Wallace did, amassing hundreds of observations to support it and publishing his ideas in the book *On the Origin of Species*. For his innumerable contributions to our understanding of natural selection, Darwin often is regarded as the father of evolutionary theory. What is amazing about this theory is that Darwin had no clue about how genetic transmission worked. All that was necessary for his theory was that somehow offspring resemble their parents. Knowing the details of genetic transmission of traits makes the theory much deeper and richer and provides much more satisfying tests of how evolutionary change works at the genetic level.

Natural selection can be defined as differential reproduction of genotypes. It simply means that individuals with certain genes produce more offspring than others; therefore, those genes increase in frequency in the next generation, as discussed in Chapter 14. Through natural selection, traits that contribute to survival and reproduction increase over time. In this way, organisms adapt to their environment.

Selection in Natural Populations. A classic example of selection in natural populations is the evolution of melanic (dark) forms of moths in association with industrial pollution, a phenomenon known as industrial melanism. Melanic phenotypes have appeared in a num-

a)
b)

Figure 24.18

(a) **Charles Darwin and**
(b) **Alfred Russel Wallace**
should be given equal credit for
developing the theory of evolu-
tion through natural selection.

ber of different species of moths found in the industrial regions of continental Europe, North America, and England. One of the best-studied cases involves the peppered moth, *Biston betularia.* The common phenotype of this species, called the *typical* form, is a greyish white color with black mottling over the body and wings.

Before 1848, all peppered moths collected in England possessed this *typical* phenotype, but in 1848, a single black moth was collected near Manchester, England. This new phenotype, called *carbonaria,* presumably arose by mutation and rapidly increased in frequency around Manchester and in other industrial regions. By 1900, the *carbonaria* phenotype had reached a frequency of more than 90 percent in several populations. High frequencies of *carbonaria* appeared to be associated with industrial regions, whereas the *typical* phenotype remained common in more rural districts. Laboratory studies by a number of

investigators, including E. B. Ford and R. Goldschmidt, demonstrated that the *carbonaria* phenotype was dominant to the *typical* phenotype. A third phenotype was also discovered, which was somewhat intermediate to *typical and carbonaria;* this phenotype, *insularia,* was produced by a dominant allele at a different locus.

H. B. D. Kettlewell investigated color polymorphism in the peppered moth, demonstrating that the increase in the *carbonaria* phenotype occurred as a result of strong selection against the *typical* form in polluted woods. Peppered moths are nocturnal; during the day they rest on the trunks of lichen-covered trees. Birds often prey on the moths during the day, but because the lichens that cover the trees are naturally grey in color, the *typical* form of the peppered moth is well camouflaged against this background (Figure 24.19a). In industrial areas, however, extensive pollution beginning with the indus-

Figure 24.19

Biston betularia, **the peppered moth, and its dark form** *carbonaria* **(a) on the trunk of a lichened tree in the unpolluted countryside and (b) on the trunk of a tree with dark bark.** On the lichened tree, the dark form of the moth is readily seen, whereas the light form is well camouflaged. On the dark tree, the dark form of the moth is well camouflaged.

a)
b)

trial revolution in the mid–nineteenth century had killed most of the lichens and covered the tree trunks with black soot. Against this black background, the *typical* phenotype was conspicuous and was readily consumed by birds. In contrast, the *carbonaria* form was well camouflaged against the blackened trees and had a higher rate of survival than did the *typical* phenotype in polluted areas (Figure 24.19b). Because *carbonaria* survived better in polluted woods, more *carbonaria* genes were transmitted to the next generation; thus, the *carbonaria* phenotype increased in frequency in industrial areas. In rural areas, where pollution was absent, the *carbonaria* phenotype was conspicuous and the *typical* form was camouflaged; in these regions the frequency of the *typical* form remained high.

Kettlewell demonstrated that selection affected the frequencies of the two phenotypes by conducting a series of mark-and-recapture experiments involving dark and light moths in smoky, industrial Birmingham, England, and in nonindustrialized Dorset. As predicted, the *typical* phenotype was favored in Dorset, and *carbonaria* was favored in Birmingham.

Fitness and Coefficient of Selection. Darwin described natural selection primarily in terms of survival, and even today, many nonbiologists think of natural selection in terms of a struggle for existence. However, what is most important in the process of natural selection is the relative number of genes that are contributed to future generations. Certainly the ability to survive is important, but survival alone does not ensure that genes are passed on; reproduction must also occur. Therefore, we measure natural selection by assessing reproduction. Natural selection is measured in terms of **Darwinian fitness,** which is defined as the relative reproductive ability of a genotype.

Darwinian fitness is often symbolized as W. Since it is a measure of the relative reproductive ability, population geneticists usually assign a fitness of 1 to a genotype that produces the most offspring. The fitnesses of the other genotypes are assigned relative to this. For example, suppose that the genotype G^1G^1 on the average produces eight offspring, G^1G^2 produces an average of four offspring, and G^2G^2 produces an average of two offspring. The G^1G^1 genotype has the highest reproductive output, so its fitness is 1 ($W_{11} = 1.0$). Genotype G^1G^2 produces on the average four offspring for the eight produced by the most fit genotype, so the fitness of G^1G^2 (W_{12}) is $4/8 = 0.5$. Similarly, G^2G^2 produces two offspring for the eight produced by G^1G^1, so the fitness of G^2G^2 (W_{22}) is $2/8 = 0.25$. Table 24.9 illustrates the calculation of relative fitness values.

Estimation of fitness values for genotypes must be done with great care. For example, equating fitness to the number of offspring is an oversimplification, because we need to know about the survival probability of those offspring as well. For example, David Lack found that starlings had an optimal number of eggs that they could successfully rear to mature offspring. If they laid too many eggs, they produced fewer successful chicks. Assigning higher fitness values to birds that laid more eggs would have been incorrect. The fitness associated with a genotype is difficult to pin down by looking at a snapshot of a part of the organism's life. Genotypes that have a high survival probability can have on average fewer offspring. Single genes can have effects on different aspects of the life cycle that affect fitness. Such **pleiotropic** effects make a big difference in the overall genotypic fitnesses. Despite these challenges, Darwinian fitness tells us how well a genotype is doing in terms of natural selection. A related measure is the **selection coefficient,** which is a measure of the relative intensity of selection against a genotype. The selection coefficient is symbolized by s and equals $1 - W$. In our example, the selection coefficients for G^1G^1 are $s = 0$; for G^1G^2, $s = 0.5$; for G^2G^2, $s = 0.75$.

Effect of Selection on Allelic Frequencies. Natural selection produces a number of different effects. At times, natural selection eliminates genetic variation, and at other times it maintains variation; it can change allelic frequencies or prevent allelic frequencies from changing; it can produce genetic divergence between populations or maintain genetic uniformity. Which of these effects occurs depends

Table 24.9 Computation of Fitness Values and Selection Coefficients of Three Genotypes

	Genotypes		
	G^1G^1	G^1G^2	G^2G^2
Number of breeding adults in one generation	16	10	20
Number of offspring produced by all adults of the genotype in the next generation	128	40	40
Average number of offspring produced per breeding adult	128/16 = 8	40/10 = 4	40/20 = 2
Fitness W (relative number of offspring produced)	8/8 = 1	4/8 = 0.5	2/8 = 0.25
Selection coefficient ($s = 1 - W$)	1 − 1 = 0	1 − 0.5 = 0.5	1 − 0.25 = 0.75

Table 24.10	General Method of Determining Change in Allelic Frequency Caused by Natural Selection		

	Genotypes		
	A^1A^1	A^1A^2	A^2A^2
Initial genotypic frequencies	p^2	$2pq$	q^2
Fitness[a]	W_{11}	W_{12}	W_{22}
Frequency after selection	p^2W_{11}	$2pqW_{12}$	q^2W_{22}
Relative genotypic frequency after selection	$P' = \dfrac{p^2W_{11}}{W^b}$	$H' = \dfrac{2pqW_{12}}{W}$	$Q' = \dfrac{q^2W_{22}}{W}$

Allelic frequency after selection $= p' = P' + \frac{1}{2}(H')$

$$q' = 1 - p'$$

Change in allelic frequency caused by selection $= \Delta p = p' = p$

[a] For simplicity, fitness in this example is considered to be the probability of survival. Change in allelic frequency caused by differences in the number of offspring produced by the genotypes is calculated in the same manner.
[b] $\overline{W} = p^2W_{11} + 2pqW_{12} + q^2W_{22}$

primarily on the relative fitness of the genotypes and on the frequencies of the alleles in the population.

The change in allelic frequency that results from natural selection can be calculated by constructing a table such as Table 24.10. This table method can be used for any type of single-locus trait, whether the trait is dominant, codominant, recessive, or overdominant. To use the table method, we begin by listing the genotypes (A^1A^1, A^1A^2, and A^2A^2) and their initial frequencies. If random mating has just taken place, the genotypes are in Hardy-Weinberg proportions, and the initial frequencies are p^2, $2pq$, and q^2. We then list the fitnesses for each of the genotypes, W_{11}, W_{12}, and W_{22}. Now, suppose that selection occurs and only some of the genotypes survive. The contribution of each genotype to the next generation is equal to the initial frequency of the genotype multiplied by its fitness. For A^1A^1 this is $p^2 \times W_{11}$. Notice that the contributions of the three genotypes do not add up to 1. We calculate the relative contributions of each genotype by dividing each by the mean fitness of the population. The *mean fitness of the population* equals $p^2W_{11} + 2pqW_{12} + q^2W_{22} = W$. The mean fitness is the average fitness of individuals in the population. After dividing the relative contributions of each genotype by the mean fitness, we have the frequencies of the genotypes after selection. We then calculate the new allelic frequency (p') from the genotypes after selection, using our familiar formula, $p' =$ (frequency of A^1A^1) + ($\frac{1}{2} \times$ frequency of A^1A^2). Finally, the change in allelic frequency resulting from selection equals $p' - p$. A sample calculation using some actual allelic frequencies and fitness values is presented in Table 24.11.

The wide range of generally unappreciated effects of natural selection discussed earlier can now be under-

stood in terms of the figures in Table 24.10. Again remembering that we will arbitrarily set the genotype with highest fitness to 1.0, we get a variety of classes of natural selection, each with its own effects by permuting all possible relationships among fitness values for the genotypes. These include the following:

1. $W_{11} = W_{12} = W_{22} = 1.0$. All fitnesses are equal, and there is no selection.

2. $W_{11} = W_{12} < 1.0$ and $W_{22} = 1.0$. The heterozygote has a fitness equal to a homozygote but less than the best fitness of the other homozygote. Natural selection is operating against a dominant allele.

3. $W_{11} = W_{12} = 1.0$ and $W_{22} < 1.0$. The heterozygote along with a homozygote has the highest fitness, which is greater than that of the other homozygote. Natural selection is operating against a recessive allele.

4. $W_{11} < W_{12} < 1.0$ and $W_{22} = 1.0$. The heterozygote has an intermediate fitness. Natural selection is operating without effects of dominance.

5. W_{11} and $W_{22} < 1.0$ and $W_{12} = 1.0$. The heterozygote has the highest fitness, and the two homozygotes have a lower fitness that may or may not be the same. Natural selection is favoring the heterozygote.

6. $W_{12} < W_{11}, W_{22} = 1.0$. The heterozygote has lower fitness than both homozygotes. Only one of the homozygotes must have a fitness equal to 1.0. Natural selection is favoring the homozygotes.

Each of the five cases of natural selection results in a characteristic pattern of change in the genetic structure of a population. Cases 2, 3, and 4 are all a type of natural selection called *directional selection* and result in the elimination or at least in a great reduction in frequency of one of the alleles. Case 5 is very different and results in no evolutionary change once a stable equilibrium has

Table 24.11 General Method of Determining Change in Allelic Frequency Caused by Natural Selection When Initial Allelic Frequencies Are $p = 0.6$ and $q = 0.4$

	Genotypes		
	A^1A^1	A^1A^2	A^2A^2
Initial genotypic frequencies	p^2	$2pq$	q^2
	$(0.6)^2 = 0.36$	$2(0.6)(0.4) = 0.48$	$(0.4)^2 = 0.16$
Fitness	$W_{11} = 0$	$W_{12} = 0.4$	$W_{22} = 1$
Frequency after selection	$p^2W_{11} =$	$2pqW_{12} =$	$q^2W_{22} =$
	$(0.36)(0) = 0$	$(0.48)(0.4) = 0.19$	$(0.16)(1) = 0.16$
Relative genotypic frequency after selection	$P' = \dfrac{p^2W_{11}}{\overline{W}^a}$	$H' = \dfrac{2pqW_{12}}{\overline{W}}$	$Q' = \dfrac{q^2W_{22}}{\overline{W}}$
	$P' = 0/0.35 = 0$	$H' = 0.19/0.35$	$Q' = 0.16/0.35$
		$= 0.54$	$= 0.46$

Allelic frequency after selection $p' = P' + \tfrac{1}{2}(H')$

$$p' = 0 + \tfrac{1}{2}(0.54) = 0.27$$
$$q' = 1 - p' = 1 - 0.27 = 0.73$$

Change in allelic frequency caused by selection $= \Delta p = p' - p$

$$\Delta p = 0.27 - 0.6 = -0.33$$

$^a\overline{W} = p^2W_1 + 2pqW_{12} = q^2W_{22}$
$\overline{W} = 0 + 0.19 + 0.16$
$\overline{W} = 0.35$

been reached. Case 6 results in what looks like a directional change in allele frequency, but the allele that is selected against depends on the initial allele frequency. We now consider how several of these cases affect the genetic structure of a population.

Selection Against a Recessive Trait. Cases 2, 3, and 4 are similar in that they result in a directed change in the allele frequency of a population. This directional effect is the one that is most often associated with natural selection. The case of the peppered moth discussed earlier falls into the category of selection against a recessive allele. We will discuss one of these cases in more detail because many important traits and most new mutations are recessive and have reduced fitness. When a trait is completely recessive (case 3), both the heterozygote and the dominant homozygote have a fitness of 1, whereas the recessive homozygote has reduced fitness, as shown here.

Genotype	Fitness
AA	1
Aa	1
Aa	$1 - s$

If the genotypes are initially in Hardy-Weinberg proportions, the contribution of each genotype to the next generation is the frequency times the fitness.

AA	$p^2 \times 1 = p^2$
Aa	$2pq \times 1 = 2pq$
aa	$q^2 \times (1 - s) = q^2 - sq^2$

The mean fitness of the population is $p^2 + 2pq + q^2 - sq^2$. Since $p^2 + 2pq + q^2 = 1$, the mean fitness becomes $1 - sq^2$, and the normalized genotypic frequencies after selection are

AA	$\dfrac{p^2}{1 - sq^2}$
Aa	$\dfrac{2pq}{1 - sq^2}$
Aa	$\dfrac{q^2 - sp^2}{1 - sq^2}$

To obtain q', the frequency after selection, we add the frequency of the *aa* homozygote and half the frequency of the heterozygote.

$$q' = \frac{q^2 - sq^2}{1 - sq^2} + \frac{1}{2} \times \frac{2pq}{1 - sq^2}$$
$$= \frac{q^2 - sq^2}{1 - sq^2} + \frac{pq}{1 - sq^2}$$
$$= \frac{q^2 - sq^2 + pq}{1 - sq^2}$$
$$= \frac{q^2 + pq - sq^2}{1 - sq^2}$$
$$= \frac{q(q + p) - sq^2}{1 - sq^2}$$

Since $(q + p) = 1$,

$$q' = \frac{q - sq^2}{1 - sq^2}$$

Therefore, the change in the frequency of a after one generation of selection is

$$\Delta q = q' - q$$

$$= \frac{q - sq^2}{1 - sq^2} - q$$

$$= \frac{q - sq^2}{1 - sq^2} - \frac{q(1 - sq^2)}{1 - sq^2}$$

$$= \frac{q - sq^2 - q(1 - sq^2)}{1 - sq^2}$$

$$= \frac{q - sq^2 - q + sq^3}{1 - sq^2}$$

$$= \frac{-sq^2 + sq^3}{1 - sq^2}$$

$$= \frac{-sq^2(1 - q)}{1 - sq^2}$$

$\Delta q = -spq^2/(1 - sq^2)$ because $1 - q = p$. When $\Delta q = 0$, no further change occurs in allelic frequencies. Notice that there is a negative sign in the equation to the left of spq^2, because the values of s, p, and q are always positive or zero, Δq is negative or zero. Thus, the value of q decreases with selection.

Selection also depends on the actual frequencies of the allele in the population. This is because the relative proportions of Aa and aa individuals at various frequencies of allele a influence how effectively selection can reduce a detrimental recessive trait. When the frequency of a recessive allele is high, many homozygous recessive individuals are present in the population and have low fitness, causing a large change in the allelic frequency. When the allelic frequency is low, however, the homozygous recessive genotype is rare, and little change in allelic frequency occurs.

Figure 24.20 shows the magnitude of change in allelic frequency for each generation in three populations with different initial allelic frequencies. Population 1 begins with allelic frequency q equal to 0.9, population 2 begins with q equal to 0.5, and population 3 begins with q equal to 0.1. In this example, the homozygous recessive genotype (aa) has a fitness of 0 and the other two genotypes (AA and Aa) have a fitness of 1 (recessive lethal condition). When the frequency of q is high, as in population 1, the change in allelic frequency is large; in the first generation, q drops from 0.9 to 0.47. However, when q is small, as in population 3, the change in q is much less; here q drops from 0.1 to

Figure 24.20

Effectiveness of selection against a recessive lethal genotype at different initial allele frequencies. The three populations had initial frequencies of 0.9, 0.5, and 0.1.

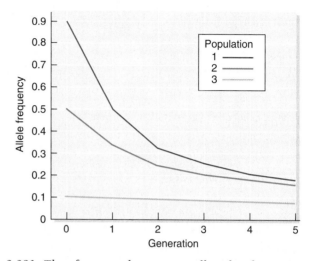

0.091. Therefore, as q becomes smaller, the change in q becomes less. Because of this diminishing change in frequency, it is almost impossible to eliminate a recessive trait from the population entirely. This is easily understood if one realizes that the final recessive alleles in a population will almost always find themselves in the heterozygote condition. However, this result applies only to completely recessive traits; if the fitness of the heterozygote is also reduced (case 4), the change in allelic frequency will be more rapid because now selection also acts against the heterozygote in addition to the homozygote. The effect of dominance on changes in allelic frequency as a result of selection is illustrated in Figure 24.21.

Figure 24.21

Fitnesses of the genotypes AA, Aa, and aa are 1, 0.5, and 0.5 for the dominant case, 1, 0.75, and 0.5 for the additive case, and 1, 1, and 0.5 for the recessive case. Frequency of the a allele is plotted.

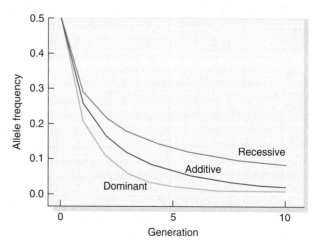

Table 24.12   Formulas for Calculating Change in Allelic Frequency After One Generation of Selection

Type of Selection	Fitnesses of Genotypes			Calculation of Change in Allelic Frequency
	A^1A^1	A^1A^2	A^2A^2	
Selection against recessive homozygote	1	1	$1 - s$	$\Delta q = \dfrac{-spq^2}{1 - sq^2}$
Selection against a dominant allele	$1 - s$	$1 - s$	1	$\Delta p = \dfrac{-spq^2}{1 - s + sq^2}$
Selection with no dominance	1	$(1 - s/2)$	$1 - s$	$\Delta q = \dfrac{-spq/2}{1 - sq}$
Selection that favors the heterozygote (overdominance)	$1 - s$	1	$1 - t$	$\Delta q = \dfrac{pq(sp - tq)}{1 - sp^2 - tq^2}$
Selection against the heterozygote	1	$1 - s$	1	$\Delta q = \dfrac{spq(q - p)}{1 - 2spq}$
General	W_{11}	W_{12}	W_{22}	$\Delta q = \dfrac{pq[p(W_{12} - W_{11}) + q(W_{22} - W_{12})]}{\overline{W}^a}$

^{a} *Note:* For calculation of $\overline{W}$ see Table 22.10.

We have discussed at length the effects of selection on a recessive trait to illustrate how the formula for change in allelic frequency can be derived from our general table method of allelic frequency change under selection. Similar derivations can be carried out for dominant traits and codominant traits (cases 2 and 4). We will not present those derivations here, but the appropriate formula for calculating changes in allelic frequency under different types of dominance are presented in Table 24.12. However, by using the table method it is possible to calculate changes in allelic frequency for any type of trait.

Heterozygote Superiority. Natural selection does not always result in a directional change in allele frequency and a decrease in genetic variation. Some forms of selection result in the maintenance of genetic variation and form the backbone of the balanced model of genetic variation discussed earlier. The simplest type of balancing selection is called **heterosis, overdominance,** or **heterozygote superiority.** An equilibrium of allelic frequencies arises when the heterozygote has higher fitness than either of the homozygotes. In this case (case 5), both alleles are maintained in the population because both are favored in the heterozygote genotype. Allelic frequencies will change as a result of selection until the equilibrium point is reached and then will remain stable. The allelic frequencies at which the population reaches equilibrium depend on the relative fitnesses of the two homozygotes. If the selection coefficient of AA is s and the selection coefficient of aa is t, it can be shown algebraically that at equilibrium

$$\hat{p} = f(A) = t/(s + t)$$

and

$$\hat{q} = f(a) = s/(s + t)$$

Notice that if selection against both homozygotes is the same (i.e., $s = t$), then the equilibrium allele frequency is 0.5. As selection against the homozygotes becomes less symmetrical the equilibrium allele frequency shifts in the direction of the most fit homozygote.

The most famous example of heterozygote superiority operating in nature is provided by human sickle-cell anemia. Sickle-cell anemia results from a mutation in the gene coding for β-hemoglobin. In some populations, there are three hemoglobin genotypes: *Hb-A/Hb-A*, *Hb-A/Hb-S*, and *Hb-S/Hb-S*. Individuals with the *Hb-A/Hb-A* genotype have completely normal red blood cells, *Hb-S/Hb-S* individuals have sickle-cell anemia, and *Hb-A/Hb-S* individuals have sickle-cell trait, a mild form of sickle-cell anemia. In an environment in which malaria is common, the heterozygotes are at a selective advantage over the two homozygotes. The mild anemia suffered by heterozygotes is enough to inhibit growth and reproduction of the malarial parasite. The heterozygotes therefore have greater resistance to malaria and thus higher fitness than do *Hb-A/Hb-A* individuals. The *Hb-S/Hb-S* individuals are at a serious selective disadvantage because they have sickle-cell anemia. As a result, in malaria-infested areas in which the *Hb-S* gene is also found, an equilibrium state is established in which a significant number of *Hb-S* alleles are found in the

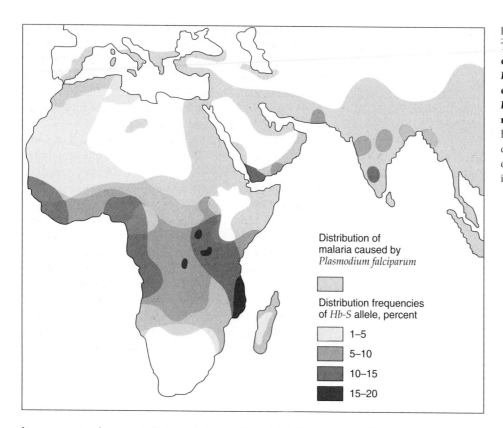

Figure 24.22

The distribution of malaria caused by the parasite *Plasmodium falciparum* coincides with distribution of the *Hb-S* allele for sickle-cell anemia. The frequency of *Hb-S* is high in areas where malaria is common because *Hb-A/Hb-S* heterozygotes are resistant to malarial infection.

Distribution of
malaria caused by
Plasmodium falciparum

Distribution frequencies
of *Hb-S* allele, percent

	1–5
	5–10
	10–15
	15–20

heterozygotes because of the selective advantage of this genotype. The distributions of malaria and the *Hb-S* allele are illustrated in Figure 24.22. Thus, despite the problems faced by *Hb-S* homozygotes, natural selection cannot eliminate this allele from the population because the allele has beneficial effects in the heterozygote state.

K E Y N O T E

Natural selection involves differential reproduction of genotypes and is measured in terms of Darwinian fitness, the relative reproductive contribution of a genotype. The effects of selection depend on the relative fitnesses of the different genotypes. Directional selection results in the directional change in allele frequency, with the disfavored allele being eliminated from the population in the cases where it is dominant or codominant but persisting in the population at low frequencies if it is invisible in the heterozygote. In either case directional selection decreases the amount of genetic variation in a population. Balancing selection, exemplified here as heterozygote superiority, results in the maintenance of genetic variation in the population.

Balance Between Mutation and Selection

We have already seen two examples where we violated the Hardy-Weinberg law assumptions more than one at a time: the combination of drift and mutation and the combination of drift and migration. We saw that mutation, migration, and small population size interact to determine the genetic structure of a population when the alleles are selectively neutral. Population genetics theory has also been extended to accommodate other violations of several assumptions simultaneously; much has been learned by taking this approach. Here, as an example, we consider the simultaneous effects of mutation and natural selection.

As we have seen, natural selection can reduce the frequency of a deleterious recessive allele. As the frequency of the allele becomes low, the change in frequency diminishes with each generation. When the allele is rare, the change in frequency is very slight. Opposing this reduction in the allele's frequency due to selection is mutation pressure, which continually produces new alleles and tends to increase the frequency. Eventually a balance, or equilibrium, is reached, in which the input of new alleles by recurrent mutation is counterbalanced by the loss of alleles through natural selection. When equilibrium is obtained, the allele's frequency remains stable, despite the fact that selection and mutation continue, unless the equilibrium is perturbed by some other process.

Consider a population in which selection occurs against a deleterious recessive allele, *a*. As we saw on pages 691–693, the amount *a* will change in one generation (Δq) as a result of selection is

$$\Delta q = -spq^2/(1 - sq^2)$$

For a rare recessive allele, q^2 will be near 0, and the denominator in this equation, $1 - sq^2$, will be approximately 1, so that the decrease in frequency caused by selection is given by

$$\Delta q = -spq^2$$

At the same time, the frequency of the *a* allele increases as a result of mutation from *A* to *a*. Provided the frequency of *a* is low, the reverse mutation of *a* to *A* essentially can be ignored. Equilibrium between selection and mutation occurs when the decrease in allelic frequency produced by selection is the same as the increase produced by mutation:

$$spq^2 = up$$

We can predict the frequency of *a* at equilibrium ($\hat{q}$) by rearranging this equation:

$$sq^2 = u$$
$$q^2 = u/s$$

and

$$\hat{q} = \sqrt{u/s}$$

If the recessive homozygote is lethal ($s = 1$), the equation becomes

$$\hat{q} = \sqrt{u}$$

As an example of the balance between mutation and selection, consider a recessive gene for which the mutation rate is 10^{-6} and *s* is 0.1. At equilibrium, the frequency of the gene will be $\hat{q} = \sqrt{10^{-6}/0.1} = 0.0032$. Most recessive deleterious traits remain within a population at low frequency because of equilibrium between mutation and selection.

For a dominant allele *A*, the frequency at equilibrium ($\hat{q}$) is

$$\hat{p} = u/s$$

If the mutation rate is 10^{-6} and *s* is 0.1, the frequency of the dominant gene at equilibrium is $10^{-6}/0.1 = 0.00001$, which is considerably less than the equilibrium frequency for a recessive allele with the same fitness and mutation rate. This is because selection cannot act on a recessive allele in the heterozygote state, whereas both the homozygote and the heterozygote for a dominant allele have reduced fitness. For this reason, detrimental dominant alleles are generally far less common than recessive ones.

Assortative Mating

A fundamental assumption of the Hardy-Weinberg law is that members of the population mate randomly. But many populations do not mate randomly for some traits, and when nonrandom mating occurs, the genotypes do not exist in the proportions predicted by the Hardy-Weinberg law. One form of nonrandom mating is **positive assortative mating,** which occurs when individuals with similar phenotypes mate preferentially. Positive assortative mating is common in natural populations. For example, humans mate assortatively for height; tall men and tall women marry more frequently and short men and short women marry more frequently than would be expected if men and women chose mates at random. **Negative assortative mating** occurs when phenotypically dissimilar individuals mate more often than randomly chosen individuals. If humans exhibited negative assortative mating for height, tall men and short women would marry preferentially and short men and tall women would marry preferentially. Positive assortative mating does not affect the allelic frequencies of a population, but it will influence the genotypic frequencies if the phenotypes for which assortative mating occurs are genetically determined. Negative assortative mating may affect both the allelic and genotypic frequencies, because types that are rare in the population may mate more frequently than those that are common.

Inbreeding

Another important departure from random mating is **inbreeding.** Inbreeding involves preferential mating between close relatives. In very small populations, even if individuals tossed a coin or used some other randomizing procedure to choose mates, their mates probably would end up being relatives. In this sense, small populations suffer the consequences of inbreeding even if there is no preferential tendency to select relatives as mates. Inbreeding often is measured in terms of the coefficient of inbreeding (*F*). The greater the value of *F*, the greater the reduction in heterozygosity relative to that expected from the Hardy-Weinberg expectation. Thus,

$$F = \frac{\text{expected heterozygosity} - \text{observed heterozygosity}}{\text{expected heterozygosity}}$$

If genotypes are in Hardy-Weinberg proportions, $F = 0$ because observed heterozygosity and expected heterozygosity are equal. However, regular systems of inbreeding exist, such as self-fertilization, sib mating, and mating between first cousins. After one generation of mating in such systems the value of *F* would be 0.5, 0.25, and 0.06, respectively.

The most extreme case of inbreeding is self-fertilization, which occurs in many plants and a few animals, such as some snails. The effects of self-fertilization are illustrated in Table 24.13. Assume that we begin with a population consisting entirely of *Aa* heterozygotes and that all individuals in this population reproduce by self-fertilization. After one generation of self-fertilization, the progeny will consist of ¼ *AA*, ½ *Aa*, and ¼ *aa*. Now only half of the population consists of heterozygotes. When this generation undergoes

Table 24.13	**Relative Genotype Distributions Resulting from Self-Fertilization over Several Generations Starting with an *Aa* Individual**		
	Frequencies of Genotypes		
Generation	***AA***	***Aa***	***aa***
0	0	1	0
1	$1/4$	$1/2$	$1/4$
2	$1/4 + 1/8 = 3/8$	$1/4$	$1/4 + 1/8 = 3/8$
3	$3/8 + 1/16 = 7/16$	$1/8$	$3/8 + 1/16 = 7/16$
4	$7/16 + 1/32 = 15/32$	$1/16$	$7/16 + 1/32 = 15/32$
5	$15/32 + 1/64 = 31/64$	$1/32$	$15/32 + 1/64 = 31/64$
n	$[1 - (1/2)^n]/2$	$(1/2)^n$	$[1 - (1/2)^n]/2$
∞	$1/2$	0	$1/2$

self-fertilization, the *AA* homozygotes will produce only *AA* progeny, and the *aa* homozygotes will produce only *aa* progeny. When the heterozygotes reproduce, however, only half of their progeny will be heterozygous like the parents, and the other half will be homozygous ($1/4$ *AA* and $1/4$ *aa*). This means that in each generation of self-fertilization, the percentage of heterozygotes decreases by 50 percent. After a large number of generations, there will be no heterozygotes and the population will be divided equally between the two homozygous genotypes. Note that the population was in Hardy-Weinberg proportions in the first generation after self-fertilization, but after further rounds the proportion of homozygotes is greater than that predicted by the Hardy-Weinberg law.

It should be noted that inbreeding has very similar effects to genetic drift in small populations. In both cases heterozygosity decreases and homozygosity increases. In the case of inbreeding in large populations, however, allele frequencies stay the same while homozygosity increases, whereas in the case of drift, allele frequency changes while homozygosity increases. Drift causes only small departures from Hardy-Weinberg; those caused by inbreeding can be extreme.

The result of continued self-fertilization is to increase homozygosity at the expense of heterozygosity. The frequencies of alleles *A* and *a* remain constant, while the frequencies of the three genotypes change significantly. When less intensive inbreeding occurs, similar but less pronounced effects occur.

K E Y N O T E

Inbreeding involves preferential mating between close relatives. Continued inbreeding increases homozygosity within a population and in most species results in reduced fitness, known as inbreeding depression.

Summary of the Effects of Evolutionary Forces on the Genetic Structure of a Population

Let us now review the major effects of the different evolutionary processes on (1) changes in allelic frequency within a population; (2) genetic divergence between populations; and (3) increases and decreases in genetic variation within populations.

Changes in Allelic Frequency Within a Population
Mutation, migration, genetic drift, and selection all have the potential to change the allelic frequencies of a population over time. However, mutation usually occurs at such a low rate that the change resulting from mutation pressure alone is usually negligible. Genetic drift produces substantial changes in allelic frequency when population size is small. Furthermore, mutation, migration, and selection may lead to equilibria where these processes continue to act, but the allelic frequencies no longer change. While negative assortative mating can lead to changes in allelic frequencies, other forms of nonrandom mating do not change allelic frequencies. All forms of nonrandom mating affect the genotypic frequencies of a population: Inbreeding leads to increases in homozygosity, and if there are deleterious recessive alleles in the population (which there almost always are), then inbreeding leads to reduced fitness or inbreeding depression.

Genetic Divergence Among Populations
Several evolutionary processes lead to genetic divergence between populations. Since genetic drift is a random process, allelic frequencies in different populations may drift in different directions, so genetic drift can produce genetic divergence among populations. Migration among populations has just the opposite effect, tending

to equalize allelic frequency differences among populations. If the population size is small, different mutations may arise in different populations, so mutation may contribute to population differentiation. Natural selection can increase genetic differences among populations by favoring different alleles in different populations, or it can prevent divergence by keeping allelic frequencies uniform among populations. Nonrandom mating, by itself, will not generate genetic differences between populations, although it may contribute to the effects of other processes by increasing or decreasing effective population size.

Increases and Decreases in Genetic Variation Within Populations

Migration and mutation tend to increase genetic variation within populations by introducing new alleles to the gene pool. Genetic drift produces the opposite effect, decreasing genetic variation within small populations through loss of alleles. Since inbreeding leads to increases in homozygosity, it also diminishes genetic variation within populations; outbreeding, on the other hand, increases genetic variation by increasing heterozygosity. Natural selection can increase or decrease genetic variation; if one particular allele is favored, other alleles decrease in frequency and can be eliminated from the population by selection. Alternatively, natural selection can increase genetic variation within populations through overdominance and other forms of balancing selection.

In natural populations, these evolutionary processes never act in isolation but combine and interact in complex ways. In most natural populations, the combined effects of these processes and their interaction determine the pattern of genetic variation observed in the gene pool over time.

The Role of Genetics in Conservation Biology

The current rate at which species are being driven to extinction is greater than it has ever been in recorded history. It is estimated that there are approximately 2 million known species and as many as 30 million that are yet to be described. As we alter the environment and reduce the amount of suitable habitat for species, their numbers frequently decline. It is important to consider the consequences on the gene pool of such species because the variability of the gene pool may affect the chances of long-term survival of the species. Many of the genetic principles and processes discussed in this chapter relate to this conservation problem. As we have seen, populations have genetic structure; conserving this structure may warrant special attention. For example, **population**

viability analysis techniques are designed to estimate how large a population must be to keep from going extinct for a particular period of time with a certain degree of certainty. If one wants to ensure that a population has the potential to evolve over long periods of time, an adequate gene pool must be maintained. Clearly, determining the genetic structure of a population and how genetic variation within populations affects the probability of extinction requires a great deal of study. The problem is particularly acute for rare and already endangered species. The effects of unintentional inbreeding of species in zoos and game management programs are diminishing as population genetics principles are being used to manage genetic structures of populations more carefully. We have seen that inbreeding, genetic drift, and selection can all decrease genetic variation, and populations may need to be maintained at certain genetic effective sizes to ensure that ample amounts of variation remain. Fortunately, we now have tools to obtain quantitative assessments of the relationships among geographic populations and of the amount of genetic variability in each. These data provide essential information for management policies. However, there is some controversy regarding the utility of genetic information in conservation biology. The central problem in most cases is loss of habitat, and unless the rate of habitat destruction can be slowed, efforts to manage the genetic structure of populations may accomplish little. Nonetheless, insights from both population ecology and genetics are necessary to better understand the best course of action for maintaining the diversity of life in our ecosystems.

KEYNOTE

Rare and endangered species risk losing genetic variability and hence the ability to adapt to changing environmental conditions. Management practices are applied in an attempt to maintain genetic diversity by keeping adequate population size and avoiding inbreeding.

Speciation

Our discussion of population genetics principles so far has considered only processes of changing allele frequencies within populations of interbreeding organisms. Population subdivision may be weak, or it may be extreme to the point that two populations never interbreed. If this occurs for a long period of time, one expects that eventually there will be fixation of different alleles in the subpopulations such that if they were to come together again, they would fail to mate or the hybrids would have low fitness. Recently, there has been great progress in identifying genes that play a role in the reproductive isolation of closely related species. Before we see how genetics provides exactly the tools we need to understand the basis for reproductive isolation, let's back up and consider some general principles about speciation.

Barriers to Gene Flow

If a species is a group of reproductively compatible organisms, then the process of speciation is likely to involve the erection of barriers to gene flow. Eventually, geographically isolated populations, by a combination of drift and natural selection, will diverge to the point that they can no longer reproduce with one another. The barriers to gene flow come in two major categories: those that result in poor fitness of hybrid offspring (postzygotic barriers) and those that keep the two species from mating in the first place (prezygotic barriers). In animals, it appears that mutations in many genes can result in infertility, especially male infertility. As a result, the earliest genetic change in our pair of isolated subpopulations results in hybrid offspring that are sterile. The adults from the two subpopulations still recognize one another and mate, but the offspring are a dead end. Thus, postzygotic isolation generally arises first. **Postzygotic isolation** may include *hybrid sterility, hybrid inviability* (the failure of the hybrids to perform well in future generations), or *hybrid breakdown*.

In the face of a postmating barrier, it seems that mating adults could increase their fitness if they could recognize the "wrong" species and avoid these unproductive matings. If the populations harbor genetic variation for mate recognition, then the alleles that allow the adults to successfully discriminate will increase in frequency. This simple model (called **reinforcement**) provides a mechanism whereby postzygotic isolation leads to **prezygotic isolation**. The genes that result in **prezygotic isolation** may keep the species apart in many different ways, including the following:

1. **Temporal isolation.** By changing the mating season or the activity periods such that they no longer overlap, the opportunity for mating is removed.
2. **Ecological isolation.** If the ecological niche of the two species is distinct, such that, for example, the two species' dietary preferences keep them geographically isolated, even on a small spatial scale, again the opportunity for mating is removed.

Consider now the cases in which the two species freely overlap and have plenty of opportunity for mating. There still are genetic means to prevent the formation of zygotes:

1. **Behavioral incompatibility.** If the two species recognize and avoid each other as mates, there is no opportunity for mating.
2. **Mechanical isolation.** The two species may not be able to discriminate one from another, but if their genitalia do not fit together, zygotes cannot be formed.

3. **Gametic isolation.** Even if they mate and gametes come in contact with one another, there still remains a highly complex process of gametic fusion that can fail. In the case of plants, pollen from the wrong species often lands on the stigma surface. There is a chemical communication between the pollen (or pollen tube) and the stigma and style that has to go correctly, or the pollen fails to germinate or the pollen tube fails to grow.

Once prezygotic isolation is partially achieved, there is a snowball effect in which the rate of divergence accelerates. Individuals who engage in interspecific matings suffer an increasing disadvantage until at last the barrier to gene flow is complete. There is good empirical support for speciation by geographical isolation, although support for other mechanisms remains controversial. The study of the genetic basis for species isolation and the mechanisms by which species differences arise remains an active field.

Genetic Basis for Speciation

Because of the argument that younger species tend to rely more on postzygotic isolation, these species are sought after by evolutionary geneticists seeking to understand the nature of the genes that partially isolate the species. It is often the case that male hybrids are sterile but female hybrids are fertile. This is especially true when the males are heterogametic (they make two different kinds of gametes, namely, X- and Y-bearing sperm), and the females are homogametic (producing only X-bearing eggs). In birds and butterflies, where males are homogametic and females are heterogametic, the pattern is reversed. The pattern is so pervasive that J. B. S. Haldane noted it, and we call the phenomenon **Haldane's rule.**

The observation of Haldane's rule immediately begs the question, what is the genetic basis for hybrid male sterility? In many species of *Drosophila*, such as *D. simulans* and *D. mauritiana*, the F_1 females are viable and fertile, whereas the F_1 males are sterile. One can backcross the females to *D. simulans*, and a few of the backcross males (who are now roughly ¾ *D. simulans*) are fertile. By doing tricks like this, or by crossing in markers and introgressing parts of the wrong species' genome in, *Drosophila* population geneticists have learned that many genes are involved in the fertility of male hybrids.

Another intriguing example of prezygotic isolation occurs in species of abalone, a subtidal marine gastropod that sheds its gametes into the sea water. Sperm and eggs of more than one species may co-occur (although some species exhibit temporal and ecological isolation), so the only way to avoid interspecific matings is for the eggs to allow penetration only by conspecific sperm. This is accomplished by means of molecules in the sperm and the egg. The sperm protein lysin is able to disaggregate the egg glycoprotein VERL in a species-specific manner.

Because the sperm and egg components must track one another and because they must be able to respond quickly if they come in contact with new species, these molecules are undergoing rapid adaptive evolution. You will learn more about the inferences of adaptive change in proteins in Chapter 25.

Summary

Population genetics is the study of the genetic structure of populations and species and how the structure changes or evolves over time. The gene pool of a population is the total of all genes within the population, and it is described in terms of allelic and genotypic frequencies. The Hardy-Weinberg law describes what happens to allelic and genotypic frequencies of a large, randomly mating population free from evolutionary processes; when these conditions are met, allelic frequencies do not change, and genotypic frequencies stabilize after one generation in the proportions p^2, $2pq$, q^2, where p and q equal the allelic frequencies of the population.

The classic, balance, and neutral mutation models have generated testable hypotheses that help explain how much genetic variation should exist within natural populations and what processes are responsible for the variation observed. Protein electrophoresis showed that most populations of plants and animals contain large amounts of genetic variation, proving that the classic model was wrong. The current view is that genetic variation is maintained in populations by a combination of forces and that some particular genes may be heavily influenced by natural selection, resulting in fixation of an advantageous allele, or by recurrent loss of deleterious mutations. Variation in other genes appears to fit the neutral model, suggesting that the variation is maintained by a balance between genetic drift and mutation.

Mutation, genetic drift, migration, and natural selection are processes that can alter allelic frequencies of a population. Although mutation is the source of all variation in a population, it usually changes allelic frequencies at a very slow rate, so slow that almost any of the other forces swamp the effects of recurrent mutation on allelic frequencies. Genetic drift, chance change in allelic frequencies due to small effective population size, leads to a loss of genetic variation within a population, genetic divergence among populations, and random change of allelic frequency within a population. Migration tends to reduce genetic divergence among populations. Natural selection is differential reproduction of genotypes. The relative reproductive contribution of genotypes is measured in terms of Darwinian fitness. The effects of natural selection depend on the fitnesses of the genotypes, the degree of dominance, and the frequencies of the alleles in the population. Nonrandom mating affects the effective population size and genotypic frequencies of a population; the allelic frequencies are unaffected, except in some cases of negative assortative mating. One type of nonrandom mating, inbreeding, leads to an increase in homozygosity.

New techniques of molecular genetics, including analysis of restriction fragment length polymorphisms and RNA and DNA sequences, have supported prior insights obtained from analyses of proteins with respect to evolutionary processes. Different parts of a gene are found to have different levels of polymorphism; the parts of the gene that have the least effect on fitness appear to evolve at the highest rates. This suggests that natural selection generally removes deleterious mutations. Exceptions occur in genes where it is advantageous to have high levels of variability, such as genes in the immune system that increase the variety of pathogens that can be recognized.

Analytical Approaches to Solving Genetics Problems

Q24.1 In a population of 2,000 gaboon vipers, a genetic difference with respect to venom exists at a single locus. The alleles are incompletely dominant. The population shows 100 individuals homozygous for the *t* allele (genotype *tt*, nonpoisonous), 800 heterozygous (genotype *Tt*, mildly poisonous), and 1,100 homozygous for the *T* allele (genotype *TT*, lethally poisonous).

a. What is the frequency of the *t* allele in the population?

b. Are the genotypes in Hardy-Weinberg equilibrium?

A24.1 This question addresses the basics of calculating allelic frequencies and relating them to the genotype frequencies expected of a population in Hardy-Weinberg equilibrium.

a. The *t* frequency can be calculated from the information given because the trait is an incompletely dominant one. There are 2,000 individuals in the population under study, meaning a total of 4,000 alleles at the *T/t* locus. The number of *t* alleles is given by

$$(2 \times tt \text{ homozygotes}) + (1 \times Tt \text{ heterozygotes})$$
$$= (2 \times 100) + (1 \times 800) = 1,000$$

This calculation is straightforward because both alleles in the nonpoisonous snakes are *t*, whereas only one of the two alleles in the mildly poisonous snakes is *t*. Since the total number of alleles under study is 4,000, the frequency of *t* alleles is $1,000/4,000 = 0.25$. This system is a two-allele system, so the frequency of *T* must be 0.75.

b. For the genotypes to be in Hardy-Weinberg equilibrium, the distribution must be $p^2\,TT + 2pq\,Tt + q^2\,tt$ genotypes, where p is the frequency of the T allele and q is the frequency of the t allele. In (a) we established that the frequency of T is 0.75 and the frequency of t is 0.25. Therefore, $p = 0.75$ and $q = 0.25$. Using these values, we can determine the expected genotype frequencies if this population is in Hardy-Weinberg equilibrium:

$$(0.75)^2\,TT + 2(0.75)(0.25)\,Tt + (0.25)^2\,tt$$

This expression gives $0.5625\,TT + 0.3750\,Tt + 0.0625\,tt$. Thus, with 2,000 individuals in the population we would expect 1,125 TT, 750 Tt, and 125 tt. These values are close to the values given in the question, suggesting that the population is indeed in genetic equilibrium.

To check this result, we should perform a chi-square analysis (see Chapter 2), using the given numbers (not frequencies) of the three genotypes as the observed numbers and the calculated numbers as the expected numbers. The chi-square analysis is as follows, where $d = $ (observed − expected):

Genotype	Observed	Expected	d	d^2	d^2/e
TT	1,100	1,125	−25	625	0.556
Tt	800	750	+50	2,500	3.334
tt	100	125	−25	625	5.000
Totals	2,000	2,000	0		8.890

Thus, the chi-square value (i.e., the sum of all the d^2/e values) is 8.89. For the reasons discussed in the text for a similar example, there is only one degree of freedom. Looking up the chi-square value in the chi-square table (Table 2.5, p. 30), we find a P value of approximately 0.0025. Therefore, about 25 times out of 10,000 we would expect chance deviations of the magnitude observed. In other words, our hypothesis that the population is in Hardy-Weinberg equilibrium is not substantiated. In this case, our guess that it was in equilibrium was inaccurate. Nonetheless, the population is not greatly removed from an equilibrium state.

Q24.2 Approximately one normal allele in 30,000 mutates to the X-linked recessive allele for hemophilia in each human generation. Assume for the purposes of this problem that one h allele in 300,000 mutates to the normal alternative in each generation. (Note that in reality it is difficult to measure the reverse mutation of a human recessive allele that is essentially lethal, such as the allele for hemophilia.) The mutation frequencies are indicated in the following equation:

$$h^+ \xrightarrow{\;u\;} h$$
$$\xleftarrow{\;v\;}$$

where $u = 10v$. What allelic frequencies would prevail at equilibrium under mutation pressures alone in these circumstances?

A24.2 This question seeks to test understanding of the effects of mutation on allelic frequencies. In this chapter, we discussed the consequences of mutation pressure. The conclusion was that if A mutates to a at n times the frequency with which a mutates back to A, then at equilibrium the value of q will be $\hat{q} = u/(u + v)$ or $\hat{q} = nv/(n + 1)v$. Applying this general derivation to this particular problem, we simply use the values given. We are told that the forward mutation rate is 10 times the reverse mutation rate, or $u = 10v$. At equilibrium the value of q will be $\hat{q} = u/(u + v)$. Since $u = 10v$, this equation becomes $\hat{q} = 10v/11v$, so $q = 10/11$, or 0.909. Therefore, at equilibrium brought about by mutation pressures, the frequency of h (the hemophilia allele) is 0.909, and the frequency of h^+ (the normal allele) is $\hat{q}$, that is, $(1 - \hat{q}) = (1 - 0.909) = 0.091$.

Questions and Problems

*24.1 In the European land snail *Cepaea nemoralis*, multiple alleles at a single locus determine shell color. The allele for brown (C^B) is dominant to the allele for pink (C^P) and to the allele for yellow (C^Y). The dominance hierarchy among these alleles is $C^B > C^P > C^Y$. In one population sample of *Cepaea*, the following color phenotypes were recorded:

Brown	236
Pink	231
Yellow	33
Total	500

Assuming that this population is in Hardy-Weinberg equilibrium (large, randomly mating, and free from evolutionary processes), calculate the frequencies of the C^B, C^Y, and C^P alleles.

24.2 Three alleles are found at a locus coding for malate dehydrogenase (MDH) in the spotted chorus frog. Chorus frogs were collected from a breeding pond, and each frog's genotype at the MDH locus was determined with electrophoresis. The following numbers of genotypes were found:

M^1M^1	8
M^1M^2	35
M^2M^2	20
M^1M^3	53
M^2M^3	76
M^3M^3	62
Total	254

a. Calculate the frequencies of the M^1, M^2, and M^3 alleles in this population.

b. Using a chi-square test, determine whether the MDH genotypes in this population are in Hardy-Weinberg proportions.

24.3 In a large interbreeding population, 81 percent of the individuals are homozygous for a recessive character. In the absence of mutation or selection, what percentage of the next generation would be homozygous recessives? Homozygous dominants? Heterozygotes?

*24.4 Let A and a represent dominant and recessive alleles whose respective frequencies are p and q in a given interbreeding population at equilibrium (with $p + q = 1$).

a. If 16 percent of the individuals in the population have recessive phenotypes, what percentage of the total number of recessive genes exist in the heterozygous condition?

b. If 1.0 percent of the individuals were homozygous recessive, what percentage of the recessive genes would occur in heterozygotes?

*24.5 A population has eight times as many heterozygotes as homozygous recessives. What is the frequency of the recessive gene?

24.6 In a large population of range cattle, the following ratios are observed: 49 percent red (RR), 42 percent roan (Rr), and 9 percent white (rr).

a. What percentage of the gametes that give rise to the next generation of cattle in this population will contain allele R?

b. In another cattle population, only 1 percent of the animals are white and 99 percent are either red or roan. What is the percentage of r alleles in this case?

24.7 In a gene pool, the alleles A and a have initial frequencies of p and q, respectively. Show that the allelic frequencies and zygotic frequencies do not change from generation to generation as long as there is no selection, mutation, or migration, the population is large, and the individuals mate at random.

*24.8 The S-s antigen system in humans is controlled by two codominant alleles, S and s. In a group of 3,146 individuals, the following genotypic frequencies were found: 188 SS, 717 Ss, and 2,241 ss.

a. Calculate the frequency of the S and s alleles.

b. Determine whether the genotypic frequencies are in Hardy-Weinberg equilibrium by using the chi-square test.

24.9 Refer to Problem 24.8. A third allele is sometimes found at the S locus. This allele S^u is recessive to both the S and the s alleles and can be detected only in the homozygous state. If the frequencies of the alleles S, s, and S^u are p, q, and r, respectively, what would be the expected frequencies of the phenotypes $S-$, Ss, $s-$, and S^uS^u?

24.10 In a large interbreeding human population, 60 percent of individuals belong to blood group O (genotype i/i). Assuming negligible mutation and no selective advantage of one blood type over another, what percentage of the grandchildren of the present population will be type O?

*24.11 A selectively neutral, recessive character appears in 40 percent of the males and in 16 percent of the females in a large, randomly interbreeding population. What is the gene's frequency? What proportion of females are heterozygous for it? What proportion of males are heterozygous for it?

24.12 Suppose you found two distinguishable types of individuals in wild populations of some organism in the following frequencies:

	Type 1	Type 2
Females	99%	1%
Males	90%	10%

The difference is known to be inherited. Are these data compatible with the trait being X linked?

*24.13 Red-green color blindness is caused by a sex-linked recessive gene. About 64 women out of 10,000 are color blind. What proportion of men would be expected to show the trait if mating is random?

24.14 About 8 percent of the men in a population are red-green color blind (because of a sex-linked recessive gene). Answer the following questions, assuming random mating in the population, with respect to color blindness.

a. What percentage of women would be expected to be color blind?

b. What percentage of women would be expected to be heterozygous?

c. What percentage of men would be expected to have normal vision two generations later?

24.15 List some of the basic differences in the classical, balance, and neutral mutation models of genetic variation.

*24.16 Two alleles of a locus, A and a, can be interconverted by mutation:

where u is a mutation rate of 6.0×10^{-7} and v is a mutation rate of 6.0×10^{-8}. What will be the frequencies of A and a at mutational equilibrium, assuming no selective difference, no migration, and no random fluctuation caused by genetic drift?

24.17

a. Calculate the effective population size (N_e) for a breeding population of 50 adult males and 50 adult females.

b. Calculate the effective population size (N_e) for a breeding population of 60 adult males and 40 adult females.

c. Calculate the effective population size (N_e) for a breeding population of 10 adult males and 90 adult females.

d. Calculate the effective population size (N_e) for a breeding population of 2 adult males and 98 adult females.

24.18 In a population of 40 adult males and 40 adult females, the frequency of allele A is 0.6 and the frequency of allele a is 0.4.

a. Calculate the 95 percent confidence limits of the allelic frequency for A.

b. Another population with the same allelic frequencies consists of only 4 adult males and 4 adult females. Calculate the 95 percent confidence limits of the allelic frequency for A in this population.

c. What are the 95 percent confidence limits of A if the population consists of 76 females and 4 males?

24.19 The land snail *Cepaea nemoralis* is native to Europe but has been accidentally introduced into North America at several localities. These introductions occurred when a few snails were inadvertently transported on plants, building supplies, soil, or other cargo. The snails subsequently multiplied and established large, viable populations in North America.

Assume that today the average size of *Cepaea* populations found in North America is equal to the average size of *Cepaea* populations in Europe. What predictions can you make about the amounts of genetic variation present in European and North American populations of *Cepaea*? Explain your reasoning.

*24.20 A population of 80 adult squirrels resides on campus, and the frequency of the Est^1 allele among these squirrels is 0.70. Another population of squirrels is found in a nearby woods, and there, the frequency of the Est^1 allele is 0.5. During a severe winter, 20 of the squirrels from the woods population migrate to campus in search of food and join the campus population. What will be the allelic frequency of Est^1 in the campus population after migration?

24.21 Upon sampling three populations and determining genotypes, you find the following three genotype distributions. What would each of these distributions imply with regard to selective advantages of population structure?

Population	AA	Aa	aa
1	0.04	0.32	0.64
2	0.12	0.87	0.01
3	0.45	0.10	0.45

24.22 The frequency of two adaptively neutral alleles in a large population is 70 percent A : 30 percent a. The population is wiped out by an epidemic, leaving only four individuals, who produce many offspring. What is the probability that the population several years later will be 100 percent AA? (Assume no new mutations occur.)

*24.23 A completely recessive gene, through changed environmental circumstances, becomes lethal in a certain population. It was previously neutral, and its frequency was 0.5.

a. What was the genotype distribution when the recessive genotype was not selected against?

b. What will be the allelic frequency after one generation in the altered environment?

c. What will be the allelic frequency after two generations?

24.24 Human individuals homozygous for a certain recessive autosomal gene die before reaching reproductive age. Despite this removal of all affected individuals, there is no indication that homozygotes occur less frequently in succeeding generations. To what might you attribute the continued appearance of these recessives?

*24.25 A completely recessive gene (Q^1) has a frequency of 0.7 in a large population, and the Q^1Q^1 homozygote has a relative fitness of 0.6.

a. What will be the frequency of Q^1 after one generation of selection?

b. If there is no dominance at this locus (the fitness of the heterozygote is intermediate to the fitnesses of the homozygotes), what will the allelic frequency be after one generation of selection?

c. If Q^1 is dominant, what will the allelic frequency be after one generation of selection?

24.26 As discussed earlier in this chapter, the gene for sickle-cell anemia exhibits heterozygote advantage. An individual who is an *Hb-A/Hb-S* heterozygote has increased resistance to malaria and therefore has greater fitness than the *Hb-A/Hb-A* homozygote, who is susceptible to malaria, and the *Hb-S/Hb-S* homozygote, who has sickle-cell anemia. Suppose that the fitness values of the genotypes in Africa are as presented here:

$Hb\text{-}A/Hb\text{-}A = 0.88$

$Hb\text{-}A/Hb\text{-}S = 1.00$

$Hb\text{-}S/Hb\text{-}S = 0.14$

Give the expected equilibrium frequencies of the sickle-cell gene (*Hb-S*).

*24.27 Achondroplasia, a type of dwarfism in humans, is caused by an autosomal dominant gene. The mutation rate for achondroplasia is about 5×10^{-5}, and the fitness of achondroplastic dwarfs has been estimated to be about 0.2, compared with unaffected individuals. What is the equilibrium frequency of the achondroplasia gene based on this mutation rate and fitness value?

24.28 To answer the following questions, consider the spontaneous mutation frequencies tabulated in Table 24.6 (p. 678).
a. In humans, why is the frequency of forward mutations to neurofibromatosis an order of magnitude larger than that for the other human diseases?
b. In *E. coli*, why is the frequency of mutations to arabinose dependence two to four orders of magnitude larger than the frequency of mutations to leucine, arginine, or tryptophan independence?
c. What factors influence the spontaneous mutation frequency for a specific trait?

24.29 The frequencies of the L^M and L^N blood group alleles are the same in each of the populations I, II, and III, but the genotypes' frequencies are not the same, as shown in the following table. Which of the populations is most likely to show each of the following characteristics: random mating, inbreeding, genetic drift? Explain your answers.

	$L^M L^M$	$L^M L^N$	$L^N L^N$
I	0.50	0.40	0.10
II	0.49	0.42	0.09
III	0.45	0.50	0.05

*24.30 DNA was collected from 100 people randomly sampled from a given human population and was digested with the restriction enzyme *Bam*HI; the fragments were separated by electrophoresis and then transferred to a membrane filter using the Southern blot technique. The blots were probed with a particular cloned sequence. Three different patterns of hybridization were seen on the blots. Some DNA samples (56 of them) showed a single band of 6.3 kb, others (6) showed a single band at 4.1 kb, and others (38) showed both the 6.3- and the 4.1-kb bands.
a. Interpret these results in terms of *Bam*HI sites.
b. What are the frequencies of the restriction site alleles?
c. Does this population appear to be in Hardy-Weinberg equilibrium for the relevant restriction sites?

*24.31 Fifty tiger salamanders from one pond in west Texas were examined for genetic variation by using the technique of protein electrophoresis. The genotype of each salamander was determined for five loci (AmPep, ADH, PGM, MDH, and LDH-1). No variation was found at AmPep, ADH, and LDH-1; in other words, all individuals were homozygous for the same allele at these loci. The following numbers of genotypes were observed at the MDH and PGM loci.

MDH Genotypes	Number of Individuals	PGM Genotypes	Number of Individuals
AA	11	DD	35
AB	35	DE	10
BB	4	EE	5

Calculate the proportion of polymorphic loci and the heterozygosity for this population.

*24.32 The success of a population depends in part on its reproductive rate, which may be affected by low genetic variability. Vyse and his colleagues have been interested in the conservation of small populations of grizzly bears, studying 304 members of 30 grizzly bear family groups in a population in northwestern Alaska. They have identified a set of polymorphic loci with these alleles, allele frequencies, and observed heterozygosities (obs. het.):

Locus G1A		Locus G10X		Locus G10C		Locus G10L	
Obs. het.: 0.776		Obs. het.: 0.783		Obs. het.: 0.770		Obs. het.: 0.651	
Allele	Freq.	Allele	Freq.	Allele	Freq.	Allele	Freq.
A194	0.398	X137	0.395	C105	0.355	L155	0.487
A184	0.240	X135	0.211	C103	0.257	L157	0.276
A192	0.211	X141	0.211	C111	0.240	L161	0.128
A180	0.086	X133	0.102	C113	0.092	L159	0.089
A190	0.036	X131	0.053	C107	0.043	L171	0.013
A200	0.016	X129	0.030	C101	0.010	L163	0.007
A186	0.007			C109	0.003		
A188	0.006						

The genotypes of a mother bear and her three cubs are shown here.

Mother	Cub #1	Cub #2	Cub #3
A184, A192	A184, A194	A184, A192	A184, A194
X135, X137	X135, X137	X133, X135	X137, X141
C105, C113	C105, C111	C105, C105	C111, C113
L155, L159	L155, L157	L159, L161	L155, L155

a. How do the observed heterozygosities compare with the expected heterozygosities? On the basis of this information, can you tell whether this grizzly bear population is in Hardy-Weinberg equilibrium?

b. What can you infer about the paternity of the mother's three cubs? How might paternity information affect the genetic variability and the effective population size in this population of grizzly bears?

24.33 What factors cause genetic drift?

*24.34 What are the primary effects of the following evolutionary processes on the gene and genotypic frequencies of a population?
a. mutation
b. migration
c. genetic drift
d. inbreeding

24.35 Explain how overdominance leads to an increased frequency of sickle-cell anemia in areas where malaria is widespread.

24.36 Since 1968, Pinter has studied the population dynamics of the montane vole, a small rodent in the Grand Teton mountains in Wyoming. For more than 25 years, severe periodic fluctuations in population density have been negatively correlated with precipitation levels: Vole density sharply declines every few years when spring precipitation is extremely high.
a. Propose several hypotheses concerning the genetic structure of the population of montane voles in two separate sampling sites if

i. there is negligible migration of voles between sampling sites
ii. here is substantial migration of voles between sampling sites
b. How would you gather data to evaluate these hypotheses?

24.37 What are some of the advantages of using DNA sequences to infer the strength of evolutionary processes?

*24.38 Multiple, geographically isolated populations of tortoises exist on the Galapagos Islands off the coast of Ecuador. Several populations are endangered, partly because of hunting and partly because of illegal capture and trade. In principle, a significant number of tortoises in captivity (in zoos or private collections) could be used to repopulate some of the endangered populations.
a. Suppose you are interested in returning a particular captive tortoise to its native subpopulation, but you have no record of its capture. How could you determine its original subpopulation?
b. A researcher planned to characterize one subpopulation of tortoises. In her first field season, she tagged and collected blood samples from all animals in the subpopulation. When she returns a year later, she cannot locate two animals. She learns that two untagged, smuggled tortoises are being held by U.S. customs officials. How can she assess whether the smuggled animals are from her field site?

25

Molecular Evolution

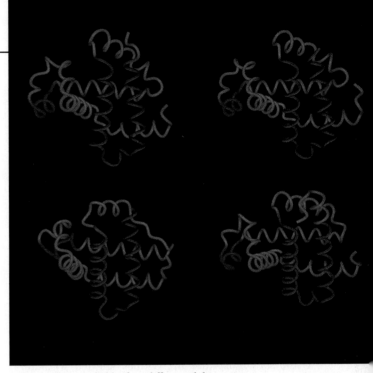

Polypeptides encoded by four different globin genes.

PRINCIPAL POINTS

- Rates of molecular evolution can be measured by comparing DNA sequences. Rates of change vary both within and between genes. Regions with the lowest rates of change usually are those that are most functionally constrained and most subject to natural selection.

- Mutations are changes in nucleotide sequences, whereas substitutions are mutations that have passed through the filter of selection. Synonymous substitution rates are a good indicator of the actual mutation rate operating within a genome. Selection acts at surprisingly small scales, for example, at the level of the choice of which synonymous codons are preferred.

- Mutations are rare events, and most changes in amino acid sequence tend to be removed through natural selection.

- Relative rate tests suggest that the molecular clocks of some genes run at a steady rate over long periods of time. However, it is unreasonable to assume that *all* lineages in a gene tree (a tree depicting the relationship of a single gene within and between species) accumulate substitutions at the same rate.

- Gene trees do not always correspond to species trees (a tree depicting the relationship of species based on morphological or paleontological analyses or molecular data from several genes) because some genetic polymorphisms within populations predate speciation events.

- Increases in the number of taxa being considered dramatically increases the possible number of phylogenetic trees that can describe the relationship of those taxa. Distance matrix (statistical analysis that groups taxa on the basis of their overall similarity), parsimony (premised on the concept that the tree that invokes the fewest number of mutations is most likely to be correct because mutations are rare events), and maximum likelihood (evaluating differences in mutation frequncies) approaches are three different ways to choose which of the many possible trees are most likely to represent the true evolutionary relationship.

- In eukaryotic organisms, genes frequently occur in multiple copies with identical or very similar sequences. A group of such genes is called a multigene family. Duplications of genes, in whole or in part, are the principal raw material from which proteins with new functions are made.

- The functional domains of many proteins correspond to regions encoded in single exons. Many genes appear to have been derived by "mixing and matching" such functional domains of already useful proteins through exon shuffling.

i IN A HOT, HARSH REGION OF THE DESERT, YOU SIFT carefully through the sediment looking for fossils. Suddenly, you glimpse a fragment of bone. Eventually, you unearth a leg bone and part of a jaw, which lab tests show to be those of an ancient hominid. Is this the distant predecessor of modern humans or merely an unrelated species that became extinct hundreds of thousands of years ago? How could you find out? In this chapter, you will learn how population geneticists can apply molecular genetic techniques to answer questions about how species evolve. Then, in the iActivity, you will have the opportunity to use some of the same tools and techniques to determine whether we are the direct descendants of Neanderthals.

While individuals are the entities affected by natural selection, it is populations and genes that change over evolutionary time. **Molecular evolution** is evolution at the molecular level of DNA and protein sequences. The study of molecular evolution uses the theoretical foundation of population genetics to address two essentially different sets of questions: How do DNA and protein molecules evolve and how are genes and organisms evolutionarily related?

Aside from the differences in the questions asked, population genetics and molecular evolution differ primarily in the time frame of their perspective. Population genetics (see Chapter 24) focuses on the changes in gene frequencies that occur from generation to generation, whereas molecular evolution typically considers the much longer time frames associated with speciation. Very small departures from the conditions needed to maintain Hardy-Weinberg equilibrium have small effects on gene frequencies in the short term but can take on great significance on evolutionary time scales. Moreover, random effects such as those associated with small amounts of sampling error along with exceptionally small differences in fitness tend to become the predominant process of genomic change when applied cumulatively over hundreds or thousands of generations.

The field of molecular evolution is multidisciplinary. It routinely invokes data and insights from genetics, ecology, evolutionary biology, statistics, and even computer science. However, before the widespread development of the tools of molecular biology in the 1970s and 1980s, researchers interested in the study of how biologically important molecules change over time had little data

available to study. The ability to clone, sequence, and hybridize DNA removed the species barrier in population genetics studies. It also opened a window on a world that had been only dimly perceived where genes evolve by the accumulation of **mutations** (see Chapter 7), **transposition** (see Chapter 7), **duplication** (see Chapter 17), and **gene conversion** (see Chapter 3). For the first time, studies of evolution had an abundance of parameters that could be measured and theories that could be tested. Molecular analyses made it clear that genomes are historical records that can be unraveled to identify the dynamics behind evolutionary processes and to reconstruct the chronology of change. The same approaches that allow these documents of evolutionary history to be put in order and deciphered also facilitate classification of the living world in true **phylogenetic relationships**—the hierarchical genealogical relationships between separated populations or species—across the vast distances of evolutionary time. Previously unimagined relationships between organisms became apparent, and even the kingdoms of life at the root of all systematics had to be rearranged.

In this chapter, we present the principles of molecular evolution studies and show how molecules with new functions arise and how the phylogenetic relationships of molecules and organisms are determined.

Patterns and Modes of Substitutions

Nucleotide Substitutions in DNA Sequences

Substitutions in Protein and DNA Sequences. An important question in the study of evolution is how the patterns and rates of substitution differ between different parts of the same gene. These studies began in earnest in the 1970s and 1980s, when the best molecular data available came from the amino acid sequences in proteins. It quickly became apparent that some amino acid differences were more likely to be observed between two **homologous** proteins—proteins in different species that share a common ancestor—than were others. Specifically, amino acids were most likely to be replaced with amino acids that had similar chemical characteristics (see the groupings in Figure 6.2, p. 113) to those of the amino acid present in the ancestral protein. This replacement bias supported two evolutionarily important principles: (1) mutations are rare events; and (2) most dramatic alterations are removed from the gene pool by natural selection.

Chemically similar amino acids tend to have similar codons (see Figure 6.7, p. 118), and fewer changes at the level of DNA are required to change one into another. For example, a leucine codon (i.e., CUU) can be changed to an isoleucine codon (i.e., AUU) by a single base pair change at the DNA level. In contrast, two base pair changes must occur to convert that same leucine codon to a codon for the chemically dissimilar asparagine (i.e., AAU). DNA polymerase error rates are typically described in terms of errors per *millions* or *billions* of replicated nucleotides. Since nucleotide and amino acid changes are such rare events, scenarios that invoked the fewest numbers of them were most likely to occur. At the same time, natural selection acting over many generations has caused most proteins to have amino acid sequences that make them optimally suited for their particular role and present environment. The more substantial an alteration to a protein's primary structure was, the more likely it was to have had a deleterious effect on its function that would not escape the scrutiny of natural selection.

Sequence Alignments. Analyzing changes at both the amino acid and nucleotide levels between two or more gene sequences begins with an alignment of all the sequences to be studied. Pairwise alignments between short, highly similar sequences can be done by hand but computer programs are usually used to make more difficult alignments. Alignments represent specific hypotheses about the evolution of two or more sequences, and the best possible alignment reflects the true ancestral relationship at each position in the sequence of a protein or gene. For instance, the alignment of two sequences

$$\text{G T A C C T}$$
$$\text{G - A T C T}$$

can be interpreted to mean the following:

1. Four of the six nucleotide positions (first, third, fifth, and sixth from the left) have not undergone any change since the pair of sequences last shared a common ancestor.
2. A substitution has occurred at one site (the fourth from the left).
3. An insertion or deletion occurred in one of the sequences (second position).

Rapid divergence or long periods of evolution often leave little in common between sequences and can make generating alignments difficult. As a result, most studies are based on approximations of the true alignment, called **optimal alignments,** in which gaps are inserted to maximize the similarity between the sequences being aligned. Gaps in alignments are necessary because gene sequences are altered not just by changes to individual nucleotides (point mutations) but also by even less common insertion and deletion events. Since it is often difficult or impossible to distinguish an insertion in one sequence from a deletion in another, such gaps often are called **indels.** The number of possible alignments between two or more sequences of even modest length is enormous. Determining which alignment is optimal typically is left to computer

algorithms that seek to maximize the number of matching amino acids or nucleotides between the sequences while invoking the smallest number of indel events possible.

Substitutions and the Jukes-Cantor Model. After two nucleotide sequences diverge from each other, each begins to accumulate nucleotide substitutions independently. The number of substitutions (K) observed in an alignment between two sequences is almost always the single most important variable in any molecular evolution analysis. If an alignment suggests that few substitutions have occurred between two sequences, then a simple count of the substitutions usually is sufficient to determine the number of substitutions that have occurred. However, even before the nucleotide sequences of any genes were available for analysis, T. Jukes and C. Cantor (1969) realized that alignments between sequences with many differences might cause a significant underestimation of the actual number of substitutions since the sequences last shared a common ancestor. Where substitutions were common, there were no guarantees that a particular site had not undergone multiple changes such as those illustrated in Figure 25.1. To address that possibility, Jukes and Cantor assumed that each nucleotide was just as likely to change into any other nucleotide. Using that assumption, they created a mathematical model in which the rate of change to any one of the three alternative nucleotides was assumed to be α, and the overall rate of substitution for any given nucleotide was 3α. In that model, if a site within a gene was occupied by a C at time 0 ($t = 0$ in Figure 25.1), then the probability (P) that that site would still be the same nucleotide at time 1 ($t = 1$) would be $P_{C(1)} = 1 - 3\alpha$. At subsequent points in time (e.g., $t = 2$) the possibility of reversions (back mutation) to C must also be considered. Specifically, if the original C changed

Figure 25.1

Two possible scenarios in which multiple substitutions at a single site would lead to underestimation of the number of substitutions that had occurred if a simple count was performed. *T* = time.

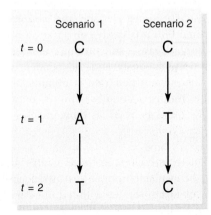

to another nucleotide (say, an "A") in that first time span, at time 2 ($t = 2$) the probability, $P_{C(2)}$, would be equal to $(1 - 3\alpha)P_{C(1)} + \alpha[1 - P_{A(1)}]$. Further expansion suggested that at any given time (t) in the future, the probability that the site would contain a C was defined by the equation:

$$P_{C(t)} = \frac{1}{4} = (\frac{3}{4})e^{-4\alpha t}$$

All that remained was for molecular biologists to determine the value for α, the rate at which nucleotide substitutions occurred.

When those data arrived 10 years later, it became clear that the model used by Jukes and Cantor was an oversimplification. For instance, **transitions** (exchanging one purine for the other or one pyrimidine for the other) were seen to accumulate at a different, faster rate than **transversions** (exchanging a purine for a pyrimidine, or vice versa). Even so, the Jukes-Cantor model provided a very useful framework for taking into account the actual number of substitutions per site (K) when multiple substitutions were possible. By manipulating the Jukes-Cantor equation, it is possible to determine that K can be calculated as

$$K = -\frac{3}{4} \ln(1 - (\frac{4}{3})p)$$

where p is the fraction of nucleotides that a simple count reveals to be different between two sequences. This equation is consistent with the idea that when two sequences have few mismatches between them, p is small and the chance of multiple substitutions at any given site is also small (e.g., when p for a stretch of 100 nucleotides is 0.02, $K = 0.02$). It also suggests that when the observed number of mismatches is large, the actual number of substitutions per site can be substantially larger than what is counted directly (e.g., when p for a stretch of 100 nucleotides is 0.50, $K = 0.82$).

Rates of Nucleotide Substitutions

The number of substitutions that two sequences have undergone since they last shared a common ancestor is a centrally important parameter to almost all molecular evolution analyses. When K is expressed in terms of the number of substitutions per site and coupled with a divergence time (T), it is easily converted into a rate (r) of substitution. Because substitutions are assumed to accumulate simultaneously and independently in both sequences, the substitution rate is obtained by simply dividing the number of substitutions between two homologous sequences by $2T$, as shown in this equation:

$$r = K/(2T)$$

Note that to estimate substitution rates, those data must always be available from at least two species. Comparisons of substitution rates within and between genes can then give valuable insights into the mechanisms

involved in molecular changes. And if evolutionary rates between several species are similar, substitution rates can give insights into the dates of evolutionary events for which no other physical evidence is available.

Variation of Evolutionary Rates Within Genes. Studies of nucleotide sequences in numerous genes have revealed that different parts of genes evolve at widely differing rates that reflect the extent that they are subject to natural selection. Recall from our discussion of molecular genetics that a typical eukaryotic gene is made up of some nucleotides that specify the amino acid sequence of a protein (coding sequences) and other nucleotides that do not code for amino acids in a protein (noncoding sequences). Noncoding sequences include introns, leader regions, trailer regions (all of which are transcribed but not translated) and 5′ and 3′ flanking sequences that are not transcribed. Additional noncoding sequences include **pseudogenes,** which are nucleotide sequences that no longer produce functional gene products because they have accumulated inactivating mutations. Even within the coding regions of a functional gene, not all nucleotide substitutions produce a corresponding change in the amino acid sequence of a protein. In particular, many substitutions occurring at the third position of triplet codons have no effect on the amino acid sequence of the protein because such changes often produce synonymous codons, one that codes for the same amino acid (see Figure 6.7, p. 118).

Synonymous and Nonsynonymous Sites. Table 25.1 shows relative rates of change in different parts of mammalian genes. Within functional genes, notice that the highest rate of change involves synonymous changes in the coding sequences. The rate of synonymous nucleotide change is about five times greater than the observed rate of nonsynonymous changes. Synonymous changes do not alter the amino acid sequence of the protein. Thus, the high rate of

Table 25.1	Relative Rates of Evolutionary Change in DNA Sequences of Mammalian Genes	
Sequence		**Nucleotide Substitutions per Site per Year ($\times 10^{-9}$)**
Functional genes		
5′ Flanking region		2.36
Leader		1.74
Coding sequence, synonymous		4.65
Coding sequence, nonsynonymous		0.88
Intron		3.70
Trailer		1.88
3′ Flanking region		4.46
Pseudogenes		4.85

evolutionary change seen there is not unexpected because these changes do not affect a protein's functioning. Consider for a moment how variation in nucleotide sequences arise. All variation must arise either from errors in DNA replication or repair processes. The enzymes responsible for DNA replication and repair are in no way capable of distinguishing between synonymous and nonsynonymous changes to DNA sequences. As a result, synonymous and nonsynonymous mutations are likely to arise with equal frequency. However, nonsynonymous changes that arise within coding sequences often are detrimental to fitness and are eliminated by natural selection, whereas synonymous mutations usually are less detrimental, so they are tolerated. That raises an interesting and subtle distinction between the use of the words *mutation* and *substitution* in molecular evolution studies. **Mutations** are changes in nucleotide sequences that occur because of mistakes in DNA replication or repair processes (see Chapter 15). **Substitutions** are mutations that have passed through the filter of selection on at least some level. Synonymous substitution rates probably are fairly reflective of the actual mutation rate operating within a genome, whereas nonsynonymous substitutions rates are not.

Flanking Regions. High rates of evolutionary change also occur in the 3′ flanking regions of functional genes (Table 25.1). Like synonymous changes, sequences in the 3′ flanking regions have no effect on the amino acid sequence of a protein and usually have little effect on gene expression. Consequently, most substitutions that occur within a 3′ flanking region are tolerated by natural selection. Rates of change in introns are also high but not as high as the synonymous changes and those in the 3′ flanking regions. Although the sequences in the introns usually are not used to spell out the sequence of amino acids in a protein, they must be properly spliced out for an mRNA to be translated into a functional protein. Some sequences within introns must be present for splicing to occur, including the 5′ and 3′ splice junctions and the branch point of the intron (see Chapter 5, pp. 98–99). And at least some introns occasionally code for protein coding regions in some tissues but not in others because of alternate splicing. As a result, not all changes in introns avoid detection by natural selection, but their overall rate of evolution is a bit lower than that seen in 3′ flanking regions and at synonymous sites within coding sequences.

Still lower rates of evolutionary change are seen in the 5′ flanking region. Although this region is neither transcribed nor translated, it does contain the promoter and other important regulatory elements for a gene; therefore, sequences in the 5′ flanking region are important for gene expression. Even subtle differences in promoter sequences, such as the TATA box, can dramatically affect how much of a particular protein is made and thus can have detrimental effects on the fitness of the organism. Natural selection usually eliminates these mutations and minimizes the observable changes in functionally important regions.

Next in evolutionary rate are the leader and trailer regions (see Table 25.1), which have somewhat lower rates than the 5′ flanking region. Although leaders and trailers are not translated, they are transcribed, and they provide important signals for processing and translation of the mRNA. Substitutions in these regions therefore are limited. The lowest rate of evolution is seen in the nonsynonymous coding sequences. Alteration of these nucleotides changes the amino acid sequence of the protein. As explained earlier, most proteins seem to have amino acid sequences that make them optimally suited for their particular role and environment, and most substitutions that cause departures from that optimum are eliminated fairly quickly by natural selection.

Pseudogenes. The highest rate of evolution seen in Table 25.1 is that of nonfunctional pseudogenes. Among human globin pseudogenes, for example, the rate of nucleotide change is approximately 5 times that observed at the nonsynonymous sites within the coding sequence of functional globin genes. The high rate of evolution observed in these sequences occurs because pseudogenes no longer code for proteins. Since further changes in these genes do not affect an organism's fitness, changes are not eliminated by natural selection. In summary, we usually observe what also makes intuitive sense: The stronger the functional constraints on a part (or the whole) of a macromolecule, the slower the rate of its evolution.

KEYNOTE

> Rates of evolution vary between different portions of a gene. Sequences with the most functional importance to an organism, such as protein-binding sites within a promoter or nucleotides that would result in an amino acid substitution, evolve at the slowest rate. Pseudogenes are inactivated versions of genes that are no longer functionally constrained and tend to evolve at the fastest rate.

Comparative Genomics. The association between low evolutionary rate and high functional significance is particularly useful in **comparative genomics,** which entails comparisons of the sequences of entire genomes of different species. Any genome project like the Human Genome Project is an ambitious attempt to compile a kind of address list of where all the genes reside within an organism's chromosomes. Such a list or map is the foundation that one day may allow researchers to repair defective genes or add or replace a desired one that is missing or lost—thereby curing or preventing genetic-based diseases such as cystic fibrosis or even cancer. However, having an address list for a large office building doesn't tell you what job each employee does, how (or if) they work, together or alone, to accomplish tasks, and whether a worker's presence saves the company or sabotages it. Numerous genome projects have been generating large amounts of nucleotide sequence information, but more than 95 percent of the nucleotides within a complex organism's genome do not appear to be functionally important. As a result, it is often difficult to tell which portions of a genome are associated with protein-coding regions and what role those proteins play. Comparison is proving to be one of the best ways to figure out the function of specific tracts of sequences within genomes. The discovery of a patch of significant sequence similarity between the genomes of two distantly related species is very suggestive of functional importance. For example, humans and mice last shared a common ancestor at least 80 to 100 million years ago. Given a pseudogene substitution rate of roughly 5×10^{-9} changes per site per year, almost half of all nucleotides that are free of selective constraint should have undergone at least one change since humans and mice diverged. Regions under functional constraint, such as nonsynonymous sites within coding regions or protein-binding sites within the promoter of a gene, accumulate changes at one-fifth that rate or less and often are easy to recognize in pairwise comparisons. Natural selection evaluates the functional consequences of an enormous number of changes on evolutionary time scales. As a result, comparative genome analysis often takes the place of the experimentally challenging and time-consuming process of saturation mutagenesis in which every position in a gene is mutated and the consequences are evaluated in a laboratory. Even better, narrowing down what an identified gene does in an organism as comparatively simple as yeast and then finding the same gene in human DNA greatly facilitates predicting the human version's function.

Comparative genomics is a powerful tool but it does have limitations. Some human genes appear to have no counterparts in simpler organisms. And remember that many proteins have multiple functions—not all of which are shared by all their homologs.

Codon Usage Bias. The effect that even tiny differences in fitness can have when subject to natural selection over the course of thousands or millions of generations is evidenced by a phenomenon known as **codon usage bias.** Take note of the slightly lower rate of evolutionary change at synonymous sites relative to that of pseudogenes in Table 25.1. This observation suggests that synonymous substitutions are not completely neutral from the perspective of natural selection and that some triplet codons may be favored over others. This

hypothesis is reinforced by the finding that synonymous codons are not used equally throughout the coding sequences of many organisms. For example, the redundancy of the genetic code allows six different codons (UUA, UUG, CUU, CUC, CUA, and CUG) to specify the amino acid leucine, but 60 percent of the leucine codons found in *Escherichia coli* are CUG and 80 percent in yeast are UUG. Since the alternative synonymous codons specify the very same amino acid, selection must be favoring some synonymous codons over others, or they would all be used equally. Remember that some synonymous codons pair with different tRNAs that carry the same amino acid. Therefore, a mutation to a synonymous codon does not change the amino acid, but it may change the tRNA used by a ribosome during translation. Studies of the different tRNAs reveal that within a cell, the amounts of the isoacceptor tRNAs (different tRNAs that accept the same amino acid) differ, and the most abundant tRNAs are those that pair with the most frequently used codons. Selection may favor one synonymous codon over another because the tRNA for the codon is more abundant and translation of mRNAs that contain that codon is more efficient. Alternatively, the bonding energy between codon and anticodon of synonymous codons may differ slightly because different bases are paired. These extremely subtle differences in translation efficiency and bonding energy appear to be subject to natural selection. This is especially true in genes that are expressed at high levels and in organisms with short generation times and large population sizes, such as bacteria, yeast, and fruit flies, where codon usage is most apparent. The existence of codon usage bias is a profound testament to the sheer power of evolutionary processes. The difference in fitness between two bacteria who are identical in every way aside from a single synonymous codon (out of approximately 1 million within their genomes) must be infinitesimally small. Still, immeasurably small as it is, it is sufficient to result in only one of the two cells' ancestors being present for counting after evolutionary time scales have passed.

Again, the tendency to use preferred rather than less-used codons is most pronounced within the genes of bacterial organisms that are expressed at the highest levels. This makes sense because the genes that are expressed the most are also the most efficient and cost effective. A large company that overpays by 1% for something that they purchase thousands of times per year is likely to be worse off than a company that saves that 1% on the commonly purchased item but overpays by 20% on something that it purchases rarely. In fact, a gene's adherence to an organism's codon usage bias has proven to be one of the most reliable indicators of the relative amount of expression of the gene during the organism's lifetime. Given that that is the case, it should also be expected that highly expressed genes use energetically inexpensive amino acids in place of energetically expensive amino acids wherever possible. The energetic cost of synthesizing each of the 20 different amino acids actually varies greatly, with glycine requiring an investment of only 11.7 adenosine triphosphate (ATP) equivalents and tryptophan requiring an average of 78.3 ATP equivalents. There are some places within a protein's primary sequence where only a tryptophan will do, but wherever it is possible to substitute a glycine (or delete the residue all together), natural selection should favor the change—especially in highly expressed genes where the cost savings would have the greatest effect. Analyses of the hundreds of completely sequenced prokaryotic genomes suggest that this is in fact the case (results for three prokaryotes are shown in Table 25.2).

KEYNOTE

> Even subtle differences in fitness, such as those associated with how efficiently one of several synonymous codons is read by ribosomes, can have significant effects on the evolution of molecules over long periods of time.

Variation in Evolutionary Rates Between Genes

Just as variation in evolutionary rates is readily apparent in comparisons of different regions within genes, striking differences in the rates of evolution between genes have been observed, even when genes from the same species are considered. As before, if stochastic factors (such as those arising from small population sizes resulting from sampling error) are ruled out, the difference must be

Table 25.2 Amino Acid Utilization in the 10 percent of Genes with the Highest Expression Levels and the 10 percent with the Lowest Expression Levels in Three Different Prokaryotic Organisms.

Amino Acid[a]	Cost	Difference from Average	*Escherichia coli* K12			*Streptococcus pneumoniae* R6			*Bacillus subtilis* subsp. *subtilis* str. 168		
			Low	High	Difference in Usage	Low	High	Difference in Usage	Low	High	Difference in Usage
Gly	11.7	−15.66	6.40	8.16	1.76	6.01	8.30	2.29	6.12	8.11	1.99
Ser	11.7	−15.66	6.65	4.99	−1.67	6.47	4.96	−1.51	6.24	5.84	−0.40
Ala	11.7	−15.66	8.48	9.76	1.27	5.85	9.80	3.95	7.40	8.70	1.29
Asp	12.7	−14.66	4.81	5.95	1.15	4.85	6.04	1.20	5.06	5.42	0.36
Asn	14.7	−12.66	4.45	3.92	−0.52	3.64	4.60	0.96	3.32	4.24	0.92
Glu	15.3	−12.06	5.23	7.10	1.87	6.54	7.91	1.37	6.71	8.11	1.40
Gln	16.3	−11.06	4.48	3.83	−0.66	4.43	3.33	−1.10	3.98	3.46	−0.52
Thr	18.7	−8.66	5.41	5.22	−0.19	4.78	5.97	1.19	4.92	5.82	0.89
Pro	20.3	−7.06	4.22	4.08	−0.15	3.17	3.72	0.55	3.87	3.62	−0.25
Val	23.3	−4.06	6.24	7.64	1.39	6.54	8.13	1.59	6.59	7.94	1.35
Cys	24.7	−2.66	1.50	0.92	−0.57	0.88	0.33	−0.54	1.02	0.52	−0.50
Arg	27.3	−0.06	5.38	5.45	0.07	4.61	4.21	−0.39	4.71	3.99	−0.72
Leu	27.3	−0.06	11.56	8.96	−2.59	12.70	7.97	−4.73	10.85	8.40	−2.45
Lys	30.3	2.94	4.34	5.66	1.32	5.96	7.16	1.19	6.25	7.55	1.30
Ile	32.3	4.94	6.72	6.03	−0.69	7.74	6.41	−1.33	6.79	7.14	0.34
Met	34.3	6.94	2.39	2.70	0.31	2.29	2.07	−0.22	2.65	2.28	−0.37
His	38.3	10.94	2.50	2.05	−0.45	2.21	1.61	−0.60	2.74	1.80	−0.93
Tyr	50	22.64	3.15	2.82	−0.34	4.44	2.96	−1.48	3.94	2.85	−1.09
Phe	52	24.64	4.35	3.71	−0.64	5.76	3.69	−2.07	5.56	3.53	−2.04
Trp	74.3	46.94	1.73	1.05	−0.69	1.13	0.81	−0.32	1.27	0.67	−0.60

[a]Amino acids are listed in terms of ascending cost of biosynthesis (shown in terms of required average number of high energy phosphate bonds from ATP to synthesize).
[b]Shown for each organism are the fraction of amino acids in the primary sequence of 10 percent of genes expressed at the lowest level and that of the 10 percent of genes expressed at the highest level.

Modeled after the work of H. Akashi and T. Gojobori, 2002. Metabolic efficiency and amino acid composition in the proteomes of *Escherichia coli* and *Bacillus subtilis. Proc. Natl. Acad. Sci. USA,* 99: 3695–3700.

attributable to one or some combination of two factors: (1) differences in mutation frequency; and (2) the extent to which natural selection affects the locus. Rigorous statistical analyses can aid in distinguishing between adaptive and random changes in nucleotide sequences. For example, the McDonald-Kreitman test (1991) compares the patterns of within-species polymorphism and between-species divergence at the synonymous and nonsynonymous sites in the coding region of a gene. If the ratio of nonsynonymous to synonymous substitutions within species is the same as that between species, then all the substitutions are likely to be neutral. If the ratios are not the same, then natural selection must be responsible—most likely by favoring nonsynonymous mutations that confer an adaptive advantage, although diversifying selection can also occur. Although some regions of genomes seem to be more prone to random changes than are others regions, synonymous substitution rates across a genome rarely differ by more than a factor of two. That difference is far from sufficient to account for the roughly thousand-fold difference in nonsynonymous substitution rates observed between different classes of mammalian genes shown in Table 25.3. As was the case for variation in observed substitution rates within genes, variation of substitution rates between genes must result largely from differences in the intensity of natural selection at each locus.

Specific examples of two classes of genes, histones and apolipoproteins, illustrate the effects of different levels of functional constraint. Histones are positively charged, essential DNA binding proteins that are present in all eukaryotes. Almost every amino acid in a histone such as histone H4 interacts directly with specific chemical residues associated with negatively charged DNA. Thus any change to the amino acid sequence of histone H4 affects its ability to interact with DNA. As a

result, histones are one of the slowest-evolving groups of proteins known, and it is possible to replace the yeast version of histone H4 with its human homolog with no effect despite hundreds of millions of years of independent evolution. Apolipoproteins, in contrast, are responsible for nonspecifically interacting with and carrying a wide variety of lipids in the blood of vertebrates. Their lipid-binding domains are made up predominantly of hydrophobic amino acids. Any similar amino acid (i.e., leucine, isoleucine, and valine) appears to function in those positions just as well as another as long as it too is hydrophobic. As a result, dozens of different versions (at the protein level) of human apolipoproteins have been characterized to date at the same time that only single versions of human histone H4 proteins have been identified.

KEYNOTE

> The rate at which genes accumulate changes reflects primarily the extent to which they are functionally constrained. Genes that are heavily subject to natural selection have much higher synonymous substitution rates than they do nonsynonymous rates.

Although amino acid substitutions within many genes generally are deleterious, it should be pointed out that natural selection favors variability within populations for some genes. For instance, it is advantageous for the genes associated with the major histocompatability complex (MHC) in mammals to differ between individuals. As a result, the rate of nonsynonymous substitutions within the MHC is greater than that of synonymous substitutions (Table 25.3). The MHC is a large multigene family whose protein products are involved with the immune system's ability to recognize foreign antigens. Within human populations, roughly 90 percent of individuals receive different sets of MHC genes from their parents, and a sample of 200 individuals can be expected to have 15 to 30 different alleles. Such high levels of diversity in this region are favored by natural selection because the number of individuals vulnerable to infection by any single virus (one that is not recognized by MHC proteins) is likely to be substantially less than it would have been if they all had similar immune systems. At the same time that host populations are driven to maintain diverse immune systems, viruses are driven to evolve rapidly. Error-prone replication, coupled with diversifying selection causes the rate of nucleotide substitutions within the influenza NS genes to be 1.9×10^{-3} per site per year, roughly one million times greater than the synonymous substitution rate for mammalian genes in Table 25.1.

Table 25.3 Relative Rates of Evolutionary Change in DNA Sequences of Different Mammalian Genes[a]

Gene	Nonsynonymous Rate	Synonymous Rate
Histone H4	0.004	1.43
Insulin	0.16	5.41
Prolactin	1.29	5.59
α-Globin	0.56	3.94
β-Globin	0.87	2.96
Albumin	0.92	6.72
α-Fetoprotein	1.21	4.90
MHC	5.10	2.40
Apolipoprotein E	0.98	4.04

[a]All rates are in nucleotide substitutions per site per year × 10^{-9}.

Rates of Evolution in Mitochondrial DNA

Organelle genomes are replicated and transmitted in a manner distinct from that of nuclear genes, and the dynamics of their substitutions are substantially different as a result. In Chapter 23, we discussed the mitochondrial genome and saw that the mammalian mitochondrial genome consists of a circular, double-stranded mitochondrial DNA (mtDNA) about 15,000 base pairs long. Human mtDNA is fairly typical and, at roughly 1/10,000 the size of the nuclear genome, encodes 2 rRNAs, 22 tRNAs, and 13 proteins. The small size of mtDNA and the discovery of its exceptionally high rate of substitution have stimulated substantial interest in its evolution.

The average synonymous substitution rate in mammalian mitochondrial genes is approximately 5.7×10^{-8} substitutions per site per year, about 10 times the average value for synonymous substitutions in nuclear genes. The rate of nonsynonymous substitution varies greatly among the different mitochondrial protein-coding genes, but in all cases that rate is much higher than the average nonsynonymous rate of nuclear genes. It is not entirely clear why animal mtDNA undergoes such rapid evolutionary change, but the reason is likely to be related to a higher error rate during mtDNA replication and repair (unlike nuclear DNA polymerases, mitochondrial DNA polymerases have no proofreading ability). Higher concentrations of mutagens such as oxygen free radicals (i.e., O_2^-) resulting from the metabolic processes carried out in mitochondria may also play a role in higher rates of substitution. It may also be that the selection pressure that normally eliminates many mutations in nuclear genes is relaxed in the mitochondria because most cells contain several dozen—each of which contains up to a dozen copies of the mtDNA molecule (with an average of two copies per mitochondria). Regardless, changes in the proteins, tRNAs, and rRNAs encoded by the mitochondrial genome appear to be less detrimental to individual fitness than are similar changes in the proteins, tRNAs, and rRNAs encoded by nuclear genes.

Mammalian mtDNA also differs from nuclear DNA in that the overwhelming majority of mtDNA is inherited clonally from the mother. Mitochondria are located in the cytoplasm, and only the mother's egg cell contributes cytoplasm to a zygote. As a consequence, mtDNA does not undergo meiosis, and all offspring should be identical to the maternal genotype for mtDNA sequences (the offspring are clones for mtDNA genes). This pattern of inheritance allows matriarchal lineages (descendants from one female) to be traced and provides a means for examining family structure in some populations. This, in conjunction with the rapid and regular rate of accumulation of nucleotide sequence differences, has allowed mtDNA to become a valuable tool for comparing closely related lineages. An example of geographic variation in mtDNA sequences of pocket gophers living in the southeastern United States is shown in Figure 25.2.

Figure 25.2

Lineage relationships among mtDNA types in pocket gophers. The lowercase letters are different mtDNA types grouped according to similarity and superimposed on a geographic map of the collection sites. The tick marks across the connecting lines are the numbers of inferred mutational steps.

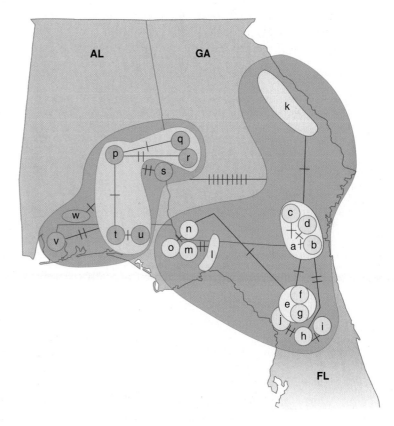

Molecular Clocks

As described earlier, the differences in the nucleotide and amino acid replacement rates between nuclear genes can be striking but are likely to result primarily from differences in the selective constraint on each individual protein. However, rates of molecular evolution for loci with similar functional constraints can be relatively constant over long periods of evolutionary time.

In fact, the very first comparative studies of protein sequences performed by Emile Zuckerkandl and Linus Pauling in the 1960s suggested that substitution rates were essentially constant within homologous proteins over many tens of millions of years. Based on these observations, they likened the accumulation of amino acid changes to the steady ticking of a clock, the **molecular clock hypothesis.** The molecular clock may run at different rates in different proteins, but the number of differences between two homologous proteins appeared to be very well correlated with the amount of time since speciation caused them to diverge independently, as shown in Figure 25.3. This observation immediately stimulated intense interest in using biological molecules in evolutionary studies. Steady rates of change between homologous sequences would facilitate not only the determination of phylogenetic relationships between species but also the times of their divergence in much the same way that radioactive decay was used to date geological times.

Despite its great promise, however, Zuckerkandl and Pauling's molecular clock hypothesis has been controversial. Classic evolutionists argued that the erratic tempo of morphological evolution was inconsistent with a steady rate of molecular change. Disagreements regarding divergence times have also placed in question the uniformity of evolutionary rates at the heart of the idea.

Relative Rate Test. Most divergence dates used in molecular evolution studies come from interpretations of the notoriously incomplete fossil record and are of questionable accuracy. To avoid any questions regarding speciation dates, Sarich and Wilson (1973) devised a simple way to estimate

Figure 25.3

The molecular clock runs at different rates in different proteins. One reason is that the neutral substitution rate differs among proteins. Fibrinogen appears to be unconstrained and has a high neutral substitution rate, whereas cytochrome c has a lower neutral substitution rate and may be more constrained. Data are from a wide variety of organisms.

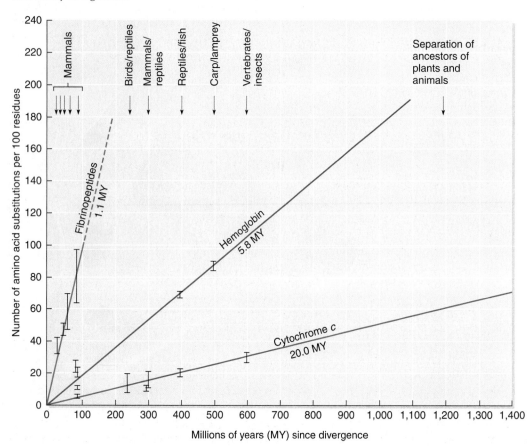

Figure 25.4

Phylogenetic tree used in a relative rate test. Species 3 is an outgroup known to have been evolving independently before the divergence of species 1 and species 2. The letter A denotes the common ancestor of species 1 and species 2.

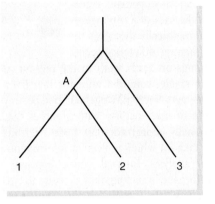

the overall rate of substitution in different lineages that does not depend on specific knowledge of divergence times. For example, to determine the *relative rate* of substitution in the lineages for species 1 and 2 in Figure 25.4, we need to have a less-related species 3 as an **outgroup**. Outgroups usually can be readily agreed upon; for instance, in this example, if species 1 and 2 are humans and gorillas, respectively, then species 3 could be another primate such as a baboon. In the evolutionary relationship portrayed in Figure 25.4, the point in time when species 1 and 2 diverged is marked with the letter A. The number of substitutions between any two species is assumed to be the sum of the number of substitutions along the branches of the tree connecting them such that

$$d_{13} = d_{A1} + d_{A3}$$
$$d_{23} = d_{A2} + d_{A3}$$
$$d_{12} = d_{A1} + d_{A2}$$

where d_{13}, d_{23}, and d_{12} are easily obtained measures of the differences between species 1 and 3, species 2 and 3, and species 1 and 2, respectively. Simple algebraic manipulation of those statements allows the amount of divergence that has taken place in species 1 and in species 2 since they last shared a common ancestor to be calculated using these equations:

$$d_{A1} = (d_{12} + d_{13} - d_{23})/2$$
$$d_{A2} = (d_{12} + d_{23} - d_{13})/2$$

By definition, the time since species 1 and species 2 began diverging independently is the same, so the molecular clock hypothesis predicts that values for d_{A1} and d_{A2} should also be the same.

Exponentially increasing amounts of DNA sequence data from a very wide variety of species are available for testing the premise of the molecular clock that the rate of evolution for any given gene is constant over time in all evolutionary lineages. Substitution rates in rats and mice have been found to be largely the same. In contrast, molecular evolution in humans and apes appears to have been only half as rapid as that which has occurred in Old World monkeys since their divergence. Indeed, relative rate tests performed on homologous genes in rats and humans suggest that rodents have accumulated substitutions at twice the rate of primates since they last shared a common ancestor during the time of the mammalian radiation 80 to 100 million years ago. The rate of the molecular clock clearly varies among taxonomic groups, and such departures from constancy of the clock rate pose a problem in using molecular divergence to date the times of existence of recent common ancestors. Before such inferences can be made, it is necessary to demonstrate that the species being examined have a uniform clock such as the one observed within rodents.

Causes of Variation in Rates. Several possible explanations have been put forward to account for the differences in evolutionary rates revealed by relative rate tests. For instance, generation times in monkeys are shorter than they are in humans, and the generation time of rodents is much shorter still. The number of germ-line DNA replications, occurring once per generation, should be more closely correlated with substitution rates than with simple divergence times. Differences may also result in part from a variety of other differences between two lineages since the time of their divergence, such as average repair efficiency, average exposure to mutagens, and the opportunity to adapt to new ecological niches and environments.

KEYNOTE

Relative rate tests suggest that substitutions do not always accumulate at the same rate in different evolutionary lineages. Primates, humans in particular, appear to be evolving more slowly at a molecular level than are other mammals since the time of the mammalian radiation 80 to 100 million years ago. Faster rates of change may result from shorter average generation times or from a variety of other factors.

Molecular Phylogeny

Because evolution can be defined as genetic change in the face of selective dynamics, genetic relationships are of primary importance in the deciphering of evolutionary relationships. The greatest promise of the molecular clock hypothesis is the implication that molecular data can be used to decipher the phylogenetic relationships among all living things. Quite simply, organisms with high degrees of molecular similarity are expected to be more closely related than those that are dissimilar. Before the tools of molecular biology were available to provide molecular data for such analyses, evolutionary biologists relied entirely on comparison of phenotypes to infer genetic similarities and differences. The underlying assumption was that if the phenotypes were similar, the genes that were responsible for the phenotypes were also similar; if the phenotypes were different, the genes were different. Originally the phenotypes examined consisted largely of gross anatomical features. Later, behavioral, ultrastructural, and biochemical characteristics were also studied. Comparisons of such traits were used successfully to construct evolutionary trees for many groups of plants and animals and are still the basis of many evolutionary studies today.

However, relying on the study of such traits has limitations. Sometimes similar phenotypes can evolve in organisms that are distantly related in a process called convergent evolution. For example, if a naive biologist tried to construct an evolutionary tree on the basis of whether wings were present or absent in an organism, he might place birds, bats, and insects in the same evolutionary group because all have wings. In this particular case, it is fairly obvious that these three organisms are not closely related; they differ in many features other than the possession of wings, and the wings themselves are very different in their design. But this extreme example shows that phenotypes can be misleading about evolutionary relationships, and phenotypic similarities do not necessarily reflect genetic similarities.

Another problem with relying on phenotypes to determine evolutionary relationships is that many organisms do not have easily studied phenotypic features suitable for comparison. For example, the study of relationships among bacteria has always been problematic because bacteria have few obvious traits that correlate with the degree of their genetic relatedness. A third problem arises when we try to compare distantly related organisms. What phenotypic features should be compared, for example, in an analysis of bacteria and mammals where so few characteristics are held in common?

Earlier in this chapter, we saw that molecular approaches can generate useful information about DNA sequences and how they evolve. Even though the relative rate of molecular evolution may vary from one lineage to another, and molecularly inferred divergence times must be treated with caution, molecular approaches to generating phylogenies usually can be relied on to group organisms correctly. Many have argued that molecular phylogenies are more reliable even when alternative data are available because the effects of natural selection generally are less pronounced at the DNA sequence level. When differences between molecular and morphological phylogenies are found, they usually create valuable opportunities to examine the effect of natural selection acting at the level of phenotypic differences.

Phylogenetic Trees

Because of the long history of evolutionary studies even before molecular data were available, the general approaches used to elucidate relationships between species are fairly well established. A central principle in all phylogenetic reconstructions is the idea of a **phylogenetic tree** that graphically describes the relationship among different species.

All living things on earth, both in the present and in the past, share a single, common ancestor that lived roughly 4 billion years ago. Every phylogenetic tree portrays at least some portion of that ancestry with *branches* that connect two (occasionally more) adjacent *nodes*. Terminal nodes indicate the taxa for which molecular information has been obtained for analysis. Internal nodes represent common ancestors before the branching that gave rise to two separate groups of organisms. Branch lengths often are scaled to reflect the amount of divergence between the taxa they connect. Where it is possible to distinguish one internal node as representing a common ancestor to all the other nodes on a tree, it is possible to make a **rooted tree.** Unrooted trees specify only the relationship between nodes and say nothing about the evolutionary path that was taken. Roots for unrooted trees usually can be determined through the use of an outgroup. As in the example of an outgroup used for the relative rate test described earlier, outgroups are taxa that have unambiguously separated the earliest from the other taxa being studied. In the case of humans and gorillas, when baboons are used as an outgroup, the root of the tree can be placed somewhere along the branch connecting baboons to the common ancestor of humans and gorillas. When only three taxa are being considered, there are three possible rooted trees but only one unrooted tree (Figure 25.5).

Number of Possible Trees. The number of possible rooted and unrooted trees quickly becomes staggering as more taxa are considered (Table 25.4). The actual number of possible rooted (N_R) and unrooted (N_U) trees for any number of taxa (n) can be determined with the following equations:

$$N_R = (2n - 3)!/[2^{n-2}(n - 2)!]$$
$$N_U = (2n - 5)!/[2^{n-3}(n - 3)!]$$

Values for n must be greater than or equal to 2 for the first equation and greater than or equal to 3 for the second equation and can be extremely large. In practice, though, the value for n is often described in terms of dozens or at most hundreds of taxa or individuals—where unimaginably large numbers of possible trees can describe the relationship between them.

Gene Versus Species Trees. A phylogenetic tree based on the divergence observed within a single homologous gene is most appropriately called a **gene tree.** This type of tree may represent the evolutionary history of a gene but not necessarily that of the species in which it is found. **Species trees**

Figure 25.5

The relationship between three taxa (labeled 1, 2, and 3) can be described by three different rooted trees but only one unrooted tree. The letter A denotes a common ancestor of all three species in the rooted tree.

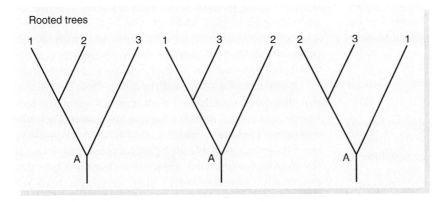

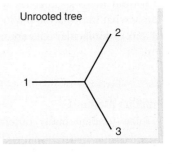

usually are best obtained from analyses that use data from multiple genes. While this may sound counterintuitive, divergence within genes typically occurs before the splitting of populations that occurs when new species are created. For the locus being considered in Figure 25.6, some individuals in species 1 may actually be more similar to individuals in species 2 than they are to other members of their own population. The differences between gene and species trees tends to be particularly important when considering loci where diversity within populations is advantageous, as in the major histocompatability locus (MHC) described earlier. If MHC alleles alone were used to determine species trees, many humans would be grouped with gorillas rather than with other humans because the polymorphism they carry is older than the time of the split in the two lineages.

KEYNOTE

Sequence polymorphisms often predate speciation events. As a result, it is possible for phylogenetic trees made from a single gene not always to reflect the relationships between species. Species trees are best constructed by considering multiple genes.

Despite the staggering number of rooted and unrooted trees that can be generated even when using a small number of taxa (Table 25.4), only one of the possible trees represents the true phylogenetic relationship between the taxa being considered. Since the true tree usually is known only when artificial data are used in computer simulations, most phylogenetic trees generated with molecular data are called **inferred trees**. Distinguishing which of all the possible trees is most likely to be the true tree can be a daunting task and is typically left

Figure 25.6

Trans-species or shared polymorphism may occur if the ancestor was polymorphic for two or more alleles and if alleles persist to the present in both species.

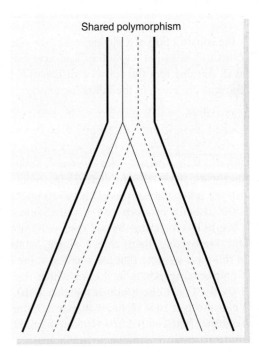

Table 25.4	Numbers of Rooted and Unrooted Trees That Describe the Possible Relationships Between Different Numbers of Taxa	
Number of Taxa	Number of Rooted Trees	Number of Unrooted Trees
2	1	1
3	3	1
4	15	3
5	105	15
10	34,459,425	2,027,025
20	8.20×10^{21}	2.22×10^{20}
30	4.95×10^{38}	8.69×10^{36}

to high-speed computers. The computer algorithms used in these searches usually use one of three different kinds of approaches: distance matrix, parsimony-based, and maximum likelihood methods. A basic understanding of the logic behind these approaches will help you understand exactly what information phylogenetic trees convey and what sort of molecular data are most useful for their generation.

Reconstruction Methods

At least three fundamentally different approaches are commonly used to determine phylogenetic relationships using molecular data. Distance matrix approaches are based on statistical principles that group things on the basis of their overall similarity to each other. This statistical approach is used for many kinds of data analysis other than just those of molecular evolution. In contrast, parsimony approaches group organisms in ways that minimize the number of substitutions that must have occurred since they last shared a common ancestor and are generally invoked only in molecular evolution studies. Maximum likelihood methods are intrinsically probabilistic/statistical and are only recently becoming feasible for typical data sets as the raw power of computers is increasing.

Distance Matrix Approaches to Phylogenetic Tree Reconstruction. The oldest distance matrix method is also the simplest of all methods for tree reconstruction. Originally proposed in the early 1960s to help with the evolutionary analysis of morphological characters, the **unweighted pair group method with arithmetic averages (UPGMA)** is largely statistically based and requires data that can be condensed to a measure of genetic distance between all the pairs of taxa being considered. To illustrate the construction of a phylogenetic tree using the UPGMA method, consider a group of four taxa called A, B, C, and D. Assume that the pairwise distances between each of the taxa are given in the following matrix:

Taxa	A	B	C
B	d_{AB}	–	–
C	d_{AC}	d_{BC}	–
D	d_{AD}	d_{BD}	d_{CD}

In this matrix, d_{AB} represents the distance (perhaps as calculated by the Jukes-Cantor model) between taxa A and B, d_{AC} is the distance between taxa A and C, and so on. UPGMA begins by clustering the two taxa with the smallest distance separating them into a single, composite taxon. In this case, assume that the smallest value in the distance matrix corresponds to d_{AB}, in which case taxa A and B are the first to be grouped together (AB). After the first clustering, a new distance matrix is computed, with the distance between the new taxon (AB) and taxa C

and D being calculated as $d_{(AB)C} = \frac{1}{2}(d_{AC} + d_{BC})$ and $d_{(AB)D} = \frac{1}{2}(d_{AD} + d_{BD})$. The taxa separated by the smallest distance in the new matrix are then clustered together to make another new composite taxa. The process is repeated until all taxa have been grouped together. If scaled branch lengths are to be used on the tree to represent the evolutionary distance between taxa, then branch points are positioned at a distance halfway between the taxa being grouped (i.e., at $d_{AB}/2$ for the first clustering).

A strength of distance matrix approaches in general is that they work equally well with morphological and molecular data as well as combinations of the two. They, like maximum likelihood analyses, also take into consideration all the data available for a particular analysis, whereas the alternative parsimony approaches described later discard many "noninformative" sites. A weakness of the UPGMA approach in particular is that it presumes a constant rate of evolution across all lineages, something that the relative rate tests tell us is not always the case. Several distance matrix-based alternatives to UPGMA such as the transformed distance method and the neighbor-joining method are more complex but capable of incorporating different rates of evolution within different lineages.

Parsimony-Based Approaches to Phylogenetic Tree Reconstruction. While the distance- and maximum likelihood-based methods of tree reconstruction are grounded in statistics, parsimony-based approaches rely more heavily on the biological principle that mutations are rare events. The word *parsimony* itself means stinginess or cheapness and refers to the fact that parsimony approaches attempt to minimize the number of mutations that a phylogenetic tree must invoke to account for the sequences of all taxa being considered. These parsimony approaches assume that the tree that invokes the fewest number of mutations is considered to be the best and that tree is deemed a tree of **maximum parsimony.**

As mentioned earlier, the parsimony-based approach does not use all sites when considering molecular data. Instead, it focuses only on positions within a multiple alignment that favors one tree over an alternative in terms of the number of substitutions they invoke. Not all positions within a multiple alignment favor one tree over an alternative from the perspective of parsimony. Consider the following alignment of four nucleotide sequences:

			Site			
	1	2	3	4	5*	6*
Sequence	G	C	G	A	T	G
	G	T	G	T	T	G
	G	T	T	G	C	A
	G	T	C	C	C	A

In such an alignment, only the fifth and sixth sites (marked with asterisks) qualify as *informative sites* from a

Figure 25.7

Three different unrooted trees describe all possible relationships between four taxa. Using the sequences (uppercase letters) and sites shown in the text, all three trees for each of the six sites are shown. Dark lines are drawn on branches along which substitutions must have occurred, and inferred ancestral states are shown in lowercase letters. The sequence for site 1 requires no substitutions regardless of which tree is used, site 2 requires one for all trees, site 3 requires two for all trees, and site 4 requires three for all trees. Only sites 5 and 6 have one tree with a fewer number of substitutions than the alternative trees; that makes them informative sites.

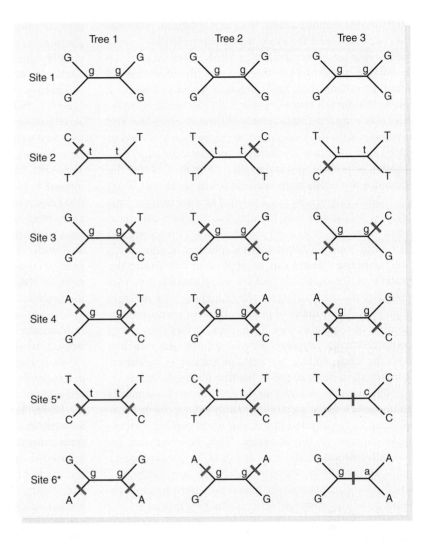

parsimony perspective. As shown in Figure 25.7, only three possible unrooted trees can be drawn that describe the relationship between four taxa. The unrooted tree that groups sequences 1 and 2 separate from sequences 3 and 4 would require only one mutation to have occurred in the branch that connects both groupings. Either of the two alternative trees that group the taxa differently would require two mutations and therefore do not represent the most parsimonious arrangement of the sequences. In contrast, all three of the possible unrooted trees for site 1 are indistinguishable from the perspective of parsimony because no mutations must be invoked for any of them. Similarly, site 2 is uninformative because one mutation occurs in all three of the possible trees. Likewise, site 3 is uninformative because all three trees require two mutations, and site 4 is uninformative because all three trees require three mutations. In general, for a site to be informative regardless of how many sequences are aligned, it has to have at least two different nucleotides, and each of these nucleotides has to be present at least twice.

Maximum parsimony trees are determined by first identifying all informative sites within an alignment and then determining which of all possible unrooted trees

invokes the fewest number of mutations for each of those sites. The tree or trees that invoke the fewest number of mutations when all sites within an alignment are considered is the most parsimonious tree. A very useful byproduct of the parsimony approach is the generation of inferred ancestral sequences at each node of a tree (Figure 25.7). These inferred ancestral sequences go a long way toward making a non-issue of the infamous "missing links" of the fossil record and, when analyzed carefully, can give remarkably clear insights into the nature of long-dead organisms and even the environment in which they lived. Of course, the parsimony approach described here assumes that all nucleotides are just as likely to mutate into any of the three alternative nucleotides. More complicated parsimony algorithms take the difference in transition and transversion frequencies into account, although none is particularly reliable when rates of substitutions between branches of a tree differ dramatically.

Maximum Likelihood Approaches to Phylogenetic Tree Reconstruction. Maximum likelihood approaches represent an alternative and purely statistically based method of phylogenetic reconstruction. With this approach, probabilities are considered for every individual nucleotide substitution in a set of sequence alignments. For instance, we know that transitions are observed roughly three times as often as transversions. In a three-way alignment where a single column is found to have a C, a T, and an A, it can be reasonably argued that a greater likelihood exists that the sequences with the C and the T are more closely related to each other than they are to the sequences with an A. Calculation of probabilities is complicated by the fact that the sequence of the common ancestor to the sequences being considered is generally not known. They are further complicated by the fact that multiple substitutions may have occurred at one or more of the sites being considered and that all sites are not necessarily independent or equivalent. Still, objective criteria can be applied to calculating the probability for every site *and* for every possible tree that describes the relationship of the sequences in a multiple alignment. The number of possible trees for even a modest number of sequences (see Table 25.4) makes this a very computationally intensive proposition, yet the one tree with the single highest aggregate probability is, by definition, the most likely to reflect the true phylogenetic tree.

The dramatic increase in the raw power of computers has begun to make maximum likelihood approaches feasible, and trees inferred in this way are becoming increasingly common in the literature. Note, however, that no one substitution model is as yet close to general acceptance and, because different models can very easily lead to different conclusions, the model used must be carefully considered and described when using this approach.

KEYNOTE

> The number of possible trees that describe the relationship between even a small number of taxa can be very large. Distance matrix and maximum likelihood methods rely on statistical relationships between taxa to group them. Parsimony approaches are more biologically based and assume that the tree that invokes the fewest number of mutations is most likely to be the best. No method can guarantee that it will yield the true phylogenetic tree, but when multiple substitutions are not likely to have occurred and evolutionary rates within all lineages are fairly equal, all three methods have been demonstrated to work well.

Bootstrapping and Tree Reliability. Obviously, longer sequence alignments require a longer time to analyze than shorter ones when the parsimony approach is used. However, because of the relationship between the number of taxa and the corresponding number of unrooted trees illustrated in Table 25.4, the addition of more sequences has a much more dramatic effect on the time required to find a

preferred tree. Once data sets involve 30 or more species, the number of possible trees is so large that it is simply not possible to examine all possible trees and assess the fit of the data to each, even when the fastest computers are used. Alternative trees are not all independent of each other, however, and many parsimony algorithms use shortcuts to avoid having to perform an exhaustive search. However, no tree reconstruction method is certain to yield the correct tree. Numerous variations on each approach have been suggested, and intensive simulation studies have been performed to compare the statistical reliability of almost all tree construction methods. The results of these simulations are easy to summarize: Data sets that allow one method to infer the correct phylogenetic relationship generally work well with all the currently popular methods. However, if many changes have occurred in the simulated data sets or rates of change vary among branches, then none of the methods works very reliably. As a general rule, if a data set yields similar trees when analyzed by two or three of the fundamentally different tree reconstruction methods, that tree can be considered to be fairly reliable.

It is also possible for portions of inferred trees to be determined with varying degrees of confidence. **Bootstrap tests** allow a rough quantification of those confidence levels. The basic approach of the bootstrap test is straightforward: A subset of the original data is drawn (with replacement) from the original data set and a tree is inferred from the new data set. In a physical sense, the process is equivalent to taking the print out of a multiple alignment; cutting it up into pieces, each of which contain a different column from the alignment; placing all those pieces into a bag; randomly reaching into the bag and drawing out a piece; copying down the information from that piece before returning it to the bag; and then repeating the drawing step until an artificial data set has been created that is as long as the original alignment. This whole process is repeated to create hundreds or thousands of resampled data sets and portions of the inferred tree that have the same groupings in many of the repetitions are those that are especially well supported by the entire original data set. Numbers that correspond to the fraction of bootstrapped trees yielding the same grouping are often placed next to the corresponding nodes in phylogenetic trees to convey the relative confidence in each part of the tree. Bootstrapping has become very popular in phylogenetic analyses even though some methods of tree inference can make it very time-consuming to perform.

Despite their often casual use in the scientific literature, bootstrap results need to be treated with some caution. First, bootstrap results based on fewer than several hundred iterations (rounds of resampling and tree-generation) are not likely to be reliable—especially when large numbers of sequences are involved. Simulation studies have also shown that bootstrap tests tend to underestimate the confidence level at high values and overestimate it at low values. And, since many trees have very large numbers of branches, there is often a significant risk of succumbing to "the fallacy of multiple tests"—some results may appear to be statistically significant by chance simply because so many groupings are being considered. Still, some studies have suggested that commonly used solutions to these potential problems (i.e., doing thousands of iterations; using a correction method to adjust for estimation biases; collapsing branches to multi-bifurcations wherever bootstrap values do not exceed a very stringent threshold) yield trees that are closer representations of the true tree than the single most parsimonious tree.

Phylogenetic Trees on a Grand Scale

One of the most striking cases in which sequence data have provided new information about evolutionary relationships is in our understanding of the primary divisions of life. In the late 1800s, biologists divided all of life into three major groups: the plants, the animals, and the protists (a catchall category for everything that did not fit into the two eukaryotic categories). As more organisms were discovered and their features examined in more detail, this simple dichotomy became unworkable. It was later recognized that organisms could be divided into prokaryotes and eukaryotes on the basis of cell structure. Several additional primary divisions of life were subsequently recognized, such as the five *kingdoms* (prokaryotes, protista, plants, fungi, and animals) proposed by R. H. Whittaker in 1959.

The Tree of Life. In the mid-1980s, RNA and DNA sequences were used to uncover the primary lines of evolutionary history among all organisms. In one study, Carl Woese, Norm Pace, and colleagues constructed an evolutionary tree of life based on the nucleotide sequences of the 16S rRNA, which all organisms (as well as mitochondria and chloroplasts) possess. As illustrated in Figure 25.8, their evolutionary tree revealed three major evolutionary groups: the **eubacteria** (the traditional prokaryotes as well as mitochondria and chloroplasts), the **archaebacteria** (including thermophilic bacteria and many other little-known organisms) and the **eukaryotes.** Eubacteria and archaebacteria, although both prokaryotic in that they had no internal membranes, were found

Figure 25.8

An evolutionary tree of life revealed by comparison of 16S rRNA sequences.
[Reprinted with permission from N. Pace, "A Molecular View of Microbial Diversity in the Biosphere," in *Science* 276:735 (1997). Copyright © 1997 American Association for the Advancement of Science.]

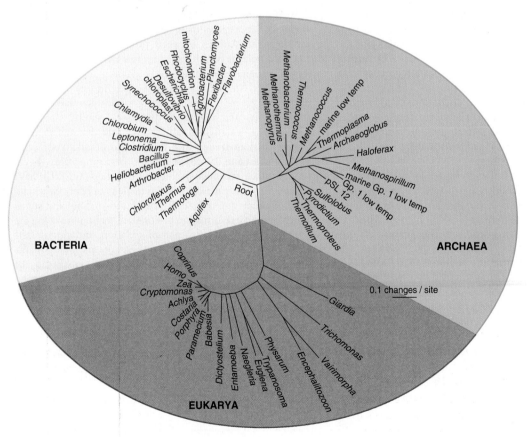

to be as different genetically as eubacteria and eukaryotes. The deep evolutionary differences that separate the eubacteria and the archaebacteria were not obvious on the basis of phenotype, and the fossil record was silent on the issue. The differences became clear only after their nucleotide sequences were compared. Sequences of other genes, including 5S rRNAs, large rRNAs, and the genes coding for some fundamentally important proteins, support the idea that three major **evolutionary domains** exist among living organisms: the Bacteria, the Archaea, and the Eukarya. Originally intended as a replacement for kingdoms, domains are now used as a higher-level rank with eukaryotes divided into four different kingdoms (protists, fungi, plants, and animals). These molecular phylogenies have led to many other surprising revelations, such as the observation that the genes of eukaryotic organelles like mitochondria and chloroplasts actually have separate, independent origins from their nuclear counterparts (Box 25.1).

Human Origins.   Another field in which DNA sequences are being used to study evolutionary relationships is human evolution. In contrast to the extensive phenotypic variation observed in size, body shape, facial features, and skin color, genetic differences among human populations are surprisingly small. For example, analysis of mtDNA sequences shows that the mean difference in sequence between two human populations is about 0.33 percent. Other primates exhibit much larger differences. For example, the two subspecies of orangutan have mtDNA sequences that differ by as much as 5 percent. This high degree of genetic similarity indicates that all human groups are closely related. Another surprising observation emerges upon careful examination of those genetic differences that do exist between different human groups: The greatest differences are not found among populations located on different continents but rather are found between human populations residing in Africa. All other human populations show fewer differences than we find among just the African populations alone. Many experts interpret these findings to mean that humans originated and experienced their early evolutionary divergence in Africa. By this theory, a small group of humans migrated out of Africa and gave rise to all other human populations only after a number of genetically differentiated populations had already evolved in Africa. This hypothesis has been called the out-of-Africa theory. Sequence data from both

---

Box 25.1   The Endosymbiont Theory

The tree of life shown in Figure 25.8 suggests that the differences between prokaryotes, eukaryotes, and the archaebacteria result from independent evolution that has been taking place for far longer than the time since plants and animals diverged (at least 1.5 billion years). Analyses such as those have also shed light on the long-standing question of how the compartmental organization of eukaryote cells could have evolved from the simpler condition still found in prokaryotes and archaebacteria. The most important clue in providing a satisfying answer to that question came with the realization that the 16S ribosomal DNA of the nucleus, mitochondria, and chloroplasts were evolving independently even before the first eukaryotes appeared. In fact, the closest living relative of mitochondria today actually appears to be the bacteria *Rickettsia prowazekii*, the causative agent of epidemic typhus. A logical inference was that mitochondria and chloroplasts were free-living organisms that at some point in the past became engulfed by a prokaryote-like organism. The endosymbiotic (endo meaning "internal," and symbiotic meaning "cooperative relationship between two or more organisms") arrangement that resulted became the eukaryotes we see today. In other words, a merger of at least two or three evolutionary lineages gave rise to significantly different new forms of life.

The endosymbiont theory was originally suggested by a pioneering physiological ecologist, A. Schimper, in the early 1880s, and then championed by G. Mereschkovsky in the early 1900s based on microscopic examinations of plants and their plastids, which Mereschkovsky described as "little green slaves." More recent molecular analyses, especially those of Lynn Margulis (1981), have led to general acceptance of this model for the origin of these eukaryotic organelles. Numerous additional similarities between prokaryotes, mitochondria, and chloroplasts corroborate the 16S rRNA-based phylogenies. For instance, all organisms (including mitochondria and chloroplasts) in the eubacteria branch in the tree of life (Figure 25.8) have circular chromosomes, similar genomic arrangements and replication processes, similar sizes, and similar drug sensitivities, all features that distinguish them from what is associated with the nucleus of eukaryotic cells. Mitochondria and chloroplasts share these properties.

In time, the endosymbionts in eukaryotic cells have become very specialized, with the nucleus (a likely endosymbiont of eukaryotic cells itself!) being the predominant site at which heritable information is stored, mitochondria being the primary site for oxidative phosphorylation, and chloroplasts being the site at which photosynthesis occurs. Many of the genes essential for organelle function have moved to the nucleus, and the relationship between organelles and their host cells has become an obligatory and elaborate one in which no unit (compartment) can live independently.

Sources: Andersson, S. G. E., Zomorodipour, A., Andersson, J. O., Sicheritz-Ponten, T., Alsmark, U. C., Podowski, R. M., Naslund, A. K., Eriksson, A. S., Winkler, H. H., Kurland, C. G. 1998. The genome sequence of *Rickettsia prowazekii* and the origin of mitochondria. *Nature* 396:133–140.

Margulis, L. 1981. *Symbiosis in Cell Evolution: Life and Its Environment in the Early Earth.* San Francisco: W. H. Freeman.

mitochondrial DNA and the nuclear Y chromosome (the male sex chromosome) are consistent with this hypothesis in that they suggest that all people alive today have mitochondria that came from a "mitochondrial Eve" and that all men have Y chromosomes derived from a "Y chromosome Adam" roughly 200,000 years ago. While the out-of-Africa theory is not universally accepted, DNA sequence data are playing an increasingly important role in the study of human evolution and in the study of the evolution of many lineages.

> *ⓘ* You have joined a team of molecular geneticists and anthropologists who have developed a technique for extracting and analyzing ancient DNA from Neanderthal fossils in the iActivity *Were Neanderthals Our Ancestors?* on your CD-ROM.

Canine Origins.   The evolution of "man's best friend" has recently been studied in a similar way in a hybrid effort of phylogenetic reconstruction and comparative genomics. Over the past several centuries, artificial selection (selective breeding) has created hundreds of different breeds of dogs having a wide variety of physical features and temperaments. The inbreeding associated with the creation of these breeds has generated many breed-specific problems such as narcolepsy, arthritis, and the various forms of cancer that also afflict humans. Comparisons of closely related breeds that differ in their prevalence of diseases are allowing researchers to track down genes responsible for many illnesses in dogs as well as their counterparts in humans. As part of those disease-gene-finding efforts, a set of researchers working with Elaine Ostrander of the Fred Hutchison Cancer Research Center in Seattle (Parker H. G. et al., 2004. Genetic structure of the purebred domestic dog; *Science* 304:1160–1164.) examined genetic variation associated with almost 100 different microsatellite markers in 85 different dog breeds in 2004. They found that dog breeds are a very real concept at a genetic level (e.g., each Saint Bernard is more closely related to other Saint Bernards than it is to any dog of a different breed). The phylogenetic trees the researchers constructed with their molecular data also revealed that all dog breeds clustered into only four different categories. The oldest (most genetically diverse) cluster (including Siberian husky, chow chow, and sharpei) have the greatest similarity to wolf DNA and appear to trace their ancestry back to Asia and Africa. Subsequent European efforts seem to have been responsible for the creation of breeds specialized for guarding (including bulldog, rottweiler, and German sheperd), hunting (including golden retriever, bloodhound, and beagle), and herding (including collie, several sheepdogs, and Saint Bernard).

Acquisition and Origins of New Functions

A long-standing question of deep interest to those who study molecular evolution is the issue of how genes with new functions arise. As early as 1932, J. B. S. Haldane suggested that new genes arose from the process of mutating redundant copies of already existing genes. Although other means such as transposition (see Chapter 8) have since been described, Haldane's argument still does a good job of describing the origin of most new genes.

Multigene Families

In eukaryotic organisms, we often find tandemly arrayed, multiple copies of genes, all having identical or very similar sequences. These **multigene families** are sets of related genes that have evolved from some ancestral gene through gene duplication. The globin gene family that encodes the proteins used to make up the oxygen-carrying hemoglobin molecule in our blood has become a classic example of such a multigene family. (The organization and expression pattern of this multigene family in humans was discussed in Chapter 21, pp. 579–580.) Briefly, the globin multigene family is composed of seven α-like genes found on chromosome 16 and six β-like genes found on chromosome 11.

Globin genes are also found in other animals, and globin-like genes are even found in plants, suggesting that this gene family is at least 1.5 *billion* years old. Almost all functional globin genes in animal species have the same general structure, consisting of three exons separated by two introns. However, the numbers of globin genes and their order vary among species, as is shown for the β-like genes in Figure 25.9. Since all globin genes have similarities in structure and sequence, it appears that an ancestral globin gene (perhaps most like the present-day myoglobin gene) duplicated and diverged to produce an ancestral α-like gene and an ancestral β-like gene. These two genes then underwent repeated, independent duplications, giving rise to the various α-like and β-like genes found in vertebrates today. Repeated gene duplication, such as that giving rise to the globin gene family, appears to be a frequent evolutionary occurrence. Indeed, the number of copies of globin gene varies even within some human populations. For example, most humans have two α-globin genes on chromosome 16 (as shown in Figure 21.8, p. 580). However, some individuals have a single α-globin gene; other individuals have three or even four copies of the α-globin gene on their chromosome 16. These observations suggest that duplication and deletion of genes in multigene families are part of a constant process that continues to operate today. Gene duplications and deletions in gene clusters often arise as a result of misalignment of sequences during crossing-over between homologous chromosomes, a process called **unequal crossing-over.**

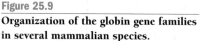

Figure 25.9

Organization of the globin gene families in several mammalian species.

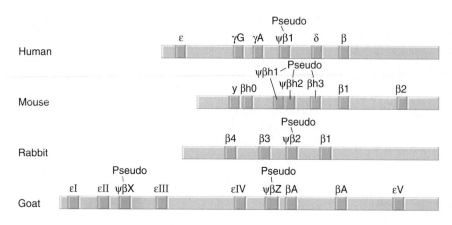

Duplications can also arise through transposition, although these duplications are likely to result in a wide dispersal of the copied genes.

Gene Duplication and Gene Conversion

Following gene duplication, one of the separate copies of a gene may undergo changes in sequence as if it were free from functional constraint as long as the other copy continues to function. As you might expect from the previous discussions in this chapter, most changes to the copy normally would have been selectively disadvantageous or even render it a nonfunctional pseudogene. On rare occasions, however, the changes may lead to subtle alterations of function or pattern of expression that are advantageous to the organism, and the change sweeps through a population. This "tinkering" approach to evolution becomes even more of a "win/no-lose" scenario when misalignments between pseudogene copies and the functional copy occur during subsequent recombination events and the inactivating changes are corrected in a process known as **gene conversion** (see Chapter 16, Box 16.1 p. 435). In this way, gene conversion events can give an organism multiple chances to create a gene with a new function from the duplicate of an already functional one. Like gene duplication, gene conversion also continues to operate to this day, although it is usually most apparent when helpful substitutions to a gene copy are "corrected." For example, the two neighboring genes on the X chromosome that allow most humans to distinguish between red and green light are 98 percent identical at the nucleotide level, and most spontaneous occurrences of deficiencies in green-color vision occur as a result of gene conversions between the two. Approximately 8 percent of human males are color blind as a result of this kind of gene conversion event.

Arabidopsis Genome Results

The extent to which organisms use gene duplication to generate proteins with new functions and thereby evolve is becoming increasingly clear as more and more genome sequencing projects are concluding. For example, with only about 125 million nucleotides in its genome, *Arabidopsis thaliana* (thale cress) was the first plant genome to be completely sequenced (see Chapter 10, pp. 253–254). Its short generation time and small size make it a favorite organism of plant geneticists, but it was a particularly appealing choice for genome sequencers because studies had indicated that its genome had undergone much less duplication than that seen in other, more commercially important plants. But when the sequencing was completed at the end of 2000, more than half of the 25,500 *Arabidopsis* genes were found to be duplicates. Phylogenetic analyses such as the distance matrix, parsimony, and maximum likelihood methods described earlier in this chapter revealed only about 11,600 distinct families of one or more genes. Even in this unusually nonredundant genome, the process of evolving through gene duplication followed by tinkering holds sway.

KEYNOTE

> Gene duplication events appear to have occurred frequently in the evolutionary history of all organisms. Copies of genes provide the raw material for evolution in that they are free to accumulate substitutions that sometimes give rise to proteins with new, advantageous functions.

Domain (Exon) Shuffling. It should also be pointed out that an increase in the number of copies of a DNA sequence can also occur in segments of a genome that are smaller than complete genes. Numerous examples of genes that contain internal duplication of one or more protein domains have been found, such as the human serum albumin gene, which is made up almost entirely of three perfect copies of a 195 amino acid domain. Elongation of a gene through internal duplication of functional domains does not lead to new proteins with significantly different functions very quickly, however. Most complex proteins are assemblages of several different protein domains that perform varied functions, such as acting as a substrate binding site or a membrane-spanning region. Perhaps not coincidentally, the beginnings and ends of exons often correspond to the beginnings and ends of domains within complex proteins. Walter Gilbert in 1978 proposed that the first genes had a limited number of protein domains within their repertoire and that most if not all of the gene families used by living things today came through **domain shuffling:** the duplication and rearrangement of those domains (usually encoded by individual exons) in different combinations. Domain (or exon) shuffling is a controversial idea that presupposes that introns were a feature of the most primitive life on earth even though they are now found all but exclusively in Eukarya rather than simpler Bacteria and Archaea domains. Still, numerous striking examples of complex genes that are made of bits and pieces of other genes are known, and it is clear that at least some genes with novel functions have been created in this way.

K E Y N O T E

> Internal duplications within genes are not uncommon, and many exons correspond to discrete functional domains within proteins. Some genes with novel functions seem to have been created through a process of domain (or exon) shuffling in which regions between and within genes are recombined in new ways.

Summary

The mathematical theory developed by population geneticists is applied to long time frames in the study of molecular evolution. This application can decipher the sometimes cryptic historical record preserved in the DNA sequences of organisms. It also provides insights into which portions of genes are functionally important, the evolutionary relationship between widely varying groups of organisms, and the mechanisms by which genes with novel functions arise.

Rates of evolution vary widely within and between genes. In most cases, the primary cause of these differences is variation in the level to which changes affect the function of genes or their proteins. Portions of a genome that have the least impact on fitness appear to evolve the fastest. Alignments between two or more sequences allow estimates to be made of the number of substitutions that have accumulated since they diverged from a single sequence in a common ancestor. Many genes accumulate substitutions at a constant rate for long periods of evolutionary time, although relative rate tests show that some

lineages (like that of humans) tend to evolve more slowly than others.

Sequence alignments also can be used as a starting point in phylogenetic reconstructions of very diverse groups of organisms. Caution must be exercised when interpreting relationships based on data from only one gene, but two fundamentally different approaches often yield trees that describe similar relationships when substitution rates are low and rates of evolution are fairly constant for all lineages. The distance matrix methods are based on statistical principles that group taxa together in a fashion that depends on their overall similarity to each other. Parsimony-based methods rely on the biological principle that mutations are rare events and the assumption that a tree that invokes the fewest number of mutations is most likely to represent the actual evolutionary relationship of sequences. Maximum likelihood approaches use the probabilities associated with any given nucleotide substitution to determine the most likely evolutionary relationship between sequences. The large number of possible trees makes phylogenetic reconstruction computationally intensive, but statistical methods are available to determine the robustness of any tree. Molecular phylogenies have numerous advantages over trees generated with morphological or data sets and have provided new insights into the very deepest branches to the tree of life.

Gene duplication, in whole or in part, seems to have played a fundamental role in the evolution of proteins with novel functions. Gene duplications often arise as a result of misalignment of sequences during crossing-over and typically result in tandemly arrayed families of genes on chromosomes. Duplicates of functional genes are free of selective constraint and can accumulate substitutions without affecting the fitness of an organism. When a duplicated gene undergoes a mutation that confers a useful, new function to an organism, it is again placed under selective constraint as the organism begins to derive an increased fitness from it. Multigene families are common in complex organisms such as eukaryotes, and more than half of all genes within any given genome are likely to have evolved in such a way.

Analytical Approaches to Solving Genetics Problems

Q25.1 Consider the following five-way multiple alignment of hypothetical homologous sequences. Generate a distance matrix that describes the pairwise relationship of all the sequences presented. Use the UPGMA method to generate a tree that describes the relationship between these sequences.

```
         10        20        30
A: GCCAACGTCC ATACCACGTT GTTTAGCACC
B: GCCAACGTCC ATACCACGTT GTCAAACACC
C: GGCAACGTCC ATACCACGTT GTTATACACC
D: GCTAACGTCC ATATCACGCT GTCATGTACC
E: GCTGGTGTCC ATATCACGTT ATCATGTACC

         40        50
A: GGTTCTCGTC CGATCACCGA
B: GGTTCTCGTC CGATCACCGA
C: GGTTCTCGTC AGGTCACCGA
D: GGTCCTCGTC AGATCCCAA
E: GGTACTCGTC CGATCACCGA
```

A25.1 A distance matrix is made by determining the number of differences observed in all possible pairwise comparisons of the sequences. The number of differences between sequence A and B (d_{AB}), for instance, is 3. The complete distance matrix is shown here:

Taxa	A	B	C	D
B	3	–	–	–
C	6	5	–	–
D	11	10	11	–
E	11	10	13	9

The smallest distance separating any of the two sequences in the multiple alignment corresponds to d_{AB}, so taxon A and taxon B are grouped together. A new distance matrix is then made in which the composite group (AB) takes their place. Distances between the remaining taxa and the new group are determined by taking the average distance between its two members (A and B) and all other remaining taxa [i.e., $d_{(AB)C} = \frac{1}{2}(d_{AC} + d_{BC})$, so $d_{(AB)C} = \frac{1}{2}(6 + 5) = 5.5$], and the resulting matrix looks like this:

Taxa	AB	C	D
C	5.5	–	–
D	10.5	11	–
E	10.5	13	9

The smallest distance separating any two taxa in this new matrix is the distance between (AB) and C, so a new combined taxon, (AB)C, is created. Another distance matrix using this new grouping then looks like this:

Taxa	(AB)C	D
D	10.75	–
E	11.75	9

In this last matrix the smallest distance is between taxa D and E ($d_{DE} = 9$), so they are grouped together as (DE). One way to symbolically represent the final clustering of taxa is ((AB)C)(DE). Alternatively, a tree such as the following can be used.

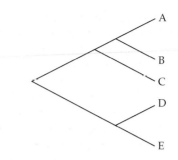

Q25.2 Using the same five sequences from Question 25.1, which positions within the alignment correspond to informative sites for parsimony analyses?

A25.2 The following positions are informative sites for parsimony analyses: 3, 14, 23, 25, 26, 27, and 41. They are the only ones that have at least two different nucleotides, with each of those nucleotides being present at least twice.

Questions and Problems

*25.1 The following sequence is that of the first 45 codons from the human gene for preproinsulin. Using the genetic code (Figure 6.7, p. 118), determine what fraction of mutations at the first, second, and third positions of these 45 codons will be synonymous.

```
ATG GCC CTG TGG ATG CGC CTC CTG CCC CTG CTG GCG
CTG CTG GCC CTC TGG GGA CCT GAC CCA GCC GCA GCC
TTT GTG AAC CAA CAC CTG TGC GGC TCA CAC CTG GTG
GAA GCT CTC TAC CTA GTG TGC GGG GAA
```

At which position is natural selection likely to have the greatest effect and nucleotides are most likely to be conserved?

*25.2 The following sequences represent an optimal alignment of the first 50 nucleotides from the human and sheep preproinsulin genes. Estimate the number of substitutions that have occurred in this region since humans and sheep last shared a common ancestor, using the Jukes-Cantor model.

*25.3 Using the alignment in Question 25.2 and assuming that humans and sheep last shared a common ancestor 80 million years ago, estimate the rate at which the sequence of the first 50 nucleotides in their preproinsulin genes have been accumulating substitutions.

*25.4 Would the mutation rate be greater or less than the observed substitution rate for a sequence of a gene such as the one shown in Question 25.2? Why?

*25.5 If the rate of nucleotide evolution along a lineage is 1.0 percent per million years, what is the rate of substitution per nucleotide per year? What would be the observed rate of divergence between two species evolving at that rate since they last shared a common ancestor?

*25.6 The average synonymous substitution rate in mammalian mitochondrial genes is approximately 10 times the average value for synonymous substitutions in nuclear genes. Why would it be better to use comparisons of mitochondrial sequences to study human migration patterns and nuclear genes when studying the phylogenetic relationships of mammalian species that diverged 80 million years ago?

*25.7 Why might mitochondrial DNA sequences accumulate substitutions at a faster rate than nuclear genes in the same organism?

*25.8 Why might substitution rates differ from one species to another, and how would such differences depart from Zuckerkandl and Pauling's assumptions for molecular clocks?

*25.9 Suppose we examine the rates of nucleotide substitution in two nucleotide sequences isolated from humans. In the first sequence (sequence A), we find a nucleotide substitution rate of 4.88×10^{-9} substitutions per site per year. The substitution rate is the same for synonymous and nonsynonymous substitutions. In the second sequence (sequence B), we find a synonymous substitution rate of 4.66×10^{-9} substitutions per site per year and a nonsynonymous substitution rate of 0.70×10^{-9} substitutions per site per year. Referring to Table 25.1 (p. 708), what might you conclude about the possible roles of sequence A and sequence B?

*25.10 What evolutionary process might explain a coding region in which the rate of amino acid replacement is greater than the rate of synonymous substitution?

*25.11 Natural selection does not always act only at the level of amino acid sequences in proteins. Ribosomal RNAs, for instance, are functionally dependent on extensive and specific intramolecular secondary structures that form when complementary nucleotide sequences within a single rRNA interact. Would the regions involved in such pairing accumulate mutations at the same rate as unpaired regions? Why?

*25.12 The following three-way alignment is of the nucleotide sequences from the beginning of the human, rabbit, and duck α-globin genes. Given that humans and rabbits are known to be more closely related to each other than they are to ducks (an outgroup), use the relative rate test to determine whether there has been a change in the rate of substitution in this region of the human and rabbit genomes since they last shared a common ancestor.

```
                    10          20          30
Human: ATGGTGCTGT CTCCTGCCGA CAAGACCAAC
Rabbit: ATGGTGCTGT CTCCCGCTGA CAAGACCAAC
Duck: ATGGTGCTGT CTGCGGCTGA CAAGACCAAC

                    40          50
Human: GTCAAGGCCC CCTGGGACAA
Rabbit: ATCAAGACTG CCTGGGAAAA
Duck: GTCAAGGGTG TCTTCTCCAA
```

*25.13 What are some of the advantages of using DNA sequences to infer evolutionary relationships?

*25.14 As suggested by the popular movie *Jurassic Park*, organisms trapped in amber have proven to be a good source of DNA from tens and even hundreds of millions of years ago. However, when using such sequences in phylogenetic analyses, it is usually not possible to distinguish between samples that come from evolutionary dead ends and those that are the ancestors of organisms still alive today. Why would the former be no more useful than simply including the DNA sequence of another living species in an analysis?

*25.15 In the phylogenetic analysis of a group of closely related organisms, the conclusions drawn from one locus were found to be at odds with those from several others. What might account for the discordant locus?

*25.16 What is the chance of randomly picking the one rooted phylogenetic tree that describes the true relationship between a group of six organisms? Are the odds better or worse for randomly picking from among all the possible unrooted trees for those organisms?

25.17 Draw all the possible unrooted trees for four species: A, B, C and D. How many rooted trees are there for the same four species?

25.18 Use the same sequence alignment provided for the Analytical Question 25.1 to generate a distance matrix, but do so by weighting transversions (As or Gs changing to Cs or Ts) twice as heavily as transitions (Cs changing to Ts, Ts changing to Cs, As changing to Gs, or Gs changing to As).

*25.19 Increasing the amount of sequence information available for analysis usually has little effect on the length of time that computer programs use to generate phylogenetic trees with the parsimony approach. Why doesn't the amount of sequence information affect the total number of possible rooted and unrooted trees?

*25.20 When bootstrapping is used to assess the robustness of branching patterns in a tree of maximum parsimony, why is it more important to use sequences that have as many informative sites as possible than simply to use longer sequences?

*25.21 What are the advantages of gene duplication (in whole or in part) in generating genes with new functions? Suggest an alternative way in which genes with new functions could arise.

Glossary

10-nm chromatin fiber The least compact form of **chromatin.** It is approximately 10 nm in diameter and has a "beads-on-a-string" morphology. *See also* **30-nm chromatin fiber.**

30-nm chromatin fiber A slightly compacted form of **chromatin,** about 30 nm in diameter, that H1 histone helps stabilize. *See also* **10-nm chromatin fiber.**

acrocentric chromosome A chromosome with the centromere near one end such that it has one long arm plus a stalk and a satellite.

activators The major class of transcription regulatory proteins in eukaryotes. Binding of these proteins to regulatory DNA sequences associated with specific genes determines the efficiency of transcription initiation. Some bacterial genes are controlled by activators. *See also* **repressors.**

adenine (A) A **purine** base found in RNA and DNA. In double-stranded DNA, adenine pairs with thymine, a **pyrimidine,** by hydrogen bonding.

allele One of two or more alternative forms of a single gene that can exist at the same **locus** in the genome. All the alleles of a gene determine the same hereditary trait (e.g., seed color), but each has a unique nucleotide sequence, which may result in different phenotypes (e.g., yellow or green seeds). *See also* **DNA polymorphism.**

allele-specific oligonucleotide (ASO) hybridization A procedure, using PCR primers, to distinguish alleles that differ by one base pair.

allelic frequency Proportion of a particular allele at a locus within a gene pool. The sum of the allelic frequencies at a given locus is 1.

allelomorph *See* **allele.**

allopolyploidy Condition in which a cell or organism has two or more genetically distinct sets of chromosomes that originate in different, though usually related, species.

alternation of generations Type of life cycle characteristic of green plants in which haploid cells **(gametophytes)** alternate with diploid cells **(sporophytes).**

alternative polyadenylation Process for generating different functional mRNAs from a single gene by cleavage and polyadenylation of the primary transcript at different **poly(A) sites.**

alternative splicing Process for generating different functional mRNAs from a single precursor mRNA (pre-mRNA) by incorporating different exons in the mature mRNA.

Ames test An assay that measures the ability of chemicals to cause mutations in certain bacteria. It can identify potential carcinogens.

amino acid Any of the small molecules, containing a carboxyl group and amino group, that are joined together to form polypeptides and proteins.

aminoacyl-tRNA A tRNA molecule covalently bound to an amino acid; also called *charged tRNA.* This complex brings the amino acid to the ribosome so that it can be used in polypeptide synthesis.

aminoacyl-tRNA synthetase An enzyme that catalyzes the addition of a specific amino acid to the tRNA for that amino acid.

amniocentesis A procedure in which a sample of amniotic sac fluid is withdrawn from the amniotic sac of a developing fetus and cells are cultured and examined for chromosomal abnormalities.

analysis of variance (ANOVA) A series of statistical procedures for determining whether differences in the **means** of a variable in two samples are significant and for partitioning the **variance** into components.

anaphase The stage in mitosis when the **sister chromatids** separate and migrate toward the opposite poles of the cell.

anaphase I The stage in meiosis I when the chromosomes in each **bivalent** separate and begin moving toward opposite poles of the cell.

anaphase II The stage in meiosis II when the **sister chromatids** are pulled to the opposite poles of the cell.

aneuploidy Any condition in which the number of chromosomes differs from an exact multiple of the normal haploid number in a cell or organism. It commonly results from the gain or loss of individual chromosomes, but also can result from the duplication or deletion of part(s) of a chromosome or chromosomes.

antibody A protein molecule that recognizes and binds to a foreign substance introduced into the organism.

anticodon A group of three adjacent nucleotides in a tRNA molecule that pairs with a **codon** in mRNA by complementary base pairing.

antigen Any large molecule that stimulates the production of specific antibodies or binds specifically to an antibody.

antiparallel In the case of double-stranded DNA, referring to the opposite orientations of the strands, with the 5′ end of one strand paired with the 3′ end of the other strand.

antisense mRNA An mRNA transcribed from a cloned gene that is complementary to the mRNA produced by the normal gene.

apoptosis Controlled process leading to cell death that is triggered by intracellular damage (e.g., DNA lesions) or by external signals from neighboring cells. Also called *programmed cell death.*

aporepressor protein An inactive repressor that is activated when bound to an effector molecule.

applied research Research done with the objective of developing products or processes that can be commercialized or at least made available to humankind for practical benefit.

archaebacteria Prokaryotes that constitute one of the three main evolutionary domains of organisms. Also called *archaeans.*

artificial selection Process for deliberating changing the phenotypic traits of a population by determining which individuals will survive and reproduce.

attenuation A regulatory mechanism in certain bacterial biosynthetic operons that controls gene expression by causing RNA polymerase to terminate transcription

autonomously replicating sequence (ARS) A specific sequence in yeast chromosomes that, when included as part of an extra-chromosomal, circular DNA molecule, confers on that molecule the ability to replicate autonomously; one type of eukaryotic **replicator.**

autopolyploidy Condition in which a cell or organism has two or more genetically distinct sets of chromosomes of the same species.

autosome A chromosome other than a **sex chromosome.**

auxotroph A mutant strain of an organism that cannot synthesize a molecule required for growth and therefore must have the molecule supplied in the growth medium for it to grow. Also called *auxotrophic mutant* or *nutritional mutant.*

auxotrophic mutant *See* **auxotroph.**

auxotrophic mutation A mutation that affects an organism's ability to make a particular molecule essential for growth. Also called *nutritional mutation.*

back mutation *See* **reverse mutation.**

bacterial artificial chromosome (BAC) A vector for **cloning** DNA fragments up to about 200 kb long in *E. coli.* A BAC contains the origin of replication of the *F* factor, a multiple cloning site, and a selectable marker.

bacteriophages Viruses that attack bacteria. Also called *phages.*

Barr body A highly condensed and transcriptionally inactive X chromosome found in the nuclei of somatic cells of female mammals. *See also* **lyonization.**

base Purine or pyrimidine component of a **nucleotide.**

base analog A chemical whose molecular structure is very similar to that of one of the bases normally found in DNA. Some chemical **mutagens,** such as 5-bromouracil (5BU), are base analogs.

base excision repair An enzyme-catalyzed process for repairing damaged DNA by removal of the altered base, followed by excision of the baseless nucleotide. The correct nucleotide then is inserted in the gap.

base-modifying agent A chemical **mutagen** that modifies the chemical structure of one or more bases normally found in DNA. Nitrous oxide, hydroxylamine, and methylmethane sulfonate are common base-modifying agents.

base pair substitution mutation A change in the genetic material such that one base pair is replaced by another base pair; for instance, an AT is replaced by a GC pair.

basic research Research done to further knowledge for knowledge's sake.

bidirectional replication Synthesis of DNA in both directions away from an **origin of replication.**

bioinformatics Application of mathematics and computer science to store, retrieve, and analyze biological data, particularly nucleic acid and protein sequence data.

biparental inheritance An uncommon phenomenon in which chloroplast genes from both parents are transmitted to progeny.

bivalent A pair of homologous, synapsed chromosomes, consisting of four **chromatids,** during the first meiotic division. *See also* **synapsis.**

bootstrap test A method for determining confidence levels attached to the branching patterns of a **phylogenetic tree** chosen by the parsimony approach.

bottleneck effect A form of **genetic drift** that occurs when a population is drastically reduced in size and some genes are lost from the gene pool as a result of chance.

branch-point sequence Specific sequence within introns of precursor mRNAs (pre-mRNAs) containing an adenylate (A) nucleotide to which the free 5′ end of an intron binds during mRNA splicing.

broad-sense heritability The proportion of the **phenotypic variance** within a population that results from genetic differences among individuals.

cAMP (cyclic AMP) Adenosine 3′,5′-monophosphate; an intracellular regulatory molecule involved in controlling gene expression and some other processes in both prokaryotes and eukaryotes.

cancer Disease characterized by the uncontrolled and abnormal division of cells and by the spread of malignant tumor cells (metastasis) to disparate sites in the organism.

5′ capping The addition of a methylated guanine nucleotide (a "cap") to the 5′ end of a **precursor mRNA (pre-mRNA)** molecule; the cap is retained on the mature mRNA molecule.

carcinogen Any physical or chemical agent that increases the frequency with which cells become cancerous.

carrier An individual who is heterozygous for a recessive mutation. A carrier usually does not exhibit the mutant phenotype.

catabolite activator protein (CAP) A regulatory protein that binds with **cAMP** at low glucose concentrations, forming a complex that stimulates transcription of the *lac* operon and some other bacterial operons.

catabolite repression The inactivation of some **inducible operons** in the presence of glucose even though the operon's inducer is present. Also called *glucose effect.*

cDNA *See* **complementary DNA.**

cDNA library Collection of cloned **complementary DNAs (cDNAs)** produced from the entire mRNA population of a cell.

cell cycle The cyclical process of growth and cellular reproduction in unicellular and multicellular eukaryotes. The cycle includes nuclear division, or **mitosis,** and cell (cytoplasmic) division, or **cytokinesis.**

cell division A process whereby one cell divides to produce two cells. *See also* **cytokinesis.**

CEN sequence Nucleotide sequence of DNA in the centromere region of chromosomes. *CEN* sequences differ among species and between chromosomes in the same species.

centimorgan (cM) The unit of distance on a **genetic map.** Equivalent to *map unit.*

centromere A specialized region of a chromosome, seen as a constriction under the microscope, that is important in the activities of chromosomes during cell division. *See also* **kinetochore.**

chain-terminating codon. *See* **stop codon.**

character *See* **hereditary trait.**

charged tRNA *See* **aminoacyl-tRNA.**

charging Addition of an amino acid to a tRNA that contains an **anticodon** for that animo acid. Also called *aminoacylation.*

checkpoints, cell-cycle Stages in the cell cycle at which progression of a cell through the cycle is blocked if there is damage to the genome or the mitotic machinery.

chiasma (*plural*, **chiasmata**) A cross-shaped structure formed during **crossing-over** and visible during the diplonema stage of meiosis.

chiasma interference *See* **interference.**

chi-square (χ^2) **test** A statistical procedure that determines what constitutes a significant difference between observed results and results expected on the basis of a particular hypothesis; a goodness-of-fit test.

chloroplasts Organelles found in green plants in which photosynthesis occurs.

chorionic villus sampling A procedure in which a sample of chorionic villus tissue of a developing fetus is examined for chromosomal abnormalities.

chromatid One of the two visibly distinct replicated copies of each chromosome that becomes visible between early prophase and metaphase of mitosis and is joined to its sister chromatid at their **centromeres.**

chromatin The DNA-protein complex that constitutes eukaryotic chromosomes and can exist in various degrees of folding or compaction.

chromatin remodeling Alteration of the structure of chromatin in the vicinity of a **core promoter** in a way that stimulates or represses transcription initiation. Remodeling is carried out by enzymes catalyzing histone acetylation or deacetylation and by nucleosome remodeling complexes.

chromosomal aberration *See* **chromosomal mutation.**

chromosomal mutation The variation from the wild-type condition in chromosome number or structure.

chromosome In eukaryotic cells, a linear structure composed of a single DNA molecule complexed with protein. Each eukaryotic species has a characteristic number of chromosomes in the nucleus of its cells. Most prokaryotic cells contain a single, usually circular chromosome.

chromosome library Collection of cloned DNA fragments produced from a particular chromosome (e.g., the human X chromosome).

chromosome theory of inheritance The theory that genes are located on chromosomes and that the transmission of chromosomes from one generation to the next accounts for the inheritance of hereditary traits.

chromosome walking A method for analyzing long segments of DNA by sequentially identifying adjacent, overlapping clones in a **genomic library.**

cis-dominant Referring to a gene or DNA sequence that can control genes on the same DNA molecule but not on other DNA molecules.

cis-trans test *See* **complementation test.**

cline A systematic change in **allelic frequencies** within a continuous population distributed over a geographic region.

clonal selection A process whereby cells that express cell-surface antibodies specific for a particular antigen are stimulated to proliferate and secrete that antibody by exposure to that antigen.

clone contig map A set of ordered, overlapping clones comprising all of the DNA of an entire chromosome, or part of a chromosome, without any gaps.

cloning (a) The production of many identical copies of a DNA molecule by replication in a suitable host; also called *DNA cloning, gene cloning,* and *molecular cloning.* (b) The generation of cells (or individuals) genetically identical to themselves and to their parent.

cloning vector A double-stranded DNA molecule that is able to replicate autonomously in a host cell and into which a DNA fragment (or fragments) can be inserted to form a recombinant DNA molecule for cloning.

coactivator In eukaryotes, a large multiprotein complex that interacts with activators bound at enhancers, general transcription factors bound near the promoter, and RNA polymerase II. These interactions help determine the efficiency of transcription of regulated genes.

coding sequence The part of an mRNA molecule that specifies the amino acid sequence of a polypeptide during translation.

codominance The condition in which an individual heterozygous for a gene exhibits the phenotypes of both homozygotes.

codon A group of three adjacent nucleotides in an mRNA molecule that specifies either one amino acid in a polypeptide chain or the termination of polypeptide synthesis.

codon usage bias A disproportionate use of one or a few synonymous codons within a codon family for a particular gene or across a genome.

coefficient of coincidence A measure of the extent of chiasma interference throughout a genetic map; ratio of the observed to the expected frequency of double crossovers. *See also* **interference.**

combinatorial gene regulation In eukaryotes, control of transcription by the combined action of several activators and repressors, which bind to particular gene regulatory sequences.

comparative genomics Comparison of the nucleotide sequences of entire genomes of different species, with the goal of understanding the functions and evolution of genes. Such comparisons can identify which genome regions are evolutionarily conserved and likely to represent functional genes.

complementary base pairs The specific A-T and G-C base pairs in double-stranded DNA. The bases are held together by hydrogen bonds between the purine and pyrimidine base in each pair.

complementary DNA (cDNA) DNA copies made from RNA templates in a reaction catalyzed by the enzyme reverse transcriptase.

complementation test A test used to determine whether two independently isolated mutations that confer the same phenotype are located within the same gene or in two different genes. Also called *cis-trans test.*

complete dominance The condition in which an allele is phenotypically expressed when one or both copies are present, so that the phenotype of the heterozygote is essentially indistinguishable from that of the homozygote.

complete medium For a microorganism, a medium that supplies all the ingredients required for growth and reproduction, including those normally produced by the wild-type organism.

complete recessiveness The condition in which an allele is phenotypically expressed only when two copies are present.

conditional mutation A mutation that results in a wild-type phenotype under one set of conditions but a mutant phenotype under other conditions. Temperature-sensitive mutations are a common type of conditional mutation.

conjugation In bacteria, process of unidirectional transfer of genetic material through direct cellular contact between a donor ("male") cell and a recipient ("female") cell.

consensus sequence The series of nucleotides found most frequently at each position in a particular DNA sequence among different species.

conservative model A model for DNA replication in which the two parental strands of DNA remain together and serve as a template for the synthesis of a new daughter double helix. The results of the Meselson-Stahl experiment did not support this model.

constitutive gene A gene whose expression is unregulated. The products of constitutive genes are essential to the normal functioning of the cell and are always produced in growing cells regardless of the environmental conditions.

constitutive heterochromatin Condensed chromatin that is always transcriptionally inactive and is found at homologous sites on chromosome pairs.

continuous trait *See* **quantitative trait.**

contributing allele An allele that affects the phenotype of a **quantitative trait.**

coordinate induction The simultaneous transcription and translation of two or more genes brought about by the action of an inducer.

core enzyme The portion of *E. coli* RNA polymerase that is the active enzyme; it is bound to the sigma factor, which directs the enzyme to the **promoter** region of genes.

core promoter In eukaryotic genomes, the gene regulatory elements closest to the transcription start site that are required for RNA synthesis to begin at the correct nucleotide.

correlation coefficient A statistical measure of the strength of the association between two variables. *See also* **regression.**

cotransduction The simultaneous transduction of two or more bacterial genes, a good indication that the bacterial genes are closely linked.

coupling In individuals heterozygous at two genetic loci, the arrangement in which the wild-type alleles of both genes are on one homologous chromosome and the recessive mutant alleles are on the other; also called *cis configuration. See also* **repulsion.**

covariance A statistical measure of the tendency for two variables to vary together; used to calculate the **correlation coefficient** between the two variables.

CpG island DNA region containing many copies of the dinucleotide CpG. Many genes in eukaryotic DNA have CpG islands in or near the promoter. Methylation of the cytosines (C) in these islands represses transcription.

crisscross inheritance Transmission of a gene from a male parent to a female child to a male grandchild.

cross *See* **cross-fertilization.**

cross-fertilization The fusion of male gametes from one individual and female gametes from another.

crossing-over The process of reciprocal chromosomal interchange that occurs frequently during meiosis and gives rise to **recombinant chromosomes.** *See also* **mitotic crossing-over.**

C value The amount of DNA found in the haploid set of chromosomes.

cyclin Any of a group of proteins whose concentrations increase and decrease in a regular pattern through the cell cycle. The cyclins act in conjunction with cyclin-dependent kinases to regulate cell cycle progression.

cyclin-dependent kinase Any of a group of protein kinases, activated by binding of specific cyclins, that regulate cell cycle progression.

cytohet The genetic condition of plant cells that contain chloroplasts derived from both parents. See also **biparental inheritance.**

cytokinesis Division of the cytoplasm following mitosis or meiosis I and II during which the two new nuclei compartmentalize into separate daughter cells.

cytological marker Any cytological feature that distinguishes one pair of homologous chromosomes from other pairs.

cytoplasmic inheritance *See* **non-Mendelian inheritance.**

cytosine (C) A **pyrimidine** found in RNA and DNA. In double-stranded DNA, cytosine pairs with guanine, a **purine,** by hydrogen bonding.

dark repair *See* **excision repair.**

Darwinian fitness (W) The relative reproductive ability of individuals with a particular genotype.

daughter chromosomes Detached sister chromatids after they separate at the beginning of mitotic anaphase or meiotic anaphase II.

deamination Removal of an amino group from a nucleotide in DNA.

degeneracy In the **genetic code,** the existence of more than one codon corresponding to each amino acid.

deletion A chromosomal mutation resulting in the loss of a segment of a chromosome and the gene sequences it contains.

deoxyribonuclease (DNase) An enzyme that catalyzes the degradation of DNA to nucleotides.

deoxyribonucleic acid (DNA) A polymeric molecule consisting of deoxyribonucleotide building blocks that in a double-stranded, double-helical form is the genetic material of most organisms.

deoxyribonucleotide Any of the nucleotides that make up DNA, consisting of a sugar (deoxyribose), a base, and a phosphate group.

deoxyribose The pentose (five-carbon) sugar found in DNA.

depurination Loss of a purine base (adenine or guanine) from a nucleotide in DNA.

determination Process early in development that establishes the fate of a cell, that is, the differentiated cell type it will become.

development Overall process of growth, differentiation, and morphogenesis by which a zygote gives rise to an adult organism. It involves a programmed sequence of phenotypic events that are typically irreversible.

diakinesis The final stage in prophase I of meiosis during which the replicated chromosomes (bivalents) are most condensed, the nuclear envelope breaks down, and the spindle begins to form.

dicentric bridge *See* **dicentric chromosome.**

dicentric chromosome A homologous chromosome pair in meiosis I in which one chromatid has two centromeres as the result of crossing-over within a paracentric inversion. As the two centromeres begin to migrate to opposite poles, a dicentric bridge stretching across the cell forms and eventually breaks.

dideoxynucleotide (ddNTP) A modified nucleotide that has a 3'-H on the deoxyribose sugar rather than a 3'-OH. A ddNTP can be incorporated into a growing DNA chain, but no further DNA synthesis can occur because no phosphodiester bond can be formed with an incoming nucleotide. *See also* **dideoxy sequencing.**

dideoxy sequencing A method for rapidly sequencing DNA molecules in which the DNA to be sequenced is used as the template for *in vitro* DNA synthesis in the presence of **dideoxynucleotides (ddNTPs).** When a ddNTP molecule is incorporated into a growing DNA chain, no further DNA synthesis occurs, generating a truncated chain in which the 5' ddNTP corresponds to the normal nucleotide that occurs at that point in the sequence.

differentiation Series of cell-specific changes in gene expression by which determined cells give rise to cell types with characteristic structures and functions.

dihybrid cross A cross between two individuals of the same genotype that are heterozygous for two pairs of alleles at two different loci (e.g., *Ss Yy* × *Ss Yy*).

dioecious Referring to plant species in which individual plants possess either male *or* female sex organs. *See also* **monoecious.**

diploid (2N) A cell or an individual with two copies of each chromosome.

diplonema The stage in prophase I of meiosis during which the **synaptonemal complex** disassembles and homologous chromosomes begin to move apart.

discontinuous trait A heritable characteristic that exhibits a small number of distinct phenotypes, which commonly are determined by variant alleles at a single locus. See also **quantitative trait.**

dispersed repeated DNA Repetitive DNA sequences that are distributed at irregular intervals in the genome.

dispersive model A model for DNA replication in which the parental double helix is cleaved into double-stranded DNA segments that act as templates for the synthesis of new double-stranded DNA segments, which are reassembled into complete DNA double helices, with parental and progeny DNA segments interspersed. The results of the Meselson-Stahl experiment did not support this model.

DNA *See* **deoxyribonucleic acid.**

DNA chip *See* **DNA microarray.**

DNA fingerprinting *See* **DNA typing.**

DNA helicase An enzyme that catalyzes unwinding of the DNA double helix at a **replication fork** during DNA replication.

DNA ligase An enzyme that catalyzes the formation of a phosphodiester bond between the 5′ end of one DNA chain and 3′ end of another DNA chain during DNA replication and DNA repair.

DNA markers Sequence variations among individuals in a specific region of DNA that are detected by molecular analysis of the DNA and can be used in genetic analysis. *See also* **gene markers.**

DNA microarray An ordered grid of DNA molecules of known sequence—*probes*—fixed at known positions on a solid substrate, either a silicon chip, glass, or, less commonly, a nylon membrane. Labeled free DNA molecules—*targets*—are added to the fixed probes to analyze identities or quantities of target molecules. DNA microarrays allow for the simultaneous analysis of thousands of DNA target molecules.

DNA molecular testing A type of genetic testing that focuses on the molecular nature of mutations associated with a particular disease.

DNA polymerase Any enzyme that catalyzes the polymerization of deoxyribonucleotides into a DNA chain. All DNA polymerases synthesize DNA in the 5′ to 3′ direction.

DNA polymerase I (DNA Pol I) One of several *E. coli* enzymes that catalyze DNA synthesis; originally called the Kornberg enzyme.

DNA polymorphism Variation in the nucleotide sequence or number of tandem repeat units at a particular locus in the genome. Most commonly, this term is used for DNA markers, variations that are located outside of genes and that are detected by molecular analysis.

DNA primase An enzyme that catalyzes formation of a short RNA primer in DNA replication.

DNA profiling *See* **DNA typing.**

DNA typing Molecular analysis of **DNA polymorphisms** to identify individuals based on the unique characteristics of their DNA. Also called *DNA fingerprinting.*

docking protein An integral protein in the membrane of the endoplasmic reticulum (ER) to which binds the complex of a growing polypeptide, **signal recognition particle (SRP),** and ribosome. This interaction facilitates binding of the polypeptide and associated ribosome to the ER.

domain shuffling Proposed mechanism for evolution of genes with new functions by the duplication and rearrangement of exons encoding protein domains in different combinations. Also called *exon shuffling.*

dominant Describing an allele or phenotype that is expressed in either the homozygous or the heterozygous state.

dominant lethal allele An allele that results in the death of an organism that is homozygous or heterozygous for the allele.

dosage compensation Any mechanism in organisms with **genotypic sex determination** for equalizing expression of genes on the sex chromosomes in males and females. *See also* **Barr body.**

Down syndrome *See* **trisomy-21.**

duplication A chromosomal mutation that results in the doubling of a segment of a chromosome and the gene sequences it contains.

effective population size The effective number of adults contributing gametes to the next generation.

effector A small molecule involved in controlling expression of a regulated gene or the activity of a protein.

endosymbiont theory The theory that mitochondria and chloroplasts originated as free-living prokaryotes that invaded primitive eukaryotic cells and established a mutually beneficial (symbiotic) relationship.

enhancer A set of gene regulatory elements in eukaryotic genomes that can act over distances up to thousands of base pairs upstream or downstream from a gene. Most enhancers bind activators and act to stimulate transcription. *See also* **silencer element.**

environmental variance (V_E) Component of the **phenotypic variance** for a trait that is due to any nongenetic source of variation among individuals in a population. V_E includes variation arising from general environmental effects, which permanently influence phenotype; special environmental effects, which temporarily influence phenotype; and family environmental effects, which are shared by family members.

epigenetic Referring to a heritable change in gene expression that does not result from a change in the nucleotide sequence of the genome.

episome In bacteria, a **plasmid** that is capable of integrating into the host cell's chromosome.

epistasis Interaction between two or more genes that controls a single phenotype. For instance, the expression of a gene at one locus can mask or suppress the expression of a second gene at another locus.

essential gene A gene that when mutated can result in the death of the organism.

EST *See* **expressed sequence tag.**

eubacteria Prokaryotes that constitute one of the three main evolutionary domains of organisms. Also known as *true bacteria* or simply *bacteria.*

euchromatin Chromatin that is condensed during mitosis but becomes uncoiled during interphase, when it can be transcribed. *See also* **heterochromatin.**

eukaryote Any organism whose cells have a membrane-bound nucleus in which the genetic material is located and membrane-bound organelles (e.g., mitochondria). Eukaryotes can be unicellular or multicellular and constitute one of the three major evolutionary domains of organisms. *See also* **prokaryote.**

euploid Referring to an organism or cell that has one complete set of chromosomes or an exact multiple of complete sets.

evolution Genetic change that takes place over time within a group of organisms.

evolutionary domains The three major lineages of organisms— **eubacteria, archaebacteria,** and **eukaryotes**—thought to have evolved from a common, single-celled ancestor.

excision repair An enzyme-catalyzed process for removal of thymine dimers from DNA and synthesis of a new DNA segment complementary to the undamaged strand.

exon A segment of a protein-coding gene and its precursor (pre-mRNA) that specifies an amino acid sequence and is retained in the functional mRNA. *See also* **intron.**

exon shuffling *See* **domain shuffling.**

expressed sequence tag (EST) A marker produced by PCR using oligonucleotide primers designed based on the sequence of a cDNA.

expression vector A cloning vector carrying a **promoter** and other sequences required for expression of a cloned gene in a host cell.

expressivity The degree to which a particular gene is expressed in the phenotype. A gene with variable expressivity can cause a range of phenotypes.

facultative heterochromatin Chromatin that may become condensed and therefore transcriptionally inactive in certain cell types, at different developmental stages, or in one member of a homologous chromosome pair.

familial trait A characteristic shared by members of a family as the result of shared genes and/or environmental factors.

fate map A diagram of an early embryo showing the cell types and tissues that different embryonic cells subsequently develop into.

F-duction Transfer of host genes carried on an F' factor in **conjugation** between an F' and an F^- cell. If the genes are different between the two cell types, the recipient becomes partially diploid for the genes on the F'.

F factor In *E. coli,* an **episome** that confers the ability to act as a donor cell in **conjugation.** Excision of an F factor from the bacterial chromosome may generate an F' factor, which carries a few host cell genes. *See also* **F-duction.**

F_1 generation The offspring that result from the first experimental crossing of two parental strains of animals or plants; the first filial generation.

F_2 generation The offspring that result from crossing F_1 individuals; the second filial generation.

fine-structure mapping Procedures for generating a high-resolution map of allelic sites within a gene.

first filial generation *See* F_1 **generation.**

first law *See* **principle of segregation.**

FISH (fluorescent in situ hybridization) A technique for detecting chromosomes carrying a particular DNA sequence by hybridizing them with a cloned DNA probe that is tagged with a fluorescent chemical. The chromosomes are treated to separate the double-stranded DNA into single strands, which base-pair with a probe whose sequence is complementary to a region of the chromosomal DNA.

fitness *See* **Darwinian fitness.**

formylmethionine (fMet) A modified form of the amino acid methionine that has a formyl group attached to the amino group. It is the first amino acid incorporated into a polypeptide chain in prokaryotes and in eukaryotic organelles.

forward mutation A point mutation in a wild-type allele that changes it to a mutant allele.

founder effect A form of **genetic drift** that occurs when a population is formed by migration of a small number of individuals from a large population.

F-pili (*singular,* **F-pilus**) Hairlike cell surface components produced by cells containing the **F factor,** which allow the physical union of F^+ and F^- cells or Hfr and F^- cells to take place. Also called *sex pili.*

frameshift mutation A mutational addition or deletion of a base pair in a gene that disrupts the normal **reading frame** of the corresponding mRNA.

frequency distribution In genetics, a graphical representation of the numbers of individuals within a population who fall within the same range of phenotypic values for a continuous **quantitative trait.** Typically, the phenotypic classes are plotted on the horizontal axis and the number of individuals in each class are plotted on the vertical axis.

functional genomics The comprehensive analysis of the functions of genes and of nongene sequences in entire genomes, including patterns of gene expression and its control.

gain-of-function mutation A mutation that confers a new property on a protein, causing a new phenotype.

gamete Mature reproductive cell that is specialized for sexual fusion. Each gamete is haploid and fuses with a cell of similar origin but of opposite sex to produce a diploid zygote.

gametogenesis The formation of male and female gametes.

gametophyte The haploid sexual generation in the life cycle of plants that produces the gametes by mitotic division of spores.

gene The physical and functional unit that helps determine the traits passed on from parents to offspring; also called *Mendelian factor.* In molecular terms, a gene is a nucleotide sequence in DNA that specifies a polypeptide or RNA. Alterations in a gene's sequence can give rise to species and individual variation.

GeneChip® array *See* **DNA microarray.**

gene conversion A nonreciprocal recombination process in which one allele in a heterozygote is changed to the other allele, thus converting a heterozygous genotype to a homozygous genotype.

gene expression The overall process by which a gene produces its product and the product carries out its function.

gene flow The movement of genes that takes place when organisms migrate and then reproduce, contributing their genes to the gene pool of the recipient population.

gene locus *See* **locus.**

gene markers Alleles that produce detectable phenotypic differences useful in genetic analysis. *See also* **DNA markers.**

gene mutation A heritable alteration in the sequence of a gene, usually from one allelic form to another, or in the sequences regulating the gene.

gene pool All of the alleles in a breeding population existing at a given time.

generalized transduction A type of transduction in which any gene may be transferred from one bacterium to another.

general transcription factor One of several proteins required for the initiation of transcription by a eukaryotic RNA polymerase.

gene regulatory elements DNA sequences that are involved in the regulation of gene expression. The regulatory elements associated with a particular gene are located within about 200 base pairs upstream of the transcription start site and also, in many eukaryotes, at much greater distances from the gene.

gene segregation *See* **principle of segregation.**

gene silencing Inactivation of a gene due to its location in the genome, DNA methylation, or **RNA interference (RNAi).** This type of gene control often represses transcription of multiple genes in a region of DNA.

genetic code The set of three-nucleotide sequences **(codons)** within mRNA that carries the information for specifying the amino acid sequence of a polypeptide.

genetic correlation Phenotypic correlation due to genetic causes such as **pleiotropy** or genetic **linkage.**

genetic counseling Evaluation of the probabilities that prospective parents will have a child who expresses a particular genetic trait (deleterious or not) and discussion with the couple of their options for avoiding or minimizing the possible risk.

genetic drift Random change in allelic frequencies within a population over time; observed most often in small populations due to sampling error.

genetic engineering Alteration of the genetic constitution of cells or individuals by directed and selective modification, insertion, or deletion of an individual gene or genes.

genetic map A representation of the relative distances separating genes on a chromosome based on the frequencies of recombination between nonallelic gene loci; also called *linkage map.* *See also* **physical map.**

genetic marker Any gene or DNA region whose sequence varies among individuals and is useful in genetic analysis, for example, in the detection of genetic recombination events.

genetic recombination A process by which parents with different alleles give rise to progeny with genotypes that differ from either parent. For example, parents with *A B* and *a b* genotypes can produce recombinant progeny with *A b* and *a B* genotypes.

genetics The science that deals with the structure and function of genes and their transmission from one generation to the next (heredity).

genetic structure of populations The patterns of genetic variation found among individuals within groups.

genetic testing Analysis to determine whether an individual who has symptoms of a particular genetic disease or is at high risk of developing it actually has a gene mutation associated with that disease.

genetic variance (V_G) Component of the **phenotypic variance** for a trait that is due to genetic differences among individuals in a population. V_G includes variation arising from the dominance effects of alleles, the additive effects of genes, and epistatic interactions among genes.

gene tree A **phylogenetic tree** based on the divergence observed within a single homologous gene. Gene trees are not always a good representation of the relationships among species because polymorphisms in any given gene may have arisen before speciation events. *See also* **species tree.**

genic sex determination System of sex determination, found primarily in eukaryotic microorganisms, in which sex is determined by different alleles at a small number of gene loci. *See also* **genotypic sex determination.**

genome The total amount of genetic material in a chromosome set; in eukaryotes, this is the amount of genetic material in the haploid set of chromosomes of the organism.

genomic imprinting Phenomenon in which the phenotypic expression of certain genes is determined by whether a particular allele is inherited from the female or male parent.

genomic library Collection of cloned DNA fragments in which every DNA sequence in the genome of an organism is represented at least once.

genomics The development and application of new mapping, sequencing, and computational procedures to analyze the entire genome of organisms.

genotype The complete genetic makeup (allelic composition) of an organism. The term is commonly used in reference to the specific alleles present at just one or a limited number of genetic loci.

genotypic frequency Percentage of individuals within a population that have a particular genotype. The sum of the genotypic frequencies at a given locus is 1.

genotypic sex determination Any system in which sex chromosomes play a decisive role in the inheritance and determination of sex. *See also* **genic sex determination.**

germ-line mutation In sexually reproducing organisms, a change in the genetic material in germ-line cells (those that give rise to gametes), which may be transmitted by the gametes to the next generation, giving rise to an individual with the mutant genotype in both its somatic and germ-line cells. *See also* **somatic mutation.**

glucose effect *See* **catabolite repression.**

Goldberg Hogness box *See* TATA **box.**

guanine (G) A **purine** found in RNA and DNA. In double-stranded DNA, guanine pairs with cytosine, a **pyrimidine,** by hydrogen bonding.

Haldane's Rule Common observation that among the offspring of crosses between two species, one sex is sterile, absent, or rare. Often, male hybrids are sterile and female hybrids are fertile.

haploid (N) A cell or an individual with one copy of each nuclear chromosome.

haploidization Formation of haploid nuclei from a diploid nucleus that undergoes mitosis without prior chromosome duplication.

haplosufficient Describing a gene that can support the normal wild-type phenotype when present in only one copy (heterozygous condition) in a diploid cell. A haplosufficient gene exhibits **complete dominance** in genetic crosses.

Hardy-Weinberg law An extension of Mendel's laws of inheritance that describes the expected relationship between gene frequencies in natural populations and the frequencies of individuals of various genotypes in the same populations.

hemizygous Possessing only one copy (allele) of a gene in a diploid cell. Usually applied to genes on the X chromosome in males with the XY genotype.

hereditary trait A characteristic that results from gene action and is transmitted from one generation to another. Also called *character.*

heritability The proportion of phenotypic variation in a population attributable to genetic factors.

hermaphroditic Referring to animal species in which each individual has both testes and ovaries and to plant species in which both stamens and pistils are on the same flower.

heterochromatin Chromatin that remains condensed throughout the cell cycle and is usually not transcribed. *See also* **euchromatin.**

heteroduplex DNA A region of double-stranded DNA with different sequence information on the two strands.

heterogametic sex In a species, the sex that has two types of sex chromosomes (e.g., X and Y) and therefore produces two kinds of gametes with respect to the sex chromosomes. In mammals, the male is the heterogametic sex.

heterogeneous nuclear RNA (hnRNA) A group of RNA molecules of various sizes that exist in a large population in the nucleus and include **precursor mRNAs (pre-mRNAs)**.

heterokaryon A cell containing two nuclei with different genotypes produced by fusing cells from different sources.

heterosis The superiority of heterozygous genotypes with respect to one or more characters compared with the corresponding homozygous genotypes in terms of growth, survival, phenotypic expression, and fertility. Also called *heterozygote superiority* or *overdominance*.

heterozygosity (H) A measure of genetic variation; with respect to a particular locus, the proportion of individuals within a population that are heterozygous at that locus.

heterozygote superiority *See* **heterosis**.

heterozygous Describing a diploid organism having different alleles of one or more genes and therefore producing gametes of different genotypes.

Hfr (high-frequency recombination) Designation for an *E. coli* cell that has an **F factor** integrated into the bacterial chromosome. When a *Hfr* cell conjugates with a recipient (F^-) cell, bacterial genes are transferred to the recipient with high frequency.

highly repetitive DNA A class of DNA sequences each of which is present in 10^5 to 10^7 copies in the haploid chromosome set.

histone One of a class of basic proteins that are complexed with DNA in **chromatin** and play a major role in determining the structure of eukaryotic nuclear chromosomes.

Holliday model One model of the molecular events in **genetic recombination** that involves an X-shaped intermediate visible in the electron microscope.

homeobox A 180-bp consensus sequence found in many genes that regulate development.

homeodomain The 60-amino acid part of proteins that corresponds to the homeobox sequence in genes. All homeodomain-containing proteins can bind to DNA and function in regulating transcription.

homeotic genes Group of genes in *Drosophila* that specify the body parts (appendages) that will develop in each segment, thus determining the identity of the segments.

homeotic mutation Any mutation that alters the identity of a particular body segment, transforming it into a copy of a different segment.

homogametic sex In a species, the sex that has one type of sex chromosome (e.g., X) and therefore produces only one kind of gamete with respect to the sex chromosomes. In mammals, the female is the homogametic sex.

homolog Each individual member of a pair of homologous chromosomes.

homologous Referring to genes that have arisen from a common ancestral gene over evolutionary time; also used in reference to proteins encoded by homologous genes.

homologous chromosomes Chromosomes that have the same arrangement of genetic loci, are identical in their visible structure, and pair during meiosis.

homozygous Describing a diploid organism having the same alleles at one or more genetic loci and therefore producing gametes of identical genotypes.

homozygous dominant A diploid organism that has the same dominant allele for a given gene locus on both members of a homologous pair of chromosomes.

homozygous recessive A diploid organism that has the same recessive allele for a given gene locus on both members of a homologous pair of chromosomes.

Human Genome Project (HGP) A project to determine the sequence of the complete 3 billion (3×10^9) nucleotide pairs of the human genome and to map all of the genes along each chromosome.

hypersensitive sites Regions of DNA around transcriptionally active genes that are highly sensitive to digestion by DNase I.

hypothetico-deductive method of investigation Research method involving making observations, forming hypotheses to explain the observations, making experimental predictions based on the hypotheses, and, finally, testing the predictions. The last step produces new observations, so a cycle is set up leading to a refinement of the hypotheses and perhaps eventually to the establishment of a law or an accepted principle.

imaginal disc In the *Drosophila* blastoderm, a group of undifferentiated cells that will develop into particular adult tissues and organs.

immunoglobulins Specialized proteins (antibodies) secreted by B cells that circulate in the blood and lymph and are responsible for humoral immune responses.

inborn error of metabolism A biochemical disorder caused by mutation in a gene encoding an enzyme in a particular metabolic pathway.

inbreeding Preferential mating between close relatives.

incomplete dominance The condition in which neither of two alleles is completely dominant to the other, so that the heterozygote has a phenotype between the phenotypes of individuals homozygous for either allele involved. Also called *partial dominance*.

indels Gaps in a sequence alignment where it is not possible to determine whether an insertion occurred in one sequence or a deletion occurred in another.

induced mutation Any mutation that results from treating a cell or organism with a chemical or physical **mutagen**.

inducer A chemical or environmental agent that stimulates transcription of specific genes.

inducible operon An **operon** whose transcription is turned on in the presence of a particular substance (inducer). The lactose (*lac*) operon is an example of an inducible operon. *See also* **repressible operon**.

induction (1) Stimulation of the synthesis of a gene product in response to the action of an inducer, that is, a chemical or environmental agent. (2) In development, the ability of one cell or group of cells to influence the developmental fate of other cells.

initiation factor Any of various proteins involved in the initiation of translation.

initiator protein A protein that binds to the **replicator**, stimulates local unwinding of the DNA, and helps recruit other proteins required for the initiation of replication.

insertion sequence (IS element) The simplest **transposable element** found in prokaryotes. An IS element contains a single gene, which encodes transposase, an enzyme that catalyzes movement of the element within the genome.

insulator A DNA regulatory element, located between a promoter and associated enhancer, that blocks the ability of activators bound at the enhancer to stimulate transcription from the promoter.

intercalating agent A chemical mutagen that can insert between adjacent nucleotides in a DNA strand.

interference Phenomenon in which the presence of one crossover interferes with the formation of another crossover nearby. Mathematically defined as 1 minus the **coefficient of coincidence**. Also called *chiasma interference*.

intergenic suppressor A mutation whose effect is to suppress the phenotypic consequences of another (primary) mutation in a different gene.

internal control region (ICR) Promoter sequence, recognized by RNA polymerase III, that is located within the gene sequence. This type of promoter is found in tRNA genes and 5S rRNA genes of eukaryotes.

interspersed repeated DNA *See* **dispersed repeated DNA.**

intragenic suppressor A mutation whose effect is to suppress the phenotypic consequences of another (primary) mutation within the same gene.

intron A segment of a protein-coding gene and its precursor mRNA (pre-mRNA) that does not specify an amino acid sequence. Introns in pre-mRNA are removed by **mRNA splicing.** *See also* **exon.**

inversion A chromosomal mutation in which a segment of a chromosome is excised and then reintegrated in an orientation 180° from the original orientation.

karyotype A complete set of all the metaphase chromatid pairs in a cell.

kinetochore Specialized multiprotein complex that assembles at the centromere of a chromatid and is the site of attachment of spindle microtubules during mitosis.

Klinefelter syndrome A human clinical syndrome that results from disomy for the X chromosome in a male, which results in a 47,XXY male. Many of the affected males are mentally deficient, have underdeveloped testes, and are taller than average.

knockout mouse A mouse in which a nonfunctional allele of a particular gene has replaced the normal alleles, thereby knocking out the gene's function in an otherwise normal individual.

lagging strand In DNA replication, the DNA strand that is synthesized discontinuously from multiple RNA primers in the direction opposite to movement of the replication fork. *See also* **leading strand** and **Okazaki fragments.**

leader sequence *See* **5′ untranslated region (5′ UTR)**

leading strand In DNA replication, the DNA strand that is synthesized continuously from a single RNA primer in the same direction as movement of the replication fork . *See also* **lagging strand.**

leptonema The first stage in prophase I of meiosis during which the chromosomes begin to coil and become visible.

lethal allele An allele whose expression results in the death of an organism.

light repair *See* **photoreactivation.**

LINEs (long interspersed elements) One class of **dispersed repeated DNA** consisting of repetitive sequences that are several thousand base pairs in length. LINEs can move in the genome by **retrotransposition.**

linkage The association of genes located on the same chromosome such that they tend to be inherited together.

linkage map *See* **genetic map.**

linked genes Genes that are located on the same chromosome and tend to be inherited together. A collection of such genes constitutes a *linkage group.*

linker *See* **restriction site linker.**

locus (*plural,* **loci**) The position of a gene on a genetic map; the specific place on a chromosome where a gene is located. More broadly, a locus is any chromosomal location that exhibits variation detectable by genetic or molecular analysis.

lod score method The lod (logarithm of odds) score method is a statistical analysis, usually performed by computer programs, based on data from pedigrees. It is used to test for linkage between two loci in humans.

long interspersed elements *See* **LINEs.**

looped domains Loops of supercoiled DNA that serve to compact the chromosomes.

loss-of-function mutation A mutation that leads to the absence or decreased biological activity of a particular protein.

Lyon hypothesis *See* **lyonization.**

lyonization A mechanism of dosage compensation, discovered by Mary Lyon, in which one of the X chromosomes in the cells of female mammals becomes highly condensed and genetically inactive.

lysogenic Referring to a bacterium that contains the genome of a temperate phage in the **prophage** state. On induction, the prophage leaves the bacterial chromosome, progeny phages are produced, and the bacterial cell lyses.

lysogenic pathway One of two pathways in the life cycle of temperate phages in which the phage genome is integrated into the host cell's chromosome and progeny phages are not formed.

lysogeny The phenomenon in which the genome of a temperate phage is inserted into a bacterial chromosome, where it replicates when the bacterial chromosome replicates. In this state, the phage genes are repressed and progeny phages are not formed.

lytic cycle Bacteriophage life cycle in which the phage takes over the bacterium and directs its growth and reproductive activities to express the phage genes and to produce progeny phages.

mapping function Mathematical formula used to correct the observed recombination frequencies for the incidence of multiple crossovers.

map unit (mu) A unit of measurement for the distance between two genes on a **genetic map.** A recombination (crossover) frequency of 1 percent between two genes equals 1 map unit. *See also* **centimorgan.**

maternal effect (a) The influence of the maternal environment (e.g., uterus size, quantity and quality of milk) on the phenotype of offspring; one of the family environmental effects that influence the variation of quantitative traits. (b) The phenotype established by expression of **maternal effect genes** in the oocyte before fertilization.

maternal effect gene A nuclear gene, expressed by the mother during oogenesis, whose product helps direct early development in the embryo.

maternal inheritance A type of uniparental inheritance in which the mother's phenotype is expressed exclusively.

mating types In lower eukaryotes, two forms that are morphologically indistinguishable but carry different alleles and will mate; equivalent to the sexes in higher organisms. *See also* **genic sex determination.**

maximum parsimony Property of the **phylogenetic tree** (or trees) that invokes the fewest number of mutations and therefore is most likely to represent the true evolutionary relationship between species or their genes.

mean The average of a set of numbers, calculated by adding all the values represented and dividing by the number of values.

meiosis Two successive nuclear divisions of a diploid nucleus, following one DNA replication, that result in the formation of haploid gametes or of spores having one-half the genetic material of the original cell.

meiosis I The first meiotic division, resulting in the reduction of the number of chromosomes from diploid to haploid.

meiosis II The second meiotic division, resulting in the separation of the chromatids and formation of four haploid cells.

meiotic tetrad The four haploid cells produced in a single meiosis. In some lower eukaryotes, these cells are contained within a common structure.

Mendelian factor *See* **gene.**

Mendelian population A group of interbreeding individuals who share a common **gene pool;** the basic unit of study in population genetics.

messenger RNA (mRNA) Class of RNA molecules that contain coded information specifying the amino acid sequences of proteins.

metacentric chromosome A chromosome with the centromere near the center such that the chromosome arms are of about equal lengths.

metaphase The stage in mitosis or meiosis during which chromosomes become aligned along the equatorial plane of the spindle.

metaphase I The stage in meiosis I when each homologous chromosome pair (bivalent) becomes aligned on the equatorial plate.

metaphase II The stage of meiosis II during which the chromosomes (each a sister chromatid pair) line up on the equatorial plate in each of the two daughter cells formed in meiosis I.

metaphase plate The plane in the cell where the chromosomes become aligned during metaphase.

metastasis The spread of malignant tumor cells throughout the body so that tumors develop at new sites.

methyl-directed mismatch repair An enzyme-catalyzed process for repairing mismatched base pairs in DNA *after* replication is completed in contrast to **proofreading,** a process for correcting mismatched base pairs *during* replication.

minimal medium For a microorganism, a medium that contains the simplest set of ingredients (e.g., a sugar, some salts, and trace elements) required for the growth and reproduction of wild-type cells.

missense mutation A **point mutation** in a gene that changes one codon in the corresponding mRNA so that it specifies a different amino acid than the one specified by the wild-type codon.

mitochondria Organelles found in the cytoplasm of all aerobic animal and plant cells in which most of the cell's ATP is produced.

mitosis The process of nuclear division in haploid or diploid cells producing daughter nuclei that contain identical chromosome complements and are genetically identical to one another and to the parent nucleus from which they arose.

mitotic crossing-over A genetic recombination that follows the rare pairing of homologous chromosomes during mitosis of a diploid cell. Also called *mitotic recombination.*

mobile genetic element *See* **transposable element.**

moderately repetitive DNA A class of DNA sequences each of which is present from a few to about 10^5 copies in the haploid chromosome set.

molecular clock hypothesis The hypothesis that for any given gene mutations accumulate at an essentially constant rate in all evolutionary lineages as long as the gene retains its original function.

molecular cloning *See* **cloning (a).**

molecular evolution Study of how genomes and macromolecules evolve at the molecular level and how genes and organisms are evolutionarily related.

molecular genetics Study of how genetic information is encoded within DNA and how biochemical processes of the cell translate the genetic information into the phenotype.

monoecious Referring to plant species in which individual plants possess *both* male and female sex organs and thus produce male and female gametes. Monoecious plants are capable of self-fertilization. *See also* **dioecious.**

monohybrid cross A cross between two individuals that are both heterozygous for the same pair of alleles (e.g., *Aa* × *Aa*). By extension, the term also refers to crosses involving the pure-breeding parents that differ with respect to the alleles of one locus (e.g., *AA* × *aa*).

monoploidy Condition in which a normally diploid cell or organism lacks one complete set of chromosomes.

monosomy A type of **aneuploidy** in which one chromosome of a homologous pair is missing from a normally diploid cell or organism. A monosomic cell is 2N − 1.

morphogen A substance that helps determine the fate of cells in early development in proportion to its concentration.

morphogenesis Overall developmental process that generates the size, shape, and organization of cells, tissues, and organs.

mRNA splicing Process whereby an intron (intervening sequence) between two exons (coding sequences) in a precursor mRNA (pre-mRNA) molecule is excised and the exons ligated (spliced) together.

multifactorial trait A characteristic determined by multiple genes and environmental factors.

multigene family A set of genes encoding products with related functions that have evolved from a common ancestral gene through gene duplication.

multiple alleles Many alternative forms of a single gene. Although a population may carry multiple alleles of a particular gene, a single diploid individual can have a maximum of only two alleles at a locus.

multiple cloning site *See* **polylinker.**

multiple-gene hypothesis for quantitative inheritance *See* **polygene hypothesis for quantitative inheritance.**

mutagen Any physical or chemical agent that significantly increases the frequency of mutational events above a spontaneous mutation rate.

mutagenesis The creation of mutations.

mutant allele Any form of a gene that differs from the wild-type allele. Mutant alleles may be **dominant** or **recessive** to wild-type alleles.

mutation Any detectable and heritable change in the genetic material not caused by genetic recombination; Mutations may occur within or between genes and are the ultimate source of all new genetic variation.

mutation frequency The number of occurrences of a particular kind of mutation in a population of cells or individuals.

mutation rate The probability of a particular kind of mutation as a function of time.

mutator gene A gene that, when mutant, increases the spontaneous mutation frequencies of other genes.

narrow-sense heritability The proportion of the **phenotypic variance** that results from the additive effects of different alleles on the phenotype.

natural selection Differential reproduction of individuals in a population resulting from differences in their **genotypes.**

negative assortative mating Preferential mating between phenotypically dissimilar individuals that occurs more frequently than expected for **random mating.**

neutral mutation A **point mutation** in a gene that changes a codon in the corresponding mRNA to that for a different amino acid but results in no change in the function of the encoded protein.

neutral mutation hypothesis The hypothesis that the extensive genetic variation observed in proteins is neutral with regard to natural selection.

nitrogenous base A nitrogen-containing **purine** or **pyrimidine** that, along with a pentose sugar and a phosphate, is one of the three parts of a **nucleotide.**

noncontributing allele An allele that has no effect on the phenotype of a **quantitative trait.**

nondisjunction A failure of homologous chromosomes or sister chromatids to separate at anaphase. See also **primary nondisjunction.**

nonhistone An acidic or neutral protein found in **chromatin.**

nonhomologous chromosomes Chromosomes that contain dissimilar genetic loci and that do not pair during meiosis.

non-Mendelian inheritance The inheritance of characters determined by genes located on mitochondrial or chloroplast chromosomes. Such extranuclear genes show inheritance patterns distinctly different from those of genes on chromosomes in the nucleus. Also called *cytoplasmic inheritance.*

nonparental ditype (NPD) One of three types of **meiotic tetrads** possible when two genes are segregating in a cross. The NPD tetrad contains four nuclei, all of which have recombinant (nonparental) genotypes, that is, two of each possible type.

nonsense codon See **stop codon.**

nonsense mutation A **point mutation** in a gene that changes an amino acid-coding codon in the corresponding mRNA to a stop codon.

nonsynonymous Referring to nucleotides in a gene that when mutated cause a change in the amino acid sequence of the encoded wild-type protein.

nontranscribed spacer (NTS) sequence Nontranscribed sequence found between adjacent ribosomal DNA (rDNA) repeat units. NTS sequences play a role in controlling transcription of rDNA.

normal distribution Common probability distribution that exhibits a bell-shaped curve when plotted graphically.

norm of reaction Range of phenotypes produced by a particular genotype in different environments.

northern blot analysis A technique for detecting specific RNA molecules in which the RNAs are separated by gel electrophoresis, transferred to a nitrocellulose filter, and then hybridized with labeled complementary probes; also called *northern blotting.* See also **Southern blot analysis.**

nuclease An enzyme that catalyzes the degradation of a nucleic acid by breaking phosphodiester bonds.

nucleic acid High-molecular-weight polynucleotide. The main nucleic acids in cells are DNA and RNA.

nucleoid Central region in a bacterial cell in which the chromosome is compacted.

nucleolus An organelle in the nucleus of eukaryotic cells that is the site of transcription of ribosomal RNA genes and assembly of the ribosomal subunits.

nucleoside A **purine** or **pyrimidine** covalently linked to a sugar.

nucleoside phosphate A nucleoside with an attached phosphate group. Also called *nucleotide.*

nucleosome The basic structural unit of eukaryotic **chromatin,** consisting of two molecules each of the four core histones (H2A, H2B, H3, and H4, the histone octamer), a single molecule of the linker histone H1, and about 180 bp of DNA.

nucleosome remodeling complex Large, multiprotein complex that uses the energy released by ATP hydrolysis to alter the position or structure of nucleosomes, thereby remodeling chromatin structure.

nucleotide The type of monomeric molecule found in RNA and DNA. Nucleotides consist of three distinct parts: a pentose (ribose in RNA, deoxyribose in DNA), a nitrogenous base (a purine or pyrimidine), and a phosphate group.

nucleotide excision repair (NER) See **excision repair.**

nucleus A discrete structure within eukaryotic cells that is bounded by a double membrane (the nuclear envelope) and contains most of the DNA of the cell.

null hypothesis A hypothesis that states there is no real difference between the observed data and the predicted data.

nullisomy A type of **aneuploidy** in which one pair of homologous chromosomes is missing from a normally diploid cell or organism. A nullisomic cell is $2N - 2$.

nutritional mutant See **auxotroph.**

nutritional mutation See **auxotrophic mutation.**

Okazaki fragments The short, single-stranded DNA fragments that are synthesized on the lagging-strand template during DNA replication and are subsequently covalently joined to make a continuous strand, the **lagging strand.**

oligomer Short DNA molecule; also called *oligonucleotide.*

oligonucleotide array See **DNA microarray.**

oncogene A gene whose protein product promotes cell proliferation. Oncogenes are altered forms of **proto-oncogenes.**

oncogenesis Formation of a tumor (cancer) in an organism.

one gene-one enzyme hypothesis The hypothesis that each gene controls the synthesis of one enzyme.

one gene-one polypeptide hypothesis The hypothesis that each gene controls the synthesis of a polypeptide chain.

oogenesis Development of female gametes (egg cells) in animals.

open reading frame In a segment of DNA, a potential protein-coding sequence identified by a start codon in frame with a stop codon.

operator A short DNA region, adjacent to the promoter of a bacterial operon, that binds repressor proteins responsible for controlling the rate of transcription of the operon.

operon In bacteria, a cluster of adjacent genes that share a common operator and promoter and are transcribed into a single mRNA. All the genes in an operon are regulated coordinately; that is, all are transcribed or none are transcribed.

optimal alignment In the comparison of nucleotide or amino acid sequences from two or more organisms, an approximation of the true alignment of sequences where gaps are inserted to maximize the similarity among the sequences being aligned. *See also* **indels.**

origin A specific site on a DNA molecule at which the double helix denatures into single strands and replication is initiated.

origin of replication A specific region in DNA where the double helix unwinds and synthesis of new DNA strands begins.

origin recognition complex (ORC) A multisubunit complex that functions as an **initiator protein** in eukaryotes.

outgroup A species that is least related to the others in a group of species because it has diverged from the group before all the others diverged from each other.

overdominance *See* **heterosis.**

ovum (*plural,* **ova**) A mature female **gamete** (egg cell); the larger of the two cells that arise from a secondary oocyte by meiosis II in the ovary of female animals.

pachynema The stage in prophase I of meiosis during which the homologous pairs of chromosomes undergo **crossing-over.**

paracentric inversion A chromosomal mutation in which a segment on one chromosome arm that does not include the centromere is inverted.

parasexual system Mechanism for formation of recombinant haploid cells that does not involve the regular alternation of meiosis and fertilization.

parental *See* **parental genotype.**

parental class *See* **parental genotype.**

parental ditype (PD) One of three types of **meiotic tetrads** possible when two genes are segregating in a cross. The PD tetrad contains four nuclei, all of which are parental genotypes, with two of one parent and two of the other parent.

parental genotype The genetic makeup (allelic composition) of individuals in the parental generation of genetic crosses. Progeny in succeeding generations may have combinations of linked alleles like one or the other of the parental genotypes or new (nonparental) combinations as the result of **crossing-over.**

partial dominance *See* **incomplete dominance.**

partial reversion A **point mutation** in a mutant allele that restores all or part of the function of the encoded protein but not the wild-type amino acid sequence.

particulate factors The term Mendel used for the entities that carry hereditary information and are transmitted from parents to progeny through the gametes. These factors are now called *genes.*

PCR. *See* **polymerase chain reaction.**

pedigree analysis Study of the inheritance of human traits by compilation of phenotypic records of a family over several generations.

penetrance The frequency with which a dominant or homozygous recessive gene is phenotypically expressed within a population.

pentose sugar A five-carbon sugar that, along with a nitrogenous base and a phosphate group, is one of the three parts of a **nucleotide.**

peptide bond A covalent bond in a polypeptide chain that joins the α-carboxyl group of one amino acid to the α-amino group of the adjacent amino acid.

peptidyl transferase Catalytic activity of an RNA component of the ribosome that forms the peptide bond between amino acids during translation.

pericentric inversion A chromosomal mutation in which a segment including the centromere and parts of both chromosome arms is inverted.

P generation The parental generation; the immediate parents of F_1 offspring.

phage Shortened form of **bacteriophage.**

phage lysate The progeny phages released after lysis of phage-infected bacteria.

phage vector A phage that carries pieces of bacterial DNA between bacterial strains in the process of transduction.

pharmacogenomics Study of how a person's unique genome affects the body's response to medicines.

phenocopy An abnormal phenotype that results from certain environmental conditions and mimics a phenotype caused by gene mutation.

phenotype The observable characteristics of an organism that are produced by the genotype and its interaction with the environment.

phenotypic correlation An association between two or more **quantitative traits** in the same individual.

phenotypic variance (V_P) A measure of all the variability for a **quantitative trait** in a population; mathematically is identical to the **variance.**

phosphate group An acidic chemical component that, along with a pentose sugar and a nitrogenous base, is one of the three parts of a **nucleotide.**

phosphodiester bond A covalent bond in RNA and DNA between a sugar of one nucleotide and a phosphate group of an adjacent nucleotide. Phosphodiester bonds form the repeating sugar-phosphate array of the backbone of DNA and RNA.

photoreactivation Repair of **thymine dimers** in DNA by exposure to visible light in the wavelength range 320–370 nm. Also called *light repair.*

phylogenetic relationship A reconstruction of the evolutionary history of groups of organisms (taxa) or genes.

phylogenetic tree A graphic representation of the evolutionary relationships among of a group of species or genes. It consists of *branches* (lines) connecting *nodes*, which represent ancestral or extant organisms. *See also* **maximum parsimony.**

physical map A representation of the physical distances, measured in base pairs, between identifiable regions or markers on genomic DNA. A physical map is generated by analysis of DNA sequences rather than by genetic recombination analysis, which is used in constructing a **genetic map.**

pistil The female reproductive organ in flowering plants. It usually consists of a pollen-receiving stigma, stalklike style, and ovary.

plaque A round, clear area in a lawn of bacteria on solid medium that results from the lysis of cells by repeated cycles of phage lytic growth.

plasmid An extrachromosomal circular, double-stranded DNA molecule that replicates autonomously from the host chromosome. Plasmids occur naturally in many bacteria and can be engineered for use as **cloning vectors.**

pleiotropic Referring to genes or mutations that result in multiple phenotypic effects.

pleiotropy The ability of a single gene to have multiple phenotypic effects.

point mutant An organism whose mutant phenotype results from an alteration of a single nucleotide pair.

point mutation A heritable alteration of the genetic material in which one base pair is changed to another.

polarity Property of certain **nonsense mutations** in a bacterial operon that abolish expression of the structural gene in which the mutation is located *and* expression of any structural genes on the operator distal side of the mutated gene.

poly(A) polymerase The enzyme that catalyzes formation of the **poly(A) tail** at the 3′ end of eukaryotic mRNA molecules.

poly(A) site In eukaryotic precursor mRNAs (pre-mRNAs), the sequence that directs cleavage at the 3′ end and subsequent addition of adenine nucleotides to form the poly-A tail, during RNA processing.

poly(A) tail A sequence of 50 to 250 adenine nucleotides at the 3′ end of most eukaryotic mRNAs. The tail is added during processing of pre-mRNA.

polycistronic mRNA An mRNA molecule, produced from a prokaryotic **operon,** that is translated into all the polypeptide encoded by the structural genes in the operon; also called *polygenic mRNA.* In eukaryotes, polygenic pre-mRNAs encoding more than one polypeptide are processed into individual mRNAs, each of which is translated into a single polypeptide.

polygene hypothesis for quantitative inheritance The hypothesis that **quantitative traits** are controlled by many genes.

polygenes Two or more genes whose additive effects determine a particular **quantitative trait.**

polygenic mRNA *See* **polycistronic mRNA.**

polylinker A region within a cloning vector that contains many different restriction sites. Also called *multiple cloning site.*

polymerase chain reaction (PCR) A method for producing many copies of a specific DNA sequence from a DNA mixture without having to clone the sequence in a host organism.

polynucleotide A linear polymeric molecule composed of **nucleotides** joined by phosphodiester bonds. DNA and RNA are polynucleotides.

polypeptide A linear polymeric molecule consisting of **amino acids** joined by peptide bonds. *See also* **protein.**

polyploidy Condition in which a cell or organism has more than two sets of chromosomes.

polyribosome (polysome) The complex between an mRNA molecule and all the ribosomes that are translating it simultaneously.

polytene chromosome A special type of chromosome representing a bundle of numerous chromatids that have arisen by repeated cycles of replication of single chromatids without nuclear division. This type of chromosome is characteristic of various tissues of Diptera.

population A specific group of individuals of the same species.

population genetics Study of the consequences of Mendelian inheritance on the population level, including the mathematical description of a population's genetic composition and how it changes over time.

population viability analysis Analysis of the survival probabilities of different genotypes in the population.

positional cloning The isolation of a gene associated with a genetic disease on the basis of its approximate chromosomal position.

position effect A change in the phenotypic effect of one or more genes as a result of a change in their position in the genome.

positive assortative mating Preferential mating between phenotypically similar individuals that occurs more frequently than expected for **random mating.**

postzygotic isolation Reduction in mating between closely related species by various mechanisms that act after fertilization, resulting in nonviable or sterile hybrids or hybrids of lowered fitness. *See also* **prezygotic isolation.**

precursor mRNA (pre-mRNA) The initial (primary) transcript of a protein-coding gene that is modified or processed to produce the mature, functional mRNA molecule.

precursor rRNA (pre-rRNA) The initial (primary) transcript produced from ribosomal DNA that is processed into three different rRNA molecules in prokaryotes and eukaryotes.

precursor tRNA (pre-tRNA) The initial (primary) transcript of a tRNA gene that is extensively modified and processed to produce the mature, functional tRNA molecule.

prezygotic isolation Reduction in mating between closely related species by various mechanisms that prevent courtship, mating, or fertilization. *See also* **postzygotic isolation.**

Pribnow box A part of the **promoter** sequence in prokaryotic genomes that is located at about 10 base pairs upstream from the transcription start site. Also called the *−10 box.*

primary nondisjunction A rare event in cells with a normal chromosome complement in which sister chromatids (in mitosis or meiosis II) or homologous chromosomes (in meiosis I) fail to separate and move to opposite poles.

primary oocytes Diploid cells that arise by mitotic division of primordial germ cells (oogonia) and undergo meiosis in the ovaries of female animals.

primary transcripts *See* **precursor mRNA, rRNA, and tRNA.**

primase *See* **DNA primase.**

primer *See* **RNA primer.**

primosome A complex of *E. coli* primase, helicase, and perhaps other proteins that functions in initiating DNA synthesis.

principle of independent assortment Mendel's second law stating that the factors (genes) for different traits assort independently of one another. In other words, genes on different chromosomes behave independently in the production of gametes.

principle of segregation Mendel's first law stating that two members of a gene pair (alleles) segregate (separate) from each other during the formation of gametes. As a result, one-half the gametes carry one allele and the other half carry the other allele.

probability The ratio of the number of times a particular event occurs to the number of trials during which the event could have happened.

proband In human genetics, an affected person with whom the study of a trait in a family begins. *See also* **proposita; propositus.**

probe array Generic term for a **DNA microarray, DNA chip,** or **GeneChip® array.**

product rule The rule that the probability of two independent events occurring simultaneously is the product of each of their probabilities.

programmed cell death *See* **apoptosis.**

prokaryote Any organism whose genetic material is not located within a membrane-bound nucleus. The prokaryotes include **eubacteria** and **archaebacteria.** *See also* **eukaryote.**

promoter A DNA region containing specific **gene regulatory elements** to which RNA polymerase binds for the initiation of transcription. *See also* **core promoter.**

promoter proximal elements Gene regulatory elements in eukaryotic genomes that are located 50–200 base pairs from the transcription start site (upstream of the TATA **box**) and help determine the efficiency of transcription.

proofreading In DNA synthesis, the process of recognizing a base pair error during the polymerization events and correcting it. Proofreading is carried out by some DNA polymerases in prokaryotic and eukaryotic cells.

prophage The genome of a temperate bacteriophage that has been integrated into the chromosome of a host bacterium in the **lysogenic pathway.** A prophase is replicated during replication of the host cell's chromosome.

prophase The first stage in mitosis or meiosis during which the replicated chromosomes condense and become visible under the microscope.

prophase I The first stage of meiosis, divided into several sub-stages, during which the replicated chromosomes condense, homologues undergo **synapsis,** and **crossing-over** occurs.

prophase II The first stage of meiosis II during which the chromosomes condense.

proportion of polymorphic loci (P) A ratio calculated by determining the number of loci with more than one allele present and dividing by the total number of loci examined.

proposita In human genetics, an affected female person with whom the study of a trait in a family begins. (*See also* **proband.**)

propositus In human genetics, an affected male person with whom the study of a trait in a family begins. (*See also* **proband.**)

protein A macromolecule composed of one or more **polypeptides.** The functional activity of a protein depends on its complex folded shape and composition.

protein array A collection of different proteins, immobilized on a solid substrate, which serve as probes for detecting labeled target proteins that bind to those affixed to the substrate. Also called *protein microarray* and *protein chip.*

proteome The complete set of proteins in a cell.

proteomics The cataloging and analysis of the proteins in a cell to determine when they are expressed, how much is made, and which proteins interact.

proto-oncogene A gene that in normal cells functions to control the proliferation of cells and that when mutated can become an **oncogene.** *See also* **tumor suppressor gene.**

prototroph A strain of an organism that is wild type for all nutritional requirements and can grow on minimal medium. *See also* **auxotroph.**

prototrophic strain *See* **prototroph.**

pseudodominance The phenotypic expression of a single recessive allele resulting from deletion of a dominant allele on the homologous chromosome.

pseudogene A nonfunctional gene that has sequence homology to one or more functional genes elsewhere in the genome.

Punnett square A matrix that describes all the possible genotypes of progeny resulting from a genetic cross.

pure-breeding strain *See* **true-breeding strain.**

purine One of the two types of cyclic nitrogenous bases found in DNA and RNA. Adenine and guanine are purines.

pyrimidine One of the two types of cyclic nitrogenous bases found in DNA and RNA. Cytosine (in DNA and RNA), thymine (in DNA), and uracil (in RNA) are pyrimidines.

quantitative genetics Study of the inheritance of complex characteristics that are determined by multiple genes.

quantitative trait A heritable characteristic that shows a continuous variation in phenotype over a range. Also called *continuous trait.*

radiation hybrid (RH) A rodent cell line that carries a small fragment of the genome of another organism, such as a human.

random mating Matings between individuals of the same or different genotypes that occur in proportion to the frequencies of the genotypes in the population.

rDNA repeat unit Set of ribosomal RNA (rRNA) genes—encoding 18S, 5.8S, and 28S rRNAs—that are located adjacent to each other and repeated many times in tandem arrays in eukaryotic genomes.

reading frame Linear sequence of codons (groups of three nucleotides) in mRNA that specify amino acids during translation beginning at a particular start codon.

recessive Describing an allele or phenotype that is expressed only in the homozygous state.

recessive lethal allele An allele that results in the death of organisms homozygous for the allele.

reciprocal cross A pair of crosses in which the genotypes of the males and females for a particular trait is reversed. In the garden pea, for example, a reciprocal cross for smooth and wrinkled seeds is smooth female × wrinkled male and wrinkled female × smooth male.

recombinant A chromosome, cell, or individual that has non-parental combinations of **genetic markers** as a result of genetic recombination.

recombinant chromosome A daughter chromosome that emerges from meiosis with an allelic composition that differs from that of either parental chromosome.

recombinant DNA Any DNA molecule that has been constructed in the test tube and contains sequences from two or more distinct DNA molecules, often from different organisms.

recombinant DNA technology A collection of experimental procedures for inserting a DNA fragment from one organism into DNA from another organism and for cloning the new recombinant DNA.

recombination *See* **genetic recombination.**

regression A statistical analysis assessing how changes in one variable are quantitatively related to changes in another variable.

regression coefficient The slope of the **regression line** drawn to show the relationship between two variables.

regression line A mathematically computed line that represents the best fit of a line to the data values for two variables plotted against each other. The slope of the regression line indicates the change in one variable (y) associated with a unit increase in another variable (x).

regulated gene A gene whose expression is controlled in response to the needs of a cell or organism.

release factor (RF) *See* **termination factor.**

replica plating Procedure for transferring the pattern of colonies from a master plate to a new plate. In this procedure, a velveteen pad on a cylinder is pressed lightly onto the surface of the master plate, thereby picking up a few cells from each colony to inoculate onto the new plate.

replication bubble A locally unwound (denatured) region of DNA bounded by replication forks at which DNA synthesis proceeds in opposite directions.

replication fork A Y-shaped structure formed when a double-stranded DNA molecule unwinds to expose the two single-stranded **template strands** for DNA replication.

replicator The entire set of DNA sequences, including the **origin of replication,** required to direct the initiation of DNA replication.

replicon A stretch of DNA in eukaryotic chromosomes extending from an origin of replication to the two termini of replication on each side of that origin. Also called *replication unit.*

replisome The complex of closely associated proteins that forms at the replication fork during DNA synthesis in bacteria.

repressible operon An **operon** whose transcription is reduced in the presence of a particular substance, often the end product of a biosynthetic pathway. The tryptophan (*trp*) operon is an example of a repressible operon. *See also* **inducible operon.**

repressor gene A regulatory gene whose product is a protein that controls the transcriptional activity of a particular operon or gene.

repressors The major class of transcription regulatory proteins in prokaryotes. Bacterial repressors usually bind to the **operator** and prevent transcription by blocking binding of RNA polymerase. In eukaryotes, repressors act in various ways to control transcription of some genes. *See also* **activators.**

repulsion In individuals heterozygous for two genetic loci, the arrangement in which each homologous chromosome carries the wild-type allele of one gene and the mutant allele of the other gene; also called *trans configuration. See also* **coupling.**

restriction endonuclease Enzyme that cleaves double-stranded DNA molecules within or near a specific nucleotide sequence (restriction site), which often is present in multiple copies with a genome. These enzymes are used in analyzing DNA and constructing recombinant DNA. Also called *restriction enzyme.*

restriction enzyme *See* **restriction endonuclease.**

restriction fragment length polymorphism (RFLP) Variation in the lengths of fragments generated by treatment of DNA with a particular restriction enzyme. RFLPs result from point mutations that create or destroy restriction enzyme cleavage sites.

restriction mapping Procedure for locating the relative positions of restriction enzyme cleavage sites in a cloned DNA fragment, yielding a restriction map of the fragment.

restriction site linker A double-stranded oligodeoxyribonucleotide about 8 to 12 base pairs long that contains the cleavage site for a specific restriction enzyme and is used in cloning cDNAs.

retrotransposition The movement of certain mobile genetic elements (retrotransposons) in the genome by a mechanism involving an RNA intermediate.

retrotransposon A type of mobile genetic element, found only in eukaryotes, that encodes **reverse transcriptase** and moves in the genome via an RNA intermediate.

retrovirus A virus with a single-stranded RNA genome that replicates via a double-stranded DNA intermediate produced by **reverse transcriptase,** an enzyme encoded in the viral genome. The DNA integrates into the host's chromosome where it can be transcribed.

reverse genetics *See* **positional cloning.**

reverse mutation A point mutation in a mutant allele that changes it back to a wild-type allele. Also called *reversion.*

reverse transcriptase An enzyme (an RNA-dependent DNA polymerase) that makes a double-stranded DNA copy of an RNA strand.

reverse transcriptase PCR (RT-PCR) A two-step method for detecting and quantitating a particular RNA in an RNA mixture by first converting the RNAs to cDNAs and then performing the **polymerase chain reaction (PCR)** using primers specific for the RNA of interest.

reversion *See* **reverse mutation.**

ribonuclease (RNase) An enzyme that catalyzes degradation of RNA to nucleotides.

ribonucleic acid (RNA) A usually single-stranded polymeric molecule consisting of ribonucleotide building blocks. The three major types of RNA in cells are **ribosomal RNA (rRNA), transfer RNA (tRNA),** and **messenger RNA (mRNA),** each of which performs an essential role in protein synthesis (translation). In some viruses, RNA is the genetic material.

ribonucleotide Any of the nucleotides that make up RNA, consisting of a sugar (ribose), a base, and a phosphate group.

ribose The pentose (five-carbon) sugar found in RNA.

ribosomal DNA (rDNA) The regions of the genome that contain the genes for rRNAs in prokaryotes and eukaryotes.

ribosomal proteins A group of proteins that along with rRNA molecules make up the ribosomes of prokaryotes and eukaryotes.

ribosomal RNA (rRNA) Class of RNA molecules of several different sizes that along with ribosomal proteins make up ribosomes of prokaryotes and eukaryotes.

ribosome A large, complex cellular particle composed of ribosomal protein and rRNA molecules that is the site of amino acid polymerization during protein synthesis (translation).

ribosome-binding site The nucleotide sequence in an mRNA molecule on which the ribosome becomes oriented in the correct reading frame for the initiation of translation.

ribozyme An RNA molecule that has catalytic activity.

RNA *See* **ribonucleic acid.**

RNA editing Unusual type of RNA processing in which the nucleotide sequence of a pre-mRNA is changed by the post-transcriptional insertion or deletion of nucleotides or by conversion of one nucleotide to another.

RNA enzyme *See* **ribozyme.**

RNA interference (RNAi) Silencing of the expression of a specific gene by double-stranded RNA whose sequence matches a portion of the mature mRNA encoded by the gene.

RNA ligase An enzyme that splices together the RNA pieces once the intervening sequence is removed from a pre-tRNA.

RNA polymerase Any enzyme that catalyzes the synthesis of RNA molecules from a DNA template in a process called **transcription.**

RNA polymerase I An enzyme in eukaryotes that catalyzes transcription of 18S, 5.8S, and 28S rRNA genes.

RNA polymerase II An enzyme in eukaryotes that catalyzes transcription of mRNA-coding genes and some snRNA genes.

RNA polymerase III An enzyme in eukaryotes that catalyzes transcription of tRNA and 5S rRNA genes and of some snRNA genes.

RNA primer A short RNA chain, produced by DNA primase during DNA replication, to which DNA polymerase adds nucleotides, thereby extending the new DNA strand.

RNA splicing *See* **mRNA splicing.**

RNA synthesis *See* **transcription.**

Robertsonian translocation A type of nonreciprocal translocation in which the long arms of two nonhomologous **acrocentric chromosomes** become attached to a single centromere.

rooted tree A **phylogenetic tree** in which one internal node is represented as a common ancestor to all the other nodes on the tree.

rRNA transcription unit *See* **ribosomal DNA.**

RT-PCR. *See* **reverse transcriptase PCR.**

sample Subset of individuals belonging to a population. Study of a sample can provide accurate information about the population if the sample is large enough and randomly selected.

sampling error Chance deviations from expected results that arise when the observed sample is small.

second law *See* **principle of independent assortment.**

secondary nondisjunction Abnormal segregation of the X chromosomes during meiosis in the progeny of females with the XXY genotype produced by a primary nondisjunction.

secondary oocyte The larger of the two daughter cells produced by unequal cytokinesis during meiosis I of a primary oocyte in the ovaries of female animals.

segmentation genes Group of genes in *Drosophila* that determine the number and organization of segments in the embryo and adult.

selection The favoring of particular combinations of genes in a given environment.

selection coefficient (s) A measure of the relative intensity of selection against a genotype; equals $1 - W$ (Darwinian fitness).

selection differential In natural and artificial selection, the difference between the mean phenotype of the selected parents and the mean phenotype of the unselected population.

selection response The amount by which a phenotype changes in one generation when natural or artificial selection is applied to a group of individuals.

self-fertilization (selfing) The union of male and female gametes from the same individual.

selfing *See* **self-fertilization.**

self-splicing The excision of introns from some pre-RNA molecules that occurs by a protein-independent reaction in certain organisms.

semiconservative model A model for DNA replication in which each daughter molecule retains one of the parental strands. The results of the Meselson-Stahl experiment supported this model.

semidiscontinuous Concerning DNA replication, when one new strand (the **leading strand**) is synthesized continuously and the other strand (the **lagging strand**) is synthesized discontinuously.

sequence tagged site (STS) A short segment of DNA of known sequence that defines a unique position in the human genome.

sex chromosome A chromosome in eukaryotic organisms that differs morphologically or in number in the two sexes. In many organisms, one sex possesses a pair of visibly different chromosomes. One is an X chromosome, and the other is a Y chromosome. Commonly, the XX sex is female and the XY sex is male.

sex-influenced trait A characteristic controlled by autosomal genes that appears in both sexes but either the frequency of its occurrence or the relationship between genotype and phenotype is different in males and females.

sex-limited trait A characteristic controlled by autosomal genes that is phenotypically exhibited in only one of the two sexes.

sex-linked *See* **X-linked.**

sexual reproduction Mode of reproduction involving the fusion of haploid gametes produced directly or indirectly by meiosis.

Shine-Dalgarno sequence A sequence in prokaryotic mRNAs upstream of the start codon that base-pairs with an RNA in the small ribosomal subunit, allowing the ribosome to locate the start codon for correct initiation of translation.

short interfering RNA (siRNA) Short double-stranded RNAs that function in gene silencing by **RNA interference (RNAi).**

short interspersed elements *See* **SINEs.**

short tandem repeat (STR) A type of **DNA polymorphism** involving variation in the number of short identical sequences (2 to 6 bp in length) that are tandemly repeated at a particular locus in the genome. Also called *microsatellite* and *simple sequence repeat.*

shuttle vector A **cloning vector** that can be introduced into and replicate in two or more host organisms (e.g., *E. coli* and yeast).

signal hypothesis The hypothesis that secreted proteins are synthesized on ribosomes that are directed to the endoplasmic reticulum (ER) by an amino terminal **signal sequence** in the growing polypeptide chain.

signal peptidase An enzyme in the cisternal space of the endoplasmic reticulum that catalyzes removal of the **signal sequence** from growing polypeptide chains.

signal recognition particle (SRP) A cytoplasmic ribonucleoprotein complex that binds to the ER **signal sequence** of a growing polypeptide, blocking further translation of the mRNA in the cytosol. *See also* **docking protein.**

signal sequence Hydrophobic sequence of 15–30 amino acids at the amino end of a growing polypeptide chain that directs the chain-mRNA-ribosome complex to the endoplasmic reticulum (ER) where translation is completed. The signal sequence is removed and degraded in the cisternal space of the endoplasmic reticulum.

signal transduction Process by which an external signal, such as a growth factor, leads to a particular cell response.

silencer element In eukaryotes, an **enhancer** that binds a repressor and acts to decrease RNA transcription rather than stimulating it, as most enhancers do.

silent mutation A **point mutation** in a gene that changes a codon in the mRNA to another codon for the same amino acid, resulting in no change in the amino acid sequence or function of the encoded protein.

simple telomeric sequences Short, tandemly repeated nucleotide sequences at or very close to the extreme ends of chromosomal DNA molecules. The same species-specific sequence is present at the ends of all chromosomes in an organism.

SINEs (short interspersed elements) One class of **dispersed repeated DNA** consisting of sequences that are 100 to 400 bp in length. SINEs can move in the genome by **retrotransposition.**

single nucleotide polymorphism (SNP) A difference in one base pair at a particular site (SNP locus) within coding or noncoding regions of the genome. SNPs that affect restriction sites cause **restriction fragment length polymorphisms (RFLPs).**

single-strand DNA-binding (SSB) protein A protein that binds to the unwound DNA strands at a **replication bubble** and prevents them from reannealing.

sister chromatids Two identical copies of a chromosome derived from replication of the chromosome during interphase of the cell cycle. Sister chromatids are held together by the replicated but unseparated centromeres.

site-specific mutagenesis Introduction of a mutation at a specific site in a particular gene by one of several *in vitro* techniques.

small nuclear ribonucleoprotein particle (snRNP) Large complex formed by small nuclear RNAs (snRNAs) and proteins in which the processing of pre-mRNA molecules occurs.

small nuclear RNA (snRNA) Class of RNA molecules, found only in eukaryotes, that associate with certain proteins to form small nuclear ribonucleoprotein particles (snRNPs).

somatic mutation In multicellular organisms, a change in the genetic material of somatic (body) cells. It may affect the phenotype of the individual in which the mutation occurs but is not passed on to the succeeding generation.

Southern blot analysis A technique for detecting specific DNA fragments in which the fragments are separated by gel electrophoresis, transferred from the gel to a nitrocellulose filter, and then hybridized with labeled complementary probes; also called *Southern blotting. See also* **northern blot analysis.**

spacer sequences Sequences located between and flanking the coding sequences in rRNA and tRNA genes that are transcribed into pre-rRNA and pre-tRNA and then removed during processing to produce mature molecules.

specialized transducing phage A temperate bacteriophage that can transfer only a certain section of the bacterial chromosome from one bacterium to another.

specialized transduction A type of transduction in which only specific genes are transferred from one bacterium to another.

species tree A **phylogenetic tree** based on the divergence observed within multiple genes. A species tree is better than a **gene tree** for depicting the evolutionary history of a group of species.

spermatogenesis Development of male gametes (sperm cells) in animals.

sperm cell A mature male **gamete,** produced by the testes in male animals. Also called **spermatozoon (plural, spermatozoa).**

spliceosome Large complex in the nucleus of eukaryotic cells that carries out **mRNA splicing.** It consists of several small nuclear ribonucleoprotein particles (snRNPs) bound to a pre-mRNA molecule.

spontaneous mutation Any mutation that occurs without the use of a chemical or physical mutagenic agent.

sporophyte The haploid asexual generation in the life cycle of plants that produces haploid spores by meiosis.

stamen The male reproductive organ in flowering plants. It usually consists of a stalklike filament bearing a pollen-producing anther.

standard deviation The square root of the **variance;** a common measure of the extent of variability in a population for **quantitative traits.**

standard error of allelic frequency A statistical measure of the amount of variation in allelic frequency among populations.

steroid hormone response element DNA sequence to which a complex of a specific steroid hormone and its receptor binds, resulting in activation of genes regulated by that hormone.

sticky ends Single-stranded ends of DNA fragments generated by cleavage of DNA with **restriction endonucleases** that make staggered, double-stranded cuts. Base pairing of complementary sticky ends on different fragments and linkage with DNA ligase produces a **recombinant DNA.**

stop codon One of three codons in mRNA for which no normal tRNA molecule exists and which signals the termination of polypeptide synthesis.

STR *See* **short tandem repeat.**

structural genomics The genetic mapping, physical mapping, and sequencing of entire genomes.

submetacentric chromosome A chromosome with the centromere nearer one end than the other such that one arm is longer than the other.

substitution A mutation that has passed through the filter of selection on at least some level.

sum rule The rule that the probability of either of two mutually exclusive events occurring is the sum of their individual probabilities.

supercoiled Referring to a double-stranded DNA molecule that is twisted in space about its own axis.

suppressor gene A gene that when mutated causes suppression of mutations in other genes.

suppressor mutation A mutation at a second site that totally or partially restores a function lost because of a primary mutation at another site.

synapsis The intimate association of replicated homologous chromosomes brought about by the formation of a zipperlike structure (the synaptonemal complex) between the homologues during prophase I of meiosis.

synaptonemal complex A complex structure that spans the region between meiotically paired (synapsed) chromosomes and facilitates **crossing-over.**

synonymous Referring to nucleotides in a gene that when mutated do not result in a change in the amino acid sequence of the encoded wild-type protein.

tandemly repeated DNA Repetitive DNA sequences that are clustered together in the genome, so that each such sequence is repeated many times in a row within a particular chromosomal region.

TATA box A part of the **core promoter** in eukaryotic genomes; it is located about 30 base pairs upstream from the transcription start point. Also called *Goldberg-Hogness box.*

tautomers Alternate chemical forms in which DNA (or RNA) bases are able to exist.

telocentric chromosome A chromosome with the centromere more or less at one end such that only one arm is visible.

telomerase An enzyme that adds short, tandemly repeated DNA sequences (**simple telomeric sequences**) to the ends of eukaryotic chromosomes. It contains an RNA component complementary to the telomeric sequence and has reverse transcriptase activity.

telomere-associated sequence Repeated, complex DNA sequence extending inward from the simple telomeric sequence at each end of a chromosomal DNA molecule.

telophase The stage in mitosis or meiosis during which the migration of the daughter chromosomes to the two poles is completed.

telophase I The stage in meiosis I when chromosomes (each a sister chromatid pair) complete migration to the poles and new nuclear envelopes form around each set of replicated chromosomes.

telophase II The last stage of meiosis II during which a nuclear membrane forms around each set of daughter chromosomes and **cytokinesis** takes place.

temperate phage A bacteriophage that is capable of following either the **lytic cycle** or **lysogenic pathway.** *See also* **virulent phage.**

temperature-sensitive mutant A strain that exhibits a wild-type phenotype in one temperature range but a defective (mutant) phenotype in another, usually higher, temperature range.

template strand DNA strand on which is synthesized a complementary DNA strand during replication or a RNA strand during transcription.

termination factor One of several proteins that recognize stop codons in mRNA and then initiate a series of specific events to terminate translation. Also called *release factor*.

terminator A DNA sequence located at the distal (downstream) end of a gene that signals the termination of transcription.

testcross A cross of an individual of unknown genotype, usually expressing the dominant phenotype, with a homozygous recessive individual to determine the unknown genotype.

testis-determining factor Gene product in placental mammals that causes embryonic gonadal tissue to develop into testes; in the absence of this factor, the gonadal tissue develops as ovaries.

tetrad analysis Technique for analyzing gene linkage and recombination in certain eukaryotic microorganisms (e.g., yeast, *Neurospora*, and *Chlamydomonas*) in which the four products of a single nucleus that has undergone meiosis are grouped together in a single structure. See also **meiotic tetrad.**

tetrasomy A type of **aneuploidy** in which a normally diploid cell or organism possesses four copies of a particular chromosome instead of two copies. A tetrasomic cell is 2N + 2.

tetratype (T) One of the three types of **meiotic tetrads** possible when two genes are segregating in a cross. The T tetrad contains two parental and two recombinant nuclei, one of each parental type and one of each recombinant type.

three-point testcross A cross between an individual heterozygous at three loci with an individual homozygous for recessive alleles at the same three loci. Commonly used in mapping linked genes to determine their order in the chromosome and the distances between them.

thymine (T) A **pyrimidine** found in DNA but not in RNA. In double-stranded DNA, thymine pairs with adenine, a **purine,** by hydrogen bonding.

thymine dimer A common lesion in DNA, caused by ultraviolet radiation, in which adjacent thymines in the same strand are linked in an abnormal way that distorts the double helix at that site.

topoisomerase Any enzyme that catalyzes the supercoiling of DNA.

totipotent Describing a cell that has the potential to develop into any cell type of the organism.

trailer sequence *See* **3′ untranslated region (3′ UTR).**

transconjugant A bacterial cell that incorporates donor DNA received during conjugation into its genome.

transcription The process for making a single-stranded RNA molecule complementary to one strand (the template strand) of a double-stranded DNA molecule, thereby transferring information from DNA to RNA. Also called *RNA synthesis*.

transcriptome The set of mRNA transcripts in a cell.

trans-dominant Referring to a gene or DNA sequence that can control genes on different DNA molecules.

transducing phage Any bacteriophage that can mediate transfer of genetic material between bacteria by **transduction.**

transducing retrovirus Retrovirus that has picked up an **oncogene** from the genome of a host cell.

transductant In bacteria, a recombinant recipient cell generated by **transduction.**

transduction A process by which bacteriophages mediate the transfer of pieces of bacterial DNA from one bacterium (the donor) to another (the recipient).

transfer RNA (tRNA) Class of RNA molecules that bring amino acids to ribosomes, where they are transferred to growing polypeptide chains during translation.

transformant In bacteria, a recombinant recipient cell generated by **transformation.**

transformation (a) In bacteria, a process in which genetic information is transferred by means of extracellular pieces of DNA. (b) In eukaryotes, the conversion of a normal cell with regulated growth properties to a cancer-like cell that can give rise to tumors.

transforming principle Term coined by Frederick Griffith for the unknown agent responsible for the change in genotype via transformation in bacteria. DNA is now known to constitute the transforming principle.

transgene A gene introduced into the genome of an organism by genetic manipulation to alter its genotype.

transgenic Referring to a cell or organism whose genotype has been altered by the artificial introduction of a different allele or gene from the same or a different species.

transition mutation A type of **base pair substitution mutation** that involves a change of one purine-pyrimidine base pair to the other purine-pyrimidine base pair (e.g., AT to GC) at a particular site in the DNA.

transition *See* **transition mutation.**

translation The process that converts the nucleotide sequence of an mRNA into the amino acid sequence of a polypeptide. Also called *protein synthesis*.

translesion DNA synthesis An inducible DNA repair process that allows the replication of DNA beyond a lesion that normally would interrupt DNA synthesis. In *E. coli*, this process is called the *SOS response*.

translocation (a) A chromosomal mutation involving a change in the position of a chromosome segment (or segments) and the gene sequences it contains. (b) In polypeptide synthesis, translocation is the movement of the ribosome, one codon at a time, along the mRNA toward the 3′ end.

transmission genetics Study of how genes are passed from one individual to another. Also called *classical genetics*.

transposable element A DNA segment that can move from one position in the genome to another (nonhomologous) position; also called *mobile genetic element*. Transposable elements are found in both prokaryotes and eukaryotes.

transposase An enzyme encoded by many types of mobile genetic elements that catalyzes the movement (transposition) of these elements in the genome.

transposition The movement of a transposable element within the genome. *See also* **retrotransposition.**

transposon (Tn) A mobile genetic element that contains a gene for transposase, which catalyzes transposition, and genes with other functions such as antibiotic resistance.

trans-splicing An unusual process found in nematodes in which a **polygenic mRNA** is cleaved at each splice site and the same short RNA sequence is added to the 3′ end of each of the resulting individual mRNAs.

transversion mutation A type of **base pair substitution mutation** that involves a change of a purine-pyrimidine base pair to a pyrimidine-purine base pair (e.g., A-T to T-A or G-C to T-A) at a particular site in the DNA.

transversion *See* **transversion mutation.**

trihybrid cross A cross between individuals of the same genotype that are heterozygous for three pairs of alleles at three different loci (e.g., *Ss Yy Cc* × *Ss Yy Cc*).

trisomy A type of **aneuploidy** in which a normally diploid cell or organism possesses three copies of a particular chromosome instead of two copies. A trisomic cell is 2N + 1.

trisomy-13 The presence of an extra copy of chromosome 13, which causes Patau syndrome in humans.

trisomy-18 The presence of an extra copy of chromosome 18, which causes Edwards syndrome in humans.

trisomy-21 The presence of an extra copy of chromosome 21, which causes Down syndrome in humans.

true reversion A **point mutation** in a mutant allele that restores it to the wild-type allele; as a result, the wild-type amino acid sequence and function of the encoded protein is restored.

true-breeding strain A strain in which mating of individuals yields progeny with the same genotype as the parents.

tumor A tissue mass composed of transformed cells, which multiply in an uncontrolled fashion and differ from normal cells in other ways as well; also called *neoplasm*. Benign tumors do not invade the surrounding tissues, whereas malignant tumors invade tissue and often spread to other sites in the body.

tumor suppressor gene A gene in normal cells whose protein product suppresses uncontrolled cell proliferation. *See also* **proto-oncogene.**

tumor virus A virus that induces cells to dedifferentiate and to divide to produce a tumor.

Turner syndrome A human clinical syndrome that results from monosomy for the X chromosome in the female, which gives a 45,X female. Affected females fail to develop secondary sexual characteristics, tend to be short, have weblike necks, have poorly developed breasts, are usually infertile, and exhibit mental deficiencies.

twin spots Two adjacent cell groups that differ in genotype and phenotype. They result from **mitotic crossing-over** within the somatic cells of a heterozygous individual.

unequal crossing-over The process of chromosomal interchange between misaligned chromosomes that may occur during meiosis.

uniparental inheritance A phenomenon, usually exhibited by mitochondrial and chloroplast genes, in which all progeny have the phenotype of only one parent.

unique sequence DNA A class of DNA sequences each of which is present in one to a few copies in the haploid chromosome set; includes most protein-coding genes. Also called *single-copy DNA*.

3′ untranslated region (3′ UTR) The untranslated part of an mRNA molecule beginning at the end of the amino acid-coding sequence and extending to the 3′ end of the mRNA.

5′ untranslated region (5′ UTR) In eukaryotes, the untranslated part of an mRNA molecule extending from the 5′ end to the first (start) codon. It contains coded information for directing initiation of protein synthesis at the translation start site.

unweighted pair group method with arithmetic averages (UPGMA) A statistically based approach used in constructing **phylogenetic trees** that groups taxa together on the basis of their overall pairwise similarities to each other. Also called *cluster analysis*.

uracil (U) A **pyrimidine** found in RNA but not in DNA.

variable number tandem repeat (VNTR) A type of **DNA polymorphism** involving variation in the number of identical sequences (7 bp to a few tens of base pairs in length) that are tandemly repeated at a particular locus in the genome. Also called a *minisatellite*.

variance A statistical measure of the extent to which values in a data set differ from the **mean.**

virulent phage A bacteriophage, such as T4, that always follows the **lytic cycle** when it infects bacteria. *See also* **temperate phage.**

visible mutation A mutation that affects the morphology or physical appearance of an organism.

VNTR *See* **variable number tandem repeat.**

wild-type allele *See* **wild-type.**

wild-type Term describing an allele or phenotype that is designated as the standard ("normal") for an organism and is usually, but not always, the most prevalent in a "wild" population of the organism; also used in reference to a strain or individual.

wobble hypothesis A proposed mechanism that explains how one **anticodon** can pair with more than one **codon.**

X chromosome A sex chromosome present in two copies in the homogametic sex (the female in mammals) and in one copy in the heterogametic sex (the male in mammals).

X chromosome-autosome balance system A genotypic sex determination system in which the ratio between the numbers of X chromosomes and number of sets of autosomes is the primary determinant of sex.

X chromosome nondisjunction Failure of the two X chromosomes to separate in meiosis so that eggs are produced with two X chromosomes or with no X chromosomes instead of the usual one X chromosome.

X-linked Referring to genes located on the X chromosome.

X-linked dominant trait A characteristic caused by a dominant mutant allele carried on the X chromosome.

X-linked recessive trait A characteristic caused by a recessive mutant allele carried on the X chromosome.

Y chromosome A sex chromosome that when present is found in one copy in the heterogametic sex, along with an X chromosome, and is not present in the homogametic sex. Not all organisms with sex chromosomes have a Y chromosome.

Y chromosome mechanism of sex determination A genotypic system of sex determination in which the Y chromosome determines the sex of an individual. Individuals with a Y chromosome are genetically male, and individuals without a Y chromosome are genetically female.

yeast artificial chromosome (YAC) A vector for cloning large DNA fragments, several hundred kilobase pairs long, in yeast. A YAC is a linear molecular with a telomere at each end, a centromere, an autonomously replicating sequence (ARS), a selectable marker, and a polylinker.

yeast two-hybrid system Experimental procedure to find genes encoding proteins that interact with a known protein. Also called *interaction trap assay*.

Y-linked trait A characteristic controlled by a gene carried on the Y chromosome for which there is no corresponding gene locus on the X chromosome. Also called *holandric* or "*wholly male*" *trait*.

zygonema The stage in prophase I of meiosis during which homologous chromosomes begin to pair in a highly specific way along their lengths.

zygote The cell produced by the fusion of a male gamete (sperm cell) and a female gamete (egg cell).

Suggested Readings

In this section are references for selected classic, historically relevant, and other research papers and reviews, and selected websites for the topics presented in the chapters. To learn more about any topic, look for general information using key words with the Google search engine (www.google.com), or for specific research and review papers using keywords at the National Library of Medicine, PubMed website (www.ncbi.nlm.nih.gov/entrez/query.fcgi?db=PubMed).

Chapter 1: Genetics: An Introduction

Genetics Review. http://www.ncbi.nlm.nih.gov/Class/MLACourse/Original8Hour/Genetics/

Sturtevant, A. H. 1965. *A history of genetics.* New York: Harper & Row.

Chapter 2: DNA: The Genetic Material

DNA Structure. http://www.johnkyrk.com/DNAanatomy.html

Introduction to DNA Structure. http://www.blc.arizona.edu/Molecular_Graphics/DNA_Structure/DNA_Tutorial.HTML

Avery, O. T., MacLeod, C. M., and McCarty, M. 1944. Studies on the chemical nature of the substance inducing transformation of pneumococcal types. Induction of transformation by a deoxyribonucleic acid fraction isolated from pneumococcus type III. *J. Exp. Med.* 79:137–158.

Blackburn, E. H. 1994. Telomeres: No end in sight. *Cell* 77:621–623.

Britten, R. J., and Kohne, D. E. 1968. Repeated sequences in DNA. *Science* 161:529–540.

Chargaff, E. 1951. Structure and function of nucleic acids as cell constituents. *Fed. Proc.* 10:654–659.

Clarke, L. 1990. Centromeres of budding and fission yeasts. *Trends Genet.* 6:150–154.

D'Ambrosio, E., Waitzikin, S. D., Whitney, F. R., Salemme, A., and Furano, A. V. 1985. Structure of the highly repeated, long interspersed DNA family (LINE or L1Rn) of the rat. *Mol. Cell. Biol.* 6:411–424.

Dickerson, R. E. 1983. The DNA helix and how it is read. *Sci. Am.* 249 (Dec):94–111.

Fraenkel-Conrat, H., and Singer, B. 1957. Virus reconstitution: Combination of protein and nucleic acid from different strains. *Biochim. Biophys. Acta* 24:540–548.

Franklin, R. E., and Gosling, R. 1953. Molecular configuration of sodium thymonucleate. *Nature* 171:740–741.

Geis, I. 1983. Visualizing the anatomy of A, B, and Z-DNAs. *J. Biomol. Struct. Dyn.* 1:581–591.

Gierer, A., and Schramm, G. 1956. Infectivity of ribonucleic acid from tobacco mosaic virus. *Nature* 177:702–703.

Greider, C. W. 1999. Telomeres do D-loop–T-loop. *Cell* 97:419–422.

Griffith, F. 1928. The significance of pneumococcal types. *J. Hyg. (Lond.)* 27:113–159.

Griffith, J. D., Corneau, L., Rosenfield, S., Stansel, R. M., Bianchi, A., Moss, H., and de Lange, T. 1999. Mammalian telomeres end in a large duplex loop. *Cell* 97:503–514.

Grosschedl, R., Giese, K., and Pagel, J. 1994. HMG domain proteins: Architectural elements in the assembly of nucleoprotein structures. *Trends Genet.* 10:94–100.

Grunstein, M. 1998. Yeast heterochromatin: Regulation of its assembly and inheritance by histones. *Cell* 93:325–328.

Hershey, A. D., and Chase, M. 1952. Independent functions of viral protein and nucleic acid in growth and bacteriophage. *J. Gen. Physiol.* 36:39–56.

Jaworski, A., Hsich, W.-T., Blaho, J. A., Larson, J. E., and Wells, R. D. 1988. Left-handed DNA in vivo. *Science* 238:773–777.

Korenberg, J. R., and Rykowski, M. C. 1988. Human genome organization: Alu, LINES, and the molecular structure of metaphase chromosome bands. *Cell* 53:391–400.

Kornberg, R. D., and Klug, A. 1981. The nucleosome. *Sci. Am.* 244 (Feb):52–64.

Kornberg, R. D., and Lorch, Y. 1999. Twenty-five years of the nucleosome, fundamental particle of the eukaryote chromosome. *Cell* 98:285–294.

Krishna, P., Kennedy, B. P., van de Sande, J. H., and McGhee, J. D. 1988. Yolk proteins from nematodes, chickens, and frogs bind strongly and preferentially to left-handed Z-DNA. *J. Biol. Chem.* 263:19066–19070.

Mason, J. M., and Biessmann, A. 1995. The unusual telomeres of *Drosophila*. *Trends Genet.* 11:58–62.

Moyzis, R. K. 1991. The human telomere. *Sci. Am.* 265 (Aug):48–55.

Olins, A. L., Carlson, R. D., and Olins, D. E. 1975. Visualization of chromatin substructure: Nu-bodies. *J. Cell Biol.* 64:528–537.

Pauling, L., and Corey, R. B. 1956. Specific hydrogen-bond formation between pyrimidines and purines in deoxyribonucleic acids. *Arch. Biochem. Biophys.* 65:164–181.

Pluta, A. F., Mackay, A. M., Ainsztein, A. M., Goldberg, I. G., and Earnshaw, W. C. 1995. The centromere: Hub of chromosomal activities. *Science* 270:1591–1594.

Pruss, D., Bartholomew, B., Persinger, J., Hayes, J., Arents, G., Moudrianakis, E. N., and Wolfe, A. P. 1996. An asymmetric model for the nucleosome: A binding site for linker histones inside the DNA gyres. *Science* 274:614–617.

Singer, M. F. 1982. SINEs and LINEs: Highly repeated short and long interspersed sequences in mammalian genomes. *Cell* 28:133–134.

Sinsheimer, R. L. 1959. A single-stranded deoxyribonucleic acid from bacteriophage ΦX174. *J. Mol. Biol.* 1:43–53.

Wang, A. H.-J., Quigley, G. J., Kolpak, F. J., Crawford, J. L., van Boom, J. H., van der Marel, G., and Rich, A. 1979. Molecular structure of a left-handed double helical DNA fragment at atomic resolution. *Nature* 282:680–686.

Wang, J. C. 1982. DNA topoisomerases. *Sci. Am.* 247 (Jul):94–109.

Watson, J. D. 1968. *The double helix.* New York: Atheneum.

Watson, J. D., and Crick, F. H. C. 1953. Genetical implications of the structure of deoxyribonucleic acid. *Nature* 171:964–969.

———. 1953. Molecular structure of nucleic acids. A structure for deoxyribose nucleic acid. *Nature* 171:737–738.

Wilkins, M. H. F., Stokes, A. R., and Wilson, H. R. 1953. Molecular structure of deoxypentose nucleic acids. *Nature* 171:738–740.

Chapter 3: DNA Replication

DNA Replication. http://users.rcn.com/jkimball.ma.ultranet/BiologyPages/D/DNAReplication.html

Andrews, B., and Measday, V. 1998. The cyclin family of budding yeast: Abundant use of a good idea. *Trends Genet.* 14:66–72.

Cimbora, D. M., and Groudine, M. 2001. The control of mammalian DNA replication: A brief history of space and timing. *Cell* 104:643–646.

Cook, P. R. 1999. The organization of replication and transcription. *Science* 284:1790–1795.

DeLucia, P., and Cairns, J. 1969. Isolation of an E. coli strain with a mutation affecting DNA polymerase. *Nature* 224:1164–1166.

Diller, J. D., and Raghuraman, M. K. 1994. Eukaryotic replication origins: Control in space and time. *Trends Biochem.* 19:320–325.

Gilbert, W., and Dressler, D. 1968. DNA replication: The rolling circle model. *Cold Spring Harbor Symp. Quant. Biol.* 33:473–484.

Greider, C. W., and Blackburn, E. H. 1996. Telomeres, telomerase and cancer. *Sci. Am.* 274 (Feb):92–97.

Huberman, J. A., and Riggs, A. D. 1968. On the mechanism of DNA replication in mammalian chromosomes. *J. Mol. Biol.* 32:327–341.

Kornberg, A. 1960. Biologic synthesis of deoxyribonucleic acid. *Science* 131:1503–1508.

Lendvay, T. S., Morris, D. K., Sah, J., Balasubramanian, B., and Lundblad, V. 1996. Senescence mutants of *Saccharomyces cerevisiae* with a defect in telomere replication identify three additional *EST* genes. *Genetics* 144:1399–1412.

Meselson, M., and Stahl, F. W. 1958. The replication of DNA in *Escherichia coli. Proc. Natl. Acad. Sci. USA* 44:671–682.

Ogawa, T., Baker, T. A., van der Ende, A., and Kornberg, A. 1985. Initiation of enzymatic replication at the origin of the *Escherichia coli* chromosome: Contributions of RNA polymerase and primase. *Proc. Natl. Acad. Sci. USA* 82:3562–3566.

Ogawa, T., and Okazaki, T. 1980. Discontinuous DNA replication. *Annu. Rev. Biochem.* 49:424–457.

Okazaki, R. T., Okazaki, K., Sakobe, K., Sugimoto, K., and Sugino, A. 1968. Mechanism of DNA chain growth. I. Possible discontinuity and unusual secondary structure of newly synthesized chains. *Proc. Natl. Acad. Sci. USA* 59:598–605.

Runge, K. W., and Zakian, V. A. 1996. *TEL2,* an essential gene required for telomere length regulation and telomere position effect in *Saccharomyces cerevisiae. Mol. Cell. Biol.* 16:3094–3105.

Shippen-Lentz, D., and Blackburn, E. H. 1990. Functional evidence for an RNA template in telomerase. *Science* 247:546–552.

Stillman, B. 1994. Smart machines at the DNA replication fork. *Cell* 78:725–728.

Taylor, J. H. 1970. The structure and duplication of chromosomes. In *Genetic organization,* E. Caspari and A. Ravin, eds., vol. 1, pp. 163–221. New York: Academic Press.

Van der Ende, A., Baker, T. A., Ogawa, T., and Kornberg, A. 1985. Initiation of enzymatic replication at the origin of the *Escherichia coli* chromosome: Primase as the sole priming enzyme. *Proc. Natl. Acad. Sci. USA* 82:3954–3958.

Zyskind, J. W., and Smith, D. W. 1986. The bacterial origin of replication, *oriC. Cell* 46:489–490.

Chapter 4: Gene Function

Beadle, G. W., and Tatum, E. L. 1942. Genetic control of biochemical reactions in *Neurospora. Proc. Natl. Acad. Sci. USA* 27:499–506.

Collins, F. 1992. Cystic fibrosis: Molecular biology and therapeutic implications. *Science* 256:774–779.

Garrod, A. E. 1909. *Inborn errors of metabolism.* New York: Oxford University Press.

Gilbert, F., Kucherlapati, R., Creagan, R. P., Murnane, M. J., Darlington, G. J., and Ruddle, F. H. 1975. Tay–Sachs and Sandhoff's diseases: The assignment of genes for hexosaminidase A and B to individual human chromosomes. *Proc. Natl. Acad. Sci. USA* 72:263–267.

Gusella, J. F., Wexler, N. S., Conneally, P. M., Naylor, S. L., Anderson, M. A., Tanzi, R. E., Watkins, P. C., Ottina, K., Wallace, M. R., Sakaguchi, A. Y., Young, A. B., Shoulson, I., Bonilla, E., and Martin, J. B. 1993. A polymorphic DNA marker genetically linked to Huntington's disease. *Nature* 306:234–238.

Guttler, F., and Woo, S. L. C. 1986. Molecular genetics of PKU. *J. Inherit. Metab. Dis.* 9 (Suppl. 1):58–68.

Ingram, V. M. 1963. *The hemoglobins in genetics and evolution.* New York: Columbia University Press.

McIntosh, I., and Cutting, G. R. 1992. Cystic fibrosis transmembrane conductance regulator and the etiology and pathogenesis of cystic fibrosis. *FASEB J.* 6:2775–2782.

Motulsky, A. G. 1973. Frequency of sickling disorders in U.S. blacks. *N. Engl. J. Med.* 288:31–33.

Neel, J. V. 1949. The inheritance of sickle-cell anemia. *Science* 110:64–66.

Pauling, L., Itano, H. A., Singer, S. J., and Wells, J. C. 1949. Sickle-cell anemia, a molecular disease. *Science* 110:543–548.

Riordan, J. R., Rommens, J. M., Kerem, B., Alon, N., Rozmahel, R., Grzelczak, Z., Zielenski, J., Lok, S., Plavsic, N., Chou, J. L., Drumm, M. L., Ianuzzi, M. C., Collins, F. S., and Tsui, L.-C. 1989. Identification of the cystic fibrosis gene: Cloning and characterization of complementary DNA. *Science* 245:1066–1073.

Rommens, J. M., Ianuzzi, M. C., Kerem, B., Drumm, M. L., Melmer, G., Dean, M., Rozmahel, R., Cole, J. L., Kennedy, D., Hidaka, N., Zsiga, M., Buchwald, M., Riordan, J. R., Tsui, L. C., and Collins, F. S. 1989. Identification of the cystic fibrosis gene: Chromosome walking and jumping. *Science* 245:1059–1065.

Scriver, C. R., and Clow, C. L. 1980. Phenylketonuria and other phenylalanine hydroxylation mutants in man. *Annu. Rev. Genet.* 14:179–202.

Scriver, C. R., and Waters, P. J. 1999. Monogenic traits are not simple: Lessons from phenylketonuria. *Trends Genet.* 15:267–272.

Srb, A. M., and Horowitz, N. H. 1944. The ornithine cycle in *Neurospora* and its genetic control. *J. Biol. Chem.* 154:129–139.

Stout, J. T., and Caskey, C. T. 1988. The Lesch-Nyhan syndrome: Clinical, molecular and genetic aspects. *Trends Genet.* 4:175–178.

Chapter 5: Gene Expression: Transcription

Eukaryotic Transcription. http://www.mun.ca/biochem/courses/3107/Topics/euk_transcription.html

Gene Expression: Transcription. http://users.rcn.com/jkimball.ma.ultranet/BiologyPages/T/Transcription.html

Baker, T. A., and Bell, S. P. 1998. Polymerases and the replisome: Machines within machines. *Cell* 92:295–305.

Barabino, S. M. I., and Keller, W. 1999. Last but not least: Regulated poly(A) tail formation. *Cell* 99:9–11.

Bogenhagen, D. F., Sakonju, S., and Brown, D. D. 1980. A control region in the center of the 5S RNA gene directs specific initiation of transcription II: The 3' border of the region. *Cell* 19:27–35.

Breathnach, R., and Chambon, P. 1981. Organization and expression of eucaryotic split genes coding for proteins. *Annu. Rev. Biochem.* 50:349–383.

Breathnach, R., Mandel, J. L., and Chambon, P. 1977. Ovalbumin gene is split in chicken DNA. *Nature* 270:314–318.

Brody, E., and Abelson, J. 1985. The "spliceosome": Yeast pre-messenger RNA associates with a 40S complex in a splicing-dependent reaction. *Science* 228:963–967.

Buratowski, S. 1994. The basics of basal transcription by RNA polymerase II. *Cell* 77:1–3.

Busby, S., and Ebright, R. H. 1994. Promoter structure, promoter recognition, and transcription activation in prokaryotes. *Cell* 79:743–746.

Cate, J. H., Yusupov, M. M., Yusupova, G. Z., Earnest, T. N., and Noller, H. F. 1999. X-ray crystal structures of 70S ribosome functional complexes. *Science* 285:2095–2104.

Cech, T. R. 1983. RNA splicing: Three themes with variations. *Cell* 34:713–716.

———. 1985. Self-splicing RNA: Implications for evolution. *Int. Rev. Cytol.* 93:3–22.

———. 1986. The generality of self-splicing RNA: Relationship to nuclear mRNA splicing. *Cell* 44:207–210.

Choi, Y. D., Grabowski, P. J., Sharp, P. A., and Dreyfuss, G. 1986. Heterogeneous nuclear ribonucleoproteins: Role in RNA splicing. *Science* 231:1534–1539.

Cook, P. R. 1999. The organization of replication and transcription. *Science* 284:1790–1795.

Cramer, P., Bushnell, D. A., Fu, J., Gnatt, A. L., Maier-Davis, B., Thompson, N. E., Burgess, R. R., Edwards, A. M., David, P. R., and Kornberg, R. D. 2000. Architecture of RNA polymerase II and implications for the transcription mechanism. *Science* 288:640–649.

Crick, F. H. C. 1979. Split genes and RNA splicing. *Science* 204:264–271.

Grabowski, P. J., Seiler, S. R., and Sharp, P. A. 1985. A multicomponent complex is involved in the splicing of messenger RNA precursors. *Cell* 42:355–367.

Green, M. R. 1986. Pre-mRNA splicing. *Annu. Rev. Genet.* 20:671–708.

———. 1991. Biochemical mechanisms of constitutive and regulated pre-mRNA splicing. *Annu. Rev. Cell Biol.* 7:559–599.

Guarente, L. 1988. UASs and enhancers: Common mechanism of transcriptional activation in yeast and mammals. *Cell* 52:303–305.

Guthrie, C. 1992. Messenger RNA splicing in yeast: Clues to why the spliceosome is a ribonucleoprotein. *Science* 253:157–163.

Guthrie, C., and Patterson, B. 1988. Spliceosomal snRNAs. *Annu. Rev. Genet.* 22:387–419.

Horowitz, D. S., and Krainer, A. R. 1994. Mechanisms for selecting 5' splice sites in mammalian pre-mRNA splicing. *Trends Genet.* 10:100–105.

Jeffreys, A. J., and Flavell, R. A. 1977. The rabbit beta-globin gene contains a large insert in the coding sequence. *Cell* 12:1097–1108.

Korzheva, N., Mustaev, A., Kozlov, M., Malhotra, A., Nikiforov, V., Goldfarb, A., and Darst, S. A. 2000. A structural model of transcription elongation. *Science* 289:619–625.

Marmur, J., Greenspan, C. M., Palecek, E., Kahan, F. M., Levine, J., and Mandel, M. 1963. Specificity of the complementary RNA formed by *Bacillus subtilis* infected with bacteriophage SP8. *Cold Spring Harbor Symp. Quant. Biol.* 28:191–199.

Narlikar, G. J., Fan, H.-Y., and Kingston, R. E. 2002. Cooperation between complexes that regulate chromatin structure and transcription. *Cell* 108:475–487.

Nilsen, T. W. 1994. RNA–RNA interactions in the spliceosome: Unraveling the ties that bind. *Cell* 78:1–4.

Nomura, M. 1973. Assembly of bacterial ribosomes. *Science* 179:864–873.

O'Hare, K. 1995. mRNA 3' ends in focus. *Trends Genet.* 11:253–257.

Orphanides, G., and Reinberg, D. 2000. A unified theory of gene expression. *Cell* 108:439–451.

Padgett, R. A., Grabowski, P. J., Konarska, M. M., and Sharp, P. A. 1985. Splicing messenger RNA precursors: Branch sites and lariat RNAs. *Trends Biochem. Sci.* (April):154–157.

Proudfoot, N., Furger, A., and Dye, A. J. 2002. Integrating mRNA processing with transcription. *Cell* 108:501–512.

Sharp, P. A. 1985. On the origin of RNA splicing and introns. *Cell* 42:397–400.

Sharp, P. A. 1994. Split genes and RNA splicing. Nobel lecture. *Cell* 77:805–815.

Simpson, L., and Thiemann, O. H. 1995. Sense from nonsense: RNA editing in mitochondria of kinetoplastid protozoa and slime molds. *Cell* 81:837–840.

Sollner-Webb, B. 1988. Surprises in polymerase III transcription. *Cell* 52:153–154.

Thompson, C. C., and McKnight, S. L. 1992. Anatomy of an enhancer. *Trends Genet.* 8:232–236.

Tilghman, S. M., Curis, P. J., Tiemeier, D. C., Leder, P., and Weissman, C. 1978. The intervening sequence of a mouse β-globin gene is transcribed within the 15S β-globin mRNA precursor. *Proc. Natl. Acad. Sci. USA* 75:1309–1313.

Tilghman, S. M., Tiemeier, D. C., Seidman, J. G., Peterlin, B. M., Sullivan, M., Maizel, J. V., and Leder, P. 1978. Intervening sequence of DNA identified in the structural portion of a mouse beta-globin gene. *Proc. Natl. Acad. Sci. USA* 78:725–729.

Weinstock, R., Sweet, R., Weiss, M., Cedar, H., and Axel, R. 1978. Intragenic DNA spacers interrupt the ovalbumin gene. *Proc. Natl. Acad. Sci. USA* 75:1299–1303.

White, R. J., and Jackson, S. P. 1992. The TATA-binding protein: A central role in transcription by RNA polymerases I, II, and III. *Trends Genet.* 8:284–288.

Woychik, N. A., and Hampsey, M. 2002. The RNA polymerase II machinery: Structure illuminates function. *Cell* 108:453–463.

Zaug, A. J., and Cech, T. R. 1986. The intervening sequence RNA of *Tetrahymena* is an enzyme. *Science* 231:470–475.

Chapter 6: Gene Expression: Translation

Translation. http://users.rcn.com/jkimball.ma.ultranet/BiologyPages/T/Translation.html

Ban, N., Nissen, P., Hansen, J., Moore, P. B., and Steitz, T. A. 2000. The complete atomic structure of the large ribosomal subunit at 2.4Å resolution. *Science* 289:905–920.

Blobel, G., and Dobberstein, B. 1975. Transfer of proteins across membranes. I. Presence of proteolytically processed and unprocessed nascent immunoglobulin light chains on membrane-bound ribosomes of murine myeloma. *J. Cell Biol.* 67:835–851.

Brenner, S., Jacob, F., and Meselson, M. 1961. An unstable intermediate carrying information from genes to ribosomes for protein synthesis. *Nature* 190:576–581.

Carter, A. P., Clemons, W. M., Brodersen, D. E., Morgan-Warren, R. J., Wimberly, B. T., and Ramakrishnan, V. 2000. Functional insights from the structure of the 30S ribosomal subunit and its interactions with antibiotics. *Nature* 407:340–348.

Crick, F. H. C. 1966. Codon–anticodon pairing: The wobble hypothesis. *J. Mol. Biol.* 19:548–555.

Crick, F. H. C., Barnett, L., Brenner, S., and Watts-Tobin, R. J. 1961. General nature of the genetic code for proteins. *Nature* 192:1227–1232.

Garen, A. 1968. Sense and nonsense in the genetic code. *Science* 160:149–159.

Khorana, H. G. 1966–67. Polynucleotide synthesis and the genetic code. *Harvey Lect.* 62:79–105.

Kozak, M. 1983. Comparison of initiation of protein synthesis in procaryotes, eucaryotes, and organelles. *Microbiol. Rev.* 47:145.

———. 1989. Context effects and inefficient initiation at non-AUG codons in eukaryotic cell-free translation systems. *Mol. Cell. Biol.* 9:5073–5080.

McCarthy, J. E. G., and R. Brimacombe. 1994. Prokaryotic translation: The interactive pathway leading to initiation. *Trends Genet.* 10:402–407.

Meyer, D. I. 1982. The signal hypothesis: A working model. *Trends Biochem. Sci.* 7:320–321.

Morgan, A. R., Wells, R. D., and Khorana, H. G. 1966. Studies on polynucleotides. LIX. Further codon assignments from amino acid incorporation directed by ribopolynucleotides containing repeating trinucleotide sequences. *Proc. Natl. Acad. Sci. USA* 56:1899–1906.

Nierhaus, K. H. 1990. The allosteric three-site model for the ribosomal elongation cycle: Features and future. *Biochemistry* 29:4997–5008.

Nirenberg, M., and Leder, P. 1964. RNA code words and protein synthesis. *Science* 145:1399–1407.

Nirenberg, M., and Matthaei, J. H. 1961. The dependence of cell-free protein synthesis in *E. coli* upon naturally occurring or synthetic polyribonucleotides. *Proc. Natl. Acad. Sci. USA* 47:1588–1602.

Nissen, P., Hansen, J., Ban, N., Moore, P. B., and Steitz, T. A. 2000. The structural basis of ribosome activity in peptide bond formation. *Science* 289:920–930.

Noller, H. F., Hoffarth, V., and Zimniak, L. 1992. Unusual resistance of peptidyl transferase to protein extraction procedures. *Science* 256:1416–1419.

Ramakrishnan, V. 2002. Ribosome structure and the mechanism of translation. *Cell* 108:557–572.

Ryan, K. R., and Jensen, R. E. 1995. Protein translocation across mitochondrial membranes: What a long, strange trip it is. *Cell* 83:517–519.

Schnell, D. J. 1995. Shedding light on the chloroplast protein import machinery. *Cell* 83:521–524.

Shine, J., and Dalgarno, L. 1974. The 3′-terminal sequence of *Escherichia coli* 16S ribosomal RNA: Complementarity to nonsense triplet and ribosome binding sites. *Proc. Natl. Acad. Sci. USA* 71:1342–1346.

Watson, J. D. 1963. The involvement of RNA in the synthesis of proteins. *Science* 140:17–26.

Zheng, N., and Gierasch, L. M. 1996. Signal sequences: The same yet different. *Cell* 86:849–852.

Chapter 7: DNA Mutation, DNA Repair, and Transposable Elements

Profiles in Science; The Barbara McClintock Papers. http://profiles.nlm.nih.gov/LL/

Transposable genetic elements. http://www.ndsu.nodak.edu/instruct/mcclean/plsc431/transelem/trans1.htm

Ames, B. N., Durston, W. E., Yamasaki, E. and Lee, F. 1973. Carcinogens are mutagens: A simple test system combining liver homogenates for activation and bacteria for detection. *Proc. Natl. Acad. Sci. USA* 70:2281–2285.

Boeke, J. D., Garfinkel, D. J., Styles, C. A., and Fink, G. R. 1985. Ty elements transpose through an RNA intermediate. *Cell* 40:491–500.

Boyce, R. P., and Howard-Flanders, P. 1964. Release of ultraviolet light–induced thymine dimers from DNA in *E. coli* K12. *Proc. Natl. Acad. Sci. USA* 51:293–300.

Cleaver, J. E. 1994. It was a very good year for DNA repair. *Cell* 76:1–4.

Cohen, S. N., and Shapiro, J. A. 1980. Transposable genetic elements. *Sci. Am.* 242 (Feb):40–49.

Devoret, R. 1979. Bacterial tests for potential carcinogens. *Sci. Am.* 241 (Aug):40–49.

Federoff, N. V. 1989. About maize transposable elements and development. *Cell* 56:181–191.

Fishel, R., Lescoe, M. K., Rao, M. R. S., Copeland, N. G., Jenkins, N. A., Garber, J., Kane, M., and Kolodner, R. 1993. The human mutator gene homolog *MSH2* and its association with hereditary nonpolyposis colon cancer. *Cell* 75:1027–1038.

Kingsman, A. J., and Kingsman, S. M. 1988. Ty: A retroelement moving forward. *Cell* 53:333–335.

Lederberg, J., and Lederberg, E. M. 1952. Replica plating and indirect selection of bacterial mutants. *J. Bacteriol.* 63:399–406.

Luria, S. E., and Delbrück, M. 1943. Mutations of bacteria from virus sensitivity to virus resistance. *Genetics* 28:491–511.

McClintock, B. 1939. The behavior in successive nuclear divisions of a chromosome broken at meiosis. *Proc. Natl. Acad. Sci. USA* 25:405–416.

———. 1950. The origin and behavior of mutable loci in maize. *Proc. Natl. Acad. Sci. USA* 36:344–355.

———. 1951. Chromosome organization and genic expression. *Cold Spring Harbor Symp. Quant. Biol.* 16:13–47.

———. 1953. Induction of instability at selected loci in maize. *Genetics* 38:579–599.

———. 1956. Controlling elements and the gene. *Cold Spring Harbor Symp. Quant. Biol.* 21:197–216.

———. 1961. Some parallels between gene control systems in maize and in bacteria. *Am. Naturalist* 95:265–277.

———. 1965. The control of gene action in maize. *Brookhaven Symp. Biol.* 18:162 ff.

———. 1984. The significance of responses of the genome to challenge. Nobel lecture. *Science* 226:792–801.

Morgan, A. R. 1993. Base mismatches and mutagenesis: How important is tautomerism? *Trends Biochem. Sci.* 18:160–163.

Setlow, R. B., and Carrier, W. L. 1964. The disappearance of thymine dimers from DNA: An error-correcting mechanism. *Proc. Natl. Acad. Sci. USA* 51:226–231.

Tessman, I., Liu, S.-K., and Kennedy, A. 1992. Mechanism of SOS mutagenesis of UV-irradiated DNA: Mostly error-free processing of deaminated cytosine. *Proc. Natl. Acad. Sci. USA* 89:1159–1163.

Chapter 8: Recombinant DNA Technology

Cloning and Molecular Analysis of Genes. http://www.ndsu.nodak.edu/instruct/mcclean/plsc431/cloning/

Polymerase Chain Reaction (PCR): Cloning DNA in the Test Tube. http://users.rcn.com/jkimball.ma.ultranet/BiologyPages/P/PCR.html

Recombinant DNA and Gene Cloning. http://users.rcn.com/jkimball.ma.ultranet/BiologyPages/R/RecombinantDNA.html

Arber, W. 1965. Host-controlled modification of bacteriophage. *Annu. Rev. Microbiol.* 19:365–378.

Arber, W., and Dussoix, D. 1962. Host specificity of DNA produced by *Escherichia coli* I. Host controlled modification of bacteriophage lambda. *J. Mol. Biol.* 5:18–36.

Boyer, H. W. 1971. DNA restriction and modification mechanisms in bacteria. *Annu. Rev. Microbiol.* 25:153–176.

Danna, K., and Nathans, D. 1971. Specific cleavage of simian virus 40 DNA by restriction endonuclease of *Haemophilus influenzae*. *Proc. Natl. Acad. Sci. USA* 68:2913–2917.

Feinberg, A. P., and Vogelstein, B. 1983. A technique for radiolabeling DNA restriction endonuclease fragments to high specific activity. *Anal. Biochem.* 132:6–13.

———. 1984. Addendum: A technique for radiolabeling DNA restriction endonuclease fragments to high specific activity. *Anal. Biochem.* 137:266–267.

Luria, S. E. 1953. Host-induced modification of viruses. *Cold Spring Harbor Symp. Quant. Biol.* 18:237–244.

Maxam, A. M., and Gilbert, W. 1977. A new method for sequencing DNA. *Proc. Natl. Acad. Sci. USA* 74:560–564.

Mullis, K. B. 1990. The unusual origin of the polymerase chain reaction. *Sci. Am.* 262 (Apr):56–65.

Mullis, K. B., and Faloona, F. A. 1987. Specific synthesis of DNA in vitro via a polymerase-catalyzed chain reaction. *Methods Enzymol.* 155:335–350.

Sambrook, J., Fritsch, E. F., and Maniatis, T. 1989. *Molecular cloning: A laboratory manual*, 2nd ed. Cold Spring Harbor, NY: Cold Spring Harbor Laboratory.

Sanger, F., and Coulson, A. R. 1975. A rapid method for determining sequences in DNA by primed synthesis with DNA polymerase. *J. Mol. Biol.* 94:441–448.

Southern, E. M. 1975. Detection of specific sequences among DNA fragments separated by gel electrophoresis. *J. Mol. Biol.* 98:503–517.

Watson, J. D., Gilman, M., Witkowski, J., and Zoller, M. 1992. *Recombinant DNA*, 2nd ed. New York: Scientific American Books, Freeman.

White, T. J., Arnheim, N., and Erlich, H. A. 1989. The polymerase chain reaction. *Trends Genet.* 5:185–188.

Chapter 9: Applications of Recombinant DNA Technology

DNA Typing and Identification. http://faculty.ncwc.edu/toconnor/425/425lect15.htm

Gene Therapy. http://www.ornl.gov/sci/techresources/Human_Genome/medicine/genetherapy.shtml

What is Genetic Testing. http://www.lbl.gov/Education/ELSI/Frames/genetic-testing-f.html

Anderson, W. F. 1992. Human gene therapy. *Science* 256:808–813.

Antonarakis, S. E. 1989. Diagnosis of genetic disorders at the DNA level. *N. Engl. J. Med.* 320:153–163.

Cavazzana-Calvo, M., Havein-Bey, S., de Saint Basile, G., Gross, F., Yvon, E., Nusbaum, P., Selz, F., Hu, C., Certain, S., Casanova, J.-L., Bousso, P., Le Deist, F., and Fischer, A. 2000. Gene therapy of human severe combined immunodeficiency (SCID)-X1 disease. *Science* 288:669–672.

Chien, C.-T., Bartel, P. L., Sternglanz, R., and Fields, S. 1991. The two-hybrid system: A method to identify and clone genes for proteins that interact with a protein of interest. *Proc. Natl. Acad. Sci. USA* 88:9578–9582.

Collins, F. 1992. Cystic fibrosis: Molecular biology and therapeutic implications. *Science* 256:774–779.

Culver, K. V., and Blaese, R. M. 1994. Gene therapy for cancer. *Trends Genet.* 10:174–178.

Eisenstein, B. I. 1990. The polymerase chain reaction: A new method of using molecular genetics for medical diagnosis. *N. Engl. J. Med.* 322:178–183.

Fields, S., and Sternglanz, R. 1994. The two-hybrid system: An assay for protein–protein interactions. *Trends Genet.* 10:286–292.

Geisbrecht, B. V., Collins, C. S., Reuber, B. E., and Gould, S. J. 1998. Disruption of a PEX1–PEX6 interaction is the most common cause of the neurological disorders Zellweger syndrome, neonatal adrenoleukodystrophy, and infantile Refsum disease. *Proc. Natl. Acad. Sci. USA* 95:8630–8635.

Gilliam, T. C., Tanzi, R. E., Haines, J. L., Bonner, T. I., Faryniarz, A. G., Hobbs, W. J., MacDonald, M. E., Cheng, S. V., Folstein, S. E., Conneally, P. M., Wexler, N. S., and Gusella, J. F. 1987. Localization of the Huntington's disease gene to a small segment of chromosome 4 flanked by *D4S10* and the telomere. *Cell* 50:565–571.

Green, E. D., and Olson, M. V. 1990. Chromosomal region of the cystic fibrosis gene in yeast artificial chromosomes: A model for human genome mapping. *Science* 250:94–98.

Harris, J. D., and Lemoine, N. R. 1996. Strategies for targeted gene therapy. *Trends Genet.* 12:400–405.

Huntington's Disease Collaborative Research Group. 1993. A novel gene containing a trinucleotide repeat that is expanded and unstable on Huntington's disease chromosomes. *Cell* 72:971–983.

Kay, M. A., and Woo, S. L. C. 1994. Gene therapy for metabolic disorders. *Trends Genet.* 10:253–257.

Kerem, B.-S., Rommens, J. M., Buchanan, J. A., Markiewicz, D., Cox, T. K., Chakravarti, A., Buchwald, M., and Tsui, L.-C. 1989. Identification of the cystic fibrosis gene: Genetic analysis. *Science* 245:1073–1080.

Knowlton, R. G., Cohen-Haguenauer, O., Van Cong, N., Frézal, J., Brown, V. A., Barker, D., Braman, J. C., Schumm, J. W., Tsui, L.-C., Buchwald, M., and Donis-Keller, H. 1985. A polymorphic DNA marker linked to cystic fibrosis is located on chromosome 7. *Nature* 318:380–385.

Koenig, M., Hoffman, E. P., Bertelson, C. J., Monaco, A. P., Feener, C., and Kunkel, L. M. 1987. Complete cloning of the Duchenne muscular dystrophy (DMD) cDNA and preliminary genomic organization of the DMD gene in normal and affected individuals. *Cell* 50:509–517.

Krings, M., Stone, A., Schmitz, R. W., Krainitzki, H., Stoneking, M., and Pääbo, S. 1997. Neanderthal DNA sequences and the origin of modern humans. *Cell* 90:19–30.

Mulligan, R. C. 1993. The basic science of gene therapy. *Science* 260:926–932.

Murray, J. M., Davies, K. E., Harper, P. S., Meredith, L., Mueller, C. R., and Williamson, R. 1982. Linkage relationship of a cloned DNA sequence on the short arm of the X chromosome to Duchenne muscular dystrophy. *Nature* 300:69–71.

Pääbo, S. 1993. Ancient DNA. *Sci. Am.* 269 (Nov):86–92.

Riordan, J. R., Rommens, J. M., Kerem, B., Alon, N., Rozmahel, R., Grzelczak, Z., Zielenski, J., Lok, S., Plavsic, N., Chou, J. L., Drumm, M. L., Ianuzzi, M. C., Collins, F. S., and Tsui, L.-C. 1989. Identification of the cystic fibrosis gene: Cloning and characterization of complementary DNA. *Science* 245:1066–1073.

Rommens, J. M., Ianuzzi, M. C., Kerem, B., Drumm, M. L., Melmer, G., Dean, M., Rozmahel, R., Cole, J. L., Kennedy, D.,

Hidaka, N., Zsiga, M., Buchwald, M., Riordan, J. R., Tsui, L.-C., and Collins, F. S. 1989. Identification of the cystic fibrosis gene: Chromosome walking and jumping. *Science* 245:1059–1065.

Rozsa, F. W., Shimizu, S., Lichter, P. R., Johnson, A. T., Othman, M. I., Scott, K., Downs, C. A., Nguyen, T. D., Polansky, J., and Richards, J. E. 1998. *GLC1A* mutations point to regions of potential functional importance on the TIGR/MYOC protein. *Mol. Vis.* 4:20.

Ryner, L. C., Goodwin, S. F., Castrillon, D. H., Anand, A., Villella, A., Baker, B. S., Hall, J. C., Taylor, B. J., and Wasserman, S. A. 1996. Control of male sexual behavior and sexual orientation in *Drosophila* by the *fruitless* gene. *Cell* 87:1079–1089.

Stafford, H. A. 2000. Crown gall disease and *Agrobacterium tumefaciens*: A study of the history, present knowledge, missing information, and impact on molecular genetics. *Botanical Rev.* 66:99–118.

Wicking, C., and Williamson, B. 1991. From linked marker to gene. *Trends Genet.* 7:288–293.

Wolfenbarger, L. L., and Phifer, P. R. 2000. The ecological risks and benefits of genetically engineered plants. *Science* 290:2088–2093.

Chapter 10: Genomics

Genome Research and Genetics News: *Nature* Genome Gateway. http://www.nature.com/genomics/

Human Genome Project Information. http://www.ornl.gov/sci/techresources/Human_Genome/home.shtml

Adams, M. D., et al. 2000. The genome sequence of *Drosophila melanogaster. Science* 287:2185–2215.

Allzadeh, A. A., et al. 2000. Distinct types of diffuse large B-cell lymphoma identified by gene expression profiling. *Nature* 403:503–511.

Arabidopsis Genome Initiative. 2000. Analysis of the genome sequence of the flowering plant *Arabidopsis thaliana. Nature* 408:796–815.

Bevan, M., and Murphy, G. 1999. The small, the large and the wild. The value of comparison in plant genomics. *Trends Genet.* 15:211–214.

Blattner, F. R., et al. 1997. The complete genome sequence of *Escherichia coli* K-12. *Science* 277:1453–1463.

Bult, C. J., et al. 1996. Complete genome sequence of the methanogenic archaeon, *Methanococcus jannaschii. Science* 273:1058–1073.

Cho, R. J., Campbell, M. J., Winzeler, E. A., Steinmetz, L., Conway, A., Wodicka, L., Wolfsberg, T. G., Gabriellan, A. E., Landsman, D., Lockhart, D. J., and Davis, R. W. 1998. A genome-wide transcriptional analysis of mitotic cell cycle. *Mol. Cell* 2:65–73.

Chu, S., DeRisi, J., Eisen, M., Mulholland, J., Botstein, D., Brown, P. O., and Herskowitz, I. 1998. The transcriptional program of sporulation in budding yeast. *Science* 282:699–705.

DeRisi, J. L., Iyer, V. R., and Brown, P. O. 1997. Exploring the metabolic and genetic control of gene expression on a genomic scale. *Science* 278:680–686.

Dib, C., Fauré, S., Fizames, C., Samson, D., Drouot, N., Vignal, A., Millasseau, P., Marc, S., Hazan, J., Seboun, E., Lathrop, M., Gyapay, G., Morissette, J., and Weissenbach, J. 1996. A comprehensive genetic map of the human genome based on 5,264 microsatellites. *Nature* 380:152–154.

Dujon, B. 1996. The yeast genome project: What did we learn? *Trends Genet.* 12:263–270.

Dunham, I., et al. 1999. The DNA chromosome of human chromosome 22. *Nature* 402:489–495.

Fleischmann, R. D., et al. Whole-genome random sequencing and assembly of *Haemophilus influenzae* Rd. *Science* 269:496–512.

Foote, S., Vollrath, D., Hilton, A., and Page, D. C. 1992. The human Y chromosome: Overlapping DNA clones spanning the euchromatic region. *Science* 258:60 66.

Fraser, C. M., et al. 1995. The minimal gene complement of *Mycoplasma genitalium*. *Science* 270:397–403.

Fraser, C. M., et al. 1998. Complete genome sequence of *Treponema pallidum,* the syphilis spirochete. *Science* 281:375–388.

Goffeau, A., et al. 1996. Life with 6000 genes. *Science* 274:546–567.

International Genome Sequencing Consortium. 2004. Finishing the euchromatic sequence of the human genome. *Nature* 431:931–945.

Klenk, H.-P., et al. 1997. The complete genome sequence of the hyperthermophilic, sulphate-reducing archaeon *Archaeoglobus fulgidus*. *Nature* 390:364–370.

Kornberg, T. B., and Krasnow, M. A. 2000. The *Drosophila* genome sequence: Implications for biology and medicine. *Science* 287:2218–2220.

Nature 15 February 2001: An issue with a special section on "The Human Genome," an analysis of the draft sequence of the human genome.

O'Brien, S. J., Wienberg, J., and Lyons, L. A. 1997. Comparative genomics: Lessons from cats. *Trends Genet.* 13:393–399.

Rubin, G. M., and Lewis, E. B. 2000. A brief history of *Drosophila*'s contributions to genome research. *Science* 287:2216–2218.

Science 16 February 2001: An issue focused on "The Human Genome," an analysis of the draft sequence of the human genome.

Tang, C. M., Hood, D. W., and Moxon, E. R. 1997. *Haemophilus* influence: The impact of whole genome sequencing on microbiology. *Trends Genet.* 13:399–404.

The *C. elegans* Sequencing Consortium. 1998. Genome sequence of the nematode *C. elegans*: A platform for investigating biology. *Science* 282:2012–2018.

Various authors. 1999. The chipping forecast. *Nat. Genet.* 21(suppl.):1–60.

White, R., and Lalouel, J.-M. 1988. Chromosome mapping with DNA markers. *Sci. Am.* 258(February):40–48.

Young, R. A. 2000. Biomedical discovery with DNA arrays. *Cell* 102:9–15.

Chapter 11: Mendelian Genetics
Basic Principles of Genetics. http://anthro.palomar.edu/mendel/mendel_1.htm

Bateson, W. 1909. *Mendel's principles of heredity.* Cambridge: Cambridge University Press.

Bhattacharyya, M. K., Smith, A. M., Ellis, T. H. N., Hedley, C., and Martin, C. 1990. The wrinkled-seed character of pea described by Mendel is caused by a transposon-like insertion in a gene encoding starch-branching enzyme. *Cell* 60:115–122.

Mendel, G. 1866. Experiments in plant hybridization (translation). In *Classic papers in genetics*, J. A. Peters, ed. 1959. Englewood Cliffs, NJ: Prentice Hall.

Peters, J. A., ed. 1959. *Classic papers in genetics.* Englewood Cliffs, NJ: Prentice Hall.

Sandler, I., and Sandler, L. 1985. A conceptual ambiguity that contributed to the neglect of Mendel's paper. *Hist. Phil. Life Sci.* 7:3–70.

Tschermak-Seysenegg, E. von. 1951. The rediscovery of Mendel's work. *J. Hered.* 42:163–171.

Chapter 12: Chromosomal Basis of Inheritance
Cell Cycle and Mitosis Tutorial. http://www.biology.arizona.edu/cell_bio/tutorials/cell_cycle/cells3.html

Meiosis Tutorial. http://www.biology.arizona.edu/cell_bio/tutorials/meiosis/main.html

Barr, M. L. 1960. Sexual dimorphism in interphase nuclei. *Am. J. Hum. Genet.* 12:118–127.

Bridges, C. B. 1916. Nondisjunction as a proof of the chromosome theory of heredity. *Genetics* 1:1–52, 107–163.

———. 1925. Sex in relation to chromosomes and genes. *Am. Nat.* 59:127–137.

Egel, R. 1995. The synaptonemal complex and the distribution of meiotic recombination events. *Trends Genet.* 11:206–208.

Lyon, M. F. 1962. Sex chromatin and gene action in the mammalian X-chromosome. *Am. J. Hum. Genet.* 14:135–148.

McClung, C. E. 1902. The accessory chromosome: Sex determinant? *Biol. Bull.* 3:43–84.

McKusick, V. A. 1965. The royal hemophilia. *Sci. Am.* 213 (Aug):88–95.

Morgan, L. V. 1922. Non criss-cross inheritance in *Drosophila melanogaster*. *Biol. Bull.* 42:267–274.

Morgan, T. H. 1910. Sex-limited inheritance in *Drosophila*. *Science* 32:120–122.

———. 1911. An attempt to analyze the constitution of the chromosomes on the basis of sex-limited inheritance in *Drosophila*. *J. Exp. Zool.* 11:365–414.

Shonn, M. A., McCarroll, R., and Murray, A. W. 2000. Requirement of the spindle checkpoint for proper chromosome segregation in budding yeast meiosis. *Science* 289:300–303.

Stern, C., Centerwall, W. P., and Sarkar, Q. S. 1964. New data on the problem of Y-linkage of hairy pinnae. *Am. J. Hum. Genet.* 16:455–471.

Sutton, W. S. 1903. The chromosomes in heredity. *Biol. Bull.* 4:231–251.

Wilson, E. B. 1905. The chromosomes in relation to the determination of sex in insects. *Science* 22:500–502.

Chapter 13: Extensions of Mendelian Genetic Principles
Gene Interactions. http://www.ndsu.nodak.edu/instruct/mcclean/plsc431/mendel/mendel6.htm

Bultman, S. J., Michaud, E. J., and Woychik, R. P. 1992. Molecular characterization of the mouse *agouti* locus. *Cell* 71:1195–1204.

Ginsburg, V. 1972. Enzymatic basis for blood groups. *Methods Enzymol.* 36:131–149.

Landauer, W. 1948. Hereditary abnormalities and their chemically induced phenocopies. *Growth Symp.* 12:171–200.

Landsteiner, K., and Levine, P. 1927. Further observations on individual differences of human blood. *Proc. Soc. Exp. Biol. Med.* 24:941–942.

Siracusa, L. D. 1994. The *agouti* gene: Turned on to yellow. *Trends Genet.* 10:423–428.

Chapter 14: Quantitative Genetics

Mapping Quantitative Trait Loci. http://www.ndsu.nodak.edu/instruct/mcclean/plsc731/quant/quant1.htm

Darwin, C. 1860. *On the origin of species by means of natural selection, or the preservation of favoured races in the struggle for life.* New York: Appleton.

Dobzhansky, T., and Pavlovsky, O. 1969. Artificial and natural selection for two behavioral traits in *Drosophila pseudoobscura. Proc. Natl. Acad. Sci. USA* 62:75–80.

East, E. M. 1910. A Mendelian interpretation of variation that is apparently continuous. *Am. Nat.* 44:65–82.

———. 1916. Studies on size inheritance in *Nicotiana. Genetics* 1:164–176.

East, E. M., and Jones, D. F. 1919. *Inbreeding and outbreeding.* Philadelphia: Lippincott.

Falconer, D. S. 1989. *Introduction to quantitative genetics.* New York: Wiley.

Hill, W. G., ed. 1984. *Quantitative genetics*, Parts I and II. New York: Van Nostrand Reinhold.

Lander, E. S., and Botstein, D. 1989. Mapping Mendelian factors underlying quantitative traits using RFLP linkage maps. *Genetics* 121:185–199.

Mather, K. 1943. Polygenic inheritance and natural selection. *Biol. Rev.* 18:32–64.

Nilsson-Ehle, H. 1909. Kreuzungsuntersuchungen an Hafer und Weizen. *Lunds Univ. Aarskr. N. F. Atd.*, Ser. 2, 5 (2):1–122.

Paterson, A. H., Lander, E. S., Hewitt, J. D., Person, S., Lincoln, S. E., and Tanksley, S. D. 1988. Resolution of quantitative traits into Mendelian factors by using a complete RFLP linkage map. *Nature* 335:721–726.

Selander, R. K., and Kaufman, D. W. 1975. Self-fertilization and genetic population structure in a colonizing land snail. *Proc. Natl. Acad. Sci. USA* 70:1186–1190.

Sokal, R. R., and Rohlf, F. J. 1981. *Biometry*, 2nd ed. San Francisco: Freeman.

Sokal, R. R., and Thoday, J. M., eds. 1979. *Quantitative genetic variation.* New York: Academic Press.

Thoday, J. M. 1961. Location of polygenes. *Nature* 191:368–370.

Thompson, J. N., Jr., and Thompson, J. M., eds. 1979. *Quantitative genetic variation.* New York: Academic Press.

Weir, B. S., Eisen, E. J., Goodman, M. M., and Namkoong, G., eds. 1988. *Proceedings of the Second International Conference on Quantitative Genetics.* Sunderland, MA: Sinauer.

Chapter 15: Gene Mapping in Eukaryotes

Gene Linkage and Genetic Maps. http://users.rcn.com/jkimball.ma.ultranet/BiologyPages/L/Linkage.html

Bateson, W., Saunders, E. R., and Punnett, R. G. 1905. Experimental studies in the physiology of heredity. *Rep. Evol. Committee R. Soc.* II:1–55, 80–99.

Blixt, S. 1975. Why didn't Mendel find linkage? *Nature* 256:206.

Creighton, H. S., and McClintock, B. 1931. A correlation of cytological and genetical crossing-over in *Zea mays. Proc. Natl. Acad. Sci. USA* 17:492–497.

Holliday, R. 1964. A mechanism for gene conversion in fungi. *Genet. Res.* 5:282–304.

Morgan, T. H. 1910. The method of inheritance of two sex-limited characters in the same animal. *Proc. Soc. Exp. Biol. Med.* 8:17.

———. 1910. Sex-limited inheritance in *Drosophila. Science* 32:120–122.

———. 1911. An attempt to analyze the constitution of the chromosomes on the basis of sex-limited inheritance in *Drosophila. J. Exp. Zool.* 11:365–414.

———. 1911. Random segregation versus coupling in Mendelian inheritance. *Science* 34:384.

Morgan, T. H., Sturtevant, A. H., Müller, H. J., and Bridges, C. B. 1915. *The mechanism of Mendelian heredity.* New York: Henry Holt.

Sturtevant, A. H. 1913. The linear arrangement of six sex-linked factors in *Drosophila* as shown by their mode of association. *J. Exp. Zool.* 14:43–59.

Sutton, W. S. 1903. The chromosomes in heredity. *Biol. Bull.* 4:231–251.

Szostak, J., Orr-Weaver, T., Rothstein, R., and Stahl, F. 1983. The double-strand break repair model for recombination. *Cell* 33:25–35.

Chapter 16: Advanced Gene Mapping in Eukaryotes

Dib, C., et al. 1996. A comprehensive genetic map of the human genome based on 5,264 microsatellites. *Nature* 380:152–154.

Ephrussi, B., and Weiss, M. C. 1969. Hybrid somatic cells. *Sci. Am.* 220 (Apr):26–35.

Holliday, R. 1964. A mechanism for gene conversion in fungi. *Genet. Res.* 5:282–304.

Kao, F., Jones, C., and Puck, T. T. 1976. Genetics of somatic mammalian cells: Genetic, immunologic, and biochemical analysis with Chinese hamster cell hybrids containing selected human chromosomes. *Proc. Natl. Acad. Sci. USA* 73:193–197.

McKusick, V. A. 1971. The mapping of human chromosomes. *Sci. Am.* 224 (Apr):104–113.

Ried, T., Baldini, A., Rand, T. C., and Ward, D. C. 1992. Simultaneous visualization of seven different DNA probes by *in situ* hybridization using combinatorial fluorescence and digital imaging microscopy. *Proc. Natl. Acad. Sci. USA* 89:1388–1392.

Chapter 17: Variations in Chromosome Structure and Number

Barr, M. L., and Bertram, E. G. 1949. A morphological distinction between neurones of the male and female, and the behavior of the nucleolar satellite during accelerated nucleoprotein synthesis. *Nature* 163:676–677.

Borst, P., and Greaves, D. R. 1987. Programmed gene rearrangements altering gene expression. *Science* 235:658–667.

Caskey, C. T., Pizzuti, A., Fu, Y.-H., Fenwick, R. G., and Nelson, D. L. 1992. Triplet repeat mutations in human disease. *Science* 256:784–789.

Dalla-Favera, R., Martinotti, S., Gallo, R., Erickson, J., and Croce, C. 1983. Translocation and rearrangements of the c-*myc* oncogene locus in human undifferentiated B-cell lymphomas. *Science* 219:963–997.

DeKlein, A., van Kessel, A. G., Grosveld, G., Bartram, C. R., Hagemeijer, A., Bootsma, D., Spurr, N. K., Heisterkamp, N., Groffen, J., and Stephenson, J. R. 1982. A cellular oncogene is translocated to the Philadelphia chromosome in chronic myelocytic leukemia. *Nature* 300:765–767.

Huntington's Disease Collaborative Research Group. 1993. A novel gene containing a trinucleotide repeat that is expanded and unstable in Huntington's disease chromosome. *Cell* 72:971–983.

Kremer, E., Pritchard, M., Lynch, M., Yu, S., Holman, K., Baker, E., Warren, S. T., Schlessinger, D., Sutherland, G. R., and Richards, R. I. 1991. Mapping of DNA instability at the fragile X to a trinucleotide repeat sequence p(CGG)n. *Science* 252:1711–1714.

Lyon, M. F. 1961. Gene action in the X-chromosomes of the mouse (*Mus musculus L*). *Nature* 190:372–373.

Penrose, L. S., and Smith, G. F. 1966. *Down's anomaly*. Boston: Little, Brown.

Richards, R. I., and Sutherland, G. R. 1992. Dynamic mutations: A new class of mutations causing human disease. *Cell* 70:709–712.

———. 1992. Fragile X syndrome: The molecular picture comes into focus. *Trends Genet.* 8:249–255.

Rowley, J. D. 1973. A new consistent chromosomal abnormality in chronic myelogenous leukemia identified by quinacrine fluorescence and Giemsa staining. *Nature* 243:290–293.

Shaw, M. W. 1962. Familial mongolism. *Cytogenetics* 1:141–179.

Tarleton, J. C., and Saul, R. A. 1993. Molecular genetic advances in fragile X syndrome. *J. Pediatr.* 122:169–185.

Verkerk, A. J. M. H., Piertti, M., Sutcliff, J. S., Fu, Y.-H., Kuhl, D. P. A., Pizzuti, A., Reiner, O., Richards, S., Victoria, M. F., Zhang, F., Eussen, B. E., van Ommen, G.-J. B., Blonden, L. A. J., Riggins, G. J., Chastain, J. L., Kunst, C. B., Galjaard, H., Caskey, C. T., Nelson, D. L., Oostra, B. A., and Warrent, S. T. 1991. Identification of a gene (*FMR-1*) containing a CGG repeat coincident with a breakpoint cluster region exhibiting length variation in fragile X syndrome. *Cell* 65:905–914.

Chapter 18: Genetics of Bacteria and Bacteriophages

Bacterial Conjugation (A History of its Discovery). http://www.mun.ca/biochem/courses/3107/Lectures/Topic/conjugation.html

Mapping within a Gene: the *rII* Locus. http://users.rcn.com/jkimball.ma.ultranet/BiologyPages/B/Benzer.html

Archer, L. J. 1973. *Bacterial transformation*. New York: Academic Press.

Benzer, S. 1959. On the topology of the genetic fine structure. *Proc. Natl. Acad. Sci. USA* 45:1607–1620.

———. 1961. On the topography of the genetic fine structure. *Proc. Natl. Acad. Sci. USA* 47:403–415.

———. 1962. The fine structure of the gene. *Sci. Am.* 206 (Jan):70–84.

Curtiss, R. 1969. Bacterial conjugation. *Annu. Rev. Microbiol.* 23:69–136.

Ellis, E. L., and Delbruck, M. 1939. The growth of bacteriophage. *J. Gen. Physiol.* 22:365–384.

Fincham, J. 1966. *Genetic complementation*. New York: W. A. Benjamin.

Hayes, W. 1968. *The genetics of bacteria and their viruses*, 2nd ed. New York: Wiley.

Hershey, A. D., and Rotman, R. 1949. Genetic recombination between host-range and plaque-type mutants of bacteriophage in single bacterial cells. *Genetics* 34:44–71.

Hotchkiss, R. D., and Gabor, M. 1970. Bacterial transformation with special reference to recombination processes. *Annu. Rev. Genet.* 4:193–224.

Jacob, F., and Wollman, E. L. 1951. *Sexuality and the genetics of bacteria*. New York: Academic Press.

Ravin, A. W. 1961. The genetics of transformation. *Adv. Genet.* 10:61–163.

Susman, M. 1970. General bacterial genetics. *Annu. Rev. Genet.* 4:135–176.

Vielmetter, W., Bonhoeffer, F., and Schutte, A. 1968. Genetic evidence for transfer of a single DNA strand during bacterial conjugation. *J. Mol. Biol.* 37:81–86.

Wollman, E. L., Jacob, F., and Hayes, W. 1962. Conjugation and genetic recombination in *E. coli K-12*. *Cold Spring Harbor Symp. Quant. Biol.* 21:141–162.

Zinder, N., and Lederberg, J. L. 1952. Genetic exchange in *Salmonella*. *J. Bacteriol.* 64:679–699.

Chapter 19: Regulation of Gene Expression in Bacteria and Bacteriophages

The Operon. http://users.rcn.com/jkimball.ma.ultranet/BiologyPages/L/LacOperon.html

Bell, C. E., Frescura, P., Hochschild, A., and Lewis, M. 2000. Crystal structure of the λ repressor C-terminal domain provides a model for cooperative operator binding. *Cell* 101:801–811.

Bertrand, K., Korn, L., Lee, F., Platt, T., Squires, C. L., Squires, C., and Yanofsky, C. 1975. New features of the structure and regulation of the tryptophan operon of *Escherichia coli*. *Science* 189:22–26.

Bertrand, K., and Yanofsky, C. 1976. Regulation of transcription termination in the leader region of the tryptophan operon of *Escherichia coli* involves tryptophan as its metabolic product. *J. Mol. Biol.* 103:339–349.

Dickson, R. C., Abelson, J., Barnes, W. M., and Reznikoff, W. S. 1975. Genetic regulation: The *lac* control region. *Science* 187:27–35.

Fisher, R. F., Das, A., Kolter, R., Winkler, M. E., and Yanofsky, C. 1985. Analysis of the requirements for transcription pausing in the tryptophan operon. *J. Mol. Biol.* 182:397–409.

Gilbert, W., Maizels, N., and Maxam, A. 1974. Sequences of controlling regions of the lactose operon. *Cold Spring Harbor Symp. Quant. Biol.* 38:845–855.

Gilbert, W., and Muller-Hill, B. 1966. Isolation of the *lac* repressor. *Proc. Natl. Acad. Sci. USA* 56:1891–1898.

Jacob, F. 1965. Genetic mapping of the elements of the lactose region of *Escherichia coli*. *Biochem. Biophys. Res. Commun.* 18:693–701.

Jacob, F., and Monod, J. 1961. Genetic regulatory mechanisms in the synthesis of proteins. *J. Mol. Biol.* 3:318–356.

Lee, F., and Yanofsky, C. 1977. Transcription termination at the *trp* operon attenuators of *Escherichia coli* and *Salmonella typhimurium*: RNA secondary structure and regulation of termination. *Proc. Natl. Acad. Sci. USA* 74:4365–4369.

Lewis, M., Chang, G., Horton, N. C., Kercher, M. A., Pace, H. C., Schumacher, M. A., Brennan, R. G., and Lu, P. 1996. Crystal structure of the lactose operon repressor and its complexes with DNA and inducer. *Science* 271:1247–1254.

Maizels, N. 1974. *E. coli* lactose operon ribosome binding site. *Nature (New Biol.)* 249:647–649.

Pabo, C. O., Sauer, R. T., Sturtevant, J. M., and Ptashne, M. 1979. The λ repressor contains two domains. *Proc. Natl. Acad. Sci. USA* 76:1608–4612.

Ptashne, M. 1967. Isolation of the λ phage repressor. *Proc. Natl. Acad. Sci. USA* 57:306–313.

————. 1984. Repressors. *Trends Biochem. Sci.* 9:142–145.

————. 1992. *A genetic switch,* 2nd ed. Oxford: Cell Press and Blackwell Scientific Publications.

Ptashne, M., and Gilbert, W. 1970. Genetic repressors. *Sci. Am.* 222 (Jun):36–44.

Winkler, M. E., and Yanofsky, C. 1981. Pausing of RNA polymerase during in vitro transcription of the tryptophan operon leader region. *Biochemistry* 20:3738–3744.

Yanofsky, C. 1981. Attenuation in the control of expression of bacterial operons. *Nature* 289:751–758.

————. 1987. Operon-specific control by transcription attenuation. *Trends Genet.* 3:356–360.

Chapter 20: Regulation of Gene Expression in Eukaryotes

Antisense RNA (*includes RNA interference*). http:// users.rcn.com/ jkimball.ma.ultranet/BiologyPages/A/ AntisenseRNA.html

Control of Gene Expression (*includes prokaryotes*). http:// web.indstate.edu/thcme/mwking/gene-regulation.html

Gene Regulation in Eukaryotes. http://users.rcn.com/ jkimball.ma.ultranet/BiologyPages/P/Promoter.html

RNAi–Interference RNA. http://fig.cox.miami.edu/~cmallery/ 150/gene/siRNA.htm

Ashburner, M. 1990. Puffs, genes, and hormones revisited. *Cell* 61:1–3.

Beermann, W., and Clever, U. 1964. Chromosome puffs. *Sci. Am.* 210 (Apr):50–58.

Blumenthal, T., and Gleason, K. S. 2003. *Caenorhabditis elegans* operons: form and function. *Nature Rev. Genet.* 4:110–118.

Carpousis, A. J., Vanzo, N. F., and Raynal, L. C. 1999. mRNA degradation. A tale of poly(A) and multiprotein machines. *Trends Genet.* 15:24–28.

Chen, C.-Y. A., and Shyu, A.-B. 1995. AU-rich elements: Characterization and importance of mRNA degradation. *Trends Biochem. Sci.* 20:465–470.

Gasser, S. M. 2001. Positions of potential: Nuclear organization and gene expression. *Cell* 104:639–642.

Gellert, M. 1992. V(D)J recombination gets a break. *Trends Genet.* 8:408–412.

Green, M. R. 1989. Pre-mRNA processing and mRNA nuclear export. *Curr. Opin. Cell Biol.* 1:519–525.

Grunstein, M. 1992. Histones as regulators of genes. *Sci. Am.* 267 (Oct):68–74B.

Hochstrasser, M. 1996. Protein degradation or regulation: Ub the judge. *Cell* 84:813–815.

Horn, P. J., and Peterson, C. L. 2002. Chromatin higher order folding: wrapping up transcription. *Science* 297:1824–1828.

Johnston, M., Flick, J. S, and Pexton, T. 1994. Multiple mechanisms provide rapid and stringent repression of *GAL* gene expression in *Saccharomyces cerevisiae*. *Mol. Cell. Biol.* 14:3834–3841.

Jones, P. A. 1999. The DNA methylation paradox. *Trends Genet.* 15:34–37.

Karlsson, S., and Nienhuis, A. W. 1985. Development regulation of human globin genes. *Annu. Rev. Biochem.* 54:1071–1078.

Keyes, L. N., Cline, T. W., and Schedl, P. 1992. The primary sex determination signal of *Drosophila* acts at the level of transcription. *Cell* 68:933–943.

Kornberg, R. D. 1999. Eukaryotic transcriptional control. *Trends Genet.* 15:M46–M49.

Oettinger, M. A., Schatz, D. G., Gorka, C., and Baltimore, D. 1990. *RAG-1* and *RAG-2*, adjacent genes that synergistically activate V(D)J recombination. *Science* 248:1517–1522.

O'Malley, B. W., and Schrader, W. T. 1976. The receptors of steroid hormones. *Sci. Am.* 234 (Feb):32–43.

Pankratz, M. J., and Jäckle, H. 1990. Making stripes in the *Drosophila* embryo. *Trends Genet.* 6:287–292.

Parthun, M. R., and Jaehning, J. A. 1992. A transcriptionally active form of GAL4 is phosphorylated and associated with GAL80. *Mol. Cell. Biol.* 12:4981–4987.

Postlethwait, J. H., and Schneiderman, H. A. 1973. Developmental genetics of *Drosophila* imaginal discs. *Annu. Rev. Genet.* 7:381–433.

Ptashne, M. 1989. How gene activators work. *Sci. Am.* 243 (Jan):41–47.

Rhodes, D., and Klug, A. 1993. Zinc fingers. *Sci. Am.* 259 (Feb):56–65.

Rivera-Pomar, R., and Jäckle, H. 1996. From gradients to stripes in *Drosophila* embryogenesis: Filling in the gaps. *Trends Genet.* 12:478–483.

Rogers, J. O., Early, H., Carter, C., Calame, K., Bond, M., Hood, L., and Wall, R. 1980. Two mRNAs with different 3′ ends encode membrane-bound and secreted forms of immunoglobin chain. *Cell* 20:303–312.

Rogers, S., Wells, R., and Rechsteiner, M. 1986. Amino acid sequences common to rapidly degraded proteins: The PEST hypothesis. *Science* 234:364–368.

Ross, J. 1996. Control of messenger RNA stability in higher eukaryotes. *Trends Genet.* 12:171–175.

Scott, M. P., Tamkun, J. W., and Hartzell III, G. W. 1989. The structure and function of the homeodomain. *Biochim. Biophys. Acta* 989:25–48.

Struhl, K. 1999. Fundamentally different logic of gene regulation in eukaryotes and prokaryotes. *Cell* 98:104.

Varshavksy, A. 1996. The N-end rule: Functions, mysteries, uses. *Proc. Natl. Acad. Sci. USA* 93:12142–12149.

Verdine, G. L. 1994. The flip side of DNA methylation. *Cell* 76:197–200.

Wilmut, I., Schnieke, A. E., McWhir, J., Kind, A. J., and Campbell, K. H. S. 1997. Viable offspring derived from fetal and adult mammalian cells. *Nature* 385:810–813.

Wolffe, A. P. 1994. Transcription: In tune with the histones. *Cell* 77:13–16.

Wolffe, A. P., and Pruss, D. 1996. Targeting chromatin disruption: Transcription regulators that acetylate histones. *Cell* 84:817–819.

Chapter 21: Genetic Analysis of Development

Zygote: A Developmental Biology Website (*by Scott Gilbert, Swarthmore College*). http://zygote.swarthmore.edu/

Albrecht, E. B., and Salz, H. K. 1993. The *Drosophila* sex determination gene *snf* is utilized for the establishment of the female-specific splicing pattern of *Sex-lethal*. *Genetics* 134:801–807.

Beachy, P. A. 1990. A molecular view of the *Ultrabithorax* homeotic gene of *Drosophila*. *Trends Genet.* 6:46–51.

Boggs, R. T., Gregor, P., Idriss, S., Belote, J. M., and McKeown, M. 1987. Regulation of sexual differentiation in *Drosophila melanogaster* via alternative splicing of RNA from the transformer. *Cell* 50:739–747.

Capel, B. 1995. New bedfellows in the mammalian sex-determination affair. *Trends Genet.* 11:161–163.

De Robertis, E. M., and Gurdon, J. B. 1977. Gene activation in somatic nuclei after injection into amphibian oocytes. *Proc. Natl. Acad. Sci. USA* 74:2470–2474.

Efstratiadis, A., Posakony, J. W., Maniatis, T., Lawn, R. M., O'Connell, C., Spritz, R. A., DeRiel, J. K., Forget, B. G., Weissman, S. M., Slighton, J. L., Blechtl, A. E., Smithies, O., Baralle, F. E., Shoulders, C. C., and Proudfoot, N. J. 1980. The structure and evolution of the human β-globin gene family. *Cell* 21:653–668.

Eicher, E. M., and Washburn, L. L. 1986. Genetic control of primary sex determination in mice. *Annu. Rev. Genet.* 20:327–360.

Gurdon, J. B. 1968. Transplanted nuclei and cell differentiation. *Sci. Am.* 219 (Dec):24–35.

Gurdon, J. B., Laskey, R. A., and Reeves, R. 1975. The developmental capacity of nuclei transplanted from keratinized skin cells of adult frogs. *J. Embryol. Exp. Morph.* 34:93–112.

Haqq, C. M., King, C.-Y., Ukiyama, E., Falsafi, S., Haqq, T. N., Donahoe, P. K., and Weiss, M. A. 1994. Molecular basis of mammalian sexual determination: activation of Müllerian inhibiting substance gene expression by SRY. *Science* 266:1494–1500.

Hodgkin, J. 1987. Sex determination and dosage compensation in *Caenorhabditis elegans*. *Annu. Rev. Genet.* 21:133–154.

———. 1989. *Drosophila* sex determination: A cascade of regulated splicing. *Cell* 56:905–906.

———. 1993. Molecular cloning and duplication of the nematode sex-determining gene *tra-1*. *Genetics* 133:543–560.

Jiménez, R., Sánchez, A., Burgos, M., and Díaz de la Guardia, R. 1996. Puzzling out the genetics of mammalian sex determination. *Trends Genet.* 12:164–166.

Kay, G. F., Barton, S. C., Surani, M. A., and Rastan, S. 1994. Imprinting and X chromosome counting mechanisms determine *Xist* expression in early mouse development. *Cell* 77:639–650.

Koopman, P., Gubbay, J., Vivian, N., Goodfellow, P., and Lovell-Badge, R. 1991. Male development of chromosomally female mice transgenic for *Sry*. *Nature* 351:117–121.

Lee, J. T., Strauss, W. M., Dausman, J. A., and Jaenisch, R. 1996. A 450 kb transgene displays properties of the mammalian X-inactivation center. *Cell* 86:83–94.

Marahrens, Y., Loring, J., and Jaenisch, R. 1998. Role of the *Xist* gene in X chromosome choosing. *Cell* 92:657–664.

Meller, V. H. 2000. Dosage compensation: Making 1X equal 2X. *Trends Cell Biol.* 10:54–59.

Meyer, B. J. 2000. Sex in the worm. Counting and compensating X-chromosome dose. *Trends Genet.* 16:247–253.

Migeon, B. R. 1994. X-chromosome inactivation: Molecular mechanisms and genetic consequences. *Trends Genet.* 10:230–235.

Page, D. C. 1985. Sex-reversal: Deletion mapping of the male-determining function of the human Y chromosome. *Cold Spring Harbor Symp. Quant. Biol.* 51:229–235.

Page, D. C., de la Chapelle, A., and Weissenbach, J. 1985. Chromosome Y–specific DNA in related human XX males. *Nature* 315:224–226.

Page, D. C., Mosher, R., Simpson, E. M., Fisher, E. M. C., Mardon, G., Pollack, J., McGillivray, B., de la Chapelle, A., and Brown, L. G. 1987. The sex-determining region of the human Y chromosome encodes a finger protein. *Cell* 51:1091–1104.

Palmer, M. S., Sinclair, A. H., Berta, P., Ellis, N. A., Goodfellow, P. N., Abbas, N. E., and Fellous, M. 1990. Genetic evidence that ZFY is not the testis-determining factor. *Nature* 342:937–939.

Penny, G. D., Kay, G. F., Sheardown, S. A., Rastan, S., and Brockdorff, N. 1996. Requirement for *Xist* in X chromosome inactivation. *Nature* 379:131–137.

Rivera-Pomar, R., and Jäckle, H. 1996. From gradients to stripes in *Drosophila* embryogenesis: Filling in the gaps. *Trends Genet.* 12:478–483.

Willard, H. F. 1996. X chromosome inactivation, *XIST*, and pursuit of the X-inactivation center. *Cell* 86:5–7.

Chapter 22: Genetics of Cancer

Oncogenes. http://users.rcn.com/jkimball.ma.ultranet/ Biology Pages/O/Oncogenes.html

Tumor suppressor genes. http://users.rcn.com/ jkimball. ma.ultranet/BiologyPages/T/TumorSuppressorGenes.html

Baltimore, D. 1985. Retroviruses and retrotransposons: The role of reverse transcription in shaping the eukaryotic genome. *Cell* 40:481–482.

Bishop, J. M. 1983. Cancer genes come of age. *Cell* 32:1018–1020.

———. 1983. Cellular oncogenes and retroviruses. *Annu. Rev. Biochem.* 52:301–354.

———. 1987. The molecular genetics of cancer. *Science* 235:305–311.

Brown, M. A., and Solomon, E. 1997. Studies on inherited cancers: Outcomes and challenges of 25 years. *Trends Genet.* 13:202–206.

Cavenee, W. K., and White, R. L. 1995. The genetic basis of cancer. *Sci. Am.* (Mar):72–79.

Fishel, R., Lescoe, M. K., Rao, M. R. S., Copeland, N. G., Jenkins, N. A., Garber, J., Kane, M., and Kolodner, R. 1994. The human mutator gene homolog MSH2 and its association with hereditary nonpolyposis colon cancer. *Cell* 75:1027–1038.

Jiricny, J. 1994. Colon cancer and DNA repair: Have mismatches met their match? *Trends Genet.* 10:164–168.

Kingston, R. E., Baldwin, A. S., and Sharp, P. A. 1985. Transcription control by oncogenes. *Cell* 41:3–5.

Levine, A. J. 1997. p53, the cellular gatekeeper for growth and division. *Cell* 88:323–331.

Rabbitts, T. H. 1994. Chromosomal translocations in human cancer. *Nature* 372:143–149.

Rebbeck, T. R., Couch, F. J., Kant, J., Calzone, K., DeShano, M., Peng, Y., Chen, K., Garber, J. E., and Weber, B. L. 1996. Genetic heterogeneity in hereditary breast cancer: Role of *BRCA1* and *BRCA2*. *Am. J. Hum. Genet.* 59:547–553.

Vousden, K. H. 2000. p53: Death star. *Cell* 103:691–694.

Weinberg, R. A. 1995. The retinoblastoma protein and cell cycle protein. *Cell* 81:323–330.

Weinstein, R. A. 1997. The cat and mouse games that genes, viruses, and cells play. *Cell* 88:573–575.

Welcsh, P. L., Owens, K. N., and King, M.-C. 2000. Insights into the functions of *BRCA1* and *BRCA2*. *Trends Genet.* 16:69–74.

Wooster, R., and Stratton, M. R. 1995. Breast cancer susceptibility: A complex disease unravels. *Trends Genet.* 11:3–5.

Zakian, V. A. 1997. Life and cancer without telomerase. *Cell* 91:1–3.

Chapter 23: Non-Mendelian Inheritance

Birky, C. W. 1978. Transmission genetics of mitochondria and chloroplasts. *Annu. Rev. Genet.* 12:471–512.

Blanchard, J. L., and Lynch, M. 2000. Organellar genes. Why do they end up in the nucleus? *Trends Genet.* 16:315–320.

Brown, M. D., Voljavec, A. S., Lott, M. T., MacDonald, I., and Wallace, D. C. 1992. Leber's hereditary optic neuropathy: A model for mitochondrial neurodegenerative diseases. *FASEB J.* 6:2791–2799.

Chiu, W.-L., and Sears, B. B. 1993. Plastome–genome interactions affect plastid transmission in Oenothera. *Genetics* 133:989–997.

Clayton, D. A. 1982. Replication of animal mitochondrial DNA. *Cell* 28:693–705.

Ephrussi, B. 1953. *Nucleo-cytoplasmic relations in microorganisms*. New York: Oxford University Press.

Freeman, G., and Lundelius, J. W. 1982. The developmental genetics of dextrality and sinistrality in the gastropod *Limnaea peregra*. *Wilhelm Roux Arch. Dev. Biol.* 191:69–83.

Grivell, L. 1983. Mitochondrial DNA. *Sci. Am.* 225 (Mar):78–89.

Gyllensten, U., Wharton, D., Josefsson, A., and Wilson, A. C. 1991. Paternal inheritance of mitochondrial DNA in mice. *Nature* 352:255–257.

Hiratsuka, J., Shimada, H., Whittier, R., Ishibashi, T., Sakamoto, M., Mori, M., Kondo, C., Honji, Y., Sun, C.-R., Meng, B.-Y., Li, Y.-Q., Kanno, A., Nishizawa, Y., Hirai, A., Shinozaki, K., and Sugiura, M. 1989. The complete sequence of the rice (*Oryza saliva*) chloroplast genome: Intermolecular recombination between distinct tRNA genes accounts for a major plastid DNA inversion during the evolution of cereals. *Mol. Gen. Genet.* 217:185–194.

Lander, E. S., and Lodish, H. 1990. Mitochondrial diseases: Gene mapping and gene therapy. *Cell* 61:925–926.

Ohyama, K., Fukuzawa, H., Kohchi, T., Sano, T., Sano, S., Shirai, H., Umesono, K., Shiki, Y., Takeuchi, M., Chang, Z., Aota, S.-I., Inokuchi, H., and Ozeki, H. 1988. Structure and organization of *Marchantia polymorpha* chloroplast genome. I. Cloning and gene identification. *J. Mol. Biol.* 203:281–298.

Umesono, K., and Ozeki, H. 1987. Chloroplast gene organization in plants. *Trends Genet.* 3:281–287.

Van Winkle-Swift, K. P., and Birky, C. W. 1978. The nonreciprocality of organelle gene recombination in *Chlamydomonas reinhardi* and *Saccharomyces cerevisiae*. *Mol. Gen. Genet.* 166:193–209.

Chapter 24: Population Genetics

Avise, J. C. 1986. Mitochondrial DNA and the evolutionary genetics of higher animals. *Phil. Trans. Roy. Soc. Lond., Ser. B* 321:325–342.

Buri, P. 1956. Gene frequency in small populations of mutant *Drosophila*. *Evolution* 10:367–402.

Clarke, C. A., and Sheppard, P. M. 1966. A local survey of the distribution of industrial melanic forms in the moth *Biston betularia* and estimates of the selective values of these in an industrial environment. *Proc. R. Soc. Lond. [Biol.]* 165:424–439.

Crow, J. F. 1986. *Basic concepts in population, quantitative, and evolutionary genetics*. New York: Freeman.

Darwin, C. 1860. *On the origin of species by means of natural selection, or the preservation of favoured races in the struggle for life*. New York: Appleton.

Dobzhansky, T. 1951. *Genetics and the origin of species*, 3rd ed. New York: Columbia University Press.

Fisher, R. A. 1930. *The genetical theory of natural selection*. Oxford: Clarendon Press.

Ford, E. B. 1971. *Ecological genetics*, 3rd ed. London: Chapman & Hall.

Gillespie, J. H. 1991. *The causes of molecular evolution*. Oxford: Oxford University Press.

Glass, B., Sacks, M. S., Jahn, E. F., and Hess, C. 1952. Genetic drift in a religious isolate: An analysis of the causes of variation in blood group and other gene frequencies in a small population. *Am. Nat.* 86:145–159.

Hardy, G. H. 1908. Mendelian proportions in a mixed population. *Science* 28:49–50.

Hartl, D. L., and Clark, A. G. 1995. *Principles of population genetics*, 3rd ed. Sunderland, MA: Sinauer.

Hedrick, P. H. 2000. *Genetics of populations*. Boston: Science Books International.

Hillis, D. M., and Moritz, C. 1990. *Molecular systematics*. Sunderland, MA: Sinauer.

Hillis, D. M., Moritz, C., Porter, C. A., and Baker, R. J. 1991. Evidence for biased gene conversion in concerted evolution of ribosomal DNA. *Science* 251:308–309.

Kettlewell, H. B. D. 1961. The phenomenon of industrial melanism in the Lepidoptera. *Annu. Rev. Entomol.* 6:245–262.

Kreitman, M. 1983. Nucleotide polymorphism at the alcohol dehydrogenase locus of *Drosophila melanogaster*. *Nature* 304:412–417.

Lehman, N., Eisenhawer, A., Hansen, K., Mech, L. D., Peterson, R. O., Gogan, P. J., and Wayne, R. K. 1991. Introgression of coyote mitochondrial DNA into sympatric North American gray wolf populations. *Evolution* 45:104–119.

Lewontin, R. C. 1974. *The genetic basis of evolutionary change*. New York: Columbia University Press.

Lewontin, R. C. 1985. Population genetics. *Annu. Rev. Genet.* 19:81–102.

Lewontin, R. C., Moore, J. A., Provine, W. B., and Wallace, B. 1981. *Dobzhansky's genetics of natural populations I–XLIII*. New York: Columbia University Press.

Li, W.-H. 1997. *Molecular evolution*. Sunderland, MA: Sinauer.

Li, W.-H., Luo, C. C., and Wu, C. I. 1985. Evolution of DNA sequences. In *Molecular evolutionary genetics*, R. J. MacIntyre, ed., pp. 1–94. New York: Plenum.

Maniatis, T., Fritsch, E. F., Lauer, L., and Lawn, R. M. 1980. The molecular genetics of human hemoglobin. *Annu. Rev. Genet.* 14:145–178.

Maynard Smith, J. 1989. *Evolutionary genetics*. Oxford: Oxford University Press.

Nei, M. 1987. *Molecular evolutionary genetics*. New York: Columbia University Press.

Nei, M., and Koehn, R. K. 1983. *Evolution of genes and proteins*. Sunderland, MA: Sinauer.

Powell, J. R. 1997. *Progress and prospects in evolutionary biology: The* Drosophila *model*. New York: Oxford University Press.

Selander, R. K., and Kaufman, D. W. 1975. Self-fertilization and genetic population structure in a colonizing land snail. *Proc. Natl. Acad. Sci. USA* 70:1186–1190.

Soulé, M. E., ed. 1986. *Conservation biology: The science of scarcity and diversity*. Sunderland, MA: Sinauer.

Stringer, C. B. 1990. The emergence of modern humans. *Sci. Am.* (Dec):98–104.

Weir, B. S. 1996. *Genetic data analysis II*. Sunderland, MA: Sinauer.

Woese, C. R. 1981. Archaebacteria. *Sci. Am.* 244:98–122.

Chapter 25: Molecular Evolution

Haldane, J. B. S. 1932. *The causes of evolution*. London: Longmans and Green.

Jukes, T. H., and Cantor, C. R. 1969. Evolution of protein molecules. In *Mammalian protein metabolism*, H. N. Munro, ed., pp. 21–123. Academic Press, New York.

Klein, J., and Figueroa, F. 1986. Evolution of the major histocompatability complex. *CRC Crit. Rev. Immunol.* 6:295–386.

Perutz, M. F. 1983. Species adaptation in a protein molecule. *Mol. Biol. Evol.* 1:1–28.

Sarich, V. M., and Wilson, A. C. 1967. Immunological time scale for hominid evolution. *Science* 158:1200–1203.

Solutions to Selected Questions and Problems

Chapter 2 DNA: The Genetic Material

2.2 a. Lived

b. Died

c. Lived

d. Died (DNA from the *IIIS* bacteria transformed the *IIR* bacteria into a virulent form.)

2.4   Parts (**a**) (**b**) and (**c**): Phage ghosts (supernatant) and progeny would have isotope. Both amino acids and nucleic acids have C, N, and H, so parental phage labeled with isotopes of C, N, or H will have labeled protein coats and DNA. Isotope would be recovered in the DNA of the progeny phage, as well as in the phage ghosts left behind in the supernatant after phage infection. This experiment would not help determine whether DNA or protein was the genetic material, as neither material is selectively labeled.

2.9   a. 3′ TCAATGGACTAGCAT 5′ (or 5′ TACGATCAGGTAACT 3′).

b. 3′ AAGAGTTCTTAAGGT 5′ (or 5′ TGGAATTCTTGA-GAA 3′).

2.10 The adenine-thymine base pair has two hydrogen bonds, while the guanine-cytosine base pair has three hydrogen bonds. Thus, the guanine-cytosine base pair requires more energy to break apart and so is harder to break apart.

2.12 Since $(G) = (C)$ and $(A) = (T)$, it follows that $(G + A) = (C + T)$ and $(G + T) = (A + C)$. Thus, (**b**) (**c**) and (**d**) are all equal to 1.

2.15 Since the DNA molecule is double-stranded, $(A) = (T)$ and $(G) = (C)$. If there are 80 T residues, there must be 80 A residues. If there are 110 G residues, there must be 110 C residues. The molecule has $(110 + 110 + 80 + 80) = 380$ nucleotides, or 190 base pairs.

2.16 Here, $(A) \neq (T)$ and $(G) \neq (C)$, so the DNA is not double-stranded. The bacterial virus appears to have a single-stranded DNA genome.

2.17 GC base pairs have three hydrogen bonds, whereas AT base pairs have two. As a consequence, GC base pairs are stronger than AT base pairs. If a double-stranded molecule in solution is heated, the thermal energy "melts" the hydrogen bonds, denaturing the double-stranded molecule into single strands. Double-stranded molecules with more GC base pairs require more thermal energy to break their hydrogen bonds, so they dissociate into single strands at higher temperatures. Put another way, the higher the GC content of a double-stranded DNA molecule, the higher its melting temperature. Reordering the molecules from lowest to highest percent GC, the melting order is (**b**) 69°, then (**a**) 73°, (**d**) 78°, (**e**) 82°, and (**c**) 84°.

2.22 a. Each base pair has 2 nucleotides, so the molecule has 200,000 nucleotides.

b. There are 10 base pairs per complete 360° turn, so there will be 100,000/10 = 10,000 complete turns in the molecule.

c. There is 0.34 nm between the centers of adjacent base pairs. There will be $100,000 \times 0.34$ nm $= 3.4 \times 10^4$ nm $= 34$ μm.

2.24 The chance of finding the sequence 5′-GUUA-3′ is $(0.30 \times 0.25 \times 0.25 \times 0.20) = 0.00375$. In a molecule 10^6 nucleotides long, there are nearly 10^6 groups of four bases: The first group of four is bases 1, 2, 3, and 4, the second group is bases 2, 3, 4, and 5, and so on. Thus, the number of times this sequence is expected to appear is $0.00375 \times 10^6 = 3,750$.

2.25 a. The sequence CGAGG in molecule 2 is complementary to the sequence GCTCC in molecule 3. These can pair to give

Each strand has two unpaired bases sticking out. These bases are complementary to each other, so that if the molecule bends, one has

a.

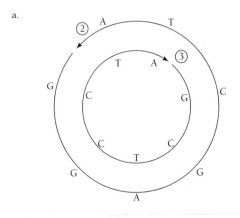

b. The sequence in molecule 3 is complementary to the sequence in molecule 4. It also has opposite polarity, so that the two strands can pair up. One has:

b.

2.29 a. Only eukaryotic chromosomes have centromeres, the sections of the chromosome found near the point of attachment of mitotic or meiotic spindle fibers. In some organisms, such as *S. cerevisiae*, they are associated with specific *CEN* sequences. In other organisms, they have a more complex repetitive structure.

b. Hexose sugars are not found in either eukaryotic or bacterial chromosomes, as pentose sugars are found in DNA.

c. Amino acids are found in proteins that are involved in chromosome compaction, such as the proteins that hold the ends of looped domains in prokaryotic chromosomes and the histone and nonhistone proteins in eukaryotic chromatin.

d. Both eukaryotic and bacterial chromosomes share supercoiling.

e. Telomeres are found only at the ends of eukaryotic chromosomes and are required for replication and chromosome stability. They are associated with specific types of sequences: simple telomeric sequences and telomere-associated sequences.

f. Nonhistone proteins are found only in eukaryotic chromosomes and have structural (higher-order packaging) and possibly other functions.

g. DNA is found in both prokaryotic and eukaryotic chromosomes (although some viral chromosomes have RNA as their genetic material).

h. Nucleosomes are the fundamental unit of packaging of DNA in eukaryotic chromosomes and are not found in prokaryotic chromosomes.

i. Circular chromosomes are found only in prokaryotes, not in eukaryotes.

j. Looping is found in both eukaryotic and prokaryotic chromosomes. In eukaryotic chromosomes, the 30-nm nucleofilament is packed into looped domains by nonhistone chromosomal proteins. In bacterial chromosomes, such as that of *E. coli*, there are about 100 looped domains, each containing about 40 kilobases of supercoiled DNA.

2.31 a. The belt forms a right-handed helix. Although you wrapped the belt around the can axis in a counterclockwise direction from your orientation (looking down at the can), the belt was winding up and around the side of can in a clockwise direction from its orientation. While the belt is wrapped around the can, curve the fingers of your right hand over the belt and use your index finger to trace the direction of the belt's spiral. Your right index finger will trace the spiral upward, the same direction your thumb points when you wrap your hand around the can. Therefore, the belt has formed a right-handed helix.

b. Three turns were present.

c. Three turns were present. The number of helical turns is unchanged, although the twist in the belt is.

d. The belt appears more twisted because the pitch of the helix was altered and the edges of the belt (positioned much like the complementary base pairs of a double helix) are twisted more tightly.

e. While twisted around the can, the length of the belt decreases by about 70 to 80 percent, depending on the initial length of the belt and the belt diameter.

f. The answer is yes. As the DNA of linear chromosomes becomes is wrapped around histones to form the 10-nm nucleofilament, it becomes supercoiled. In much the same manner as you must add twists to the belt for it to lie flat on the surface of

the can, supercoils must be introduced into the DNA for it to wrap around the histones.

g. Topoisomerases increase or reduce the level of negative supercoiling in DNA. For linear DNA to be packaged, negative supercoils must be added.

2.32 All 16 yeast centromeres of have similar but not identical DNA sequences called *CEN* sequences. Each contains three sequence domains, called centromere DNA elements (CDEs). CDEII, a 76–86-bp region that is >90% AT, is flanked by CDEI, a conserved RTCACRTG sequence (R = A or G), and CDEIII, a 25-bp AT-rich conserved domain. The CDEs interact with kinetochore proteins that mediate the attachment of centromeres to spindle fiber microtubules. In one model, CDEI binds centromere-binding factor I (CBFI), CDEII is wrapped once around a histone octamer, and CDEIII binds the protein complex CBF3. In turn, a linker protein may attach CBF3 to the spindle microtubule.

2.34 You would find them in unique-sequence DNA.

2.36 a. These findings support the current view that telomeres are specialized chromosome structures with two distinct structural components: simple telomeric sequences and telomere-associated sequences. They show that functional genes do not reside in the telomeric region, consistent with the view that telomeres are heterochromatic and have special protective functions in chromosomes. They add significantly to our knowledge of the structure of telomeric and near-telomeric regions. For example, they document the considerable distance over which the telomere-associated sequences are found, about 36 kb, and give a sense of the number, size, and density of genes in the region near this telomere.

b. At least in this region, *Alu* sequences are found more often in AT-rich areas. These areas are not as gene rich as adjacent GC-rich areas. Thus, this class of moderately repetitive sequences and the genes in this area appear to have a nonrandom distribution.

Chapter 3 DNA Replication

3.2 Key: $^{15}N-^{15}N$ DNA = HH; $^{15}N-^{14}N$ DNA = HL; $^{14}N-^{14}N$ DNA = LL.

a. Generation 1: all HL; 2: $\frac{1}{2}$ HL, $\frac{1}{2}$ LL; 3: $\frac{1}{4}$ HL, $\frac{3}{4}$ LL; 4: $\frac{1}{8}$ HL, $\frac{7}{8}$ LL; 6: $\frac{1}{32}$ HL, $\frac{31}{32}$ LL; 8: $\frac{1}{128}$ HL, $\frac{127}{128}$ LL.

b. Generation 1: $\frac{1}{2}$ HH, $\frac{1}{2}$ LL; 2: $\frac{1}{4}$ HH, $\frac{3}{4}$ LL; 3: $\frac{1}{8}$ HH, $\frac{7}{8}$ LL; 4: $\frac{1}{16}$ HH, $\frac{15}{16}$ LL; 6: $\frac{1}{64}$ HH, $\frac{63}{64}$ LL; 8: $\frac{1}{256}$ HH, $\frac{255}{256}$ LL.

3.4 a. Establishing that DNA replication is semiconservative *does not* ensure that it is semidiscontinuous. For example, if the old strands were completely unwound and replication were initiated from the 3′ end of each, it could proceed continuously in a 5′→3′ direction along each strand. Alternatively, if DNA polymerase were able to synthesize DNA in both the 3′→5′ and 5′→3′ directions, DNA replication could proceed continuously on both DNA strands.

b. Establishing that DNA replication is semidiscontinuous *does* ensure that it is semiconservative. In the semidiscontinuous model, each old separated strand serves as a template for a new strand. This is the essence of the semiconservative model.

c. Semiconservative DNA replication is ensured by two enzymatic properties of DNA polymerase: It synthesizes just one new strand from each "old" single-stranded template and it can synthesize new DNA in only one direction (5′→3′).

3.5 DNA can be synthesized in vitro using DNA polymerase I; dATP, dGTP, dCTP, and dTTP; magnesium ions; and a fragment of double-stranded DNA to serve as a template.

3.6 The primary evidence that the Kornberg enzyme is not the main enzyme for DNA synthesis in vivo stems from an analysis of the growth and biochemical phenotypes of the mutants *polA1* and *polAex1*. The mutant *polA1* lacks 99 percent of polymerase activity, but is nonetheless able to grow, replicate its DNA, and divide. The conditional mutant *polAex1* retains most of the polymerizing activity at the restrictive temperature 42°C, but is still unable to replicate its chromosomes and divide (it has lost the enzyme's $5' \rightarrow 3'$ exonuclease activity).

3.9 None is an analog for adenine, B and D are analogs of thymine, C is an analog of cytosine, and A is an analog of guanine.

3.13 Since a replication fork moves at a rate of 10^4 bp per minute and each replication has two replication forks moving in opposite directions, in one replicon, replication occurs at a rate of 2×10^4 bp/minute. Assume for the purposes of this calculation only that DNA replication is distributed among similarly sized replicons initiating replication at the same time. Since all the DNA replicates in 3 minutes, the number of replicons in the diploid genome is

$$\frac{4.5 \times 10^8 \text{ bp}}{3 \text{ minutes}} \times \frac{1 \text{ replicon}}{2 \times 10^4 \text{ bp/minute}} = 7{,}500 \text{ replicons.}$$

3.14 DNA ligase catalyzes the formation of a phosphodiester bond between the 3'-OH and the 5'-monophosphate groups on either side of a single-stranded DNA gap, sealing the gap. See Figure 11.7. Temperature-sensitive ligase mutants would be unable to seal such gaps at the restrictive (high) temperature, leading to fragmented lagging strands and presumably cell death. If a biochemical analysis were performed on DNA synthesized after *E. coli* were shifted to a restrictive temperature, there would be an accumulation of DNA fragments the size of Okazaki fragments. This would provide additional evidence that DNA replication must be discontinuous on one strand.

3.15 Assume the amount of a gene's product is directly proportional to the number of copies of the gene present in the *E. coli* cell. Assay the enzymatic activity of genes at various positions in the *E. coli* chromosome during the replication period. Then, some genes (those immediately adjacent to the origin) will double their activity very shortly after replication begins. Relate the map position of genes having doubled activity to the amount of time that has transpired since replication was initiated. If replication is bidirectional, there should be a doubling of the gene products both clockwise and counterclockwise from the origin.

3.17 Clearly DNA replication in the Jovian bug does not occur as it does in *E. coli*. Assuming that the double-stranded DNA is antiparallel as it is in *E. coli*, the Jovian DNA polymerases must be able to synthesize DNA in the $5' \rightarrow 3'$ direction (on the leading strand) as well as in the $3' \rightarrow 5'$ direction (on the lagging strand). This is unlike any DNA polymerase on earth.

3.19 a. The DNA endonuclease encoded by the *ter* gene recognizes sequences at *cos* sites appearing just once within a λ genome. It makes a staggered cut at these sites to produce the unit-length linear DNA molecules that are packaged.

b. The *ter* enzyme produces complementary ("sticky") 12-base-long, single-stranded ends. After λ infects *E. coli*, these ends pair and gaps in the phosphodiester backbone are sealed by DNA ligase to produce a closed circular molecule. This molecule recombines into the *E. coli* chromosome if the lysogenic pathway is followed, or replicates using rolling circle replication if a lytic pathway is followed.

3.20 a. Since M13 has a closed circular genome with $(A) \neq (T)$ and $(G) \neq (T)$, it must have a single-stranded DNA genome. Bidirectional replication would require the initial synthesis of a complementary strand. To produce many phage, many rounds of bidirectional replication would be necessary. However, upon completion of replication and prior to packaging, the nongenomic strand of the resulting double-stranded molecules would need to be selectively degraded.

b. To produce single-stranded molecules with the same sequence and base composition as the packaged M13 genome, rolling circle replication must use a complementary template. Therefore, DNA polymerase must initially synthesize the genome's complementary strand to make a double-stranded molecule. Then, a nuclease could nick the genomic strand to create a displacement fork. Continuous rolling circle replication using the intact complementary strand as the leading strand template and without discontinuous replication on the displaced genomic strand will generate single-stranded M13 genomes. To prevent concatamer formation, the newly replicated DNA must be cleaved by an endonuclease after exactly one genome has been replicated. To form a closed circle, the molecule's ends would need to be ligated to each other.

3.21 Multiple DNA polymerases have been identified in all cells: There are five in prokaryotes and 15 or more in eukaryotes. All DNA polymerases synthesize DNA from a primed strand in the $5' \rightarrow 3'$ direction using a template. In both eukaryotes and prokaryotes, certain DNA polymerases are used for replication, while others are used for repair. Prokaryotes and eukaryotes differ in how many polymerases they use and how they use them in each of these processes.

In *E. coli*, DNA polymerase I and III function in DNA replication. Both have $3' \rightarrow 5'$ exonuclease activity that is used in proofreading. DNA polymerase III is the main synthetic enzyme and can exist as a core enzyme with three polypeptides or as a holoenzyme with an additional six different polypeptides. DNA polymerase I consists of one polypeptide. Unlike DNA polymerase III, it has the $5' \rightarrow 3'$ exonuclease activity needed to excise RNA from the 5' end of Okazaki fragments. DNA polymerases II, IV, and V function in DNA repair.

In eukaryotes, nuclear DNA replication requires three DNA polymerases: Pol α/primase, Pol δ, and Pol ε. After primase initiates new strands in replication by making about 10 nucleotides of an RNA primer, Pol α extends them by adding about 30 nucleotides of DNA. The RNA/DNA primers are extended by Pol δ and Pol ε on the leading and lagging strand. However, it is not clear which enzyme synthesizes which strand. Other DNA polymerases function in DNA repair and mitochondrial DNA replication.

3.23 Assuming cells spend 4 hours in G_2, there are 4.5 hours from the last 30 minutes of S to metaphase in M. Late-replicating chromosomal regions can be identified by adding ^{3}H thymidine to the medium, waiting 4.5 hours, and then preparing a slide of metaphase chromosomes. Chromosomal regions displaying silver grains are late replicating, as cells that were at earlier stages of S when the ^{3}H was added will be unable to reach metaphase in 4.5 hours.

3.26 a. Harlequin chromosomes are prepared by allowing tissue culture cells to undergo two rounds of DNA replication in the presence of BUdR. After BUdR (a base analog that replaces T) becomes incorporated in the DNA, metaphase chromosomes stained with Giemsa stain and a fluorescent dye have one darkly stained and one lightly stained chromatid. Since BUdR-DNA stains less intensely than T-DNA, the presence of two differently stained chromatids indicates that one chromatid has two BUdR-strands and the other has one BUdR- and one T-labeled strand.

This supports the semiconservative model of DNA replication. After one round of replication, each DNA molecule has one (new) BUdR- and one (old) T-labeled strand. After two rounds of replication, each of these molecules produces one DNA molecule with two BudR-labeled strands (the light chromatid) and one DNA molecule with one BUdR- and one T-labeled strand (the dark chromatid).

b. Chromosomes with chromatids containing segments of T-labeled DNA and BUdR-labeled DNA have had a sister-chromatid exchange (mitotic crossing-over).

3.28 Telomerase synthesizes the simple-sequence telomeric repeats at the ends of chromosomes. The enzyme is made up of both protein and RNA, and the RNA component has a base sequence that is complementary to the telomere repeat unit. The RNA component is used as a template for the telomere repeat, so that if the RNA component were altered, the telomere repeat would be as well. Thus, the mutant in this question is likely to have an altered RNA component.

Chapter 4 Gene Function

4.4 A double homozygote should have PKU, but not AKU. The PKU block should prevent most homogentisic acid from being formed, so it could not accumulate to high levels and cause AKU.

4.6 Autosomes are chromosomes that are found in two copies in both males and females. That is, an autosome is any chromosomes except the X and Y chromosomes. Since individuals have two of each type of autosome, they have two copies of each gene on an autosome. The forms of the gene, or the alleles at the gene, can be the same or different. They can have either two normal alleles (homozygous for the normal allele), one normal allele and one mutant allele (heterozygous for the normal and mutant alleles), or two mutant alleles (homozygous for the mutant allele). A recessive mutation is one that exhibits a phenotype only when it is homozygous. Therefore, an autosomal recessive mutation is a mutation that occurs on any chromosome except the X or Y and that causes a phenotype only when homozygous. Heterozygotes exhibit a normal phenotype.

Of the diseases discussed in this chapter, many are autosomal recessive. For example, phenylketonuria, albinism, Tay–Sachs disease, and cystic fibrosis are autosomal recessive diseases. Heterozygotes for the disease allele are normal, but homozygotes with the disease allele are affected. For phenylketonuria and albinism, homozygotes are affected because they lack a required enzymatic function. In these cases, heterozygotes have a normal phenotype because their single normal allele provides sufficient enzyme function.

Parents contribute one of their two autosomes to their gametes, so that each offspring of a couple receives an autosome from each parent. If in a particular conception each of two heterozygous parents contributes a chromosome with the normal allele, the offspring will be homozygous for the normal allele and be normal. If in a particular conception one of the two heterozygous parents contributes a chromosome with the normal allele and the other parent contributes a chromosome with the mutant allele, the offspring will be heterozygous but be normal. If in a particular conception each parent contributes the chromosome with the mutant allele, the offspring will be homozygous for the mutant allele and develop the disease. Therefore, heterozygous parents can have both normal and affected children. Since each conception is independent, two heterozygous parents can have all normal, all affected, or any mix of normal and affected children.

4.7 A genetic disease such as sickle-cell anemia is caused by a change in DNA that alters levels or forms of one or more gene products. In turn, the altered forms or levels of gene products result in changes in cellular functions, which in turn cause a disease state. The examples given in this chapter demonstrate that genetic diseases can be associated with mutations in single genes that affect their protein products. For example, sickle-cell anemia is caused by mutations in the gene for β-globin. Mutations lead to amino acid substitutions that cause the β-globin polypeptide to fold incorrectly. This in turn leads to sickled red blood cells and anemia. The environment can have a significant effect on disease severity and many genetic diseases are treatable. For example, PKU can be treated by altering the diet. Unlike diseases caused by an invading microorganism or other external agent that are subject to the defenses of the human immune system and that generally have short-lived clinical symptoms and treatments, genetic diseases are caused by heritable changes in DNA that are associated with chronically altered levels or forms of one or more gene products.

4.10 Wild-type T4 will produce progeny phages at all three temperatures. Consider what will happen under each model if *E. coli* is infected with a doubly mutant phage (one step is cold sensitive, one step is heat sensitive), and the growth temperature is shifted between 17°C and 42°C during phage growth. Suppose model 1 is correct and cells infected with the double mutant are first incubated at 17°C and then shifted to 42°C. Progeny phages will be produced and the cells will lyse, as each step of the pathway can be completed in the correct order. In model 1, the first step, *A* to *B*, is controlled by a gene whose product is heat sensitive but not cold sensitive. At 17°C, the enzyme works, and *A* will be converted to *B*. While phage are at 17°C, the second, cold-sensitive step of the pathway prevents the production of mature phage. However, when the temperature is shifted to 42°C, the accumulated *B* product can be used to make mature phage so that lysis will occur. Under model 1, a temperature shift performed in the reverse direction does not allow for growth. When *E. coli* cells are infected with a doubly mutant phage and placed at 42°C, the heat-sensitive first step precludes the accumulation of *B*. When the culture is shifted to 17°C, *B* can accumulate, but now the second step cannot occur, so no progeny phage can be produced. Therefore, if model 1 is correct, lysis will be seen only in a temperature shift from 17°C to 42°C. If model 2 is correct, growth will be seen only in a temperature shift from 42°C to 17°C. Hence, the correct model can be deduced by performing a temperature shift experiment in each direction and observing which direction allows progeny phage to be produced.

4.11 A strain blocked at a later step in the pathway accumulates a metabolic intermediate that can "feed" a strain blocked at an earlier step. It secretes the metabolic intermediate into the medium, thereby providing a nutrient to bypass the earlier block of another strain. Consequently, a strain that feeds all others (but itself) is blocked in the last step of the pathway, while a strain that feeds no others is blocked in the first step of the pathway. Mutant *a* is blocked in the earliest step in the pathway because it cannot feed any of the others. Mutant *c* is next because it can supply the substance *a* needs but cannot feed *b* or *d*. Mutant *d* is next, and mutant *b* is last in the pathway because it can feed all the others. The pathway is $a \rightarrow c \rightarrow d \rightarrow b$.

4.12 One approach to this problem is to try to sequentially fit the data to each pathway, as if each were correct. Check where each mutant *could* be blocked (remember, each mutant carries only one mutation), whether the mutant would be able to grow if supplemented with the single nutrient that is listed, and whether the mutant would not be able to grow if supplemented with the "no growth" intermediate. It will not be possible to fit the data for mutant 4 to pathway a, the data for mutants 1 and 4 to pathway b, or data for mutants 3 and 4 to pathway c. The data for all mutants can be fit only to pathway d. Thus, d must be the correct pathway.

A second approach to this problem is to realize that in any linear segment of a biochemical pathway (a segment without a branch), a block early in the segment can be circumvented by any metabolites that normally appear later in the same segment. Consequently, if two (or more) intermediates can support growth of a mutant, they normally are made after the blocked step in the same linear segment of a pathway. From the data given, compounds D and E both circumvent the single block in mutant 4. This means that compounds D and E lie after the block in mutant 4 on a linear segment of the metabolic pathway. The only pathway where D and E lie in an unbranched linear segment is pathway d. Mutant 4 could be blocked between A and E in this pathway. Mutant 4 cannot be fit to a single block in any of the other pathways that are shown, so the correct pathway is pathway d.

4.13 If the enzyme that catalyzes the d → e reaction is missing, the mutant strain should accumulate d and be able to grow on minimal medium to which e is added. In addition, it should not be able to grow on minimal medium or on minimal medium to which X, c, or d is added but should grow if Y is added. Therefore, plate the strain on these media and test which allow for growth of the mutant strain and which intermediate is accumulated if the strain is plated on minimal medium.

4.14 a. In each of these diseases, the lack of an enzymatic step leads to the toxic accumulation of a precursor or its byproduct. The proposed treatments are ineffective because they do not prevent the accumulation of the toxic precursor. Administering purines will worsen Lesch-Nyhan syndrome, since the disease results from purine accumulation.

b. The loss of 25-hydroxycholecalciferal 1 hydroxylase should lead to increased serum levels of 25-hydroxycholecalciferol, the precursor it acts upon. Since administration of the end product of the reaction, 1,25-dihydroxycholecalciferol (vitamin D), is an effective treatment, this disease is unlike those in (a). It appears that this disease is caused by the loss of the reaction's end product and not the accumulation of its precursor.

4.15 a. Since normal parents have affected offspring, the disease appears to be recessive. However, since patients with 50

percent of GSS activity have a mild form of the disease, individuals may show mild symptoms if they are heterozygous (mutant/+) for a mutation that eliminates GSS activity. In a population, individuals having the disease may not all show identical symptoms, and some may have a more severe disease form than others. The severity of the disease in an individual will depend on the nature of the person's GSS mutation and, possibly, whether they are heterozygous or homozygous for the disease allele. The alleles discussed here appear to be recessive.

b. Patient 1, with 9 percent of normal GSS activity, has a more severe form of the disease, whereas patient 2, with 50 percent of normal GSS activity, has a less severe form of the disease. Thus, increased disease severity is associated with less GSS enzyme activity.

c. The two different amino acid substitutions may disrupt different regions of the structure of the enzyme (consider the effect of different amino acid substitutions on the function of hemoglobin, discussed in the text). As amino acids vary in their polarity and charge, different amino acid substitutions within the same structural region could have different chemical effects on protein structure. This, too, could lead to different levels of enzymatic function. (For a discussion of the chemical differences between amino acids, see Chapter 6.)

d. By analogy with the disease PKU discussed in the text, 5-oxoproline is produced only when a precursor to glutathione accumulates in large amounts due to a block in a biosynthetic pathway. When GSS levels are 9 percent of normal, this occurs. When GSS levels are 50 percent of normal, there is sufficient GSS enzyme activity to partially complete the pathway and prevent high levels of 5-oxoproline.

e. The mutations are allelic (in the same gene), since both the severe and the mild forms of the disease are associated with alterations in the same polypeptide that is a component of the GSS enzyme. (Note that while the data in this problem suggest that the GSS enzyme is composed of a single polypeptide, they do not exclude the possibility that GSS has multiple polypeptide subunits encoded by different genes.)

f. If GSS is normally found in fetal fibroblasts, one could, in principle, measure GSS activity in fibroblasts obtained via amniocentesis. The GSS enzyme level in cells from at-risk fetuses could be compared to that in normal control samples to predict disease due to inadequate GSS levels. Some variation in GSS level might be seen, depending on the allele(s) present. Since more than one mutation is present in the population, it is important to devise a functional test that assesses GSS activity, rather than a test that identifies a single mutant allele.

4.19 a. From Figure 4.12, Hb-Norfolk affects the α-chain whereas Hb-S affects the β-chain of hemoglobin. Since each chain is encoded by a separate gene, there remains one normal allele at the genes for each of α- and β-chains in a double heterozygote. Thus, some normal hemoglobin molecules form and double heterozygotes do not have severe anemia. However, unlike double heterozygotes for two different, completely recessive mutations that lie in one biochemical pathway, these heterozygotes exhibit an abnormal phenotype. This is because some mutations in the α- and β-chains of hemoglobin show partial dominance. In particular, Hb-S/+ heterozygotes show symptoms of anemia if there is a sharp drop in oxygen tension, so these double heterozygotes exhibit mild anemia.

b. Both Hb-C and Hb-S affect the sixth amino acid of the β-chain. The Hb-C mutation alters the normal glutamate to lysine, while the Hb-S mutation alters it to valine. Since both mutations affect the β-chain, no normal hemoglobin molecules are present. According to the text, only one type of β-chain is found in any one hemoglobin molecule. Therefore, an Hb-C/Hb-S heterozygote has two types of hemoglobin: those with Hb-C β-chains and those with Hb-S β-chains.

4.23 a. In Caucasians, PKU occurs in about 1 in 12,000 births while CF occurs in about 1 in 2,000 births. In African-Americans and Asians, the CF frequency is 1 in 17,000 and 1 in 90,000, respectively. Given their relative frequencies in Caucasians, the choice of which diseases have mandated testing is not based on disease frequency alone.

b. The Guthrie test is a simple clinical screen for phenylalanine in the blood. A drop of blood is placed on a filter paper disc and the disc is then placed on a solid culture medium containing *B. subtilis* and β-2-thienylalanine. The β-2-thienylalanine normally inhibits the growth of *B. subtilis*, but the presence of phenylalanine prevents this inhibition. Therefore, the amount of growth of *B. subtilis* is a measure of the amount of phenylalanine in the blood. The test provides an easy, relatively inexpensive means to reliably quantify blood phenylalanine levels, making it an effective preliminary screen for PKU in newborn infants.

c. Mandated diagnostic testing requires a highly accurate test—one that has very low false-positive and false-negative rates—as misdiagnosis of a genetic disease in a genetically normal individual has significant potential for emotional distress in the family of the misdiagnosed child and misdiagnosis of an affected individual as normal may delay necessary therapeutic treatment. A set of mutations with a range of different disease phenotypes may make it difficult to employ a single easy-to-use test. For example, different mutations may make it impossible to use just one DNA-based test and non-DNA-based tests that are effective at diagnosing severe disease phenotypes may not be equally effective at diagnosing mild disease forms, as they may give results that overlap with those from normal individuals.

d. Testing for PKU in newborns is essential for early intervention to prevent the toxic accumulation of phenylketones and the resulting neurological damage in early infancy. Unless it is documented that intervention in newborns is critical for CF disease management, testing for CF in newborns is less critical. Testing is warranted to confirm a diagnosis when severe CF symptoms are apparent in a newborn.

4.25 a. Tests can be DNA based and determine the genotype of a parent or fetus or they can be biochemically based and determine some aspect of the individual's physiology. For example, the Guthrie test determines the relative amount of phenylalanine in a drop of blood to assess whether an individual has PKU; enzyme assays can determine whether a person has a complete or partial enzyme deficiency; gel electrophoresis can determine whether a person has an altered α- or β-globin that might be associated with anemia. DNA-based tests assess the presence or absence of a specific mutation, and are normally employed only when there is already suspicion that an individual may carry that mutation (e.g., the couple has already had an affected offspring). Biochemical tests typically focus on assessing gene function, so they are often used in screens. However, they may not provide detailed information about which gene or biochemical step is affected and

require that the biochemical activity be present in the tested cell population, such as cells obtained from an amniocentesis.

b. Lesch–Nyhan syndrome is caused by an X-linked mutation and the affected son's X came from his mother, so use a DNA-based test to evaluate whether Mrs. Chávez is a carrier for the same mutation as her son. Tay–Sachs disease is caused by an autosomal recessive mutation and each parent contributed one of their autosomes to the affected son, so use a DNA-based test to evaluate whether each parent is heterozygous for an allele present in the affected son. If any of the tested parents do not carry the mutation present in their affected son, that son has a new mutation.

c. Each conception has produced a female carrier. Mr. and Mrs. Lieberman should be relieved, as their daughter is heterozygous for a recessive disease and will not develop Tay–Sachs disease. However, Mr. and Mrs. Chávezes' concerns have not been resolved as the female carrier of an X-linked recessive disorder may be symptomatic due to random X-chromosome inactivation. If the X chromosome bearing the normal allele is inactivated in most cells, the mutant allele will be expressed and their daughter will develop Lesch–Nyhan syndrome. If the X chromosome bearing the mutant allele is inactivated in most cells, the normal allele will be expressed and their daughter will be normal. In this case, if the mutation is not new, she would be like her carrier mother. The fetal cells tested from amniocentesis or chorionic villus sampling may not be representative of the X chromosome inactivation pattern of all fetal cells, so it is not possible to infer from these tests alone whether Mr. and Mrs. Chávezes' daughter will develop Lesch–Nyhan disease.

Chapter 5 Gene Expression: Transcription

5.1 While both DNA and RNA are composed of linear polymers of nucleotides, their bases and sugars differ. DNA contains deoxyribose and thymine, while RNA contains ribose and uracil. Their structures also differ. DNA is frequently double stranded, while RNA is usually single stranded. Single-stranded RNAs are capable of forming stable, functional, and complex stem–loop structures, such as those seen in tRNAs. Double-stranded DNA is wound in a double helix and packaged by proteins into chromosomes, either as a nucleoid body in prokaryotes or within the eukaryotic nucleus. After being transcribed from DNA, RNA can be exported into the cytoplasm. If it is mRNA, it can be bound by ribosomes and translated. Eukaryotic RNAs are highly processed before being transported out of the nucleus. DNA functions as a storage molecule, while RNA functions either as a messenger (mRNA carries information to the ribosome), or in the processes of translation (rRNA functions as part of the ribosome, tRNA brings amino acids to the ribosome), or in eukaryotic RNA processing (snRNA functions within the spliceosome).

Both DNA polymerases and RNA polymerases catalyze the synthesis of nucleic acids in the $5' \rightarrow 3'$ direction. Both use a DNA template and synthesize a nucleic acid polynucleotide that is complementary to the template. However, DNA polymerases require a $3'$-OH to add onto, while RNA polymerases do not. That is, RNA polymerases can initiate chains without primers, while DNA polymerases cannot. Furthermore, RNA polymerases usually require specific base-pair sequences as signals to initiate transcription.

5.3 Both eukaryotic and *E. coli* RNA polymerases transcribe RNA in a $5' \rightarrow 3'$ direction, using a $3' \rightarrow 5'$ DNA template strand. There are many differences between the enzymes, however. In *E. coli*, a single RNA polymerase core enzyme is used to transcribe genes. In eukaryotes, there are three types of RNA polymerase molecules: RNA polymerase I, II, and III. RNA polymerase I synthesizes 28S, 18S, and 5.8S rRNA and is found in the nucleolus. RNA polymerase II synthesizes hnRNA, mRNA, and some snRNAs and is nuclear. RNA polymerase III synthesizes tRNA, 5S rRNA, and some snRNAs and also is nuclear.

Each RNA polymerase uses a unique mechanism to identify those promoters at which it initiates transcription. In prokaryotes such as *E. coli*, a sigma factor provides specificity to the sites bound by the four-polypeptide core enzyme, so that it binds to promoter sequences. The holoenzyme loosely binds a sequence about 35 bp before transcription initiation (the −35 region), changes configuration, and then tightly binds a region about 10 bp before transcription initiation (the −10 region) and melts about 17 bp of DNA around this region. The two-step binding to the promoter orients the polymerase on the DNA and facilitates transcription initiation in the 5′→3′ direction. After about eight or nine bases are formed in a new transcript, sigma factor dissociates from the holoenzyme, and the core enzyme completes the transcription process. Although the principles by which eukaryotic RNA polymerases bind their promoters are similar in that they use a set of ancillary protein factors—transcription factors—the details are quite different. In eukaryotes, each of the three types of RNA polymerases recognizes a different set of promoters by using a polymerase-specific set of transcription factors, and the mechanisms of interaction are different.

5.7 a. b. There are multiple 5′-AG-3′ sequences in each strand, and transcription may proceed in either direction. Determine the correct initiation site by locating the −10 and −35 consensus sequences recognized by RNA polymerase and σ⁷⁰. Good −35 (TTGACA) and −10 (TATAAT) consensus sequences are found on the top strand, starting at the 8th and 32nd bases from the 5′ end, respectively, indicating that the initiation site is the 5′-AG-3′ starting at the 44th base from the 5′ end of that strand.

  c. Transcription proceeds from left to right in this example.

  d. the bottom (3′→5′) strand

  e. the top (5′→3′) strand

5.9 a. *E. coli* promoters vary with the type of sigma factor that is used to recognize them. More than four types of promoters exist, each having different recognition sequences. Most promoters have −35 and −10 sequences that are recognized by σ⁷⁰. Other promoters have consensus sequences that are recognized by different sigma factors, which are used to transcribe genes needed under altered environmental conditions, such as heat shock and stress (σ³²), limiting nitrogen (σ⁵⁴), or when cells are infected by phage T4 (σ²³).

  b. Although there is one core RNA polymerase enzyme, different RNA polymerase holoenzymes are formed from different sigma factors. Promoter recognition is determined by the sigma factor.

  c. Utilizing different sigma factors allows for a quick response to altered environmental conditions (for example, heat shock, low N₂, phage infection) by the coordinated production of a set of newly required gene products.

5.10 RNA polymerase I transcribes the major rRNA genes that code for 18S, 5.8S, and 28S rRNAs; RNA polymerase II transcribes the protein-coding genes to produce mRNA molecules and some snRNAs; RNA polymerase III transcribes the 5S rRNA genes, the tRNA genes, and some small nuclear RNAs.

In the cell, the 18S, 5.8S, 28S, and 5S rRNAs are structural and functional components of the ribosome, which functions during translation. The mRNAs are translated to produce proteins. The tRNAs bring amino acids to the ribosome to donate to the growing polypeptide chain during protein synthesis. Small nuclear RNAs function in nuclear processes such as RNA splicing and processing.

5.18 In eukaryotes, 18S, 28S, and 5.8S rRNA genes are transcribed from the rDNA into a single pre-rRNA molecule, which is then processed by removing the internal and external spacer sequences, leading to the production of the mature rRNAs. (See Figure 5.19 in text.) The eukaryotic 5S rRNA is transcribed separately to produce mature rRNA molecules that need no further processing. Eukaryotic 5S rRNAs are imported into the nucleolus, where they are assembled with the other mature rRNAs and the ribosomal protein subunits to produce functional ribosomal subunits. Some *Tetrahymena* pre-rRNAs have self-splicing introns (group I) in the 28S rRNA. Remember, the removal of the intron in the 28S rRNA via self-splicing is a process that is separate from the cleavage of spacer sequences from the mature rRNAs. (See Figure 5.14 in text.)

Protein-coding eukaryotic transcripts synthesized by RNA polymerase II are extensively processed. A 5′-5′ bonded m⁷Gppp cap is added to the 5′ end of the nascent transcript when the chain is 20–30 nucleotides long. A poly(A) tail of variable length can be added to the 3′ end of the transcript, and spliceosomes remove intronic sequences. The site of poly(A) addition, as well as splice-site selection, can be regulated, so that more than one alternatively processed mature mRNA is sometimes produced from a single precursor mRNA transcript. Some mRNAs are posttranscriptionally edited: Nucleotides are inserted, deleted, or converted from one base to another.

5.21 A recessive lethal is a mutation that causes death when it is homozygous—that is, when only mutant alleles are present. Heterozygotes for such mutations can be viable. Recessive lethal mutations result in death because some essential function is lacking. Neither copy of the gene functions, so the organism dies.

  a. Deletion of the U1 genes will be recessive lethal, as U1 snRNA is essential for the identification of the 5′ splice site in RNA splicing. Incorrect splicing would lead to nonfunctional gene products for many genes, a nonviable situation.

  b. This mutation would prevent U1 from base-pairing with 5′ splice sites and thus, by the same reasoning as in part a, would be recessive lethal.

  c. If a deletion within intron 2 did not affect a region important for its splicing (for example, the branch point or the regions near the 5′ or 3′ splice sites), it would have no effect on the mature mRNA produced. Consequently, such a mutation would lack a phenotype if it were homozygous. However, if the splicing of intron 2 were affected and the mRNA altered, such a mutation, if homozygous, could result in the production of only nonfunctional hemoglobin, leading to severe anemia and death.

  d. The deletion described would affect the 3′ splice site of intron 2, leading to, at best, aberrant splicing of that intron. If the mutation were homozygous, only a nonfunctional protein would be produced, resulting in severe anemia and death.

5.22 1 (5′ m⁷G cap) + 100 (exon 1) + 50 (exon 2) + 25 (exon 3) + 200 (poly(A) tail) = 376 bases.

5.23 The first two bases of an intron are typically 5′-GU-3′ which are essential for base pairing with the U1 snRNA during spliceosome assembly. A GC-to-TA mutation at the initial base pair of the first intron impairs base pairing with the U1 snRNA, so that the 5′ splice site of the first intron is not identified. This causes the retention of the first intron in the *tub* mRNA and a longer mRNA transcript in *tub/tub* mutants. When the mutant *tub* mRNA is translated, retention of the first intron could result in the introduction of amino acids not present in the *tub*⁺ protein or, if the intron contained a chain termination (stop) codon, premature translation termination and the production of a truncated protein. In either case, a nonfunctional gene product is produced.

The *tub* mutation is recessive because the single *tub*⁺ allele in a *tub/tub*⁺ heterozygote produces mRNAs that are processed normally, and when these are translated, enough normal (*tub*⁺) product is produced to obtain a wild-type phenotype. Only the *tub* allele produces abnormal transcripts. When both copies of the gene are mutated in *tub/tub* homozygotes, no functional product is made and a mutant, obese phenotype results.

5.26 a. 3
   b. 1, 2, 3, 4
   c. 3
   d. 1, 2
   e. 1 (note that some tRNAs and rRNAs have introns)
   f. 4
   g. 1

Chapter 6 Gene Expression: Translation

6.1 b

6.3 a. The hemoglobin will dissociate into its four component subunits, because the heat will destabilize the ionic bonds that stabilize the quaternary structure of the protein. An individual subunit's tertiary structure may also be altered, because the thermal energy of the heat may destabilize the folding of the polypeptide.

   b. The protein will denature. Its tertiary structure is destabilized by heating, so it does not retain a pattern of folding that allows it to be soluble.

   c. The protein will denature when its tertiary structure is destabilized by heating. Unlike albumin, RNase will renature if cooled slowly and will reestablish its normal, functional tertiary structure.

   d. It is likely that the meat proteins will be denatured when their tertiary and quaternary structures are destabilized by the acid conditions of the stomach. Then the primary structure of the polypeptides will be destroyed as they are degraded into their amino-acid components by proteolytic enzymes in the digestive tract.

   e. Valine is a neutral, nonpolar amino acid, unlike the acidic glutamic acid. (See Figure 6.2, p. 113.) A change in the chemical properties of the sixth amino acid may alter the function of the hemoglobin molecule by affecting multiple levels of protein structure. Since it is an amino-acid substitution, it changes the primary structure of the β polypeptide. This change could affect local interactions between amino acids lying near it and, in doing so, alter the secondary structure of the β polypeptide. It could also affect the folding patterns of the protein and alter the tertiary structure of the β polypeptide. Finally, the sixth amino-acid residue is known to be important for interactions between the subunits of hemoglobin molecules (see Figures 4.10–4.12, pp. 77–78), because some mutations which alter that amino acid result in sickle-cell anemia. Thus, this change could alter the quaternary structure of hemoglobin.

6.4 a. The primary structure, or amino-acid sequence, of the prion protein would be unchanged, as the disease is caused, not by a mutation, but rather by misfolding of the prion protein. One misfolded protein can convert a normally folded protein to the misfolded state, so the misfolded proteins are infectious. The secondary structure is affected, as α-helical regions are misfolded into β-pleated sheets. This is likely to lead to an altered tertiary structure that results in the formation of amyloid.

   b. If a genetic mutation led to an amino-acid substitution, it would affect the primary structure of the prion protein. A particular amino-acid substitution in the prion protein could make it more susceptible to being misfolded and lead, as in (a), to changes in its secondary and tertiary structures.

6.6 Multiple lines of evidence support the view that the rRNA component of the ribosome serves more than a structural role. First, the 3′ end of the 16S rRNA is important for identifying where the small ribosomal subunit should bind the mRNA. It has a sequence that is complementary to the Shine–Dalgarno sequence, the ribosome-binding site (RBS) in the mRNA. Mutational analyses demonstrated that the 3′ end of the 16S rRNA must base pair with this mRNA sequence for correct initiation of translation. Second, the 23S rRNA is required for peptidyl transferase activity. Evidence that the peptidyl transferase consists entirely of RNA comes from studies of the atomic structure of the large ribosomal subunit and is supported by experiments which show that peptidyl transferase activity remains following the depletion of the 50S subunit proteins, but not after the digestion of rRNA with ribonuclease T1.

6.9 Figure 6.7, p. 118, and Table 6.1, p. 119, aid in answering this question. The answer is given in the following table:

Amino Acid	tRNAs needed	Rationale
Ile	1	3 codons can use 1 tRNA (wobble)
Phe	1	2 codons can use 1 tRNA (wobble)
Tyr	1	" " " " " " "
His	1	" " " " " " "
Gln	1	" " " " " " "
Asn	1	" " " " " " "
Lys	1	" " " " " " "
Asp	1	" " " " " " "
Glu	1	" " " " " " "
Cys	1	" " " " " " "
Trp	1	1 codon
Met	2	Single codon, but need 1 tRNA for initiation and 1 tRNA for elongation
Val	2	4 codons: 2 can use 1 tRNA (wobble)
Pro	2	" " " " " " "
Thr	2	" " " " " " "
Ala	2	" " " " " " "
Gly	2	" " " " " " "
Leu	3	6 codons: 2 can use 1 tRNA (wobble)
Arg	3	" " " " " " "
Ser	3	" " " " " " "
Total	**32**	**61 codons**

6.10 Since a dipeptide is formed, translation initiation is not affected, nor is the first step of elongation—the binding of a charged tRNA in the A site and the formation of a peptide bond. However, since *only* a dipeptide is formed, it appears that translocation is inhibited.

6.14 A eukaryotic mRNA is modified to contain a 5′ 7-methyl-G cap and a 3′ poly(A) tail. The 5′ cap is required early in translation initiation—it binds to the eIF-4F complex just prior to the binding of a complex of the 40S ribosomal subunit, the initiator Met-tRNA, and other eIF proteins. Transcription initiation is stimulated by the looping of the poly(A) tail close to the 5′ end. This occurs when the poly(A) binding protein (PAB) binds to eIF-4G, which is part of the eIF-4F complex.

6.17 Determine the expected amino acids in each case by calculating the expected frequency of each kind of triplet codon that might be formed and inferring from these what types and frequencies of amino acids would be used during translation.

a. 4 A : 6 C gives $2^3 = 8$ codons—specifically, AAA, AAC, ACC, ACA, CCC, ACA, CAC, and CAA. Since there is 40% A and 60% C,

$$P(\text{AAA}) = 0.4 \times 0.4 \times 0.4 = 0.064, \text{ or } 6.4\% \text{ Lys}$$
$$P(\text{AAC}) = 0.4 \times 0.4 \times 0.6 = 0.096, \text{ or } 9.6\% \text{ Asn}$$
$$P(\text{ACC}) = 0.4 \times 0.6 \times 0.6 = 0.144, \text{ or } 14.4\% \text{ Thr}$$
$$P(\text{ACA}) = 0.4 \times 0.6 \times 0.4 = 0.096, \text{ or } 9.6\% \text{ Thr}$$
$$(24\% \text{ Thr total})$$
$$P(\text{CCC}) = 0.6 \times 0.6 \times 0.6 = 0.216, \text{ or } 21.6\% \text{ Pro}$$
$$P(\text{CCA}) = 0.6 \times 0.6 \times 0.4 = 0.144, \text{ or } 14.4\% \text{ Pro}$$
$$(36\% \text{ Pro total})$$
$$P(\text{CAC}) = 0.6 \times 0.4 \times 0.6 = 0.144, \text{ or } 14.4\% \text{ His}$$
$$P(\text{CAA}) = 0.6 \times 0.4 \times 0.4 = 0.096, \text{ or } 9.6\% \text{ Gln}$$

b. 4 G : 1 C gives $2^3 = 8$ codons—specifically, GGG, GGC, GCG, GCC, CGG, CGC, CCC, and CCG. Since there is 80% G and 20% C,

$$P(\text{GGG}) = 0.8 \times 0.8 \times 0.8 = 0.512, \text{ or } 51.2\% \text{ Gly}$$
$$P(\text{GGC}) = 0.8 \times 0.8 \times 0.2 = 0.128, \text{ or } 12.8\% \text{ Gly}$$
$$(64\% \text{ Gly total})$$
$$P(\text{GCG}) = 0.8 \times 0.2 \times 0.8 = 0.128, \text{ or } 12.8\% \text{ Ala}$$
$$P(\text{GCC}) = 0.8 \times 0.2 \times 0.2 = 0.032, \text{ or } 3.2\% \text{ Ala}$$
$$(16\% \text{ Ala total})$$
$$P(\text{CGG}) = 0.2 \times 0.8 \times 0.8 = 0.128, \text{ or } 12.8\% \text{ Arg}$$
$$P(\text{CGC}) = 0.2 \times 0.8 \times 0.2 = 0.032, \text{ or } 3.2\% \text{ Arg}$$
$$(16\% \text{ Arg total})$$
$$P(\text{CCC}) = 0.2 \times 0.2 \times 0.2 = 0.008, \text{ or } 0.8\% \text{ Pro}$$
$$P(\text{CCG}) = 0.2 \times 0.2 \times 0.8 = 0.032, \text{ or } 3.2\% \text{ Pro}$$
$$(4\% \text{ Pro total})$$

c. 1 A : 3 U : 1 C gives $3^3 = 27$ different possible codons. Of these, one will be UAA, a chain-terminating codon. Since there is 20% A, 60% U, and 20% C, the probability of finding this codon is $0.6 \times 0.2 \times 0.2 = 0.024$, or 2.4%. All of the remaining 26 (97.6%) codons will be sense codons. Proceed in the same manner as in (a) and (b) to determine their frequency, and determine the kinds of amino acids expected. To take the frequency of nonsense codons into account, divide the frequency of obtaining a particular amino acid considering all 27 possible codons by the frequency of obtaining a sense codon. This gives

$$(0.8/0.976)\% = 0.82\% \text{ Lys}$$
$$(3.2/0.976)\% = 3.28\% \text{ Asn}$$
$$(12.0/0.976)\% = 12.3\% \text{ Ile}$$
$$(9.6/0.976)\% = 9.84\% \text{ Tyr}$$
$$(19.2/0.976)\% = 19.67\% \text{ Leu}$$
$$(28.8/0.976)\% = 29.5\% \text{ Phe}$$
$$(4.0/0.976)\% = 4.1\% \text{ Thr}$$
$$(0.8/0.976)\% = 0.82\% \text{ Gln}$$
$$(3.2/0.976)\% = 3.28\% \text{ His}$$
$$(4.0/0.976)\% = 4.1\% \text{ Pro}$$
$$(12.0/0.976)\% = 12.3\% \text{ Ser}$$

It is likely that the chains produced would be relatively short, due to the chain-terminating codon.

d. 1 A : 1 U : 1 G : 1 C will produce $4^3 = 64$ different codons, all possible in the genetic code. The probability of each codon is $\frac{1}{64}$, so there will be a $\frac{3}{64}$ chance of a codon being chain terminating. With those exceptions, the relative proportion of amino-acid incorporation is dependent directly on the codon degeneracy for each amino acid. Inspecting the table of the genetic code in Figure 6.7 and taking the frequency of nonsense codons into account yields the following table:

Amino Acid	Number of Codons	Frequency
Trp	1	$1/61 = 1.64\%$
Met	1	1.64%
Phe	2	$2/61 = 3.28\%$
Try	2	3.28%
His	2	3.28%
Gln	2	3.28%
Asn	2	3.28%
Lys	2	3.28%
Asp	2	3.28%
Glu	2	3.28%
Cys	2	3.28%
Ile	3	$3/61 = 4.92\%$
Val	4	$4/61 = 6.56\%$
Pro	4	6.56%
Thr	4	6.56%
Ala	4	6.56%
Gly	4	6.56%
Leu	6	$6/61 = 9.84\%$
Arg	6	9.84%
Ser	6	9.84%

6.18 The minimum word size must be able to uniquely designate 20 amino acids, so the number of combinations must be at least 20. The following table gives the number of combinations as a function of the word size?

Word Size	Number of Combinations
a. 5	$2^5 = 32$
b. 3	$3^3 = 27$
c. 2	$5^2 = 25$

6.20 a. 3'-TAC AAA ATA AAA ATA AAA ATA AAA ATA...-5' (The first fMet or Met is removed following translation of the mRNA.)

b. 5'-ATG TTT TAT TTT TAT TTT TAT TTT TAT...-3'

c. 3'-AAA-5' is the anticodon for Phe, and 3'-AUA-5' is the anticodon for Tyr

6.22 In population A, the codons that can be produced encode Lys (AAA, AAG), Arg (AGG, AGA), Glu (GAG, GAA), and Gly (GGA, GGG). All of these are sense codons, so long polypeptide chains containing these amino acids will be synthesized. In population B, the codons that can be produced encode Lys (AAA), Asn (AAU), Ile (AUA, AUU), Tyr (UAU), Leu (UUA), Phe (UUU), and stop (UAA). The frequency of the stop codon will be ($\frac{1}{4} \times \frac{3}{4} \times \frac{3}{4}$) = $\frac{9}{64}$ = 0.14, or 14%. Thus, the polypeptides formed in population B will, on average, be shorter than those formed in population A. If a stop codon appears about 14 percent of the time, there will be, on average, 1/0.14 = 7.14 codons from one stop codon to the next. On average, six sense codons will lie in between the stop codons, so polypeptides will be synthesized that are about six amino acids long.

6.24 The anticodon 5'-GAU-3' recognizes the codon 5'-AUC-3', which encodes Ile. The mutant tRNA anticodon 5'-CAU-3' would recognize the codon 5'-AUG-3', which normally encodes Met. The mutant tRNA would therefore compete with tRNA.Met for the recognition of the 5'-AUG-3' codon, and if successful, insert Ile into a protein where Met should be. Since a special tRNA.Met is used for initiation, only AUG codons other than the initiation AUG will be affected. Thus, this protein will have four different N-terminal sequences, depending on which tRNA occupies the A site in the ribosome when the codon AUG is present there:

```
Met-Val-Ser-Ser-Pro-Ile-Gly-Ala-Ala-Ile-Ser
Met-Val-Ser-Ser-Pro-Met-Gly-Ala-Ala-Ile-Ser
Met-Val-Ser-Ser-Pro-Ile-Gly-Ala-Ala-Met-Ser
Met-Val-Ser-Ser-Pro-Met-Gly-Ala-Ala-Met-Ser
```

6.26 Rewriting the sequences to readily visualize the codons shows that mutants *a, b, c, d,* and *f* are point mutations in which one base has been substituted for another and that mutant *e* is a deletion of one base that causes a frameshift mutation. The following proteins are produced (alterations to the normal sequence are underlined):

```
Normal: Met-Phe-Ser-Asn-Tyr-...-Met-Gly-Trp-Val
     a: Met-Phe-Ser-Asn
     b: Starts at later AUG to give Met-Gly-Trp-Val
     c: Met-Phe-Ser-Asn-Tyr-...-Met-Gly-Trp-Val
     d: Met-Phe-Ser-Lys-Tyr-...-Met-Gly-Trp-Val
     e: Met-Phe-Ser-Asn-Ser-...-Trp-Gly-Gly-Cys...
        (no stop codon, protein continues)
     f: Met-Phe-Ser-Asn-Tyr-...-Met-Gly-Trp-Val-Trp...
        (no stop codon, protein continues)
```

6.30 a. If the primary mRNA for this gene is 250 kb, it must be substantially processed by RNA splicing (removing introns) and polyadenylation to a smaller mature mRNA.

b. A 1,480-amino-acid protein requires 1,480 × 3 = 4,440 bases of protein-coding sequence. This leaves 6,500 − 4,440 = 2,060 bases of 5' untranslated leader and 3' untranslated trailer sequence in the mature mRNA—about 32%.

c. The ΔF508 mutation could be caused by a DNA deletion for the three base pairs encoding the mRNA codon for phenylalanine. This codon is UUY (Y = U or C), and the DNA sequence of the nontemplate strand is 5'-TTY-3'. The segment of DNA containing these bases would be deleted in the appropriate region of the gene.

d. If positioned at random and solely within a gene's coding region (that is, not in 3' or 5' untranslated sequences or in intronic sequences), a deletion of three base pairs results either in an mRNA missing a single codon or an mRNA missing bases from two adjacent codons. If three of the six bases from two adjacent codons were deleted, the remaining three bases would form a single codon. In this case, an incorrect amino acid might be inserted into the polypeptide at the site of the left codon, and the amino acid encoded by the right codon would be deleted. If the 3' base of the left codon were deleted, it would be replaced by the 3' base of the right codon. Since the code is degenerate and wobble occurs in the 3' base, this type of deletion might not alter the amino acid specified by the left codon. The adjacent amino acid would still be deleted, however.

6.31 Some genes can inhibit the activity of others. An increase in an enzyme's activity will be seen if actinomycin D blocks the transcription of a gene that codes for an inhibitor of the enzyme's activity.

Chapter 7 DNA Mutation, DNA Repair, and Transposable Elements

7.1 b

7.3 e. none of the above (all are classes of mutation)

7.4 c. The key to this answer is the word "usually." The other choices might apply rarely, but not usually.

7.6 a. If the normal codon is 5'-CUG-3', the anticodon of the normal tRNA is 5'-CAG-3'. If a mutant tRNA recognizes 5'-GUG-3', it must have an anticodon that is 5'-CAC-3'. The mutational event was a CG-to-GC transversion.

b. Presumably Leu

c. Val

d. Leu

7.8 Acridine is an intercalating agent and so can be expected to induce frameshift mutations 5BU is incorporated into DNA in place of T. During DNA replication, it is likely to be read as C by DNA polymerase because of a keto-to-enol shift. This results in point mutations, usually TA-to-CG transitions. Considering these expectations, we find that *lacZ-1* would probably result in a completely altered amino-acid sequence after some point, although it might be truncated (due to the introduction—out of frame—of a nonsense codon). In either case, it would be very likely to have a different molecular weight and charge and so migrate differently than the wild-type would during gel electrophoresis (see Figure 4.9, p. 77). *lacZ-2* is likely to contain a single amino-acid difference, due to a missense mutation, although it, too, could contain a nonsense codon. A missense mutation might lead to the protein's having a different charge, while a nonsense codon would lead to a truncated protein that would have a lower molecular weight. Both would migrate differently during gel electrophoresis.

7.9 a. Six

b. Three, since the UGG codon would be replaced by UAG, a nonsense (chain termination) codon, to give 5'-AUG-ACC-CAU-UAG-...-3'.

7.11 a. UAG: Gln (CAG), Lys (AAG), Glu (GAG), Leu (UUG), Ser (UCG), Trp (UGG), Tyr (UAC, UAU), chain terminating (UAA)

b. UAA: Lys (AAA), Gln (CAA), Glu (GAA), Leu (UUA), Ser (UCA), Tyr (UAC, UAU), chain terminating (UGA, UAG)

c. UGA: Arg (AGA, CGA), Gly (GGA), Leu (UUA), Ser (UCA), Cys (UGC, UGU), Trp (UGG), chain terminating (UAA)

7.13

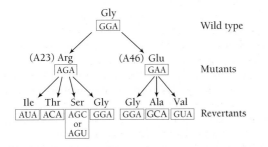

7.18 a. In its normal state, 5-bromouracil is a T analog that can base pair with A. In its rare state, it resembles C and can base pair with G. It will induce an AT-to-GC transition as follows:

$$\text{A-T} \rightarrow \text{A-5BU} \rightarrow \text{G-5BU} \rightarrow \text{G-C}$$

b. Nitrous acid can deaminate C to U, resulting in a CG-to-TA transition.

7.21 The absence of dGTP leads to a block in polymerization after the first two bases:

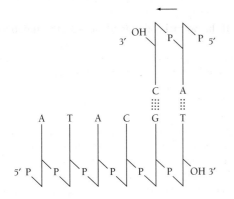

7.23 Pretreatment of the template with HNO₂ deaminates G to X, C to U, and A to H. X will still pair with C, but U pairs with A and H pairs with C, causing "mutations" in the newly synthesized strand:

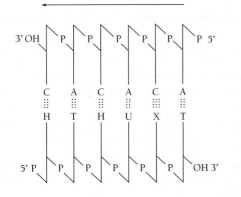

7.27 Nitrous acid deaminates C to make it U. U pairs with A, so treatment with nitrous acid leads to CG-to-TA transitions. Analyze how this treatment would affect the codons of this protein. Use N to represent any nucleotide and Y to represent a pyrimidine (U or C). Then the codons for Pro are CCN, the (relevant) codons for Ser are UCN, the codons for Leu are CUN, and the codons for Phe are UUY. (Nucleotides are written in the 5′→3′ direction, unless specifically noted otherwise.)

The codon CCN for Pro would be represented by CCN in the nontemplate DNA strand. Deamination of the 5′-C would lead to a nontemplate strand of UCN and a template strand of 3′-AGN-5′. This would produce a UCN codon encoding Ser. Deamination of the middle C would lead to a nontemplate strand of CUN and a template strand of 3′-GAN-5′, producing a CUN codon encoding Leu.

Further treatment of either mutant would result in deamination of the remaining C and a template strand of 3′-AAN-5′. This would result in a UUN codon. Since we are told that Phe is obtained, N must be C or U, and the template strand must be 3′-AAA-5′ or 3′-AAG-5′.

To explain why further treatment with nitrous acid has no effect, observe that nitrous acid acts via deamination and that T has no amine group. If the template strand were 3′-AAA-5′, the nontemplate strand would have been TTT. Since T cannot be deaminated, nitrous acid will have no effect on the nontemplate strand.

7.28 Use the revertant frequencies under "none" to estimate the spontaneous reversion frequency. *ara-1*: BU and AP, but not HA or a frameshift, can revert *ara-1*. Both BU and AP cause CG-to-TA and TA-to-CG transitions, while HA causes only CG-to-TA transitions. If HA cannot revert *ara-1*, it must require a TA-to-CG transition to be reverted and must be caused by a CG-to-TA transition. *ara-2*: BU, AP, and HA, but not a frameshift, can revert *ara-2*. Since HA causes only CG-to-TA transitions, *ara-2* must have been caused by a TA-to-CG transition. *ara-3*: By the same logic as that for *ara-2*, *ara-3* must have been caused by a TA-to-CG transition. Provided that this is a representative sample, mutagen X appears to cause both TA-to-CG and CG-to-TA transitions. It does not appear to cause frameshift mutations.

7.36 Since introns are spliced out only at the RNA level, a transposition event that results in the loss of an intron (such as that used by Ty elements) indicates that the transposition occurred via an RNA intermediate. Thus, A is likely to move via an RNA intermediate. The lack of intron removal during B transposition suggests that it uses a DNA→DNA transposition mechanism (either conservative or replicative, or some other mechanism).

Chapter 8 Recombinant DNA Technology

8.2 Examples of methods that utilize the hydrogen bonding in complementary base pairing include (1) the ligation of a DNA fragment with sticky ends into a site with complementary sticky ends in a cloning vector, (2) annealing of a labeled nucleic acid probe with a single-stranded DNA or RNA molecule attached to a membrane during screens of libraries and Southern and northern hybridizations, (3) annealing of an oligo(dT) primer to a poly(A) tail during the synthesis of cDNA from mRNA, and (4) annealing of a primer to a template during PCR. In each case, base pairing allows for nucleotides to interact in a sequence specific manner essential for the procedure's success. For example, detection of membrane-bound single-stranded nucleic acid sequences in hybridization of probes to Southern blots requires complementary base pairing between a bound strand of DNA and complementary sequences in the probe.

8.4 The average length of the fragments produced indicates how often, on average, the restriction site appears. If the DNA is composed of equal amounts of A, T, C, and G, the chance of finding one specific base pair (A-T, T-A, G-C, or C-G) at a particular site is $1/4$. The chance of finding two specific base pairs at a site is $(1/4)^2$. In general, the chance of finding n specific base pairs at a site is $(1/4)^n$. Here, $1/4{,}096 = (1/4)^6$, so the enzyme recognizes a six base-pair site.

8.7 If the enzyme is not inactivated, the restriction enzyme produced by the *hsdR* gene will cleave any DNA transformed into *E. coli* with the appropriate recognition sequence. This will make it impossible to clone DNA with the recognition sequence that is not methylated at the A in this sequence.

8.12 Use an expression vector. Expression vectors have the signals necessary for DNA inserts to be transcribed and for these transcripts to be translated. In prokaryotes, the vector should have a prokaryotic promoter sequence upstream of the site where the cDNA is inserted, and possibly, a terminator sequence downstream of this site. In eukaryotes, a eukaryotic promoter would be needed, and a poly(A) site should be provided downstream of the site where the cDNA is inserted. If the cDNA lacked a start codon, a start AUG codon embedded in a Kozak consensus sequence would be needed upstream of the site where the cDNA is inserted so that the transcript can be efficiently translated. In the event that the cDNA lacked a start codon, care must be taken during the design of the cloning steps to insure that the open reading frame (ORF) of the cDNA is in the same reading frame with the start codon provided by the vector.

8.13 It would be preferable to use cDNA. Human genomic DNA contains introns, while cDNA synthesized from cytoplasmic poly(A)+ mRNA does not. Prokaryotes do not process eukaryotic precursor mRNAs having intron sequences, so genomic clones will not give appropriate translation products. Since cDNA is a complementary copy of a functional mRNA molecule, the mRNA transcript will be functional, and when translated human (pro-)insulin will be synthesized.

8.14 If genomic DNA had been used, there could be concerns that an intron in the genomic DNA was not removed, since *E. coli* does not process RNAs as eukaryotic cells do. However, the cDNA is a copy of a mature mRNA, so this should not be a concern. There are other potential concerns however. First, depending on the nature of the sequence inserted, a fusion protein with β-galactosidase may have been produced, and not just human insulin. In pUC19, the polylinker is within the β-galactosidase gene. Sequences inserted into the polylinker, if inserted in the correct reading frame (the same one as used for β-galactosidase), will be translated into a β-galactosidase-fusion protein. If this was acceptable, it would be important to insure that only the ORF (the open reading frame) of the insulin gene is properly inserted into the polylinker of the pUC19 vector. In order for the insulin ORF to be properly inserted, it must be inserted in the correct reading frame, so that premature termination of translation does not occur, and the correct polypeptide is produced. One could not use a complete copy of the human mRNA transcript for insulin. If transcribed, it would have features of eukaryotic transcripts but not features required for prokaryotic translation. Indeed, some of its 5′ UTR and 3′ UTR sequences may interfere with prokaryotic transcription and translation. For example, it will lack a Shine-Dalgarno sequence to

specify where translation should initiate and identify the first AUG codon. In the pUC19 vector, a Shine-Dalgarno sequence is supplied after the promoter for the *lacZ* gene, since without an insert in the polylinker, β-galactosidase is produced. However, the cDNA may have 5′ UTR sequences which interfere with translation initiation in prokaryotes, or which contain stop codons, terminating translation of the β-galactosidase fusion protein. Second, the cDNA may encode a protein that is post-translationally processed to become human insulin. The protein produced in *E. coli* may not be processed.

8.16 A genomic library made in a plasmid vector is a collection of plasmids that have different yeast genomic DNA sequences into them. Like two volumes of a book series, two plasmids in the library will have identical vector sequences, but different yeast DNA inserts. Such a library is made as follows:

1. Isolate high-molecular-weight yeast genomic DNA by isolating nuclei, lysing them, and gently purifying their DNA.

2. Cleave the DNA into fragments that are 5–10 kb, an appropriate size for insertion into a plasmid vector. This can be done by cleaving the DNA with *Sau*3A for a limited time (i.e., performing a *partial* digest), and then selecting fragments of an appropriate size by either sucrose density centrifugation or agarose gel electrophoresis.

3. Digest a plasmid vector such as pUC19 with *Bam*HI. This will leave sticky ends that can pair with those left by *Sau*3A.

4. Mix the purified, *Sau*3A-digested yeast genomic DNA with the plasmid vector and DNA ligase.

5. Transform the recombinant DNA molecules into *E. coli*.

6. Recover colonies with plasmids by plating on media with ampicillin (pUC19 has a gene for resistance to this antibiotic), and with X-gal (to allow for blue-white selection to identify plasmids with inserts). Each colony will have a different yeast DNA insert, and all of the colonies comprise the yeast genomic library.

In a BAC vector, much larger DNA fragments—200 to 300 kb in size—would be used.

8.17 From the text, $N = \ln(1 - p)/\ln(1 - f)$, where N is the necessary number of recombinant DNA molecules, p is the probability of including one particular sequence, and f is the fractional proportion of the genome in a single recombinant DNA molecule. Here, $p = 0.90, f = (2 \times 10^5)/(3 \times 10^9)$, so $N = 34{,}538$.

8.20 a. As in Question 16.16, $N = [\ln(1 - p)]/[\ln(1 - f)]$. Here, $p = 0.95$. For (i), $f = (7 \times 10^3)/(3 \times 10^9)$; for (ii), $f = (1 \times 10^6)/(3 \times 10^9)$. The number of clones required is (i) 1.28×10^6 plasmids and (ii) 8.99×10^3 YACs.

b. By screening libraries with larger average inserts, fewer clones must be screened. This advantage must be evaluated relative to the added difficulty of analyzing the larger inserts of positive clones. For example, restriction mapping 1 Mb is substantially more difficult than restriction mapping 7 kb. Since a single gene ranges in size from hundreds of base pairs to hundreds of kilobase pairs, an essential question to consider is how the cloned sequences will be analyzed and used once identified.

c. Genomic clones provide for the analysis of gene structure: intron/exon boundaries, transcriptional control regions, and polyadenylation sites. This analysis is important for evaluating how a gene's expression is controlled. Since mutations in regulatory regions can affect the expression of the gene, analysis of these regions is important for understanding the molecular basis of a mutation.

8.22 She should clone the genes by complementation. First, cross the mutants into a genetic background to allow transformants to be selected using the *ura3* marker. Then, transform each mutant with a library containing wild-type sequences that has been made in a vector having the *URA3* gene. Plate the transformants at an elevated, restrictive temperature on media that also selects for *URA3*. Colonies that grow have a plasmid that complements the cell division mutation—they are able to overcome the functional deficit of the mutation because the plasmid has provided a copy of the wild-type gene—and also provides *URA3* function. Purify the plasmid from these colonies and characterize the cloned gene.

8.24 Compare the amino acid sequence to the genetic code, and design a "guessmer"—a set of oligonucleotides which could code for this sequence. Here, the guessmer would have the sequence 5'-ATG TT(T or C) TA(T or C) TGG ATG AT((T, C, or A) GG(A, G, T, or C) TA(T or C)-3', and be composed of 96 different oligonucleotides. Synthesize and then label these oligonucleotides, and use them as a probe to screen a cDNA library as described in Figure 8.10.

8.27 Construct a map stepwise, considering the relationship between the fragments produced by double digestion and the fragments produced by single-enzyme digestion. Start with the larger fragments. The 1,900-bp fragment produced by digestion with both A and B is a part of the 2,100-bp fragment produced by digestion with A, and the 2,500-bp fragment produced by digestion with B. Thus, the 2,500-bp and 2,100-bp fragments overlap by 1,900 bp, leaving a 200-bp A-B fragment on one side and a 600-bp A-B fragment on the other. One has:

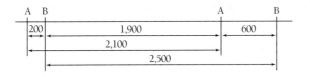

The map is extended in a stepwise fashion, until all fragments are incorporated into the map. The restriction map is:

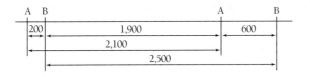

8.29 a. The lane with genomic DNA will have a smear: there are many *Eco*RI sites in a genome and the distances between these sites will vary. The smear reflects the large number and many different sizes of *Eco*RI fragments. Since *Eco*RI recognizes a 6-bp site, the average size will be about 4,096 bp (assume the genome is 25 percent A, G, C and T), and more intense staining will be seen around this size. The pUC19 plasmid has a single *Eco*RI restriction site into which the 10-kb insert has been cloned, so the lane with plasmid DNA will have two bands: the genomic DNA insert at 10 kb, and the plasmid DNA at 3 kb.

b. The probe will specifically detect the 10-kb *Eco*RI fragment, so signal will be seen in each lane at 10 kb.

8.30 The gel is soaked in an alkaline solution to denature the DNA to single-stranded form. It must be bound to the membrane in single-stranded form so that the probe can bind in a sequence-specific manner using complementary base pairing.

8.31 a. She should see a 2.0-kb band, as the 2.0-kb probe is a single-copy genomic DNA sequence.

b. LINE sequences are moderately repetitive DNA sequences, which may be distributed throughout the genome. Since the LINE sequence has an internal *Eco*RI site, each LINE element in the genomic DNA will be cut by *Eco*RI during preparation of the Southern blot. When the blot is incubated with the probe, both fragments will hybridize to the probe. The size of the fragments produced from each LINE element will vary according to where the element is inserted in the genome, and where the adjacent *Eco*RI sites are. Hence there will be many different-sized bands seen on the genomic Southern blot.

c. As in (b), there will be many different-sized bands on the genomic Southern blot. The sizes of the bands seen reflect the distances between *Eco*RI sites that flank a LINE element. All of the bands will be larger in size than the element, as the element is not cleaved by *Eco*RI. Counting the number of bands can give an estimate of the number of copies of the element in the genome.

d. Since the heterozygote has one normal chromosome 14, the probe will bind to the 3.0-kb *Eco*RI fragment derived from the normal chromosome 14. If the translocation is a reciprocal translocation, the remaining chromosome 14 is broken in two, and attached to different segments of chromosome 21. Since chromosome 14 has a breakpoint in the 3.0-kb *Eco*RI fragment, the 3.0-kb fragment is now split into two parts, each attached to a different segment of chromosome 21. Consequently, the 3.0-kb probe spans the translation breakpoint and will bind to two different fragments, one from each of the translocation chromosomes. The sizes of the fragments are determined by where the adjacent *Eco*RI sites are on the translocated chromosomes. Thus, the blot will have three bands, one of which is 3.0 kb.

e. Since the *TDF* gene is on the Y chromosome, no signal should be seen in a Southern blot prepared with DNA from a female having only X chromosomes.

8.32 a. *Not*I recognizes an 8-bp site, while *Bam*HI recognizes a 6-bp site. 8-bp sites appear about $\frac{1}{16}$ less frequently than 6-bp sites (See Question 8.6).

b. There are many *Bam*HI fragments in the BAC DNA insert, while there are fewer fragments in each *Not*I fragment. Digesting first with *Not*I allows regions of the BAC to be evaluated in an orderly, systematic manner and allows for the *Bam*HI fragments containing the gene to be more precisely identified and purified.

c. The 47-kb *Not*I fragment contains the gene, since it is the only *Not*I fragment that has sequences hybridizing to the cDNA.

d. The 10.5, 8.2, 6.1, and 4.1 *Bam*HI fragments contain the gene, since they hybridize to the cDNA probe.

e. The RNA-coding region is about 28.9 kb. It is larger than the cDNA since genomic DNA contains intronic sequences.

8.34 If the same gene functions in the brain, the gene be transcribed into a precursor mRNA, processed to a mature mRNA and then translated to produce the functional enzyme. Thus, transcripts for the gene should be found in the brain. To address this issue, label the cloned DNA, and use it to probe a northern blot having mRNA isolated from brain tissue. If the mRNA is rare, it may be prudent to use mRNA isolated from a specific region of the brain, such as the hypothalamus. An alternate, quite sensitive approach would be to sequence the cloned DNA, analyze the sequence to identify the coding region, and then design PCR primers that could be used to amplify cDNA made from mRNA isolated from various brain regions. To perform this RT-PCR (reverse transcriptase PCR), isolate mRNA, reverse transcribe it into cDNA, and then perform PCR. Obtaining an RT-PCR product in such an investigation would provide evidence that the gene is transcribed in the brain. In this alternative method, it would be important to be sure that no genomic DNA was present in the PCR amplification mixture, as the gene for the enzyme would be found in genomic DNA in both tissues.

8.35 a. Since both cDNAs hybridize to the same bands on a genomic Southern blot, they are copies of mRNAs transcribed from the same sequences. Therefore, it is likely that they are from the same gene.

b. Different bands on the northern blot indicate that the primary mRNA for this gene may be processed differently in brain and liver tissue. For example, it is possible that the 0.8-kb size difference between the two bands reflects a 0.8-kb intron that is spliced out in brain tissue that is not spliced out in liver tissue.

c. The two cDNAs are copies of mRNAs found in two different tissues. The northern blot indicates that there are some differences between the mRNAs in the different tissues. Thus, it is not surprising that the restriction maps are not identical. Note that the ends of the restriction maps are identical (the same EcoRI–HindIII and BamHI fragments), while the internal regions are not (the brain cDNA lacks the 0.5-kb EcoRI–HindIII fragment and some of each adjoining fragment).

d. The genomic Southern blot gives an indication of the gene organization at the DNA level, while the cDNA maps give an indication of the structure of the mRNA transcript(s). When the cDNA is used to probe genomic DNA sequences, it will hybridize to any sequences that are transcribed. Since restriction sites in the genome do not delineate where the transcribed regions are, the probe will hybridize to genomic DNA fragments that are only partly transcribed. That is, the probe will hybridize to transcribed sequences that are "connected to" nontranscribed sequences. Thus, the large (7.8-, 7.4-, 6.1-, and 3.6-kb) bands reflect the parts of the cDNA that hybridize to genomic DNA fragments that are only partly transcribed. Since they are the same size fragments that appear in the liver cDNA, the smaller fragments (2.0, 1.4, and 1.3 kb) represent fragments that are entirely transcribed. Based on this data, a possible gene organization is illustrated below:

Possible Gene Structure

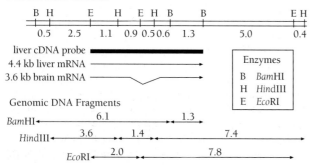

8.36 The primer will anneal to the fragment, and be extended at its 3′ end in four separate reactions, each with small amounts of a different dideoxynucleotide. In each reaction, some chains will be prematurely terminated when the dideoxynucleotide is incorporated. By using labeled dNTP precursors, all extension products will be labeled and be observed as distinct, labeled bands after separation on a denaturing polyacrylamide gel and signal detection.

Gel Analysis
fragment: 3′-GATCCAAGTCTACGTATAGGCC-5′

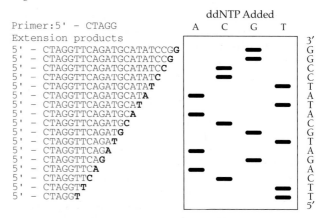

8.42 The insert Katrina sequenced was obtained from genomic DNA, while the inserts Marina sequenced were obtained from PCR. *Taq* DNA polymerase introduces errors during PCR, so that individual double-stranded molecules that are amplified during PCR may have small amounts of sequence variation. If PCR products are sequenced directly, the amount of variation is small enough that it may not be noticed—at a particular position in the sequence, only a very small number of molecules have an error. However, when PCR products are cloned, each independently isolated plasmid has an insert derived from a different double-stranded DNA PCR product, so that errors will be apparent.

Chapter 9 Applications of Recombinant DNA Technology

9.3 A knockout mouse is a mouse in which the gene has been made nonfunctional, for example, by deleting part of it. Under the first hypothesis, heterozygotes develop the disease because they have only half the normal dose of the gene's function. If this hypothesis is correct, then heterozygous mice having a knockout allele and a normal allele will also have only half of the normal dose of the gene's function, and so these mice should develop symptoms of the disease. If they do not develop symptoms of the disease, this hypothesis would not be supported. The second hypothesis, that the missense mutation alters the protein to interfer with a normal process, should then be investigated further.

9.6 A SNP is a single nucleotide polymorphism. Since a single base change can alter the site recognized by a restriction endonuclease, a SNP can also be a RFLP, or restriction fragment length polymorphism. Since simple tandem repeats (STRs) and variable number of tandem repeats (VNTRs) are based on tandemly repeated sequences (1 to 4 base-pair repeats for STRs, 5 to 10s of base pairs for VNTRs), they will not usually be SNPs.

9.7 If an individual is homozygous for an allele at an STR, all of their gametes have the same STR allele. The STR cannot be used as a marker to distinguish the recombinant- and parental-type gametes of the individual, and so will not be useful for mapping studies. In a population, individuals will be heterozygous more often for STRs with more alleles and higher levels of heterozygosity. The recombinant- and parental-type offspring may be distinguished in individuals heterozygous for an STR, making crosses informative for mapping studies.

If an STR has few alleles and a low heterozygosity, many individuals in a population will share the same STR genotypes. Therefore, there will be many individuals in the population who, by chance alone, will share the same genotype as a test subject and the STR will not be very useful for DNA fingerprinting studies.

9.8 a. The probe hybridizes to the same genomic region in each of the 10 individuals. Different patterns of hybridizing fragments are seen because of polymorphism of the *Eco*RI sites in the region. If a site is present in one individual but absent in another, different patterns of hybridizing fragments are seen. This provides evidence of restriction fragment length polymorphism. To distinguish between sites that are invariant and those that are polymorphic, analyze the pattern of bands that appear. Notice that the sizes of the hybridizing bands in individual 1 add up to 5 kb, the size of the band in individual 2 and the largest hybridizing band. This suggests that there is a polymorphic site within a 5.0-kb region. This is indicated in the diagram below, where the asterisk over site *b* depicts a polymorphic *Eco*RI site:

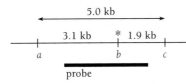

Notice also that the size of the band in individual 3 equals the sum of the sizes of the bands in individual 4. Thus, there is an additional polymorphic site in this 5.0-kb region. Since the 1.9-kb band is retained in individual 4, the additional site must lie within the 3.1-kb fragment. This site, denoted *x*, is incorporated into the diagram below. Notice that, because the 1.0-kb fragment flanked by sites *a* and *x* is not seen on the Southern blot, the probe does not extend into this region.

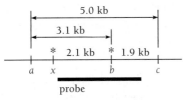

Depending on whether *x* and/or *b* are present, one will see either 5.0-, 3.1- and 1.9-kb, 2.1- and 1.9-kb, or 4.0-kb bands. In addition, if an individual has chromosomes with different polymorphisms, one can see combinations of these bands. Thus, individual 5 has one chromosome that lacks sites *x* and *b* and

one chromosome that has site *b*. The chromosomes in each individual can be tabulated as follows:

Individual	Sites on Each Homolog	Homozygote or Heterozygote?
1	*a, b, c*	homozygote
2	*a, c*	homozygote
3	*x, c*	homozygote
4	*x, b, c*	homozygote
5	*a, c/a, b, c*	heterozygote
6	*x, c/a, b, c*	heterozygote
7	*a, b, c/x, b, c*	heterozygote
8	*a, c/x, c*	heterozygote
9	*a, c/x, b, c*	heterozygote
10	*x, c/x, b, c*	heterozygote

b. Since individual 1 is homozygous, chromosomes with sites at *a*, *b*, and *c* will be present in all of the offspring, giving bands at 3.1 and 1.9 kb. Individual 6 will contribute chromosomes of two kinds, one with sites at *x* and *c* and one with sites at *a*, *b*, and *c*. Thus, if this analysis is performed on their offspring, two equally frequent patterns will be observed: a pattern of bands at 3.1 and 1.9 kb and a pattern of bands at 4.0, 3.1, and 1.9 kb. This is just like the patterns seen in the parents.

9.9 Chromosomes bearing CF mutations have a shorter restriction fragment than chromosomes bearing wild-type alleles. Both parent lanes (M and P) have two bands, indicating that each parent has a normal and a mutant chromosome. The parents are therefore heterozygous for the CF trait. The fetus lane (F) shows only one (lower molecular weight) band. The size of the band indicates that the fetus has only mutant chromosomes. The intensity of the band is about twice that of the same-sized band in the parent lanes. This is because the diploid genome of the fetus has two copies of the fragment, while the diploid genome of each parent only has one. Since the fetus is homozygous for the CF trait, it will have CF. (Do not confuse the black bands with the open boxes, which represent the origin, the location where the DNA was loaded before electrophoresis.)

9.10 a. Use the PCR-RFLP method: Isolate genomic DNA from the parkinsonian individual, and use PCR to amplify the 200-bp segment of exon 4; purify the PCR product, digest it with *Tsp*45I, and resolve the digestion products by size using gel electrophoresis. The normal allele will contain the *Tsp*45I site, and so produce 120- and 80-bp fragments. The mutant allele will not contain the *Tsp*45I site, and so produce only a 200-bp fragment.

b. Homozygotes for the normal allele will have 120- and 80-bp fragments; homozygotes for the mutant allele will have a 200-bp fragment; heterozygotes will have 200-, 120-, and 80-bp fragments.

c. Use RT-PCR to amplify a DNA copy of the mRNA, and digest the RT-PCR product with *Tsp*45I. First isolate RNA from the tissue. Then make a single-stranded cDNA copy using reverse transcriptase and an oligo(dT) primer. Then amplify exon 4 of the cDNA using PCR, digest the product with *Tsp*45I, and separate the digestion products by size using gel electrophoresis. If a 200-bp fragment is identified in a heterozygote, then the mutant allele is transcribed. If only 120- and 80-bp fragments are identified, then the mutant allele is not transcribed. Note that to accurately assess expression of either allele, it is essential that the RT-PCR reaction is performed on a purified RNA template without contaminating genomic DNA.

9.14 James and Susan Scott are not the parents of "Ronald Scott." There are several bands in the fingerprint of the boy that are not present in either James or Susan Scott and thus could not have been inherited from either of them (e.g., bands a and b in the figure below). In contrast, whenever the boy's DNA exhibits a band that is missing from one member of the Larson couple, the other member of the Larson couple has that band (e.g., bands c and d). Thus, there is no band in the boy's DNA that he could not have inherited from one or the other of the Larsons. These data thus support an argument that the boy is, in fact, Bobby Larson. These data should be used together with other, non-DNA-based evidence to support the claim that the boy is Bobby Larson.

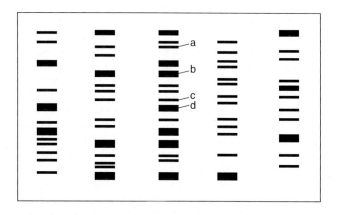

9.15 a. The PCR method requires very small (nanograms) amounts of template DNA, and if the primers are designed to amplify only small regions, the DNA can be partially degraded. In contrast, VNTR methods require larger amounts (micrograms) of intact DNA, as restriction digests are used to produce relatively large (kb-size fragments) that are then detected by Southern blotting. Some of the DNA samples used in forensic analysis are found in crime scenes and may be stored for years, so that they may often be degraded and not be present in large amounts. STR methods can still be used on such samples, while VNTR methods cannot.

b. Multiplexing PCR reactions insures that (1) the different STR results obtained in the reaction are all derived from a single DNA sample (laboratory labeling and pipetting errors are minimized), and (2) limited amounts of DNA samples are used efficiently.

c. P(random match) = (0.112 × 0.036 × 0.081 × 0.195) = 6.4 × 10^{-5}. About 1 person in 15,702 would be misidentified by chance alone using just these four markers.

d. P(random match) = (0.112 × 0.036 × 0.081 × 0.195 × 0.062 × 0.075 × 0.158 × 0.065 × 0.067 × 0.085 × 0.089 × 0.028 × 0.039) = 1.7 × 10^{-15}. About 1 person in 5.94 × 10^{14} would be misidentified by chance alone using all 13 markers.

9.17 Use the reverse ASO method: see text for details.

9.20 Several different methods can be helpful to orient a chromosomal walk that has proceeded through several steps, or which includes one or more jumps. After restriction mapping is used to establish how the clones of the walk overlap, DNA markers such as RFLPs that have been positioned relative to each other using genetic mapping studies may be used to identify the orientation of a walk. If a walk has extended far enough, and includes two RFLP markers that have been previously oriented relative to the gene you are seeking, the orientation of the walk can be determined by using Southern blotting to identify which clone contains each marker. An alternative method is to use *in situ* hybridization to chromosomes. Probes can be made from unique sequences at the ends of a partially completed walk, and hybridized to chromosomes. The orientation of the walk relative to the centromere and telomere of the chromosome can then be determined.

9.23 Use an interaction trap assay (the yeast two-hybrid system). Fuse the *fru* coding region (obtained from a cDNA) to the sequence of the Gal4p BD, and cotransform this plasmid into yeast with a plasmid library containing the Gal4p AD sequence, which is fused to protein sequences encoded by different cDNAs from the *Drosophila* brain. Purify colonies that express the reporter gene (see text Figure 9.15, p. 230). In these colonies, the transcription of the reporter gene was activated when the AD and BD domains were brought together by the interactions of the *fru* protein with an unknown protein encoded by one of the brain cDNAs. Isolate and characterize the brain cDNA found in these yeast colonies.

9.25 Isolate RNA from the livers of the alcohol-fed and control rats. Measure the levels of mRNA for alcohol dehydrogenase by either (1) separating the RNA by size using gel electrophoresis, preparing a northern blot, and hybridizing it with a probe made from a cDNA for the alcohol dehydrogenase gene; or (2) using RT-PCR or quantitative RT-PCR.

Chapter 10 Genomics
10.3 a. In a random sequence that is 25% each A, G, C, and T, the chance of finding a 6-bp site is (1/4)^6 = 1/4,096, and the chance of finding an 8-bp site is (1/4)^8 = 1/65,536. In such a sequence, *Apa*I, *Hind*III, *Sac*I, and *Ssp*I should produce fragments that average 4,096 bp in size, and *Srf*I and *Not*I should produce fragments that average 65,536 bp in size.

b. i. The large variation in average fragment sizes when one restriction enzyme is used to cleave different genomes could reflect (1) the nonrandom arrangements of base pairs in the different genomes (e.g., there is variation in the frequencies of certain sequences that are part of the restriction site in the different genomes), and/or (2) the different base compositions of the genomes (e.g., genomes that are rich in A-T base pairs should have fewer sites for enzymes recognizing sites containing only G-C base pairs).

ii. The large variation in fragment sizes when the same genome is cut with different enzymes that recognize sites having the same length could reflect (1) the nonrandom arrangement of base pairs *in that genome*, and/or (2) the base composition *of that genome*.

iii. If the sequence of *Mycobacterium tuberculosis* was random and contained 25% each of A, G, T, and C, enzymes recognizing a 6-bp site should produce fragments that are about 16-fold smaller than enzymes recognizing an 8-bp site. That this is not the case here suggests that at least one of these assumptions is incorrect. Two possibilities are that the genome of *Mycobacterium tuberculosis* is very rich in G-C base pairs and poor in A-T base pairs, and/or that there is a nonrandom arrangement of base pairs so that 5′-AA-3′, 5′-TT-3′, 5′-AT-3′ and/or 5′-TA-3′ sequences are rare. (The data given for *Sac*I suggest that the sites 5′-AG-3′ and 5′-CT-3′, which are part of the *Hind*III site, are not rare.)

c. Choose an enzyme that cuts infrequently, so that it becomes tractable to construct a map from the fragments that are produced, but not so infrequently that there is little information gained from the map. The answer will depend in part on the size of the fragment being mapped and the objectives of constructing the map. For example, if a 4 Mb region of the *Caenorhabditis elegans* genome is being mapped, cleavage with *Hind*III would produce about 1,333 fragments while cleavage with *Srf*I would produce about 4 fragments. It will be difficult to construct a map from 1,333 fragments, while the map constructed from just four fragments might not be very helpful. Cleavage with *Not*I would produce about 15 fragments, a number more tractable for constructing a map and which might provide the desired organizational information. Reasonable choices to consider are: for *E. coli*, *Apa*I, *Srf*I, or *Not*I; for *M. tuberculosis*, *Hind*III or *Ssp*I; for *A. thaliana* and *C. elegans*, *Not*I; for *S. cerevisiae*, *D. melanogaster*, *M. musculus*, and *H. sapiens*, *Srf*I or *Not*I.

10.4 a. Synthesize an oligonucleotide of, say, 30 bases, that is either $(AT)_N$ or $(CAG)_N$. Label one end of the oligonucleotide with ^{32}P, using polynucleotide kinase. Then use this as a probe in a colony hybridization to screen the colonies. Sequence the inserts of the positive clones using universal primer sites near the multiple cloning site of the plasmids.

b. After sequencing the positive clones, design primers (based on the unique sequences flanking the repetitive DNA sequences of each insert) for PCR. Have these synthesized and use them in PCR amplifications of genomic DNA. The sizes of the PCR products will indicate the length of the amplified segment. If an STS is polymorphic, multiple alleles corresponding to different lengths of repeats will be found in different individuals. Polymorphic STSs can be used as markers in mapping experiments. The genotypes of individuals in a cross would be determined using PCR, and linkage between two different STSs or between an STS and a phenotype (e.g., disease) can be assessed.

c. $960 \times 8,203 = 7,874,880$ PCR reactions.

d. (i) Under this scheme, the 96 wells on each plate are pooled into 12 (rows) + 8 (columns), or 20 samples. For 10 plates, there would be 200 samples. **(ii)** For 8,203 STSs, there would be $200 \times 8,203 = 1,640,600$ PCRs. **(iii)** Plate II, row F, column 6. **(iv)** It is probably a false-positive result, an artifact.

e. STS markers that are shared by a set of YACs define the regions of the YACs that overlap. Identify which YACs share the same markers and then draw the contigs. One possible order is shown in Figure 10.A. The exact order of some sets of markers

cannot be determined from the data given. Some sets of markers may be in reversed order relative to each other: 2631 and 63; 210 and 6892; 5192 and 719; 6193 and 991; 3097, 4630, and 522.

f. Multiple pooling schemes are possible, but not all insure that positive YACs can be identified unambiguously. Try combining similarly positioned wells in each of the 10 different microtiter plates: combine IA1, IIA1, … XA1 into pool A1, combine IB1, IIB1, … XB1 into pool B1, etc., to give 96 pools. Also use the Plate-Row pools described in the question. This would require $96 + 80 = 176$ pools ($176 \times 8,203 = 1,443,728$ PCRs), almost the same as the 200 pools previously described.

10.5 a. After a large number of BAC clones are typed using DNA fingerprinting (see Chapter 9, p. 224) and STS mapping, the overlaps between BAC clones are used to assemble BAC clone contigs. Chimeric BAC clones can be identified because they will be assembled into multiple clone contigs that derive from different parts of the genome, leading to consistencies in assembled clone-contig maps. FISH (see Chapter 16, p. 445) can also be used to verify that a fluorescently labeled BAC clone hybridizes to a single chromosomal site.

b. A laborious but precise approach is to sequence each BAC and use computerized algorithms to search for the sequences of the STSs relative to the entire BAC sequence. A faster but approximate approach is to generate a restriction map of each BAC using rare-cutting restriction enzymes and identify which restriction fragments contain a STS. This can be done by testing each restriction digest fragment for the presence of an STS using PCR or using a labeled STS as a probe in a Southern blot hybridization.

10.7 a. Approximately, $500 \times (13,543,099 + 10,894,467) = 1.22 \times 10^{10}$ nucleotides.

b. If a plasmid with a 2-kb insert has a unique sequence at one end but repetitive sequence at the other end, it will not be possible to continue to assemble the sequence past this plasmid, since many clones in the library have the same repetitive sequence and they come from all over the genome. Since many repetitive sequences are about 5 kb in length, sequencing plasmids with 10-kb inserts circumvents this problem. Some of the 10-kb inserts will have a sequence at one end that overlaps with the unique sequence in the plasmid with the 2-kb insert, as well as unique sequence at their other end that lies past the repetitive element, and which can be assembled with unique sequence from other plasmids.

c. The sequence of the central region is obtained from the sequence of overlapping clones during sequence assembly.

10.8 Repetitive sequences pose at least two problems for sequencing eukaryotic genomes. Highly repetitive sequences associated with centromeric heterochromatin consist of short, simple repeated sequences. These are unclonable, making it

Figure 10.A

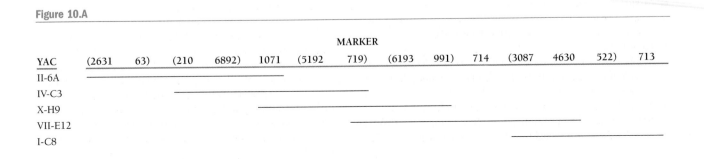

YAC	(2631	63)	(210	6892)	1071	(5192	719)	(6193	991)	714	(3087	4630	522)	713
II-6A														
IV-C3														
X-H9														
VII-E12														
I-C8														

impossible to obtain the complete genome sequence of organisms with them. More complex repetitive sequences such as those found within euchromatic regions can be cloned and sequenced. However, since they can originate from different genomic locations and a shotgun sequencing approach only provides short sequences, the assembly of overlapping sequences can be ambiguous. Some of the ambiguities can be resolved by comparing these sequences to overlapping sequences generated from sequencing clones with larger inserts.

10.9 Sequencing of archaeon genomes has shown that their genes are not uniformly similar to those of the Bacteria or the Eukarya. While most of the archaeon genes involved in energy production, cell division, and metabolism are similar to their counterparts in Bacteria, the genes involved in DNA replication, transcription, and translation are similar to their counterparts in Eukarya.

10.12 a. Since prokaryotic ORFs should reside in transcribed regions, they should follow a bacterial promoter containing consensus sequences recognized by a sigma factor. For example, promoters recognized by σ^{70} would contain -35 (TTGACA) and -10 (TATAAT) consensus sequences. Within the transcribed region, but before the ORF, there should be a Shine-Dalgarno sequence (UAAGGAGG) used for ribosome binding. Nearby should be an AUG (or GUG, in some systems) initiation codon. This should be followed by a set of in-frame, sense codons. The ORF should terminate with a stop (UAG, UAA, UGA) codon.

b. Eukaryotic introns are transcribed but not translated sequences in the RNA-coding region of a gene. They will be spliced out of the primary mRNA transcript before it is translated. If not accounted for, they could introduce additional amino acids, frameshifts, and chain-termination signals.

c. The small, average size of exons relative to the range of sizes for introns makes it challenging to predict whether a region with only a short set of in-frame codons is used as an exon. Such regions could have arisen by chance or be the remnants of exons that are no longer used due to mutation in splice site signals.

d. Eukaryotic introns typically contain a GU at their 5′ ends, an AG at their 3′ ends, and a YNCURAY branch-point sequence 18 to 38 nucleotides upstream of their 3′ ends. To identify eukaryotic ORFs in DNA sequences, scan sequences following a eukaryotic promoter for the presence of possible introns by searching for sets of these three consensus sequences. Then try to translate sequences obtained if potential introns are removed, testing whether a long ORF with good codon usage can be generated. Since alternative mRNA splicing exists at many genes, more than one possible ORF may be found in a given DNA sequence.

10.14 a. Comparison of cDNA and genomic DNA sequences can define the structure of transcription units by elucidating the location of the intron-exon boundaries, poly(A) sites, and the approximate locations of promoter regions. Comparison of different full-length cDNAs representing the same gene can identify the use of alternative splice sites, alternative poly(A) sites, and alternative promoters.

b. The analysis of full-length cDNAs provides information about an entire open reading frame, information about the site at which transcription starts and where the promoter lies, and the location of the poly(A) site. Partial-length cDNAs might provide some but not all of this information. While multiple ESTs could be compared and assembled to obtain more information, assembling ESTs is challenging as alternative splice sites, alternative promoters and/or alternative poly(A) sites can be used.

c. Genes are not uniformly distributed among different chromosomes, and some chromosomes have more genes than others. While consistent with the finding that chromosomes have gene-rich regions and gene deserts, more data is needed to infer the relationship between the density of genes on a chromosome and how gene-rich it is. For example, a chromosome with many small genes could still have regions classified as gene deserts.

d. Two possible explanations are that (1) some regions of the genome sequence are incorrectly assembled (e.g., due to the large numbers of repetitive sequences they contain), so that the cDNAs are unable to be mapped to just one region, and (2) some of the genes are in regions that have not yet been assembled (e.g., because they are difficult to clone or sequence using current technologies). As the genome sequence is revised, these issues should be resolved.

10.15 One approach is to use model organisms (e.g., transgenic mice) that have been developed as models to study a specific human disease. Expose them and a control population to specific environmental conditions, and then simultaneously assess disease progression and alterations in patterns of gene expression using microarrays. This would provide a means to establish a link between environmental factors and patterns of gene expression that are associated with disease onset or progression.

10.18 Isolate DNA from the blood of the patient and label it with a red fluorescent dye and label DNA from a normal individual with a green fluorescent dye. Prepare a DNA chip consisting of oligonucleotides that collectively represent the entirety of the normal dystrophin gene, and hybridize the chip with the labeled DNAs. The normal, green-labeled DNA will hybridize on all sites. The normal sequences in the patient's red-labeled DNA will hybridize to the oligonucleotides on the chip but mutated sequences will not. Consequently, normal hybridization will be seen as a yellow (red + green) spot, while a mutation will be recognized as a green spot. Identify the region of the gene corresponding to the oligonucleotide spots that have green hybridization signals to determine the site(s) of the mutation.

10.19

S, F, C	Aligning DNA sequences within databases to determine the degree of matching
F, C	Annotating sequences within a sequenced genome
F	Characterizing the transcriptome and proteome present in a cell at a specific developmental stage or in a particular disease state
C	Comparing the overall arrangements of genes and non-gene sequences in different organisms to understand how genomes evolve
F	Describing the function of all genes in a genome
F, C	Determining the functions of human genes by studying their homologs in nonhuman organisms
F	Developing a comprehensive two-dimensional polyacrylamide gel electrophoresis map of all proteins in a cell
S	Developing a physical map of a genome
S, F, C	Developing DNA microarrays (DNA chips)
F, C	Identification of homologs to human disease genes in organisms suitable for experimentation

_____S_____ Identifying a large collection of simple tandem repeat, or microsatellite sequences to use as DNA markers within one organism

_S, F, C_ Identifying expressed sequence tags (ESTs)

_F, C_ Making gene knockouts and observing the phenotypic changes associated with them

_____S_____ Mapping a gene in one organism using the lod score method

_____S_____ Sequencing individual BAC or PAC clones aligned in a contig using a shotgun approach

_____S_____ Using oligonucleotide hybridization analysis to type an SNP

10.21 a. On chromosome V and the X chromosome, genes are distributed uniformly. However, especially on chromosome V, conserved genes are found more frequently in the central regions. In contrast, inverted and tandem-repeat sequences are found more frequently on the arms. It appears that at least on chromosome V, there is an inverse relationship between the frequency of inverted and tandem-repeats and the frequency of conserved genes.

b. Since there are fewer conserved genes on the arms, there appears to be a greater rate of change on chromosome arms than in the central regions.

c. Yes, since increased meiotic recombination provides for greater rates of exchange of genetic material on chromosomal arms.

Chapter 11 Mendelian Genetics

11.1 a. Let R represent red and r represent yellow. The cross $RR \times rr$ gives all Rr. The F_1 are all red.

b. The F_2 is obtained from selfing the F_1. $Rr \times Rr$ gives $3/4$ $R-$: $1/4$ rr. The F_2 are $3/4$ red and $1/4$ yellow.

c. $Rr \times RR$ gives all $R-$. The fruits all red.

d. $Rr \times rr$ gives $1/2$ Rr : $1/2$ rr. The fruits are $1/2$ red and $1/2$ yellow.

11.3 In Mendelian monohybrid crosses, F_2 plants display phenotypic ratios that are $3/4$ dominant : $1/4$ recessive. Since the F_2 ratio here is 3 colored : 1 colorless, we can infer that colored is dominant to colorless. Let C represent colored and c represent colorless. The F_2 has a 1 CC : 2 Cc : 1 cc genotypic ratio, so there are two types of colored plants, CC and Cc. If a CC plant is picked and selfed, only colored plants will be seen in its offspring. In contrast, if a Cc plant is picked and selfed, both colored and colorless plants will be seen in the offspring. To satisfy the conditions of the problem, a Cc plant must be picked. Since the F_2 colored plants are present in a 1 CC : 2 Cc ratio, the chance of picking a Cc plant is $2/3$.

11.4 a. Parents are Rr (rough) and rr (smooth); F_1 are Rr (rough) and rr (smooth).

b. $Rr \times Rr \rightarrow 3/4$ $R-$ (rough) and $1/4$ rr (smooth).

11.6 To obtain a 3 purple : 1 white ratio, the selfed plant must have been heterozygous, and purple (P) must be dominant to white (p). The purple-flowered progeny of a Pp heterozygote have two genotypes, PP and Pp, and are present in a 2 Pp : 1 PP ratio. Since only PP plants breed true and these are $1/3$ of the purple progeny, $1/3$ of the purple progeny will breed true.

11.7 Black is dominant to brown. Let B represent the black allele and b represent the brown allele. Then female X is Bb, female Y is BB, and the male is bb.

11.10

Parents	Progeny		Female Parent
Female × Male	Grey	White	Genotype
grey × white	81	82	Gg
grey × grey	118	39	Gg
grey × white	74	0	GG
grey × grey	90	0	GG or Gg ($G-$)

11.11 The farmer now has only black babbits, so must breed animals that are either BB or Bb. His initial pair gave both black and white progeny and is not true breeding. To obtain white offspring from them, each must be heterozygous with a Bb genotype. The unsold black babbit offspring should therefore have a 1 BB : 2 Bb ratio.

a. To obtain a white offspring from a cross of two black parents, both parents must be Bb and a bb offspring must be produced. The chance of picking a Bb individual from among the F_1 offspring is $2/3$. The chance that a bb offspring is produced from a cross of two Bb individuals is $1/4$.

P(white offspring) = P(both F_1 babbits are Bb and a bb offspring is produced)

= P(both F_1 babbits are Bb)
× P(bb offspring)

= ($2/3 \times 2/3$) × ($1/4$)

= $1/9$

b. If he crosses an F_1 male (Bb or BB) to the parental female (Bb), two types of crosses are possible. The crosses and their probabilities are (1) Bb (F_1 male) × Bb (parental female), $P = 2/3 \times 1 = 2/3$ and (2) Bb (F_1 male) × Bb (parental female), $P = 1/3 \times 1 = 1/3$. Only the first cross can produce white progeny, $1/4$ of the time. Using the product rule, the chance that this strategy will yield white progeny is

$P = 2/3$ (chance of $Bb \times Bb$ cross) × $1/4$ (chance of bb offspring) = $1/6$.

c. While it is more work initially, a productive long-term strategy is to remate the initial two black babbits (both are known to be Bb) to obtain a white male offspring [$P = 1/4$ (white bb) × $1/2$ (male) = $1/8$]. Since only the fertility of white females and not that of white males is affected, retain this male and breed it back to its mother. This cross would be $Bb \times bb$ and give $1/2$ white (bb) and $1/2$ black (Bb) offspring. Use the progeny of this cross to develop a "breeding colony" consisting of black (Bb) females and white (bb) males. These would consistently produce half white and half black offspring.

11.14 Try fitting the data to a model in which catnip sensitivity/insensitivity is controlled by a pair of alleles at one gene. Since sensitivity is seen in all of the progeny of the initial mating between catnip-sensitive Cleopatra and catnip-insensitive Antony, hypothesize that sensitivity is dominant. Let S represent the sensitive allele, and s represent the insensitive allele. Then the initial cross is $S- \times ss$, and the progeny are Ss. If two of the Ss kittens mate, a 3 sensitive ($S-$) : 1 insensitive (ss) progeny ratio is expected. In the mating with Augustus, the cross would be $Ss \times ss$, and should give a 1 Ss (sensitive) : 1 ss (insensitive) progeny ratio. The observed progeny ratios are not far off from these expectations.

An alternative hypothesis is that sensitivity (s) is recessive and insensitivity (S) is dominant. For Antony and Cleopatra to have sensitive (*ss*) offspring, they would need to be *Ss* and *ss*, respectively. When two of their sensitive (*ss*) progeny mate, only sensitive (*ss*) offspring should be produced. Since this is not observed, this hypothesis does not explain the data.

11.15 **a.** *Ww Dd* × *ww dd*

  b. *Ww dd* × *ww dd*

  c. *ww DD* × *WW dd*

  d. *Ww Dd* × *ww dd*

  e. *Ww Dd* × *Ww dd*

11.16 To determine the desired probabilities in the cross *Aa Bb Cc* × *Aa Bb Cc*, consider each gene separately and then use the product rule.

  a. In the cross *Aa* × *Aa*, there is a ¾ chance of obtaining an *A*– individual. Similarly, the chance of obtaining a *B*– individual from the cross *Bb* × *Bb* is ¾, and the chance of obtaining a *C*– individual from the cross *Cc* × *Cc* is ¾. Using the product rule, the probability of obtaining a phenotypically *A B C* (*A*– *B*– *C*–) offspring is ¾ × ¾ × ¾ = ²⁷/₆₄.

  b. There is a ¼ chance of obtaining an *AA* offspring from the cross *Aa* × *Aa*. This is also the probability for obtaining a *BB* or *CC* offspring from a *Bb* × *Bb* or *Cc* × *Cc* cross, respectively. Using the product rule, the probability of obtaining an *AA BB CC* offspring is ¼ × ¼ × ¼ = ¹/₆₄.

11.20 **a.** The cross is *aa BB CC* × *AA bb cʰcʰ*. The F₁ trihybrids are all *Aa Bb Ccʰ* and are agouti and black. A branch diagram will show that the F₂ consists of ²⁷/₆₄ agouti and black; ⁹/₆₄ agouti, black, Himalayan; ⁹/₆₄ agouti, brown; ⁹/₆₄ black; ³/₆₄ agouti, brown, Himalayan; ³/₆₄ black, Himalayan; ³/₆₄ brown; and ¹/₆₄ brown, Himalayan.

  b. F₂ animals that are not Himalayan, black, and agouti are *A*– *B*– *C*–. Among the *A*– animals, ⅔ are *Aa*. Among the *B*– animals, ⅓ are *BB*. Among the *C*– animals, ⅔ are *Ccʰ*, so the proportion of *Aa BB Ccʰ* animals is ⅔ × ⅓ × ⅔ = ⁴/₂₇.

  c. From the cross *Aa Bb Ccʰ* × *Aa Bb Ccʰ*, ¼ of the progeny will be *bb* and show brown pigment. This will be the case regardless of whether the animals are pigmented over their entire body or are Himalayan. Thus, ¼ of the Himalayan mice will show brown pigment.

  d. From the cross *Aa Bb Ccʰ* × *Aa Bb Ccʰ*, ¾ of the progeny will be *B*– and show black pigment. This will be the case regardless of whether the animals are agouti or nonagouti. Thus, ¾ of the agouti mice will show black pigment.

11.24 Mating type C is determined only by the genotype *aa bb*. Thus, C must be genotype *aa bb*. Crosses of the other strains to C, then, are testcrosses and the progeny ratios indicate the genotypes of the strains. Therefore, A is *Aa Bb*, B is *aa Bb*, and D is *Aa bb*.

11.25 **a.** The initial cross is *Ww Rr* × *W r*. The progeny females result from a queen's egg (¼ *W R*, ¼ *W r*, ¼ *w R*, ¼ *w r*) being fertilized by the drone's sperm (all *W r*). These will be workers and will be ½ *W*– *Rr* (black-eyed, wax sealers) and ½ *W*– *rr* (black-eyed, resin sealers).

  b. Males arise solely from unfertilized eggs and receive chromosomes only from their mother. The progeny males will be ¼ *W R* (black-eyed, wax sealers), ¼ *W r* (black-eyed, resin sealers), ¼ *w R* (white-eyed, wax sealers), ¼ *w r* (white-eyed, resin sealers).

  c. The egg fertilized by the mutation-bearing sperm results in a *Cc* female (Madonna). Since fertilization occurs in flight, males that fertilize a queen must be *C*. Hence, Madonna's first generation arises from the cross *Cc* × *C*. There is a ½ chance of

obtaining daughters that are *Cc*. Since a *Cc* daughter can also only be fertilized by a *C* male, the chance of her having a *Cc* daughter is also ½. The chance of Madonna having a *Cc* granddaughter, who will produce ½ wingless males, is thus ½ × ½ = ¼.

  d. The chance that the F₄ generation great-great-granddaughter will be heterozygous is (½)⁴ = ¹/₁₆.

11.26 **a.** The mother must be heterozygous *Aa* in order to have children that exhibit the recessive trait.

  b. The father must be homozygous *aa*, since he expresses the trait.

  c. Since the cross is *Aa* (mother) × *aa* (father), all offspring receive the recessive *a* allele from their father. If a child receives the recessive *a* allele from their mother, they will be affected and homozygous *aa* (II.2, II.5). If a child receives the normal *A* allele from their mother, they will be normal and heterozygous *Aa* (II.1, II.3, II.4).

  d. In the cross *Aa* × *aa*, the prediction is that ½ of the progeny will be *Aa* (normal) and ½ will be *aa* (express the trait). There are five children, two affected and three normal. Thus, the ratio fits as well as it could for five children.

11.29 **a.** It is uncertain whether the brother of the man's wife's paternal grandmother had Gaucher disease. If this distant relative had the disease, a disease allele might have been passed on to the man's wife. Therefore, in a worst-case scenario, this distant relative would have had the disease. Under this scenario, the pedigree is as shown here:

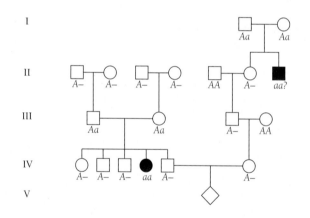

In this pedigree, the man is IV.5, his affected sister is IV.4, his wife is IV.6, and the brother of his wife's paternal grandmother is II.7. Since II.7 is affected but his parents are not, the disease must be a recessive trait, and each of his parents must be heterozygous.

  b. For the couple IV.5 and IV.6 to have an affected child (V.1), both IV.5 and IV.6 must give V.1 a recessive *a* allele. Since the trait is recessive and IV.5 and IV.6 are not affected, we know that IV.5 and IV.6 are *A*–. We must calculate the chance that they are *Aa* and that both pass on the *a* allele.

IV.5 has an affected sister, so his parents must both be *Aa*, and there is a ⅔ chance that he is *Aa*. Therefore, there is a ⅔ × ½ = ⅙ chance that IV.5 will pass the *a* allele to V.1.

$$P(\text{IV.6 is } Aa) = P(\text{III.3 is } Aa \text{ and III.3 passed } a \text{ to IV.6})$$
$$= P[(\text{II.6 was } Aa \text{ and II.6 passed } a \text{ to III.3})$$
$$\textit{and } (\text{III.3 passed } a \text{ to IV.6})]$$
$$= [(\tfrac{2}{3} \times \tfrac{1}{2}) \times \tfrac{1}{2}] = \tfrac{1}{6}.$$

Therefore, there is a $1/6 \times 1/2 = 1/12$ chance that IV.6 will pass the a allele to V.1. In this worst-case scenario, the chance that both parents will pass on an a allele and have an affected child is $1/12 \times 1/3 = 1/36$. If the brother of the wife's paternal grandmother did not have the disease, IV.6 would be AA insuring that V.1 will be $A-$ and phenotypically normal.

11.30 The F_1 cross is $a^+/a\ b^+/b\ c^+/c\ d^+/d \times a^+/a\ b^+/b\ c^+/c\ d^+/d$.

a. A colorless F_2 individual would result if an individual has an a/a, b/b, and/or c/c genotype. This would consist of many possible genotypes. Rather than identify all of these combinations, use the fact that the proportion of colorless individuals $= 1 -$ the proportion of pigmented individuals. The proportion of pigmented individuals $(a^+/-\ b^+/-\ c^+/-)$ is $3/4 \times 3/4 \times 3/4 = 27/64$. The chance of not obtaining this genotype is $1 - 27/64 = 37/64$.

b. A brown individual is $a^+/-\ b^+/-\ c^+/-\ d/d$. The proportion of brown individuals is $3/4 \times 3/4 \times 3/4 \times 1/4 = 27/256$.

11.32 a. Since any of the normal alleles a^+, b^+, or c^+ is sufficient to catalyze the reaction leading to color, in order for color to fail to develop, all three normal alleles must be missing. That is, the colorless F_2 must be $a/a\ b/b\ c/c$. The chance of obtaining such an individual is $1/4 \times 1/4 \times 1/4 = 1/64$.

b. Now, colorless F_2 are obtained if either one or both steps of the pathway are blocked. That is, colorless F_2 are obtained in either of the following genotypes: $d/d\ -/-\ -/-\ -/-$ (the first or both steps blocked) or $d^+/-\ a/a\ b/b\ c/c$ (second step blocked). The chance of obtaining such individuals is $(1/4 \times 1 \times 1 \times 1) + (3/4 \times 1/4 \times 1/4 \times 1/4) = 67/256$.

11.33 a. $1/2\ w/w^+\ bw/bw^+\ st/st^+$, fire-red-eyed daughters; $1/2\ w/Y\ bw/bw^+\ st/st^+$, white-eyed sons

b. $w/w^+\ se/se^+\ bw/bw^+$ and $w^+/Y\ se/se^+\ bw/bw^+$, all fire-red eyes

c. $w/w^+\ v/v^+\ bw/bw$ and $w^+/Y\ v/v^+\ bw/bw$, all brown eyes

d. $1/4\ w^+/w$ or w^+/Y, $bw/bw^+\ st/st^+$, fire-red eyes; $1/4\ w^+/w$ or w^+/Y, $bw/bw\ st/st^+$, brown eyes; $1/4\ w^+/w$ or w^+/Y, $bw/bw^+\ st/st$, scarlet eyes; $1/4\ w^+/w$ or w^+/Y, $bw/bw\ st/st$, (the color of 3-hydroxykynurenine plus the color of the precursor to biopterin, or colorless $=$ white)

11.34 As is described in Figure 7.24, p. 162, the spotted-kernel phenotype arises from the transposition of Ds elements (incomplete transposons incapable of transposition themselves) out of the C gene during kernel development. This means that the mutation in colorless strain B is caused by an insertion of a Ds element in the C gene. It is able to transpose because the purple strain C has Ac elements—full-length transposons—that can activate Ds transposition.

Spotted kernels are not seen in any of the other crosses, so Ds elements are not transposed in them. There are two different explanations for this. First, since spotted kernels are not seen in the A × C cross, the colorless phenotype in strain A is not caused by the insertion of a Ds element. (For example, it could result from just a point mutation in the C gene.) Second, since spotted kernels are not seen in the B × D cross, the D strain lacks Ac elements. For these reasons, the crosses A × C, A × D, and B × D all show only expected Mendelian patterns of inheritance.

Chapter 12 Chromosomal Basis of Inheritance

12.1 c

12.3 c

12.6 a. Yes, providing that the species has a sexual mating system in its life cycle. Meiosis can be initiated only in diploid cells. If a sexual mating system exists, two haploid cells can fuse to produce a diploid cell, which can then go through meiosis to produce haploid progeny. The fungi *Neurospora crassa* and *Saccharomyces cerevisiae* exemplify this positioning of meiosis in a life cycle.

b. No, because a diploid cell cannot be formed in a haploid individual and meiosis can be initiated only in a diploid cell.

12.8 c. For example, in an organism with a haploid life cycle, gametes and somatic cells are both 1N.

12.10 a. Metaphase: Metaphase in mitosis, metaphase I and metaphase II in meiosis.

b. Anaphase: Anaphase in mitosis, anaphase I and anaphase II in meiosis.

12.14 a. The chance that a gamete would have a particular maternal chromosome is $1/2$. Applying the product rule, the chance of obtaining a gamete with all three maternal chromosomes is $(1/2)^3 = 1/8$.

b. The set of gametes with some maternal and paternal chromosomes is composed of all gametes *except* those that have only maternal or only paternal chromosomes. That is P(gamete with both maternal and paternal chromosomes) $= 1 - P$ (gamete with only maternal chromosomes or gamete with only paternal chromosomes). From (a), the chance of a gamete having chromosomes from only one parent is $1/8$. Using the sum rule, P(gamete with both maternal and paternal chromosomes) $= 1 - (1/8 + 1/8) = 3/4$.

12.15 Since the cells are normal and diploid, chromosomes should exist in pairs. There are pairs of medium and long chromosomes, leaving one short and one long chromosome. These could be members of a heteromorphic pair such as the X and Y chromosomes of a male mammal.

12.17 False. Genetic diversity in the male's sperm is achieved during meiosis, when there is crossing-over between nonsister chromatids and independent assortment of the males' maternal and paternal chromosomes. These processes make it very unlikely that any two sperm cells are genetically identical.

12.19 a. $17 + 26 = 43$ chromosomes.

b. Similar chromosomes pair in meiosis. The pairing pattern seen in the hybrid indicates that some of the chromosomes in these two species share evolutionary similarity, while others do not. Unpaired chromosomes will not segregate in an orderly manner, giving rise to unbalanced meiotic products with either extra or missing chromosomes. This can lead to sterility for two reasons. First, meiotic products that are missing chromosomes may not have genes necessary to form gametes. Second, even if gametes are able to form, a zygote generated from them will not have a complete chromosome set from the hybrid, the red, or the arctic fox. The zygote will be an aneuploid with missing or extra genes, causing it to be infertile.

12.20 The chance of a particular paternal chromosome being present in a gamete is $1/2$. Using the product rule, the chance of all five paternal chromosomes being in one gamete is $(1/2)^5 = 1/32$.

12.23 Fathers always give their X chromosome to their daughters, so the woman must be heterozygous for the color-blindness trait and is c^+c. Her husband received his X chromosome from his mother and has normal color vision, so he is c^+Y. The cross is therefore $c^+c \times c^+Y$. All daughters will receive the paternal X bearing the c^+ allele and have normal color vision. Sons will receive the maternal X, so half will be cY and be color blind, and half will be c^+Y and have normal color vision.

12.25 a. The parental cross is $ww\ vg^+vg^+ \times w^+Y\ vgvg$. This produces F_1 males that are $wY\ vg^+vg$ (white, long wings) and F_1 females that are $w^+w\ vg^+vg$ (red, long wings).

b. In both males and females, the F_2 will be $^3/_8$ white, long; $^3/_8$ red, long; $^1/_8$ white, vestigial; $^1/_8$ white, vestigial.

c. If the F_1 males are crossed back to the female parent, the cross is $ww\ vg^+vg^+ \times wY\ vg^+vg$. All the progeny are white, long. If the F_1 females are crossed back to the male parent, the cross is $w^+w\ vg^+vg \times w^+Y\ vgvg$. Male progeny: $^1/_4$ white, vestigial; $^1/_4$ white, long; $^1/_4$ red, vestigial; $^1/_4$ red, long. All female progeny are red, half are long and half are vestigial.

12.28 The crisscross inheritance pattern (father to daughter) suggests an X-linked trait. Man A marries a normal woman and all his daughters have the trait, so the trait must be dominant. Let X^B be the defective enamel allele and X^b be the normal allele. Man A is X^BY and his wife is X^bX^b, so all of their daughters are X^BX^b. As heterozygotes, they have defective enamel and 50 percent of their offspring receive the X^B allele and are affected. The sons inherit the mother's X^b allele, so they are normal and transmit only the normal allele.

12.30 a. The unaffected parents have offspring affected with an autosomal recessive disorder, so both must be heterozygous. If c^+ is the normal allele and c the affected allele, the cross is $c^+c \times c^+c$ and there is a $^1/_4$ chance of having a cc offspring. Each conception is independent, so the probability that their next child has cystic fibrosis is $^1/_4$.

b. Unaffected offspring are expected in a $1\ c^+c^+ : 2\ c^+c$ ratio, so there is a $^2/_3$ chance that an unaffected child is heterozygous.

12.32 a. Since only a single trait is being followed, consider just that part of the cross: $AA \times aY$. The progeny will be Aa or AY, so the probability of obtaining an $A-$ individual in the F_1 is 1.

b. From (a), there is no chance $(P = 0)$ of obtaining an a male individual in the F_1.

c. The F_1 progeny will be $A-\ Bb\ Cc\ Dd$. Half will be female, so $P = ^1/_2$.

d. Two, $Aa\ Bb\ Cc\ Dd$ females and $AY\ Bb\ Cc\ Dd$ males.

e. For the X chromosome trait, the F_1 cross is $AY \times Aa$. Half of the female offspring ($^1/_4$ of the total) will be heterozygous Aa individuals. For each of the autosomal traits, $^1/_2$ of the offspring will be heterozygous (e.g., $Bb \times Bb$ gives $^1/_2\ Bb$ offspring). Therefore, the chance that an F_2 individual will be heterozygous at all four traits is $^1/_4 \times ^1/_2 \times ^1/_2 \times ^1/_2 = ^1/_{32}$.

f. At any autosomal gene, the chance of obtaining either type of homozygote (e.g., BB, bb) is $^1/_4$ and the chance of obtaining a heterozygote (e.g., Bb) is $^1/_2$. At the A/a gene, the cross is $Aa \times AY$, so the chance of obtaining an AY male or an aY male is $^1/_4$, and the chance of obtaining an $A-$ female is $^1/_2$. Using the product rule, the chance of obtaining (1) an $A-bb\ CC\ dd$ (female) is $P = (^1/_2 \times ^1/_4 \times ^1/_4 \times ^1/_4) = ^1/_{128}$; (2) an $aY\ BB\ Cc\ Dd$ (male) is $P = (^1/_4 \times ^1/_4 \times ^1/_2 \times ^1/_2) = ^1/_{64}$; (3) an $AY\ bb\ CC\ dd$ (male) is $P = (^1/_4 \times ^1/_4 \times ^1/_4 \times ^1/_4) = ^1/_{256}$; and (4) an $aa\ bb\ Cc\ Dd$ (female) is $P = (0 \times ^1/_4 \times ^1/_2 \times ^1/_2) = 0$.

12.33 Primary nondisjunction of sex chromosomes in a ww female produces two types of eggs: ww eggs having two X chromosomes and eggs lacking an X chromosome. Red-eyed males have w^+-bearing and Y-bearing sperm. As shown in the Punnett

square here, the only viable and fertile offspring produced from this cross are wwY females:

	Sperm	
	w^+	Y
ww	www^+ Usually dies	wwY white ♀
O	w^+O Sterile red ♂	YO Dies

Eggs (labeling the left column)

The wwY females are the consequence of primary nondisjunction. They have XY (wY) and X (w) gametes resulting from normal disjunction and, less frequently, XX (ww) and Y gametes resulting from secondary nondisjunction. The results of backcrossing a wwY female to a w^+Y male are shown in the following Punnett square:

		Sperm	
		w^+	Y
Normal X segregation	wY	w^+wY red ♀	wYY white ♂
	w	w^+w red ♀	wY white ♂
Secondary nondisjunction	ww	w^+ww Triplo-X; usually dies	wwY white ♀
	Y	w^+Y red ♂	YY dies

Eggs (labeling left)

12.37 This problem raises the issue that the precise mode of inheritance of a trait often cannot be determined when a pedigree is small and the trait's frequency in a population is unknown. For example, pedigree A could easily fit an autosomal dominant trait (AA and Aa = affected): The affected father would be heterozygous for the trait (Aa), the mother would be unaffected (aa), and half of their offspring would be affected ($A-$). However, pedigree A could also fit an autosomal recessive trait (aa = affected): The father would be homozygous for the trait (aa), the mother would be heterozygous (Aa), and half of their offspring would be affected (aa). Furthermore, pedigree A could fit an X-linked recessive trait: The mother would be heterozygous (X^AX^a), the father would be hemizygous (X^aY), and half of the progeny would be affected (either X^aX^a or X^aY). An X-linked dominant trait would not fit the pedigree because it would require all the daughters of the affected father to be affected (because they all receive their father's X). Pedigrees B and C can be solved by similar analytical reasoning.

	Pedigree A	Pedigree B	Pedigree C
Autosomal recessive	Yes	Yes	Yes
Autosomal dominant	Yes	Yes	No
X-linked recessive	Yes	Yes	No
X-linked dominant	No	No	No

12.39 a. Y-linked inheritance can be excluded because females are affected. X-linked recessive inheritance can also be excluded, as an affected mother (I-2) has a normal son (II-5). Autosomal recessive inheritance can also be excluded since two affected parents, II-1 and II-2, have unaffected offspring.

b. The two remaining mechanisms of inheritance are X-linked dominant and autosomal dominant. Genotypes can be assigned to all members of the pedigree that satisfy either inheritance mechanism. Of these two, X-linked dominant inheritance may be more likely, as II-6 and II-7 have only affected daughters, suggesting crisscross inheritance. If the trait were autosomal dominant, half of the daughters and half of the sons should be affected.

12.42 Answer (a) is false because an affected father who is heterozygous should only have half affected children. Answer (b) is false because an affected mother who is heterozygous should have half affected offspring, regardless of sex type. Answer (c) is false because two heterozygous parents should have $1/4$ of their offspring be homozygous for the recessive, normal allele. Answer (d) is true. However, if the mutation were newly arisen in either the child or his or her parents, his or her grandparent could have been unaffected.

12.43 Answer (a) is true because two affected individuals will always have affected children ($aa \times aa$ can give only aa offspring). Answer (b) is false because an autosomal trait is inherited independent of sex type. Answer (c) need not be true, as the trait could be masked by normal dominant alleles through many generations before two heterozygotes marry and produce affected, homozygous offspring. Answer (d) could also be true. If the trait is rare, then it is likely that an unaffected individual marrying into the pedigree is homozygous for a normal allele. Since the trait is recessive, and the children receive the dominant, normal allele from the unaffected parent, the children will be normal. Answer (d) would not be true if the unaffected individual was heterozygous. In this case, half of the children would be affected.

12.45 Since hemophilia is an X-linked trait, the most likely explanation is that random inactivation of X chromosomes (lyonization) produces individuals with different proportions of cells with a functioning allele. This in turn leads to different amounts of clotting factor being made. Normal clotting times would be expected in females whose X^h chromosome was very frequently inactivated. In these individuals, most cells have a functioning h^+ allele and near-normal amounts of clotting factor are made. In contrast, a clotting time consistent with clinical hemophilia would be seen in a woman having the X^{h^+} chromosome inactivated, say, 90 percent of the time. In these individuals, only 10 percent of the cells have a functioning h^+ allele, and very little clotting factor would be made.

Chapter 13 Extensions of Mendelian Genetic Principles

13.2 Six genotypes are possible: w^1/w^1, w^1/w^2, w^1/w^3, w^2/w^2, w^2/w^3, w^3/w^3.

13.6 The woman's genotype is $I^A I^B$ and the man's genotype is $I^A i$. Their offspring have four equally likely genotypes ($I^A I^A$, $I^A I^B$, $I^A i$, $I^B i$) and three phenotypes: A ($P = 1/2$), AB ($P = 1/4$) and B ($P = 1/4$).

a. $P = 1/2 \times 1/2 = 1/4$.

b. $P = 0$, as there is no chance of producing a group O child.

c. P [(first child is male and AB) and (second child is male and B)] = ($1/2 \times 1/4$) × ($1/2 \times 1/4$) = $1/64$.

13.8 Blood type O, because the genotype is i/i.

13.10 The cross $C^R/C^W \times C^R/C^W$ gives a 1 C^W/C^W : 2 C^R/C^W : 1 C^R/C^R progeny ratio, so half of the progeny resemble the parents in coat color.

13.12 The parental cross is $F/F \ G^N/G^N \times f/f \ G^O/G^O$.

a. The F_1 is $F/f \ G^N/G^O$ and has fuzzy skin and round leaf glands.

b. Since the alleles at the G gene show incomplete dominance, there will be a modified 9:3:3:1 ratio in the F_2. The progeny will be $3/16$ fuzzy with oval glands ($F/- \ G^O/G^O$), $6/16$ fuzzy with round glands ($F/- \ G^N/G^O$), $3/16$ fuzzy with no glands ($F/- \ G^N/G^N$), $1/16$ smooth with oval glands ($f/f \ G^O/G^O$), $2/16$ smooth with round glands ($f/f \ G^O/G^N$), and $1/16$ smooth with no glands ($f/f \ G^N/G^N$).

c. The offspring of the cross $F/f \ G^N/G^O \times f/f \ G^O/G^O$ are $1/4$ each of fuzzy with oval glands ($f/f \ G^O/G^O$), fuzzy with round glands ($f/f \ G^O/G^N$), smooth with oval glands ($F/f \ G^O/G^O$), and smooth with round glands ($F/f \ G^N/G^O$).

13.18 The 9:7 F_2 ratio is a modified 9:3:3:1 ratio obtained from the F_1 cross $A/a \ B/b \times A/a \ B/b$. The $9/16$ colored plants are $A/- \ B/-$ and will show the genotypic ratio 1 $A/A \ B/B$: 2 $A/a \ B/B$: 2 $A/A \ B/b$: 4 $A/a \ B/b$. Since both A and B are required for color, only if a true-breeding $A/A \ B/B$ plant is selfed will there be "no segregation of the two phenotypes among its progeny." The $A/A \ B/B$ plants are $1/9$ of the colored plants, so $P = 1/9$.

13.19 The 9:7 ratio in the F_2 is a modified 9:3:3:1 ratio, where the $A/- \ B/-$ genotypes are "runner" and the $A/- \ b/b$, $a/a \ B/-$, and $a/a \ b/b$ genotypes are "bunch." This is an example of duplicate recessive epistasis: Recessive alleles at either of the genes block (are epistatic to) the "runner" phenotype, resulting in the "bunch" phenotype.

13.20 a. The cross is $A/a \ B/b \times A/a \ B/b$, which gives $9/16$ $A/- \ B-$, $3/16$ $A/- \ b/b$, $3/16$ $a/a \ B/-$, and $1/16$ $a/a \ b/b$. The $A/- \ b/b$, $a/a \ B/-$, and $a/a \ b/b$ individuals are deaf because they are homozygous for either one or both recessive alleles. Only the $A/- \ B/-$ individuals can hear. Therefore, the phenotypic ratio is 9 hearing rabbits : 7 deaf rabbits.

b. This alleles show duplicate recessive epistasis. Homozygous recessive alleles at either of two genes block hearing, and are epistatic to the dominant alleles at the other gene.

c. The cross is $a/a \ B/b \times A/a \ B/b$, which gives $5/8$ deaf progeny ($1/8$ $A/a \ b/b$ + $1/2$ $a/a \ -/-$) and $3/8$ hearing progeny ($A/a \ B/-$).

13.23 Let C/c represent alleles at the locus controlling the pigment production and Y/y represent the alleles at the yellow/agouti locus. The 3 colored : 1 albino progeny ratio indicates that $C/-$ individuals produce pigment while c/c individuals do not and are albino. The 2 yellow : 1 agouti progeny ratio is a modified monohybrid cross ratio indicating recessive lethality: Y/Y individuals die, Y/y are yellow, and y/y are agouti. Since albino mice don't express alleles at the agouti locus, c/c is epistatic to alleles at the Y/y locus and $c/c \ Y/y$ and $c/c \ y/y$ individuals are albino.

a. First, infer the partial genotypes: yellow mice are $C/- \ Y/y$ and albino mice are $c/c \ -/y$. Then determine the complete genotypes from the progeny ratios for each trait. A 1 colored : 1 albino progeny ratio is expected from a $C/c \times c/c$ cross. A 2 yellow : 1 agouti progeny ratio is expected from a $Y/y \times Y/y$ cross. Therefore, the parental genotypes were $C/c \ Y/y \times c/c \ Y/y$.

b. The cross is $C/c \ Y/y \times C/c \ Y/y$, and, since Y/Y progeny are inviable, will produce a 1 albino : 2 yellow : 1 agouti phenotypic ratio. None of the yellow mice will be true breeding, as they are all Y/y.

13.25 a. The three complementation groups identify three genes.

b. A, F, and D have defects in one gene; B and G are in a second gene; C and E are in a third gene.

13.26 a. Y/Y R/R (crimson) × y/y r/r (white) gives a Y/y R/r magenta-rose F_1. Selfing the F_1 gives an F_2 that is $\frac{1}{16}$ crimson (Y/Y R/R), $\frac{1}{8}$ orange-red (Y/Y R/r), $\frac{1}{16}$ yellow (Y/Y r/r), $\frac{1}{8}$ magenta (Y/y R/R), $\frac{1}{4}$ magenta-rose (Y/y R/r), $\frac{1}{8}$ pale yellow (Y/y r/r), and $\frac{1}{4}$ white (y/y $-/-$). A backcross of the F_1 to the crimson parent will give $\frac{1}{4}$ crimson (Y/Y R/R), $\frac{1}{4}$ magenta-rose (Y/y R/r), $\frac{1}{4}$ magenta (Y/y R/R), and $\frac{1}{4}$ orange-red (Y/Y R/r).

b. Y/Y R/r (orange-red) × Y/y r/r (pale yellow) gives $\frac{1}{4}$ orange-red (Y/Y R/r), $\frac{1}{4}$ magenta-rose (Y/y R/r), $\frac{1}{4}$ yellow (Y/Y r/r), and $\frac{1}{4}$ pale yellow (Y/y r/r).

c. Y/Y r/r (yellow) × y/y R/r (white) gives $\frac{1}{2}$ magenta-rose (Y/y R/r) and $\frac{1}{2}$ pale yellow (Y/y r/r).

13.29 a. In order to be black using this pathway, individuals must have all three normal alleles: They must be $A/- B/- C/-$. The F_1 is the trihybrid A/a B/b C/c that when selfed, gives black $A/- B/- C/-$ individuals in ($\frac{3}{4} \times \frac{3}{4} \times \frac{3}{4}$) = $\frac{27}{64}$ of the progeny. The remaining $1 - \frac{27}{64} = \frac{37}{64}$ of the progeny are colorless, having a/a, b/b, and/or c/c genotypes. A 27 black : 37 colorless progeny ratio is expected in the F_2.

b. In order to be black using this pathway, individuals must have the A and B functions, but not the inhibitor function provided by C: They must be $A/- B/- c/c$. The chance of obtaining this genotype in the F_2 progeny is $\frac{3}{4} \times \frac{3}{4} \times \frac{1}{4} = \frac{9}{64}$. The remaining $1 - \frac{9}{64} = \frac{55}{64}$ of the progeny will be colorless. A 9 black : 55 colorless progeny ratio is expected in the F_2.

c. Here, the ratio of black to colorless in the F_2 can be used to distinguish between hypotheses concerning the two pathways proposed in (a) and (b). Evaluate whether the F_2 results fit either pathway using a chi-square test.

13.30 A single p^+ allele provides 50 percent of the enzyme activity seen in a p^+/p^+ homozygote. Since p^+ is dominant (i.e., $P-$ C/C plants are purple), this appears to be enough activity to provide for a wild-type phenotype. If a plant with less than 50 percent of normal activity does not synthesize enough purple pigment for a wild-type phenotype (e.g., 25% of normal activity gives a light-purple flower) and a plant with more than 100 percent of normal activity produces noticeably darker purple pigmentation, the phenotypes in Table 13.A should be seen.

13.35 Hornless is a sex-influenced trait. In males, H/H and H/h are horned, and h/h is hornless. In females, H/H is horned. The cross is H/H W/W ♂ × h/h w/w ♀. The F_1 is H/h W/w — horned white males and hornless white females. Interbreeding the F_1 gives the following F_2:

	Male	**Female**
$\frac{3}{16}$ H/H $W/-$	horned, white	horned, white
$\frac{6}{16}$ H/h $W/-$	horned, white	hornless, white
$\frac{3}{16}$ h/h $W/-$	hornless, white	hornless, white
$\frac{1}{16}$ H/H w/w	horned, black	horned, black
$\frac{2}{16}$ H/h w/w	horned, black	hornless, black
$\frac{1}{16}$ h/h w/w	hornless, black	hornless, black

In sum, the ratios for males are $\frac{9}{16}$ horned white : $\frac{3}{16}$ hornless white : $\frac{3}{16}$ horned black : $\frac{1}{16}$ horned black. The ratios for females are $\frac{3}{16}$ horned white : $\frac{9}{16}$ hornless white : $\frac{1}{16}$ horned black : $\frac{3}{16}$ hornless black.

13.37 Initially, use the information in Question 13.35 to infer partial genotypes from phenotypes:

Individual	Phenotype	Inferred Partial Genotype
Male parent	horned white male	$H/- W/-$
Ewe A	hornless black female	$-/h$ w/w
Ewe A offspring	horned white female	H/H $W/-$
Ewe B	hornless white female	$-/h$ $W/-$
Ewe B offspring	hornless black female	$-/h$ w/w
Ewe C	horned black female	H/H w/w
Ewe C offspring	horned white female	H/H $W/-$
Ewe D	hornless white female	$-/h$ $W/-$
Ewe D offspring #1	hornless black male	h/h w/w
Ewe D offspring #2	horned white female	H/H $W/-$

Then compare the offspring to their parents. Since Ewe D's male offspring is h/h w/w, both Ewe D and the male parent must have at least one recessive allele at each gene. Since Ewe A and Ewe D have H/H offspring, Ewe A and Ewe D must each have an H allele. Since Ewe B has a w/w offspring, she must have a w allele. Therefore, the male parent and Ewe D are H/h W/w, Ewe A is H/h w/w, Ewe B is either H/h W/w or h/h W/w, and Ewe C is H/H w/w.

13.38 c.

Table 13.A

Genotype	Percent of +/+ Activity	Percent of +/+ Activity When Mixed 50:50 with +/+ Extract	(A) Homozygote Phenotype	(B) Heterozygote Phenotype	(C) Hemizygote Phenotype	(D) Allele Classification
p^+/p^+	100	100	purple	purple	purple	wild type
p^1/p^1	20	60	light purple	purple	very light purple	hypomorph
p^2/p^2	0	50	white	purple	white	amorph
p^3/p^3	300	200	very dark purple	dark purple	dark purple	hypermorph
p^4/p^4	0	5	white	very light purple	white	antimorph
p^5/p^5	0	50	red	reddish purple	red	neomorph

Chapter 14 Quantitative Genetics

14.1 a. The mean of a sample is obtained by summing all the individual values and dividing by the total number of those values. The mean head width is 25.21/8 = 3.15 cm, and the mean wing length is 281.7/8 = 35.21 cm.

The standard deviation equals the square root of the variance (s^2). The variance is computed by summing the squares of the differences between each measurement and the mean value, and dividing this sum by the number of measurements minus one:

$$s_{\text{head width}} = \sqrt{s^2} = \sqrt{\frac{\Sigma (x_i - \bar{x})^2}{n - 1}}$$

$$= \sqrt{\frac{1.70}{7}} = \sqrt{0.24} = 0.49 \text{ cm}$$

$$s_{\text{wing length}} = \sqrt{s^2} = \sqrt{\frac{\Sigma (x_i - \bar{x})^2}{n - 1}}$$

$$= \sqrt{\frac{413.35}{7}} = \sqrt{59.05} = 7.68 \text{ cm}$$

b. The correlation coefficient, r, is calculated from the covariance, cov, of two sets of data. Let head width be represented by x, and wing length be represented by y. r is defined as

$$r = \frac{cov_{xy}}{s_x s_y} = \frac{\dfrac{\Sigma x_i y_i - n\bar{x}\bar{y}}{n - 1}}{s_x s_y}$$

The first factor ($\Sigma x_i y_i$) is obtained by taking the sum of the products of the individual measurements of head width and wing length for each duck. The next factor is the product of the number of individuals and the means of these two sets of measurements. The difference between these values is then divided by ($n - 1$), and then by the products of the standard deviations of each measurement. Thus,

$$r = \frac{\dfrac{913.50 - 8 \times 3.15 \times 35.21}{7}}{0.49 \times 7.68} = \frac{3.74}{3.76} = 0.99$$

c. Head width and wing length show a strong positive correlation, nearly 1.0. This means that ducks with larger heads will almost always have longer wings, and ducks with smaller heads will almost always have shorter wings.

14.2 When given a series of data, the best first step is to graph the data. We can determine the minimum and maximum values, and then create histograms for each series of data using different bin sizes (e.g., by 1, by 2, by 5, etc.) to get a feel for the distribution of the data. Some of these sample histograms can be found in the graphs presented here, along with notes on interpretation and the sources of the original data. One final note is that while many times data from a particular sample do not appear to have the bell-shaped distribution characteristic of a quantitative trait, if we know that we are dealing with a quantitative trait we often assume that the data are normally distributed so that we can apply certain statistical techniques in analyzing the data.

a.

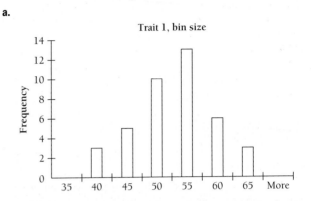

These data appear to be normally distributed, so we could assume that they are representative of the phenotypic data we would see for a quantitative trait. In fact, these are 40 sample values taken from a normal distribution with $\mu = 50$ and $s^2 = 5$.

b.

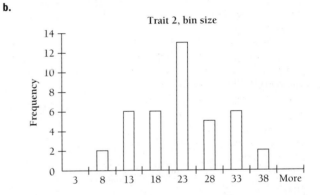

These data appear to have a more pronounced peak in the middle, without the "shoulders" next to the peak, such as those seen in (a). We could conclude that these data are not representative of a quantitative trait. In fact, these data are 10 sample values from a normal distribution, with $\mu = 10$ and $s^2 = 2$, 20 values from $\mu = 20$ and $s^2 = 2$, and 10 values from $\mu = 30$ and $s^2 = 2$. This is what we might expect to see from a simple additive Mendelian character with some environmental variance.

c.

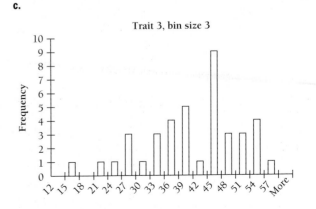

These data have a strong peak, but they do not have the characteristic shape of a normal distribution. You can see that there are no shoulders to the peak and that the peak trails off to the left, but not to the right. In fact, these data include 10 sample values from a normal distribution with $\mu = 25$ and $s^2 = 5$, 20 values from $\mu = 42$ and $s^2 = 5$, and 10 values from $\mu = 55$ and $s^2 = 5$, something we might expect to see from a trait showing simple Mendelian inheritance with a small degree of dominance and substantial environmental variance.

14.3 The degree of phenotypic variability is related to the degree of genetic variability. Since each pure-breeding parent is homozygous for the genes (however many there are) that control the size character, the variation seen within parental lines is due only to the environmental variation present. A cross of two pure-breeding strains will generate an F_1 heterozygous for those loci controlling the size trait, but genetically as homogeneous as each of the parents. Therefore, the only variation we would expect to see in the F_1 is that caused by the environment, and the F_1 should show no greater variability than do the parents.

14.4 a. Since the cross is $AA\ BB \times aa\ bb$, the F_1 genotype will be $Aa\ Bb$. Since capital alleles additively determine height, and individuals with four capital alleles have a height of 50 cm, while individuals with no capital alleles have a height of 30 cm, each capital allele appears to confer $(50 - 30)/4 = 5$ cm of height over the 30-cm base. $Aa\ Bb$ individuals with two capital alleles should have an intermediate height of 40 cm.

b. Any individuals with two capital alleles will show a height of 40 cm. Thus, $Aa\ Bb$, $AA\ bb$, and $aa\ BB$ individuals will be 40 cm high.

c. In the F_2, $1/16$ of the progeny are $AA\ bb$, $4/16$ are $Aa\ Bb$, and $1/16$ are $aa\ BB$. Thus, $6/16 = 3/8$ of the progeny will be 40 cm high.

d. In answering this question, we have assumed that the A and B loci assort independently and that each locus and each allele contribute equally to the phenotype.

14.10 Internode length shows the characteristics of a quantitative trait. These characteristics include F_1 progeny that show a phenotype intermediate between the two parental phenotypes, and an F_2 showing a range of phenotypes with extremes in the range of the two parents, some of which have all the original parental alleles.

14.13 In order to see transgressive segregation, at least one of the parents must have some alleles that are "opposite" in effect of the expected direction. For example, imagine we are looking at height. If we assume that there are six loci that contribute to height, and that capital alleles contribute a 5-cm increase over a base height of 1 meter, a cross between an $AA\ BB\ CC\ DD\ EE\ FF$ individual (160 cm) and an $aa\ bb\ cc\ dd\ ee\ ff$ individual (100 cm) will produce an F_2 with extreme individuals only as tall and as short as the original parents. If, however, the original parents have the genotypes $AA\ BB\ CC\ DD\ EE\ ff$ (150 cm) and $aa\ bb\ cc\ dd\ ee\ FF$ (110 cm), segregation in the F_2 can produce an $AA\ BB\ CC\ DD\ EE\ FF$ genotype (160 cm) and an $aa\ bb\ cc\ dd\ ee\ ff$ genotype (100 cm), which are taller and shorter than the original lines used. In this case, the taller parent has "shorter" alleles at one locus and vice versa.

A more extreme case to consider is where the parents have the same phenotype, but produce segregating offspring. Imagine, for example, if an $AA\ BB\ CC\ dd\ ee\ ff$ (130 cm) individual were crossed with an $aa\ bb\ cc\ DD\ EE\ FF$ (130 cm) individual. Their F_1 offspring would again be 130 cm ($Aa\ Bb\ Cc\ Dd\ Ee\ Ff$), but in the F_2, individuals from 160 cm ($AA\ BB\ CC\ DD\ EE\ FF$) to 100 cm ($aa\ bb\ cc\ dd\ ee\ ff$) could be seen!

14.14 The F_1 is $A/a\ B/b\ C/c\ D/d\ E/e$, and so it is greyish-brown. The F_2 phenotypes are determined by the number of capital alleles contributed from each F_1 parent. The easiest way to proceed is to try to find the proportions of individuals with the light tan pigmentation (3 or 4 capital alleles) and the whitish blue pigmentation (0 or 1 capital alleles), and the proportion of greyish brown offspring will be the rest. Start off by determining the chance of obtaining 0, 1, 2, or 3 capital alleles in a gamete from each F_1 parent and then determining how these combinations make the desired genotypes.

Since the parent is heterozygous at all five loci, the chance of obtaining any specified set of five alleles in one gamete is $(1/2)^5$. The chance of obtaining a particular number of capital alleles from an F_1 is the number of ways in which that number of alleles can be obtained, multiplied by $(1/2)^5$. There is one way to obtain 0 capital alleles, 5 ways to obtain 1 capital allele, 10 ways of obtaining two capital alleles, and 10 ways of obtaining three capital alleles. With this information, we can tabulate the ways in which the progeny classes we are interested in can be formed:

| F₁ Gamete #1 | | F₁ Gamete #2 | | F₂ Progeny | | |
Capital Alleles	Gamete Fraction	Capital Alleles	Gamete Fraction	Capital Alleles	F₂ Fraction	Phenotype
0	$(1/2)^5$	0	$(1/2)^5$	0	$(1/2)^{10}$	whitish blue
0	$(1/2)^5$	1	$5(1/2)^5$	1	$5(1/2)^{10}$	whitish blue
1	$5(1/2)^5$	0	$(1/2)^5$	1	$5(1/2)^{10}$	whitish blue
0	$(1/2)^5$	2	$10(1/2)^5$	2	$10(1/2)^{10}$	light tan
1	$5(1/2)^5$	1	$5(1/2)^5$	2	$25(1/2)^{10}$	light tan
2	$10(1/2)^5$	0	$(1/2)^5$	2	$10(1/2)^{10}$	light tan
0	$(1/2)^5$	3	$10(1/2)^5$	3	$10(1/2)^{10}$	light tan
1	$5(1/2)^5$	2	$10(1/2)^5$	3	$50(1/2)^{10}$	light tan
2	$10(1/2)^5$	1	$5(1/2)^5$	3	$50(1/2)^{10}$	light tan
3	$10(1/2)^5$	0	$(1/2)^5$	3	$10(1/2)^{10}$	light tan

In the F_2, $11(1/2)^{10} = {}^{11}/_{1,024}$ will be bluish white and $165(1/2)^{10} = {}^{165}/_{1,024}$ will be light tan. The remaining $[1 - 176(1/2)^{10}] = {}^{848}/_{1,024}$ F_2 progeny will be greyish brown.

There is another method to solve this problem: Use the coefficients of the binomial expansion to determine the proportion of progeny with different numbers of capital and lowercase alleles. Let n = total number of alleles, s = number of capital alleles, t = number of lowercase alleles, a = chance of obtaining a capital allele, b = chance of obtaining a lowercase allele, and $x! = (x)(x - 1)(x - 2)\ldots(1)$, with $0! = 1$. Then the chance p of obtaining progeny with a specified number of each type of allele is given by

$$p(s,t) = \frac{n!}{s!t!}a^s b^t$$

$$p(0,10) = \frac{10!}{0!10!}\left(\frac{1}{2}\right)^0\left(\frac{1}{2}\right)^{10} = \frac{1}{1024}$$

$$p(1,9) = \frac{10!}{1!9!}\left(\frac{1}{2}\right)^1\left(\frac{1}{2}\right)^9 = \frac{10}{1024}$$

$\left.\begin{array}{c} \\ \\ \end{array}\right\}\ \dfrac{11}{1024}$ whitish blue

$$p(2,8) = \frac{10!}{2!8!}\left(\frac{1}{2}\right)^2\left(\frac{1}{2}\right)^8 = \frac{45}{1024}$$

$$p(3,7) = \frac{10!}{3!7!}\left(\frac{1}{2}\right)^3\left(\frac{1}{2}\right)^7 = \frac{120}{1024}$$

$\left.\begin{array}{c} \\ \\ \end{array}\right\}\ \dfrac{165}{1024}$ light tan

$$1 - \frac{11 + 165}{1024} = \frac{848}{1024}\ \text{greyish brown}$$

14.15 a. From the data that are given, it appears that some proportion of cases of AD can be attributed to genetic factors. Multiple genes that increase the risk for AD have been identified, some of which appear to act in a dose-dependent manner. Thus, it could be that a number of different genes contribute to the onset of AD, with some having a greater contribution factor than others. This is somewhat similar to how polygenic traits control a phenotype, since there, alleles at multiple genes contribute in an additive, dose-dependent fashion to the phenotype.

b. Consider two explanations: First, if AD can be caused by environmental agents, mutation, and/or a combination of both environmental agents and mutation, the presence of AD in both twins could be due to the presence of one or more abnormal alleles in both and/or the exposure of both twins to adverse environmental conditions. The presence of AD in only one twin may be due to differences in the exposure of that twin to a contributing or causative environmental agent(s). Second, the presence of a particular allele or a specific mutation may only increase the risk of disease, and not determine its occurrence, since an allele's penetrance may be strongly affected by the environment. In the case of AD, the environmental factors may not be clear cut or even small in number. There may be multiple environmental factors, some of which may be complex or subtle.

14.18 SHR rats will continue to respond to salt by developing hypertension. Since the strain is inbred, any variation in blood pressure will result from the amount of exposure to salt, and not from genetic variation. Therefore, heritability for this population will be zero. Similarly, the inbred TIS rats would also have a heritability of zero (and retain a low blood pressure).

14.20 a. The narrow-sense heritability of the number of triradii will equal the slope, b, of the regression line of the mean offspring phenotype on the mean parental phenotype.

$$b = \frac{cov_{xy}}{(s_x)^2} = \frac{\dfrac{\Sigma x_i y_i - n\overline{x}\overline{y}}{n - 1}}{\dfrac{\Sigma(x_i - \overline{x})^2}{n - 1}} = \frac{\Sigma x_i y_i - n\overline{x}\overline{y}}{\Sigma(x_i - \overline{x})^2}$$

For this data set, x is the mean number of triradii in the parents and y is the mean number of triradii in the offspring. Using either a calculator or a spreadsheet or statistics program, you can find the following:

$$\Sigma x_i = 111,\ \overline{x} = 11.1$$
$$\Sigma y_i = 108.5,\ \overline{y} = 10.85$$
$$\Sigma(x_i - \overline{x})^2 = 51.4$$
$$\Sigma x_i y_i = 1,257.5$$
$$b = \frac{1257.5 - 10 \times 11.1 \times 10.85}{51.4} = \frac{53.15}{51.4} = 1.04$$

b. A slope of 1.04 indicates that additive genetic variation is responsible for essentially all of the observed variation in phenotype. Note that the estimate obtained for h^2 is greater than one, showing that methods for estimating narrow-sense heritability can overestimate the amount of additive genetic variation among individuals.

14.22 The selection differential (S) equals $14.3 - 9.7 = 4.6$ cm. The response to selection (R) equals $13 - 9.7 = 3.3$ cm. The narrow-sense heritability H_N^2 equals $R/S = 3.3/4.6 = 0.72$.

14.26 A response to selection depends on (a) variation on which selection can act and (b) a high narrow-sense heritability so that the selected individuals produce similar offspring. The narrow-sense heritability for each of the traits is V_A/V_P: 0.165 for body length, 0.061 for antenna bristle number, and 0.144 for egg production. The amount of raw variation is also greatest for body length. Thus, body length will respond most to selection, and antenna bristle number will respond least to selection.

14.27 Assume that there are multiple loci that contribute equally to fruit weight and days to first flower. In order to recover the cultivated phenotype most quickly from selection after crossing it with the wild genotype, we would like to find the cross where most of the variation is due to additive effects. A quick way to assess this is to look at the phenotype of the F_1: If most of the variation is due to additive effects, the phenotype of the F_1 will be intermediate to both parents. If the F_1 is closer in phenotype to one parent or the other, this can be taken as an indicator that that parent harbors some nonadditive variation. Using this criterion for both traits, crosses 2 and 4 appear to be the best initial crosses to work with.

Chapter 15 Gene Mapping in Eukaryotes

15.2 In a chi-square test of these data under the hypothesis of independent assortment, $\chi^2 = 16.1$ and $P < 0.01$: There is $< 1\%$ likelihood of observing this much deviation from the expected values by chance alone. Linkage might seem reasonable initially, but it is inconsistent with the minority classes not being reciprocal classes (both carry the a/a phenotype). Consider the segregation at each locus: The $B/- : b/b$ ratio is about 1:1 (203:197), while the $A/- : a/a$ ratio is not (240:160). The large deviation, then, is due to a reduced number of a/a

individuals. This reduction should be confirmed in other crosses that test segregation at the *A/a* locus. In corn, further evidence might show up as a class of ungerminated seeds or of seedlings that die early.

15.5 If *m* were X linked, the parental cross could be written: $X^m/X^m\ vg^+/vg^+$ ♀ × $X^{m^+}/Y\ vg/vg$ ♂. The F_1 males would be $X^m/Y\ vg^+/vg$, and thus would have maroon eyes. Since the observed F_1 are all wild type, *m* cannot be X linked. If *vg* and *m* are linked, the crosses could be written $vg^+\ m/vg^+\ m$ ♀ × $vg\ m^+/vg\ m^+$ ♂. This would produce an F_1 that is $vg^+\ m/vg\ m^+$ and all wild type. The progeny of an $F_1 \times F_1$ cross are diagrammed below, recognizing that *Drosophila* females, but not males, exhibit recombination:

		Gametes of $vg^+\ m/vg\ m^+$ ♂	
		$vg^+\ m$ (parental)	$vg\ m^+$ (parental)
Gametes of $vg^+\ m/vg\ m^+$ ♀	$vg^+\ m$ (parental)	$vg^+\ m/vg^+\ m$ maroon	$vg^+\ m/vg\ m^+$ wild type
	$vg\ m^+$ (parental)	$vg\ m^+/vg^+\ m$ wild type	$vg\ m^+/vg\ m^+$ vestigial
	$vg^+\ m^+$ (recombinant)	$vg^+\ m^+/vg^+\ m$ wild type	$vg^+\ m^+/vg\ m^+$ wild type
	$vg\ m$ (recombinant)	$vg\ m/vg^+\ m$ maroon	$vg\ m/vg\ m^+$ vestigial

No recombinants are produced in the male parent, so it is impossible to obtain a $vg\ m/vg\ m$ animal from this cross. Since a double mutant was found in the F_2 progeny, the genes cannot be linked. Finding a double mutant is enough evidence to conclude that the genes assort independently, as this is the only way to obtain $vg\ m$ gametes from *both* parents. Since *m* is not on the X or chromosome 2, it must be on chromosome 3 or 4.

15.6 The multiple crossovers that occur over distant intervals result in large recombination frequencies being less accurate measures of map distance than the small recombination frequencies observed between close neighbors. Therefore, construct a map starting with the smallest recombination frequencies, working upward.

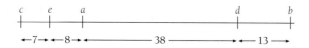

Due to the effects of multiple crossovers, recombination frequencies between loci are not strictly additive. Although recombination frequency will not exceed 50 percent, map distances can exceed 50 map units. The map distance between *c* and *b* is 66 map units (= 7 + 8 + 38 + 13), but *c* and *b* show 50 percent recombination.

15.10 In this testcross, the genotypes of the dihybrid's gametes determine the progeny phenotypes. The $a\ b^+/a^+\ b$ parent has 90 percent parental type ($a\ b^+$, $a^+\ b$) and 10 percent recombinant ($a\ b$, $a^+\ b^+$) gametes, giving 45% $a\ b^+$, 45% $a^+\ b$, 5% $a\ b$, and 5% $a^+\ b^+$ offspring.

15.11 The genes are 7 mu apart, so the female has 7% recombinant (3.5% each $a\ b$, $a^+\ b^+$) gametes and 93% parental (46.5% each $a^+\ b$, $a\ b^+$) gametes. The wild-type male has either $a^+\ b^+$ or Y gametes.

    a. $P = 0.035(a^+\ b^+) + 0.465(a\ b^+) = 0.50$

    b. $P = 1$, as all daughters receive their father's X, which is carrying $a^+\ b^+$.

15.14 Infer the recombination frequency between two genes from the map distance between them. For example, *a* and *b* are 20 mu apart, so $A\ B/a\ b$ will have 20% recombinant (10% each $A\ b$, $a\ B$) and 80% parental (40% each $A\ B$, $a\ b$) gametes. Then, since each chromosome pair segregates independently, use the product rule and multiply together the probabilities of obtaining alleles from each chromosome.

    a. $P(A\ B\ C\ D\ E\ F) = 0.40 \times 0.45 \times 0.35 = 0.063$, or 6.3 percent.

    b. $P(A\ B\ C\ d\ e\ f) = 0.40 \times 0.05 \times 0.35 = 0.007$, or 0.7 percent.

    c. $P(A\ b\ c\ D\ E\ f) = 0.10 \times 0.05 \times 0.15 = 0.00075$, or 0.075 percent.

    d. $P(a\ B\ C\ d\ e\ f) = 0.10 \times 0.05 \times 0.35 = 0.00175$, or 0.175 percent.

    e. $P(a\ b\ c\ D\ e\ F) = 0.40 \times 0.05 \times 0.15 = 0.003$, or 0.3 percent.

15.15 With respect to the *D/d* and *P/p* loci, there are 95 percent parental type progeny and 5 percent recombinants. The *H/h* locus assorts independently.

    a. 47.5 percent each *D P h* and *d p h*; 2.5 percent each *D p h* and *d P h*.

    b. 23.75 percent each *d P H*, *d P h*, *D p H*, and *D p h*; 1.25 percent each *D P H*, *D P h*, *d p H*, and *d p h*.

15.19 The double-crossover classes will always be the least frequent.

    a. *F m W*, *f M w*

    b. *M f W*, *m F w*

    c. *F w M*, *f W m*

15.21 a. The crosses are (the correct order of the loci is not yet determined):

    P: $+ + dp/+ + dp$ ♀ × $b\ hk + /b\ hk +$ ♂

    F_1 testcross: $+ + dp/b\ hk +$ ♀ × $b\ hk\ dp/b\ hk\ dp$ ♂

Tabulate the data to include parental, single-crossover (SCO) and double-crossover (DCO) classes (determined from their frequency):

Phenotype	Gamete Genotype	Number	Class
dumpy	$+ + dp$	305	parental
black, hooked	$b\ hk +$	301	parental
hooked, dumpy, black	$b\ hk\ dp$	171	SCO
wild type	$+ + +$	169	SCO
dumpy, hooked	$+ hk\ dp$	21	SCO
black	$b + +$	19	SCO
hooked	$+ hk +$	8	DCO
dumpy, black	$b + dp$	6	DCO

Compare the DCO to parental classes to determine the correct gene order: *b* is in the middle. Rewrite the F_1 trihybrid using the correct gene order ($dp + +/+ b\ hk$) and determine which SCOs belong to each interval: The *dp b hk* and $+ + +$ are SCOs between *dp* and *b*, while the *dp + hk* and $+ b +$ are SCOs between *b* and *hk*. Calculate the recombination frequencies (RFs):

$$RF(dp - b) = [(171 + 169 + 6 + 8)/1,000] \times 100\% = 35.4\%$$
$$RF(b - hk) = [(21 + 19 + 6 + 8)/1,000] \times 100\% = 5.4\%$$

The map distances are: $dp - b$, 35.4 mu; $b - hk$, 5.4 mu; $dp - hk$, 40.8 mu.

b. Interference = 1 − coefficient of coincidence

= 1 − frequency observed DCO/ frequency expected DCO

= 1 − (14/1,000)/0.354 × 0.054)

= 1 − 0.73 = 0.27

15.25 Start by considering a simpler cross between a female heterozygous for an X-linked lethal (*l*) and a normal male: P: $l/+\ ♀ × +/Y\ ♂$; F_1: $+/Y\ ♂$, $l/Y\ ♂$ (dead), $l/+\ ♀$, $+/+\ ♀$. Half of the male progeny are not recovered due to the *l* allele, and the *l*-bearing chromosome is recovered only in the female offspring where it is masked by the dominant wild-type allele. Here, half of the male progeny are also not recovered due to the *l* allele, so each of the four classes seen represents one of the two reciprocal classes of progeny recovered in a three-point cross. The classes not seen bear the *l* allele. Tabulate the data from the male progeny into parental, single-crossover (SCO) and double-crossover (DCO) classes (determined from their frequency), including the third (*l/+*) locus:

Gamete Genotype	Number	Class
a + +	405	parental
+ *b* +	44	SCO
+ + +	48	SCO
a *b* +	2	DCO

Comparison of the parental and DCO classes indicates that the correct order is $a - b - l$. Since one of the parental-type chromosomes is $a + +$, the other is its reciprocal, $+ b\ l$. This means the heterozygous female was $a + +/+ b\ l$. The 44 $+ b +$ progeny result from single crossovers between b and l, the 48 $+ + +$ progeny result from single crossovers between a and b, and the 2 $a\ b +$ progeny result from crossovers in both intervals. For each class, the progeny are half of the total crossovers, as only one of the two reciprocal events in each class is viable. Since half of the progeny in each class are not recovered, use these numbers to estimate recombination frequencies (RF) and construct a map:

$$RF(a - b) = [(48 + 2)/499] × 100\% = 10\%$$
$$RF(b - l) = [(44 + 2)/499] × 100\% = 9.2\%$$

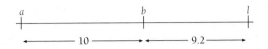

15.26 Since the male parent is triply recessive, the phenotypes of both male and female progeny are determined by the female's gametes. The map distances between the loci give the frequency of recombinants (i.e., crossovers) in each gene interval. There are 14% recombinants in the $a - c$ interval (7% each $a\ c$, + +), and 12% recombinants in the $c - b$ interval (6% each + +, $c\ b$). These recombinants are distributed between both single- and double-crossover classes.

The coefficient of coincidence gives the percentage of expected double crossovers that are observed. The expected double crossover frequency is $(0.12 × 0.14) × 100\% = 1.68\%$. Since the coefficient of coincidence is 0.3, only 30% of expected double crossovers are observed, or $1.68\% × 0.30 = 0.50\%$ (0.25% each of $a\ c\ b$ and + + +).

The remaining recombinants will be single-crossover classes. In calculating this frequency, we must account for the double crossovers that result from crossovers in both gene intervals. The frequency of single crossovers in each gene interval equals the difference between the frequency of crossovers in that interval (inferred from a map distance) and the frequency of double crossovers. There will be 14% − 0.5% = 13.5% single crossovers between a and c (6.75% each $a\ c +$, + + b), and 12% − 0.5% = 11.5% single crossovers between c and b (5.75% each $a + +$, + $c\ b$).

The remaining progeny [100% − (13.5% + 11.5% + 0.5%) = 74.5%] will be parental types (37.25 percent each $a + b$, + $c +$). Therefore, the types of progeny are

Genotype	Percent	Number
$a + b$	37.25	745
+ $c +$	37.25	745
$a\ c +$	6.75	135
+ + b	6.75	135
$a + +$	5.75	115
+ $c\ b$	5.75	115
$a\ c\ b$	0.25	5
+ + +	0.25	5

15.29 a. Consider two genes at a time (see table at bottom of page):

Gene Pair	# Parental-Type Progeny	# Recombinant-Type Progeny	Recombination Frequency	Linked?
a, b	902	98	9.8	yes
a, c	973	27	2.7	yes
a, d	957	43	4.3	yes
a, e	497	503	50.0	no
b, c	875	125	12.5	yes
b, d	945	55	5.5	yes
b, e	497	503	50.0	no
c, d	930	70	7.0	yes
c, e	498	502	50.0	no
d, e	496	504	50.0	no

The genes *a*, *b*, *c*, and *d* are linked because the recombination frequencies between these genes is less than 50 percent. Gene *e* is unlinked to the other four genes—it is either on a separate chromosome or far from the other loci. Develop a map starting with the smallest distances, as they are the most accurate:

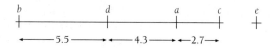

b. Rewrite the cross using the correct gene order: b^+ *d* a^+ *c*/*b* d^+ *a* c^+; *e*/e^+ × *b* *d* *a* *c*/*b* *d* *a* *c*; *e*/*e*. A b^+ d^+ a^+ c^+ fly is obtained from a triple crossover: There must be crossovers between *b* and *d*, *d* and *a*, and *a* and *c*. The reciprocal products of the triple crossover are b^+ d^+ a^+ c^+ and *b* *d* *a* *c*. Among these, half will be e^+ and half will be *e*. Thus,

$$P(b^+\ d^+\ a^+\ c^+\ e^+) = P\ (\text{receiving one of the two triple-}$$
$$\text{crossover products and }e^+)$$
$$= \tfrac{1}{2} \times (0.055 \times 0.043 \times 0.027) \times \tfrac{1}{2}$$
$$= 1.6 \times 10^{-5}$$

15.31 First, use the chi-square test to evaluate the hypothesis that there is no relationship between chestnut coat color and class. Further investigation into the potential linkage of this coat color gene and a hypothetical class gene is warranted if we can reject this hypothesis. The assumptions in this initial chi-square test involve the genotypes of the horses bred to Sharpen Up, as stated in the problem, and the hypothesis of the chi-square test, that chestnut coat color and class are unrelated.

From the initial hypothesis of *no relationship between class and chestnut coat color*, the likelihood of Sharpen Up siring a classy horse is uniform with regard to its coat color. There were 83 classy horses produced from a total of $367 + 260 = 627$ progeny, so the chance of Sharpen Up siring a classy horse, independent of its coat color, is $83/627 = 13.24\%$. To perform the chi-square test, we need to compare the expected and observed number of classy chestnut and classy bay horses.

Assumption I: Sharpen Up is mated equally frequently to homozygous bay, heterozygous bay/chestnut, and homozygous chestnut mares. Chestnut is recessive to bay, so let *c* represent chestnut and *C* represent bay. Sharpen Up is chestnut (*cc*), so the crosses and their progeny are (1) $cc \times CC \rightarrow$ all *C*– (bay); (2) $cc \times Cc \rightarrow \tfrac{1}{2}$ *Cc* (bay), $\tfrac{1}{2}$ *cc* (chestnut); and (3) $cc \times cc \rightarrow$ all *cc* (chestnut). If each cross is equally likely ($P = \tfrac{1}{3}$), the expected number of chestnut offspring, rounding up or down to the nearest whole horse, is $627 \times [(\tfrac{1}{3} \times 0) + (\tfrac{1}{3} \times \tfrac{1}{2}) + (\tfrac{1}{3} \times 1)] = 314$. The remaining $627 - 314 = 313$ offspring are expected to be bay. Using assumption I and assuming that the frequency of classy offspring is uniform (13.24 percent) with respect to their coat color, the expected number of classy chestnut progeny is $314 \times 0.1324 = 42$, and the expected number of classy bay progeny is $313 \times 0.1324 = 41$. The observed numbers of classy horses were 45 chestnut and 38 bay. For these values, $\chi^2 = [(45 - 42)^2/42 + (38 - 41)^2/41] = 0.43$, df $= 1$, $0.50 < P < 0.70$. Under assumption I, then, the hypothesis that chestnut coat color and class are unrelated is accepted as possible.

Assumption II: Sharpen Up is mated equally frequently to heterozygous bay/chestnut and chestnut mares. These crosses and their progeny are (1) $cc \times Cc \rightarrow \tfrac{1}{2}$ *Cc* (bay), $\tfrac{1}{2}$ *cc* (chestnut); and (2) $cc \times cc \rightarrow$ all *cc* (chestnut). If each cross is equally likely ($P = \tfrac{1}{2}$), the expected number of chestnut offspring is $627 \times [(\tfrac{1}{2} \times \tfrac{1}{2}) + (\tfrac{1}{2} \times 1)] = 470$. The expected number of bay offspring is $627 - 470 = 157$. Using assumption II and assuming that the frequency of classy offspring is uniform (13.24 percent) with respect to their coat color, the classy progeny are expected to be $470 \times 0.1324 = 62$ chestnut and $157 \times 0.1324 = 21$ bay. The observed numbers of classy horses were 45 chestnut and 38 bay. For these values, $\chi^2 = [(45 - 62)^2/62 + (38 - 21)^2/21] = 18.4$, df $= 1$, $P < 0.001$. Under assumption II, then, the hypothesis that chestnut coat color and class are unrelated is rejected as being unlikely. It would be reasonable to consider the hypothesis that a gene closely linked to chestnut/bay coat color might contribute to class.

Notice that the evidence for a relationship between chestnut coat color and class hinges on knowing what alleles at the chestnut/bay gene were present in the mares bred to Sharpen Up. This information is available (although not in this problem). Additional assumptions required to specifically test for linkage to a class gene might include assumptions about the number of alleles in the population of horses, the dominance relationships between them, and which alleles reside on the same homolog with the chestnut allele.

15.32 About seven percent of the meioses will produce a^+ *b* or *a* b^+ chromosomes. In the Holliday model for reciprocal recombination, a physical exchange between two loci will result in the genetic exchange of outside chromosome markers about half of the time, depending on whether the Holliday intermediate is cleaved to produce a patched duplex (no exchange of outside chromosome markers), or a spliced duplex (exchange of outside markers). See Box Figure 15.1.

Chapter 16 Advanced Gene Mapping in Eukaryotes

16.3 The eight progeny classes appearing in four frequencies indicate that there are three linked genes, so solve for the map distances as in a three-point testcross.

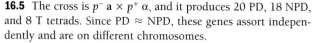

16.5 The cross is p^- **a** × p^+ α, and it produces 20 PD, 18 NPD, and 8 T tetrads. Since PD ≈ NPD, these genes assort independently and are on different chromosomes.

16.7 a. Consider one pair of loci at a time and identify what types of tetrads are produced. Then determine map distances by applying the formula map distance = [(½ T + NPD)/total number of asci] × 100%. The distances are: *a–b*, 19.7 mu; *b–c*, 11 mu; *a–c*, 14 mu. Thus, the gene order is *a–c–b*.

b. Since we know the gene order and intergene distances, we need only to determine whether *a* or *b* is nearer the centromere. Consider each locus individually to identify the number of second-division (MII) segregation patterns. Then determine gene–centromere map distances by applying the formula: Map distance = [(½ × MII-type patterns)/(total number of asci)] × 100%. The distances are: *a*, 20.7 mu; and *b*, 6 mu. Construct the map using the distances from smaller intervals, as these are more accurate due to the effects of multiple crossovers in larger intervals.

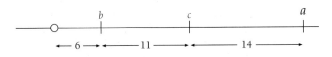

16.8 a. *c/c d⁺/d⁺* and *c⁺/c⁺ d/d*

b. The diploid *c/c⁺ d/d⁺* produces equal numbers of PD tetrads (4 *c d⁺* and 4 *c⁺ d* spores, all mutant) and NPD tetrads (4 *c⁺ d⁺* and 4 *c d* spores, 4 wild-type and 4 mutant), so *c* and *d* are not linked.

c. The pathway is Y → X → Z. Mutant *c* is blocked in the synthesis of X from Y. Mutant *d* is blocked before the synthesis of Y.

16.10 The cross is *w/Y* (white ♂) × *w^ch/w^ch* (cherry ♀), so the female progeny are *w/w^ch* heterozygotes. Normal mitotic division in a *w/w^ch* cell, diagrammed in the top half of Figure 16.A, generates two identical *w/w^ch* daughter cells. As shown in the bottom half of the figure, when a single mitotic crossover occurs between the *w* locus and its centromere, two genetically different daughter cells are produced after cell division: *w/w* and *w^ch/w^ch*.

The alleles at the *white* locus control the amount of pigment deposition in the cells of the eye (Chapter 13). Phenotypically white homozygotes (*w/w*) have 0.44% of wild-type pigment levels, whereas cherry-colored homozygotes (*w^ch/w^ch*) have 4.1% of wild-type pigment levels. The *white* alleles show partial dominance, so *w/w^ch* heterozygotes should have about 2% of wild-type pigment levels and show an intermediate pink color. If mitotic recombination occurs during the development of the eye, neighboring *w/w* and *w^ch/w^ch* daughter cells will be produced. After these and the surrounding *w/w^ch* cells divide by mitosis to form the adult eye, twin spots will be seen: In a pink background (*w/w^ch*), there will be neighboring white (*w/w*) and cherry (*w^ch/w^ch*) twin spots.

16.13 The two homologs of each chromosome assort independently during haploidization. Here, three blocks of genes segregate independently, identifying three linkage groups: *ad w pu* and *ad⁺ w⁺ pu⁺* segregate independently of *y bi⁺* and *y⁺ bi*, which segregate independently from *sm phe* and *sm⁺ phe⁺*.

16.16 a. GB4 has [(3 × 10⁹ bp/genome × 0.32 genome)/(25 × 10⁶ bp/segment)] ≈ 38 segments. G3 has ~200 segments.

b. The two markers may reside on segments derived from different chromosomes that, by chance alone, reside in the same hybrid. Their coexistence in one or even several hybrids is insufficient to indicate linkage without calculating the likelihood that this occurred by chance.

c. To resolve the relative order of a set of markers, overlapping DNA segments bearing marker subsets must be employed. GB4 has larger DNA segments, so it is useful for mapping markers that span larger distances, at least 1 MB. However, it will not resolve the order of more closely spaced markers, as they will most often lie within the same DNA segment. In contrast, since it has smaller DNA segments, G3 will resolve markers that are at least 250,000 bp apart. The differing properties of the two panels make them complementary: GB4 provides long-range map continuity, while G3 gives higher local resolution.

d. The two panels have human DNA segments with different average lengths, and this determines their capacity to resolve close markers. Even though the GB4 panel has a similar total number of hybrids containing each marker, it fails to resolve the markers. All three markers are simultaneously present in the majority of positive hybrids and so these markers are relatively close together, probably within a 1-MB interval. The G3 panel is able to resolve them as hybrids when? just one or two markers are obtained. There are more hybrids positive for only *A* and *B* than for *A* and *C*, and there are no hybrids positive for *B* and *C* or *A, B,* and *C*. This is consistent with a hypothesis that *A* and *B* are closer together than *A* and *C*, and their order is *B–A–C*.

16.18 A graph of θ versus lod score reveals a maximum lod score of 4.01 at a map distance of about 25 mu. Since the lod score is greater than 3, the marker is linked to the *waf* gene.

Figure 16.A

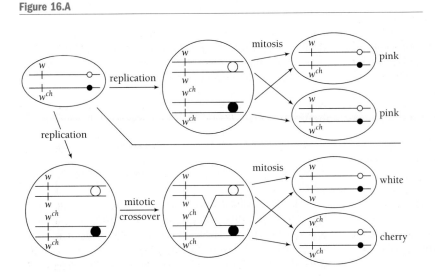

However, the marker is not closely linked. Using the estimate from humans that 1 mu corresponds on average to 1 Mb, the marker is about 25 Mb away from the *waf* gene.

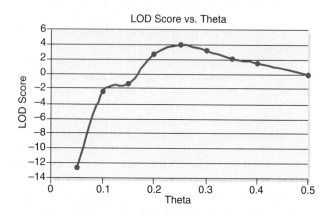

16.20 The crosses involve alleles at just one gene, and in each case, tetrads containing doubly mutant and/or completely normal spores are recovered due to events associated with intragenic recombination. In the *a1 a2⁺ × a1⁺ a2* cross, each allele shows a 2:2 segregation pattern, and the middle two spores are recombinants that result from cleaving a Holliday intermediate between the *a1* and *a2* sites to produce spliced duplexes. For the *a1 a3⁺ × a1⁺ a3* and the *a2 a3⁺ × a2⁺ a3* crosses, there is evidence that the segregation of one of the alleles in the tetrad has resulted from gene conversion caused by mismatch repair of heteroduplex DNA. The *a1 a3⁺ × a1⁺ a3* cross shows 2:2 segregation of *a3⁺* : *a3* and 3:1 segregation of *a1⁺* : *a1* resulting from gene conversion of an *a1* allele to *a1⁺*. In the *a2 a3⁺ × a2⁺ a3* cross, the *a2* allele segregates in a Mendelian fashion while the *a3* allele segregates 3:1 *a3⁺* : *a3* as a result of gene conversion of one *a3* allele to *a3⁺*.

16.23 Since both *ry²⁰⁶* and *ry²⁰⁹* are point mutations, rare reversion to *ry⁺* could produce *kar ry⁺ ace* and *kar⁺ ry⁺ ace⁺* chromosomes. The lack of flanking marker exchange in these chromosomes is also consistent with a *ry⁺* allele arising from gene conversion by mismatch repair. The *ry²⁰⁶* allele on the *kar ry²⁰⁶ ace* chromosome was converted to *ry⁺* using the normal sites present in the *ry* locus on the *kar⁺ ry²⁰⁹ ace⁺* chromosome. The *ry²⁰⁹* allele on the *kar⁺ ry²⁰⁹ ace⁺* chromosome was converted to *ry⁺* using the normal sites present in the *ry* locus on the *kar ry²⁰⁶ ace* chromosome. (See Box Figure 16.1.) The chromosomes where the flanking *kar* and *ace* marker loci are exchanged (*kar⁺ ry⁺ ace, kar ry⁺ ace⁺*) result from intragenic recombination: a Holliday intermediate forming between the *ry²⁰⁶* and *ry²⁰⁹* sites was cleaved to produce spliced duplexes.

Chapter 17 Variations in Chromosome Structure and Number

17.1 **a.** pericentric inversion [*D ○ E F* inverted]

   b. nonreciprocal translocation [*B C* moved from left to right arm]

   c. tandem duplication [*E F* duplicated]

   d. reverse tandem duplication [*E F* duplicated]

   e. deletion [*C* deleted]

17.2 A pericentric inversion includes the centromere, while a paracentric inversion lies wholly within one chromosomal arm. See Figure 17.8 in the text.

17.4 **a.** This is paracentric inversion, because the centromere is not included in the inverted DNA segment.

   b.

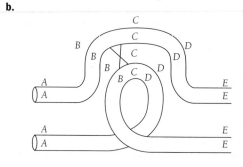

   c. A crossover between B and C results in the following chromosomes:

A B C D E   (normal order)

A B C D A   (dicentric, duplication for A, deletion for E)

E B C D E   (acentric, duplication for E, deletion for A)

A D C B E   (inverted order)

17.7 One series of sequential inversions is

$$a \to c \to e \to d$$
$$\downarrow$$
$$b$$

The regions inverted in each step are illustrated here.

a. A B C D E F G H I

c. A B F E D C G H I

e. A B F E H G C D I

b. H E F B A G C D I   d. A B F C G H E D I

17.9 **a.** The following diagram shows a normal chromosome bearing *w⁺*, the *w^M4*-associated inversion and a second inversion found in a *w⁺* revertant. The genes *a, b, c,* and *d* are inserted near the breakpoints of the different inversions to help visualize them. Euchromatin is represented by a thin line, centromeric heterochromatin by a thick line, and the centromere by an open circle. The brackets delineate inverted regions.

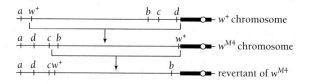

The mottled-eye phenotype is associated with chromosomal rearrangements induced on a *w⁺*-bearing chromosome that place the *w⁺* gene near heterochromatin. When a rearrangement is heterozygous with a *w* allele, only the *w⁺* gene on the rearranged chromosome can provide for normal

eye pigmentation. The mottled appearance of the eye indicates that it functions in some, but not all cells. This suggests that the DNA sequence of the *white* gene is unaltered. It is more likely an epigenetic phenomenon caused by a position effect, a phenotypic change due to inactivation of the w^+ allele by neighboring heterochromatin.

b. The second inversion that occurs on the w^{M4} chromosome repositions the w^+ gene to a euchromatic location. This supports the view that the mottled-eye phenotype is caused by a position effect.

17.10 a. Parents of Rec(8) individuals are heterozygous for a pericentric inversion with breakpoints at 8p23.1 and 8q22.1. Rec(8) offspring with 8q duplication and 8p deletion probably arose from a single crossover within the pericentric inversion. Such an event is diagrammed in Figure 17.10 in the text.

b. As shown in Figure 17.10 in the text a single crossover between two nonsister chromatids in an inversion heterozygote results in four products: two have the noncrossover chromosomes (one normal-ordered and one inverted) and two are duplication/deletion products. Here, the product with 8q duplication and 8p deletion contribute to a viable zygote with Rec(8) syndrome. The product with 8q deletion and 8p duplication is not discussed in the problem. It may be that zygotes with this product do not survive. In this case, $\frac{1}{3}$ of the surviving zygotes have Rec(8) syndrome. Of the $\frac{2}{3}$ normal zygotes, $\frac{1}{2}$ carry the chromosome 8 inversion.

c. The phenotypes of Rec(8) individuals could vary for one or a combination of reasons. (1) There could be several different chromosome 8 inversions in the population that vary slightly in their inversion breakpoints. The Rec(8) individuals resulting from single crossovers in inversion heterozygotes would differ symptomatically due to variation in genes that are duplicated and deleted or due to differences in gene activation or gene inactivation. (2) There may be a position effect. (3) The genetic background could vary. The phenotypic effects of gene deletion or duplication could depend on genetic interactions with other genes in the genome. In this case, alleles inherited from the father that are different from those inherited from the mother and grandmother could contribute to the phenotype. (4) Environmental effects could exacerbate the effects of the deleted and duplicated region. These effects may not be uniform, and so could contribute to the observed phenotypic variability. Since many of the symptoms associated with Rec(8) syndrome are developmental abnormalities, variation in the environment during fetal development may contribute to phenotypic variability. (5) There may be other, cytologically invisible mutations associated with the Rec(8) individuals that could strongly affect their phenotype.

d. The child has the chromosome 8 inversion, but not the duplication/deletion chromosome that results from a single crossover in an inversion heterozygote; she is an atypical Rec(8) individual. There are several explanations for why some of her symptoms overlap with those of Rec(8) syndrome. She may have an additional mutation near one of the Rec(8) breakpoints, in a region that is duplicated or deleted in Rec(8) syndrome, or in a gene that interacts with genes in the duplicated or deleted regions. Alternatively, it is possible that the inversion disrupts the function of a gene or genes at one or both breakpoints, and that normally, the inversion is an asymptomatic condition. In this case, the inversion chromosome (in the child's mother and grandmother) would bear a recessive mutation. If she had a new allelic mutation, or her

paternally contributed chromosome had an allelic mutation, she would be affected. This could also explain why she has only some of the symptoms of Rec(8) syndrome; she would have fewer genes affected than would most Rec(8) individuals.

Small deletions would be cytologically invisible, as would point mutations. Thus, the explanations given above could not be evaluated solely by karyotype analysis. FISH, DNA marker, and/or DNA sequence analyses (see text Chapters 16 and 17) could to be used to evaluate the integrity of the chromosomal regions near the breakpoints.

17.11 a. The irradiated chromosome has a paracentric inversion. Single crossovers produce dicentric chromosomes and fragments; a four-strand double crossover produces dicentric chromosomes with two bridges and two fragments.

b. The bridge chromosome would arise by a single crossover within the inversion loop. See Figure 17.9 in the text.

17.13 a. Mr. Lambert is heterozygous for a pericentric inversion of chromosome 6. Relative to the centromere of the normal chromosome 6, one of the breakpoints is within the fourth light band up from the centromere, while the other is in the sixth dark band below the centromere. Mrs. Lambert's chromosomes are normal.

b. When Mr. Lambert's number 6 chromosomes paired during meiosis, they formed an inversion loop that included the centromere. Crossing-over occurred within the loop, and gave rise to the partially duplicated, partially deficient chromosome 6 that the child received.

c. The child's abnormalities stem from having three copies of some and only one copy of other chromosome 6 regions. The top part of the short arm is duplicated, and there is a deficiency of the distal part of the long arm in this case.

d. The inversion appears to cover more than half of the length of chromosome 6, so crossing-over will occur in this region in the majority of meioses. In the minority of meioses where there is a two-strand double crossover inside the loop, or where crossing-over occurs outside the loop, and in the cases where a crossover has occurred within the loop but the child receives a noncrossover chromosome, the child can be normal. There is significant risk of abnormality, so fetal chromosomes should be monitored.

17.15

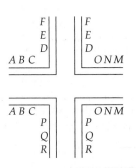

17.17 a. Mr. Denton has normal chromosomes. Mrs. Denton is heterozygous for a balanced reciprocal translocation between chromosomes 6 and 12. Most of the short arm of chromosome 6 has been reciprocally translocated onto the long arm of chromosome 12. The breakpoints appear to be in the first thick, dark band just above the centromere of 6 and in the third dark band below the centromere of 12.

b. The child received a normal chromosome 6 and a normal chromosome 12 from his father. In prophase I of meiosis in Mrs. Denton, chromosome 6 and 12 and the reciprocally translocated 6 and 12 paired to form a cruciform-like structure. Segregation of adjacent, nonhomologous centromeres to the same pole ensued, so that the child received a gamete containing a normal 6 and one of the translocation chromosomes. See Figure 17.12 in the text for an illustration of adjacent-1 segregation.

c. The child has a normal chromosome 6 and a normal chromosome 12 from Mr. Denton. The child also has a normal chromosome 6 from Mrs. Denton. However, the child also has one of the translocation chromosomes from Mrs. Denton. With this chromosome, the child is partially trisomic as well as partially monosomic. It has three copies of part of the short arm of chromosome 6 and only one copy of most of the long arm of chromosome 12. This abnormality in gene dosage is the cause of its physical abnormality.

d. Segregation of adjacent homologous centromeres to the same pole is relatively rare. The segregation pattern seen in this child (adjacent-1 segregation) and alternate segregation (see Figure 17.12) are more common. About half the time, when alternate segregation occurs the gamete will have a complete haploid set of genes, and the embryo should be normal. However, half of the gametes resulting from alternate segregation will be translocation heterozygotes.

e. Prenatal monitoring of fetal chromosomes could be done, and given the severity of the abnormalities (high probability of miscarriage and multiple congenital abnormalities), therapeutic abortion of chromosomally unbalanced fetuses would be a consideration.

17.19　a. First, the pedigree is consistent with an X-linked recessive trait such as fragile X syndrome. Second, in fragile X-syndrome, normal transmitting males carry a premutation that is passed to their daughters, and the sons of these daughters frequently show mental retardation. Here, the daughters of individual I-1 all have sons that have mental retardation, but neither he, nor his son's children show mental retardation. Therefore, individual I-1 could have an X chromosome bearing a premutation that is passed to his daughters (but not his sons). During DNA replication in his daughters, the CGG triplet repeat in the *FMR-1* gene is amplified to generate a full mutation. Mental retardation is seen in their offspring when the X-chromosome bearing the full mutation is transmitted.

b. Culture cells from the affected individuals, and examine chromosomes cytologically to determine whether a fragile site is present. Use PCR with primers that flank the CGG repeat in the *FMR-1* gene to evaluate the size of the repeat. Individuals with fragile X syndrome will exhibit a fragile site at Xp27.3, and have 200 to 1,300 copies of the CGG repeat.

c. (i) I-1, II-3, II-8, II-11; (ii) I-1; (iii) I-1, II-3, II-8, II-11, III-5, III-6, III-13, III-14, IIII-15, III-21; (iv) III-7, III-8, III-20, III-26, III-27

d. In females, one X chromosome is inactivated. In some cells, the fragile X chromosome will be inactivated while the other X chromosome will have a normal *FMR-1* gene. This could underlie the less severe phenotype seen in females.

17.22　a. 45

b. 47

c. 23

d. 69

e. 48

17.23　b. trisomic

17.24　a. The cross can be written as $X^+X^+ \times X^cY$. The X^cO child received its father's X^c. A chromosomally normal X-bearing

sperm fertilized a nullo-X egg, so nondisjunction occurred in the mother.

b. The X^+O child received its mother's X^+. The chromosomally normal X-bearing egg was fertilized by a nullo-X, nullo-Y sperm, so nondisjunction occurred in the father.

17.25 This problem considers what happens when a chromosome is lost at the very first mitotic division (and only at that division).

a. The cross is y/y $+/+$ (female) $\times$ $+/Y$ pal/pal (male), with progeny $y/+$ $pal/+$ (daughters) and y/Y $pal/+$ (sons). The paternally contributed X is found only in the $y/+$ $pal/+$ daughter, so we need only consider the consequence of its loss in daughters. If a paternally contributed X chromosome (+) is lost during the first mitotic division in a $y/+$ $pal/+$ zygote, one daughter cell will lose an X chromosome and be y $pal/+$. The other daughter cell will have two X chromosomes and be $y/+$ $pal/+$. The cell with two X chromosomes would be female (XX) and produce nonyellow cells ($y/+$), while the cell with one X chromosome would be male (XO) and produce yellow cells (y). The animal will be a mosaic with cells of two sex types that are marked by yellow (male) or grey (female) cuticle.

b. The cross is $+/+$ eye/eye (female) $\times$ pal/pal $+/+$ (male), with progeny $pal/+$ $eye/+$ (daughters and sons). The paternally contributed fourth chromosome is +. If it is lost during the first mitotic division in a $pal/+$ $eye/+$ zygote, one daughter cell will lose a fourth chromosome and be $pal/+$ eye. The other daughter cell will have two fourth chromosomes and be normal, while the cell with one fourth chromosome will be eye. The animal will be a mosaic with some cells that are haploid for the fourth chromosome and some cells that are diploid for the fourth chromosome. If a patch of haplo-4 cells forms an eye during development, the eye will be reduced in size.

c. The cross is $+/+$ e/e (female) $\times$ pal/pal $+/+$ (male), with progeny $pal/+$ $e/+$. The paternally contributed third chromosome is +. If it is lost during the first mitotic division in a $pal/+$ $e/+$ zygote, one daughter cell will lose a third chromosome, and be $pal/+$ e. This cell is inviable, and so will not be recovered in the organism, should the organism survive. Consequently, if the organism survives, it will be phenotypically normal ($pal/+$ $e/+$).

17.29　a. The genotype of the F_1 peas will be AA aa.

b. If we label the four alleles in the F_1 A^1, A^2, a^1, and a^2, there are six possible gametes—A^1A^2, A^1a^1, A^1a^2, A^2a^1, A^2a^2, a^1a^2 —giving ⅙ AA, ⁴⁄₆ Aa, and ⅙ aa. As shown in the following Punnett square, selfing the F_1 gives a phenotypic ratio of 35 $A- : 1$ aa.

	⅙ AA	⁴⁄₆ Aa	⅙ aa
⅙ AA	¹⁄₃₆ $AAAA$	⁴⁄₃₆ $AAAa$	¹⁄₃₆ $AAaa$
⁴⁄₆ Aa	⁴⁄₃₆ $AAAa$	¹⁶⁄₃₆ $AAaa$	⁴⁄₃₆ $Aaaa$
⅙ aa	¹⁄₃₆ $AAaa$	⁴⁄₂₆ $Aaaa$	¹⁄₃₆ $aaaa$

17.33 The initial allopolyploid will have 17 chromosomes. After doubling, the somatic cells will have 34 chromosomes.

Chapter 18 Genetics of Bacteria and Bacteriophages

18.1 For a recipient to be converted to a donor, a complete F factor must be transferred. In $F^+ \times F^-$ crosses, only the F factor is transferred, and this occurs relatively quickly. In $Hfr \times F^-$ crosses, transfer starts at the origin within the F element and then must proceed through the bacterial chromosome before reaching the *F* factor. For transfer of the entire *F* factor, the whole chromosome would have to be transferred. This would take about 100 minutes, and usually the conjugal unions break apart before then.

18.2

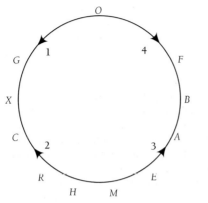

18.5 Strain *A* is *thy⁻ leu⁺* while strain *B* is *thy⁺ leu⁻*. DNA from *A* can transform *B* if DNA from *A* can transform the *leu⁻* allele of *B* to *leu⁺*. Test this by adding DNA from *A* to a leucine-fortified culture of *B*. Incubate long enough for transformation to occur, and then plate the potentially transformed *B* cells on minimal medium or on medium supplemented only with thymine. This selects for *leu⁺* transformants.

18.6 a. Initially select for *c⁺ str^R* recombinants by plating the progeny on minimal medium without compound C, but supplemented with streptomycin and compounds A, B, D, E, F, G, and H. To assess the complete genotype of the *c⁺ str^R* recombinants, replica plate them onto different minimal media supplemented with streptomycin and all but two of the compounds (compound C and one other). For example, to test if a *c⁺ str^R* colony was also *a⁺*, replica plate it onto a medium that lacked compound A, but was supplemented with streptomycin and B, D, E, F, G, and H. If the colony were able to grow on this medium, it would be *a⁺ c⁺ str^R*. If it were unable to grow, it would be *a c⁺ str^R*.

b. Strain 1: Since no *c⁺* recombinants are ever obtained, strain 1 is unable to transfer *c⁺*. This means it is either (1) *F⁻*; (2) *Hfr* but with the F factor inserted either far from *c⁺* or close to it but in an orientation so that genes are transferred in a direction opposite to *c⁺*; or (3) *F′*, with *c⁺* in the bacterial chromosome. It should not be *F⁺* because then, at a very low frequency, some *c⁺* recombinants would be obtained.

Strain 2: Since *c⁺* recombinants are obtained at 6 minutes, and *g⁺*, *h⁺*, *a⁺*, and *b⁺* recombinants are obtained at subsequent time intervals, strain 2 is *Hfr*. The genes are transferred in the sequence *c⁺*, *g⁺*, *h⁺*, *a⁺* and *b⁺*. From the times of transfer of the genes, the map position of the genes is as follows: origin (0)–*c⁺*(6)–*g⁺*(8)–*h⁺*(11)–*a⁺*(14)–*b⁺*(16). The location of genes *d⁺*, *e⁺*, and *f⁺* cannot be determined precisely; because they are not transferred in an *Hfr* × *F⁻* cross, they are either far away from the F factor insertion site, or close to it but near the fertility genes, which are rarely transferred by an *Hfr* strain. When the recombinants obtained from the strain 2 × *F⁻* mating at the 16-minute time period are crossed to an *amp^R F⁻* strain, *c⁺* is not transferred. If these recombinants cannot conjugate with *F⁻*, this indicates that although strain 2 is fertile, it did not transfer a complete F factor. Therefore, it must be *Hfr*.

Strain 3: Strain 3 transfers *c⁺* within 1 minute and *g⁺* by 3 minutes. From analysis of the strain 2 × *F⁻* cross, we knew

that these genes are 2 minutes apart. These data support this conclusion. Since no other recombinants are obtained, no other genes are transferred. This suggests that strain 3 is *F′*, and that the segment of DNA containing *c⁺* and *g⁺* is in the *F′* factor. If this is the case, the complete F factor will be transferred in a strain 3 × *F⁻* cross if the mating is allowed to proceed long enough. This is observed: *c⁺* recombinants from the strain 3 × *F⁻* cross obtained at 16 minutes are able to transfer *c⁺* to an *F⁻ amp^R* strain. Therefore, strain 3 is *F′*.

c. Information known with certainty is diagrammed here. The location of genes in strains 1 and 3 is inferred from crosses with strain 2. The location of genes *d⁺*, *e⁺*, and *f⁺* is unknown.

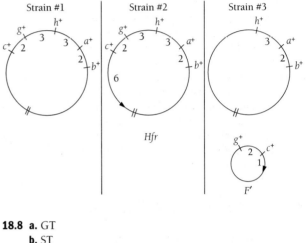

18.8 a. GT
b. ST
c. ST
d. GT
e. GT
f. B
g. B
h. B
i. N

18.10 Closer genes have a higher cotransduction frequency. The *pryD* and *cmlB* genes show the highest cotransduction frequency, so they are the closest together and (c) is eliminated. The genes *aroA* and *pyrD* show the lowest cotransduction frequency, so they are the farthest apart and (b) is eliminated. The *aroA* and *cmlB* genes show an intermediate cotransduction frequency, as would be expected if *cmlB* is between *aroA* and *pyrD*. Thus, the correct answer is (a), *aroA–cmlB–pyrD*.

18.12 Since relatively closer loci show a relatively higher cotransduction frequency, pairs of loci can be ordered in terms of their proximity. The order (closest together to farthest apart) is *cheB–eda, cheA–eda, cheA–supD, cheB–supD, eda–supD*. A gene order consistent with these relationships is *eda–cheB–cheA–supD*.

18.13 The genetic distance is 0.07 mu. The plaques produced on *E. coli* K12(λ) are *r⁺*, while those on *E. coli* B may be either *r⁺* or *r⁻*. Thus, the total number of progeny can be inferred from the number of plaques formed on *E. coli* B: Total number of progeny in 1 mL = (dilution factor) × (# progeny phage/mL) = 1,000 × (672/0.1) = 6.72 × 10⁶/mL. Since *E. coli* B is coinfected with *rIIx* and *rIIy*, the only way to obtain an *r⁺* progeny phage is to have a crossover within the *rII* locus. The progeny resulting from a crossover would be ½ *r⁺* and ½ *rIIxy* recombinants. The number of recombinant phage is twice the number of r^+ phage, which can be assayed for by growth on *E. coli* K12(λ): Number of recombinant progeny in 1 mL = 2 × (number of r^+ phage/mL) = 2 × (470/0.2) = 4,700/mL. The map distance between *rIIx* and *rIIy* is [4,700/(6.72 × 10^6)] × 100% – 0.07 mu.

18.15 Analyze the data as you would a set of two-factor crosses. Two maps are compatible with the data:

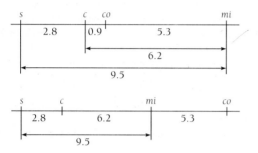

18.17 Identify the region missing in each deletion mutant by applying two principles: (1) If no r^+ recombinants are obtained, the deletion removes the site of the point mutant. (2) If r^+ recombinants are obtained, the site of the point mutation is not within the boundaries of the deletion.

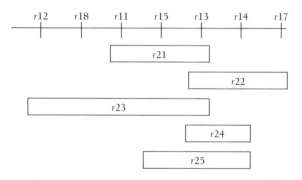

18.18 Apply two principles to delineate the region where a point mutant lies: (1) If a point mutant can recombine with a deletion mutant, it must lie outside of the deleted region. (2) If a point mutant cannot recombine with a deletion mutant, it must lie within the deleted region. Use the complementation data to determine the positions of the *A* and *B* cistrons: If two mutants are unable to grow on *E. coli* K12(λ), they do not complement each other and cannot together provide the functions to complete the *rII* pathway. Both mutants must be defective in the same function and either the *rIIA* or *rIIB* function is missing.

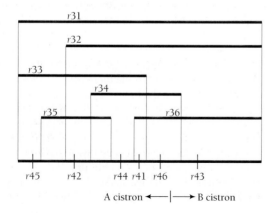

A cistron ←—|—→ B cistron

18.21 a. If DNA transduced into a particular *leu* mutant recombines with its chromosome to produce a *leu*$^+$ recombinant, the transduced DNA must contain a wild-type site that can replace the mutated *leu* site. Therefore, pairs of mutants that produce *leu*$^+$ recombinants affect different sites, while pairs of mutants unable to produce *leu*$^+$ recombinants affect one or more common sites. The sites may be single or multiple base-pair regions. Mutants that affect more than one site must be deletions. Deletions can be recognized by identifying mutants unable to produce *leu*$^+$ recombinants with pairs of mutants that lie in different sites. For example, mutants 3 and 8 can recombine to produce *leu*$^+$ recombinants, so they lie in different sites. Mutant 7 cannot recombine with either mutant 3 or mutant 8, so it must delete both sites. Using this logic, mutants 2, 4, and 7 must be deletions. In the map here, these deletions are shown by open boxes.

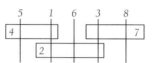

b. A site may be one or more base pairs. To address if a site is a point mutation, check if a mutant can be reverted. Point mutants, but not deletions, can be reverted.

c. This analysis does not address how many cistrons are present in this region. For this, complementation tests are needed.

18.24 There are at least two options. First, the enzyme could be composed of multiple polypeptide subunits. The mutants affect different genes, each of which encodes a polypeptide that is part of the multimeric enzyme. Second, the polypeptide for the enzyme is modified before it becomes active as an enzyme. One mutant affects the gene for the polypeptide that will be modified to become the enzyme; the second mutant affects the gene for a modifier protein that is required to make the enzymatic polypeptide functional (e.g., it could affect a protease that cleaves a proenzyme to make an active enzyme; it could affect a kinase that phosphorylates a nonactive form of the enzyme to make it active.)

18.26 a. Since all nine very early mutants can be reverted, all are point mutants.

b. All nine very early mutants fail to complement each other (none produce enough virus in pairwise coinfections to be considered positive), so they affect one function.

c. Eight of the mutants (*B2, B21, B27, B28, B32, 901, LB2, D*) are able to recombine with each other, and so affect different sites. Mutant *c75* fails to recombine with mutant *D*, so these two mutants may affect the same site.

d. Mutant *c75* may be a mutant affecting multiple points. It reverts at a lower frequency than the others do and also shows inconsistent recombination relative to the other mutants.

e. The mutants incompletely block viral growth, so some virus is produced by each mutant. *I* compares the amount of virus produced by coinfection of two mutants to the sum of the amounts produced by individual infections. If two viruses are blocked in the same function, a coinfection should produce low amounts of virus similar to the sum of two separate infections and *I* will be about 1.0. If two mutants are blocked in different functions, coinfection allows for complementation, as the function blocked in one mutant is provided by the other mutant. Substantial viral growth will occur and I should be much larger than 1.

At the restrictive temperature of 39°C, neither parent nor doubly mutant recombinants can grow. Only wild-type recombinants can grow. These are half of the recombinants, so doubling the amount of virus produced at 39°C will estimate the number of recombinants between the two mutant sites. At the permissive temperature of 34°C, mutant and wild-type viruses can grow, so the amount of virus produced at 34°C measures the total amount of virus produced by coinfecting the two mutant strains. RF is calculated by doubling the amount of virus produced by coinfecting two mutants at 39°C and then dividing this number by the amount of virus produced by coinfecting the mutants at 34°C, so RF estimates recombination between two mutants.

f.

g. Mutant $c75$'s reversion rate is less than the other mutants, suggesting that it is more complex than a simple point mutant. If it affected multiple sites, it would have recombination data inconsistent with the other mutants.

18.27 a. Cross each mutation to the wild-type, deep-red strain and note the progeny phenotypes. A dominant mutation, by definition, appears in a heterozygote, so if the progeny have brownish eyes, the mutation is dominant. If they have deep-red eyes, the mutation is recessive.

b. Set up pairs of crosses between the mutants to perform complementation tests. Mutations that affect the same gene function will produce brown-eyed progeny when crossed and belong to the same complementation group. Mutations that affect different functions will produce deep-red-eyed progeny when crossed and belong to different complementation groups. Counting the number of different complementation groups will give the number of genes that are affected.

c. Allelic mutations are those that are members of the same complementation group, as determined in (b).

d. One could determine if a particular mutant is allelic to a known eye-color gene by performing complementation tests between it and mutants at all known eye-color genes. This would involve crossing the mutant to mutants from a collection of strains with known eye-color mutations and observing the progeny of each cross. If the progeny have a mutant eye color, one would infer that the mutations carried in the two strains are allelic. However, this would be a tremendous amount of work. There are many eye color mutations and this would require a large number of crosses. It would be faster to first determine which of the six mutations are allelic, and then choose a representative allele from each complementation group and identify its chromosomal location using a set of two- and/or three-point mapping crosses. Once this is done, examine published genetic maps of the *Drosophila* genome (visit the site http://www.flybase.org) and ask if any known eye-color mutations lie in the same region. Then, obtain strains with these eye-color mutations and perform complementation tests between

these mutant strains and a representative mutant from each complementation group identified with the new eye-color mutations.

Chapter 19 Regulation of Gene Expression in Bacteria and Bacteriophages

19.2 Allolactose and tryptophan are effector molecules that regulate the *lac* and *trp* operons, respectively. Effectors cause allosteric shifts in repressor proteins to alter their affinity for operator sites in DNA. When allolactose binds to the *lac* repressor, it loses its affinity for the *lac* operator, inducing transcription at the *lac* operon. When tryptophan interacts with the *trp* aporepressor, it is converted to an active repressor that can bind the *trp* operator, repressing transcription at the *trp* operon.

19.5 A constitutive phenotype results from a $lacI^-$ or $lacO^c$ mutation.

19.7 a. A missense mutation results in partial or complete loss of β-galactosidase activity, but no loss of permease and transacetylase activities.

b. Unless the nonsense mutation is very close to the normal chain-terminating codon for β-galactosidase, it is likely to have polar effects. Therefore, permease and transacetylase activities would be lost in addition to the loss of β-galactosidase activity.

19.9 The partial diploid genotype is $lacI^+ \, lacO^c \, lacP^+ \, lacZ^+$ $lacY^-/lacI^+ \, lacO^+ \, lacP^- \, lacZ^- \, lacY^+$. (Only one $lacI^+$ gene is required, so one of the repressor genes may be $lacI^-$.)

19.11 The answer is given in Table 19.A.

19.14 The CAP, in a complex with cAMP, is required to facilitate RNA polymerase binding to the *lac* promoter. The RNA polymerase binding occurs only in the absence of glucose and only if the operator is not occupied by repressor (i.e., lactose is also absent). A mutation in the CAP gene, then, would render the *lac* operon incapable of expression because RNA polymerase would be unable to recognize the promoter.

19.16 a. A DNase protection experiment is an in vitro method to identify DNA sites that are bound by a protein. After a purified protein is allowed to bind a DNA segment, the complex is treated with DNase and sequences unprotected by protein binding are digested. Then, the sequence of the protected region is determined. DNase protection experiments defined the location of the operator, promoter and CAP-cAMP-binding site.

b. See Figure 19.14, p. 528.

c. i. This deletion disrupts the operator, so the operon would be expressed constitutively.

ii. A -12 transversion alters the -10 promoter consensus sequence, possibly decreasing the efficiency of transcription initiation. The operon may still be coordinately induced, but there will be diminished levels of β-galactosidase, permease, and transacetylase activity.

iii. A -69 transversion alters the consensus sequence for the CAP-binding site. If CAP-cAMP is unable to bind the CAP site, RNA polymerase will not be recruited to the promoter and the operon will not be coordinately induced.

iv. A $+28$ transition alters the Shine-Dalgarno sequence, and by affecting translation initiation, could result in diminished or absent β-galactosidase, and, due to polar

Table 19.A

	Genotype	Inducer Absent		Inducer Present	
		β Galactosidase	Permease	β-Galactosidase	Permease
a.	I^+ P^+ O^+ Z^+ Y^+	−	−	+	+
b.	I^+ P^+ O^+ Z^- Y^+	−	−	−	+
c.	I^+ P^+ O^+ Z^+ Y^-	−	−	+	−
d.	I^- P^+ O^+ Z^+ Y^+	+	+	+	+
e.	I^S P^+ O^+ Z^+ Y^+	−	−	−	−
f.	I^+ P^+ O^c Z^+ Y^+	+	+	+	+
g.	I^S P^+ O^c Z^+ Y^+	+	+	+	+
h.	I^+ P^+ O^c Z^+ Y^-	+	−	+	−
i.	I^{-d} P^+ O^+ Z^+ Y^+	+	+	+	+
j.	$\dfrac{I^-\ P^+\ O^+\ Z^+\ Y^+}{I^+\ P^+\ O^+\ Z^-\ Y^-}$	−	−	+	+
k.	$\dfrac{I^-\ P^+\ O^+\ Z^+\ Y^-}{I^+\ P^+\ O^+\ Z^-\ Y^+}$	−	−	+	+
l.	$\dfrac{I^S\ P^+\ O^+\ Z^+\ Y^-}{I^+\ P^+\ O^+\ Z^-\ Y^+}$	−	−	−	−
m.	$\dfrac{I^+\ P^+\ O^c\ Z^-\ Y^+}{I^+\ P^+\ O^+\ Z^+\ Y^-}$	−	+	+	+
n.	$\dfrac{I^-\ P^+\ O^c\ Z^+\ Y^-}{I^+\ P^+\ O^+\ Z^-\ Y^+}$	+	−	+	+
o.	$\dfrac{I^S\ P^+\ O^+\ Z^+\ Y^+}{I^+\ P^+\ O^c\ Z^+\ Y^+}$	+	+	+	+
p.	$\dfrac{I^{-d}\ P^+\ O^+\ Z^+\ Y^-}{I^+\ P^+\ O^+\ Z^-\ Y^+}$	+	+	+	+
q.	$\dfrac{I^+\ P^-\ O^c\ Z^+\ Y^-}{I^+\ P^+\ O^+\ Z^-\ Y^+}$	−	−	−	+
r.	$\dfrac{I^+\ P^-\ O^+\ Z^+\ Y^-}{I^+\ P^+\ O^c\ Z^-\ Y^+}$	−	+	−	+
s.	$\dfrac{I^-\ P^-\ O^+\ Z^+\ Y^+}{I^+\ P^+\ O^+\ Z^-\ Y^-}$	−	−	−	−
t.	$\dfrac{I^-\ P^+\ O^+\ Z^+\ Y^-}{I^+\ P^-\ O^+\ Z^-\ Y^+}$	−	−	+	−

effects, diminished or absent expression of permease and transacetylase.

v. A +9 transition alters the operator. It could either have no effect, cause the repressor to have more affinity for the operator (preventing coordinate induction in a cis-dominant manner), or cause the repressor to have less affinity for the operator (leading to constitutive expression).

d. None of the mutants will prevent catabolite repression.

19.19 For a wild-type *trp* operon, the absence of tryptophan results in antitermination; that is, the structural genes are transcribed and the tryptophan biosynthetic enzymes are made. This occurs because a lack of tryptophan results in the absence of, or at least at a very low level of, Trp-tRNA.Trp. In turn, this causes the ribosome translating the leader sequence to stall at the Trp codons (see Figure 19.17a, p. 531). When the ribosome is stalled at the Trp codons, the RNA being synthesized just ahead of the ribosome by RNA polymerase assumes a particular secondary structure. This favors continued transcription of the structural genes by the polymerase. If the two Trp codons were mutated to stop codons, then the mutant operon would function constitutively in the same way as the wild-type operon in the absence of tryptophan. The ribosome would stall in the same place, and antitermination would result in transcription of the structural genes.

For a wild-type *trp* operon, the presence of tryptophan turns off transcription of the structural genes. This occurs because the presence of tryptophan leads to the accumulation of Trp-tRNA.Trp, which allows the ribosome to read the two Trp codons and stall at the normal stop codon for the leader sequence. When stalled in that position, the antitermination signal cannot form in the RNA being synthesized; instead, a termination signal is formed, resulting in the termination of transcription. In a mutant *trp* operon with two stop codons instead of the Trp codons, the stop codons cause the ribosome to stall, even though tryptophan and Trp-tRNA.Trp are present. This results in an antitermination signal and transcription of the structural genes.

In sum, in both the presence and the absence of tryptophan, the mutant *trp* operon will not show attenuation. The structural genes will be transcribed in both cases and the tryptophan biosynthetic enzymes will be synthesized.

19.20 1: If the aporepressor cannot bind to tryptophan, it will not be converted to an active repressor when tryptophan is present. This will lead to constitutive expression of tryptophan synthetase: In medium without tryptophan, expression will be the same as in the wild type; In medium with tryptophan, expression will be reduced only through attenuation, and so it will be about seventy-fold more than in the wild type.

2: The *trp* operon will exhibit constitutive expression, so mutant 2 will show the same expression patterns as mutant 1.

3: The *trpE* gene is the first gene transcribed in the *trp* operon (see Figure 19.15, p. 529). A nonsense mutation could have a polar effect, leading to diminished or absent translation of *trpB* and *trpA*, which encode tryptophan synthetase. Therefore, in medium without tryptophan where the operon is not repressed, mutant 3 would produce diminished levels of tryptophan synthetase compared to wild-type cells. In medium with tryptophan, the levels will be the same as in wild-type cells (very low).

4: Trp-tRNA.Trp molecules are needed to attenuate transcription at the *trp* operon. Therefore, if the levels of Trp-tRNA.Trp are always low, transcription of the *trp* operon will not be attenuated even when tryptophan levels are high. Therefore, in medium with tryptophan, mutant 4 will have about eight- to ten-fold higher levels of tryptophan synthetase than will wild-type cells. In medium without tryptophan, attenuation does not occur, so mutant 4 will have levels of tryptophan synthetase that are similar to the wild type.

5: In mutant 5, the 3:4 attenuator structure shown in Figure 19.17, p. 531 will not form, so attenuation will not occur when tryptophan levels are high. Tryptophan synthetase levels will be the same as in mutant 4.

19.24 The *cI* gene product is a repressor protein that acts to keep the lytic functions of the phage repressed when λ is in the lysogenic state. A *cI* mutant strain would lack the repressor and be unable to repress lysis, so that the phage would always follow a lytic pathway.

19.26 See table below.

Chapter 20 Regulation of Gene Expression in Eukaryotes

20.2 In prokaryotes, a nonsense mutation in an upstream gene of a polygenic transcript can prevent reinitiation of translation at downstream genes, leading to polarity. The polygenic transcripts of eukaryotic operons are processed by 5′ trans-splicing, which adds a capped, spliced leader RNA, and 3′ cleavage and polyadenylation, so that each gene is separately translated. Consequently, nonsense mutations in an upstream gene will not affect translation of downstream genes, and they will not show polarity.

20.5 a. The disappearance of the 4-kb band indicates that the promoter region has a DNase I hypersensitive site—a less highly coiled site where DNA is more accessible to DNase I for digestion. Since DNase I digestion produces a 2-kb band, the site lies near the middle of the 4-kb *Eco*RI fragment. The 3-kb band does not diminish in intensity except at the highest DNase I concentration, so the region of the gene containing the 3-kb *Eco*RI fragment is more highly coiled by nucleosomes.

b. Following the ecdysone pulse, increased concentrations of DNase I lead to the disappearance of both the 4- and 3-kb bands, indicating that DNase I has increased access and the gene is less tightly coiled during transcription. The appearance of low molecular weight digestion products indicates that DNase I cannot access some regions of the 4- and 3-kb *Eco*RI fragments. These regions may be bound by proteins such as general transcription factors.

20.7 a. Histones repress gene expression, so if they are present on DNA, promoter-binding proteins cannot bind promoters and transcription cannot occur.

b. Histones will compete more strongly for promoters than will promoter-binding proteins, so transcription will not occur.

c. If promoter-binding proteins are already assembled on promoters, nucleosomes will be unable to assemble on these sites, so transcription will occur.

d. Enhancer-binding proteins will help promoter-binding proteins to bind promoters even in the presence of histones, so transcription will occur.

20.9 The final product of an rRNA gene is an rRNA molecule. Therefore, a large number of genes are required to produce the large number of rRNA molecules required for ribosome biosynthesis. In contrast, ribosomal proteins are the end products of the translation of mRNAs. The mRNAs can be read over and over to produce the large number of ribosomal protein molecules required for ribosomal biosynthesis.

20.11 The data indicate that the synthesis of ovalbumin is dependent upon the presence of the hormone estrogen. These data do not address the mechanism by which estrogen achieves its effects. Theoretically, it could act (1) to increase transcription of the ovalbumin gene by binding to an intracellular receptor

Mutant	Molecular Phenotype	Lytic Growth	Lysogenic Growth	Inducible by UV Light
1	The Cro protein is unable to bind DNA.	no	yes	no
2	The N protein does not function.	no	no	no
3	The cII protein does not function.	yes	no	yes
4	The Q protein does not function.	no	yes	no
5	P_{RM} is unable to bind RNA polymerase.	yes	no	yes

that, as an activated complex, stimulates transcription at the ovalbumin gene; (2) to stabilize the ovalbumin precursor mRNA; (3) to increase the processing of the precursor ovalbumin mRNA; (4) to increase the transport of the processed ovalbumin mRNA out of the nucleus; (5) to stabilize the mature ovalbumin mRNA once it has been transported into the cytoplasm; (6) to stimulate translation of the ovalbumin mRNA in the cytoplasm; or (7) to stabilize (or process) the newly synthesized ovalbumin protein. Experiments in which the levels of ovalbumin mRNA were measured have shown that the production of ovalbumin mRNA is primarily regulated at the level of transcription.

20.16 a. In fragile X syndrome, the expanded CGG repeat results in hypermethylation and transcriptional silencing. A CAG codon specifies glutamine, so in Huntington disease, the expanded CAG repeat results in the inclusion of a polyglutamine stretch within the huntingtin protein. This causes it to have a novel, abnormal function.

b. A heterozygote with a CGG repeat expansion near one copy of the *FMR-1* gene will still have one normal copy of the *FMR-1* gene. The normal gene can produce a normal product, even if the other is silenced. (The actual situation is made somewhat more complex by the process of X inactivation in females, but in general one would expect that a mutation that caused transcriptional silencing of one allele would not affect a normal allele on a homolog.) In contrast, a novel, abnormal protein is produced by the CAG expansion in the disease allele in Huntington disease. Since the disease phenotype is due to the presence of the abnormal protein, the disease trait is dominant.

c. Transcriptional silencing may require significant amounts of hypermethylation, and so require more CGG repeats for an effect to be seen. In contrast, protein function may be altered by a stretch of more than 36 glutamines.

20.20 a. In bacterial operons, a common regulatory region controls the production of single mRNA from which multiple protein products are translated. These products function in a related biochemical pathway. Here, two proteins that are involved in the synthesis and packaging of acetylcholine are both produced from a common primary mRNA transcript.

b. Unlike the proteins translated from an mRNA synthesized from a bacterial operon, the protein products produced at the *VAChT/ChAT* locus are not translated sequentially from the same mRNA. Here, the primary mRNA appears to be alternatively processed to produce two distinct mature mRNAs. These mRNAs are translated starting at different points, producing different proteins.

c. At least two mechanisms are involved in the production of the different ChAT and VAChT proteins: alternative mRNA processing and alternative translation initiation. After the first exon, an alternative 3′ splice site is used in the two different mRNAs. In addition, different AUG start codons are used.

20.21 a. Four different protein isoforms differing in their C-terminus are produced.

b. The cDNA structures indicate that alternative mRNA splicing is used to generate the different protein isoforms. Specifically, alternative 5′ splice sites are used: The last exon of cDNA 4 contains a 5′ splice site that is used by cDNAs 1, 2, and 3; the last exon of cDNA 1 contains a 5′ splice site that is used by cDNAs 2 and 3; and the last exon of cDNA 2 contains a 5′ splice site that is used by cDNA 3.

Chapter 21 Genetic Analysis of Development

21.3 This experiment demonstrates the phenomenon of *determination* and the point at which it occurs during development. The tissue taken from the blastula or gastrula has not yet been committed to its final differentiated state in terms of its genetic programming; that is, it has not yet been *determined*. Thus, when the tissue is transplanted into the host, it adopts the fate of nearby tissues and differentiates in the same way that they do. Presumably, cues from the tissue surrounding the transplant determine its fate. In contrast, tissues in the neurula stage are stably determined. By the time the neurula developmental stage has been reached, a developmental program has been set. In other words, the fate of neurula tissue transplants is *determined*. Upon transplantation, they will differentiate according to their own set genetic program. Tissue transplanted from a neurula to an older embryo cannot be influenced by the surrounding tissues. It will develop into the tissue type for which it has been determined, in this case, an eye.

21.6 a. Based on the work of Wilmut and his colleagues, the nose cells would first be dissociated and grown in tissue culture. The cells would be induced into a quiescent state (the G_0 phase of the cell cycle) by reducing the concentration of growth serum in the medium. Then they would be fused with enucleated oocytes from a donor female and allowed to grow and divide by mitosis to produce embryos. The embryos would be implanted into a surrogate female. After the establishment of pregnancy, its progression would need to be maintained.

b. While the nuclear genome would generally be identical to that in the original nose cell, cytoplasmic organelles presumably would derive from those in the enucleated oocyte. Therefore, the mitochondrial DNA would not derive from the original leader. In addition, because telomeres in an older individual are shorter, one might expect the telomeres in the cloned leader to be those of an older individual.

c. In mature B cells, DNA rearrangements at the heavy- and light-chain immunoglobulin genes have occurred. One would expect the cloned leader to be immunocompromised, as he would be unable to make the wide spectrum of antibodies present in a normal individual.

d. It is likely that the cloning process will be very inefficient, with most clones dying before or soon after birth. The surviving clones are also likely to differ in body shape and personality, and are unlikely to be normal since a nucleus donated from the differentiated nose cell is unlikely to be completely reprogrammed. One suggestion is for the government to hire a good plastic surgeon to alter the appearance of a good actor able to assume the role of the totalitarian leader.

e. There is no way to predict the psychological profile of the cloned leader based on his genetic identity. Even identical twins, who are genetically more identical than such a clone, do not always share behavioral traits.

21.13 Experiment A results in all of the DNA becoming radioactively labeled. The distribution of radioactive label throughout the polytene chromosomes indicates that DNA is a fundamental and major component of these chromosomes. The even distribution of label suggests that each region of the chromosome has been replicated to the same extent. This provides support for the contention that band and interband regions are the result of different types of packaging, not of different amounts of DNA replication. Experiment B results in the radioactive labeling of RNA molecules. The finding that label is found first in puffs indicates that these are sites of transcriptional activity that arise from molecules that are in the process of being synthesized. The later appearance of label in the cytoplasm reflects the completed RNA molecules that have been processed and transported into the cytoplasm, where they will be translated. Experiment C provides additional support for the hypothesis that transcriptional activity is associated with puffs. The inhibition of RNA transcription by actinomycin D blocks the appearance of signal over puffs, indicating that it blocks the incorporation of ^{3}H-uridine into RNA in puffed regions. The fact that the puffs are much smaller indicates that the puffing process itself is associated with the onset of transcriptional activity for the genes in a specific region of the chromosome.

21.17 a. Each cell should exhibit green fluorescence, since the gene is constitutively expressed.

b. About half the cells will exhibit green fluorescence. If more than one *Xic* is present, X inactivation will occur on one *Xic*-containing chromosome. Either the X chromosome or the *Xic*-bearing autosome will be inactivated, at random. If the X chromosome is inactivated, the *gfp* gene will not be expressed.

c. The *Xist* gene on the autosome is being expressed. Since the cell exhibits green fluorescence, the X chromosome with the *gfp* gene is not inactivated and the *Xic*-bearing autosome is inactivated. The *Xist* gene on the autosome is transcribed and its RNA coats the autosome to trigger the methylation of histone H3. This initiates chromatin remodeling to silence genes on the *Xic*-bearing autosome.

21.18 a. The X:A ratio is detected by interactions of the protein products of three X-linked numerator genes (*sis-a*, *sis-b*, *sis-c*) and one autosomal denominator gene (*dpn*). The numerator gene products can form either homodimers or heterodimers with the denominator gene product. When the X:A ratio is 2:2, an excess of numerator gene products leads to the formation of many homodimers. These serve as transcription factors to activate *Sxl* transcription from P$_E$. When the X:A ratio is 1:2, most numerator subunits are found in heterodimers, so *Sxl* transcription is not activated. Therefore, activation of *Sxl* at P$_E$ by the homodimers serves to detect the X:A ratio and leads to the early sex-specific synthesis of SXL protein.

b. Transcription at P$_E$ is essential to generate a functional SXL protein in animals with an X:A ratio of 2:2. It is not used in individuals with an X:A ratio of 1:2, so these animals will be unaffected, and differentiate as males. In individuals with an X:A ratio of 2:2, SXL initiates a cascade of alternative mRNA splicing at *Sxl, tra,* and *dsx* that leads to the implementation of female differentiation. If there is no transcription from P$_E$, no functional SXL protein will be produced in these animals, and a default set of splice choices at *Sxl, tra,* and *dsx* will be used. In principle this would lead to male differentiation in individuals with an X:A ratio of 2:2. However, SXL also prevents the translation of *msl-2* transcripts so that dosage compensation does not normally occur in individuals with an X:A ratio of 2:2. Without SXL, *msl-2* transcripts will be translated so dosage compensation will occur, leading to four doses of X-linked gene products. The imbalance in X and autosomal gene product dosage is likely to be lethal to these animals.

c. This *tra* mutant will eliminate functional TRA protein. Since functional TRA is not normally present in animals with an X:A ratio of 1:2, this mutation will have no affect on these animals—they will differentiate normally into males. TRA is normally present in individuals with an X:A ratio of 2:2, where it functions to regulate alternative splicing at *dsx* and produce DSX-F, which implements female differentiation by repressing male-specific gene expression. Without TRA, default splicing will occur at *dsx* and produce DSX-M, which implements male differentiation by repressing female-specific gene expression. Thus, animals with an X:A ratio of 2:2 will be males.

d. Animals with knockout mutations at *dsx* will produce neither DSX-M, which represses female-specific gene expression, nor DSX-F, which represses male-specific gene expression. Therefore, neither male- nor female-specific gene expression will be repressed, and both male and female differentiation pathways will proceed.

21.21 Preexisting, maternally packaged mRNAs that have been stored in the oocyte are recruited into polysomes as development begins following fertilization.

21.24

Mutant	Class
a	segmentation gene (segment polarity)
b	maternal effect gene (anterior-posterior gradient)
c	segmentation gene (gap)
d	homeotic (transforms cell fate from eye to wing)
e	segmentation gene (gap)

21.26 Preexisting mRNA that was made by the mother and packaged into the oocyte prior to fertilization is translated up to the gastrula stage. After gastrulation, new mRNA synthesis is necessary for the production of proteins needed for subsequent embryonic development.

Chapter 22 Genetics of Cancer

22.5 If FeSV contributed to the feline sarcoma, FeSV should be found in the neoplastic tissues (muscle and bone marrow). The Southern blot provides this evidence: A 1.2-kb DNA fragment hybridizes to the *fes* cDNA probe in the lanes with DNA from muscle and bone marrow, and in the control lane with FeSV cDNA. The size difference between the 1.2-kb hybridizing fragment and the 1.0-kb *fes* proto-oncogene HindIII-cut cDNA probe reflects their different origins. The *fes* proto-oncogene is found normally in a cat, while the FeSV *fes* oncogene is found in a retrovirus. The size of the fragment in the retrovirus may reflect a polymorphic HindIII site and/or a gene rearrangement. The *fes* proto-oncogene normally functions in the cat, so it is DNA should be present in all tissues. The 3.4-kb DNA fragment is found in all of the cat tissues, so it is likely to be the genomic sequence. Since the *fes* proto-oncogene cDNA has a 1.0-kb HindIII fragment, the mRNA of this gene is very likely spliced to remove a 2.4-kb intron.

22.6 The high degree of conservation of proto-oncogenes suggests that they function in normal, essential, conserved cellular processes. Given the relationship between oncogenes and proto-oncogenes, it also suggests that cancer occurs when these processes are not correctly regulated.

22.7 a. Proto-oncogenes encode a diverse set of gene products that includes growth factors, receptor and nonreceptor protein kinases, receptors lacking protein kinase activity, membrane-associated GTP-binding proteins, cytoplasmic regulators involved in intracellular signaling, and nuclear transcription factors. These gene products all function in intercellular and intracellular pathways that regulate cell division and differentiation.

b. In general, mutations that activate a proto-oncogene convert it into an oncogene. Since (i), (iii), and (viii) cause a decrease in gene expression, they are unlikely to result in an oncogene. Since (ii) and (vii) could activate gene expression, they could result in an oncogene. Mutations (iv), (v), and (vi) cannot be predicted with certainty. The deletion of a $3'$ splice site acceptor would alter the mature mRNA and possibly the protein produced, and may or may not affect the function and regulation of the protein. Similarly, it is difficult to predict the effect of a nonspecific point mutation or a premature stop codon. The text presents examples in which these types of mutations have caused the activation of a proto-oncogene and resulted in an oncogene.

22.8 a. Increased transcription of the mRNA will lead to increased levels of the growth factor, which in turn will stimulate fibroblast growth and division.

b. Constitutive activation of a nonreceptor tyrosine kinase could lead to aberrant, unregulated phosphorylation and activation of many different proteins, including growth factor receptors, that are involved in signaling cascades used in regulating cellular growth and differentiation.

c. When the growth factor EGF binds its membrane-bound receptor, it stimulates its autophosphorylation, which allows for Grb2 to bind and recruit SOS. SOS displaces GDP from Ras, a membrane-associated G protein, so that it can bind GTP. GTP-bound Ras recruits and activates Raf-1 to initiate the cytoplasmic MAP kinase signaling cascade. In turn, this signaling cascade activates transcription factors such as Elk-1 to induce transcription of cell-cycle specific target genes. Therefore, if a membrane-associated G protein were unable to hydrolyze GTP, it would be constitutively active and lead to constant expression of genes needed for cell cycle progression.

22.12 One hypothesis is that the proviral DNA has integrated near the proto-oncogene, and the expression of the proto-oncogene has come under the control of promoter and enhancer sequences in the retroviral LTR. This could be assessed by performing a whole-genome Southern blot analysis to determine whether the organization of the genomic DNA sequences near the proto-oncogene has been altered.

22.13 Tumor growth induced by transforming retroviruses results either from the activity of a single viral oncogene or from the activation of a proto-oncogene caused by the nearby integration of the proviral DNA. The oncogene can cause abnormal cellular proliferation via the variety of mechanisms discussed in the text. The expression of a proto-oncogene, normally tightly regulated during cell growth and development, can be altered if it comes under the control of the promoter and enhancer sequences in the retroviral LTR.

DNA tumor viruses do not carry oncogenes. They transform cells through the action of one or more genes within their genomes. For example, in a rare event, the DNA virus can be integrated into the host genome, and the DNA replication of the host cell may be stimulated by a viral protein that activates viral DNA replication. This would cause the cell to move from the G_0 to the S phase of the cell cycle.

For both transducing retroviruses and DNA tumor viruses, abnormally expressed proteins lead to the activation of the cell from G_0 to S and abnormal cell growth.

22.14 Experimentally fuse cells from the two cell lines and then test the resultant hybrids for their ability to form tumors. If the uncontrolled growth of the tumor cell line was caused by a mutated pair of tumor suppressor alleles, then the normal alleles present in the normal cell line would "rescue" the tumor cell line defect. The hybrid line would grow normally and be unable to form a tumor. If the uncontrolled growth of the tumor cell line was caused by an oncogene, the oncogene would also be present in the hybrid cell line. The hybrid line would grow uncontrollably and form a tumor.

22.15 Hereditary cancer is associated with the inheritance of a germ-line mutation; sporadic cancer is not. Consequently, hereditary cancer runs in families. For some cancers, both hereditary and sporadic forms exist, with the hereditary form being much less frequent. For example, retinoblastoma occurs when both normal alleles of the tumor suppressor gene *RB* are inactivated. In hereditary retinoblastoma, a mutated, inactive allele is transmitted via the germ line. Retinoblastoma occurs in cells of an *RB/rb* heterozygote when an additional somatic mutation occurs. In the sporadic form of the disease, retinoblastoma occurs when both alleles are inactivated somatically.

22.18 a. Studies of hereditary forms of cancer have led to insights into the fundamental cellular processes affected by cancer. For example, substantial insights into the important role of DNA repair and the relationship between the control of the cell cycle and DNA repair have come from analyses of the genes responsible for hereditary forms of human colorectal cancer. For breast cancer, studying the normal functions of the *BRCA1* and *BRCA2* genes promises to provide substantial insights into breast and ovarian cancers.

b. Genetic predisposition for cancer refers to the presence of an inherited mutation that, with additional somatic mutations during the individual's life span, can lead to cancer. For diseases such as retinoblastoma, a genetic predisposition has been associated with the inheritance of a recessive allele of the *RB* tumor suppressor gene. Retinoblastoma occurs in *RB/rb* individuals when the normal allele is mutated in somatic cells and the pRB protein no longer functions. Because somatic mutation is likely, the disease appears dominant in pedigrees.

Although there is a substantial understanding of the genetic basis for cancer and the genetic abnormalities present in somatic cancerous cells, there are also substantial environmental risk factors for specific cancers. Environmental risk factors must be investigated thoroughly when a pedigree is evaluated for a genetic predisposition for cancer.

22.22 Apoptosis is programmed, or suicidal, cell death. Cells targeted for apoptosis are those that have high levels of DNA damage and so are at a greater risk for neoplastic transformation. (During the development of some tissues in multicellular organisms, cell death via apoptosis is a normal process.) Apoptosis is regulated by p53, among other proteins. In cells with large amounts of DNA damage, p53 accumulates and functions as a transcription factor to activate transcription of DNA repair genes and *WAF1*, whose product, p21, leads cells to arrest in G_1. If very high levels of DNA damage exist, p53 does not induce DNA repair genes and *WAF1*, but activates the *BAX* gene, whose product blocks the BCL-2 protein from repressing the apoptosis pathway. By blocking BCL-2 function, the apoptosis pathway is activated.

22.25 Tumors result from multiple mutational events that typically involve both the activation of oncogenes and the inactivation of tumor suppressor genes. The analysis of hereditary adenomatous polyposis, an inherited form of colorectal cancer, has shown that the more differentiated cells found in benign, early-stage tumors are associated with fewer mutational events, while the less differentiated cells found in malignant and metastatic tumors are associated with more mutational events. Although the path by which mutations accumulate varies between tumors, additional mutations that activate oncogenes and inactivate tumor suppressor genes generally lead to the breaking down of the multiple mechanisms that regulate growth and differentiation.

22.26 a. The fact that some translocations are found as the only cytogenetic abnormality in certain cancers probably means that they are a key event in tumor formation. It does not necessarily mean that they are the primary cause of the tumor or the first of many mutational events.

b. A chimeric fusion protein may have different functional properties than do either of the two proteins from which it derives. If it results in the activation of a proto-oncogene product into a protein that has oncogenic properties or if it results in the inactivation of a tumor suppressor gene product, it could play a key role in the genetic cascade of events leading to tumor formation.

c. Before drawing conclusions as to whether these chromosomal aberrations inactivate the function of tumor suppressor genes, or activate quiescent proto-oncogenes, it is necessary to have additional molecular information on the effects of the translocation breakpoints on specific transcripts. Finding that the translocation breakpoints result in a lack of gene transcription or in transcripts that encode nonfunctional products would support the hypothesis that the translocation inactivated a tumor suppressor gene. Finding that the translocation breakpoints result in activation of gene transcription or in the production of an active fusion protein would support the hypothesis that the translocation activated a previously quiescent proto-oncogene.

d. One hypothesis is that the various fusion proteins that result from different translocations involving the *EWS* gene somehow result in the transcription activation of different proto-oncogenes, and this leads to the different sarcomas that are seen. (Sarcomas are cancers found in tissues that include muscle, bone, fat, and blood vessels.)

e. If translocation breakpoints are conserved within a tumor type, molecular-based diagnostics can be developed to identify the breakpoints relatively quickly from a tissue biopsy. For example, if the genes at the breakpoints have been cloned, PCR methods can be used to address whether the gene is intact or disrupted, using the DNA from cells of a tumor biopsy. Primers can be designed to amplify different segments of the normal gene. Then, PCR reactions containing these primers and either normal, control DNA or tumor-cell DNA can be set up to determine if each segment of a candidate gene is intact (a PCR product of the expected size is obtained) or disrupted (no PCR product will be obtained, because the gene has been rearranged).

Such molecular analyses would provide fast, accurate tumor diagnosis. If the different tumor types respond differentially to different regimens of therapeutic intervention, then a more rapid, unequivocal diagnosis of a particular tumor type

should allow for the earlier prescription of a more optimized regime of therapeutic intervention. In addition, understanding the nature of the normal gene products of the affected genes may allow for the development of sarcoma-specific therapies.

22.28 Direct-acting carcinogens are chemicals that bind to DNA and act as mutagens. Procarcinogens are chemicals that must be converted by normal cellular enzymes to become active carcinogens. These products, most of which also bind DNA and act as mutagens, are referred to as ultimate carcinogens.

Chapter 23 Non-Mendelian Inheritance

23.1 a. The endosymbiont theory holds that ancestral eukaryotic cells were anaerobic organisms that lacked mitochondria and chloroplasts. They formed symbiotic relationships with a purple nonsulfur photosynthetic bacterium whose oxidative phosphorylation activities benefited them and then, as atmospheric oxygen increased due to photosynthetic activity, whose photosynthetic activity was lost. Over time, the eukaryotic cell became dependent on the intracellular bacterium for survival, and mitochondria were formed. Chloroplasts were formed through a similar symbiotic relationship when eukaryotic cells ingested an oxygen-producing photosynthetic cyanobacterium.

b. During the evolution of mitochondria and chloroplasts, some of their ancestral genes were transferred to the nuclear genome. Without these genes, they can no longer survive as independent organisms.

c. Photosynthetic activity led to increased atmospheric O_2, so the photosynthetic activity provided by the purple nonsulfur photosynthetic bacterium was no longer essential and was lost.

d. Like many bacterial genomes, many mitochondrial and all chloroplast genomes are circular. Like bacterial genomes, all mitochondrial and chloroplast genomes are supercoiled and packaged without histones in nucleoid bodies. Translation in mitochondria and chloroplasts is not identical to that in prokaryotes, but some features are reminiscent of prokaryotes, including the use of 70S ribosomes with 50S and 30S subunits, a 30S subunit having a 16S rRNA, fMet-tRNA as an initiator codon, and the sensitivity to inhibitors of bacterial but not cytoplasmic ribosome function.

23.4 Multiple approaches are possible. You could purify mitochondria, isolate DNA from them, and use CsCl density gradient centrifugation to determine if this DNA has an identical density to that of the minor peak. You could isolate the minor peak and visualize the DNA under the electron microscope. If the molecules are circular, they are unlikely to be nuclear fragments. You could grow the yeast in the presence of an intercalating agent such as acridine and see whether this treatment causes the minor peak species to disappear. If it does, the minor peak is organellar in origin. You could isolate the minor peak, label it (by nick translation, for example), and hybridize this labeled DNA to DNA within suitably prepared yeast cells. Then, using the electron microscope, you could determine whether the label is found over the nucleus or the mitochondria. Finally, you could isolate the minor peak DNA and study its sequence similarity with mitochondrial DNA from other yeast.

23.6 Mitochondrial transcription in animals is unlike nuclear or prokaryotic transcription in that each strand is transcribed into a single RNA that is then processed to produce the RNAs for individual genes. Most of the genes encoding mRNAs and rRNAs are separated by tRNA genes. The mRNAs and rRNAs are produced as the tRNA genes are recognized and cut out by specific enzymes. The mRNAs also lack 5′ caps, introns, and chain-terminating codons (mRNAs end in U or UA; polyadenylation completes the missing parts of a UAA stop codon).

23.8 a. Selective use of antibiotics that inhibit translation by either mitochondrial or cytoplasmic ribosomes has shown that components 2, 5, and 6 of cytochrome oxidase are synthesized on mitochondrial ribosomes, while components 1, 3, 4, and 7 are synthesized on cytoplasmic ribosomes. See Figure 23.5 (p. 636) and the accompanying text.

b. Compare cytochrome oxidase activity in cybids made with platelets from diseased individuals and in cybids made with platelets from age-matched control individuals. It is important to assess several different enzyme activities associated with mitochondrial proteins to ensure that the deficits in cytochrome oxidase are specific.

c. As discussed in the text (p. 647), the cells of individuals with diseases resulting from mitochondrial DNA defects have a mixture of mutant and normal mitochondria; that is, they show heteroplasmy. Thus, assays in cybids are measurements of the enzyme activity present in a population of mitochondria in a cell. It would be unlikely that each of the mitochondria of an affected individual has an identical defect.

23.9 The two codes differ in codon designation. The mitochondrial code has more extensive wobble, so fewer tRNAs are needed to read all possible sense codons. As a consequence, fewer mitochondrial genes are necessary. The advantage is that fewer tRNA genes need be present than for cytoplasmic tRNAs.

23.11 a. Rubisco is the first enzyme used in the pathway for fixation of carbon dioxide in photosynthesis.

b. Determine which subunit is made in of plant cells grown in tissue culture with an antibiotic that selectively inhibits translation by either mitochondrial (e.g., neomycin) or cytoplasmic (e.g., cycloheximide) ribosomes.

c. Since the small subunit is encoded by the nuclear genome, a mutation in one of the two gene copies will eliminate about half of the small subunit produced, decreasing functional rubisco levels by about half. These plants might show phenotypic effects (e.g., slower growth), making the mutation dominant. Homozygous mutants lack functional rubisco, so would be unable to carry out photosynthesis. Such mutants would die soon after seed germination, making the mutation a recessive lethal one.

d. Each chloroplast contains multiple cpDNA copies with different species having different copy numbers. In species with a high copy number (e.g., *Chlamydomonas*), a mutation in one copy would have a negligible effect on the amount of functional rubisco in the cell, and could have little phenotypic effect. If the cpDNA copy number were small, a phenotypic effect might be seen. Plants with the cpDNA mutation could also have cells with different numbers of mutation-bearing cpDNAs and show variegation.

23.12 Traits showing non-Mendelian inheritance exhibit differences in reciprocal crosses that are unrelated to sex, cannot be mapped to nuclear linkage groups, produce progeny ratios atypical of Mendelian segregation, and are indifferent to nuclear substitution.

23.15 a. The *tudor* mutation is a maternal effect mutation. Homozygous *tudor* mothers give rise to sterile progeny, regardless of their mate.

b. The grandchildless phenotype results from the absence of some maternally packaged component in the egg needed for the development of the F_1's germ line.

23.16 a. If the normal cytoplasm is [N] and the male-sterile cytoplasm is [Ms], the cross is [Ms] *rf/rf* ♀ × [N] *Rf/Rf* ♂, and the F_1 would be [Ms] *Rf/rf* and is male-fertile.

b. The cross is [Ms] *Rf/rf* ♀ × [N] *rf/rf* ♂. Half of the progeny would be [Ms] *Rf/rf* and be male fertile, and half would be [Ms] *rf/rf* and be male sterile.

23.17 a. Parents: [*poky*]F ♀ × [N]+ ♂; progeny: [*poky*]F and [*poky*]+ in equal numbers. Standard [*poky*] ([*poky*]+) is found among the offspring, so gene F must not effect a permanent alteration of [*poky*] cytoplasm. The 1:1 ratio of [*poky*] to *fast*-[*poky*] indicates that all progeny have the [*poky*] mitochondrial genotype (by maternal inheritance) and that the F gene must be a nuclear gene segregating according to Mendelian principles. Therefore, the [*poky*] progeny are [*poky*]+, and the *fast*-[*poky*] are [*poky*]F.

b. Parents: [N]+ ♀ × [*poky*]F ♂; progeny [N]+ and [N]F in equal numbers. These two genotypes are phenotypically normal and indistinguishable.

23.19 ½ *petite*, ½ wild type (*grande*)

23.24 If the male lethality is caused by a sex-linked, male-specific lethal mutation (*l*), the cross can be written as *l/l* ♀ × +/Y ♂, giving *l/+* (♀) and *l/*Y (dead ♂) progeny. A cross of the F_1 females to normal males (*l/+* ♀ × +/Y ♂) will give a 2:1 ratio of females to males (¼ *l/+* ♀, ¼ +/+ ♀, ¼ +/Y ♂, and ¼ *l/*Y dead ♂). If the male lethality is caused by a maternally inherited cytoplasmic factor lethal to males, then the F_1 females will receive this factor in cytoplasm from their mother, so they (like their mothers) should have only female offspring when mated to wild-type males.

23.27 Draw out two pedigrees to illustrate the lineage of Carlos. In one, include Mr. and Mrs. Escobar, Mr. and Mrs. Sanchez, their murdered children, and Carlos. In the other, include Mr. and Mrs. Mendoza and Carlos. Analyze each pedigree to determine who could have contributed mitochondrial DNA to Carlos.

a. Mitochondrial RFLP data can be helpful to trace the maternal line of descent. Carlos Mendoza will have inherited his mitochondrial DNA from his mother, and she will have inherited it from her mother. If Mrs. Escobar and Mrs. Mendoza have different mitochondrial RFLPs, it can be determined which of them contributed mitochondria to Carlos.

b. Only Carlos and individuals who might have maternally contributed his mitochondria (Mrs. Escobar and Mrs. Mendoza) need to be tested. The potential grandfathers need not be tested. Mrs. Sanchez also need not be tested: She may have given mitochondria to Carlos's father, but the father would not have passed them on to Carlos.

c. If Mrs. Mendoza and Mrs. Escobar do not differ in mitochondria RFLPs, the data will not be helpful. If the mitochondrial RFLPs do differ, and Carlos matches Mrs. Mendoza, the case should be dismissed. If Carlos matches Mrs. Escobar, then the Escobar and Sanchez couples are indeed the grandparents, and the Mendozas have claimed a stolen child.

23.30 The F_1 snail gives sinistral offspring when selfed, so it is *dd*. Therefore, both parents had a *d* allele. Since the F_1 has a dextral pattern, its maternal parent was *Dd*. The paternal parent could have been either *Dd* or *dd*.

Chapter 24 Population Genetics

24.1 Equate the frequency of each color with the frequency expected in Hardy-Weinberg equilibrium, letting $p = f(C^B)$, $q = f(C^P)$, and $r = f(C^Y)$.

Brown: $f(C^B C^B) + f(C^B C^P) + f(C^B C^Y) = p^2 + 2pq + 2pr = 236/500 = 0.473$

Pink: $f(C^P C^P) + f(C^P C^Y) = q^2 + 2qr = 231/500 = 0.462$

Yellow: $f(C^Y C^Y) = r^2 = 33/500 = 0.066$

Now solve for p, q, and r, knowing that $p + q + r = 1$:

$$r^2 = 0.066, \text{ so } r = \sqrt{0.066} = 0.26$$

There are two approaches to solving for q. First, because $q^2 + 2qr = 0.462$, we can substitute $r = 0.26$, giving $q^2 + 2q(0.26) = 0.462$. Recognize this as a quadratic equation, set it equal to 0, and solve for q; that is, solve the equation $q^2 + 0.52q - 0.462 = 0$

Solving the quadratic equation for q, we have

$$q = \frac{-0.52 \pm \sqrt{(0.52)^2 - 4(1)(-0.462)}}{2(1)} = 0.467$$

A second approach to solve for q is to realize that

$$q^2 + 2qr = 0.462$$
$$r^2 = 0.066$$

Adding left and right sides of the equations together, we have

$$q^2 + 2qr - r^2 = 0.066 + 0.467$$
$$(q + r) = 0.528$$
$$q = 0.726 - r = 0.726 - 0.26 = 0.467$$

Since $p + q + r = 1$,

$$p = 1 - (q + r) = 1 - (0.26 + 0.467) = 0.273$$

24.4 a. $\sqrt{0.16} = 0.40 = 40\%$ = frequency of recessive alleles: $1 - 0.4 = 0.6 = 60\%$ = frequency of dominant alleles; $2pq = (2)(0.4)(0.6) = 0.48$ = probability of heterozygous diploids. Then, $(0.48)/[(2 \times 0.16) + 0.48] = 0.48/0.80 = 60\%$ of recessive alleles are heterozygotes.

b. if $q^2 = 1\% = 0.01$, then $q = 0.1, p = 0.9$, and $2pq = 0.18$ heterozygous diploids. Therefore, $(0.18)/[0.18 + 2 (0.01)] = 0.18/0.20 = 0.90 = 90\%$ of recessive alleles in heterozygotes.

24.5 $2pq/q^2 = 8$, so $2p = 8q$; then $2(1 - q) = 8q$, and $2 = 10q$, or $q = 0.2$.

24.8 a. Let p equal the frequency of S and q equal the frequency of s. Then,

$$q = \frac{2(188)SS + 717Ss}{2(3,146)} = \frac{1,093}{6,292} = 0.1737$$

$$p = \frac{717Ss + 2(2,241)ss}{2(3,146)} = \frac{5,199}{6,292} = 0.8263$$

b.

Class	Observed	Expected	d	d^2/e
SS	188	95	+93	91.0
Ss	717	903	−186	38.3
ss	2,241	2,148	+93	4.0
	3,146	3,146	0	133.3

There is only one degree of freedom because the three genotypic classes are completely specified by two allele frequencies: p and q (df = number of phenotypes − number of *alleles* = 3 − 2 = 1). The X^2 value of 133.3, for one degree of freedom, gives $P < 0.0001$. Therefore, the distribution of gene types differs significantly from that expected if the population were in Hardy-Weinberg equilibrium.

24.11 Since the frequency of the trait is different in males than it is in females, the character might be caused by a sex-linked recessive gene. If the frequency of this gene is q, females would occur with the character at a frequency of q^2, and males with the frequency of q. The frequency in males is 0.4 and thus we may predict that the frequency in females is $(0.4)^2 = 0.16$ if this is a sex-linked gene. This result fits the data. Therefore, the frequency of heterozygous females is $2pq = 2(0.6)(0.4) = 0.48$. For sex-linked genes, no heterozygous males exist.

24.13 64/10,000 are color blind, that is, $0.0064 = q^2$, so $q = 0.08$ = probability of color-blind male.

24.16

$$q = \frac{u}{u + v} = \frac{6 \times 10^{-7}}{(6 \times 10^{-7}) + (6 \times 10^{-8})}$$

$$= \frac{6 \times 10^{-7}}{(6 \times 10^{-7}) + (0.6 \times 10^{-7})} = \frac{6}{6.6} = 0.91$$

Thus, the frequencies are 0.0081 *AA*, 0.1638 *Aa*, and 0.8281 *aa*.

24.20 $p'_x = mp + (1 + m)p$

Here,

$p'_x = [20/(20 + 80)](0.50) + \{1 - [20/(20 + 80)]\}$
$(0.70) = 0.66$

24.23 a. When selectively neutral, the genes distribute themselves according to the Hardy-Weinberg law, so 0.25 are *AA*, 0.5 are *Aa*, and 0.25 are *aa*.

b. $q = 0.33$

c. $q = 0.66$

24.25 a. $q = 0.63$

b. $q = 0.64$

c. $q = 0.66$

24.27 $q = u/s = (5 \times 10^{-5})/0.8 = 0.0000625$

24.30 a. The data fit the idea that a single *Bam*HI site varies. The probe is homologous to a region wholly within the 4.1-kb piece bounded on one end by the variable *Bam*HI site and on the other end by a constant site. When the variable site is present, the hybridized fragment is 4.1 kb. When the variable site is absent, the fragment extends to the next constant *Bam*HI site and is 6.7 kb long. People with only 4.1- or only 6.7-kb bands are homozygotes; people with both are heterozygotes.

b. The + allele of the variable site is present in $2(6) + 38 = 50$ chromosomes, and the − allele is present in $2(56) + 38 = 150$ chromosomes. Thus, $f(+)$ is 0.25 and $f(-)$ is 0.75.

c. If the population is in Hardy-Weinberg equilibrium, we would expect $(0.25)^2$ or 0.0625 of the sample to show only the 4.1-kb band. This would be 6.25 individuals. We observed 6. We expect $(0.75)^2$ or 0.5625 to be homozygous for the 6.7-kb band, which is 56.25 individuals. We saw 56. Finally, we would expect $2(0.25)(0.75)$ or 0.375 to be heterozygotes, or 37.5 individuals. We observed 38. The observed numbers are so close to the expected that a chi-square test is unnecessary.

24.31 To calculate the expected heterozygosity in the nucleotide sequence, use the formula presented in the text:

$$H_{nuc} = \frac{n(\Sigma c_i) - \Sigma c_i^2}{j(\Sigma c_i)(n-1)}$$

Here, n equals the number of homologous DNA molecules examined. In this example, 10 armadillos were examined, each having two homologous chromosomes, so $n = 20$. The quantity j equals the number of nucleotides in the restriction site. In this example, HindIII recognizes a 6-bp sequence, so $j = 6$. For each restriction site, c_i represents the number of molecules in the sample that were cleaved at that restriction site. Here, a single restriction site was cleaved at 10 of 20 sites, so $\Sigma c_i = 10$. (Individuals with two bands had sites on both chromosomes cleaved, individuals with three bands had a site on one but not the other chromosome cleaved, and individuals with one band had no sites cleaved.) Therefore, we obtain

$$H_{nuc} = [20(10) - 10^2]/[6(10)(19)] = 100/1{,}140 = 0.088$$

24.32 a. The expected heterozygosity is $1 - $ (frequency of expected homozygotes). If the frequency of alleles in the population are $p_1, p_2, p_3, \ldots, p_n$, the expected frequency of homozygotes is $p_1^2 + p_2^2 + p_3^2 + \cdots p_n^2$. For locus *G1A*, the expected frequency of homozygotes is $(0.398)^2 + (0.240)^2 + (0.211)^2 + (0.086)^2 + (0.036)^2 + (0.016)^2 + (0.007)^2 + (0.006)^2 = 0.270$, and the expected heterozygosity is $1 - 0.270 = 0.730$. The expected heterozygosity for the other loci are *G10X*, 0.741; *G10C*, 0.740; and *G10L*, 0.662. These are approximately the observed frequencies of heterozygosities. Since the numbers and types of different heterozygotes are given, it is not possible to use the chi-square test to directly evaluate whether the population is in Hardy-Weinberg equilibrium. The population appears to be close to Hardy-Weinberg equilibrium.

b. The three cubs of the mother show evidence of multiple paternity. For each of the loci *G10X* and *G10L*, three alleles present in the cub must have been contributed paternally (*G10X*: X133 or X137, X141; *G10L*: L155, L157, L161). This could have happened only if the cubs were sired by at least two different fathers. Multiple paternity within one set of cubs would tend to increase the genetic variability in the population because it would allow a larger number of males to contribute gametes seen in the next generation. Since $N_e = (4 \times N_f \times N_m)/(N_f + N_m)$, a larger N_m will tend to increase the effective population size.

24.34 a. Mutation leads to change in allelic frequencies within a population if no other forces are acting and so introduces genetic variation. If population size is small, mutation may lead to genetic differentiation between populations.

b. Migration increases the population size and has the potential to disrupt a Hardy-Weinberg equilibrium. It can increase genetic variation and may influence the evolution of allelic frequencies within populations. Over many generations, migration reduces divergence between populations and equalizes allelic frequencies between populations.

c. Genetic drift produces changes in allelic frequencies within a population. It can reduce genetic variation and increase the homozygosity within a population. Over time, it leads to genetic change. When several populations are compared, genetic drift can lead to increased genetic differences between populations.

d. Inbreeding increases the homozygosity within a population and decreases genetic variation.

24.38 In both instances, protein electrophoresis, RFLP analyses, and DNA sequence analysis of specific genes could be used to gather information on the genotype of the captured individuals and member of each island population. In (a), the captured individual should be returned to the subpopulation from which it shows the least genetic variation. In (b), evaluate the genotype of the two missing tortoises using DNA from the previously collected blood samples and compare these genotypes with the genotypes of the two captured tortoises. If a captured animal was taken from the field site, its genotype will exactly match a genotype obtained from one of the blood samples.

Chapter 25 Molecular Evolution

25.1 Each of the three codon positions can change to three different nucleotides, so a total of 135 substitutions must be considered for each position. At the first position, only 9 of the 135 possible changes have no effect on the amino acid sequence of the polypeptide (0.07 are synonymous). Every change at the second codon position results in an amino acid substitution (0.00 are synonymous). A total of 98 changes at the third codon position have no effect on the amino acid sequence of the protein (0.73 are synonymous). Natural selection is much more likely to act on mutations that change amino acid sequences, such as the second and first codon positions, and those are the ones where the least change is likely to be seen as sequences diverge.

25.2 $K = -3/4 \ln[1 - 4/3(p)]$
$K = -3/4 \ln[1 - 4/3(0.12)]$
$K = -3/4 \ln(1 - 0.16)$
$K = -3/4(-0.17)$
$K = 0.13$

25.3 $r = K/2T$
$r = 0.13/[2(80{,}000{,}000)]$
$r = 8.1 \times 10^{-10}$

25.4 Mutation rates would be greater than or equal to the substitution rate for any locus. Mutations are any nucleotide changes that occur during DNA replication or repair, whereas substitutions are mutations that have passed through the filter of selection. Many mutations are eliminated through the process of natural selection.

25.5 1×10^{-8} substitutions/nucleotide/year. 2×10^{-8} substitution/nucleotide/year.

25.6 The high substitution rate in mammalian mitochondrial genes is useful when determining the relationships between evolutionary closely related groups of organisms such as members of a single species. When longer divergence times are involved such as those associated with the mammalian radiation, more slowly evolving nuclear loci are more convenient to study because multiple substitutions are less likely to have occurred.

25.7 Mitochondria have higher concentrations of potentially mutagenic oxygen free radicals than those found in the nucleus. Mitochondrial genes may also be under less functional constraints because of their high copy number inside each cell. Also, mitochondrial DNA polymerases and repair processes may be less efficient than those of the nucleus.

25.8 Generation times of the organisms may be significantly different now or have been different at some point since they diverged. Average repair efficiency, average exposure to mutagens, and the need to adapt to new ecological niches and environments may also differ between the two species.

25.9 Sequence A is a pseudogene, and sequence B is a functioning gene.

25.10 Diversifying selection is the evolutionary process.

25.11 Regions involved in base pairing would not accumulate substitutions as quickly as those that are not. If the secondary structure of an RNA molecule is under selective constraint, then the only changes that would be found would be those for which a compensatory change also occurred in its complimentary sequence.

25.12 $d_{HR} = 6$; $d_{HD} = 10$; $d_{RD} = 11$

$$d_{A1} = (d_{HR} + d_{HD} - d_{RD})/2$$
$$d_{A2} = (d_{HR} + d_{RD} - d_{HD})/2$$
$$d_{A1} = (6 + 10 - 11)/2$$
$$d_{A2} = (6 + 11 - 10)/2$$
$$d_{A1} = 2.5$$
$$d_{A2} = 3.5$$

Since d_{A1} and d_{A2} are not equal, there has been a difference in the rate of substitution in this region of the human and rabbit genomes since they last shared a common ancestor. The human linage appears to have accumulated substitutions at a slower rate.

25.13 Molecular changes underlie all heritable morphological differences. Disadvantages associated with morphological characters include the possibilities of convergent evolution, little basis of difference, and little basis of comparison.

25.14 The appeal of analyzing ancient DNA samples is that they might allow ancestral sequences to be determined (and not just inferred). However, it is almost impossible to prove that an ancient organism is from the same lineage as an extant species. Increasing the number of taxa in any analysis increases the robustness of any phylogenetic inferences, but extant taxa are almost invariably easier to obtain.

25.15 Divergence within genes typically occurs before the splitting of populations that occurs when new species are created. Preexisting polymorphisms such as those seen in the major histocompatibilty locus could account for such a discordance.

25.16 These chances are related to the total number of rooted and unrooted trees that can be generated using six taxa.

$$N_{rooted} = (2n - 3)!/[2^{n-2}(n - 2)!]$$
$$N_{rooted} = 9!/[16 \times (4)!]$$
$$N_{rooted} = 362,80/384$$
$$N_{rooted} = 945$$
$$N_{unrooted} = (2n - 5)!/[2^{n-3}(n - 3)!]$$
$$N_{unrooted} = 7!/[8 \times (3)!]$$
$$N_{unrooted} = 5,040/48$$
$$N_{unrooted} = 105$$

Since there are only 105 possible unrooted trees and 945 possible rooted trees, choosing a correct unrooted tree is more likely.

25.19 The equations used to determine how many rooted and unrooted trees can describe the relationship between taxa have no parameter that considers the amount of data associated with each taxon but do consider the number of taxa.

25.20 Informative sites are the only sites that are considered in parsimony analyses.

25.21 Most new genes arise from the process of mutating redundant copies of already existing genes. Copies of genes are free to accumulate substitutions, whereas the original version remains under selective constraint. One alternative might be to have tracts of noncoding sequences undergo random changes until they code for something that provides a selective advantage. The chance of a sequence randomly accumulating mutations that give it an open reading frame and appropriate promoter elements all at the same time is extremely small.

Credits

Text and Illustration Credits

Figure 1.6: From Robert H. Tamarin, *Principles of Genetics.* Copyright © 1996 McGraw Hill. Used by permission of McGraw Hill Companies, Inc.

Figure 1.8: Peregrine Publishing

Figure 1.9: Peregrine Publishing

Figure 2.19: Reprinted from *Genes IV* by Benjamin Lewin. Copyright © 1990 with permission from Excerpta Medica, Inc.

Figure 2.21a, b and c: Reprinted from James Watson et al., *Molecular Biology of the Gene,* 5/e, fig 7.21 and 7.29. Copyright © 2003 Benjamin/Cummings. Reprinted by permission of Pearson Education, Inc.

Figure 2.24a and c: Reprinted from James Watson et al., *Molecular Biology of the Gene,* 5/e, figure 7.21 and 7.23b. Copyright © 2003 Benjamin/Cummings. Reprinted by permission of Pearson Education, Inc.

Figure 2.24b: Figure 21.4 p. 421 from Hartwell, Genetics: From Genes to Genomes, © 2003.

Figure 2.26: From Pluta et al., *Science,* Vol. 270, 1995, pp. 1591–94. Copyright © 1995 American Association for the Advancement of Science. Reprinted by permission.

Figure 2.31: Reprinted from *Cell,* Vol. 97, Greider, p. 419–422, copyright 1999 with permission from Excerpta Medica, Inc.

Figure 3.4: From *Molecular Biology of the Gene,* Fifth Edition by Watson et. al., Copyright © 2004. Reproduced by permission of Pearson Education.

Figure 3.6: From *Molecular Biology of the Cell,* Second Edition by Bruce Alberts. Copyright © 1989. Reproduced by permission of Routledge, Inc., part of The Taylor & Francis Group.

Figure 3.8: From *Molecular Biology of the Cell,* Second Edition by Bruce Alberts. Copyright © 1989. Reproduced by permission of Routledge, Inc., part of The Taylor & Francis Group.

Figure 3.16 From *Molecular Biology of the Gene,* Fifth Edition by Watson, et. al., Copyright © 2004. Reproduced by permission of Pearson Education.

Figure 5.7 From *Molecular Biology of the Gene,* Fifth Edition by Watson, et. al., Copyright © 2004. Reproduced by permission of Pearson Education.

Figure 5.11: Reprinted from *Cell,* Vol. 87, N. Proudfoot. "Ending the Message is Not So Simple" pp. 779–781, copyright 1996 with permission by Excerpta Medica, Inc.

Figure 5.14: From Maizels and Winer, "RNA Editing" *Nature,* Vol. 334, 1988, p. 469. Copyright © 1988 Macmillan Magazines Limited. Reprinted by permission.

Figure 6.4: © Irving Geis. Rights owned by Howard Hughes Medical Institute. Not to be used without permission.

Figure 6.13: Peregrine Publishing

Figure 7.16: From Brooker, *Genetics: Analysis and Principles,* p. 474. Copyright © 1998. Reprinted by permission of Pearson Education, Inc.

Figure 7.26: "Ty-transposable element of yeast" adapted from Watson by permission of Gerald B. Fink. Reprinted by permission of Pearson Education, Inc.

Figure 9.9a: Roza et al., *Molecular Vision,* Vol. 4, No. 20, 1998, figure 3a. Used with permission.

Figure 9.10: From *Biochemistry* by Donald Voet and Judith G. Voet. © 1990 by Donald Voet and Judith G. Voet. Reprinted by permission of John Wiley & Sons, Inc.

Figure 9.13: From Johnston et al., *Molecular Cell Biology,* Vol. 14, pp. 3834–3841, 1994. Copyright © 1994 American Society for Microbiology. Used with permission.

Figure 9.14: Reprinted from *Cell,* Vol. 87, Ryner et al., pp. 1079–1089, Fig. 2B, copyright 1996 with permission from Excerpta Medica, Inc.

Figure 9.16: From *Recombinant DNA* by J. D. Watson, M. Gillman, J. Witkowski, and M. Zoller. © 1983, 1992 by J. D. Watson, M. Gillman, J. Witkowski, and M. Zoller. Used with the permission of W.H. Freeman and Company.

Figure 9.17: From *Genetics* by Robert Weaver and Philip Hedrick, © 1989. Reprinted by permission of McGraw Hill Companies, Inc.

Figure 9.18: From *Recombinant DNA* by J. D. Watson, M. Gillman, J. Witkowski, and M. Zoller. © 1983, 1992 by J. D. Watson, M. Gillman, J. Witkowski, and M. Zoller. Used with the permission of W. H. Freeman and Company.

Figure 10.1: Reprinted from *Trends in Genetics,* Vol. 15, No. 2, R. K. Wilson, "How the Worm was Won: The *C. elegans* Genome Sequencing Project," pp. 51–58, copyright 1991 with permission from Elsevier Science.

Figure 10.2: From "DNA Mapping" *Time,* July 3, 2000, p. 71. © 2000 Time Inc. Reprinted by permission.

Figure 10.3: Reprinted with permission from Fleischmann et al., *Science,* July 28, 1995, Vol. 269, p. 507. Copyright © 1995 American Association for the Advancement of Science.

Figure 10.7: Reprinted from James Watson et al., *Molecular Biology of the Gene,* 5/e, figure 7.2. Copyright © 2003 Benjamin/Cummings. Reprinted by permission of Pearson Education, Inc.

Figure 10.9: Reprinted from *Trends in Genetics,* Vol. 12, No. 7B. Dujon, Fig. 3, p. 267, copyright 1966 with permission from Elsevier Science.

Figure 10.10: Copyright © Stanford University. Used with permission.

Figure 10.11a: Reprinted with permission from Chu et al., *Science,* Vol. 282, No. 699, figure 1. Copyright © 1998 American Association for the Advancement of Science.

Figure 10.11b: From http://orca/ucsc.edu/mathdocs/microarray/madn/microarray_ exp.jpg

with permission from Warren & Nelson (JAMA, 2-16-94, 271; 536–542) © 1994, American Medical Association 8.18a: National Library of Medicine 8.18b: M. Coleman/Visuals Unlimited 8.21a, b: Dr. Laird Jackson, Thomas Jefferson University Hospital, Division of Medical Genetics 8.22a, b: Dr. Laird Jackson, Thomas Jefferson University Hospital, Division of Medical Genetics

Chapter 18 Opener: © Dr. Dennis Kunkel/Phototake 9.1: Custom Medical Stock Photo, Inc. 9.4: Micrograph courtesy of David P. Allison, Biology Division, Oak Ridge National Laboratory. 9.11a,b: Courtesy of Dr. Harold W. Fisher, University of Rhode Island 9.13: Bruce Iverson 9.18: Courtesy of Gunther S. Stent, University of California, Berkeley 9.19: Courtesy of Dr. D. P. Snustad, Department of Genetics and Cell Biology, College of Biological Sciences, University of Minnesota

Chapter 19 Opener: © Ken Eward/ Science Source/Photo Researchers, Inc.

Chapter 20 Opener: Courtesy of Dr. Michael J. Kruhlak (NCIC) and Dr. Michael J. Hendzel (University of Alberta) and reprinted with the permission of cellnucleus.com 20.5: BioInfo Bank, RIKEN Tsukuba Institute

Chapter 21 Opener: Courtesy of Stephen Paddock, James Langeland, Peter DeVries, and Sean B. Carroll of the Howard Hughes Medical Institute at the University of Wisconsin (*Bio-Techniques,* Jan. 1993) 21.1: Courtesy of Dr. Bill Stark, Marcey, D. and Stark, W. S. The morphology, physiology and neural projections of supernumerary compound eyes in Drosophila melanogaster. Developmental Biology, 1985, 107, 180–197. 21.2: Edward T. Kipreos, Department of Cellular Biology, University of Georgia 21.3: Courtesy of Dr. Marty Yanofsky, Center for Molecular Agriculture, Department of Biology, University of California, San Diego 21.4a: Dr. Charles Kimmel, University of Oregon 21.4b: ZFIN, The Zebrafish Model Organism Database 21.7a,b: Z. Legacy Images Resource Centers

Chapter 22 Opener: Courtesy of Y. Cho et al., *Science* 265: 346–55 22.1: SIU/Visuals Unlimited 22.9: Custom Medical Stock Photo, Inc.

Chapter 23 Opener: Courtesy of Dr. Roderrick Capaldi & Daciana Margineantu 23.2: CNRI/Science Photo Library/PhotoResearchers, Inc. 23.8a: Color-Pic, Inc. 23.12: M.I. Walker/ Science Source/ Photo Researchers, Inc. 23.16: Kim Taylor/Bruce Coleman Inc.

Chapter 24 Opener: © Dieter and Mary Plage/Bruce Coleman, Inc. 24.1a: CORBIS 24.1b: Hildegard Adler, photographer; courtesy of James F. Crow 24.1c: UPI/CORBIS 24.5: The Field Museum, Neg #CSA 118, Chicago 24.8: Chip Clark 24.16a: Kevin Byron/Bruce Coleman Inc. 24.16b: W. Perry Conway/Tom Stack & Associates 24.17: Cliff B. Frith/Bruce Coleman Inc. 24.18a: AKG/Photo Researchers, Inc. 24.18b: The Granger Collection 24.19a, b: Breck P. Kent

Chapter 25 Opener: Photo courtesy of Mike Raymer

Index

Page numbers in *italics* indicate material in figures and tables.

A antigen, 342–343, *343*
*Aat*II, 195
Abalone, prezygotic isolation in, 698
abdominal-A gene, 596–597
abdominal-B gene, 596–597
Abelson murine leukemia
    virus, *614*
abl oncogene, 464, *614*
ABO blood group, 341–343, *342–343*,
    346, 358, 682–683, *683*
Abortion, spontaneous, 454, 473
Abscisic acid, 556, *557*
Accessory chromosome, 313
Ac element, 159–162, *162–163*
Acentric fragment, 455–456, 459
Acetylation, of histones, 547, *548*
N-Acetylhexosaminidase A,
    74, 76, *76*
O-Acetylhomoserine, 72, *72*
Achondroplasia, 291, 362, *678*
Acid phosphatase deficiency, *74*
Acquired immunodeficiency
    syndrome, 610
Acrocentric chromosome, 301
Activator, 94, 544, 551, 567
    activation of transcription by,
    549–550, *549*
ada gene, 150
Adaptation, 687, 712
    mutation versus, 134–136, *135*
Adapter, 187
Addition mutation, 116, *116*, 140,
    142, 146
Additive genetic variance, 386–387
Adenine, 2, 13, 20, *20*, 22, 23, *23*, 26,
    26, *141*
Adenoma
    class I, 625, *625*
    class II, 625, *625*
    class III, 625, *625*
Adenosine deaminase, 231
Adenylate cyclase, 526, *527*, 554
Adenylate/uridylate-rich element (ARE),
    563–564
Adjacent 1 segregation, 462, *463*
Adjacent 2 segregation, 462, *463*
A-DNA, 25–26, *26–27*
Aflatoxin, 626
agamous mutant, in *Arabidopsis
    thaliana*, 576, *577*

Agarose gel electrophoresis, of DNA,
    166, 192, *193–194*, 196, *196*
    preparative, 194
Age of onset, 359–360
Agglutination reaction, 342, *343*
Agouti coat pattern, 350–351,*352*, 356
Agrobacterium tumefaciens, 233–234,
    234–235, 256
AIDS, 610
Alanine, *113*
Albinism, 75, 666
    among Hopi Indians, 668–669, *669*
    human, 289–290, *290*
    tyrosinase-negative, 666
Albumin gene, *712*
Alcohol dehydrogenase gene, in *Droso-
    phila melanogaster*, 675–676, *676*
Alcoholism, nature-versus-nurture
    debate, 363
Aldosterone, 554
Aleurone layer, 556, *558*
Alkaptonuria, 68, *68*, *74*
Alkylating agent, 144–146, *145*, 626
    repair of alkylation damage, 150
Allele, 3, 213–214, 276, 278, 286, 339
    contributing, 376
    fixed, 683–684, *683*
    genetic symbols for, 317
    lethal, 339, 356–357, 363–364
    multiple. *See* Multiple alleles
    mutable, 159
    mutant, 315, 317, 341, *341*
    noncontributing, 376
    null, 259
    stable, 159
    unstable, 158
    wild-type, 280–281, 317, 341, *341*
Allele-specific oligonucleotide (ASO)
    hybridization, 211, 215–216, *216*
    analysis of single-nucleotide polymor-
    phisms, 215–216, *216*
    in DNA molecular testing, 221
    reverse, 221
Allelic frequency, 5, 657, 660–663, 696
    calculation of
    by gene counting, 660, *662*
    from genotype frequency,
    660–661, *662*
    estimation from Hardy-Weinberg
    law, 668–669

forces that change, 676–696
    genetic drift, 679–685, 699
    migration, 685–687, *686*, 699
    mutation, 677–679, *678*, 699
    natural selection, 687–694,
    690–691, 699
    Hardy-Weinberg law, 663–669
    hemoglobin variants among
    Nigerians, *662*
    with multiple alleles, 661
    variation in space and time, 669–670,
    670
    for X-linked alleles, 661–663, 667
Allelomorph, 286
Allolactose, 517, *517*, 519, 521, *521*
Allopolyploidy, 473
Allosteric shift, 521
Alpha complementation, 181, *181*
Alpha-fetoprotein gene, *712*
α-helix, 112, 114, *114*
Alternate segregation, 462, *463*
Alternation of generations, 312, *313*
Alu family, 36–37, 165
Ames, Bruce, 147
Ames test, 147–148, *147*, 626
Amino acid, 111–112
    abbreviations for, *113*
    acidic, *113*
    basic, *113*
    energetic cost of synthesis of,
    711, *711*
    essential, 73
    neutral, nonpolar, *113*
    neutral, polar, *113*
    peptide bond formation, 112, *114*
    in protein synthesis, 112–126
    structure of, 112, *112–113*
Amino-acid biosynthesis operons,
    528–532, *528–532*
Aminoacyl-tRNA, 119–120, *120–121*
    binding to ribosome, 123, *124–125*
Aminoacyl-tRNA synthetase, 119–120,
    120–121
Amniocentesis, 80–81, *81*, 219
Amoeba proteus, C value of, *31*
Amplification, 200
Amplimer, 200
α-Amylase, 556, *558*
Analysis of variance (ANOVA), 379,
    382–383

Anaphase
  meiosis I, *307*, 308–309, *311*
  meiosis II, *307*, 309, *311*
  mitotic, 302, *303–304*, *305*, *311*
Anastasia (missing Romanov), 635–636
Ancient DNA, 203, 227
Andalusian blue fowl, 346
Anemia. *See* specific types of anemia
Aneuploidy, 319, 453, 467
  generation of, 467
  in humans, 468–471, *468–471*
  meiosis and, 468, *468*
  sex chromosome, 322, *323*
  types of, 467–468, *467*
Anfinsen, Christian, 115
Angelman syndrome, 561
Angiosperm, inheritance of plastids
  in, 649
Angiotensin receptor, *616*
Anhidrotic ectodermal dysplasia, 325
Animal
  cloning of, 577–579, *579*
  meiosis in, 310–311, *312*
  polyploidy in, 472–473
  steroid hormone regulation of gene
    expression in, 553–556, *553*,
    555–556
Animal breeding, 2, 6, 390
Animal cell, 9, *11*
  cytokinesis in, 305, *306*
  meiosis in, *307*
  mitosis in, *303*
Aniridia, 465, 678
Annealing, of DNA ends, 179, *180*
Annotation, of genome sequence,
  251, 258
ANOVA. *See* Analysis of variance
Ant, chromosome number in, *313*
Antennapedia complex (ANT-C),
  596, *598*
Antennapedia (*Antp*) gene, 596–597
Anthocyanin, 157–159
Antibiotic resistance, 149, 180
  in *Chlamydomonas reinhardtii*, 641
Antibody, 342, 573, 581, 600. *See also*
  Immunoglobulin
Antibody probe, screening cDNA library
  for specific clone, 188, *188*
Anticodon, 105, 119, 710
Antigen, 342, 573, 581
  cellular, 342
Antisense RNA, 236, 565
Antitermination signal, 530–531, *531*
Antiterminator, 534
APC gene, *619*, 625, *625*
Ape, molecular evolution in, 715
Apolipoprotein, evolution of, 712, *712*
Apoptosis, 623
Aporepressor protein, 530
Applied research, 6–7
Apurinic site, 140
Arabidopsis thaliana
  Arabidopsis 2010 Project, 253–254
  floral development in, 576, *577*
  genome analysis in, *256*
  genome sequence of, 244, 253–254, 724

homeotic genes in, 597
  as model organism for research, *10*,
    576, 577
Arber, Werner, 177
Archaea, 11
  gene density in, 255, *256*
  genome sequence of, 251–252
  genome size in, 255, *256*
  tree of life, 721, *721*
Archeoglobus fulgidus, chromosome of, 28
ARE. *See* Adenylate/uridylate-rich element
Arginine, *113*
Aristapedia mutation, 596, *598*
armadillo gene, 595
ARS. *See* Autonomously replicating
  sequence
Artificial chromosome, 182–183
Artificial selection, 390
Asbestos, 626
Ascospore, 69–70, *69*, *430*, 431, 641, 643
Ascus, 69–70, *69*, *430*, 431–432, 641, 643
Ashkenazi Jews, 75, 219
ASO hybridization. *See* Allele-specific
  oligonucleotide hybridization
Asparagine, *113*
Aspartame (NutraSweet), 73
Aspartic acid, *113*
Aspergillus nidulans
  colony color pattern in, 440–442,
    440–442
  gene mapping in, 429
  mitotic recombination in, 438–442,
    440–442
  as model organism for research, 438
Assortative mating, 658, 695
  negative, 695
  positive, 695
Aster (mitotic), 304
Ataxia, intermittent, *74*
Ataxia-telangiectasia, 154
ATP
  production in mitochondria, 633
  use in protein synthesis, 120, *120*
ATPase, 633
Attardi, G., 634
Attenuation, 515, 530–532, *531–532*, 536
  in amino-acid biosynthetic operons,
    532
  molecular model for, 530–532,
    531–532
  tRNA and, 532
Attenuator, 515, 531–532
Attenuator site, 529, *529*
att lambda site, 497
Autonomous element, 158–159, 165
Autonomously replicating sequence
  (ARS), 58, 183
Autoradiogram, 188, *188*, 191
  of dideoxy sequencing gel, *199*
Autoradiography, 188, *188*, 191
Autosomal trait
  dominant, 300
  recessive, 300
Autosome, 301, 313
Auxin, 556, *557*
Auxotroph, 70, *70*, 148, 149, 483

Avery, Oswald, transformation
  experiment, 16, *16*, 490
Avian erythroblastosis virus, *611*
Avian myeloblastosis virus, *614*
Azidothymidine, 143
Azo dye, 626
AZT, 143

BAC. *See* Bacterial artificial chromosome
Bacillus amyloliquefaciens, 648
Bacillus subtilis
  C value of, *31*
  transformation in, 490, *491*
Back-fat thickness, in pigs, *391*
Back mutation. *See* Reverse mutation
Bacteria, 11. *See also* Prokaryote
  conjugation in, 481, 507
  gene mapping in
    by conjugation, 484–490, *484–489*
    by transduction, 492–499, *492–498*
    by transformation, 490–492, *491*
  genetically engineered, 233
  genetic analysis of, 482–483, *483*
  transduction in, 481, 507
  transformation in, 481, 507
Bacterial artificial chromosome (BAC),
  175, 183
  clone contig maps of human genome,
    247–249
Bacterial colony, 482, *483*
Bacterial lawn, 493
Bacteriophage, 492–494
  chromosomes of, 37
  gene mapping in, 481, 499–507, *500*
  helper, 499
  host range gene of, 499–500, *500*
  lysogenic cycle of, 493–494
  lytic cycle of, 493, *493*
  plaque phenotype, 499–500, *500*
  regulation of gene expression in, 515
  replication in, 54–57, *55–56*
  temperate, 494
  transducing, 481, 492–499, *492–498*
  virulent, *17*, 493
Bacteriophage lambda, 492
  chromosome of, 27, *28*
  C value of, *31*
  early transcription events in,
    533–534, *535*
  genetic map of, *533*
  λ*d gal⁺*, 497–499, *498*
  life cycle of, 493–494, *495*
  lysogenic cycle of, 56–57, 493, *495*,
    534, *535*
  lytic cycle of, 28, 56–57, *56*, 493–494,
    495, 497–499, *498*, 534–535, *535*
  regulation of gene expression in, 515,
    533–536, *535*
  replication in, 54, 56
  specialized transduction by, 497–499,
    498
  structure of, *492*
Bacteriophage lambda operator, 534, *535*
Bacteriophage lambda promoters, 534, *535*
Bacteriophage lambda repressor, 494,
  534–535, *535*

Bacteriophage P1, transduction in *Escherichia coli*, 495, *496*
Bacteriophage P22, transduction in *Escherichia coli*, 494
Bacteriophage ΦX174, *27*
  chromosome of, 27
  replication in, 54
Bacteriophage T1, resistance in *Escherichia coli*, 149
Bacteriophage T2, 17, *17*, 492
  chromosome of, 27
  genetic analysis of, 499–500, *500*
  Hershey-Chase experiment with, 16–19, *18*
  life cycle of, 17–18, *17*, 493, *493*
  plaques of, *494*
  spontaneous mutation frequency at specific loci, *678*
  structure of, *492*
Bacteriophage T4, 492
  chromosome of, 27
  C value of, *31*
  evidence that genetic code is triplet code, 115–116, *116–117*
  host range properties of, 501
  plaque morphology in, 501, *501*
  resistance in *Escherichia coli*, 135–136, *135*
  rII mutants of, 115, 481
    complementation tests, 503–507, *506*
    deletion mapping, *502*, 503, *504–506*
    evidence that genetic code is triplet code, 115–116, *116–117*
    fine-structure mapping, 501–507
    recombination analysis of mutants, 501–503, *502*
  thymidylate synthetase gene of, 98
Bacteriophage T6, 27, 492
*Bam*HI, 177, *178–179*, 179, 184–185, 187, *187*, 192, *193*, 194, 673, *674*
Banana, polyploidy in, 473
B antigen, 342–343, *343*
Bar-eye trait, in *Drosophila melanogaster*, 408–409, *409*, 458, *458*
barnase gene, 648
Barnett, Leslie, 115
barnstar gene, 648
Barr, Murray, 324
Barr body, 34, 324, *324*
Base. *See* Nitrogenous base
Base analog, 47, *47*, 143, *144*
Base excision repair, 150
Base-modifying agent, 144–146, *145*
Base-pair substitution, 133, 136, *137*, 140. *See also* Nucleotide substitution
Basic research, 6–7
Bateson, William, 286
Baur, Erwin, 357
BAX gene, 623
B cells, 581–582
  development of, assembly of antibody genes during, 583–585, *583–584*
BCL-2 protein, 623
BCR-ABL gene, 464

B-DNA, 25–26, *26–27*
Beadle, George, 4, 69–72, *71–72*, 160
Bean, seed weight in, 375, 377–378, *378*
Becker muscular dystrophy, *74*
Beet, E.A., 77
Behavioral incompatibility, between species, 698
Behavioral trait, 363
Bender, Welcome, 595
Benign tumor, 606
Benzer, Seymour, 119, 501–507
Berg, Paul, 2, 176
Berger, Susan, 97
β-pleated sheet, 114
*Bgl*II, *178*
bicoid (*bcd*) gene, 594, *594–595*
Bidirectional replication, 51, 54, *55*, 57
Binomial distribution, 680
Biochemical genetics, 69
Biochemical pathway, genetic dissection of, 69–72, *71*
Bioinformatics, 258
Biotechnology, 212, 232–233
Biparental inheritance, in *Chlamydomonas reinhardtii*, 647
Bipolar disorder, 355
Bird, sex chromosomes in, 326
Birth weight, human, 374, *374*, 377
Bishop, J. Michael, 614
Biston betularia. See Peppered moth
bithorax complex (*BX-C*), 596, *597*, *599*
Bivalent, 308
Blackburn, Elizabeth, 59
BLAST (database), 7–8, 258
Blastoderm
  cellular, *592–593*, 593
  syncytial, *592*, 593
Blobel, Günther, 127
Blood group, 340, 342
  ABO, 341–343, *342–343*, 346, 358, 682–683, *683*
  Bombay, 343
  M-N, 346–347, 664
Bloom syndrome, *154*
Blue mussel, leucine aminopeptidase in, 667, 669–670, *670*
Blue-white colony screen, 181, *181*
Body color
  in *Drosophila melanogaster*, 341, 410, 437–438, *438*
  in mouse, 356–357, *357*
Body length, in salamanders, 379, *379*, *381*
Body size, in *Drosophila melanogaster*, 391
Body weight
  in cattle, *391*
  heritability of, 389
  in mouse, *391*, *393*
  in poultry, *391*, 393, *393*
Bombay blood type, 343
Bootstrapping, 720–721
Borrelia burgdorferi
  chromosomes of, 28–29
  C value of, *31*
  genome sequence of, 482

Bottleneck, 681–683
Bouquet (telomeres), 306
Boveri, Theodor, 4, 312
Boyce, R.P., 150
Boyer, Herbert, 2
Brachydactyly, 288, *288*, 291, 359
Bradyrhizobium japonicum, genome of, 255, *256*
Branch diagram, 278–279, *279*
  of dihybrid cross, 284, *284*
Branch-point sequence, 99, *100*
BRCA1 gene, 219, 261, 345, *619*, 623
BRCA2 gene, 219, 261, *619*, 623
Breast cancer, 219, 261, *619*, 623
Brenner, Sidney, 115, 253
Bridges, Calvin, 317–319
Bristle number, in *Drosophila melanogaster*, 391, 437–438, *438*
  QTL for, 397
Broad-sense heritability, 373, 387–388, 397
5-Bromodeoxyuridine (BUdR), 47, *47*
5-Bromouracil (5BU), 143, *144*
Brown, Pat, 261
*Bss*HII, 246
*Bst*XI, 177, *178*
5BU. *See* 5-Bromouracil
BUdR. *See* 5-Bromodeoxyuridine
Buri, P., 681, *681*
Burkitt lymphoma, 462, 464, 610
Burnham, C.R., 160

CAAT box, 94
Caenorhabditis elegans
  C value of, *31*
  development in, 575–576, *576*
  genome analysis in, 256
  genome sequence of, 244, 253
  hermaphrodites in, 325–326
  as model organism for research, 5, 9, 10, 253, *253*
  operons in, 544–545, *545*
  RNA interference in, 565–566
  sex determination in, 325–326
Café-au-lait spot, 359, *359*
Cairns, John, 48
Calcitonin, 561, *562*
Calcitonin gene (*CALC*), 561–562, *562*
Calcitonin gene-related peptide, 562, *562*
Calico cat, 324–325, *325*, 578, *579*
cAMP. *See* Cyclic AMP
Cancer, 607
  breast, 219, 261, *619*, 623
  cell cycle and, 605, 607–608, *608*, 627
  cervical, 618
  chromosomal mutations in, 454, 462–465
  colorectal, *619*, 624–625
  familial (hereditary), 606, 609
  gene expression in, 261
  genes and, 610–624
  as genetic disorder, 609
  genetics of, 605–627
  hereditary disposition for, 621
  induction by retroviruses, 605
  kidney, *619*

Cancer, *Continued*
 liver, 626
 multistep nature of, 606, 625, *625*, 627
 mutator genes and, 152, 610, 624, 627
 oncogenes and, 610–618
 ovarian, 219, 261, *619*, 623
 retroviruses and, 618, 627
 skin, 626
 sporadic, 606, 609
 telomerase and, 606, 624, 627
 thyroid, 626
 tumor suppressor genes and, 610,
  618–623, *619*, 627
 two-hit mutation model for, 605–606,
  619–621, *620–621*
 viruses and, 605, 609–610
Candidate gene, for cystic fibrosis, 224
Cantor, C., 707
CAP. *See* Catabolite activator protein
Cap-binding protein (CBP), 122
5′ Capping, of mRNA, 97, *97*
Capping enzyme, 97
CAP site, 525–526, *528*
Capture array, 263–264
carbonaria phenotype, in peppered
  moth, 688–689, *688*
Carcinogen, 606, 609, 625, 627
 chemical, 626
 direct-acting, 626
 radiation, 626
 screening for, 147–148, *147*
 ultimate, 626
Carrier, W., 150
Carrier detection, 67, 79–81, 219
Carrot, regeneration of plants from
  mature single cells, 577, *578*
Castle, William Ernest, 356, 663
Castle-Hardy-Weinberg law, 663
Cat
 calico, 324–325, *325*, 578, *579*
 chromosome number in, *313*
 cloning of, 578, *579*
 coat color in, 324–325, *325*, 578, *579*
Catabolite activator protein (CAP),
  525–527, *526–527*
Catabolite repression, 525–526,
  526–527
 in yeast *GAL* gene system, 552–553
Cataract, 74
Cattle
 body weight in, *391*
 milk production in, 360, 386, *391*,
  393, *393*
caudal (*cad*) gene, 594
Cause-effect relationship, 380
Cavalli-Sforza, Luca, 485
CBP. *See* Cap-binding protein
Cdk. *See* Cyclin-dependent kinase
cDNA. *See* Complementary DNA
Cech, Tom, 100
Celera Genomics, 254
Cell cycle, 30–31, 58, 302, *302*, 605
 in cancer cells, 605, 607–608,
  608, 627
 molecular control of, 607, *608*
Cell-cycle checkpoint, 605, 607, *608*

Cell division, 302
 regulation in normal cells,
  607–608, *609*
Cell-free protein-synthesizing system, 117
Cell plate, 305
Cellular antigen, 342
Cellular blastoderm, 592–593, *593*
Cellular oncogene, 615, *615*
Cellular proto-oncogene. *See*
  Proto-oncogene
Cellular respiration, 633
CEN sequence, 34, *35*
Centimorgan (cM), 405, 411
Central dogma, 88
Central Park jogger case, 227
Centre d'Etude du Polymorphisme
  Humain (CEPH), 444
Centriole, 9, *11*, 304
Centromere, 14, 30, 301, *303*,
  304–306, 309
 calculating gene-centromere distance
  with ordered tetrads, 429,
  432–434, *433*
 DNA of, 34–35, *35*
 human, 35
 of *Saccharomyces cerevisiae*,
  34–35, *35*
 of *Schizosaccharomyces pombe*, 35
Cepaea nemoralis. *See* Snail
CEPH. *See* Centre d'Etude du
  Polymorphisme Humain
Cervical cancer, 618
Cesium chloride density gradient
  centrifugation, 45–47, *46*
CF. *See* Cystic fibrosis
CFTR. *See* Cystic fibrosis transmem-
  brane conductance regulator
C gene, 159
CGG repeat, 466
Chain-terminating codon. *See* Stop
  codon
Chance, laws of, 279
Chaperone, 115, 554
 histone, 61–62
Character, 272
Chargaff, Erwin, 23
Chargaff's rules, 23
Charged tRNA. *See* Aminoacyl-tRNA
Chase, Martha, 16–19, *18*
Checkpoint, cell-cycle, 605, 607, *608*
Chemical(s), effect on gene expression,
  360–362
Chemical carcinogen, 626
Chemical mutagen, 143–147,
  144–146
Chiasma, 307–308, *308*, 405, 407
Chicken. *See also* Poultry
 Andalusian blue, 346
 chromosome number in, *313*
 comb shape in, 348–349, *349–351*
 as model organism for research, 9, *10*
 plumage color in, 346, *347*
Chimpanzee, chromosome number
  in, *313*
Chinese hamster ovary cells, replication
  in, 47, *47*

Chinese hamster somatic cell tissue
  culture, spontaneous mutation
  frequency at specific loci, *678*
Chi-square test, 286–287, *286–287*,
  410–411, *410*, 668
Chi-square value, 287, 411
Chlamydomonas reinhardtii, *645*
 chloroplast genome of, 637
 erythromycin resistance in, 631,
  645–647, *646*
 life cycle of, 431
 mating type in, 431
 as model organism for research,
  9, *10*
 non-Mendelian inheritance in,
  645–647, *646*
 random-spore analysis in, 432
 tetrad analysis in, 430–437
Chloramphenicol, inhibition of mitochondrial protein synthesis, 635
Chloride channel, defective, 79, *80*
Chlorophyll, 636, 640–641, *642*
Chloroplast, 9, *11*, 636
 DNA of, 3, 631–632, 651–652
  gene organization of, 637, *638*
  genetic map of, *638*
  replication of, 636–637
  size of, 637
 genetic code in, 637
 genome of, 636–637, 651
  structure of, 636–637
 non-Mendelian inheritance in
  Chlamydomonas, 645–647, *646*
 origin of, 632, 722
 shoot variegation in four o'clocks,
  639–641, *640–642*
 structure of, 636, *637*
 translation in, 637
Chorionic villus sampling, 81,
  81, 219
Chromatid, 13, 302
 sister. *See* Sister chromatids
Chromatin, *11*, 30–31
 structure of, 31–33, *32–33*
Chromatin fiber
 10-nm, 32, 547, *548*
 30-nm, 33, *33*, 547, *548*
Chromatin remodeling, 543, 547–550,
  548, 574, 591
Chromocenter, 454, *455*
Chromosomal mutation, 133–134,
  453–474
 in cancer cells, 454, 462–465, 610
 developmental disorders and, 454
 spontaneous abortions and, 454
 types of, 454
 variations in chromosome number,
  466–474
 variations in chromosome structure,
  454–466, 474
Chromosome, 3, 9, 13. *See also*
  Meiosis; Mitosis
 accessory, 313
 acrocentric, 301
 artificial, 182–183
 of bacteriophage, 37

cellular reproduction and, 300–312, 329
circular, 28, *28*
daughter, 302, *303*, 305
dicentric, 459
DNA in, *1*, 13–14, 27–37, 305, *305*
  of eukaryotes, 29–37, *31–35*, 300–302,
    300–301
  genetic symbols for, 317
  harlequin, 47, *47*
  homologous, 300, 407
  metacentric, 301, *301*, 305
  metaphase, *305*
  nonhomologous, 300
  Philadelphia, 464
  polytene, 454, *455*, 580–581, *581*
  of prokaryotes, 28–29, *29*, 36
  proof that DNA is genetic material,
    14–19, *15–18*
  recombinant, 308
  separation of individual chromosomes,
    184, 247
  sex. *See* Sex chromosome
  structure of, variations in, 453–466
  submetacentric, 301, *301*, 305
  telocentric, 301
  viral, 27–28, *27*
Chromosome arm, 302
Chromosome banding, 301–302, *301*,
  454, *455*
Chromosome jumping, 223
Chromosome library, 183, 185
Chromosome number, 29, 309, 312
  variations in, 453, 466–474
    changes in complete sets of
      chromosomes, 472–474, *472*
    changes in one or a few chromo-
      somes, 467–471, *467–471*, 474
    in various organisms, *313*
Chromosome puff, 580–581, *581*
Chromosome-specific library, 166
Chromosome theory of inheritance, 4,
  299, 312–320, 329
Chromosome walking, 223, *223*
Chronic myelogenous leukemia (CML),
  462, 464, 610
Cigarette smoking, 469, 626
cII gene, 534, *535*
cIII gene, 534
Cis-acting element, 94
cis configuration of mutations, 411
Cis-dominance, 519
cis-trans test. *See* Complementation test
Cistron, 507
Citrullinemia, 74
Classical genetics, 2
Classical model, for genetic
  variation, 672
Cleft lip, 360
Cleft palate, 360
Clethrionomys gapperi. See Red-backed
  vole
Cline, 669
Clonal selection, 581
Clone, 176
Clone contig map, 246, 248–249
  of human genome, 246–247, 248

Cloned DNA sequence, 176
  restriction mapping of, 192–195,
    193–195
  sequencing of DNA, 176, 197–200,
    198–200
  Southern blot analysis of,
    195–197, *196*
  subcloning segment of, 202–203
Cloning, 175–183
  of animals, 577–579, *579*
    problems with, 578–579, *579*
  of carrot plant from mature single cell,
    577, *578*
  of cat, 578, *579*
  cloning vectors, 179–183
  molecular, 176
  positional, 222, 619, 622
  of quantitative trait loci, 397
  restriction enzymes, 176–179,
    177–180
  of sheep, 577–578, *579*
Cloning vector, 175–176, 179–183
  plasmid, 175, 179–182, *180–181*
Closed promoter complex, 90, *91*
Clubfoot, 360
CML. *See* Chronic myelogenous
  leukemia
Coactivator, activation of transcription
  by, 549–550, *549*
Coat color
  in cats, 324–325, *325*, 578, *579*
  in horses, 346, 353, *354*
  in labrador retrievers, 351, *352*
  in mice, 350–351, *352*
  in rabbits, 360, *362*
Cockayne syndrome, *154*
Coding single nucleotide
  polymorphism, 214
Codominance, 214, 339, 346–347
Codon, 5, 105, 111, 115–116,
  116–117
  anticodon recognition, 119
  initiator, 111, 120–121
  sense, 118
  start, 123
  stop, 111, 119, *124*, 125–126,
    126, 136
Codon usage bias, 710–711
Coefficient of coincidence, 417–418
Coefficient of selection, 689, *689*
Cohen, Stanley, 2
Coincidence, coefficient of, 417–418
Cointegrate, 156, *158*
Cointegration, 156
Colinearity rule, 597
Collins, Francis, 244, 254
Colony, bacterial, 482, *483*
Colony color pattern, in *Aspergillus
  nidulans*, 440–442, *440–442*
Color blindness, 327, 444, 667
Colorectal cancer, 619, *624–625*
Color pattern, of Cuban tree snail, *672*
Combinatorial gene regulation,
  550–551, *551*
Comb shape, in chickens, 348–349,
  349–350

Common ancestor, 719
Common family environmental
  effect, 387
Comparative genome analysis, 709
Comparative genomics, 243, 245,
  264–265, 709–710
Compartmentalization, of eukaryotic
  cells, 546, 722
Competent cell, 490
Complementary base pairing, 13,
  23–24, *25–26*
  errors in, 140, *141*
  wobble in, 118–119, *118–119*, *142*
Complementary DNA (cDNA), 166
  cloning of, 187, *187*
Complementary DNA (cDNA)
  library, 175, 183, 185–187,
    186–187, 204
  construction of, 185–187, *186–187*
  screening for specific clone,
    187–189, *188*
Complementary gene action, 353
Complementation group, 505
Complementation test, 340–341,
  340–341, 482, 507
  identifying genes in libraries,
    190, *192*
  in *rII* mutants of bacteriophage T4,
    503–507, *506*
  in *Saccharomyces cerevisiae*, 507
Complete dominance, 339, 346, 348
Complete medium, 70, 483
Complete penetrance, 358
Complete recessiveness, 346
Complete transcription initiation
  complex, 94, *95*
Composite transposon, 155–156, *157*
Concatamer, DNA, 56, *57*
Conditional mutation, 148–149
Conformation, of protein, 115
Conidia, 69, *69*, 431, *431*, 641
Conifer, paternal inheritance of plastids
  in, 649
Conjugation, in bacteria, 481, 507
  discovery of, 484
  gene mapping by, 484–490,
    484–489
Consensus sequence, 89
Conservation biology, 227, 697
Conservative replication, 44–47, *44*
Conservative transposition, 157
Constitutional thrombopathy, 327
Constitutive gene, 516
Constitutive heterochromatin, 34
Contact inhibition, 607
Contig, 246
Continuous trait, 373, 397
  inheritance of, 375–376
  nature of, 374–375, *374*
Contributing allele, 376
Controlled cross, 641
Controlling element. *See* Transposable
  element
Convergent evolution, 716
Coordinate induction, 517
Core promoter, 94

Corn
  association of recombination with chromosome exchange, 408, *408*
  base composition of DNA from, *23*
  C value of, *31*
  ear length in, 383–384, *384*
  hybrid seed production in, 648
  kernel color in, 6, 157–162, *159, 162–163*
  as model organism for research, 9, *10*
  spontaneous mutation frequency at specific loci, *678*
  teosinte branched 1 QTL, 397
  transposable elements in, 157–162, *159, 162–163*
Correlation, 373, 380–381
  genetic, 392–394, *393–394*
  negative, 380, *382*
  positive, 380, *382*
Correlation coefficient, 380–381, *381–382*
Correns, Carl, 286
cos sequence, 56–57, *56*
Cotranduction, 496–497
Cotransductant, 496
Cotransformation, determining gene order from, 491, *491*
Cotton, chromosome number in, *313*
Coupled transcription and translation, 96, *96*, 530
Coupling of alleles, 411
Covariance, 380
CpG island, 224, 558, 560
CPSF protein, 97, *98*
Creighton, Harriet, 160, 408
Cremello horse, 346
Crick, Francis H.C., 22–23, *22*, 88, 115, 119
Cri-du-chat syndrome, 457, *457*
Criminal investigation. *See* DNA forensics
Crisscross inheritance, 316
Crithridia fasiculata, RNA editing in, 101, *101*
cro gene, 534–535, *535*
Cross, 274, 278
Cross-fertilization, 274
Crossing-over, 5, *307–308*, 308–309, 405, 407, 419
  association with recombination, 408–410, *408–409*
  double, 413, 416
  in inversion heterozygote, 459–460, *460–461*
  in mitosis, 429, 437–443
  multiple, 419
  single, 413
  unequal, 723
Crossover frequency, 411
Crown gall disease, 233–234, *234*
CstF protein, 97, *98*
Cuban tree snail, color patterns of, *672*
Cuenot, Lucien, 356
Culture medium, 483
Cut-and-paste transposition. *See* Conservative transposition
C value, 29–30
C value paradox, 30, 255

Cyclic AMP (cAMP), 525–527, *526–527*, 554
Cyclin, 605, 607, *608*, 621, *621*
Cyclin-dependent kinase (Cdk), 58, 605, 607, *608*, 621, *621*
Cystathionine, 72, *72*
Cysteine, *113*
Cystic fibrosis (CF), 74, 79–80, *79–80*, 218
  gene therapy for, 232
Cystic fibrosis (CF) gene
  candidate gene, 224
  carrier detection, 219
  cloning of, 222–224
    between flanking markers, 223, *223*
  defects in cystic fibrosis, 224
  finding chromosomal location of, 222–223
  identifying gene in cloned DNA, 224
  knockout mice for, 213
  RFLP markers linked to, 222
Cystic fibrosis transmembrane conductance regulator (CFTR), 79, *80*
Cytochrome *c*, evolution of, *714*
Cytochrome oxidase
  non-Mendelian inheritance of, 633, 635, *636*
  subunit III in protozoans, 101, *101*
Cytochrome P450, genetic tests for variations in, 263
Cytogenetic map, 246
Cytohet, 647
Cytokinesis, 302, *303*, 305, *307*, 309
  in animal cell, 305, *306*
  in plant cell, 305, *306*
Cytokinin, 556, *557*
Cytological marker, 408
Cytolytic virus, 611
Cytoplasm, *11*
Cytoplasmic genes, 632
Cytoplasmic male sterility, 647–648, *648*
Cytosine, 2, 13, 20, *20*, 22, 23, *23*, 26, *26, 141*
  deamination of, 140, *143*
  methylation of, 557, *559*
Cytoskeleton, *11*

Dalgarno, Lynn, 121
Danio rerio. See Zebrafish
Dark repair. *See* Nucleotide excision repair
Darwin, Charles, 5, 687, *688*
Darwinian fitness, 689, *689*
Darwin's finch, *657*
Data mining, 254
Daughter chromosome, 302, *303*, 305
Dauphin (son of Louis XVI and Marie Antoinette), 227–228
Davis, Bernard, 484
DCC gene, *619*, 625, *625*
DCP1 gene, 564
DDT resistance, in insects, *677*
deadpan gene, 588, *588*
Deaminating agent, 144–146, *145*
Deamination, of nitrogenous base, 140–141

Decapping, of mRNA, 564
deformed (Dfd) gene, 596–597
Degeneracy, of genetic code, 118
Degradation control
  mRNA, *546*, 564, *564*, 567
  proteins, *546*, 564–565
Degrees of freedom, 287
Delbrück, Max, 135
Deletion, 116, 133, 140, *142*, 146, 453–457, *455–457*, 503, 676
  changing cellular proto-oncogenes into oncogenes, 617
  fragile sites and, 465–466
  genetic diseases due to, 457, *457*
  induced, 455
Deletion mapping, 456
  in *Drosophila melanogaster*, 456–457, *456*
  of *rII* region of bacteriophage T4, 502, 503, *504–506*
DeLucia, Peter, 48
Demerec, Milislav, 160
Deoxyribonuclease (DNase), 16, 233
Deoxyribonucleic acid. *See* DNA
Deoxyribonucleotide, 21
Deoxyribose, 13, 20, *20*, 22
Depurination, 140–141
Descriptive science, 244
Determination, 573, 575, 600
Determined cell, 575
Development, 357–358, 573–601
  in *Arabidopsis thaliana*, 576, *577*
  basic events of, 574–575
  in *Caenorhabditis elegans*, 575–576, *576*
  constancy of DNA in genome during, 577–579, *577–579*
  definition of, 574
  in Diptera, 580–581, *581*
  in *Drosophila melanogaster*, 261, 573, 574–576, 591–599, *592–600*, 601
  in frogs, 394
  gene expression during, 573–601
  model organisms for genetic analysis, 575–577, *576*
  in mouse, 576
  in *Saccharomyces cerevisiae*, 575
  in zebrafish, 576, *577*
Developmental abnormalities, 454
  in cloned mammals, 579
Developmental genetics, 574–601
Developmental potential, 575
Development rate, in frogs, 391, *393*
Deviation squared (d^2), 287, 411
Deviation value (*d*), 287
de Vries, Hugo, 286
Diakinesis, 308
Dicentric bridge, 459
Dicentric chromosome, 459
Dicer (enzyme), 565–566
Dicot, 233
Dideoxynucleotide, 198, *198*
Dideoxy sequencing of DNA, 197–199, *198–199*
Differentiation, 357, 573, 575, 600, 607
Digoxigenin labeling, of DNA, 191

Dihybrid cross, 282–286, *282–286*
  branch diagram of, 284, *284*
Dihydrouridine, *106*
Dimethylguanosine, *106*
dinB gene, 50
Dioecious plant, 326
Diplococcus pneumoniae, spontaneous
    mutation frequency at specific
    loci, *678*
Diploid (2N), 29, 278, 299–300, *300*,
    306, *467*
Diploidization, 440
Diplonema, 308, *308*
Diptera, development in, 580–581, *581*
Direct-acting carcinogen, 626
Directional selection, 690
Disaccharide intolerance I, 74
Discontinuous trait, 373–374, *374*
Disjunction, 305
Dispersed (interspersed) repeated
    DNA, 36–37
Dispersive replication, 44–47, *44*
Displacement loop. *See* D-loop
Distance matrix approach, to phyloge-
    netic tree reconstruction, 705,
    718, 725
D-loop, 35, *35*, 633, *634*
DNA, 2, 13–14
  A-DNA, 25–26, *26–27*
  agarose gel electrophoresis of, 166,
    192, *193–194*, 196, *196*
    preparative, 194
  ancient, 203, 227
  antiparallel strands in, 23
  assembling into nucleosomes, 61–62,
    61
  base composition of, 23, *23*
  B-DNA, 25–26, *26–27*
  centromeric, 34–35, *35*
  of chloroplast. *See* Chloroplast, DNA of
  in chromosomes, *1*, 13–14, 27–37,
    305, *305*
  circular, 28, *28*
    replication of, 54, *55*
  cloning of. *See* Cloning
  compared to RNA, 22
  complementary. *See* Complementary
    DNA
  composition of, 20–27, *20–26*
  concatameric, *56*, *57*
  constancy in genome during develop-
    ment, 577–579, *577–579*
  dispersed (interspersed) repeated, 36–37
  DNase-hypersensitivity site, 547
  double helix, *3*, 13, 22–24, *22–25*
  genetic variation measured at DNA
    level, 673–676, *674–676*
  hemimethylation of, 561
  heteroduplex, 490, *491*
  highly repetitive, 36
  hydrogen bonds in, 23, *25–26*
  inverted repeats, 155
  labeling of
    nonradioactive, 191, 216
    radioactive, 191, *191*
  linker, 32, *32*

long terminal repeats, 162–164, *163*
looped domains of, 29, *31*, 33, *33*
loss in antibody-producing cells,
    581–585, *582–584*
major groove of, 24, *25*
methylation of, 544, 557–561, *559*,
    625, *625*, 646
minor groove of, 24, *25*
of mitochondria. *See* Mitochondria,
    DNA of
moderately repetitive, 36
molecular evolution, 705–725
mutations in. *See* Mutation
nicked, 29, 54–55, *55*
nontemplate strand of, 89, *89*
nucleotide substitutions. *See*
    Nucleotide substitution
polarity of, 21, *21*
proof that it is genetic material,
    14–19, *15–18*, 36
proviral, 611, *612–613*
recombinant. *See* Recombinant DNA
    technology
recombination. *See* Recombination
relaxed, 29, *30–31*, 54
repetitive-sequence, 14, 36–37
replication of. *See* Replication
short tandem repeats, 211, 216–217,
    218, 224–227, 676
single nucleotide polymorphisms. *See*
    Single nucleotide polymorphism
with sticky ends, 28, *28*, *56*, *57*,
    179, *180*
structure of, *3*, 13, 20–27, *20–26*, 36
sugar-phosphate backbone of, 23
supercoiled, 29–30, *30–31*
    negative supercoiling, 29, 54
    positive supercoiling, 29
tandemly repeated, 35–36
T-DNA, 233–234, *234–235*
telomeric, 34–35
template strand of, 49, 89, *89–90*
terminal inverted repeats, *157*
transcription of, 87–108. *See also*
    Transcription
transformation with. *See*
    Transformation
translesion synthesis of, 152–153
unique-sequence, 36–38
variable-number tandem repeats, 211,
    217–218, 224–226
X-ray diffraction studies of, 23, *24*
Z-DNA, 25–26, *26–27*
DNA-binding domain, 549–550, *550*, 554
DNA-binding protein, 200, *616*
DNA chip. *See* DNA microarray
DNA fingerprinting. *See* DNA typing
DNA forensics, 225–227
  applications of PCR, 203
  Central Park jogger case, 227
  Green River murders, 226–227
  Narborough murders, 226, *226*
dna genes, 50–51
DNA gyrase, 50, 52, 54
DNA helicase, 50–51, *51–52*, 150
DNA length polymorphism, 676

DNA ligase, *50*, 52, *53–54*, 152, 175,
    179, *186*, 187
DNA marker, 214, 245, 406
  genetic mapping in humans, 444
DNA methylase, 559
DNA microarray, 216, 261, 262
  analysis of *Drosophila* development,
    598–599
  analysis of single-nucleotide polymor-
    phisms, 216, *217*
DNA molecular testing, 211, 218–222
  availability of, 222
  concept of, 218–219
  using PCR approaches, 221–222, *221*
  using RFLP analysis, 219–221, *220*
DNA polymerase, 43, 47–50, *49*, 62
  of eukaryotes, 59
  $3' \rightarrow 5'$ exonuclease activity of, 50, 62,
    150
  $5' \rightarrow 3'$ exonuclease activity of, 48,
    52
  Klenow fragment, 191
  proofreading activities of, 50, 62, 150
  repair, 150
  roles of, 48–50, *49*
  Taq polymerase, *201*, 202
  for translesion DNA synthesis, 153
  Vent polymerase, 202
DNA polymerase α, 59
DNA polymerase δ, 59
DNA polymerase ε, 59
DNA polymerase I, 47–48, 50, *50*, 52,
    53, *186*, 187
  Klenow fragment of, 191
  mutant enzyme, 48
DNA polymerase II, 50
DNA polymerase III, 50, *50*, 52, *53–54*,
    55, 150, *152*
  holoenzyme, 50
DNA polymerase IV, 50
DNA polymerase V, 50
DNA polymorphism, 211
  classes of, 214–218
  definition of, 213–214
  short tandem repeats, 211, 216–217,
    218, 224–227, 676
  single-nucleotide polymorphisms,
    214–216, *215–216*
  variable-number tandem repeats, 211,
    217–218, 224–226
DNA primase, 43, *50*, 51–52, *53*
DNA probe
  labeling of, 191, *191*
  screening genomic library for specific
    DNA sequence, 189–190, *189*
DNA profiling. *See* DNA typing
DNA repair, 133, 149–153, *151–153*,
    166, 624. *See also* specific
    repair systems
  defects in genetic diseases, 153,
    153–154
  direct correction (direct reversal) of
    damage, 149–150
  DNA polymerase I in, 48
  involving excision of nucleotides,
    150–152

DNase. *See* Deoxyribonuclease
DNase-hypersensitivity site, 547
DNA sequencing
    analysis of DNA sequences, 199–200
    automated, 199, *200*, 247–248
    dideoxy method of, 197–199,
        198–199
    identification of genetic variation, 676
    generating sequence of genome,
        247–249
    PCR-based, 248
DNA typing, 7, 203, 211, *211*
    applications of, 227
    crime scene investigations, 225–227
    in paternity case, 224–225, *225*
DNA virus
    double-stranded DNA, 27
    single-stranded DNA, 27
    tumor virus, 610, 618
Dobberstein, B., 127
Docking protein, 127, *127*
Dog
    breeding of, 723
    canine origins, 723
    chromosome number in, *313*
    C value of, *31*
    evolution under domestication,
        392, *392*
Dolly (cloned sheep), 577–578, *579*
Domain (evolutionary), 722
Domain shuffling, 724–725
Dominance, 276, 363
    codominance, 214, 339, 346–347
    complete, 339, 346, 348
    incomplete, 339, 346–347, *347*
    molecular explanation of, 347
    partial. *See* Dominance, incomplete
    pseudodominance, 456
Dominance variance, 386–387
Dominant epistasis, 351–355, *353*
Dominant trait, 4, 276, 278
    autosomal, 300
    general characteristics of, 291
    in humans, 290–291, *290*
    lethal, 356–357
    X-linked, 300, 327, *329*
Donor site, 159
Dosage compensation, 574, 586, 600
    in *Drosophila melanogaster*, 590–591
    for X-linked genes, 323–326, *324*
    in mammals, 586
Double crossover, 413, 416
    four-strand, 413
    three-strand, 413
    two-strand, 413
Double helix, 3, *13*, 22–24, *22–25*
Double monosomic, 467, 468
Doubly tetrasomic, 467, 468
Down syndrome, 363. *See also*
    Trisomy-21
    familial, 470–471
Drosophila melanogaster
    alcohol dehydrogenase gene of,
        675–676, *676*
    association of recombination with
        chromosomal exchange, 408–409, *409*

bar-eye in, 408–409, *409*, 458, *458*
base composition of DNA from, *23*
body color in, 341, 410,
    437–438, *438*
body size in, 391
bristle number in, *391*, 437–438, *438*
    QTL for, 397
chromosome number in, 300, *313*
chromosomes of, *314*
C value of, *31*
deletion mapping in, 456–457, *456*
development in, 261, 573, 574–576,
    591–599, *592–600*, 601
    microarray analysis of, 598–599
dosage compensation in, 590–591
eye color in, 164, *164*, 314–320, *316*,
    318, 320–321, 344, *344–345*,
    406–409, *407*, *409*, 465, 681, *681*
eye shape in, 501
fecundity in, 393
fruitless gene in, 228–229, *229*
genetic drift in, 681, *681*
genetic map of, *8*
genome analysis in, 256–257
genome of, 255
genome sequence of, 244, 253
homeotic genes in, 574, 593, 595–600,
    597–600
imaginal discs of, 593, *594*
intersex flies, 325
linkage studies in, 406–407, *407*
maternal effect genes in, 593–595,
    594–595
mitotic recombination in, 437–438,
    438–439
as model organism for research, 6, 9,
    10, 253, *314*, 575–576, *576*
polytene chromosomes in, 454, *455*
replication in, 57, *57*
segmentation genes in, 574, 593,
    595, *596*
sex chromosomes of, 314, *314–315*
sex determination in, 325–326, *325*,
    586–590, *587–590*
sex linkage in, 314–317, *315–316*,
    318, 357
sexual behavior in, 228–229, *229*
spontaneous mutation frequency at
    specific loci, *678*
starvation resistance in, *393*
telomeres of, 35
transposons in, 164, *164*
twin spots in, 437–438, *438–439*
wing morphology in, 390, *390*,
    405–407, *407*, 410
X chromosome of, 456, *456*,
    574, 591
Drosophila pseudoobscura, phototaxis
    in, 392, *392*
Ds element, 159–162, *162–163*
dsx gene (*doublesex*), 587,
    589–590, *590*
Dt element, 157–158
Duchenne muscular dystrophy, 74, 219,
    327, 360, 445
Dukepoo, Frank C., 668

Dunkers
    ABO blood group among, 682–683, *683*
    allelic frequencies among, 682
Duplicate dominant epistasis, 354–355
Duplicate recessive epistasis, 353, *356*
Duplication, 133, 453–454, 457–458,
    457–458
    epistasis involving duplicate genes,
        353–355, *356*
    reverse tandem, 458, *458*
    tandem, 458, *458*
    terminal tandem, 458, *458*
Dwarfism, 362
Dyad, 307, 309

Ear length, in corn, 383–384, *384*
East, Edward M., 378–379, 383
Ecdysone, 581, 598–599
Ecological isolation, 698
*Eco*RI, 177, *178*, 179, *180*, 192,
    193, 194
Edible vaccine, 236
Edwards syndrome. *See* Trisomy-18
Effective population size, 680–681
Effector, 516
Egg, 306, 311, *312*
Egg production, in poultry, *391*,
    393, *393*
Egg weight, in poultry, *391*, 393, *393*
Electrophoresis. *See also* Agarose gel
    electrophoresis
    of hemoglobin, 77, *77*
    of proteins, finding proportion of
        polymorphic loci, 672–673,
        673, 699
Electroporation
    transformation of bacteria, 490
    with recombinant DNA, 181
    transformation of plant cells, 234
Elongation factor
    EF-G, 123
    EF-Ts, 123, *124*
    EF-Tu, 123, *124*
Embryonic development, in *Drosophila
    melanogaster*, 591–593, *592*
Embryonic hemoglobin, 580, *580*
Emerson, Rollins, 160, 383
Enamel hypoplasia, hereditary, 327, *329*
Endangered species, conservation of,
    658, 679, 697
Endoplasmic reticulum, 9
    protein sorting in cells, 127, *127*
    rough, *11*
    smooth, *11*
Endoreduplication, 454
Endosperm, 472, 556, *558*
Endosymbiont theory, 632, 722
Engineered transformation, 490
engrailed gene, 595
Enhancer, 94, 543, 550–551, *551*, 567
Entrez (database searching), 8
env gene, 611, *612–613*, 614
Environment
    chemical mutagens in, 147–148
    genotype-by-environment interaction,
        385, 386

Environmental effect
  common family, 387
  on development, 573, 578, *579*
  on gene expression, 339, 357–364
  general, 387
  on phenotype, 272, *272*, 373, 377
  special, 387
Environmental variance, 385–386
Enzyme
  deficiencies in humans, 73–75, *74–76*
  gene control of, 67–78
  RNA, 100, 123
  temperature-sensitive, 360, *362*
Ephrussi, Boris, 643
Epigenetic phenomenon
  position effect, 465
  X inactivation, 324
Epiloia, *678*
Episome, 157, 485
Epistasis, 339, 348–355, *352–356*, 374
  dominant, 351–355, *353*
  duplicate dominant, 354–355
  duplicate recessive, 353, *356*
  involving duplicate genes, 353–355, *356*
  recessive, 350–353, *352*
Equilibrium density gradient
    centrifugation, 45–47, *46*
erbA oncogene, *614*
erbB oncogene, *614*
ERK protein, 617, *617*
Erythromycin resistance, in
    Chlamydomonas reinhardtii,
    645–647, *646*
Escherichia coli
  bacteriophages of, 492–494, *492*, *494*
  base composition of DNA from, 23
  cell-free protein-synthesizing system
    from, 117
  chromosome of, 29, *29*
  conjugation in, 484–490, *484–489*
  C value of, *31*
  excision repair in, 150
  foodborne, 227
  genetic analysis of, 482–483, *483*
  genetic map of, 489–490, *489*
  genome analysis in, 256–257
  genome sequence of, 244, 246, 251,
    482, 490, 516
  IS elements in, 155
  lac operon of, 515–528, *516–528*, 536
  Meselson-Stahl experiment on,
    45–47, *46*
  as model organism for research, 6,
    11, *12*
  phage T4 resistance in, 135–136, *135*
  plasmid cloning vectors for, 180
  production of recombinant proteins
    in, 232
  regulation of gene expression in,
    515–533
  replication in, 48–51, *50*
  resistance mutants in, 149
  resistance to phage T1, 149
  SOS response in, 152–153
  spontaneous mutation frequency at
    specific loci, *678*

transduction in, 494–499, *496*, *498*
transformation in, 490
trp operon of, 528–532, *528–532*, 536
Essential amino acid, 73
Essential gene, 339, 356–357, 363–364
EST. *See* Expressed sequence tag
EST1 gene, 60
Esterase, in mouse, 684
Esterase 4F, in prairie vole, *671*
Estrogen, 554–555, *555*
Ethical, legal, and social issues, Human
    Genome Project, 244, 264–265
Ethidium bromide, 194, 643
Ethylene, 556, *557*
Eubacteria, 721, *721*
Euchromatin, 34, 465
Euglena, base composition of DNA
    from, 23
Eukaryote, 1, 9
  chromosome mutations in, 454–474
  chromosomes of, 29–37, *31–35*,
    300–302, *300–301*
  evolutionary tree of life, 721–722, *721*
  gene density in, 255–257, *256*
  gene mapping in, 405–419, 429–446
    random-spore analysis, 432
    tetrad analysis, 429–437, *433–436*
  genome sequences of, 252–255
  genome size in, 255–257, *256*
  mRNA of, 95–102, *96*
    production of mature mRNA, 97–99,
    97–100
  mutation rate in, 140
  operons in, 544–545, *545*
  protein secretion in, 127, *127*
  regulation of gene expression in,
    543–568
  replication in, 43, 47, *47*, 57–60,
    57–60
  ribosomes of, 88, 102–103
  RNA polymerase of, 93
  rRNA of, 103–104, *104*
  transcription in, 93–106
  translation in, 111
  transposable elements in, 134, 153,
    157–165, *159*, *162–164*
Eukaryotic cell, *11*
  cell cycle, 302, *302*
Euploidy, 466–467
even-skipped gene, 595, *596*
Evolution, 390
  convergent, 716
  molecular, 705–725. *See also*
    Molecular evolution
Excision repair, 150–151
Exon, 87, 97–98, 706
Exon shuffling, 724–725
Expressed sequence tag (EST), 247
Expression vector, 182, 188, *188*, 232
Expressivity, 339, 358–359, *358*, 364, 374
  variable, *358–359*, 359
External transcribed spacer, 104, *104*
Extinction, 697
Extrachromosomal genes, 632
Extranuclear genes, 632
  organization of, 632–639

Eye color
  in *Drosophila melanogaster*, 164, *164*,
    314–320, *316*, *318*, *320–321*, 344,
    344–345, 406–409, *407*, *409*, 465,
    681, *681*
  in humans, 373
Eye shape, in *Drosophila melanogaster*, 501

Facial hair, distribution of, 360
Factor VIII, 233
Facultative heterochromatin, 34
Familial adenomatous polyposis (FAP),
    619, 625, *625*
Familial trait, 389
Fanconi anemia, *154*
FAP. *See* Familial adenomatous polyposis
Farabee, W., 288
Fas gene, 566
Fate map, 575–576
F-duction, 487
Fecundity
  in *Drosophila melanogaster*, 393
  in milkweed bugs, *391*, 393
Feline leukemia virus, *31*, 610
Feline sarcoma virus, 614
Fetal analysis, 67, 80–81, *81*
Fetal hemoglobin, 580, *580*
F factor, 481, 484–485, *485–486*, 489
  excision of, 485
  $F^+ \times F^-$ cross, 485
  IS elements in, 157
  replication of, 54
F' factor, 487, *487*
F_1 generation, 4, 275, *275–277*
F_2 generation, 4, 275, *275–277*
Fibrinopeptides, evolution of, *714*
Field horsetail, chromosome number
    in, *313*
Fields, Stanley, 229
Filterable agent, 494
Fine-structure mapping, of *rII* region of
    bacteriophage T4, 501–507
Fire, A., 565
First-division segregation, 432–434, *433*
First filial generation. *See* F_1 generation
FISH. *See* Fluorescent in situ
    hybridization
Fisher, Sir Ronald, 379, 658, *658*, 679
FIS protein, *50*
Fitness, 689, *689*
  Darwinian, 689, *689*
  mean fitness of the population,
    690–691
Fixed allele, 683–684, *683*
Flanking region, evolution in, *708*, 709
Flavr Savr tomato, 236
Floral development, in *Arabidopsis
    thaliana*, 576, *577*
Floral traits, in monkeyflower, 395–397,
    395–396
Flow cytometry, separation of individual
    chromosomes by, 184, 247
Flower
  imperfect, 326
  perfect, 326
  structure of, 311–312, *313*

Flower color
   in garden pea, 274, *274*, 277,
      285–286, *285*
   in snapdragon, 346
   in sweet pea, 353
Flower length, in tobacco, 378–379
Flower position, in garden pea, 274,
   274, 277
Fluctuation test, 135–136, *135*
Fluorescent in situ hybridization
   (FISH), 445, *445*
   FISH maps, 246–247
FMR-1 gene, 466, 560
fms oncogene, *614*
Food
   from genetically modified crops, 227,
      234, 236
   pathogen detection in, 227
Forensic science. *See* DNA forensics
Fork diagram. *See* Branch diagram
Formylmethionine, 121
Forward mutation, 677
fos oncogene, *614*
Founder effect, 681–683
Four o'clock, shoot variegation in, 631,
   639–641, *640–642*
F-pilus, *485*
Fraenkel-Conrat, Heinz, 20
Fragile site, 465–466, *466*
Fragile site mental retardation. *See*
   Fragile X syndrome
Fragile X syndrome, 465–466, *466*, 560
Frameshift mutation, 115–116, *116*,
   137, 138
Franklin, Rosalind, 23–24, *24*
Freeman, G., 650
Frequency distribution, 377–378
Frequency histogram, 378, *378*
Frog
   development in, *391*, *393*, 394
   size at metamorphosis, *391*, *393*
Fructose intolerance, 74
Fruit, seedless, 473
Fruit color
   in summer squash, 351–352,
      353–354
   in tomato, 397
Fruit fly. *See Drosophila melanogaster*
fruitless gene, 228–229, *229*
Fruit shape
   in shepherd's purse, 354–355
   in summer squash, 349, *351*
Fugu rubripes. See Pufferfish
Functional genomics, 243, 245,
   257–265. *See also* Proteome;
      Transcriptome
FUN gene, 258–259, *259*
Fur color. *See* Coat color
fushi tarazu gene, 595
fw2.2, QTL in tomato, 397

gag gene, 611, *612–613*, 614
Gain-of-function mutation, 290
Gal4p protein, 552–553, *552*
Galactosemia, 74

β-Galactosidase, 517–518, *517*,
   521–522, 528
   alpha fragment of, 181, *181*
   omega fragment of, 181, *181*
GAL genes, of *Saccharomyces cerevisiae*,
   229–230, *230*
   glucose repression of *GAL1*,
      228, *228*
   regulation of, 551–553, *552*
Gallus. See Chicken
Galton, Francis, 375
Gamete, 278, 306, 310–311, *312*
Gametic isolation, 698
Gametogenesis, 306
Gametophyte, 311–312, *313*
Ganglioside, 76, *76*
GAP. *See* GTPase activating protein
Gap genes, 595, *596*
Garden pea
   flower color in, 274, *274*, 277,
      285–286, *285*
   flower position in, 274, *274*, 277
   Mendel's experiments with, 3–4, *3*,
      272–286
   as model organism for research, 9, *10*
   pod traits in, 274, *274*, 277
   procedure for crossing, 273–274, *273*
   seed traits in, *271*, 274–276, *274–277*,
      280–286, *280–285*
   stem height in, 274, *274*, 277
   wrinkled-pea phenotype, 280–281
Garrod, Archibald, 68
Garter snake, speed and neurotoxin
   resistance in, 393, *394*
Gaucher disease, 74
G banding, *301*, 302
GC box, 94
Gefter, Martin, 48–49
Gehring, Walter, 595
GenBank, 8
Gene, 3, 271, 278, 286, 507
   cancer and, 610–624
   constitutive, 516
   essential, 339, 356–357, 363–364
   homologous, 258
   housekeeping, 94, 516
   identifying in DNA sequences, 258, *259*
   inducible, 516, *516*
   linked, 406
   in meiosis, *322*
   protein-coding. *See* Protein-coding gene
   regulated, 516
   reporter, 229–230, *230*
   syntenic, 406
Gene amplification, changing
   cellular proto-oncogenes into
   oncogenes, 618
GeneChip array. *See* DNA microarray
Gene conversion, 706, 724
Gene counting, 660, *662*
Gene density, 243, 255–256, *256*
   in eukaryotes, 255–257, *256*
   in prokaryotes, 255, *256*
Gene desert, 256
Gene dosage, 574

Gene duplication, 723–725, *724*.
   See also Duplication
Gene expression, 4–5, *4*. *See also*
   Proteome; Transcriptome
   analysis of expression of individual
      genes, 228–229
   in cancer cells, 261
   in cloned mammals, 579
   describing patterns of, 260–264, *262*
   environmental effects on, 339, 357–363
   gene control of proteins, 67–81
   in plants, hormonal control of,
      556–558, *557*
   regulation of, 5
      in bacteriophage, 515, 533–536, *535*
      in eukaryotes, 543–568
      in prokaryotes, 515–533, *544*
      in *Saccharomyces cerevisiae*, 261
      in tissues during development,
         573–601
   transcription, 87–108
   translation, 111–128
Gene flow, 658, 685–687, *686*
Gene frequency. *See* Allelic frequency
Gene gun, 234
Gene interactions, 348–356, 363
   epistasis. *See* Epistasis.
   producing new phenotypes, 348–349,
      349–351
Gene knockout, 259, *260*
   in mice, 213
   in *Saccharomyces cerevisiae*, 259, *260*
Gene locus. *See* Locus
Gene mapping, 213, 419
   in *Aspergillus nidulans*, 429
   in bacteriophage, 481, 499–507, *500*
   calculating map distance, 418–419, *418*
   calculating recombination frequencies
      for genes, 416–417, *416–417*
   coincidence, 417–418
   by conjugation, 484–490, *484–489*
   deletion mapping, *502*, 503, *504–506*
   establishing gene order, 415–416,
      415–416
   in eukaryotes, 405–419, 429–446
      gene-centromere distance, 429,
         432–434, *433*
      random-spore analysis, 432
      tetrad analysis, 429–437, *433–436*
   fine-structure mapping, 501–507
   of genome, 245
   in humans, 443–446, *443–446*
      high-density genetic maps, 444, *445*
      lod score method, 444
      by recombination analysis, 443–444,
         443
   interference, 417–418
   intergenic, 481, 501
   intragenic, 481, 501–507
   linkage detection through testcrosses,
      410–413, *410*
   by transduction, 492–499, *492–498*
   by transformation, 490–492, *491*
   using three-point testcross, 413, *415*
   using two-point testcross, 412, *412*

Gene mutation, 134
Gene pool, 657–658, 699
General environmental effect, 387
Generalized transduction, 494–497, 496–497
General transcription factor, 94, 95, 549, 549
Gene regulatory element, 88
Gene-rich regions, 256
Gene segregation, 278
 in meiosis, 309, 310
 in mitosis, 305–306
Gene silencing, 543–544, 557–561, 558, 567, 586
 RNA interference, 544, 565–566, 566
Gene therapy, 211–212, 231–232
 germ-line, 231
 somatic, 212, 231
Genetically modified organism (GMO), detection of, 227
Genetic code, 5, 111, 115–119, 118
 characteristics of, 118–119
 in chloroplasts, 637
 comma free, 118
 deciphering of, 117–118
 degeneracy of, 118
 in mitochondria, 631, 635
 nonoverlapping, 118
 redundancy in, 710
 start and stop signals in, 118–119
 triplet nature of, 115–116, 116–117
 universality of, 118
Genetic correlation, 392–394, 393–394
 negative, 393, 393
 positive, 393, 393
 traits in humans, domesticated animals, and natural populations, 393
Genetic counseling, 67, 79–81, 80
Genetic database, 7–8
Genetic disease, 6, 67, 81
 distribution in humans, 666
 DNA molecular testing for, 218–222
 from DNA replication and repair mutations, 153, 153–154
 enzyme deficiency-related, 73–76, 74–76
 molecular testing for, 211
 mtDNA defects, 647
 multiple alleles in, 345
 prenatal diagnosis of, 80–81, 81
 screening for, 261
 testing for, 264
Genetic drift, 657–658, 664, 696
 alterations in allelic frequency, 679–685, 699
 balance between migration and genetic drift, 686
 balance between mutation and genetic drift, 684–685, 685
 bottlenecks, 681–683
 effective population size and, 680–681
 effects of, 683–684, 683
 founder effects, 681–683
Genetic engineering, 176, 203, 212
 of plants, 233–236
  applications of, 234–236, 235

Geneticist, 5–11
Genetic linkage. See Linkage
Genetic map, 8, 8, 406, 410. See also Gene mapping
 of bacteriophage lambda, 533
 of chloroplast DNA, 638
 concept of, 411–412
 of Drosophila melanogaster, 8
 of Escherichia coli, 489–490, 489
 generation of, 412–413
 high-density, 214
  in humans, 444, 445
 of mtDNA, 633–634, 634
 of rII region of phage T4, 501–507, 502
Genetic marker, 406
Genetic material, 13
 characteristics of, 14
 search for, 14–20, 15–18
 of viruses, 13, 19–20, 19
Genetic mosaic, 324
Genetic privacy, 264–265
Genetic recombination. See Recombination
Genetics
 basic concepts of, 2–5
 biochemical, 69
 classical, 2
 definition of, 2
 developmental, 574–601
 Mendelian, 271–291
  extensions of, 339–364
 modern, 2
 molecular, 1, 6, 658
 population, 1, 6, 657–699
 quantitative, 1, 6, 373–397, 658
 subdisciplines of, 6
 transmission, 1, 6, 658
Genetics research, 2, 5–11
 applied, 6–7
 basic, 6–7
 model organisms for, 6, 9, 10
Genetic structure, of population, 375, 658–663
 variation in space and time, 669–670, 670
Genetic switch, 534–535
Genetic symbols, 289, 289, 317
Genetic testing, 218–219
 purpose of human testing, 219
Genetic variance, 385–387, 669
 additive, 386–387
Genetic variation, 657, 659, 699
 classical model for, 672
 at DNA level
  DNA length polymorphism and microsatellites, 676
  DNA sequence variation, 676
 increases and decreases within populations, 697
 measurement of
  at DNA level, 673–676, 674–676
  at protein level, 671–673, 672–673
 in natural populations, 670–676
 neutral mutation model for, 672–673

sources of, 5, 134
 in space and time, 669–670, 670
 transposable elements and, 153
Gene tree, 705, 716–718
Genic male sterility, 648
Genic sex determination, 320, 326
Genome, 2, 27, 300
 gene mapping of, 245
 generating sequence of, 247–249
 insights from analysis of, 255–257, 256
 of mitochondria. See Mitochondria, genome of
 physical mapping of, 246–247, 248
Genome sequence
 annotation of, 251, 258
 of Arabidopsis thaliana, 244, 253–254, 724
 assigning gene function experimentally, 259, 260
 of Borrelia burgdorferi, 482, 490
 of Caenorhabditis elegans, 244, 253
 describing patterns of gene expression, 260–264, 262
 of Drosophila melanogaster, 244, 253
 of Escherichia coli, 244, 246, 251, 482, 516
 of eukaryotes, 252–255
 of Haemophilus influenzae, 249, 251, 252
 of Helicobacter pylori, 482
 of humans, 254–255, 255
 identifying genes in DNA sequence, 258, 260
 of Methanococcus jannaschii, 251–252, 482
 of mouse, 244, 254, 255
 of prokaryotes, 251–252, 252
 of rat, 254, 255
 of Saccharomyces cerevisiae, 244, 252–253, 258, 259
 sequence similarity searches, 258–259
 of Treponema pallidum, 482
Genome sequencing, 243–244, 265
 assembling and finishing sequence, 250–251
 mapping approach to, 243, 245–249, 248
 whole-genome shotgun approach to, 243, 245, 249–250, 250, 254
Genome size, 255–256, 256
 in eukaryotes, 255–257, 256
 in prokaryotes, 255, 256
Genomic imprinting, 544, 560–561, 560
Genomic library, 166, 183–185, 184, 204
 construction of, 184–185, 184
 screening for specific clone, 189–190, 189
Genomics, 2, 243–265
 comparative, 243, 245, 264–265, 709–710
 functional, 243, 245, 257–265
 structural, 243, 245–257, 265
Genotype, 3, 271–272, 272, 278
 differential reproduction of, 687
 genotype-by-environment interaction, 385, 386

Genotypic frequency, 657, 659–660
    calculation of, 659–660, 662
    calculation of allelic frequency from, 660–661, 662
    Hardy-Weinberg law, 663–669
Genotypic ratio, 4, 284
Genotypic sex determination, 320–326
Geographical isolation, 698
Geographic variation, in allelic frequency, 669–670, 670
Germination
    in jewelweed, 391, 393
    of seeds, 556, 558
Germ-line gene therapy, 231
Germ-line mutation, 136
giant gene, 595
Giant sequoia, chromosome number in, 313
Gibberellic acid, 557
Gibberellin, 556, 557–558
Giemsa stain, 302
Gierer, A., 19
Gilbert, Walter, 197, 725
Glass, William, 682
Glaucoma, open-angle, DNA molecular testing, 221, 221
GLC1A gene, 221
Gleevec. See ST1571
Globin gene family
    α-globin gene, 78, 458, 579–580, 580
        evolution of, 712
    β-globin gene, 78, 98, 220, 220, 458, 579–580, 580
        evolution of, 712
        nucleotide heterozygosity in, 675
    evolution of, 723–724, 724
    during human development, 579–580, 580
Glucocorticoid, 554–555, 555–556
Glucose effect, 525–526, 526–527
    in yeast GAL gene system, 552–553
Glucose-6-phosphate dehydrogenase deficiency, 74
Glucose repression, of yeast GAL1 gene, 228, 228
Glutamic acid, 113
Glutamine, 113
Glycine, 113
Glycogen storage disease, 74
Glycolysis, 633
Glycoproteins, 127
Glycosylase, 150
Glycosyltransferase, 343, 343
GMO. See Genetically modified organism
Goldberg-Hogness box. See TATA box
Goldfish, chromosome number in, 313
Golgi apparatus, 11
Goodness-of-fit test, 287
gooseberry gene, 595, 596
Gout, 360
G_0 phase, 302
G_1 phase, 30, 58, 61, 302, 302, 607
G_1-to-S checkpoint, 607, 608, 621, 621, 623
G_2 phase, 30, 58, 302, 302, 607

G_2-to-M checkpoint, 607, 608
G protein, membrane-associated, proto-oncogene products, 616–617, 617
Grandparental phenotype, 406
Granum, 636, 637
Grasshopper, sex chromosomes of, 313–314
Grb2 protein, 617, 617
Green River murders, 226–227
Greider, Carol W., 59
Greying, in horses, 353, 354
Griffith, Frederick, transformation experiment, 15–16, 15, 490
Grizzly bear, genetic variability in, 636
Growth factor, 607–608, 609, 617
    proto-oncogene products, 616–617, 616
Growth hormone
    bovine, 233
    human, 233
Growth hormone gene, 675
Growth-inhibitory factor, 608, 609
GTP, in translation, 120, 122, 123, 124
GTPase activating protein (GAP), 617, 617
Guanine, 2, 13, 20, 20, 22, 23, 23, 26, 26, 141
Guessmer, 192
Guide RNA, 102
Guo, S., 565
Gurdon, John, 577
Guthrie test, 73, 219

H4 gene region, 675
H19 gene, 560–561, 560
HaeII, 178
HaeIII, 178
Haemophilus influenzae, genome sequence of, 249, 251, 251
Hairy ears trait, 328
hairy gene, 595
Haldane, J.B.S., 419, 658, 658, 698
Haldane's rule, 698
H antigen, 343, 343
Haploid (N), 29, 278, 299–300, 300, 306
Haploidization, 440
Haploid segregant, 440
Haploidy, 472
Haplosufficient gene, 347
Hardy, Godfrey H., 663
Hardy-Weinberg equilibrium, 664, 666
Hardy-Weinberg law, 657, 659, 663–669
    assumptions of, 663–664
    derivation of, 665–666, 665
    estimation of allelic frequencies from, 668–669
    extensions to loci with more than two alleles, 661, 667
    extensions to sex-linked alleles, 667, 667
    forces that change gene frequencies in populations, 676–690
    historical aspects of, 663
    predictions of, 664–665
    testing for Hardy-Weinberg proportions, 667–668
Harlequin chromosome, 47, 47

Harris, Henry, 618
Hayes, William, 485
Head width, in salamanders, 381
Heavy chain, 582, 583
    constant region of, 582, 583
    recombination in heavy chain genes, 584–585, 584
    variable region of, 582, 583
Height. See Stature
Helicobacter pylori
    C value of, 31
    genome sequence of, 482
Helix-turn-helix motif, 549, 550, 597
Helper phage, 499
Helper virus, 614
Hemizygote, 315
Hemoglobin. See also Globin gene family
    changes during human development, 579–580, 580
    electrophoresis of, 77, 77
    embryonic, 580, 580
    evolution of, 714
    fetal, 580, 580
    gene control of protein synthesis, 76–79, 77–80
    mouse, 684
    structure of, 77, 77, 114, 115
    variants of, 78–79, 78
        genotypic and allelic frequencies among Nigerians, 662
Hemoglobin A, 77, 579–580, 580
Hemoglobin A2, 580
Hemoglobin C, 78–79, 78
Hemoglobin S, 77–78, 77–78, 219–220, 220, 693–694, 694
Hemolytic anemia, 74
Hemophilia, 165, 218, 357, 444
Hemophilia A, 327, 328
Herbicide tolerance, in plants, 234, 235
Hereditary disposition, for cancer, 621
Hereditary nonpolyposis colon cancer (HNPCC), 152, 154, 624
Hereditary trait, 272
Heredity, infectious, 631, 649–650, 649
Heritability, 385–390
    broad-sense, 373, 387–388, 397
    calculation of, 389–390
    limitations to estimates of, 388–389
    narrow-sense, 373, 387–388, 391
    from parent-offspring regression, 389–390, 390
    of traits in humans, domesticated animals, and natural populations, 391
Hermaphrodite, in Caenorhabditis elegans, 325–326
Herrick, J., 77
Hershey, Alfred D., 16–19, 18
Hershey-Chase bacteriophage experiments, 16–19, 18
Herskowitz, Ira, 261
Heteroallelic mutation, 502
Heterochromatin, 34, 465, 544, 557
    constitutive, 34
    facultative, 34
Heteroduplex DNA, 490, 491
Heterogametic sex, 299, 314, 325–326

Heterogeneous nuclear RNA (hnRNA), 98, 563
Heterokaryon, 438–439
Heterologous probe, identifying DNA sequences in libraries, 190
Heteroplasmy, 647, 649
Heterosis. *See* Heterozygote superiority
Heterozygosity, 671–672
  loss of, 621
  nucleotide, 675, *675*
Heterozygote, 3, 276, *276*, 278, 280
  carrier detection, 67, 79–81, 219
  deletion, 455–456
  duplication, 458
  genetic symbols for, 317
  inversion, 459–460, *460–461*
  translocation, 462, *463*
Heterozygote superiority, 648, 693–694, *694*
HEXA gene, 357
Hexosaminidase A, 357
Hfr strain, 481, 485–487, 489
  Hfr × *F⁻* mating, 485, *486*
  in interrupted-mating experiments, 487–489, *488–489*
  production of, 485, *486*
HFT lysate. *See* High-frequency transducing lysate
HGP. *See* Human Genome Project
HGPRT. *See* Hypoxanthine-guanine phosphoribosyl transferase
*Hha*I, 177, *178*
High-frequency recombination strain. *See Hfr* strain
High-frequency transducing (HFT) lysate, *498*, 499
Highly repetitive DNA, 36
Himalayan rabbit, 360, *362*
*Hind*III, *178*
*Hinf*I, 177
Histidine, *113*
Histogram, frequency, 378, *378*
Histone, 13, 31–33, *31*, 547
  acetylation of, 547, *548*, 574, 591
  deacetylation of, 547, *548*
  evolutionary conservation of, 32
  methylation of, 557, 586
  nucleosome assembly, 61–62, *61*
  repression of gene activity by, 547, 567
  synthesis of, 61
Histone acetyl transferase, 547, *548*, 591
Histone chaperone, 62
Histone deacetylase, 547, *548*, 550, 557
Histone genes, 712, *712*
Histone methyl transferase, 557
HIV. *See* Human immunodeficiency virus
hMLH1 gene, 151–152, 624
hMSH2 gene, 151–152, 624
HNPCC. *See* Hereditary nonpolyposis colon cancer
hnRNA. *See* Heterogeneous nuclear RNA
HO endonuclease, 548
Holandric trait. *See* Y-linked trait
holE gene, 50
Holley, Robert, 117
Homeobox, 574, 597

Homeodomain, 597
Homeotic genes
  in *Drosophila melanogaster*, 574, 593, 595–600, *597–600*
  in plants, 597
  in vertebrates, 597, *600*
Homeotic mutation, 595
Homoallelic mutation, 502
Homocysteine, 72, *72*
Homogametic sex, 299, 314, 325–326
Homogentisic acid, 68, *68*
Homolog, 300, 550
Homologous chromosomes, 300, 407
Homologous genes, 258
Homologous proteins, 706
Homozygote, 3, 276, *276*, 278, 280
Hopi Indians, albinism among, 668–669, *669*
Horiuchi, Takashi, 251
Hormone, 553
Horns, in sheep, 360
Horse
  chromosome number in, *313*
  coat color in, 346, 353, *354*
  cremello, 346
  C value of, *31*
  greying in, 353, *354*
  palomino, *339*, 346
  sorrel, 346
Host range, of bacteriophage, 499–501, *500*
Hot spot, mutational, 141, 503, *506*
Housekeeping gene, 94, 516
Howard-Flanders, P., 150
Hox genes, 597. *See also* Homeotic genes
*Hpa*II, 178, *178*, 558–559, *559*
hPMS1 gene, 151–152, 624
hPMS2 gene, 151–152, 624
H-*ras* oncogene, *614*
Hsp90 chaperone, 554
Hubby, John, 671
HUGO. *See* Human Genome Organization
Human
  aneuploidy in, 468–471, *468–471*
  base composition of DNA from, 23
  birth weight in, 374, *374*, 377
  chromosome number in, 300, *313*
  C value of, *31*
  development of, hemoglobin types and, 579–580, *580*
  DNA repair in, 151–152
  dominant traits in, 290–291, *290*
  eye color in, 373
  gene density in, 256–257
  gene mapping in, 443–446, *443–446*
    high-density genetic maps, 444, *445*
    by lod score method, 444
    by recombination analysis, 443–444, *443*
  genetic diseases in. *See* Genetic disease
  isolation of CF gene, 222–224
  karyotype of, 301, *301*
  Mendelian genetics in, 288–291
  molecular evolution in, 715
  mtDNA of, 633
  mutation rate in, 140

  origins of, 722–723
  physical mapping of genes, 445–446, *445–446*
  polyploidy in, 473
  quantitative trait loci in, 397
  recessive traits in, 289–290, *290*
  retrotransposons in, 165
  sex determination in, 585–586
  sex-linked traits in, 326–330
  spontaneous mutation frequency at specific loci, 678
  X-linked genes in, 443–444, *443*
Human genome
  clone contig map of, 246–247, *248*
  generating sequence of, 247–249
  genome sequence of, 254–255, *255*
  genome size in, 256–257
  nucleotide heterozygosity in, 675, *675*
  physical map of, 246–247
  radiation hybrid map of, 247
  short tandem repeats in, 216–217
Human Genome Organization (HUGO), 244
Human Genome Project (HGP), 244–245, 254, 265, 405, 444
  ethical, legal, and social issues in, 244, 264–265
  goals of, 244
Human Genome Sequencing Project Consortium, 254
Human immunodeficiency virus (HIV), 163, 610–611
  C value of, *31*
  genome of, 611
Human papillomavirus, 618
Human Proteome Organization (HUPO), 263
Human remains, identification of, 227–228, 635–636
Humulin, 7, 233
hunchback (*hb*) gene, 595
Huntington disease, 218, 357, 466, *678*
HUPO. *See* Human Proteome Organization
Hybrid breakdown, 698
Hybrid inviability, 698
Hybridization (nucleic acid), 445
Hybrid seed, production of, 647–648, *648*
Hybrid sterility, 698
Hydrocortisone, 554
Hydrogen bonds
  in DNA, 23, 25–26
  in proteins, 112
Hydroxylamine, 144–146, *145*
Hydroxylaminocytosine, *145*
Hydroxylating agent, 144–146, *145*
Hypersensitivity site, 547
Hypostasis, 349
Hypothesis, 6
  null, 286–287
Hypothesis-driven science, 244
Hypothetico-deductive method of investigation, 6
Hypoxanthine, 144, *145*
Hypoxanthine-guanine phosphoribosyl transferase (HGPRT), 74, 75

ICR. *See* Internal control region
Ideogram, 302
I gene, 341–343, *342–343*
Igf2 gene, *560, 560*
IHF protein, *50*
III^Glc, *526*
Imaginal disc, 593, *594*
Immortal cells, 624
Immunoglobulin, *391, 393,* 573,
　　581–585. *See also* Heavy chain;
　　Light chain
　antigen-binding site on, *582,* 583
　structure of, *582*
Immunoglobulin A, 583, *584,* 585
Immunoglobulin D, 583, *584,* 585
Immunoglobulin E, 583, *584,* 585
Immunoglobulin G, *582,* 583, 585
Immunoglobulin M, 583, *584,* 585
Immunoglobulin genes, 573
　assembly from segments during B
　　cell development, 583–585,
　　583–584, 600
Imperfect flower, 326
Inborn errors of metabolism, 68, 82
Inbreeding, 658, 695–696, *696*
Incomplete dominance, 339,
　　346–347, *347*
Incomplete penetrance, 358–359, *358*
Indel, 707
Independent assortment, principle of, 4,
　　271, 282–286, *282–286,* 320, 322
Indoleacetic acid, 557
Induced mutation, 140–148, *143–147*
　chemical mutagens, 143–147, *144–146*
　radiation-induced, 141–143, *143*
Inducer, 516
Inducible gene, 516, *516*
Induction, 516
　for cell determination, 575
　coordinate, 517
Industrial melanism, 687–689, *688*
Infantile amaurotic idiocy. *See*
　　Tay-Sachs disease
Infectious heredity, 631, 649–650, *649*
Inferred ancestral sequence, 719
Inferred tree, 717
Infinitely many alleles model,
　　684–685, *685*
Informative site, 718–719
Ingram, V.M., 77
Inheritance
　chromosome theory of, 4, 299,
　　312–320, 329
　crisscross, 316
　maternal. *See* Maternal inheritance
　uniparental, 639
Initial committed complex, *95*
Initiation codon, 111, 120–121
Initiation complex
　30S, 121, *122*
　70S, 121, *122, 124*
Initiation factor, 120, 123
　eIF-4F, *122*
　IF1, 121, *122*
　IF2, 121, *122*
　IF3, 121, *122*

Initiator protein (replication), 51, *51*
Initiator tRNA, 121, *122,* 635
The Innocence Project, 227
Inosine, *106*
Inr element, 94
Insect
　DDT resistance in, 677
　sex chromosomes of, 313–314
Insertion, 133, 676
Insertional mutagenesis, 165, 618
Insertion sequence (IS), 134, 166
　characteristics of, 155, *155–156*
　in *F* factor, 157
　γ-δ, 157
　insertion of, 155, *156*
　IS*1,* 155, *155*
　IS*2,* 155, 157
　IS*3,* 157
　IS*10R,* 155
　IS module, 155
　in transposons, 155–156
insularia phenotype, in peppered moth,
　　688–689, *688*
Insulator, *560, 560*
Insulin, genetically engineered, 7, 233
Intelligence, 363
Interaction trap assay. *See* Yeast
　　two-hybrid system
Interaction variance, 387
Intercalating agent, 146, *146,* 643
Interference, 417–418
Intergenic mapping, 481, 501
Intergenic region, 545
Intergenic suppressor, 138–139, *139*
Internal control region (ICR), 88, 105
Internal transcribed spacer, 104, *104*
Interphase, 302, *303–304*
Interrupted-mating experiment,
　　487–489, *488–489*
Intersex individual, 325
Intestinal polyposis, *678*
Intragenic mapping, 481, 501–507
Intragenic suppressor, 138–139
Intron, 87, 97–98, 545
　group I, 100
　in mtDNA, 635
　in proto-oncogenes, 615
　self-splicing, 88, 100, *101*
　in tRNA genes, 105
Inversion, 453–454, 459–460, *459–461*
　paracentric, 459, *459*
　pericentric, 459, *459*
　position effect, 465
Inversion loop, 459–460, *460–461*
Invertebrate, genome analysis in, *256*
Inverted repeat, DNA, 155
Ionizing radiation
　as carcinogen, 626
　induction of mutations by,
　　141–143, *143*
IQ (intelligence quotient), 363
IS. *See* Insertion sequence
Island population, genetic drift in
　　humans, 682
Isoacceptor tRNA, 710
Isoleucine, *113*

Jacob, François, 5, 487, 518, 520–525,
　　520–525
Jaenisch, Rudolph, 579
Jeffreys, Alec J., 217, 226, *226*
Jewelweed
　germination time in, *391, 393*
　seed weight in, *393*
Johannsen, W.L., 286, 375
Jukes, T., 707
Jukes-Cantor model, of nucleotide
　　substitutions, 707–708, *707*

Karpechenko, 473
Karyokinesis. *See* Mitosis
Karyotype, 299, 301, *301,* 323
Kaufman, Thomas, 595
Kearns-Sayre syndrome, 647
Kemphues, K., 565
Kendrew, John, 115
Kennedy disease, 466
Kernel color
　in corn, 6, 157–162, *159, 162–163*
　in wheat, 375–376, *376,* 386
Ketoacidosis, 74
Kettlewell, H.B.D., 688–689
Khorana, Ghobind, 117
Kidney cancer, *619*
Kidney tubular acidosis with deafness, 74
Killer yeast, 649–650, *649*
Kimura, Motoo, 672, 683–684
Kinetochore, 34–35, *35,* 304–305
Kinetochore microtubule, *303*
Kingdom, 721–722
Klenow fragment, 191
Klinefelter syndrome, 322, *323,* 468
Knippers, Rolf, 48
"Knock-down," 566
Knudson, Alfred, 620
Koehn, Richard K., 667
Kornberg, Arthur, 47
Kornberg, Tom, 48
Kornberg enzyme. *See* DNA
　　polymerase I
Kozak, Barbara, 123
Kozak sequence, 123
K-*ras* oncogene, *614*
Kreitman, Martin, 676
Krüppel gene, 595, *596*

labial (lab) gene, 596–597
Labrador retrievers, coat color in,
　　351, *352*
lacA gene, 518–519, *518*
lacI gene, 519, 520–525, *521,* 528
　lacI⁻ mutants, 520, 523, *523–524*
　lacI⁻ᵈ mutants, 524
　lacI^Q mutants, 524
　lacI^S mutants, 523–524, *525*
　lacI^SQ mutants, 524–525
　mutations in, 519–525
　promoter region of, 526–527, *527*
lac operator, 519–520, *519–522,*
　　527, 528
　lacO^c mutants, 519, 521, *522,*
　　528, *528*
　mutations in, 519

lac operon, of *Escherichia coli*, 515–528, 516–528, 536
  cells grown in absence of lactose, 520, 522
  cells grown in presence of lactose, 521–522
  experimental evidence for regulation of *lac* genes, 517–520, 518–519
  lactose as carbon source, 516–517, 517
  molecular details of regulation, 526–528, 527–528
  mutations affecting regulation of gene expression, 519–520, 519
  mutations in protein-coding genes of, 518–519, 518
  negative control of, 520–525
  operon model for *lac* genes, 520–525, 520–525
  positive control of, 525–526, 526–527
  regulatory sequences of, 527–528, 528
lac promoter, 519–520, 522, 527, 528
  mutations in, 520
lac repressor, 515, 515, 519–525, 520–525, 527
  molecular model for, 521
  promoter region of gene for, 526–527, 527
lacY gene, 518–519, 518
lacZ gene, 181, 181, 518–519, 518
Lack, David, 689
Lactase deficiency, intestinal, 74
Lactose, as carbon source for *Escherichia coli*, 516–517, 517
Lactose permease, 517–518, 517, 521–522
Lamarckism, 135
Landsteiner, Karl, 340–341
Large-scale protein array, 264
Larva, 591, 592
Leader region, 528, 529, 531, 532
  evolution in, 708, 709
  trp mRNA, 530–532, 531–532
Leader sequence, 95, 545
Leber's hereditary optic neuropathy (LHON), 647
Leder, Philip, 98, 117
Lederberg, Esther, 483
Lederberg, Joshua, 483–484, 484, 494
Leigh's necrotizing encephalopathy, 74
Leishmania tarentolae, RNA editing in, 101, 101
Leptonema, 306
Lesch-Nyhan syndrome, 74–75, 75
Lethal allele, 339, 356–357, 363–364
  dominant, 356–357
  recessive, 356–357
  sex-linked, 357
Leucine, 113
Leucine aminopeptidase, in blue mussel, 667, 669–670, 670
Leucine zipper motif, 549, 550
Leukemia, 626
  chronic myelogenous, 462, 464, 610
Leukoplast, 640–641, 642
Lewis, Edward, 340, 504, 595
Lewontin, Richard, 671

lexA gene, 152
LFT lysate. *See* Low-frequency transducing lysate
LHON. *See* Leber's hereditary optic neuropathy
Life cycle
  of bacteriophage lambda, 493–494, 495
  of bacteriophage T2, 17, 17, 493, 493
  of *Chlamydomonas reinhardtii*, 431
  of *Neurospora crassa*, 69–70, 69, 431–432, 431, 641
  of retrovirus, 611, 612
  of *Saccharomyces cerevisiae*, 430, 430, 643
Li-Fraumeni syndrome, 619
Light chain, 582, 583
  constant region of, 582, 583
  J$_{kappa}$ segment of, 584
  kappa, 583, 583
  lambda, 583
  recombination in light chain genes, 583–584, 583
  variable region of, 582, 583
Light repair, 150
Lilium formosanum, C value of, 31
Limnaea peregra, shell coiling in, 650–651, 650–651
Linear tetrad, 429
LINEs (long-interspersed elements), 36, 38, 166
  in humans, 165
  L1 element, 165
  LINE-1 family, 36
Linkage, 406, 419
  genetic correlation and, 392–393
  human genes, 429
Linkage group, 406
Linkage map. *See* Gene mapping; Genetic map
Linked genes, 406
  in *Drosophila melanogaster*, 406–407, 407
Linker DNA, 32, 32
Little, Clarence Cook, 356
Liver cancer, 626
Liver cells, 472
Locus, 8, 213–214, 278
Locusta migratoria, genome of, 255
Lod score method, 444
Long interspersed elements. *See* LINEs
Long terminal repeat (LTR), 162–164, 163
Looped domain, of DNA, 29, 31, 33, 33
Loss-of-function mutation, 259, 281, 289
Loss of heterozygosity, 621
Low-frequency transducing (LFT) lysate, 499
LTR. *See* Long terminal repeat
Lundelius, J., 650
Luria, Salvador, 135
L virus, 649, 649
Lymphocytes, 581
Lymphoma, Burkitt, 462, 464, 610

Lyon, Mary, 324
Lyon hypothesis, 324
Lyonization, 324–325, 324
Lysin, 698
Lysine, 113
Lysine intolerance, 74
Lysogen, 497
Lysogenic cycle
  of bacteriophage, 493–494
  of bacteriophage lambda, 56–57, 493, 495, 534, 535
Lysogeny, 494, 495
Lysosomal storage disease, 75
Lysosome, 11, 75
Lytic cycle, 17
  of bacteriophage, 493, 493
  of bacteriophage lambda, 28, 28, 56–57, 56, 493–494, 495, 497–499, 498, 534–535, 535
  of bacteriophage T2, 17–18, 17

MacLeod, Colin M., 16
Major histocompatibility complex (MHC) genes, evolution of, 712, 712, 717
Malaria, sickle-cell anemia and, 693–694, 694
Male pseudohermaphroditism, 74
Male sterility
  cytoplasmic, 647–648, 648
  genetic engineering approach to, 648
  genic, 648
Malignant tumor, 607
Mammals
  cloning of, 577–579, 579
    problems with, 578–579, 579
  dosage compensation for X-linked genes, 586
  sex determination in, 320–325, 585–586
Mammogram, 606, 606
Map distance, 405, 417, 429–430
  calculation of, 418–419, 418
  from transduction experiments, 497
MAP kinase cascade, 617, 617
Maple sugar urine disease, 74
Mapping function, 405, 419, 419
Mapping panel, 444
Map unit, 8, 405, 411, 502
Marfan syndrome, 291
Margulis, Lynn, 722
Marker, 406
Marker-based mapping, identifying QTL, 395–397
Mate recognition, 698
Maternal age, trisomy-279 and, 469, 469
Maternal effect, 387, 631–632, 650–652
Maternal effect gene, in *Drosophila melanogaster*, 593–595, 594–595
Maternal inheritance, 631, 639, 651–652, 713
  exceptions to, 648–649
Maternal lineage, 635–636
MAT gene, 430–431, 430, 575

Mating
  assortative, 658, 695
  negative assortative, 658
  random, 664
Mating type
  in *Chlamydomonas reinhardtii*, 431
  in *Neurospora crassa*, 69–70, *69*,
    431–434, *431*, *433*
  in *Saccharomyces cerevisiae*, 326, 430,
    430, 548, 575
Mating-type switch, 548
Maximum likelihood approach, to
    phylogenetic tree reconstruction,
    705, 719–720, 725
Maximum parsimony, 718–719
McCarty, Maclyn, 16
McClintock, Barbara, 6, 158–161,
    160, 408
McClung, Clarence E., 313
McDonald-Kreitman test, 712
McGinnis, W., 595
M checkpoint, 607, *608*
McKusick, Victor A., 7
Mdm2 protein, 622–623, *622*
Mealworm, sex chromosomes of, 314
Mean, 378–379, *378*
Mean fitness of the population,
    690–691
Measles virus, C value of, *31*
Mechanical isolation, 698
Mediator Complex, 549, *550*
Megagametophyte, *313*
Meiocyte. *See* Spermatocyte, primary
Meiosis, 299, 306–312, *307*,
    310–312, 407
  in aneuploid, 468, *468*
  in animals, *307*, 310–311, *312*
  compared to mitosis, *310–311*
  gene segregation in, 309, *310*
  meiosis I, 306–309, *307*
    nondisjunction at, *319*, 467, 472
  meiosis II, *307*, 309
    nondisjunction at, *319*, 467, 472
  parallel behavior of genes and
    chromosome in, *322*
  in plants, 311–312, *313*
Meiospore, 306
Meiotic tetrad, 430. *See also* Tetrad
    analysis
Melanin, 75, 289
Melanism, industrial, 687–689, *688*
Melanoma, *619*, 626
Mello, C., 565
Mendel, Gregor Johann, 2, 272
  experiments with garden pea, 3–4,
    3, 272–286
  portrait of, *273*
  rediscovery of Mendel's principles, 286
Mendelian genetics, 271–291
  extensions of, 339–364
  in humans, 288–291
Mendelian population, 658. *See also*
    Population
Mendelian ratio, modified, 348–356, 363
Mendel's first law, 3–4, 271, 275–282,
    275–281, 320

Mendel's second law, 4, 271, 282–286,
    282–286, 320, *322*
Meningioma, *619*
Mental retardation, 363
Mereschkovsky, G., 722
Merodiploid, 487
MERRF disease, 647
Meselson, Matthew, 45–47, *46*
Meselson-Stahl experiment, 45–47, *46*
Messenger RNA (mRNA), 4–5
  antisense, 236
  central dogma, 88
  construction of cDNA library, 175,
    184–187, *186–187*
  decapping of, 564
  of eukaryotes, 95–102, *96*
    production of mature mRNA, 97–99,
    97–100
  genetic code, 115–119, *118*
  gradients in developing *Drosophila*,
    594–595
  northern blot analysis of, 197
  polycistronic, 97, 515–516, 518–521,
    518, *521*
  precursor. *See* Precursor mRNA
  processing of pre-mRNA to mature
    mRNA, 98–99, *99–100*, 107
  of prokaryotes, 96, *96*
  purification from cellular RNAs, 186
  RNA editing, 101–102, *101*
  splicing of. *See* Precursor mRNA
  stability of, 544, 564, *564*
  stored, inactive, 563
  structure of, 26–27
  synthesis of, 87–88. *See also*
    Transcription
    in eukaryotes, 93
    in mitochondria, 633–635
  synthetic, 117
  trailer sequence of, 96
  transcriptome, 260–261, *262*
  in translation, 119–126, *120–126*.
    See also Translation
  3′ untranslated region of, 96, 563
  5′ untranslated region of, 95, 96
Messenger RNA (mRNA) degradation
    control, 546, 564, *564*, 567
  deadenylation-dependent decay
    pathway, 564
  deadenylation-independent decay
    pathway, 564
Messenger RNA (mRNA) translation
    control, 544, 546, *546*, 563
Messenger RNA (mRNA) transport
    control, *546*, 563, 567
Metabolic pathway. *See* Biochemical
    pathway
Metabolism, inborn errors of, 68, 82
Metacentric chromosome, 301, *301*, 305
Metamorphosis, in frogs, *391*, *393*
Metaphase
  meiosis I, *307*, 308–309, *310*
  meiosis II, *307*, 309, *311*
  mitotic, 302, *303–305*, 304–305, *310*
Metaphase chromosome, *305*
Metaphase plate, 304, 308–309

Metastasis, 607, *625*
Methanococcus jannaschii
  chromosomes of, 28
  C value of, *31*
  genome sequence of, 251–252, 482
Methanosarcina acetivorans, genome of,
    255, 256
Methionine, *113*
  biosynthetic pathway for, 72, *72*
Methylation
  of DNA, 544, 557–561, *559*, 625,
    625, 646
  of histones, 557, 586
5-Methylcytosine, 557, *559*
  deamination of, 141, *143*
Methyl-directed mismatch repair,
    151–152, *152*
O^6-Methylguanine, *145*, 146
O^6-Methylguanine methyltransferase, 150
Methylguanosine, 97, *106*
Methylinosine, *106*
Methylmethane sulfonate (MMS),
    145, 146
MHC genes. *See* Major histocompatibility
    complex genes
Microgametophyte, *313*
Microsatellite. *See* Short tandem repeat
Microtus ochrogaster. *See* Prairie vole
Miescher, Friedrich, 14
MIG1 gene, 553
Migration, 657–658, 685–687, *686*, 699
  balance between migration and
    genetic drift, 686
Milk
  recombinant proteins secreted into,
    232, *232*
  yield in cattle, 360, 386, *391*, *393*, 393
Milkweed beetle, phosphoglucomutase
    of, 661
Milkweed bug
  fecundity in, *391*, *393*
  wing length in, *391*, *393*
Mimulus lewisii. *See* Monkeyflower
Minimal medium, 70, 483
Minimal transcription initiation
    complex, 95
Minisatellite. *See* Variable-number
    tandem repeat
−10 box, 89
Mirabilis jalapa. *See* Four o'clock
Mismatch repair, 152, 624
  by DNA polymerase proofreading, 150
  methyl-directed, 151–152, *152*
Missense mutation, 136, *137*, 214
"Missing link," 719
Mitchell, Herschel, 641
Mitchell, Mary, 641
Mitochondria, 9, 11, 631
  DNA of, 3, 631–632, *633*, 651–652
    defects in human genetic diseases, 647
    evolution of, 713, *713*
    exceptions to maternal inheritance,
    648–649
    genetic map of, 633–634, *634*
    investigating genetic relationships by
    mtDNA analysis, 635–636

mutations in, 647
nucleotide heterozygosity in, *675*
in pocket gophers, 713, *713*
polymorphisms in, 635
in primates, 722
replication of, 633, *634*, 713
transcription of, 633–635
functions of, 632–633
genetic code in, 118, 631, 635
genome of, 632–636, 651
cytoplasmic male sterility, 648
gene content, 633–635
size of, 633
structure of, 633
origin of, 632, 722
petite mutants of *Saccharomyces cerevisiae*, 643–645, *644*
[*poky*] mutant of *Neurospora*, 641–643, *643*
structure of, 632
translation in, 635, *636*
"Mitochondrial Eve," 723
Mitosis, 299, 302–306, *302–306*
compared to meiosis, *310–311*
crossing-over in, 429, 437–443
gene segregation in, 305–306
Mitotic recombination, 437–443
in *Aspergillus nidulans*, 438–442, *440–442*
discovery of, 437–438, *438–439*
in *Drosophila melanogaster*, 437–438, *438–439*
in human tumor, 442–443
Mitotic spindle, *303*, 304, 650
Mito Tracker Red, *631*
mle (*maleness*) gene, 591
MMS. *See* Methylmethane sulfonate
M-N blood group, 346–347, 664
Model organisms, 6, 9, *10*
Moderately repetitive DNA, 36
Modern genetics, 2
mof (*males absent on the first*) gene, 591
Molecular chaperone. *See* Chaperone
Molecular clock, 705, 714–715, *714–715*
causes of variation in rates, 715
relative rate test, 714–715, *715*
Molecular cloning, 176
Molecular evolution, 705–725. *See also* Nucleotide substitution
acquisition and origins of new functions, 723–725
comparative genomics, 264–265, 709–710
definition of, 706
molecular clocks, 714–715, *714–715*
molecular phylogeny, 715–723
neutral theory of, 684
patterns and modes of substitutions, 706–715
rates of, 724
variation in rates between genes, 711–712, *712*
variation in rates within genes, 708
Molecular genetics, 1, 6, 658
Molecular marker. *See* DNA marker

Molecular phylogeny, 715–723, 725
Molecular testing. *See* DNA molecular testing
Monarch butterfly, gene flow in, 686–687, *686*
Monkeyflower, floral traits in, 395–397, *395–396*
Monocot, 234
Monod, Jacques, 5, 518, 520–525, *520–525*
Monoecious plant, 326
Monohybrid cross, 275–282, 286
Monolayer, 607
Monomer, 20
Monoploidy, 453, 472, *472*
Monosomy, 467–468, *467*
double monosomic, *467*, 468
Morgan, Thomas Hunt, 314, 344, 357, 406–407, *407*, 411, 458
Morphogen, gradients in developing *Drosophila*, 594–595
Morphogenesis, 573, 575
Morton, Newton, 444
Mosaic
genetic, 324
twin spots in *Drosophila melanogaster*, 437–438, *438–439*
Mosquito, chromosome number in, *313*
Mouse
body color in, 356–357, *357*
body weight in, *391*, *393*
chromosome number in, *313*
coat color in, 350–351, *352*
C value of, *31*
development in, 576
esterase in, 684
genome analysis in, *256*
genome sequence of, 244, 254, *255*
hemoglobin in, 684
homeotic genes of, *600*
knockout, 213, 623
as model organism for research, 9, *10*, 576
sex determination in, 585–586
site-specific mutagenesis in, 212–213
spontaneous mutation frequency at specific loci, *678*
tail length in, *393*
TP53 gene in, 213, 623
Mouse mammary tumor virus, 610
M phase, 30, 58, 302, *302*
M protein. *See* Lactose permease
mRNA. *See* Messenger RNA
msl (*male-specific lethal*) gene, 591
*Msp*I, 558–559, *559*
mtDNA. *See* Mitochondria, DNA of
Müller, Hermann Joseph, 141
Mullis, Kary, 2, 200
Multifactorial trait, 375
Multigene family, 458, 705, 712, 723–725, 724
Multiple alleles, 341–346, *341–346*, 363
allelic frequency with, 661, 667
in genetic diseases, 345
relating to molecular genetics, 344–345

Multiple allelic series, 341
Multiple cloning site. *See* Polylinker
Multiple crossovers, 419
Multiplex PCR, 221–222
Murine sarcoma virus, *614*
Muscular dystrophy
Becker, 74
Duchenne, *74*, 219, 327, 360, 445
Mus musculus. *See* Mouse
Mutable allele, 159
Mutagen, 115, 133, 143–147, *143–146*, 609
base analog, 143, *144*
in environment, 147–148
Mutagenesis, 140
insertional, 165, 618
site-specific, 146–147, 211–213, *213*
Mutant, 2
nutritional, 69–70, *71*, 148, *149*
temperature-sensitive, 48
Mutant allele, 315, 317, 341, *341*
Mutation, 5, 133, 166, 657, 705–706
adaptation versus, 134–136, *135*
advantageous, 677
alterations in allelic frequencies, 677–679, *678*, 699
balance between mutation and genetic drift, 684–685, *685*
balance between mutation and selection, 694–695
chromosomal. *See* Chromosomal mutation
compared to nucleotide substitution, 709
complementation test, 340–341, *340–341*
consequence to organism, 133
definitions of, 136–140, *137*
detection of, 148–149, *149*
detrimental, 677
gain-of-function, 290
gene, 134
heteroallelic, 502
homeotic, 595
homoallelic, 502
induced, 5
loss-of-function, 259, 281, 289
neutral, 677
spontaneous, 5
transposition-related, 157–158
unit of, 503
Mutation frequency, 136
Mutation rate, 136, 140, 677, 678
Mutator gene, 157–158, 605–606, 624, 627
cancer and, 152, 610
Mutator mutation, 150
mut genes, 150–152, *152*, 624
M virus, 649, *649*
myb oncogene, *614*
myc oncogene, 464, *614*, 618
Mycoplasma genitalium, genome of, 255, *256*
Myoclonic epilepsy with ragged-red fiber (MERRF) disease, 647
Myotonic dystrophy, 466, 678

NADH-dehydrogenase, 633
nanos (*nos*) gene, 594–595
Narborough murders, 226, *226*
Narrow-sense heritability, 373,
    387–388, 391
Nathans, Daniel, 177
National Center for Biotechnology
    Information (NCBI), 7
Natural selection, 390, 657–658,
    687–694
  against recessive trait, 691–693,
    692–693
  balance between mutation and
    selection, 694–695
  definition of, 687
  directional, 690
  effect on allelic frequencies, 689–691,
    690–691, 699
  fitness and coefficient of selection,
    689, *689*
  heterozygote superiority, 693–694, *694*
  in natural populations, 687–689, *688*
  response to, 373, 390–394
    estimation of, 390–392, *392*
  selection coefficient, 689
Natural transformation, 490
Nature-versus-nurture debate,
    362–363, 377
NCBI. *See* National Center for
    Biotechnology Information
Neel, J.V., 77
Negative assortative mating, 695
Negative correlation, 380, *382*
Neisseria meningitidis, C value of, *31*
Nematode. See also *Caenorhabditis elegans*
  chromosome number in, *313*
N-end rule, 565
Neo-Darwinian theory, *659*
Neomycin, inhibition of mitochondrial
    protein synthesis, 635
Neoplasm, 606
Nephroblastoma, *619*
Neufield, Peter, 227
Neurofibroma, 359, *359*, 619
Neurofibromatosis, 165, 359, *359*,
    619, 678
Neurospora crassa
  Beadle and Tatum experiments with,
    69–72, *71–72*
  calculating gene-centromere distance
    in, 432–434, *433*
  chromosome number in, *313*
  genome analysis in, *256*
  life cycle of, 69–70, *69*, 431–432,
    431, 641
  mating type in, 69–70, *69*, 431–434,
    431, *433*
  as model organism for research, 9, *10*
  nutritional mutants of, 69–70, *71*
  [*poky*] mutant of, 641–643, *643*
  random-spore analysis in, 432
  spontaneous mutation frequency at
    specific loci, *678*
  tetrad analysis in, 430–437
Neurotoxin resistance, in garter snakes,
    393, *394*

Neutral mutation, 136–137
Neutral mutation model, for genetic
    variation, 672–673
Neutral *petite*, 644–645, *644*
Neutral theory of molecular
    evolution, 684
Newborn screening, 219
NF genes, *619*
N gene, 534, *535*
Nicholas, Tzar of Russia, 635–636
Niemann-Pick disease, *74*
Nilsson-Ehle, Hermann, 375–376
Nirenberg, Marshall, 117
Nitrogenous base, 2, 13, 20, *20*
  base analogs, 47, *47*, 143, *144*
  depurination and deamination of,
    140–141
  tautomeric forms of, 140, *141*
Nitrosamine, 626
Nitrous acid, 144, *145*
Noah's ark blot, 224
Noller, Harry, 123
Nonautonomous element, 158–159, 165
Non-coding single nucleotide
    polymorphism, 214
Noncomposite transposon, 156, *157*
Noncontributing allele, 376
Nondisjunction, 318
  at meiosis I, *319*, 467, 472
  at meiosis II, *319*, 467, 472
  primary, 318, *320*
  secondary, 319–320, *321*
  of X chromosome, 317–320, *319–320*
Nonhistone chromosomal protein,
    31–33, 547
Nonhomologous chromosomes, 300
Nonhomologous recombination, 153
Non-Mendelian inheritance, 631–652
  contrasts to, 650–651
  examples of, 639–640
  rules of, 639
Nononcogenic retrovirus, 611, *613*
Nonparental ditype tetrad, 434–437,
    434–436
Nonpermissive host, 499
Nonreciprocal translocation, *461*, 462
Nonsense codon. *See* Stop codon
Nonsense mutation, 136, *137–138*
  polar effects of, 518–519, *518*
Nonsense suppressor, 139–140, *139*
Nonsynonymous site, 676, 708–709,
    708, 712–713, *712*
Nontranscribed spacer sequence, 104, *104*
Normal distribution, 378–379, *378*
Normal transmitting male, 465
Norm of reaction, 363, 374–375
Northern blot analysis, 197, 224
*Not*I, 177, *178*, 246
NS gene, influenza virus, 712
Nuclear division. *See* Mitosis
Nuclear envelope, 9, *11*, *303*, 305,
    307, 309
  mRNA transport control, 563
Nuclear *petite*, 643–644, *644*
Nuclear pore, *11*, 563
Nuclease, 16

Nucleic acid, 14. *See also* DNA; RNA
Nuclein, 14
Nucleoid, 28, 637
Nucleoid region, *11*, 631, 633
Nucleolus, 9, *11*, *303*, 305
Nucleoside, 21, *21–22*
Nucleoside phosphate, 21
Nucleosome, 32, *32–33*, 547
  assembly of, 61–62, *61*
Nucleosome remodeling complex,
    547–548, *548*
Nucleotide, 2, 13, 20–21, *21–22*
Nucleotide excision repair, 150–151, *151*
Nucleotide heterozygosity, 675, *675*
Nucleotide substitution, 706–714. *See
    also* Base-pair substitution
  compared to mutation, 709
  Jukes-Cantor model of, 707–708, *707*
  in mtDNA, 713, *713*
  multiple substitutions at one site,
    707–708, *707*
  rates of, 708
    codon usage bias, 710–711
    in flanking regions, *708*, 709
    in pseudogenes, *708*, 709
    synonymous and nonsynonymous
      sites, 708–709, *708*
    variation in evolutionary rates
      between genes, 711–712, *712*
    variation in evolutionary rates
      within genes, 709
  sequence alignments, 707
  substitutions in protein and DNA
    sequences, 706–707
Nucleus, cell, 9, *11*
Null allele, 259
Null hypothesis, 286–287
Nullisomy, 467, *467*
Null mutation, 158
Nutritional mutant, 148, *149*, 483.
    See also Auxotroph
  of *Neurospora crassa*, 69–70, *71*

Observation, 6
O gene, 534, *535*
Okazaki, Reiji, 52
Okazaki, Tuneko, 52
Okazaki fragment, 52–53, *53–54*
Oligo(dT) primer, 186, *186*
Oligomer, 24
Oligonucleotide array, 216
Oligonucleotide primer, for PCR,
    200–202, *202*, 248
Oligonucleotide probe, identifying
    genes or cDNA in libraries,
    190–192
Oliver, C.P., 501
OMIM (Online Mendelian Inheritance
    in Man), 7
Oncogene, 464, 605–606, 610–618
  cellular, 615, *615*
  changing cellular proto-oncogenes
    into oncogenes, 617–618
  retroviruses and, 610–618
  viral, 605, 610, 614, *614–615*
Oncogenesis, 607

Oncogenic retrovirus, 611, *612*, 614
One gene-one enzyme hypothesis, 4, 67, 69–72, 81
One gene-one polypeptide hypothesis, 4, 67, 73, 81
On the Origin of Species (Darwin), 687
Oocyte, 308
  mRNA stored in, 563
  primary, 311, *312*
  secondary, 311, *312*
Oogenesis, 311, *312*
Oogonia, *312*
  primary, 311
  secondary, 311
Open-angle glaucoma, 221, *221*
Open promoter complex, 90, *91*
Open reading frame (ORF), 119, 251, 258, *259*
  analysis of DNA sequences, 199–200
  of mtDNA, 634
  searching genome for, 258, *259*
Operator, 515, 519, 536
Operon, 5, 515–516, 536, 544
  in eukaryotes, 544–545, *545*
  Jacob and Monod model for *lac* genes, 520–525, *520–525*
  repressible, 529
Optimal alignment, 707
Oral contraceptives, 469
Orange bread mold. See *Neurospora crassa*
ORC. *See* Origin recognition complex
Ordered tetrad, 432
ORF. *See* Open reading frame
Organellar genes, 632
Origin, on *F* factor, 485
Origin of replication, 51, 57
Origin recognition complex (ORC), 58
Orotic aciduria, 74
Orphan gene family, 258–259, *259*
Oryza sativa. See Rice
Osteogenesis imperfecta, 359, *678*
Osteoporosis, 360
Osteosarcoma, *631*
Ostrander, Elaine, 723
Outgroup, 715, *715*
Out-of-Africa theory, 722–723
Ovarian cancer, 219, 261, *619*, 623
Overdominance, 693
Ovum, 311, *312*
Oxygen free radicals, 713

p14 protein, *623*
p16 gene, *619*
p21 protein, *622*, 623
p53 protein, 622–623, *622*
  function of, 622–623, *622*
PABP. *See* Poly(A) binding protein
Pace, Norm, 721
Pachynema, 308
Pair-rule genes, 595, *596*
Palomino horse, *339*, 346
PAN1 gene, 564
Panaxia dominula. See Scarlet tiger moth
PAR. *See* Pseudoautosomal region
Paracentric inversion, 459, *459*

Paramecium, as model organism for research, 9, *10*
Parasegment, in *Drosophila* development, 593, *593*
Parasexual system, 442–443
Parental, 406, *407*
Parental class. *See* Parental
Parental ditype tetrad, 434–437, *434–436*
Parental generation. *See* P generation
Parental genotype. *See* Parental
Parent-offspring regression, heritability from, 389–390, *390*
Parsimony approach, to phylogenetic tree reconstruction, 705, 718–719, *719*, 725
Partial diploid, 519
Partial dominance. *See* Incomplete dominance
Partial reversion, 138
Particulate factors, 276
Partitioning, of variance, 383
Patau syndrome. *See* Trisomy-13
Paternal inheritance, of plastids in conifers, 649
Paternity case, DNA typing in, 224–225, *225*
  exclusion result, 225
  inclusion result, 225
Pattern baldness, 360, *361*
Pauling, Linus, 77, 714
Pause signal, transcription, 530
PAX6 gene, 465
PCR. *See* Polymerase chain reaction
Pearson, Karl, 375
Pedigree analysis, 271, 288–289, *289*, 429, 443
  dominant trait, 290, *290*
  for genetic counseling, 80
  recessive trait, 289–290, *290*
  sex-linked trait, 326
  symbols used in, 289, *289*
  X-linked dominant trait, *329*
  X-linked recessive trait, 328
P element, in *Drosophila melanogaster*, 164, *164*
Pelger anomaly, *678*
Penetrance, 339, 358–359, *358*, 364, 374
  complete, 358
  incomplete, 358–359, *358*
Penn, Keone, 231
Pentose, 20, *20*
Peppered moth
  carbonaria phenotype in, 688–689, *688*
  industrial melanism in, 687–689, *687*
  insularia phenotype in, 688–689, *688*
  typical phenotype in, 688–689, *688*
Peptide bond, 111–112, *114*
  formation of, 123, *124–125*
Peptidyl transferase, 123, *124–125*, 126
Peptidyl-tRNA, 123, *124*
Perfect flower, 326
Pericentric inversion, 459, *459*
Permissive host, 499
Peroxin, 230

Peroxisome, *11*, 230
Perutz, Max, 115
petite mutant, of *Saccharomyces cerevisiae*, 643–645, *644*
  characteristics of, 643
  neutral *petite*, 644–645, *644*
  nuclear *petite*, 643–644, *644*
  suppressive *petite*, 644–645, *644*
PEX genes, 230
P gene, 534, *535*
P generation, 275, *275–277*
Phage. *See* Bacteriophage
Phage ghost, 18
Phage lysate, 17, 493
Phage vector, 492
Pharmacogenomics, 261–263
Pharming, 236
Phaseolus vulgaris. See Bean
Phenocopy, induced by chemicals, 361–362
Phenocopying agent, 362
Phenotype, 3, 271–272, *272*, 278
  of continuous trait, 374
  determining evolutionary relationships from, 716
  environmental effect on, 272, *272*
  epistasis and, 349–355, *352–356*
  gene interactions, 348–356
  producing new phenotypes, 348–349, *349–351*
Phenotypic correlation, 392
Phenotypic ratio, 4
  3:1, 276–280
  9:7, 353, 356
  15:1, 355, 375–376
  9:3:4, 350–351, *352*
  12:3:1, 351–352
  1:1:1:1, 284
  9:3:1:1, 282, 284
  9:3:3:1, 348
  27:9:9:9:3:3:3:1, 285, *285*
Phenotypic rule, 286, *286*
Phenotypic structure, of population, 375
Phenotypic variance, 385, 397
  components of, 385–387, *385*
Phenotypic variation, 272, *272*
  environmental contribution to, 339, 373, 377, 386
  genetic contribution to, 339, 373, 377, 386
Phenylalanine, *113*
Phenylalanine hydroxylase, 73, 220
Phenylalanine tRNA, of yeast, *106*
Phenylalanine-tyrosine metabolic pathway, 68, *68*
Phenylketonuria (PKU), 73–75, *74*, 360–361, 363
  DNA molecular testing, 220–221
  newborn testing for, 73, 219
Philadelphia chromosome, 464
Phocomelia, 362
α-Phosphate, 191
Phosphate group, 13, 20
Phosphodiesterase, 526, 527
Phosphodiester bond, 21, *21*, 48, *90*
  2′-5′ bond, 99

Phosphoglucomutase, of milkweed beetles, 661
Photolyase, 150
Photoreactivation repair, 150
Photosynthesis, 636–637
Phototaxis, in *Drosophila pseudoobscura*, 392, *392*
phr gene, 150
Phylogenetic relationships, 706, 715–723
Phylogenetic tree, 705, 716–718
  branches of, 716
  gene versus species trees, 716–718
  on grand scale, 721–723, *721*
  inferred, 717
  nodes of, 716
  number of possible trees, 716, 717
  reconstruction methods
    bootstrapping, 720–721
    distance matrix approach, 718, 725
    maximum likelihood approach, 719–720, 725
    parsimony approach, 718–719, *719*, 725
  rooted, 716, 717
  tree of life, 721–722, *721*
  unrooted, 716, 717
Phylogeny, molecular, 715–723, 725
Physical mapping
  deletion mapping in *Drosophila melanogaster*, 456–457, *456*
  of genome, 246–247, *248*
  of human genes, 246, 445–446, *445–446*
PIC. *See* Preinitiation complex
Pig, back-fat thickness in, *391*
Pistil, 311, *313*
Pisum sativum. *See* Garden pea
Pitchfork, Colin, 226
PKU. *See* Phenylketonuria
Plant. *See also* specific plants
  cytoplasmic male sterility in, 647–648
  dioecious, 326
  edible vaccines from, 236
  genetic engineering of, 233–236
    applications of, 234–236, *235*
  genome analysis in, 256
  genome sequence of, 253–254
  herbicide tolerance in, 234, *235*
  homeotic genes in, 597
  hormonal control of gene expression in, 556–558, *557*
  meiosis in, 311–312, *313*
  monoecious, 326
  polyploidy in, 472–473
  sex chromosomes of, 326
  transformation of plant cells, 233–234, *234*
Plant breeding, 2, 6–7, 390
Plant cell, 9, *11*
  cytokinesis in, 305, *306*
Plaque, phage, 493, *494*, 499–501, *500–501*
Plasma cells, 582
Plasma membrane, 9, *11*

Plasmid, 28, 485
  cloning vectors, 175, 179–182, *180–181*
    recircularization of, 181–182
  pUC19, 180–182, *180–181*, 195
Plasmodesmata, *11*
Platelet-derived growth factor, 233, 616, *616*
Pleiotropy, 73, 374, 392, 689
Plumage color, in chickens, 346, *347*
Poaching investigation, 227
Pocket gopher, mtDNA in, 713, *713*
Pod traits, in garden pea, 274, *274*, 277
Point mutation, 133–134, 503
  changing cellular proto-oncogenes into oncogenes, 617
  types of, 136–138, *137*
[*poky*] mutant, of *Neurospora crassa*, 641–643, *643*
Polar body, 459
  first, 311, *312*
  second, 311, *312*
Polar cytoplasm, in *Drosophila* development, 592–593, *592*
Polarity
  of DNA, 21, *21*
  of nonsense mutations, 518–519, *518*, *521*
Polar mutation, 518–519, *518*
pol genes, 48, 50, 611, *612–613*, 614
Poliovirus, 19
Poly(A), 117
Poly(A) binding protein (PABP), 123, 564
Poly(AC), 117
Poly(A) polymerase, 97, *98*
Poly(A) site, 97
Poly(A) tail, of mRNA, 97, *98*, 123, 186, *186*, 545, *545*, 561–563, *562*
Poly(C), 117
Polycistronic mRNA, 97, 515–516, 518–521, *518*
Polycyclic aromatic hydrocarbon, 626
Polygalacturonase, 236
Polygene, 376
Polygene hypothesis, for quantitative inheritance, 375–376
Polygenic inheritance, 373
Polylinker, *180–181*, 181–182
Polymerase chain reaction (PCR), 2, 176, 200–204, *201*
  advantages and limitations of, 202
  applications of, 202–203, 211–236
  in DNA molecular testing, 221–222, *221*
  multiplex PCR, 221–222
  PCR-RFLP analysis method, 215, *215*
  procedure for, 200–202, *201*
  real-time RT-PCR, 203
  reverse-transcriptase PCR, 203
Polymorphic loci, 672, 699
  proportion of, 671–672
Polymorphism, DNA. *See* DNA polymorphism
Polynucleotide, 21, *21*
Polyp, colonic, 625, *625*
Polypeptide, 111–112
Polypeptide hormone, mechanism of action of, *553*, 554

Polyploidy, 453, 472–473, *472*
  in animals, 472–473
  with even number of chromosome sets, 472
  with odd number of chromosome sets, 472
  in plants, 472–473
Polyribosome, 125, *125*
Polysome. *See* Polyribosome
Polytene chromosome, 454, *455*, 580–581, *581*
  during development in Diptera, 580–581, *581*
Poly(U), 117
Population, 377
  allelic frequencies in. *See* Allelic frequency
  genetic divergence among, 696–697
  genetic structure of, 658–663
    variation in space and time, 669–670, *670*
  genetic variation in, 670–676
  genotype frequencies in, 659–660
Population genetics, 1, 6, 657–699
  questions studied in, 659
Population size, 681, *681*, 684, *684*
  effective, 680–681
  infinite, 664
Population viability analysis, 697
Porphyria
  acute intermittent, 74
  congenital erythropoietic, 74
Positional cloning, 222, 619, 622
Position effect, 465
  telomere, 557
Positive assortative mating, 695
Positive correlation, 380, *382*
Posttranscriptional control, 561–565, 567
Postzygotic barrier, 698
Postzygotic isolation, 698
Potato, chromosome number in, *313*
Pott, Sir Percival, 626
Poultry. *See also* Chicken
  body weight in, *391*, 393, *393*
  egg production in, *391*, *393*
  egg weight in, *391*, *393*
Prader-Willi syndrome, 457, 561
Prairie vole, esterase 4F in, *671*
Precursor mRNA (pre-mRNA), 87, 94
  alternative polyadenylation sites, 561–562, *562*
  alternative splicing of, 228–229, *229*, 588
  5′ capping of, 97, *97*
  introns of, 98–99, *99–100*
  poly(A) tail of, 97, *98*, 123, 186, *186*, 545, *545*, 561–563, *562*
  processing to mature mRNA, 98–99, *99–100*, 107
  self-splicing introns, 88, 100, *101*
  trans-splicing of, 545, *545*
Precursor rRNA (pre-rRNA), 88, 103–104
Precursor tRNA (pre-tRNA), 105
Prediction, 6
Preinitiation complex (PIC), 94, *95*
Premutation, 465–466

Prenatal diagnosis, 80–81, *81*, 219
Preparative agarose gel
  electrophoresis, 194
Prereplicative complex, 58
Prezygotic barrier, 698
Prezygotic isolation, 698
Pribnow, David, 89
Pribnow box, 89
Primary nondisjunction, 318, *320*
Primary structure, of proteins, 111–112,
  706–707, 711–712, *711*
Primate, mtDNA of, 722
Primosome, 51
Principle of uniformity in F₁, 275
Privacy, genetic, 264–265
Probability (P), 279, 287
Proband, 288
Probe array, 216, 260–261, *262*
proboscipedia (*Pb*) gene, 596–597
Procarcinogen, 626
Product rule, 279
Proflavin, 115–116
Progesterone, 554
Programmed cell death, 623
Prokaryote, 1, 9, 11
  chromosomes of, 28–29, *29*, 36
  gene density in, 255, *256*
  genome sequences of, 251–252, *252*
  genome size in, 255, *256*
  mRNA of, 96, *96*
  regulation of gene expression in,
    515–533, 544
  replication in, 43, 62
  ribosomes of, 88, 102
  rRNA of, 103, *103*
  transcription in, 87–92
  translation in, 111
  transposable elements in, 134, 153,
    155–157, *155–157*
Prokaryotic cell, *11*
Prolactin gene, *712*
Proline, *113*
Prometaphase, 304
Promoter, 89–92, *89–91*, 94, 526–527,
  527, 536, 543, 551
  core, 94
  strength of, 92
Promoter complex
  closed, 90, *91*
  open, 90, *91*
Promoter proximal element, 94
Proofreading activities
  of DNA polymerase, 150
  in replication, 50, 62
  of RNA polymerase, 92
Prophage, 493–494, 497, 533
Prophase
  meiosis I, 306, *307*, *310*, 407
  meiosis II, *307*, 309
  mitotic, 302, *303–304*, 304, *310*
Proportion of polymorphic loci,
  671–672
Proposita, 288
Propositus, 288
Protein
  conformation of, 115
  C-terminal end of, 112, *114*

domain shuffling in, 724–725
electrophoresis of, finding proportion of
  polymorphic loci, 672–673, *673*, 699
functional domains of, 706
gene control of protein synthesis,
  76–79, *77–80*
genetic variation at protein level,
  671–673, *672–673*
homologous, 706
isoforms of, 561
as molecular clock, 714–715, *714–715*
molecular evolution, 705–725
N-terminal end of, 112, *114*
primary structure of, 111–112,
  706–707, 711–712, *711*
proteome, 263–264
quaternary structure of, 115
secondary structure of, 112, 114, *114*
sorting in cell, 111, 127, *127*
structure of, 111–113, *112–114*
synthesis of. *See* Translation
tertiary structure of, 114–115, *114*
ubiquitination of, 564–565
Protein array, 263–264
  large-scale, 264
Protein chip. *See* Protein array
Protein-coding gene, 88
  mitochondrial, 633–634
  mutation in, 134, *134*
  transcription in eukaryotes, 93–102
  translation of, 111–128
Protein degradation control, *546*, 564–565
Protein kinase, proto-oncogene
  products, 616, *616*
Protein microarray. *See* Protein array
Protein-protein interactions
  analysis of, 229–231, *230*
  yeast two-hybrid system, 229–230, *230*
Proteolysis, 564
Proteome, 243, 260, 263–265
Proto-oncogene, 464, 605, 614–618,
  615–617, 627
  changing cellular proto-oncogenes
    into oncogenes, 617–618
  protein products of, 615–617, *616*
Protoperithecia, 641
Prototroph, 70, 483
Protozoa, genome analysis in, *256*
Proviral DNA, 611, *612–613*
Provirus, 605
Pseudoautosomal region (PAR),
  Y chromosome, 308
Pseudodominance, 456
Pseudogene, evolution in, *708*, 709
Pseudouridine, *106*
*Pst*I, 178–179, *179*
Puberty, female, 308
PubMed, 7
Pufferfish, 257
  genome of, *256*, 257
Punnett, R., 276
Punnett square, 276, *277*
  for dihybrid cross, 282–283, *282–283*
Pupa, 591, *592*
Pure-breeding strain, 274, 276
Purine, 20–21, *20*, 22, 23, *23*
  depuration, 140–141

Pyridoxine dependency with seizures, 74
Pyrimidine, 20–21, *20*, 22, 23, *23*
  thymine dimer, 142–143, *143*, 150

Q gene, 534, *535*
QTL. *See* Quantitative trait loci
Quantitative genetics, 1, 6, 373–397, 658
  polygene hypothesis for, 375–376
  questions studied in, 375
  statistical tools for, 377–383
Quantitative trait, 374
Quantitative trait loci (QTL), 373,
  394–397, *395–396*
  cloning of, 397
  in humans, 397
  marker-based mapping of, 395–397
Quaternary structure, of proteins, 115
Quenching, 553

Rabbit
  fur color in, 360, *362*
  Himalayan, 360, *362*
Radiation
  as carcinogen, 626
  induction of mutations by, 133,
    141–143, *143*
Radiation hybrid (RH), 446, *446*
  radiation hybrid map, 246–247
Radioactive labeling, of DNA, 191, *191*
Radiologist, 626
Raf-1 protein, 617, *617*
raf oncogene, *614*
Raly gene, 356–357
Randolph, Lowell, 160
Random copolymer, 117
Random genetic drift. *See* Genetic drift
Random mating, 664
Random-primer method, radioactive
  labeling of DNA, 191, *191*
Random-spore analysis, 432
RAP1 gene, 557
Raphanobrassica, 473
ras oncogene, 617, 625, *625*
Ras protein, 617, *617*
Rat
  C value of, *31*
  genome analysis in, *256*
  genome sequence of, 254, 255
RB gene, 442–443, 619–622,
  619–622
  function of, 621–622, *621*
RBS. *See* Ribosome-binding site
Reading frame, 116, 137–138
  open. *See* Open reading frame
Real-time reverse transcriptase PCR, 203
recA gene, 152
Recessive epistasis, 350–353, *352*
Recessive trait, 4, 276, 278, 281
  autosomal, 300
  complete recessiveness, 346
  general characteristics of, 290
  in humans, 289–290, *290*
  lethal, 356–357
  pedigree analysis of, 289–290, *290*
  selection against, 691–693,
    692–693
  X-linked, 300, 327, *328*

Reciprocal cross, 275–276, 316, *318*
Reciprocal translocation, *461*, 462
Recombinant, 405–406, *407*
Recombinant chromosome, 308
Recombinant DNA, 2, 176
Recombinant DNA library
  finding specific clone in, 187–192,
    188–189
  heterologous probes to identify DNA
    sequences, 190
  identifying genes by complementation
    test, 190, *192*
  oligonucleotide probes to identify
    genes of cDNA, 190–192
  preparation of, 183–187
Recombinant DNA molecule, 175
Recombinant DNA technology, 6–7, *7*
  applications of, 211–236
  cloning, 176–183
  DNA sequencing, 197–200, *198–200*
  finding specific clone in DNA library,
    187–192
  molecular analysis of cloned DNA,
    192–197
  polymerase chain reaction, 200–203, *201*
  recombinant DNA libraries, 183–187
  techniques in, 175–204
Recombinant vaccine, 233
Recombination, 5, 308, 405–406
  association with chromosomal
    exchange, 408–410, *408–409*
  mitotic. *See* Mitotic recombination
  nonhomologous, 153
  somatic, 583
  unit of, 503
Recombination analysis, gene mapping
    in humans, 443–444, *443*
Recombination frequency, 411–412
  calculation of, 416–417, *416–417*
  for genes located far apart on same
    chromosome, 413, *414*
Red-backed vole, transferrin in, 668
Redundancy, in genetic code, 710
Regeneration, of carrot plants from
    mature single cells, 577, *578*
Regression, 381–382, *382*
Regression coefficient, 382
Regression line, 381–382, *382*
Regulated gene, 516
Regulation cascade model, for sex deter-
    mination in *Drosophila*, 587, *587*
Regulatory promoter element, 551
Reinforcement, 698
Relative rate test, 705, 714–715, *715*
  causes of variation in rates, 715
Release factor, 111, 126
  eRF1, 126
  RF1, 126
  RF2, 126
  RF3, 126
Repetitive-sequence DNA, 36–37
Replica plating, 148–149, *149*, 189, 483
Replication, 43–62
  assembly of DNA into nucleosomes,
    61–62, *61*
  in bacteriophage, 54–57, *55–56*

bidirectional, 51, 54, *55*, 57
chain elongation step in, 48, *49*, *53*
of chloroplast DNA, 636–637
of circular DNA, 54, *55*
conservative, 44–47, *44*
direction of, 48, *49*, 50, 52, *53*
dispersive, 44–47, *44*
DNA polymerase. *See* DNA polymerase
in *Drosophila melanogaster*, 57, *57*
errors in, 50, 140, *141–142*
in *Escherichia coli*, 48–51, *50*
in eukaryotes, 43, 47, *47*, 57–60,
  57–60
initiation of, 51, *51*, *53*, 58
lagging strand in, 52, *53*, 55, 59
leading strand in, 52, *53*, 59
Meselson-Stahl experiment on,
  45–47, *46*
molecular model of, 50–57, *50–56*
of mtDNA, 633, *634*, 713
in prokaryotes, 43, 62
proofreading in, 50, 62
rate of, 57
RNA primer, 43, 51–52, *51*, *53*, 59
rolling circle, 54–57, *55–56*
in *Saccharomyces cerevisiae*, 58
semiconservative, 43–47, *44*, *46–47*, 62
semidiscontinuous, 43, 52–54, *53–54*,
  56, 62
of telomeric DNA, 59–60, *59–60*
template strand, 51
Replication fork, 51–54, *51*,
  53–54, 57
Replication unit. *See* Replicon
Replicative senescence, 624
Replicative transposition, 156, *158*
Replicator, 51, *51*, 58
Replicator selection, 58
Replicon, 57–58, *58*
Replisome, 52, *54*
Reporter gene, 229–230, *230*
Repressible operon, 529
Repressor, 515, 519, 544
  blocking transcription with, 550
  eukaryotic, 550–551, 567
  translational, 595
Repressor gene, 519
Repressor protein, 494
Repulsion of alleles, 411
Research. *See* Genetics research
Resistance mutation, 149
Resolvase, 156
Response to selection. *See* Natural
  selection, response to
Restorer of fertility (Rf) gene, 648, *648*
Restriction endonuclease. *See*
  Restriction enzyme
Restriction enzyme, 175–179,
  177–180, 673
  cloning using, 176–179, *177–180*
  general properties of, 176–177
  partial digestion with, 184, *184*
  rare cutters, 246
Restriction fragment length
  polymorphism (RFLP), 211, 214
  DNA molecular testing, 219–221, *220*

estimation of genetic variation,
  674–675, *674–675*
PCR-RFLP analysis, 215, *215*
Restriction map, 192
Restriction mapping, 246
  of cloned DNA sequence, 176,
    192–195, *193–195*
Restriction site, 176–177, *177*
  frequency of occurrence in DNA,
    177–178, *179*
  rotational symmetry in, 177, *177*
  single-nucleotide polymorphisms that
    alter, 214–215, *215*
Restriction site linker, 187, *187*
Retinoblastoma, 442–443, 619–622,
  619–622, 678
  bilateral, 442, 619
  hereditary, 442, 619, *620*, 621
  sporadic, 442, 619–620, *620*
  unilateral, 442, 619
Retrotransposition, 164
Retrotransposon, 164
  in humans, 165
Retrovirus, 163, 605
  cancer-inducing, 605, 618, 627
  life cycle of, 611, *612*
  nononcogenic, 611, *613*
  oncogenes and, 610–618
  oncogenic, 611, *612*, 614
  structure of, 610–611, *610*
  transducing, 605, 614–615,
    614, 618
Reverse ASO hybridization, 221
Reverse mutation, 115–116, *116–117*,
  133, 138–140, 677
  partial reversion, 138
  true reversion, 138
Reverse tandem duplication,
  458, *458*
Reverse transcriptase, 164–165,
  611, *613*
  construction of cDNA, *186*, 187
  telomerase, 60
Reverse transcriptase PCR
  (RT-PCR), 203
  real-time RT-PCR, 203
Reverse transcription, 60
Reversion. *See* Reverse mutation
Revertant, 115–116, *116–117*
Reyes, Matias, 227
RFLP. *See* Restriction fragment length
  polymorphism
R group, 112, *112–113*
RH. *See* Radiation hybrid
Rheumatoid arthritis, 360
Rhoades, Marcus, 157, 160
Rho-dependent terminator, 92
Rho-independent terminator, 92, *93*
Ribonuclease. *See* RNase
Ribonucleic acid. *See* RNA
Ribonucleotide, 21
Riboprobe, 182
Ribose, 13, 20, *20*, 22, 26
Ribosomal DNA (rDNA), 103
Ribosomal DNA (rDNA) repeat unit,
  104, *104*

Ribosomal protein, 102, 105
  mitochondrial, 635
Ribosomal RNA (rRNA), 4–5, 102
  central dogma, 88
  of chloroplast, 637
  of eukaryotes, 103–104, 104
  mitochondrial, 635, 641–642
  pre-rRNA, 88
  of prokaryotes, 103, 103
  5S, 105
  16S, 121, 122
    evolutionary tree of life from,
      721–722, 721
  23S, 123
  structure of, 26–27
  synthesis of, 87–88, 93, 103–105, 103
Ribosomal RNA (rRNA) genes, 37
  of chloroplast, 631
  of mitochondria, 631
Ribosomal RNA (rRNA) transcription
    unit, 103
Ribosome, 5, 5, 9, 11, 88
  A site on, 122–123, 122, 124–125, 125
  E site on, 122–125, 122
  of eukaryotes, 88, 102–103
  mitochondrial, 635
  of prokaryotes, 88, 102
  P site on, 122–123, 122, 124–125,
    125–126
  structure of, 102–103, 102
  subunits of, 102, 102, 111
  in translation, 111–128, 120–126
Ribosome-binding assay, 117
Ribosome-binding site (RBS),
    121–122, 122
Ribothymidine, 106
Ribozyme, 100, 123
Ribulose bisphosphate carboxylase/
    oxygenase (rubisco), 637
Rice
  chloroplast genome of, 637
  genome of, 256
Richardson, C.C., 48
Ridgway, Gary, 226
rII region, of bacteriophage T4, 481
  complementation tests on,
    503–507, 506
  deletion mapping of, 502, 503, 504–506
  evidence that genetic code is triplet
    code, 115–116, 116–117
  fine-structure mapping of, 501–507
  recombination analysis of mutants,
    501–503, 502
RISC, 565–566, 566
RNA, 2, 13–14
  antisense, 236, 565
  compared to DNA, 22
  composition of, 20–21, 20–22
  double-stranded, 27, 565
  guide, 102
  messenger. See Messenger RNA
  northern blot analysis of, 197
  ribosomal. See Ribosomal RNA
  small nuclear. See Small nuclear RNA
  structure of, 13, 20–21, 20–22,
    26–27, 36

synthesis of. See Transcription
  telomerase, 59–60, 60
  transfer. See Transfer RNA
RNA editing, 101–102, 101
RNA enzyme, 100, 123
RNAi. See RNA interference
RNA-induced silencing complex
    (RISC), 565–566, 566
RNA interference (RNAi), 544,
    565–566, 566
RNA lariat structure, 99, 100
RNA ligase, 105
RNA polymerase, 4, 4, 87–90, 89–91, 527
  core enzyme, 89, 92
  of eukaryotes, 93
  holoenzyme, 89, 91, 623
  proofreading activities of, 92
  sigma factor, 87, 89, 91, 92
RNA polymerase I, 87, 93, 104
RNA polymerase II, 87, 93–94, 93, 95,
    105, 545, 550, 611
RNA polymerase III, 87–88, 93, 105
RNA primer, replication, 43, 51–52, 51,
    53, 59
RNA processing control, 546, 561–562,
    562, 567
RNase, 16
RNase III, 103
RNase H, 186, 187
RNA virus, 19–20, 19, 26
  double-stranded RNA, 27
  single-stranded RNA, 27
  tumor virus, 610–618
Roberts, Richard, 97
Robertsonian translocation, 470–471, 470
Rolling circle replication, 54–57,
    55–56
Romanov family (Russian rulers),
    635–636
Rooted phylogenetic tree, 716, 717
Rough endoplasmic reticulum, 11
Roundup Ready soybean, 234
Rous, F. Peyton, 611
Rous sarcoma virus, 610–611, 612,
    614–615
rRNA. See Ribosomal RNA
rrn region, 103
RT-PCR. See Reverse transcriptase PCR
Rubella, 362
Rubin, G.M., 164
Rubisco, 637
runt gene, 595
Russell, Lillian, 324

Saccharomyces cerevisiae
  centromeres of, 34–35, 35
  chromosome number in, 300, 313
  cloning genes of, 190, 192
  complementation tests in, 507
  C value of, 31
  development in, 575
  GAL genes of, 228–230, 228, 230
    glucose repression of GAL1, 228, 228
    regulation of, 551–553, 552
  gene expression in, 261
  gene knockouts in, 259, 260

genome analysis in, 256
  genome sequence of, 244, 252–253,
    258, 259
  killer yeast, 649–650, 649
  life cycle of, 430, 430, 643
  mating type in, 326, 430, 430, 575
  mating-type switch in, 548
  as model organism for research, 6, 9,
    10, 252–253, 253, 575
  mRNA degradation in, 564
  mtDNA of, 633
  petite mutants of, 643–645, 644
  phenylalanine tRNA of, 106
  random-spore analysis in, 432
  replication in, 58
  sporulation in, 261, 262
  telomeres of, gene silencing at,
    557, 558
  tetrad analysis in, 430–437
Saccharomyces pombe, centromeres of, 35
Sager, Ruth, 645
Salamander
  body length in, 379, 379, 381
  head width in, 381
SalI, 178
Salmonella typhimurium
  Ames test, 147–148, 147
  spontaneous mutation frequency at
    specific loci, 678
Sample, 377
Sampling error, 680, 682
Sanger, Frederick, 197
SAR. See Scaffold-associated region
Sau3A, 178, 184–185
SBE. See Starch-branching enzyme
Scaffold, chromosome, 14, 33, 34,
    305, 305
Scaffold-associated region (SAR), 33, 33
Scanning model, for initiation of
    translation, 123
Scarlet tiger moth, spot pattern of,
    660, 660
Scheck, Barry, 227
Schimper, A., 722
Schramm, G., 19
Schwannoma, 619
SCID. See Severe combined
    immunodeficiency
Scott, Matthew, 595
Secondary nondisjunction, 319–320, 321
Secondary structure, of proteins, 112,
    114, 114
Second-division segregation, 432–434,
    433
Second filial generation. See F_2 generation
Second messenger, 554
Seed, 313
  germination of, 556, 558
  hybrid, production of, 647–648, 648
Seedless fruit, 473
Seed traits, in garden pea, 271,
    274–276, 274–277, 280–286,
    280–285
Seed weight
  in bean, 375, 377–378, 378
  in jewelweed, 393

Segment, in *Drosophila* development, 593, *593*

Segmentation genes, in *Drosophila melanogaster*, 574, 593, 595, *596*

Segment polarity genes, 595, *596*

Segregation, principle of, 3–4, 271, 275–282, *275–281*, 320

Selander, Robert K., 684

Selectable marker, 180

Selected marker, 495, *497*

Selection, 5
  artificial, 390
  natural. *See* Natural selection
  response to, 373, 390–394
  estimation of, 390–392, *392*

Selection coefficient, 689

Selection differential, 391

Selection response (R), 391–392

Selector genes. *See* Homeotic genes

Self-fertilization, 273, 473, 695–696, *696*

Selfing, 274

Self-splicing introns, 88, 100, *101*

Semiconservative replication, 43–47, *44*, *46–47*, 62

Semidiscontinuous replication, 43, 52–54, *53–54*, 56, 62

Semisterility, 462

Sense codon, 118

Sequence similarity search, 258–259

Sequence-tagged site (STS), 245, 247
  construction of clone contig maps, 247, *248*

Sequencing ladder, 199, *199*

Serine, *113*

Serine-protein kinase, *616*

Setlow, R., 150

Severe combined immunodeficiency (SCID), gene therapy for, 231

Sex chromosome, 299–301, 312–314. *See also* X chromosome; Y chromosome
  in birds, 326
  in *Caenorhabditis elegans*, 325–326
  in *Drosophila melanogaster*, 325–326, *325*
  in plants, 326

Sex combs reduced (Scr) gene, 596–597

Sex determination, 299, 329, 573, 600
  in *Caenorhabditis elegans*, 325–326
  in *Drosophila melanogaster*, 325–326, *325*, 586–590, *587–590*
  genic, 320, 326
  genotypic, 320–326
  in mammals, 320–325, 585–586
  X chromosome-autosome balance system of, 325–326, 586–590, *587–590*
  Y chromosome mechanism of, 320–325, 585

Sexduction, 487

Sex factor F. See F factor

Sex-influenced trait, 360, *361*

Sex-limited trait, 360

Sex linkage, 299, 314–317, *315–316*, *318*
  Hardy-Weinberg law for, 667, *667*

Sex-linked trait, 299, 317, 329–330. *See also* X-linked trait; Y-linked trait

in humans, 326–330
recessive lethal, 357

Sex reversal, 585–586

S9 extract, 147–148

Sexual behavior, in *Drosophila melanogaster*, 228–229, *229*

Sexual reproduction, 311

SH-2/3 protein, *616*

Sharp, Philip, 97

Sheep
  cloning of, 577–578, *579*
  horns in, 360
  production of recombinant proteins in, 232, *232*

Shell coiling pattern, in *Limnaea peregra*, 650–651, *650–651*

Shell color, in snail, 374

Shepherd's purse, fruit shape in, 354–355

Shine, John, 121

Shine-Dalgarno sequence. *See* Ribosome-binding site

Shoot variegation, in four o'clocks, 631, 639–641, *640–642*

Short interfering RNA (siRNA), 544, 565–566, *566*

Short interspersed elements. *See* SINEs

Short tandem repeat (STR), 211, 216–217, *218*
  DNA typing with, 224–225
  estimation of genetic variation, 676
  forensic applications of, 226–227

Shotgun approach, to genome sequencing, 243, 245, 249–250, *250*, 254

Shuttle vector, 175, 182

Sickle-cell anemia, 77–78, *77–78*, 218
  DNA molecular testing, 219–220, *220*
  gene therapy for, 231
  malaria and, 693–694, *694*
  newborn screening for, 219

Sickle-cell trait, 77–78, 693

Sigma factor, 87, 89, *91*, 92

Signal hypothesis, 127, *127*

Signaling cascade, 616–617, *617*

Signal peptidase, 127, *127*

Signal recognition particle (SRP), 127, *127*

Signal sequence, 111, 127, *127*, 584

Signal transducer, 608

Signal transduction, 554, 608, *609*, 616–617, *617*

Silencer element, 551

Silent mutation, 137, *137*, 214

Simian virus 40, C value of, *31*

Simple sequence repeat. *See* Short tandem repeat

Simple telomeric sequences, 35, *36*

SINEs (short interspersed elements), 36, 38, 166
  Alu family, 36–37, 165
  in humans, 165

Singer, B., 20

Single crossover, 413

Single-nucleotide polymorphism (SNP), 211, 214–216, *215–216*
  ASO hybridization analysis of, 215–216, *216*
  coding, 214
  DNA microarrays to detect, 216, *217*
  non-coding, 214
  PCR-RFLP analysis of, 215, *215*
  Southern blot analysis of, 214–215, *215*
  that alter restriction sites, 214–215, *215*

Single orphan, 259, *259*

Single-strand DNA-binding (SSB) protein, 50, 52, 53, 55, *55*

Sinsheimer, Robert, 27

SIR genes, 557

siRNA. *See* Short interfering RNA

sis genes (*sisterless*), 588, *588*

Sister chromatids, 30, 302, 304, 306, 309

Site-specific mutagenesis, 146–147, 211–213, *213*

Skin cancer, 626

Slope, of regression line, 382, *383*

*Sma*I, *178–179*, 179, 187

Small nuclear ribonucleoprotein particle (snRNP), 87, 99, *100*

Small nuclear RNA (snRNA), 5, 93, 563
  central dogma, 88
  structure of, 26–27

Small nuclear RNA (snRNA) genes, 105

Smith, Hamilton O., 177

Smoking, 469, 626

Smooth endoplasmic reticulum, *11*

Snail
  shell coiling in, 650–651, *650–651*
  shell color in, 374, *672*

Snapdragon
  aurea, 357
  flower color in, 346

SNP. *See* Single-nucleotide polymorphism

snRNA. *See* Small nuclear RNA

snRNP. *See* Small nuclear ribonucleo-protein particle

Snurp. *See* Small nuclear ribonucleo-protein particle

Solenoid model, for 288-nm chromatin fiber, 33, *33*

Somatic gene therapy, 212, 231

Somatic mutation, 136

Somatic recombination, 583

Sorrel horse, 346

SOS protein, 617, *617*

SOS response, 152–153

Southern blot analysis, 224
  procedure for, 195–197, *196*
  of sequences in genome, 195–197, *196*
  of single-nucleotide polymorphisms, 214–215, *215*

Soybean, Roundup Ready, 234

Spacer sequence, 104, *104*

Special environmental effect, 387

Specialized transduction, 494, 497–499, *498*

Speciation, 697–699
  barriers to gene flow, 698
  genetic basis for, 698–699

Species tree, 705, 716–718
Speed, in garter snake, 393, *394*
Sperm, 306, 311, *312*
Spermatid, 311, *312*
Spermatocyte
  primary, 311, *312*
  secondary, 311, *312*
Spermatogenesis, 311, *312*
Spermatogonia, *312*
  primary, 311, *312*
  secondary, 311, *312*
Sphagnum moss, chromosome number
  in, *313*
S phase, 30, 58, 302, *302*, 607
Spindle apparatus, *307*
Spinobulbar muscular atrophy, 466
Spliced leader-RNA, 545, *545*
Spliceosome, 87, 99, *100*
Spliceosome retention, 563
Spontaneous mutation, 140–141,
  141–142
Spore, plant, 312, *313*
Sporogenesis, 306
Sporophyte, 311–312, *313*
Sporulation, in *Saccharomyces cerevisiae*,
  261, *262*
Spot pattern, of scarlet tiger moth,
  660, *660*
Spradling, A.G., 164
src oncogene, 611, 614–616, *614–615*
SRP. *See* Signal recognition particle
SRY gene, 327, 573, 585–586, 600
SSB protein. *See* Single-strand DNA-
  binding protein
ST1571, 462–464
Stable allele, 159
Stadler, Lewis, 160
Stahl, Frank, 45–47, *46*
Stamen, 311, *313*
Standard deviation, 378–379, *379*
Standard error, 680–681
Starch-branching enzyme (SBE), 281
Starfish, chromosome number in, *313*
START, in yeast, 607, *608*
Start codon, 123
Starvation resistance, in *Drosophila
  melanogaster*, 393
Statistical analysis, 286–287, *286–287*,
  377–383
Stature, 362–363, 388
  heritability of, *391*
Stem cells, 607
Stem height, in garden pea, 274, *274*, 277
Stern, Curt, 408–409, *409*, 437
Steroid hormone
  control of chromosome puffing, 581
  mechanism of action of, *553*,
    554–556, *556*
  proteins induced by, 554, *555*
  regulation of gene expression by, 543,
    553–556, *553*, *555–556*, 586
  structure of, *554*
Steroid hormone receptor, 554–555, *556*
Steroid hormone response element, 555
Stevens, Nettie, 313
Steward, Frederick, 577

Stop codon, 111, 119, *124*, 125–126,
  126, 136
STR. *See* Short tandem repeat
Strawberry, polyploidy in, 472
Streptococcus pneumoniae
  Avery's transformation experiment
    with, 16, *16*
  Griffith's transformation experiment
    with, 15–16, *15*
Streptomycin, inhibition of mitochondrial
  protein synthesis, 635
Stroma, of chloroplast, 636, *637*
Structural genomics, 243, 245–257, 265
STS. *See* Sequence-tagged site
Sturtevant, Alfred, 344, 411, 458
Submetacentric chromosome, 301,
  301, 305
Sulston, John, 575
Summer squash
  fruit color in, 351–352, *353–354*
  fruit shape in, 349, *351*
Sum rule, 279
Supercoiled DNA, 29–30, *30–31*
  negative supercoiling, 29, 54
  positive supercoiling, 29
Suppressive *petite*, 644–645, *644*
Suppressor mutation, 133, 138–140
  intergenic suppressor, 138–139, *139*
  intragenic suppressor, 138–139
Sutton, Walter, 4, 312
Sweet pea, flower color in, 353
SWI/SNF, 548
Sxl gene, 574, 587–588, *587*, *589*, 591
SYBR green, 203
Synapsis, 306, *307*, 308
Synaptonemal complex, 306, 308
Syncytial blastoderm, 592, *593*
Syncytium, multinucleate, in *Drosophila*
  development, 592, *593*
Syngamous mating, 645
Synonymous site, 676, 705, 708–710,
  708, 712–713, *712*
Syntenic gene, 406
Systemic lupus erythematosus, 360

TAF. *See* TBP-associated factor
Tail length, in mouse, 393
tailless gene, 595
Tandem duplication, 458, *458*
Tandemly repeated DNA, 35–36
Taq DNA polymerase, *201*, 202
Target site, for IS element, 155, *156*
Target-site duplication, 155, *156*
TATA-binding protein (TBP), 95
TATA box, 94, 95, 551, 709
Tatum, Edward, 4, 69–72, *71–72*,
  484, *484*
Tautomer, 140, *141*
Tautomeric shift, 140, *141*
Tay-Sachs disease, 74, 75–76, *76*,
  218–219, 357
TBP. *See* TATA-binding protein
TBP (*ter* binding protein), *50*
TBP-associated factor (TAF), 95
T cells, 581
TDF gene, 585

T-DNA, 233–234, *234–235*
TEL genes, 60
Telocentric chromosome, 301
Telomerase, 43, 59–60, *60*
  in cancer cells, 606, 624, 627
Telomere, 14, 306, 454
  DNA of, 34–35
    replication of, 59–60, *59–60*
  of *Drosophila melanogaster*, 35
  length of, 60
  of *Saccharomyces cerevisiae*, gene
    silencing at, 557, *558*
  shortening, cancer and, 624
  simple telomeric sequences, 35, *36*
  telomere-associated sequence, 35
  of *Tetrahymena*, 35, 59, *60*
Telomere-associated sequence, 35
Telomere position effect, 557
Telophase
  meiosis I, *307*, 309
  meiosis II, *307*, 309, *311*
  mitotic, 302, *303–304*, 305, *311*
Temperate phage, 494
Temperature effect, on gene expression,
  360, *362*
Temperature-sensitive mutant, 48, 149
Template strand, replication, 49, 51
Temporal isolation, 698
Temporal variation, in allelic frequency,
  669–670, *671*
Tenebrio molitor. *See* Mealworm
teosinte branched 1 QTL, in corn, 397
TEP1 gene, 261, *262*
ter gene, 57
Terminal inverted repeat, *157*
Terminally differentiated cell, 607
Terminal tandem duplication, 458, *458*
Termination factor, 126
Terminator (transcription), 89, *90*, 92
  Rho-dependent, 92
  Rho-independent, 92, *93*
Terminator sequence, 92
Tertiary structure, of proteins,
  114–115, *114*
Testcross, 271, 279–280, *280–281*
  gene mapping through, 410–411, *410*
Testis-determining factor, 321, 585, 600
Testosterone, 554, *555*, 586
Tetrad
  linear, 429
  ordered, 432
  unordered, 431
Tetrad analysis, 429–437, *433–436*
  calculating gene-centromere distance,
    429, 432–434, *433*
  mapping two linked genes, 434–437,
    434–436
  of mitochondrial mutants
    of *Neurospora crassa*, 641
    of *Saccharomyces cerevisiae*, 643–645
Tetrahymena
  genetic code in, 118
  genome of, 256
  as model organism for research, *10*
  self-splicing introns of, 100, *101*
  telomeres of, 35, 59, *60*

Tetraploid (4N), 472–473, *472*
Tetrasomy, *467*, 468
  double tetrasomic, *467*, 468
Tetratype tetrad, 434–437, *434–436*
T-even phages, 492
Thalidomide, 362
Thermal cycler, 202–203
Thermoplasma acidophilum, genome of, 255, *256*
Thiomargarita namibiensis, 11
Three-point testcross, gene mapping with, 413, *415*
Threonine, *113*
Thylakoid, 636, *637*
Thymidylate synthetase gene, of bacteriophage T4, 98
Thymine, 2, 13, 20, *20*, 22, 23, *23*, *25*, 141
Thymine dimer, 142–143, *143*
  repair of, 150
Thyroid cancer, 626
Thyroxine receptor, *616*
Ti plasmid, transformation of plant cells, 233–234, *234*
Tissue growth factor-beta, 233
Tissue plasminogen activator, 233
Titer, 482
TLC1 gene, 60
t-loop, 35, *36*
TMV. *See* Tobacco mosaic virus
Tn. *See* Transposon
Toad, chromosome number in, *313*
Tobacco
  chloroplast genome in, 637
  chromosome number in, *313*
  flower length in, 378–379
Tobacco mosaic virus (TMV), RNA as genetic material in, 19–20, *19*
Toes, webbed, 327
Tomato
  chromosome number in, *313*
  Flavr Savr, 236
  fruit color in, 397
  QTL *fw2.2*, 397
Tonoplast, *11*
Topoisomerase, 29, *50*, 54
Totipotent cells, 573–574
TP53 gene, *619*, 622–623, *622; 625*
  genetics of, 622
  in mice, 213
tra gene (*transformer*), 587, 588–589, *590*
Trailer sequence, 96
  evolution in, *708*, 709
Trait, 3
Transacetylase, 517–518, *517*, 521–522
trans configuration, 411
Transconjugant, 484
Transcription, 4, *4*, 87–108
  antitermination signal, 530–531, *531*
  central dogma, 88
  coupled transcription and translation, 96, *96*, 530
  direction of, 89, *89–90*
  divergent, 552
  elongation stage of, *91*, 92

in eukaryotes, 93–106
initiation of, 87, 89–92, *89–91*, 94–95, *95*
  of mtDNA, 633–635
  of non-protein-coding genes, 102–106
  pause signals, 530
  of polytene chromosomes, 581, *581*
  in prokaryotes, 87–92
  of protein-coding genes, 93–102
  rate of, 92
  regulation of, 228, *228*
  reverse, 60
  by RNA polymerase III, 105
  RNA synthesis, 88–89
  of rRNA genes, 103–105, *103*
  termination of, 92, *93*
Transcriptional control, 543, 546, *546*, 567
  by activators and coactivators, 549–550, *549*
  blocking transcription with repressors, 550
  chromatin remodeling, 547–548, *548*
  combinatorial gene regulation, 550–551, *551*
  GAL genes in *Saccharomyces cerevisiae*, 551–553, *552*
  by steroid hormones, 553–556, *553*, *555–556*
  transcription initiation, 546–556
Transcriptional "fingerprint," 505
Transcription factor, 88, 95, 543, 550, 594–595, *616*, 623
  E2F, 621, *621*
  general, 94, *95*, 549, *549*
Transcriptome, 243, 260–261, *262*, 265
Transducing phage, 492–499, *492–498*
  specialized, 497
Transducing retrovirus, 605, 614–615, *614*, 618
Transductant, 492, 494
Transduction
  in bacteria, 481, 507
  in *Escherichia coli*, 494–499, *496*, *498*
  gene mapping in bacteria, 492–499, *492–498*
  generalized, 494–497, *496–497*
  specialized, 494, 497–499, *498*
Transfection, 231
Transferrin, in red-backed vole, 668
Transfer RNA (tRNA), 4–5
  adding amino acids to, 119–120, *120–121*
  attenuation and, 532
  central dogma, 88
  of chloroplast, 637
  cloverleaf structure of, 105, *106*
  initiator, 121, *122*
  isoacceptor, 710
  mitochondrial, 635
  modified bases in, *106*
  pre-tRNA, 105
  structure of, 26–27
  synthesis of, 87–88, 93
  in translation, 111–128, *120–126*
  tRNA.fMet, 121–122, *122*, 635

tRNA.Met, 121
Transfer RNA (tRNA) genes, 103, 105, 631
  redundancy in, 139–140
  suppressor mutations, 139–140, *139*
  of *Xenopus laevis*, 105
Transformant, 490, *491*
Transformation
  Avery's experiment on, 16, *16*
  in *Bacillus subtilis*, 490, *491*
  in bacteria, 481, 507
  of cells, 606–607
  engineered, 490
  in *Escherichia coli*, 490
  gene mapping in bacteria, 490–492, *491*
  Griffith's experiment on, 15–16, *15*
  introduction of recombinant DNA into bacteria, 181
  natural, 490
  of plant cells, 233–234, *234*
Transformed cells, 606–607
Transforming principle, 16, *16*
Transformylase, 121
Transfusion, blood type and, 342
Transgene, 231
Transgenic cell, 231
Transgenic mammal, production of recombinant proteins in, 232, *232*
Transition mutation, 136, *137*, 141, 143–144, *144–145*, 708
Translation, 5, 88, 111–128
  central dogma, 88
  in chloroplasts, 637
  coupled transcription and translation, 96, *96*, 530
  elongation of polypeptide chain, 111, 123–125, *124–125*, 128
  in eukaryotes, 111
  initiation of, 111, 120–123, *122*, 128
    scanning model, 123
  in mitochondria, 635, *636*
  in prokaryotes, 111
  synonymous codons, 710
  termination of, 111, 126, *126*, 128
Translational control, 544, 546, *546*, 563
Translational repressor, 595
Translesion DNA synthesis, 152–153
Translocation, 408, 453–454, 460–462, *461*, *463*
  nonreciprocal interchromosomal, *461*, 462
  nonreciprocal intrachromosomal, *461*, 462
  position effect, 465
  reciprocal interchromosomal, *461*, 462
  Robertsonian, 470–471, *470*
Translocation step, in translation, 123–125, *124*
Transmission genetics, 1, 6, 658
Transposable element, 6, 134, 153–166
  in corn, 157–162, *159*, *162–163*
  in eukaryotes, 134, 153, 157–165, *159*, *162–164*
  general features of, 153–155
  in prokaryotes, 134, 153, 155–157, *155–157*

Transposase, 134, 155–156
Transposition, 134, 156–157, *158*, 161
  conservative, 157
  cut-and-paste. *See* Transposition,
    conservative
  replicative, 156, *158*
Transposon (Tn), 6, 134, 166
  autonomous elements, 158–159
  characteristics of, 155–157, *157*
  composite, 155–156, *157*
  in *Drosophila melanogaster*, 164, *164*
    Drosophila telomeres, 35
  nonautonomous element, 158–159
  noncomposite, 156, *157*
  in plants, 158–162, *159, 162–163*
  Tn3, 156, *157*
  Tn10, 155–156, *157*
  transposition of, 156–157, *158*
  wrinkled-pea phenotype, 281
Transversion mutation, 136, *137*, 708
Tree of life, 721–722, *721*
Tree snail, Cuban, color patterns
    of, *672*
Treponema pallidum, genome sequence
    of, 482
Trihybrid cross, 285–286, *285–286*
Triplet repeat amplification, *466, 560*
Triploid (3N), 472–473, *472*
Trisomy, 467–468, *467–468*
Trisomy-13, *468*, 471, *471*
Trisomy-18, *468*, 471, *471*
Trisomy-21, 468–470, *469–471*
  maternal age and, 469, *469*
Tristan da Cunha, genetic drift in
    human population, 682
tRNA. *See* Transfer RNA
trp genes, 529–530, *529*, 532
trp operator, *529*
trp operon, of *Escherichia coli*, 528–532,
    528–532, 536
  attenuation in, 530–532, *531–532*
  cells grown in limited tryptophan,
    530, *531*
  cells grown in presence of tryptophan,
    529–530, *531*
  organization of tryptophan biosynthesis
    genes, 529, *529*
  regulation of, 529–532
trp promoter, *529*
trp repressor, 530
Trp-tRNA, in attenuation, 532
True-breeding strain, 274, 276
True reversion, 138
Trypanosoma brucei, RNA editing in,
    101, *101*
Tryptophan, *113*
Tubulin, 304
Tumor, 606
  benign, 606
  malignant, 607
  mitotic recombination in retinoblastoma,
    442–443
Tumor suppressor gene, 605–606,
    618–623, *619*, 627
  BRCA genes, 623
  cancer and, 610

  identification of, 619
  RB gene, 619–622, *619–622*
  TP53 gene, 622–623, *622*
Tumor virus, 610
  DNA virus, 610, 618
  retroviruses. *See* Retrovirus
  RNA virus, 610–618
Turner syndrome, 321–322,
    323, 468
Twin spots, in *Drosophila melanogaster*,
    437–438, *438–439*
Two-hit mutation model, for cancer,
    605–606, 619–621, *620–621*
Two-point testcross, gene mapping
    with, 412, *412*
Ty element, in yeast, 162–164, *163*
typical phenotype, in peppered moth,
    688–689, *688*
Tyrosinase, 75
Tyrosinase-negative albinism, 666
Tyrosine, *113*
Tyrosinemia, 74
Tyrosine protein kinase, 616, *616*

UAS. *See* Upstream activator sequence
Ubiquitin, 564–565
Ultimate carcinogen, 626
Ultrabithorax (Ubx) gene, 596
Ultraviolet light
  as carcinogen, 626
  induction of mutations by, 141–143, *143*
umuDC gene, 50
Unequal crossing-over, 723
Unidentified reading frame (URF), 634
Uniparental inheritance, 639
  in *Chlamydomonas reinhardtii*,
    646–647, *646*
Unique-sequence DNA, 36–38
Unit of mutation, 503
Unit of recombination, 503
Universal donor, 342
Universal recipient, 342
Universal sequencing primer, 198
Unordered tetrad, 431
Unrooted phylogenetic tree,
    716, *717*
Unselected marker, 495, *497*
Unstable allele, 158
3′ Untranslated region, 96, 563
5′ Untranslated region, 95, *96*
Unweighted pair group method with
    arithmetic averages (UPGMA), 718
UPGMA. *See* Unweighted pair group
    method with arithmetic averages
Upstream activator sequence (UAS),
    229–230, *230*
  for *GAL*, 552, *552*
Upstream repressing sequence (URS),
    for galactose, 553
Uracil, 13, 20, *20*, 22, 23, 89
URF. *See* Unidentified reading frame
URS. *See* Upstream repressing
    sequence
U-tube experiment, discovery of bacter-
    ial conjugation, 484, *484*
uvr genes, 150–151, *151*

Vaccine
  edible, 236
  recombinant, 233
Valine, *113*
Variable expressivity, *358–359*, 359
Variable-number tandem repeat
    (VNTR), 211, 217–218
  DNA typing with, 224–225
  forensic applications of, 226
  multicopy loci, 218
  unique loci, 218
Variance, 378–379, *378*. *See also* specific
    types of variance
  partitioning of, 383
Varmus, Harold, 614
Venter, J. Craig, 244, 249, 254
Vent polymerase, 202
Vertebrate
  genome analysis in, 256
  homeotic genes of, 597, *600*
VHL gene, *619*
Victoria, Queen of England, 327, *328*
Viral oncogene, 605, 610, 614, *614–615*
Virulent phage, *17*, 493
Virus
  cancer and, 605, 609–610
  chromosomes of, 27–28, *28*
  DNA, 27
  genetic material of, 13
  helper, 614
  RNA, 19–20, *19*, 26
Visible mutation, 148
VNTR. *See* Variable-number
    tandem repeat
Vogelstein, Bert, 625
von Ehrenstein, G., 119
von Hippel-Lindau syndrome, *619*
von Tschermak, Erich, 286

WAF1 gene, *622*, 623
Wallace, Alfred Russel, 687, *688*
Watson, James, 22–23, *22*
Watts-Tobin, R., 115
W chromosome, 326
Webbed toes, 327
Weinberg, R.A., 614
Weinberg, Wilhelm, 663
Weisblum, B., 119
Wheat
  chromosome number in, *313*
  kernel color in, 375–376, *376*, 386
  polyploidy in, 472–473
Whitefish embryo, mitosis in, *304*
Whittaker, R.H., 721
Whole-genome shotgun approach, to
    genome sequencing, 243, 245,
    249–250, *250*, 254
Wigler, M., 614
Wildlife crimes, forensic analysis in, 227
Wild type, 314
Wild-type allele, 280–281, 317,
    341, *341*
Wilkins, Maurice H.F., 23–24, *24*
Wilms tumor, *619*
Wilmut, Ian, 577
Wilson, Edmund B., 313

Wing length
  in *Drosophila melanogaster*, 390, *390*
  in milkweed bug, *391, 393*
Wing morphology, in *Drosophila melanogaster*, 405–407, *407*, 410
Winter, Johnny and Edgar, *290*
Wobble hypothesis, 118–119, *118–119, 142*
Woese, Carl, 721
Wollman, Elie, 487
Woolf, Charles M., 668
Woolly hair, 290, *291*
Wright, Sewall, 658, *658*, 679
Wrinkled-pea phenotype, in garden pea, 280–281
WT1 gene, *619*

Xanthine, 144, *145*
X chromosome, *299*, 301, 313–314, *314–315*
  abnormal numbers of, 322–324, *324*
  of *Drosophila melanogaster*, 456, *456*, 574, 591
  fragile X syndrome, 465–466, *466*, 560
  inactivation of, 34, 324–325, *324*, 574, 578, 586
  nondisjunction of, 317–320, *319–320*
X chromosome-autosome balance system, of sex determination, 325–326, 586–590, *587–590*
X-controlling element, 586
Xenopus laevis
  C value of, *31*
  tRNA genes of, 105

Xeroderma pigmentosum, 153, *153–154*
X-gal, 181
X inactivation center, 586
XIST gene, 574, 586
X-linked alleles, allelic frequency, 661–663
X-linked trait, 317
  dominant, 300, 327, *329*
  dosage compensation
    in *Caenorhabditis elegans*, 326
    in *Drosophila*, 325
    in mammals, 323–325, *324*, 586
    extension of Hardy-Weinberg law to, 677, *677*
    in humans, 443–444, *443*
  recessive, 300, 327, *328*
XO female. *See* Turner syndrome
X-ray diffraction studies, on DNA, 23, *24*
X rays
  as carcinogen, 626
  induction of mutations by, 141–143
XRN1 gene, 564
XX male, 585
XXX (triplo-X) female, 319, 322–323, 468
XXXX female, 468
XXXXY male, 322, 468
XXY male. *See* Klinefelter syndrome
XXYY male, 322, 468
XY female, 585
XYY male, 322, 468

YAC. *See* Yeast artificial chromosome
Y chromosome, *299*, 301, 308, 314, *314–315*, 573, 585, 600
  abnormal numbers of, 322–324, *324*
  pseudoautosomal regions of, 308
"Y chromosome Adam," 723
Y chromosome mechanism, for sex determination, 320–325
Yeast. See also *Saccharomyces cerevisiae*
  Ty element in, 162–164, *163*
Yeast artificial chromosome (YAC), 175, 182–183, *183*
Yeast two-hybrid system, 229–230, *230*
Y-linked trait, 300, 327–328
Yule, G.U., 663

Z chromosome, 326
Z-DNA, 25–26, *26–27*
Zea mays. *See* Corn
Zeatin, 557
Zebrafish
  C value of, *31*
  development in, 576, *577*
Zellweger syndrome, 230–231
Zinc finger motif, 549, *550, 555*
Zinder, Norton, 494
Zoo blot, 224
Zuckerkandl, Emile, 714
Zygonema, 306
Zygote, 278, 300, 573–574, *592*